Stellar Evolution Physics
Volume 1: Physical Processes in Stellar Interiors

This volume describes the microscopic physics operating in stars and demonstrates how stars respond from formation, through hydrogen-burning phases, up to the onset of helium burning. Intended for beginning graduate students and senior undergraduates with a solid background in physics, it illustrates the intricate interplay between the microscopic physical processes and the stars' macroscopic responses. The volume examines the gravitationally contracting phase which carries the star from formation to the core hydrogen-burning main sequence, through the main sequence phase, through shell hydrogen-burning phases as a red giant, up to the onset of core helium burning. Particular emphasis is placed on describing the gravothermal responses of stars to nuclear transformations in the interior and energy loss from the surface, responses which express the very essence of stellar evolution. The volume is replete with many illustrations and detailed numerical solutions to prepare the reader to program and calculate evolutionary models.

The processes in this volume are added to in Volume 2 of *Stellar Evolution Physics: Advanced Evolution of Single Stars* (ISBN 978-1-107-01657-6), which explains the microscopic physics operating in stars in advanced stages of their evolution and describes how they respond to this microphysics. *Stellar Evolution Physics* is also available as a 2-volume set (ISBN 978-1-107-60253-3). Taken together, the two volumes will prepare a graduate student for professional-level research in this key area of astrophysics.

Icko Iben, Jr. is Emeritus Distinguished Professor of Astronomy and Physics at the University of Illinois at Urbana-Champaign, where he also gained his MS and PhD degrees in Physics and where a Distinguished Lectureship in his name was established in 1998. He initiated his teaching career at Williams College (1958–61), engaged in astrophysics research as a Senior Research Fellow at Cal Tech (1961–4), and continued his teaching career at MIT (1964–72) and Illinois (1972–99). He has held visiting professorships at over a dozen institutions, including Harvard University, the University of California at Santa Cruz, the University of Bologna, Italy, and Niigata University, Japan. He was elected to the US National Academy of Sciences in 1985, and his awards include the Russell Lectureship of the American Astronomical Society (1989), the George Darwin Lectureship (1984) and the Eddington Medal (1990) of the Royal Astronomical Society, and the Eminent Scientist Award of the Japan Society for the Promotion of Science (2003–4).

Stellar Evolution Physics

Volume 1: Physical Processes in Stellar Interiors

ICKO IBEN, JR.

University of Illinois at Urbana-Champaign

Shaftesbury Road, Cambridge CB2 8EA, United Kingdom

One Liberty Plaza, 20th Floor, New York, NY 10006, USA

477 Williamstown Road, Port Melbourne, VIC 3207, Australia

314–321, 3rd Floor, Plot 3, Splendor Forum, Jasola District Centre, New Delhi – 110025, India

103 Penang Road, #05–06/07, Visioncrest Commercial, Singapore 238467

Cambridge University Press is part of Cambridge University Press & Assessment,
a department of the University of Cambridge.

We share the University's mission to contribute to society through the pursuit of
education, learning and research at the highest international levels of excellence.

www.cambridge.org
Information on this title: www.cambridge.org/9781107016569

First published 2013

A catalogue record for this publication is available from the British Library

Library of Congress Cataloging-in-Publication data
Iben, Icko, 1931–
Stellar evolution physics / Icko Iben, Jr.
p.   cm.
Includes bibliographical references and index.
ISBN 978-1-107-01656-9 (Hardback)
1. Stars–Evolution.   2. Stellar dynamics.   I. Title.
QB806.I24 2012
523.8´8–dc23

2012019504

ISBN   978-1-107-01656-9   Hardback

Also available as part of a two-volume set, *Stellar Evolution Physics* ISBN 978-1-107-60253-3

Contents

Preface

The bright stars in the familiar constellations of the Milky Way have intrigued mankind for millennia. Over the past several centuries we have obtained by observations a quantitative understanding of the intrinsic global and surface characteristics of these stars, and over the past century we have learned something about their internal structure and the manner in which they change with time. An awareness that one kind of star can transform into another kind of star and an appreciation of how this transformation is achieved have been accomplishments of the last half of the twentieth century. One of the objectives of this monograph is to describe some of the transformations and to understand how they come about.

The microscopic and macroscopic physics that enters into the construction of the equations of stellar structure and evolution is described in many other monographs and texts. For highly personal reasons, this physics is nonetheless developed here in some detail. My undergraduate and graduate training was in physics, but I did not fully appreciate the beauty of physics until, just prior to my second year of college teaching, during an enforced sedentary period occasioned by a collision between myself on a bicycle and an automobile, I discovered the book *Frontiers of Astronomy* by Fred Hoyle and became entranced with the idea that the evolution of stars could be understood by applying the principles of physics. During my next two years of teaching, I embarked on a self study course heavily influenced by the vivid discription of physical processes in stars by Arthur S. Eddington in his book *The Internal Constitution of the Stars* and by the straightforward description of how to construct solutions to the equations of stellar structure by Martin Schwarzschild in his book *The Structure and Evolution of the Stars*. These books taught me that stars provide a context for understanding physics on many different levels.

Stars are fascinating objects in their own right and the marvelous transformations which they experience as they evolve provide a captivating story worth telling over and over again. In this book, the distinction between the microscopic physics that occurs in stars and the macroscopic response of the stars to the operation of this physics is highly blurred. In short, physics is physics and every effort has been made to understand the relationship between the microscopic and macroscopic aspects of this physics. As Eddington put it, "It would be hard to say whether the star or the electron is the hero of our epic."

In the world of single stars and of stars in wide binaries, focus has been placed on how a star of low to intermediate mass reaches the main sequence, then evolves successively through the main sequence phase, into a red giant, through the core helium-burning phase, into and through the shell helium- and hydrogen-burning asymptotic giant branch phase, and finally into a white dwarf. Attention has also been given to the evolution of stars massive enough to experience central and shell carbon burning, experience a type II supernova explosion, and then evolve into a neutron star or black hole.

A basic objective has been to provide some feeling for what stars really are – not just how big, how bright, and how old they are, but what is going on inside them. How hot are they inside, how dense are they, what sources of pressure do they rely on to balance gravity, what are their energy sources, what are the dominant modes of energy transport, what has been the extent of nuclear transformations in their interiors, and so forth, are questions that have been deemed of central importance. Although one needs to understand physics deeply in order to address these questions properly, and although one can solve the equations based on this physics, many aspects of the solution must be described in a very qualitative way. In an effort to offset the impression of remoteness between the input physics and the set of numbers which constitutes a model star, some partial solutions have been constructed which attempt to tie together in a simple and straightforward way the macroscopic characteristics of stars with the microscopic physics occurring in their interiors.

Just as it has benefited from other disciplines, the science of stellar evolution has contributed to other disciplines. For example, much of the light coming from distant galaxies is the sum of the contributions of individual stars which are themselves unresolvable, but knowledge of whose characteristics is essential for understanding the structure and evolution of these galaxies. By the same token, understanding cosmology requires understanding the characteristics of galaxy evolution. Thus, to fully understand both galaxy evolution and cosmology, one needs also to understand stellar evolution, and so the colors (surface temperatures), luminosities, and evolutionary time scales which characterize different evolutionary stages receive careful attention.

The heavy elements which are essential for the formation of terrestial planets and for the existence of life on such planets have been constructed in stars and ejected by them into the interstellar medium out of which successive generations of stars are born. Hence, to truly understand the origins of the Solar System and of life, one needs to understand the manner in which stars contribute to the build-up of the heavy elements in the Galaxy. A description of element building in stars is therefore a central theme.

The logic of the presentation, including the grouping of the observational evidence, has evolved through almost five decades of emersion in theoretical model building and comparison with the observations. The observations have, in some instances, eliminated possible theoretical paths, made multiple by uncertainties in and incompleteness of the input physics, and, in others, it has opened up new possibilities for theoretical exploration not previously envisioned.

One of the greatest pleasures in writing this book has been to experience moments of revelation when insight comes for the first time. At such moments, I recall the words of T. S. Eliot in *Four Quartets* to the effect that it is in the nature of mankind to explore, and that the end result of all our exploring is to return to the place from which we started and to understand it for the first time. The exact quote, near the end of the fourth quartet (*Little Gidding*), is:

> "We shall not cease from exploration
> And the end of all our exploring
> Will be to arrive where we started
> And know the place for the first time."

An even greater pleasure would be realized if perusal of this book were to inspire some in future generations to contribute to the description of the physics that constitutes stellar evolution in simple and transparent ways, free of jargon and complex mathematics. I would encourage those so motivated to actually construct a stellar evolution code to explore a phase of evolution not yet properly addressed. One of the most exciting moments in my life occurred when in 1963, after almost a year of effort as a senior research fellow at Cal Tech, an evolution program I was constructing finally worked. The music and words of Franz Joseph Haydn's *The Creation* exploded in my brain: "The heavens are telling the glory of God, the wonder of His work displays the firmament."

Due to its length, the book is broken into two volumes and the text is divided into six parts, three parts in the first volume and three in the second volume. In Part I, a qualitative description of single and binary star evolution is followed by a quantitative description of the global characteristics (luminosity, radius, and mass) of stars provided by the observations. In Part II, the basic physical processes occurring in main sequence stars are outlined and the stage is set for constructing realistic mathematical models. Equations of state are constructed by exploring the consequences of momentum and energy balance under conditions of thermodynamic equilibrium and simple stellar models are constructed by demanding only a balance between gravity forces inward and pressure gradient forces outward and a one-parameter relationship between pressure and density. Estimates of the rates of hydrogen-burning reactions and of cross sections for photon–matter interactions are obtained. Part II ends with a description of the equations necessary for constructing theoretical models and of methods for solving the equations. In Part III, solutions of the equations as applied to the evolution of 1 $M_\odot$, 5 $M_\odot$, and 25 $M_\odot$ models approaching the main sequence are presented. Solar models are constructed and a discussion is given of how the initial discrepancy between a predicted and detected solar neutrino flux forced scientists to rethink neutrino physics and resolve the discrepancy, thus vindicating the basic validity of the input physics on which astrophysicists had relied for decades in constructing stellar models. Part III ends with the solutions of the equations of stellar structure as applied to the evolution of 1 $M_\odot$, 5 $M_\odot$, and 25 $M_\odot$ models through all phases of hydrogen burning until the onset of helium burning in the hydrogen-exhausted core.

The second volume begins with a description, in Part IV, of physical processes that are particularly important during advanced phases of evolution. These processes are diffusion, heat conduction by electrons, neutrino losses by beta-decay reactions of the sort occurring in terrestial laboratories and of a sort uniquely occurring in stars, and helium-burning reactions. In Part V, models of 1 $M_\odot$ and 5 $M_\odot$ are evolved through the core helium-burning phase and into the thermally pulsing AGB (TPAGB) phase, with a special chapter devoted to the production of s-process elements. The 25 $M_\odot$ model is evolved through the core helium-burning phase and into the shell carbon-burning phase. In the sixth and final part, the 1 $M_\odot$ TPAGB model is subjected to wind mass loss, evolved through the planetary nebula phase, and, after an exploration of the physics of solids and liquids, is evolved into the white dwarf stage through phases of liquefaction and solidification.

The presentation in the book assumes a solid grounding in basic physics at the advanced undergraduate and early graduate levels and a firm grasp of calculus. The reader would benefit from having had an elementary course in astronomy, but this is not an essential requirement. The content and level of presentation of Volume 1 are suitable for a one-semester beginning graduate-level course which is also within the capabilities of well-prepared senior undergraduate physics students. Volume 2 discusses somewhat more sophisticated physical processes, and carries the discussion of stellar evolution to more advanced stages than described in the first volume. It could form the basis of a second one-semester graduate level course in which students might be encouraged to write and/or make use of a stellar evolution code to construct their own evolutionary models.

My original intent was to provide a definitive account of the evolution from start to finish of all stars whose evolution is essentially quasistatic. To my surprise, I have found that there remain as many unanswered questions regarding this evolution as there have been answers, a discovery that makes me quite sanguine about the future of the field. The story which I tell is far from complete and there are mountains of rewarding work remaining for future generations to tackle.

Bibliography and references

Arthur S. Eddington, *The Internal Constitution of the Stars* (Cambridge: Cambridge University Press), 1926.

T. S. Eliot, *Four Quartets*, fourth quartet, *Little Gidding*, written in 1942.

Franz Joseph Haydn, *The Creation*, written between October, 1796 and April, 1798.

Fred Hoyle, *Frontiers of Astronomy* (New York: Harper & Row, Inc.), 1955;
    Signet Science Library Books (New York: New American Library of World Literature, Inc.), 1957.

Martin Schwarzschild, *Structure and Evolution of the Stars* (Princeton: Princeton University Press), 1958.

PART I

INTRODUCTION AND OVERVIEW

1 Qualitative description of single and binary star evolution

Although locally conspicuous, stars are but one of many forms in which matter and energy in the Universe can manifest themselves; and there is a continuous interaction between these other forms and the stars. Stars are made out of diffuse interstellar matter that gathers itself into the condensations seen as giant molecular clouds; the stars lose mass as they evolve, returning to the interstellar medium material which they have enriched in heavy elements, thus causing a gradual change in the composition and cooling characteristics of the interstellar medium. Radiation from the stars interacts with interstellar matter in the immediate environment, changing the thermodynamic characteristics of this matter. Ejection at high velocity of stellar envelope material, such as occurs in supernova explosions, makes another important contribution to the energetics of the interstellar medium. The change in the composition, thermodynamic, and dynamical characteristics of the interstellar medium then alters the character and dynamics of the star formation process. Thus, although one may concentrate on the evolutionary behavior of stars as if they were isolated entities, stars are actually in a state of interaction with their environment, both feeding and feeding upon the matter in this environment. There is, however, a degree of asymmetry in the interaction. Although one cannot understand the evolution of the interstellar medium without taking into account the influence of stars, one can understand the evolution of a star without worrying about how it was made. Thus, in this book, stars are viewed as more or less isolated entities, their existence being accepted as a given and the circumstances that lead to their formation being examined only to the extent that they offer insight as to an appropriate initial configuration with which to begin evolutionary model calculations.

The growth in our understanding of stellar evolution has been a joint venture between observational astronomers, who have been aided immensely over the years by the development of more powerful telescopes and detectors at all wavelengths, and computational astronomers, who have been aided by the development of ever more powerful digital computers. It is safe to say that each new advance in detector or in computer technology and in marketing this technology has led to an understanding of at least one facet of stellar evolution which eluded understanding prior to this new advance. In this sense, the science of stellar evolution owes much of its success to engineers in both the academic and industrial sectors, and to the business community.

A vast amount and variety of observational and theoretical information goes into providing the foundation on which the edifice of stellar evolution is built. From the Sun and from stars near enough to permit a determination of distance by trigonometry, it has been established that the bulk of the stars lie within a well-defined band or main sequence of low dispersion in a Hertzsprung–Russell (HR) diagram. The HR diagram can take many forms. From an observer's perspective, color and visual magnitude can be a convenient coordinate pair; from a theoretician's perspective, bolometric luminosity and surface temperature are the logical coordinate pair. The orbital characteristics of wide non-interacting binary systems in which at least one of the components is a main sequence star establish a relatively well-defined correlation between the luminosity and mass for a star in the main sequence band. Analysis of the strengths of lines of different atomic and molecular species in stellar spectra demonstrates that hydrogen and helium are by far the dominant constituents of matter at the surfaces of main sequence stars, and permits quantitative estimates of the relative abundances of the heavier elements. Finally, solutions of the equations of stellar evolution for stellar models which are initially of homogeneous composition (with the distribution of elements like that seen at the surfaces of main sequence stars) and which are burning hydrogen in their cores mimic in fair quantitative detail the observed properties of real main sequence stars, thereby solidifying the identification of the main sequence as the residence of core hydrogen-burning stars.

Among the visually brightest stars in the solar vicinity are many which are intrinsically much brighter and considerably cooler at their surfaces than main sequence stars of the same mass. These include stars known as red giants and supergiants, blue supergiants, and RR Lyrae and Cepheid variables. There are others which are intrinsically fainter, but have higher surface temperatures than main sequence stars of the same mass. These include white dwarfs and some subdwarf O and B (sdO and sdB) stars. Still others exhibit at their surfaces abundance distributions which are significantly different from those in main sequence stars. Examples are: red (super) giant carbon stars which show spectral lines of molecules containing the element technicium (all isotopes of which are radioactive with half lives much shorter than the ages of the stars); Wolf–Rayet stars which appear to lie near the main sequence in the luminosity–surface temperature plane but whose spectra show unusual abundances of helium, carbon, and nitrogen; FG Sge stars which have been observed to alter their surface abundances in real time; and hydrogen-deficient supergiant R CrB stars and helium-rich PG 1159 white dwarfs. Still others are both brighter and hotter at their surfaces than are main sequence stars of the same mass. Examples are the central stars of planetary nebulae and some sdO and sdB stars.

White dwarfs and red giants define distinct, well populated sequences in the HR diagram. A typical white dwarf in the observed sample is roughly the size of the Earth and one hundred times less luminous than the Sun. A typical red giant in the nearby sample is one hundred times larger and one hundred times brighter than the Sun. White dwarfs can be easily detected from the Earth only within 100 parsecs of the Sun (a parsec, or pc for short, is approximately 3.26 light years), but bright giants can be observed throughout the Galaxy. Altogether in the entire Galaxy, red giants are roughly one hundred times less numerous than white dwarfs, and white dwarfs are roughly ten times less numerous than main sequence stars more massive than 0.1 $M_\odot$ (one tenth of a solar mass). Hence, a random sampling of stars chosen from within a sphere of radius one hundred pc about the Sun will turn up a few red giants and a modest number of white dwarfs relative to the number of main sequence stars, and a random sampling of visually bright stars will turn up many giants and intrinsically bright main sequence stars, but very few white dwarfs, if any.

In the next two sections, properties of theoretical models of evolving stars are used to interpret the observed characteristics of real single (Section 1.1) and binary (Section 1.2) stars. Although making use of theoretical results before describing these results might seem to be putting the cart before the horse, the objective has been to emphasize that a full appreciation of the luminous objects in the sky that we gaze at in awe and wonder requires the knowledge gained by staring into the bowels of a computer, as the observational astronomer Olin Eggen once characterized my approach to astronomy. It should be noted that, until he died, Olin and I remained good friends and collaborators. In a more serious vein, the following two sections are recommended reading for those who have had an elementary course in astronomy. Those who have not had such a course might find it more profitable to skip immediately to Chapter 2 and then return to Sections 1.1 and 1.2 in this chapter.

1.1   On the evolutionary status of real single stars

The relative characteristics of stars in stellar aggregates both in our Galaxy and in several nearby galaxies, especially in the nearby Magellanic Clouds, have played crucial roles in helping us to determine the internal evolutionary state of many of the stellar types which are not main sequence stars. This determination has been accomplished by comparing the observed properties of stars in these aggregates with theoretical stellar models which have been constructed by taking into account changes in composition wrought by nuclear transformations in the interior. Basic to the exercise is the fact that, if the aggregate is far enough away from us, the distance between stars in the aggregate is sufficiently small compared with the distance from us to the center of the aggregate that the apparent luminosity of one star relative to the apparent luminosity of another is essentially identical with the ratio of the intrinsic (absolute) luminosities of the two stars. The second basic element in the exercise is the assumption that the main sequence defined by stars in the cluster is roughly the same as the main sequence defined by stars in our vicinity for which trigonometric parallaxes can be obtained; this permits an estimate of the distance to the cluster and therefore an estimate of the absolute brightness of each star. The third basic element is the assumption that all of the stars in an aggregate are of essentially the same age and initial composition; if the period of time over which the stars in the aggregate have been formed is small compared with the time which has elapsed since star formation in the aggregate ceased, then the only attribute which differentiates one star from another at birth is stellar mass. Since initially more massive stars evolve more rapidly than less massive ones, the most massive cluster stars which are still burning nuclear fuel will be in the most advanced evolutionary state prior to becoming white dwarfs, neutron stars, or black holes; since the lifetime in phases more advanced than the main sequence phase is short compared with the lifetime of the main sequence phase, the distribution in the color–magnitude plane of the stars in the aggregate which are not on the main sequence effectively defines the evolutionary path of an individual star of an initial mass which slightly exceeds the initial mass of the most luminous stars still on the main sequence. This path may be compared with the path obtained by solving the equations of stellar evolution.

In this way, it is learned from the globular clusters in our Galaxy that, after spending approximately 10^{10} yr converting hydrogen into helium on the main sequence, stars of near solar mass exhaust hydrogen at the center and evolve into red giants which are approximately 2000 times brighter than the Sun before igniting helium in a hot white-dwarf-like core composed of nearly pure helium. The helium nuclei have average energies of $\frac{3}{2} kT$, where k is Boltzmann's constant and T is the temperature, but the electrons are forced by the Pauli exclusion principle into higher energy states of average energy much larger than kT, so the electrons, which are said to be degenerate, provide most of the pressure support for the core. The ignition of helium is initially explosive, with the explosion continuing until the kinetic energy imparted to the degenerate electrons by the nuclear energy released as helium is fused into carbon exceeds the degeneracy energy of the electrons, forcing the hydrogen-exhausted core to expand. In response, the envelope of the star contracts. The process continues until the model star has reached a new equilibrium configuration in which helium burns quiescently in a core in which electrons are not degenerate and hydrogen burns in a shell. The new equilibrium configuration lasts for approximately 10^8 yr, or approximately as long as it takes a precursor to ascend the red giant branch.

A quiescent helium-burning model evolves within a region in the HR diagram defined by horizontal branch (HB) stars in globular clusters. Stars on this branch are approximately twenty-five times more luminous than the most luminous main sequence stars in the clusters and the spread in surface temperatures along observed HBs, when compared with model evolutionary tracks, demonstrates that approximately $0.2 \pm 0.1 \, M_\odot$ is lost by precursor red giants in a stochastic wind, the physics of which is not fully understand. The identification of HB stars as core helium and shell hydrogen burning stars is strengthened by the fact that the number of HB stars and the number of ascending red giant stars are comparable, in agreement with the theoretical result that the lifetimes of the two phases are comparable. The near constancy of the luminosity of HB models is a consequence of the fact that the mass of the hot white dwarf core of the model which ignites helium explosively is essentially independent of the initial mass of a main sequence precursor.

If their evolution carries them into a pulsational instability strip, the location of which can be estimated from theory and determined from observation, HB model stars pulsate acoustically as RR Lyrae stars. Such stars are typically of about the same surface temperature as the Sun. For stars pulsating in the fundamental radial mode, typical pulsation periods are in the range 6–9 hours and, for stars pulsating in the first overtone, periods are typically in the range 10–17 hours. The eponymous variable star RR Lyra is a first overtone pulsator with a period of 13.4 hours. By comparing observed properties with the results of pulsation calculations, one can corroborate that the masses of HB stars in the instability strip are less massive than their main sequence precursors by an amount which varies between about $0.1 \, M_\odot$ and $0.2 \, M_\odot$.

After hydrogen is exhausted at the centers of model stars initially more massive than about $2 \, M_\odot$, the hydrogen-exhausted core contracts and heats rapidly enough that helium is ignited at the center before electrons have become degenerate. The transition to a new equilibrium configuration proceeds in a way which is qualitatively similar to the way in which low-mass models reach the HB branch except that the location in the HR diagram of the quiescent core helium burning and shell hydrogen burning model depends strongly on the mass of the main sequence precursor. In fact, the models define a core helium-burning band which is roughly parallel to the hydrogen-burning main sequence band and can be identified with a similar band defined by nearby intrinsically bright stars such as Rigel, Canopus, and Capella. Just as along the hydrogen-burning main sequence band, the luminosity and surface temperature in the core helium-burning band increase with increasing model mass.

The instability strip defined by pulsation theory passes through the theoretical core helium-burning band defined by evolutionary calculations. Observational counterparts are the Cepheids which, in our Galaxy, have radial pulsation periods between about two and thirty days, peaking in frequency at about six days. The locations of both the band and the strip are functions of the composition, and theory predicts that the functional dependences are such that the intersection between the band and the strip occurs at lower luminosities and lower periods as the metallicity is decreased. Since lifetime increases with decreasing luminosity, one infers that the relative number of stars in an aggregate which are Cepheids should increase with decreasing metallicity. The mean metallicity of stars decreases in going from our Galaxy to the Large Magellanic Cloud (LMC) and then to the Small Magellanic Cloud (SMC). The decrease in the average pulsation period and the increase in the relative number of Cepheids in passing from our Galaxy to the Large and Small Magellanic Clouds is consistent with these predictions.

After exhausting helium at the center, stars of near solar mass evolve again into a red giant configuration, alternately burning hydrogen and helium in separate shells above a hot white-dwarf-like core in which the electrons are degenerate. The bare nuclei in the core are either a mixture of carbon and oxygen (stars less massive than $\sim 8\, M_\odot$) or oxygen and neon (stars of mass in the range 8–10 $M_\odot$). The asymptotic giant branch (AGB) phase, as this phase is called, lasts for a few times 10^5 years (for stars initially of intermediate mass, 2–10 $M_\odot$) to a few times 10^6 years (for stars initially of low mass $\gtrsim 2\, M_\odot$).

AGB stars experience thermonuclear flashes which are also called thermal pulses. Most of the time a model burns hydrogen quiescently in a shell, depositing the resultant hot helium ashes onto a layer of increasing mass above the electron-degenerate carbon–oxygen (CO) core. The helium layer becomes dense and heats as it is compressed, and, once it becomes hot enough, helium is ignited and helium burning begins to produce energy more rapidly than this energy can be carried away by photon diffusion. A thermonuclear runaway (helium shell flash) ensues and continues until the extra pressure built up by heat deposition in the burning region has forced a significant expansion of matter in the helium layer and in overlying layers. With expansion comes cooling; temperatures at the base of the hydrogen-containing portion of the model drop to such low values that hydrogen ceases to burn and the rate at which helium burning produces energy comes into equilibrium with the rate at which photon diffusion can carry energy outward. The model embarks on a phase of quiescent helium burning which lasts about one-tenth as long as the preceeding hydrogen-burning phase.

Eventually, the temperatures in the transition layer between hydrogen-rich matter and helium- and carbon-rich matter decrease to such an extent that helium burning ceases. The decrease in the outward flux of helium-burning energy is compensated for by the reinstitution of contraction and release of gravothermal energy in immediately overlying hydrogen-rich layers, and the concomitant heating in these layers means that hydrogen burning recommences. The duration of the thermonuclear runaway which punctuates each cycle is only a few dozen years and the time between thermal pulses for a typical AGB star of low mass is of the order of 100 000 yr, so the probability of detecting a real star in the process of undergoing the first stages of a thermal pulse is not very large. However, there are several stars which may be identified as being in the process of relaxing after the thermonuclear runaway has taken place. Stars undergoing these alternating hydrogen- and helium-burning episodes are called TPAGB (thermally pulsing AGB) stars.

The lifetimes of real stars in the AGB phase can be estimated by comparing the distribution in luminosity achieved by real AGB stars with that predicted by model AGB stars. If no mass loss were to occur to terminate the ascent of the AGB branch, a model star of near solar mass would achieve a luminosity which is $\sim 3 \times 10^4$ times the Sun's luminosity as its carbon–oxygen core grows to contain most of the mass of the model. Since the maximum luminosity of a globular cluster AGB star is only a few thousand times the Sun's luminosity, one infers that mass loss must abstract from real AGB stars most of their remaining hydrogen-rich envelope before the carbon–oxygen core exceeds about 0.5–0.6 $M_\odot$, or after only about 10^6 years of evolution along the AGB. The short time scale for the mass-loss phase and the large amount of mass loss have prompted the name superwind. The atoms in the shell of matter that is lost by the AGB star are eventually excited into fluorescence by photons from the still bright stellar remnant which contracts to high surface temperature before cooling into a white dwarf. The fluorescing shell is known as a planetary nebula.

The Magellanic Clouds contain aggregates also called globular clusters, but these aggregates typically have only a few times 10^4 stars, over ten times fewer stars than are found in most galactic globular clusters, and their ages span a much larger range. Thus, there are clusters containing AGB stars whose progenitors are of near solar mass, as in the galactic globular clusters, but there are also clusters containing AGB stars whose progenitors have masses near 5 $M_\odot$. One may therefore learn how AGB evolution depends on progenitor mass. The most important lesson is that real AGB stars deriving from intermediate-mass stars expel their hydrogen-rich envelopes within only a few times 10^5 yr after arriving on the AGB; therefore, the mass of the white dwarf remnant of such stars is only $\sim$0.05–0.1 $M_\odot$ larger than the mass of the CO or ONe core at the beginning of the thermally-pulsing AGB phase. Since, as is learned from model studies, this largest initial CO core mass is about $\sim 1.1 M_\odot$, and since an electron-degenerate configuration can burn carbon explosively only if its mass exceeds the Chandrasekhar mass ($\sim 1.4\ M_\odot$), it has been learned that all single stars initially less massive than about 8 $M_\odot$ become CO white dwarfs; they do not explode as supernovae. Single stars of initial mass in the range $\sim$8–10 $M_\odot$ burn carbon non-explosively during the AGB phase and evolve into white dwarfs with substantial ONe cores topped by CO shells. More massive single stars ultimately develop cores composed of iron peak elements which collapse into neutron stars or black holes; the envelopes of such stars are ejected in type II supernova explosions.

As AGB stars evolve to larger luminosities, their envelopes begin to pulsate acoustically, with pulsation periods which can be typically between several hundreds of days and about two thousand days and with amplitudes up to several magnitudes. It is suspected that shock waves formed during pulsations inflate the atmosphere into an extended envelope in which grains can form, and that radiation pressure on the grains converts the envelope into an escaping superwind. It is found empirically that, for periods larger than about 400 days, mass loss from the surface occurs typically at a rate of $10^{-4}\,M_\odot\,\mathrm{yr}^{-1}$. The huge, rapidly expanding OH/IR sources, which emit maser light of 18 cm wavelength from the OH molecule and are strong sources of infrared emission, contain hidden in their interiors an object which produces energy at a rate typical of an AGB star and are themselves composed of matter expelled by this interior object. It is evident that the central object was once an AGB star or is still an AGB star in the process of losing mass and that OH/IR sources are destined to become planetary nebulae, forced into fluorescence by radiation from the central object which has maintained a high luminosity but has contracted into a compact object of high surface temperature.

The evolution of the stellar remnant can be understood with the help of stellar models and the properties of large samples of planetary nebulae and white dwarfs. After the ejection of most of its hydrogen-rich envelope is completed, the remnant star, with a fuel-exhausted core of mass between about $0.6\,M_\odot$ and something short of $1.4\,M_\odot$, rapidly evolves into a compact object of size comparable to that of the Earth. For approximately $10\,\mathrm{yr}$ ($M_{\mathrm{core}} \sim 1.1\,M_\odot$) to $10^4\,\mathrm{yr}$ ($M_{\mathrm{core}} \sim 0.6\,M_\odot$), it burns hydrogen or helium, or both, in a thin layer near its surface, remaining bright enough and hot enough (surface temperature higher than $30\,000\,\mathrm{K}$) to cause the fluorescence of ejected material, which is seen as a planetary nebula 3000–30 000 times more luminous than the Sun. Following the extinction of all nuclear fuels, the remnant cools along the white-dwarf sequence in the HR diagram at nearly constant radius, but at steadily decreasing luminosity and surface temperature. It requires approximately $10^8\,\mathrm{yr}$ for the luminosity to drop to about $10^{-2}\,L_\odot$ (one one-hundredth of the Sun's luminosity), and another 10^{10} years to drop two more decades in luminosity to $10^{-4}\,L_\odot$.

Another lesson learned from Magellanic Cloud AGB stars, in conjunction with model stars, is that the ^{12}C made in the interior during a thermal pulse is dredged to the surface following the pulse. Also made in the interior during a pulse and brought to the surface after a pulse are about 200 varieties of neutron-rich isotopes called s-process isotopes. The neutrons necessary for neutron capture on the seed nucleus ^{56}Fe and on its progeny are a consequence of reactions between ^{4}He and ^{13}C (in AGB stars of small core mass) or of reactions between ^{4}He and ^{22}Ne (in AGB stars of both large and small core mass). The most striking demonstration of carbon production and dredge-up is the fact that, in Magellanic Cloud clusters of intermediate age (3×10^8–$3 \times 10^9\,\mathrm{yr}$), there is a fairly sharp demarcation in luminosity between AGB stars which are also carbon stars and those which are not. The carbon stars are brighter and, at their surfaces, the abundance of carbon exceeds the abundance of oxygen, in contrast with the normal situation during less advanced evolutionary stages when oxygen is more abundant than carbon. This agrees qualitatively with the facts that AGB model stars: brighten as their CO core grows during the quiescent hydrogen-burning phase; do not dredge up freshly synthesized material until the mass of the CO core exceeds a critical value; and, depending on initial composition and total mass, may not develop a surface ratio of carbon to oxygen greater than one until several dozens of carbon dredge-up episodes have occurred. Theory and observation combine to demonstrate that AGB stars are the source of most of the ^{12}C and s-process elements in the Universe.

Dramatic demonstrations of the occurrence of neutron-capture nucleosynthesis and dredge-up are the presence of unstable technetium at the surfaces of several galactic AGB stars and the presence of unusually strong lines of ZrO in the spectra of long period variables in the Magellanic Clouds which are also AGB stars. The long period variable AGB stars have large core masses and their paucity, relative to their precursors which have passed through the Cepheid instability strip, reinforces the inference from the Magellanic Cloud globular clusters that AGB stars deriving from intermediate-mass main sequence stars eject a nebular shell within roughly 10^5 yr after reaching the thermally-pulsing phase. Their short lifetimes as AGB stars is further strengthened by the fact that these stars are not carbon stars, even though the presence of strong ZrO lines indicates that dredge-up is occurring; models suggest that typically 10^6 yr of dredge-up is necessary to develop the carbon-star syndrome.

Since stars take time to form, the assumption that all stars in an aggregate are of the same age is obviously inappropriate for a sufficiently young aggregate. This lesson is most clearly demonstrated by the properties of stars in young open clusters and in superclusters and groups in the disk of our Galaxy. Instead of finding just one cluster locus of low dispersion, one sometimes finds evidence for two or more bursts of star formation occurring at distinctly different times. In the very youngest of open clusters it is clear that the lowest mass stars in clusters have not yet reached the main sequence, but are still gravitationally contracting toward the main sequence; the dispersion about the mean cluster locus suggests, on comparison with models of contracting stars, that a burst of star formation lasts for of the order of 10^7 years.

Galactic globular clusters (halo clusters) and the less tightly bound open clusters (disk clusters) occupy distinctly different locations in our Galaxy and the stars in the two cluster types have quite different metallicities (roughly solar for the disk clusters and 10–100 times less than solar for the halo clusters). The striking dichotomy in abundances has led to the concept of two distinct populations having quite different histories. The conditions at the time of the formation of the metal-poor (population II) globular clusters must have been quite different from the conditions at the time of the formation of the metal-rich (population I) open clusters. Comparison with theoretical stellar models shows that the disk clusters, which orbit the galactic center in nearly circular paths, range in age from 10^7 yr to 10^{10} yr. The age of the oldest white dwarfs in our vicinity, as estimated by comparing with theoretical white dwarf cooling models, is about 9×10^9 yr, and this is close to the age of the oldest known disk cluster, NGC 188. The globular clusters, which traverse highly elliptical orbits within a huge halo of diameter about 30 kpc, have a mean age near 14×10^9 yr. These differences in ages, based on comparisons between theoretical stellar models and real stars, offer intriguing clues as to how galaxies evolve.

In disk clusters, the brightness of the brightest red giants relative to the brightness of the brightest main sequence stars undergoes a dramatic change with increasing cluster age. In the youngest clusters, such as the Pleiades and Alpha Persei, the two types of star are of comparable luminosity. In the oldest clusters, such as NGC 188 and M67, the brightest red giants are much brighter than the brightest main sequence stars, as is also the case in the halo globular clusters. Comparison with models shows that this change in morphology has to do with whether or not, after a star leaves the main sequence, electrons in its hydrogen-exhausted core become degenerate before helium is ignited in the core. If they do not, then ascent along the giant branch is aborted at a luminosity not much above that which the star had on leaving the main sequence. If they do, then the giant branch extends to high luminosity as the electron-degenerate core grows in mass to about $0.45\,M_\odot$, when conditions become ripe for helium to ignite explosively in the core. Models of a solar-like initial abundance show that the transition between the two types of behavior occurs at a critical stellar mass which is somewhere between $2\,M_\odot$ and $2.5\,M_\odot$, the imprecision being a consequence of the uncertainty in the extent to which matter is mixed outward beyond the boundary of the formal convective core found in these models while on the main sequence.

Since the luminosity of a main sequence star is not overly sensitive to this uncertainty in the degree of convective overshoot, there is the possibility of ascertaining, by comparison with the observations, the value of the critical mass, thereby learning something about the physics of convective overshoot. The mass of the most luminous main sequence star in the Hyades cluster is about $2.3\,M_\odot$ and the brightest red stars are the population I analogues of HB stars in galactic globular clusters, namely core helium-burning, shell hydrogen-burning stars. They are called clump stars because, instead of being distributed in the HR diagram over a wide range in surface temperature, they form a nearly constant luminosity peak in the distribution along the first red giant branch. The larger opacity in the metal-rich population I disk stars relative to the metal-poor population II globular cluster stars is responsible for the different HR-diagram morphology of the clump and HB stars. The Hyades cluster does not exhibit an extended giant branch and one may infer only that the critical mass for metal-rich stars is less than about $2.5\,M_\odot$, the mass of the progenitors of the clump stars, and that, therefore, the degree of convective overshoot may be quite modest.

Within the Galaxy there are of the order of 10^4 Cepheids, luminous stars (typically 10^3–$10^4\,L_\odot$) of low surface temperature (typically 5000–6000 K) which pulsate at periods of a few days to several tens of days. They are rare stars, but are conspicuous because of their large-amplitude variability. Only about a dozen disk clusters contain a Cepheid. The distances of the parent clusters can be estimated by the main sequence fitting technique and so both the luminosity and the radius of each Cepheid can be estimated. On comparing with evolutionary models of the appropriate mass and composition which pass near the location of the Cepheids in the temperature–luminosity plane, it is evident that Cepheids are in the core helium-burning stage of evolution. A quantitative fit between the estimated luminosity of each Cepheid and the luminosity–mass relationship given by evolutionary calculations allows one to determine an evolutionary mass for each Cepheid. A pulsation mass can be assigned as well by making use of a relationship between mass, radius, and period obtained by theoretical pulsation calculations (which do not require a specification of the evolutionary state of the interior) and fitting with the luminosity and radius estimated from the observations. On first inspection, the evolutionary masses do not agree in detail with the pulsation masses. That is, the mass–luminosity relationship obtained by using the theory of evolution differs from the mass–luminosity relationship obtained by combining pulsation theory with the luminosities estimated by the main sequence fitting technique. Similarly, the period–luminosity relationship obtained by combining the results of evolution and pulsation theory calculations differs from that obtained by adopting the luminosities estimated by main sequence fitting. A resolution of these discrepancies can be achieved by adjusting the luminosities of the relevant clusters by approximately 20% (accomplished by increasing the estimated distances of the clusters by 10%) or by assuming that the radiative opacities in the envelopes of the theoretical models have been underestimated by a considerable factor. Since the early ladders in the cosmological distance scale involve knowing the distances to Cepheids in nearby external galaxies, it is of importance to clear up the current uncertainties.

An understanding of the late stages of evolution of massive stars was immensely aided by the discovery in the late 1960s of a pulsar in the Crab nebula, which is the ejectum of a supernova recorded by Chinese court astrologers in the eleventh century (SN 1054), and by the occurrence in the Large Magellanic Cloud of a type II supernova (SN 1987a), which is known to have as a precursor a luminous $20\,M_\odot$ star evolving between the main sequence and the giant branch. The properties of the pulsar in the Crab nebula and of the hundreds of other known pulsars are most easily interpreted in terms of the theoretical construct known as a neutron star, invented independently by Lev Landau and by Walter Baade and Fritz Zwicky two years after James Chadwick discovered the neutron in 1932. In 1939, Oppenheimer and Volkov constructed model stars made of neutrons, demonstrating that such objects have radii which are a 1000 times smaller than the radii of white dwarfs and have binding energies of the order of 10^{53} erg.

The period and rate of period change of the Crab pulsar, coupled with the rate of energy emission of the Crab nebula, may be used to support the picture that the pulsar is a rotating neutron star, beaming photons and relativistic particles through a narrow cone. The 33 ms period of the signal is too short to be understood in terms of global, low mode radial oscillations of a white dwarf and it is smaller than the rotation period of a white dwarf rotating at breakup velocity, so the signal is obviously not from a white dwarf. The period is too long to be understood in terms of low mode radial oscillations of a neutron star, but an interpretation of the observed period as the rotation period of a neutron star rotating roughly 30 times slower than the breakup period is, in retrospect, a natural one to make, even though the necessity of assuming that the energy is emitted in a relatively narrow beam to account for the pulsing character of the signal was initially widely decried as highly unlikely. Even more persuasive is the fact that the observed rate of change of the period of the signal is such that, when interpreted in terms of the slowing down of a neutron star, the inferred rate of decrease of the rotational kinetic energy of such a star of mass $\sim 1.4\,M_\odot$ and radius ~ 10 km is $\sim 10^{38}$ erg s^{-1}, which is about the rate at which the Crab nebula is emitting light energy. The picture that has developed is that the pulsar beams out into the nebula a stream of relativistic electrons, which emit synchrotron radiation as they spiral down in the magnetic field threading the nebula.

Baade and Zwicky suggested that the source of the energy released by a supernova as visible light (typically $\sim 10^{49}$ erg) could be the gravitational potential energy released as the fuel-exhausted core of a massive star collapsed to neutron star dimensions. It has since been realized that much more of the energy released in a supernova explosion goes into the kinetic energy of the ejected material ($\sim 10^{51}$ erg) than into visible light, and that at least some of the energy emitted as visible light is a consequence of decays of radioactive elements made and ejected in the explosion. More important, the total amount of energy in forms detected prior to the occurrence of SN 1987a is several orders of magnitude smaller than the binding energy of neutron stars, models of which have now reached a very high level of sophistication, with superfluidity, pion condensates, and strange matter being seriously considered characteristics.

Thus, the detection of a seven-second burst of neutrinos from SN 1987a by a detector in Japan and by one in the US Midwest, and the inference from this detection of a total energy release close to 10^{53} erg, the binding energy of a neutron star, were most exciting discoveries. The detectors were designed primarily to test the possibility that protons might be unstable (on time scales of 10^{31} yr or so); the first direct verfication of a more than 50-year-old prediction about neutron star binding energies was in this respect serendipitous. The occurrence of SN 1987a at precisely the moment when a highly sensitive neutrino detector was available and at such a fortuitously short distance ($\sim$50 kpc) from this detector was an astonishing coincidence in several other respects. Intrinsically bright supernovae occur in galaxies the size of our own only once every $\sim$30 yr. Andromeda, at a distance of 700 kpc, is the closest such galaxy. The Large Magellanic Cloud (LMC) has ten times fewer stars than are contained in our Galaxy and in Andromeda and one might therefore anticipate a supernova event only once every 300 yr in the LMC. Further, as its intrinsic visual brightness was less by several orders of magnitude than that of typical supernova, SN 1987a would have gone undetected (visually as well as in neutrinos) had it occurred in Andromeda, some 12 times further from us than the LMC.

Finally, had the supernova not occurred in the nearby LMC, it would probably also not have been discovered that the precursor was a luminous blue star (luminosity $\sim$100 000 $L_\odot$ and surface temperature $\sim$10 000K) of mass about 20 $M_\odot$. As it turns out, records of the color and brightness of the precursor star existed, thus providing the first direct evidence for the long-standing theoretical prediction that stars of mass as large as 20 $M_\odot$ develop a fuel-exhausted core of 1–2 $M_\odot$ which collapses to neutron-star dimensions.

A class of objects known as Wolf–Rayet (WR) stars has provided insight into the evolutionary behavior of very massive stars and into the manner in which these stars contribute to the enrichment of heavy elements in the Galaxy. Massive ($>40\,M_\odot$) main sequence stars are known to lose mass at very high rates, typically $\dot{M} \sim 10^{-6}\,M_\odot\,\mathrm{yr}^{-1}$ or higher. The mechanism driving mass loss is thought to be momentum transferred from photons to matter through resonant absorption by several atomic species in several spectral lines, but the theory is very complex. Mass-loss rates appear to vary considerably from one star to the next, even among stars of similar luminosity and surface temperature, but the average mass-loss rate increases roughly linearly both with luminosity and with radius. The main sequence lifetime predicted by evolutionary models which neglect mass loss decreases inversely with luminosity. Thus, the product of mass-loss rate and main sequence lifetime increases with increasing initial mass and one can imagine a critical mass such that stars initially more massive than this critical mass will lose most of their hydrogen envelopes during the main sequence phase, eventually exposing layers which have undergone extensive nuclear processing.

The WR stars (classified into subgroups WN and WC) are thought to be just such objects. The spectra of WN stars show strong lines of nitrogen, and, as both carbon and oxygen are converted into nitrogen during hydrogen burning, this is an indication that layers which were once engaged in hydrogen burning have been exposed. This does not mean that WN stars have completed the main sequence phase; stellar models constructed with mass loss taken into account show that hydrogen burns in a convective core which steadily shrinks in mass, leaving behind more and more highly processed material which no longer burns hydrogen. Thus, highly processed material may be exposed without disturbing the continued burning of hydrogen in the core.

The spectra of WC stars show strong lines of carbon, suggesting that layers which have experienced a degree of helium burning, which converts helium into carbon and oxygen, have been exposed. These stars clearly can no longer be in the main sequence phase, and are probably in the core helium-burning phase. However, stellar models constructed by taking mass loss into account, but employing otherwise standard surface boundary conditions, lie far to the blue of the main sequence, in contrast with the WC stars themselves, which are near the main sequence. The solution to this puzzle is probably to be found in the construction of models with unbound, expanding envelopes, rather than with standard static envelopes. In such models, the photosphere can actually be in the expanding envelopes (i.e., in the wind) with the "real surface" of the star at a much smaller distance from the center, consistent with the results of standard calculations. The quite strong winds supported by WR stars ($\dot{M} \sim 10^{-5}$–$10^{-4} \, M_\odot \, \mathrm{yr}^{-1}$) make this solution quite probable.

In any case, the WR stars demonstrate that initially very massive stars lose a large fraction of their initial mass in stellar winds while burning hydrogen and helium at their centers and make contributions to the enrichment of the interstellar medium in helium and in CNO elements (carbon, nitrogen, and oxygen) even before experiencing supernova explosions.

1.2   Close binary stars and evolutionary scenarios

In comparison with the theory and understanding of single star evolution, the theory and understanding of close binary star evolution is exceedingly rudimentary. There is a vast array of systems in which at least one of the components is a highly evolved star (white dwarf, neutron star, helium star, or black hole) at a distance from its companion which is smaller than what must have been the distance between the two components in the main sequence precursor system. In many instances, mass is now being actively exchanged by the pair and, in many other instances, it is clear that considerable mass has been lost from the primordial system in establishing the current system.

Cataclysmic variables are classical examples. They consist of a relatively massive ($\sim 1 \, M_\odot$) white dwarf which is accreting material from a less massive main sequence companion. The large mass of the white dwarfs in observed systems is primarily a consequence of observational selection. In any case, in the typical observed system, the large mass of the white dwarf tells us that the mass of the progenitor star must have been close to the maximum mass of a single star which can become a CO white dwarf ($\sim 8 \, M_\odot$), and that, therefore, most of the initial mass of the system has been blown away. The maximum radius of a metal-rich main sequence star of $8 \, M_\odot$ is about six times the radius of the Sun ($6 \, R_\odot$), which means that the orbital diameter of the precursor system must have been larger than $\sim 16 \, R_\odot$, quite large compared with the current separation (~ 1–$2 \, R_\odot$) of the components of the cataclysmic variable. Obviously, considerable orbital shrinkage has taken place in the past. Precisely when and how this has occurred are questions which have been actively debated.

A combination of theory and observation suggests that mass transfer in cataclysmic variables takes place because (1) the donor slightly overfills its "Roche lobe", a teardrop-shaped geometrical figure enclosing the donor and having the property that any particle outside of the figure along the line joining the centers of the two components is gravitationally attracted more to the stellar component outside of this figure than to the one inside of it, and (2) some agency abstracts orbital angular momentum from the system, forcing the donor to remain in contact with its Roche lobe. In long period cataclysmic variables (orbital period typically $\sim$3–20 hr), the donor loses spin angular momentum via a magnetic stellar wind (MSW) similar to the wind which abstracts angular momentum from the Sun. Tidal forces keep the spin rate of the donor locked into synchronism with the orbital rotation rate, and hence the spin angular momentum lost by the donor is continually resupplied by a contribution from the orbital angular momentum, which decreases. The dependence of the Roche-lobe geometry on the stellar mass ratio and on the orbital angular momentum is such that the loss of orbital angular momentum forces the donor to maintain contact with its Roche lobe, even as it loses mass and shrinks in size. The time scale for angular momentum loss of a typical long period cataclysmic variable via a MSW is thought to be of the order of 10^9 yr. In short period cataclysmics (orbital period $\sim$80 min–2 hr), the driving agency which maintains the donor in Roche-lobe contact is the loss of orbital angular momentum via gravitational wave radiation (GWR) on a time scale of 10^{10} yr.

The cataclysmic variable class of stars owes its name to the fact that many representatives have been discovered by monitoring classical nova outbursts which reach intrinsic luminosities similar to those of AGB stars and remain bright for a few months to a few years. By comparing observed characteristics with models of accreting white dwarfs, one may infer that these outbursts are consequences of a thermonuclear runaway which is initiated when enough (typically $10^{-5}\,M_\odot$) hydrogen-rich material has been transferred from the main sequence component to the white-dwarf component. The hydrogen-burning runaway is triggered near the base of the accreted layer, where electrons are degenerate, and it develops into a (sometimes hydrodynamical) explosive event. The spectra of many novae suggest abundances of CNO elements which are much larger than those found at the surfaces of main sequence stars and this is most easily explained as a consequence of mixing of material from the underlying CO white dwarf into the accreted layer and of nuclear processing in this layer during the nova outburst. It is thought that essentially all of the accreted layer, as well as the matter mixed up from the interior, is lost during an outburst that is characterized by a spectrum exhibiting strong CNO features, and one has the picture that the white-dwarf component actually loses mass over the course of many cycles of quiet accretion interrupted by outburst. Still other novae exhibit evidence in their spectra for overabundances of neon and yet heavier elements, and this suggests that the composition of the underlying core includes these elements. Theoretical models of evolving binary stars suggest that cores composed of O and Ne (ONe white dwarfs) must be of mass larger than $1.1\,M_\odot$ and can be made only in a close binary from a component with an initial main sequence mass in the narrow range $\sim$8–10 $M_\odot$. This last statement leads naturally to the question of how cataclysmic variables are formed.

The answer is illustrative of a very basic characteristic of the theory of close binary star evolution in its current state of development. It is not a theory in the ordinary sense, consisting of a set of equations which summarize the necessary physics and which can be solved to provide numerical models. Rather, the theory is a crude picture, or "scenario," which makes use of broad principles of energy and angular momentum conservation and makes use of results of simple one-dimensional experiments to gauge the response of a spherically symmetrical model star to accretion of matter and to loss of matter. The alternative is to attempt to solve an extremely complex problem in three dimensional fluid dynamics with self gravity, turbulent viscosity, radiative and convective mass transport, and still other phenomena, suitable equations for which do not yet exist.

The first episode of mass transfer begins when, in consequence of the exhaustion of a nuclear fuel at its center, the initally more massive component tries to become a red giant, but on the way encounters its Roche lobe. As it expands outside of this lobe, matter flowing from its surface at first forms a disk about its companion. Thereafter, mass flows through the disk onto the surface of the companion. It is reasonable to assume that, at least during the initial phase of mass transfer, the total mass and orbital angular momentum of the system is conserved. The orbital angular momentum of the system depends on the mass ratio of the two components and on their separation in such a way that, as the mass donor loses mass, the orbital separation must decrease, forcing the donor to overfill its Roche lobe still further and at a more rapid rate, thus causing the mass-transfer rate to increase exponentially. The rate of transfer becomes so large that the disk is converted into a thick, hot, expanding layer which entirely encases the accretor. The faster the rate of mass transfer, the more rapidly the accreted layer expands. This situation must persist at least as long as the mass of the donor is larger than the mass of the accretor. However, if the initial mass ratio is large, long before such a situation is achieved, the accreted layer will have expanded to fill the Roche lobe of the accretor and the situation arises that both stars overfill their Roche lobes simultaneously. Matter has nowhere to go but out of the system.

At this point, the theory becomes fully a matter of scenario building. By now, the concepts of Roche lobes and of mass transfer have lost their usefulness and it becomes convenient to think in terms of a "common envelope". Matter in this expanding envelope is supplied by the internally more evolved component, which is still attempting to grow in consequence of interior structural changes forced by nuclear transformations at the edge of a compact, hydrogen-exhausted core, and both the underlying main sequence accretor and the compact core of the donor can be thought of as two parts of an "eggbeater" which engages in a dissipative, frictional interaction with the matter in the common envelope. The ultimate source of energy released in this frictional interaction is the energy of orbital motion of the stellar cores – the helium core of the primary and the main sequence secondary which supports a hot accreted layer – and some of the energy released goes into driving common-envelope matter out of the system. The process is envisioned to continue until essentially all of the initial hydrogen-rich envelope of the evolved component has been stripped off, forcing the remaining matter above the hydrogen-burning shell to contract to the dimensions of the remnant helium core. A reduction in orbital energy (an increase in orbital binding energy) during the common envelope phase means a reduction in the separation of the stellar cores during this phase, and this explains how the components of the system emerging from the common envelope phase – the hot helium core of the initial primary and the original main sequence component, perhaps slightly enhanced in mass – can be much closer together than were the components of the initial system.

The remnant helium star contracts, heating still further, and it eventually ignites helium at the center, adopting the characteristics of a subdwarf O or B star (an sdO or sdB star), depending upon how much hydrogen it has retained at its surface. Subdwarfs lie in the HR diagram between the main sequence and the region of white dwarfs. The sdO stars are in general hotter ($T_e > 40\,000$ K) than the sdB stars and their spectra show an overabundance of helium. The binary UU Sge, which is at the center of the planetary nebula Abell 63, is a good example of a system in this stage of evolution. UU Sge is an eclipsing binary with a period of 11.2 hr. One component is a main sequence star of mass about $0.7\,M_\odot$ and the other is an sdO star of mass about $0.8\,M_\odot$. The nebula is matter which was presumably once in a common envelope and the sdO star is hot enough ($\sim35\,000$ K) to explain the fluorescence of the nebula. One may estimate the mass of the precursor of the sdO star to be about $5\,M_\odot$.

Helium burning continues in the helium star until most of its interior has been converted into carbon and oxygen. The compact remnant then evolves into a CO white dwarf. In this way, the system has evolved into the immediate precursor of a cataclysmic variable. Examples of such a system are MT Serpentis and V 471 Taurus. MT Serpentis is the binary central star of the planetary nebula Abell 41. Its orbital period is 2hr 40min, one component is a near main sequence star of mass $\sim0.2\,M_\odot$, and the other component is a hot white dwarf with a surface temperature near $60\,000$ K and its mass is perhaps near $0.6\,M_\odot$. One may estimate the mass of the main sequence progenitor of the white dwarf to be $\sim3\,M_\odot$.

V 471 Taurus is in the Hyades supercluster, either at the center of the Hyades cluster or at a distance about 10 pc beyond the center of the cluster. The white dwarf component and its cooler near main sequence companion have about the same mass ($\sim0.7\,M_\odot$), and the radius of the cool component is about half the radius of its associated Roche lobe. The mass of the main sequence progenitor of the white dwarf is estimated to be between $2\,M_\odot$ and $4\,M_\odot$. The cool component spins at a period close to the orbital period of 12.5 hr and exhibits many signs of magnetic activity. It belongs to the class known as VAC (very active chromosphere) stars. Several lines of evidence indicate that it supports a stellar wind, and it is natural to assume that the same mechanism which maintains the cool component of a cataclysmic variable in Roche-lobe contact is forcing the components of V 471 Taurus closer together and that this system will ultimately become a cataclysmic variable when the cool component fills its Roche lobe. The direct evidence for mass loss is of a spectroscopic nature and the indirect evidence for mass loss is that the white dwarf component is brighter by a factor of about 30 than theoretical models would suggest a white dwarf of age $\sim10^9$ yr to be. Accretion of matter onto the WD from a wind emitted by the cool companion could account for this excess brightness. An accretion rate of only $10^{-13}\,M_\odot$ yr^{-1}, small compared with the expected wind mass-loss rate of $\sim10^{-9}\,M_\odot$ yr^{-1}, would be sufficient.

A potentially puzzling feature of the VAC component is that, if it is indeed 10 pc beyond the center of the cluster, it is currently brighter and larger than a main sequence star of the same mass, even though it is old enough to have evolved off the main sequence. Furthermore, this component varies in brightness by almost a factor of two on a fifty year time scale. It is possible that there is sufficient asynchronism between the orbital period and the spin period of the VAC star that tidally induced heating contributes significantly to the emitted light and/or that magnetic dynamo activity is particularly strong in this star, due in part to the action of tidal forces.

The degree of orbital shrinkage which occurs during the common envelope phase can be only crudely estimated. Invoking energy conservation and assuming that the energy needed to eject the matter in the common envelope is precisely equal to the change in orbital binding energy leads to an estimate of the degree of shrinkage, but some of the frictional energy generated will go into the directed kinetic energy of ejected material and some will be lost by radiation, so this estimate is only a lower limit if no other source of energy can be tapped to eject the envelope. Such an additional source may be available if the initial binary system is so wide that the primary does not fill its Roche lobe until it has become an AGB star. Then, it is possible that the gravitational potential about the AGB star, modified by the presence of a companion, triggers the planetary nebula ejection mechanism, so that the system thereby partially circumvents the common envelope process; in this case, the estimate of shrinkage based on the assumption of precise conversion from orbital binding energy into the energy necessary to drive common envelope material to infinity at zero velocity and zero temperature is an overestimate. The best one can do at the moment is to formalize the relationship between initial and final orbital separation by means of an arithmetic algorithm expressing the conservation of energy, parameterizing the uncertainty as to the exact flow of energy among different forms. Values for the parameters in the algorithm can be set by comparing with those individual systems for which orbital elements and global properties of components are known with some accuracy. Conflicting results for different systems then leads to a reassessment of the algorithm and to attempts to incorporate into this algorithm more subtle and realistic physical principles.

The properties of the class of stars known as Algols demonstrate that the initial mass transfer between close components does not always lead to a common envelope phase in which most of the primary's initial mass is lost from the system. In an Algol system, the more massive component lies on the main sequence and is brighter and bluer than its companion which, despite its smaller mass, has a more highly evolved interior structure, with an electron degenerate core made of helium and a hydrogen-burning shell. The companion, lying between the main sequence and the extended giant branch, is called a subgiant. Models of single stars show that the radius of an isolated subgiant does not depend sensitively on the stellar mass but varies roughly as the fourth power of the mass of the helium core. In typical Algols, the proximity of the subgiant to the main sequence suggests a core mass of the order of only $0.2 \, M_\odot$ or less. This estimate of core mass is possible because the subgiant fills its Roche lobe. A good example is, of course, the eponymous star, Algol itself. The period of this system is 1.87 days and the orbital separation is about $14 \, R_\odot$. The mass, luminosity, and radius of the main sequence component are $3.7 \, M_\odot$, $160 \, L_\odot$, and $2.7 \, R_\odot$, respectively. These same quantities for the Roche-lobe filling subgiant component are $0.8 \, M_\odot$, $6 \, L_\odot$, and $3.6 \, R_\odot$. The mass of the helium core of the subgiant can be estimated to be about $0.19 \, M_\odot$.

The simplest explanation of the Algol configuration is that the more evolved star was initially the more massive component of the primordial binary, that it filled its Roche lobe shortly after leaving the main sequence, and that essentially all of the mass which it lost was accreted by its companion. This is a remarkable conclusion. Assuming that the orbital angular momentum of the stellar system is conserved, the radius of the Roche lobe about the donor decreases until the initial component mass ratio is reversed. This means that mass transfer will proceed on a time scale intermediate between the thermal time scale of the donor envelope and the dynamical time scale of this envelope. Somehow, a common envelope which is driven from the system is avoided. Once the mass of the subgiant donor becomes less than about eight-tenths of the mass of the accretor, a solution for the system may be found on the assumptions that orbital angular momentum is conserved during subsequent mass transfer and that the agency which maintains the subgiant in contact with its Roche lobe is the expansion which accompanies growth of the hydrogen-exhausted core. However, the subgiant in most Algol systems is relatively close to the main sequence, so the mass of the helium core has not increased much beyond its value at the termination of the main sequence phase. Further, the subgiant is spinning about its own axis at a frequency synchronous with the orbital frequency and it shows signs of vigorous magnetic activity – it is a VAC star. These facts suggest that orbital angular momentum is being lost from the system via an MSW on a time scale which could be less than the time scale for the growth of the hydrogen-exhausted core of the subgiant, and that the MSW could be the primary agency driving mass transfer, rather than the growth of the helium core.

Some close members of the class of binaries known as RS Canum Venaticorum (RS CVn) stars will undoubtedly evolve into Algols. In RS CVn systems the brighter, cooler, and generally more massive component is a subgiant or giant which does not fill its Roche lobe and the dimmer component is a main sequence star. The cooler component shows strong signs of magnetic activity, including indications of solar "spots" (in analogy with sunspots) and strong X-ray emission, and is rotating rapidly, with a spin period close to the orbital period. This means that an MSW is operating and will force the primary in many systems to come into contact with its Roche lobe. Those systems in which the primary is much more massive than its companion or is a giant with a large core mass and a deep convective envelope, or both, when it fills its Roche lobe may experience a common-envelope episode and evolve into the short period central star of a planetary nebula and thence into a cataclysmic variable. Those in which the primary is of mass comparable to the mass of the secondary and is a subgiant with a small core mass and a shallow convective envelope when it fills its Roche lobe may evolve into Algols.

A nice example of a system of the latter type is UX Ari, only 50 pc from the Sun. Component masses are approximately the same at 1.07 $M_\odot$ and 0.93 $M_\odot$, and the slightly more massive component is redder and three times brighter than its main sequence companion. The orbital period of the system is 6.44 days and the orbital separation is about 18.5 $R_\odot$. The radius of the brighter, subgiant component is about 6 $R_\odot$, slightly smaller than the 7 $R_\odot$ radius of its Roche lobe. Comparing with theoretical models, one can deduce that the mass of the helium core of the subgiant component is about 0.17 $M_\odot$, close to the canonical value for Algol systems. The system should very shortly (in, say, 10^8 yr) evolve into an Algol system.

Thus far, attention has been focussed on systems in which the more evolved component either is or will become a white dwarf. However, if its mass is initially or grows to be in the range $\sim$10–40 $M_\odot$, that component will probably evolve into a neutron star, and if its mass exceeds $\sim$40 $M_\odot$, it may evolve into a black hole (the initial mass which separates neutron star and black hole precursors continues to elude theoretical determination). Several classes of system originate in close binaries in which one or both components evolve into a neutron star or black hole, both of which one may call relativistic remnants. These include high mass X-ray binaries (HMXBs), Be/X-ray stars, low mass X-ray binaries (LMXBs), some runaway OB stars, binary radio pulsars, and perhaps the majority of single radio pulsars.

Most members of these systems show above average peculiar space velocities which are a consequence of the recoil of the remnant system (whether it remains bound or becomes unbound) to the mass ejected in the supernova explosion which accompanies the formation of the relativistic component. Bound systems with relativistic components can exhibit large eccentricities (if the mass lost in the explosion is comparable to the mass remaining in the system after the explosion) or have almost zero eccentricity (if the companion to the relativistic remnant donates mass efficiently).

The optical companion of the relativistic star in the brightest HMXBs is a fairly evolved main sequence star (typically of mass $\sim$10–30 $M_\odot$) which does not quite (but almost) fill(s) its Roche lobe and supports a radiative stellar wind. The relativistic component can accrete matter from the wind at a rate up to $\sim$10^{-8} yr^{-1} and can emit energy at X-ray wavelengths at a rate up to 3×10^4 $L_\odot$, although typical rates are ten times less than this. The origin of the X-rays is, of course, the release of gravitational potential energy as matter flows onto the neutron star surface where it becomes thermalized at temperatures up to 1–2×10^7 K. Comparison with models of accreting neutron stars suggests that hydrogen and helium are being converted into iron peak elements in subsurface layers as rapidly as these elements are being accreted at the surface.

LMXBs are similar to cataclysmic variables (CVs) in that the lighter component can be a low-mass main sequence star filling or nearly filling its Roche lobe and, in this case, a MSW undoubtedly plays a role in the mass-transfer process. However, the donor in an LMXB can also frequently be a subgiant (which is not the case among CVs) and there are no counterparts among LMXBs of short period CVs (80 min to 2 hr) in which the donor is a completely convective main sequence star of mass less than $\sim$0.3 $M_\odot$. The first of these differences may be due to the fact that the relativistic component in an LMXB (of mass $\sim$1.4 $M_\odot$ if a neutron star and $\sim$10 $M_\odot$ if a black hole) can accrete stably from subgiant donors of mass larger than would be the case if its mass were similar to that of a white dwarf in a typical CV. The second difference may be related to the fact that a neutron-star component is spun up as it accretes matter and can evolve into a pulsar. The intense radiation from the pulsar can destroy the donor if it is a (loosely bound) main sequence star. This kind of evolution may account for single millisecond pulsars. In contrast, as it evolves to larger radii, the subgiant donor in long period LMXBs moves away from the neutron star component and, as the neutron star evolves into a pulsar, the dense helium core of the subgiant survives as a helium white dwarf. Binary millisecond pulsars are almost certainly products of this kind of evolution.

LMXBs experience bursts of X-ray emission, which, when compared with theoretical models, can be identified as thermonuclear in origin. A burst releases typically 10^{39} erg in 1–10 s and the time between bursts is from hours to days. This contrasts with classical CVs, which experience nova outbursts lasting months to years and for which the time between outbursts may be estimated to be of the order of centuries to tens of centuries. The differences in time scales are due to the fact that, for a given accretion rate, the much larger gravitational force acting on matter near the surface of an accreting neutron star produces a much larger rate of release of gravitational potential energy by matter being compressed in subsurface layers than in the case in accreting white dwarfs. This in turn leads to a much larger rate of increase in subsurface temperatures so that conditions for a thermonuclear runaway are achieved much sooner and after the accretion of much less material in the accreting neutron-star case (only $\sim 10^{-13}$ $M_\odot$ instead of $\sim 10^{-5}$ $M_\odot$). Another striking contrast between thermonuclear outbursts experienced by LMXBs and by classical CVs is that very little matter can be ejected by LMXB bursters. The reason is very simple. The amount of energy released by converting one gram of hydrogen into one gram of iron is only one percent of the rest mass energy contained in that gram, while the energy required to impart escape velocity to one gram of material at the surface of a neutron star is about 10 percent of the rest mass energy.

The flux of X-ray energy from the neutron star in LMXBs is so large that, if it were not shadowed from this radiation by the accretion disk, the donor would absorb energy at a rate which is several orders of magnitude larger than the rate at which it produces energy by nuclear burning in its interior. There is the distinct possibility that the X-ray energy which does strike the donor causes it to form a hot corona which emits a strong wind. This opens the possibility that the neutron star can accrete matter from the donor wind and that the donor need not precisely fill its Roche lobe for mass transfer to occur. Since most of the mass lost in a wind escapes the system, this means that the lifetime of a typical LMXB may be considerably shorter than that of a typical CV at periods larger than 3 hr.

The evolution of wide binaries into evolved systems with either a relativistic component or a white dwarf component are similar in the sense that mass loss from the primary (planetary nebula ejection for low and intermediate mass primaries and a supernova explosion for high mass primaries) leads to a widening of the orbit. The minimum mass of an isolated (single or in a wide binary) progenitor of a neutron star is ~ 10–11 $M_\odot$. Since the baryonic mass of a neutron star is about 1.5 $M_\odot$, one may anticipate the loss of at least ~ 8–9 $M_\odot$ of matter from a wide binary system during the supernova explosion which leaves the neutron star as a remnant. In order for the system to remain bound following the explosion, the mass of the remnant system, a neutron star and the original main sequence secondary, must be larger than the mass of material that is ejected in the supernova explosion. Thus, the mass of the secondary, main sequence star must be at least as large as 7–8 $M_\odot$ and typically even larger.

On the other hand, in a close binary, when it fills its Roche lobe after having developed a non-degenerate helium core, the primary will continue to lose mass until its helium core (capped by a thin hydrogen-rich layer) is exposed. Whether the mass lost by the primary is transferred conservatively to the secondary or whether this mass is lost from the system through a common envelope depends on the details (the initial mass ratio, whether or not the primary develops a deep convective envelope before filling its Roche lobe, and so on). In any case, the mass of the helium-star remnant of the primary is substantially smaller than that of the initial primary, and this reduces considerably the minimum mass of the secondary in order for the system to remain bound after the formation of the neutron star.

It is for this reason that one expects binaries beginning as two massive main sequence stars to survive the first supernova explosion as a runaway OB star (orbited by a neutron star) which may evolve into a persistent HMXB or transient Be/X-ray source.

It is also this first mass-transfer/mass-loss episode prior to the supernova explosion that makes the formation of LMXBs possible. In rough approximation, the mass M_{He} of the helium star into which the primary evolves is related to the mass M_{MS} of the initial primary by $M_{He} \sim 0.1\, M_{MS}^{1.4}$. For example, the helium-star remnant of a primary of mass $M_{MS} \sim 11\, M_\odot$ has a mass $M_{He} \sim 2.9\, M_\odot$, and, when the helium star explodes, the mass lost from the system is only $M_{lost} \sim 1.4\, M_\odot$, allowing the stellar system to remain bound even when the mass of the secondary is as small as $0.1\, M_\odot$ or even smaller. Similarly, if the secondary is of mass $\sim 1\, M_\odot$, the system will remain bound even when the initial mass of the primary is as large as $M_{MS} \sim 13.7\, M_\odot$, corresponding to a helium star mass of $M_{He} \sim 3.9\, M_\odot$. In summary, LMXBs with main sequence donors which support a MSW (i.e., of mass in the range 0.3–$1.0\, M_\odot$) come from systems in which the primary is a star of mass in the range 11.9–$13.7\, M_\odot$. The large initial primary-to-secondary mass ratio ensures that the mass-transfer episode will be highly non-conservative and that orbital shrinkage in a common envelope episode will be sufficient to drive systems close enough that the potential donor moves into (near) Roche-lobe contact by the action of an MSW.

In the dense stellar environment of globular clusters, there are other channels for the formation of LMXBs. If the distribution of primordial binaries over orbital parameters in such clusters is not radically different from the distribution in the field, then one expects the presence of many close binary systems in which both components are main sequence stars of mass less than $0.8\, M_\odot$, the maximum initial mass of a star which is still on the main sequence. Presumably there are also neutron stars in globular clusters, left over from the time when such a cluster was in the active phase of star formation and was still making massive stars which evolve into type II supernovae. Numerical experiments which examine the consequence of bombarding a binary system with another star whose mass is larger than that of either component in the binary show that, for sufficiently small impact parameters, the more massive star will replace one (usually the lightest) of the original binary components. Thus, three body collisions in the high stellar density environment of a globular cluster may occasionally produce a binary consisting of a neutron star and a low-mass main sequence star. In consequence of an MSW, GWR, or a hardening collision with another star, this binary may evolve into an LMXB.

Another possible way of forming LMXBs is by two-body interactions. On passing very close to an unattached main sequence star, a neutron star will raise tides in this relatively large star and tidal torques will induce differential motions which result in frictional dissipation; the ultimate source of energy for raising the tides and for supplying the energy lost by dissipation is the relative kinetic energy of the interacting pair. A close enough passage (distance of closest approach less than about three times the radius of the main sequence star) can lead to sufficient energy transfer into raising tides and into heat that the interacting pair becomes bound. Estimates indicate that the probability of forming an LMXB by this means is large enough to account for the majority of the observed LMXBs in clusters.

It has been suggested that some precursor LMXB binaries have been formed in globular clusters and that either they have escaped from the cluster or the parent cluster has been disrupted after too many close passages through the galactic bulge, where most of the LMXBs not currently in clusters are found. A very small minority of astronomers believes that exchange collisions between main sequence binaries and single neutron stars and tidal captures of main sequence stars by neutron stars, even at the comparatively low stellar densities in the bulge, can account for the LMXBs there.

Some stars appear to be born close enough to each other that, thanks to the action of a MSW or of GWR, or both, both stars have been brought into contact with their Roche lobes before either has left the main sequence. Such stars are known as W Ursa Majoris (W UMa) stars and no fully satisfactory theoretical model to account for their properties has yet been devised. Since both stars fill their Roche lobes, they are highly distorted and deeply eclipsing, and it is therefore difficult to define and assign individual luminosities and surface temperatures. Nevertheless, attempts can be made and it appears that, relative to main- sequence stars of comparable mass, the more massive component in a W UMa system is underluminous for its mass and the less massive component is overluminous. Furthermore, the less massive component is about the same size as a main sequence star of the same mass. Typical component masses are less than 1 $M_\odot$ and typical surface temperatures are 5000 K. Actually, both components have about the same surface temperature, which can be interpreted as the temperature of a common envelope that is not expanding outward away from the system and is therefore not to be confused with the common envelope invoked to understand the formation of evolved binaries such as cataclysmic variables. Both components must be forced by tidal torques to spin at the orbital frequency and a strong MSW must be operating to abstract orbital angular momentum from the system. Orbital periods can be short enough that angular momentum loss by GWR also plays a role in the evolution of the system.

A possible interpretation of the observed characteristics of W UMa systems is that (1) in response to orbital angular momentum loss by an MSW from the common envelope, the lighter component is transferring mass to its companion, causing the rate of nuclear energy production in the interior of this companion to increase, and (2) some of the energy produced in the interior of the more massive component flows into the envelope of the lighter component before emerging from its surface, and the extra flow of energy through this envelope forces it to swell to a size larger than the envelope of a main sequence star of the same mass. It is difficult to see how this scenario could be viable unless, prior to the initiation of mass transfer, the presently lighter component has developed significant inhomogeniety in composition due to the conversion of hydrogen into helium, perhaps even to the extent of having developed a hydrogen-exhausted core. Otherwise, one might expect the direction of mass transfer to oscillate back and forth. In this connection, it is worth pointing out that low mass stars do not evolve rapidly away from the main sequence toward the giant branch until hydrogen has been exhausted over a mass slightly larger than ten percent of the initial mass of the star, so a star in the observationally defined main sequence band can still contain a highly evolved core.

Circumstantial evidence for this picture of W UMa evolution is provided by the simultaneous existence in old galactic disk clusters of W UMa systems and of stars known as blue stragglers. Blue stragglers are main sequence stars of mass significantly larger than the masses of stars near the dominant cluster turnoff in the HR diagram. Stars defining the turnoff are, of course, in the process of leaving the main sequence on the way to the giant branch, and the only way to account for the presence of blue stragglers in terms of single-star evolution is to suppose that the stragglers were formed at times substantially later than when the bulk of the cluster stars (including turnoff stars) were formed. The difference in formation times demanded by this picture amounts to many galactic rotation periods (i.e., many times 2×10^8 yr) and it is difficult to see how enough gas could be retained by a cluster for this length of time.

On the other hand, the picture that has been painted here to describe the evolution of W UMa stars provides an attractive alternative explanation for the blue stragglers. Suppose that all stars in a given old cluster, including many close binaries, were formed over an interval of time short compared with the age of the cluster. Among the binaries formed there should be some which are close enough (and cool enough) that an MSW will force the more massive component into contact with its Roche lobe in a time short compared with the main sequence lifetime of this component but long compared with the time required for a substantial chemical inhomogeneity to develop in its interior. Once Roche-lobe contact is made, the system develops into a W UMa star. If the sum of the masses of the component stars is greater than the mass of a turnoff star, and if the process of mass transfer can continue until the donor no longer exists and on a time scale short compared with the lifetime of a main sequence star of mass equal to the total mass of the initial system, then the result will be a single blue straggler. A very nice example of a system which may be far advanced in the merger process is the star AW UMa in the group HR 1614. Both components of this system fill their Roche lobes and component masses are approximately 1.3 $M_\odot$ and 0.1 $M_\odot$, respectively. The more massive component is on the main sequence, well to the blue of the group turnoff. That is, it is already a blue straggler, albeit in a binary. The age of the group is about 5×10^9 yr and the mass of a star at turnoff is about 1.2 $M_\odot$. One may estimate that the merger of the two components of AW UMa will be completed in another 10^8 yr, the result being a single blue straggler approximately two times brighter than stars near cluster turnoff.

Having examined the characteristics of several classes of close binaries in which only one of the components is an evolved star, and having understood these characteristics at least qualitatively as consequences of (or, in several instances, as a prelude to) a Roche-lobe overflow episode, it is reasonable to next examine the characteristics of several binaries containing two evolved stars, anticipating that these characteristics can be understood as consequences of a second Roche-lobe overflow event, when the initially lighter component exhausts a nuclear fuel at its center and grows to fill its Roche lobe. Among the many exotic systems whose past history one would like to understand is the famous binary pulsar PSR 1913+16, (probably) two neutron stars which are orbiting each other once every eight hours in a highly eccentric orbit ($e = 0.62$) and semimajor axis of $\sim 3 R_\odot$. The pulsar component (of mass $\sim 1.44 M_\odot$) is surely a neutron star and the other, unseen component is of mass ($\sim 1.38 M_\odot$) so close to the Chandrasekhar mass, the upper limit on the mass of a white dwarf, that it is also probably a neutron star. The main sequence progenitors of both components must have been relatively massive ($> 10 M_\odot$) and, therefore, the orbital separation of the initial main sequence precursor pair must have been larger than $20 R_\odot$ (and probably much larger). Once again, it is clear that considerable orbital shrinkage must have accompanied the formation of one or both of the compact stellar remnants, implying at least one episode of mass loss involving a common envelope. One can actually go a step further, using the algorithm for orbital shrinkage developed for understanding cataclysmic variable formation and using the theorem relating the mass lost in a supernova explosion to the mass of the remnant binary system, to construct detailed scenarios. One such scenario begins with two main sequence stars of mass $20\,M_\odot$ and $17\,M_\odot$ at an orbital separation of $600\,R_\odot$. Models suggest that the primary will fill its Roche lobe when it has developed a helium core of mass about $5\,M_\odot$ and the orbital shrinkage algorithm then suggests that the semimajor axis of the system will be reduced by a factor of about 4 in a common envelope event. The first star, now of mass $5\,M_\odot$, eventually explodes as a supernova, leaving a neutron star remnant. The orbital separation increases in consequence of the explosive loss of mass from the system. Then the second star develops a helium core of mass about $4\,M_\odot$ and grows to fill its Roche lobe. Another common envelope episode occurs, leaving a $4\,M_\odot$ helium star which grows to fill its Roche lobe. Another common envelope episode occurs, leaving a highly evolved remnant in orbit about the first neutron star at a distance of about $2\,R_\odot$. Finally the second star explodes, leaving a second neutron star, and the average orbital separation approximately doubles to about $4\,R_\odot$. The loss of angular momentum due to GWR over the course of about 4×10^9 yr then shrinks the average size of the orbit to the current $3\,R_\odot$ average separation.

The observed rate of decrease in the orbital period of the current system provides a quantitative demonstration that the gravitational wave radiation predicted by Einstein's general theory of relativity occurs at precisely the predicted rate. In approximately one billion years, the components will be brought so close to one another (less than 20 km if the unseen component is a neutron star) that a merger must occur. Whether the consequences of that merger will be the formation of a black hole or will be an explosion of the magnitude of those that accompany the formation of a neutron star has yet to be deduced from theory.

If two neutron stars can form in a tight orbit from an initially much wider and much more massive main sequence pair, it seems reasonable to suppose that two white dwarfs can also form in a tight orbit from an initial pair of low or intermediate mass main sequence components. There are now several established cases of white dwarfs in orbits with periods of about a day or longer. The discovery of these relatively close white dwarf pairs has been a response to theoretical predictions of their occurrence. These systems must emit gravitational wave radiation, but the systems found so far are not close enough for a merger to occur until a time much longer than the age of the Universe has elapsed.

Nevertheless, there are many reasons to suspect that a good fraction of all low-mass main sequence binaries evolve into pairs of helium white dwarfs which are close enough that loss of orbital angular momentum due to the emission of gravitational waves will lead to mergers in a much shorter time. One can explore the possible consequence of such mergers by means of numerical experiments in which fuel of a given type is accreted rapidly onto an initially cold white dwarf model. For example, the rapid accretion of helium onto a helium white dwarf model results in the ignition of helium at the base of the accreted layer; helium burning works its way toward the center of the model in a series of about a dozen flashes and then the model embarks on a long phase of quiescent core helium burning. From this experiment, one can infer that the merger of two real helium white dwarfs may produce an sdO or sdB star. Whether the spectrum of the real merger will be hydrogen-rich (sdB) or helium-rich (sdO) depends, among other things, on how much of the hydrogen remaining near the surfaces of the precursor white dwarfs survives burning during the merger and diffuses to the surface of the final product. The expected frequency of occurrence of mergers of helium white dwarfs is, if anything, somewhat large compared with the frequency of occurrence of sdO and sdB stars. This is welcome news, as it is very difficult to understand how single stars could evolve into sdO or sdB stars at the observed frequency.

Further evidence for the merger scenario is the star Eridani B, one of the five nearby white dwarfs for which a mass can be determined because it is in a binary system (it is actually in a triple system, with a distant third component) with well-determined orbital parameters. The mass of Eridani B is $0.43 \pm 0.02 \, M_\odot$, and this is significantly smaller than $\sim 0.55 \, M_\odot$, the mass of the lightest white dwarf which a single star evolving off the main sequence within the lifetime of the Universe can produce. A straightforward scenario to explain the facts begins with the assumption that Eridani B was once itself a binary system consisting of two low-mass main sequence stars at a separation such that the more massive component filled its Roche lobe after developing an electron-degenerate helium core of mass about $0.2 \, M_\odot$. If the initial primary–secondary mass ratio was not too large (the RS CVn star UX Ari is a good example of a possible precursor), the system might be expected to have evolved into an Algol system (systems such as TW Dra, AW Peg, and RY Aqr are good examples). Eventually, the initial primary would have lost most of its hydrogen-rich envelope and shrunk within its Roche lobe to become a helium white dwarf of mass about $0.2 \, M_\odot$. The more massive component would exhaust hydrogen over a region of order $0.2 \, M_\odot$ ($\sim 10 \, \%$ of its total mass), swell to fill its Roche lobe, and begin sending matter over to its compact white dwarf companion. Because of the large component mass ratio, the large amount of gravitational potential energy liberated as matter accelerated onto the surface of the white dwarf, and the long thermal time scale of the WD, a common envelope would form and the final result would be two close helium white dwarfs of comparable mass. The emission of GWR might then bring the lighter of the pair into Roche-lobe contact, with a merger and transformation into an sdB star being the immediate result. After $\sim 10^8$ yr of nuclear burning, the sdB star would evolve into a white dwarf, reaching the luminosity of Eridani B in another 10^8 yr.

One can envision scenarios which produce much more spectacular results than a single white dwarf of mass smaller than can be produced by a single star, or even two neutron stars in a tight orbit. A particularly exotic scenario is one which produces a type Ia supernova. Type Ia supernovae are a rather uniform class, characterized by a maximum luminosity of about $10^{10} \, L_\odot$, a total integrated light intensity of about 10^{51} erg, a lack of evidence for either hydrogen or helium in their spectra, evidence for outflowing matter with velocities $\sim 10^4$ km s^{-1}, and a late-time light curve which exhibits an exponential decline with a half life of about 77 days, suggesting that energy release by some radioactive species made in the initial explosion may be responsible for powering the light output. Type Ia supernovae occur in both spiral and elliptical galaxies, but the fact that they occur in elliptical galaxies is particularly instructive because type II supernovae do not occur in elliptical galaxies. The absence of massive star presursors of type II supernovae indicates that star formation in ellipticals has ceased, implying an upper limit on the mass of stars which are currently still burning nuclear fuel. From the distribution of stars in the HR diagram, one infers that nuclear burning stars in elliptical galaxies are in fact very old, of mass less than $1\,M_\odot$. Hence, the current stellar content of elliptical galaxies consists of (1) low mass stars which require of the order of a Hubble time ($\sim 14 \times 10^{10}$ yr) or more to evolve into white dwarfs and (2) the compact remnants of initially more massive stars: white dwarfs, neutron stars, or black holes.

Some of the compact remnants must be in binary systems. The double star scenario to account for type Ia supernovae begins with two intermediate-mass main sequence stars which interact in a series of four common envelope events to produce two close CO white dwarfs. If the mass of each of the precursor main sequence stars is in the range 5–8 $M_\odot$, then the total mass of the white dwarf pair will be larger than the Chandrasekhar mass, which is about $1.4\,M_\odot$ and which is the maximum mass for which a static model of a single white dwarf can be constructed. If the initial system is close enough, then the white dwarf pair will be close enough that GWR will drive the two together until the lighter of the two fills its Roche lobe and a merger will result.

What happens next is still by no means clear. However, the consequences of one-dimensional hydrodynamical experiments are very suggestive as to the possible outcome. Adding mass to a model CO white dwarf of mass slightly less than the Chandrasekhar mass causes this model to contract and heat until carbon is ignited explosively at the center. Enough of the energy released by carbon burning goes into heat that the products of carbon burning also ignite. A nuclear-burning front develops and works its way outward through the model. In the front, as it passes through the inner half of the model, nucleosynthesis proceeds all the way to the formation of iron-peak elements. Particularly large quantities of the isotope ^{56}Ni are made, as this is the isotope with the largest binding energy per nucleon under conditions in the deep interior of the white dwarf model. At densities found in terrestrial laboratories, ^{56}Ni is unstable to decay into ^{56}Co, and thence into ^{56}Fe, but at densities as large as 10^9 g cm^{-3} the electron Fermi energy is larger than the energy involved in the beta decays, so these decays cannot occur. Most of the nuclear energy released by nucleosynthesis in the front is converted into kinetic energy and matter left behind by the front expands outward at velocities of about 10^4 km s^{-1}. As the front passes through the outer half of the model, it weakens, and nucleosynthesis does not extend as far as the iron peak elements. As matter expands outward, densities decline until the ^{56}Ni in the model begins to decay into ^{56}Co and ^{56}Co then decays into ^{56}Fe, producing a light curve with a 77 day half life. Thus, almost all of the characteristics of Type Ia supernovae are reproduced by the model. Not only is there no hydrogen or helium to see, but the energetics and time scales fit extremely well with the observations. It is even possible to match the complicated spectra seen as the matter in the model expands to such low densities that products of nucleosynthesis made in the interior are revealed. If real type Ia supernovae produce as much iron as the models which match their properties do, then it is clear that type Ia supernovae are the major source of iron peak elements in the Galaxy.

Presented with the remarkable coincidence between the properties of exploding white dwarf models and the observed characteristics of Type Ia supernovae, it is disappointing to realize that convincing direct evidence does not yet exist for a link between binary stars and the formation of an electron-degenerate configuration of mass greater than the Chandrasekhar mass. Part of the problem is, of course, that white dwarfs are in general so dim that one cannot search much of the Galaxy for them. This contrasts with the situation with regard to evolved binary systems containing a pulsar or a mass-accreting X-ray binary, as such systems can be seen throughout the Galaxy. Another problem is that the lifetime of a pair of degenerate dwarfs prior to merger is in general very short. This life-time depends on the fourth power of the orbital separation and is approximately 10^{10} yr when the orbital period is 1 day. Suppose that white dwarf pairs of characteristics appropriate for producing supernovae were formed with initial periods between one hour and one day at a rate that would explain the observed rate of occurrence of type Ia supernovae in external galaxies similar to our own. By examining the time evolution of envisioned precursor systems formed at this rate and in this distribution, one may show that only one out of perhaps 2000 white dwarfs might be expected to currently be in a supernova-precursor system. Given that relevant properties of only a few thousand white dwarfs are known, a detailed search for immediate precursors among the known white dwarfs would not appear to be particularly promising. Nonetheless, the hope is there, and the search must continue.

In summary, mankind has learned a tremendous amount about single and binary star evolution, but there is still a long way to go before the whole story has been understood. The most pressing problem is to raise the level of the theory of binary star evolution above its current rudimentary state. This means facing up to the task of setting up the fully three-dimensional gravo-fluid dynamics equations and solving them with some precision. There will, as always, be some of the physics which must remain parameterized, and in this case, an important parameter will be the one which characterizes turbulent viscosity. Because of this, those who attempt solutions should also develop an intimate familiarity with individual binary systems, as these will provide the essential clues for choosing appropriate parameter values in the theory and for selecting from among many effects those which are the really important ones for influencing the course of evolution.

Bibliography and references

Walter Baade & Fritz Zwicky, *Phys. Rev.*, **45**, 138, 1934.
Richard L. Bowers & Terry Deeming, *Astrophysics I. Stars*, (Boston: James Bartlett), 1984.
James Chadwick, *Nature*, **129**, 312, 1932.
Icko Iben, Jr., *Quart. J. Roy. Astr. Soc.*, **26**, 1, 1985.
Icko Iben, Jr., *ApJ*, **353**, 215, 1990.
Icko Iben, Jr., *ApJS*, **76**, 55, 1991.
Icko Iben, Jr., Physics Reports, **250**, 1, 1995.
Icko Iben, Jr. & Mario Livio, PASP, **105**, 1373, 1993.
Icko Iben, Jr. & Alvio Renzini, Physics Reports, **105**, 330, 1984.

Icko Iben, Jr., & Alexander V. Tutukov, *ApJ*, **511**, 324, 1999.

Icko Iben, Jr. & Alexander V. Tutukov, *Sky and Telescope*, **94**, No. 6, 36, December, 1997; **95**, No. 1, 42, January, 1998.

Icko Iben, Jr. & Alexander V. Tutukov, APJS, **54**, 335, 1984; **58**, 661, 1985; **105**, 145, 1996.

Icko Iben, Jr., Alexander V. Tutukov, & Lev Yungelson, *ApJS*, **100**, 217 & 233, 1995.

William J. Kaufmann, III, *Universe* (New York: Freeman), second edition, 1988; third edition, 1991.

Lev Landau, private communication, 1934.

J. Robert Oppenheimer & G. R. Volkov, *Phys. Rev.*, **55**, 324, 1939.

Cecilia Payne-Gaposhkin, *Stars and Clusters* (Cambridge, Mass: Harvard University Press.) 1979.

Deana Prialnik, *An Introduction to the Theory of Stellar Evolution* (Cambridge: Cambridge University Press), 2000.

Frank H. Shu, *The Physical Universe*, (Mill Valley, Ca.: University Science Books), 1982.

2   Quantitative foundations of stellar evolution theory

The characteristics of the Sun and of other bright stars for which distances can be estimated constitute the major observational foundation of the disciplines of stellar structure and stellar evolution. The Sun's basic global characteristics, which provide natural units for cataloguing the global characteristics of other stars, are descibed in Section 2.1. Properties of some bright stars in familiar constellations and of some nearby stars for which masses have been estimated are described in Section 2.2, with several of the more familiar stars being shown in the Hertzsprung–Russell (HR) diagram, where a measure of intrinsic brightness (luminosity) is plotted against surface temperature (color). It is evident that nearby and/or visually bright stars form distinctive sequences in the HR diagram.

Mass and luminosity estimates for stars in relatively wide binary systems as well as for those in close, but detached systems are presented in Section 2.3. Comparing these estimates in a mass–luminosity (ML) diagram, one may infer that, for systems in which proximity does not imply mass transfer between components or significant tidal interaction, the relationship between mass and luminosity for either component is not greatly affected by the presence of a companion. Comparing the locations of individual stars in the HR and ML diagrams, one can infer that stars probably evolve between sequences in the HR diagram.

In Section 2.4, the evolution of the interior and global characteristics of theoretical stellar models is sketched and, in Section 2.5, the theoretical results are employed to interpret the significance of the different branches defined in the HR diagram by nearby and/or visually bright stars and to identify the evolutionary status of familiar stars in the night sky.

2.1   Observed properties of the Sun

Although there is more known quantitatively about the Sun than about all other stars put together, this knowledge tells us very little directly about the future evolution of the Sun and does not provide very strong constraints on the theory of stellar structure and evolution. One can learn more about the past and future history of the Sun and about the principles underlying stellar evolution by studying other stars. However, the Sun does provide the basic yardsticks for classifying observable properties of other stars and provides clues as to the essential physics occurring within it and within other stars and as to the approximations one might introduce in constructing a first-order theory of stellar structure and evolution.

The most directly observable global characteristics of the Sun are its radius and luminosity. From measurements of the Sun's angular size coupled with the mean distance $d_\odot$ to the Sun measured by a variety of methods to be

$$d_\odot = 1.496 \times 10^{13} \text{ cm}, \qquad (2.1.1)$$

the mean radius of the Sun is determined to be

$$R_\odot = 6.96 \times 10^{10} \text{ cm}. \qquad (2.1.2)$$

The luminosity of the Sun, defined as the time-averaged rate at which the Sun radiates energy, is given by multipying the flux of energy reaching the Earth by the surface area of a sphere having a radius $d_\odot$ and is found to be

$$L_\odot = 3.86 \times 10^{33} \text{ erg s}^{-1}. \qquad (2.1.3)$$

Fluctuations about this value on the time scale of human civilization may have been no more than a fraction of a percent.

The constant G in Isaac Newtons's law of gravitational attraction (*Principia Mathematica*, 1687) is

$$G = 6.673 \times 10^{-8} \text{ cm}^3 \text{ g}^{-1} \text{ s}^{-2}, \qquad \cdot \qquad (2.1.4)$$

not unlike the estimate first obtained by Henry Cavendish over 100 years after the appearance of the *Principia*. Cavendish reported his results as an estimate of the Earth's mass and mean density and it was not until 1873 that A. Cornu and J. B. Baile (1873) undertook to evaluate G explicitly from Cavendish's data. It was about 20 years later that, on reporting results of his own experiments to determine G directly, C. Vernon Boys (1894) emphasized that the fundamental result of Cavendish's experiments and of others like it is a determination of G, rather than a determination of the mass and density of the Earth. An extensive history of these matters is provided by John Henry Poynting in *The Mean Density of the Earth* (1984). Using the length of the year and measurements of the Earth's orbital path about the Sun, it follows from Newton's theory of gravitation that the mass of the Sun is

$$M_\odot = 1.991 \times 10^{33} \text{ g}. \qquad (2.1.5)$$

Radioactive dating techniques applied to meteoritic matter lead to an estimated age of the Solar System, and therefore of the Sun, of

$$t_\odot = 4.6 \times 10^9 \text{ yr}. \qquad (2.1.6)$$

Several additional characteristics of the Sun provide the foundation for the assumptions and approximations which are useful for constructing an initial theory of stellar structure and evolution. If one does not look too closely at the detail provided by solar telescopes, and ignores what one might call the weather above its relatively stable photosphere, the Sun appears to be quite spherical. Direct visual observation of the rising and setting Sun without the aid of a telescope demonstrates sphericity at wavelengths accessible to the human eye. This evidence suggests that it is quite reasonable to construct stellar models which are spherically symmetric.

The Sun emits an evaporative wind, but the mass-loss rate associated with this wind is only of the order of $10^{-14} M_\odot$ yr^{-1}. Thus, in good approximation, the Sun is of constant mass, and one may adopt as an initial working hypothesis that other single stars are also nearly of constant mass. It turns out that this is a safe hypothesis for most stars during most nuclear-burning stages of their lives, but it is dramatically inappropriate for very massive and luminous stars which support strong radiative winds and for low mass and intermediate mass stars which, during a brief stage near the end of their nuclear burning lives, become very luminous and spatially large and support strong evaporative winds or, if they also pulsate acoustically, form shock-inflated photospheres and lose mass due to radiation pressure on graphite grains.

The fact that, as determined from radioactive dating of fossils, life has existed on the Earth for $\gtrsim 3.6 \times 10^9$ years, coupled with the fact that, for life to exist as we know it, the mean temperature on the surface of the Earth cannot have varied much outside of the range 0–100 C, tells us that the Sun's luminosity cannot have varied by more than perhaps 25% over this time. In first approximation, then, one may construct models which do not require a consideration of dynamical motions and evolve quasistatically. Finally, the Sun rotates very slowly about its axis approximately once a month, suggesting that the dynamical effects of rotation in the Sun's interior are not very pronounced. Thus, one can speak initially of spherically symmetric, constant-mass models in hydrostatic or near hydrostatic equilibrium.

The number of baryons in the Sun is

$$N_{\text{baryons}} = \frac{M_\odot}{M_{\text{proton}}} = \frac{1.991 \times 10^{33} \text{ g}}{1.67 \times 10^{-24} \text{ g}} \sim 1.2 \times 10^{57}. \tag{2.1.7}$$

Combining this number with the solar age and the solar luminosity, and assuming that the Sun's luminosity has been roughly constant over its lifetime, one has that, averaged over the Sun's lifetime, approximately 0.3 MeV has been liberated for every baryon. This is persuasive evidence that nuclear energy is being liberated in the Sun's interior and that nuclear energy is the dominant contributor to the outward flow of energy that determines the run of temperature in the stellar interior. The generation of nuclear energy implies that nuclear transformations are occurring and one can anticipate that these transformations are perhaps the primary driver of stellar evolution, at least until all nuclear fuels have been exhausted.

More information about the interior comes from a consideration of the spectral distribution of light from the surface. Perhaps the most striking aspect of this distribution is the fact that it approximates that of a black body, demonstrating that matter and radiation are in near thermodynamic equilibrium in the photosphere, despite the fact that the mean free path of a typical photospheric photon is comparable to the thickness of the photosphere. Given this fact, one may expect that, in the deep interior where the mean free path of a typical photon is many orders of magnitude smaller than a temperature or pressure scale height, local thermodynamic equilibrium prevails to a very good approximation. This circumstance simplifies enormously the calculation of energy transport in the interior.

The flux of radiation emitted by a laboratory furnace containing black body radiation at temperature T is given by

$$\frac{c}{4} a T^4 = \sigma T^4, \tag{2.1.8}$$

where c is the velocity of light, σ is the Stefan–Boltzmann constant given by experiment to be

$$\sigma = 5.6704 \times 10^{-5} \text{erg}^{-1} \text{K}^{-4} \text{cm}^{-2}, \tag{2.1.9}$$

and aT^4 is the energy density of the radiation field in erg cm^{-3}. This same result (see Section 4.11) can also be obtained from a consideration of the thermodynamics of electromagnetic radiation coupled with an experimental determination of Boltzmann's constant,

$$k = 1.38045 \times 10^{-16} \text{erg K}^{-1}, \tag{2.1.10}$$

which has the significance that the average energy of a particle in a perfect gas is $\frac{3}{2} kT$.

The flux of energy from the Sun's photosphere is given by the Sun's luminosity divided by its surface area. Thus, using eq. (2.1.8) for flux in terms of temperature, one can define the Sun's surface or "effective" temperature $T_e = T_\odot$ by the equation

$$\sigma T_\odot^4 = \frac{L_\odot}{4\pi R_\odot^2}. \tag{2.1.11}$$

Using in this equation the radius and luminosity given, respectively, by eqs. (2.1.2) and (2.1.3) produces

$$T_\odot = 5783 \text{ K}, \tag{2.1.12}$$

or

$$kT_\odot \sim 0.5 \text{ eV}. \tag{2.1.13}$$

A similar temperature follows from the fact that, for a black body,

$$(\lambda T)_{\text{max intensity}} = 0.288 \text{ cm K}, \tag{2.1.14}$$

where the wavelength λ is taken at the peak in the best fit black body curve to the solar intensity versus wavelength distribution. Since, for the Sun, the peak occurs in the yellow at $\sim$5000 Å, this gives

$$T_\odot \sim 5760 \text{ } K. \tag{2.1.15}$$

The fact that the surface temperatures estimated in two quite different ways agree so well reinforces the surmises that (1) near the surface, matter and radiation are approximately in thermodynamic equilibrium, even though the radiation field there is clearly not isotropic, and (2) in the deep interior, thermodynamic equilibrium is an even better approximation.

The presence and strength of both emission and absorption lines at wavelengths corresponding to transitions between energy levels in known atomic species demonstrates that matter at the surface of the Sun is gaseous and permits additional estimates of photospheric temperatures to be made. Because energy is flowing outward, it is clear that temperature decreases outwards; nevertheless, the mean temperature where most of the absorption lines are formed is relatively close to the surface temperature T_e. Line strengths may also be used in conjunction with simple atmospheric models to estimate the relative abundances of the various atomic species. In this way it is found that the most abundant element in the Sun is hydrogen (ionization energy = 13.6 eV) and that helium (ionization energies of 24 eV and 54 eV) is also prominent. However, since kT_e is much less than the ionization energy for these two elements, a quantitative estimate of their relative abundances cannot be made from an analysis of photospheric lines. On the other hand, many of the elements heavier than helium have low enough first ionization potentials relative to kT_e to permit accurate abundance estimates relative to hydrogen.

Quantitative solar abundance determinations began with the work of H. N. Russell (1929), who estimated abundances of 56 elements in the Sun's photosphere, concluding that "The calculated abundance of hydrogen in the Sun's atmosphere is almost incredibly great." At the time, it was difficult to imagine a solar distribution of abundances so different from the distribution of abundances in terrestial matter, and Russell conservatively underestimated the hydrogen abundance as being only 60% "by volume" of solar matter. Subsequent estimates have supported the "incredibility" of the hydrogen abundance in the Sun, as well as in other stars.

Estimated abundances of a selection of species are compared in Table 2.1.1. In this table, Z is the atomic number of the element listed in the first column and A is the atomic mass of the dominant isotope of this element. The third through fifth columns give the logarithm to the base 10 of the abundance by number found in three determinations: (1) Russell (1929); (2) Anders & Grevesse (1989), and (3) Asplund, Grevesse & Sauval (2005). It has become conventional for the logarithm of the hydrogen abundance to be set arbitrarily at 12.

The abundance of helium in Table 2.1.1 has not been obtained by spectroscopic analysis of the Sun's spectrum. Instead, it is the result of the analysis of the spectra of other stars for which the photospheric value of Fe/H is similar to the Sun's and for which the surface temperature is large enough to permit a reliable estimate of the He/H ratio. Other abundance estimates enclosed in parentheses in columns 4 and 5 are also indirect estimates. Semicolons next to estimates in column 3 are indications of uncertainty and a question mark beside the estimate of the nitrogen abundance indicates concern that nitrogen might not be responsible for the lines used to make the abundance estimate. This is ironic since, among the CNO elements, the best agreement between Russell's estimate and the estimate published 76 years later is for nitrogen.

A major difference between the two modern sets of abundance estimates is that the earlier one makes use of one-dimensional hydrostatic model atmospheres whereas the later one makes use of three-dimensional hydrodynamical model atmospheres. It is sobering to note that the total abundance by number of CNO elements in the later set is smaller than in the earlier set by about 40% and that the Fe abundance is also smaller by approximately 40%. In the 1989 determination, the total abundance by mass of elements heavier than helium is approximately 1.92% of the sum of all abundances by mass, including hydrogen and helium, whereas it is only 1.22% in the 2005 determination. Put in terms of

		log($N_\odot$) R29	log($N_\odot$) AG89	log($N_\odot$) AGS05	log(N_{cc}) AGS05	log(N_{ss}) Cam73	Ionization potential (eV)
Element	Z(A)						
H	1(1)	11.5::	12.0	12.0	8.25	12.0	13.527
He	2(4)	—	(10.99)	(10.93)	1.29	10.8	24.46
Li	3(7)	2.0:	1.16	1.05	3.25	3.19	5.363
Be	4(9)	1.8	1.15	1.38	1.38	1.41	9.28
B	5(11)	—	(2.6)	2.70	2.75	4.04	8.257
C	6(12)	7.4	8.56	8.39	7.40	8.57	11.217
N	7(14)	7.6?	8.05	7.78	6.25	8.07	14.48
O	8(16)	9.0:	8.93	8.66	8.39	8.83	13.550
C+N+O	—	9.0:	9.12	8.88	8.44	9.07	
F	9(19)	—	4.56	4.56	4.43	4.89	17.34
Ne	10(20)	—	(8.09)	(7.84)	−1.06	8.03	21.47
Na	11(23)	7.2	6.33	6.17	6.27	6.03	5.12
Mg	12(24)	7.8	7.58	7.53	7.53	7.52	7.61
Al	13(27)	6.4	6.47	6.37	6.43	6.42	5.96
Si	14(28)	7.4	7.55	7.51	7.51	7.50	8.12
S	16(32)	5.7:	7.21	7.14	7.16	7.20	10.30
Ar	18(40)	—	(6.56)	(6.18)	−0.45	7.07	15.68
Ca	20(40)	6.7	6.36	6.31	6.29	6.36	6.09
Ti	22(48)	5.2	4.99	4.90	4.89	4.94	6.07
Cr	24(52)	5.7	5.67	5.64	5.63	5.60	6.74
Fe	26(56)	7.3	7.67	7.45	7.45	7.42	7.83
Ni	28(58)	6.0	6.25	6.23	6.19	6.18	7.61

Table 2.1.1 Abundances of elements in the Sun's photosphere

the abundances by mass of hydrogen (X), helium (Y), and heavy elements (Z), the values $X = 0.707$, $Y = 0.274$, and $Z = 0.0192$ in the 1989 estimate become $X = 0.739$, $Y = 0.249$, and $Z = 0.0122$ in the 2005 estimate. These differences strongly influence interpretations of the results of neutrino detection experiments and results of solar oscillation observations when these results are compared, respectively, with the neutrino fluxes and the oscillation spectrum produced by theoretical solar models (see Chapter 10).

The sixth column in Table 2.1.1 gives the value of log(N) in stony (C1) meteorites, as quoted by Asplund *et al.* (2005). With the exception of very volatile elements, meteorites of the C1 variety are thought to have the same relative abundances as those in the original solar nebula out of which their parent bodies, the planetesimals, were formed. Comparing columns five and six in Table 2.1.1, it appears that the relative abundances of many non-volatile elements in the solar photosphere are very much the same as in the primitive solar nebula.

For comparison, the "Solar System" abundances in the seventh column of Table 2.1.1 are the result of an attempt by A. G. W. Cameron (1973) to reproduce abundances in the primitive solar nebula. These estimates rely on relative abundances in C1 meteorites for many non-volatile elements, on abundance ratios in low energy (10 MeV) cosmic rays from solar flares, as well as on solar photospheric determinations for many elements.

Close inspection of photospheric layers reveals vertical and horizontal structure which is time dependant. From the point of view of its relevance to internal structure, one of the most important features is granulation. Granules have characteristic horizontal dimensions of about 100 km and persist for times of about 5–20 min. Matter appears to be welling up within the granules and descending in the interstices bordering the granules, the whole phenomenon being a form of turbulent convection. Stellar evolution calculations show that energy flow by convection is one of the dominant forms of energy transport in other stars and many questions of astrophysical importance hinge upon its efficiency. Unfortunately, a completely satisfactory theory for this efficiency does not yet exist and it remains important to study the solar case both theoretically and observationally in an effort to develop a theory which can be applied reliably to other stars.

In typical convective regions, elements are thoroughly mixed on time scales short compared with time scales for nuclear transformations. The low abundance of Li at the solar surface compared to that in Solar System material suggests that, over the course of the Sun's evolution, the convective region beginning in the solar photosphere has extended inward to depths large enough for Li to burn. As detailed in Chapters 6, 9, and 10, this means that convection has extended in the past (and may even now extend) to depths where temperatures are of the order of 1–2 million degrees Kelvin. The fact that Li has not completely disappeared at the surface means that, during most of the Sun's lifetime, the base of the solar convective envelope has not extended to temperatures significantly larger than this.

Given the fact that temperatures much larger than 2×10^6 K are required for the CNO elements to experience significant nuclear transformations over 4.6×10^9 yr, one might infer that the relative abundances of these elements in the solar photosphere represent the relative abundances in the primitive solar nebula. On the other hand, if diffusion at the base of the convective envelope occurs on a time scale comparable to the Sun's age, the total abundance of CNO elements in the solar photosphere is smaller than the total abundance in the primitive solar nebula. The fact that the abundances of CNO elements in C1 meteorites are much smaller than in the photosphere of the present Sun may be attributed to the volatility of these elements in meteorites.

Another important characteristic of the Sun is its time-dependent magnetic-field structure. Among the many conspicuous manifestations of surface magnetic activity are the sunspots, whose number and location vary with an approximately 11 year period. The spots appear at the beginning of a cycle at about 30 degrees latitude and migrate toward the equator. Spot movement has been used historically to infer that the Sun rotates with about a 24 day period near the poles and with about a 30 day period near the equator.

The magnetic field structure is highly variable. There is a global dipolar component which alters polarity every 11 years. Near the surface, the strength of the non-dipolar magnetic field can be very intense, with root mean square intensities near 1000 gauss, suggesting magnetic field energy densities comparable with gas kinetic energy densities there. Theories of dynamo action have been constructed to account for the Sun's activity cycle.

Some theories place the seat of the dynamo just below the base of the convective envelope, across which one might expect a gradient in the angular momentum per unit mass that is associated with rotation.

Even though it occurs at what would seem to be a very modest rate, mass loss from the Sun has exerted and still exerts a not inconsequential influence on the human psyche and on human communication (by radio waves). One of the more spectacular evidences for mass loss are the aurora borealis and aurora australis phenomena, which are the consequence of atoms in the Earth's atmosphere being ionized by collisions with impinging solar wind particles. The visible light emitted in the auroral displays is due to transitions from highly excited electronic states populated by recombination.

The flux of particles from the Sun is highly variable with time as well as with direction. The wind seems to break out along magnetic flux tubes through "holes" in the solar corona and vary in strength with the solar activity cycle. The corona is a tenuous plasma which extends outward from the Sun's surface by several solar radii and is thought to be heated by Alfvén waves arising as a consequence of the movement of the footprints of the magnetic field in response to turbulent convective motions in the convective envelope; the matter in the envelope is tied to the field by virtue of being considerably ionized. Temperatures in the corona reach millions of degrees. The mechanism which accelerates the wind particles must occur in the corona, as both helium and hydrogen are completely ionized in the wind; and it must be electromagnetic in origin, as the ratio of helium to hydrogen in the wind is on the average about half what the ratio is found to be in other astrophysical sites where the ratio can be accurately estimated (such as at the surfaces of OB stars, in HII regions, and in planetary nebulae); furthermore, the relative abundances of carbon, nitrogen, and oxygen in the wind are the same as the relative abundances of these elements at the stellar surface, as deduced from spectral line strengths. On emerging from the influence of the Sun's magnetic field, wind particles have velocities of approximately 400 km s^{-1} and protons in the solar wind reach the Earth with kinetic energies of the order of 800 eV, far more energy than is necessary to ionize nitrogen and oxygen atoms in the Earth's atmosphere.

In addition to mass, the solar wind also carries off angular momentum. Being ionized, the wind plasma is coupled to the Sun's magnetic field, the dipole component of which is itself rooted in the rotating Sun. The result is a forward drag on the wind particles in a direction perpendicular to their radial motion as they make their way out through the field. The angular momentum imparted to the wind results in a slowing down of the rotation rate of the Sun itself. Theoretical estimates of the interaction between wind, magnetic field, and solar rotation give a time scale for the loss of solar angular momentum of about 5×10^9 years (Weber & Davis, 1967).

Estimates of stellar spin rates in stars in galactic clusters of different ages shows that the time scale for angular momentum losses for other solar-type stars is similar to that inferred for the Sun and, as importantly, that the time scale depends on the rotation rate. Because of its dependence on rotation rate, angular momentum loss by a "magnetic stellar wind" (MSW), plays an important role in close binary systems in which one or other of the components displays magnetic activity, supports a particle wind, and is tidally constrained to rotate at the orbital frequency (see Section 1.2).

Two additional characteristics of the Sun have been used to help constrain theoretical models of the solar interior: acoustical oscillations and the production of neutrinos close to the solar center. The Sun oscillates at very low amplitude at a very large number of frequencies, but the spectrum of frequencies is sharply peaked at about 5 minutes. Most of the oscillations are non-radial, some in modes which involve the entire Sun and have a finite amplitude even at the center. The excitation mechanism may involve turbulent motions in the convective envelope, and this view is encouraged by the fact the persistence time scales for features such as the granules are close to 5 minutes.

In the last decades of the twentieth century, scientists succeeded in determining the flux of electron neutrinos from the Sun. Electron neutrinos are emitted in several of the nuclear reactions that occur in the deep interior of the Sun. Using a detector in the Homestake gold mine in South Dakota, Raymond Davis determined initial upper limits on the neutrino flux which were considerably smaller than theoretical predictions at the time. Ultimately, finite counting rates were obtained in the Homestake mine experiment, as well as in the Kamiokande experiment in Japan, establishing that the flux of electron neutrinos is approximately half that suggested by solar models then currently in vogue. In recent years it has come to be understood that the discrepancy is not due to inadequacy of the solar models, but is due to the fact that, of the three flavors of neutrinos, at least two of them have a finite mass, allowing oscillations between the flavors to occur. Thus, for example, if only two neutrino types were involved and the oscillation length involving the two types were small compared to the distance from the Sun to the Earth, the flux of electron neutrinos at the Earth would be only half of the total flux of neutrinos. Another factor to consider is that the electron neutrino can interact with electrons in the Sun, while the other neutrinos presumably do not. The interaction imparts an effective mass to the neutrino. Thus, on its way from the solar center to the solar surface, an electron neutrino acquires an effective mass which can affect the probability of its conversion into another flavor neutrino. This theme is explored at length in Chapter 10.

2.2   Nearby stars in the Hertzsprung–Russell diagram

Another major observational cornerstone of the disciplines of stellar structure and evolution is the set of stars which are near enough for a determination of distance by trigonometry, using the Earth's orbit as a baseline. For these, once an analysis of the spectral distribution has led to an estimate of the surface temperature and allowance has been made for the departure of the spectral distribution from a black body distribution, intrinsic or "absolute" luminosities can be estimated. Table 2.2.1 contains estimated global properties of 21 stars chosen from a group of the 100 optically brightest stars, as seen from the Earth. Data from C. W. Allen (*Astrophysical Quantities*, Athlone: University of London, 1973) have been subjected to conversions from visual absolute magnitude M_V and color B-V to bolometric luminosity and surface temperature. The conversions are not intended to be definitive.

The luminosities and surface temperatures of the stars cited in Table 2.2.1 are shown in the upper half of Fig. 2.2.1. This figure is called a theoretical Hertzsprung–Russell

Star	constellation	$\log L/L_\odot$	$\log T_e$	$R/R_\odot$	distance(pc)
Rigel	β Ori	4.94	4.03	85.5	250
Deneb	α Cyg	4.89	3.95	117	500
Betelgeuse	α Ori	4.87	3.51	870	200
Antares	α Sco	4.41	3.52	489	130
Spica	α Vir	4.07	4.28	9.98	80
Canopus	α Car	3.83	3.92	39.6	60
Polaris	α UMi	3.75	3.72	91	24
Mirfak	α Per	3.63	3.80	54.7	160
Achernar	α Eri	3.43	4.20	6.91	39
Mira	o Cet	3.25	3.45	177	40
Aldebaran	α Tau	2.67	3.54	60	21
Regulus	α Leo	2.57	4.11	3.88	26
Algol	β Per	2.43	4.10	3.45	32
Formalhaut	α PsA	2.33	4.15	2.44	7
Capella	α Aur	2.21	3.71	16	14
Vega	α Lyr	1.87	4.01	2.74	8
Castor	α Gem	1.69	3.98	2.56	14
Pollux	β Gem	1.67	3.62	13.4	11
Sirius	α CMa	1.50	4.01	1.80	2.7
Altair	α Aql	1.01	3.89	1.77	5
Procyon	α CMi	0.85	3.82	1.99	3.5

Table 2.2.1 A selection from among the 100 brightest stars

(HR) diagram, to distinguish it from observational HR diagrams in which stars are located according to their visual absolute magnitude M_V and spectral type or according to M_V and color B-V (color–magnitude diagram). References to the original works by Hertzsprung and Russell have all but disappeared from the literature. Chandrasekhar (*The Study of Stellar Structure*, 1939) refers to a review by E. Hertzsprung (1911) summarizing the original work done by Hertzsprung in the years 1905–9 and refers to H. N. Russell by citing a text book (*Astronomy II*, by Russell, Dugan, and Stewart, 1927, see p. 910). There are two articles by Russell in the magazine *Popular Astronomy* (1914 and 1914).

Many of the stars in the lower half of Fig. 2.2.1 are selected from a set of nearby binary stars analysed by P. van de Kamp and M. D. Worth (1971) and from a set of visual binaries analysed by D. M. Popper (1980). Characteristics of the stars in these two sets are listed in Tables 2.2.2 and 2.2.3, respectively, where P_{orb} is in years, A is in astronomical units ($1 AU = 1.5 \times 10^{13}$ cm), and L, M, and R are in solar units. Six of the stars in Table 2.2.2 are also in Table 2.2.3; a comparison of the characteristics assigned in the two tables gives an indication of the accuracy of the assignments.

In Fig. 2.2.1, lines of constant radius permit one to see that, based on size, nearby and optically bright stars may be grouped into three broad classes – stars which define a band passing through the Sun's position, with radii varying from about 10 $R_\odot$ at the bright end

Table 2.2.2 Binaries within 5.2 pc (van de Kamp & Worth, 1971) (luminosities are visual)						
Star	P_{orb} (years)	A (AU)	L_1 ($L_\odot$)	M_1 ($M_\odot$)	L_2 ($L_\odot$)	M_2 ($M_\odot$)
α Cen	79.92	23.1	1.3	1.06	0.36	0.87
Sirius (αCMa)	50.09	19.9	23	2.20	2.7/−3	0.94
Luyten 726-8	25	5.7	6/−5	0.15	4/−5	0.15
61 Cygni	720	84.2	0.083	0.58	0.040	0.57
Procyon	40.65	15.9	7.6	1.78	5/−4	0.65
σ 2398	453	55.1	0.0028	0.41	0.0013	0.41
Krüger 60	44.6	9.5	0.0017	0.27	0.00044	0.16
Ross 614	16.5	3.9	0.0004	0.15	2/−5	0.07
40 Eridani	247.9	33.6	0.0027	0.42	0.00063	0.20
70 Ophiuchi	87.85	23.3	0.44	0.95	0.083	0.69

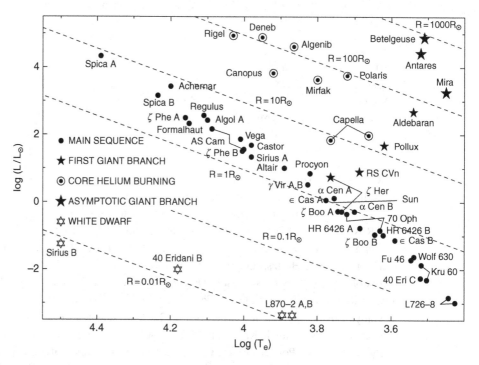

Fig. 2.2.1 Luminosity versus surface temperature for a selection of stars, some near, some visibly bright, some in binaries

to about 0.1 $R_\odot$ at the dim end (main sequence stars); stars with radii ranging from approximately 10 $R_\odot$ to as large as 10^3 $R_\odot$, with surface temperatures ranging from ∼10^4 K (blue giants) to as low as 3×10^3 K (red giants and supergiants); and very compact objects with radii of the order of 0.01 $R_\odot$ (white dwarfs). A further subdivision based on an understanding of evolutionary paths of model stars is made in Section 2.5.

Star	P_{orb}	A	M_1	$\log L_1$	$\log T_{e1}$	R_1	M_2	$\log L_2$	$\log T_{e2}$	R_2
α CMa	50.1	19.9	2.20	1.37	3.975	1.68	0.94			
α CMi	40.6	15.9	1.77	0.82	3.816	2.06	0.65			
ζ Her	34.4	13.2	1.25	0.74	3.765	2.24	0.70	−0.28	3.735	0.79
α Cen	79.9	23.6	1.14	0.12	3.755	1.27	0.93	−0.28	3.700	0.94
γ Vir	171.4	39.9	1.08	0.52	3.826	1.35	1.08	0.52	3.826	1.35
ϵ Cas	480	69.7	0.91	0.06	3.777	0.98	0.56	−1.13	3.590	0.59
ζ Boo	152	33.5	0.90	−0.27	3.745	0.77	0.72	−0.96	3.645	0.55
70 Oph	88.1	22.4	0.84	−0.35	3.721	0.79	0.61	−0.84	3.631	0.68
HR 6426	42.1	13.3	0.78	−0.77	3.685	0.58	0.54	−0.98	3.622	0.60
Wolf 630	144.5	26.0	0.42	−1.62	3.538	0.43	0.42	−1.62	3.538	0.60
Fu 46	13.0	4.7	0.30	−1.71	3.546	0.37	0.30	−1.73	3.528	0.39
Kr 60	44.4	9.5	0.28	−1.86	3.518	0.35	0.16	−2.30	3.503	0.23
o Eri B,C	252	33.5	0.43	−2.00	4.180	1.48/−2	0.16	−2.25	3.520	0.22
L726-8	26.5	5.4	0.11	−2.83	3.443	0.16	0.11	−2.98	3.425	0.15

Table 2.2.3 Visual binaries (Popper, 1980)

The density of stars in the HR diagram does not reflect the actual space density of stars of various sorts. There is an extremely high degree of observational selection. The apparent luminosity of a star in a given wave length band is smaller than its intrinsic luminosity in that band by a factor which is proportional to the inverse square of its distance from us. Thus, at any given apparent luminosity, the volume of space sampled for stars of a given intrinsic luminosity in a given wave length band is proportional to the three-halves power of that intrinsic luminosity. For example, white dwarfs can be seen out to only 100 pc or so, but luminous giants can be seen throughout much of the Galaxy. Hence, although it would appear from Fig. 2.2.1 that there are many more giants than white dwarfs, the actual space density of white dwarfs far exceeds that of giants. Along the main sequence in Fig. 2.2.1, there appear to be as many stars brighter than the Sun's luminosity as there are stars less bright, but, in fact, the space density of stars of a particular intrinsic luminosity decreases rapidly with increasing luminosity along the main sequence. Main sequence stars of mass smaller than the Sun's mass outnumber white dwarfs in the solar vicinity by about ten to one, and white dwarfs outnumber giants by about one hundred to one.

2.3 Mass–luminosity relationships

Of prime importance for an interpretation of the location of stars in the HR diagram is a knowlege of the masses of the stars in this diagram. Ironically, a reliable estimate of a star's mass requires that the star be in a binary system, even though it would be desirable to know the masses of single stars whose properties are not influenced by the presence of a companion. That is, it would be desirable not to have to worry about the effects of spin

| **Table 2.3.1** Spectroscopic binaries (Popper, 1980) | | | | | | | | | |
Star	P_{orb} (years)	A (AU)	M_1 ($M_\odot$)	$\log L_1$	$\log T_{e1}$	R_1 ($R_\odot$)	M_2 ($M_\odot$)	$\log L_2$	$\log T_{e2}$	R_2 ($R_\odot$)
ADS10598	46.1	16.8	1.11	−0.03	3.745	1.02	1.11	−0.03	3.745	1.02
δ Equ	5.71	4.28	1.19	0.32	3.794	1.22	1.19	0.32	3.794	1.22
α Aur	104.0	38.3	2.55	1.84	3.765	8.13	2.65	1.98	3.662	14.8
α Vir	4.0	6.55	10.8	4.33	4.390	8.13	6.8	3.16	4.235	4.17
12 Per	331.0	62.5	1.19	0.38	3.782	1.38	1.04	0.23	3.770	1.23

up and spin down associated with tidal forces and with the effects of mass and radiative energy transfer between components.

Fortunately, nature does provide systems of short enough orbital period that this period can be determined on the time scale of a human life but long enough that the influence of the companion is relatively minimal. Examples are the visual binary stars whose characteristics are listed in Tables 2.2.2 and 2.2.3 and by the set of spectroscopic binary stars whose characteristics, as given by Popper (1980), are listed in Table 2.3.1.

The luminosities and masses of many of the stars cited in Tables 2.2.2, 2.2.3, and 2.3.1 are plotted in the mass–luminosity (ML) diagram of Fig. 2.3.1. All but six of the stars represented are main sequence stars. Four of the exceptions are white dwarfs (Sirius B, Eridani B, and L870-2 A,B) and two (Capella A and B) lie near the red giant branch in the HR diagram.

To enlarge the data base, estimates of the masses, luminosities, and radii of the components of 34 eclipsing binaries, as given by Popper (1980), are presented in Tables 2.3.2 and 2.3.3. In these tables, orbital period is in days and orbital separation A is in solar radii. Although, in all cases, the components are quite close to each other, they comfortably underfill their Roche lobes. Understanding the significance of this fact requires a small digression. For any binary system one may construct a set of gravitational equipotential surfaces. At distances from the stars large compared with the orbital size, these surfaces are nearly spheres, with values of the potential essentially equal to those of a single star of mass equal to the sum of the masses of the two components. Close to the two stars the potential surfaces transform into detached figures of revolution. There is one unique surface which is essentially a figure eight rotated about the line joining the two stellar centers. The figure thus consists of two hollow tear drops joined at a point that, for equal mass components, is halfway between the two stellar centers. The tear drops are called Roche lobes and have the property that, when either component fills its lobe, it will begin to transfer matter to its companion. Inspection of the entries in Tables 2.3.2 and 2.3.3 shows that, on average, stellar radii are less than one-sixth of the orbital separation. Thus, no mass transfer occurs because of Roche-lobe filling. Further, tidal forces on each component due to the presence of a companion varies roughly as the cube of the ratio of stellar radius to the orbital separation. Thus, on average, tidal forces are only a fraction of a percent. These features suggest that a serious error is not being made in treating each component as a

Star	P_{orb} (days)	A ($R_\odot$)	M_1 ($M_\odot$)	$\log L_1$	$\log T_{e1}$	R_1 ($R_\odot$)	M_2 ($M_\odot$)	$\log L_2$	$\log T_{e2}$	R_2 ($R_\odot$)
ζ Phe	1.67	11.1	3.92	2.50	4.160	2.85	2.54	1.51	4.005	1.85
χ² Hya	2.27	13.5	3.61	2.49	4.063	4.39	2.64	1.64	4.005	2.16
AS Cam	3.43	17.3	3.31	2.17	4.088	2.70	2.51	1.56	4.001	2.00
V451 Oph	2.20	12.4	2.78	1.86	4.015	2.65	2.36	1.54	3.985	2.12
RX Her	1.78	10.4	2.75	1.79	4.015	2.44	1.96	1.48	3.985	1.96
AR Aur	4.13	18.3	2.48	1.67	4.048	1.83	2.29	1.46	3.997	1.83
β Aur	3.96	12.5	2.35	1.56	3.955	2.49	2.27	1.56	3.995	2.49
SZ Cen	4.11	18.0	2.28	1.60	3.885	3.62	2.32	1.79	3.878	4.55
EE Peg	2.63	12.1	2.08	1.28	3.927	2.05	1.32	0.40	3.806	1.29
V624 Her	3.90	16.5	2.1	1.47	3.890	3.0	1.8	1.16	3.882	2.2
805 Aql	2.41	11.9	2.06	1.26	3.915	2.10	1.75	0.86	3.855	1.75
RR Lyn	9.95	32.3	2.00	1.26	3.880	2.50	1.55	0.93	3.850	1.93
WW Aur	2.52	12.2	1.98	1.15	3.910	1.89	1.82	1.065	3.890	1.89
CM Lac	1.60	8.66	1.88	1.095	3.935	1.59	1.47	0.68	3.885	1.42
RS Cha	1.67	9.19	1.86	1.21	3.885	2.28	1.82	1.12	3.863	2.28
MY Cyg	4.00	16.3	1.81	1.03	3.848	2.20	1.78	1.02	3.845	2.20
V477 Cyg	2.35	10.9	1.78	1.08	3.940	1.52	1.34	0.41	3.820	1.20

Table 2.3.2 Detached eclipsing binaries (Popper, 1980)

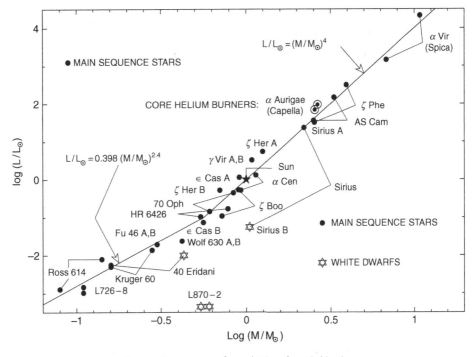

Fig. 2.3.1   Luminosity versus mass for a selection of stars in binaries

		Table 2.3.3	Detached eclipsing binaries (Popper 1980)							
Star	P_{orb} (days)	A $(R_\odot)$	M_1 $(M_\odot)$	$\log L_1$	$\log T_{e1}$	R_1 $(R_\odot)$	M_2 $(M_\odot)$	$\log L_2$	$\log T_{e2}$	R_2 $(R_\odot)$
EI Cep	8.44	26.5	1.68	1.12	3.840	2.54	1.78	1.15	3.827	2.80
XY Cet	2.78	12.6	1.76	1.00	3.875	1.88	1.63	0.94	3.860	1.88
ZZ Boo	4.99	18.6	1.72	0.94	3.824	2.22	1.72	0.94	3.824	2.22
TX Her	2.06	9.95	1.62	0.80	3.863	1.58	1.45	0.60	3.827	1.48
CW Eri	2.73	11.6	1.52	0.92	3.830	2.11	1.28	0.57	3.820	1.48
RZ Cha	2.83	12.2	1.51	0.86	3.800	2.26	1.51	0.86	3.800	2.26
BK Peg	5.49	18.4	1.27	0.43	3.785	1.48	1.43	0.69	3.780	2.03
DM Vir	4.67	16.9	1.46	0.64	3.800	1.76	1.46	0.64	3.800	1.76
CD Tau	3.44	13.4	1.40	0.625	3.798	1.74	1.31	0.56	3.798	1.61
TV Cet	9.10	21.8	1.39	0.58	3.820	1.50	1.27	0.365	3.803	1.26
BS Dra	3.36	13.3	1.37	0.50	3.808	1.44	1.37	0.50	3.808	1.44
VZ Hya	2.90	11.4	1.23	0.46	3.812	1.35	1.12	0.24	3.798	1.12
UX Men	4.18	14.4	1.17	0.31	3.785	1.28	1.11	0.29	3.782	1.28
WZ Oph	4.18	14.4	1.12	0.35	3.785	1.34	1.12	0.35	3.785	1.34
UV Leo	0.6	3.73	0.99	0.09	3.768	1.08	0.92	0.09	3.768	1.08
YY Gem	0.81	3.88	0.59	−1.13	3.576	0.62	0.59	−1.13	3.576	0.62
CM Dra	1.27	3.80	0.24	−2.16	3.500	0.25	0.21	−2.16	3.500	0.235

single star. Data describing the components of several eclipsing binaries are presented in the ML diagram of Fig. 2.3.1.

Those stars in Fig. 2.3.1 which lie within the main sequence band in the HR diagram of Fig. 2.2.1 define a distribution which, at a given luminosity, has a remarkably low dispersion. All of the main sequence stars shown in Fig. 2.3.1 have luminosities which are within a factor of approximately two in luminosity of the curve which has been drawn through the distribution. This curve, which is intended primarily as a guide to the eye rather than as a fit to the distribution, consists of two straight-line segments. The upper segment is described by

$$\frac{L}{L_\odot} = \left(\frac{M}{M_\odot}\right)^{4.0} \tag{2.3.1}$$

and the lower one is described by

$$\frac{L}{L_\odot} = 0.398 \left(\frac{M}{M_\odot}\right)^{2.4}. \tag{2.3.2}$$

The segments intersect at $L = 0.1\ L_\odot$ and $M = 0.562\ M_\odot$.

It is interesting that several giants for which masses have been determined (e.g., Capella A and B, also known as α Aurigae) reside close to the mass–luminosity relationship defined by main sequence stars. A possible interpretation of this fact is that these stars have lost very little mass either during their main sequence phase or during their evolution away from the main sequence. This interpretation presupposes that the main sequence is the locus of stars in early stages of evolution after forming out of the interstellar medium, that there is an end to the main sequence phase, and that stars then evolve onto other sequences. This interpretation is now accepted as the correct one, but, beginning in the early part of the twentieth century, speculations as to evolutionary paths in the HR diagram provoked animated discussions of competing scenarios. A very popular idea in some quarters was that stars arrive on the upper main sequence as massive, luminous objects, and that they then descend the main sequence, becoming less luminous as they lose mass.

2.4 Evolutionary paths of theoretical models in the HR diagram

Instead of speculating about possible scenarios for evolution in the HR diagram, it is more constructive to proceed directly to a description of the results of calculations of stellar model evolution. In Fig. 2.4.1 are tracks in the HR diagram of three stellar models that are of constant mass (1 $M_\odot$, 5 $M_\odot$, and 25 $M_\odot$) and are initially of a homogeneous composition similar to the composition of Solar System material (Iben, 1985). Each initial model is located at the lower end of the heavy portion of the evolutionary track next to the mass designator for the track. Over most of the interior of each model, most atoms are completely ionized and the relationship between gas pressure and temperature is proportional to the product of the number density of particles (including free electrons) and the temperature. In the most massive model, photons contribute to the pressure by an amount proportional to the fourth power of the temperature. At any point, the balance between the pressure gradient and the gravitational pull of matter interior to this point maintains near hydrostatic equilibrium.

As hydrogen is converted into helium in central regions, each model evolves upward to larger luminosities along the first heavy portion of the track until hydrogen is exhausted at the center of the model. The primary mechanism for the gradual change in interior characteristics which leads to the increase in luminosity is the decrease in particle densities in regions where four protons and four electrons are being converted into one helium nucleus and two electrons. In order for pressure balance to be maintained, the temperature must increase as particle density decreases. The increasing temperature leads to an increased rate of nuclear energy generation due to an increase in the frequency with which nuclear reactions occur. The duration of this first, or core hydrogen-burning, phase depends strongly on the mass of the model, varying from about 10^{10} yr for the 1 $M_\odot$ model, through about 6.5×10^7 yr for the 5 $M_\odot$ model, and to about 6×10^6 yr for the 25 $M_\odot$ model.

In the 5 $M_\odot$ and 25 $M_\odot$ models, hydrogen burns primarily by the carbon–nitrogen–cycle (CN-cycle) reactions, with carbon and nitrogen acting as catalysts for the conversion of hydrogen into helium. The fluxes generated by CN-cycle reactions are so large that, over a large region that extends beyond where most of the nuclear energy is produced, convection rather than radiation is the principle agent for carrying energy outward. In the convective core, hydrogen and helium are thoroughly mixed so that, when hydrogen vanishes at the

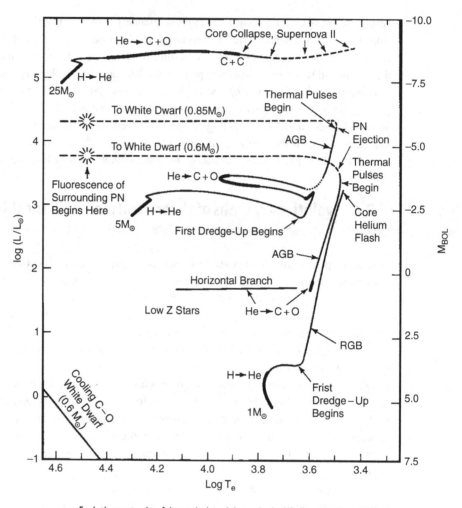

Fig. 2.4.1 Evolutionary tracks of theoretical model stars in the HR diagram (Iben, 1985)

center, it vanishes over the entire core. Robbed of a source of nuclear energy, the helium core contracts and heats; the release of gravitational potential energy in the region just outside the core heats matter there and hydrogen burning by CN-cycle reactions in a shell takes over as the primary source of energy escaping the model. The duration of this shell-forming phase is only a few percent of the duration of the core hydrogen-burning phase.

The helium core in the two more massive shell hydrogen-burning models continues to contract and heat until the rate at which the so-called triple alpha reactions release nuclear energy becomes large enough to stabilize the core while the matter in the core still behaves like a nearly perfect gas. A triple alpha reaction consists of the formation of an unstable ^{8}Be nucleus during a rare close collision of two ^{4}He nuclei (α particles), followed by the radiative capture by ^{8}Be of another alpha particle to form an excited ^{12}C nucleus which decays to the ground state, releasing nuclear energy. As the abundance of ^{12}C builds up, nuclear energy released in the reaction ^{12}C$(\alpha, \gamma)^{16}$O contributes to the flux of energy generated by helium burning. This flux is large enough that, once again, a convective core is formed. Although the duration of the core helium-burning phase is determined by the rate of the triple alpha reaction, most of the energy released by the model continues to be energy from the hydrogen-burning shell. The lifetime of the core helium-burning phase varies from about 1.6×10^6 yr for the 5 $M_\odot$ model to about 7×10^5 yr for the 25 $M_\odot$ model. Put another way, the core helium-burning lifetime of the 5 $M_\odot$ model is about one-fourth of its core hydrogen-burning lifetime, while the core helium-burning lifetime of the 25 $M_\odot$ model is about one-tenth of its core hydrogen-burning lifetime.

During the core helium-burning phase, the 25 $M_\odot$ model moves at a nearly constant rate along the thick portion of the track in Fig. 2.4.1 labeled He $\rightarrow$ C + O. The track of the 5 $M_\odot$ model is slightly more complex, consisting of two main phases of comparable lifetime, the second being the thick portion of the track in Fig. 2.4.1 labeled He $\rightarrow$ C + O and the first being the short thick segment at lower surface temperature.

After helium is exhausted at the center, the carbon–oxygen (CO) core of the 25 $M_\odot$ model contracts and heats at a rate controlled by the loss of energy to electron neutrino–antineutrino pairs produced by the annihilation of electron–positron pairs. The rate of contraction and heating are temporarily reduced as carbon-burning reactions are ignited. Densities have become large enough in the core that electrons there are degenerate, so that the pressure exerted by electrons dominates the pressure exerted by nuclei and the total pressure is thus nearly independent of the temperature. During the carbon-burning phase, which lasts for only about 250 yr, the core is essentially uncoupled from the rest of the model, with the radiant luminosity at the surface of the model being due entirely to energy produced in the hydrogen- and helium-burning shells outside of the core.

The decoupling between the core and the rest of a model star persists during all subsequent nuclear burning phases which proceed on a dynamical time scale until the core consists of iron-peak elements, for which the binding energy per nucleon is at a maximum. As the core continues to contract and heat, iron-peak nuclei are broken into alpha particles by gamma-capture reactions, alpha particles are broken into nucleons, and electron capture converts protons into neutrons. The core collapses into a neutron star and the outer portions of the star are ejected in a complicated process involving neutrino deposition and shock waves. The observational result is presumably a type II supernova. A thorough discussion of these matters is given by W. D. Arnett in *Supernovae and Nucleosynthesis* (1996). From the point of view of element production in the Universe and interaction with the interstellar medium, the last phases of evolution of massive stars are of the highest relevance. From the point of view of understanding sequences in the HR diagram, the phases are of such a short duration as to be essentially irrelevant.

After the core helium-burning phase, the evolution of the 5 $M_\odot$ model star is quite different from that of the 25 $M_\odot$ model star. At any given temperature, the density in the CO core of the 5 $M_\odot$ model is much larger than in the CO core of the 25 $M_\odot$ model. So much larger, in fact, that electrons become degenerate in the core before temperatures large enough to initiate carbon burning are attained. Energy flows by electron conduction out of the core and energy escapes the model star in the form of neutrinos generated by the plasma process. As the cooling core shrinks to dimensions comparable to that of a white dwarf, the envelope of the 5 $M_\odot$ model acquires the dimensions of a giant. Nuclear burning that takes place alternately in a hydrogen-burning shell and in a helium-burning shell is responsible for the luminosity of the model. First, the hydrogen-burning shell deposits helium onto a layer of almost pure helium which grows in mass. As it grows in mass, the helium layer is compressed and heated until, when the mass of the layer reaches a critical value, helium ignites in the layer and helium burning evolves into a thermonuclear runaway known as a helium shell flash. The energy injected into the helium layer causes it to heat, heating increases the pressure, and the layer pushes outward matter which lies above it, including matter in the hydrogen-burning shell. Propelled outward to lower densities and temperatures, matter in the hydrogen-burning shell becomes less dense and cools. Hydrogen burning is replaced by helium burning as the major source of model luminosity. The thermonuclear flash mitigates and a phase of quiescent helium burning ensues. During this second quiescent nuclear burning phase, the location of the weakly burning hydrogen shell source remains nearly fixed in the mass coordinate, while burning at the base of the helium layer adds carbon and oxygen to the CO core. As the mass of the helium layer decreases and the base of the helium layer and the location of the hydrogen-burning shell source approach each other in mass, helium burning dies out and hydrogen burning takes over once again as the model's primary source of luminosity. Another long phase of hydrogen burning continues until a new helium layer attains the critical mass for igniting helium. The entire cycle is known as a thermal pulse cycle and the model is said to be in the thermally pulsing asymptotic giant branch (TPAGB) phase.

Only a few million years elapse from the time when helium vanishes at the center and the time when the thermally pulsing phase begins and, due to the development of a superwind, this latter phase lasts for of the order of a million years or even less. The story of the superwind is not one established by stellar interior model calculations, but one which has been built up over many years of observationally based estimates of mass-loss rates from TPAGB stars and from the characteristics and statistics of planetary nebulae which are the end product of TPAGB evolution. The mechanism of superwind mass loss is thought to involve the production of grains from the carbon formed in helium shell flashes and dredged to the surface during thermal pulses (see Chapter 18 in Volume 2), inflation of the stellar photosphere by shocks produced during acoustical pulsations of the envelope, and radiation pressure on the grains in the inflated photosphere.

In the current context, suffice it to say that, after only a few million years as an AGB star, the real counterpart of the 5 $M_\odot$ model sheds most of its hydrogen-rich envelope, and its CO core evolves in a matter of only about 1000 years into a white dwarf of mass about 0.9 $M_\odot$. Devoid of a source of tappable nuclear energy, the white dwarf remnant cools at first rapidly and then ever more slowly, reaching a luminosity of $10^{-2} \, L_\odot$ in about 10^8 yr, a luminosity of $10^{-3} \, L_\odot$ in about 10^9 yr, and so on.

What remains to be described is the nature of hydrogen-burning reactions in low mass stars and of the advanced evolutionary stages of the 1 $M_\odot$ model. Mean temperatures in the deep interior of core hydrogen-burning models decrease with decreasing mass. This is basically because less massive stars exert smaller gravitational forces so that smaller pressure forces, hence smaller temperatures, are required to balance the gravitational forces. At temperatures at and below those at the center of the 1 $M_\odot$ model, the so-called pp reaction which converts two protons into a deuterium nucleus with the emission of a positron occurs more rapidly than do the CN-cycle reactions. The positron annihilates with an electron with the emission of one or more gamma rays. Shortly after it is formed, the deuterium nucleus captures another proton to become a ^{3}He nucleus, with the emission of a gamma ray. Next, two ^{3}He nuclei react to form a ^{4}He nucleus with the emission of two protons. The net result is the same as in the catalytic CN cycle: four protons and four electrons become one ^{4}He nucleus and two electrons, with the liberation of approximately the same energy.

The difference between the two nuclear burning processes is that the pp reaction, which is the rate-controlling reaction in the pp chain (the pp and subsequent reactions), is much less temperature-sensitive than the rate-controlling reaction (^{14}N$(p, \gamma)^{15}$O reaction) in the CN-cycle set of reactions. This has the consequence that the energy generated by the pp chain is spread out over a much larger fraction of the model mass than is the case for the energy generated by the CN-cycle reactions. This, in turn, has the consequence that, when the pp chain dominates in energy production, the fluxes near the model center are not large enough to force energy to flow by convection. Thus, the 1 $M_\odot$ model does not possess a convective core and, when hydrogen vanishes at the center, it does not vanish at the same time over a large region about the center. The result is that the development of a hydrogen-burning shell is a much more gradual process in the 1 $M_\odot$ model than in the more massive models and the configurational changes that occur as the model makes its way to becoming a giant are much more measured. Along the first two thirds of the heavy portion of the evolutionary track in the HR diagram of Fig 2.4.1, hydrogen continues to burn at the model center; along approximately the last third of the track, hydrogen burns in a thick shell which gradually narrows as the base of the shell moves outward in mass. Once hydrogen is exhausted over the inner approximately 13% of the model mass, the helium core contracts rapidly, the hydrogen-burning shell becomes quite narrow in the mass coordinate, and the envelope expands to giant dimensions in response to an increasing flux of energy from the hydrogen-burning shell.

Electrons in the helium core become quite degenerate. Thanks to efficient conduction by electrons, the core becomes nearly isothermal. But, because of plasma-induced neutrino emission, which occurs at a larger rate, the larger the density, the maximum temperature in the core is not at the model center. The helium core grows in mass due to the deposition of helium ash by the hydrogen-burning shell. As it becomes more massive, the core contracts, and, in response to the release of gravitational potential energy, it also becomes hotter. Ultimately, when the mass of the helium core reaches about 0.45 $M_\odot$, the maximum temperature in the core is close to 8×10^6 K and a series of thermonuclear helium shell flashes occur. At the end of this series of flashes, the electron degeneracy of the core has been "lifted," the core has expanded until its diameter is roughly one-tenth that of the Sun, and the radius of the model (initially of solar composition) has been reduced by a factor of about ten from its value of over 100 $R_\odot$ at the start of the flashing phase. The entire duration of the phase during which the helium core grows in mass from 0.13 $M_\odot$ to 0.45 $M_\odot$ is about 10^9 yr, about 10% of this time being spent at luminosities larger than those which prevail during the following quiescent core helium-burning phase. The location of this latter phase, which lasts for about 10^8 yr, is denoted by the small heavy portion of the track in Fig. 2.4.1 labeled He $\rightarrow$ C + O.

During the quiescent helium-burning phase, helium is converted into carbon and oxygen in a convective core, and, just as in more massive models, most of the model luminosity is produced in a thin hydrogen-burning shell, even though the evolutionary time scale for the entire model is set by the rate of the triple alpha process. When helium is exhausted at the center, evolution is a scaled down version of the evolution of the 5 $M_\odot$ model. At the end of the TPAGB phase, the mass of the electron-degenerate CO core, at $M_{CO} \sim 0.55$ $M_\odot$, is smaller than that of the final TPAGB descendant of the 5 $M_\odot$ model. The maximum luminosity of the central star of the ejected envelope is given approximately by (Bohdan Paczyński, 1970)

$$\frac{L}{L_\odot} = 60\,000 \left(\frac{M_{CO}}{M_\odot} - 0.52 \right), \tag{2.4.1}$$

so the luminosity of the central star descendent of the lower mass model is smaller than that of the more massive model. Finally, the duration of the associated planetary nebula phase is much longer, being of the order of 10 000 yr.

If, in the HR diagram, one now connects the thick portions of the three model tracks that are associated with the core hydrogen-burning phase of evolution and connects those that are associated with the core helium-burning phase of evolution, the resulting two bands define where real counterparts of the theoretical models should lie. The core hydrogen-burning band can be unambiguously identified with the main sequence band defined by observed stars (the filled circles in Fig. 2.2.1). The core helium-burning band can be identified with a band in Fig. 2.2.1 extending from RS CVn to Rigel and Deneb (the filled circles surrounded by open circles in Fig. 2.2.1). This identification was not understood until, in the middle 1960s, models of evolving intermediate mass stars in the core helium-burning phase were constructed.

The third sequence unambigously relating observed and model stars is the white dwarf sequence. This sequence is uniquely located in the HR diagram defined by observed stars with radii of the order of 0.01 $R_\odot$ (e.g., Sirius B, Eridani B, and L870-2 A and B in Fig. 2.2.1). Comparing the locations of the four white dwarfs in the HR and ML diagrams, it is evident that the more massive the white dwarf, the smaller is its radius. This inverse relationship between radius and mass is a natural consequence of the relationship between pressure and density in an electron-degenerate gas, as is demonstrated in Chapter 5 with very simple models. The final phase of evolution of model stars initially less massive than about 10 $M_\odot$ is a white dwarf of mass in the range 0.5–1.4 $M_\odot$, with the peak in the final mass distribution being in the range 0.5–0.6 $M_\odot$, exactly as is observed.

The theoretical model results show that the traditional red giant branch is actually populated by two distinct types of star: (1) low mass, lower luminosity red giants with electron-degenerate cores which are composed primarily of helium and which burn hydrogen in a shell, reaching a maximum luminosity $L \sim 2 \times 10^3$ $L_\odot$ before igniting helium in their cores; and (2) similar luminosity AGB stars which have electron-degenerate cores which are composed of carbon and oxygen and which, after an initial phase of helium shell burning, alternately burn hydrogen in a shell and helium in a shell, eject their hydrogen-rich envelope in a superwind and become white dwarfs. After completing the first (type I) red giant phase, models of initial mass ≤ 2 $M_\odot$ burn helium in central regions and hydrogen in a shell for approximately 10^8 years, and then experience the second (type II) red giant (AGB) phase at luminosities somewhat larger than those of the first red giant branch, before ejecting their envelopes and becoming CO white dwarfs of mass in the range 0.55–0.65 $M_\odot$.

Models of initial mass in the range 2.25–10 $M_\odot$ experience a very brief first red giant phase during which they ignite helium in a non-electron-degenerate core and evolve quickly to the core helium-burning band, blueward of the red giant branch. They then evolve to the second red giant, or AGB, branch at higher luminosities than those stars which experience an extended first giant branch. Models of initial mass in the range 2.25–8 $M_\odot$ become CO white dwarfs in the mass range 0.65–1.1 $M_\odot$. During the first part of the AGB phase, models of initial mass in the range 8–10 $M_\odot$ convert their helium- and hydrogen-exhausted core into an oxygen–neon (ONe) core and, after a TPAGB phase terminated by a superwind, evolve into ONe white dwarfs.

2.5 The evolutionary status of familiar stars

Who among us has not gazed at the night sky on a clear, moonless night and experienced a sense of awe on contemplating the nature of the specks of light that shine at us from the deep reaches of space? These specks are objects that have come to be understood as bodies like our Sun whose inner workings obey the laws of physics derived in consequence of several centuries of experimentation in terrestial laboratories combined with theoretical formulations which summarize and integrate these laws mathematically. Thanks to the development of high-speed computers with large memories, it has been possible to construct theoretical models of stars which are solutions of equations which incorporate these laws. On comparing the theoretically predicted locations of model stars in the HR diagram, Fig. 2.4.1, with the locations of observed stars as seen through telescopes, Fig. 2.2.1, it has been possible to infer the evolutionary status of observed stars.

Thus, for example, one can state that our Sun, at an age of 4.6 billion years, is approximately half way through its core hydrogen-burning phase of evolution. Two constellations in the southern sky related by legend are Orion the Hunter and Taurus the Bull. Aldebaran, the brightest star in Taurus, is a red giant star, with an interior structure very much like the structure which our Sun will achieve after another six billion years or so of evolution when (1) its hydrogen-exhausted core will have contracted to approximately the size of the Earth, even though the core contains half its mass and hence is over 100 000 times more massive than the Earth, (2) it will be shining 1000–2000 times more brightly than it does now, burning hydrogen in a thin shell at the base of a giant envelope, and (3) its radius will become over 100 times larger than it is now, certainly engulfing Mercury and perhaps even reaching as far as the Earth.

The two intrinsically brightest stars in Orion are Betelgeuse and Rigel. Both of these stars are approximately 10 times as massive as the Sun, both are about 100 000 times as bright as the Sun in total rate of energy emission, but they are in two quite different stages of evolution. Rigel is burning helium in its hydrogen-exhausted core, with most of its luminosity being due to hydrogen burning in a thin shell at the base of an envelope the radius of which is about 100 times larger than the Sun's radius. Betelgeuse is in a more advanced stage of evolution, having converted helium into carbon and oxygen over an inner fuel-exhausted region of mass approximately equal to the Sun's mass. This inner region is very compact, even smaller than the size of the Earth, and yet the envelope of the star has a radius which is nearly 1000 times larger than the Sun's radius. Betelgeuse is burning hydrogen and helium alternately in two thin shells and is manufacturing and dredging into its envelope carbon and s-process elements. Within another million years or so, it will develop a "superwind" which ejects into interstellar space essentially its entire hydrogen-rich, carbon-rich, and s-process-rich envelope, leaving a compact remnant which in less than 1000 years will evolve into a white dwarf. Rigel will ultimately evolve into a star very much like Betelgeuse.

Thus, both Rigel and Betelgeuse will have enriched the interstellar medium with elements produced in their interiors and their remnants will join the white dwarf graveyard of "dead" stars which houses the remains of 99 out of every 100 stars that have formed since the Big Bang and have completed their nuclear burning lives. After having completed its red giant branch phase of evolution, our Sun will first adopt a configuration very much like that of Rigel, except on a more modest scale, and then during about 100 million years evolve into a configuration very much like that of Betelgeuse, but on a more modest scale. During this stage, its radius will extend almost to and perhaps slightly beyond the Earth's orbit. Finally, after approximately a million years in this configuration, the Sun will eject its envelope (which may contain the debris of human civilization) and become a white dwarf.

Since the identification of hydrogen-burning nuclear energy sources in the fourth decade of the twentieth century, it has been understood that the main sequence in the HR diagram is made up of stars which are burning hydrogen at their centers, that many low luminosity red giants are burning hydrogen in a shell outside of an electron-degenerate helium core which is not burning helium, and that the white dwarf branch is made up of stars which have completed their nuclear burning phase of evolution and become compact objects in which the force of gravity is balanced by the pressure of degenerate electrons. Stars like the Sun, Spica, Achernar, Regulus, Algol, Formalhaut, Vega, Castor, Sirius A, Altair, and Procyon are main sequence stars. Stars like Aldebaran, Pollux, and RS CVn are ordinary low-mass red giants and Sirius B, the companion of Sirius A, is a white dwarf.

Since the middle 1960s it has been understood that there is a fourth branch, the core helium-burning branch, which is populated by stars which are burning helium in a convective core and hydrogen in a shell. Examples are Rigel, Deneb, Algenib, Canopus, Mirfak, and Capella A and B, forming a sequence which is roughly parallel to the main sequence branch in the HR diagram and running into the red giant branch at about 10 times the luminosity of the Sun. The hottest and brightest of these stars were formerly classified as blue giants, but their interior structure was not understood.

During the last third of the twentieth century, it became apparent that there is a fifth branch, the asymptotic giant branch, which consists of stars which are burning helium and hydrogen alternately in shells outside of a hot white-dwarf-like core which is composed of carbon and oxygen if the initial mass is less than 8 $M_\odot$ and is composed of oxygen and neon if the initial mass is in the range 8–10 $M_\odot$. Luminosities are typically an order of magnitude larger than luminosities of stars on the low-mass red giant branch. In addition to Betelgeuse, good examples are Antares and Mira. It is now clear that stars of initial mass less than 10 solar masses evolve through the branches in the order (1) main sequence, (2) red giant with a helium core (electron degenerate if initial mass is less than 2 $M_\odot$), (3) core helium-burning, (4) asymptotic giant with an electron-degenerate CO or ONe core, and (5) white dwarf. Put another way, one can say that such stars evolve through phases which are (1) like the Sun and Spica, (2) like Pollux and Aldebaran, (3) like Capella A and B and Rigel, (4) like Mira and Betelgeuse, and (5) like Sirius B, the companion of Sirius A.

Two additional primary avenues of stellar evolution require mention. Single stars initially more massive than about 10 times the mass of the Sun continue to burn nuclear fuels until they form a core composed of iron-peak elements. The core collapses into a neutron star or black hole, depending upon the mass of the initial main sequence precursor, and the heavy-element-rich envelope is ejected into the interstellar medium, appearing as a type II supernova. These stars are the source of perhaps half of all of the iron and of most of the other elements heavier than helium in the Universe other than the nitrogen, carbon, and s-process elements produced by less massive stars. The neutron star remnants occupy approximately one tenth of the graveyard of "dead" stars and the black holes occupy one one hundredth.

The label "dead" has here been applied to the compact objects into which nuclear burning stars ultimately evolve for poetic effect. The label does such stars a serious injustice. Although they are not burning a nuclear fuel, they are releasing from their surfaces energy that is produced in most interesting ways: in the case of white dwarfs, by cooling of heavy ions in their interiors; in the case of neutron stars known as millisecond pulsars, by the interaction of energetic electrons with a magnetic field which transforms rotational kinetic energy into beamed radiation at radio wavelengths. In the interiors of these stars, which are very much alive, conditions are so far from those achievable in terrestial laboratories that considerable imagination must be exercised in contemplating their behavior, and the exercise leads to a more complete understanding of the physics of matter.

The last avenue of evolution which deserves mention is that followed by close binary stars which can transfer mass back and forth and evolve into configurations not accessible to single stars. Among the more interesting final products are pairs of close binary white dwarfs of combined mass larger than 1.4 times the Sun's mass which, on being drawn into contact by gravitational wave radiation, merge and perhaps explode as type Ia supernovae, producing of the order of half of the iron peak elements in the Universe.

Hopefully, the reader has been persuaded that we now understand in some detail the interior properties and the past and future history of familiar stars such as the Sun, Sirius, Achernar, Betelgeuse, Capella A and B, and others that can be seen by the human eye, adding depth to their appreciation of the beauty of the night sky. Other stars, including neutron stars and mass-transferring binary stars, which radiate predominantly at wavelengths outside of the optical wave length band, but can be studied with special detectors, tell stories as fascinating as those told by the stars whose radiation is accessible to the human eye.

Bibliography and references

Hans Alfvén, *Cosmical Electrodynamics*, (London: Oxford University Press), 1950.

C. W. Allen, *Astrophysical Quantities* (Athlone: University of London), 1973, third edn.

Edward Anders & N. Grevesse, *Geochim. Cosmochim. Acta*, **53**, 197, 1989.

W. David Arnett, *Supernovae and Nucleosynthesis* (Princeton: Princeton University Press), 1996.

M. Asplund, N. Grevesse & A. J. Sauval, in *Cosmic Abundances as Records of Stellar Evolution and Nucleosynthesis*, eds. F. N. Bash & T. G. Barnes, ASP Conf. Ser., Vol. **30**, 2005.

Ludwig Boltzmann, *Ann. der Physik und Chemie*, **22**, 291, 1884.

C. Vernon Boys, *Proc. Roy. Inst. of G.B.*, **14**, 353, 1894

Alastair G. W. Cameron, *Space Sci. Rev.*, **15**, 121, 1973.

Henry Cavendish, *Phil. Trans. Roy. Soc.*, **88**, 469, 1798.

Donald D. Clayton, *Principles of Stellar Evolution and Nucleosynthesis*, (New York: McGraw Hill), 1968.

Subrahmanyan Chandrasekhar, *The Study of Stellar Structure*, (Chicago: University of Chicago Press), 1939.

A. Cornu & J. B. Baile, *Comptes Rendus Acad. Sci., Paris*, **76**, 954, 1873.

Ejnar Hertzsprung, *Potsdam Publishing*, **63**, 1, 1911.

Icko Iben, Jr., *Quart. J. Roy. Astr. Soc.*, **26**, 1, 1985.

James Kaler, *Cambridge Encyclopedia of Stars* (Cambridge: Cambridge University Press), 2006.

Peter van de Kamp & M. D. Worth, *AJ.*, **76**, 1129, 1971.

Susan A. Lamb, W. Michael Howard, & Icko Iben, Jr., *ApJ*, **207** 209, 1975.

Isaac Newton, *Principia Mathematica*, 1687.

Bohdan Paczyñski, *Acta Astron.*, **20**, 47 287, 1970.

Daniel M. Popper, *Ann. Rev. Astr. & Ap.*, **18**, 115, 1980.

John Henry Poynting, *The Mean Density of the Earth* (1893 Adams Prize Essay, University of Cambridge), (Exeter St. Strand: Charles Griffen & Co.), 1894.

Henry Norris Russell, *Popular Astronomy*, **22**, 275, 1914; **22**, 331, 1914.

Henry Norris Russell, *ApJ*, **70**, 11, 1929.

H. N. Russell, R. S. Dugan, and J. Q. Stewart, *Astronomy II*, (Boston: Ginn & Co.), 1927.

Joseph Stefan, *Sitz. Berichte der Kais. Acad. der Wiss.*, **79**, 391, 1879.

E. J. Weber & Lyman Davis, Jr., *ApJ.*, **148**, 217, 1967.

PART II

BASIC PHYSICAL PROCESSES IN STELLAR INTERIORS

3 Properties of and physical processes in main sequence stars – order of magnitude estimates

Considerable insight into the interior characteristics of main sequence stars can be obtained by means of back of the envelope estimates. Such estimates are vital, for, while being engrossed in the construction of numerical solutions of the rigorous equations of stellar structure, one can lose sight of the basic physics underlying the equations.

In Section 3.1, the balance between the pressure gradient force and the gravitational force is used in conjunction with the equation of state for a perfect gas to estimate interior temperatures in homogeneous main sequence stars, emphasizing that the thermal energy of a particle in the deep interior of the star is comparable to the gravitational potential energy of a particle near the surface. In Section 3.2, the effects on the equation of state of electrostatic forces, electron degeneracy, and radiation pressure are examined, with the conclusion that the first two effects become important in stars less massive than the Sun and the third effect becomes important in stars considerably more massive than the Sun. In Section 3.3, theorems relating the binding energy of a star with the kinetic energy of material particles in the star and to the overall gravitational binding energy of the star are constructed.

In Section 3.4, three modes of energy transport – radiation, convection, and conduction – are explored. In the discussion of radiative flow, emphasis is placed on the microscopic physical processes involved in estimating the radiative opacity and a theorem relating stellar luminosity to stellar mass, mean molecular weight, and mean interior opacity is constructed. The discussion of convection begins with a criterion for instability against convection, develops the mixing length algorithm for estimating the convective flux, and demonstrates that, in convective regions which appear in the deep interior, the temperature gradient is very close to the adiabatic gradient, whereas, in convective regions which appear near the surface, the temperature gradient can be very superadiabatic. In the discussion of conduction, a diffusion equation is introduced and a characteristic cross section for electron–ion scattering under non electron-degenerate conditions is estimated. It is shown that the effective opacity is the geometric mean of the radiative opacity and the conductive opacity and it is found that electron conduction is not important in main sequence stars.

In Section 3.5, the physics involved in determining hydrogen-burning energy-generation rates is outlined and main sequence lifetimes based on these rates are estimated. In the final section, Section 3.6, it is shown that, in regions where matter is in a highly ionized state, the fact that the gravitational force on electrons is three orders of magnitude larger than the gravitational force on heavy ions, coupled with the fact that charge neutrality must prevail over regions containing a macroscopic number of particles, implies the existence of a static electric field the strength E of which is, in order of magnitude, related to the gravitational acceleration g by $E \sim M_H g/e$, where e is the charge of an electron and M_H is the mass of a hydrogen atom.

3.1 Particle numbers and separations, pressures and temperatures

For simplicity, assume that a star is spherically symmetric and homogeneous in composition. Defining the mass of an individual particle as μM_H, where $M_H = 1.6735 \times 10^{-24}$ g is the mass of the hydrogen atom and μ is called the molecular weight, one has that the total number of particles in a star of mass M is

$$N = \frac{M}{\mu M_H} = \frac{1.19 \times 10^{57}}{\mu} \frac{M}{M_\odot}, \tag{3.1.1}$$

where $M_\odot = 1.991 \times 10^{33}$ g is the mass of the Sun. If one defines the mean volume occupied by a particle to be $(4\pi/3)\, r_0^3$, then $r_0^3\, N = R^3$, where R is the radius of the star. Combining with eq. (3.1.1), one has that

$$r_0 = \frac{R}{N^{1/3}} = 0.66 \times 10^{-8}\ \mathrm{cm}\ \mu^{1/3} \left(\frac{M_\odot}{M}\right)^{1/3} \frac{R}{R_\odot} = 1.25\, a_0\, \mu^{1/3} \left(\frac{M_\odot}{M}\right)^{1/3} \frac{R}{R_\odot}, \tag{3.1.2}$$

where $a_0 = 0.529177 \times 10^{-8}$ cm is the Bohr orbital radius of an electron in the ground state of the hydrogen atom, and $R_\odot$ is the radius of the Sun ($= 6.96 \times 10^{10}$ cm).

Equation (3.1.2) shows that, in contrast with the situation at stellar surfaces where atoms can be found in many different states of excitation, highly excited, bound atomic states of light atoms do not exist in the deep interior. Another way of seeing this is to note that the mean density in the interior, defined by

$$\bar{\rho} = \frac{M}{(4\pi/3)\, R^3} = 1.41\ \mathrm{g\ cm^{-3}}\ \frac{M}{M_\odot} \left(\frac{R_\odot}{R}\right)^3, \tag{3.1.3}$$

is comparable to that of liquids and solids at the Earth's surface.

Since their properties do not change perceptibly over human lifetimes, main sequence stars are obviously not evolving on a dynamical time scale and one may assume that inward gravitational forces are balanced by outward pressure forces. Locally,

$$\frac{dP}{dr} = -g\rho = -\frac{GM(r)}{r^2}\, \rho, \tag{3.1.4}$$

where P is the pressure, g is the acceleration of gravity at a distance r from the center, G is Newton's gravitational constant ($= 6.673 \times 10^{-8}$ dyne cm^2 g^{-2}), $M(r)$ is the mass contained within a sphere of radius r, and ρ is the density at point r. Equation (3.1.4), which embodies both the principle of the conservation of momentum and the action of the inverse square law of gravity, is the basic equation of stellar structure. It is a magical equation the simplicity of which is deceiving: it appears to be a local equation, but it is the consequence of the attraction of all of the particles in the star. That is, for a spherical distribution of mass, the gravitational forces exerted by particles outside of the sphere of radius r cancel each other, and the forces exerted by particles inside of this sphere act as if they were concentrated in a point at the center, as first shown by Isaac Newton (*Principia Mathematica*, 1687). Setting in eq. (3.1.4) $r \sim R/2$, $M(r) \sim M/2$, $dP/dr \sim -P_c/R$, and $\rho \sim \rho_c/2$ gives

$$\frac{P_c}{\rho_c} \sim \frac{GM}{R}. \tag{3.1.5}$$

In this equation, P_c and ρ_c are, respectively, the pressure and density at the center of the star.

Further progress requires the choice of an equation of state. It is clear from observed spectral characteristics that matter at stellar surfaces is gaseous. Both a relatively high surface temperature and a low surface density contribute to this circumstance. From the fact that stars emit energy, it follows that temperature must increase inward and, even though density also increases inward, there is a chance that matter in the interior is also gaseous. If one supposes that the equation of state is that of a perfect gas, then

$$P = P_{\text{gas}} = \frac{\rho}{\mu M_H} kT, \tag{3.1.6}$$

where k is Boltzmann's constant $(= 1.38065 \times 10^{-16}$ erg K$^{-1})$ and T is the temperature. Inserting eq. (3.1.6) into eq. (3.1.5) yields

$$kT_c \sim \frac{GM}{R} \mu M_H. \tag{3.1.7}$$

This last equation makes the remarkable statement that the kinetic energy of a particle at the center of a main sequence star is approximately equal to the gravitational potential energy of a particle at its surface (i.e., the energy required to remove the particle to infinity). Inserting values for fundamental constants in eq. (3.1.7) gives

$$T_c(K) \sim 23.1 \times 10^6 \, \mu \, \frac{M}{R} \frac{R_\odot}{M_\odot}. \tag{3.1.8}$$

Another way of obtaining a similar result is to make use of a virial theorem (eq. (3.3.12)) which states that the kinetic energy content of a bound system of particles is approximately half of the gravitational binding energy of this system. Setting

$$N \frac{3}{2} k\tilde{T} \sim \frac{1}{2} \frac{GM^2}{R} \tag{3.1.9}$$

defines $\tilde{T}$ as *one measure* of the mean temperature. From eqs. (3.1.9), (3.1.7), and (3.1.1), it follows that

$$\tilde{T} \sim \frac{1}{3} T_c. \tag{3.1.10}$$

The quantity kT, which, at any point in the stellar interior, provides a measure of the energy of a typical particle $\left(E \sim \frac{3}{2} kT\right)$ and of a typical photon (E $\sim 2.7 \, kT$), may be written as $kT = 86.164\ T_6$ eV, where T_6 is the temperature in units of 10^6 K and the unit eV is an electron volt ($= 1.60206 \times 10^{-12}$erg). The interior temperatures predicted by eqs. (3.1.7)–(3.1.10) show that typical values of kT are much larger than the ionization potentials of hydrogen and helium (in the terrestial laboratory, the ionization potential of hydrogen is 13.6 eV and the first and second ionization potentials of helium are 24 eV and 54 eV, respectively). Therefore, these two elements, which are the main constituents of the interstellar matter out of which stars are born, are completely ionized over most of the interior of a main sequence star. Other light elements are also quite highly ionized.

Defining the abundance by mass of hydrogen as X and that of helium as Y, one has

$$\frac{\rho}{\mu M_H} = \frac{\rho X}{M_H}(1+1) + \frac{\rho Y}{4M_H}(1+2) + \sum \frac{\rho X_i}{A_i M_H}(1+Z_i), \qquad (3.1.11)$$

where μ is defined as the molecular weight and the sum in the last term on the right of the equation extends over constituents of atomic mass (nucleon number) $A_i > 4$, degree of ionization Z_i, and abundance by mass X_i. Since the number of protons in a light nucleus is close to the number of neutrons in that nucleus (and assuming complete ionization), $\sum X_i(1+Z_i)/A_i \approx (1/2)\sum X_i = Z/2$, where $Z = 1 - X - Y$ is the abundance by mass of elements heavier than helium, not to be confused with the atomic numbers Z_i. Thus,

$$\frac{1}{\mu} \approx 2X + \frac{3}{4}Y + \frac{1}{2}Z, \qquad (3.1.12)$$

Inserting typical values $X = 0.71$, $Y = 0.27$, and $Z = 0.02$ for population I stars, eq. (3.1.12) gives $\mu \sim 0.61$, so

$$T_c(K) \sim 14.1 \times 10^6\ \bar{\mu}\ \frac{M}{M_\odot}\frac{R_\odot}{R}, \qquad (3.1.13)$$

where $\bar{\mu} = \mu/0.61$. For the Sun, this last estimate is within $\sim 10\%$ of values obtained by constructing very elaborate inhomogeneous models which take into account the change in composition variables in central regions due to the nuclear transformations discussed in Section 3.5 (see also Chapter 10).

Another interesting feature of homogeneous stars governed by the perfect gas law is that the ratio of mean density to the cube of the mean temperature is inversely related to the square of the mass of the star. Using eq. (3.1.3) for the mean density $\bar{\rho}$ and setting

$$\bar{T} \sim \frac{T_c}{2}, \qquad (3.1.14)$$

where T_c is given by eq. (3.1.8), one has

$$\left\langle \frac{\rho}{T^3} \right\rangle \sim \frac{\bar{\rho}}{\bar{T}_6^3} = \frac{0.00091}{\mu^3}\left(\frac{M_\odot}{M}\right)^2 = \frac{0.0040}{\bar{\mu}^3}\left(\frac{M_\odot}{M}\right)^2. \qquad (3.1.15)$$

3.2 Departures from a classical perfect gas: electrostatic interactions, electron degeneracy, and radiation pressure

It is instructive to examine to what extent the description of the interior of a main sequence star as a classical, perfect gas of material particles is justified. At least three phenomena are worth considering: the reduction in pressure due to the electrostatic attraction between ions and electrons; the enhancement of pressure due to the quantum mechanical exclusion principle acting on electrons; and the contribution of photons to the pressure.

3.2.1 Electrostatic forces

Because interior densities are not unlike those of terrestial solids and of solids comprising other earthlike planets in which electrostatic forces balance the gravitational force, it is reasonable to ask how important electrostatic interactions are in stars. For simplicity, consider a star made of pure, completely ionized, hydrogen for which $\mu = 1/2$. A measure of the average separation between nuclei is given by the quantity a_s in the equation

$$a_s^3 \, \frac{\rho}{M_H} = 1, \tag{3.2.1}$$

from which it follows that

$$a_s = \frac{1.186 \times 10^{-8}}{\rho_0^{1/3}} \text{ cm}, \tag{3.2.2}$$

where ρ_0 is the density in units of g cm^{-3}. If all of the nuclei were arranged in a lattice in a uniform sea of electrons, the electrostatic binding energy per nucleon between shielded nuclei and electrons would be of the order of

$$E_{\text{Coulomb}} \sim -\frac{e^2}{2a_s} = -9.725 \times 10^{-12} \, \rho^{1/3} \text{ erg} = -6.1 \, \rho_0^{1/3} \text{ eV}, \tag{3.2.3}$$

where e is the charge of an electron (4.80286×10^{-10} statcoulomb). For every proton there is an electron, so the ratio of the electrostatic interaction energy to thermal energy per electron–proton pair is

$$\frac{E_{\text{Coulomb}}}{E_{\text{thermal}}} \sim -\frac{e^2}{2a_s} \div \left(2 \times \frac{3}{2} kT\right) = -0.0235 \left(\frac{\rho_0}{T_6^3}\right)^{1/3}, \tag{3.2.4}$$

where T_6 is the temperature in units of 10^6 K. At the center of a 1 $M_\odot$ pure hydrogen star, eq. (3.1.8) suggests a central temperature of about 11.5×10^6 K, and, assuming a central density of, say, 100 g cm^{-3}, eq. (3.2.4) gives $E_{\text{Coulomb}}/E_{\text{thermal}} \sim 0.01$. The effect is small enough that one can be assured that nuclei are not ordered in a lattice and that both electrons and nuclei move freely past one another very much as in a perfect gas.

An estimate of the effect that is more representative of the whole star is worth attempting. Defining a quantity α by

$$\left\langle \left(\frac{\rho}{T^3}\right)^{1/3} \right\rangle = \frac{1}{\alpha} \left(\frac{\bar{\rho}}{T_c^3}\right)^{1/3}, \tag{3.2.5}$$

where brackets denote an average over the star and $\bar{\rho}$ and T_c are defined by eqs. (3.1.3) and (3.1.8), respectively, one has

$$\left\langle \left(\frac{\rho_0}{T_6^3}\right)^{1/3} \right\rangle \sim \frac{0.0969}{\alpha} \left(\frac{M_\odot}{M}\right)^{2/3}. \tag{3.2.6}$$

Using eq. (3.2.6) in eq. (3.2.4) gives

$$\left\langle \frac{E_{\text{Coulomb}}}{E_{\text{thermal}}} \right\rangle \sim -\frac{0.00228}{\alpha} \left(\frac{M_\odot}{M}\right)^{2/3}. \tag{3.2.7}$$

Setting $\alpha = 1/2 - 1/3$, as suggested by eqs. (3.1.14) and (3.1.10), it is evident that, over most of a star of mass 1 $M_\odot$ or more, Coulomb interaction energies are over two orders of magnitude smaller than thermal energies. As shown in Section 4.17, by an argument that leads to eq. (4.17.31), the contribution to the pressure by Coulomb forces is equal to $1/3\, U_{\text{Coulomb}}$, where U_{Coulomb} is the Coulomb energy per unit volume. Since $P_{\text{thermal}} = 2/3\, U_{\text{thermal}}$, where P_{thermal} and U_{thermal} are, respectively, the thermal pressure and the thermal energy per unit volume, one has that

$$\frac{P_{\text{Coulomb}}}{P_{\text{thermal}}} = \frac{1}{2} \frac{U_{\text{Coulomb}}}{U_{\text{thermal}}} = \frac{1}{2} \frac{E_{\text{Coulomb}}}{E_{\text{thermal}}}. \tag{3.2.8}$$

From eq. (3.2.7), it is evident that, over most of the interior of main sequence stars as massive as the Sun, the reduction in the pressure due to Coulomb forces is, at most, a fraction of a percent of the pressure given by the perfect gas law.

However, in lower main sequence stars, Coulomb interactions can affect the pressure by several percent. For example, as discussed in Section 5.6, in a star of mass $M \sim \frac{1}{9} M_\odot$, the central density and temperature are of the order of 400 g cm^{-3} and 4.6×10^6 K, respectively. Under these conditions, eqs. (3.2.4) and (3.2.8) suggest a $\sim 2\,\%$ effect on the pressure by Coulomb interactions.

3.2.2 The exclusion principle and electron degeneracy

The Heisenberg uncertainty principle suggests that the minimum momentum p of an electron is given by

$$p_{\text{min}} \sim \frac{\hbar}{a_s}, \tag{3.2.9}$$

where $\hbar = h/2\pi$, h is Planck's constant (6.626×10^{-27} erg s), and a_s is the average separation between electrons. Therefore, the average energy of an electron is at least as large as

$$E_{\min} = \frac{\hbar^2}{2m_e a_s^2} = \frac{\hbar^2}{2m_e a_0^2}\left(\frac{a_0}{a_s}\right)^2 = 13.6 \, \text{eV}\left(\frac{a_0}{a_s}\right)^2, \qquad (3.2.10)$$

where m_e is the mass of an electron (0.9109×10^{-27} g) and a_0 is the Bohr radius of the hydrogen atom. Again simplifying to stars made of pure hydrogen and using a_s from eq. (3.2.2), one has that, per electron–proton pair, the additional energy due to the uncertainty principle relative to the perfect gas component of kinetic energy is

$$\frac{E_{\min}}{3\,kT} = 0.0105 \, \frac{\rho_0^{2/3}}{T_6}. \qquad (3.2.11)$$

At the center of a 1 $M_\odot$ hydrogen star with $T_c \sim 11.5$ and $\rho_c \sim 100$, eq. (3.2.10) gives $E_{\min}/E_{\text{thermal}} \sim 0.02$. Approximating $\langle \rho^{2/3}/T \rangle$ by $\sim (2\text{–}3)\,(\bar{\rho}^{2/3}/T_c)$, one has that, over much of the star,

$$\left\langle \frac{E_{\min}}{3\,kT} \right\rangle \sim \frac{0.0023\text{–}0.0034}{\mu}\left(\frac{M_\odot}{M}\right)^{1/3}\frac{R_\odot}{R}. \qquad (3.2.12)$$

Thus, over most of the interiors of main sequence stars as massive as the Sun, the classical kinetic energy of an electron is at least two orders of magnitude larger than the minimum required by the uncertainty principle.

This does not mean, however, that the effect on the pressure is also less than a percent or so. In determining the effect on the pressure of crowding electrons together, what is important is not the uncertainty principle, but the Pauli exclusion principle. That is, in doing the statistical mechanics of electrons, what counts is not the minimum product $\Delta r \Delta p \gtrsim \hbar$ defined by the uncertainty principle, but rather the size of a unit cell in phase space and the requirement that only two electrons of opposite spin can occupy such a cell. The unit cell size is $h = 2\pi\hbar$, and, as described in an elementary fashion in Section 5.8, when electron crowding is large enough, electrons are distributed throughout the momentum space $p = 0$ to $p = p_F$, where

$$p_F \sim \left(\frac{3}{4\pi}\right)^{1/3}\frac{h}{a_s} \sim 3.9\, p_{\min}, \qquad (3.2.13)$$

p_F is called the Fermi momentum, and $p_{\min}$ is given by eq. (3.2.9). The net effect is that the pressure due to electron crowding, normally referred to as electron degeneracy, is significantly larger than might have been guessed from considerations based on the minimum momentum given by the uncertainty principle.

When electron crowding is of the order of that encountered in solar-type stars, the situation is as described exhaustively in Section 4.5 for weak degeneracy. There it is shown that electron degeneracy increases the electron pressure at the center of the Sun by about 4% above that given by the perfect gas law. In stars on the lower main sequence, pressure due to electron degeneracy can become comparable to the thermal pressure. For example, at the center of the star of mass $M \sim (1/9)M_\odot$ discussed in Section 5.6 and 6.12, $P_e/n_e kT \sim 1.7$, where P_e is the actual pressure exerted by electrons and $n_e kT$ is the pressure which would be exerted by the electrons if they were not partially degenerate.

3.2.3 Radiation pressure

In relatively massive main sequence stars, the contribution of photons to the pressure is much more important than the negative contribution of Coulomb interactions and the positive contribution of electron degeneracy. The average temperature gradient in a main-sequence star ($\sim 1\,°$K in 100 meters) is small compared with what can normally be achieved in a terrestial laboratory. The mean free path of a typical photon is of the order of one centimeter (see Section 3.4). Hence, at any point in the star, the radiation field is in thermodynamic equilibrium with the gas and the radiation intensity is highly isotropic. This means that the radiation field can be described to a high degree of accuracy by a black body distribution. In a black body, the energy density of photons is given by $U_{rad} = aT^4$, where the radiation constant is $a = 7.5675 \times 10^{-15}$ erg cm^{-3} K^{-4}, and U_{rad} is in units of energy per unit volume. The related pressure is

$$P_{rad} = \frac{1}{3} U_{rad} = \frac{1}{3} aT^4. \tag{3.2.14}$$

The ratio of radiation pressure to particle, or gas pressure (in the perfect gas approximation) is

$$\frac{P_{rad}}{P_{gas}} = 3.05 \times 10^{-23} \frac{\mu\, T^3}{\rho} = 3.05 \times 10^{-5} \frac{\mu\, T_6^3}{\rho_0}. \tag{3.2.15}$$

Defining $\bar{\alpha}$ by

$$\left\langle \frac{T^3}{\rho} \right\rangle = \bar{\alpha}^3 \frac{T_c^3}{\bar{\rho}}, \tag{3.2.16}$$

and using eqs. (3.1.8) and (3.1.13) one has

$$\left\langle \frac{P_{rad}}{P_{gas}} \right\rangle \sim 0.269\, \bar{\alpha}^3\, \mu^4 \left(\frac{M}{M_\odot} \right)^2 = 0.0372\, \bar{\alpha}^3\, \bar{\mu}^4 \left(\frac{M}{M_\odot} \right)^2. \tag{3.2.17}$$

With $\bar{\alpha} \sim 1/3 - 1/4$, the numerical coefficient in the middle term of eq. (3.2.17) is in the range 0.01–0.004 and the numerical coefficient in the third term is in the range 0.0014–0.0006. Thus, in stars of solar mass or less, radiation pressure is of negligible importance; in stars ten times more massive than the Sun, it contributes several percent to the pressure; and, in very massive stars, it can be the dominant contributor to the total pressure.

Neglecting all forms of pressure other than gas pressure and radiation pressure, so that

$$P = P_{gas} + P_{rad}, \tag{3.2.18}$$

and defining a quantity β by

$$\beta = \frac{P_{gas}}{P}, \tag{3.2.19}$$

one has that

$$P = \frac{P_{\text{gas}}}{\beta} = \frac{P_{\text{rad}}}{1 - \beta}. \tag{3.2.20}$$

All previous order of magnitude relationships can be appropriately modified by substituting $\mu \beta / k$ for μ / k. For example, eq. (3.1.7) becomes

$$kT_c \sim \frac{GM}{R} \beta_c \mu M_H, \tag{3.2.21}$$

where β_c is the value of β at the center, and eq. (3.2.15) becomes

$$\left\langle \frac{P_{\text{rad}}}{P_{\text{gas}}} \right\rangle = \frac{1 - \bar{\beta}}{\bar{\beta}} \sim 0.269 \, \bar{\alpha}^3 \, \mu^4 \left(\frac{M}{M_\odot} \right)^2 \beta_c^3 = 0.0372 \, \bar{\alpha}^3 \, \beta_c^3 \, \bar{\mu}^4 \left(\frac{M}{M_\odot} \right)^2, \tag{3.2.22}$$

where $\bar{\beta}$ is an appropriate average of β. Setting $\bar{\mu} = 1.0$, $\beta_c \sim \bar{\beta}$, and $\bar{\alpha} \sim 1/4$, eq. (3.2.22) gives for homogeneous main sequence stars of masses 10, 50, and 100 $M_\odot$, respectively, $\bar{\beta} \sim 0.95$, 0.68, and 0.53.

For even more massive stars in which radiation pressure dominates gas pressure, it makes sense to repeat the exercise which leads to eq. (3.1.7), this time assuming that radiation is the only source of pressure. From eq. (3.1.4), it follows that

$$\frac{P_c}{R} \sim \frac{aT_c^4/3}{R} \sim G \frac{M/2}{(R/2)^2} \, 2 \, \frac{3}{4\pi} \frac{M}{R^3}, \tag{3.2.23}$$

or

$$T_c^4 \sim \frac{9G}{\pi a} \frac{M^2}{R^4}. \tag{3.2.24}$$

Numerically,

$$T_c(\text{K}) \sim 54.2 \times 10^6 \left(\frac{M}{M_\odot} \right)^{1/2} \frac{R_\odot}{R}. \tag{3.2.25}$$

3.3 Virial theorems relating kinetic, gravitational binding, and net binding energies

For main sequence stars, thanks to the dominance of classical gas pressure and radiation pressure, there exist two simple relationships between bulk properties of the star: (1) a relationship between the binding energy of the star and the total thermal energy of the material particles in the star, and (2) a relationship between the binding energy, the gravitational binding energy, and the relative importance of gas pressure to radiation pressure in the star. The two relationships are called virial theorems (from the Latin *vis*, *viris*, pertaining to strength). Derivation of the relationships begins by multiplying both the leftmost and rightmost sides of eq. (3.1.4) by $4\pi r^3 dr = r\, dV = r\, dM(r)/\rho$. Integration by parts of the leftmost side of the resulting equation yields

$$3 \int_0^s P \mathrm{d}V - 4\pi R^3 P_s = \int_0^s \frac{GM(r)\mathrm{d}M(r)}{r} = -\Omega. \tag{3.3.1}$$

Here, integration is from the center (0) to the surface (s), P_s is the pressure at the surface, R is the stellar radius, and $-\Omega = |\Omega|$ is the gravitational binding energy of the star. Since the average value of P in the interior is much, much larger than P_s, the second term on the leftmost side of eq. (3.3.1) is minuscule compared with the first term. Thus,

$$\Omega = -3 \int_0^s P \mathrm{d}V. \tag{3.3.2}$$

For a gas consisting of ions and electrons at non-relativistic energies, $P_{gas} = (2/3)\, U_{gas}$, where U_{gas} is the kinetic energy density of particles in units of energy per volume. Using this relationship and eq. (3.2.14), and neglecting all sources of pressure other than gas pressure and radiation pressure, one has

$$P = P_{gas} + P_{rad} = \frac{2}{3} U_{gas} + \frac{1}{3} U_{rad}, \tag{3.3.3}$$

and eq. (3.3.2) becomes

$$2E_{gas} + E_{rad} = -\Omega = |\Omega|, \tag{3.3.4}$$

where

$$E_{gas} = \frac{3}{2} \int P_{gas}\, \mathrm{d}V = \int U_{gas}\, \mathrm{d}V \tag{3.3.5}$$

is the total kinetic energy of particles and

$$E_{rad} = 3 \int P_{rad}\, \mathrm{d}V = \int U_{rad}\, \mathrm{d}V \tag{3.3.6}$$

is the total energy of radiation.

The binding energy E_{bind} of a star may be defined as the energy required to place all of the particles in the star at rest at infinite distances from one another. That is, the binding energy is the difference between the gravitational binding energy of the star and the total thermal energy content of the star:

$$E_{bind} = |\Omega| - (E_{gas} + E_{rad}). \tag{3.3.7}$$

Then, using eq. (3.3.4), it follows that

$$E_{bind} = E_{gas}. \tag{3.3.8}$$

Thus, the binding energy, or the energy required to convert the star into an infinitely diffuse object at zero temperature, is equal to the kinetic energy of the material particles in the star. In this form of the virial theorem, the two types of energy being compared are of equal strength. The most interesting feature of the relationship is that it is entirely independent of the radiation energy content of the star.

Replacing P by P_{gas}/β in eq. (3.3.2) gives

$$\Omega = -3\,\frac{1}{\tilde{\beta}}\int P_{gas}\,dV = -\frac{2}{\tilde{\beta}}\,E_{gas},\tag{3.3.9}$$

where $\tilde{\beta}$ is yet another weighted average of β defined by

$$\tilde{\beta} = \int U_{gas}\,dV \div \int \frac{1}{\beta}\,U_{gas}\,dV.\tag{3.3.10}$$

Thus, as long as material particles have non-relativistic kinetic energies,

$$E_{gas} = -\frac{\tilde{\beta}}{2}\,\Omega = \frac{\tilde{\beta}}{2}\,|\Omega|.\tag{3.3.11}$$

For low and intermediate mass main sequence stars in which radiation is not an important contributor to pressure,

$$E_{gas} \sim -\frac{1}{2}\,\Omega = \frac{1}{2}\,|\Omega|.\tag{3.3.12}$$

In this second virial theorem, the strengths of the two types of energy which are being compared differ by a factor of two. Equation (3.3.12) is equivalent to the virial theorem in classical mechanics in its most general form: under the influence of an inverse square law of force between particles, the average kinetic energy of a system of particles is equal to minus one-half of the average potential energy of the system. Making use of eq. (3.3.8), one can also write eq. (3.3.12) as

$$E_{bind} \sim -\frac{1}{2}\,\Omega = \frac{1}{2}\,|\Omega|,\tag{3.3.13}$$

which is also true in classical mechanics for bound systems. It is remarkable that, although the formulations pertinent to stellar structure make central use of thermodynamics and the existence of pressure forces between particles, the results are the same as in classical mechanics which makes no mention of thermodynamics or pressure forces.

Returning to the more general case where

$$E_{gas} = E_{bind} = \frac{\tilde{\beta}}{2}\,|\Omega|,\tag{3.3.14}$$

and making use of the relationship

$$E_{thermal} = E_{gas} + E_{rad},\tag{3.3.15}$$

it follows from eqs. (3.3.4), (3.3.7), and (3.3.11) that

$$E_{rad} = 2\left(\frac{1}{2}\,|\Omega| - E_{bind}\right)\tag{3.3.16}$$

and

$$E_{rad} = 2\left(E_{thermal} - \frac{1}{2}\,|\Omega|\right).\tag{3.3.17}$$

Combining eqs. (3.3.16) and (3.3.14), one has that

$$E_{\rm rad} = \left(1 - \tilde{\beta}\right) \frac{1}{2} |\Omega|. \qquad (3.3.18)$$

Finally, eqs. (3.3.18) and (3.3.14) give

$$\frac{E_{\rm rad}}{E_{\rm gas}} = 2 \, \frac{1 - \tilde{\beta}}{\tilde{\beta}}. \qquad (3.3.19)$$

This last equation shows that the global ratio of photon energy to particle kinetic energy increases monotonically with decreasing $\tilde{\beta}$.

As stellar mass is increased and radiation pressure grows in importance, the ratios

$$\frac{E_{\rm gas}}{|\Omega|} = \frac{E_{\rm bind}}{|\Omega|} = \frac{\tilde{\beta}}{2} \qquad (3.3.20)$$

become smaller and smaller. Ultimately, in very, very massive main sequence stars (say, $M \sim 10^6 - 10^8 \, M_\odot$, if one can imagine the formation of such stars), the total thermal energy content (particles plus radiation) and the gravitational binding energy become nearly identical. In such stars, at least three additional effects must be taken into account: electrons become relativistic; electron–positron pairs are formed; and the effects of general relativity (in which the space-time metric is determined by the distribution of mass) become important.

As the mass and interior temperatures of a model are increased, the number densities of charged leptons of each charge increase and eventually become comparable with the number density of photons (see Section 4.10). The rest mass of charged leptons can therefore exceed the rest mass of nucleons. Since the relationship between pressure and energy density for the leptons becomes identical to that for radiation ($U = 3 \, P$), eq. (3.3.8) is replaced by

$$E_{\rm bind} = E_{\rm baryon \, gas}, \qquad (3.3.21)$$

the equivalent of $\tilde{\beta}$ in eq. (3.3.11) tends toward zero, and the binding energy becomes a very small fraction of the gravitational binding energy. This last circumstance allows general relativity to play a role in determining the stability of a hypothetical supermassive star (R. P. Feynman 1963, private communication; Iben, 1963).

3.4 Energy transport: radiation, convection, and conduction

Energy can be carried outwards through a star by photons, by bulk convective motions, and by conduction. In main sequence stars, radiation and convection are the primary modes of transport.

3.4.1 Radiative flow and mass–luminosity relationships

To the extent that the mean free path of a typical photon is small compared with a temperature scale height, one may treat radiative flow as a diffusive phenomenon in which the rate of energy transport is related to the local temperature gradient in a fashion which depends only on the local composition and the local thermodynamic conditions. The simplest way to derive the equation of radiative transport is to proceed by analogy with the elementary treatment of diffusion and conduction in the kinetic theory of gases. One begins with

$$F_{\text{rad}} = -\frac{1}{3}\frac{dU_{\text{rad}}}{dr}\, l_{\text{ph}}\, c, \tag{3.4.1}$$

where F_{rad} is the flux of energy via photons, l_{ph} is the mean free path of a typical photon, and c is the speed of light. A mean free path is defined by setting the attenuation dI of a beam of radiation of initial intensity I passing through a thin slab of material of thickness dx equal to $I\,(dx/l_{\text{ph}})$. Defined in this way,

$$\frac{1}{l_{\text{ph}}} \equiv n\sigma \equiv \kappa\rho, \tag{3.4.2}$$

where σ is the cross section of absorbers (or scatterers) which are at a number density n in the slab, and κ is the opacity. Using eqs. (3.2.14) and (3.4.2), eq. (3.4.1) becomes

$$F_{\text{rad}} = -\frac{4ac}{3}\frac{T^3}{\kappa\rho}\frac{dT}{dr}. \tag{3.4.3}$$

Exactly the same result follows from considerations of momentum conservation. Consider a spherical shell of matter into which radiation flows from below (at a flux F_+) and from above (at a flux F_-). The momentum flux associated with an energy flux F is just F/c and the rate at which the shell of thickness dr absorbs momentum from the radiation field is $(F_+/c - F_-/c)\kappa\,\rho\,dr$. But this is a force per unit area and is therefore equivalent to a pressure difference. Thus,

$$\frac{1}{c}(F_+ - F_-)\kappa\rho = -\frac{dP_{\text{rad}}}{dr}. \tag{3.4.4}$$

Differentiating the first and third terms in eq. (3.2.14), noting that

$$F_+ - F_- = F_{\text{rad}} = \frac{L}{4\pi r^2}, \tag{3.4.5}$$

and setting the results in eq. (3.4.4), one recovers eq. (3.4.3).

A very far-reaching theorem can be derived if it is assumed that radiative diffusion is the dominant form of energy transport through most of the star. Then,

$$F_{\text{rad}} = F_{\text{tot}} = \frac{L(r)}{4\pi r^2} = -\frac{4ac}{3}\frac{T^3}{\kappa\rho}\frac{dT}{dr}, \tag{3.4.6}$$

where $L(r)$ is the total rate at which energy flows through a spherical shell at radius r and F_{tot} is the total flux. In the spirit of the approximations made up to this point, eq. (3.4.6) is equivalent to the order of magnitude equality

$$\frac{L}{R^2} \sim ac \left\langle \frac{T^3}{\kappa\rho} \right\rangle \frac{T_c}{R}, \tag{3.4.7}$$

where L is the surface luminosity and brackets denote an average.

A similar result is achieved by comparing the time t_{diff} required for energy to diffuse from the interior to the surface with the time t_{leak} it would take all of the radiation energy in the star to leak out at the rate L. Supposing that photons diffuse outward in a random walk of step size $\bar{l}_{ph}$, it follows that

$$t_{diff} \sim \frac{R^2}{c \, \bar{l}_{ph}}. \tag{3.4.8}$$

The leakage time scale is given in order of magnitude by

$$t_{leak} \sim \frac{a\bar{T}^4 R^3}{L}. \tag{3.4.9}$$

Setting $\bar{l}_{ph} = 1/\bar{\kappa}\bar{\rho}$, $\bar{T} \propto T_c$, and equating t_{diff} with t_{leak} leads to roughly the equivalent of eq. (3.4.7).

To progress further, in eq. (3.4.7) one may set $\langle T^3/\kappa\rho\rangle = \langle T^3/\rho\rangle \div \bar{\kappa}$, where $\bar{\kappa}$ is a measure of the mean opacity, $\langle T^3/\rho\rangle$ is given by eq. (3.2.16), and T_c is given by eq. (3.1.8). The result is

$$L = \bar{\alpha}^3 \frac{\mu^4}{\bar{\kappa} \, (\text{cm}^2 \, \text{g}^{-1})} \left(\frac{M}{M_\odot} \right)^3 L_0, \tag{3.4.10}$$

where

$$L_0 = ac \left(\frac{GM_H}{k} \right)^4 M_\odot^3 = 197 L_\odot, \tag{3.4.11}$$

$L_\odot$ is the Sun's luminosity (3.86×10^{33} erg s^{-1}), and $\bar{\kappa}$ is in units of cm^2 g^{-1}. Equation (3.4.10) may be written as

$$\frac{L}{L_\odot} \sim (24.6\text{--}7.30) \frac{\mu^4}{\bar{\kappa}} \left(\frac{M}{M_\odot} \right)^3 = (3.41\text{--}1.01) \frac{\bar{\mu}^4}{\bar{\kappa}} \left(\frac{M}{M_\odot} \right)^3. \tag{3.4.12}$$

Remarkably, the mass–luminosity relationship given by eqs. (3.4.10) and (3.4.12) has been derived without any reference to the nature of the energy sources in the star. This suggests that the actual energy sources in main sequence stars are highly sensitive to the temperature or to the density, or to both. For a given choice of stellar composition, mass, and radius, the mean values of temperature and density in the interior are determined; therefore, the opacity (which is a function of composition, density, and temperature) is also determined. Thus, not only is the rate at which energy flows from the surface fixed by the mass and radius of the star, but the mean interior structure variables are constrained as well. Interior energy sources can adjust to meet the demand only if they are very sensitive to those alterations in the structure variables which are consistent with the imposed constraints; that is, they must be sensitive to small changes in the temperature and/or in the density. This interpretation must not be carried too far, as the rates at which nuclear reactions proceed (Section 3.5) do help determine the precise structure of the star and, hence, the average opacity. Therefore, hidden in $\bar{\kappa}$ are effects of nuclear reaction rates.

The quantitative estimate of the mass–luminosity relationship given by eq. (3.4.12), when compared with the empirical relationship, $L/L_\odot \sim (M/M_\odot)^4$, which holds for main sequence stars of mass in the range 1–10 $M_\odot$ (see Chapter 2), suggests that the opacity in such stars is of order unity and decreases with increasing mass. Since interior temperatures increase and densities decrease with increasing stellar mass along the main sequence, one may infer that, in general, opacities decrease with increasing temperatures and decreasing densities. A preliminary estimate of opacities based on atomic physics and an awareness of interior temperatures and densities strengthens this inference. Given that a mean opacity is of order unity, one has from eqs. (3.4.8) and (3.4.9) that the energy diffusion time scale, $t_{\text{diff}} = \bar{\kappa}\rho R^3/cR \sim \bar{\kappa}\, M/cR \sim 3 \times 10^4\, \bar{\kappa}\, (M/M_\odot)\,(R_\odot/R)$ yr, is of the order of 10^4–10^5 yr for stars like the Sun.

For very massive stars in which β is small, a different mass–luminosity relationship is more appropriate. In such stars, $dP/dr \sim dP_{\text{rad}}/dr$ and eqs. (3.1.4) and (3.4.6) give

$$L = \frac{4\pi cGM}{\kappa_s} = \frac{1.30 \times 10^4}{\kappa_s(\text{cm}^2\,\text{g}^{-1})}\,\frac{M}{M_\odot}\,L_\odot, \qquad (3.4.13)$$

where κ_s is the opacity near the surface ($M(r) \to M$) in units of cm^2 g^{-1}. The luminosity given by eq. (3.4.13) when κ_s is set equal to the electron scattering opacity (eqs. (3.4.17) and (3.4.19)) is known as the Eddington luminosity.

3.4.2 Opacity sources

There are four major sources of opacity: electron scattering, inverse bremsstrahlung (free–free absorption), the photoelectric effect (bound–free absorption), and transitions between bound atomic levels (bound–bound absorption). Electron scattering, which provides a lower bound to the opacity, is the process most easy to estimate.

Consider a beam of radiation hitting an electron. Let the amplitude of the electric field be E_{electric} and the acceleration of the electron be a. Then, $m_e a = -eE_{\text{electric}}$, and, from classical electromagnetic theory, the rate at which the electron emits energy is

$$\dot{E}_{\text{emit}} = \frac{2}{3}\frac{e^2}{c^3}a^2 = \frac{2}{3}\frac{e^2}{c^3}\left(\frac{eE_{\text{electric}}}{m_e}\right)^2. \qquad (3.4.14)$$

The rate at which the electron absorbs energy is

$$\dot{E}_{\text{absorb}} = c\,\frac{E^2_{\text{electric}}}{4\pi}\,\sigma_{\text{es}}, \qquad (3.4.15)$$

where σ_{es} is the electron scattering cross section. Setting $\dot{E}_{emit} = \dot{E}_{absorb}$ gives

$$\sigma_{es} = \frac{8\pi}{3} \left(\frac{e^2}{m_e c^2} \right)^2 = \frac{8\pi}{3} r_{el}^2 = 0.665246 \times 10^{-24} \text{ cm}^2, \qquad (3.4.16)$$

where $r_{el} = 2.81794 \times 10^{-13}$ cm is the classical radius of the electron. The quantity σ_{es} is called the Thomson cross section σ_{Th} after the nineteenth century British physicist J. J. Thomson who discovered the electron in 1897.

If electron scattering were the only source of opacity,

$$\kappa = \kappa_{es} = \frac{n_e \sigma_{es}}{\rho} = \frac{1}{\mu_e} \frac{\sigma_{es}}{M_H}, \qquad (3.4.17)$$

where n_e is the electron density and μ_e (called the electron molecular weight) is given by

$$\frac{1}{\mu_e} = \frac{M_H n_e}{\rho} = X + \frac{2}{4} Y + \sum_i X_i \left(\frac{Z_i}{A_i} \right) \approx \frac{1 + X}{2}. \qquad (3.4.18)$$

Thus, the electron scattering opacity (at temperatures typical of main sequence stars) is

$$\kappa_{es} = 0.20 \, (1 + X) \text{ cm}^2 \text{ g}^{-1}. \qquad (3.4.19)$$

It is ironic that it has been possible to obtain a correct first estimate for the *scattering* cross section of photons by assuming that these photons are actually *absorbed* and then *reradiated*.

The next most easily estimated process is that in which photons are absorbed by "free" electrons which, though unbound, are deflected by the much heavier positively charged nuclei. An interaction between an electron and another more massive charged particle is necessary because a truly free electron cannot absorb a photon and still conserve energy and momentum. The process is the inverse of the "bremsstrahlung" (radiation-braking) process whereby an electron radiates energy as it accelerates and then decelerates about a positive ion.

To proceed, consider a small cube of matter immersed in a radiation bath and suppose that a detailed balance exists between the rate at which electrons in the cube emit bremsstrahlung radiation and the rate at which photons entering the cube from the radiation bath are absorbed. The volume must be imagined small enough that it is "optically thin", so that essentially all of the radiation which is emitted in the cube by bremsstrahlung escapes the cube. The rate at which photon energy enters through the six sides of the cube is

$$\left(\frac{dE}{dt} \right)_{absorb} = 6 \, \sigma T^4 = 6 \, \frac{c}{4} \, a T^4, \qquad (3.4.20)$$

where σ is the Stefan–Boltzmann constant (5.6704×10^{-5} erg s^{-1} K^{-4}) and σT^4 is the radiation flux in any direction in a black body.

An electron which has a speed v_0 when it is far from a nucleus of charge Ze can come within a distance b of this nucleus before receding again. In first approximation,

$$\frac{1}{2} m_e v_0^2 = \frac{Ze^2}{b}. \qquad (3.4.21)$$

The energy radiated by the electron is, in the classical approximation,

$$\Delta E = \frac{2}{3}\frac{e^2}{c^3}\int a^2 dt = \frac{2}{3}\frac{e^2}{c^3}\int \frac{dv}{dt}dv = \frac{2}{3}\frac{e^2}{c^3}\int \frac{Ze^2}{m_e r^2}dv, \tag{3.4.22}$$

where r is the distance between the electron and the nucleus.

The rate at which bremsstrahlung radiation is emitted by the cube is equal to the rate of collisions between electrons and nuclei per unit volume times the energy lost per collision, averaged over all collisions, and may be written as

$$\left(\frac{dE}{dt}\right)_{emit} = \langle n_e\,\sigma_{coll}\,v\,\Delta E\,n_Z\rangle, \tag{3.4.23}$$

where σ_{coll} is the collision cross section and n_Z is the number density of ionized nuclei. At this point, all further pretense at rigor is abandoned and all numerical factors are assumed to be of the order of unity. For example, in eq. (3.4.22), the integral is replaced by $(Ze^2 v_0/m_e b^2)$; in eq. (3.4.23), σv is replaced by $b^2 v_0$ from eq. (3.4.21) and ΔE is replaced by $(e^4/m_e c^3)(v_0/b^2)$. Further, n_e is set equal to $n_Z \sim (\rho/M_H)$ and v_0 is set equal to $(kT/m_e)^{1/2}$. Then, setting $(dE/dt)_{emit} = (dE/dt)_{absorb}$, it follows that

$$\kappa_{ff} \sim \frac{k}{a M_H^2}r_0^2\frac{\rho}{T^3} = 540\frac{\rho_0}{T_6^3}\ cm^2\ g^{-1}. \tag{3.4.24}$$

A more careful attention to coefficients (especially noting that the integral in eq. (3.4.22) has been considerably overestimated) shows that, in completely ionized hydrogen- and helium-rich matter, as given by the development in Section 7.6 culminating in eq. (7.6.44),

$$\kappa_{ff} \sim 41\frac{\rho_0}{T_6^{7/2}}\ cm^2\ g^{-1}. \tag{3.4.25}$$

Bound–free absorption becomes important when densities are small enough that, for a substantial number of nuclei, at least the ground electron–nucleus state exists, and temperatures are small enough that this state can be significantly populated. Bound–bound absorption becomes important when densities are small enough that, again for a substantial number of nuclei, several bound electron–nucleus states exist and temperatures are small enough that these states can be significantly populated. At this juncture, no attempt is made to estimate the contributions to the opacity of bound–free and bound–bound transitions. These contributions are strong functions of the composition and a proper calculation requires a detailed estimate of the energy level population, a fully quantum mechanical estimate of cross sections, and an integration over the photon distribution (see Chapter 7).

3.4.3 Convection

Dynamical characteristics of granulations at the solar surface reveal that energy is carried outward by convection in subphotospheric layers of the Sun, and that the process involves the continuous formation and dissipation of evanescent "cells". Hot matter rises upward at slow speeds within a cell and cold matter sinks at higher velocities at cell boundaries. A similar phenomenon is exhibited by oatmeal cooking in a pot.

Full blown turbulent convection is a very complex phenomenon and sophisticated methods have been developed to deal with it more rigorously, but, from the point of view of developing an intuitive understanding of the effect of convection, it is difficult to improve upon the simple treatment which replaces the evanescent cells and the associated complicated mathematics with a picture of clumps of matter moving up and down in an idealized, symmetric way.

The most commonly used criterion for determining whether or not convection occurs in layers of homogeneous composition is the one developed at the beginning of the nineteenth century by Karl Schwarzschild. Suppose that a first approximation to the local structure variables exists and that a first estimate of the gradients of these variables can be made. For example, one could construct a model assuming that energy transport is everywhere by photon diffusion. That is, the temperature gradient at any point in the model could be approximated by setting $F_{rad} = F_{tot}$ in eq. (3.4.3), where F_{tot} is the total energy flux:

$$\frac{dT}{dr} = \left(\frac{dT}{dr}\right)_{rad} \equiv -\frac{3}{4ac} \frac{\kappa\rho}{T^3} F_{tot}. \tag{3.4.26}$$

Next, ask what would happen to a hypothetical clump of matter in the model that, due to a statistical fluctuation, acquires a temperature larger than the ambient mean temperature. Since it remains in pressure equilibrium with its surroundings (pressure equilibrium is reached on a dynamical time scale), the clump expands and thereupon experiences a buoyant force which causes it to move upward against gravity. If transfer of thermal energy through the walls of the clump proceeds initially on a time scale long compared with the time scale for upward motion, the clump continues for a while to expand adiabatically. If the adiabatic temperature gradient is larger (in absolute value) than the temperature gradient of the ambient matter, the clump eventually reaches a point where it is again in thermal as well as in pressure equilibrium with ambient matter. Having a finite velocity, the clump continues to rise, but now, since its density decreases less rapidly than does the density of ambient matter, it experiences a drag force impeding its progress upward and eventually reversing its direction of motion. The hypothetical clump oscillates back and forth without contributing to the net flow of energy outward. Thus, a criterion for stability against convection (the Schwarzschild criterion) is

$$-\left(\frac{dT}{dr}\right)_{ad} > -\left(\frac{dT}{dr}\right)_{rad}, \tag{3.4.27}$$

where $(dT/dr)_{ad}$ is the adiabatic gradient obtained from an equation of state and $(dT/dr)_{rad}$ is given by eq. (3.4.26).

Combining eq. (3.1.4) with eq. (3.4.26), one can define a logarithmic radiative gradient V_{rad} by

$$V_{rad} \equiv \left(\frac{d\log T}{d\log P}\right)_{rad} = \frac{3}{4ac} \frac{\kappa P}{gT^4} F_{tot}$$

$$= \frac{3}{4ac} \frac{\kappa P}{T^4} \frac{L}{4\pi GM}, \tag{3.4.28}$$

where the second line follows from the first by noting that $F_{tot}/g = L/4\pi GM$. Similarly, a logarithmic adiabatic gradient can be defined as

$$V_{ad} \equiv \left(\frac{d \log T}{d \log P} \right)_{ad}. \tag{3.4.29}$$

In terms of these two gradients, the stability criterion takes the form

$$V_{rad} < V_{ad}. \tag{3.4.30}$$

If eq. (3.4.27) (or eq. (3.4.30)) is not satisfied, the hypothetical clump continues to rise, but thermal equilibrium with the ambient material is eventually achieved as a consequence of a transfer of thermal energy from the clump to the ambient medium. The clump has carried energy outward and deposited it along its path. For every clump that, due to a statistical fluctuation, acquires a temperature larger than the ambient mean temperature, there will be another one which acquires a lower temperature and sinks; as it sinks, it acquires energy from its surrounding and comes into thermal equilibrium.

To estimate the flux of energy carried by convection, one may postulate the existence of a characteristic distance, or "mixing length", l_{mix}, over which a perturbed element persists and delivers energy to its surroundings. As a measure of the rate of flow of energy one may adopt

$$F_{conv} \sim \rho \, C_P \left(\left| \frac{dT}{dr} - \left(\frac{dT}{dr} \right)_{ad} \right| l_{mix} \right) v_{conv}, \tag{3.4.31}$$

where dT/dr is the ambient temperature gradient, C_P is the specific heat at constant pressure, and v_{conv} is the mean speed of a clump. The speed v_{conv} may be estimated by equating the mean kinetic energy of a clump to the work done by the buoyant force over a distance l_{mix}:

$$\rho v_{conv}^2 = \langle \delta \rho \rangle \, g \, l_{mix} \sim \left\langle \left| \left(\frac{d\rho}{dr} \right) - \left(\frac{d\rho}{dr} \right)_{ad} \right| \right\rangle l_{mix} \, g \, l_{mix}. \tag{3.4.32}$$

Since clumps and the ambient medium are in pressure balance,

$$\left(\frac{d \log_e \rho}{dr} \right) - \left(\frac{d \log_e \rho}{dr} \right)_{ad} \sim \left(\frac{d \log_e T}{dr} \right) - \left(\frac{d \log_e T}{dr} \right)_{ad}. \tag{3.4.33}$$

Equations (3.4.31)–(3.4.33) give

$$F_{conv} \sim (\rho C_P T) \, g^{1/2} \left\langle \left| \left(\frac{d \log_e T}{dr} \right) - \left(\frac{d \log_e T}{dr} \right)_{ad} \right|^{3/2} \right\rangle l_{mix}^2. \tag{3.4.34}$$

Using the gradient V_{ad} given by eq. (3.4.29) and the fact that the pressure scale height H_P is given by

$$\frac{1}{H_P} = -\frac{1}{P} \frac{dP}{dr} = -\frac{g\rho}{P}, \tag{3.4.35}$$

eq. (3.4.34) can also be written as

$$F_{\text{conv}} \sim \rho \, C_P T \left(\frac{P}{\rho}\right)^{1/2} (V - V_{\text{ad}})^{3/2} \left(\frac{l_{\text{mix}}}{H_P}\right)^2. \tag{3.4.36}$$

where

$$V = \left(\frac{\text{d} \log T}{\text{d} \log P}\right) \tag{3.4.37}$$

is the ambient logarithmic gradient of temperature with respect to pressure. Similarly, eq. (3.4.32) for the convective speed can be written as

$$v_{\text{conv}} \sim \left(\frac{P}{\rho}\right)^{1/2} (V - V_{\text{ad}})^{1/2} \left(\frac{l_{\text{mix}}}{H_P}\right), \tag{3.4.38}$$

and, thus,

$$F_{\text{conv}} \sim \rho \, C_P T \, v_{\text{conv}} \, (V - V_{\text{ad}}) \left(\frac{l_{\text{mix}}}{H_P}\right). \tag{3.4.39}$$

These equations constitute the mixing-length algorithm for convection. The choice of l_{mix}/H_P as the parameter in the algorithm is motivated by the thought that postulated clumps cannot retain their identity over a distance larger than a scale height and by the fact that, in the context of a star, there is often no other natural length scale present, particularly if a convective zone is much larger than a pressure scale height. It is important to recognize that this formulation is really just an algorithm and not a theory. If any stellar evolutionary consequences hinge delicately on the choice of l_{mix}/H_P, recourse must be had to more sophisticated treatments. Either that, or one may experiment with a range of choices of l_{mix}/H_P and be content with an estimate of the range of possibily acceptable answers.

The speed of sound in a perfect gas is

$$v_{\text{sound}} = \left(\frac{\text{d}P}{\text{d}\rho}\right)_{\text{ad}}^{1/2}, \tag{3.4.40}$$

so one can also express v_{conv} as

$$v_{\text{conv}} \sim (V - V_{\text{ad}})^{1/2} \left(\frac{l_{\text{mix}}}{H_P}\right) \left(\frac{\text{d} \log \rho}{\text{d} \log P}\right)_{\text{ad}}^{1/2} v_{\text{sound}}. \tag{3.4.41}$$

Using this in eq. (3.4.36) gives

$$F_{\text{conv}} \sim \rho \, C_P T \, v_{\text{sound}} \left(\frac{\text{d} \log \rho}{\text{d} \log P}\right)_{\text{ad}}^{1/2} (V - V_{\text{ad}}) \left(\frac{l_{\text{mix}}}{H_P}\right). \tag{3.4.42}$$

Since one may reasonably insist that $v_{\text{conv}} \leq v_{\text{sound}}$, eq. (3.4.41) with $v_{\text{conv}} = v_{\text{sound}}$ is a particularly useful constraint, placing a cap on an acceptable estimate of $V - V_{\text{ad}}$.

Estimating V with the mixing-length algorithm is a straightforward operation. Note first that

$$F_{rad} = \frac{V}{V_{rad}} F_{tot} = K\,V, \tag{3.4.43}$$

where

$$F_{tot} = \frac{L(r)}{4\pi r^2} \tag{3.4.44}$$

and

$$K = \frac{4ac}{3\kappa\rho}\frac{T^4}{H_P} = \frac{F_{tot}}{V_{rad}}. \tag{3.4.45}$$

Then, defining J as the coefficient of $(V - V_{ad})^{3/2}$ in eq. (3.4.36),

$$J = \frac{F_{conv}}{(V - V_{ad})^{3/2}} = \rho\,C_P T \left(\frac{P}{\rho}\right)^{1/2}\left(\frac{l_{mix}}{H_P}\right)^2, \tag{3.4.46}$$

one has

$$F_{tot}(r) = F_{conv} + F_{rad} = J(V - V_{ad})^{3/2} + K\,V. \tag{3.4.47}$$

From the leftmost equalities in eqs. (3.4.43) and (3.4.47), it follows that

$$F_{conv} = \left(1 - \frac{V}{V_{rad}}\right)F_{tot}, \tag{3.4.48}$$

and that, therefore,

$$\left(1 - \frac{V}{V_{rad}}\right)F_{tot} = J\,(V - V_{ad})^{3/2}. \tag{3.4.49}$$

At any point in a model star under construction, one first calculates V_{rad} from eq. (3.4.28) and V_{ad} from thermodynamics and an equation of state. If $V_{ad} > V_{rad}$, one adopts $V = V_{rad}$. If $V_{ad} < V_{rad}$, one finds J from eq. (3.4.46) and solves eq. (3.4.49) for V, finding $V_{ad} < V < V_{rad}$.

To progress further, it is necessary to compare F_{tot} and J numerically. At any point in a model star,

$$F_{tot} = \frac{L(r)}{4\pi r^2} = 6.34 \times 10^{10}\ \text{erg cm}^{-2}\ \text{s}^{-1}\ \frac{L(r)}{L_\odot}\left(\frac{R_\odot}{r}\right)^2. \tag{3.4.50}$$

In a convective region in which the perfect gas approximation is adequate, $P/\rho = kT/\mu M_H$, $\left(d\log\rho/d\log P\right)_{ad} = 3/5$, and $C_P = (5/2)\,kT/(\mu M_H)$, so

$$J = 1.875 \times 10^{21}\rho_0 \left(\frac{T_6}{\mu}\right)^{3/2}\left(\frac{l_{mix}}{H_P}\right)^2\ \text{erg cm}^{-2}\ \text{s}^{-1}. \tag{3.4.51}$$

In central regions of main sequence stars, $F_{tot}/J \ll 1$ and, provided that $V_{ad} \stackrel{>}{\sim} V_{rad}$, it is evident from eq. (3.4.49) that $(V - V_{ad}) \ll 1$. One can therefore replace V by V_{ad} on the left hand side of eq. (3.4.49) and solve to find

$$(V - V_{ad}) \sim \left(1 - \frac{V_{ad}}{V_{rad}}\right)^{2/3} \left(\frac{F_{tot}}{J}\right)^{2/3}$$

$$= 0.458 \times 10^{-8} \frac{(L(r)/L_\odot)^{2/3} (R_\odot/r)^{4/3}}{\rho_0^{2/3} (T_6/\mu) (l_{mix}/H_P)^{4/3}} \left(1 - \frac{V_{ad}}{V_{rad}}\right)^{2/3} \ll 1. \qquad (3.4.52)$$

That is, in convective cores of main sequence stars, the temperature gradient is essentially the adiabatic gradient. Inserting eq. (3.4.52) into eq. (3.4.41) shows that the estimated convective speed is typically very small compared with the sound speed, which can be of the order of 10^8 cm s^{-1}. This means that the time scale for mixing in a convective core is much larger than a dynamical time scale determined by dividing the radius of the core by a characteristic sound speed in the core. Typically, the convective mixing time scale is measured in months as compared with a dynamical time scale which is measured in tens of seconds. Although long compared with a dynamical time scale, the mixing time in the convective cores of main sequence stars of intermediate mass is quite short compared with the time scale for exhausing hydrogen in the core (see Section 3.5), and this fact prolongs the main sequence lifetimes of such stars over what would be the case in the absence of mixing. In subsurface layers of low mass main sequence stars and of red giants, in regions where the adiabatic gradient is small and the opacity is large because atoms are only partially ionized, $V_{ad} \ll V_{rad}$, $J/F_{tot} \ll 1$, and

$$V_{rad} \stackrel{>}{\sim} V \gg V_{ad}. \qquad (3.4.53)$$

That is, in convective envelopes of stars with low surface temperatures, the temperature gradient can be highly superadiabatic. As shown by eq. (3.4.41), the estimated convective speed in such regions can exceed the sound speed. This causes no practical problems in determining stellar structure, however, since, as is evident from eq. (3.4.48), the convective flux is formally very small compared with the radiative flux. On the other hand, one may guess that the time scale for mixing in a convective envelope is of the order of the distance between the base of the convective envelope and the stellar surface divided by a characteristic sound speed in the interior of the envelope.

3.4.4 Heat conduction by electrons

Heat conduction consists of the transfer of energy from material particles at higher temperatures to adjacent material particles at lower temperatures. Because, when the degree of electron degeneracy is small, an average electron moves faster than an average ion by the square root of the mass ratio between ions and electrons, it makes sense to concentrate only on electrons for an order of magnitude estimate. The starting point for the discussion is a diffusion equation which, for non-degenerate electrons, is

$$F_{\text{cond}} \sim -\frac{1}{3} \frac{\mathrm{d}}{\mathrm{d}T} \left(\frac{3}{2}kT\right) \frac{\mathrm{d}T}{\mathrm{d}r} \, n_{\text{e}} \, v_0 \, l_{\text{ei}} = -K_0 \frac{\mathrm{d}T}{\mathrm{d}r}, \tag{3.4.54}$$

where F_{cond} is an energy flux,

$$l_{\text{ei}} = \frac{1}{\sigma_{\text{ei}} \, n_{\text{e}}} \tag{3.4.55}$$

is a mean free path for an electron (thought of as being scattered only by heavy ions), σ_{ei} is a mean cross section for scattering,

$$v_0 = \left(\frac{3kT}{m_{\text{e}}}\right)^{1/2} \tag{3.4.56}$$

is the velocity of a typical electron, and K_0 is the conductivity. For electron–ion collisions, a very crude guess for the cross section is $\sigma_{\text{ei}} \sim b^2$, where b is the closest distance of approach between an average electron and an ion, given by

$$\frac{Ze^2}{b} = \frac{1}{2} \frac{m_{\text{e}}}{2} v_0^2 = \frac{3}{4} kT. \tag{3.4.57}$$

Thus,

$$K_0 \sim \frac{1}{3} \frac{3k}{2} \left(\frac{3kT}{m_{\text{e}}}\right)^{1/2} \left(\frac{3kT}{4Ze^2}\right)^2 = \frac{9\sqrt{3}}{32} \frac{1}{Z^2} \frac{ck}{r_{\text{el}}^2} \left(\frac{kT}{m_{\text{e}}c^2}\right)^{5/2}, \tag{3.4.58}$$

where $r_{\text{el}} = e^2/m_{\text{e}}c^2$ is the classical radius of the electron.

In order to compare the effectiveness of conduction with the effectiveness of radiation in transporting energy, one can define a conductive opacity κ_{c} by

$$F_{\text{cond}} = -\frac{4ac}{3} \frac{T^3}{\kappa_{\text{c}}\rho} \frac{\mathrm{d}T}{\mathrm{d}r} = -K_0 \frac{\mathrm{d}T}{\mathrm{d}r}, \tag{3.4.59}$$

or

$$\kappa_{\text{c}} = \frac{4ac}{3} \frac{T^3}{\rho} \frac{1}{K_0}. \tag{3.4.60}$$

Adopting K_0 from eq. (3.4.58), one has

$$\kappa_{\text{c}} \sim 3.23 \times 10^4 \, Z^2 \frac{T_6^{1/2}}{\rho_0} \, \text{cm}^2 \, \text{g}^{-1}. \tag{3.4.61}$$

From conservation of energy, and using eqs. (3.4.59) and (3.4.61), it follows that

$$F_{\text{tot}} - F_{\text{conv}} = F_{\text{cond}} + F_{\text{rad}} = -\frac{4ac}{3} \frac{T^3}{\kappa_{\text{eff}} \, \rho} \frac{\mathrm{d}T}{\mathrm{d}r}, \tag{3.4.62}$$

where

$$\frac{1}{\kappa_{\text{eff}}} = \frac{1}{\kappa_{\text{c}}} + \frac{1}{\kappa_{\text{rad}}}, \tag{3.4.63}$$

and κ_{rad} is the radiative opacity. For typical values of ρ and T in main sequence stars, κ_c given by eq. (3.4.61) is much larger than the radiative opacity κ_{rad}. Hence, $\kappa_{eff} \approx \kappa_{rad}$ in such stars.

As described in Chapter 13 in Volume 2, the situation is significantly altered in regions of electron degeneracy which prevail in the fuel-depleted cores of red giants, asymptotic giant branch stars and in white dwarfs. There are two basic reasons for this result, both due to the fact that, under electron-degenerate conditions, electrons are forced by the exclusion principle into higher energy states with much larger velocities than under non-degenerate conditions. Larger velocities mean (1) a more rapid transfer process and (2) smaller Coulomb cross sections and, hence, a longer mean free path and a more effective transfer process. These two effects combine to make the conductive opacity more than competitive with the radiative opacity.

3.4.5 Summary

To summarize the development in this section, it is convenient to replace the relationship between the temperature gradient and the ambient energy flux by a relationship between the temperature gradient and the pressure gradient. From eqs. (3.4.28) and (3.4.61) it follows that, if convection is not important, the logarithmic temperature–pressure gradient is

$$V = \frac{d \log T}{d \log P} = V_{rad+cond} = \frac{3}{4ac} \frac{\kappa_{eff}}{T^4} \frac{L}{4\pi GM} = \frac{3}{4ac} \frac{\kappa_{eff} \, r^2}{GMT^4} \, F_{tot}, \qquad (3.4.64)$$

where κ_{eff} is given by eq. (3.4.63).

If $V_{rad+cond} > V_{ad}$, where V_{ad} is the adiabatic logarithmic gradient, it is necessary to take convection into account. Then, the logarithmic gradient is given by

$$V = V_{rad+cond} \left(1 - \frac{F_{conv}}{F_{tot}} \right). \qquad (3.4.65)$$

The crude approximation to F_{conv} developed in eqs. (3.4.31)–(3.4.42) shows that the convective flux is a function of the temperature–pressure gradient V which is sought and, so, an estimate of V requires the solution of a transcendental equation (eq. (3.4.49)). However, when $V_{rad+cond} \gg V_{ad}$, $V \to V_{ad}$.

3.5 Nuclear energy-generation rates and evolutionary time scales

The fact that the Earth is approximately 4.6×10^9 yr old, coupled with the fact that life has existed on Earth for most of this time, could be interpreted to mean that the luminosity of the Sun has not varied much from its present value over the past 4.6 billion years. That is, since a major constituent of living things is water, the mean temperature on the Earth's surface must have remained between 273 K and 373 K. Assuming that the mean properties of the Earth's atmosphere (particularly its albedo and the extent of the greenhouse effect) have remained fairly constant since the emergence of life, the permissible temperature range corresponds to a variation of at most a factor of three in the Sun's luminosity. Hence, the total energy emitted by the Sun since the formation of the planetary system is of the order of

$$E_{\text{emitted}} = L_\odot \, t_{\text{Earth}} \sim 5.6 \times 10^{50} \text{ erg,} \tag{3.5.1}$$

which, dividing by the number of nucleons in the Sun as given by eq. (3.1.1), means that about 0.47×10^{-6} erg or 0.29 MeV has been produced per nucleon.

Since hydrogen and helium, with ionization potentials much smaller than 0.29 MeV, are the dominant constituents of the matter out of which stars are born, it is obvious that a liberation of chemical energy has contributed essentially nothing to the luminosity of the Sun over most of its lifetime. The total gravitational binding energy of the Sun is of the order of

$$E_{\text{bind}} \sim \frac{G M_\odot^2}{R_\odot} = 3.80 \times 10^{48} \text{ erg,} \tag{3.5.2}$$

or about 0.002 MeV per nucleon, and it is obvious that the release of gravitational binding energy has also contributed essentially nothing to the luminosity of the Sun over most of its lifetime.

The total rest mass energy of the Sun is

$$E_{\text{rest}} = M_\odot \, c^2 = 1.79 \times 10^{54} \text{ erg} \tag{3.5.3}$$

or 940 MeV per nucleon. The maximum amount of energy that could be achieved by the conversion of all of the original hydrogen and helium into iron peak elements (for which the nuclear binding energy per particle is at or near maximum) is about

$$E_{\text{nuc}} \sim 1.7 \times 10^{52} \text{ erg,} \tag{3.5.4}$$

or about 8.8 MeV per nucleon. This is far more energy than is necessary to account for the luminosity of the Sun over most of its lifetime until now, but it demonstrates that nuclear energy is the probable source of this luminosity and that only a modest fraction of the total store is required to account for the luminosity history of the Sun up to the present time.

The minimum amount of energy released by nuclear reactions comes from the conversion of a proton into deuterium and a positron, with the subsequent annihilation of the positron with an electron. This process releases into the star about 0.6 MeV per proton (see Chapter 6). However, deuterium has a very small binding energy and it is very "fluffy" (its wave function extends considerably beyond the range of nuclear forces) so that, if conditions are suitable for forming deuterium from two protons (a process that involves weak interactions), they are even more suitable for capture by a deuterium nucleus of a proton with the release of a gamma ray of energy ~ 5.5 MeV. Thus, the entire process of converting three protons into a ^{3}He nucleus and destroying an electron releases about 2.2 MeV per nucleon. Reactions between helium nuclei and an additional proton-capture reaction and electron-capture reaction result in an overall conversion of four protons and four electrons into a ^{4}He nucleus and two electrons, releasing altogether about 6 MeV per initial proton. It is evident from these energetics that, up to the current epoch, only a fraction (~ 0.05–0.13) of the initial hydrogen in the Sun can have been converted into an isotope of helium.

The reason that nuclear reactions occur so slowly in main sequence stars is that the typical kinetic energy of a collision between nuclei is small compared with the repulsive electrical potential between nuclei at separations comparable to the range of nuclear forces, namely ~ 1 Fermi ($= 10^{-13}$ cm). At a separation of 1 Fermi, the repulsive electrical potential of two point charges of charge $Z_1 e$ and $Z_2 e$ is

$$V_{\text{Coul}} = 1.44 \, Z_1 \, Z_2 \quad \text{MeV}. \tag{3.5.5}$$

This "Coulomb barrier" is three orders of magnitude larger than typical kinetic energies of particles at the centers of main sequence stars. This means that only a very small fraction of all nuclei actively participate in nuclear reactions, and these nuclei must have energies reaching well into the high energy tail of the Maxwell–Boltzmann distribution.

From a solution of the Schrödinger equation for two point positive charges which are plane waves at large separations in a box of size 1 cm^3, the probability density at separations of the order of a few Fermi or less is

$$|\psi|^2 \approx 2\pi \eta \exp(-2\pi \eta), \tag{3.5.6}$$

where

$$\eta = \frac{Z_1 Z_2 e^2}{\hbar v} \tag{3.5.7}$$

and v is the relative velocity of the two point charges when they are far from one another. If n_1 is the total number density of nuclei of type 1 and dn_2 is the number density of particles of type 2 with a velocity between v and $v + dv$ *relative* to a particle of type 1, then the number density of type 2 particles which overlap with particles of type 1 within a nuclear volume of size V_{nuc} is

$$n_1 \, dn_2 \, |\psi|^2 \, V_{\text{nuc}}. \tag{3.5.8}$$

The differential rate at which nuclear reactions take place may now be written as

$$dR \, (\text{ reactions cm}^{-3} \, \text{s}^{-1}) = n_1 \, dn_2 \, |\psi|^2 \, V_{\text{nuc}} \, \frac{1}{\tau_{\text{react}}}, \tag{3.5.9}$$

where τ_{react} is the time (τ_{react}^{-1} is the probabilty per unit time) for a reaction to take place once the nuclei are overlapping. The differential reaction rate can also be written as

$$dR = n_1 \, dn_2 \, \sigma(E) \, v, \tag{3.5.10}$$

where $\sigma(E)$ is the cross section for the process as a function of the relative energy E of the particles and v is the relative speed. Energy and speed are, of course, related by $E = \frac{1}{2} \mu v^2$, where $\mu = m_1 m_2/(m_1 + m_2)$ is the reduced mass of the two-nucleus system. Thus,

$$\sigma(E) = \frac{|\psi|^2 \, V_{\text{nuc}}}{\tau_{\text{react}}} \frac{1}{v}. \tag{3.5.11}$$

Using eqs. (3.5.6) and (3.5.7) as well as $E = \frac{1}{2} \mu v^2$, eq. (3.5.11) may also be written as

$$\sigma(E) = \frac{S_0(E)}{E} \exp(-2\pi \eta), \tag{3.5.12}$$

where

$$S_0(E) = \frac{\pi Z_1 Z_2 e^2 \mu}{\hbar} \frac{V_{\text{nuc}}}{\tau_{\text{react}}}, \tag{3.5.13}$$

is called a center of mass cross section factor.

One might expect that $V_{\text{nuc}}/\tau_{\text{react}}$ does not vary strongly with E for values of E which are relevant in a star (typical kinetic energies are very small compared with V_{Coul} and with the energies of motion of nucleons in a nucleus). In fact, nuclear reaction experiments show that, in many instances, reaction cross sections can be characterized by eq. (3.5.12), where $S_0(E)$ is a fairly slowly varying function of energy when $E \ll V_{\text{Coul}}$. Thus, at low energies, $V_{\text{nuc}}/\tau_{\text{react}}$, which characterizes the excited compound nucleus formed by the temporary coalescence of the reacting nuclei, is relatively independent of the precise manner in which this compound nucleus is formed.

To find the total reaction rate under stellar conditions, it is necessary to integrate over the Maxwell–Boltzmann distribution function (see Chapter 6)

$$\mathrm{d}n_2 = \frac{n_2}{(2\pi \mu kT)^{3/2}} \exp\left(-\frac{E}{kT}\right) 4\pi p^2 \, \mathrm{d}p, \tag{3.5.14}$$

where n_2 is the total number density of particles of type 2 and $p^2 = 2\mu E$. The reaction rate (reactions $\text{cm}^{-3}\,\text{s}^{-1}$) becomes

$$R = n_1 \int \frac{S_0(E)v}{E} \exp(-2\pi \eta) \frac{n_2}{(2\pi \mu kT)^{3/2}} \exp\left(-\frac{E}{kT}\right) 4\pi p^2 \, \mathrm{d}p, \tag{3.5.15}$$

and, if $S_0(E) \sim S_0$, where S_0 is a constant,

$$R = n_1 n_2 \frac{8\pi \mu S_0}{(2\pi \mu kT)^{3/2}} \int \exp\left(-\frac{E}{kT} - \frac{\alpha}{\sqrt{E}}\right) \mathrm{d}E, \tag{3.5.16}$$

where

$$\alpha = 2\pi Z_1 Z_2 \frac{e^2}{\hbar c} \left(\frac{\mu c^2}{2}\right)^{1/2}. \tag{3.5.17}$$

A final transformation gives

$$R = n_1 n_2 \frac{4 S_0}{(2\pi \mu kT)^{1/2}} \int \exp\left(-x - \frac{\lambda}{x^{1/2}}\right) \mathrm{d}x, \tag{3.5.18}$$

where

$$\lambda = \frac{\alpha}{\sqrt{kT}} = 2\pi Z_1 Z_2 \frac{e^2}{\hbar c} \left(\frac{\mu c^2}{2kT}\right)^{1/2}. \tag{3.5.19}$$

Integration gives (see Chapter 6)

$$R \approx n_1 n_2 \frac{4 S_0}{(2\pi \mu kT)^{1/2}} \exp(-3x_0) \left(\frac{4\pi}{3} x_0\right)^{1/2}, \tag{3.5.20}$$

where

$$x_0 = \left[\left(\frac{\mu c^2}{2kT}\right)^{1/2} \pi Z_1 Z_2 \frac{e^2}{\hbar c}\right]^{2/3} \tag{3.5.21}$$

After some rearrangement, one finds

$$R = \bar{S}_0 \left(\frac{\nu}{T^{1/3}}\right)^2 n_1 n_2 \exp\left(-\frac{\nu}{T^{1/3}}\right), \tag{3.5.22}$$

where

$$\frac{\nu}{T^{1/3}} = 3 \left[\left(\frac{\mu c^2}{2kT}\right)^{1/2} \pi Z_1 Z_2 \frac{e^2}{\hbar c}\right]^{2/3} \tag{3.5.23}$$

and

$$\bar{S}_0 = \frac{S_0}{\mu c} \frac{8}{9\pi \sqrt{3}} \frac{1}{Z_1 Z_2} \frac{\hbar c}{e^2}. \tag{3.5.24}$$

The rate at which energy is generated (in erg g^{-1} s^{-1}) by a reaction which liberates an energy ϵ_{12} is

$$\epsilon = \epsilon_{12} \frac{R}{\rho} \propto \frac{\rho}{T^{2/3}} \exp\left(-\frac{\nu}{T^{1/3}}\right). \tag{3.5.25}$$

If one approximates

$$\epsilon \approx A \rho T^s, \tag{3.5.26}$$

then

$$s = -\frac{2}{3} + \frac{1}{3} \frac{\nu}{T^{1/3}}. \tag{3.5.27}$$

Two cases are to be emphasized. When the two interacting particles are protons and T is given in units of 10^6 K, $\nu = 33.84$. At temperatures of $\sim 15 \times 10^6$ K, as near the center of the Sun, $s \sim 4$. For reactions between protons and ^{14}N nuclei, $\nu = 152.3$, and, at temperatures of $\sim 20 \times 10^6$ K, $s \sim 18$. Thus, as anticipated, the energy sources in main sequence stars are very temperature sensitive and therefore concentrated toward the center, the degree of concentration being much larger in stars massive enough to burn nitrogen than in low mass stars in which energy generation relies on the occurrence of the pp reaction.

A star of near solar mass leaves the main sequence after it has burned approximately 13% of its original store of hydrogen (see Chapter 11). Thus, the lifetime of the main sequence phase is given by

$$t_{MS}(\text{yr}) \sim \frac{0.13 M \epsilon_0}{L_{MS}} \sim 10^{10} \frac{M}{M_\odot} 1.3X \frac{L_\odot}{L_{MS}}, \tag{3.5.28}$$

where ϵ_0 is the average energy per unit mass liberated by hydrogen burning ($\sim 6 \times 10^{18}$ erg g^{-1}), X is the initial abundance by mass of hydrogen, and L_{MS} is the average main sequence luminosity. Making use of the fact that the observed main sequence mass–luminosity relationship is

$$\frac{L_{MS}}{L_\odot} \sim \left(\frac{M}{M_\odot} \right)^4, \tag{3.5.29}$$

it follows that

$$t_{MS}(M) \sim 10^{10} \text{ yr} \left(\frac{M_\odot}{M} \right)^3. \tag{3.5.30}$$

In words, the main sequence lifetime of a 10 $M_\odot$ star is about a thousand times shorter than the main sequence lifetime of a 1 $M_\odot$ star.

The dependence of theoretical lifetime on mass was used by Edwin E. Salpeter (1955) to estimate the mass dependence of the birthrate function for stars in the galactic disk. The observations provide the current distribution of stars on the main sequence as a function of mass – $dn(M)/dM$, but the birthrate $(d/dt)(dn(M, t)/dM)$ is proportional to $dn(M)/dM \div t_{MS}(M)$. Salpeter found that

$$\frac{dn(M, t)}{dM} \underset{\propto}{\sim} \frac{1}{M^{2.3}}. \tag{3.5.31}$$

A comparison of eqs. (3.5.1) and (3.5.2) suggests that the time required to contract to the main sequence by a star which is initially too cool to burn nuclear fuel is several orders of magnitude smaller than the main-sequence lifetime of the star. The contraction lifetime can be approximated by dividing the gravitational binding energy on reaching the main sequence (eq. (3.5.2)) by the main-sequence luminosity, assuming that the luminosity during the contraction phase was not too different from the present luminosity. This gives a gravitational contraction, or Kelvin–Helmholtz, time scale of

$$t_{KH}(\text{yr}) \sim 3 \times 10^7 \frac{M^2}{RL} \frac{R_\odot L_\odot}{M_\odot^2}. \tag{3.5.32}$$

For a star like the Sun, the lifetime of the pre-main sequence gravitational contraction phase is roughly 300 times shorter than the lifetime of the main sequence phase.

3.6　The static electrical field

A very fundamental characteristic of ionized matter in a gravitational field that significantly enriches our understanding of how gravity works in stars is the presence of a static electrical field which exerts a force on electrons and ions alike that is comparable in magnitude to the force exerted on ions by the gravitational field (Milne, 1924; Eddington, 1926). That an electrical force must exist is clear from the fact that the inward gravitational force on an electron is three orders of magnitude smaller than the gravitational force on an ion and yet, given that charge neutrality must prevail over regions large enough to contain many particles, there must be an inward force on electrons other than gravity which prevents a macroscopic separation of electrons and ions. If this force is electrical, the electrical field must be directed outward and exert an inward force on an electron comparable to the gravitational force on an ion. This being the case, the electrical force on an ion is directed outward and is comparable in magnitude to the gravitational force on the ion, so that the net inward force on an ion is significantly smaller than the force exerted by gravity alone.

These remarkable inferences are easily demonstrated. Consider, for simplicity, the case of pure, completely ionized, hydrogen in an inwardly directed gravitational field of magnitude g and an outwardly directed electric field of strength E. The pressure-gradient force on each type of particle must balance the sum of the gravitational and electrical forces on each type. Each electron has a charge $-e$ and mass m_e and each proton has a charge $+e$ and mass M_p. The net outward pressure force on an electron is

$$F_e = -\frac{1}{n_e}\frac{dP_e}{dr} = -m_e g - eE \tag{3.6.1}$$

and the net outward pressure force on a proton is

$$F_p = -\frac{1}{n_p}\frac{dP_p}{dr} = -M_p g + eE. \tag{3.6.2}$$

Note that $dP/dr = dP_e/dr + dP_p/dr = -\left(n_e m_e + n_p M_p\right) g = -\rho g$, in accord with eq. (3.1.4), and independent of the value of E. To avoid charge separation, the forces F_e and F_p must be equal, with the result that

$$eE = \frac{1}{2}(M_p - m_e)g, \tag{3.6.3}$$

and

$$F_e = F_p = -\frac{1}{2}(M_p + m_e)g. \tag{3.6.4}$$

Thus, the existence of a gravitational field leads to the development of a macroscopic static electrical field of a magnitude such that the electron and the proton share equally in responding to the force of gravity.

This contrasts with the case of completely neutral hydrogen in a gravitational field where the electron and proton are locked into a bound state by the force of an atomic electrical field many orders of magnitude larger than the gravitational force which is exerted on the neutral atom. The contrast between the two cases is, incidentally, a nice illustration of why calling the gravitational force weak relative to the electrical force is not always an appropriate way of characterizing the forces.

Bibliography and references

Ludwig Boltzmann, *Vorlesungen über Gastheorie*, Vol. 1, 1896.

C. Vernon Boys, *Proc. Roy. Inst. of G.B.*, **14**, 353, 1894.

Henry Cavendish, *Phil. Trans. Roy. Soc.*, **88**, 469, 1798.

A. Cornu & J. B. Baile, *C.R. Acad. Sci., Paris*, **76**, 954, 1873.

Charles-Augustin de Coulomb, Premier et Second Mémoires sur l'Electricité et le Magnétisme, Historie de l'Académie Royale des Sciences, p. 569–577 & 578–611, 1785.

Paul Adrien Maurice Dirac, *Proc. Roy. Soc. (London)*, **112**, 661, 1926.

Arthur S. Eddington, *The Internal Constitution of the Stars* (Cambridge: Cambridge University Press), 1926.

Enrico Fermi, *Zeits. für Physik*, **36**, 902, 1926.

Werner Heisenberg, *Zeits. für Physik*, **25**, 691, 1927.

Icko Iben, Jr., *ApJ*, **138**, 1090, 1963.

James Clerk Maxwell, *Phil. Mag.*, **19**, 31, 1860.

E. A. Milne, *Proc. Cambridge Phil. Soc.*, **22**, 1924.

Isaac Newton, *Principia Mathematica*, 1687.

Wolfgang Pauli, *Zeits. für Physik*, **31**, 765, 1925.

Max Planck, *Ann. der Physik*, **4**, 553, 1901.

F. K. Richtmeyer, E. H. Kennard, & T. Lauritsen, *Introduction to Modern Physics* (New York: McGraw-Hill), fifth edition, 1955.

Edwin E. Salpeter, *ApJ*, **121**, 161, 1955.

Karl Schwarzschild, *Göttinger Nachrichten*, No. **1**, 41, 1906.

Joseph John Thomson, *Phil. Mag.*, **44**, 293, 1897.

4 Statistical mechanics, thermodynamics, and equations of state

Given that the central feature of stellar structure is a balance between pressure forces and gravitational forces, it is essential to understand quantitatively, at every point in a star, how pressure P is related to temperature T, density ρ, and composition. The relationship is known as an equation of state (EOS) and the primary objective of this chapter is to explore the EOS for conditions frequently found in stars.

From the perspective of stellar evolution, the relationship between the internal energy density U and T, ρ, and composition is of central importance, for, as is developed at length in Chapter 8, it is the quantity

$$\epsilon_{\mathrm{grav}} = -\left[\frac{\mathrm{d}U}{\mathrm{d}t} + P\,\frac{\mathrm{d}}{\mathrm{d}t}\left(\frac{1}{\rho}\right)\right], \tag{4.0.1}$$

where U is in units of erg g^{-1} when P is in units of dyne cm^{-2} and ρ is in units of g cm^{-3}, which expresses how gravothermal energy contributes to the ambient flow of energy and plays a key role in influencing changes in both local and global stellar characteristics. Since P and U are intimately related, both the EOS and the relationship between U and T, ρ, and composition are most conveniently found at the same time.

The basic concept of an irreducible unit cell in position–momentum phase space is presented in Section 4.1 and the connection between statistical mechanics and thermodynamics is developed in Section 4.2 in the context of a system of particles which obey the Pauli exclusion principle (Fermi–Dirac statistics). A general algorithm for finding an EOS and the energy density is presented in Section 4.3 and applied to a non-degenerate ionic gas in Section 4.4.

Equations of state for free electrons with non-relativistic energies and for different degrees of degeneracy are developed in Sections 4.5 (weak degeneracy), 4.6 (strong degeneracy), and 4.7 (intermediate degeneracy). The EOS for relativistically degenerate electrons is discussed in Sections 4.8 and 4.9, and an EOS for electron–positron pairs at high temperatures is discussed in Section 4.10.

The statistics applicable to a system of indistinguishable particles which do not obey the exclusion principle (Bose–Einstein statistics) is developed in Section 4.11 and applied to the radiation field. In Section 4.12, Maxwell–Boltzmann statistics is discussed and the connection between entropy and the probability of the most probable distribution is presented.

In the outer envelopes and atmospheres of most stars, atoms are only partially ionized and, in order to construct an EOS, it is necessary to determine the degree of ionization for all species of ion. Degrees of ionization depend upon partition functions (which depend upon the occupancy of excited states) as well as upon ionization potentials. In Section 4.13, the Saha equation which describes the equilibrium ratio of free protons and hydrogen atoms in a gas of pure hydrogen is derived under various assumptions regarding the partition function for the atom and, in Section 4.14, several thermodynamic properties of pure hydrogen are derived in the approximation that the partition function is unity. In Section 4.15, the Saha equations which give the ratios of equilibrium abundances of two adjacent states of ionization for an arbitrary assemblage of multi-electron atoms are derived and applied to a gas of pure helium. In Section 4.16, the thermodynamic properties of a gas of partially ionized hydrogen and helium and of photons are presented.

A pedagogical discussion of the effects of Coulomb forces between charged particles on the EOS for a configuration of heavy ions and free electrons in various stages of degeneracy is given in Section 4.17, with the general result that the negative pressure P_{Coulomb} and the negative energy per unit volume $\bar{U}_{\text{Coulomb}}$ due to these forces are related by $P_{\text{Coulomb}} = \frac{1}{3}\,\bar{U}_{\text{Coulomb}}$.

Given that heavy ions are in the gaseous state in the interiors of evolutionary models presented in this volume and in all but the last chapter of Volume 2, a discussion of the physics of solids and liquids is postponed to the last chapter of Volume 2. In Section 21.4, concepts involved in understanding the solid state configuration and the transition from solid to liquid are discussed and, in Section 21.5, ingredients of an equation of state for solid and liquid phases based on rigorous treatments in the literature are presented and algorithms based on these ingredients for an EOS useful in calculations of the evolution of white dwarfs are given.

4.1 Quantum-mechanical wave functions and the unit cell in phase space

Distributions with respect to energy and momentum are important characteristics of systems of particles in thermal equilibrium. In order to find these distributions it is first necessary to specify the *states* which a particle may occupy. For example, for a gas of N identical particles in a cubic box of volume $V = L^3$, where L is the length of a side, states are characterized by specific values of linear momentum and of energy which are of a particularly simple form.

Neglecting interactions between particles, states characterized by a sharp value of momentum $\mathbf{p}$ can be described by wave functions of the form

$$\psi_i = \frac{1}{\sqrt{V}} \exp\left(\sqrt{-1}\,\frac{\mathbf{p}_i}{\hbar} \cdot \mathbf{r}\right), \tag{4.1.1}$$

where $\hbar$ is Planck's (Max Planck, 1901) constant divided by 2π. For material particles with kinetic energies much less than their rest mass energies, these functions are solutions of the Schrödinger equation. They are also approximations to solutions of the Dirac equation for electrons and are solutions of Maxwell's equations for photons. The components of the momentum vector $\mathbf{p}_i$ satisfy

$$\frac{p_{ix}}{\hbar}L_x = 2\pi m_{ix}, \quad \frac{p_{iy}}{\hbar}L_y = 2\pi m_{iy}, \quad \frac{p_{iz}}{\hbar}L_z = 2\pi m_{iz}, \tag{4.1.2}$$

where m_{ix}, m_{iy}, and m_{iz} are integers which are components of a vector $\mathbf{m}_i$, and L_x, L_y, and L_z are cube edges, each of length L. The absolute value of the momentum vector is thus

$$p_i = \frac{h}{L} m_i = \frac{h}{L} \sqrt{m_{ix}^2 + m_{iy}^2 + m_{iz}^2}. \tag{4.1.3}$$

In first approximation, eqs. (4.1.1)–(4.1.3) hold for all particles, with or without mass. Total energy and momentum are related by

$$\epsilon_i^2 = (cp_i)^2 + (Mc^2)^2, \tag{4.1.4}$$

where M is the rest mass of the particle, kinetic energy ϵ_i' is given by

$$\epsilon_i' = \epsilon_i - Mc^2, \tag{4.1.5}$$

and velocity $\mathbf{v}$ and momentum $\mathbf{p}$ are related by

$$\frac{\mathbf{v}}{c} = \frac{c \, \mathbf{p}}{\epsilon}. \tag{4.1.6}$$

The relationship between energy and the quantum vectors $\mathbf{m}_i$ depends on the type of particle. In the case of a gas of material particles at non-relativistic energies, eqs. (4.1.3)–(4.1.5) give

$$\epsilon_i' \sim \frac{p_i^2}{2M} = \frac{\hbar^2}{2M} \left(\frac{2\pi}{L} \right)^2 \left(m_{ix}^2 + m_{iy}^2 + m_{iz}^2 \right) = \frac{1}{2M} \frac{h^2}{L^2} m_i^2. \tag{4.1.7}$$

When typical particles have energies much larger than their rest mass,

$$\epsilon_i' \sim \epsilon_i \sim cp_i = \frac{ch}{L} \sqrt{m_{ix}^2 + m_{iy}^2 + m_{iz}^2} = \frac{ch}{L} |m_i|. \tag{4.1.8}$$

In general, energy is a very incomplete description of state. In the current example, if the absolute values j, k, l of the three components of a particular vector are all different, there are 47 additional vectors one can construct with this same set of integers. That is, if one vector has components j, k, l, five other vectors may be constructed by permuting the integers, making a total of six vectors. Next, let the x component be negative and again permute the three integers to construct six more vectors. Then, let the y component be negative and permute the three integers to construct six more vectors. Continuing in this way, a total of 48 different states having the same energy, i.e., the same $(j^2 + k^2 + l^2)$, can be constructed. If no other set of integers gives the same energy, the degeneracy is said to be 48-fold.

Thus, in constructing a distribution function for any system, it is essential to think of states as being characterized not only by their energy, but also by all other physically meaningful attributes. In the case at hand, this means specification of the vectors $\mathbf{p}$ and $\mathbf{m}$. In more complex systems made of particles with structure, other quantum characteristics such as spin, orbital angular momentum, polarization, vibrational quantum number, and so forth must be included in the specification of a single state. For example, in specifying a state for free electrons, spin is included, and in specifying a state for photons, polarization is included. Insisting on a complete set ensures that all possible configurations are included in whatever counting schemes are adopted for estimating the probabilities of various distributions.

When $\mathbf{p}$ (equivalently $\mathbf{m}$) is large enough, the spacing between adjacent states in momentum and energy becomes small enough that one may treat the states as being essentially continuously distributed. Thus, one may approximate the number of states with components of the momentum vector in the ranges $p_x \rightarrow p_x + \mathrm{d}p_x$, $p_y \rightarrow p_y + \mathrm{d}p_y$, and $p_z \rightarrow p_z + \mathrm{d}p_z$ by

$$g_i \rightarrow \mathrm{d}g = \mathrm{d}m_x \mathrm{d}m_y \mathrm{d}m_z = \left(\frac{L}{2\pi}\right)^3 \frac{\mathrm{d}p_x \mathrm{d}p_y \mathrm{d}p_z}{\hbar^3} = V \frac{\mathrm{d}p_x \mathrm{d}p_y \mathrm{d}p_z}{h^3}. \tag{4.1.9}$$

Equation (4.1.9), which holds for material particles, whatever the relationship between energy and momentum, as well as for massless photons and (nearly massless) neutrinos, conveys the remarkable and profound lesson that the size of the smallest statistical cell in the six-dimensional position–momentum phase space is the cube of Planck's constant h.

By asking how N particles are distributed among the "non-interacting" particles in thermodynamic equilibrium, one encounters an apparent paradox: the only way to achieve equilibrium is through interactions, but, in constructing particle states, it has been assumed that there are no interactions between different particles in the box or between particles in the box and particles in the walls of the box. For the present, assume that the states with sharply defined momenta are adequate approximations insofar as counting states is concerned, and that the interactions which must occur are small perturbations that act primarily to endow each state with an energy width which is small compared with the energy of the state.

4.2 The connection between thermodynamics and Fermi–Dirac statistics for particles which obey the Pauli exclusion principle

If particles obey the Pauli exclusion principle (Wolfgang Pauli, 1925), no two particles may occupy the same state. One way of estimating the probability that N_i particles will be found in a group of g_i states which are closely related in energy is to assume that the *a priori* probability of finding a particle in this group is proportional to the number of member states which can be filled. In the present instance, think of the particles as darts which are thrown randomly, one at a time, at the entire ensemble of states $\sum g_i$. The *a priori* probability of hitting (filling) one of the states in the group g_i with the first particle is proportional to g_i. Once one of the g_i states is filled, the exclusion principle reduces the number of available states from g_i to $g_i - 1$, so that the probability of filling a second level among the g_i states is proportional to $g_i - 1$. Continuing in this way, it follows that the probability of finding N_i particles in the g_i group of states is proportional to

$$g_i(g_i - 1)(g_i - 2) \cdots (g_i - N_i + 1) = \frac{g_i!}{(g_i - N_i)!} \tag{4.2.1}$$

Another way of achieving the same result is to argue that the probability of finding N_i particles in g_i states is proportional to the number of *distinct ways* in which these particles can be distributed among the g_i states. Thus, there are g_i ways in which the first particle may be placed in a state, $g_i - 1$ ways in which the second may be placed, and $g_i - (N_i - 1)$ ways in which the last particle may be placed, leading again to eq. (4.2.1).

Indistinguishability is taken into account by noting that any permutation among the N_i particles in the states g_i gives the same result insofar as the total energy of the g_i states is concerned. Since the number of permutations is $N!$, the final probability is proportional to

$$P_i \propto \frac{g_i!}{N_i!(g_i - N_i)!}. \tag{4.2.2}$$

The probability $P_i(N_i, N_j, \dots)$ of finding N_i particles in states g_i, N_j particles in states g_j, and so on, is equal to the product of all of the individual probabilities. That is,

$$P_i(N_i, N_j, \dots) \propto \prod_{i=1}^{\infty} P_i \propto \prod_{i=1}^{\infty} \frac{g_i!}{N_i!(g_i - N_i)!}. \tag{4.2.3}$$

If the walls of the box are impermeable, and are perfect insulators,

$$N = \sum_{i=1}^{\infty} N_i = \text{constant, and} \tag{4.2.4}$$

$$E = \sum_{i=1}^{\infty} N_i \epsilon_i = \text{constant,} \tag{4.2.5}$$

where E is the total energy of the system.

Using the method of Lagrange multipliers, one may maximize the logarithm of the probability described by eq. (4.2.3), subject to the constraints represented by eqs. (4.2.4) and (4.2.5). The result is

$$\sum_{i=1}^{\infty} \frac{\partial \log_e P_i}{\partial N_i} \delta N_i - \alpha \sum_{i=1}^{\infty} \delta N_i - \beta \sum_{i=1}^{\infty} \epsilon_i \delta N_i = 0, \tag{4.2.6}$$

where α and β are parameters to be determined.

The first term on the left of eq. (4.2.6) can be evaluated with the help of what is called Stirling's approximation:

$$n! \to \left(\frac{n}{e}\right)^n \sqrt{2\pi n} \text{ as } n \to \infty, \tag{4.2.7}$$

or, equivalently,

$$\log_e n! \sim n (\log_e n - 1) + \frac{1}{2} (\log_e n + \log_e 2\pi). \tag{4.2.8}$$

The approximation was published originally by Abraham de Moivre in *Doctrine of Chances* (1733) as $n! \sim \text{constant } n^{n+1/2} e^{-n}$, and the constant, estimated numerically by de Moivre, was determined by James Stirling to be $\sqrt{2\pi}$. The essence of the approximation can be found in a very elementary way beginning with the observation that, when n is an integer, $\log_e n! = \log_e 1 + \log_e 2 + \log_e 3 + \cdots + \log_e n$. Drawing a smooth curve through the points in a graph of $\log_e j$ versus j, where j is an integer, it is clear that a lower limit to the sum is given by the area under the curve. That is, $\log_e n! \gtrsim \int_1^n \log_e x \, dx = (x \log_e x - x)_1^n = n \log_e n - n + 1$. The exact value of $\log n!$ is the area under the staircase in the $\log x$ versus x diagram constructed by inserting another set of points in the diagram at $(x, \log_e(x+1)) = (1, \log_e 2), (2, \log_e 3)$, etc., and drawing straight lines between adjacent points in the entire ensemble of points. The difference in area under the staircase and the area under the smooth curve passed through the first set of points is only slightly overestimated by summing the areas of a series of triangles, resulting in $(1/2) \sum_{j=2}^n (\log_e j - \log_e (j-1)) = (1/2) \log_e n$. Altogether, $\log_e n! \gtrsim n (\log_e n - 1) + 1 + (1/2) \log_e n$, which differs from the approximation given by eq. (4.2.8) by only $(1 - \log_e \sqrt{2\pi}) = 0.0811$.

This elaborate discussion of the niceties of Stirling's approximation has been presented primarily as a curiosity. For an n which is even a tiny fraction of Avogadro's number ($N_0 = 6.02486 \times 10^{23}$), $(1/2) \log_e n$ plus either $\log_e \sqrt{2\pi}$ or 1 is negligible compared with the first term on the right side of eq. (4.2.8), so that

$$\log_e n! \sim n (\log_e n - 1) \qquad (4.2.9)$$

is entirely adequate for the purposes of statistical mechanics.

Using eq. (4.2.9) in eq. (4.2.3), one obtains

$$\log_e P_i = g_i (\log_e g_i - 1) - N_i (\log_e N_i - 1) - (g_i - N_i)(\log_e(g_i - N_i) - 1), \qquad (4.2.10)$$

from which it follows that

$$\frac{\partial \log_e P_i}{\partial N_i} = \log_e(g_i - N_i) - \log_e N_i = \log_e \left(\frac{g_i - N_i}{N_i} \right). \qquad (4.2.11)$$

Using this result in eq. (4.2.6), one has that

$$\sum_{i=0}^\infty \delta N_i \left[\log_e \left(\frac{g_i - N_i}{N_i} \right) - \alpha - \beta \epsilon_i \right] = 0. \qquad (4.2.12)$$

Equation (4.2.12) can be satisfied only if every expression in square brackets vanishes. Thus,

$$\log_e \left(\frac{g_i - N_i}{N_i} \right) = \alpha + \beta \epsilon_i \qquad (4.2.13)$$

or

$$N_i = \frac{g_i}{\exp(\alpha + \beta \epsilon_i) + 1}. \qquad (4.2.14)$$

Equation (4.2.14) describes what is known as the Fermi–Dirac distribution function (Fermi, 1926; Dirac, 1926). The quantity

$$f_i = \frac{1}{\exp(\alpha + \beta \epsilon_i) + 1} \qquad (4.2.15)$$

can be interpreted as the probability that a state in the group g_i is occupied. It is also an occupation number which gives the fraction of states g_i that are occupied.

For g_i, one may adopt the continuum approximation given by eq. (4.1.9), and replace ϵ_i by the continuous variable ϵ. The parameters α and β are determined, in principle, by applying the constraints of eqs. (4.2.4) and (4.2.5). Since, in the current example, energy depends only on the absolute value of the momentum, one can integrate $dp_x dp_y dp_z = p^2 dp d\Omega$ over all solid angles $d\Omega$ to obtain $4\pi p^2 dp$. Then, eqs. (4.2.14), (4.2.4), and (4.2.5) give

$$N = \sum N_i \rightarrow \frac{4\pi V}{h^3} \int_0^\infty \frac{p^2 dp}{\exp(\alpha + \beta\epsilon) + 1}, \text{ and} \tag{4.2.16}$$

$$E = \sum N_i \epsilon_i \rightarrow \frac{4\pi V}{h^3} \int_0^\infty \frac{\epsilon p^2 dp}{\exp(\alpha + \beta\epsilon) + 1}. \tag{4.2.17}$$

By considering two different kinds of gas in the same box and repeating the steps in the derivation, it is evident that, whereas each gas is characterized by a different value of α, β is the same for both gases. That is, particles are not exchanged between gases, so that a constraint of the form defined by eq. (4.2.4) holds for each gas separately, but there is only one energy constraint which involves all of the particles. It is convenient to choose, as one of the gases, a system for which $f(\epsilon)$ is small for all values of ϵ, so that the integer 1 in the denominator of the integrand in eqs. (4.2.16) and (4.2.17) can be neglected relative to the exponential term, and to choose the system such that $\epsilon = p^2/2M$. These choices imply a rarified gas and temperatures such that, in general, $p \ll Mc$. There is a large body of experimental evidence that provides a relationship between bulk properties of such gases which may be used to determine β.

For the rarified component

$$N \rightarrow \frac{4\pi V}{h^3} \int_0^\infty \exp\left(-\alpha - \frac{\beta p^2}{2M}\right) p^2 dp \tag{4.2.18}$$

$$= \frac{4\pi V}{h^3} e^{-\alpha} \left(\frac{2M}{\beta}\right)^{3/2} \int_0^\infty e^{-x^2} x^2 dx \tag{4.2.19}$$

$$= \left(\frac{2\pi M}{h^2 \beta}\right)^{3/2} V e^{-\alpha}. \tag{4.2.20}$$

Similarly, for the kinetic energy E' of the rarified component,

$$E' = E - N Mc^2 \rightarrow \frac{4\pi V}{h^3} \int_0^\infty \exp\left(-\alpha - \beta\frac{p^2}{2M}\right) \frac{p^2}{2M} p^2 dp \tag{4.2.21}$$

$$= -\frac{\partial N}{\partial \beta} \tag{4.2.22}$$

$$= \frac{3}{2\beta} N. \tag{4.2.23}$$

There is a logical inconsistency in allowing integrations to extend to ∞ when it has been assumed that $p \ll Mc$; in practice, however, the contributions from those parts of the integrands violating $p \ll Mc$ are totally negligible.

To find β, one may first invoke the first law of thermodynamics for gases:

$$\delta Q = P\delta V + \delta E, \tag{4.2.24}$$

where δQ is the amount of heat that enters through the walls of the box, δE is the change in the internal energy of the gas and $P\delta V$ is the work done by the gas in changing the size of the box. Next, one may return to the statistical-mechanics formulation in terms of discrete states and consider the variation of all terms in eq. (4.2.5) that occur in consequence of slowly varying the size of the box and allowing heat to be exchanged between particles which are in the box and particles which are in the walls of the box. One has

$$\delta E = \sum_{i=1}^{N} N_i \frac{\partial \epsilon_i}{\partial V} \delta V + \sum_{i=1}^{N} \epsilon_i \delta N_i. \tag{4.2.25}$$

On setting $\delta V = 0$ in eqs. (4.2.25) and (4.2.24), it follows that

$$\delta Q = \sum_{i=1}^{N} \epsilon_i \delta N_i, \tag{4.2.26}$$

and, on setting $\delta Q = 0$, it follows that the work done by particles in expanding the box is

$$P\delta V = -\sum_{i=1}^{N} N_i \frac{\partial \epsilon_i}{\partial V} \delta V, \tag{4.2.27}$$

so that

$$P = -\sum_{i=1}^{N} N_i \frac{\partial \epsilon_i}{\partial V} = -\frac{1}{V} \sum_{i=1}^{N} N_i \epsilon_i \frac{\partial \log \epsilon_i}{\partial \log V}. \tag{4.2.28}$$

To find the derivatives of ϵ_i for particles in a rarified gas at low temperatures, note from eq. (4.1.7) that

$$\epsilon_i' \propto V^{-2/3}, \tag{4.2.29}$$

where $V = L^3$. Since ϵ_i and ϵ_i' differ only by a constant,

$$\frac{\partial \epsilon_i}{\partial V} = \frac{\partial \epsilon_i'}{\partial V} = -\frac{2}{3} \frac{\epsilon_i'}{V} \tag{4.2.30}$$

and

$$P = \frac{2}{3} \frac{E'}{V}. \tag{4.2.31}$$

Comparing eq. (4.2.31) with eq. (4.2.23), one has that

$$PV = \frac{N}{\beta}. \tag{4.2.32}$$

It is known from experiment that, for a mole of any sufficiently rarified gas, the law

$$PV = R_0 T \tag{4.2.33}$$

holds, where the gas constant R_0 has the value $8.316\,62 \times 10^7$ erg, independent of the type of gas. A mole (gram molecular weight) is an amount of gas the mass of which in grams is equal to the molecular weight of the constituent particles (e.g., $\equiv 32$ g in the case of molecular oxygen O_2). A gas constant per particle can be defined as

$$k = \frac{R_0}{N_A} = 1.38044 \times 10^{-16} \text{erg K}^{-1}, \tag{4.2.34}$$

where k is Boltzmann's constant, N_A is Avogadro's number, and the choice $N_A = 6.024\,62 \times 10^{23}$ mol^{-1} has been adopted. Boltzmann's constant can be estimated in a variety of ways, one of which is described in Section 4.9, and eq. (4.2.34) can be viewed as a way of estimating Avogadro's number, which has been assigned values ranging from

$$N_A = 6.024\,86 \times 10^{23} \text{mol}^{-1} \tag{4.2.35a}$$

(Leighton, 1964) to

$$N_A = 6.022\,141\,99 \times 10^{23} \text{mol}^{-1} \tag{4.2.35b}$$

(Mohr & Taylor, 2002). Whatever its precise value, it has been established that, for a rarified gas, k in the equation

$$P = \frac{N}{V} kT = n\, kT, \tag{4.2.36}$$

where

$$n = \frac{N}{V} \tag{4.2.37}$$

is the particle number density (particles per unit volume), is a constant.

Comparing eqs. (4.2.32) and (4.2.36), one has that, for any other gas in thermal equilibrium with the rarified gas,

$$\beta = \frac{N_A}{R_0} \frac{1}{T} = \frac{1}{kT}. \tag{4.2.38}$$

Having determined β as a function of temperature, the appropriate value of α for any gas can be found from eq. (4.2.16). An explicit example is that given by eq. (4.2.20) for a rarified gas at low temperature.

Another way of arriving at the Fermi–Dirac distribution function is to consider the consequences of assuming detailed balance and, instead of hiding the fact of interactions, to employ the necessary existence of these interactions in a fundamental way. The rate at which particles in states of energy ϵ_1 and ϵ_2 interact to form particles in states of energy ϵ_1' and ϵ_2' may be written as

$$R_{1,2 \to 1',2'} = C_{1,2 \to 1',2'} f(\epsilon_1) f(\epsilon_2) \left[1 - f\left(\epsilon_1'\right)\right] \left[1 - f\left(\epsilon_2'\right)\right], \tag{4.2.39}$$

where $C_{1,2 \to 1',2'}$ is a rate constant and $f(\epsilon)$ is the probability that a state at energy ϵ is occupied. The factors $1 - f(\epsilon)$ follow from the Pauli exclusion principle. Conservation of energy gives $\epsilon_1 + \epsilon_2 = \epsilon_1' + \epsilon_2'$. In a similar fashion,

$$R_{1,2 \leftarrow 1',2'} = C_{1,2 \leftarrow 1',2'} f\left(\epsilon_1'\right) f\left(\epsilon_2'\right) \left[1 - f(\epsilon_1)\right] \left[1 - f(\epsilon_2)\right]. \tag{4.2.40}$$

The rate constants C may be thought of as matrix elements of a Hermitian operator. Hence, $C_{1,2\to1',2'} = C_{1,2\leftarrow1',2'}$. Setting $\epsilon_1' = \epsilon_1 + x$, $\epsilon_2' = \epsilon_2 - x$, and $R_{1,2\leftarrow1',2'} = R_{1,2\to1',2'}$,

$$
f(\epsilon_1)f(\epsilon_2)\left[1 - f(\epsilon_1 + x)\right]\left[1 - f(\epsilon_2 - x)\right]
$$
$$
= f(\epsilon_1 + x)f(\epsilon_2 - x)\left[1 - f(\epsilon_1)\right]\left[1 - f(\epsilon_2)\right]. \tag{4.2.41}
$$

This can be rewritten as

$$
\left(\frac{1}{f(\epsilon_1)} - 1\right)\left(\frac{1}{f(\epsilon_2)} - 1\right) = \left(\frac{1}{f(\epsilon_1 + x)} - 1\right)\left(\frac{1}{f(\epsilon_2 - x)} - 1\right). \tag{4.2.42}
$$

Setting

$$
\omega(\epsilon) = \frac{1}{f(\epsilon)} - 1, \tag{4.2.43}
$$

and expanding about $x = 0$ gives

$$
\omega(\epsilon_1)\,\omega(\epsilon_2) = \omega(\epsilon_1 + x)\,\omega(\epsilon_2 - x) \tag{4.2.44}
$$

$$
= \omega(\epsilon_1)\,\omega(\epsilon_2)\left(1 + \left[\left(\frac{\partial\log_e \omega(\epsilon)}{\partial\epsilon}\right)_{\epsilon=\epsilon_1}\right.\right.
$$
$$
\left.\left. - \left(\frac{\partial\log_e \omega(\epsilon)}{\partial\epsilon}\right)_{\epsilon=\epsilon_2}\right]x\right) + O(x^2). \tag{4.2.45}
$$

Since this result must be independent of x, the expression in square brackets vanishes, giving

$$
\frac{\partial\log_e \omega(\epsilon)}{\partial\epsilon} = \beta, \tag{4.2.46}
$$

where β is independent of ϵ, and

$$
\omega(\epsilon) = \frac{1}{f(\epsilon)} - 1 = \exp(\alpha + \beta\epsilon), \tag{4.2.47}
$$

where α is a constant of integration. This is the continuum equivalent of eq. (4.2.15), and therefore gives the continuum version of the distribution function described by eq. (4.2.14).

4.3 Calculation of pressure and energy density

In all but one of the instances discussed in this chapter (see Section 4.17), the equation of state, which is the relationship between pressure, temperature, density, and composition, follows from a theorem derived by considering specular reflections of particles off a wall. The momentum imparted in a single reflection by a particle of initial momentum p_x and velocity v_x in the x direction to a wall perpendicular to the x axis is $2p_x$. The quantity $v_x\,dn_x/2$ is the flux of particles with velocity in the range $v_x \to v_x + dv_x$, where n_x is the *number density* of particles moving in the $\pm x$ direction. The pressure P on the wall is the rate at which momentum is imparted to the wall by all particles hitting the wall. Thus,

$$P = \text{Pressure} = \int_{-\infty}^{\infty} 2p_x v_x \frac{1}{2} dn_x = \int_{-\infty}^{\infty} p_x v_x dn_x = \langle v_x p_x \rangle n. \qquad (4.3.1)$$

In eq. (4.3.1), brackets denote an average over the momentum distribution, and $n = N/V$ is the total number density. Since $\langle v_x p_x \rangle = \langle v_y p_y \rangle = \langle v_z p_z \rangle$,

$$P = \frac{1}{3} \langle vp \rangle n. \qquad (4.3.2)$$

Another way to obtain this result is to take eq. (4.2.28) rather than eq. (4.3.1) as the starting point. From eqs. (4.1.4) and (4.1.3), one has that

$$\epsilon_i^2 = (cp_i)^2 + (Mc^2)^2 = \frac{\gamma_i^2}{V^{2/3}} + (Mc^2)^2, \qquad (4.3.3)$$

where $\gamma_i = chm_i$ and $V = L^3$. Differentiation gives

$$\epsilon_i \frac{\partial \epsilon_i}{\partial V} = -\frac{1}{3} \frac{\gamma_i^2}{V^{5/3}} = -\frac{1}{3} \frac{1}{V} (cp_i)^2. \qquad (4.3.4)$$

Inserting into eq. (4.2.28), and using eq. (4.1.6), it follows that

$$P = -\sum_{i=1}^{N} N_i \frac{\partial \epsilon_i}{\partial V} = \frac{1}{3} \sum_i n_i \frac{cp_i}{\epsilon_i} cp_i = \frac{1}{3} \sum_i n_i v_i p_i, \qquad (4.3.5)$$

which is equivalent to eq. (4.3.2). Although the two approaches in this instance give the same result, the second approach is more fundamental, being applicable to systems in which the energy is not simply kinetic energy (see Section 4.17).

The kinetic energy per unit volume,

$$\bar{U}' = \frac{E'}{V} = \sum_i n_i \epsilon_i', \qquad (4.3.6)$$

is related to the pressure in a straightforward way, as may be seen by making use of eq. (4.1.4) to write eq. (4.3.5) as

$$P = \frac{1}{3} \sum_i n_i \frac{\epsilon_i^2 - (Mc^2)^2}{\epsilon_i} = \frac{1}{3} \sum_i n_i \frac{(\epsilon_i - Mc^2)(\epsilon_i + Mc^2)}{\epsilon_i} = \frac{1}{3} \sum_i n_i \epsilon_i' \frac{\epsilon_i + Mc^2}{\epsilon_i}. \qquad (4.3.7)$$

For non-relativistic particles, $\epsilon_i \to Mc^2$ and $(\epsilon_i + Mc^2)/\epsilon_i \to 2$, so

$$P \to \frac{2}{3} \sum_i n_i \epsilon_i' = \frac{2}{3} \bar{U}'. \qquad (4.3.8)$$

For relativistic particles, $Mc^2/\epsilon_i \to 0$ and $(\epsilon_i + Mc^2)/\epsilon_i \to 1$, so

$$P \to \frac{1}{3} \sum_i n_i \epsilon_i' = \frac{1}{3} \bar{U}' = \frac{1}{3} \frac{E'}{V}. \qquad (4.3.9)$$

At this point, a word about notation with regard to energy and energy density is appropriate. The words energy density are commonly used to denote both energy per unit volume and energy per unit mass. The convention adopted here is that E is energy, U is energy per unit mass, and $\bar{U}$ is energy per unit volume. Thus,

$$U = \frac{\bar{U}}{\rho} \text{ and } \bar{U} = \frac{E}{V} = \rho\, U. \qquad (4.3.10)$$

Superscript primes are attached when energy is kinetic energy.

4.4 Equation of state for non-degenerate, non-relativistic ions

Over most of the interior of most stars, including the electron-degenerate cores of evolved stars and white dwarfs, ions move at velocities significantly smaller than the speed of light, and are separated from one another by distances much larger than the deBroglie wavelength $\hbar/\langle p \rangle$, where $\langle p \rangle$ is a mean momentum (Louis deBroglie, 1924). Thus, $\epsilon = p^2/2M$, and the occupation number $f(\epsilon)$ given by eq. (4.2.15) (eq. (4.2.47) in the continuum approximation) is small compared to 1. For protons, ^{3}He nuclei, and other nuclei of spin 1/2 in the gaseous state,

$$dn(p_x, p_y, p_z) = 2\, \frac{dp_x dp_y dp_z}{h^3}\, \exp\left(-\alpha - \frac{p^2}{2MkT}\right) \qquad (4.4.1)$$

is an excellent approximation to the differential number density. The factor of 2 in this equation comes from the fact that particles can be in either of two spin states.

Integrating over all angles, eq. (4.4.1) gives

$$n = \frac{8\pi}{h^3} e^{-\alpha} \int_0^\infty \exp\left(-\frac{p^2}{2MkT}\right) p^2 dp \qquad (4.4.2)$$

$$= \frac{8\pi}{h^3} e^{-\alpha} (2MkT)^{3/2} \int_0^\infty e^{-x^2} x^2 dx, \qquad (4.4.3)$$

and eqs. (4.3.5) and (4.4.1) together give

$$P = \frac{1}{3} \frac{8\pi}{h^3} e^{-\alpha} \int_0^\infty \exp\left(-\frac{p^2}{2MkT}\right) v\, p^3 dp \qquad (4.4.4)$$

$$= \frac{1}{3} \frac{8\pi}{h^3} e^{-\alpha} \frac{(2MkT)^{5/2}}{M} \int_0^\infty e^{-x^2} x^4 dx. \qquad (4.4.5)$$

With the definition

$$I_m(\lambda) = \int_0^\infty x^m e^{-\lambda x^2} dx, \tag{4.4.6}$$

it follows that

$$I_0(\lambda) = \frac{\sqrt{\pi}}{2} \lambda^{-1/2}, \tag{4.4.7}$$

and

$$I_{m+2}(\lambda) = -\frac{1}{\lambda} \frac{dI_m(\lambda)}{d\lambda}. \tag{4.4.8}$$

Using these results,

$$n = \frac{8\pi}{h^3} e^{-\alpha} (2MkT)^{3/2} \frac{\sqrt{\pi}}{4} \tag{4.4.9}$$

$$= 2 e^{-\alpha} \left(\frac{2\pi MkT}{h^2}\right)^{3/2}, \tag{4.4.10}$$

and

$$P = 2e^{-\alpha} kT \left(\frac{2\pi MkT}{h^2}\right)^{3/2}. \tag{4.4.11}$$

Eliminating α from eqs. (4.4.11) and (4.4.10) gives

$$P = nkT, \tag{4.4.12}$$

in agreement with the development in Section 4.2 (eq. (4.2.36)).

Defining the mean momentum $\langle p \rangle$ by

$$\langle p \rangle = \frac{1}{n} \int_0^\infty p \, dn = \frac{2}{\pi} (2\pi MkT)^{1/2}, \tag{4.4.13}$$

it follows that

$$\lambda_M \equiv \frac{2}{\pi} \frac{h}{\langle p \rangle} = \left(\frac{h^2}{2\pi MkT}\right)^{1/2} \tag{4.4.14}$$

is a measure of the deBroglie wave length of a typical particle of mass M. Rewriting eq. (4.4.10) as

$$e^{-\alpha} = \frac{1}{2} \lambda_M^3 \, n, \tag{4.4.15}$$

it is evident that $e^{-\alpha}$ is identical to the number of particles (of spin in a given direction) in a box with sides of length equal to the deBroglie wavelength of a typical particle.

Quantitatively,

$$\lambda_M = \frac{1.739\,50 \times 10^{-10}}{T_6^{1/2}} \left(\frac{M_H}{M}\right)^{1/2} \text{cm}, \tag{4.4.16}$$

where T_6 is the temperature in units of 10^6 K and $M_H = 1.673\,53 \times 10^{-24}$ g is the mass of the hydrogen atom. Setting

$$n = \frac{\rho}{M} = \frac{1}{\mu_M} \frac{\rho}{M_H}, \tag{4.4.17}$$

where ρ is the density in units of g cm^{-3} and

$$\mu_M = \frac{M}{M_H} \tag{4.4.18}$$

is the molecular weight (number of nucleons per particle) of particles of mass M, one has

$$e^{-\alpha} = \frac{1.572\,576 \times 10^{-6}}{\mu_M^{5/2}} \frac{\rho}{T_6^{3/2}}. \tag{4.4.19}$$

Under most conditions in nuclear burning stars as well as in typical white dwarfs, $e^{-\alpha}$ for ions is much smaller than 1, verifying the statement made at the outset of this section that the average distance between ions is large compared with the deBroglie wavelength. Thus, quantum effects for ions in the gaseous state are typically unimportant, the second term in the denominator of eq. (4.2.15) may be neglected, and eq. (4.4.12) may be generalized to give

$$P_{\text{ions}} = \sum_i n_i\, kT = \frac{1}{\mu_{\text{ions}}} \frac{\rho}{M_H}\, kT, \tag{4.4.20}$$

where μ_{ion} is the average molecular weight of ions. Since relativistic effects are also negligible, the relationship between the kinetic energy per unit volume and pressure is given by eq. (4.3.8), namely,

$$\bar{U}'_{\text{ions}} = \frac{E'_{\text{ions}}}{V} = \frac{3}{2}\, P_{\text{ions}}. \tag{4.4.21}$$

4.5 Equation of state for weakly degenerate, non-relativistic electrons

In contrast with the case for ions, the deBroglie wavelength of a typical electron in the interior of even a main sequence star can be a significant fraction of the separation between electrons. As long as $\alpha > 0$, it is possible to expand the occupation number given by eq. (4.2.15) in powers of $e^{-(\alpha+\beta\epsilon)}$:

$$f(\epsilon) = \frac{e^{-(\alpha+\beta\epsilon)}}{1 + e^{-(\alpha+\beta\epsilon)}} = \sum_{m=1}^{\infty} (-1)^{m+1} e^{-m(\alpha+\beta\epsilon)}. \tag{4.5.1}$$

The electron number density is then given by

$$n_e = \frac{8\pi}{h^3} \int_0^\infty f(\epsilon) p^2 \mathrm{d}p = \frac{8\pi}{h^3} \int_0^\infty \sum_{m=1}^\infty (-1)^{m+1} e^{-m(\alpha+\beta\epsilon)} p^2 \mathrm{d}p \qquad (4.5.2)$$

$$= \frac{8\pi}{h^3} \sum_{m=1}^\infty (-1)^m e^{-m\alpha} \int_0^\infty e^{-m\beta\epsilon} p^2 \mathrm{d}p. \qquad (4.5.3)$$

Assuming that electron energies are not relativistic, setting $x^2 = \beta\epsilon = p^2/2m_e kT$, and making use of eqs. (4.4.6)–(4.4.8), one finds

$$n_e = \frac{8\pi}{h^3} (2m_e kT)^{3/2} \sum_{m=1}^\infty (-1)^{m+1} e^{-m\alpha} \int_0^\infty e^{-mx^2} x^2 \mathrm{d}x \qquad (4.5.4)$$

$$= 2 \left(\frac{2\pi m_e kT}{h^2} \right)^{3/2} \sum_{m=1}^\infty (-1)^{m+1} \frac{e^{-m\alpha}}{m^{3/2}}. \qquad (4.5.5)$$

Defining the deBroglie wavelength of a typical electron to be

$$\lambda_e = \left(\frac{h^2}{2\pi m_e kT} \right)^{1/2} = \frac{0.745383 \times 10^{-8}}{T_6^{1/2}} \text{ cm}, \qquad (4.5.6)$$

it follows from eq. (4.5.5) that

$$e^{-\alpha} - \frac{1}{2^{3/2}} e^{-2\alpha} + \cdots = \frac{1}{2} \lambda_e^3 n_e, \qquad (4.5.7)$$

or

$$e^{-\alpha} = \frac{1}{2} \left(\lambda_e^3 n_e \right) \left(1 + \frac{1}{2^{3/2}} \frac{1}{2} \left(\lambda_e^3 n_e \right) \right) + O\left[\left(\lambda_e^3 n_e \right)^3 \right]. \qquad (4.5.8)$$

Defining the electron molecular weight μ_e by

$$n_e = \frac{\rho}{\mu_e M_H} \qquad (4.5.9)$$

and a degeneracy parameter δ by

$$\delta = \frac{\rho}{\mu_e T_6^{3/2}}, \qquad (4.5.10)$$

one has that

$$\frac{1}{2} \lambda_e^3 n_e = \frac{0.207\,066 \times 10^{-24} \text{ cm}^3}{T_6^{3/2}} \, n_e = 0.123\,730\,\delta, \qquad (4.5.11)$$

and

$$e^{-\alpha} = 0.123\,730\,\delta + 0.005\,412\,59\,\delta^2 + O(\delta^3)$$

$$= 0.123\,730\,\delta\,(1 + 0.043\,745\,\delta) + O(\delta^3). \tag{4.5.12}$$

Pressure may be calculated by combining eqs. (4.3.2) and (4.5.1) to obtain

$$P_e = \frac{1}{3}\frac{8\pi}{h^3}\sum_{m=1}^{\infty}(-1)^{m+1}e^{-m\alpha}\int_0^{\infty}e^{-m(\alpha+\beta\epsilon)}\,vp^3\mathrm{d}p \tag{4.5.13}$$

$$= \frac{1}{3m_e}\frac{8\pi}{h^3}\sum_{m=1}^{\infty}(-1)^{m+1}e^{-m\alpha}\int_0^{\infty}e^{-mp^2/2m_ekT}\,p^4\mathrm{d}p \tag{4.5.14}$$

$$= \frac{1}{3m_e}\frac{8\pi}{h^3}\sum_{m=1}^{\infty}(-1)^{m+1}e^{-m\alpha}\left(\frac{2m_ekT}{m}\right)^{5/2}\int_0^{\infty}e^{-x^2}x^4\mathrm{d}x \tag{4.5.15}$$

and then using eqs. (4.4.6)–(4.4.8) to find

$$P_e = 2\frac{h^2}{2\pi m_e}\left(\frac{2\pi m_ekT}{h^2}\right)^{5/2}\sum_{m=1}^{\infty}(-1)^{m+1}\frac{e^{-m\alpha}}{m^{5/2}}. \tag{4.5.16}$$

Equations (4.5.5) and (4.5.16) together give

$$\frac{P_e}{n_ekT} = \frac{\Sigma_{5/2}(\alpha)}{\Sigma_{3/2}(\alpha)}, \tag{4.5.17}$$

where

$$\Sigma_k(\alpha) = \sum_{m=1}^{\infty}(-1)^{m+1}\frac{e^{-m\alpha}}{m^k}. \tag{4.5.18}$$

Explicitly,

$$\frac{P_e}{n_ekT} = \left[\left(1 - \frac{1}{2^{5/2}}e^{-\alpha}\right) + \left(\frac{1}{3^{5/2}} - \frac{1}{4^{5/2}}e^{-\alpha}\right)e^{-2\alpha} + \cdots\right]$$

$$\div\left[\left(1 - \frac{1}{2^{3/2}}e^{-\alpha}\right) + \left(\frac{1}{3^{3/2}} - \frac{1}{4^{3/2}}e^{-\alpha}\right)e^{-2\alpha} + \cdots\right] \tag{4.5.19}$$

$$= 1 + \frac{1}{2^{5/2}}e^{-\alpha} - \left(\frac{2}{3^{5/2}} - \frac{1}{2^4}\right)e^{-2\alpha} + \cdots \tag{4.5.20}$$

$$= 1 + 0.176\,776\,695\,e^{-\alpha} - 0.065\,800\,06\,e^{-2\alpha} + \cdots. \tag{4.5.21}$$

Using eq. (4.5.12) in eq. (4.5.21) gives

$$\frac{P_e}{n_ekT} = 1 + 0.021\,8726\,\delta - 0.000\,050\,529\,\delta^2 + O(\delta^3)$$

$$= 1 + 0.021\,8726\,\delta\,(1 - 0.002\,310\,\delta) + O(\delta^3). \tag{4.5.22}$$

At the center of the Sun, where $\rho \sim 150$ gm cm^{-3}, $T_6 \sim 15$, and $\mu_e \sim 1.4$, one has $\delta \sim 1.844$, $\lambda_e^3 n_e/2 \sim 0.228$, $e^{-\alpha} \sim 0.247$, and $P_e/n_e kT \sim 1 + 0.0403 - 0.0001 + \cdots = 1.0402$. Thus, even at the comparatively low density at the Sun's center, the exclusion principle plays a non-negligible role in the equation of state for electrons. The occupation number for states at energies small compared with kT is $f(0) \sim 0.20$. Occupation numbers for states at energies kT, $3kT/2$, and $2kT$ are 0.08, 0.05, and 0.03, respectively.

Since, for $\epsilon = 0$, the expansion of f undertaken in eq. (4.5.1) is not formally convergent when $\alpha < 0$, one might anticipate that the approximation explored in this section reaches its limit when $\alpha \to 0$. At this limit, the sums in eq. (4.5.10) converge very slowly towards their limiting values, even when terms are grouped as in eq. (4.5.20). Setting

$$S_k = \sum_{m=1}^{\infty} (-1)^{m+1} \frac{1}{m^k}, \tag{4.5.23}$$

one obtains $S_{3/2} \sim 0.764\,906$ by explicitly adding 26 terms and then integrating to approximate the remainder of the sum, and $S_{5/2} \sim 0.867\,171$ by explicitly adding 20 terms and integrating to approximate the remainder of the sum. A machine calculation including 500 terms gives

$$S_{3/2} = 0.765\,102\,370\,347\,702 \tag{4.5.24}$$

and

$$S_{5/2} = 0.867\,199\,799\,793\,070. \tag{4.5.25}$$

Using these sums in eq. (4.5.17), one has that, when $\alpha = 0$,

$$\left(\frac{P_e}{n_e kT} \right)_{\alpha=0} \sim 1.133\,443. \tag{4.5.26}$$

Equations (4.5.5), (4.5.6), (4.5.23), and (4.5.24) together yield, when $\alpha = 0$,

$$\frac{1}{2}\lambda_e^3 n_e = 0.123\,730 \frac{\rho_0}{\mu_e T_6^{3/2}} = 0.123\,730\,\delta = S_{3/2} = 0.765\,102, \tag{4.5.27}$$

or

$$\delta_{\alpha=0} = 6.1836. \tag{4.5.28}$$

One may think of the condition $\alpha = 0$ as defining the borderline between very degenerate and not very degenerate electrons. At this value of α, the occupation number for electrons at zero kinetic energy is $f = 1/2$; said another way, the average number of electrons in a cube one deBroglie wavelength on a side, or in a sphere one deBroglie wavelength in diameter, is essentially unity; said in still another way, the average distance between an electron and its nearest neighbor is one deBroglie wavelength. These are all indications that quantum mechanics plays an important, but not completely dominant, role in determining the distribution of electrons in phase space.

With the value of δ given by eq. (4.5.28), the first two terms in the expansion of $e^{-\alpha}$ in terms of δ (eq. (4.5.12)) give $\alpha = 0.028\,65$ instead of 0.0, showing that the omitted higher order terms make a positive contribution to $e^{-\alpha}$. By adding a cubic term to eq. (4.5.12) of the just the right magnitude to give $\alpha = 0.0$ when $\delta = 6.182$, one obtains

$$e^{-\alpha} \sim 0.123\,730\,\delta\,(1 + 0.043\,745\,\delta\,(1 + 0.022\,09\,\delta)). \qquad (4.5.29)$$

After this lengthy and detailed exploration, it is worth remembering that, since it has been assumed that electrons have non-relativistic energies, the very simple relationship expressed by eq. (4.3.8) still holds. That is,

$$\bar{U}'_e = \frac{E'_e}{V} = \frac{3}{2} P_e, \qquad (4.5.30)$$

where E'_e is the total electron kinetic energy, $\bar{U}'_e$ is the electron kinetic energy per unit volume, and P_e is the electron pressure.

4.6 Equation of state for strongly degenerate, non-relativistic electrons

The discussion of weak degeneracy has demonstrated that, at a given density, the occupation numbers of states with the lowest energies increase with decreasing temperature. As temperature is lowered, it is clear that, at some critical value of kT, α must become negative. Then, it is customary to write

$$f(\epsilon) = \frac{1}{e^{(\epsilon - \epsilon_F)/kT} + 1}, \qquad (4.6.1)$$

where ϵ_F is called the Fermi energy. In this section, the convention will be adopted that both ϵ and ϵ_F are kinetic energies. When $\epsilon = \epsilon_F$, the occupation number is 1/2. As kT is decreased, a larger and larger fraction of the electron population shifts into states with energy smaller than ϵ_F, until, ultimately, as $kT \to 0$, all states with energy smaller than the Fermi energy are filled and none with energy larger than the Fermi energy are occupied.

Of course, the Fermi energy itself changes as temperature is changed, and it is instructive to see analytically how it does so in the limit that $kT/\epsilon_F \ll 1$. It is convenient to divide all integrals into three parts. Defining

$$x = \frac{\epsilon - \epsilon_F}{kT} > 0, \qquad (4.6.2)$$

one can write

$$f(\epsilon) = f_{\epsilon > \epsilon_F} = \frac{e^{-(\epsilon - \epsilon_F)/kT}}{1 + e^{-(\epsilon - \epsilon_F)/kT}} = \frac{e^{-x}}{1 + e^{-x}} = \sum_{m=1}^{\infty} (-1)^{m+1} e^{-mx}. \qquad (4.6.3)$$

When

$$\bar{x} = \frac{\epsilon_F - \epsilon}{kT} = -x > 0, \qquad (4.6.4)$$

one can write

$$f(\epsilon) = f_{\epsilon < \epsilon_F} = \frac{1}{1 + e^{-(\epsilon_F - \epsilon)/kT}} = \frac{1}{1 + e^{-\bar{x}}} = 1 - \Delta f_{\epsilon < \epsilon_F}, \qquad (4.6.5)$$

where

$$\Delta f_{\epsilon < \epsilon_F} = e^{-\bar{x}}/[1 + e^{-\bar{x}}] = \sum_{m=1}^{\infty} (-1)^{m+1} e^{-m\bar{x}}. \qquad (4.6.6)$$

Using these functions,

$$n_e = \frac{8\pi}{h^3} \int_0^{\infty} f(\epsilon) p^2 dp = \frac{8\pi}{h^3} \left(\int_0^{p_F} p^2 dp - \int_0^{p_F} \Delta f_{\epsilon < \epsilon_F} \; p^2 dp + \int_{p_F}^{\infty} f_{\epsilon > \epsilon_F} \; p^2 dp \right), \qquad (4.6.7)$$

where p_F is the value of the electron momentum when $\epsilon = \epsilon_F$. Converting coordinates from momentum to kinetic energy in the non-relativistic approximation, $p^2 = 2m_e\epsilon$, one obtains

$$n_e = \frac{4\pi}{h^3} (2m_e)^{3/2}$$
$$\times \left(\int_0^{\epsilon_F} \epsilon^{1/2} d\epsilon - \int_0^{\epsilon_F} \sum_{m=1}^{\infty} (-1)^{m+1} e^{-m\bar{x}} \epsilon^{1/2} d\epsilon + \int_{\epsilon_F}^{\infty} \sum_{m=1}^{\infty} (-1)^{m+1} e^{-mx} \epsilon^{1/2} d\epsilon \right). \qquad (4.6.8)$$

Converting to the coordinates x and $\bar{x}$, one can write

$$n_e = \frac{4\pi}{h^3} (2m_e)^{3/2} (I_1 + \Delta I_1 + I_2), \qquad (4.6.9)$$

where

$$I_1 = \frac{2}{3} \epsilon_F^{3/2}, \qquad (4.6.10)$$

$$\Delta I_1 = -kT \int_0^{\epsilon_F/kT} \sum_{m=1}^{\infty} (-1)^{m+1} e^{-m\bar{x}} (\epsilon_F - \bar{x}kT)^{1/2} d\bar{x}, \qquad (4.6.11)$$

and

$$I_2 = kT \int_0^{\infty} \sum_{m=1}^{\infty} (-1)^{m+1} e^{-mx} (\epsilon_F + xkT)^{1/2} dx. \qquad (4.6.12)$$

Expanding $\epsilon^{1/2} = (\epsilon_F + xkT)^{1/2}$ in powers of x,

$$\epsilon^{1/2} = \sum_{k=0}^{\infty} \frac{1}{k!} \left(\frac{d^k \epsilon^{1/2}}{dx^k} \right)_{x=0} x^k = \sum_{k=0}^{\infty} a_k x^k, \qquad (4.6.13)$$

and inserting this into eq. (4.6.12) yields

$$I_2 = kT \int_0^\infty \sum_{m=1}^\infty (-1)^{m+1} e^{-mx} \sum_{k=0}^\infty a_k x^k dx. \tag{4.6.14}$$

Similarly, eq. (4.6.11) becomes

$$\Delta I_1 = -kT \int_0^{\epsilon_F/kT} \sum_{m=1}^\infty (-1)^{m+1} e^{-m\bar{x}} \sum_{k=0}^\infty (-1)^k a_k \bar{x}^k d\bar{x}. \tag{4.6.15}$$

If it is assumed that $\epsilon_F/kT \gg 1$, the upper limit on the integral in ΔI_1 may be extended to ∞. Then, terms in even powers of k in eq. (4.6.15) and (4.6.14) cancel each other and those in odd powers add. Thus,

$$\Delta I_1 + I_2 = 2kT \int_0^\infty \sum_{m=1}^\infty (-1)^{m+1} e^{-mx} \sum_{k=1}^\infty a_{2k-1} x^{2k-1} dx \tag{4.6.16}$$

$$= 2kT \sum_{m=1}^\infty (-1)^{m+1} \sum_{k=1}^\infty a_{2k-1} \int_0^\infty e^{-mx} x^{2k-1} dx \tag{4.6.17}$$

$$= 2kT \sum_{m=1}^\infty (-1)^{m+1} \sum_{k=1}^\infty a_{2k-1} \frac{(2k-1)!}{m^{2k}}. \tag{4.6.18}$$

Since

$$a_k = \frac{1}{k!} \left(\frac{d^k \epsilon^{1/2}}{dx^k} \right)_{x=0}, \tag{4.6.19}$$

$$\frac{\Delta I_1 + I_2}{2kT} = \left(\frac{d\epsilon^{1/2}}{dx} \right)_{x=0} \sum_{m=1}^\infty (-1)^{m+1} \frac{1}{m^2} + \left(\frac{d^3 \epsilon^{1/2}}{dx^3} \right)_{x=0} \sum_{m=1}^\infty (-1)^{m+1} \frac{1}{m^4} + \cdots . \tag{4.6.20}$$

Using

$$S_2 = \sum_{m=1}^\infty (-1)^{m+1} \frac{1}{m^2} = \frac{\pi^2}{12} = 0.822\,467\,033, \tag{4.6.21}$$

$$S_4 = \sum_{m=1}^\infty (-1)^{m+1} \frac{1}{m^4} = \frac{7\pi^4}{720} = 0.947\,032\,829, \tag{4.6.22}$$

$$\left(\frac{d\epsilon^{1/2}}{dx} \right)_{x=0} = \frac{1}{2} \frac{kT}{\epsilon_F^{1/2}}, \tag{4.6.23}$$

and

$$\left(\frac{d^3 \epsilon^{1/2}}{dx^3} \right)_{x=0} = \frac{3}{8} \frac{(kT)^3}{\epsilon_F^{5/2}}, \tag{4.6.24}$$

in eq. (4.6.20), using eq. (4.6.10), and summing up according to eq. (4.6.9), one has finally that

$$n_e = \frac{8\pi}{3} \left(\frac{2m_e \epsilon_F}{h^2} \right)^{3/2} \left(1 + \frac{\pi^2}{8} \left(\frac{kT}{\epsilon_F} \right)^2 + \frac{7\pi^4}{640} \left(\frac{kT}{\epsilon_F} \right)^4 + \cdots \right) \tag{4.6.25}$$

$$= \frac{8\pi}{3} \left(\frac{2m_e \epsilon_F}{h^2} \right)^{3/2} \left(1 + 1.2337 \left(\frac{kT}{\epsilon_F} \right)^2 + 1.0654 \left(\frac{kT}{\epsilon_F} \right)^4 + \cdots \right). \tag{4.6.26}$$

At zero temperature, $\epsilon_F = \epsilon_{F0}$, where

$$\epsilon_{F0} = \left(\frac{3n_e}{8\pi} \right)^{2/3} \frac{h^2}{2m_e} = \left(\frac{3n_e}{8\pi} \right)^{2/3} \frac{1}{2} \left(\frac{h}{m_e c} \right)^2 (m_e c^2) \tag{4.6.27}$$

$$= 0.506\,245 \, (m_e c^2) \left(\frac{\rho_6}{\mu_e} \right)^{2/3} = 0.258\,6906 \text{ MeV} \left(\frac{\rho_6}{\mu_e} \right)^{2/3}, \tag{4.6.28}$$

and ρ_6 is the density in units of 10^6 g cm^{-3}. Additional useful relationships are

$$\frac{\epsilon_{F0}}{kT} = 3001.977 \left(\frac{\rho_6}{\mu_e T_6^{3/2}} \right)^{2/3} = 0.300\,1977 \, \delta^{2/3} = \left(\frac{\delta}{6.0798} \right)^{2/3}, \tag{4.6.29}$$

where δ is given by eq. (4.5.10). Equations (4.6.25) and (4.6.27) can be solved to find ϵ_F in terms of ϵ_{F0}:

$$\epsilon_F = \epsilon_{F0} \left(1 - \frac{\pi^2}{12} \left(\frac{kT}{\epsilon_{F0}} \right)^2 - \frac{\pi^4}{80} \left(\frac{kT}{\epsilon_{F0}} \right)^4 - \cdots \right). \tag{4.6.30}$$

To the extent that ρ_6 is small compared with 1, the non-relativistic approximation for electron energy is adequate. For example, when $\rho_6/\mu_e = 0.01$, $\epsilon_{F0} = 0.0235 \, m_e c^2$ and the approximation is a reasonable one. When $\rho_6/\mu_e = 0.1$, $\epsilon_{F0} = 0.109 \, m_e c^2$ and the approximation begins to break down.

It is interesting that, had the approximation not been made of extending to ∞ the upper limit on the integral in eq. (4.6.15), the result of formally evaluating the integral at the correct upper limit would have added

$$\Delta I_{1,\,add} = \epsilon_F^{1/2} \sum_{m=1}^{\infty} (-1)^{m+1} e^{-m\epsilon_F/kT}$$

$$\times \left[\frac{1}{m} (1-1)^{1/2} - \frac{1}{m^2} \frac{kT}{\epsilon_F} \frac{1}{2} \frac{1}{(1-1)^{1/2}} - \frac{1}{m^3} \left(\frac{kT}{\epsilon_F} \right)^2 \frac{1}{4} \frac{1}{(1-1)^{3/2}} - \cdots \right],$$

where a term of the form $(1-1)^n$ denotes the expansion of $(1-x)^n$ in powers of x evaluated at $x = 1$. In short, the terms do not converge and the formal, asymptotic, expansion gives an indeterminate result.

The calculation of pressure begins with

$$P_e = \frac{1}{3}\frac{8\pi}{h^3} \int_0^\infty f(\epsilon)\, vp\, p^2 dp = \frac{8\pi}{h^3}(2m_e)^{3/2} \int_0^\infty f(\epsilon)\epsilon^{3/2} d\epsilon \qquad (4.6.31)$$

and proceeds in exactly the same way as in the calculation of n_e. The penultimate result is

$$P_e = \frac{8\pi}{3h^3}(2m_e)^{3/2}\left(I_1' + \Delta I_1' + I_2'\right), \qquad (4.6.32)$$

where

$$I_1' = \frac{2}{5}\epsilon_F^{5/2} \qquad (4.6.33)$$

and

$$\frac{\Delta I_1' + I_2'}{2kT} = \left(\frac{d\epsilon^{3/2}}{dx}\right)_{x=0}\sum_{m=1}^\infty (-1)^{m+1}\frac{1}{m^2} + \left(\frac{d^3\epsilon^{3/2}}{dx^3}\right)_{x=0}\sum_{m=1}^\infty (-1)^{m+1}\frac{1}{m^4} + \cdots . \tag{4.6.34}$$

With

$$\left(\frac{d\epsilon^{3/2}}{dx}\right)_0 = \frac{3}{2}kT\epsilon_F^{1/2}, \qquad (4.6.35)$$

and

$$\left(\frac{d^3\epsilon^{3/2}}{dx^3}\right)_0 = -\frac{3}{8}\frac{(kT)^3}{\epsilon_F^{3/2}}, \qquad (4.6.36)$$

and using eq. (4.6.21) and (4.6.22), one obtains

$$P_e = \frac{16\pi}{15}\left(\frac{2m_e\epsilon_F}{h^2}\right)^{3/2}\epsilon_F\left(1 + \frac{5\pi^2}{8}\left(\frac{kT}{\epsilon_F}\right)^2 - \frac{7\pi^4}{384}\left(\frac{kT}{\epsilon_F}\right)^4 + \cdots\right) \qquad (4.6.37)$$

$$= \frac{16\pi}{15}\left(\frac{2m_e\epsilon_F}{h^2}\right)^{3/2}\epsilon_F\left(1 + 6.169\left(\frac{kT}{\epsilon_F}\right)^2 - 1.776\left(\frac{kT}{\epsilon_F}\right)^4 + \cdots\right). \qquad (4.6.38)$$

Combining eqs. (4.6.37) and (4.6.25) gives the electron equation of state,

$$P_e = \frac{2}{5}n_e\,\epsilon_F\left(1 + \frac{5}{8}\pi^2\left(\frac{kT}{\epsilon_F}\right)^2 - \cdots\right) \div \left(1 + \frac{1}{8}\pi^2\left(\frac{kT}{\epsilon_F}\right)^2 + \cdots\right) \qquad (4.6.39)$$

$$= \frac{2}{5}n_e\,\epsilon_F\left(1 + \frac{\pi^2}{2}\left(\frac{kT}{\epsilon_F}\right)^2 - \frac{11\pi^4}{120}\left(\frac{kT}{\epsilon_F}\right)^4 + \cdots\right). \qquad (4.6.40)$$

Using eq. (4.6.30) to eliminate ϵ_F in favor of ϵ_{F0}, it follows that

$$P_e = \frac{2}{5} n_e \epsilon_{F0} \left(1 + \frac{5\pi^2}{12} \left(\frac{kT}{\epsilon_{F0}} \right)^2 - \frac{\pi^4}{16} \left(\frac{kT}{\epsilon_{F0}} \right)^4 + \cdots \right). \tag{4.6.41}$$

At zero temperature, eqs. (4.6.41) and (4.6.29) give

$$P_e = \frac{2}{5} \left(\frac{3}{8\pi} \right)^{2/3} \left(\frac{\rho}{\mu_e M_H} \right)^{5/3} \frac{h^2}{2m_e}$$

$$= 0.990\,647 \times 10^{13} \left(\frac{\rho}{\mu_e} \right)^{5/3} \text{ dyne cm}^{-2}. \tag{4.6.42}$$

Electron kinetic energy density per unit volume and pressure are, once again, related by eq. (4.3.8), but, for completeness, the relationship is repeated here:

$$U'_e = \frac{3}{2} P_e. \tag{4.6.43}$$

4.7 Equation of state for non-relativistic electrons of intermediate degeneracy

The analytic relationships between thermodynamic quantities which have been derived in the previous two sections are fully applicable only in the domains of weak degeneracy ($\alpha \geq 0$) and of strong degeneracy ($\epsilon_F/kT \gg 1$), respectively. In the region of intermediate degeneracy, say ($0 < -\alpha, \epsilon_F/kT < 3$), numerical calculations are necessary to obtain relationships in which confidence can be placed.

In Table 4.7.1 are results of such calculations which cover the complete range of non-relativistic degeneracy (Iben, 1968). The quantities α, ϵ_F/kT, $\delta = \rho/(\mu_e T_6^{3/2})$, and $P_e/(n_e kT)$ have been described in previous sections and, of course,

$$\epsilon_F = -\alpha\, kT. \tag{4.7.1}$$

The quantity in the fifth column of Table 4.7.1 gives a measure of the deBroglie wavelength $\bar{\lambda}_e$ of a typical electron divided by the average distance between electrons. The quantities in the last two columns of Table 4.7.1 are of relevance to electron screening; they are discussed in Section 4.17.

The entry $\alpha = \epsilon_F/kT = 0.0000$ in Table 4.7.1 occurs at $\delta = 6.185$ rather than at the value $\delta = 6.1836$ found in Section 4.5. This difference of 0.02% is an indication of the accuracy of the numerical calculations.

The approximations for weak degeneracy derived in Section 4.5 and those for strong degeneracy derived in Section 4.6 can be used as guides for producing analytic expressions which relate thermodynamic variables explicitly to the degeneracy parameter $\delta = \rho/(\mu_e T_6^{3/2})$ and which approximate the tabular data in the region of intermediate degeneracy.

Table 4.7.1 Properties of a non-relativistic electron gas						
$e^{-\alpha}$	$\frac{\epsilon_F}{kT}$	$\frac{\rho}{\mu_e T_6^{3/2}}$	$\frac{P_e}{n_e kT}$	$\frac{\lambda_e}{r_{0e}}$	$\frac{1}{n_e}\left|\frac{\partial n_e}{\partial \alpha}\right|$	$\frac{R_D}{r_{0i}}\frac{1}{T_6^{1/4}}$
0.01	−4.3026	0.08055	1.002	0.127	0.9965	1.0655
0.03	−3.5066	0.2400	1.005	0.181	0.9896	0.8892
0.06	−2.8134	0.4751	1.010	0.228	0.9797	0.7950
0.10	−2.3026	0.7812	1.017	0.268	0.9671	0.7332
0.20	−1.6094	1.513	1.033	0.331	0.9382	0.6599
0.30	−1.2040	2.203	1.048	0.373	0.9127	0.6226
0.40	−0.9163	2.856	1.062	0.404	0.8899	0.5986
0.60	−0.5108	4.063	1.088	0.450	0.8508	0.5683
0.80	−0.2231	5.169	1.112	0.481	0.8183	0.5491
1.0	0.0000	6.185	1.133	0.506	0.7906	0.5355
1.25	0.2231	7.354	1.158	0.530	0.7612	0.5231
1.5	0.4055	8.430	1.181	0.556	0.7362	0.5136
2.0	0.6931	10.36	1.221	0.581	0.6956	0.5000
2.5	0.9163	12.05	1.257	0.600	0.6638	0.4905
3.0	1.0986	13.57	1.288	0.617	0.6379	0.4831
4.0	1.3863.	16.21	1.342	0.641	0.5979	0.4727
5.0	1.609	18.46	1.388	0.658	0.5679	0.4652
8.0	2.079	23.78	1.494	0.690	0.5091	0.4512
10.0	2.303	26.56	1.549	0.704	0.4834	0.4553
20.0	2.996	36.21	1.734	0.738	0.4234	0.4289
30.0	3.401	42.51	1.852	0.753	0.3790	0.4206
50.0	3.912	51.09	2.009	0.769	0.3416	0.4111
100.0	4.605	63.77	2.233	0.786	0.2997	0.3999
250.0	5.521	82.20	2.544	0.800	0.2566	0.3869
500.0	6.215	97.29	2.788	0.808	0.2308	0.3784
1000.0	6.908	113.3	3.038	0.815	0.2095	0.3707
3000.0	8.006	140.4	3.442	0.823	0.1825	0.3599
10^4	9.210	172.4	3.894	0.829	0.1597	0.3496
3×10^4	10.309	203.6	4.312	0.833	0.1432	0.3413
10^5	11.513	239.7	4.774	0.834	0.1287	0.3333
3×10^5	12.612	274.4	5.199	0.838	0.1177	0.3267
10^6	13.816	314.2	5.668	0.839	0.1076	0.3202
10^7	16.118	395.3	6.569	0.842	0.0925	0.3093
10^8	18.421	482.4	7.475	0.842	0.0810	0.3000
10^9	20.723	575.2	8.384	0.844	0.0721	0.2920
10^{10}	23.026	673.3	9.296	0.844	0.0650	0.2849

As a first example, consider ϵ_F/kT. Equations (4.6.29) and (4.6.30) give

$$\frac{\epsilon_F}{kT} \sim \left(\frac{\delta}{6.0798}\right)^{2/3}\left[1 - 0.822\,467\left(\frac{6.0798}{\delta}\right)^{4/3} - 1.217\,614\left(\frac{6.0798}{\delta}\right)^{8/3}\right] \quad (4.7.2)$$

$$= \left(\frac{\delta}{6.0798}\right)^{2/3} - \left(\frac{4.5349}{\delta}\right)^{2/3} - \left(\frac{6.7088}{\delta}\right)^2. \qquad (4.7.2a)$$

These equations give values for ϵ_F/kT which differ from the tabular values by -2.3%, -0.1%, and $+0.17\%$ for $\delta = 18.46$, 26.56, and 36.21, respectively. For values of δ larger than these, the values given by the analytic approximation are as accurate as the numerically derived values.

The relationship between $e^{-\alpha}$ and δ given by eq. (4.5.30), derived as it is from expansions which do not converge when $\alpha < 0$, is in remarkably good agreement with the numerically derived relationship for values of $-\alpha$ as large as 25, corresponding to $\epsilon_F/kT > 2$. For example, eq. (4.5.30) underestimates tabular values for ϵ_F/kT by 0.1%, 0.3%, 0.8%, 1.5%, and 3% for $\delta = 7.354$, 10.36, 13.57, 18.46, and 26.56, respectively. By modifying eq. (4.5.30) to read

$$\frac{\epsilon_F}{kT} = -\alpha = \log_e \, [0.123\,730\,\delta \, (1 + 0.043\,745\,\delta \, (1 + 0.020\,130\,\delta \, (1 + 0.015\,754\delta)\,)\,)\,], \tag{4.7.3}$$

ϵ_F/kT is fitted exactly at $\delta = 6.182$ and at $\delta = 26.56$, and the fit to tabular values over the entire range $\delta < 30$ is accurate to within a small fraction of a percent.

For calculating ϵ_F/kT in stellar models, it is reasonable to adopt eq. (4.7.3) for all $\delta < 30$ and eq. (4.7.2) for all $\delta \geq 30$. Alternatively, if, for some application, exact continuity is required, one could restrict the use of eq. (4.7.3) to some favorite choice of $\delta \leq \delta_1$ and eq. (4.7.2) to some favorite choice of $\delta \geq \delta_2$ and use, for $\delta_1 < \delta < \delta_2$,

$$\frac{\epsilon_F}{kT} = g_1 \cos^2 \phi + g_2 \sin^2 \phi, \tag{4.7.4}$$

where

$$\phi = \frac{\pi}{2} \frac{\delta - \delta_1}{\delta_2 - \delta_1}, \tag{4.7.5}$$

and g_1 and g_2 are the estimates of ϵ_F/kT given by eqs. (4.7.3) and (4.7.2), respectively.

Either algorithm is adequate for use in evolution calculations for stars of all masses and for evolutionary stages up to and including the helium-burning stage. In stars initially more massive than 2.5 $M_\odot$, helium ignition occurs when temperatures reach the order of $T_6 \sim 100$, densities reach the order of 10^4 g cm^{-3}, and electrons are weakly degenerate $(\delta \sim 5, \epsilon_F/kT \leq 1)$. In stars initially less massive than ~ 2 $M_\odot$, helium ignition occurs when densities are several times 10^5 g cm^{-3}, temperatures are $T_6 \sim 80$, and electrons are strongly degenerate $(\delta \gtrsim 200, \epsilon_F/kT \gtrsim 10)$.

The analytic approximations for $P_e/(n_e kT)$ derived in Section 4.5 understandably agree well with the numerically derived estimates for $\alpha > 0$. However, the agreement extends into the domain of large, negative α. Equation (4.5.22) gives values for $P_e/(n_e kT)$ which are smaller than tabular values by 0.05%, 0.24%, 0.48%, and 2.0% for $\delta = 13.57, 26.56, 36.21,$ and 63.77, respectively. When $\delta = 63.77$, $-\alpha = \epsilon_F/kT = 4.6$, and the agreement between analytic and numerical estimates is nothing short of astonishing, given the fact that the analytical estimates have been constructed by manipulating series expansions which are violently non-convergent for negative α.

That a finite set of terms selected from a non-convergent series expansion nevertheless provides a good approximation to the function from which the series has been derived is not self evident. Integrations over momentum performed on terms of the series do not succeed in producing convergence. For example, for $\alpha < 0$ and $\epsilon = 0$, terms proportional to $e^{-n(\alpha + \epsilon/kT)}$ in the series expansion of f in eq. (4.5.1) are larger than unity and the series does not formally converge. For modest positive values of $-\alpha$, integration of $e^{-n\epsilon/kT}$ over $d^3\mathbf{p}$ reduces the first few terms (see eqs. (4.5.5) and (4.5.9)) to less than unity, but, far enough into the series, individual terms increase without limit. It appears that the non-convergent series may be related to the class of asymptotic series which have the property that a judicious choice of the number of included terms can give a good approximation to the function from which the series has been derived (see, e.g., Whittaker & Watson, *A Course in Modern Analysis*, 1952).

An improvement of the agreement for large δ is achieved by adding to eq. (4.5.22) a cubic term such that analytic and tabular values of $P_e/(n_e kT)$ match precisely at $\alpha = 26.56$. The result,

$$\frac{P_e}{n_e kT} = 1 + 0.021\,8726\,\delta\,(1 - 0.002\,310\,\delta\,(1 - 0.003\,9151\,\delta)), \qquad (4.7.6)$$

provides accuracy to within a fraction of a percent over the entire range $\delta \leq 64$, corresponding to $\epsilon_F/kT \leq 4.6$. In the region of strong non-relativistic degeneracy, eq. (4.6.41) pertains. Replacing kT/ϵ_{F0} in this equation with its equivalent in terms of δ as given by eq. (4.6.29), one has

$$\frac{P_e}{n_e kT} = \frac{2}{5}\left(\frac{\delta}{6.0798}\right)^{2/3}$$

$$\times \left[1 + 4.112\,335\left(\frac{6.0798}{\delta}\right)^{4/3} - 6.088\,068\left(\frac{6.0798}{\delta}\right)^{8/3}\right] + \cdots \quad (4.7.7)$$

$$= \left(\frac{\delta}{24.0325}\right)^{2/3} + \left(\frac{12.8266}{\delta}\right)^{2/3} - \left(\frac{10.1942}{\delta}\right)^2 + \cdots. \qquad (4.7.8)$$

For $\delta = 26.56$, these equations give a value of $P_e/(n_e kT)$ that is 2.2% less than the numerical estimate in Table 4.7.1. However, for $\delta = 36.21, 51.09, 63.77,$ and 82.20, the analytic estimates of $P_e/(n_e kT)$ agree with the numerical estimates to an accuracy of 0.1%, 0.1%, 0.06%, and 0.02%, respectively.

For stellar model calculations, one may adopt eq. (4.7.6) for all $\delta \leq 40$–60 and either eq. (4.7.7) or eq. (4.7.8) for all $\delta \geq 40$–60. Again, if strict continuity is required, one may blend the analytic approximations by employing an algorithm similar to that described by eqs. (4.7.4) and (4.7.5), but using $\delta_1 \sim 40$ and $\delta_2 \sim 60$.

The entries in the fifth column of Table 4.7.1 show that, with increasing degeneracy, the ratio between the deBroglie wavelength of a typical electron and the average separation between adjacent ions (as well as the average separation between adjacent electrons) approaches a limiting value. This is basically a statement of consistency with the Heisenberg uncertainty principle.

A measure of the average distance between two adjacent electrons can be obtained by supposing that space is divided into cubes, with 27 electrons being found in a set of 27 cubes. By placing the electrons at the centers of the cubes, it is evident that

$$n_e \, (2r_{0e})^3 = 1, \tag{4.7.9}$$

where $2r_{0e}$ is shortest length of any side of the cube. With this placement, $2r_{0e}$ is the distance from one electron to its six nearest neighbors, $\sqrt{2} \, 2r_{0e}$ is the distance to its 12 second nearest neighbors, and $\sqrt{3} \, 2r_{0e}$ is the distance to its eight third nearest neighbors. A very similar result follows from the definition

$$n_e \, \frac{8\pi}{3} \, r_{0e}^3 = 1. \tag{4.7.10}$$

Assuming a body centered cubic distribution, $\sqrt{3} \, r_{0e}$ is the distance from one electron to its eight nearest neighbors, and $2r_{0e}$ is the distance to its six second nearest neighbors. In the case of a face centered cubic, $\sqrt{2} \, r_{0e}$ is the distance from one electron to its 12 nearest neighbors and $2r_{0e}$ is the distance to its six second nearest neighbors. In a gaseous configuration, no one of these prescriptions is better than another and, for lack of a more cogent argument, aesthetics suggests the choice of $r_e = 2r_{0e}$ given by eq. (4.7.10) as a measure of the average distance between adjacent electrons. Adopting eq. (4.7.10) as a model, a measure of the average separation between adjacent ions in a completely ionized medium consisting entirely of ions of type i and associated electrons is $r_i = 2r_{0i}$, where r_{0i} is given by

$$n_e \, \frac{8\pi}{3} \, (r_{0i})^3 = Z_i, \tag{4.7.11}$$

or

$$r_i = 2 \, r_{0i} = Z_i^{1/3} \, r_e = 2 \, Z_i^{1/3} \left(\frac{3}{8\pi} \, \frac{\mu_e M_H}{\rho} \right)^{1/3}. \tag{4.7.12}$$

The definition of an average or typical deBroglie wavelength λ_e is, to some extent, also a matter of taste. In Sections 4.4 and 4.6, it was convenient to define, for non-degenerate electrons (see eqs. (4.4.14) and (4.5.6)),

$$\lambda_e = \frac{2}{\pi} \, \frac{h}{\langle p \rangle} = \frac{h}{\sqrt{2\pi m_e kT}}, \tag{4.7.13}$$

where $\langle p \rangle$ is the average of the electron momentum p over the Maxwell–Boltzmann distribution. However, for electrons of arbitrary degeneracy, the average of p^2 over the Fermi distribution is more easily related to n_e and P_e than is the average of p, so in this section the definition

$$\lambda_e = \frac{h}{\sqrt{\langle p^2 \rangle}} \tag{4.7.14}$$

is used. For non-relativistic electrons,

$$\langle p^2 \rangle = \frac{\int f \, p^2 \, d^3\mathbf{p}}{\int f \, d^3\mathbf{p}} = 3m_e \frac{P_e}{n_e} = 3m_e \, kT \frac{P_e}{n_e kT}, \tag{4.7.15}$$

and so

$$\lambda_e = \frac{h}{\sqrt{3m_e kT}} \left(\frac{n_e kT}{P_e} \right)^{1/2}. \tag{4.7.16}$$

With this definition, the deBroglie wavelength for non-degenerate electrons, for which $P_e = n_e kT$, is

$$\lambda_e = \frac{h}{\sqrt{3m_e kT}}, \tag{4.7.17}$$

larger by $(2\pi/3)^{1/2} = 1.4472$ than given by eq. (4.7.13).

In applications to screening and conductivity, it is convenient to deal with $\lambdabar$ defined by

$$\lambdabar_e = \frac{\lambda_e}{2\pi}, \tag{4.7.18}$$

and the quantities

$$\frac{\lambdabar_e}{r_{0e}} = \frac{\hbar}{m_e c} \left(\frac{1}{3} \frac{m_e c^2}{kT} \right)^{1/2} \left(\frac{n_e kT}{P_e} \right)^{1/2} \left(\frac{8\pi}{3} \frac{\rho}{\mu_e M_H} \right)^{1/3} \tag{4.7.19}$$

$$= 0.293\,691 \left(\frac{n_e kT}{P_e} \right)^{1/2} \delta^{1/3} \tag{4.7.20}$$

and

$$\frac{\lambdabar_e}{r_{0i}} = \left(\frac{1}{Z_i} \right)^{1/3} \frac{\lambdabar_e}{r_{0e}} \tag{4.7.21}$$

are useful.

For non-degenerate electrons,

$$\frac{\lambdabar_e}{r_{0e}} = 0.293\,691 \, \delta^{1/3}. \tag{4.7.22}$$

Early entries in column five of Table 4.7.1 are consistent with this. Using eqs. (4.6.40) and (4.6.26) in eq. (4.7.20), one has that, for strongly degenerate electrons,

$$\frac{\lambda_e}{r_{0e}} = \frac{1}{2\pi} \left(\frac{5}{3}\right)^{1/2} \left(\frac{8\pi}{3}\right)^{2/3} = 0.847\,534. \tag{4.7.23}$$

The last entries in column five of Table 4.7.1 are asymptotic to $\lambda_e/r_{0e} = 0.848$. This limiting value, which is valid also for relativistically degenerate electrons, is equivalent to

$$\sqrt{\langle p^2 \rangle} \, r_{0e} = 1.18 \, \hbar, \tag{4.7.24}$$

completing the demonstration that the constancy of λ_e/r_{0e} for large degeneracy is consistent with the Heisenberg uncertainty principle.

4.8 Equation of state for relativistically degenerate electrons at zero temperature

In regions within stars where most atoms are completely ionized, the approximations given in the previous section are acceptable as long as the density is small compared with 10^6 g cm^{-3} and the temperature is small compared with 10^9 K. Thus, they are quite adequate approximations in the interiors of core hydrogen-burning and core helium-burning stars and they are reasonably adequate approximations over most of the interiors of the electron-degenerate cores which low mass stars develop along the first giant branch and over most of the interiors of the electron-degenerate carbon–oxygen cores which stars of low and intermediate mass develop along the asymptotic giant branch.

However, when densities exceed 10^5 g cm^{-3}, it is imperative that relativistic effects be taken into account. To explore the relativistically correct equation of state, the electron momentum p is defined in units of $m_e c$, the electron energy ϵ is defined in units of $m_e c^2$, and the electron rest mass is included in ϵ. In these dimensionless units, the total electron energy and the electron momentum are related by

$$\epsilon^2 = p^2 + 1 \tag{4.8.1}$$

and the electron kinetic energy ϵ' is related to the total electron energy and electron momentum by

$$\epsilon' = \epsilon - 1 = \sqrt{1 + p^2} - 1. \tag{4.8.2}$$

For $p < 1$, eq. (4.8.2) can be written as

$$\epsilon' = \frac{1}{2} p^2 \left(1 - \frac{1}{4} p^2 + \cdots \right) \tag{4.8.3}$$

In the dimensionless units, eq. (4.6.1) is still valid if ϵ_F is interpreted as the total electron Fermi energy ($\epsilon_F = \epsilon_F' + 1$). At $T = 0$, $f(\epsilon) = 1$ for all $\epsilon < \epsilon_F = \epsilon_{F0}$ and $f(\epsilon) = 0$ for all $\epsilon > \epsilon_F = \epsilon_{F0}$, where ϵ_{F0} is ϵ_F at $T = 0$ and $p_{F0} = \sqrt{\epsilon_{F0}^2 - 1}$. Hence,

$$n_e = \frac{8\pi}{3} \left(\frac{m_e c}{h}\right)^3 \int_0^{p_{F0}} p^2 dp = \frac{8\pi}{\lambda_C^3} \int_0^{p_{F0}} p^2 dp, \tag{4.8.4}$$

where λ_C is the Compton wavelength of the electron,

$$\lambda_C = \frac{h}{m_e c} = 2.426\,309 \times 10^{-10} \text{ cm}. \tag{4.8.5}$$

Performing the integration in eq. (4.8.4) gives

$$n_e = \frac{8\pi}{\lambda_C^3} \frac{p_{F0}^3}{3}, \tag{4.8.6}$$

which defines the Fermi momentum at $T = 0$ as

$$p_{F0} = \lambda_C \left(\frac{3}{8\pi} n_e\right)^{1/3} = \lambda_C \left(\frac{3}{8\pi} \frac{\rho}{\mu_e M_H}\right)^{1/3} = 1.006\,2199 \left(\frac{\rho_6}{\mu_e}\right)^{1/3}, \tag{4.8.7}$$

where μ_e is the number of nucleons per electron, M_H is the mass of the hydrogen atom, and ρ_6 is the density in units of 10^6 g cm^{-3}. Had one chosen to set $n_e = N_A (\rho/\mu_e)$, where N_A is Avogadro's number, the coefficient in eq. (4.8.7) would be 1.008 8376.

Using eq. (4.3.2) and the properties of $f(\epsilon)$ at $T = 0$ ($f(\epsilon) = 1$ for $\epsilon \leq \epsilon_{F0}$ and $f(\epsilon) = 0$ for $\epsilon > \epsilon_{F0}$), one has for the pressure

$$P_e = \frac{8\pi}{3} \frac{m_e c^2}{\lambda_C^3} \int_0^{p_{F0}} \left(\frac{v}{c} p\right) p^2 \, dp. \tag{4.8.8}$$

Since $v/c = p/\epsilon$ and $p\, dp = \epsilon\, d\epsilon$,

$$P_e = \frac{8\pi}{3} \frac{m_e c^2}{\lambda_C^3} \int_0^{p_{F0}} \frac{p^4}{(p^2 + 1)^{1/2}} \, dp. \tag{4.8.9}$$

When $p < 1$, $(p^2 + 1)^{-1/2}$ can be expanded in powers of p^2, so that eq. (4.8.9) becomes

$$P_e = \frac{8\pi}{3} \frac{m_e c^2}{\lambda_C^3} \int_0^{p_{F0}} p^4 \, dp \left(1 - \frac{1}{2}p^2 + \frac{3}{8}p^4 - \frac{5}{16}p^6 + \frac{35}{128}p^8 - \cdots\right), \tag{4.8.10}$$

or

$$P_e = \frac{8\pi}{3} \frac{m_e c^2}{\lambda_C^3} \left(\frac{1}{5} p_{F0}^5 - \frac{1}{14} p_{F0}^7 + \frac{1}{24} p_{F0}^9 - \frac{5}{176} p_{F0}^{11} + \frac{35}{1664} p_{F0}^{13} - \cdots\right). \tag{4.8.11}$$

Combining eqs. (4.8.6), (4.8.9), and (4.8.11) gives

$$P_e = \frac{1}{5}\, n_e\, p_{F0}^2\, m_e c^2 \left(1 - \frac{5}{14}\, p_{F0}^2 + \frac{5}{24}\, p_{F0}^4 - \frac{25}{176}\, p_{F0}^6 + \frac{175}{1664}\, p_{F0}^8 - \cdots \right). \quad (4.8.12)$$

From this result, it is evident that the non-relativistic relationship between pressure and the Fermi momentum is a good approximation only to the extent that

$$\frac{5}{14}\, p_{F0}^2 = \left(\frac{\rho/\mu_e}{4.5990 \times 10^6 \text{ g cm}^{-3}} \right)^{2/3} \ll 1. \quad (4.8.13)$$

When $\rho/\mu_e = 10^5$ g cm^{-3}, $p_{F0} = 0.467$ and the first order relativistic correction to the electron pressure is $(5/14)\, p_{F0}^2 = 0.0779$, or roughly 8%. Additional terms in the series reduce the overall correction to roughly 7%.

The total electron energy per unit volume is

$$\bar{U}_e = 8\pi\, \frac{m_e c^2}{\lambda_C^3} \int_0^{p_{F0}} p^2\, dp\, (p^2 + 1)^{1/2}, \quad (4.8.14)$$

which, when $p_{F0} < 1$, can be written as

$$\bar{U}_e = 8\pi\, \frac{m_e c^2}{\lambda_C^3} \int_0^{p_{F0}} p^2\, dp \left(1 + \frac{1}{2}p^2 - \frac{1}{8}p^4 + \frac{1}{16}p^6 - \frac{5}{128}p^8 + \frac{7}{256}p^{10} + \cdots \right). \quad (4.8.15)$$

Performing the integrations and using eq. (4.8.6), eq. (4.8.14) becomes

$$\bar{U}_e = n_e\, m_e c^2 \left(1 + \frac{3}{10}\, p_{F0}^2 - \frac{3}{56}\, p_{F0}^4 + \frac{1}{48}\, p_{F0}^6 - \frac{15}{1408}\, p_{F0}^8 + \frac{21}{3328}\, p_{F0}^{10} + \cdots \right). \quad (4.8.16)$$

The electron kinetic energy per unit volume is

$$\bar{U}_e' = \bar{U}_e - n_e\, m_e c^2, \quad (4.8.17)$$

so

$$\bar{U}_e' = n_e\, m_e c^2$$
$$\times \frac{3}{5}\, \frac{p_{F0}^2}{2} \left(1 - \frac{5}{28}p_{F0}^2 + \frac{5}{72}p_{F0}^4 - \frac{25}{704}p_{F0}^6 + \frac{35}{1664}p_{F0}^8 - \cdots \right). \quad (4.8.18)$$

Finally, combining eqs. (4.8.13) and (4.8.18),

$$\frac{P_e}{\bar{U}_e'} = \frac{2}{3}\, \frac{\left(1 - \frac{5}{14}\, p_{F0}^2 + \frac{5}{24}\, p_{F0}^4 - \frac{25}{176}\, p_{F0}^6 + \frac{175}{1664}\, p_{F0}^8 - \cdots \right)}{\left(1 - \frac{5}{28}\, p_{F0}^2 + \frac{5}{72}\, p_{F0}^4 - \frac{25}{704}\, p_{F0}^6 + \frac{35}{1664}\, p_{F0}^8 - \cdots \right)}, \quad (4.8.19)$$

reconfirming the non-relativistic result that $P_e = (2/3)\, \bar{U}_e'$ when $p_{F0} \ll 1$.

To obtain a more general solution, eq. (4.8.9) may be rewritten as

$$P_e = \frac{8\pi}{3} \frac{m_e c^2}{\lambda_C^3} \int_1^{\epsilon_{F0}} p^3 \, d\epsilon.$$ (4.8.20)

Integration gives

$$P_e = \frac{8\pi}{3} \frac{m_e c^2}{\lambda_C^3} \left(\frac{1}{4} \epsilon_{F0} \, p_{F0}^3 - \frac{3}{8} \epsilon_{F0} \, p_{F0} + \frac{3}{8} \log_e (\epsilon_{F0} + p_{F0}) \right)$$ (4.8.21)

$$= \frac{8\pi}{3} \frac{m_e c^2}{\lambda_C^3} \frac{1}{4} \epsilon_{F0} \, p_{F0}^3 \left[1 - \frac{3}{2 p_{F0}^2} \left(1 - \frac{\log_e (\epsilon_{F0} + p_{F0})}{\epsilon_{F0} \, p_{F0}} \right) \right].$$ (4.8.22)

Combining eqs. (4.8.22) and (4.8.6), one has

$$P_e = \frac{1}{4} n_e \, m_e c^2 \, \epsilon_{F0} \left[1 - \frac{3}{2 p_{F0}^2} \left(1 - \frac{\log_e (\epsilon_{F0} + p_{F0})}{\epsilon_{F0} \, p_{F0}} \right) \right].$$ (4.8.23)

Similarly, the total electron energy per unit volume is

$$\bar{U}_e = 8\pi \frac{m_e c^2}{\lambda_C^3} \int_1^{\epsilon_{F0}} \epsilon^2 \, p \, d\epsilon$$ (4.8.24)

$$= 8\pi \frac{m_e c^2}{\lambda_C^3} \left(\frac{1}{4} \epsilon_{F0} \, p_{F0}^3 + \frac{1}{8} \epsilon_{F0} \, p_{F0} - \frac{1}{8} \log_e (\epsilon_{F0} + p_{F0}) \right)$$ (4.8.25)

$$= 8\pi \frac{m_e c^2}{\lambda_C^3} \frac{1}{4} \epsilon_{F0} \, p_{F0}^3 \left[1 + \frac{1}{2 p_{F0}^2} \left(1 - \frac{\log_e (\epsilon_{F0} + p_{F0})}{\epsilon_{F0} \, p_{F0}} \right) \right].$$ (4.8.26)

Then, using eq. (4.8.6),

$$\bar{U}_e = \frac{3}{4} n_e \, m_e c^2 \, \epsilon_{F0} \left[1 + \frac{1}{2 p_{F0}^2} \left(1 - \frac{\log_e (\epsilon_{F0} + p_{F0})}{\epsilon_{F0} \, p_{F0}} \right) \right].$$ (4.8.27)

$$= \frac{3}{8} n_e \, m_e c^2 \, \frac{\epsilon_{F0}}{p_{F0}^2} \left(\epsilon_{F0}^2 + p_{F0}^2 - \frac{\log_e (\epsilon_{F0} + p_{F0})}{\epsilon_{F0} \, p_{F0}} \right),$$ (4.8.28)

Finally, using eqs. (4.8.17), (4.8.23), and (4.8.27),

$$\frac{P_e}{\bar{U}'_e} = \frac{1}{3} \frac{1 - \frac{3}{2 p_{F0}^2} \left(1 - \frac{\log_e (\epsilon_{F0} + p_{F0})}{\epsilon_{F0} \, p_{F0}} \right)}{1 + \frac{1}{2 p_{F0}^2} \left(1 - \frac{\log_e (\epsilon_{F0} + p_{F0})}{\epsilon_{F0} \, p_{F0}} \right) - \frac{4}{3 \epsilon_{F0}}}.$$ (4.8.29)

This last equation reconfirms that, when $\epsilon_{F0} \gg 1$, $P_e = (1/3) \, \bar{U}'_e$.

When the density is small, as measured by eq. (4.8.13), eqs. (4.8.12) and (4.8.15) show more transparently than do eqs. (4.8.23) and (4.8.27) how relativistic effects increase in importance with increasing density and, when the density is large, eqs. (4.8.23) and (4.8.27) offer more insight.

It is instructive to examine the relative contributions of the different terms in square brackets in eqs. (4.8.23) and (4.8.27) for various values of p_F. For example, when $p_F = 1$, $\epsilon_F = \sqrt{2}$, the value of the bracketed term in eq. (4.8.23) for P_e is

$$1 - \frac{3}{2p_{F0}^2} \left(1 - \frac{\log_e (\epsilon_{F0} + p_{F0})}{\epsilon_{F0}\, p_{F0}}\right) = 1 - 1.5\,(1. - 0.6232) = 1 - 0.5652 = 0.4348,$$

and the value of the bracketed term in eq. (4.8.27) for U_e is

$$1 + \frac{1}{2p_{F0}^2} \left(1 - \frac{\log_e (\epsilon_{F0} + p_{F0})}{\epsilon_{F0}\, p_{F0}}\right) = 1 + 0.5\,(1. - 0.6232) = 1 + 0.1884 = 1.1884.$$

The denominator of the second fraction on the right hand side in eq. (4.8.29) is $1.1884 - 0.9428 = 0.2456$, giving $3P_e/\bar{U}_e' = 1.770$. When $\epsilon_{F0} = 2$, $p_{F0} = \sqrt{3}$, the value of the bracketed term in eq. (4.8.23) for P_e is

$$1 - \frac{3}{2p_{F0}^2} \left(1 - \frac{\log_e (\epsilon_{F0} + p_{F0})}{\epsilon_{F0}\, p_{F0}}\right) = 1 - 0.5\,(1. - 0.3802) = 1 - 0.3099 = 0.6901,$$

and the value of the bracketed term in eq. (4.8.27) for U_e is

$$1 + \frac{1}{2p_{F0}^2} \left(1 - \frac{\log_e (\epsilon_{F0} + p_{F0})}{\epsilon_{F0}\, p_{F0}}\right)$$
$$= 1 + 0.16667\,(1. - 0.3802) = 1 + 0.1033 = 1.1033.$$

The denominator of the second fraction on the right hand side in eq. (4.8.29) is $1.1033 - 0.6667 = 0.4366$, giving $3P_e/\bar{U}_e' = 1.581$.

In order to demonstrate explictly that, for $p_{F0} \ll 1$, the two solutions given by eqs. (4.8.12) and (4.8.23), are equivalent, it is necessary to expand the quantities $\epsilon_{F0}\, p_{F0}$ and $\log_e (\epsilon_{F0} + p_{F0})$ in powers of p_{F0} out to the fifth power. Explicitly,

$$\epsilon_{F0}\, p_{F0} - \log (\epsilon_{F0} + p_{F0}) = \frac{2}{3}\, p_{F0}^3 \left(1 - \frac{3}{10}\, p_{F0}^2 + \frac{9}{56}\, p_{F0}^4 - \frac{5}{48}\, p_{F0}^6 + \cdots\right).$$

$$(4.8.30)$$

Insertion of this result into eq. (4.8.23) reproduces eq. (4.8.12). Insertion into eq. (4.8.27) reproduces eq. (4.8.16).

In Table 4.8.1 are shown several characteristics of the electron Fermi sea as a function of the electron density when $kT = 0$. In this table,

$$P_e^{NR} = \frac{2}{5}\, n_e\, \frac{p_{F0}^2}{2}\, m_e c^2 \tag{4.8.31}$$

is the electron pressure in the non-relativistic approximation, and P_e is better given by eq. (4.8.12) when $p_{F0} < 0.4$ and is better given by eq. (4.8.23) when $p_{F0} > 0.4$. The fact that P_e^R/P_e^{NR} decreases with increasing degeneracy is responsible for the characterization of

Table 4.8.1 Characteristics of the Fermi sea at $T = 0$						
$\frac{\rho}{\mu_e}$ (g cm^{-3})	p_{F0}	$\frac{p_{F0}^2}{2}$	$\epsilon_{F0}{}'$	$(\frac{v}{c})_F^2$	$\frac{P_e}{P_e^{NR}}$	$\frac{3P_e}{\bar{U}_e'}$
10^2	0.046445	0.001079	0.001078	0.002152	0.99923	1.99923
10^3	0.100622	0.005062	0.005050	0.010023	0.99660	1.99641
10^4	0.215577	0.023237	0.022971	0.044441	0.98385	1.98382
10^5	0.464448	0.107856	0.102593	0.177437	0.93142	1.93165
10^6	1.006220	0.506239	0.418619	0.503100	0.76678	1.76878
10^7	2.155775	2.323683	1.376419	0.822926	0.49374	1.50053
10^8	4.644476	10.78560	3.750911	0.955696	0.25810	1.26662
10^9	10.06220	50.62393	9.111769	0.990220	0.12478	1.13069
10^{10}	21.55775	232.36829	20.58093	0.997853	0.05805	1.06585

the relativistic EOS as a softer equation of state than the non-relativistic EOS. The last two columns show that, in first approximation,

$$\frac{3P_e}{\bar{U}_e'} \sim 1 + \frac{P_e}{P_e^{NR}}. \tag{4.8.32}$$

4.9 Equation of state for relativistically degenerate electrons at finite temperatures

When the temperature is finite, but $\frac{kT}{\epsilon_F} \ll 1$, by inserting $f(\epsilon)$ into the integrand of eq. (4.8.4), replacing p_{F0} by ∞ as an integration limit, and, using the appropriate series expansions for $f(\epsilon)$ developed in Section 4.6, one finds

$$\frac{\lambda_C^3}{8\pi} n_e = \frac{p_F^3}{3} - kT \sum_{m=1}^{\infty} (-1)^{m+1} \int_0^{\epsilon_F/kT} e^{-m\bar{x}} (p\epsilon) \, d\bar{x}$$

$$+ kT \sum_{m=1}^{\infty} (-1)^{m+1} \int_0^{\infty} e^{-mx} (p\epsilon) dx, \tag{4.9.1}$$

where x and $\bar{x}$ are given by eqs. (4.6.2) and (4.6.4), respectively. Setting

$$p\epsilon = \sum_{k=0}^{\infty} a_k x^k = \sum_{k=0}^{\infty} (-1)^k a_k \bar{x}^k, \tag{4.9.2}$$

where

$$a_k = \frac{(kT)^k}{k!} \left[\frac{d^k (p\epsilon)}{d\epsilon^k} \right]_{\epsilon = \epsilon_F}, \tag{4.9.3}$$

and extending the upper limit on the first integral in eq. (4.9.1) to ∞, one has

$$\frac{\lambda_C^3}{8\pi} n_e = \frac{p_F^3}{3} + 2kT \sum_{m=1}^{\infty} (-1)^{m+1} \sum_{k=1}^{\infty} a_{2k-1} \int_0^{\infty} e^{-mx} x^{2k-1} \, dx \qquad (4.9.4)$$

$$= \frac{p_F^3}{3} + 2kT \sum_{m=1}^{\infty} (-1)^{m+1} \sum_{k=1}^{\infty} a_{2k-1} \frac{(2k-1)!}{m^{2k}}. \qquad (4.9.5)$$

The first and third derivatives in eq. (4.9.3) are, respectively,

$$\frac{d(p\epsilon)}{d\epsilon} = \frac{\epsilon}{p} \epsilon + p = \frac{1 + 2p^2}{p} \qquad (4.9.6)$$

and

$$\frac{d^3(p\epsilon)}{d\epsilon^3} = \frac{3}{p^5}. \qquad (4.9.7)$$

Inserting these into eq. (4.9.3) and the resulting a_ks into eq. (4.9.5), one obtains

$$\frac{\lambda_C^3}{8\pi} n_e = \frac{p_F^3}{3} + 2 \sum_{m=1}^{\infty} (-1)^{m+1} \left[\left(\frac{kT}{m}\right)^2 \left(\frac{1 + 2\,p_F^2}{p_F}\right) + \left(\frac{kT}{m}\right)^4 \frac{3}{p_F^5} + \cdots \right]. \qquad (4.9.8)$$

Using eqs. (4.6.21) and (4.6.22), one obtains

$$\frac{\lambda_C^3}{8\pi} n_e = \frac{p_F^3}{3} \left[1 + 6\,S_2 \left(\frac{kT}{p_F^2}\right)^2 \left(1 + 2\,p_F^2\right) + 18\,S_4 \left(\frac{kT}{p_F^2}\right)^4 + \cdots \right] \qquad (4.9.9)$$

$$= \frac{p_F^3}{3} \left[1 + \frac{\pi^2}{2} \left(\frac{kT}{p_F^2}\right)^2 \left(1 + 2\,p_F^2\right) + \frac{7\pi^4}{40} \left(\frac{kT}{p_F^2}\right)^4 + \cdots \right]. \qquad (4.9.10)$$

Comparing eqs. (4.9.5) and (4.9.10), one has

$$R^3 = \left(\frac{p_{F0}}{p_F}\right)^3 = 1 + \frac{\pi^2}{2} \left(\frac{kT}{p_F^2}\right)^2 \left(1 + 2\,p_F^2\right) + \frac{7\pi^4}{40} \left(\frac{kT}{p_F^2}\right)^4 + \cdots . \qquad (4.9.11)$$

Noting that $p_F^2 = (\epsilon_F - 1)(\epsilon_F + 1) = 2\epsilon_F' \left(1 + \epsilon_F'/2\right)$, eq. (4.9.10) can be written as

$$\frac{\lambda_C^3}{8\pi} n_e = \frac{p_F^3}{3} \left[1 + \frac{\pi^2}{8} \left(\frac{kT}{\epsilon_F'}\right)^2 \frac{1 + 4\epsilon_F' \left(1 + \epsilon_F'/2\right)}{\left(1 + \epsilon_F'/2\right)^2} \right.$$

$$\left. + \frac{7\pi^4}{640} \left(\frac{kT}{\epsilon_F'}\right)^4 \frac{1}{\left(1 + \epsilon_F'/2\right)^4} + \cdots \right]. \qquad (4.9.12)$$

In the non-relativistic limit, when $\epsilon_F' = \epsilon_F - 1 \ll 1$, this last equation becomes eq. (4.6.25) (ϵ_F in eq. (4.6.25) corresponds to ϵ_F' in the notation of this section).

In a similar fashion, one can find the pressure in terms of p_F and ϵ_F. Inserting appropriate series expansions for $f(\epsilon)$ into the integrand of eq. (4.9.9) and replacing p_{F0} by ∞ as an integration limit, one finds

$$\frac{3\lambda_C^3}{8\pi} \frac{P_e}{m_e c^2} = I_P + I_{PT}, \tag{4.9.13}$$

where

$$I_P = \int_0^{p_F} \frac{p^4}{\epsilon} \, dp = \frac{1}{4} \epsilon_F \, p_F^3 \left[1 - \frac{3}{2 p_F^2} \left(1 - \frac{\log_e (\epsilon_F + p_F)}{\epsilon_F \, p_F} \right) \right] \tag{4.9.14}$$

$$I_{PT} = 2kT \sum_{m=1}^{\infty} (-1)^{m+1} \sum_{k=1}^{\infty} b_{2k-1} \frac{(2k-1)!}{m^{2k}}, \tag{4.9.15}$$

and

$$b_k = \frac{(kT)^k}{k!} \left(\frac{d^k (p^3)}{d\epsilon^k} \right)_{\epsilon = \epsilon_F}. \tag{4.9.16}$$

For $p_F < 1$, one can also write

$$I_P = \frac{1}{5} \, p_F^5 \left(1 - \frac{5}{14} p_F^2 + \frac{15}{72} p_F^4 - \frac{5}{24} p_F^6 + \frac{175}{1664} p_F^8 - \cdots \right). \tag{4.9.17}$$

Using

$$\frac{d(p^3)}{d\epsilon} = 3 \, p \, \epsilon \tag{4.9.18}$$

and

$$\frac{d^3 (p^3)}{d\epsilon^3} = \left(6 - \frac{3}{p^2} \right) \frac{\epsilon}{p} \tag{4.9.19}$$

in eq. (4.9.16), inserting the resulting b_ks into eq. (4.9.15), and using eqs. (4.6.21) and (4.6.22), one obtains

$$I_{PT} = \frac{\pi^2}{2} \, (kT)^2 \, p_F \, \epsilon_F + \frac{7\pi^4}{120} \, (kT)^4 \left(2 - \frac{1}{p_F^2} \right) \frac{\epsilon_F}{p_F} + \cdots . \tag{4.9.20}$$

Equation (4.9.5) permits one to write

$$\frac{3\lambda_C^3}{8\pi} \frac{P_e}{m_e c^2} = \frac{P_e}{m_e c^2} \, p_{F0}^3 \, n_e, \tag{4.9.21}$$

and combining this with eqs. (4.9.11) and (4.9.13), one has

$$P_e = \frac{m_e c^2}{R^3} \, n_e \left(\frac{I_P + I_{PT}}{p_F^3} \right). \tag{4.9.22}$$

Combining with eqs. (4.9.14) and (4.9.20) gives

$$P_e = \frac{1}{4R^3} n_e \, m_e c^2 \, \epsilon_F \left[1 - \frac{3}{2p_F^2} \left(1 - \frac{\log_e (\epsilon_F + p_F)}{\epsilon_F \, p_F} \right) \right.$$
$$\left. + 2\pi^2 \left(\frac{kT}{p_F^2} \right)^2 \left[1 - \frac{7\pi^2}{60} \left(\frac{kT}{p_F^2} \right)^2 \left(1 - 2p_F^2 \right) + \cdots \right].$$ (4.9.23)

Note that eq. (4.9.23) reduces to eq. (4.8.23) when $T = 0$.

Similar arithmetic produces an expression for the total energy density. Inserting appropriate series expansions for $f(\epsilon)$ into the integrand of eq. (4.8.14) and replacing p_{F0} by ∞ as an integration limit, one obtains

$$\frac{\lambda_C^3}{8\pi} \frac{\bar{U}_e}{m_e c^2} = I_E + I_{ET},$$ (4.9.24)

where

$$I_E = \int_0^{p_F} \epsilon \, p^2 \, dp = \frac{1}{4} \epsilon_F \, p_F^3 + \frac{1}{8} \epsilon_F \, p_F - \frac{1}{8} \log_e (\epsilon_F + p_F)$$ (4.9.25)

$$= \frac{1}{4} \epsilon_F \, p_F^3 \left[1 + \frac{1}{2p_F^2} \left(1 - \frac{\log_e (\epsilon_F + p_F)}{\epsilon_F \, p_F} \right) \right],$$ (4.9.26)

$$I_{ET} = 2kT \sum_{m=1}^{\infty} (-1)^{m+1} \sum_{k=1}^{\infty} c_{2k-1} \frac{(2k-1)!}{m^{2k}},$$ (4.9.27)

and

$$c_k = \frac{(kT)^k}{k!} \left(\frac{d^k (\epsilon^2 \, p)}{d\epsilon^k} \right)_{\epsilon = \epsilon_F}.$$ (4.9.28)

When $p_F < 1$, one has also that (see eq. (4.8.16))

$$I_E = \frac{1}{3} p_F^3 \left(1 + \frac{3}{10} p_F^2 - \frac{3}{56} p_F^4 + \frac{1}{48} p_F^6 - \frac{15}{1408} p_F^8 + \frac{21}{3328} p_F^{10} + \cdots \right).$$ (4.9.29)

Using

$$\frac{d(\epsilon^2 \, p)}{d\epsilon} = (1 + 3 \, p^2) \frac{\epsilon}{p}$$ (4.9.30)

and

$$\frac{d^3 (\epsilon^2 \, p)}{d\epsilon^3} = \left(6 - \frac{3}{p^2} + \frac{3}{p^4} \right) \frac{\epsilon}{p}$$ (4.9.31)

in eq. (4.9.28), inserting the resulting c_ks into eq. (4.9.27), and using eqs. (4.6.21) and (4.6.22), one obtains

$$I_{ET} = \frac{\pi^2}{6} (kT)^2 (1 + 3\, p_F^2) \frac{\epsilon_F}{p_F} + \frac{7\pi^4}{120} (kT)^4 \left(2 - \frac{1}{p_F^2} + \frac{1}{p_F^4}\right) \frac{\epsilon_F}{p_F} + \cdots . \quad (4.9.32)$$

Combining eqs. (4.9.5), (4.9.24), and (4.9.11), one has

$$\bar{U}_e = \frac{3}{R^3} m_e c^2 n_e \left(\frac{I_E + I_{ET}}{p_F^3}\right). \quad (4.9.33)$$

Combining with eqs. (4.9.26) and (4.9.32), one has

$$\bar{U}_e = \frac{3}{4R^3} n_e\, m_e c^2\, \epsilon_F \left[1 + \frac{1}{2p_F^2} \left(1 - \frac{\log_e(\epsilon_F + p_F)}{\epsilon_F\, p_F}\right)\right]$$

$$+ \frac{2\pi^2}{3} \left(\frac{kT}{p_F^2}\right)^2 \left[(1 + 3p_F^2) + \frac{7\pi^2}{20} \left(\frac{kT}{p_F^2}\right)^2 (1 - p_F^2 + 2\, p_F^4) + \cdots\right] + \cdots . \quad (4.9.34)$$

Note that eq. (4.9.34) reduces to eq. (4.8.27) when $T = 0$.

In order to evaluate the T-dependent expansions, it is necessary to find p_F and ϵ_F as functions of the electron density n_e. The process begins by solving eq. (4.8.7) for p_{F0}. The next step is to solve eq. (4.9.11) for p_F in terms of p_{F0} and to then use $\epsilon_F = \sqrt{1 + p_F^2}$. The most direct approach is to rewrite eq. (4.9.11) as

$$R^3 = \left(\frac{p_{F0}}{p_F}\right)^3 = 1 + \frac{\pi^2}{2} \left(\frac{kT}{p_{F0}^2}\right)^2 \left(R^4 + 2\, p_{F0}^2\, R^2\right) + \frac{7\pi^4}{40} \left(\frac{kT}{p_{F0}^2}\right)^4 R^8 + \cdots, \quad (4.9.35)$$

and solve numerically for R. From eqs. (4.8.7) and (4.5.10) one has

$$p_{F0} = 0.010\,062\,199\, \delta^{1/3}\, T_6^{1/2} \quad (4.9.36)$$

and, expressing kT in units of $m_e\, c^2$, one has

$$\frac{p_{F0}^2}{kT} = 0.600\,4805\, \delta^{2/3}. \quad (4.9.37)$$

Inserting these relationships in eq. (4.9.35) gives

$$R^3 = 1 + \frac{13.685\,85}{\delta^{4/3}} R^2 \left(R^2 + 2p_{F0}^2\right) + \frac{131.1118}{\delta^{8/3}} R^8 + \cdots, \quad (4.9.38)$$

an equation which can be solved by the Newton–Raphson method.

A starting value for R can be obtained by setting $R = 1 + y$, expanding all powers of R to the second power in y, and solving for y. The result is

$$\frac{p_{F0}}{p_F} = 1 + \frac{\pi^2}{6} \left[1 + 2\, p_{F0}^2 \right] \left(\frac{kT}{p_{F0}^2} \right)^2 + \frac{\pi^4}{18} \left(\frac{31}{20} + 4 p_{F0}^2 + 2 p_{F0}^4 \right) \left(\frac{kT}{p_{F0}^2} \right)^4 + \cdots ,$$

(4.9.39)

or

$$\frac{p_{F0}}{p_F} = 1 + \frac{4.561\,95}{\delta^{4/3}} \left[1 + 2\, p_{F0}^2 \right] + \frac{64.515\,35}{\delta^{8/3}}$$

$$\times \left[1 + 2.580\,645\, p_{F0}^2 + 1.290\,323\, p_{F0}^4 \right] + \cdots .$$

(4.9.40)

Inverting eq. (4.9.39) gives

$$\frac{p_F}{p_{F0}} = 1 - \frac{\pi^2}{6} \left(1 + 2\, p_{F0}^2 \right) \left(\frac{kT}{p_{F0}^2} \right)^2 - \frac{\pi^4}{54} \left(\frac{83}{20} + 10 p_{F0}^2 + 4 p_F^4(0) \right) \left(\frac{kT}{p_{F0}^2} \right)^4 + \cdots ,$$

(4.9.41)

or

$$\frac{p_F}{p_{F0}} = 1 - \frac{4.561\,95}{\delta^{4/3}} \left(1 + 2\, p_{F0}^2 \right) - \frac{57.578\,21}{\delta^{8/3}}$$

$$\times \left(1 + 2.409\,639\, p_{F0}^2 + 0.963\,855\, p_{F0}^4 \right) + \cdots .$$

(4.9.42)

With p_F (and R) calculated numerically as a function of electron density and of temperature, the electron pressure can then be found from eq. (4.9.23) and the electron energy density can be found from eq. (4.9.34). Thus, the electron pressure is

$$P_e = \frac{1}{4R^3} n_e\, m_e c^2\, \epsilon_F \left[1 - \frac{3}{2 p_F^2} \left(1 - \frac{\log_e (\epsilon_F + p_F)}{\epsilon_F\, p_F} \right) \right]$$

$$+ \frac{1}{4R^3} n_e\, m_e c^2\, \epsilon_F\, 54.743\,42\, \frac{R^2}{\delta^{2/3}} \left[1 - 3.193\,366 \frac{R^2}{\delta^{2/3}} \left(1 - 2 p_F^2 \right) + \cdots \right]$$

(4.9.43)

and the average energy of an electron is

$$\frac{\bar{U}_e}{n_e\, m_e c^2} = \frac{3}{4} \frac{1}{R^3} \epsilon_F \left[1 + \frac{1}{2 p_F^2} \left(1 - \frac{\log_e (\epsilon_F + p_F)}{\epsilon_F\, p_F} \right) \right]$$

$$+ \frac{3}{4} \frac{1}{R^3} \epsilon_F\, 18.247\,807 \frac{R^2}{\delta^{4/3}} \left[\left(1 + 3 p_F^2 \right) + 9.580\,098 \frac{R^2}{\delta^{4/3}} \right.$$

$$\left. \times \left(1 - p_F^2 + 2\, p_F^4 \right) + \cdots \right].$$

(4.9.44)

In performing numerical calculations, it is recommended that, when $p_{F0} < 0.4$, the series expressions for I_P and I_E given, respectively, by eqs. (4.9.17) and (4.9.29) be used instead of the exact expressions for these quantities given, respectively, by eqs. (4.9.14) and (4.9.26). That is, in eq. (4.9.43), the quantity

$$\left[1 - \frac{3}{2p_F^2}\left(1 - \frac{\log_e(\epsilon_F + p_F)}{\epsilon_F \, p_F}\right)\right]$$

is to be replaced by

$$\frac{4p_F^2}{5\epsilon_F}\left(1 - \frac{1}{14}p_F^2 + \frac{5}{24}p_F^4 - \frac{25}{176}p_F^6 + \frac{175}{1664}p_F^8 + \cdots\right)$$

and, in eq. (4.9.44), the quantity

$$\left[1 + \frac{1}{2p_F^2}\left(1 - \frac{\log_e(\epsilon_F + p_F)}{\epsilon_F \, p_F}\right)\right]$$

is to be replaced by

$$\frac{4}{3\epsilon_F}\left(1 + \frac{3}{10}p_F^2 - \frac{3}{56}p_F^4 + \frac{1}{48}p_F^6 - \frac{15}{1408}p_F^8 + \frac{21}{3328}p_F^{10} + \cdots\right)$$

Another way of estimating the equation of state and other thermodynamic relationships for a degenerate electron gas has been devised by Eggleton, Faulkner, and Flannery (1973). Their algorithm produces variables as functions of the temperature and a parameter f which is related to the degeneracy parameter $\psi = -\alpha = \frac{\epsilon_F}{kT}$ by

$$\psi = \log_e f + 2\left[\sqrt{1+f} - \log_e\left(1 + \sqrt{1+f}\right)\right] \qquad (4.9.45)$$

and is not to be confused with the occupation number f. Given the temperature and electron density as input, expressions (4.7.2) and (4.7.3) may be used to estimate ψ and eq. (4.9.45) may then be solved for f. For $\psi < -2$,

$$f \sim 4\,e^{\psi - 2} \qquad (4.9.46)$$

is a good starting approximation for f in terms of ψ. For $\psi > 4$,

$$f \sim 1 + \left(\frac{\psi}{2}\right)^2 \qquad (4.9.47)$$

is a good starting approximation. For values of ψ in the range $-2 < \psi < 4$,

$$\log_{10} f \sim -0.354\,68 + 0.355\,19\,\psi - 0.023\,433\,\psi^2. \qquad (4.9.48)$$

When introduced into eq. (4.9.45), values of f obtained from eq. (4.9.48) lead to values of ψ which differ from the input choice of ψ by less than 1%. The Eggleton, Faulkner, and Flannery algorithm gives the electron density, pressure, energy density, and entropy as functions of f and T. Beginning with the estimates of f given by eqs. (4.9.46)–(4.9.48), the algorithm may be used to find that f which reproduces the input electron density exactly. In regions where the series expansions in kT/p_F^2 constructed in this chapter are valid, the Eggleton, Faulkner, & Flannery algorithm gives results that match those given by the analytic approximations to a high degree of accuracy (often to within a fraction of a percent) and it has the virtue of being applicable over a region much larger than the realm of applicability of each of the analytic approximations.

However, its range of applicability is limited to temperatures and densities where positrons do not contribute comparably with electrons in supplying pressure. From Fig. 1 in Eggleton et al., the border of the region of inapplicability is defined by the density and temperature pairs given approximately by $(\log \rho_0, \log T_9) = (-10, -0.7)$, $(0, 0.0)$, and $(10, 0.6)$, where ρ_0 is in units of g cm^{-3} and T_9 is in units of 10^9 K. A power law approximation to the border is $\rho_0 \sim T_9^{10}$. At any given temperature T_9, the density must be larger than given by the equation for the border, and, at any density ρ_0, the temperature must be smaller than given by the border equation.

4.10 High temperatures and electron–positron pairs

Photons more energetic than $2m_e c^2$ can interact with the electric fields of charged particles to produce real electron–positron pairs. Electron–positron pairs, in turn, annihilate to produce photons which are deposited locally in and diffuse slowly out of the star. But $e^- - e^+$ pairs also annihilate to produce neutrino–antineutrino pairs which, except under extreme conditions found in the collapsing neutron-rich cores of massive stars, escape directly from the star. It is of interest to determine where in the density–temperature plane positrons form at sufficient number densities that associated neutrino losses sensibly affect the course of evolution of a star. As shown in Chapter 20 in Volume 2, in massive stars, large neutrino-loss rates first arise during the carbon-burning phase at temperatures such that kT is an order of magnitude smaller than $m_e c^2$ and the number abundance of positrons is substantially smaller than the number abundance of electrons contributed by heavy ions.

Consider a system consisting only of completely ionized nuclei, electrons, and positrons in a box of size V at temperature T. Number densities and numbers of the three types of particle are represented, respectively, by $n_n = N_n/V$, $n_e = N_-/V$, and $n_+ = N_+/V$. Conservation of charge requires that

$$\sum_i N_{-i} - \sum_j N_{+j} = A_Z N_n = N_Z, \qquad (4.10.1)$$

where A_Z is the number of protons in an average nucleus, N_{-i} is the number of electrons in the ith group of electron states at energy ϵ_i, N_{+j} is the number of positrons in the jth group of states at energy ϵ_j, and N_Z is the number of free electrons contributed by the ionized nuclei. Conservation of energy requires that

$$\sum_i N_{-i}\epsilon_i + \sum_j N_{+j}\epsilon_j = \text{constant}. \qquad (4.10.2)$$

Lepton rest mass must be included in the energies ϵ_i and ϵ_j. The probability that the ith group of electron states contains N_{-i} particles is given by eq. (4.2.2) and the probability that the jth group of positron states contains N_{+j} particles is also given by eq. (4.2.2).

Maximizing the total probability of the lepton system subject to the constraints given by eqs. (4.10.1) and (4.10.2) proceeds along exactly the same lines as the development in Section 4.2, with the number distribution for electrons being given by eq. (4.2.14) and the number distribution for positrons being given by

$$N_{+j} = \frac{g_j}{\exp(-\alpha + \beta\epsilon_j) + 1}. \tag{4.10.3}$$

Since the number of electrons must exceed the number of positrons, $\alpha < 0$. Because rest mass is included in the definition of energy, the chemical potential α defined in this section differs from the quantity α calculated in previous sections when energy is taken to be kinetic energy rather than total energy. That is, α (without rest mass) $= \alpha$ (with rest mass)$+ m_e c^2/kT$. Thus, α in eqs. (4.2.18)–(4.2.23) and as used in Sections 4.4–4.9 is larger than α defined in this section by $m_e c^2/kT$. With the current definition, it follows that a value of $-\alpha$ larger than $-\alpha = m_e c^2/kT$ is required for electrons to be incipiently degenerate.

Before examining exact solutions, it is instructive to examine the characteristics of approximate solutions under various limiting conditions. Keeping just the first terms in Taylor expansions of the occupation numbers in eqs. (4.10.3) and (4.2.14) and converting to continuous distributions, one has

$$n_{\pm} = \frac{N_{\pm}}{V} \sim e^{\pm\alpha} \frac{8\pi}{h^3} \int_0^\infty e^{-\epsilon/kT} p^2 dp = e^{\pm\alpha} I, \tag{4.10.4}$$

where

$$I = \frac{8\pi}{h^3} \int_0^\infty e^{-\epsilon/kT} p^2 dp = 8\pi \left(\frac{kT}{ch}\right)^3 \int_0^\infty e^{-\sqrt{(m_e c^2/kT)^2 + x^2}} x^2 dx \tag{4.10.5}$$

and

$$n_- = \frac{1}{V}\sum_i N_{-i}, \text{ and } n_+ = \frac{1}{V}\sum_j N_{+j}. \tag{4.10.6}$$

From eq. (4.10.4), it follows that

$$n_- n_+ \sim I^2. \tag{4.10.7}$$

Since the integral I is a function only of the temperature and increases monotonically with temperature, it is evident that the product of the positron and electron number abundances increases with increasing temperature and is independent of the number abundance of heavy ions.

Coupling eqs. (4.10.1) and (4.10.3) gives

$$n_- - n_+ \sim I(e^{-\alpha} - e^{+\alpha}) = n_Z = \frac{N_Z}{V}, \tag{4.10.8}$$

which has the solution

$$e^{\mp\alpha} = \sqrt{1 + \left(\frac{n_Z}{2I}\right)^2} \pm \frac{n_Z}{2I}. \tag{4.10.9}$$

The ratio of positrons to electrons contributed by heavy ions is just

$$\frac{n_+}{n_Z} = \frac{n_+}{n_- - n_+} = \frac{e^\alpha}{e^{-\alpha} - e^\alpha} = \frac{\sqrt{1 + (n_Z/2I)^2} - (n_Z/2I)}{2(n_Z/2I)}$$

$$= 0.5 \left(\sqrt{\left(\frac{2I}{n_Z}\right)^2 + 1} - 1 \right). \tag{4.10.10}$$

When $2I/n_Z \gg 1$ (relatively large temperatures, small nuclear matter densities), eq. (4.10.10) gives

$$\frac{n_+}{n_Z} = \frac{n_- - n_Z}{n_Z} \sim \frac{I}{n_Z} \gg 1. \tag{4.10.11}$$

In words, the electron and positron number abundances are essentially equal, with the number abundance of electron–positron pairs far exceeding the number abundance of electrons contributed by heavy ions. The number abundance of $e^- - e^+$ pairs is due to a balance between pair production by photons and pair annihilation into photons and this balance is essentially unaffected by the presence of electrons whose numbers balance the numbers of protons in nuclei.

To explore this situation more quantitatively, suppose that kT is small enough that, for most leptons,

$$\epsilon = m_e c^2 + \frac{p^2}{2m_e}. \tag{4.10.12}$$

Then,

$$I = \frac{8\pi}{h^3} e^{-m_e c^2/kT} \int_0^\infty \exp\left(\frac{-p^2}{2m_e kT}\right) p^2 \mathrm{d}p = 2\left(\frac{2\pi m_e kT}{h^2}\right)^{3/2} e^{-m_e c^2/kT}$$

$$= 1.52714 \times 10^{29} \, T_9^{3/2} \, e^{-5.930/T_9}. \tag{4.10.13}$$

The number abundance of electrons balancing protons is

$$n_Z = n_- - n_+ = 3.00 \times 10^{23} \, \frac{\rho_0}{(\mu_e/2)} \, \mathrm{cm}^{-3}, \tag{4.10.14}$$

where the numerical coefficient of the variable term on the far right of the equation is a compromise between the two approximations $n_e \sim (\rho_0/\mu_e) \, N_A$ and $n_e \sim (\rho_0/\mu_e) \, (1/M_H)$. Here, μ_e is the electron molecular weight and the matter density ρ_0 is expressed in units of g cm^{-3}.

Noting from eq. (4.10.11) that $n_\pm \sim I$, combining eqs. (4.10.13) and (4.10.14) gives

$$\frac{n_\pm}{n_Z} \sim 1.7947 \times 10^4 \, \frac{(\mu_e/2)}{\rho_0} \, T_9^{3/2} \, e^{-5.930/T_9}. \tag{4.10.15}$$

Rewriting eq. (4.10.15) as

$$\frac{m_e(n_+ + n_-)}{2 M_p n_Z} \sim 10 \, \frac{(\mu_e/2)}{\rho_0} \, T_9^{3/2} \, e^{-5.930/T_9}, \tag{4.10.16}$$

one has the remarkable result that situations can be envisioned in which the lepton mass density exceeds the heavy ion mass density.

It is much more frequently the case that the number abundance of $e^- - e^+$ pairs is significantly influenced by the existence of the electrons provided by heavy ions. At any given temperature, as the number abundance of electrons which balance the protons in nuclei is increased, the relative number abundance of positrons decreases. Solving eq. (4.10.10) in the limit that $n_Z/2I \gg 1$ (low temperatures, large nuclear matter densities),

$$\frac{n_+}{n_Z} = \left(\frac{I}{n_Z} \right)^2 \ll 1. \tag{4.10.17}$$

Thus, even when electrons are quite far from being degenerate, the formation of electron–positron pairs can be dramatically depressed.

The expressions for the number density of electrons given by eqs. (4.10.4) and (4.10.5) are valid only if $-\alpha$ does not exceed

$$|\alpha|_{\max} = \frac{m_e c^2}{kT}. \tag{4.10.18}$$

When this condition is violated, electrons are degenerate in the sense that, at zero electron kinetic energy, the occupation number is larger than $f = \frac{1}{2}$. At the degeneracy border,

$$n_\pm = \frac{8\pi}{h^3} \int \frac{p^2 \, dp}{e^{\pm m_e c^2 / kT} \, e^{\sqrt{(m_e c^2)^2 + (cp)^2}/kT} + 1}. \tag{4.10.19}$$

Setting $n_- - n_+$ given by eq. (4.10.19) equal to the proton number density given by eq. (4.10.14) when $\mu_e = 2$ produces the relationship between temperature and density described by the curve labeled DEGENERACY BORDER in Fig. 4.10.1. The ratio along the degeneracy border of the positron number density n_+ to n_Z, the number density of free electrons provided by ions, is described by the curve labeled n_+/n_Z, the scale being given along the right hand vertical axis of the figure. The curves are results of exact integrations. In the lower left hand quadrant of Fig. 4.10.1, electron velocities are not very relativistic and the slope of the degeneracy border is distinctly larger than the slope of the border in the upper right hand quadrant of the figure where electron and positron velocities become quite relativistic. The break in the slope occurs in the neighborhood of $\log \rho_6 \sim -0.3$ and $\log T_9 \sim 0$. Comparing with the curve labeled n_+/n_Z, it is apparent that, as long as lepton velocities are not relativistic, electron–positron pairs are a very minor lepton constituent whereas, in the region where lepton velocities are decidedly relativistic, the number

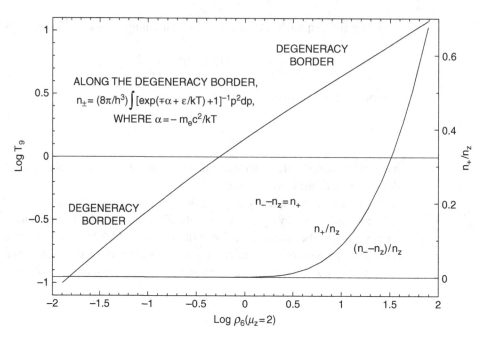

Fig. 4.10.1 Temperature and positron/proton ratio versus density at the electron-degeneracy border ($\alpha = -m_e c^2/kT$)

abundance of $e^- - e^+$ pairs relative to the abundance of electrons contributed by heavy ions increases rapidly with increasing temperatures and densities along the degeneracy border.

In the non-relativistic approximation, the number density of electrons given by eq. (4.10.19) is

$$
n_- = \frac{8\pi}{h^3} \int \frac{p^2\, dp}{e^{(p^2/2m_e kT)} + 1}
$$

$$
= \frac{8\pi}{h^3} \int e^{-(p^2/2m_e kT)} \left(1 - e^{-(p^2/2m_e kT)} + e^{-2(p^2/2m_e kT)} + \cdots \right) p^2\, dp
$$

$$
= \frac{8\pi}{h^3} (2m_e kT)^{3/2} \sum_{n=1}^{\infty} (-1)^{n+1} \frac{1}{n^{3/2}} \int_0^{\infty} e^{-x^2} x^2\, dx
$$

$$
= 2 \left(\frac{2\pi m_e kT}{h^2} \right)^{3/2} \sum_{n=1}^{\infty} (-1)^{n+1} \frac{1}{n^{3/2}} \tag{4.10.20}
$$

and the number abundance of positrons is given by

$$
n_+ = \frac{8\pi}{h^3} \int \frac{p^2\, dp}{e^{(2m_e c^2/kT)}\, e^{(p^2/2m_e kT)} + 1}
$$

$$
= 2 \left(\frac{2\pi m_e kT}{h^2} \right)^{3/2} \sum_{n=1}^{\infty} (-1)^{n+1} \frac{1}{n^{3/2}} \exp\left\{ -n \left(\frac{2m_e c^2}{kT} \right) \right\}. \tag{4.10.21}
$$

Thus, in first approximation,

$$\frac{n_+}{n_Z} \sim \frac{n_+}{n_-} = \left(\sum_{n=1}^{\infty} (-1)^{n+1} \frac{1}{n^{3/2}}\right)^{-1} \exp\left(-\frac{2m_e c^2}{kT}\right) \sim 1.291 \exp\left(-\frac{11.86}{T_9}\right).$$

$$(4.10.22)$$

This relationship gives $n_+/n_Z \sim 0.91 \times 10^{-5}$ for $T_9 = 1$ and $n_+/n_Z \sim 1.2 \times 10^{-3}$ for $T_9 = 1.7$. So, along the degeneracy border, the importance of $e^- - e^+$ pairs increases dramatically over only a very small range in temperature.

Moving beyond the degeneracy border into regions of increasing electron degeneracy, as the degree of degeneracy increases, it becomes appropriate to set

$$\alpha = -\frac{m_e c^2 + \epsilon_F}{kT}. \tag{4.10.23}$$

where

$$\epsilon_F = \sqrt{(m_e c^2)^2 + (c p_F)^2} - m_e c^2 \tag{4.10.24}$$

is the Fermi kinetic energy determined in the standard way when positrons are neglected. In first approximation, the standard approach gives the electron number density as

$$n_e \sim n_Z \sim \frac{8\pi}{h^3} \frac{p_F^3}{3} = \frac{8\pi}{3} \left(\frac{2m_e \epsilon_F}{h^2}\right)^{3/2}. \tag{4.10.25}$$

The positron number density (using eq. 4.10.3 for the occupation number) is then given by

$$n_+ \sim \frac{8\pi}{h^3} \int_0^{\infty} \frac{p^2 dp}{\exp\left([\epsilon_F + m_e c^2]/kT + [m_e c^2 + p^2/2m_e]/kT\right) + 1} \tag{4.10.26}$$

$$\sim \frac{8\pi}{h^3} \exp\left(-[\epsilon_F + 2m_e c^2]/kT\right) \int_0^{\infty} \exp\left(-p^2/2m_e kT\right) p^2 dp \tag{4.10.27}$$

$$= 2\left(\frac{2\pi m_e kT}{h^2}\right)^{3/2} \exp\left(-\frac{\epsilon_F + 2m_e c^2}{kT}\right). \tag{4.10.28}$$

Using eq. (4.10.25), one has finally that, when $kT/\epsilon_F \ll 1$, but $\epsilon_F \gg m_e c^2$,

$$n_+ \sim n_e \frac{3}{4} \sqrt{\pi} \left(\frac{kT}{\epsilon_F}\right)^{3/2} \exp\left(-\frac{\epsilon_F + 2m_e c^2}{kT}\right). \tag{4.10.29}$$

In a similar way one can derive that, when $kT/\epsilon_F \ll 1$, but $\epsilon_F \gg m_e c^2$,

$$n_+ \sim n_e \, 9 \left(\frac{kT}{c p_F}\right)^3 \exp\left(-\frac{\epsilon_F + 2m_e c^2}{kT}\right). \tag{4.10.30}$$

In stark contrast with the situation in regions of electron degeneracy which are characterized by a paucity of positrons relative to electrons is the situation encountered at the beginning of this section where it is shown that the number abundance of electron–positron pairs can exceed the number abundance of electrons contributed by heavy ions even to the extent that pairs contribute more to the pressure than do the heavy ions and associated electrons. In even starker contrast are situations in which the total rest mass of pairs is larger than the rest mass of heavy ions. Such situations occur at temperatures large enough that lepton kinetic energies are highly relativistic and pairs behave very much like photons. Then, $kT \gg m_e c^2$, $\epsilon \to cp$, and eq. (4.10.5), which is the result of keeping just the first term in the expansion of the occupation numbers in powers of $e^{-\epsilon/kT}$, becomes

$$I \to \frac{8\pi}{h^3} \int_0^\infty e^{-cp/kT} p^2 \mathrm{d}p = 8\pi \left(\frac{kT}{hc}\right)^3 \int_0^\infty e^{-x} x^2 \mathrm{d}x = 16\pi \left(\frac{kT}{hc}\right)^3, \quad (4.10.31)$$

giving

$$n_\pm \to e^{\pm\alpha} 16\pi \left(\frac{kT}{hc}\right)^3. \quad (4.10.32)$$

In this approximation, the pressures due to electrons P_- and to positrons P_+ are

$$P_\pm \sim \frac{1}{3} \frac{8\pi}{h^3} e^{\pm\alpha} \int_0^\infty e^{-\epsilon/kT} \, vp \, p^2 \mathrm{d}p \to 16\pi e^{\pm\alpha} \left(\frac{kT}{hc}\right)^3 kT, \quad (4.10.33)$$

so that

$$P_\pm = n_\pm kT. \quad (4.10.34)$$

The corresponding energy densities (energy per unit volume) are

$$\bar{U}_\pm = 48\pi \, e^{\pm\alpha} \left(\frac{kT}{hc}\right)^3 kT = 3P_\pm. \quad (4.10.35)$$

Had the occupation numbers been expanded in powers of $\exp(\mp\alpha - \epsilon/kT)$, and it were still assumed that $\epsilon = cp$, $e^{\pm\alpha}$ in eq. (4.10.32) would be replaced by

$$r = \sum_{k=1}^\infty (-1)^{k+1} e^{\pm k\alpha} \frac{1}{k^3} \to 0.901\,542\,675 \text{ as } \alpha \to 0. \quad (4.10.36)$$

and $e^{\pm\alpha}$ in eqs. (4.10.33)–(4.10.34) would be replaced by

$$r' = \sum_{k=1}^\infty (-1)^{k+1} e^{\pm k\alpha} \frac{1}{k^4} \to \frac{7\pi^4}{720} = 0.947\,032\,829 \text{ as } \alpha \to 0. \quad (4.10.37)$$

Thus, for example, with $\alpha = 0$,

$$\bar{U}_\pm = \frac{7\pi^5}{15} \left(\frac{kT}{hc}\right)^3 kT = 3P_\pm, \text{ and } P_\pm = 1.050\,458\,126 \, n_\pm \, kT. \quad (4.10.38)$$

Comparing with eq. (4.11.14) in the next section (4.11), it is evident that, at high enough temperatures, leptons behave very similarly to photons, a result that could have been anticipated from the fact that, when $kT \gg m_e c^2$, the energy–momentum relationship for leptons is essentially the same as for photons. A minor numerical difference is occasioned by the fact that there is a minus 1 in the denominator of the photon occupation number while there is a plus 1 in the occupation numbers for electrons and positrons, making the numerical coefficients for number abundances for leptons smaller by a factor of 3/4 than the corresponding quantities for photons and the coefficients for pressure and for energy densities for leptons smaller by a factor of 7/8 than the corresponding quantities for photons.

Thus, in the limiting case that $\alpha \to 0$, $P_+/P_{\rm rad} = P_-/P_{\rm rad} = 7/8$ and

$$\beta = \frac{P_{\rm gas}}{P_{\rm total}} = \frac{P_n + P_- + P_+}{P_n + P_- + P_+ + P_{\rm rad}} \to \left(1 + \frac{P_{\rm rad}}{P_- + P_+}\right)^{-1} \sim \left(1 + \frac{1}{7/4}\right)^{-1} = \frac{7}{11}.$$
(4.10.39)

In writing this equation, it has been assumed that the number abundance of $e^- - e^+$ pairs far exceeds the number abundance of heavy ions and that neutrinos formed by interactions between photons and charged particles do not come into thermal equilibrium before escaping from the region under consideration, so that $P_{\rm tot} \sim P_+ + P_- + P_{\rm rad}$. However, in the cores of massive stars collapsing into neutron stars, the mean free paths of neutrinos become small compared with the linear dimensions of the core and during a phase in the early universe when neutrinos have nowhere to escape, neutrinos do come into local thermodynamic equilibrium. In these cases, since neutrinos are fermions, the number abundance of $\nu_e - \bar{\nu}_e$ pairs is the same as that of $e^- - e^+$ pairs , so $\beta = P_{\rm gas}/P_{\rm total} = 7/18$.

Having explored the consequences of $e^- - e^+$ pair production both when lepton kinetic energies are non-relativistic and when they are highly relativistic, it is appropriate to explore consequences at intermediate values of $kT/m_e c^2$. Converting eqs. (4.10.3) and (4.2.14) to their continuum forms, eq. (4.10.4) is replaced by

$$n_\pm = \frac{8\pi}{h^3} \int_0^\infty \frac{1}{e^{\mp\alpha + \epsilon/kT} + 1} \, p^2 dp = \frac{8\pi}{h^3} \int_0^\infty \frac{e^{-(\epsilon/kT \mp \alpha)}}{1 + e^{-(\epsilon/kT \mp \alpha)}} \, p^2 dp. \qquad (4.10.40)$$

Carrying out the full Taylor expansions of the occupation numbers on the assumption that $\epsilon/kT \mp \alpha > 0$, one has

$$n_\pm = \frac{8\pi}{h^3} \sum_{n=1}^\infty (-1)^{n+1} \int_0^\infty e^{-n(\epsilon/kT \mp \alpha)} \, p^2 dp$$

$$= \frac{8\pi}{h^3} \sum_{n=1}^\infty (-1)^{n+1} e^{\pm n\alpha} \int_0^\infty e^{-n\epsilon/kT} \, p^2 dp, \qquad (4.10.41)$$

where the first term in the series reproduces eq. (4.10.4).

Continuing the development,

$$n_\pm = \frac{8\pi}{h^3} \sum_{n=1}^{\infty} (-1)^{n+1} e^{\pm n\alpha} \int_0^\infty e^{-n\sqrt{(m_e c^2)^2 + (cp)^2}/kT} \, p^2 dp$$

$$= \frac{8\pi}{h^3} \sum_{n=1}^{\infty} (-1)^{n+1} e^{\pm n\alpha} \int_0^\infty e^{-n(m_e c^2/kT)\sqrt{1+(p/m_e c)^2}} \, p^2 dp$$

$$= 8\pi \left(\frac{m_e c}{h}\right)^3 \sum_{n=1}^{\infty} (-1)^{n+1} e^{\pm n\alpha} \int_0^\infty e^{-n(m_e c^2/kT)\sqrt{1+y^2}} \, y^2 dy$$

$$= 8\pi \left(\frac{m_e c}{h}\right)^3 \sum_{n=1}^{\infty} (-1)^{n+1} e^{\pm n\alpha} \, J_n, \tag{4.10.42}$$

where

$$J_n = \int_0^\infty e^{-n(m_e c^2/kT)\sqrt{1+y^2}} \, y^2 dy. \tag{4.10.43}$$

In these equations, $y = p/m_e c$, and each J_n integral, which is a function only of $n\, m_e c^2/kT$, looks innocent enough. However, exact integration involves hyberbolic functions and asymptotic expansions of Bessel functions, as described by Subrahmanyan Chandrasekhar in *An Introduction to the Study of Stellar Structure* (1939). Relevant properties of hyperbolic functions are described by H. B. Dwight in *Tables of Integrals and Other Mathematical Data* (1934). Derivations of asymptotic series expansions for Bessel functions and analysis of their properties are given by Whittaker & Watson in *A Course in Modern Analysis* (1952).

Setting

$$y = \frac{p}{m_e c} = \cosh\theta = \frac{e^\theta + e^{-\theta}}{2}, \quad \sqrt{1+y^2} = \sinh\theta = \frac{e^\theta - e^{-\theta}}{2}, \quad \text{and } z = \frac{m_e c^2}{kT}, \tag{4.10.44}$$

one has

$$J_n = \int_0^\infty e^{-n(m_e c^2/kT)\sqrt{1+y^2}} \, y^2 dy = \int_0^\infty e^{-nz\sinh\theta} \, \sinh^2\theta \cosh\theta \, d\theta. \tag{4.10.45}$$

Using the identity

$$\sinh^2\theta \cosh\theta = \frac{1}{4} (\cosh 3\theta - \cosh\theta), \tag{4.10.46}$$

one obtains

$$J_n = \int_0^\infty e^{-nz\sinh\theta} \frac{1}{4} (\cosh 3\theta - \cosh\theta) d\theta = \frac{1}{4} (K_3(nz) - K_1(nz)) = \frac{K_2(nz)}{nz}, \tag{4.10.47}$$

where the K_νs are Bessel functions and the last equality follows from recurrence relations for these functions.

When $kT/m_e c^2 \ll 1$, the Bessel functions can be represented by the asymptotic series

$$K_\nu(nz) = \sqrt{\frac{\pi}{2}} \frac{e^{-nz}}{\sqrt{nz}} \left[1 + \frac{4\nu^2 - 1}{1! \, 8nz} + \frac{(4\nu^2 - 1^2)(4\nu^2 - 3^2)}{2! \, (8nz)^2} + \cdots \right], \qquad (4.10.48)$$

and when $kT/m_e c^2 \gg 1$,

$$K_\nu(nz) = \frac{(\nu - 1)!}{2} \left(\frac{2}{z} \right)^\nu. \qquad (4.10.49)$$

Using eq. (4.10.49) in eq. (4.10.47) and then in eq. (4.10.42), one recovers the result for $n_\pm$ already obtained in the extreme relativistic approximation, as described by eqs. (4.10.32) and (4.10.35). This is also true for $P_\pm$ and $\bar{U}_\pm$.

In the region where eq. (4.10.48) is a satisfactory approximation,

$$n_\pm = 8\pi \left(\frac{m_e c}{h} \right)^3 \sqrt{\frac{\pi}{2}} \sum_{n=1}^{\infty} (-1)^{n+1} e^{\pm n\alpha} \left(\frac{1}{n} \frac{kT}{m_e c^2} \right)^{3/2} e^{-n(m_e c^2/kT)} S_n, \qquad (4.10.50)$$

or

$$n_\pm = 2 \left(\frac{2\pi m_e kT}{h^2} \right)^{3/2} \sum_{n=1}^{\infty} (-1)^{n+1} \left(\frac{1}{n} \right)^{3/2} e^{-n(\mp\alpha + m_e c^2/kT)} S_n, \qquad (4.10.51)$$

where

$$S_n = 1 + \frac{16 - 1^2}{1!} \left(\frac{1}{8n} \frac{kT}{m_e c^2} \right) + \frac{(16 - 1^2)(16 - 3^2)}{2!} \left(\frac{1}{8n} \frac{kT}{m_e c^2} \right)^2 + \cdots$$

$$= 1 + \frac{15}{8} \frac{1}{n} \frac{kT}{m_e c^2} + \frac{105}{128} \frac{1}{n^2} \left(\frac{kT}{m_e c^2} \right)^2 - \frac{945}{6144} \frac{1}{n^3} \left(\frac{kT}{m_e c^2} \right)^3 + \cdots, \qquad (4.10.52)$$

Keeping just the first term in the sum over n in eq. (4.10.51) and setting $S_n = 1$ in this term, one recovers the result of inserting I from eq. (4.10.13) into the expression for $n_\pm$ given by eq. (4.10.4), namely

$$n_\pm \sim 8\pi \left(\frac{m_e c}{h} \right)^3 \sqrt{\frac{\pi}{2}} \left(\frac{kT}{m_e c^2} \right)^{3/2} e^{\pm\alpha - m_e c^2/kT}$$

$$= e^{\pm\alpha} \, 2 \left(\frac{2\pi m_e kT}{h^2} \right)^{3/2} e^{-m_e c^2/kT}. \qquad (4.10.53)$$

Although the sum n_- for electrons in eq. (4.10.56) diverges when $-\alpha \geq m_e c^2/kT$, the sum n_+ for positrons remains valid. In this case, the techniques for finding the equation of state for degenerate electrons outlined in Sections 4.3–4.9 may be used to find the chemical potential for electrons and this potential may be inserted as a good approximation in the sum for n_+ in eq. (4.10.51). But remember that α as used in Sections 4.4–4.9 is larger than α defined in this section by $m_e c^2/kT$.

Both care and judgement must be exercised in determining when to truncate the series for S_n given by eq. (4.10.52). The series may be rewritten as

$$S_n = s_1 + \sum_{j=2}^{\infty} s_j, \qquad (4.10.54)$$

where

$$s_1 = 1 \text{ and } s_j = s_{j-1} \frac{16 - (2j - 3)^2}{j - 1} \frac{1}{8n} \frac{kT}{m_e c^2}, \quad \text{for } j \geq 2 . \qquad (4.10.55)$$

The asymptotic nature of the series is obvious from the recurrence relationship given by eq. (4.10.55). For any given choice of the quantity

$$\lambda = 8n \frac{m_e c^2}{kT} \qquad (4.10.56)$$

there exists a j_{max} such that, for $j \geq j_{max}$, the value of $|s_j|$ increases with increasing j. The larger λ is, the larger is j_{max}. Setting the coefficient of s_{j-1} in eq. (4.10.55) equal to unity and solving for j_{max}, one obtains

$$j_{max} = \frac{12 + \lambda}{8} \left(1 + \sqrt{1 - \frac{4(\lambda - 7)}{(12 + \lambda)^2}} \right) . \qquad (4.10.57)$$

For example, for $n = 1$ and $T_9 = 1.5$, $\lambda = 31.6$, and $j_{max} = 10.76 \sim 11$. The question arises as to how many of these terms should be retained. The first ten terms in the sum give

$$S_1 = 1 + 0.474\,283 + 0.052\,4871 - 0.004\,978\,75 + 0.001\,298\,74 - 0.000\,533\,840$$

$$+ 0.000\,295\,390 - 0.000\,204\,143 + 0.000\,168\,631$$

$$- 0.000\,161\,735 + \cdots = 1.522\,65 + \cdots . \qquad (4.10.58)$$

Adding more terms simply results in oscillations of increasing amplitude. Common sense suggests compromising on an acceptable accuracy and terminating the series when this accuracy has been achieved. In the case at hand, insisting on four digit accuracy means truncation of the series at the sixth term. Retaining just three terms gives $S_1 = 1.527$ compared with the value $S_1 = 1.523$ achieved by retaining ten terms.

Consolidating the coefficients in front of the sums in eq. (4.10.51), one obtains

$$n_{\pm} = 1.527\,14 \times 10^{29} \, T_9^{3/2} \sum_{n=1}^{\infty} (-1)^{n+1} \left(\frac{1}{n} \right)^{3/2} e^{-n(\mp\alpha + m_e c^2 / kT)} S_n . \qquad (4.10.59)$$

The number abundance of protons in nuclei remains in the form given by eq. (4.10.14),

$$n_Z = n_- - n_+ \sim 3.00 \times 10^{29} \left(\frac{2}{\mu_e} \rho_6 \right), \qquad (4.10.60)$$

where ρ_6 is the density in units of 10^6 g cm^{-3}. For every set of choices of T_9 and α, $n_{\pm}$ can be determined from eq. (4.10.59) and, with the additional choice of the molecular weight μ_e, the baryonic mass density ρ_6 can be determined from eq. (4.10.60).

Keeping just the first ($n = 1$) term for n_+ and for n_- in eq. (4.10.59), and solving for $e^{\pm\alpha}$, one obtains the generalization of eq. (4.10.9),

$$e^{\mp\alpha} = \sqrt{1 + x^2} \pm x, \qquad (4.10.61)$$

where

$$x = 0.982\,23 \left(\rho_6 / T_9^{3/2} \right) \exp\left(5.93 / T_9 \right) S_1^{-1}. \qquad (4.10.62)$$

Equation (4.10.10) becomes

$$\frac{n_+}{n_Z} = 0.5 \left(\sqrt{\frac{1}{x^2} + 1} - 1 \right). \qquad (4.10.63)$$

In this approximation, the first order inclusion of relativistic effects has been accomplished by multiplying the quantity I defined by eq. (4.10.13) by the series S_1, the first three terms of which are

$$S_1 = 1 + 0.316\,189\,T_9 \left(1 + 0.073\,777\,T_9 \right) + \cdots. \qquad (4.10.64)$$

The quantity x is just $n_Z/2I$ divided by S_1. It is apparent that treating electrons and positrons relativistically has the effect of increasing their number abundances at any temperature compared with number abundances obtained in the non-relativistic approximation.

As described in Fig. 4.10.1, when electrons are degenerate, the number abundance of electron–positron pairs becomes appreciable only when temperatures are large enough for the electrons and positrons to have highly relativistic velocities. However, as shown at the beginning of this section, at any given temperature, as the baryonic density is decreased and electrons become increasingly less degenerate, the abundance of electron–positron pairs increases rapidly with respect to the abundance of electrons provided by heavy ions. The transition in behavior as one moves away from the borderline for electron degeneracy is described in Fig. 4.10.2 for the case $T_9 = 1.5$. In this figure, the curve with positive slope gives the ratio of $-\alpha$ to $m_e c^2 / kT$ (right hand scale) and the solid curve of negative slope gives n_+/n_Z (left hand scale) as determined by exact integrations of eq. (4.10.40). Circles along the curve for n_+/n_Z show where calculations have been performed. The dotted curve of negative slope is given by eqs. (4.10.62)–(4.10.64) and its proximity to the solid curve of negative slope demonstrates the extent to which the approximation provided by these equations is adequate.

The information provided in Fig. 4.10.2 for $T_9 = 1.5$ is repeated and extended in Fig. 4.10.3 where results of exact solutions of eq. (4.10.40) are shown for temperatures $T_9 = 0.5, 0.75, 1.0, 1.5, 3.0$, and 7.0. The unnumbered curves with positive slope give the ratio of $-\alpha$ to $m_e c^2 / kT$ (right hand scale), the temperature associated with each curve increasing from left to right at any given value of the ratio. The numbered curves of negative slope give n_+/n_Z (left hand scale), with numbers giving the temperature in units of T_9. Circles along these curves show where calculations were performed. A portion of the degeneracy border is described by the dotted curve (left hand scale). The approximations

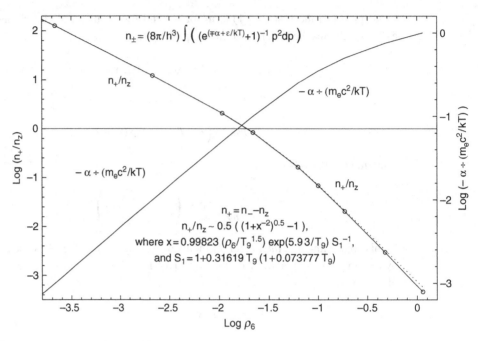

Fig. 4.10.2 Positron/proton ratio versus density when $T_9 = 1.5$; $n_z = n_- - n_+ = $ number density of protons in nuclei $= \sum n_i z_i$

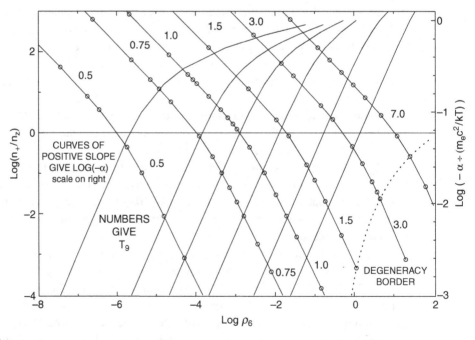

Fig. 4.10.3 Positron/proton ratio versus density and temperature; $n_z = n_- - n_+ = $ number density of protons in nuclei $= \sum n_i z_i$

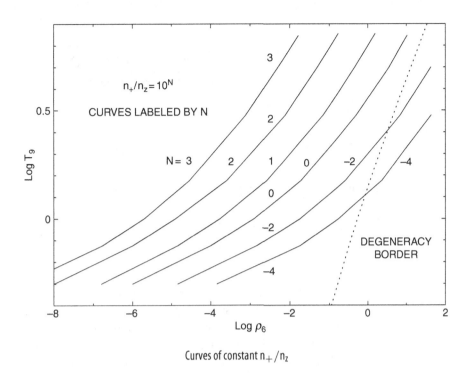

Fig. 4.10.4   Curves of constant n_+/n_Z

given by eqs. (4.10.62)–(4.10.64) reproduce the results for temperatures other than $T_9 = 1.5$ as well as in the $T_9 = 1.5$ case described in Fig. 4.10.2.

Figure 4.10.4 has been constructed by first determining, for each selection of the integer N in the equation $n_+/n_Z = 10^N$, the intersection of curves of negative slope in Fig. 4.10.3 with the line defined by $n_+/n_Z = 10^N$ and then passing a cubic spline through the resulting points. The procedures for constructing cubic splines are presented in Section 7.14, eqs. (7.14.41)–(7.14.90). The dotted curve in Fig. 4.10.4 is the electron-degeneracy border. It is a bit sobering to realize that, for temperatures larger than $T_9 = 0.1$, the region in the density–temperature plane where free electrons provided by heavy ions are not degenerate and are more abundant than the electrons which accompany positrons is extremely small (bounded by the degeneracy border and the curve for $N = 0$).

The same sort of calculations which lead to eqs. (4.10.40)–(4.10.52) for number abundances can be performed to find that the pressures due to positrons and electrons are given by

$$P_\pm = 2 \left(\frac{2\pi m_e kT}{h^2} \right)^{3/2} kT \sum_{n=1}^{\infty} (-1)^{n+1} \left(\frac{1}{n} \right)^{5/2} e^{-n(\mp\alpha+m_e c^2/kT)} S_n . \quad (4.10.65)$$

Dividing eq. (4.10.65) by eq. (4.10.51) gives

$$\frac{P_\pm}{n_\pm} = kT \frac{\sum_{n=1}^{\infty} (-1)^{n+1} e^{-n(\mp\alpha+m_e c^2/kT)} S_n/n^{5/2}}{\sum_{n=1}^{\infty} (-1)^{n+1} e^{-n(\mp\alpha+m_e c^2/kT)} S_n/n^{3/2}} . \quad (4.10.66)$$

or

$$\frac{P_\pm}{n_\pm} = kT \; \frac{1 - e^{-(\pm\alpha + m_e c^2/kT)} \; (S_2/S_1)/2^{5/2} + \cdots}{1 - e^{-(\pm\alpha + m_e c^2/kT)} \; (S_2/S_1)/2^{3/2} + \cdots} \tag{4.10.67}$$

Approximating still further, one has

$$\frac{P_\pm}{n_\pm} = kT \left(1 + \frac{1}{2^{5/2}} \, e^{-(\mp\alpha + m_e c^2/kT)} \; (S_2/S_1) + \cdots \right) \tag{4.10.68}$$

or, using eq. (4.10.52) to estimate S_2/S_1,

$$\frac{P_\pm}{n_\pm} = kT \left(1 + \frac{1}{2^{5/2}} \, e^{\pm\alpha - m_e c^2/kT} \left(1 + \frac{15}{16} \frac{kT}{m_e c^2} + \frac{135}{512} \left(\frac{kT}{m_e c^2} \right)^2 + \cdots \right) + \cdots \right). \tag{4.10.69}$$

It is quite remarkable that, even in regions far from where electrons are degenerate, the pressure per particle is not the same for positrons as it is for electrons, with the pressure per electron being larger than the pressure per positron. Intuition suggests that, because the two types of particle are identical in all respects except for the sign of their electric charge and their helicity, the pressures per particle should be essentially identical for the two types of lepton. However, the fact that the pressure due to the more numerous electrons is larger than the pressure due to the less numerous positrons is consistent with the expectation that the Pauli exclusion principle should increase energies and hence pressures over and above what would prevail if this principle were not having effects which are proportional to the number abundances of the particles. It is nevertheless somewhat astonishing to recognize the power which the principle exerts. Intuition is somewhat molified by the fact that, in regions above the x-axis in Fig. 4.10.3, where $n_\pm \gg n_Z$ and $e^{\pm\alpha} \sim 1$, the pressures per particle are basically identical, with

$$\frac{P_\pm}{n_\pm} \sim kT \left(1 + 0.176\,778 \, e^{-5.930/T_9} \; (1 + 0.158\,09 \, T_9) + \cdots \right). \tag{4.10.70}$$

In the same approximation that gives particle number abundances described by eq. (4.10.51) and pressures described by eq. (4.10.65), lepton kinetic energy densities per unit volume are given by

$$\bar{U}'_\pm = 2 \left(\frac{2\pi m_e kT}{h^2} \right)^{3/2} \left(\frac{2z}{\pi} \right)^{1/2} \sum_{n=1}^{\infty} (-1)^{n+1} \frac{e^{\pm\alpha n}}{n} \left\{ \frac{3K_3(nz) + K_1(nz)}{4} - K_2(nz) \right\}. \tag{4.10.71}$$

where $z = m_e c^2/kT$ and the functions $K_\nu(nz)$ are described by eq. (4.10.48). Writing

$$K_1(nz) = \left(\frac{\pi}{2z} \right)^{1/2} \frac{e^{-nz}}{n^{1/2}} \, R_n, \tag{4.10.72}$$

where

$$R_n = 1 + \frac{3}{8}\frac{1}{nz} - \frac{15}{128}\left(\frac{1}{nz}\right)^2 + \frac{945}{6144}\left(\frac{1}{nz}\right)^3 + \cdots, \tag{4.10.73}$$

$$K_2(nz) = \left(\frac{\pi}{2z}\right)^{1/2}\frac{e^{-nz}}{n^{1/2}}S_n, \tag{4.10.74}$$

where S_n is given by eq. (4.10.52), and

$$K_3(nz) = \left(\frac{\pi}{2z}\right)^{1/2}\frac{e^{-nz}}{n^{1/2}}T_n, \tag{4.10.75}$$

where

$$T_n = 1 + \frac{35}{8}\frac{1}{nz} + \frac{945}{128}\left(\frac{1}{nz}\right)^2 + \frac{103\,95}{6144}\left(\frac{1}{nz}\right)^3 + \cdots, \tag{4.10.76}$$

one has

$$\frac{3T_n + R_n}{4} - S_n = \frac{3}{2}\frac{1}{nz}U_n, \tag{4.10.77}$$

where

$$U_n = \left(1 + \frac{25}{12}\left(\frac{1}{nz}\right) + \frac{5985}{6144}\left(\frac{1}{nz}\right)^2 + \cdots\right). \tag{4.10.78}$$

Kinetic energy densities per unit volume are therefore

$$\bar{U}'_\pm = 2\left(\frac{2\pi m_e kT}{h^2}\right)^{3/2}\frac{3}{2}kT\sum_{n=1}^{\infty}(-1)^{n+1}\frac{e^{n(\pm\alpha - m_e c^2/kT)}}{n^{5/2}}U_n. \tag{4.10.79}$$

Derivatives of U with respect to P, T, and ρ are relevant for questions of stability against convection and dynamical motions. When the chemical potential is small and the abundances of positrons and electrons are comparable, it transpires that there are regions in the density–temperature plane where the adiabatic temperature–pressure and adiabatic pressure-density derivatives satisfy, respectively,

$$V_{\mathrm{ad}} = \left(\frac{d\log T}{d\log P}\right)_{\mathrm{adiabatic}} < 0.25, \text{ and } \Gamma_{\mathrm{ad}} = \left(\frac{d\log P}{d\log \rho}\right)_{\mathrm{adiabatic}} < \frac{4}{3}. \tag{4.10.80}$$

In the non-relativistic approximation described by eqs. (4.10.4)–(4.10.15), derivatives can be given in closed form. Setting

$$x = \frac{nz}{2I} \text{ and } y = \frac{x}{\sqrt{1+x^2}}, \tag{4.10.81}$$

one has from eq. (4.10.9) that

$$e^{-\alpha} + e^{\alpha} = 2\sqrt{1+x^2}, \quad e^{-\alpha} - e^{\alpha} = 2x, \quad \text{and} \quad \frac{e^{-\alpha} - e^{\alpha}}{e^{-\alpha} + e^{\alpha}} = \frac{x}{\sqrt{1+x^2}} = y. \quad (4.10.82)$$

From eq. (4.10.8), one has

$$\frac{dI}{I} + \frac{-e^{-\alpha} - e^{\alpha}}{e^{-\alpha} - e^{\alpha}} \, d\alpha = \frac{dn_Z}{n_Z} = \frac{d\rho}{\rho}, \quad (4.10.83)$$

where the fact that $n_z \propto \rho$ has been used. Using eqs. (4.10.82) and (4.10.13), this can be written as

$$d\alpha = \frac{e^{-\alpha} - e^{\alpha}}{e^{-\alpha} + e^{\alpha}} \left(\frac{dI}{I} - \frac{d\rho}{\rho} \right) = y \left\{ \left(\frac{3}{2} + \frac{m_e c^2}{kT} \right) \frac{dT}{T} - \frac{d\rho}{\rho} \right\}. \quad (4.10.84)$$

From the first term of eq. (4.10.79), and with the help of eqs. (4.10.8) and (4.10.13), one has that

$$\bar{U}'_{\text{lep}} = (n_- + n_+) \frac{3}{2} kT = I \left(e^{-\alpha} + e^{\alpha} \right) \frac{3}{2} kT, \quad (4.10.85)$$

the differential of which is

$$\frac{d\bar{U}'_{\text{lep}}}{\bar{U}'_{\text{lep}}} = \frac{dI}{I} + \frac{-e^{-\alpha} + e^{\alpha}}{e^{-\alpha} + e^{\alpha}} \, d\alpha + \frac{dT}{T} = \left(\frac{5}{2} + \frac{m_e c^2}{kT} \right) \frac{dT}{T} - y \, d\alpha.$$

$$= \left(\frac{5}{2} + \frac{m_e c^2}{kT} \right) \frac{dT}{T} - y^2 \left\{ \left(\frac{3}{2} + \frac{m_e c^2}{kT} \right) \frac{dT}{T} - \frac{d\rho}{\rho} \right\}, \quad (4.10.86)$$

which can also be written as

$$d\bar{U}'_{\text{lep}} = \frac{\bar{U}'_{\text{lep}}}{1 + x^2} \left\{ \left(\frac{5}{2} + \frac{m_e c^2}{kT} + x^2 \right) \frac{dT}{T} + x^2 \frac{d\rho}{\rho} \right\}. \quad (4.10.87)$$

The change in the lepton kinetic energy per gram due to small changes in temperature and density is given by

$$dQ_{\text{lep}} = d \left(\frac{\bar{U}'_{\text{lep}}}{\rho} \right) + P'_{\text{lep}} \, d \left(\frac{1}{\rho} \right) = \frac{1}{\rho} dU'_{\text{lep}} + \left(P_{\text{lep}} + U'_{\text{lep}} \right) d \left(\frac{1}{\rho} \right)$$

$$= \frac{1}{\rho} \left(d\bar{U}'_{\text{lep}} - \frac{5}{3} \bar{U}'_{\text{lep}} \frac{d\rho}{\rho} \right) = \frac{1}{\rho} \bar{U}'_{\text{lep}} \left(\frac{d\bar{U}'_{\text{lep}}}{\bar{U}'_{\text{lep}}} - \frac{5}{3} \frac{d\rho}{\rho} \right). \quad (4.10.88)$$

Inserting eq. (4.10.87) into eq. (4.10.88) gives

$$dQ_{\text{lep}} = \frac{\left(\bar{U}'_{\text{lep}} / \rho \right)}{1 + x^2} \left\{ \left(\frac{5}{2} + \frac{m_e c^2}{kT} + x^2 \right) \frac{dT}{T} - \left(\frac{5}{3} + \frac{2}{3} x^2 \right) \frac{d\rho}{\rho} \right\}. \quad (4.10.89)$$

For every positron made, the rest mass energy of leptons is increased by $2m_ec^2$, so

$$dQ_{rest} = 2\,m_ec^2\,d\left(\frac{n_+}{\rho}\right) = 2m_ec^2\left(\frac{n_+}{\rho}\right)\left(\frac{dn_+}{n_+} - \frac{d\rho}{\rho}\right). \tag{4.10.90}$$

From eqs. (4.10.4), (4.10.13), and (4.10.84), one has that

$$\frac{dn_+}{n_+} = \left(\frac{3}{2} + \frac{m_ec^2}{kT}\right)\frac{dT}{T} + d\alpha$$

$$= \left(\frac{3}{2} + \frac{m_ec^2}{kT}\right)\frac{dT}{T} + y\left\{\left(\frac{3}{2} + \frac{m_ec^2}{kT}\right)\frac{dT}{T} - \frac{d\rho}{\rho}\right\}$$

$$= (1 + y)\left(\frac{3}{2} + \frac{m_ec^2}{kT}\right)\frac{dT}{T} - y\,\frac{d\rho}{\rho}. \tag{4.10.91}$$

Hence

$$dQ_{rest} = 2m_ec^2\left(\frac{n_+}{\rho}\right)\left\{(1 + y)\left(\frac{3}{2} + \frac{m_ec^2}{kT}\right)\frac{dT}{T} - y\frac{d\rho}{\rho}\right\}. \tag{4.10.92}$$

The contribution of heavy ions to the energy budget is given by

$$dQ_{ions} = d\left(\frac{\bar{U}'_{ions}}{\rho}\right) + P_{ions}\,d\left(\frac{1}{\rho}\right) = \frac{1}{\mu_n M_H}\left\{d\left(\frac{3}{2}\,kT\right) + \rho kT\,d\left(\frac{1}{\rho}\right)\right\}$$

$$= \frac{kT}{\mu_n M_H}\left(\frac{3}{2}\frac{dT}{T} - \frac{d\rho}{\rho}\right) = \frac{\bar{U}'_{ions}}{\rho}\left(\frac{dT}{T} - \frac{2}{3}\frac{d\rho}{\rho}\right), \tag{4.10.93}$$

and the contribution of photons to the energy budget is given by

$$dQ_{rad} = d\left(\frac{aT^4}{\rho}\right) + \frac{1}{3}\,aT^4\,d\left(\frac{1}{\rho}\right) = aT^4\left(4\,\frac{dT}{T} - \frac{4}{3}\frac{d\rho}{\rho}\right)$$

$$= \frac{U_{rad}}{\rho}\left(4\,\frac{dT}{T} - \frac{4}{3}\frac{d\rho}{\rho}\right). \tag{4.10.94}$$

Note that, in writing $U_{rad} = aT^4$, the convention established in Section 4.4 (eq. (4.4.21)) that U without a bar is energy per unit mass has been violated.

For adiabatic changes, one has that

$$dQ_{tot} = dQ_{lep} + dQ_{rest} + dQ_{ions} + dQ_{rad} = \frac{1}{\rho}\left(Q_T\frac{dT}{T} - Q_\rho\frac{d\rho}{\rho}\right) = 0. \tag{4.10.95}$$

From eqs. (4.10.89) and (4.10.92)–(4.10.94), Q_T and Q_ρ are given, respectively, by

$$Q_T = \frac{\bar{U}'_{lep}}{1 + x^2}\left(\frac{5}{2} + \frac{m_ec^2}{kT} + x^2\right) + 2m_ec^2\,n_+(1 + y)\left(\frac{3}{2} + \frac{m_ec^2}{kT}\right) + \bar{U}'_{ions} + 4\,U_{rad}$$

$$\tag{4.10.96}$$

and

$$Q_\rho = \frac{\bar{U}'_{\text{lep}}}{1+x^2}\left(\frac{5}{3}+\frac{2}{3}x^2\right) + 2m_ec^2\,n_+\,y + \frac{2}{3}\,\bar{U}'_{\text{ions}} + \frac{4}{3}\,U_{\text{rad}}. \tag{4.10.97}$$

When positrons are much less abundant than ions, $x \to \infty$, $y \to 1$, $n_+ \to 0$, $Q_T \to \bar{U}'_{\text{lep}} + \bar{U}'_{\text{ions}} + 4U_{\text{rad}}$, $Q_\rho \to (2/3)\left(\bar{U}'_{\text{lep}} + U'_{\text{ions}}\right) + (4/3)\,U_{\text{rad}}$, and

$$\left(\frac{d\log T}{d\log \rho}\right)_{\text{ad}} = \frac{Q_\rho}{Q_T} \to \frac{2}{3}\frac{\left(\bar{U}'_{\text{lep}} + \bar{U}'_{\text{ions}}\right) + 2U_{\text{rad}}}{\left(\bar{U}'_{\text{lep}} + \bar{U}'_{\text{ions}}\right) + 4U_{\text{rad}}} = \frac{2}{3}\frac{\bar{U}'_{\text{gas}} + 2U_{\text{rad}}}{\bar{U}_{\text{gas}} + 4U_{\text{rad}}}. \tag{4.10.98}$$

This gives the well-known results that, under adiabatic changes, when $P_{\text{gas}} \gg P_{\text{rad}}$, $\rho \propto T^{3/2}$, and when $P_{\text{rad}} \gg P_{\text{gas}}$, $\rho \propto T^3$.

At the other extreme, $x \to 0$, $y \to 0$, $n_+ \sim n_- \gg n_{\text{ions}}$, $Q_T \to \bar{U}'_{\text{lep}}(5/2 + m_ec^2/kT) + 2m_ec^2\,n_+(3/2 + m_ec^2/kT) + U'_{\text{ions}} + 4U_{\text{rad}}$, $Q_\rho \to (5/3)\,\bar{U}'_{\text{lep}} + \bar{U}'_{\text{ions}} + U_{\text{rad}}$, and

$$\left(\frac{d\log T}{d\log \rho}\right)_{\text{ad}} \to \frac{(5/3)\,\bar{U}'_{\text{lep}} + (2/3)\,\bar{U}'_{\text{ions}} + (4/3)\,U_{\text{rad}}}{\bar{U}'_{\text{lep}}\,(5/2 + m_ec^2/kT) + 2m_ec^2\,n_+ + \bar{U}'_{\text{ions}} + 4U_{\text{rad}}}. \tag{4.10.99}$$

Discarding the ion contribution to the derivatives and making use of the fact that, in the current approximation, $\bar{U}'_{\text{lep}} = 2n_+\frac{3}{2}kT$, one obtains

$$\left(\frac{d\log T}{d\log \rho}\right)_{\text{ad}} \to \frac{5n_+kT + (4/3)U_{\text{rad}}}{5n_+[(3/2)kT + m_ec^2] + 4U_{\text{rad}}}. \tag{4.10.100}$$

In the adopted approximation, derivatives of pressure with respect to T and ρ follow directly from derivatives with respect to energy density. Thus, for example, from eq. (4.10.87) one has that

$$dP_{\text{lep}} = \frac{P_{\text{lep}}}{1+x^2}\left\{\left(\frac{5}{2} + \frac{m_ec^2}{kT} + x^2\right)\frac{dT}{T} + x^2\frac{d\rho}{\rho}\right\}. \tag{4.10.101}$$

For the total pressure differential one has

$$dP = dP_{\text{lep}} + dP_{\text{ions}} + dP_{\text{rad}}$$

$$= \frac{P_{\text{lep}}}{1+x^2}\left\{\left(\frac{5}{2} + \frac{m_ec^2}{kT} + x^2\right)\frac{dT}{T} + x^2\frac{d\rho}{\rho}\right\} + P_{\text{ions}}\left(\frac{dT}{T} + \frac{d\rho}{\rho}\right) + P_{\text{rad}}\,4\,\frac{dT}{T}, \tag{4.10.102}$$

or

$$dP = \left\{\frac{P_{\text{lep}}}{1+x^2}\left(\frac{5}{2} + \frac{m_ec^2}{kT} + x^2\right) + P_{\text{ions}} + 4\,P_{\text{rad}}\right\}\frac{dT}{T} + \left(P_{\text{lep}}\,y^2 + P_{\text{ions}}\right)\frac{d\rho}{\rho}. \tag{4.10.103}$$

In the region of most interest, where $n_+ \sim n_- \gg n_Z$, so that $x \ll 1$ and $y \ll 1$,

$$\left(\frac{d \log P}{d \log \rho}\right)_{ad} \to \frac{1}{P} \left\{ P_{lep} \left(\frac{5}{2} + \frac{m_e c^2}{kT}\right) + P_{ions} + 4P_{rad} \right\} \left(\frac{d \log T}{d \log \rho}\right)_{ad} + \frac{P_{ions}}{P}.$$

(4.10.104)

Since, in the region of interest, eq. (4.10.100) is applicable and $P_{ions} \ll P_{lep} + P_{rad}$,

$$\left(\frac{d \log P}{d \log \rho}\right)_{ad} \to \frac{P_{lep} (5/2 + m_e c^2/kT) + 4P_{rad}}{P_{lep} + P_{rad}} \frac{5n_+ kT + 4P_{rad}}{5n_+[(3/2)kT + m_e c^2] + 12P_{rad}}.$$

(4.10.105)

$$= \frac{2n_+ [(5/2)kT + m_e c^2] + 4P_{rad}}{2n_+ kT + P_{rad}} \frac{5n_+ kT + 4P_{rad}}{5n_+[(3/2)kT + m_e c^2] + 12P_{rad}}.$$

(4.10.106)

For $n_+ = 0$, this gives

$$\left(\frac{d \log P}{d \log \rho}\right)_{ad} = \frac{4}{3}$$

(4.10.107)

and, for $P_{rad} = 0$, it gives

$$\left(\frac{d \log P}{d \log \rho}\right)_{ad} = \frac{1 + (5/2)(kT/m_e c^2)}{1 + (3/2)(kT/m_e c^2)},$$

(4.10.108)

which is less than 4/3 as long as $kT/m_e c^2 < 2/3$. Thus, to the extent that the non-relativistic treatment of leptons is adequate, it is evident that, in regions where $n_+ \sim n_- \gg n_Z$, dynamical instability can be expected. This situation is encountered during late stages of the evolution of massive stars, as first described by William A. Fowler & F. Hoyle (1964).

For the adiabatic temperature–pressure gradient, one has

$$V_{ad} = \left(\frac{d \log T}{d \log P}\right)_{ad} = P \left(\frac{P_{lep}}{1 + x^2} \left(\frac{5}{2} + \frac{m_e c^2}{kT} + x^2\right) + P_{ions} + 4P_{rad}\right)^{-1}$$

$$\times \left(1 - \frac{P_{lep} y^2 + P_{ions}}{P} \left(\frac{d \log \rho}{d \log P}\right)_{ad}\right),$$

(4.10.109)

which, in the region of interest, becomes

$$V_{ad} = \frac{P_{lep} + P_{rad}}{P_{lep} (5/2 + m_e c^2/kT) + 4P_{rad}}.$$

(4.10.110)

The extension of these results to include first order relativistic effects is reasonably straightforward, the only complication being the difference between the functions

$U_1 = 1 + 0.351\,32\,T_9\,(1 + 0.078\,850\,T_9) + \cdots$ and $S_1 = 1 + 0.316\,189\,T_9\,(1 + 0.073\,777\,T_9) +$
$\cdots$. Equation (4.10.84) becomes

$$d\alpha = y \left\{ \left(\frac{3}{2} + \frac{m_e c^2}{kT} + \frac{d \log S_1}{d \log T} \right) \frac{dT}{T} - \frac{d\rho}{\rho} \right\}, \tag{4.10.111}$$

eq. (4.10.101) becomes

$$dP_{\text{lep}} = \frac{P_{\text{lep}}}{1 + x^2} \left\{ \left(\frac{5}{2} + \frac{m_e c^2}{kT} + \frac{d \log S_1}{d \log T} + x^2 \right) \frac{dT}{T} + x^2 \frac{d\rho}{\rho} \right\}, \tag{4.10.112}$$

and eq. (4.10.87) becomes

$$\frac{d\bar{U}_{\text{lep}}}{\bar{U}_{\text{lep}}} = \frac{1}{1 + x^2} \left\{ \frac{5}{2} + \frac{m_e c^2}{kT} + \frac{d \log U_1}{d \log T} + x^2 \left(1 + \frac{d \log U_1}{d \log T} - \frac{d \log S_1}{d \log T} \right) \right\}$$

$$\times \frac{dT}{T} + y^2 \frac{d\rho}{\rho}. \tag{4.10.113}$$

Since S_1 is so similar to U_1, it is reasonable to approximate

$$\frac{d\bar{U}_{\text{lep}}}{\bar{U}_{\text{lep}}} \sim \frac{1}{1 + x^2} \left\{ \frac{5}{2} + \frac{m_e c^2}{kT} + \frac{d \log S_1}{d \log T} + x^2 \right\} \frac{dT}{T} + y^2 \frac{d\rho}{\rho}. \tag{4.10.114}$$

Finally,

$$dQ_{\text{rest}} = 2 m_e c^2 \left(\frac{n_+}{\rho} \right) \left\{ (1 + y) \left(\frac{3}{2} + \frac{m_e c^2}{kT} + \frac{d \log S_1}{d \log T} \right) \frac{dT}{T} - y \frac{d\rho}{\rho} \right\}. \tag{4.10.115}$$

Thus, the derivatives of number abundances, pressures, and energy densities established
in the non-relativistic approximation can be converted to derivatives in the first order relativistic approximation simply by adding $d \log S_1 / d \log T$ to the quantity $m_e c^2 / kT$ wher-
ever this quantity appears in an expression multiplying dT/T. Adiabatic pressure–density
and temperature–pressure derivatives follow in a straightforward fashion.

In all examples discussed thus far, it has been assumed that a reasonable approximation
to the chemical potential α can be found by ignoring in all expansions terms which involve
powers of $e^{\pm\alpha}$ greater than the first. This limitation is easily remedied. Calling $\alpha_0 = \alpha$
as given by eqs. (4.10.61) and (4.10.62) and setting $\alpha_1 = \alpha_0 + \Delta\alpha_0$, it follows from eqs.
(4.10.59) and (4.10.60) that

$$\Delta\alpha_0 = \sum_{n=2}^{\infty} (-1)^{n+1} \frac{1}{n^{3/2}} e^{-n m_e c^2 / kT} \left(e^{-n\alpha_0} - e^{n\alpha_0} \right) S_n$$

$$\div \sum_{n=1}^{\infty} (-1)^{n+1} \frac{1}{n^{1/2}} e^{-n m_e c^2 / kT} \left(e^{-n\alpha_0} + e^{n\alpha_0} \right) S_n. \tag{4.10.116}$$

Then, setting $\alpha_2 = \alpha_1 + \Delta\alpha_1$, it follows that

$$\Delta\alpha_1 \frac{1}{S_1} \sum_{n=1}^{\infty} (-1)^{n+1} \frac{1}{n^{1/2}} e^{-(n-1)m_e c^2/kT} \left(e^{-n\alpha_1} + e^{n\alpha_1}\right) S_n$$

$$= \left[(e^{-\alpha_1} - e^{\alpha_1}) - 2x\right] + \frac{1}{S_1} \sum_{n=2}^{\infty} (-1)^{n+1} \frac{1}{n^{3/2}} e^{-(n-1)m_e c^2/kT} \left(e^{-n\alpha_1} - e^{n\alpha_1}\right) S_n,$$

$$(4.10.117)$$

where x is given by eq. (4.10.62). The next iteration for α is obtained by replacing 2 by 3 and 1 by 2 in eq. (4.10.117), and so forth. In practical applications, the sums must be limited to a finite number of terms, say three, and iterations are to be continued until the absolute value of $\Delta\alpha_j$ in the jth iteration becomes suitably small. The penalty to be paid for refining the estimate of α is that all estimates of thermodynamic variables and of their derivatives must be repeated by retaining more than just the first term in relevant expansions.

One additional refinement is to be noted. In regions where $e^- e^+$ number abundances are substantially larger than the number abundance of ions, the mass density of leptons may be large enough that it is a poor approximation to equate the mass density ρ with just the mass density of ions and associated electrons, as has been assumed in derivations of derivatives in this section. Then, one must write

$$\rho = n_{\text{ions}} \left(M_{\text{ions}} + Z m_e c^2\right) + n_+ \, 2m_e c^2, \qquad (4.10.118)$$

where M_{ions} is the mass of an average ion and Z is the number of electrons contributed by an average ion. Given the added complication of implementing this modification, one should check to see whether or not general relativistic effects are also of importance before incorporating the modification (see, e.g., Iben, 1963).

4.11 Indistinguishable particles, Bose–Einstein statistics, and the electromagnetic radiation field

Bose–Einstein statistics (S. N. Bose, 1924; A. Einstein, 1924) describes gases of indistinguishable particles of integer spin to which the Pauli exclusion principle does not apply. There are no restrictions on the number of particles in a state, and counting the number of distinct ways of placing N_i particles in g_i states is somewhat more subtle than in the case of Fermi–Dirac statistics. In constructing the general result, it is useful to illustrate the steps in the argument with a concrete example. To this end, a simple system that consists of two particles (labeled 1 and 2) that can occupy any of three different states (labeled a, b, and c) is chosen. The different arrangements are shown in Table 4.11.1.

The first step is to form a linear array containing both particles and states in some conveniently prescribed order. As in the top left line of the illustrative example, all of the states are placed at the left of the line in the order a, b, c, ..., g_i. All of the particles are placed

Table 4.11.1 Arrangements of three states and two particles

1:	a	b	c	1	2		1′:	a	c	1	2	b
2:	a	b	1	c	2		2′:	a	c	2	b	1
3:	a	b	1	2	c		3′:	a	c	b	1	2
4:	a	1	b	2	c		4′:	a	1	c	b	2
5:	a	1	2	b	c		5′:	a	1	2	c	b
6:	a	1	b	c	2		6′:	a	1	c	2	b
$\bar{1}$:	a	b	c	2	1		$\bar{1}$′:	a	c	2	1	b
$\bar{2}$:	a	b	2	c	1		$\bar{2}$′:	a	c	1	b	2
$\bar{3}$:	a	b	2	1	c		$\bar{3}$′:	a	c	b	2	1
$\bar{4}$:	a	2	b	1	c		$\bar{4}$′:	a	2	c	b	1
$\bar{5}$:	a	2	1	b	c		$\bar{5}$′:	a	2	1	c	b
$\bar{6}$:	a	2	b	c	1		$\bar{6}$′:	a	2	c	1	b

at the right of the same line in the order $1, 2, \ldots, N_i$. In this first line and in all subsequent distributions, it is understood that a particle occupies the closest state to its left. In the example, both particles occupy state c in the order 1, 2.

Maintaining all letters and all numbers *in the same order*, move numbers and groups of numbers so as to construct all possible particle-in-state combinations. This is illustrated in lines 1–6 in the top left quadrant of Table 4.11.1. The array of six arrangements is actually the total number of *different* arrangements if the particles are indistinguishable and the only thing that matters in describing the macroscopic state of the system is the number of particles in each state. Note that, under these conditions, exactly the same result follows if either state b or state c had been permanently placed in the preferred position at the leftmost spot of the initial line.

Next, permute the $(g_i - 1)$ states, *carrying along with each state the particles in unchanged order in that state*. Each of the $(g_i - 1)!$ permutations leads to a distribution which has the same physical content as the distribution from which it has been derived, namely, the same *number* of particles per state. In the example, this procedure converts the lines in the upper left hand quadrant of Table 4.11.1 into the lines in the upper right hand quadrant.

Return now to the initial configuration and permute the N_i particles among themselves. For each of the $N_i!$ permutations, again move numbers and groups of numbers so as to construct all possible particle-in-state combinations, maintaining all letters and all numbers in the same order. In the illustrative case, the permutations reduce to a simple switch $1, 2 \rightarrow 2, 1$, and the lines in the upper left hand quadrant of Table 4.11.1 are transformed into the lines in the lower left hand quadrant.

Finally, for each permutation of the numbered particles, permute the $(g_i - 1)$ states, carrying along the particles in each state in unchanged order. In the explicit example, this procedure converts the lines in the lower left hand quadrant of Table 4.11.1 into the lines in the lower right hand quadrant.

The total number of constructed arrangements is the number of permutations $((g_i - 1) + N_i)!$ one can produce from $(g_i - 1) + N_i$ objects. By construction, this total number is also equal to $(g_i - 1)! N_i! P_i$, where P_i is the number of arrangements which differ with regard to the *numbers* of particles in states, independent of the *order* in which the particles are arranged within the states. In the concrete illustration, $((g_i - 1) + N_i)! = ((3 - 1) + 2)! = 4! = 24$, $(g_i - 1)! = 2! = 2$, $N_i! = 2! = 2$, and $P_i = 24/(2 \cdot 2) = 6$. The same value of P_i is achieved by placing any other of the g_i states in the preferred place at the very left in the initial array.

Hence, it has been established that

$$P_i = \frac{(g_i + N_i - 1)!}{(g_i - 1)! N_i!} \tag{4.11.1}$$

is the total number of distinct ways in which N_i indistinguishable particles can be arranged in g_i cells when there is no limit on the number of particles which can occupy a cell. Proceeding as in Section 4.2, one maximizes the logarithm of the function $P = \prod_{i=1}^{\infty} P_i$, subject to the constraints expressed in eqs. (4.2.4) and (4.2.5), to obtain

$$N_i = \frac{g_i}{\exp(\alpha + \beta \epsilon_i) - 1}. \tag{4.11.2}$$

This is known as the Bose–Einstein distribution function. By considering a gas obeying this distribution to be in equilibrium with a non-degenerate Fermi–Dirac gas, it follows that $\beta = 1/kT$.

The quantity

$$f_i = \frac{1}{\exp(\alpha + \beta \epsilon_i) - 1} \tag{4.11.3}$$

is the occupation number, giving the number of particles per state; as it has the potential of becoming larger than unity for a range of energies, f_i cannot be interpreted as a probability of occupation.

If the particles are photons, conservation of photon number is not a constraint, so $\alpha = 0$. With $cp = h\nu$, the number density of states (number of states per unit volume) for photons moving in the solid angle $d\Omega$ is

$$dg = 2V \frac{(h\nu/c)^2 d(h\nu/c) d\Omega_\nu}{h^3} = \frac{2}{c^3} \nu^2 d\nu d\Omega, \tag{4.11.4}$$

where the factor of 2 comes from the two possible polarization states. The number density of photons is

$$n_{\text{ph}} = \frac{2}{c^3} \int_0^\infty \frac{\nu^2 d\nu d\Omega}{\exp(h\nu/kT) - 1}. \tag{4.11.5}$$

Integrating over angles and expanding the denominator,

$$n_{\text{ph}} = \frac{8\pi}{c^3} \int_0^\infty \sum_{m=1}^\infty e^{-mh\nu/kT} \nu^2 d\nu \tag{4.11.6}$$

$$= \frac{8\pi}{c^3} \sum_{m=1}^\infty \left(\frac{kT}{mh}\right)^3 \int_0^\infty x^2 e^{-x} dx \tag{4.11.7}$$

$$= 16\pi \left(\frac{kT}{ch}\right)^3 \sum_{m=1}^\infty \frac{1}{m^3} \tag{4.11.8}$$

$$= 16\pi \left(\frac{kT}{ch}\right)^3 \times 1.201\,975. \tag{4.11.9}$$

Similarly, the energy per unit volume is

$$\bar{U}_{\text{ph}} = \frac{8\pi}{c^3} \int_0^\infty \frac{h\nu\, \nu^2 d\nu}{\exp(h\nu/kT) - 1} \tag{4.11.10}$$

$$= \frac{8\pi h}{c^3} \int_0^\infty \sum_{m=1}^\infty e^{-mh\nu/kT} \nu^3 d\nu \tag{4.11.11}$$

$$= \frac{8\pi h}{c^3} \sum_{m=1}^\infty \left(\frac{kT}{mh}\right)^4 \int_0^\infty x^3 e^{-x} dx \tag{4.11.12}$$

$$= 48\pi \left(\frac{kT}{ch}\right)^3 kT \sum_{m=1}^\infty \frac{1}{m^4} \tag{4.11.13}$$

$$= 48\pi \left(\frac{kT}{ch}\right)^3 kT \frac{\pi^4}{90} = 48\pi \left(\frac{kT}{ch}\right)^3 kT \times 1.082\,323. \tag{4.11.14}$$

From experiment, it has been found that a black body (a cavity containing electromagnetic radiation in equilibrium with the cavity walls at temperature T) radiates energy through a hole in the cavity wall at the rate

$$F = \sigma T^4, \tag{4.11.15}$$

where σ is the Stefan–Boltzmann constant. According to the 1998 CODATA set of recommended values (see Mohr & Taylor, 2002),

$$\sigma = 5.6704 \times 10^{-5} \text{ erg s}^{-1} \text{ cm}^{-2} \text{ K}^{-4}. \tag{4.11.16}$$

The energy density in the electromagnetic field can be written from eq. (4.11.14) as

$$\bar{U}_{\text{ph}} = aT^4, \tag{4.11.17}$$

where

$$a = \frac{48\pi^5}{90} \left(\frac{k}{ch}\right)^3 k = \frac{8\pi^5}{15} \left(\frac{k}{ch}\right)^3 k, \tag{4.11.18}$$

and it is easily shown that the flux of energy in any direction is

$$F \, (= \sigma T^4) = \frac{c}{4} U_{\text{ph}} \left(= \frac{c}{4} a T^4\right). \tag{4.11.19}$$

Hence,

$$a = \frac{8\pi^5}{15} \left(\frac{k}{ch}\right)^3 k = \frac{4}{c}\sigma = 7.5658 \times 10^{-15} \text{ erg cm}^{-3} \text{ K}^{-4}. \tag{4.11.20}$$

Equation (4.11.20) provides an important relationship between the fundamental constants k, c, and h.

From eqs. (4.11.9) and (4.11.14), it follows that the average energy of a photon is

$$\langle h\nu \rangle = \frac{\bar{U}_{\text{ph}}}{n_{\text{ph}}} = \frac{\left(\sum_{m=1}^{\infty} \frac{1}{m^4}\right)}{\left(\sum_{m=1}^{\infty} \frac{1}{m^3}\right)} \, 3kT = 2.701\,362\,kT. \tag{4.11.21}$$

This says that the average photon has roughly twice as much energy as a particle with mass. However, since the number of photons increases with temperature, this comparison is not particularly illuminating.

Since, for photons, $vp = cp = h\nu$, it follows from eqs. (4.3.2) and (4.11.17) that the pressure exerted by radiation is

$$P = P_{\text{rad}} = \frac{1}{3}\langle h\nu \rangle n_{\text{ph}} = \frac{1}{3}\bar{U}_{\text{ph}} = \frac{1}{3}a T^4. \tag{4.11.22}$$

It is instructive to differentiate the integrand of eq. (4.11.10) to determine where the maximum in the energy density distribution with regard to photon frequency occurs. The result is

$$3\left[1 - \exp\left(-\frac{h\nu}{kT}\right)\right] = \frac{h\nu}{kT}. \tag{4.11.23}$$

The most important lesson of this equation is that the value of $h\nu/kT$ at maximum $dU_\nu/d\nu$ is a universal constant, independent of the temperature. The non-trivial solution of eq. (4.11.23) Is

$$\left(\frac{h\nu}{kT}\right)_{\text{at max } dU_\nu/d\nu} = 2.821\,439. \tag{4.11.24}$$

Converting the energy distribution with regard to frequency into a distribution with regard to wavelength (using $\lambda \nu = c$), one finds that the maximum in the latter distribution, $dU_\lambda/d\lambda$, occurs when

$$5\left[1 - \exp\left(-\frac{hc}{\lambda kT}\right)\right] = \frac{hc}{\lambda kT},$$
(4.11.25)

which has the solution

$$\left(\frac{hc}{\lambda kT}\right)_{\text{at max } dU_\lambda/d\lambda} = 4.965\,114.$$
(4.11.26)

This equation can be written as

$$\sigma_{\lambda T} = (\lambda T)_{\text{at max } dU_\lambda/d\lambda} = \frac{1}{4.965\,114}\frac{hc}{k} = 0.289\,782 \text{ cm K},$$
(4.11.27)

where the constant on the right-most side of eq. (4.11.27) is determined empirically, setting another important relationship between the fundamental constants h, c, and k.

On comparing eqs. (4.11.20) and (4.11.27), it is clear that Boltzmann's constant k can be determined directly from experiments with black body radiators. That is,

$$k = \frac{15}{2\pi^5}\left(\frac{ch}{k}\right)^3\frac{\sigma}{c} = \frac{15}{2\pi^5}(4.965\,114\,\sigma_{\lambda T})^3\frac{\sigma}{c} = 2.999\,850\,\sigma_{\lambda T}^3\frac{\sigma}{c}.$$
(4.11.28)

Using the value of σ from eq. (4.11.16), the value of $\sigma_{\lambda T}$ from eq. (4.11.27), and $c = 2.997\,930 \times 10^{10}$ cm s^{-1} in eq. (4.11.28) gives

$$k = 1.3803 \times 10^{-16} \text{ erg K}^{-1}.$$
(4.11.29)

4.12 Maxwell–Boltzmann statistics and entropy

For complicated systems of Fermi–Dirac particles of a multiplicity of types, when the probability of any state being occupied (eq. (4.2.12)) is small, it is practical to use Maxwell–Boltzmann statistics (J. C. Maxwell, 1860; L. Boltzmann, 1872), which gives essentially the same end result as Fermi–Dirac statistics, but is simpler to work with. When there is no restriction on the number of particles in a state, a logical extension of eq. (4.2.1) is

$$g_i\ g_i\ g_i \cdots g_i = g_i^{N_i},$$
(4.12.1)

giving

$$P_i \propto \frac{g_i^{N_i}}{N_i!}$$
(4.12.2)

as a logical extension of eq. (4.2.2). In this way of counting, highest weight has been given to the *a priori* probability of occupying any state within a cell g_i, rather than to the details of how the particles are distributed among the states within the cells.

By the theory of permutations and combinations, one has that

$$\sum_{i=1}^{k} \frac{g_1^{N_1}}{N_1!} \frac{g_2^{N_2}}{N_2!} \cdots \frac{g_k^{N_k}}{N_k!} = \frac{1}{N!}(g_1 + g_2 + \cdots + g_k)^N, \tag{4.12.3}$$

where the sum is over all values of $N_1, N_2, \ldots, N_k$ such that $\sum_{i=1}^{k} N_i = N$. If one normalizes the g_is so that the total a priori probability is unity, then

$$\sum_{i=1}^{k} g_i = 1, \text{ and} \tag{4.12.4}$$

$$P = \prod_{i=1}^{k} P_i = N! \prod_{i=1}^{k} \frac{g_i^{N_i}}{N_i!} \tag{4.12.5}$$

is the probability for realizing a given distribution.

Using the approximation given by eq. (4.2.9) for the logarithm of a factorial, one obtains

$$\log_e P \sim \sum_{i} \left[N_i \log_e g_i - N_i(\log_e N_i - 1) \right] = \sum_{i} N_i \log_e \left(\frac{g_i}{N_i} \right) + \sum_{i} N_i \tag{4.12.6}$$

and

$$\delta \log_e P = \sum_{i} \delta N_i \log_e \left(\frac{g_i}{N_i} \right). \tag{4.12.7}$$

Maximizing the total probability subject to the constraints expressed in eqs. (4.2.4) and (4.2.5) gives

$$\sum_{i} \delta N_i \left[\log_e \left(\frac{g_i}{N_i} \right) - (\alpha + \beta \epsilon_i) \right] = 0, \tag{4.12.8}$$

which has the solutions

$$\frac{g_i}{N_i} = \exp(\alpha + \beta \epsilon_i), \tag{4.12.9}$$

or

$$N_i = \frac{g_i}{\exp(\alpha + \beta \epsilon_i)}, \tag{4.12.10}$$

for all i. Equation (4.12.10) describes the Maxwell–Boltzmann, or classical distribution function. Equations (4.12.9) and (4.12.10) are identical to eqs. (4.2.13) and (4.2.14) when the occupation numbers are very small compared to unity and the Fermi–Dirac occupation number, eq. (4.2.15), reduces to the Maxwell–Boltzmann occupation number,

$$f_i = \frac{1}{\exp(\alpha + \beta \epsilon_i)}. \tag{4.12.11}$$

Just as in the case of Fermi–Dirac statistics, as described at the end of Section 4.2, the Maxwell–Boltzmann distribution can also be obtained by considering the consequences of detailed balance. In this case, one may set

$$R_{1,2 \to 1',2'} = Cf(\epsilon_1)f(\epsilon_2), \tag{4.12.12}$$

and

$$R_{1,2 \leftarrow 1',2'} = Cf(\epsilon_1')f(\epsilon_2'), \tag{4.12.13}$$

where $f(\epsilon)$ is the probability that a state at energy ϵ is occupied and $\epsilon_1 + \epsilon_2 = \epsilon_1' + \epsilon_2'$. As no restriction has been placed on the number of particles in a state, the factors $1 - f(\epsilon_i)$ do not enter the rate equations.

As before, one sets $R_{1,2 \leftarrow 1',2'} = R_{1,2 \to 1',2'}$, $\epsilon_1' = \epsilon_1 + x$, and $\epsilon_2' = \epsilon_2 - x$ to obtain

$$f(\epsilon_1)f(\epsilon_2) = f(\epsilon_1 + x)f(\epsilon_2 - x). \tag{4.12.14}$$

Expanding about $x = 0$ gives

$$f(\epsilon_1)f(\epsilon_2) = f(\epsilon_1)f(\epsilon_2)\left(1 + \left[\left(\frac{\partial \log_e f}{\partial \epsilon}\right)_{\epsilon=\epsilon_1} - \left(\frac{\partial \log_e f}{\partial \epsilon}\right)_{\epsilon=\epsilon_2}\right]x\right)$$
$$+ O(x^2). \tag{4.12.15}$$

This is true for all x only if

$$\frac{\partial \log_e f(\epsilon)}{\partial \epsilon} = -\beta. \tag{4.12.16}$$

Integration of eq. (4.12.16) leads to eq. (4.12.11) again.

It is interesting to explore the content of the probability function, eq. (4.12.5), in the light of the result obtained by maximizing this function subject to the constraints of constant total particle number, constant total energy, and constant volume. Inserting eq. (4.12.9) into eq. (4.12.7) gives

$$\delta \log_e P_{\max} = \sum_i \delta N_i (\alpha + \beta \epsilon_i) = \beta \sum_i \epsilon_i \delta N_i, \tag{4.12.17}$$

where $P_{\max}$ is the probabilty of the most probable distribution and use has been made of the fact that $\sum_i \delta N_i = 0$. From eqs. (4.2.38) and (4.2.26), it then follows that

$$k\,\delta \log_e P_{\max} = \frac{\delta Q}{T}, \tag{4.12.18}$$

which demonstrates that the quantity $\delta Q/T$ is a perfect differential. From thermodynamics, of course, $\delta Q/T$ is known as the differential of the entropy S, and eq. (4.12.18) shows that, to within an arbitrary constant,

$$S = k\,\log_e P_{\max}. \tag{4.12.19}$$

Equations (4.12.17)–(4.12.19) are also rigorously true for Fermi–Dirac statistics. Thus, the entropy of a system of particles at a specific temperature and density is a measure of the most probable distribution of particles among energy states available at that temperature and density.

Next, one may rewrite eq. (4.12.6) as

$$\log_e P = \sum_i g_i \exp(-\alpha - \beta\epsilon_i)(\alpha + \beta\epsilon_i), \tag{4.12.20}$$

where the constant $\sum_i N_i = N$ has been omitted. Making the transition to the continuum, and selecting, for definiteness, a non-degenerate gas of neutral hydrogen atoms, it follows that

$$S_H = k \int_0^\infty V \frac{4\pi}{h^3} p^2 dp \, e^{-\alpha} \exp\left(-\frac{p^2}{2M_H kT}\right)\left(\alpha + \frac{p^2}{2M_H kT}\right). \tag{4.12.21}$$

Converting to the dimensionless variable $x = \sqrt{p^2/2M_H kT}$ and using the results of integrations appearing and evaluated in eqs. (4.4.2)–(4.4.11), one obtains

$$S_H = kV \left(\frac{2\pi M_H kT}{h^2}\right)^{3/2} e^{-\alpha}\left(\alpha + \frac{3}{2}\right). \tag{4.12.22}$$

In eq. (4.12.22), the expression between k and the quantity in parentheses containing α is just the number of hydrogen atoms N_H in the volume V. To see this, in eq. (4.4.10) set $n = N_H/V$, and, since hydrogen atoms in the ground state have zero spin, drop the factor of 2. Thus,

$$n_H = \frac{N_H}{V} = \left(\frac{2\pi M_H kT}{h^2}\right)^{3/2} e^{-\alpha}, \tag{4.12.23}$$

so that

$$S_H = kN_H \log_e\left[\frac{1}{n_H}\left(\frac{2\pi M_H kT}{h^2}\right)^{3/2}\right] + \frac{3}{2}kN_H. \tag{4.12.24}$$

Since only derivatives of entropy enter into useful thermodynamic relationships, one could renormalize eq. (4.12.24) in such a way that

$$S_H = kN_H \log_e\left(\frac{T^{3/2}}{\rho}\right), \tag{4.12.25}$$

but, for comparison with the entropies of other types of particle (see Section 5.4), it is prudent to maintain the more elaborate expression in eq. (4.12.24). From either equation, it follows that

$$T \, dS_H = kN_H\left(\frac{3}{2}dT - T\frac{d\rho}{\rho}\right) = d\left(N_H \frac{3}{2}kT\right) + \frac{N_H}{V} kT \, dV = dE + P dV. \tag{4.12.26}$$

This marvelous result demonstrates exquisitely the deep concordance between statistical mechanics and thermodynamics.

4.13   Ionization equilibrium and the Saha equation for pure hydrogen

In the outer layers of main sequence stars and giants, hydrogen and helium are not completely ionized and it is necessary to take this into account in determining an appropriate equation of state. In classical pulsational variables such as Cepheids and RR Lyrae stars, it is the thermodynamic and opacity properties of the helium ionization zone (and, to some extent, also those of the hydrogen ionization zone) which are responsible for driving the pulsation. In addition, the opacity of stellar material can be strongly influenced by transitions from bound atomic levels to the continuum and the reverse, as well as by transitions between bound levels in regions where the lighter elements are fully ionized, but heavy ones such as abundant iron are not. Thus, it is important to be able to calculate the probabilities with which various atomic configurations occur for a wide variety of conditions. In this section, in order to highlight the major principles involved, the ionization state of pure hydrogen is examined as a function of density and temperature, and it is shown that a detailed consideration of the excitation state of the hydrogen atom and a consideration of Coulomb interactions between atoms, ions, and electrons are not necessary to construct a reasonably adequate equation of state.

Consider a gas consisting of N_e free electrons, N_f free protons, and N_b hydrogen atoms in thermal equilibrium in a box of volume V. In the Maxwell–Boltzmann approximation, one maximizes the probability

$$P(N_e, N_f, N_b) \propto \prod_{i=1}^{\infty} \frac{g_{ei}^{N_{ei}}}{N_{ei}!} \prod_{j=1}^{\infty} \frac{g_{fj}^{N_{fj}}}{N_{fj}!} \prod_{k=1}^{\infty} \frac{g_{bk}^{N_{bk}}}{N_{bk}!}, \tag{4.13.1}$$

subject to the restrictions that

$$N_{electrons} = \sum_i N_{ei} + \sum_k N_{bk} = \text{constant}, \tag{4.13.2}$$

$$N_{protons} = \sum_j N_{fj} + \sum_k N_{bk} = \text{constant}, \tag{4.13.3}$$

and

$$E = \sum_i N_{ei}\epsilon_{ei} + \sum_j N_{fj}\epsilon_{fj} + \sum_k N_{bk}(\epsilon_{bk} - \chi) = \text{constant}. \tag{4.13.4}$$

Here, $N_{electrons}$ is the total number of electrons, regardless of whether the electrons are free or in an atom, and $N_{protons}$ is the total number of protons, regardless of whether the protons are free or in an atom. In eq. (4.13.4), χ is the ionization energy of an isolated hydrogen atom in free space and ϵ_{bk} consists of the kinetic energy of the atom plus the excitation energy of the electron in the atom. Ultimately, integration over the continuous kinetic energy distribution and a summation over bound states is performed.

Proceeding with the method of Lagrange multipliers, as in Section 4.2, eq. (4.13.1) is differentiated with respect to the N_i. From this is subtracted the result of differentiating eqs. (4.13.2)–(4.13.4) with respect to the N_i and multiplying, respectively, by α_e (eq. (4.13.2)), α_f (eq. (4.13.3)), and β (eq. (4.13.4)). Setting the total sum equal to zero, one has

$$\sum_i \delta N_{ei}[\log_e(N_{ei}/g_{ei}) + \beta\epsilon_{ei} + \alpha_e] + \sum_j \delta N_{fj}[\log_e(N_{fj}/g_{fj}) + \beta\epsilon_{fj} + \alpha_f]$$

$$+ \sum_k \delta N_{bk}[\log_e(N_{bk}/g_{bk}) + \beta(\epsilon_{bk} - \chi) + \alpha_e + \alpha_f] = 0, \tag{4.13.5}$$

with the solutions

$$N_{ei} = g_{ei}\exp(-\alpha_e - \beta\epsilon_{ei}), \tag{4.13.6}$$

$$N_{fj} = g_{fj}\exp(-\alpha_f - \beta\epsilon_{fj}), \text{ and} \tag{4.13.7}$$

$$N_{bk} = g_{bk}\exp[-\alpha_e - \alpha_f - \beta(\epsilon_{bk} - \chi)]. \tag{4.13.8}$$

Thus,

$$N_e = \sum_i N_{ei} = e^{-\alpha_e}\sum_i g_{ei}e^{-\beta\epsilon_{ei}}, \tag{4.13.9}$$

$$N_f = \sum_j N_{fj} = e^{-\alpha_f}\sum_j g_{fj}e^{-\beta\epsilon_{fj}}, \text{ and} \tag{4.13.10}$$

$$N_b = \sum_k N_{bk} = e^{-\alpha_e - \alpha_f}e^{\beta\chi}\sum_k g_{bk}e^{-\beta\epsilon_{bk}}. \tag{4.13.11}$$

The αs may be eliminated from the last three equations to give

$$\frac{N_e N_f}{N_b} = \frac{\exp(-\beta\chi)}{\sum_k g_{bk}\exp(-\beta\epsilon_{bk})}\sum_i g_{ei}\exp(-\beta\epsilon_{ei})\sum_j g_{fj}\exp(-\beta\epsilon_{fj}). \tag{4.13.12}$$

The same result could have been obtained by replacing the condition expressed by either eq. (4.13.2) or eq. (4.13.3) with the condition that the number of free electrons equals the number of free protons ($\sum_i N_{ei} - \sum_j N_{fj} = 0$).

In the non-relativistic approximation,

$$\sum_i g_{ei}e^{-\beta\epsilon_{ei}} \rightarrow 2V\frac{4\pi}{h^3}\int_0^\infty \exp\left(-\frac{p^2}{2m_e kT}\right)p^2 dp = 2V\left(\frac{2\pi m_e kT}{h^2}\right)^{3/2}, \tag{4.13.13}$$

$$\sum_j g_{fj}e^{-\beta\epsilon_{fj}} \rightarrow 2V\frac{4\pi}{h^3}\int_0^\infty \exp\left(-\frac{p^2}{2m_f kT}\right)p^2 dp = 2V\left(\frac{2\pi m_f kT}{h^2}\right)^{3/2}, \tag{4.13.14}$$

where the factor of 2 after each equal sign takes into account the two spin states of the free particles. For the bound atoms,

$$\sum_k g_{bk} e^{-\beta \epsilon_{bk}} \rightarrow V \frac{4\pi}{h^3} \int_0^\infty \exp(-p^2/2m_b kT) p^2 dp \sum_k g_{bk}{}' \exp(-\epsilon_{bk}{}'/kT)$$

$$= V \left(\frac{2\pi m_b kT}{h^2}\right)^{3/2} \sum_k g_{bk}{}' \exp(-\epsilon_{bk}{}'/kT), \tag{4.13.15}$$

where the integral takes into account the motion of the bound atom as a whole, $g_{bk}{}'$ is the statistical weight of a bound atomic state, and $\epsilon_{bk}{}'$ is the energy of this state above the ground level.

Using eqs. (4.13.13)–(4.13.15), and neglecting the difference between m_f and m_b, eq. (4.13.12) becomes

$$\frac{N_e N_f}{N_b} = 4V \left(\frac{2\pi m_e kT}{h^2}\right)^{3/2} \frac{\exp(-\chi/kT)}{\sum_k g_{bk}{}' \exp(-\epsilon_{bk}{}'/kT)}. \tag{4.13.16}$$

This equation, which gives the ratio of ions (of any species) in two adjacent states of ionization as a function of the temperature and the electron density ($n_e = N_e/V$) is one form of what is called the Saha equation in honor of M. N. Saha, who was the first to recognize that the presence and strengths of the spectral lines associated with any given element are more an indication of the temperature than of the abundance of that element (Saha, 1920; 1921).

When atoms are far apart, so that many bound levels exist, the energy of states above the ground level are given, in first approximation, by

$$\epsilon_{bk}{}' = \chi \left(1 - \frac{1}{k^2}\right), \tag{4.13.17}$$

where k is an integer, and the statistical weight of the kth state is given by

$$g_{bk}{}' = 4k^2. \tag{4.13.18}$$

The factor of 4 in eq. (4.13.18) comes from the fact that there are four electron–proton spin states, one of total spin zero (statistical weight = 1) and three of spin 1 (statistical weight 3). Thus, in first approximation,

$$\frac{N_e N_f}{N_b} = V \exp\left(-\frac{\chi}{kT}\right) \left(\frac{2\pi m_e kT}{h^2}\right)^{3/2} \frac{1}{\Sigma}, \tag{4.13.19}$$

where

$$\Sigma = 1 + \exp\left(-\frac{\chi}{kT}\right) \sum_{k=2}^{k_{max}} k^2 \exp\left(\frac{\chi}{kT} \frac{1}{k^2}\right). \tag{4.13.20}$$

The sum in eq. (4.13.20), known as a partition function, is terminated at some maximum value $k = k_{max}$ determined by insisting that the spatial extent of the corresponding wave

function be smaller than the average separation between particles. The same considera-
tion means that the effective ionization potential is reduced to $\chi' = \left(1 - 1/k_{\max}^2\right)\chi$. This
nicety will, for the moment, be ignored.

Defining number densities by $n_e = N_e/V$, $n_f = N_f/V$, and $n_b = N_b/V$, eq. (4.13.19) can
be written as

$$\frac{n_e n_f}{n_b} = \exp\left(-\frac{\chi}{kT}\right) \left(\frac{2\pi m_e kT}{h^2}\right)^{3/2} \frac{1}{\Sigma}. \tag{4.13.19a}$$

The number density of protons, whether the proton is free or in an atom, is $\sim \rho/M_H$, so
one can write

$$n_f = n_e = c\frac{\rho}{M_H}, \tag{4.13.21}$$

$$n_b = (1 - c)\frac{\rho}{M_H}, \tag{4.13.22}$$

where c is the degree of ionization. Combining eqs. (4.13.19a)–(4.13.22) gives

$$\frac{c^2}{1 - c} = \frac{M_H}{\rho}\left(\frac{2\pi m_e kT}{h^2}\right)^{3/2} \exp\left(-\frac{\chi}{kT}\right) \frac{1}{\Sigma} = \frac{\Lambda_0}{\Sigma} = \Lambda, \tag{4.13.23}$$

where

$$\Lambda_0 = \frac{M_H}{\rho}\left(\frac{2\pi m_e kT}{h^2}\right)^{3/2} \exp\left(-\frac{\chi}{kT}\right) \tag{4.13.24}$$

$$= 4.04\frac{T_6^{3/2}}{\rho_0} \exp\left(-\frac{0.158}{T_6}\right). \tag{4.13.25}$$

The solution of eq. (4.13.23) is

$$c = \frac{\Lambda}{2}\left(\sqrt{1 + \frac{4}{\Lambda}} - 1\right), \tag{4.13.26}$$

showing that the degree of ionization increases monotonically with increasing Λ. For small
Λ ($\ll 4$),

$$c \sim \sqrt{\Lambda}, \tag{4.13.27}$$

and, for large Λ ($\gg 4$),

$$c \sim 1 - \frac{1}{\Lambda}. \tag{4.13.28}$$

At constant density, the degree of ionization increases with increasing temperature because
the density of ionizing photons increases with increasing temperature. At constant temperature, the degree of ionization increases with decreasing density because the number of
continuum levels per proton increases with decreasing density. Calling c_0 the degree of
ionization when Λ is approximated by Λ_0 (set $\Sigma = 1$ in eq. (4.13.23)), it is evident from

Table 4.13.1 Ionization properties of pure hydrogen under solar-like conditions

ρ_0	T_6	χ/kT	Λ_0	c_0	k_{max}	Σ	c	R_D/a_0	χ_D/χ
1.0	3.447	0.0458	24.63	0.962	1	1.00	0.962	1.00	0.112
0.1	1.600	0.0988	73.86	0.987	1	1.00	0.987	4.76	0.580
1.0/−2	0.743	0.213	208.5	0.995	2	4.41	0.980	10.4	0.807
1.0/−3	0.345	0.458	515.5	0.998	3	9.82	0.982	22.3	0.910
1.0/−4	0.160	0.988	960.4	0.999	4	13.0	0.987	47.9	0.953
1.0/−5	0.0743	2.128	971.3	0.999	7	19.2	0.981	103.5	0.981
1.0/−6	0.0345	4.584	263.4	0.996	10	5.42	0.980	223.2	0.991
1.0/−7	0.0160	9.875	4.194	0.834	15	1.07	0.826	523.7	0.996
3.3/−8	1.11/−2	14.29	8.82/−2	0.256	18	1.0016	0.256	1362	0.999
1.0/−8	7.48/−3	21.13	1.70/−4	0.013	22	1.00	0.013	9025	1.000

eq. (4.13.26) that $c < c_0$. The obvious reason for this is that, the larger the number of bound levels that are taken into account, the larger is the number of atoms which are not ionized.

As an example, suppose that

$$\frac{\rho_0}{T_6^3} = \frac{100}{(16)^3}, \tag{4.13.29}$$

which approximates the situation over much of the interior of the Sun. Then,

$$\Lambda_0 \sim \frac{165}{T_6^{3/2} \exp(0.158/T_6)} \tag{4.13.30}$$

and c_0 as given by setting $\Lambda = \Lambda_0$ in eq. (4.13.26) is a first approximation to the degree of ionization. The values of Λ_0 and c_0 for several values of density and temperature related by eq. (4.13.29) are given in Table 4.13.1. Densities are in units of g cm^{-3} and temperatures are in units of 10^6 K; an entry $x.y/-z$ is to be interpreted as $x.y$ times 10^{-z}. At densities less than unity and at temperatures larger than 35 000 K, hydrogen is highly ionized. Not until densities decrease below 10^{-6} g cm^{-3} and temperatures drop below $\sim 2 \times 10^4$ K do neutral atoms become prominent.

Entries in columns 6–10 in Table 4.13.1 describe how the fact that ions and electrons are not isolated from one another affects the degree of ionization. From eq. (3.2.1), the average separation between nucleons in the case of pure hydrogen is of the order of

$$r_{sep} \sim \frac{2.24 \, a_0}{\rho_0^{1/3}}, \tag{4.13.31}$$

where a_0 is the Bohr radius of the free hydrogen atom. If one assumes that, when $r_{sep} < 2 a_0$, all electron levels are in the continuum, hydrogen is completely ionized for $\rho > 1.4$ g cm^{-3}, the mean density of the Sun, independent of the temperature. Thus, over most of the mass of the Sun, hydrogen is completely ionized.

The Bohr radius of the kth level of the free hydrogen atom is

$$a_k = k^2 a_0. \tag{4.13.32}$$

For illustrative purposes, suppose that, for densities less than 1.4 g cm^{-3}, all bound levels for which $2\,a_k < r_{\text{sep}}$ exist. This condition determines

$$k_{\text{max}} < \left(\frac{1.4}{\rho_0}\right)^{1/6}. \tag{4.13.33}$$

Values of k_{max}, Σ, and the degree of ionization c as calculated using $\Lambda = \Lambda_0/\Sigma$ in eq. (4.13.26) are shown, respectively, in the sixth through eighth columns of Table (4.13.1).

As one moves outward from the largest temperatures and densities (smaller than $\rho = 1.4$ g cm^{-3}) to smaller temperatures and densities, although the number of bound levels increases, the partition function does not change monotonically, but, beginning and ending at unity, passes through a maximum. The upshot is that, both at the largest and smallest densities and temperatures in the chosen range, the degree of ionization c_0 calculated by approximating Σ by 1 is very close to the value obtained by using a value of Σ which takes excited bound states into account. However, at intermediate temperatures and densities, taking excited states into account reduces the estimate of the degree of ionization by up to 1.6%. The fraction of un-ionized hydrogen atoms remains typically at about 1–2% throughout the region of intermediate temperatures and densities. The increase in the fraction of un-ionized atoms to $\sim 4\%$ at $\rho \sim 1$ g cm^{-3} is a considerable overestimate, given the fact that the fraction rigorously goes to zero at $\rho \geq 1.4$ g cm^{-3}. The appropriate action to take is to devise an interpolation scheme to repace the discontinuous, step-wise change Σ with a continuous one.

Analytical estimates of the order of magnitude of Σ are possible in two limiting situations. From eq. (4.13.20), it follows that

$$1 + \exp\left(-\frac{\chi}{kT}\right)\sum_{k=2}^{k_{\text{max}}} k^2 < \Sigma < 1 + \exp\left(-\frac{3}{4}\frac{\chi}{kT}\right)\sum_{k=2}^{k_{\text{max}}} k^2. \tag{4.13.34}$$

Approximating the sums by integrals (when k_{max} is larger than 3), one has

$$\Sigma = 1 + \exp\left(-\sigma\frac{\chi}{kT}\right)\frac{k_{\text{max}}^3 - 8}{3}, \tag{4.13.35}$$

where $3/4 < \sigma < 1$. For small values of χ/kT (such that $\exp(-\sigma\chi/kT) \sim 1$, independent of σ) and large values of k_{max},

$$\Sigma \sim 1 + \frac{k_{\text{max}}^3 - 8}{3}. \tag{4.13.36}$$

For sufficiently large values of χ/kT, $\exp[\sigma(\chi/kT)] \gg (k_{\text{max}}^3 - 8)/3$ and $\Sigma \sim 1$.

As noted earlier, the presence of surrounding ions and electrons decreases the effective ionization potential. In the simplistic model which envisions a reduction in the number of possible bound levels due to the proximity of surrounding particles, the ionization potential decreases from χ to

$$\chi' \sim \chi \left(1 - \frac{1}{(k_{max} + 1)^2} \right). \tag{4.13.37}$$

As is evident from eq. (4.13.24), the effect is to increase the value of Λ_0 and to thereby increase the degree of ionization. For conditions examined in Table 4.13.1, the effect on the degree of ionization is less than a fraction of a percent, even at the largest density.

Another way to estimate the amount by which the ionization energy is reduced due to screening by free electrons and protons is to adopt classical Debye–Hückel theory (see Section 4.17), which estimates the average electrical potential about an ion due to the presence of other charged particles, and to then solve for the ground state of the screened hydrogen atom. As follows from eqs. (4.17.14) and (4.17.10), the electrostatic potential about a proton in a pure hydrogen mixture is

$$\phi(r) = \frac{e}{r} \exp\left(-\frac{r}{R_D}\right), \tag{4.13.38}$$

where the Debye radius R_D is given by

$$\frac{1}{R_D^2} = 8\pi \frac{e^2}{kT} n_e. \tag{4.13.39}$$

In terms of the Bohr radius of the ground state of the isolated hydrogen atom, this translates into

$$\frac{a_0}{R_D} = 0.840 \sqrt{\frac{\rho c}{T_6}}, \tag{4.13.40}$$

where c is the degree of ionization.

The next step is to look for solutions of the Schrödinger equation for the screened atom:

$$H\psi = \left[-\frac{\hbar^2}{2m_e}\nabla^2 - \frac{e^2}{r}\exp\left(-\frac{r}{R_D}\right) \right]\psi = E\psi. \tag{4.13.41}$$

One could solve this equation numerically, but more insight can be achieved by adopting a trial function

$$\psi_\alpha(r) = \frac{1}{(\pi\alpha^3)^{1/2}} \exp\left(-\frac{r}{\alpha}\right) \tag{4.13.42}$$

and minimizing the integral

$$E(\alpha) = \langle \psi_\alpha | H | \psi_\alpha \rangle = \frac{1}{\pi\alpha^3} \int_0^\infty \exp\left(-\frac{r}{a}\right)\left[H \exp\left(-\frac{r}{a}\right) \right] 4\pi r^2 dr \tag{4.13.43}$$

with respect to α. The result is

$$\frac{\alpha_R}{a_0}\left(1 + \frac{3\alpha_R}{2R_D} \right) = \left(1 + \frac{\alpha_R}{2R_D} \right)^3, \tag{4.13.44}$$

where α_R is the value of α at the minimum, and

$$E(\alpha_R) = -\frac{e^2}{2\alpha_R}\left(1 - \frac{\alpha_R}{2R_D}\right)\left(1 + \frac{\alpha_R}{2R_D}\right)^{-3} \tag{4.13.45}$$

is the minimum energy.

For small α_R/R_D,

$$\alpha_R \to a_0 \tag{4.13.46}$$

and

$$E(\alpha_R) \to -\frac{e^2}{2a_0} + \frac{e^2}{R_D}. \tag{4.13.47}$$

Thus, for $R_D \gg a_0$, the ionization energy in the Debye–Hückel approximation is

$$\chi_D = -E(\alpha_R) \to \chi - \frac{e^2}{R_D} = \chi\left(1 - \frac{2a_0}{R_D}\right) \tag{4.13.48}$$

$$= \chi\left(1 - 1.68\sqrt{\frac{\rho c}{T_6}}\right). \tag{4.13.49}$$

Values of R_D and χ_D/χ are given, respectively, in the last two columns of Table 4.13.1. The effect on the degree of ionization is very small. In conclusion, a quite reasonable first approximation to the degree of ionization, insofar as it affects the equation of state, is given by setting Λ in eq. (4.13.26) equal to Λ_0 given by eq. (4.13.24).

Taking the effects of screening into account, the energy per gram of the gas becomes (setting $\rho V = 1$ g) approximately

$$E = \frac{1}{M_H}\left[(1 + c)\frac{3}{2}kT + (1 - c)\left(-\chi' + \sum_{k=2}^{k_{max}} g_{bk}'\epsilon_{bk}'/\Sigma\right)\right], \tag{4.13.50}$$

where χ' may be chosen from eq. (4.13.37) or eq. (4.13.47), g_{bk}' is given by eq. (4.13.18), and ϵ_{bk}' and Σ are given, respectively, by eqs. (4.13.17) and (4.13.20), with χ replaced by χ'.

4.14 Thermodynamic properties of partially ionized hydrogen

Since the complications introduced in Section 4.13 have only a minor effect on the degree of ionization and on the equation of state, it is worthwhile to explore the thermodynamic characteristics of a partially ionized pure hydrogen gas in the approximation that the partition function given by eq. (4.13.20) reduces to just the first term. Setting $\Sigma = 1$, eq. (4.13.23) simplifies to

$$\frac{c^2}{1 - c} = \frac{M_H}{\rho}\left(\frac{2\pi m_e kT}{h^2}\right)^{3/2}\exp\left(-\frac{\chi}{kT}\right). \tag{4.14.1}$$

For the pressure P, one has

$$P = (n_e + n_f + n_b) kT = (2n_e + n_b) kT \tag{4.14.2}$$

$$= \{2c + (1 - c)\} \frac{\rho}{M_H} kT = (1 + c) \frac{\rho}{M_H} kT, \tag{4.14.3}$$

and, for energy per gram, one has

$$U = \frac{1}{\rho} \left((n_e + n_f + n_b) \frac{3}{2} kT + n_f \chi \right) \tag{4.14.4}$$

$$= \frac{1}{M_H} \left((1 + c) \frac{3}{2} kT + c\chi \right). \tag{4.14.5}$$

Taking derivatives of eqs. (4.14.1) and (4.14.3) while holding the temperature constant, it follows that

$$\left(\frac{\partial \rho}{\partial P} \right)_T = \left[1 + \frac{c (1 - c)}{2} \right] \frac{\rho}{P} \tag{4.14.6}$$

and

$$\left(\frac{\partial c}{\partial P} \right)_T = -\frac{c (1 - c^2)}{2} \frac{1}{P}. \tag{4.14.7}$$

Similarly, holding the pressure constant, it follows that

$$\left(\frac{\partial \rho}{\partial T} \right)_P = -\left[1 + \frac{c (1 - c)}{2} \left(\frac{5}{2} + \frac{\chi}{kT} \right) \right] \frac{\rho}{T} \tag{4.14.8}$$

and

$$\left(\frac{\partial c}{\partial T} \right)_P = \frac{c (1 - c^2)}{2} \left(\frac{5}{2} + \frac{\chi}{kT} \right) \frac{1}{T}. \tag{4.14.9}$$

Using these relationships, one has that, for one gram of matter,

$$\delta Q = T \, \delta S = \delta U + P \delta V = \delta U - \frac{P}{\rho} \frac{\delta \rho}{\rho} \tag{4.14.10}$$

$$= \frac{kT}{M_H} (1 + c) \left[\frac{5}{2} + \frac{c (1 - c)}{2} \left(\frac{5}{2} + \frac{\chi}{kT} \right)^2 \right] \frac{\delta T}{T}$$

$$- \frac{kT}{M_H} (1 + c) \left[1 + \frac{c (1 - c)}{2} \left(\frac{5}{2} + \frac{\chi}{kT} \right) \right] \frac{\delta P}{P}. \tag{4.14.11}$$

Setting $\delta P = 0$ in eq. (4.14.10) shows that the specific heat per gram at constant pressure is

$$C_P = \frac{k}{M_H} (1 + c) \left[\frac{5}{2} + \frac{c (1 - c)}{2} \left(\frac{5}{2} + \frac{\chi}{kT} \right)^2 \right], \tag{4.14.12}$$

and setting $\delta Q = 0$ in the same equation shows that the logarithmic gradient of temperature with respect to pressure at constant entropy is

$$V_{ad} = \left(\frac{d\log T}{d\log P}\right)_{ad} = \frac{2 + c\,(1-c)\,(5/2 + \chi/kT)}{5 + c\,(1-c)\,(5/2 + \chi/kT)^2}. \qquad (4.14.13)$$

Recall from Section 3.4 (eq. (3.4.30)) that the criterion for stability against convection is $V_{rad} < V_{ad}$, where V_{rad} is the logarithmic gradient of temperature with respect to pressure if the energy flow is carried entirely by radiation (see eq. (3.4.28)). From eq. (4.14.13), it follows that, when hydrogen is mostly in the atomic state or is almost completely ionized, $V_{ad} = 0.4$ and that, in regions of partial ionization, V_{ad} drops below 0.4. In typical situations, $c \sim (1-c) \sim 1/2$ when $\chi/kT \sim 10$ and, then, $V_{ad} \sim 0.116$. Further, in regions of partial ionization, the bound atoms of hydrogen are highly excited. Although, as shown in Section 4.13, this high degree of excitation does not have an important effect on either the state of ionization or the equation of state, the presence of many different occupied levels means that photons of energies covering a wide range can be absorbed in bound–bound and bound–free transitions. This in turn means that the radiative opacity and the related V_{rad} are large. It is thus the combined effect of an unusually small V_{ad} and an unusually large V_{rad} that leads to the formation of convective regions in the hydrogen-rich, low-temperature envelopes of low mass main sequence stars and red giants.

In some contexts, it is useful to choose density and temperature rather than pressure and temperature as the independent variables. For example, in the standard mixing length treatment of convective flow, the specific heat at constant volume plays a role. Again manipulating eqs. (4.14.1) and (4.14.3), one finds that

$$\left(\frac{\partial c}{\partial T}\right)_{\rho} = \frac{c\,(1-c^2)}{2 + c\,(1-c)}\left(\frac{3}{2} + \frac{\chi}{kT}\right)\frac{1}{T}. \qquad (4.14.14)$$

Then, from eqs. (4.14.5) and (4.14.10), it follows that the specific heat per gram at constant volume is

$$C_V = \frac{k}{M_H}\,(1+c)\left[\frac{3}{2} + \frac{c\,(1-c)}{2 + c\,(1-c)}\left(\frac{3}{2} + \frac{\chi}{kT}\right)^2\right]. \qquad (4.14.15)$$

In stellar evolution calculations, in situations in which particle creation and destruction can occur, it proves convenient to introduce a creation–destruction potential $\bar{\mu}_i$ for each type of particle i which can be created or destroyed. As discussed in Sections 8.1 and 8.2, the gravothermal energy-generation rate ϵ_{grav} can be written as the sum of two terms:

$$\epsilon_{grav} = \epsilon_{gth} + \epsilon_{cdth}, \qquad (4.14.16)$$

where

$$\epsilon_{gth} = -\left(\frac{dU}{dt}\right)_{Y_i} + \frac{P}{\rho^2}\frac{d\rho}{dt} \qquad (4.14.17)$$

and

$$\epsilon_{\text{cdth}} = -\sum_i \bar{\mu}_i \frac{dY_i}{dt}. \tag{4.14.18}$$

In eq. (4.14.17), the derivative of E is taken with all of the particle abundances by number Y_i held constant and, in eq. (4.14.18), $\bar{\mu}_i$ is the creation–destruction potential for the ith particle.

It is instructive to anticipate the nature of the $\bar{\mu}_i$s in the current context where Y_e, Y_f, and Y_b are, respectively, the number abundances of free electrons, free protons, and hydrogen atoms. Using eqs. (4.14.3) and (4.14.4) in eq. (4.14.10), it follows that

$$\epsilon_{\text{grav}} = -\frac{dQ}{dt} = -\frac{kT}{M_H}(1+c)\frac{d}{dt}\left(\log_e \frac{T^{3/2}}{\rho}\right) - \frac{kT}{M_H}\left(\frac{3}{2} + \frac{\chi}{kT}\right)\frac{dc}{dt}, \tag{4.14.19}$$

Noting that $Y_f = 1 - Y_b = Y_e = c$, one may write eq. (4.14.19) as

$$\epsilon_{\text{grav}} = -N_A(Y_b + Y_f + Y_e)kT\frac{d}{dt}\log_e\left(\frac{T^{3/2}}{\rho}\right) - N_A\left(\frac{3}{2}kT + \chi\right)\frac{d(Y_b + Y_f + Y_e)}{dt}, \tag{4.14.20}$$

where $1/M_H$ has been replaced by Avogadro's number N_A to facilitate comparison with the development in Chapter 8. From the definitions in eqs. (4.14.16)–(4.14.18), it follows that

$$\bar{\mu}_b = \bar{\mu}_f = \bar{\mu}_e = +N_A\left(\frac{3}{2}kT + \chi\right). \tag{4.14.21}$$

One can also write

$$\epsilon_{\text{cdth}} = +N_A\left(\frac{3}{2}kT + \chi\right)\frac{dY_b}{dt} \tag{4.14.22}$$

and

$$\epsilon_{\text{cdth}} = -N_A\left(\frac{3}{2}kT + \chi\right)\frac{dY_e}{dt}. \tag{4.14.23}$$

One interpretation of eq. (4.14.22) is that, whenever a bound atom is added, the ionization energy χ is released and shared with all of the other particles as kinetic energy; at the same time an electron and a free proton are removed for a net decrease of one particle, the kinetic energy of which is also shared with all of the other particles. An interpretation of eq. (4.14.23) is that, whenever an electron is added, the ionization energy must be supplied by all of the other particles; at the same time, there is a net gain of one particle which acquires a kinetic energy at the expense of the kinetic energy of all of the other particles.

This example is instructive and worth recalling in Sections 8.1 and 8.2, but in evolutionary calculations, a more practical approach is to use, for example, eq. (4.14.11), which gives

$$
\epsilon_{\text{grav}} = \frac{kT}{M_{\text{H}}} (1 + c) \times \left\{ -\left[\frac{5}{2} + \frac{c(1-c)}{2} \left(\frac{5}{2} + \frac{\chi}{kT} \right)^2 \right] \frac{1}{T} \frac{dT}{dt} \right.
$$
$$
\left. + \left[1 + \frac{c(1-c)}{2} \left(\frac{5}{2} + \frac{\chi}{kT} \right) \right] \frac{1}{P} \frac{dP}{dt} \right\}. \tag{4.14.24}
$$

Equation (4.14.24) may also be written as

$$
\epsilon_{\text{grav}} = -C_P \frac{dT}{dt} + C_T \frac{dP}{dt}, \tag{4.14.25}
$$

where C_P is given by eq. (4.14.12) and

$$
C_T = \frac{kT}{M_{\text{H}}} (1 + c) \left[1 + \frac{c(1-c)}{2} \left(\frac{5}{2} + \frac{\chi}{kT} \right) \right] \frac{1}{P}. \tag{4.14.26}
$$

Another approach is to begin with eq. (4.14.19) and to find dc/dt in terms of derivatives of T and ρ as independent variables. From eq. (4.14.1),

$$
\left(\frac{\partial c}{\partial \rho} \right)_T = -\frac{c(1-c)}{2-c} = -\frac{c(1-c^2)}{2+c(1-c)}, \tag{4.14.27}
$$

and, coupling with eq. (4.14.14), one has that

$$
\delta c = \frac{c(1-c^2)}{2+c(1-c)} \left[-\frac{\delta \rho}{\rho} + \left(\frac{3}{2} + \frac{\chi}{kT} \right) \frac{\delta T}{T} \right], \tag{4.14.28}
$$

or

$$
\delta c = \frac{c(1-c^2)}{2+c(1-c)} \left[\delta \left(\log_e \frac{T^{3/2}}{\rho} \right) + \frac{\chi}{kT} \frac{\delta T}{T} \right]. \tag{4.14.29}
$$

Joining eq. (4.14.19) with

$$
\frac{dc}{dt} = \frac{c(1-c^2)}{2+c(1-c)} \left[\frac{d}{dt} \left(\log_e \frac{T^{3/2}}{\rho} \right) + \frac{\chi}{kT} \frac{1}{T} \frac{dT}{dt} \right], \tag{4.14.30}
$$

one has

$$
\epsilon_{\text{grav}} = -\frac{kT}{M_{\text{H}}} (1 + c) \left[\left(1 + g(c, T) \right) \frac{d}{dt} \left(\log_e \frac{T^{3/2}}{\rho} \right) + g(c, T) \frac{\chi}{kT} \frac{1}{T} \frac{dT}{dt} \right], \tag{4.14.31}
$$

where

$$g(c, T) = \frac{c\,(1-c)}{2 + c\,(1-c)}\left(\frac{3}{2} + \frac{\chi}{kT}\right). \tag{4.14.32}$$

4.15 The generalized Saha equations, with application to pure helium

The Saha equation for hydrogen (eq. (4.13.16)) can be easily generalized to different species of atoms in different stages of ionization. Let $N_{\mu\sigma(i\mu\sigma)}$ represent the number density of atoms of the μth kind of element in the σth stage of ionization and in the $(i\mu\sigma)$th state, and let $\chi_{\mu\sigma}$ be the energy required to ionize the atom σ times. The constraints are:

$$N_e - \sum_\mu \sum_\sigma \sum_{(i\mu\sigma)} \sigma N_{\mu\sigma(i\mu\sigma)} = 0, \tag{4.15.1}$$

relating the number of free electrons to the number of ionized species;

$$N_n = \sum_\mu \sum_\sigma \sum_{(i\mu\sigma)} N_{\mu\sigma(i\mu\sigma)} = \text{constant}, \tag{4.15.2}$$

expressing the fact that the total number of nuclei, whether in bound atoms or not, is constant; and

$$E = \sum_\mu \sum_\sigma \sum_{(i\mu\sigma)} \left[\epsilon_{\mu\sigma(i\mu\sigma)} + \chi_{\mu\sigma}\right] N_{\mu\sigma(i\mu\sigma)} + \sum_i \epsilon_{ei} N_{ei} = \text{constant}, \tag{4.15.3}$$

expressing the constancy of the energy. In eq. (4.15.3) energy is measured relative to the state of complete neutrality, rather than to the state of complete ionization as was done in treating pure hydrogen.

Adopting Maxwell–Boltzmann statistics, the probability for the $(i\mu\sigma)$th state (see eq. (4.13.1)) is proportional to

$$P_{\mu\sigma(i\mu\sigma)} \frac{g_{\mu\sigma(i\mu\sigma)}^{N_{\mu\sigma(i\mu\sigma)}}}{N_{\mu\sigma(i\mu\sigma)}!}. \tag{4.15.4}$$

The total probability is the product over all i, μ, and σ of all $P_{\mu\sigma(i\mu\sigma)}$ times the first product on the right hand side of eq. (14.13.1). Using the method of Lagrange multipliers, as in Section 4.13, maximizing the total probability, subject to the conditions expressed by eqs. (4.15.1)–(4.15.3), gives

$$N_{\mu\sigma(i\mu\sigma)} = g_{\mu\sigma(i\mu\sigma)} \exp(\sigma\alpha_e - \alpha_n) \exp[-\beta(\epsilon_{\mu\sigma(i\mu\sigma)} + \chi_{\mu\sigma})] \tag{4.15.5}$$

and

$$N_{ei} = g_{ei} \exp(-\alpha_e - \beta\epsilon_{ei}), \tag{4.15.6}$$

where α_e, α_n, and β are the multipliers. Summing up over $(i\mu\sigma)$, eq. (4.15.5) yields

$$N_{\mu\sigma} = B_{\mu\sigma} \exp(\sigma\alpha_e - \alpha_n - \beta\chi_{\mu\sigma}), \tag{4.15.7}$$

where $N_{\mu\sigma}$ is the number of atoms of the μth kind of element in the σth stage of ionization, and

$$B_{\mu\sigma} = \sum_{(i\mu\sigma)} g_{\mu\sigma(i\mu\sigma)} \exp[-\beta\epsilon_{\mu\sigma(i\mu\sigma)}]. \tag{4.15.8}$$

The partition functions $B_{\mu\sigma}$ are sometimes referred to as Zustand (German for status or condition) functions.

Summing over ei, eq. (4.15.6) yields

$$N_e = 2V \left(\frac{2\pi m_e kT}{h^2}\right)^{3/2} \exp(-\alpha_e). \tag{4.15.9}$$

Finally, combining eqs. (4.15.7) and (4.15.9) gives

$$\frac{n_{\mu\sigma} n_e}{n_{\mu(\sigma-1)}} = 2 \left(\frac{2\pi m_e kT}{h^2}\right)^{3/2} \frac{B_{\mu\sigma}}{B_{\mu(\sigma-1)}} \exp(-\beta\bar{\chi}_{\mu\sigma}), \tag{4.15.10}$$

where $n_{\mu\sigma} = N_{\mu\sigma}/V$, $n_e = N_e/V$, and $n_{\mu(\sigma-1)} = N_{\mu(\sigma-1)}/V$ are number densities and

$$\bar{\chi}_{\mu\sigma} = \chi_{\mu\sigma} - \chi_{\mu(\sigma-1)} \tag{4.15.11}$$

is the energy required to abstract one electron from an atom in the $(\sigma - 1)$th state of ionization.

As an example, apply eqs. (4.15.10) and (4.15.11) to a gas of pure helium. Let n_0, n_1, and n_2 be, respectively, the number densities of neutral, singly ionized, and doubly ionized helium. Then,

$$\frac{n_1 n_e}{n_0} = 2 \left(\frac{2\pi m_e kT}{h^2}\right)^{3/2} \frac{B_1}{B_0} \exp\left(-\frac{\chi_1}{kT}\right) \tag{4.15.12}$$

and

$$\frac{n_2 n_e}{n_1} = 2 \left(\frac{2\pi m_e kT}{h^2}\right)^{3/2} \frac{B_2}{B_1} \exp\left(-\frac{\chi_2}{kT}\right). \tag{4.15.13}$$

Given the experience with hydrogen in Section 4.13, it is reasonable to assume that, for the purpose of estimating the degree of ionization, one can neglect all excited states in the partition functions. Since it has spin zero, the helium nucleus (the doubly ionized atom) has a statistical weight of $g_2 = 1$. For the singly ionized atom with one electron in the ground state, $g_1 = 2$, and for the neutral atom with two electrons in the singlet (spin zero) ground state, $g_0 = 1$. Thus,

$$\frac{B_1}{B_0} \sim \frac{g_1}{g_0} = \frac{2}{1} \tag{4.15.14}$$

and

$$\frac{B_2}{B_1} \sim \frac{g_2}{g_1} = \frac{1}{2}. \tag{4.15.15}$$

Defining concentration factors c by

$$n_0 = c_0 \, n_{\mathrm{He}}, \quad n_1 = c_1 \, n_{\mathrm{He}}, \text{ and } n_2 = c_2 \, n_{\mathrm{He}}, \qquad (4.15.16)$$

where

$$n_{\mathrm{He}} = \frac{\rho}{M_{\mathrm{He}}} \qquad (4.15.17)$$

is the number density of helium nuclei regardless of whether or not they are in atoms, number conservation gives

$$c_0 + c_1 + c_2 = 1 \qquad (4.15.18)$$

and the electron density is given by

$$n_{\mathrm{e}} = (c_1 + 2 \, c_2) \, n_{\mathrm{He}}. \qquad (4.15.19)$$

Combining eqs. (4.15.12)–(4.15.19), one has

$$\frac{c_1}{c_0} \, (c_1 + 2 \, c_2) = 2 \left(\frac{2\pi m_e kT}{h^2} \right)^{3/2} 2 \, \exp\left(-\frac{\chi_1}{kT} \right) \frac{M_{\mathrm{He}}}{\rho}, \qquad (4.15.20)$$

and

$$\frac{c_2}{c_1} \, (c_1 + 2 \, c_2) = 2 \left(\frac{2\pi m_e kT}{h^2} \right)^{3/2} \frac{1}{2} \, \exp\left(-\frac{\chi_2}{kT} \right) \frac{M_{\mathrm{He}}}{\rho}. \qquad (4.15.21)$$

For any choice of ρ and T, the three equations (4.15.20), (4.15.21), and (4.15.18) can be solved for the concentration factors.

Making use of the fact that the electron pressure is given by

$$P_{\mathrm{e}} = \frac{k\rho T}{M_{\mathrm{He}}} \, (c_1 + 2c_2), \qquad (4.15.22)$$

eqs. (4.15.20) and (4.15.21) may be written as

$$c_1 = \frac{P_{\mathrm{e}} A_1}{P_{\mathrm{e}} A_1 + \left(P_{\mathrm{e}}^2 + A_1 A_2 \right)} \qquad (4.15.23)$$

and

$$c_2 = \frac{A_1 A_2}{A_1 A_2 + \left(P_{\mathrm{e}}^2 + A_1 P_{\mathrm{e}} \right)}, \qquad (4.15.24)$$

where

$$A_1 = 1.333 \, T^{5/2} \exp\left(-\frac{24.580 \text{ eV}}{kT} \right), \qquad (4.15.25)$$

and

$$A_2 = 0.3332 \, T^{5/2} \exp\left(-\frac{54.503 \text{ eV}}{kT} \right). \qquad (4.15.26)$$

When P_e is interpreted as the total electron pressure, rather than being restricted by eq. (4.15.22), eqs. (4.15.23) and (4.15.24) are actually valid for an arbitrary composition mixture.

For a gas of pure helium, however, it is instructive to use eq. (4.15.22) in combination with eqs. (4.15.23) and (4.15.24) to write

$$P_e = \frac{k\rho T}{M_{He}} \frac{A_1 P_e + A_1 A_2 + A_1 A_2}{A_1 P_e + P_e P_e + A_1 A_2}. \tag{4.15.27}$$

Equations (4.15.25) and (4.15.26) give $A_2/A_1 = \exp\{-(1.386\,294 + 0.345\,843/T_6)\}$, and, at temperatures $T_6 < 0.377$, A_2/A_1 is always smaller than 0.1. For much lower temperatures such that $A_2 \ll A_1$, it is clear from eq. (4.15.24) that $c_2 \ll 1$. Equations (4.15.22) and (4.15.27) then give

$$P_e = c_1 \frac{k\rho T}{M_{He}} = \frac{A_1}{2} \left\{ \sqrt{1 + \frac{4}{A_1} \frac{k\rho T}{M_{He}}} - 1 \right\}. \tag{4.15.28}$$

When $A_1 < k\rho T/M_H$,

$$c_1 = \sqrt{A_1 \frac{M_{He}}{k\rho T}} \left\{ 1 - \sqrt{A_1 \frac{M_H}{k\rho T}} + \frac{1}{2} A_1 \frac{M_H}{k\rho T} + \cdots \right\}, \tag{4.15.29}$$

where M_{He} has been approximated by $4M_H$. For intermediate temperatures such that c_2 remains small compared with unity, but A_1 is large compared with $k\rho T/M_H$,

$$c_1 = 1 - \frac{1}{A_1} \frac{k\rho T}{M_{He}} + 2 \left(\frac{1}{A_1} \frac{k\rho T}{M_{He}} \right)^2 + \cdots. \tag{4.15.30}$$

Since the concentration factors must satisfy $c_i \leq 1$, it is evident that as $T \to \infty$, P_e can increase no faster than linearly with T. But, at high enough temperatures, A_1 and A_2 increase in proportion to $T^{5/2}$. Equations (4.15.19) and (4.15.21) show that, at such temperatures,

$$c_2 \to 0 \quad \text{and} \quad c_3 \to 1, \tag{4.15.31}$$

and eqs. (4.15.22) and (4.15.27) show that

$$P_e \to 2 \frac{k\rho T}{M_{He}} = 2 \frac{\rho}{M_{He}} kT. \tag{4.15.32}$$

When, as in the envelopes of most stars, helium is much less abundant than hydrogen, eqs. (4.15.22) and (4.15.23) demonstrate that, for any given density and temperature, the degree of ionization of helium is smaller than in a pure helium gas. This occurs because the free electrons provided by hydrogen reduce the phase space available for free electrons from helium.

4.16 Thermodynamic properties of hydrogen- and helium-rich matter in stellar envelopes

Most stars possess low-temperature, low-density envelopes made up predominantly of hydrogen and helium. To incorporate these envelopes in stellar evolution models, it is necessary that, among other things, specific heats, adiabatic and radiative gradients, molecular weight, and density- and pressure-scale heights be known as functions of gas pressure and temperature. In stellar envelope construction, it has been customary to include hydrogen and helium in the stages H, H^+, He, He^+, He^{++}, and one or more hypothetical metals in two stages M, M^+. For simplicity, only one metal of ionization potential = 7.5 eV is included here, following the treatment of Pierre Demarque (1960). At the low temperatures in the envelopes of very light stars, hydrogen is predominantly in molecular form. The molecule H_2 is therefore explicitly included in the following, which is a rendering of the discussion by Iben (1963), corrected for typographical errors.

The procedure chosen for calculating the specific heat and other pertinent quantities is a generalization of the method given by D. Barbier (1958) for the special case of hydrogen in the stages H and H^+. Let

$$n_X = \frac{\rho X}{M_H}, \; n_Y = \frac{\rho Y}{4M_H}, \text{ and } n_M = \alpha \, n_X \qquad (4.16.1)$$

represent, respectively, the number of hydrogen nuclei, the number of helium nuclei, and the number of hypothetical metal nuclei per cm^3. From the Solar System abundances given by L. H. Aller (*The Abundances of the Elements*, 1961), one finds that, for a hypothetical metal having a number abundance given by the total number abundances of Fe, Si, Mg, and Ni, $\alpha = (199 \, Z/X) \times 10^{-3}$, where X and Z are, respectively, the abundances by mass of hydrogen and the hypothetical heavy element. The abundance by mass of helium satisfies $Y = 1 - X - Z$. Let

$$n_H = c_a \, n_X, \; n_{H^+} = c_1 \, n_H, \; n_{H_2} = c_{aa} \, n_X,$$

$$n_{He^+} = c_2 \, n_Y, \; n_{He^{++}} = c_3 \, n_Y, \; n_{He} = (1 - c_2 - c_3) \, n_Y,$$

$$n_{M^+} = c_M \, \alpha \, n_X \qquad (4.16.2)$$

represent, respectively, the number per cm^3 of neutral hydrogen atoms, ionized hydrogen atoms, hydrogen molecules, singly ionized helium atoms, doubly ionized helium atoms, neutral helium atoms, and singly ionized metal atoms. Number conservation gives

$$c_a + c_1 + 2c_{aa} = 1. \qquad (4.16.3)$$

In statistical equilibrium the concentration factors c_i are related by the Saha equations:

$$c_1 P_e = c_a A_1, \tag{4.16.4}$$

$$c_2 = \frac{P_e A_2}{P_e^2 + A_2(A_3 + P_e)}, \tag{4.16.5}$$

$$c_3 = \frac{A_3 c_2}{P_e}, \tag{4.16.6}$$

$$c_M = \frac{A_M}{A_M + P_e}, \tag{4.16.7}$$

and

$$\frac{c_a^2}{c_{aa}} = \frac{K_V}{n_X kT}, \tag{4.16.8}$$

where P_e is the electron pressure,

$$A_i = B_i T^{5/2} \exp\left(-\frac{\chi_i}{kT}\right), \tag{4.16.9}$$

$$B_1 = B_3 = 0.333\,20, \quad B_2 = 1.3332, \quad B_M = 0.833\,24, \tag{4.16.10}$$

and

$$\chi_1 = 13.595 \text{ eV}, \quad \chi_2 = 24.580 \text{ eV}, \quad \chi_3 = 54.503 \text{ eV}, \quad \text{and} \quad \chi_M = 7.5 \text{ eV}. \tag{4.16.11}$$

The molecular dissociation parameter K_V is taken from Vardya (1960, eq. (6.16)):

$$\log_{10} K_V = \Theta \left\{ (-0.003\,268\,766\,\Theta + 0.056\,191\,27)\,\Theta - 4.925\,164 \right\} + 12.533\,51, \tag{4.16.12}$$

where

$$\Theta = \frac{0.005\,040\,39}{T_6}. \tag{4.16.13}$$

In stellar envelope regions, if electron degeneracy is neglected, gas pressure, density, temperature, and molecular weight are related by the equation of state:

$$P_g = \rho \frac{kT}{\mu M_H}, \tag{4.16.14}$$

where the overall molecular weight μ satisfies

$$\frac{1}{\mu} = (1 + c_1 - c_{aa})\,X + (1 + c_2 + 2c_3)\,\frac{Y}{4} + (1 + c_M)\,\alpha\,X. \tag{4.16.15}$$

The electron pressure is given by

$$P_e = \rho \frac{kT}{M_H} \left[c_1 X + (c_2 + 2c_3) \frac{Y}{4} + c_M \alpha X \right] \qquad (4.16.16)$$

$$= \mu P_g \left[c_1 X + (c_2 + 2c_3) \frac{Y}{4} + c_M \alpha X \right]. \qquad (4.16.17)$$

Derivatives of the Saha equations and the equation of state may be solved to give

$$T \frac{dc_i}{dT} = U_{iT} - U_{iP} \frac{1}{V}, \qquad (4.16.18)$$

where

$$V = \frac{P}{T} \frac{dT}{dP}, \quad U_{iT} = U_{i1} - U_{i2} U_T, \quad U_{iP} = U_{i2} U_P, \text{ and } i = 1, 2, 3, M \text{ and } aa, \qquad (4.16.19)$$

$$U_{11} = \beta \left[2c_{aa} \left\{ \xi_{aa} + \mu \frac{P}{P_e} \left[\frac{Y}{4}(U_{21} + 2U_{31}) + \alpha X U_{M1} \right] \right\} + (1 - c_1 + 2c_{aa})\xi_1 \right], \qquad (4.16.20)$$

$$U_{12} = \beta \left[2c_{aa} \left\{ \xi_{aa} + \mu \frac{P}{P_e} \left[\frac{Y}{4}(U_{22} + 2U_{32}) + \alpha X U_{M2} \right] \right\} + (1 - c_1 + 2c_{aa}) \right], \qquad (4.16.21)$$

$$\beta = c_1 [1 + 2c_{aa}(1 - c_1 \mu P/P_e)]^{-1}, \qquad (4.16.22)$$

$$U_{21} = c_2 [\xi_2(1 - c_2 - c_3) - \xi_3 c_3], \qquad (4.16.23)$$

$$U_{22} = c_2(1 - c_2 - 2c_3), \qquad (4.16.24)$$

$$U_{31} = c_3 [\xi_2(1 - c_2 - c_3) + \xi_3(1 - c_3)], \qquad (4.16.25)$$

$$U_{32} = c_3(2 - c_2 - 2c_3), \qquad (4.16.26)$$

$$U_{M2} = c_M(1 - c_M), \quad U_{M1} = \xi_M U_{M2}, \qquad (4.16.27)$$

$$U_{aa1} = \frac{1}{2} \left[c_a \xi_1 - \frac{c_a + c_1}{c_1} U_{11} \right], \qquad (4.16.28)$$

$$U_{aa2} = \frac{1}{2} \left[c_a - \frac{c_a + c_1}{c_1} U_{12} \right], \qquad (4.16.29)$$

$$\frac{1}{U_P} = 1 + \mu \left\{ \left(\frac{P_g}{P_e} - 1 \right) \left[X\, U_{12} + \frac{Y}{4}\, (U_{22} + 2U_{32}) + \alpha X\, U_{M2} \right] + X\, U_{aa2} \right\},$$

(4.16.30)

$$U_T = \mu U_P \left\{ \left(\frac{P_g}{P_e} - 1 \right) \left[X\, U_{11} + \frac{Y}{4}\, (U_{21} + 2U_{31}) + \alpha X\, U_{M1} \right] + X\, U_{aa1} \right\},$$

(4.16.31)

$$\xi_i = 2.5 + \frac{\chi_i}{kT}$$

(4.16.32)

for $i = 1, 2, 3, 4$, and

$$\xi_{aa} = \frac{T}{K_V} \frac{dK_V}{dT} = 2.5 + \frac{D}{kT}$$

(4.16.33)

$$= 2.302\,585\, \Theta \left\{ (0.009\,806\,298\, \Theta - 0.112\,3825)\, \Theta + 4.925\,164 \right\},$$

(4.16.34)

where D is the binding energy minus the excitation energy of the hydrogen molecule (Vardya, 1960, eqs. (3.7) and (3.15)). In general, above 20 000 K, hydrogen no longer exists in molecular form, and, by setting $c_{aa} = 0$, the equations are considerably simplified.

The total kinetic and binding energy per gram associated with any given configuration may be written as

$$U = \frac{n_X}{\rho} \left[c_{aa} \left(\frac{3}{2} kT - D - 2\xi_1 \right) + c_a \left(\frac{3}{2} kT - \xi_1 \right) + c_1 \left(2 \times \frac{3}{2} kT \right) \right]$$

$$+ \frac{n_Y}{\rho} \left[\frac{3}{2} kT + c_2 \left(\frac{3}{2} kT + \xi_2 \right) + c_3 \left(2 \times \frac{3}{2} kT + \xi_2 + \xi_3 \right) \right]$$

$$+ \frac{n_M}{\rho} \left[\frac{3}{2} kT + c_M \left(\frac{3}{2} + \xi_M \right) \right] + \text{constant}.$$

(4.16.35)

The constant defines the zero-point energy and may be chosen arbitrarily. The heat per gram required to alter a given configuration is given by

$$dQ = dU - \frac{P}{\rho} \frac{d\rho}{\rho}.$$

(4.16.36)

After differentiation and rearrangement, one finds

$$\frac{dU}{dT} = \frac{k}{\mu M_H} \left\{ \frac{3}{2} + \mu \left[\left(\sum_i \Gamma_i T \frac{dc_i}{dT} \right) + c_{aa} \left(-\frac{1}{k} \frac{dD}{dT} \right) \right] \right\},$$

(4.16.37)

where

$$\Gamma_1 = X\, (\xi_1 - 1), \quad \Gamma_2 = \frac{Y}{4}(\xi_2 - 1), \quad \Gamma_3 = \frac{Y}{4}\, (\xi_2 + \xi_3 - 2),$$

$$\Gamma_M = \alpha X\, (\xi_M - 1), \quad \text{and} \quad \Gamma_{aa} = -X.$$

(4.16.38)

Differentiation of the equation of state gives

$$\frac{P}{\rho^2}\frac{\mathrm{d}\rho}{\mathrm{d}T} = \frac{k}{\mu M_H}\left\{\frac{1}{V} - 1 - \mu T\left[X\frac{\mathrm{d}c_1}{\mathrm{d}T} + \frac{Y}{4}\left(\frac{\mathrm{d}c_2}{\mathrm{d}T} + 2\frac{\mathrm{d}c_2}{\mathrm{d}T}\right) + \alpha X\frac{\mathrm{d}c_M}{\mathrm{d}T}\right]\right\}$$

(4.16.39)

$$= \frac{k}{\mu M_H}\left\{\frac{1}{V}\left(\sum_i \lambda_i U_{iP}\right) - \left[1 + \mu X\left(\sum_i \lambda_i U_{iT}\right)\right]\right\},$$

(4.16.40)

where $\lambda_1 = X$, $\lambda_2 = Y/4 = \lambda_3/2$, $\lambda_{aa} = -X$, and $\lambda_M = \alpha X$, and, finally,

$$\frac{\mathrm{d}Q}{\mathrm{d}T} = \frac{k}{\mu M_H}$$

(4.16.41)

$$\times\left\{\frac{5}{2} + \mu\left[\left(\sum_i \xi' U_{iT}\right) + c_{aa}\left(-\frac{1}{k}\frac{\mathrm{d}D}{\mathrm{d}T}\right)\right] - \frac{1}{V}\left[1 + \mu\left(\sum_i \xi'_i U_{iP}\right)\right]\right\},$$

(4.16.42)

where

$$\xi'_1 = X\xi_1,\ \xi'_2 = \xi_2\frac{Y}{4},\ \xi'_3 = (\xi_2 + \xi_3)\frac{Y}{4},$$

$$\xi'_M = \alpha X\,\xi_M,\ \xi'_{aa} = -X\,\xi_{aa},$$

$$U'_{ij} = U_{ij}\ \text{for}\ i = 1, 2, 3, M,\quad\text{and}\quad U'_{aaj} = -U_{aaj}.$$

(4.16.43)

For a change at constant pressure, $1/V_P = 0$. The specific heat at constant pressure is therefore

$$\left(\frac{\mathrm{d}Q}{\mathrm{d}T}\right)_P = C_P = \frac{k}{\mu M_H}\left\{\frac{5}{2} + \mu\left[\left(\sum_i \xi' U_{iT}\right) + c_{aa}\left(-\frac{1}{k}\frac{\mathrm{d}D}{\mathrm{d}T}\right)\right]\right\} = \frac{k}{\mu M_H}\bar{C}_P.$$

(4.16.44)

For an adiabatic change, $\left(\frac{\mathrm{d}Q}{\mathrm{d}T}\right)_{\mathrm{ad}} = 0$, so

$$\frac{1}{V_{\mathrm{ad}}} = \frac{\bar{C}_P}{\left[1 + \mu\left(\sum_i \xi'_i U_{iP}\right)\right]}.$$

(4.16.45)

A change at constant density gives

$$\frac{1}{V_\rho} = \frac{\left[1 + \mu\left(\sum_i \lambda'_i U_{iT}\right)\right]}{\left[1 + \mu\left(\sum_i \lambda'_i U_{iP}\right)\right]},$$

(4.16.46)

and a specific heat at constant volume,

$$C_V = C_P\left(1 - \frac{V_{\mathrm{ad}}}{V_\rho}\right).$$

(4.16.47)

The local density-scale height at any point in the stellar envelope may be written as

$$\frac{1}{H_{\text{dens}}} = -\frac{1}{\rho}\frac{d\rho}{dx} = \frac{1}{H_{\text{pres}}}\left(1 + \mu\sum_i \lambda_i U'_{iT}\right)(V_\rho - V_t),$$ (4.16.48)

where $1/H_{\text{pres}} = -(1/P)\,dP/dx = g\rho/P$ and V_t is the logarithmic gradient to be associated with the average medium at the given point. In a similar fashion one may obtain the speed of sound:

$$v_s = \sqrt{\left(\frac{dP}{d\rho}\right)_{\text{ad}}} = \left(\frac{kT}{M_H}\right)^{1/2}\left\{\left[1 + \mu\left(\sum_i \lambda_i U'_{iT}\right)\right](V_\rho - V_{\text{ad}})\right\}^{-1/2}.$$ (4.16.49)

The preceeding equations are easily generalized to include the effect of radiation pressure.

4.17 The effect of Coulomb interactions on the equation of state for a gas

The Coulomb forces between charged particles lead to a reduction in energy per unit volume and to a reduction in pressure. The reduction in energy can be understood from elementary considerations. The energy of a completely ionized monatomic gas in volume V may be written as

$$E = E_{\text{kinetic}} + E_{\text{Coulomb}},$$ (4.17.1)

where

$$E_{\text{kinetic}} = \sum_{j=1}^{N_e} N_j \epsilon_j^{\text{free}} + \sum_{l=1}^{N_{\text{ions}}} N_l \epsilon_l^{\text{free}}$$ (4.17.2)

and

$$E_{\text{Coulomb}} = \frac{1}{2}\left\langle\sum_j\sum_{k\neq j}\frac{e_j e_k}{r_{jk}}\right\rangle + \frac{1}{2}\left\langle\sum_l\sum_{m\neq l}\frac{q_l q_m}{r_{lm}}\right\rangle + \left\langle\sum_j\sum_l\frac{e_j q_l}{r_{jl}}\right\rangle.$$ (4.17.3)

In eq. (4.17.2), ϵ_j^{free} is the average kinetic energy of an electron in the jth set of free states in which there are, on average, N_j electrons and ϵ_l^{free} is the average kinetic energy of an ion in the lth set of free states in which there are, on average, N_l ions. In eq. (4.17.3), e_j and e_k are the charges of the jth and kth electrons, r_{jk} is the distance between the two electrons, q_l and q_m are the charges of the lth and mth ions, and r_{lm} is the distance between the ions l and m. The double summations are over all particles and the brackets denote a time average. There are $N_e(N_e - 1)$ terms in the first double sum, $N_{\text{ions}}(N_{\text{ions}} - 1)$ terms in the second double sum, and $N_e N_{\text{ions}}$ terms in the third double sum.

Since $e_j = e_k = -e$, $q_j = q_k = Z_{ions}e$, and $N_e = Z_{ions}N_{ions}$ one can write eq. (4.17.3) as

$$E_{Coulomb} = \frac{1}{2}N_e^2\, e^2 \left\langle\!\left\langle \frac{1}{r_{jk}} \right\rangle\!\right\rangle + \frac{1}{2}N_{ions}^2\, Z_{ions}^2\, e^2 \left\langle\!\left\langle \frac{1}{r_{lm}} \right\rangle\!\right\rangle - N_e\, N_{ions}\, Z_{ions}\, e^2 \left\langle\!\left\langle \frac{1}{r_{jl}} \right\rangle\!\right\rangle,$$

(4.17.4)

or

$$E_{Coulomb} = N_e^2\, e^2 \left[\frac{1}{2}\left(\left\langle\!\left\langle \frac{1}{r_{jk}} \right\rangle\!\right\rangle + \left\langle\!\left\langle \frac{1}{r_{lm}} \right\rangle\!\right\rangle \right) - \left\langle\!\left\langle \frac{1}{r_{jl}} \right\rangle\!\right\rangle \right],$$

(4.17.5)

where the double brackets denote averages over time and particles and $N(N-1)$ has been approximated by N^2. Since charges of like sign repel and charges of opposite sign attract, $\langle\!\langle r_{jl}^{-1} \rangle\!\rangle$ is larger on average for the last term in eq. (4.17.5) than the sum of the first two terms, demonstrating that the net Coulomb energy is negative. Finding the actual averages of the reciprocal separations is not straightforward.

In the case of a relatively rarified gas, a simple, classical estimate of $U_{Coulomb}$ can be made using the Debye–Hückel (1923) approximation to the average electrical potential about any given charged particle. In the Debye–Hückel picture, particles of type j and charge $Z_j e$ are distributed about a particle of type i according to the Maxwell–Boltzmann prescription

$$n_j(r) = n_{0j} \exp\left(-\frac{Z_j e \phi_i(r)}{kT} \right),$$

(4.17.6)

where $n_j(r)$ is the number density of type-j particles in the close neighborhood of a type-i particle, n_{0j} is the mean density of type-j particles averaged over a volume containing many such particles, and ϕ_i is the potential field about the ith particle due both to this particle *and* to all of the others. The summation in eq. (4.17.6) is over all types of particle, including electrons, for which $Z_j = -1$.

With this prescription, Poisson's equation for the electrical potential in the vicinity of the ith particle is

$$\nabla^2 \phi_i(r) = -4\pi \left(\sum_j (n_{0j} Z_j e)\, \exp\left(-\frac{\phi_i(r) Z_j e}{kT} \right) \right) - 4\pi Z_i e\, \delta_i(r),$$

(4.17.7)

where $\delta_i(r)$ is a delta function. At large r, the exponential term may be expanded in a Taylor series and use can be made of the fact that

$$\sum_j n_{0j} Z_j e = 0$$

(4.17.8)

to obtain

$$\nabla^2 \phi_i(r) \sim K^2 \phi_i(r),$$

(4.17.9)

where

$$K^2 = \frac{1}{R_D^2} = \frac{4\pi}{kT} \sum_j n_{0j}(Z_j e)^2 \tag{4.17.10}$$

defines the Debye radius R_D.

The relevant solution of eq. (4.17.9) is

$$\phi_i(r) = C_i \frac{\exp(-Kr)}{r} = C_i \frac{\exp(-r/R_D)}{r}, \tag{4.17.11}$$

where C_i is a constant to be determined. Very close to the origin,

$$\nabla^2 \phi_i(r) = -4\pi Z_i e\, \delta_i(r), \tag{4.17.12}$$

for which a solution is

$$\phi_i(r) = \frac{Z_i e}{r}. \tag{4.17.13}$$

Thus, an approximate, continuous function which has the correct behavior at $r \to 0$ and at $r \gtrsim R_D$ is

$$\phi_i(r) = \frac{Z_i e}{r} \exp(-Kr) = \frac{Z_i e}{r} \exp\left(-\frac{r}{R_D}\right), \tag{4.17.14}$$

which may be rewritten as

$$\phi_i(r) \sim \frac{Z_i e}{r} + \phi_i^{\text{others}}(r), \tag{4.17.15}$$

where

$$\phi_i^{\text{others}} = \frac{Z_i e}{r} \left(\exp\left(-\frac{r}{R_D}\right) - 1\right). \tag{4.17.16}$$

When $r/R_D \ll 1$,

$$\phi_i^{\text{others}}(r) \sim -\frac{Z_i e}{R_D}. \tag{4.17.17}$$

The total Coulomb interaction energy per unit volume may now be approximated by

$$\bar{U}_{\text{Coulomb}} \sim \frac{1}{2} \sum_i Z_i e\, \phi_i^{\text{others}}(0)\, n_{0i}, \tag{4.17.18}$$

where the summation extends over all ions and electrons. To the extent that eq. (4.17.17) is a reasonable approximation,

$$\bar{U}_{\text{Coulomb}} \sim \frac{1}{2} \sum_i Z_i e \left(-\frac{Z_i e}{R_D}\right) n_{0i} \tag{4.17.19}$$

$$= -\frac{1}{2} \sum_i (Z_i e)^2 n_{0i} \left(\frac{4\pi}{kT} \sum_j (Z_j e)^2 n_{0j}\right)^{1/2} \tag{4.17.20}$$

$$= -\left(\frac{\pi}{kT}\right)^{1/2} \left(\sum_j (Z_j e)^2 n_{0j}\right)^{3/2}. \tag{4.17.21}$$

The Coulomb energy per unit volume predicted by eq. (4.17.21) times the volume behaves as $E_{\text{Coulomb}} = V \, \bar{U}_{\text{Coulomb}} \propto 1/\sqrt{TV}$, so

$$E = E_{\text{kinetic}} + E_{\text{Coulomb}} = \frac{3}{2} NkT + \frac{B}{\sqrt{VT}}, \tag{4.17.22}$$

where $B < 0$ is a constant. The fact that the Coulomb binding energy is negative agrees with the conclusion reached from eq. (4.17.5) that the energy of attraction between electrons and ions is larger than the energy of repulsion between charges of like sign. The fact that the Coulomb binding energy increases with decreasing temperature is consistent with the intuitive expectation that the smaller particle kinetic energies are, the greater is the time which oppositely charged particles spend near each other rather than far from each other.

Making use of eq. (4.17.22), the first and second laws of thermodynamics give

$$dQ = T \, dS(V, T) = PdV + dE = P \, dV + \frac{3}{2} \, Nk \, dT - \frac{1}{2} \, \frac{B}{\sqrt{VT}} \left(\frac{dV}{V} + \frac{dT}{T} \right)$$

$$= \left(P - \frac{1}{2} \, \frac{B}{\sqrt{VT}} \, \frac{1}{V} \right) dV + \left(\frac{3}{2} \, Nk - \frac{1}{2} \, \frac{B}{\sqrt{VT}} \, \frac{1}{T} \right) dT. \tag{4.17.23}$$

It follows that

$$\left(\frac{\partial S(V, T)}{\partial T} \right)_V dT = \left(\frac{3}{2} \, Nk - \frac{1}{2} \, \frac{B}{\sqrt{VT}} \, \frac{1}{T} \right) \frac{dT}{T} \tag{4.17.24}$$

$$= \frac{3}{2} \, Nk \, d \log_e T + \frac{1}{3} \, \frac{B}{\sqrt{V}} \, d \left(\frac{1}{T^{3/2}} \right), \tag{4.17.25}$$

or

$$S(V, T) = \frac{3}{2} \, \log_e T + \frac{1}{3} \, \frac{B}{\sqrt{VT}} \, \frac{1}{T} + g(V), \tag{4.17.26}$$

where $g(V)$ is a constant of integration. Differentiation of eq. (4.17.26) with respect to volume gives

$$\left(\frac{\partial S(V, T)}{\partial V} \right)_T = -\frac{1}{6} \, \frac{B}{(VT)^{3/2}} + \frac{dg(V)}{dV}. \tag{4.17.27}$$

Since, from eq. (4.17.23), it is also true that

$$\left(\frac{\partial S(V, T)}{\partial V} \right)_T = \left(P - \frac{1}{2} \, \frac{B}{\sqrt{VT}} \, \frac{1}{V} \right) \frac{1}{T}, \tag{4.17.28}$$

it follows that

$$\frac{P}{T} = \frac{1}{3} \, \frac{B}{(VT)^{3/2}} + \frac{dg(V)}{dV}. \tag{4.17.29}$$

Since, when $B = 0$, $P/T = Nk/V$, one has that

$$\frac{dg(V)}{dV} = \frac{Nk}{V}.$$ (4.17.30)

Therefore, in general,

$$P = \frac{NkT}{V} + \frac{1}{3}\frac{B}{\sqrt{VT}}\frac{1}{V} = P_{\text{kinetic}} + \frac{1}{3}\frac{E_{\text{Coulomb}}}{V} = P_{\text{kinetic}} + \frac{1}{3}\bar{U}_{\text{Coulomb}}.$$ (4.17.31)

Thus, the reduction in energy density due to a net electrical attraction between charged particles translates into a negative contribution to the pressure, the ratio between this negative contribution to the pressure and the negative contribution to the energy density being the same as the ratio between the positive contribution to the pressure by the radiation field and the positive contribution to the energy density by the radiation field.

In summary, the pressure exerted by a completely ionized, non-degenerate, non-relativistic gas is:

$$P \sim \sum_i n_{0i}\, kT - \frac{1}{3}\left(\frac{\pi}{kT}\right)^{1/2}\left(\sum_j (Z_j e)^2\, n_{0j}\right)^{3/2}$$ (4.17.32)

$$= \sum_i n_{0i}\, kT\left(1 - \frac{\pi^{1/2}}{3}\left(\frac{e^2}{kT}\right)^{3/2}\frac{\left(\sum Z_i^2\, n_{0i}\right)^{3/2}}{\sum n_{0i}}\right).$$ (4.17.33)

For pure, fully ionized hydrogen,

$$\sum n_{0i} = \sum Z_i^2 n_{0i} = \frac{2\rho}{M_{\text{H}}}$$ (4.17.34)

and eq. (4.17.24) gives

$$P \sim \frac{2\rho}{M_{\text{H}}}kT\left(1 - \frac{\pi^{1/2}}{3}\left(\frac{e^2}{kT}\right)^{3/2}\left(\frac{2\rho}{M_{\text{H}}}\right)^{1/2}\right)$$ (4.17.35)

$$= P_{0\text{H}}\left(1 - 0.0155\left(\frac{\rho}{T_6^3}\right)^{1/2}\right),$$ (4.17.36)

where

$$P_{0\text{H}} = 2\frac{\rho}{M_{\text{H}}}kT$$ (4.17.37)

is the pressure when Coulomb interactions are neglected. For pure, fully ionized helium,

$$P \sim P_{0\text{He}}\left(1 - 0.038\left(\frac{\rho}{T_6^3}\right)^{1/2}\right),$$ (4.17.38)

where

$$P_{0\text{He}} = 3\,\frac{\rho}{4M_\text{H}}\,kT. \tag{4.17.39}$$

Near the Sun's center, where, say, $\rho_0 \sim 150$, $T_6 \sim 15$, and hydrogen and helium abundances by mass are comparable, $P \sim P_0(1 - 0.0033)$, where P_0 is the pressure in the absence of Coulomb and electron-degeneracy effects. Comparing with the estimates in Section 4.5, it is evident that the negative contribution to pressure due to Coulomb interactions at the Sun's center is only $\sim 7\%$ of the positive contribution to pressure due to electron degeneracy.

4.17.1 Modifications when electrons are modestly degenerate

When electrons are modestly degenerate, it is necessary to modify eq. (4.17.7) to take into account that the distribution function for electrons is given by Fermi–Dirac statistics rather than by Maxwell–Boltzmann statistics. Thus, one writes

$$\nabla^2\phi_i(r) = -4\pi\left(\overset{\text{ions only}}{\sum_j}(n_{0j}Z_je)\,\exp\left(-\frac{\phi_i(r)Z_je}{kT}\right)\right) + 4\pi e\,n_e(r) - 4\pi Z_ie\,\delta_i(r), \tag{4.17.40}$$

where

$$n_e(r) = \int\frac{d^3\mathbf{p}}{h^3}\left(\exp\left(\frac{\epsilon - e\phi_i(r)}{kT} + \alpha\right) + 1\right)^{-1}. \tag{4.17.41}$$

To first order in $e\phi_i(r)/kT$, one has

$$n_e(r) = n_e - \frac{dn_e}{d\alpha}\,\frac{e\phi_i(r)}{kT}, \tag{4.17.42}$$

where n_e is the average electron density. Keeping just the first term in the expansion of the exponents for the ions in eq. (4.17.40), the development proceeds exactly as before, the only modification being that the Debye radius is now given by

$$\frac{1}{R_\text{D}^2} = K^2 = \frac{4\pi e^2}{kT}\left(\overset{\text{ions only}}{\sum_j}n_{0j}Z_j^2 + \left|\frac{dn_e}{d\alpha}\right|\right), \tag{4.17.43}$$

where it is recognized that $dn_e/d\alpha$ is a negative quantity. The behavior of $-n_e^{-1}\,dn_e/d\alpha$ as a function of degeneracy is shown in the sixth column of Table 4.7.1.

It is instructive to write

$$\frac{1}{R_\text{D}^2} = \frac{1}{R_\text{I}^2} + \frac{1}{R_\text{E}^2}, \tag{4.17.44}$$

where

$$\frac{1}{R_I^2} = \frac{4\pi e^2}{kT} \left(\overset{\text{ions only}}{\underset{j}{\sum}} n_{0j} Z_j^2 \right) \tag{4.17.45}$$

and

$$\frac{1}{R_E^2} = \frac{4\pi e^2}{kT} n_e \left(\frac{1}{n_e} \left| \frac{dn_e}{d\alpha} \right| \right). \tag{4.17.46}$$

The most interesting aspect of this representation is that the screening of the electric field of an ion is not properly described as due to the build-up of negative charges around an ion. It is the consequence of the polarization of the entire aggregate of charges, positive as well as negative. As shown by entries in the sixth column of Table 4.7.1, the quantity $(1/n_e) |dn_e/d\alpha|$ decreases rapidly with increasing degeneracy. The interpretation is that, with increasing degeneracy, the sea of electrons becomes more rigid, resisting polarization around positive ions. Yet, even when electrons are no longer a factor (in the weak screening approximation), the electrical field of any ion selected for individual consideration is effectively neutralized at large enough distances by the electric field produced by surrounding positive charges.

It is of interest to compare the magnitude of the screening radius with the average interionic separation. Combining eqs. (4.7.12) and (4.17.43) gives

$$\frac{R_D}{2r_{0i}} = \frac{1}{2} \left(\frac{8\pi}{3} \right)^{1/3} \left(\frac{kT}{4\pi e^2} \right)^{1/2} \left(\frac{M_H}{\rho} \right)^{1/6} \left(\frac{\mu_e}{Z_i^2} \right)^{1/6} \left(\mu_e \sum_j Y_j Z_j^2 + \frac{1}{n_e} \left| \frac{dn_e}{d\alpha} \right| \right)^{-1/2} \tag{4.17.47}$$

$$= 0.763\,58 \, \frac{T_6^{1/2}}{\rho^{1/6}} \left(\frac{\mu_e}{Z_i^2} \right)^{1/6} \left(\mu_e \sum_j Y_j Z_j^2 + \frac{1}{n_e} \left| \frac{dn_e}{d\alpha} \right| \right)^{-1/2}, \tag{4.17.48}$$

where Y_i is the abundance by number of the ith ion. The last column in Table 4.7.1 shows how the quantity $(R_D/r_{0i}) (1/T_6^{1/4})$ depends on the degree of degeneracy for a gas of completely ionized helium. At the center of the electron-degenerate core of a low mass red giant about to experience the helium core flash, $R_D/r_{0i} \sim 1.0$. After degeneracy is lifted, $R_D/r_{0i} \sim 1.7$.

An analytic representation of the quantity $(1/n_e) |dn_e/d\alpha|$ can be constructed from formulae in Sections 4.5–4.7. Differentiation of eq. (4.6.25) gives, for strongly but non-relativistically degenerate electrons,

$$\frac{1}{n_e} \left| \frac{dn_e}{d\alpha} \right| = \frac{3}{2} \frac{kT}{\epsilon_F} - \frac{\left(\frac{\pi^2}{4} \left(\frac{kT}{\epsilon_F} \right)^3 + \frac{7\pi^4}{180} \left(\frac{kT}{\epsilon_F} \right)^5 \right)}{\left(1 + \frac{\pi^2}{8} \left(\frac{kT}{\epsilon_F} \right)^2 + \frac{7\pi^4}{720} \left(\frac{kT}{\epsilon_F} \right)^4 \right)} + \cdots. \tag{4.17.49}$$

$$= \frac{3}{2} \left(\frac{kT}{\epsilon_F} \right) \left(1 - \frac{\pi^2}{6} \left(\frac{kT}{\epsilon_F} \right)^2 - \frac{11\pi^4}{2160} \left(\frac{kT}{\epsilon_F} \right)^4 + \cdots \right). \tag{4.17.50}$$

Using eq. (4.6.32), this becomes

$$\frac{1}{n_e}\left|\frac{dn_e}{d\alpha}\right| = \frac{3}{2}\left(\frac{kT}{\epsilon_{F0}}\right)\left(1 - \frac{\pi^2}{12}\left(\frac{kT}{\epsilon_{F0}}\right)^2 - \frac{59\pi^4}{2160}\left(\frac{kT}{\epsilon_{F0}}\right)^4 + \cdots\right), \qquad (4.17.51)$$

and using eq. (4.6.31), one has

$$\frac{1}{n_e}\left|\frac{dn_e}{d\alpha}\right| = \frac{3}{2}\left(\frac{6.0798}{\delta}\right)^{2/3}\left(1 - 0.822\,467\left(\frac{6.0798}{\delta}\right)^{4/3} - 2.660\,71\left(\frac{6.0798}{\delta}\right)^{8/3}\right). \tag{4.17.52}$$

At $\delta = 63.77$, eq. (4.17.54) overestimates the numerical estimate in Table 4.7.1 by 0.18%. Differentiation of eq. (4.5.5) gives, for weakly degenerate electrons,

$$\frac{1}{n_e}\left|\frac{dn_e}{d\alpha}\right| = \frac{\sum_{m=1}^{\infty}(-1)^{m+1}e^{-m\alpha}/m^{1/2}}{\sum_{m=1}^{\infty}(-1)^{m+1}e^{-m\alpha}/m^{3/2}} \tag{4.17.53}$$

$$= 1 - \frac{1}{2^{3/2}}\,e^{-\alpha} + \left(\frac{2}{3^{3/2}} - \frac{1}{2^3}\right)e^{-2\alpha} + \cdots \tag{4.17.54}$$

$$= 1 - 0.353\,553\,391\,e^{-\alpha} + 0.259\,900\,179\,e^{-2\alpha} + \cdots. \tag{4.17.55}$$

Using eq. (4.5.11), this becomes

$$\frac{1}{n_e}\left|\frac{dn_e}{d\alpha}\right| = 1 - 0.043\,7452\,\delta + 0.002\,065\,20\,\delta^2 + \cdots$$

$$= 1 - 0.043\,7452\,\delta\,(1 - 0.047\,2098\,\delta) + \cdots. \tag{4.17.56}$$

Equation (4.17.56) overestimates the numerical result in Table 4.7.1 at $\delta = 6.182$ by 2.26%. By adding a cubic term, so that

$$\frac{1}{n_e}\left|\frac{dn_e}{d\alpha}\right| = 1 - 0.043\,7452\,\delta\,(1 - 0.047\,2098\,\delta\,(1 - 0.036\,6725\,\delta)), \tag{4.17.57}$$

the numerical estimates are fitted tolerably well for $\delta \leq 6.182$. For values of δ between about 6 and 60, the derivative of $(1/n_e)\,|dn_e/d\alpha|$ with respect to $\log\delta$ is nearly a constant and equal to about $-1/2$. More precisely, this derivative, as determined by the numerical results in Table 4.7.1 at $\delta = 6.182$ and $\delta = 63.77$ is $-0.484\,663$. A reasonable algorithm for stellar evolution calculations is to use eq. (4.17.56) for $\delta \leq 6.182$, eq. (4.17.54) for $\delta \geq 63.77$, and

$$\frac{1}{n_e}\left|\frac{dn_e}{d\alpha}\right| = 0.7906 - 0.484\,663\,\log\left(\frac{\delta}{6.182}\right) \tag{4.17.58}$$

for $6.182 < \delta < 63.77$.

4.17.2 When electrons are significantly degenerate

The Debye–Hückel treatment is strictly applicable only when electrons are only modestly degenerate and $U/kT \ll 1$. When electrons are quite degenerate, they are resistant to concentration about nuclei and, in first approximation, form a uniform density fluid within which the nuclei are immersed. This being the case, a radius R_i is defined about every nucleus such that

$$Z_i = n_e \frac{4\pi}{3} R_i^3. \tag{4.17.59}$$

Within the spherical volume of radius R_i, the total electrical charge is zero. The electrostatic energy of the volume is

$$E_i = \int_0^{R_i} \frac{e^2 N_e(r)}{r} \, dN_e(r) - \int_0^{R_i} \frac{Z_i e^2 N_e(r)}{r} \, dn_e(r), \tag{4.17.60}$$

where

$$N_e(r) = n_e \frac{4\pi}{3} r^3. \tag{4.17.61}$$

Integration gives

$$E_i = \frac{3}{5} \frac{e^2 Z_i^2}{R_i} - \frac{3}{2} \frac{e^2 Z_i^2}{R_i} = -\frac{9}{10} \frac{e^2 Z_i^2}{R_i}. \tag{4.17.62}$$

Using eq. (4.17.59) to replace R_i in eq. (4.17.62) gives

$$
\begin{aligned}
E_i &= -\frac{9}{10} \left(\frac{4\pi}{3}\right)^{1/3} e^2 \, Z_i^{5/3} \, n_e^{1/3} \\
&= -\frac{9}{5} \left(\frac{4\pi}{3}\right)^{1/3} \frac{e^2}{2a_0} \frac{a_0}{M_H^{1/3}} Z_i^{5/3} \left(\frac{\rho}{\mu_e}\right)^{1/3} \\
&= -1.293\,66 \, \frac{e^2}{2a_0} Z_i^{5/3} \left(\frac{\rho}{\mu_e}\right)^{1/3},
\end{aligned}
\tag{4.17.63}
$$

where a_0 is the Bohr radius of the hydrogen atom and $e^2/2a_0$ is the binding energy of the electron in the ground state of the hydrogen atom. The total Coulomb interaction energy per unit volume between ions and electrons is given by

$$\bar{U}_{\text{Coulomb}} = \sum_{i=\text{all nuclei}} n_i \, E_i = -\left(\frac{\rho}{M_H}\zeta\right) \left(1.293\,66 \, \frac{e^2}{2a_0}\right) \left(\frac{\rho}{\mu_e}\right)^{1/3}, \tag{4.17.64}$$

where

$$\zeta = \sum_{i=\text{all nuclei}} Y_i Z_i^{5/3}, \tag{4.17.65}$$

and Y_i is the abundance by number of the ith nucleus.

One can also write eq. (4.17.64) as

$$\bar{U}_{\text{Coulomb}} = -\left(\frac{3}{2}\frac{\rho}{\mu_n M_H} kT\right)\gamma, \tag{4.17.66}$$

where

$$\gamma = 0.136\,17\left(\frac{\mu_n\,\zeta}{\mu_e^{1/3}}\right)\left(\frac{\rho^{1/3}}{T_6}\right) \tag{4.17.67}$$

is the ratio of the nucleus–electron binding energy to the kinetic energy of nuclei and $\mu_n = (\sum_i Y_i)^{-1}$ is the nuclear molecular weight.

When electrons are highly degenerate, the contribution to the pressure due to the electrostatic interactions between nuclei and electrons is

$$P_{\text{Coulomb}} = \frac{1}{3}\bar{U}_{\text{Coulomb}} = -\frac{\rho}{\mu_n M_H} kT\,\frac{1}{2}\gamma, \tag{4.17.68}$$

and the total pressure due to nuclei and electrons in the gaseous phase is

$$P_e + P_{\text{nuclei}} + P_{\text{Coulomb}} = P_e + \frac{\rho}{\mu_n M_H} kT\left(1 - \frac{1}{2}\gamma\right). \tag{4.17.69}$$

It is instructive to compare P_{Coulomb} with P_e in the limit $kT \ll \epsilon_F$. In the non-relativistic approximation, P_e is given by eq. (4.8.22), so

$$\frac{P_{\text{Coulomb}}}{P_e} \sim -\frac{3}{2}\left(\frac{4\pi}{3}\right)^{1/3}\frac{e^2}{m_e c^2}\frac{\mu_e n_e^{1/3}}{p_F^2(0)}\,\zeta. \tag{4.17.70}$$

Using eq. (4.8.7) for p_{F0}, this becomes

$$\frac{P_{\text{Coulomb}}}{P_e} \sim -\frac{4\pi}{2^{1/3}}\frac{e^2/m_e c^2}{\lambda_C}\frac{\mu_e}{\lambda_C n_e^{1/3}}\zeta$$

$$= -2^{2/3}\frac{e^2}{\hbar c}\frac{\mu_e}{\lambda_C n_e^{1/3}}\zeta = -3.5603\left(\frac{\mu_e}{\rho_0}\right)^{1/3}\mu_e\,\zeta$$

$$= -0.035\,603\left(\frac{\mu_e}{\rho_6}\right)^{1/3}\mu_e\,\zeta. \tag{4.17.71}$$

A more general expression follows from the use of eqs. (4.8.22) and (4.8.7):

$$\frac{P_{\text{Coulomb}}}{P_e} = -\frac{6}{5}\left(\frac{4}{9\pi}\right)^{1/3}\frac{e^2}{\hbar c}\frac{p_{F0}}{\sqrt{1+p_{F0}^2}}\frac{\mu_e\zeta}{f(0)} \tag{4.17.72}$$

$$= -0.004\,5628\,\frac{p_{F0}}{\sqrt{1+p_{F0}^2}}\frac{\mu_e\zeta}{f(0)}, \tag{4.17.73}$$

where

$$f(0) = 1 - \frac{3}{2p_{F0}^2} \left(1 - \frac{\log_e [\epsilon_{F0} + p_{F0}]}{\epsilon_{F0} \, p_{F0}} \right). \qquad (4.17.74)$$

From these equations it is evident that, for densities of 10^6 gm cm^{-3} and larger, $|P_{\text{Coulomb}}|$ is much smaller than P_e. At densities of terrestrial solids, $|P_{\text{Coulomb}}|$ and P_e are of comparable size. However, at such densities, the assumption of complete ionization is not valid, and these equations are not appropriate discriptions of the pressure.

Bibliography and references

Lawrence H. Aller, *The Abundances of the Elements* (New York: Interscience Publishing), 1961.

Daniel Barbier, *Handbuch der Physik* (Berlin: Springer), **50**, 274, 1958.

Niels Bohr, *Phil. Mag., Ser. 6*, **26**, 153, 1913.

Ludwig Boltzmann, *Akad. Wiss. (Wien) Sitzb.*, II Abt. **66**, 275, 1872.

Ludwig Boltzmann, *Vorlesungen über Gastheorie*, (Leipzig: Barth), Vol. 1, 1896

Satyendra Nath Bose, *Zeits. für Physik*, **26**, 178, 1924.

Louis deBroglie, *Phil. Mag.*, **47**, 446, 1924.

Subrahmanyan Chandrasekhar, *An Introduction to the Study of Stellar Structure* (Chicago: University of Chicago Press), 1939.

Peter Debye & Erich Hückel, *Zeits. für Physik*, **24**, 185, 1923.

Pierre Demarque, *ApJ*, **132**, 366, 1960.

Paul Adrien Maurice Dirac, *Proc. Roy. Soc. (London)*, **112**, 661, 1926.

H. B. Dwight, *Tables of Integrals and Other Mathematical Data* (New York: MacMillan), 1934; fourth edition, 1947.

Peter P. Eggleton, John Faulkner, & Brian P. Flannery, *A&Ap*, **23**, 325, 1973.

Albert Einstein, *Ann. Phys.*, **17**, 132 & 891, 1905.

Albert Einstein, *Berl. Ber.*, **261**, 1924.

Enrico Fermi, *Zeits. für Physik*, **36**, 902, 1926.

William A. Fowler & F. Hoyle, *ApJS*, **9**, 201, 1964.

Werner Heisenberg, *Zeits. für Physik*, **25**, 691, 1927.

Icko Iben, Jr., *ApJ*, **138**, 452 & 1090, 1963; **154**, 557, 1968.

Robert B. Leighton, *Principles of Modern Physics*, (New York: McGraw-Hill), 1964.

James Clerk Maxwell, *Phil. Mag.*, **19**, 31, 1860.

Peter J. Mohr & Barry N. Taylor, *Physics Today*, BG6-BG13, August, 2002.

Abraham de Moivre, *Doctrine of Chances*, 1733.

Wolfgang Pauli, *Zeits. für Physik*, **31**, 765, 1925.

Max Planck, *Ann. d. Physik*, **4**, 553, 1901.

Ernest Rutherford, *Phil. Mag.*, **21**, 669, 1911.

Meghnad N. Saha, *Phil. Mag.*, **40**, 472, 809, 1920.

Meghnad N. Saha, *Proc. Roy. Soc. (London)*, **99**, 135, 1921.

Erwin Schrödinger, *Ann. Der Physik*, **79**, 361, 489, 1926; **81**, 109, 1926.

Joseph John Thomson, *Phil. Mag.*, **14**, 293, 1897

M. S. Vardya, *ApJS*, **42**, 281, 1960.

E. T. Whittaker & G. N. Watson, *A Course in Modern Analysis* (Cambridge: Cambridge University Press), 1902; fourth edition, 1952.

5 Polytropes and single zone models

Polytropes are simple but pedagogically very useful models of self gravitating spheres. They were invented and explored in the last quarter of the nineteenth century, long before the development of theories for describing energy generation and energy flow in the stellar interior. The models follow when an exact balance between an outward pressure gradient force and the inward gravitational force is assumed and a parameterized power-law relationship between pressure and density is adopted. Solutions require no explicit use of a temperature-dependent equation of state, a law of energy transport, or a law of energy generation. The only equations to be solved are Poisson's equation for the gravitational field and the pressure balance equation. By varying the parameters in the power law, one can obtain plausible zeroth order models of various classes of real stars. A discussion of the contributions to the subject by Lane, Ritter, Emden, Kelvin, and others is given at the end of Chapter IV of Subrahmanyan Chandrasekhar's book *Stellar Structure* (1939). It is revealing that the construction of complete polytropes did not begin until 1878, five years after Cornu & Baile in 1873 determined explicitly for the first time the value of the gravitational constant G from results of Cavendish's 1798 tortional balance experiments. Arthur S. Eddington, who pioneered the development of the theory of radiative energy transport in stellar interiors, made imaginative use of polytropes and argued that one particular polytrope, that of index $N = 3$, provides an approximate description of the observed properties of main sequence stars. His discussion of polytropes and of the index $N = 3$ model in the book *The Internal Constitution of the Stars* (1926) is well worth study, even though many of the details of the theory of radiative transport given by him have long since been replaced by more sophisticated treatments.

Polytropes and one zone models are presented in this book because they offer a means of exploring the physics of stellar structure in a way which does not require extensive calculations with a powerful computer, making them useful instructional tools for students equiped with only a hand-held electronic calculator or even with only a slide rule. Using polytropes, theorems can be constructed which confirm relationships inferred intuitively in Chapter 3, insight into the structure of main sequence stars can be obtained, and several very important quantitative truths about white dwarfs and neutron stars can be determined. Because of their extreme simplicity, one zone models provide even more transparent insight into these same matters.

In Section 5.1, the basic polytropic structure equation is derived, physical variables are related to dimensionless quantities, and boundary conditions are set. Solutions are discussed in Section 5.2 as functions of the polytropic index N; two of three possible analytic solutions are derived ($N = 0$ and $N = 1$), properties of two particularly important numerical solutions (for $N = 3/2$ and $N = 3$) are given, a general theorem relating the gravitational binding energy to the mass, radius, and index N of a polytropic model is derived, and some of the properties of the third ($N = 5$) analytic solution are presented.

In Section 5.3, additional internal properties of polytropes are described which follow when an equation of state and an energy-generation law are adopted. In particular, the relationship between interior temperatures and global characteristics is examined as a function of polytropic index, and it is shown that, although it is very sensitive to the temperature dependence of the nuclear fuel, the effective mass fraction of the nuclear burning core of a homogeneous model is nearly independent of the polytropic index. As a fraction of total model mass, the mass of the central nuclear burning region is smaller, the steeper the temperature dependence. In Section 5.4, the relationship between luminosity and mass in polytropes burning a nuclear fuel in central regions is examined as a function of position in the model. This dependence is required for an analysis of the mass of convective cores in Section 7.13.

Properties of zero age main sequence models derived from evolutionary calculations are presented in Section 5.5 and compared with properties of polytropes. In Section 5.6, models in adiabatic equilibrium are discussed as approximations to very low mass main sequence stars and as an illustration of two universal characteristics of stars: (1) the differences in the temperature- and density-dependences of electron-degenerate matter and non-electron-degenerate matter is responsible for the existence of a maximum interior temperature and (2) as a star evolves, its global entropy decreases with time.

In Section 5.7, the radius–mass relationship for low mass white dwarfs approximated by index $N = 3/2$ polytropes is compared with the relationship predicted by more realistic models of such white dwarfs, and the properties of the $N = 3$ polytrope are used to find the formal maximum, or Chandrasekhar mass, of a cold, non-rotating white dwarf. It is also shown that, as the maximum mass is approached, although the gravitational binding energy tends toward infinity, the binding energy (gravitational binding energy minus the electron kinetic energy) remains finite and approaches the total rest mass energy of the electrons in the white dwarf.

One zone models are utilized in Sections 5.8 and 5.9 to demonstrate in remarkably transparent ways several fundamental properties of stars in various stages of evolution. In Section 5.8, a one zone model is used to show that the radius of a very low mass white dwarf varies inversely with the one third power of the mass, that the average separation between adjacent electrons decreases as the Chandrasekhar mass is approached, and that, before igniting a nuclear fuel, a stellar core first heats as it shrinks, and then, after exhausting the nuclear fuel, shrinks further until igniting another nuclear fuel or, after achieving a maximum temperature, evolves into a white dwarf which both shines and shrinks because nuclei in its interior cool.

In Section 5.9, a one zone model is used to show that neutron stars – stars of near solar mass in which degenerate baryons are the primary source of pressure – are smaller than white dwarfs of the same mass by roughly the ratio of the electron mass to the proton mass. The densities and particle kinetic energies in neutron stars are so large that (1) only electrons and protons coexist with neutrons, with neutrons being by far the dominant component, (2) in the pressure balance equation, it is necessary to take general relativity into account, and (3) in the construction of an equation of state at the highest densities, it is necessary to take nuclear forces into account.

5.1 The basic structure equation when pressure is proportional to a fixed power (1+1/N) of the density

It proves convenient to focus on the gravitational potential. Consider a collection of point masses and define a potential function $\phi(\mathbf{r}_p)$ by

$$\phi(\mathbf{r}_p) = G \sum_{\text{all } i} \frac{m_i}{|\mathbf{r}_p - \mathbf{r}_i|}, \tag{5.1.1}$$

where m_i is the mass of a particle at a distance $\mathbf{r}_i - \mathbf{r}_p$ from the point $\mathbf{r}_p$ at which the potential is evaluated, and the sum extends over all of the particles in the system.

Taking the gradient of $\phi(\mathbf{r}_p)$ with respect to $\mathbf{r}_p$, one has

$$\nabla_p \phi(\mathbf{r}_p) = -G \sum_{\text{all } i} m_i \frac{\mathbf{r}_p - \mathbf{r}_i}{|\mathbf{r}_p - \mathbf{r}_i|^3} = -G \sum_{\text{all } i} m_i \frac{\hat{\mathbf{r}}_{pi}}{|\mathbf{r}_p - \mathbf{r}_i|^2}, \tag{5.1.2}$$

where $\hat{\mathbf{r}}_{pi}$ is a unit vector in the direction $\mathbf{r}_p - \mathbf{r}_i$. This equation demonstrates that the gravitational force per unit mass on a test particle at position $\mathbf{r}_p$ is given by the gradient of the chosen potential.

Next, consider an arbitrary closed surface. At any point on the surface, construct the dot product of $\nabla\phi(\mathbf{r}_p)$ and the vector $d\mathbf{S}_p$ which is perpendicular to the surface element of area dS_p and is directed outward from the enclosed volume. Then, form the integral of this dot product over the entire surface:

$$\oint \nabla_p \phi(\mathbf{r}_p) \cdot d\mathbf{S}_p = -G \oint \sum_{\text{all } i} m_i \frac{\hat{\mathbf{r}}_{pi}}{|\mathbf{r}_p - \mathbf{r}_i|^2} \cdot d\mathbf{S}_p = -G \sum_{\text{all } i} m_i \oint \frac{\hat{\mathbf{r}}_{pi} \cdot d\mathbf{S}_p}{|\mathbf{r}_p - \mathbf{r}_i|^2}. \tag{5.1.3}$$

Note that

$$\frac{\hat{\mathbf{r}}_{pi} \cdot d\mathbf{S}_p}{|\mathbf{r}_p - \mathbf{r}_i|^2} = d\Omega_{pi} \tag{5.1.4}$$

is, in absolute value, the solid angle subtended by the area $|(\hat{\mathbf{r}}_{pi} \cdot d\mathbf{S}_p|$ at the ith particle. If a mass point lies outside of the enclosed volume, for every positive value of $d\Omega_{pi}$ there is another of value $d\Omega_{p'i} = -d\Omega_{pi}$; if a mass point is inside the enclosed volume, all values of $d\Omega_{pi}$ are positive and integrate to 4π. Thus,

$$\oint \nabla_p \phi(\mathbf{r_p}) \cdot d\mathbf{S}_p = -4\pi G \sum_{i \text{ inside}} m_i. \tag{5.1.5}$$

The summation extends *only* over the particles enclosed by the surface.

The right hand side of eq. (5.1.5) may be converted into an integral by making use of the fact that

$$\sum_{i \text{ inside}} m_i = \int \rho \, d\tau, \tag{5.1.6}$$

where ρ is the mass density, $d\tau$ is a volume element, and the integral extends over the volume enclosed by the surface defined in eq. (5.1.3). The surface integral is related by Green's theorem to an integral of the Laplacian over the volume enclosed by this surface:

$$\oint \nabla_p \phi \cdot d\mathbf{S}_p = \int \nabla^2 \phi \, d\tau. \tag{5.1.7}$$

Joining eqs. (5.1.5)–(5.1.7), it follows that

$$\int (\nabla^2 \phi + 4\pi G\rho) \, d\tau = 0, \tag{5.1.8}$$

which can, in general, be true only if the integrand vanishes. Thus, one has finally that

$$\nabla^2 \phi + 4\pi G\rho = 0, \tag{5.1.9}$$

which is Poisson's equation for the gravitational potential. In spherical coordinates, if mass is distributed in a spherically symmetric way, eq. (5.1.9) reduces to

$$\frac{d^2\phi}{dr^2} + \frac{2}{r}\frac{d\phi}{dr} + 4\pi G\rho = 0. \tag{5.1.10}$$

Equation (5.1.10) can also be obtained by observing that, for a spherically symmetric distribution of mass,

$$\frac{dM(r)}{dr} = 4\pi \rho \, r^2, \tag{5.1.11}$$

where $M(r)$ is the mass contained within the sphere of radius r, and that the acceleration due to gravity is directed inward and has the magnitude

$$g = -\frac{d\phi}{dr} = \frac{GM(r)}{r^2}. \tag{5.1.12}$$

Differentiating eq. (5.1.12) once and using eq. (5.1.11) and eq. (5.1.12) again, one recovers eq. (5.1.10).

The pressure balance equation (eq. (3.1.4)) becomes

$$\frac{dP}{dr} = -g\rho = \frac{d\phi}{dr}\rho. \tag{5.1.13}$$

Polytropes are special solutions of the pressure balance equation when it is assumed that

$$P = K\rho^\gamma, \tag{5.1.14}$$

where K and γ are constants. It will become apparent that, for any choice of $\gamma > 6/5$, K and the pressure and density distributions throughout the model are fixed by the choice of the model mass M and radius R. Conversely, if K and γ are fixed, M and R are determined.

Inserting P from eq. (5.1.14) into eq. (5.1.13) and integrating yields

$$\frac{K\gamma}{\gamma - 1} \rho^{\gamma - 1} = \phi + \text{constant}. \tag{5.1.15}$$

Choosing the constant to be zero defines the potential to be zero at the model surface, where the density vanishes. By setting

$$\gamma = 1 + \frac{1}{N}, \tag{5.1.16}$$

one obtains

$$\rho = \left(\frac{\phi}{K(N + 1)}\right)^N. \tag{5.1.17}$$

Equations (5.1.14), (5.1.16), and (5.1.17) give

$$\frac{P}{\rho} = \frac{\phi}{N + 1}. \tag{5.1.18}$$

Replacing ρ in eq. (5.1.10) by its value in terms of ϕ as given by eq. (5.1.17), one obtains

$$\frac{d^2\phi}{dr^2} + \frac{2}{r}\frac{d\phi}{dr} + 4\pi G \left(\frac{\phi}{K(N + 1)}\right)^N = 0, \tag{5.1.19}$$

a differential equation involving just the single dependent variable ϕ.

A solution of eq. (5.1.19) is known as a polytrope of index N. Solutions for several values of N provide good representations of certain classes of real stars. For example, if one assumes a perfect gas equation of state, a polytrope of index $N = 3/2$ corresponds to a star in which entropy per gram is independent of position in the star. Over most of the interior of main–sequence stars less massive than ~ 0.3–$0.5\ M_\odot$, energy flow is by convection, with $V \sim V_{ad}$ (see Chapter 3). Thus, such stars, which are also the companions of white dwarfs in short period cataclysmic variables, are describable by $N = 3/2$ polytropes. Low mass white dwarfs, in which the electrons are non-relativistically degenerate, are also describable by $N = 3/2$ polytropes. Index $N = 3$ polytropes, in which the ratio of gas pressure to radiation pressure is everywhere the same, provide rough representations of homogeneous main sequence stars of mass $\geq M_\odot$. The upper limit on the mass of a cold white dwarf can be derived by using an $N = 3$ polytrope in which pressure is given by relativistic electrons. Clearly, the exploration of polytropic structures is worthwhile.

Equation (5.1.19) can be made dimensionless by choosing the constants λ and ϕ_c in the variables

$$z = \frac{r}{\lambda} \quad \text{and} \quad u = \frac{\phi}{\phi_c} \tag{5.1.20}$$

in such a way that

$$\frac{4\pi G\lambda^2}{\phi_c} \left(\frac{\phi_c}{K(N+1)}\right)^N = 1. \tag{5.1.21}$$

Then,

$$\frac{d^2u}{dz^2} + \frac{2}{z}\frac{du}{dz} + u^N = 0. \tag{5.1.22}$$

At the center, where $z = 0$, one may choose

$$u = 1 \text{ and } \frac{du}{dz} = 0. \tag{5.1.23}$$

The choice $u = 1$ at the center means that ϕ_c is the gravitational potential at the center. The choice $du/dz = 0$ at the center follows from eq. (5.1.12) when it is recognized that, near the center,

$$M(r) \sim \rho_c \frac{4\pi}{3} r^3, \tag{5.1.24}$$

where ρ_c is the density at the center.

In general, u can change sign at some value of z. But, as is evident from eq. (5.1.17), only those portions of the solution where $u \geq 0$ are physically meaningful. Thus, the surface of a satisfactory model may be chosen to be at the point where u first vanishes. At this point, one may define quantities z_s and $(du/dz)_s$ such that

$$z_s = z(u = 0) \text{ and } \left(\frac{du}{dz}\right)_s = \left(\frac{du}{dz}\right)_{u=0}. \tag{5.1.25}$$

The corresponding physical variables of the model are

$$r(u = 0) = R \text{ and } M(u = 0) = M(R), \tag{5.1.26}$$

where R and $M(R)$ are, respectively, the radius and mass of the physical model.

With these definitions, one can determine the scaling factors in eq. (5.1.20). Thus, for example,

$$\lambda = \frac{R}{z_s}. \tag{5.1.27}$$

Using eqs. (5.1.20), (5.1.27), and (5.1.12), one has that, at any point in the model,

$$\frac{d\phi}{dr} = \frac{\phi_c}{\lambda}\frac{du}{dz} = \frac{\phi_c z_s}{R}\frac{du}{dz} = -\frac{GM(r)}{r^2}. \tag{5.1.28}$$

Setting $r = R$ and $M(r) = M(R)$ in this equation, one obtains

$$\phi_c = \frac{GM(R)}{R}\frac{1}{[-z_s \, (du/dz)_s]}. \tag{5.1.29}$$

The factor K in eq. (5.1.14) can now be determined as a function of model properties by using eqs. (5.1.27) and (5.1.29) in eq. (5.1.21) to obtain

$$[K(N+1)]^N = 4\pi G \left(\frac{R}{z_s}\right)^2 \phi_c^{N-1}. \tag{5.1.30}$$

Using eq. (5.1.30) in eq. (5.1.17) gives

$$\rho_c = \frac{\phi_c}{4\pi G} \left(\frac{z_s}{R}\right)^2, \tag{5.1.31}$$

where ρ_c is the density at the center. Then, employing eq. (5.1.29) in eq. (5.1.31) gives

$$\rho_c = \frac{M(R)}{4\pi R^3} \left(-\frac{z_s}{(du/dz)_s}\right) = \frac{1}{3} \bar{\rho} \left(-\frac{z_s}{(du/dz)_s}\right), \tag{5.1.32}$$

where

$$\bar{\rho} = \frac{M(R)}{4\pi R^3/3} \tag{5.1.33}$$

is the mean density.

From eq. (5.1.18), the central pressure is given by

$$P_c = \frac{\phi_c}{N+1}\rho_c. \tag{5.1.34}$$

Using eqs. (5.1.29) and (5.1.32) to replace ϕ_c and ρ_c in this equation and multiplying by the volume, one obtains

$$\frac{4\pi}{3} R^3 P_c = e(N) \frac{GM^2(R)}{R}, \tag{5.1.35}$$

where

$$e(N) = \frac{1}{3(N+1)} \left(-\frac{1}{(du/dz)_s}\right)^2. \tag{5.1.36}$$

Since, in a real star, pressure is proportional primarily to the kinetic energy density of particles and radiation, eq. (5.1.35) shows that, for a given N, the kinetic energy content of a model star is proportional to its gravitational binding energy, as measured roughly by GM^2/R.

5.2 Several properties of solutions as functions of N

Except when $N = 0$, 1, and 5, solutions of the polytropic equation can be obtained only by numerical integration. None of the three analytic solutions has much practical use, but finding them is simple. In the $N = 0$ and $N = 1$ cases, solutions are made easier by the choice of a modified independent variable $\bar{u}$ related to u by

$$u = \frac{\bar{u}}{z}. \tag{5.2.1}$$

With this substitution, eq. (5.1.22) becomes

$$\frac{d^2\bar{u}}{dz^2} + z\left(\frac{\bar{u}}{z}\right)^N = 0. \tag{5.2.2}$$

Setting $N = 0$,

$$\frac{d^2\bar{u}}{dz^2} = -z, \tag{5.2.3}$$

which has the solution

$$\bar{u} = zu = -\frac{z^3}{6} + Az + B, \tag{5.2.4}$$

where A and B are constants. Since $u = 1$ at $z = 0$, it follows that $B = 0$ and $A = 1$. Thus,

$$u = 1 - \frac{z^2}{6}, \tag{5.2.5}$$

and, at the surface,

$$z_s = \sqrt{6} \quad \text{and} \quad \left(\frac{du}{dz}\right)_s = -\frac{z_s}{3} = -\frac{\sqrt{6}}{3}. \tag{5.2.6}$$

Internal physical characteristics of the model are related to global characteristics by (see eqs. (5.1.29), (5.1.32), and (5.1.34))

$$\phi_c = \frac{1}{2}\frac{GM(R)}{R}, \tag{5.2.7}$$

$$\rho_c = \bar{\rho}, \quad \text{and} \tag{5.2.8}$$

$$\frac{4\pi}{3}R^3 P_c = \frac{1}{2}\frac{GM^2(R)}{R}. \tag{5.2.9}$$

In words, (1) the work required to carry a particle from the center to the surface is half of the work required to carry the particle from the surface to infinity, (2) the model is of constant density, and (3) the kinetic energy content of the model is comparable to the gravitational binding energy as measured in units of $GM^2(R)/R$.

Similar results follow from the choice $N = 1$. In this case,

$$\frac{d^2\bar{u}}{dz^2} + \bar{u} = 0 \tag{5.2.10}$$

and the general solution is

$$\bar{u} = zu = A\sin z + B\cos z. \tag{5.2.11}$$

The boundary conditions demand that $B = 0$ and $A = 1$. Thus,

$$u = \frac{\sin z}{z} \tag{5.2.12}$$

and, at the surface,

$$z_s = \pi \quad \text{and} \quad \left(\frac{du}{dz}\right)_s = -\frac{1}{\pi}. \tag{5.2.13}$$

Internal physical characteristics of the $N = 1$ model are related to global characteristics by

$$\phi_c = \frac{GM(R)}{R}, \tag{5.2.14}$$

$$\rho_c = \frac{\pi^2}{3}\, \bar{\rho}, \quad \text{and} \tag{5.2.15}$$

$$\frac{4\pi}{3} R^3\, P_c = \frac{\pi^2}{6} \frac{GM^2(R)}{R}. \tag{5.2.16}$$

In words, (1) the work required to carry a particle from the center to the surface is equal to the work required to carry the particle from the surface to infinity, (2) the model is centrally condensed, with the central density being approximately three times larger than the mean density, and (3) the kinetic energy content of the model is, as in the $N = 0$ case, comparable to the gravitational binding energy as measured in units of $GM^2(R)/R$.

For all other values of N except $N = 5$, solutions require numerical integration. Tables describing solutions for a variety of indices may be found in Chapter II of R. Emden's book *Gaskugeln – Anwendungen der Mechanische Wärmtheorie* (1907), three of which (for $N = 1.5, 2,$ and 3) are reproduced in Chapter IV of Eddington's book *Internal Constitution of the Stars*. Results of higher accuracy and finer zoning are given in *Mathematical Tables of the British Association for the Advancement of Science, Volume II - Emden Functions* (1932). They were obtained by J. C. P. Miller and D. H. Sadler using a quadrature scheme due to J. R. Airey.

Of the 50 points given in the British Association table for $N = 1.5$, 38 are presented in Table 5.2.1. In this table, $r \propto z$, $\rho \propto u^{3/2}$, $P \propto u^{5/2}$, and $\bar{\rho}(r)/\rho_c \propto -(z/3)\, dz/du$, where $\bar{\rho}(r) = 3/4\pi r^3\, M(r)$, and $M(r) \propto -z^2\, du/dz$. For a perfect gas equation of state, $T \propto u$, where T is the temperature.

Structure variables for the $N = 1.5$ polytrope are displayed as functions of radius in Fig. 5.2.1 and as functions of mass in Fig. 5.2.2. The relationships $P(r)/P_c = u^{2.5}$ and $\rho(r)/\rho_c = u^{1.5}$ are exact, but the relationship $T(r)/T_c = u$ is exact only in the limit of a perfect gas of uniform composition. The relationship between $M(r)/M(R)$ and r/R shown in the figures follows from eq. (5.1.28), which gives

$$\frac{M(r)}{M(R)} = \left(\frac{r}{R}\right)^2 \frac{du/dz}{(du/dz)_s}, \tag{5.2.17}$$

and from eqs. (5.1.20) and (5.1.27), which give

$$\frac{r}{R} = \frac{z}{z_s}. \tag{5.2.18}$$

		Table 5.2.1 Structure variables in a polytrope of index $N = 3/2$				
z	u	$u^{3/2}$	$u^{5/2}$	$-\dfrac{du}{dz}$	$-\dfrac{z}{3}\dfrac{dz}{du}$	$-z^2\dfrac{du}{dz}$
0.00	1.00000	1.00000	1.00000	0.000000	1.0000000	0.00000000
0.10	0.99834	0.99750	0.99584	0.033283	1.0015002	0.00033283
0.20	0.99335	0.99005	0.97347	0.066268	1.0060160	0.0026507
0.30	0.98510	0.97773	0.96317	0.098660	1.013581	0.0088794
0.40	0.97365	0.96074	0.93542	0.13018	1.02426	0.020828
0.50	0.95910	0.93929	0.90087	0.16054	1.03813	0.040136
0.60	0.94159	0.91367	0.86030	0.18952	1.05531	0.068226
0.70	0.92125	0.88424	0.81461	0.21686	1.0760	0.10626
0.80	0.89828	0.85136	0.76476	0.24238	1.1002	0.15512
0.90	0.87285	0.81547	0.71178	0.26589	1.1283	0.21537
1.00	0.84517	0.77699	0.65669	0.28726	1.1604	0.28726
1.10	0.81547	0.73640	0.60051	0.30636	1.1969	0.37069
1.20	0.78398	0.69415	0.54420	0.32311	1.2380	0.46528
1.30	0.75093	0.65072	0.48865	0.33747	1.2841	0.57033
1.40	0.71656	0.60657	0.43465	0.34942	1.3356	0.68486
1.50	0.68112	0.56210	0.38285	0.35896	1.3929	0.80766
1.60	0.64485	0.51783	0.33392	0.36614	1.4566	0.93731
1.70	0.60797	0.47405	0.28821	0.37102	1.5273	1.0722
1.80	0.57072	0.43116	0.24607	0.37368	1.6056	1.2107
1.90	0.53331	0.38946	0.20770	0.37425	1.6923	1.3510
2.00	0.49594	0.34925	0.17321	0.37283	1.7881	1.4913
2.10	0.45880	0.31077	0.14258	0.36958	1.8940	1.6299
2.20	0.42208	0.27421	0.11574	0.36464	2.0111	1.7649
2.30	0.38592	0.23975	0.092524	0.35818	2.1405	1.8948
2.40	0.35049	0.20749	0.072724	0.35036	2.2834	2.0181
2.50	0.31589	0.17755	0.056085	0.34134	2.44135	2.1338
2.60	0.28225	0.14995	0.042325	0.33131	2.6159	2.2396
2.70	0.24966	0.12475	0.031144	0.32042	2.8088	2.3358
2.80	0.21819	0.10192	0.022238	0.30884	3.0221	2.4213
2.90	0.18791	0.081456	0.015306	0.29673	3.2577	2.4955
3.00	0.15886	0.063316	0.010058	0.28425	3.5180	2.5583
3.10	0.13107	0.047450	0.0062191	0.27154	3.8054	2.6095
3.20	0.10455	0.033806	0.0035345	0.25875	4.1224	2.6496
3.30	0.079315	0.022337	0.0017717	0.24601	4.4714	2.6790
3.40	0.055344	0.013020	0.00072058	0.23344	4.8549	2.6986
3.50	0.032617	0.0058903	0.00019212	0.22119	5.2745	2.7096
3.60	0.011091	0.0011680	0.000012955	0.20939	5.7309	2.7137
3.65375	0.00000	0.00000	0.00000	0.203301	5.99070	2.71406

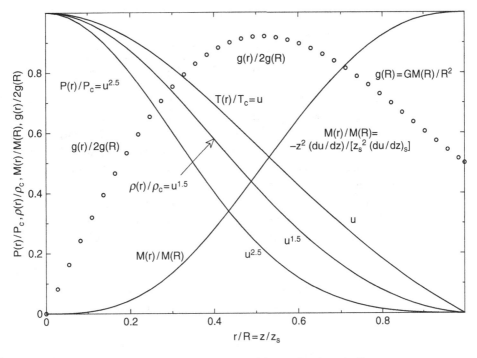

Fig. 5.2.1   Structure variables in an $N = 1.5$ polytrope as functions of radius

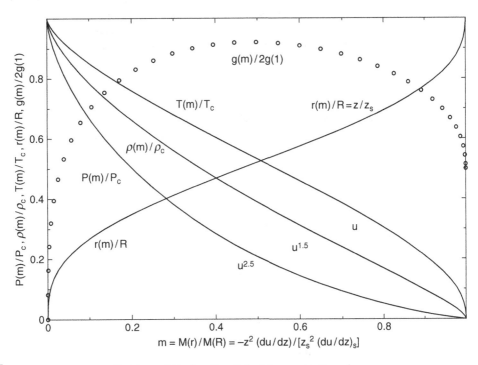

Fig. 5.2.2   Structure variables in an $N = 1.5$ polytrope as functions of mass

Equations (5.2.17) and (5.2.18) together give

$$\frac{M(r)}{M(R)} = \left(\frac{z}{z_s}\right)^2 \frac{du/dz}{(du/dz)_s} = \frac{z^2 \, du/dz}{(z^2 \, du/dz)_s}.$$ (5.2.19)

The circles in Figs. 5.2.1 and 5.2.2 show the ratio of the local gravitational acceleration $g(r)$ to twice the surface gravity g_s, where $g_s = (\phi_c/R) \, (du/dz)_s = GM(R)/R^2$. The circles are plotted at the points given explicitly in Table 5.2.1. Any discontinuities that can be discerned in the slopes of the curves for the structure variables can be understood as consequences of the spacing between points and of a plotting routine which passes straight lines between available points.

Of the 70 points given in the British Association Tables for $N = 3$, 49 are presented in Table 5.2.2. Out of curiosity, four points from Emden's book, at $z = 0.25, 0.75, 1.25$, and at 1.75, have been included. In Table 5.2.2, $\rho \propto u^3$, $P \propto u^4$, and all other variables are as in Table 5.2.1.

Structure variables for the $N = 3$ polytrope are shown in Figs. 5.2.3 and 5.2.4 as functions of radius and mass, respectively. The circles show the gravitational acceleration (relative to seven times the gravitational acceleration at the surface) at the points given explicitly in Table 5.2.2.

Comparisons between Figs. 5.2.3 and 5.2.1 show that structure variables are more concentrated toward the center in the $N = 3$ polytrope than in the $N = 1.5$ polytrope. The same conclusion follows from comparisons between Figs. 5.2.4 and 5.2.2. Another way of illustrating the difference in mass concentrations follows from the definition in atomic physics of the probability distribution of an electron in a stationary state. In the simplest case of the ground state of a single electron atom, this distribution is given by the function

$$dP(r) = 4\pi r^2 \, |\psi(r)|^2 \, dr,$$ (5.2.20)

where $\psi(r)$ is the wave function of the electron and $dP(r)$ may be interpreted as the probability that the electron is in the interval dr at the distance r from the central nucleus. The quantity

$$\rho_P(r) = |\psi(r)|^2$$ (5.2.21)

may be interpreted as a probability density. In a spherically symmetric model star, the quantity

$$dM(r) = 4\pi r^2 \, \rho(r) \, dr,$$ (5.2.22)

gives the contribution to the mass of the star of matter actually in the interval dr at the distance r from the center. Differentiating eqs. (5.2.19) and (5.2.18) yields

$$\frac{dM(r)}{dr} = \frac{dM(r)}{dz} \frac{dz}{dr} = \frac{M(R)}{(z^2 \, du/dz)_s} \frac{d}{dz} \left(z^2 \frac{du}{dz}\right) \frac{z_s}{R}.$$ (5.2.23)

			Table 5.2.2 Structure variables in a polytrope of index $N = 3$		
z	u	u^3	u^4	$-\dfrac{du}{dz}$	$-\dfrac{z}{3}\dfrac{dz}{du}$	$-z^2\dfrac{du}{dz}$
0.0	1.00000	1.00000	1.00000	0.00000	1.0000	0.00000
0.10	0.99834	0.99502	0.99336	0.033234	1.0030	0.00033
0.20	0.99337	0.98025	0.97375	0.065874	1.01203	0.00263
0.25	0.98975	0.96960	0.95966	0.08204	1.01576	0.00513
0.30	0.98520	0.95625	0.94210	0.097354	1.02718	0.00876
0.40	0.97396	0.92389	0.89983	0.12716	1.04857	0.02035
0.50	0.95984	0.88429	0.84878	0.15484	1.07638	0.03871
0.60	0.94307	0.83876	0.79101	0.18004	1.11087	0.06481
0.70	0.92392	0.78869	0.72869	0.20249	1.15231	0.09922
0.75	0.91355	0.76242	0.69650	0.21270	1.17536	0.11964
0.80	0.90267	0.73551	0.66393	0.22203	1.20105	0.14210
0.90	0.87962	0.68058	0.59865	0.23857	1.25749	0.19324
1.00	0.85506	0.62515	0.53454	0.25213	1.32207	0.25213
1.10	0.82929	0.57032	0.47296	0.26279	1.39530	0.31797
1.20	0.80259	0.51699	0.41493	0.27069	1.47770	0.38979
1.25	0.78897	0.49111	0.38747	0.27370	1.52235	0.42766
1.30	0.77524	0.46591	0.36119	0.27603	1.56990	0.46648
1.40	0.74746	0.41761	0.31215	0.27902	1.67253	0.54688
1.50	0.71950	0.37247	0.26800	0.27991	1.78628	0.62980
1.60	0.69154	0.33072	0.22871	0.27896	1.91190	0.71412
1.70	0.66376	0.29244	0.19411	0.27640	2.05017	0.79879
1.75	0.64996	0.27458	0.17847	0.27460	2.12430	0.84096
1.80	0.63631	0.25764	0.16394	0.27249	2.20194	0.88286
1.90	0.60930	0.22621	0.13783	0.26745	2.36807	0.96548
2.00	0.58285	0.19800	0.11541	0.26149	2.54948	1.04596
2.10	0.55703	0.17284	0.096275	0.25481	2.74714	1.12371
2.20	0.53191	0.15049	0.080047	0.24758	2.96205	1.19827
2.30	0.50753	0.13073	0.066350	0.23994	3.19523	1.26929
2.40	0.48393	0.11333	0.054843	0.23203	3.44776	1.33652
2.50	0.46113	0.098053	0.045215	0.22396	3.72075	1.39981
2.60	0.43914	0.084683	0.037187	0.21584	4.01534	1.45907
2.70	0.41796	0.073013	0.030516	0.20772	4.33269	1.51430
2.80	0.39759	0.062850	0.024988	0.19969	4.67398	1.56555
2.90	0.37802	0.054017	0.020419	0.19178	5.04044	1.61289
3.00	0.35923	0.046356	0.016652	0.18405	5.43331	1.65645
3.10	0.34120	0.039722	0.013553	0.17652	5.85383	1.69638
3.20	0.32391	0.033985	0.011008	0.16922	6.30329	1.73285
3.30	0.30735	0.029033	0.0089231	0.16217	6.78295	1.76604
3.40	0.29147	0.024762	0.0072175	0.15538	7.29412	1.79615
3.50	0.27626	0.021085	0.0058249	0.14885	7.83809	1.82336
3.60	0.26169	0.017922	0.0046900	0.14258	8.41617	1.84787
3.80	0.23437	0.012873	0.0030171	0.13086	9.67984	1.88956

			Table 5.2.2 (Cont.)			
z	u	u^3	u^4	$-\dfrac{\mathrm{d}u}{\mathrm{d}z}$	$-\dfrac{z}{3}\dfrac{\mathrm{d}z}{\mathrm{d}u}$	$-z^2\dfrac{\mathrm{d}u}{\mathrm{d}z}$
4.00	0.20928	0.0091663	0.0019183	0.12017	11.0955	1.92272
4.20	0.18623	0.0064592	0.0012029	0.11047	12.6733	1.94866
4.50	0.15507	0.0037289	0.00057823	0.09762	15.3659	1.97678
4.80	0.12748	0.0020717	0.00026410	0.086590	18.4779	1.99503
5.00	0.11082	0.0013610	0.00015082	0.080126	20.8006	2.00315
5.20	0.095388	0.00086792	0.000082789	0.074292	23.3314	2.00885
5.50	0.074286	0.0004100	0.0000305	0.066585	27.5350	2.01411
6.00	0.043738	0.0000837	0.0000037	0.056044	35.6863	2.01758
6.50	0.017866	0.0000057	0.0000001	0.047768	45.3578	2.01821
6.80	0.004678	0.0000000	0.0000000	0.043647	51.9318	2.01824
6.89685	0.00000	0.000000	0.000000	0.042430	54.1825	2.01824

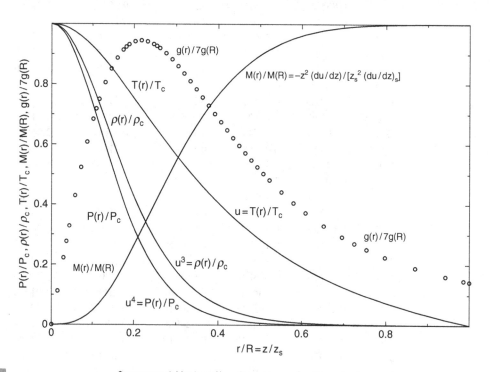

Fig. 5.2.3   Structure variables in an $N = 3$ polytrope as functions of radius

From eq. (5.1.22) one has that

$$\frac{\mathrm{d}}{\mathrm{d}z}\left(z^2\,\frac{\mathrm{d}u}{\mathrm{d}z}\right) = z^2\left(\frac{\mathrm{d}^2u}{\mathrm{d}z^2} + \frac{2}{z}\,\frac{\mathrm{d}u}{\mathrm{d}z}\right) = -z^2\,u^N, \qquad (5.2.24)$$

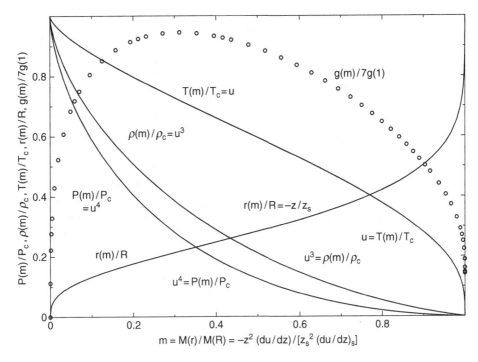

Fig. 5.2.4 Structure variables in an $N = 3$ polytrope as functions of mass

and so

$$\frac{\mathrm{d}M(r)}{\mathrm{d}r} = -\frac{M(R)}{R} \frac{z^2 \, u^N}{z_\mathrm{s} \, (\mathrm{d}u/\mathrm{d}z)_\mathrm{s}}. \qquad (5.2.25)$$

Figure 5.2.5 compares the differential contribution to mass as a function of r for the $N = 1.5$ and $N = 3$ polytropes. For a given mass $M(R)$ and radius R, the $N = 3$ polytrope is roughly twice as compact as the $N = 1.5$ polytrope.

Table 5.2.3 gives global characteristics of polytropes for other choices of N. The entries for N other than 0, 1, and 5 have been taken or derived from information in Table 4, p. 96, of Chandrasekhar's 1939 book *Stellar Structure*. Most of the entries in the Chandrasekhar book have, in turn, been taken from the British Association Tables with a few from Emden's 1907 book *Gaskugeln*.

From column 6 in Table 5.2.3, it is evident that, as N increases, more work is required to carry a particle from the center to the surface, as measured by ϕ_c, relative to the work required to carry the particle from the surface to infinity, as measured by $GM(R)/R$, and the central concentration, $\rho_\mathrm{c}/\bar{\rho}$, increases much more rapidly than does the potential at the center.

The last column in Table 5.2.3 gives the ratio of the gravitational binding energy $-\Omega$, defined in Section 3.2 by eq. (3.3.9), to the quantity $GM^2(R)/R$. The dependence of this

Table 5.2.3 Several properties of polytropes as functions of N

N	z_s	$\left(\dfrac{du}{dz}\right)_s$	$\left(z^2\dfrac{du}{dz}\right)_s$	$(5-N)z_s$	$\dfrac{\phi_c}{GM/R}$	$\rho_c/\bar{\rho}$	$e(N)$	$\dfrac{|\Omega|}{GM^2/R}$
0.0	$\sqrt{6}$	$-\dfrac{\sqrt{6}}{3}$	$-2\sqrt{6}$	$5\sqrt{6}$	$\dfrac{1}{2}$	1	$\dfrac{1}{2}$	$\dfrac{3}{5}$
0.5	2.7528	-0.49976	-3.7871	12.388	0.7269	1.8361	0.8897	0.667
1.0	π	$-\dfrac{1}{\pi}$	$-\pi$	4π	$\dfrac{\pi^2}{3}$	1	$\dfrac{\pi^2}{6}$	$\dfrac{3}{4}$
1.5	3.6538	-0.20330	-2.7141	12.788	1.3462	5.9907	3.2260	0.857
2.0	4.3529	-0.12725	-2.4111	13.059	1.8054	11.403	6.8619	1.000
2.5	5.3553	-0.076264	-2.1872	13.388	2.4485	23.407	16.375	1.200
3.0	6.8969	-0.042430	-2.0182	13.794	3.3174	54.185	46.293	1.500
3.25	8.0189	-0.030322	-1.9498	14.033	4.1127	88.153	85.305	1.714
3.5	9.5358	-0.020791	-1.8906	14.304	5.0438	152.88	171.36	2.000
4.0	14.9716	$-8.018/-3$	-1.7972	14.972	8.3305	622.42	1037.0	3.000
4.5	31.8365	$-1.715/-3$	-1.7378	15.918	18.320	6189.7	20618	6.000
5.0	∞	0.0	$-\sqrt{3}$	$\dfrac{32\sqrt{3}}{\pi}$	∞	∞	∞	∞

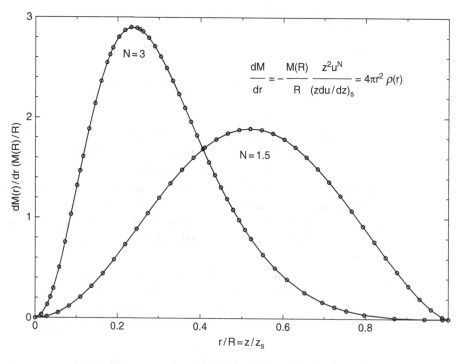

$$\frac{dM}{dr} = -\frac{M(R)}{R}\,\frac{z^2 u^N}{(z\,du/dz)_s} = 4\pi r^2\,\rho(r)$$

Fig. 5.2.5 Differential mass versus radius distributions in $N = 1.5$ and $N = 3$ polytropes

ratio on N is easy to derive. From the definition in eq. (3.3.1) of Ω as an integral over mass, one has

$$
\begin{aligned}
\Omega = -\int_0^{M(R)} \frac{GM(r)\,\mathrm{d}M(r)}{r} &= -\frac{1}{2}\int_0^{M^2(R)} \frac{G\,\mathrm{d}[M(r)]^2}{r} \\
&= -\frac{1}{2}\frac{GM^2(R)}{R} - \frac{1}{2}\int_0^R \frac{GM(r)}{r^2}M(r)\,\mathrm{d}r \\
&= -\frac{1}{2}\frac{GM^2(R)}{R} + \frac{1}{2}\int_0^R \frac{\mathrm{d}\phi}{\mathrm{d}r}M(r)\,\mathrm{d}r \\
&= -\frac{1}{2}\frac{GM^2(R)}{R} - \frac{1}{2}\int_0^{M(R)} \phi\,\mathrm{d}M(r),
\end{aligned}
\tag{5.2.26}
$$

where integrals extend from the center to the surface and use has been made of the fact that ϕ vanishes at the surface and $M(r) \to 0$ as $r \to 0$. Next, by inserting the polytropic relationship between P, ρ, and ϕ given by eq. (5.1.18) into the rightmost integral of eq. (5.2.26), one finds that

$$
\Omega = -\frac{1}{2}\frac{GM^2(R)}{R} - \frac{(N+1)}{2}\int P\mathrm{d}V,
\tag{5.2.27}
$$

where $\mathrm{d}V = 4\pi r^2 \mathrm{d}r$ is a differential volume element. Replacing the integral in eq. (5.2.27) by its equivalent

$$
\int P\mathrm{d}V = -\frac{\Omega}{3},
\tag{5.2.28}
$$

as given by eq. (3.3.2), one has finally that

$$
\Omega = -\frac{3}{5-N}\frac{GM^2(R)}{R},
\tag{5.2.29}
$$

or

$$
\Omega = -3.8006 \times 10^{48}\ \mathrm{erg}\ \frac{3}{5-N}\frac{\bar{M}^2}{\bar{R}},
\tag{5.2.30}
$$

where $\bar{M} = M/M_\odot$ and $\bar{R} = R/R_\odot$. Equation (5.2.29) confirms the intuitive guess made in Chapter 3 that $-\Omega$ is of the order of $GM^2(R)/R$. From the reasoning in Section 3.3 that leads to eqs. (3.3.11)–(3.3.13), it follows that the binding energy of a polytrope of index N is

$$
E_{\mathrm{bind}} = \frac{\tilde{\beta}}{2}|\Omega| = \frac{\tilde{\beta}}{2}\frac{3}{5-N}\frac{GM^2(R)}{R},
\tag{5.2.31}
$$

where $\tilde{\beta}$ is a weighted average of P_{gas}/P defined by eq. (3.3.10).

To complete the discussion of polytropic solutions, it is of interest to examine the character of the $N = 5$ polytrope. The analytic solution in this case is

$$
u = \frac{1}{\left(1 + \frac{1}{3}z^2\right)^{1/2}},
\tag{5.2.32}
$$

as may be verified by insertion into eq. (5.1.22). The solution extends to infinity, with both z_s and R being infinite. However, the ratio $\lambda = R/z_s$ is, by construction, finite and determined by the choice of boundary conditions not involving R. From eqs. (5.1.29) and (5.1.31), it follows that

$$\lambda = \frac{R}{z_s} = \left(\frac{M(R)}{4\pi\rho_c} \left| \frac{1}{z_s^2 (du/dz)_s} \right| \right)^{1/3}. \tag{5.2.33}$$

For large z,

$$-z^2 \frac{du}{dz} \rightarrow \sqrt{3}, \tag{5.2.34}$$

so that

$$\lambda = \frac{R}{z_s} = \left(\frac{M(R)}{4\pi\sqrt{3}\,\rho_c} \right)^{1/3}. \tag{5.2.35}$$

Using eqs. (5.2.34) and (5.2.35) in eq. (5.1.29), one obtains

$$\phi_c = \frac{GM(R)}{\lambda} \left| \frac{1}{z_s^2 (du/dz)_s} \right| = \frac{GM(R)}{\sqrt{3}} \left(\frac{4\pi\sqrt{3}\,\rho_c}{M(R)} \right)^{1/3}. \tag{5.2.36}$$

From eqs. (5.1.34) and (5.2.36), it follows that

$$P_c = \frac{GM(R)}{6\sqrt{3}} \left(\frac{4\pi\sqrt{3}\rho_c}{M(R)} \right)^{1/3} \rho_c. \tag{5.2.37}$$

From eqs. (5.1.14) and (5.1.16), $P = K\rho_c^{1+1/N}$, and from eq. (5.1.17), $\phi = K\,(N+1)\,\rho^{1/N}$, so, from both eqs. (5.2.36) and (5.2.37), for the case $N = 5$, one has

$$K = \frac{P_c}{\rho_c^{6/5}} = \frac{\phi_c}{6\,\rho_c^{1/5}} = \frac{GM(R)}{6\sqrt{3}\,\rho_c^{1/5}} \left(\frac{4\pi\sqrt{3}\rho_c}{M(R)} \right)^{1/3}. \tag{5.2.38}$$

Thus, once the central density and mass of the configuration are chosen, all physical relationships in the model are determined.

Since its radius is infinite, the $N = 5$ polytrope is of limited utility in the context of stellar structure. However, the characteristics of the solution provide closure to the discussion of polytropes for which one can define a finite gravitational potential energy, as expressed by eq. (5.2.29), and they pose an interesting question. Two secure results are that, as shown in the third and fourth columns of Table 5.2.3, the surface derivative $-(du/dz)_s$ and the quantity $-(z^2 du/dz)_s$ decline monotonically as N increases from 0 to 5, to zero in the first instance, and by a factor of $2\sqrt{2} = 2.8284$ from $2\sqrt{6}$ to $\sqrt{3}$ in the second instance.

The interesting question is, how does z_s approach infinity as N approaches 5? Equation (5.2.29) may be written as

$$\Omega = -\frac{3}{5-N} \frac{GM^2(R)}{z_s} \frac{z_s}{R} = -\frac{G\,M^2(R)}{\lambda} \left(\frac{3}{(5-N)z_s} \right), \tag{5.2.39}$$

Since λ may be chosen to be finite (e.g., as given by eq. (5.2.35) when $N = 5$) and since, for a finite $M(R)$, Ω as defined by the first equality in eq. (5.2.26) is finite, eq. (5.2.39) implies that the quantity $(5-N)z_s$ approaches a finite limiting value as $N \rightarrow 5$. What is this limiting value? The fifth column in Table 5.2.3 shows that, for $N = 1 \rightarrow 4.5$, the product $(5-N)z_s$ is remarkably stable, increasing monotonically from 12.247 at $N = 0$ to 15.918 at $N = 4.5$. In the $(N-5)z_s$ versus N plane, the slope of a curve passing through points for $N < 4.5$ is everywhere positive, and an eye fit extrapolation yields $(5-N)z_s \rightarrow \stackrel{>}{\sim} 17$ as $N \rightarrow 5$.

Fortunately, an exact value for the integral over mass in eq. (5.2.26) exists for the $N = 5$ case. Using eqs. (5.2.19), (5.2.25), and (5.2.18), one has that

$$
\Omega = -\int_0^{M(R)} \frac{G M(r) \mathrm{d} M(r)}{r} = -G \int_0^{z_s} M(R) \frac{z^2 \mathrm{d} u/\mathrm{d} z}{z_s^2 (\mathrm{d} u/\mathrm{d} z)_s} \left(-\frac{M(R)}{R} \frac{z^2 u^N}{z_s (\mathrm{d} u/\mathrm{d} z)_s} \right) \frac{1}{z} \mathrm{d} z
$$

$$
= G \left(\frac{M(R)}{z_s^2 (\mathrm{d} u/\mathrm{d} z)_s} \right)^2 \frac{z_s}{R} \int_0^{z_s} z^3 \frac{\mathrm{d} u}{\mathrm{d} z} u^N \, \mathrm{d} z. \tag{5.2.40}
$$

Setting $N = 5$ and $(z^2 \mathrm{d} u/\mathrm{d} z)_s = -\sqrt{3}$, and using eq. (5.2.32) for u and its derivative, eq. (5.2.40) becomes

$$
\Omega_{N=5} = -G \frac{z_s}{R} \frac{M^2(R)}{3} \frac{1}{3} \int_0^{z_s} \frac{z^4 u^5 \, \mathrm{d} z}{(1 + z^2/3)^{3/2}} = -G \frac{z_s}{R} \frac{M^2(R)}{9} \int_0^{z_s} \frac{z^4}{(1 + z^2/3)^4} \, \mathrm{d} z. \tag{5.2.41}
$$

From # 124.4 in H. B. Dwight's *Tables of Integrals and Other Mathematical Data, 1934* (MacMillan, fourth edition, 1947), one finds that

$$
\int_0^\infty \frac{z^4}{(1 + z^2/3)^4} \, \mathrm{d} z = 3^4 \frac{1}{48\sqrt{3}} \frac{\pi}{2} = 9 \frac{\sqrt{3}\,\pi}{32}. \tag{5.2.42}
$$

Combining eqs. (5.2.42) and (5.2.41) yields

$$
\Omega_{N=5} = -\frac{\sqrt{3}\,\pi}{32} \frac{z_s}{R} G M^2(R) = -\frac{\sqrt{3}\,\pi}{32} \frac{G M^2(R)}{\lambda}. \tag{5.2.43}
$$

Comparing with eq. (5.2.39), one can conclude that, as $N \rightarrow 5$,

$$
(5-N)z_s \rightarrow \frac{32\sqrt{3}}{\pi} = 17.6425, \tag{5.2.44}
$$

consistent with the lower limit derived by extrapolation of results for $N < 4.5$.

5.3 Additional properties of solutions when an equation of state and a law of nuclear energy generation are assumed

In order to relate polytropes to realistic stellar configurations, it is necessary to invoke an equation of state. If one assumes that only gas pressure in the non-relativistic approximation and radiation pressure are important,

$$P = P_{\text{gas}} + P_{\text{rad}} = \frac{2}{3}\bar{U}_{\text{gas}} + \frac{1}{3}\bar{U}_{\text{rad}} = \frac{k}{\mu M_{\text{H}}}\rho T + \frac{1}{3}aT^4. \tag{5.3.1}$$

Defining total energies E by

$$E = \int \bar{U}\,\mathrm{d}\tau, \tag{5.3.2}$$

one has that the total thermal energy content of a model is

$$E_{\text{thermal}} = E_{\text{gas}} + E_{\text{rad}} = 3\left(1 - \frac{\bar{\beta}}{2}\right)\bar{P}\frac{4\pi}{3}R^3, \tag{5.3.3}$$

where $\bar{P}$ is the volume-weighted mean pressure and $\bar{\beta}$ is defined by

$$\bar{\beta}\bar{P} = \frac{\int \beta P\,\mathrm{d}V}{(4\pi R^3/3)}. \tag{5.3.4}$$

It has been shown in Section 3.3 (eq. (3.3.9)) that $E_{\text{thermal}} = -(\tilde{\beta}/2)\,\Omega$, where $\tilde{\beta}(<1)$ is a weighted value of β. Thus, using eqs. (5.1.35), (5.2.19), and (5.3.4),

$$\frac{P_c}{\bar{P}} = 2\left(1 - \frac{\bar{\beta}}{2}\right)(5 - N)\left(\frac{e(N)}{\tilde{\beta}}\right) = \frac{(2 - \bar{\beta})}{3\tilde{\beta}}\frac{5 - N}{N + 1}\left(\frac{1}{(\mathrm{d}u/\mathrm{d}z)_s}\right)^2. \tag{5.3.5}$$

Values of $P_c/\bar{P}$ for $\tilde{\beta} = 1$ are in the second column of Table 5.3.1.

From eqs. (5.1.18) and (5.1.20), one has

$$\frac{P}{\rho} = \frac{P_c}{\rho_c}u. \tag{5.3.6}$$

Assuming that the gas pressure is that of an ideal non-degenerate, non-relativistic perfect gas and that the composition is homogeneous,

$$\frac{T}{T_c} = \frac{\beta}{\beta_c}u, \tag{5.3.7}$$

where $\beta = P_{\text{gas}}/P$ and T_c and β_c are, respectively, T and β at the center. Using P_{gas} and P_{rad} as given by eq. (5.3.1), one has

$$\frac{P_{\text{rad}}}{\beta^3 P_{\text{gas}}} = \frac{1 - \beta}{\beta^4} = \frac{a}{3}\left(\frac{\mu M_{\text{H}}}{k}\right)^4\frac{P^3}{\rho^4}. \tag{5.3.8}$$

N	$\frac{P_c}{P}$	$f(N)$	$\frac{kT_c}{\beta_c \Phi_s}$	$\frac{\bar{T}/\bar{\beta}}{T_c/\beta_c}$	$\Gamma(N)$	$\left(\frac{L}{\epsilon_c M}\right)_{s=4}$	$\left(\frac{L}{\epsilon_c M}\right)_{s=16}$
0.0	$\frac{5}{2}$	24	$\frac{1}{2}$	$\frac{2}{5}$	$\frac{1}{8}\sqrt{\frac{\pi}{6}}$	0.0976	0.0183
0.5	4.0038	48.405	0.48459	0.45857	0.11700	0.113	0.0226
1.0	$\frac{2\pi^2}{3}$	$8\pi^2$	$\frac{1}{2}$	$\frac{1}{2}$	$\frac{1}{4\sqrt{\pi}}$	0.123	0.0260
1.5	11.291	115.10	0.53847	0.53059	0.16326	0.126	0.0287
2.0	20.586	156.96	0.60179	0.55389	0.18378	0.125	0.0308
2.5	40.937	205.11	0.69957	0.57178	0.20260	0.122	0.0323
3.0	92.586	260.68	0.85435	0.58524	0.21959	0.117	0.0334
3.25	149.28	291.84	0.96769	0.59050	0.22726	0.115	0.0338
3.5	257.04	325.71	1.12087	0.59479	0.23438	0.112	0.0340
4.0	1037.0	403.74	1.66610	0.60020	0.24656	0.106	0.0342
4.5	10309	502.44	3.33100	0.60044	0.25499	0.0994	0.0338
5.0	∞	648	∞	$\frac{3\pi}{16}$	$\frac{1}{4}\sqrt{\frac{\pi}{3}}$	0.0910	0.0325

Table 5.3.1 Additional approximate properties of polytropes as functions of N

Using eqs. (5.1.29), (5.1.31), and (5.1.33) in eq. (5.3.8) one finds

$$\frac{1-\beta_c}{\beta_c^4} = \frac{4\pi}{3}\, a\, G^3 \left(\frac{M_H}{k}\right)^4 \frac{1}{(N+1)^3 z_s^4 (du/dz)_s^2}\, \mu^4\, M^2(R). \tag{5.3.9}$$

Since

$$\frac{4\pi}{3}\, a\, G^3 \left(\frac{M_H}{k}\right)^4 M_\odot^2 = 0.805\,68, \tag{5.3.10}$$

one has that

$$\frac{1-\beta_c}{\beta_c^4} = 0.805\,68\, \frac{1}{f(N)}\, \mu^4 \left(\frac{M(R)}{M_\odot}\right)^2 = 0.111\,55\, \bar{\mu}^4\, \frac{1}{f(N)} \left(\frac{M(R)}{M_\odot}\right)^2, \tag{5.3.11}$$

where

$$f(N) = (N+1)^3\, z_s^4 (du/dz)_s^2 \tag{5.3.12}$$

and

$$\bar{\mu} = \frac{\mu}{0.61}. \tag{5.3.13}$$

Values of $f(N)$ are given in the third column of Table 5.3.1.

Since $P/\rho \propto u$ and $\rho \propto u^N$, one has from eq. (5.3.8) that, away from the center, β varies according to

$$\frac{1-\beta}{\beta^4} = \frac{1-\beta_c}{\beta_c^4}\, u^{3-N}. \tag{5.3.14}$$

Thus, for an $N=3$ polytrope, β is everywhere the same.

From the theorem expressed by eq. (3.3.2),

$$-\Omega = 3 \int_0^{M(R)} \frac{P}{\rho} dM(r) = \frac{3k}{\mu M_H} \int_0^{M(R)} \frac{T}{\beta} dM(r) = \frac{3}{\tilde{\beta}} \frac{k\bar{T}}{\mu M_H} M(R) = \frac{2}{\tilde{\beta}} E_{gas},$$

$$(5.3.15)$$

where $\bar{T} = \int T dM(r)/M(R)$ is the mass-weighted mean temperature and $\tilde{\beta}$ is an average defined by the equation. Note that this result has already been derived in Chapter 3, eq. (3.3.9).

Using eqs. (5.1.33) and (5.1.29), one can also write

$$\frac{1}{\beta_c} \frac{kT_c}{\mu M_H} = \frac{P_c}{\rho_c} = \frac{\phi_c}{N+1} = \frac{1}{N+1} \frac{GM(R)}{R} \left(\frac{1}{-(z \, du/dz)_s} \right),$$

$$(5.3.16)$$

or

$$\frac{\beta_c}{kT_c} \Phi_s = (N+1) \left(-z \frac{du}{dz} \right)_s,$$

$$(5.3.17)$$

where

$$\Phi_s = \mu M_H \frac{GM(R)}{R}$$

$$(5.3.18)$$

is the gravitational potential energy of an average particle at the surface, defined as the energy required to remove the particle to infinity and not to be confused with the surface potential $\phi_s = 0$. Numerically,

$$T_c = 14.1135 \times 10^6 \, \text{K} \, \beta_c \, \bar{\mu} \left((N+1) \left(-z \frac{du}{dz} \right)_s \right)^{-1} \frac{M(R)}{M_\odot} \frac{R_\odot}{R},$$

$$(5.3.19)$$

where $\bar{\mu} = \mu/0.61$. Equation (5.3.17) is a generalization of eq. (3.1.7) and eq. (5.3.19) is the extension of eq. (3.1.13). Values of $kT_c/\beta_c \Phi_s$ are given in the fourth column of Table 5.3.1.

Using eqs. (5.3.15), (5.2.19), and (5.3.16), one has also that

$$\frac{\bar{T}}{T_c} = \frac{N+1}{5-N} \left(-z \frac{du}{dz} \right)_s \frac{\tilde{\beta}}{\beta_c}.$$

$$(5.3.20)$$

For $N = 0$ and $N = 1$, eqs. (5.3.16) and (5.3.17) give

$$kT_c = 0.5 \, \beta_c \left(\mu M_H \frac{GM(R)}{R} \right),$$

$$(5.3.21)$$

and eq. (5.3.20) gives

$$\frac{\bar{T}}{T_c} = \frac{2}{5} \frac{\tilde{\beta}}{\beta_c} \quad \text{for} \ N = 0,$$

$$(5.3.22)$$

and

$$\frac{\bar{T}}{T_c} = \frac{1}{2} \frac{\tilde{\beta}}{\beta_c} \quad \text{for} \ N = 1.$$

$$(5.3.23)$$

Values of $\bar{T}/\tilde{\beta}$ divided by T_c/β_c for other values of N are given in the fifth column of Table 5.3.1.

The last three columns in Table 5.3.1 have to do with an approximate theorem regarding nuclear energy generation. The derivation makes use of the fact that, near the origin, the leading term in a series expansion for the variable u is an exponential. Inserting

$$u = \exp\left(-\frac{z^2}{6}\right) \sum_{n=1}^{\infty} a_n z^n \qquad (5.3.24)$$

into eq. (5.1.22) and solving for the coefficients a_n, one finds

$$u = \exp\left(-\frac{z^2}{6}\right)\left[1 + \left(\frac{N}{120} - \frac{1}{72}\right)z^4 + O(z^6)\right]. \qquad (5.3.25)$$

The explicit terms in eq. (5.3.25) match the exact solutions to a fraction of a percent for values of $z < 1$. For example, for $N = 0$ and $z = 1$, the approximation gives a value which is larger than the exact solution by 0.167%. For $N = 3$ and $z = 1$, it gives a value larger by 0.098% than the numerical solution in Table 5.2.2.

The luminosity of the polytrope is given by integrating from $r = 0$ to $r = R$ the differential

$$dL(r) = \epsilon \, dM(r), \qquad (5.3.26)$$

where ϵ is the energy generation rate. From eqs. (5.2.25) and (5.2.18), one has that

$$dM(r) = -\frac{M(R)}{z_s^2 (du/dz)_s} z^2 u^N \, dz, \qquad (5.3.27)$$

and so

$$dL(r) = -\frac{M(R)}{z_s^2 (du/dz)_s} \epsilon \, z^2 u^N \, dz. \qquad (5.3.28)$$

To make further progress, approximate the energy-generation rate by a power law

$$\epsilon = A \, \rho^{1+k} \, T^s, \qquad (5.3.29)$$

where A, k, and s are constants. Since $\rho = \rho_c u^N$ and $P/\rho = (P_c/\rho_c)u$, and using eq. (5.3.7), one can write

$$L(R) = -\frac{M(R)}{z_s^2 (du/dz)_s} \epsilon_c \int_0^{z_s} u^{N(1+k)} \left(u\frac{\beta}{\beta_c}\right)^s z^2 u^N dz, \qquad (5.3.30)$$

where ϵ_c is the value of ϵ at the origin, and $L(R)$ is the total luminosity of the model.

The next step is to insert u from eq. (5.3.25) into eq. (5.3.30) and integrate over z from zero to infinity, assuming that the exponential term eliminates most of the error committed by extending the integration past $z = z_s$. Strictly speaking, the factor of $(\beta/\beta_c)^s$ should be retained in the integrand of eq. (5.3.30), but this would require a numerical integration. Instead, β/β_c in the integrand is here approximated by unity. However, to determine ϵ_c, one should still use the central temperature given by eqs. (5.3.19) and (5.3.18), with β_c given by eq. (5.3.9).

Assuming that $[1 + Cz^4]^{6\alpha} \sim [1 + 6\alpha\, Cz^4]$, where C is the coefficient of z^4 in eq. (5.3.25), eq. (5.3.30) becomes

$$L(R) \sim -\frac{M(R)}{z_s^2 (du/dz)_s}\, \epsilon_c \int_0^\infty e^{-\alpha z^2} z^2 \, dz \left[1 + 6\,\alpha \left(\frac{N}{120} - \frac{1}{72}\right) z^4 + \cdots\right], \quad (5.3.31)$$

where

$$\alpha = \frac{N(2+k) + s}{6}. \quad (5.3.32)$$

To perform the necessary integrals, one may use the facts that

$$\int_0^\infty e^{-\alpha x^2} \, dx = \frac{\sqrt{\pi}}{2} \frac{1}{\alpha^{1/2}} \quad (5.3.33)$$

and

$$\int_0^\infty x^{2n}\, e^{-\alpha x^2} \, dx = (-1)^n \frac{\partial^n}{\partial \alpha^n} \left(\frac{\sqrt{\pi}}{2} \frac{1}{\alpha^{1/2}}\right), \quad (5.3.34)$$

to obtain

$$\frac{L(R)}{\epsilon_c M(R)} \sim \Gamma(N) \frac{1}{\alpha^{3/2}} \left(1 + \frac{3N - 5}{16} \frac{1}{\alpha} + \cdots\right), \quad (5.3.35)$$

where

$$\Gamma(N) = -\frac{\sqrt{\pi}}{4 z_s^2 (du/dz)_s}. \quad (5.3.36)$$

Values of $\Gamma(N)$ are given in the third from last column in Table 5.3.1. Inserting α as a function of N, k, and s from eq. (5.3.32), one has

$$\frac{L(R)}{\epsilon_c M(R)} \sim \Gamma(N) \left(\frac{6}{N(2+k) + s}\right)^{3/2} \left(1 + \frac{3}{8} \frac{3N - 5}{N(2+k) + s}\right). \quad (5.3.37)$$

An effective mass-fraction $\Delta M_{\text{eff}}/M(R)$ may be defined by

$$\left(\frac{1}{2} \epsilon_c \Delta M_{\text{eff}}\right) \frac{1}{M(R)} = \left(\frac{1}{2}\epsilon_c \frac{\Delta M_{\text{eff}}}{M(R)}\right) = \frac{L(R)}{M(R)} = \epsilon_{\text{Average}}. \quad (5.3.38)$$

This mass fraction, which is twice the average energy-generation rate divided by the energy-generation rate at the center, is a measure of the fraction of the mass of the model star over which nuclear energy generation is important. Values of $L(R)/\epsilon_c M(R)$ when $k = 0$ are given in the last two columns of Table 5.3.1 for $s = 4$ (when pp reactions dominate energy production) and $s = 16$ (when CN-cycle reactions dominate energy production), respectively (see Chapter 6). Three lessons are contained in these last two columns. First, the production of nuclear energy is strongly concentrated toward the center. Second, for a given value of s, the degree of concentration is remarkably insensitive to variations in N. Third, the larger the temperature sensitivity of the nuclear reaction rate (the larger s), the smaller is the effective mass fraction (the larger is the central concentration) of the nuclear energy source. In Section 6.12, a more comprehensive discussion of the ratio $\Delta M_{\mathrm{eff}}/M$ is given that takes into account the details of the pp-chain and CN-cycle reactions and examines the variation with stellar mass in the effectiveness of the two energy sources in zero age main sequence stars.

5.4 Luminosity as a function of position in core nuclear burning models

In Section 7.13, using an analytic approximation to opacities at high densities and temperatures presented in Section 7.12, the conditions necessary for the existence of convective cores in main sequence stars are explored. Estimating the existence and size of such cores relative to the size of the energy-producing region requires knowledge of the ratio of luminosity to mass as a function of position within the model. Integrating eqs. (5.3.27) and (5.3.28) from 0 to r and dividing, one has that

$$\frac{L(r)}{M(r)} = \frac{\int_0^z \epsilon \, x^2 \, u^N(x) \, dx}{\int_0^z x^2 \, u^N(x) \, dx}. \tag{5.4.1}$$

Setting

$$\epsilon \sim \epsilon_c \, u^{N(1+k)+s}, \tag{5.4.2}$$

it follows that

$$\frac{L(r)}{M(r)} = \epsilon_c \frac{\int_0^z x^2 \, dx \, u^{N(2+k)+s}(x)}{\int_0^z x^2 \, dx \, u^N(x)}. \tag{5.4.3}$$

Inserting the approximation to u given by eq. (5.3.25) and keeping just the first term in the expansion of the exponential in u, one obtains the approximation

$$\frac{L(r)}{M(r)} \sim \epsilon_c \frac{\int_0^z x^2 \, dx \left(1 - \frac{1}{6} \left[N(2+k)+s\right] x^2 + \cdots\right)}{\int_0^z x^2 \, dx \left(1 - \frac{1}{6} N x^2 + \cdots\right)}. \tag{5.4.4}$$

Performing the integrations, one obtains

$$\frac{L(r)}{M(r)} \sim \epsilon_c \frac{\left(1 - \frac{1}{10} \left[N(2+k)+s\right] z^2 + \cdots\right)}{\left(1 - \frac{1}{10} N z^2 + \cdots\right)}. \tag{5.4.5}$$

Expanding in a Taylor series converts eq. (5.4.5) into

$$\frac{L(r)}{M(r)} \sim \epsilon_c \left(1 - \frac{1}{10}\left[N(1+k)+s\right]z^2 + \cdots\right). \tag{5.4.6}$$

Equations (5.4.5) and (5.4.6) are acceptable approximations as long as the terms proportional to z^2 are small compared with unity. When conditions are such that CN-cycle reactions are the dominant contributors to energy generation, the applicability of the equations is limited to z such that the mass contained in the sphere of radius z is small compared with the effective mass of the nuclear energy generating region. For example, choosing $k=0$, $s=17$, and $N=3/2$ (as is the case if the core is convective), the second term in the numerator of eq. (5.4.5) is <0.5 only for $z < 0.5$. From Table 5.2.1 and eq. (5.2.19), one can determine that this corresponds to less than $\sim$2% of the mass of the model, whereas eq. (5.3.38) and Table 5.3.1 show that the effective mass of the energy-generating region is of the order of 7% of the mass of the model. Outside of the rather small region where eqs. (5.4.5) and (5.4.6) are applicable, another approach is required.

If the polytropic index of the model is constant, $M(r)$ is always given by eq. (5.2.19) and, for $z < 1$, the required value of $z^2\, du/dz$ is tolerably well approximated by differentiating eq. (5.3.25) to obtain

$$z^2\frac{du}{dz} \sim -m(z) = -\frac{z^3}{3}\exp\left(-\frac{z^2}{6}\right)\left[1 - \frac{1}{10}\left(N - \frac{5}{3}\right)z^2\left(1 - \frac{1}{12}z^2\right) + \cdots\right].$$
$$\tag{5.4.7}$$

For $z < 1$, using eq. (5.4.7) in eq. (5.2.19) gives a result which agrees with the exact answer to roughly 2%. For example, for $N=3$, the approximation gives $M(r)/M(R) = 0.123$ compared with the value 0.125 which follows from the appropriate entries in Tables 5.2.2 and 5.2.3. For larger values of z, one may use in eq. (5.2.19) values of $z^2\, du/dz$ found by interpolation in British Association tables.

Series expansions of the probability integral exist which allow one to construct analytic approximations to $L(r)/M(r)$ which are valid for values of z larger than those for which eqs. (5.4.5) and (5.4.6) are adequate approximations. They are # 590 and # 591 in H. B. Dwight, *Tables of Integrals and Other Mathematical Data* 1934, fourth edition, 1947). In the series

$$\int_0^x e^{-t^2}\, dt = x\left(1 - \frac{1}{1!3}x^2 + \frac{1}{2!5}x^4 - \frac{1}{3!7}x^6 + \cdots\right), \tag{5.4.8}$$

the ratio of the $(n+1)$th term to the nth term is $a_n/a_{n-1} = -[(2n-1)/(2n+1)](x^2/n)$. Thus, the series is convergent for all x, but, for a desired accuracy, the number of terms which must be retained increases with increasing x. Another approximation is the asymptotic expansion

$$\int_0^x e^{-t^2}\, dt \sim \frac{\sqrt{\pi}}{2} - \frac{1}{2}\frac{e^{-x^2}}{x}\left(1 - \frac{1}{2x^2} + \frac{1\cdot 3}{2^2 x^4} - \frac{1\cdot 3\cdot 5}{2^3 x^6} + \cdots\right). \tag{5.4.9}$$

The ratio of the nth term to the $(n-1)$th term in the series in square brackets is $a_n/a_{n-1} = -(2n-1)/(2x^2)$. This means that, for a value of $2x^2$ several times larger than unity, successive terms in the series at first decrease and then increase. The error committed by truncating the asymptotic expansion is less than the last term retained (see, e.g., Whittaker & Watson, *Modern Analysis*, 1952). If, for example, $2x^2 = 10$, the series looks like $1 - 0.1 + 0.03 - 0.015 + 0.0105 - 0.009\,45 + 0.010\,395$, and so forth, so the best one can do is state that the series has the value $\sim 0.926 \pm 0.01$.

Integrals relevant in the present context are related. Defining

$$J_{2n}(\alpha, x) = \int_0^x t^{2n}\, e^{-\alpha t^2}\, dt, \tag{5.4.10}$$

and differentiating with respect to α, one has that

$$\frac{\partial J_{2n}(\alpha, x)}{\partial \alpha} = - \int_0^x t^{2(n+1)}\, e^{-\alpha t^2}\, dt = -J_{2(n+1)}(\alpha, x). \tag{5.4.11}$$

One may also write eq. (5.4.10) as

$$J_{2n}(\alpha, x) = \frac{1}{\alpha^{n+1/2}} \int_0^{\alpha^{1/2}x} y^{2n}\, e^{-y^2}\, dy. \tag{5.4.12}$$

Differentiating this second formulation with repect to α, one obtains

$$\frac{\partial J_{2n}(\alpha, x)}{\partial \alpha} = -\frac{n+1/2}{\alpha^{n+3/2}} \int_0^{\alpha^{1/2}x} y^{2n}\, e^{-y^2}\, dy + \left[\frac{1}{\alpha^{n+1/2}} \left(y^{2n}\, e^{-y^2} \right)^{y=\alpha^{1/2}x} \right] \frac{1}{2} \frac{1}{\alpha^{1/2}}\, x, \tag{5.4.13}$$

where the first term comes from differentiating the coefficient of the integral in eq. (5.4.51) and the second term comes from evaluating the integrand at $y = \alpha^{1/2}x$ and multiplying by the result of differentiating the upper limit of the integral. Using eq. (5.4.11) in eq. (5.4.12) gives

$$\frac{\partial J_{2n}(\alpha, x)}{\partial \alpha} = \frac{1}{2\alpha} \left[-(2n+1)\, J_{2n}(\alpha, n) + x^{2n+1}\, e^{-\alpha x^2} \right]. \tag{5.4.14}$$

Comparing eq. (5.4.11) with eq. (5.4.14) leads to the recurrence relationship

$$\begin{aligned}
J_{2(n+1)}(\alpha, x) &= \frac{1}{2\alpha} \left[(2n+1)\, J_{2n}(\alpha, x) - x^{2n+1}\, e^{-\alpha x^2} \right] \\
&= \frac{1}{2\alpha} \left[(2n+1)\, J_{2n}(\alpha, x) - \frac{x}{\alpha^n} (\alpha x^2)^n\, e^{-\alpha x^2} \right]. \tag{5.4.15}
\end{aligned}$$

Adopting the series approximation given by eq. (5.4.8), one has

$$J_0(\alpha, x) = \int_0^x e^{-\alpha t^2}\, dt = \frac{1}{\alpha^{1/2}} \int_0^{\sqrt{\alpha}x} e^{-y^2}\, dy = x \sum_{k=0}^{\infty} (-1)^k \frac{(\alpha x^2)^k}{k!(2k+1)}. \tag{5.4.16}$$

Adopting the asymptotic expansion given by eq. (5.4.9), one has

$$J_0(\alpha, x) = \int_0^x e^{-\alpha t^2}\, dt = \frac{\sqrt{\pi}}{2\alpha^{1/2}} \left(1 - \frac{1}{\sqrt{\pi}} \frac{e^{-\alpha x^2}}{\alpha^{1/2}x} S_0(\alpha, x) \right), \tag{5.4.17}$$

where

$$S_0(\alpha, x) = 1 - \frac{1}{2\alpha x^2} + \frac{1 \cdot 3}{(2\alpha x^2)^2} - \frac{1 \cdot 3 \cdot 5}{(2\alpha x^2)^3} + \cdots . \tag{5.4.18}$$

From eq. (5.4.15), it follows that

$$J_2(\alpha, x) = \frac{1}{2\alpha} \left(J_0(\alpha, x) - x\, e^{-\alpha x^2} \right). \tag{5.4.19}$$

Formulae for $J_{2n}(\alpha, x)$ for larger values of n can be found by successive applications of eq. (5.4.15). Of particular interest here is the integral $J_6(\alpha, x)$, which can be written as

$$J_6(\alpha, x) = \frac{1 \cdot 3 \cdot 5}{(2\alpha)^3} \left[J_0(\alpha, x) - x\, e^{-\alpha x^2} \left(1 + \frac{2}{3} \alpha x^2 + \frac{2}{3} \frac{2}{5} (\alpha x^2)^2 \right) \right]. \tag{5.4.20}$$

Relationships between $J_{2n}(\alpha, x)$ and $J_0(\alpha, x)$ for $n = 2$ and $n \geq 4$ follow by induction from eq. (5.4.20).

Adopting u in the form given by eq. (5.3.25), assuming that

$$u(x)^p \sim e^{-px^2/6} \left[1 + p \left(\frac{N}{120} - \frac{1}{72} \right) x^4 \right] \tag{5.4.21}$$

is an adequate approximation, and making use of the definition of $J_{2n}(\alpha, x)$ given by eqs. (5.4.10) and (5.4.12), one obtains from eq. (5.4.3), after some manipulation,

$$\frac{L(r)}{M(r)} \sim \epsilon_c \frac{J_2(\alpha, z) + \alpha(N/20 - 1/12)\, J_6(\alpha, z)}{J_2(\bar{\alpha}, z) + \bar{\alpha}(N/20 - 1/12)\, J_6(\bar{\alpha}, z)}, \tag{5.4.22}$$

where α is given by eq. (5.3.32) and

$$\bar{\alpha} = \frac{N}{6}. \tag{5.4.23}$$

One can, of course, also use eq. (5.2.19) and eq. (5.3.30) (with $\beta = \beta_c$ in the integrand set equal to unity) to write

$$\frac{L(r)}{M(r)} \sim \epsilon_c \frac{J_2(\alpha, z) + \alpha(N/20 - 1/12)\, J_6(\alpha, z)}{-z^2\, (du/dz)}, \tag{5.4.24}$$

where the denominator can be taken directly from British Association Tables or from the approximation given by eq. (5.4.7).

It is worth demonstrating that, to lowest order in z^2, eq. (5.4.22) is identical with eq. (5.4.5). Keeping just the first term in both the numerator and the denominator in the ratio on the right hand side of eq. (5.4.22) and using eq. (5.4.19) to relate $J_2(\alpha, z)$ to $J_0(\alpha, z)$, one has that

$$\frac{L(r)}{M(r)} \to \epsilon_c \frac{J_2(\alpha, z)}{J_2(\bar{\alpha}, z)} = \epsilon_c \frac{J_0(\alpha, z) - z\, e^{-\alpha z^2}}{J_0(\bar{\alpha}, z) - z\, e^{-\bar{\alpha} z^2}} \frac{\bar{\alpha}}{\alpha}. \tag{5.4.25}$$

Using the convergent series expansion for $J_0(\alpha, x)$ given by eq. (5.4.15) and expanding the exponentials using $e^{-x} = 1 - x + \frac{1}{2} x^2 + \cdots$, one has that

$$\frac{L(r)}{M(r)} \to \epsilon_c \frac{z \left[1 - \alpha z^2/3 + (\alpha z^2)^2/10 - \cdots\right] - z \left[1 - \alpha z^2 + (\alpha z)^2/2 - \cdots\right]}{z \left[1 - \bar{\alpha} z^2/3 + (\bar{\alpha} z^2)^2/10 - \cdots\right] - z \left[1 - \bar{\alpha} z^2 + (\bar{\alpha} z)^2/2 - \cdots\right]} \frac{\bar{\alpha}}{\alpha},$$

(5.4.26)

or

$$\frac{L(r)}{M(r)} \sim \epsilon_c \frac{1 - (3/5)\alpha z^2 + \cdots}{1 - (3/5)\bar{\alpha} z^2 + \cdots}.$$

(5.4.27)

Replacing α and $\bar{\alpha}$ in eq. (5.4.27) by their equivalents given, respectively, by eq. (5.3.32) and eq. (5.4.23), one recovers eq. (5.4.5).

Provided that enough terms are included in the relevant series, the accuracy of the approximations given by eq. (5.4.22) and (5.4.24) is larger than the accuracy given by eq. (5.4.25), and the accuracy of eq. (5.4.25) is larger than that given by eq. (5.4.27). To demonstrate the kind of accuracy achieved, begin with the relationship

$$M(r) = -\frac{M(R)}{z_s^2 (du/dz)_s} \left[J_2(\bar{\alpha}, z) + \bar{\alpha} \left(\frac{N}{20} - \frac{1}{12} \right) J_6(\bar{\alpha}, z) + \cdots \right],$$

(5.4.28)

simplify to

$$M(r) = -\frac{M(R)}{z_s^2 (du/dz)_s} \left(J_2(\bar{\alpha}, z) + \cdots \right),$$

(5.4.29)

and use eqs. (5.4.19) and (5.4.16) to give

$$M(r) = -\frac{M(R)}{z_s^2 (du/dz)_s} \frac{1}{2\bar{\alpha}} z$$
$$\times \left[\left(1 - \frac{\bar{\alpha} z^2}{3} + \frac{(\bar{\alpha} z^2)^2}{10} - \frac{(\bar{\alpha} z^2)^3}{42} + \cdots \right) - \exp(-\bar{\alpha} z^2) + \cdots \right].$$

(5.4.30)

For an $N = 3/2$ polytrope, $\bar{\alpha} = 1/4$, and $z_s^2 (du/dz)_s = -2.714$, so

$$[M(r)]_{N=3/2} = \frac{M(R)}{2.714} 2 z \left[\left(1 - \frac{z^2}{12} + \frac{z^4}{160} - \frac{z^6}{2688} + \cdots \right) - \exp\left(-\frac{z^2}{4} \right) + \cdots \right].$$

(5.4.31)

For $z = 1$, eq. (5.4.31) gives $M(r)/M(R) = 0.1059$, compared with $M(r)/M(R) = 0.1058$ which follows from eq. (5.2.19) and column 7 in Table 5.2.1. For $z = 2$, the two procedures give $M(r)/M(R) = 0.455$ and $M(r)/M(R) = 0.549$, respectively. For $z = 2$, however, eqs. (5.2.19) and (5.4.7) produce $M(r)/M(R) = 0.527$, demonstrating that, for z not much larger than unity, eq. (5.4.24) is a better choice than eq. (5.4.22) for estimating analytically the mass required in the ratio of luminosity to mass.

For large values of αz^2, the luminosity is best estimated by choosing eq. (5.4.17) for $J_0(\alpha, z)$. When CN-cycle reactions dominate nuclear energy production, $\alpha \sim 3$. With this value of α, the quantity $S_0(3, z)$ given by eq. (5.4.18) is

$$S_0(3, z) = 1 - \frac{1}{6z^2} + \frac{3}{36z^4} - \frac{15}{216z^6} + \frac{105}{1296z^8} - \cdots$$
$$= 1 - \frac{0.166\,67}{z^2} \left(1 - \frac{0.5}{z^2} + \frac{0.4167}{z^4} - \frac{0.4861}{z^6} + \cdots \right). \qquad (5.4.32)$$

When $z = 1$,

$$S_0(3, 1) = 1 - 0.166\,67 \begin{pmatrix} 0.5 \pm 0.5 \\ 0.9167 \pm 0.4167 \\ 0.4306 \pm 0.4861 \end{pmatrix}, \qquad (5.4.33)$$

where the values in the three rows of the vector obtain when, respectively, two, three, and four terms are retained in the sum. It is clear that, on this evidence alone, one can say only that the coefficient of $0.166\,67$ in $S_0(3, 1)$ is in the range defined by 0.75 ± 0.25, and that, therefore,

$$S_0(3, 1) = 1 - 0.166\,67\,(0.75 \pm 0.25) = 0.8750\,(1 \pm 0.0476). \qquad (5.4.34)$$

From eqs. (5.4.17) and (5.4.34) one has that

$$J_0(3, 1) = \frac{\sqrt{\pi}}{2 \cdot 3^{1/2}} \left(1 - \frac{1}{\sqrt{3\pi}}\,e^{-3}\,S_0(3, 1) \right) = 0.511\,66\,[1. - 0.016\,22\,S_0(3, 1)]$$
$$= 0.5044\,(1 \pm 0.000\,685). \qquad (5.4.35)$$

Thus, despite the 5% uncertainty in $S_0(3, 1)$, there is less than a 0.1% uncertainty in $J_0(3, 1)$. From eqs. (5.4.19) and (5.4.35) one has

$$J_2(3, 1) = \frac{1}{2 \cdot 3} \left(J_0(3, 1) - e^{-3} \right) = \frac{1}{6} \left[0.5044\,(1 \pm 0.000\,685) - 0.049\,79 \right]$$
$$= 0.075\,77\,(1 \pm 0.000\,865). \qquad (5.4.36)$$

From eqs. (5.4.20) and (5.4.35) one has

$$J_6(3, 1) = \frac{15}{216} \left[J_0(3, 1) - e^{-3} \left(1 + 2 + \frac{36}{15} \right) \right]$$
$$= 0.069\,44\,[0.5044\,(1 \pm 0.000\,783) - 0.2687]$$
$$= 0.016\,35\,(1 \pm 0.0017). \qquad (5.4.37)$$

Writing the numerator in the ratio on the right hand side of eq. (5.4.24) as

$$J_L(\alpha, z) = J_2(\alpha, z) + \alpha \left(\frac{N}{20} - \frac{1}{12} \right) J_6(\alpha, z), \qquad (5.4.38)$$

eqs. (5.4.36) and (5.4.37) give

$$J_L(3, 1) = J_2(3, 1) + 3 \left(\frac{N}{20} - \frac{1}{12} \right) J_6(3, 1)$$

$$= 0.075\,77\,(1 \pm 0.000\,865) + 0.002\,453\,(1 \pm 0.0017) \left(N - \frac{5}{3} \right). \qquad (5.4.39)$$

When $N = \frac{3}{2}$,

$$J_L(3, 1) = 0.074\,13\,(1 \pm 0.000\,94). \qquad (5.4.40)$$

Thus, when $\alpha = 3$, $z = 1$, and $N = 1.5$, $M(r)/M(R) \sim (z^2 du/dz)/(z^2 du/dz)_s = 0.1058$ and, from eq. (5.4.24),

$$\left(\frac{L(r)}{M(r)} \right)_{\alpha=3, z=1, N=1.5} \sim \epsilon_c \frac{J_L(3, 1)}{-z^2(du/dz)} \sim \epsilon_c \frac{0.074\,13}{0.2873} = 0.258\,\epsilon_c. \qquad (5.4.41)$$

Proceeding in the same way for the case $z = 2$, one finds

$$S_0(3, 2) = 1 - 0.041\,666 + 0.005\,208 - 0.001\,085 + 0.000\,3165 - 0.000\,1167 + \cdots$$

$$= 0.962\,55 \pm 0.000\,12,$$

$$J_0(3, 2) = 0.511\,66 \left[1 - 1.001 \times 10^{-6} \right],$$

$$J_2(3, 2) = \frac{1}{6} \left(J_0(3, 2) - 2\,e^{-12} \right) = \frac{1}{6} \left(0.511\,66 - 1.2288 \times 10^{-5} \right) = 0.085\,28,$$

$$J_6(3, 2) = 0.069\,444\,(0.085\,277 - 0.000\,5825)$$

$$= 0.059\,22\,(1 - 0.006\,83) = 0.005\,882, \text{ and}$$

$$J_L(3, 2) = 0.085\,28 + 0.000\,8822 \left(N - \frac{5}{3} \right). \qquad (5.4.42)$$

When $N = 3/2$,

$$J_L(3, 2) = 0.084\,71. \qquad (5.4.43)$$

Altogether, when $\alpha = 3$, $z = 2$, and $N = 1.5$, $M(r)/M(R) = 0.549$ and

$$\left(\frac{L(r)}{M(r)} \right)_{\alpha=3, z=2, N=1.5} \sim \epsilon_c \frac{0.0847}{1.4913} = 0.0568\,\epsilon_c. \qquad (5.4.44)$$

Inspection of the details of the calculation of each quantity J_{2n} in eq. (5.4.42) shows that, for $\alpha z^2 \geq 12$, terms in each J_{2n} involving $e^{-\alpha z^2}$ are essentially negligible relative to the main term. Thus, for large αz^2, one can approximate

$$J_0(\alpha, z_s) \sim \frac{1}{2} \sqrt{\frac{\pi}{\alpha}}, \tag{5.4.45}$$

$$J_2(\alpha, z_s) \sim \frac{1}{2\alpha} J_0(\alpha, z_s) \sim \frac{1}{4\alpha} \sqrt{\frac{\pi}{\alpha}}, \tag{5.4.46}$$

and

$$J_6(\alpha, z_s) \sim \frac{15}{8\alpha^3} J_0(\alpha, z_s) \sim \frac{15}{16} \frac{1}{\alpha^3} \sqrt{\frac{\pi}{\alpha}}, \tag{5.4.47}$$

so that

$$J_L(\alpha, z_s) = \frac{\sqrt{\pi}}{4\alpha^{3/2}} \left[1 + \frac{3}{16} \frac{1}{\alpha} \left(N - \frac{5}{3} \right) \right]. \tag{5.4.48}$$

Inserting this in eq. (5.4.24), one obtains for the overall luminosity to mass ratio

$$\frac{L(R)}{M(R)} \sim -\epsilon_c \frac{1}{z_s^2 (du/dz)_s} \frac{\sqrt{\pi}}{4\alpha^{3/2}} \left[1 + \frac{3}{16} \frac{1}{\alpha} \left(N - \frac{5}{3} \right) \right]. \tag{5.4.49}$$

Not surprisingly, this approximation is the same as that given by eqs. (5.3.35) and (5.3.37).

5.5 Polytropic characteristics of zero age main sequence models

Comparisons of properties of polytropic models with properties of real stars can be very rewarding, even though the ultimate test of the relationship between the theoretical and observational worlds must be based on comparisons between observations and realistic evolutionary models which have been constructed using the full panoply of input physics, including a realistic equation of state (Chapter 4), realistic nuclear energy-generation rates not approximated by power laws (Chapter 6), and realistic opacities (Chapter 7). To acquire a feeling for differences in the inferences achieved in making the two types of comparison, it is of interest at this juncture to compare some characteristics of realistic evolutionary models which have just entered the main sequence phase with characteristics of polytropic models.

In Table 5.5.1 are shown several characteristics of a selection of models constructed in the middle 1960s. The models are a consequence of evolutionary calculations that begin during the gravitational contraction stage which precedes the hydrogen-burning main sequence and the models have been somewhat arbitrarily selected at times when the overall gravitational energy-generation rate has been reduced to a relative minimum compared with the overall nuclear energy-generation rate. Because this minimum is not precisely defined by the calculations, the variation with model mass of any given quantity can be uneven.

Table 5.5.1 Properties of initial main sequence models from evolutionary calculations (Iben, 1967)

$\dfrac{M}{M_\odot}$	$\dfrac{R}{R_\odot}$	$\dfrac{\rho_c}{\text{g cm}^{-3}}$	$\dfrac{T_c}{10^6\,\text{K}}$	$\dfrac{\rho_c}{\rho_{Av}}$	$\dfrac{kT_c}{\Phi_s}$	$\tilde{M}^2\dfrac{\rho_c}{T_c^3}$	$\dfrac{\epsilon_{Av}}{\epsilon_c}$
15	4.40	6.17	34.4	24.9	0.72	3.42	0.033
9	3.40	10.5	31.0	32.5	0.83	2.85	0.033
5	2.40	21.4	27.3	41.9	0.93	2.63	0.030
3	1.75	40.4	24.1	51.2	1.00	2.60	0.030
2.25	1.45	58.8	22.2	56.5	1.02	2.72	0.035
1.5	1.18	87.7	18.8	68.1	1.05	2.97	0.06
1.25	1.08	93.5	16.6	66.8	1.02	3.19	0.09
1	0.87	90.0	13.9	42.0	0.86	3.35	0.12
0.5	0.41	73.6	9.01	7.22	0.53	2.53	0.12
0.25	0.236	155	7.46	5.78	0.50	2.33	0.12

The initial abundances by mass of elements in the pre-main sequence models are $X = 0.708$, $Y = 0.272$, $X(^{12}C) = 0.003\,61$, $X(^{14}N) = 0.001\,20$, $X(^{16}O) = 0.0108$, and $Z = 0.02$. In zero age main sequence models in which CN-cycle reactions make a major contribution to the nuclear energy-generation rate, ^{12}C has been effectively converted into ^{14}N, so the abundance by mass of nitrogen in nuclear burning regions has become $X(^{14}N) = 0.0054$.

The central concentration of the main sequence models, displayed in column 5 of Table 5.5.1, lies in the range associated with polytropic models of index $1.5 \gtrsim N \gtrsim 3.1$, as shown in column 6 of Table 5.2.3. The central concentration of the 0.25 $M_\odot$ and 0.5 $M_\odot$ models is close to that of an index $N = 3/2$ polytrope, consistent with the fact that the 0.25 $M_\odot$ model is completely convective below the photosphere and the fact that the 0.50 $M_\odot$ model has a very deep convective envelope. The central concentrations of the main sequence models of mass 1–3 $M_\odot$ correspond to those of polytropes in the approximate range $2.75 \gtrsim N \gtrsim 3.1$. The fact that the central concentrations of models of mass $M \geq 1.5$ decrease with increasing model mass is associated with the fact that these models all possess convective cores and that, relative to the mass of the model, the mass of the convective core increases with increasing model mass.

The ratio of the kinetic energy of a particle at the center to the gravitational potential of an average particle at the surface is measured by the quantity

$$\frac{kT_c}{\Phi_s} = kT_c \div \left(\mu\, M_H \frac{GM_\odot}{R_\odot} \frac{\bar{R}}{\bar{M}} \right) = k\, T_c \frac{1}{\mu} \frac{R_\odot}{GM_H M_\odot} \frac{\bar{R}}{\bar{M}}$$

$$= \frac{1}{\mu} \frac{T_c}{23.1} \frac{\bar{R}}{\bar{M}}, \qquad (5.5.1)$$

where μ is the molecular weight, $\bar{M}$ and $\bar{R}$ are, respectively, M and R in solar units, and, in the last equality, T_c is in units of 10^6 K. For the main sequence models, $\mu = 0.61$, and as demonstrated by the entries in column 6 of Table 5.5.1, kT_c/Φ_s is constant to within a factor of about 2 over the entire range of masses examined. This near constancy is shared with polytropes of index $N \gtrsim 3.25$, as shown by the entries for $kT_c/\beta_c\Phi_s$ in column 4 of Table 5.3.1.

The next to last column of Table 5.5.1 gives the quantity $\tilde{M}^2\,\rho_c/T_c^3$, where $\tilde{M}^2 = 10^{-2}M^2$. Remarkably, over an almost two order of magnitude range in mass,

$$\frac{\rho_c}{T_c^3} = 288\,(1 \pm 0.19)\,\left(\frac{M_\odot}{M}\right)^2.$$ (5.5.2)

That is, for zero age main sequence models, the relationship $\rho_c/T_c^3 \propto 1/M^2$ is an exceedingly good approximation. Since $\rho_{Av} \propto M/R^3$, $\rho_{Av}/T_c^3 \propto \left(1/M^2\right)\left(\Phi_s/kT_c\right)^3$, suggesting that $\rho_c/\rho_{Av} \propto \left(kT_c/\Phi_s\right)^3$. In fact, for the models with $M \geq 1\,M_\odot$,

$$\frac{\rho_c}{\rho_{Av}}\left(\frac{\Phi_s}{kT_c}\right)^3 = 59\,(1 \pm 0.135).$$ (5.5.3)

For the 0.5 $M_\odot$ and 0.25 $M_\odot$ models, respectively, this quantity is 48.5 and 46.2.

The ratio of the average nuclear energy-generation rate ϵ_{Av} to the nuclear energy-generation rate at the center ϵ_c is displayed for the main sequence models in the last column of Table 5.5.1. The ratio follows the pattern established with polytropic models in Section 5.3: namely, the ratio is 0.032–0.039 for $1.5 \gtrsim N \gtrsim 3$ if CN-cycle burning dominates in producing nuclear energy and is $\sim$0.12 when burning by the pp chains dominates.

5.6 Models in adiabatic equilibrium: existence of a maximum temperature and decrease of entropy with time

In the interiors of very low mass main sequence stars, because densities are large and temperatures are low, the opacity over much of the star is expected to be large enough that energy flow by convection is more efficient than energy flow by radiation, with the actual temperature gradient being close to the adiabatic temperature gradient. Thus, it is reasonable to explore the properties of adiabatic models as first approximations to the structure of very low mass main sequence stars.

Neglecting radiation pressure and assuming that particle velocities are not relativistic, pressure P and $\bar{U}$, the energy per unit volume, are related by $P = (2/3)\,\bar{U}$. From the first law of thermodynamics,

$$dQ = PdV + d(\bar{U}V)$$ (5.6.1)

or

$$dQ = P\,d\left(\frac{1}{\rho}\right) + d\left(\frac{\bar{U}}{\rho}\right),$$ (5.6.2)

where dQ is the heat entering a parcel of gas of volume V, ρ is the density, and in eq. (5.6.2) a parcel of gas of fixed mass ($\rho V = 1$ g) has been focussed upon. Differentiating, and making use of the adopted relationship between P and $\bar{U}$,

$$dQ = \frac{1}{\rho}\left(-\frac{5}{3}\bar{U}\frac{d\rho}{\rho} + d\bar{U}\right). \tag{5.6.3}$$

Assuming adiabaticity, or no heat exchange between parcels, $dQ = 0$, and

$$\frac{d\bar{U}}{\bar{U}} = \frac{dP}{P} = \frac{5}{3}\frac{d\rho}{\rho}, \tag{5.6.4}$$

or

$$P = K\rho^{5/3} = K\rho^{1+2/3}, \tag{5.6.5}$$

where K is a constant. Thus, assuming adiabaticity leads to a polytrope of index $N = 3/2$.

Next, assume that the heavy nuclei obey a perfect gas equation of state and that the electrons are in an arbitrary state of non-relativistic degeneracy. That is (see Chapter 4),

$$P = P_{\text{nuc}} + P_{\text{e}} = \frac{k}{\mu_{\text{n}}M_{\text{H}}}\rho T + \frac{1}{3}\langle v_{\text{e}}p_{\text{e}}\rangle n_{\text{e}}, \tag{5.6.6}$$

where

$$n_{\text{e}} = \frac{\rho}{\mu_{\text{e}}M_{\text{H}}} = \frac{8\pi}{h^3}(2m_{\text{e}}kT)^{3/2}\,F_{1/2}(\alpha), \tag{5.6.7}$$

$$F_{1/2}(\alpha) = \int_0^\infty \frac{x^2\,dx}{\exp(\alpha + x^2) + 1}, \tag{5.6.8}$$

and $x^2 = p_{\text{e}}^2/2m_{\text{e}}kT$. Since $v_{\text{e}}p_{\text{e}} = p_{\text{e}}^2/m_{\text{e}}$,

$$P_{\text{e}} = \frac{1}{3m_{\text{e}}}\frac{8\pi}{h^3}(2m_{\text{e}}kT)^{5/2}\,F_{3/2}(\alpha), \tag{5.6.9}$$

where

$$F_{3/2}(\alpha) = \int_0^\infty \frac{x^4\,dx}{\exp(\alpha + x^2) + 1}. \tag{5.6.10}$$

The functions $F_{1/2}(\alpha)$ and $F_{3/2}(\alpha)$ are known as Fermi–Dirac integrals.

Dividing eq. (5.6.9) by eq. (5.6.7) and dividing again by kT, one obtains a quantity

$$f(\alpha) = \frac{P_{\text{e}}}{n_{\text{e}}kT} = \frac{2}{3}\frac{F_{3/2}(\alpha)}{F_{1/2}(\alpha)} \tag{5.6.11}$$

which is a function of α alone. Thus,

$$P_{\text{e}} = n_{\text{e}}kT\,f(\alpha) = \frac{\rho}{\mu_{\text{e}}M_{\text{H}}}kT\,f(\alpha). \tag{5.6.12}$$

Combining eqs. (5.6.6) and (5.6.12), one has that, for fixed μ_e and μ_n, the quantity

$$\frac{P}{\rho T} = \frac{k}{M_H} \left(\frac{1}{\mu_n} + \frac{f(\alpha)}{\mu_e} \right)$$

(5.6.13)

is a function only of α. Equation (5.6.7) shows that the quantity

$$\delta = \frac{\rho}{\mu_e T_6^{3/2}} = 10^9 \frac{\rho}{\mu_e T^{3/2}}$$

(5.6.14)

is also a function only of α. Hence, one can write

$$K = \frac{P}{\rho^{5/3}} = \frac{T}{\rho^{2/3}} \frac{k}{M_H} \left(\frac{1}{\mu_n} + \frac{f(\alpha)}{\mu_e} \right),$$

(5.6.15)

or

$$K = \frac{1}{(\mu_e \delta)^{2/3}} \frac{k \times 10^6}{M_H} \left(\frac{1}{\mu_n} + \frac{f(\alpha)}{\mu_e} \right) = 0.824\,99 \times 10^{14} \frac{1}{(\mu_e \delta)^{2/3}} \left(\frac{1}{\mu_n} + \frac{f(\alpha)}{\mu_e} \right).$$

(5.6.16)

Useful approximations to $f(\alpha)$ as a function of δ are given by eqs. (4.7.6)–(4.7.8). Another approximation (Henyey, LeLevier, & Levee, 1959),

$$f(\alpha) \sim 1 + \frac{0.021\,893\,\delta}{(1 + 0.006\,05\,\delta)^{1/3}},$$

(5.6.17)

has the virtue of being exact in the limits of both $T \to \infty$ and $T \to 0$, giving $P_e = n_e k T$ in the first limit and giving, in the second limit (see eq. (4.6.42)),

$$P_e = K_1 \rho^{5/3},$$

(5.6.18)

where, as given by eq. (4.6.42),

$$K_1 = \frac{1}{5} \left(\frac{3}{8\pi} \right)^{2/3} \frac{h^2}{m_e} \left(\frac{1}{\mu_e M_H} \right)^{5/3}$$

$$= \frac{0.991\,39 \times 10^{13}}{\mu_e^{5/3}} \text{ dyne cm}^{-2} (\text{g cm}^{-3})^{-5/3}.$$

(5.6.19)

To complete the story, one may establish explicit relationships between the global properties of the adiabatic model (mass and radius) and the central density and temperature. The choice $N = 3/2$ in eqs. (5.1.21), (5.1.27), and (5.1.29), gives

$$\left(K_1 \frac{5}{2} \right)^{3/2} = 4\pi G \left(\frac{R}{z_s} \right)^2 \left(-\frac{GM(R)}{R} \frac{z_s}{(z^2 du/dz)_s} \right)^{1/2}.$$

(5.6.20)

Inserting appropriate values from Table 5.2.3, $z_s = 3.65375$ and $(z \mathrm{d}u/\mathrm{d}z)_s = -0.74282$, one has

$$R = 2.35728 \, \frac{K_1}{G} \, \frac{1}{M^{1/3}(R)}, \tag{5.6.21}$$

or

$$\frac{R}{R_\odot} = 4.0345 \times 10^{-15} \, K_1 \left(\frac{M_\odot}{M(R)} \right)^{1/3}. \tag{5.6.22}$$

Then, using the numerical value of K_1 given by eq. (5.6.14), one has

$$\frac{R}{R_\odot} = \frac{0.33285}{(\mu_e \delta)^{2/3}} \left(\frac{1}{\mu_n} + \frac{f(\alpha)}{\mu_e} \right) \left(\frac{M_\odot}{M(R)} \right)^{1/3}. \tag{5.6.23}$$

Since, for $N = 3/2$, $\rho_c/\bar{\rho} = 5.9908$ (Table 5.2.3), and $\rho_\odot = 1.410 \text{ g cm}^{-3}$,

$$\rho_c \, (\text{g cm}^{-3}) = 8.447 \, \frac{M(R)}{M_\odot} \left(\frac{R_\odot}{R} \right)^3. \tag{5.6.24}$$

Finally, the central temperature in units of 10^6 K is given by

$$T_{c,6} = \left(\frac{\rho_c}{\mu_e \delta} \right)^{2/3}. \tag{5.6.25}$$

For every choice of δ, the value of $f(\alpha)$ may be obtained from eq. (4.7.6) or eq. (4.7.8), or from eq. (5.6.14), and values of R, ρ_c, and $T_{c,6}$ follow, respectively, from eqs. (5.6.23), (5.6.24), and (5.6.25).

Interior properties of an adiabatic model of mass $M = (1/9) \, M_\odot$ and solar-like composition are displayed in Table 5.6.1. Successive columns in this table give, respectively, the electron-degeneracy parameter δ, model radius, central density, central temperature, $P_e/n_e kT$, ϵ_F/kT, a measure $\bar{S}_n$ of the contribution of nuclei to the entropy, and a measure $\bar{S}_e$ of the contribution of electrons to the entropy.

The entries in Table 5.6.1 show general trends which are independent of the effective polytropic index: (1) for radii large enough that electrons are not degenerate, a smaller radius means larger interior densities and temperatures; (2) for radii small enough that degenerate electrons provide the main contribution to the pressure, although a smaller radius still means larger interior densities, it also means lower interior temperatures; and (3) there is a unique minimum radius characterizing a completely electron-degenerate configuration at zero temperature. This behavior actually encapsulates the essence of the evolution of most stars from birth in a giant molecular cloud to old age as a white dwarf. Beginning as a large configuration at temperatures too low for nuclear reactions to occur, a star contracts so that, by the release of gravitational potential energy, it can supply the radiant energy escaping from its surface. As the star contracts, interior temperatures become high enough for nuclear reactions to occur and stabilize the star at radii such that nuclear reactions supply the radiant energy being lost at the surface. Ultimately, when nuclear fuel is exhausted,

Table 5.6.1 Properties of a homogeneous star in adiabatic equilibrium; model mass $= \frac{1}{9} M_\odot$ and composition $(X,Y,Z) = (0.71, 0.27, 0.02)$

δ	$\frac{R}{R_\odot}$	$\frac{\rho_c}{\text{g cm}^{-3}}$	$\frac{T_c}{10^6 \text{ K}}$	$\left(\frac{P_e}{n_e kT}\right)_c$	$\left(\frac{\epsilon_F}{kT}\right)_c$	$\bar{S}_n$	$\bar{S}_e$
1	1.03138	0.8561	0.8122	1.0218	-2.046	16.373	4.6006
2	0.67229	3.0909	1.2042	1.0435	-1.309	15.680	3.9182
5	0.36847	18.774	2.1762	1.1081	-0.263	14.764	3.0334
10	0.24427	64.442	3.1195	1.2139	0.6443	14.071	2.3904
20	0.17240	183.30	4.0368	1.4157	1.7554	13.378	1.7838
30	0.14229	326.02	4.5225	1.6054	2.5584	12.972	1.4550
40	0.12567	473.28	4.6775	1.7814	3.2179	12.685	1.2356
60	0.10736	759.10	4.8911	2.1732	4.4096	12.279	1.0234
80	0.098307	988.60	4.8150	2.5105	5.4190	11.991	0.8518
100	0.094840	1101.0	4.4584	2.8323	6.3359	11.768	0.7449
200	0.080716	1786.0	3.8775	4.2643	10.185	11.075	0.4753
500	0.072489	2465.8	2.6100	7.6511	18.868	10.159	0.2601
1000	0.069190	2835.6	1.8047	12.0626	29.992	9.4656	0.1642
2000	0.067214	3093.1	1.2047	19.0959	47.636	8.7725	0.1036
5000	0.065731	3307.1	0.6839	35.1301	87.769	7.8562	0.0562
10000	0.065081	3407.2	0.4395	55.7476	139.33	7.1631	0.0390
∞	0.064033	3577.2	0.0000	∞	∞	0.0000	0.0000

the star contracts again, releasing gravitational potential energy to satisfy energy losses from the surface until pressure due to degenerate electrons prevents further contraction; then, surface energy losses drain the residual thermal energy from the star which cools to zero temperature, finally unable to shine.

A fundamental truth of stellar evolution is expressed in the final two columns of Table 5.6.1: the entropy of a star decreases as it evolves. In calculating the entropy, it has been assumed that both nuclei and electrons are non-relativistic. From the first law of thermodynamics and the non-relativistic relationship between pressure and energy density, one has

$$T \, dS = dQ = P \, d\left(\frac{1}{\rho}\right) + dU = (P_n + P_e) \, d\left(\frac{1}{\rho}\right) + \frac{3}{2} d\left(\frac{P_n}{\rho} + \frac{P_e}{\rho}\right), \quad (5.6.26)$$

or

$$dS = -\frac{5}{2} \frac{P_n + P_e}{\rho T} \frac{d\rho}{\rho} + \frac{3}{2} \frac{1}{\rho T} (dP_n + dP_e), \quad (5.6.26a)$$

where the subscripts n and e refer, respectively, to nuclei and electrons. It is clear that the contributions of nuclei and electrons are independent of one another and that one may write

$$dS = dS_n + dS_e,$$

where

$$dS_n = -\frac{5}{2} \frac{P_n}{\rho T} \frac{d\rho}{\rho} + \frac{3}{2} \frac{1}{\rho T} dP_n = \frac{3}{2} \frac{P_n}{\rho T} d\log_e\left(\frac{P_n}{\rho^{5/3}}\right), \tag{5.6.27}$$

and

$$dS_e = \frac{3}{2} \frac{P_e}{\rho T} d\log_e\left(\frac{P_e}{\rho^{5/3}}\right). \tag{5.6.28}$$

For the nuclei, the first term on the rightmost side of eq. (5.6.6) may be used to obtain

$$dS_n = \frac{3}{2} \frac{k}{\mu_n M_H} d\log_e\left(\frac{T}{\mu_n \rho^{2/3}}\right), \tag{5.6.29}$$

or

$$S_n = \frac{3}{2} \frac{k}{\mu_n M_H} \log_e\left(\frac{T}{\mu_n \rho^{2/3}}\right) + \text{constant}. \tag{5.6.30}$$

For the electrons, one may use eqs. (5.6.7)–(5.4.10) to obtain

$$\frac{3}{2} \frac{P_e}{\rho T} = \frac{k}{\mu_e M_H} \frac{F_{3/2}}{F_{1/2}} \tag{5.6.31}$$

and

$$\frac{P_e}{\rho^{5/3}} = \frac{1}{3m_e} \left(\frac{8\pi}{h^3}\right)^{2/3} \frac{F_{3/2}}{(\mu_e M_H F_{1/2})^{5/3}} \propto \frac{F_{3/2}}{(\mu_e F_{1/2})^{5/3}}, \tag{5.6.32}$$

giving

$$dS_e = \frac{k}{\mu_e M_H} \frac{F_{3/2}}{F_{1/2}} d\log_e\left(\frac{F_{3/2}}{(\mu_e F_{1/2})^{5/3}}\right). \tag{5.6.33}$$

Differentiating the logarithmic term in eq. (5.6.33) and rearranging gives

$$dS_e = \frac{k}{\mu_e M_H} \left[\frac{dF_{3/2}}{F_{1/2}} + \frac{5}{3} F_{3/2} d\left(\frac{1}{F_{1/2}}\right) - \frac{F_{3/2}}{F_{1/2}} \frac{d\mu_e}{\mu_e}\right]. \tag{5.6.34}$$

By examining the expansions in Chapter 4 for the Fermi–Dirac functions, one can establish that

$$\frac{dF_{3/2}}{d\alpha} = -\frac{3}{2} F_{1/2}. \tag{5.6.35}$$

Using this result in eq. (5.6.34), holding μ_e constant, and integrating by parts, one finds

$$S_e = \frac{k}{\mu_e M_H} \left(\frac{5}{3} \frac{F_{3/2}}{F_{1/2}} + \alpha\right) + \text{constant}. \tag{5.6.36}$$

Using eq. (5.6.11), this can also be written as

$$S_e = \frac{k}{\mu_e M_H} \left(\frac{5}{2} f(\alpha) + \alpha \right) + \text{constant.} \tag{5.6.37}$$

or

$$S_e = \frac{k}{\mu_e M_H} \left(\frac{5}{2} \frac{P_e}{n_e kT} + \alpha \right) + \text{constant.} \tag{5.6.38}$$

When electron degeneracy is small, $P_e/(n_e kT) \to 1$ and (see eqs. (4.5.6) and (4.5.8)) $e^{-\alpha} \to -(1/2)\left(h^2/2\pi m_e kT\right)^{3/2} n_e$. So, dispensing with the constant in eq. (5.6.38), one obtains, in the limit of weak electron degeneracy,

$$S_e \to \frac{k}{\mu_e M_H} \left[\frac{5}{2} - \log_e \frac{1}{2} \left(\frac{h^2}{2\pi m_e kT} \right)^{3/2} n_e \right]. \tag{5.6.39}$$

Using the normalization in eq. (5.6.39) as a guide, one can write, by inspection, the contribution of protons to the entropy as

$$S_p = \frac{k}{\mu_p M_H} \left[\frac{5}{2} - \log_e \frac{1}{2} \left(\frac{h^2}{2\pi M_H kT} \right)^{3/2} n_p \right], \tag{5.6.40}$$

where μ_p and n_p are, respectively, the molecular weight and the number density of protons. Similarly, for alpha particles,

$$S_\alpha = \frac{k}{\mu_\alpha M_H} \left[\frac{5}{2} - \log_e \left(\frac{h^2}{2\pi M_\alpha kT} \right)^{3/2} n_\alpha \right], \tag{5.6.41}$$

where μ_α, M_α, and n_α are, respectively, the molecular weight, the mass, and the number density of alpha particles. The factor of $1/2$ in the logarithmic term is absent in eq. (5.6.41) because alpha particles have no spin. Altogether,

$$S_n = S_p + S_\alpha. \tag{5.6.42}$$

Again dispensing with the constant, eq. (5.6.38) can also be written as

$$S_e = \frac{k}{\mu_e M_H} \left(\frac{5}{2} \frac{P_e}{n_e kT} - \frac{\epsilon_F}{kT} \right), \tag{5.6.43}$$

where ϵ_F is the electron Fermi energy. This formulation is, of course, most meaningful when electrons are degenerate. In the limit of strong non-relativistic electron degeneracy, as $T \to 0$,

$$f(\alpha) = \frac{P_e}{n_e kT} \to \frac{2}{3} \left(\frac{3}{5} n_e \epsilon_F(0) \right) \div n_e kT = \frac{2}{5} \frac{\epsilon_F(0)}{kT}. \tag{5.6.44}$$

Since, as $T \to 0$, $\alpha = -\epsilon_F/kT \to -\epsilon_F(0)/kT$, one has that $S_e \to 0$ as $T \to 0$. The quantity $\bar{S}_e$ in the last column of Table 5.6.1 is the quantity in parentheses in eq. (5.6.43).

The fact that the electron entropy vanishes at absolute zero temperature is not just an artifact achieved by construction. It is a consequence of the normalization of probabilities that quantum physics introduces into thermodynamics through Planck's constant, and it is an example of Nernst's theorem (sometimes called the third law of thermodynamics) at work.

Manipulation of eqs. (5.6.40)–(5.6.42) for the chosen composition gives

$$S_n \sim S_p + S_\alpha = \frac{k}{\mu_n M_H} [16.3734 - \log_e \delta]. \tag{5.6.45}$$

The quantity $\bar{S}_n$ in the next to the last column of Table 5.6.1 is the quantity in square brackets in eq. (5.6.45). As $T \to 0$, S_n given by eq. (5.6.45) formally tends toward $-\infty$. This is a consequence of not taking into account the fact that matter solidifies when the kinetic energies of nuclei become smaller than the Coulomb binding of the nuclei to electrons and the lowest energy configuration becomes a lattice, with nuclei at lattice sites oscillating with amplitudes smaller than the mean separation between nuclei. When solidification is properly taken into account, $S_n \to 0$ as $T \to 0$. For example, in the Debye formulation described in Section 21.4 in Volume 2, when $T \ll \Theta_{\text{Debye}}$, the entropy of a solid configuration reduces to

$$S_n = \frac{k}{M_H} \left[\frac{4}{5} \pi^4 \left(\frac{T}{\Theta_{\text{Debye}}} \right)^3 \right], \tag{5.6.46}$$

where $\Theta_{\text{Debye}} = h\nu_{\text{max}}/k$ and ν_{max} is the frequency with which nuclei oscillate in the Coulomb potential provided by free electrons.

5.7 White dwarf properties revealed by polytropes: radius–mass relationships and the maximum mass

The $N = 3/2$ and $N = 3$ polytropes are marvelous tools for deriving quantitatively two very basic characteristics of white dwarfs: (1) for small masses, radius and mass are related by $R \propto 1/M^{1/3}$, and (2) there exists a maximum mass which a cold non-rotating white dwarf can have; if the nucleons in the white dwarf outnumber electrons two to one, this mass is not much larger than the mass of the Sun. The first result has historical importance in that it demonstrated the possibility of estimating the mass of a white dwarf which is not in a close binary system but is close enough to the Earth that its distance can be estimated trigonometrically. Given this distance, from its apparent luminosity, the intrinsic luminosity of the white dwarf can be estimated, and, in conjuction with its observed surface temperature, its radius and hence its mass can be estimated. The second result was stated first by Chandrasekhar as a theorem which shows that, as the limiting mass is approached, the radius of the model vanishes. On hearing of this result, Eddington is said to have remarked that he would rather invent new physics than accept the result. Of course, both great men were right: the theorem is absolutely true if only degenerate electrons contribute to the pressure and nuclear reactions and weak interactions are not taken into account. In the real world, the existence of these reactions and interactions would cause a fuel-exhausted star of the Chandrasekhar mass to collapse into a neutron star.

5.7.1   Consequences of the $N = 3/2$ polytropic approximation

Polytropes of index $N = 3/2$ provide insight into the structure of real white dwarfs which are cool enough that the contribution of ions to the pressure is not very important and are of mass small enough that electrons do not have highly relativistic energies. Then, $P \sim P_e \overset{\sim}{\propto} \rho^{5/3}$ and eq. (5.6.20) for the relationship between the radius and mass applies. This equation can be recast to read

$$R_{N=3/2} = \frac{2.5\,K_1}{(4\pi)^{2/3}G}\, z_s \left(-z^2 \frac{du}{dz}\right)_s^{1/3} \frac{1}{M^{1/3}}, \tag{5.7.1}$$

where K_1 is given by eliminating the density from the rightmost identity in eq. (5.6.18):

$$K_1 = K_{WD,N=3/2} = \frac{1}{5}\left(\frac{3}{8\pi}\right)^{2/3} \frac{h^2}{m_e M_H^{5/3}} \left(\frac{1}{\mu_e}\right)^{5/3}. \tag{5.7.2}$$

Inserting eq. (5.7.2) into eq. (5.7.1) and rearranging gives

$$R_{WD,N=3/2} = \left[\left(\frac{3}{32\sqrt{\pi}}\right)^{2/3} z_s \left(-z^2 \frac{du}{dz}\right)_s^{1/3}\right] \frac{1}{\mu_e^{5/3}}\, \lambda_C N_0^{1/3} \left(\frac{N_0 M_H}{M_{WD}}\right)^{1/3}, \tag{5.7.3}$$

where

$$\lambda_C = \frac{h}{m_e c} = 2.426\,26 \times 10^{-10}\ \text{cm} \tag{5.7.4}$$

is the Compton wavelength of the electron, and

$$N_0 = \left(\frac{c\hbar}{G M_H^2}\right)^{3/2} = 2.200 \times 10^{57} \tag{5.7.5}$$

is a characteristic number which is related to what may be called the gravitational fine structure constant, $G M_H^2/\hbar c$ and is approximately twice the number of nucleons in the Sun. The product $\lambda_C N_0^{1/3}$ is a characteristic length which is approximately 4.5% of the radius of the Sun:

$$\lambda_C N_0^{1/3} = 3.1566 \times 10^9\ \text{cm} = 0.045\,34 R_\odot. \tag{5.7.6}$$

The total number of nucleons in the Sun is

$$N_\odot = \frac{M_\odot}{M_H} = 1.19 \times 10^{57}, \tag{5.7.7}$$

so the mass of a star composed of N_0 nucleons is

$$M_0 = N_0\,M_H = \frac{N_0}{N_\odot}\,M_\odot = 1.8494\,M_\odot. \tag{5.7.8}$$

Using eq. (5.7.8) in the last term eq. (5.7.3) and evaluating the term in square brackets in this equation, eq. (5.7.3) becomes

$$R_{\mathrm{WD},N=3/2} = \frac{0.8815}{\mu_e^{5/3}} \lambda_C N_0^{1/3} \left(\frac{M_\odot}{M_{\mathrm{WD}}}\right)^{1/3},$$ (5.7.9)

or, using eq. (5.7.7) and $R_\odot = 6.96 \times 10^{10}$ cm,

$$R_{\mathrm{WD},N=3/2} = \frac{0.0400 \, R_\odot}{\mu_e^{5/3}} \left(\frac{M_\odot}{M_{\mathrm{WD}}}\right)^{1/3}.$$ (5.7.10)

For most white dwarfs, which are made of carbon and oxygen or of oxygen and neon, $\mu_e = 2$, so

$$R_{\mathrm{WD},N=3/2} = 0.012\,60 \, R_\odot \left(\frac{M_\odot}{M_{\mathrm{WD}}}\right)^{1/3}.$$ (5.7.11)

By calculating the electron pressure following the methodology of Section 4.8, where the first order relativistic correction to the relationship between velocity and momentum is taken into account at zero temperature, one can demonstrate that eqs. (5.7.9)–(5.7.11) overestimate the radius of a white dwarf. Keeping the first two terms in the expansion for v in the expression

$$v = \frac{c^2 p}{\sqrt{(m_e c^2)^2 + (cp)^2}} = \frac{p}{m_e}\left[1 - \frac{1}{2}\left(\frac{p}{m_e c}\right)^2 + \frac{1}{8}\left(\frac{p}{m_e c}\right)^4 + \cdots\right],$$ (5.7.12)

and inserting into eq. (4.3.2) for the pressure, gives

$$P_e = \frac{1}{3}\frac{8\pi}{h^3}\int_0^{p_F} (vp)\, p^2 \mathrm{d}p = \frac{8\pi}{15h^3}\frac{1}{m_e}p_F^5\left[1 - \frac{5}{14}\left(\frac{p_F}{m_e c}\right)^2 + \cdots\right]$$

$$= \frac{1}{5}m_e c^2 \, n_e \left(\frac{3}{8\pi}\lambda_C^3 \, n_e\right)^{2/3}\left[1 - \frac{5}{14}\left(\frac{3}{8\pi}\lambda_C^3 \, n_e\right)^{2/3} + \cdots\right].$$ (5.7.13)

The last identity in this equation has been achieved by making use of the fact that the Fermi momentum p_F is given to first order by

$$n_e = \frac{\rho}{\mu_e M_H} = \frac{8\pi}{h^3}\int_0^{p_F} p^2 \mathrm{d}p = \frac{8\pi}{3}\frac{p_F^3}{h^3},$$ (5.7.14)

so that

$$\frac{p_F}{m_e c} = \left(\frac{3}{8\pi}\lambda_C^3 \, n_e\right)^{1/3}.$$ (5.7.15)

These results are consistent with the development given in Section 4.8 by eqs. (4.8.8)–(4.8.12).

It is clear from eq. (5.7.13) that relativistic effects reduce the pressure from that given in the pure non-relativistic approximation. The equation of state becomes softer, with the consequence that the radius of a white dwarf model constructed with the proper equation of state throughout is smaller than that given by an $N = 3/2$ polytrope. The degree to which relativistic effects reduce the pressure at a given density is measured by the quantity

$$\alpha_r = \frac{5}{14} \left(\frac{p_F}{m_e c} \right)^{2/3} = \frac{5}{14} \left(\frac{3}{8\pi} \lambda_C^3 n_e \right)^{2/3} = 0.3616 \left(\frac{\rho_6}{\mu_e} \right)^{2/3}. \tag{5.7.16}$$

A value of $\alpha_r = 0.1$ corresponds to a roughly 10% relativistic correction to the equation of state for electrons and $\alpha_r = 0.2$ corresponds to a roughly 20% correction. This interpretation cannot be continued much beyond 30% since the expansion which produces eq. (5.7.16) is valid only if $(p_F/m_e c)^2 = (14/15)\alpha_r < 1$. Inserting numbers, one obtains

$$\frac{\rho_6}{\mu_e} = 4.599 \, \alpha_r^{3/2}. \tag{5.7.17}$$

The central density of the $N = 3/2$ polytrope is obtained by inserting R from eq. (5.7.10) into eq. (5.6.24). The result is

$$\rho_c = 1.32 \times 10^5 \mu_e^5 \left(\frac{M}{M_\odot} \right)^2 \text{ g cm}^{-3}$$

$$= 4.22 \times 10^6 \left(\frac{M}{M_\odot} \right)^2 \text{ g cm}^{-3} \text{ when } \mu_e = 2. \tag{5.7.18}$$

Solving eqs. (5.7.17) and (5.7.18) for $M/M_\odot$ in terms of α_r, one obtains

$$\frac{M_\alpha}{M_\odot} = \frac{5.90 \, \alpha_r^{3/4}}{\mu_e^2} = 1.48 \, \alpha_r^{3/4} \text{ when } \mu_e = 2. \tag{5.7.19}$$

The non-relativistic equation of state overestimates the pressure at the center of the $N = 3/2$ polytrope by approximately $\alpha_r \times 100\%$. When $\mu_e = 2$, eq. (5.7.19) gives $M_\alpha = 0.47 \, M_\odot$ for $\alpha_r = 0.1$ and $M_\alpha = 0.66 \, M_\odot$ for $\alpha_r = 0.2$. These results suggest that the $N = 3/2$ polytrope will resemble closely only the lowest mass helium white dwarfs in close binaries. On the other hand, since the mean density of the $N = 3/2$ polytrope is ~ 6 times smaller than the central density, one might guess that, even if the equation of state overestimates pressure by 10–20% at the center, the error committed over most of the white dwarf is much smaller than this, and that the radius given by eq. (5.7.10) might be close to that of properly constructed white dwarf models considerably more massive than, say, $0.2 \, M_\odot$.

In Table 5.7.1, radii of $N = 3/2$ polytropic models are compared with radii of cold white dwarf models constructed by Hamada and Salpeter (1962) using the exact form of the electron equation of state.

Values of α_r which follow from eq. (5.7.19) are given in the last column of Table 5.7.1. All things considered, one can be reasonably pleased that, up to $M \sim 0.6 \, M_\odot$, the $N = 3/2$ polytropic radii are not radically larger than the radii of the realistic models, despite the fact that the assumption of non-relativistic degeneracy is violated significantly at the center of the polytropic models over the last part of this mass range.

Table 5.7.1 Radii of $N = 3/2$ polytropes and of Hamada–Salpeter models				
$\frac{M}{M_\odot}$	$R_{N=3/2}$	R_{HS}	$\frac{R_{N=3/2}}{R_{HS}}$	α_r
0.2	0.0215	0.0200	1.07	0.069
0.4	0.0171	0.0153	1.12	0.175
0.6	0.0149	0.0124	1.20	0.300
0.8	0.0136	0.0103	1.32	0.441
1.0	0.0126	0.0080	1.60	0.593

The relationship between radius and mass for cold white dwarf models constructed with a proper electron equation of state is roughly linear over the mass range $\sim 0.4\ M_\odot \rightarrow 1.2\ M_\odot$, and the expression

$$\frac{R_{WD}}{R_\odot} \sim 0.0195 - 0.0115\ \frac{M_{WD}}{M_\odot} \tag{5.7.20}$$

fits the numerical results to better than about 3% over this range (see Fig. 1 in Iben (1982)).

5.7.2 The Chandrasekhar mass limit from the $N = 3$ polytrope

Using the relativistically correct electron equation of state, as model mass is increased beyond $\sim 1.2\ M_\odot$, model radius declines ever more rapidly with respect to mass, formally going to zero at the mass named after its discoveror, Subrahmanyan Chandrasekhar. Its traditionally quoted value is

$$M_{Chandrasekhar} = \frac{5.76}{\mu_e^2} M_\odot = 1.44\ M_\odot \text{ when } \mu_e = 2. \tag{5.7.21}$$

White dwarfs with masses up to at least $\sim 1.3\ M_\odot$ have been detected in nature, and it is certain that white dwarfs of mass nearly as large as the Chandrasekhar limit are formed. The Chandrasekhar mass can be found by going to the limit of extreme relativistic degeneracy for the electrons. How the radius of the model goes to zero at this mass is shown in Section 5.8. Setting $v = c$ in eq. (4.3.2) for the electron pressure and using the relationship between density and the Fermi momentum given by eq. (4.8.7), one has that

$$P_e = \frac{8\pi}{3h^3} \int_0^{p_F} (cp)\ p^2\ dp = c\ \frac{8\pi}{3h^3}\ \frac{p_F^4}{4} = K_2\ \rho^{4/3}, \tag{5.7.22}$$

where

$$K_2 = K_{\mathrm{WD},N=3} = \frac{c}{4}\left(\frac{3h^3}{8\pi}\right)^{1/3}\left(\frac{1}{\mu_e M_{\mathrm{H}}}\right)^{4/3} = \frac{1}{4}\left(\frac{3}{8\pi}\right)^{1/3}\frac{c\,h}{M_{\mathrm{H}}^{4/3}}\left(\frac{1}{\mu_e}\right)^{4/3}$$

$$= \frac{1.231\times10^{15}}{\mu_e^{4/3}}\frac{\mathrm{dyne}}{\left(\mathrm{g\ cm}^{-3}\right)^{4/3}}. \tag{5.7.23}$$

When $N = 3$, eqs. (5.1.20), (5.1.21), and (5.1.29) give

$$4\pi G\left(\frac{R}{z_s}\right)^2\frac{1}{(4K_2)^3}\left(-\frac{GM}{R}\frac{1}{z_s\,(du/dz)_s}\right)^2 = 1. \tag{5.7.24}$$

Radius R disappears from this equation and, using eq. (5.7.23) for K_2, one obtains a unique mass:

$$M_{\mathrm{Chandrasekhar}} = -\frac{4}{\sqrt{\pi}}\left(z^2\frac{du}{dz}\right)_s\left(\frac{K_2}{G}\right)^{3/2}. \tag{5.7.25}$$

Using $G = 6.673\times10^{-8}$ in cgs units, K_2 from eq. (5.7.23), and $\left(z^2\,du/dz\right)_s$ for $N = 3$ from Table 5.2.3, eq. (5.7.25) comes within 0.5% of being the same as eq. (5.7.21).

It is interesting that, although the gravitational binding energy of a model of mass near the Chandrasekhar limit tends towards infinity, the kinetic energy of electrons also tends towards infinity in such a way that the binding energy asymptotically becomes equal to the total rest mass energy of the electrons in the model. To see this, set $R^3 P_s = 0$ and $P = P_e$ in eq. (3.3.1) to obtain

$$\int 3P_e dV = -\Omega. \tag{5.7.26}$$

From the first equalities in eqs. (5.7.13) and (5.7.12), one has

$$P_e = \frac{1}{3}\int v_e p_e\,dn_e = \frac{1}{3}m_ec^2\int\frac{p^2}{\epsilon}\,dn_e = \frac{1}{3}m_ec^2\int\frac{\epsilon^2-1}{\epsilon}\,dn_e, \tag{5.7.27}$$

where, now,

$$\epsilon = \sqrt{1+p^2} \tag{5.7.28}$$

is the (relativistic) electron energy in units of m_ec^2, and

$$p = \frac{p_e}{m_ec}. \tag{5.7.29}$$

Inserting P_e from eq. (5.7.27) into eq. (5.7.26) gives

$$m_ec^2\int\int\left(\epsilon-\frac{1}{\epsilon}\right)\,dn_e\,dV + \Omega = 0. \tag{5.7.30}$$

The kinetic energy of one electron is $m_e c^2 (\epsilon - 1)$, so the kinetic energy of all of the electrons in the white dwarf is

$$E_e = m_e c^2 \int \int (\epsilon - 1) \, dn_e \, dV. \qquad (5.7.31)$$

It follows from eqs. (5.7.30) and (5.7.31) that the binding energy is

$$E_{\text{bind}} = |\Omega| - E_e = m_e c^2 \int \int \left(1 - \frac{1}{\epsilon}\right) dn_e \, dV = N_e \, m_e c^2 \left(1 - \left\langle \frac{1}{\epsilon} \right\rangle\right), \qquad (5.7.32)$$

where N_e is the number of electrons in the white dwarf and $\langle 1/\epsilon \rangle$ is an average over the model. The limiting value

$$E_{\text{bind}} \to N_e \, m_e c^2 \qquad (5.7.33)$$

follows from the fact that, formally, $\langle 1/\epsilon \rangle \to 0$ as the mass of the white dwarf approaches the Chandrasekhar mass and stellar radius shrinks to zero.

The limiting situation is not encountered in the real world since, as the limit is approached, several pieces of vitally important physics which have been neglected come into play and alter the equations which are relevant. For example, when the electron Fermi energy exceeds 10.42 MeV, ^{16}O is converted into ^{16}N and, when it exceeds 13.37 MeV, ^{12}C is converted into ^{12}B. More important, as kinetic energies of particles exceed the average binding energies in nuclei, the nuclei begin to disintegrate. In general, one expects that, as more mass is added, the erstwhile white dwarf collapses into a more compact object in which pressure is supplied by nucleons and electrons, with the the most stable final configuration resembling a neutron star. That is, collapse continues until typical separations become of the order of the Compton wavelength of nucleons or smaller. At the end of the collapse, the pressure supplied by every kind of particle becomes proportional to the number density of that kind of particle. Thus, for example, as the radius of the theoretical configuration stabilizes, whatever the final mix of particles, and in the context of Newtonian gravity, the binding energy approaches the total rest mass energy of the particles:

$$E_{\text{bind}} = |\Omega| - \sum_i E_i = \sum_i N_i M_i c^2 \left(1 - \left\langle \frac{1}{\epsilon_i} \right\rangle\right) \to \sum_i N_i M_i c^2, \qquad (5.7.34)$$

where the sum is over all types of particle. This gives $E_{\text{bind}} = 1.8 \times 10^{34} (M_{\text{max}}/M_\odot)$ erg for a neutron star of mass M_{max}. Before this happens, general relativity enters the picture and alters the relationship between baryonic mass, gravitational mass, and binding energy.

5.8 Insights from one zone models: white dwarf radius versus mass, heating and cooling in a low mass star as functions of radius, and compression and ion cooling as luminosity sources in white dwarfs

In Chapter 3, one zone models were constructed to explore in a qualitative fashion some of the internal characteristics of main sequence stars. In this section and the next, one zone models are constructed to demonstrate in an intuitive fashion the essence of relationships which have been derived rigorously in previous sections with polytropes. For example, it is possible to demonstrate the dependence of radius on mass for low mass white dwarfs and to demonstrate the formal existence of a maximum white dwarf mass. A single zone model can also be used to demonstrate the nature of the evolution of a very low mass main sequence star. When the radius of the model is large enough that the kinetic energy of its constituents is larger than the kinetic energy of electrons due to the exclusion principle, the model heats as it contracts. A maximum temperature is reached and, as the model continues to contract, the temperature declines as thermal kinetic energy is replaced by the kinetic energy of degenerate electrons as the dominant source of pressure.

A measure of the average distance between adjacent electrons is obtained by imagining that electrons are fixed at the corners of a cubic lattice, adjacent corners of which are separated by a distance $d = 2r$. Surrounding each electron by a sphere of radius r, one has that

$$N_e \frac{4\pi}{3} r^3 \sim \frac{4\pi}{3} R^3, \tag{5.8.1}$$

where

$$N_e = \frac{M}{\mu_e M_H} = \frac{N_\odot}{\mu_e} \frac{M}{M_\odot}, \tag{5.8.2}$$

with $N_\odot$ given by eq. (5.7.7), is the total number of electrons in the white dwarf.

An approximation to the average electron Fermi momentum p_F in the interior follows by applying a basic tenet of statistical mechanics, to wit, $N = g \int d^3 p \, dV / h^3$, where g is a statistical weight. In the current context,

$$1 = 1 \left(\frac{4\pi}{3} p_F^3 \frac{4\pi}{3} r^3 \right) \frac{1}{h^3}, \tag{5.8.3}$$

or

$$p_F = \left(\frac{3}{4\pi} \right)^{2/3} \frac{h}{r} = 2.418 \frac{\hbar}{r} = 4.836 \, p_{min}, \tag{5.8.4}$$

where, as given by the Heisenberg uncertainty principle,

$$p_{min} \sim \frac{\hbar}{d} = \frac{\hbar}{2r} = \frac{m_e c}{2\pi} \left(\frac{\lambda_C}{2r} \right) \tag{5.8.5}$$

is the minimum momentum of an electron confined to a region of dimension $d = 2r$.

Assuming that electrons are non-relativistically degenerate, the average value of p^2 over the Fermi distribution is $\langle p^2 \rangle = (3/5) \, p_F^2 = (3.746 \, p_{min})^2$ and the typical kinetic energy of an electron is

$$\langle E \rangle = \frac{\langle p^2 \rangle}{2m_e} = \frac{3}{5} \frac{p_F^2}{2m_e} = \frac{(3.746 \, p_{min})^2}{2m_e}. \tag{5.8.6}$$

Combining eqs. (5.8.6) and (5.8.4) gives

$$\langle E \rangle \sim \frac{3}{5} \left(\frac{3}{4\pi} \right)^{4/3} \frac{h^2}{2m_e r^2}, \tag{5.8.7}$$

which can be written as

$$\langle E \rangle \sim f_{\text{Pauli}}^2 \frac{h^2}{2m_e r^2} = \frac{1}{2} m_e c^2 \left(f_{\text{Pauli}} \frac{\lambda_C}{r} \right)^2, \tag{5.8.8}$$

where

$$f_{\text{Pauli}} = \sqrt{ \frac{3}{5} \left(\frac{3}{4\pi} \right)^{4/3} } = \sqrt{0.088\,86} = 0.2981. \tag{5.8.9}$$

In this formulation, the momentum of an average interior electron is given by

$$p_{\text{average}} = \sqrt{\langle p^2 \rangle} = f_{\text{Pauli}} \frac{h}{r} = 1.873 \frac{\hbar}{r} = 3.746\, p_{\text{min}}. \tag{5.8.10}$$

The final step in constructing the one zone model is to equate the total internal kinetic energy with the global binding energy of the model. When radiation pressure can be neglected, this latter is proportional to the gravitational binding energy of the model. Thus,

$$E_{\text{kinetic}} = \langle E \rangle \, N_e \sim \frac{|\Omega|}{2} \sim \alpha \frac{1}{2} \frac{GM^2}{R}, \tag{5.8.11}$$

where α is a parameter to be determined. Equation (5.2.31) shows that for an index $N = 3/2$ polytrope, $\alpha = 6/7$, and that for an index $N = 3$ polytrope, $\alpha = 3/2$.

On the right hand side of eq. (5.8.11), one may use eq. (5.8.2) to replace M in terms of N_e, μ_e, and M_H, and use eq. (5.8.1) to replace R in terms of r and N_e. Thus,

$$f_{\text{Pauli}}^2 \frac{h^2}{2m_e r^2} N_e \sim \frac{\alpha}{2} \frac{G(\mu_e N_e M_H)^2}{N_e^{1/3} r}. \tag{5.8.12}$$

Rearrangement yields

$$r \sim \frac{2\pi}{\alpha} \frac{f_{\text{Pauli}}^2}{\mu_e^2} \frac{h}{m_e c} \frac{c\hbar}{GM_H^2} \frac{1}{N_e^{2/3}}. \tag{5.8.13}$$

Using eqs. (5.7.4) and (5.7.5), one has

$$r = \frac{2\pi}{\alpha} \frac{f_{\text{Pauli}}^2}{\mu_e^2} \lambda_C \left(\frac{N_0}{N_e} \right)^{2/3}, \tag{5.8.14}$$

and using eqs. (5.7.5) and (5.8.2) then gives

$$r = \frac{3.013\pi}{\alpha} \frac{f_{\text{Pauli}}^2}{\mu_e^{4/3}} \left(\frac{M_\odot}{M} \right)^{2/3} \lambda_C. \tag{5.8.15}$$

Inserting f_{Pauli} from eq. (5.8.8) and evaluating the numerical factor gives

$$r = \frac{0.8411}{\alpha} \frac{1}{\mu_e^{4/3}} \left(\frac{M_\odot}{M}\right)^{2/3} \lambda_C. \tag{5.8.16}$$

showing that the average distance between adjacent electrons is of the order of the Compton wavelength of the electron.

Using eq. (5.8.16) in eq. (5.8.1) gives

$$R = N_e^{1/3} r = \left(\frac{M}{\mu_e M_H}\right)^{1/3} \frac{0.8411}{\alpha} \frac{1}{\mu_e^{4/3}} \left(\frac{M_\odot}{M}\right)^{2/3} \lambda_C$$

$$= \left(\frac{M_\odot}{M_H}\right)^{1/3} \lambda_C \frac{0.8411}{\alpha} \frac{1}{\mu_e^{5/3}} \left(\frac{M_\odot}{M}\right)^{1/3}. \tag{5.8.17}$$

Finally,

$$R = \frac{0.009\,788}{\alpha} R_\odot \left(\frac{2}{\mu_e}\right)^{5/3} \left(\frac{M_\odot}{M}\right)^{1/3}$$

$$= 0.011\,42\, R_\odot \left(\frac{2}{\mu_e}\right)^{5/3} \left(\frac{M_\odot}{M}\right)^{1/3} \quad \text{when } \alpha = \frac{6}{7}. \tag{5.8.18}$$

The final numerical coefficient obtained by choosing α as given by the index $N = 3/2$ polytrope is smaller (by about 10%) than the coefficient in eq. (5.7.11) given rigorously by the polytrope, which means, ironically, that the heuristic result matches the result of using a relativistically correct equation of state, Table 5.7.1, better than the rigorous result.

In a next exercise, assume again that ion pressure can be neglected, but allow for an arbitrary degree of electron degeneracy. The kinetic energy of a typical particle is given by

$$E = \sqrt{\left(m_e c^2\right)^2 + (cp)^2} - m_e c^2. \tag{5.8.19}$$

Assume in this expression that

$$p = f\,\frac{h}{r}, \tag{5.8.20}$$

where f is a parameter to be determined. In the limit of extreme degeneracy,

$$p_{average} = \frac{3}{4}\, p_F = \frac{3}{4} \left(\frac{3}{4\pi}\right)^{2/3} \frac{h}{r} = 0.289\,\frac{h}{r} = f_{ER}\,\frac{h}{r}, \tag{5.8.21}$$

which defines an $f = f_{ER}$ that is almost identical with f_{Pauli} given by eqs. (5.8.9) in the non-relativistic approximation.

Casting it as a balance between the total internal kinetic energy and the global binding energy, the consequence of the local balance between the pressure gradient force and the gravitational force can be expressed by

$$N_e \left(\sqrt{\left(m_e c^2\right)^2 + \left(f\,\frac{ch}{r}\right)^2} - m_e c^2\right) \sim \frac{3}{4} \frac{G\,(N_e M_H \mu_e)^2}{r\, N_e^{1/3}}, \tag{5.8.22}$$

where the factor of 3/4 on the right hand side of the equation comes from eq. (5.2.31) for an $N = 3$ polytrope when $\tilde{\beta} = 1$. Dividing through by $m_e c^2$ and using eqs. (5.7.4) and (5.7.5),

$$\sqrt{1 + \left(f\,\frac{\lambda_C}{r}\right)^2} - 1 \sim \frac{3}{8\pi}\left(\frac{N_e}{N_0}\right)^{2/3}\mu_e^2\,\frac{\lambda_C}{r}. \tag{5.8.23}$$

When $(1/f)\,(r/\lambda_C) < 1$, the equation is best rewritten as

$$\sqrt{1 + \left(\frac{1}{f}\,\frac{r}{\lambda_C}\right)^2} - \frac{1}{f}\,\frac{r}{\lambda_C} \sim \frac{3}{8\pi}\left(\frac{N_e}{N_0}\right)^{2/3}\mu_e^2\,\frac{1}{f}. \tag{5.8.24}$$

Expanding in powers of $(1/f)\,(r/\lambda_C)$ gives

$$1 - \frac{1}{f}\,\frac{r}{\lambda_C} + \frac{1}{2}\,\frac{1}{f^2}\left(\frac{r}{\lambda_C}\right)^2 + \cdots \sim \frac{3}{8\pi}\left(\frac{N_e}{N_0}\right)^{2/3}\mu_e^2\,\frac{1}{f}. \tag{5.8.25}$$

It is clear from all three equations (5.8.23)–(5.8.25) that, as N_e increases, r/λ_C decreases. For very small r,

$$\frac{r}{\lambda_C} \sim f\left(1 - \frac{3}{8\pi}\left(\frac{N_e}{N_0}\right)^{2/3}\mu_e^2\,\frac{1}{f}\right), \tag{5.8.26}$$

from which it follows that, as

$$N_e \to N_{e,\mathrm{max}} = \left(f\,\frac{8\pi}{3}\,\frac{1}{\mu_e^2}\right)^{3/2} N_0, \tag{5.8.27}$$

$$\frac{r}{\lambda_C} \to 0. \tag{5.8.28}$$

That is, as N_e is increased to a critical value $N_{e,\mathrm{max}}$, the separation between electrons becomes vanishingly small. The formal limiting mass given by the single zone model is

$$M_{\mathrm{max}} = \mu_e M_H N_{e,\mathrm{max}} = 2\left(\frac{2\pi}{3}\,f\right)^{3/2}\left(\frac{2}{\mu_e}\right)^2 M_H N_0$$

$$= 1.737\left(\frac{2}{\mu_e}\right)^2 M_\odot \quad \text{when } f = f_{\mathrm{ER}} = 0.289. \tag{5.8.29}$$

The maximum mass estimated by the one zone model is approximately 20% larger than the maximum mass estimated by use of an $N = 3$ polytrope (eq. (5.7.28)).

Transparent demonstration of how the white dwarf radius goes to zero as its mass approaches a limiting value is a major achievement of the single zone model. Of course, the argument breaks down when electron kinetic energies approach binding energies of nucleons in nuclei and average separations between particles become considerably smaller than the Compton wavelength of an electron. But, as emphasized in Section 5.7, this is also true for the derivation involving the $N = 3$ polytrope.

It is interesting to examine quantitatively how the radius approaches zero as the limiting mass is approached. Using eqs. (5.7.9) and (5.7.14) to find N_e/N_0 and eq. (5.7.7) for $\lambda_C N_0^{1/3}$, the predicted model radius given by eqs. (5.8.1), (5.8.26), and (5.8.27) is

$$R = r\, N_e^{1/3} \sim \lambda_C\, f \left(1 - \left(\frac{M}{M_{\max}}\right)^{2/3}\right) N_e^{1/3} \tag{5.8.30}$$

$$= \lambda_C\, N_0^{1/3}\, f \left(1 - \left(\frac{M}{M_{\max}}\right)^{2/3}\right) \left(\frac{N_e}{N_0}\right)^{1/3} \tag{5.8.31}$$

$$= 0.0293\, R_\odot\, f \left(1 - \left(\frac{M}{M_{\max}}\right)^{2/3}\right) \left(\frac{2}{\mu_e}\frac{M}{M_\odot}\right)^{1/3} \tag{5.8.32}$$

$$= 0.00847\, R_\odot \left(1 - \left(\frac{M}{M_{\max}}\right)^{2/3}\right) \left(\frac{2}{\mu_e}\frac{M}{M_\odot}\right)^{1/3} \quad \text{when } f = f_{\mathrm{ER}} = 0.289. \tag{5.8.33}$$

While the dependence on mass given by eq. (5.8.33) may be trusted for a mass very near the critical mass, for masses far from the upper limit for observed white dwarfs, the radius given by it is, perversely, much smaller than the radius given by exact models. For example, for a cold white dwarf of mass $M = M_\odot$, eq. (5.8.33) gives $R = 0.0090\, f\, R_\odot = 0.0026\, R_\odot$, compared with the correct $R = 0.0080\, R_\odot$.

5.8.1 The temperature maximum and the nature of energy sources in cooling white dwarfs and completely convective low mass stars

A final exercise elucidates the relationship between the mean interior temperature, the mean interior molecular weight, and the radius of a star which relies primarily on gravitational potential energy and internal thermal energy as sources of luminosity. It is, therefore, useful in understanding the behavior of pre-main sequence stars and cooling white dwarfs. It is also applicable to a completely convective low mass star whether or not it is burning a nuclear fuel.

The internal kinetic energy of the single zone model may be taken to consist of two parts: (1) a component which expresses the effect of the Heisenberg exclusion principle on the kinetic energy of free electrons and does not depend on the temperature and (2) a component which expresses the thermal kinetic energy of non-degenerate nuclei and the thermal kinetic energy of free electrons over and above the degeneracy-related kinetic energy. Thus, in the single zone model,

$$E_{\mathrm{kinetic}} \sim N_e \left(\frac{1}{2m_e}\left(f\,\frac{h}{r}\right)^2 + \alpha\,\frac{3}{2}\,kT\right), \tag{5.8.34}$$

or

$$E_{\text{kinetic}} \sim \frac{1}{2} \, (N_e \, m_e c^2) \left(\left(f \, \frac{\lambda_C}{r} \right)^2 + 3 \, \alpha \, \frac{kT}{m_e c^2} \right), \tag{5.8.35}$$

where α, the number of free particles per electron which contribute to the thermal energy, is given by

$$\alpha = 1 + \frac{\mu_e}{\mu_n}, \tag{5.8.36}$$

when electrons are not degenerate, and by

$$\alpha = \frac{\mu_e}{\mu_n}, \tag{5.8.37}$$

when electrons are degenerate. Here, μ_n is the nuclear molecular weight (i.e., the number of nucleons per nucleus).

The net binding energy may be written as

$$E_{\text{bind}} \sim \frac{3}{7} \, \frac{G(N_e \mu_e M_H)^2}{N_e^{1/3} r}, \tag{5.8.38}$$

or, using eqs. (5.7.4) and (5.7.5) to introduce N_0 and λ_C,

$$E_{\text{bind}} \sim \frac{1}{2} \, (N_e \, m_e c^2) \, \frac{3}{7\pi} \, \mu_e^2 \left(\frac{N_e}{N_0} \right)^{2/3} \frac{\lambda_C}{r}. \tag{5.8.39}$$

The balance between the internal kinetic energy and the global binding energy is achieved by equating the expressions on the right hand sides of eqs. (5.8.35) and (5.8.39). Dividing the result by $\frac{1}{2} \, (N_e m_e c^2)$ and rearranging, one obtains

$$3\alpha \, \frac{kT}{m_e c^2} \sim \frac{\lambda_C}{r} \left[\frac{3}{7\pi} \, \mu_e^2 \left(\frac{N_e}{N_0} \right)^{2/3} - f^2 \, \frac{\lambda_C}{r} \right], \tag{5.8.40}$$

where N_e is given by eq. (5.8.2) and N_0 is given by eq. (5.7.5).

Equation (5.8.40) describes the evolution of a completely convective low mass star even more succinctly and in greater generality than does Table 5.6.1 which describes the evolution of an adiabatic model in which the molecular weight is constant in time. Beginning with a large enough radius, the internal temperature is too small for nuclear reactions to be important. Since the model radiates energy from its surface, it must rely on the release of gravitational potential energy to supply the lost energy. Hence, its radius must decrease and, from eq. (5.8.40), its mean internal temperature must rise. If it ignites a nuclear fuel, the product $\mu_e^2 N_e^{2/3} \propto \mu_e^{4/3}$ increases as μ_e, the number of nucleons per electron, increases, and the mean interior temperature continues to rise while the radius changes on a nuclear burning time scale rather than on a Kelvin–Helmholtz time scale. Once the nuclear fuel is exhausted, μ_e and N_e cease to change, and the release of gravitational potential energy again becomes the primary source of energy loss from the surface. The release of gravitational potential energy, however, requires the model to shrink until, eventually, the quantity in square brackets in eq. (5.8.40) decreases with increasing λ_C/r more rapidly than its multiplicant λ_C/r increases. The mean temperature reaches a maximum and thereafter decreases.

The maximum value of $\alpha\, kT$ occurs when

$$f^2 \left(\frac{\lambda_C}{r}\right)_{\max T} = \frac{1}{2}\, \frac{3}{7\pi}\, \mu_e^2 \left(\frac{N_e}{N_0}\right)^{2/3} = \frac{3}{14\pi}\, \mu_e^{4/3} \left(\frac{M}{M_0}\right)^{2/3}, \qquad (5.8.41)$$

and, at the maximum mean temperature,

$$\left(3\,\alpha\,\frac{3kT}{m_e c^2}\right)_{\max} \sim \left(\frac{3}{14\pi}\, \mu_e^2 \left(\frac{N_e}{N_0}\right)^{2/3}\right)^2 \frac{1}{f^2} = \left(\frac{0.045\,28}{f}\right)^2 \left(\mu_e^2\, \frac{M}{M_\odot}\right)^{4/3},$$

$$(5.8.42)$$

or

$$\left(\frac{kT}{m_e c^2}\right)_{\max} \sim \frac{1}{\alpha_{\max}} \left(\frac{0.026\,14}{f}\right)^2 \left(\mu_e^2\, \frac{M}{M_\odot}\right)^{4/3}, \qquad (5.8.43)$$

where $\alpha_{\max}$ is the value of α when, for electrons, the exclusion-principle related kinetic energy and the thermal energy are comparable. To accommodate this comparability, set $\alpha_{\max} \sim 0.5 + \mu_e/\mu_n$. Inserting $f = 0.315$ from eq. (5.8.29) and setting $m_e c^2/k = 5.93 \times 10^9$ K, eq. (5.8.43) becomes

$$\langle T \rangle_{\max} \sim \frac{1}{0.5 + \mu_e/\mu_n} \left(\mu_e^2\, \frac{M}{M_\odot}\right)^{4/3} 4.08 \times 10^7 \text{ K}, \qquad (5.8.44)$$

where angle brackets have been placed around T to emphasize that the temperature estimate applies to a typical nucleon in the model and not to a nucleon at the model center where the temperature is of the order of two times larger than the average temperature.

Applying eq. (5.8.44) to the $M = (1/9)\, M_\odot$ model of Section 5.6, for which $\mu_e = 1.17$ and $\mu_n = 1.27$, one obtains a typical interior temperature of

$$\langle T \rangle_{\max} \sim 2.3 \times 10^6 \text{ K}, \qquad (5.8.45)$$

which compares with the maximum central temperature of $\sim 5 \times 10^6$ K for the adiabatic model (see Table 5.6.1).

For a $0.5\, M_\odot$ model of the same composition,

$$\langle T \rangle_{\max} \sim 17 \times 10^6 \text{ K}. \qquad (5.8.46)$$

Since the central temperature is roughly two times larger than the mean temperature, this means that, long before the maximum temperature is reached, interior temperatures become high enough for hydrogen to ignite, with the release of nuclear energy replacing the release of gravitational potential energy as the primary luminosity source. Realistic evolutionary models show that the time required by a star of initial mass $0.5\, M_\odot$ to become a helium star is much longer than a Hubble time.

A more relevant initial model to consider is the hot helium white dwarf below the hydrogen-burning shell of a low mass star on the first red giant branch. In the hydrogen-exhausted core, $\mu_e = 2$, $\mu_n = 4$, and, as M_{He} approaches $0.45\, M_\odot$,

$$\langle T \rangle_{\max} \sim 0.9 \times 10^8 \text{ K}, \qquad (5.8.47)$$

indicating that interior temperatures can become high enough for helium to be ignited under electron-degenerate conditions, thus initiating a helium core flash. This result anticipates the fact that realistic evolutionary models of initial mass in the 1–2 $M_\odot$ range develop electron-degenerate helium cores which ignite helium in a flash when the core mass reaches approximately 0.45 $M_\odot$. The injection of nuclear energy raises the temperature to the extent that electron degeneracy is lifted and the helium core is transformed into a helium star which burns helium into carbon and oxygen quiescently as hydrogen continues to burn in a shell just above the outer edge of the helium-exhausted core, adding to the mass of the hydrogen-free core.

After the exhaustion of helium in central regions, the star expands, and if it is in a binary and fills its Roche lobe for the first time, its hydrogen-rich envelope will be lost and the remnant core will evolve quickly into a hot hybrid white dwarf consisting of a carbon-oxygen core in which $\mu_n \gtrsim 12$ and a helium envelope in which $\mu_n = 4$. If the star is single or in a wide binary, helium burning continues until the hot white dwarf core, of mass $\sim$0.55 $M_\odot$, consists almost entirely of carbon and oxygen, and a superwind ejects the remaining hydrogen-rich envelope into a shell which is caused by radiation from the compact remnant central star to fluoresce for $\sim 10^4$ yr as a planetary nebula.

During the subsequent cooling phase, λ_C/r in the remnant central white dwarf approaches an asymptote given by setting to zero the term in square brackets in eq. (5.8.40):

$$\frac{\lambda_C}{r_{\text{final}}} = \frac{1}{f^2} \frac{3}{7\pi} \mu_e^2 \left(\frac{N_e}{N_0}\right)^{2/3}, \tag{5.8.48}$$

where r_{final} is the final mean separation of electrons. Comparing the separations r given by eqs. (5.8.41) and (5.8.48), it is evident that the white dwarf shrinks by a factor of two during the cooling process.

Using the notation of eq. (5.8.48), eq. (5.8.40) may be written as

$$3\alpha \frac{kT}{m_e c^2} \sim \left(f \frac{\lambda_C}{r_{\text{final}}}\right)^2 \left[\frac{r_{\text{final}}}{r}\left(1 - \frac{r_{\text{final}}}{r}\right)\right], \tag{5.8.49}$$

As r_{final}/r approaches unity, the relationship between temperature and radius is controlled primarily by the second factor in square brackets. Making use of the fact that $N_e/N_0 = M/(\mu_e\, 1.8494\, M_\odot)$, and setting $f = 0.315$, one obtains

$$f \frac{\lambda_C}{r_{\text{final}}} = \frac{0.090\,54}{f} \left(\mu_e^2 \frac{M}{M_\odot}\right)^{2/3} = 0.2874 \left(\mu_e^2 \frac{M}{M_\odot}\right)^{2/3}. \tag{5.8.50}$$

Using this and $\alpha = \mu_e/\mu_n$ in eq. (5.8.49), one obtains

$$\frac{kT}{m_e c^2} \sim 0.027\,54\, \mu_n\, \mu_e^{5/3} \left(\frac{M}{M_\odot}\right)^{4/3} \frac{r_{\text{final}}}{r}\left(1 - \frac{r_{\text{final}}}{r}\right), \tag{5.8.51}$$

which shows explicitly that, the more massive the nuclei are, the larger is the maximum possible temperature. The final numerical result for the mean interior temperature during the cooling phase is

$$\langle T \rangle \sim 5.19 \times 10^8 \text{ K } \mu_n \left(\frac{\mu_e}{2}\right)^{5/3} \left(\frac{M}{M_\odot}\right)^{4/3} \frac{r_{\text{final}}}{r} \left(1 - \frac{r_{\text{final}}}{r}\right). \tag{5.8.52}$$

It is interesting to compare the rate at which the thermal energy of nuclei is released during cooling with the rate of release of gravitational binding energy. Equations (5.8.35) and (5.8.46) give

$$E_{\text{bind}} \sim \frac{1}{2} \left(N_e \, m_e c^2\right) \left(f \frac{\lambda_C}{r_{\text{final}}}\right)^2 \frac{r_{\text{final}}}{r}. \tag{5.8.53}$$

Since the number of nucleons is related to the number of electrons by $N_{\text{nucelons}} = \alpha N_e$, eq. (5.8.45) gives for the thermal energy of the nuclei:

$$E_{\text{thermal}} = \alpha N_e \frac{3}{2} kT = \frac{1}{2} \left(N_e \, m_e c^2\right) \left(f \frac{\lambda_C}{r_{\text{final}}}\right)^2 \frac{r_{\text{final}}}{r} \left(1 - \frac{r_{\text{final}}}{r}\right). \tag{5.8.54}$$

The binding energy increases at the rate

$$\frac{dE_{\text{bind}}}{dt} = A \frac{d}{dt}\left(\frac{r_{\text{final}}}{r}\right), \tag{5.8.55}$$

whereas the thermal energy changes at the rate

$$-\frac{dE_{\text{thermal}}}{dt} = \left(2\frac{r_{\text{final}}}{r} - 1\right) A \frac{d}{dt}\left(\frac{r_{\text{final}}}{r}\right), \tag{5.8.56}$$

where A is the common constant in eqs. (5.8.49) and (5.8.50). Conservation of energy requires that the gravitational binding energy changes at the rate

$$\frac{d|\Omega|}{dt} = \frac{dE_{\text{bind}}}{dt} + \frac{dE_{\text{thermal}}}{dt} = 2\left(1 - \frac{r_{\text{final}}}{r}\right) A \frac{d}{dt}\left(\frac{r_{\text{final}}}{r}\right). \tag{5.8.57}$$

As $r \to r_{\text{final}}$,

$$-\frac{dE_{\text{thermal}}}{dt} \to \frac{dE_{\text{bind}}}{dt} = A \frac{d}{dt}\left(\frac{r_{\text{final}}}{r}\right) \tag{5.8.58}$$

and

$$\frac{d|\Omega|}{dt} \to 0. \tag{5.8.59}$$

In other words, as the white dwarf radius approaches its asymptote, the luminosity is described by

$$L = \frac{dE_{\text{bind}}}{dt} \sim -\frac{dE_{\text{thermal}}}{dt}, \tag{5.8.60}$$

showing that cool white dwarfs shine because they release the thermal energy of nuclei and not because they release gravitational binding energy. That is, the binding energy increases primarily because the temperatures of the nuclei decrease and not because the white dwarf shrinks. On the other hand, eq. (5.8.52) demonstrates that the white dwarf shrinks because it cools.

White dwarfs lose energy not only by radiating photons from their surfaces at rates which are determined by the temperature gradient and the opacity in subsurface layers. They also radiate neutrinos from their interiors due to various weak interaction processes described in Chapter 15 in Volume 2. The net result is that, during the very early phases of cooling at high temperatures, the maximum temperature in a white dwarf is found at a watershed point off center; energy flows inward from this point to accomodate neutrino losses from the interior, and energy flows outward to account for the photon losses from the surface. Examples of the temperature development in realistic models of mass $0.6 M_\odot$ (Figs. 2 and 3 in Iben and Tutukov, 1984) show that, despite the complications, the mean interior temperature decreases monotonically as evolution progresses. The mean interior temperature given by the single zone model does not rely on the details of energy loss mechanisms and therefore corresponds to the mean temperature found in the realistic model calculations. After about 3×10^7 yr of evolution, neutrino losses become unimportant and the white dwarf thereafter possesses an electron-degenerate core which, because of the large conductivity of degenerate electrons, is essentially isothermal. The temperature of the isothermal core may be identified with the mean temperature of the single zone model.

One last inference concerning white dwarfs can be constructed with very little additional work. The luminosity of a white dwarf with an isothermal core may be written as

$$L = \sigma \, T_e^4 = a \, T^\alpha = -b \, \frac{dT}{dt}, \tag{5.8.61}$$

where the first equality defines the white dwarf surface temperature T_e, the second equality is an assumption relating the temperature T of the isothermal core to T_e, and the third equality follows from eq. (5.8.60) and the first equality in eq. (5.8.54). As shown in Section 8.4, the relationship between the surface temperature and an interior temperature not far below the photosphere satisfies

$$T^4 = \frac{2}{3} T_e^4 \left(\tau + \frac{2}{3} \right), \tag{5.8.62}$$

where $\tau = \int \kappa \, dr$ is the optical depth between the photosphere and the isothermal core. Assuming that τ is a constant, one may guess $\alpha \sim 4$ in eq. (5.8.57). Solving eq. (5.8.58) with all coefficients and powers assumed to be constants gives

$$L = L_0 \left(1 + \frac{(\alpha - 1) \, a \, T_0^{\alpha-1}}{b} \, (t - t_0) \right)^{-\alpha/(\alpha-1)}, \tag{5.8.63}$$

where t_0 is the time when $T = T_0$. Differentiation gives

$$\frac{d \log L}{d \log t} \sim -\frac{\alpha}{\alpha - 1} \sim -\frac{4}{3}. \tag{5.8.64}$$

Referring to Fig. 2.2.1, one can surmise that 40 Eridani B has been a white dwarf roughly four times longer than Sirius B and that the pair L870-2 A,B have been white dwarfs roughly ten times longer than 40 Eridani B.

5.9 One zone models of neutron stars, neutron star composition, and neutron star masses

A star of initial mass less than ~ 11 $M_\odot$ evolves ultimately into a white dwarf that is supported against gravity primarily by the pressure of degenerate electrons, whereas a star initially more massive than ~ 11 $M_\odot$ develops an electron-degenerate core of iron-peak isotopes which, on reaching a mass of the order of the Chandrasekhar mass, collapses into a neutron star that is supported against gravity primarily by the pressure of degenerate neutrons. The one zone model for white dwarfs teaches that, for stars of near solar mass, the typical separation between pressure-supplying particles is of the order of the Compton wavelength of the pressure-supplying particles. Since the Compton wavelength of a neutron is ~ 1800 times smaller than the Compton wavelength of an electron, it follows that the radius of a neutron star is of the order of 1800 times smaller than the radius of a white dwarf of the same mass. That is, for a given stellar mass,

$$\frac{R_{\text{neutron star}}}{R_{\text{white dwarf}}} \sim \frac{m_{\text{electron}}}{m_{\text{neutron}}}. \tag{5.9.1}$$

By analogy with the white dwarf case, one can write at once for a neutron star containing N_{NS} nucleons:

$$R_{\text{NS}} = r_{\text{NS}}\, N_{\text{NS}}^{1/3} \sim \left[\lambda_{\text{C,neutron}} \left(\frac{N_0}{N_{\text{NS}}} \right)^{2/3} \right] N_{\text{NS}}^{1/3} = \lambda_{\text{C,neutron}}\, N_0^{1/3} \left(\frac{N_0}{N_{\text{NS}}} \right)^{1/3}, \tag{5.9.2}$$

where

$$\lambda_{\text{C,neutron}} = \frac{h}{m_{\text{neutron}} c} = 1.3205 \times 10^{-13}\ \text{cm} \tag{5.9.3}$$

is the Compton wavelength of a proton and N_0 is given by eq. (5.7.5). In the $N = 1.5$ polytropic approximation, the neutron star radius is given by setting $\mu_e = 1$ in eq. (5.7.10) (one mass-supplying nucleon per one pressure-supplying particle) and multiplying by $\lambda_{\text{C, neutron}}/\lambda_{\text{C}} = m_e/m_{\text{neutron}}$. Thus,

$$R_{\text{NS}, N=3/2} = \frac{\lambda_{\text{C,neutron}}}{\lambda_{\text{C}}}\, 0.0400\, R_\odot \left(\frac{M_\odot}{M} \right)^{1/3}, \tag{5.9.4}$$

or

$$R_{\text{NS}, N=3/2} \sim 2.177 \times 10^{-5} \left(\frac{M_\odot}{M} \right)^{1/3} R_\odot = 15.15 \left(\frac{M_\odot}{M} \right)^{1/3}\ \text{km}. \tag{5.9.5}$$

A typical neutron star mass is of the order of $\sim 1.4\ M_\odot$, so one can anticipate the radius of a typical neutron star to be of the order of

$$R_{\mathrm{NS}} \sim 13.5\ \mathrm{km}, \tag{5.9.6}$$

comparable with the linear dimensions of a mid-sized city containing one hundred thousand people.

The radius given by eq. (5.9.6) is astonishingly close to radii given by realistic models of neutron stars which, apart from being multiple zone models, take into account the facts that general relativity modifies the pressure balance equation, that weak interactions influence the composition of matter as a function of the density, and that nuclear forces modify the equation of state. The experience gained by comparing white dwarf model radii in the $N = 3/2$ polytropic approximation with radii given by exact solutions (see Table 5.7.1), coupled with the fact that the single zone white dwarf approximation produces slightly smaller radii than does the polytropic approximation, suggests that eq. (5.9.6) does not overestimate the typical neutron-star radius by more than, say, 50%.

The gravitational potential energy of a particle in much of the interior of a neutron star is not negligible compared with its rest mass energy, implying that general relativity must be taken into account in determining the force of gravity at any point. For example, at the surface of a neutron star with radius given by eq. (5.9.5), one has

$$\frac{GM M_{\mathrm{H}}}{R}\frac{1}{M_{\mathrm{H}}c^2} = \frac{GM}{Rc^2} = 0.087\left(\frac{M}{M_\odot}\right)^{2/3} = 0.11 \ \text{ for } M = 1.4\ M_\odot. \tag{5.9.7}$$

In spherical symmetry, the general relativistic equation for pressure balance is (e.g., eq. (11.1.13) in Steven Weinberg, *Gravitation and Cosmology*, 1972):

$$\frac{\mathrm{d}P(r)}{\mathrm{d}r} = -G\left(1 - 2\frac{GM(r)}{rc^2}\right)^{-1}\left(M(r) + \frac{4\pi r^3 P(r)}{c^2}\right)\frac{1}{r^2}\left(\rho(r) + \frac{P(r)}{c^2}\right). \tag{5.9.8}$$

The consequent increase in the pressure gradient means that, for a given baryonic mass, the radius of the model becomes smaller than it would otherwise have been. Another effect is a transformation of the baryonic mass of the model into an observable gravitational mass given by $M_{\mathrm{observed}}\ c^2 = M_{\mathrm{baryonic}}\ c^2 - E_{\mathrm{bind}}$, where E_{bind} is the net binding energy of the model.

The radii of realistic models of neutron stars are sensitive to the fact that, despite their name, neutron stars can contain particles other than neutrons, sometimes at substantial abundances. In particular, because a neutron can decay into a proton, an electron, and an antineutrino, neutron stars contain electrons and protons. In free space, if the energy of an electron exceeds the difference between the rest mass energy of a neutron and the rest mass energy of a proton, electron capture on the neutron can occur. At high densities, the situation is slightly more complicated.

To estimate the equilibrium abundances of neutrons, protons, and electrons at finite densities and temperatures, one may invoke the formalism of Section 4.2. The constraints are

$$\alpha_{ep} \left(\sum_i n_{ei} - \sum_j n_{pj} \right) = 0, \tag{5.9.9}$$

$$\alpha_{pn} \left(\sum_j n_{ej} + \sum_k n_{nk} \right) = \text{constant}, \tag{5.9.10}$$

and

$$\beta \left(\sum_i n_{ei} \epsilon_{ei} + \sum_j n_{pj} \epsilon_{pj} + \sum_k n_{nk} \epsilon_{nk} \right) = \text{constant}, \tag{5.9.11}$$

where the quantities n_{st} are number densities and α_{ep}, α_{pn}, and β are Lagrange multipliers. Maximizing probabilities leads to relationships between number densities and statistical weights g_{st} of the form

$$n_{ei} = \frac{g_{ei}}{\exp\left(-\alpha_{ep} + \epsilon_{ei}/kT\right) + 1} = \frac{g_{ei}}{\exp\left((\epsilon_{ei} - \epsilon_{Fe})/kT\right) + 1}, \tag{5.9.12}$$

$$n_{pj} = \frac{g_{pj}}{\exp\left(\alpha_{ep} - \alpha_{pn} + \epsilon_{pi}/kT\right) + 1} = \frac{g_{pj}}{\exp\left((\epsilon_{pj} - \epsilon_{Fp})/kT\right) + 1}, \tag{5.9.13}$$

and

$$n_{nk} = \frac{g_{nk}}{\exp\left(-\alpha_{pn} + \epsilon_{nk}/kT\right) + 1} = \frac{g_{nk}}{\exp\left((\epsilon_{nk} - \epsilon_{Fn})/kT\right) + 1}. \tag{5.9.14}$$

Comparing the two forms for each distribution in eqs. (5.9.12)–(5.9.14) gives the Fermi energy for each type of particle in terms of the Lagrangian multipliers α_{st}. Thus,

$$\frac{\epsilon_{Fe}}{kT} = -\alpha_{ep}, \quad \frac{\epsilon_{Fp}}{kT} = \alpha_{ep} + \alpha_{pn}, \quad \text{and} \quad \frac{\epsilon_{Fn}}{kT} = \alpha_{pn}, \tag{5.9.15}$$

from which it follows that

$$\frac{\epsilon_{Fe}}{kT} + \frac{\epsilon_{Fp}}{kT} = \frac{\epsilon_{Fn}}{kT}. \tag{5.9.16}$$

The total number density of electrons is given by

$$n_e = \frac{8\pi}{h^3} \int_0^\infty \frac{p^2 \, dp}{\exp\left((\epsilon - \epsilon_{Fe})/kT\right) + 1}, \tag{5.9.17}$$

and similarly for protons and neutrons. The development in Section 4.9 which leads to eq. (4.9.10) shows that the final result of integration is a function only of the Fermi momentum given by $p_{Fe} = c^{-1}\sqrt{\epsilon_{Fe}^2 - m_e c^2}$ and of kT. Since $n_p = n_e$, it follows that, regardless of temperature,

$$p_{Fp} = p_{Fe}, \tag{5.9.18}$$

so that, since $\epsilon = \sqrt{(cp)^2 + (mc^2)^2} = c\sqrt{p^2 + (mc)^2}$, eq. (5.9.16) may be written as

$$\sqrt{p_{Fe}^2 + (m_e c)^2} + \sqrt{p_{Fe}^2 + (M_p c)^2} = \sqrt{p_{Fn}^2 + (M_n c)^2}. \tag{5.9.19}$$

In retrospect, one could argue that eqs. (5.9.16)–(5.9.19) are intuitively obvious. The argument is justified when $kT = 0$, for then one may state that (1) conservation of energy requires that $\epsilon_{Fe} + \epsilon_{Fp} = \epsilon_{Fn}$, (2) a neutron cannot decay into a proton and an electron because all accessible states are filled, (3) an electron and a proton cannot combine because all accessible neutron states are filled, and (4) electron and proton Fermi momenta must be equal because the number abundance of each type of particle is precisely proportional to the cube of its Fermi momentum. However, when $kT \neq 0$, transitions between particle types can occur because states are only partially filled for a range of energies of the order of kT below the Fermi energy and only partially empty for a range of energies of the order of kT above the Fermi energy. Hence, the relationship given by eq. (5.9.16) requires mathematical justification, as does the fact that eq. (5.9.18) holds independent of temperature, leading to eq. (5.9.19).

The minimum p_{Fe} required for neutrons to be formed is given by setting $p_{Fn} = 0$ in eq. (5.9.19), resulting in

$$\sqrt{(cp_{Fe})^2 + (m_e c^2)^2} + \sqrt{(cp_{Fe})^2 + (M_p c^2)^2} = M_n c^2. \tag{5.9.20}$$

When this requirement is fulfilled, electrons are relativistic but nucleons are not, and a series of steps,

$$\sqrt{(cp_{Fe})^2 + (m_e c^2)^2} = M_n c^2 - M_p c^2 \sqrt{1 + \left(\frac{p_{Fe}}{M_p c}\right)^2} \sim (M_n c^2 - M_p c^2) - \frac{p_{Fe}^2}{2M_p},$$

$$(cp_{Fe})^2 + (m_e c^2)^2 = (M_n c^2 - M_p c^2)^2 - 2(M_n c^2 - M_p c^2)\frac{p_{Fe}^2}{2M_p} + \left(\frac{p_{Fe}^2}{2M_p}\right)^2,$$

and

$$(cp_{Fe})^2 \left(1 + \frac{M_n - M_p}{M_p} - \frac{1}{4}\left(\frac{p_{Fe}}{M_p c}\right)^2\right) = (M_n c^2 - M_p c^2)^2 - (m_e c^2)^2,$$

leads ultimately to

$$cp_{Fe} \sim \sqrt{\frac{(M_n c^2 - M_p c^2)^2 - (m_e c^2)^2}{1 + (M_n - M_p)/M_p}} = 1.1869 \text{ MeV}. \tag{5.9.21}$$

The corresponding proton mass density is $\rho_p = 1.232 \times 10^7$ g cm^{-3}. So, the first result regarding the composition of a neutron star is that, in its outer portions, a neutron star contains no neutrons!

When

$$cp_{Fe} = M_n c^2 - M_p c^2 = 1.293 \text{ MeV}, \tag{5.9.22}$$

another series of steps leads to

$$cp_{Fn} \sim \left(\frac{M_n}{M_n - M_p} \right)^{1/2} m_e c^2 = 13.77 \text{ MeV}. \tag{5.9.23}$$

When $kT/(\epsilon_{Fn} - 1) \ll 1$, the number density of neutrons is given by

$$n_n = \frac{8\pi}{3} \left(\frac{p_{Fn}}{h} \right)^3, \tag{5.9.24}$$

and similar relationships hold for electron and proton number abundances. This means that

$$\frac{n_n}{n_p} = \frac{n_n}{n_e} = \left(\frac{p_{Fn}}{p_{Fe}} \right)^3. \tag{5.9.25}$$

Thus, when $cp_{Fe} = 1.293$ MeV, $cp_{Fn} = 13.77$ MeV, and $n_n/n_p = n_n/n_e = 1208$. Incredibly, over an interval of only $\Delta(cp_{Fe}) \sim 0.1$ MeV, matter goes from being neutron free to being completely dominated by neutrons.

The mass density of neutrons is given by

$$\rho_n = \frac{8\pi}{3} \left(\frac{p_{Fn}}{h} \right)^3 M_n = \frac{8\pi}{3} \left(\frac{cp_{Fn}}{M_n c^2} \right)^3 \frac{M_n}{\lambda_{C,n}^3}$$

$$= 6.118 \times 10^{15} \text{ g cm}^{-3} \left(\frac{cp_{Fn}}{939.565 \text{ MeV}} \right)^3, \tag{5.9.26}$$

and the average separation between neutrons is given approximately by

$$2r_n = \left(\frac{3}{\pi} \frac{M_n}{\rho_n} \right)^{1/3} = \frac{1.1695 \times 10^{-8} \text{ cm}}{[\rho_n(\text{g cm}^{-3})]^{1/3}}. \tag{5.9.27}$$

For $cp_{Fe} = 1.293$ MeV, the neutron mass density is $\rho_n = 1.926 \times 10^{10}$ g cm^{-3} and the average separation between neutrons is $2r_n \sim 4.363 \times 10^{-12}$ cm, or roughly 15 times the diameter of a nucleon.

Values of cp_{Fn}, n_n/n_e, ρ_n, and $2r_n$ for other values of cp_{Fe} are displayed in Table 5.9.1. As a function of cp_{Fe}, the neutron–proton ratio passes through a broad peak, reaching a maximum of approximately 7400 at $cp_{Fe} \sim 2.5$ MeV, where the neutron mass density is $\rho_n \sim 8.5 \times 10^{11}$ g cm^{-3} and the average distance between adjacent neutrons is $2r_n \sim 12 \times 10^{-13}$ cm, or roughly four times the diameter of a nucleon.

To determine more precisely where the maximum occurs, differentiate with respect to p_{Fe} the ratio p_{Fn}/p_{Fe} defined by eq. (5.9.19) and set the result equal to zero. This leads to

$$\left(\frac{p_{Fn}}{p_{Fe}} \right)^2 = 2 + \frac{M_p c}{\sqrt{(p_{Fe})^2 + (m_e c)^2}} \frac{1 + (m_e/M_p)^2 + 2(p_{Fe}/M_p c)^2}{\sqrt{1 + (p_{Fe}/M_p c)^2}}$$

$$\sim \frac{M_p c}{\sqrt{(p_{Fe})^2 + (m_e c)^2}} \sim \frac{M_p c}{p_{Fe}} \left(1 - \frac{1}{2} \left(\frac{m_e c}{p_{Fe}} \right)^2 \right), \tag{5.9.28}$$

Table 5.9.1 Fermi momenta, neutron–proton ratio = neutron–electron ratio, neutron mass density, and neutron separation

$\frac{cp_{Fe}}{MeV}$	$\frac{cp_{Fn}}{MeV}$	$\frac{n_n}{n_p} = \frac{n_n}{n_e}$	$\frac{\rho_n}{\text{g cm}^{-3}}$	$\frac{\text{separation}}{10^{-13}\text{ cm}}$
1.1869	0	0	0	∞
1.2	4.764	62.56	7.974×10^{8}	126.1
1.3	14.03	1257	2.038×10^{10}	42.82
1.4	19.31	2624	5.311×10^{10}	31.11
2.0	38.13	6930	4.089×10^{11}	15.76
2.3	44.77	7374	6.618×10^{11}	13.42
2.5	48.71	7399	8.526×10^{11}	12.33
3.0	57.45	7024	1.399×10^{12}	10.46
5.0	83.99	4740	4.370×10^{12}	7.153
10.0	128.7	2132	1.572×10^{13}	4.669
30.0	236.0	486.9	9.698×10^{13}	2.545
100.0	454.2	93.70	6.911×10^{14}	1.323
300.0	876.7	24.96	7.531×10^{15}	0.5966
1000.0	2177	10.32	7.612×10^{16}	0.2760
∞	∞	8	∞	0

where the approximate equalities on the second line take advantage of the numerical results in Table 5.9.1. Expanding eq. (5.9.19) gives, in the same approximation,

$$\frac{p_{Fn}^2}{2M_p c} \sim \sqrt{p_{Fe}^2 + (m_e c)^2} - Qc$$

$$\sim p_{Fe}\left(1 + \frac{1}{2}\left(\frac{m_e c}{p_{Fe}}\right)^2\right) - Qc, \qquad (5.9.29)$$

where $Q = M_n - M_p$. Combining the final forms of eqs. (5.9.29) and (5.9.28) yields

$$p_{Fe}^2 - 2\,p_{Fe}\,Qc + \frac{3}{2}\,(m_e c)^2 = 0, \qquad (5.9.30)$$

which has the solution

$$cp_{Fe} = Qc^2\left(1 \pm \sqrt{1 - \frac{3}{2}\left(\frac{m_e}{Q}\right)^2}\right) = 2.424 \text{ MeV}, \qquad (5.9.31)$$

where the final numerical value follows when the plus sign is chosen. Using this value for cp_{Fe} in the first form of eq. (5.9.28) gives

$$p_{Fn} = 47.26 \text{ MeV}\ \ \text{and}$$

$$\left(\frac{n_n}{n_p}\right)_{max} = \left(\frac{p_{Fn}}{p_{Fe}}\right)_{max}^3 = 7406. \qquad (5.9.32)$$

In order to compare with analytic estimates of the maximum neutron–proton ratio in the literature, expand eq. (5.9.31) in powers of m_e^2/Q^2 to get

$$p_{Fe} \sim 2Q \left(1 - \frac{3}{8} \frac{m_e^2}{Q^2}\right),\tag{5.9.33}$$

then use this result in eq. (5.9.28) to find

$$p_{Fn}^2 \sim 2M_p Q \left(1 - \frac{1}{2} \frac{m_e^2}{Q^2}\right).\tag{5.9.34}$$

Equations (5.9.33) and (5.9.34) give

$$\left(\frac{p_{Fn}}{p_{Fe}}\right)^2 \sim \frac{M_p}{2Q} \left(1 + \frac{1}{4} \frac{m_e^2}{Q^2}\right),\tag{5.9.35}$$

or

$$\left(\frac{n_n}{n_p}\right)_{max} = \left(\frac{p_{Fn}}{p_{Fe}}\right)^3_{max} \sim \left(\frac{M_p}{2Q}\right)^{3/2} \left(1 + \frac{3}{8} \frac{m_e^2}{Q^2}\right) = 7316,\tag{5.9.36}$$

which differs from eq. (5.9.32) by about 1%.

At $cp_{Fe} \sim 30$ MeV, $p_{Fn} \sim (1/4)M_n c$, the neutron density ρ_n approaches a value defined by

$$\rho_{nuc} = \frac{M_p A}{(4\pi/3)\,(1.5 \times 10^{-13}\text{ cm } A^{1/3})^3} = 1.2 \times 10^{14}\text{ g cm}^{-3},\tag{5.9.37}$$

which is typical of terrestial nuclei, and the separation between neutrons is roughly equal to the diameter of a nucleon. Even under these conditions, protons constitute no more than $\sim$0.2% of the total nucleon abundance. It is not until cp_{Fn} reaches about half of the rest mass energy of the neutron that the abundance of protons becomes as large as 1% of the neutron abundance.

At this point, with the neutron density approaching 10^{15} g cm^{-3}, the separation between neutrons is several times smaller than nucleon dimensions, and it is obvious that the equation of state requires a consideration of nuclear forces. Entries in the last three rows of Table 5.9.1 are primarily of academic interest, the factor of 8 in the last row following by inspection from eqs. (5.9.19) and (5.9.25) by setting $p_{Fn} = p_{Fe} + p_{Fp} = 2\,p_{Fp} \gg M_n c^2$. With $\rho_n > \rho_{nuc}$, many effects involving elementary particle physics become relevant and require consideration. A good discussion of the issues involved in constructing realistic equations of state and realistic models of neutron stars is given in *Black Holes, White Dwarfs, and Neutron Stars* by Stuart L. Shapiro & Saul A. Teukolsky (1983).

The first detailed examination of neutron star structure using the general relativistic pressure balance equation was made by J. R. Oppenheimer and G. R. Volkoff (1939), who found that, if interactions between nucleons other than those implied by the exclusion principle are neglected, the maximum gravitational mass of a stable model is $\sim 0.7\ M_\odot$, with baryonic mass being somewhat larger than this. Early observations of radio pulsars in binary systems produced gravitational masses of real neutron stars in the range 1–2 $M_\odot$, as described by Paul C. Joss & Saul A. Rappaport (1984). Although there are large error bars associated with any individual estimate of mass, the fact that the mean estimated gravitational mass is significantly larger than 0.7 $M_\odot$ implies that, at small nucleon separations, repulsive nuclear forces come into play, leading to the prediction of a maximum baryonic mass considerably larger than the value obtained when repulsive forces are neglected.

Subsequent dynamical measurements of component masses in binary pulsars have continued to reduce the error bars associated with the mass estimates. From estimates of masses of a sample of more than 50 neutron stars, Thorsett & Chakabaraty (1999) identify 26 estimates which are of particulary high precision and describe the associated mass distribution by a Gaussian characterized by $M = 1.35 \pm 0.04\ M_\odot$. In most cases, the white dwarf in the binary system is of such a low mass that it must be the helium core remnant of a low mass star which filled its Roche lobe early on in its first ascent of the red giant branch and one may guess that mass transfer from the red giant precursor, although sufficient to lead to the spin up of the neutron star to millisecond periods, was minimal (say, $\leq 0.1\ M_\odot$) and that most of the mass from the red giant was lost from the system in a common envelope event. Thus, one may surmise that the typical baryonic mass of the neutron star remnant of a massive star is slightly larger than the Chandrasekhar mass which is, in turn, of the order of the mass of the nickel–iron core of an initially massive star just before this core experiences a dynamical collapse.

More recently, it has been discovered by Demorest, Pennuci, Ransom, Roberts, & Hessels (2010) that the 3.1 ms radio pulsar in the binary system J1614-2230 has a gravitational mass $M = 1.97 \pm 0.04\ M_\odot$. The binary has an orbital period of 8.7 days and the white dwarf component has a mass $M = 0.500 \pm 0.006\ M_\odot$. The high precision achieved is due to the fact that the general relativistic time delay identified by Shapiro (1964) is very large. The inclination of the orbit is such ($87.17^0 \pm 0.02^0$) that the delay which the X-ray beam from the pulsar experiences on passing the white dwarf component when the white dwarf is effectively exactly between the neutron star and the Earth reaches a maximum of 50 μs.

Since its mass is larger than the maximum mass of a helium white dwarf which can be formed in a binary system, the white dwarf component of J1614-2230 must be the progeny of an intermediate mass or fairly massive low mass AGB star with an electron-degenerate CO core. This means that, immediately after the supernova event which produced the precursor of the current neutron star, the orbital separation between the precursor of the present neutron star and the precursor of the white dwarf must have been large enough to accomodate the AGB precursor of the white dwarf. The very short orbital period of the current binary demonstrates that, when the white dwarf precursor filled its Roche lobe, a common envelope was formed, the most likely consequence being that most of the envelope of the precursor was lost from the system. The large mass of the current neutron star compared with the mass of typical neutron stars could be interpreted to mean that the initially formed neutron star accreted considerable mass from its AGB companion during the common envelope episode which led to orbital shrinkage and spinup of the neutron star. It is more likely that the precursor of the original neutron star had a mass placing it on the borderline between stars which form neutron stars and those which form black holes. In any case, it has been established that neutron stars of baryonic mass considerably larger than the Chandrasekhar mass can be formed.

That there is a maximum baryonic mass for neutron stars is indicated by the evidence for the existence of black holes as remnants of initially very massive stars. For example, Geis & Bolton (1986) find for the visible and invisible components of Cygnus X-1 most probable masses of $(33 \pm 9)M_\odot$ and $(16 \pm 5)M_\odot$, respectively. The discovery of a neutron star with a baryonic mass as large as ~ 2 $M_\odot$ has implications for the equation of state in central regions of neutron stars where densities are larger than terrestial nuclear densities. As discussed by Ösel, Psaltis, Ransom, Demorest, & Alfor (2010), among other possibilities, a large hyperonic component and a large component of strongly interacting strange quarks are ruled out, leaving repulsive nuclear forces as the most likely agents for preventing collapse to a black hole of neutron star configurations of baryonic mass at least as massive as 2 $M_\odot$.

Bibliography and references

J. R. Airey, *Phil. Mag.*, **6**, XXII, 658, 1911.

Wilhelm Anderson, *Zeits. für Physik*, **56**, 851, 1929.

Henry Cavendish, *Phil. Trans. Roy. Soc.*, **88**, 469, 1798.

Subrahmanyan Chandrasekhar, *ApJ*, **74**, 81, 1931.

Subrahmanyan Chandrasekhar, *An Introduction to the Study of Stellar Structure* (Chicago: University of Chicago Press), 1939.

A. Cornu, & J. B. Bailes, *Comptes Rendu, Acad. Sci.*, Paris, **76**, 954, 1873.

P. B. Demorest, T. Pennuci, S. M. Ransom, M. S. E. Roberts, & J. W. T. Hessels, *Nature*, **467**, 1081, 2010.

H. B. Dwight, *Tables of Integrals and Other Mathematical Data* (New York: MacMillan), 1934; fourth edition, 1947.

Arthur S. Eddington, *The Internal Constitution of the Stars* (Cambridge: Cambridge University Press), 1926.

Robert Emden, *Gaskugeln – Anwendungen der Mechanische Wärmtheorie* (Berlin: Tuebner), 1907.

Geis, D. R., & Bolton, C. T., *ApJ*, **304**, 371, 1986.

T. Hamada & Edwin E. Salpeter, *ApJ*, **161**, 587, 1962.

L. G. Henyey, Robert LeLevier, & R. D. Levee, *ApJ*, **129**, 2, 1959; **130**, 344, 1959.

Icko Iben, Jr., *ARAA*, **5**, 571, 1967.

Icko Iben, Jr., *ApJ*, **259**, 244, 1982.

Icko Iben, Jr. & Alexander V. Tutukov, *ApJ*, **282**, 615, 1984.

Paul C. Joss & Saul A. Rappaport, *ARAA*, **22**, 537, 1984.

J. C. P. Miller & D. H. Sadler, *Mathematical Tables of the British Association for the Advancement of Science, Vol. II - Emden Functions* (Edinburgh: Neill) 1932.

J. Robert Oppenheimer & G. R. Volkov, *Phys. Rev.*, **55**, 374, 1939.

F. Ösel, D. Psaltis, S. Ransom, P. Demorest, & M. Alfor, *ApJ*, **724**, L199, 2010.

Irwin I. Shapiro, *Phys. Rev. Lett.*, **13**, 789, 1964.

Stuart L. Shapiro & Saul A. Teukolsky, *Black Holes, White Dwarfs, and Neutron Stars* (New York: Wiley), 1983.

Edmund C. Stoner, *Phil. Mag.*, **9**, 944, 1930.

Edmund C. Stoner & Frank Tyler, *Phil. Mag.*, **11**, 986, 1931.

S. E. Thorsett & D. Chakabaraty, *ApJ*, **512**, 288, 1999.

Steven Weinberg, *Gravitation and Cosmology*, (New York: Wiley & Sons), 1972.

E. T. Whittaker & G. N. Watson, *Modern Analysis* (Cambridge: Cambridge University Press), 1902; fourth edition, 1952.

6

Hydrogen-burning reactions and energy-generation rates

Stars spend most of their nuclear burning lives on the main sequence converting hydrogen into helium in central regions. After leaving the main sequence, single stars and stars in wide binaries continue to burn hydrogen in a shell as helium is converted into carbon and oxygen in the hydrogen-exhausted core. The lifetime of a star in the core helium-burning phase, which in intermediate mass stars is typically 10–30% of the main sequence lifetime, is determined by the rate of helium burning, but hydrogen-burning reactions contribute most of the light emitted by the star. The time spent in more advanced stages of nuclear burning is quite small compared with that spent during the main hydrogen- and helium-burning phases. Thus, over most of a star's nuclear burning lifetime, hydrogen-burning reactions are the major contributors to the stellar luminosity.

In population I stars of mass smaller than $\sim 2\,M_\odot$, the reactions which dominate energy production during the main sequence phase are those in the so-called pp chains, which are initiated by the transformation of two protons into a deuterium nucleus. The reactions which follow this initial pp-chain reaction terminate with the formation of ^{3}He (at low temperatures) or with the formation of ^{4}He (at higher temperatures). These subsequent reactions release considerably more energy than is released in the pp reaction itself, but the overall rate of energy release is nevertheless controlled by the pp reaction since it is, by far, the slowest reaction in the chains.

During the main sequence phase in population I stars of mass larger than $\sim 2\,M_\odot$ and during the shell hydrogen-burning phase in red giants and asymptotic giant branch stars, hydrogen burning proceeds primarily by the so-called CN cycle which converts hydrogen into helium by a series of reactions in which isotopes of carbon and nitrogen act as catalysts. At sufficiently high temperatures, other reactions convert most of the oxygen with which the star is born slowly into nitrogen.

At interaction energies typically found in stars, the cross section for the reaction which converts two protons into a deuterium nucleus with the emission of an electron is so small that it cannot be measured in the terrestial laboratory. The energetics of the reaction are examined in Section 6.1. In Section 6.2–6.5, the ingredients of a theoretical calculation of the cross section at energies relevant to stellar interiors are explored in detail. In Section 6.2, it is shown that the matrix element involved in determining the reaction probability can be written as the product of two factors, one factor being given by weak interaction theory, the other factor being an overlap integral between the wave function of the initial di-proton state and the wave function of the deuteron. In Section 6.3, the weak interaction coupling constant is estimated from the observed properties of the ^{6}He$(e^-, \bar{\nu}_e)^6$Li decay. In Section 6.4, considerations involved in estimating the nuclear matrix element are explored and, in Section 6.5, the cross section is determined as a function of the kinetic energy of the interacting protons. In Section 6.6, the overall rate of the pp reaction is obtained by averaging over a Maxwell–Boltzmann energy distribution and it is concluded that, given that the Sun is approximately 4.6×10^9 years old, average temperatures in the energy-producing region of the Sun over this interval of time have been of the order of 12×10^6 K.

The fact that life on Earth has existed over much of the lifetime of the Sun suggests that the Sun has shone at roughly its present rate over this lifetime and that other sources of energy generation, more powerful than the pp reaction, must be at work in the Sun. It is fairly obvious that, since the Coulomb barriers are essentially the same, if a proton can interact with another proton to form an unstable di-proton, a proton can also interact with a deuteron to form a stable ^{3}He nucleus, and since the weak interaction need not be invoked to establish a stable product nucleus, the cross section for the $d(p, \gamma)^3$He reaction must be much larger than the cross section for the $p(p, e^+ \nu_e)d$ reaction. Once ^{3}He is formed, a whole host of nuclear reactions can take place which produce ^{4}He as a final product.

Cross sections for most of the reactions that make up this host, which constitute the pp chains, can be measured in the laboratory, but at energies large compared with those relevant in stellar interiors. In Section 6.7, general considerations involved in extrapolating from nuclear cross sections found in laboratory experiments to cross sections relevant at the much smaller interaction energies typically available in stars are discussed. Properties of reactions in the pp chains are reviewed in Section 6.8. In addition to the deuteron and ^{3}He, isotopes of Li, Be, and B are involved and, interestingly, ^{4}He, the final product of the pp-chain reactions, is itself an active participant. The concept of local equilibrium is introduced in Section 6.9, prescriptions for finding equilibrium abundances for isotopes in the pp chains are provided, and pp-chain energy-generation rates relevant in different temperature regimes are given. In Section 6.10, CN cycle and CNO bi-cycle reactions are described. Cross sections for these reactions are based on extrapolations of cross sections found in laboratory experiments.

The effects of electrostatic screening in a stellar plasma are explored in Section 6.11. The concentration of electrons about positively charged nuclei effectively decreases the Coulomb barrier between interacting nuclei and enhances the associated reaction rates. In main sequence stars more massive than $\sim 2\ M_\odot$, the enhancements are only a few percent, but in stars less massive than $\sim 0.5\ M_\odot$, enhancements can be by factors of 3–10.

Polytropes of index $N = 3$ were first championed as standard models by Arthur S. Eddington (*The Internal Constitution of the Stars*, 1926). As nuclear energy sources had not yet been explicitly identified, Eddington's confidence was based entirely on the relationship between radius and mass for observed intermediate mass main sequence stars. In Section 6.12, theoretical and experimentally based energy-generation rates are used to determine the luminosities of polytropic models with mass–radius ratios similar to those of real zero age main sequence stars. It is found that the luminosity–mass ratio for zero age main sequence stars of mass in the mass range 1–9 $M_\odot$ is approximated by polytropes of index ~ 2.85–3.15. For main sequence stars substantially less massive than the Sun, it is expected that convection is the dominant mode of energy transport in the interior and that such stars should therefore be reasonably well approximated by polytropes of index $N = 3/2$. A final exercise in Section 6.12 shows that a relatively good match with the relationship between luminosity and mass for main sequence stars of mass $M \sim \frac{1}{9} M_\odot$ is given by $N = 3/2$ polytropes in which electrons are quite degenerate and temperatures are so low that hydrogen burning does not proceed beyond the formation of ^{3}He.

6.1 The nature and energetics of the pp reaction

The basic reaction that controls the rate at which energy is emitted by the Sun transforms two protons into a deuteron, with the emission of a positron (anti-electron) and an electron neutrino:

$$p + p \rightarrow d + e^+ + \nu_e. \tag{6.1.1}$$

Nucleons are thought to consist of three valence quarks in a sea of quark–antiquark pairs (mostly pions), and reaction (6.1.1) may be viewed as the transformation of one up quark into a down quark with the emission of a W^+ vector boson which then decays into a positron and an electron neutrino. Here, a much more elementary description is adopted in which baryons are treated as elementary particles and the reaction is interpreted simply as a conversion of a proton into a neutron with the emission of two leptons. The discussion is purposely cast in an elementary fashion, with the aim of providing an understanding of the basic physics involved rather than an accurate quantitative estimate of the reaction cross section. Rather sweeping simplifications are made, particularly with regard to the central nuclear matrix element, but it is hoped that the pedagogical gain justifies the sacrifice of rigor. This is done because the pp reaction controls the rate at which the Sun shines to make our planet habitable for us and for other forms of water-based life and it is deeply satisfying to understand in the simplest terms possible the basic physics of the fundamental reaction that makes our existence possible.

The nuclear energy released in the pp reaction is converted into positron energy $E_{\bar{e}}$ (rest mass $m_e c^2$ plus kinetic energy E'_{e^+}) and neutrino kinetic energy E_{ν_e}. That is,

$$E_{e^+} + E_{\nu_e} = m_e c^2 + E'_{e^+} + E_{\nu_e} = 2M_p c^2 - M_d c^2, \tag{6.1.2}$$

where M_p and M_d are the mass of the proton and the mass of the deuteron, respectively. From experiment, it is known that the binding energy of the deuteron is 2.23 MeV $= M_p c^2 + M_n c^2 - M_d c^2$ and that the difference in mass of the neutron and proton is $M_n c^2 - M_p c^2 = 1.29$ MeV. Hence

$$E_{e^+} + E_{\nu_e} = (M_p + M_n - M_d)c^2 - (M_n - M_p)c^2 = 0.94 \text{ MeV}, \tag{6.1.3}$$

and, since $m_e c^2 = 0.511$ MeV,

$$E'_{e^+} + E_{\nu_e} = 0.43 \text{ MeV}. \tag{6.1.4}$$

Approximately 60% of the kinetic energy is taken, on the average, by the neutrino, leaving 40%, or 0.17 MeV, for the kinetic energy of the average positron. Having a cross section with matter of the order of 10^{-48} cm^2, the neutrino escapes from the star. The positron is rapidly thermalized by scattering from other charged particles and then either annihilates directly with an electron or, if the density is small enough, combines with an electron to first form a positronium atom. In either case, electron–positron annihilation produces either two or three gamma rays of total energy 1.02 MeV. If positronium is formed in the singlet ground state, annihilation into two gamma rays occurs in 1.25×10^{-10} s; if it is formed in the triplet ground state, annihilation into three gamma rays occurs in 1.38×10^{-7} s. The gamma rays are rapidly thermalized by Compton scattering from electrons. Altogether, the amount of nuclear and lepton rest-mass energy that is converted locally into thermal energy is

$$E_{\text{heat}} = 1.19 \text{ MeV}. \tag{6.1.5}$$

6.2 Ingredients of the pp-reaction probability

To estimate the cross section for the pp reaction, one may begin with a simplified version of Fermi's beta decay theory, which envisions a point-like interaction of strength g_{weak} (see Chapters 14 and 15 in Volume 2). Employing the "golden rule" for first order transitions (see Section 7.1), one may approximate the probability $d\omega_{\text{pp}}$ of converting the unbound di-proton into a deuteron and creating a positron with momentum in the range $p_e \rightarrow p_e + dp_e$ and a neutrino with momentum in the range $p_\nu \rightarrow p_\nu + dp_\nu$ as

$$d\omega_{\text{pp}} = \frac{2\pi}{\hbar} g_{\text{weak}}^2 \left| \int \psi_d^* \psi_e^* \psi_\nu^* \psi_{\text{pp}} \, d\tau_{\text{nuc}} \right|^2 \bar{\Sigma}_{\text{pp}} \frac{d\rho_f}{dE_f}, \tag{6.2.1}$$

where ψ_d, ψ_e, ψ_ν, and ψ_{pp} are, respectively, the spatial wave functions of the deuteron, the positron, the neutrino, and the di-proton, $\bar{\Sigma}_{\text{pp}}$ is a statistical factor to be evaluated later, and $d\rho_f$ is the density of states for the final system of positron and neutrino. There is a spin change because the di-proton state which has finite amplitude at zero separation must be of spin zero, and the only bound state of the deuteron has spin unity. Because beta decays occur in which the nuclear spin change is zero, and because the coupling constant for such decays is different than for decays with unit spin change, it is to be emphasized that the coupling constant g_{weak} in eq. (6.2.1) is appropriate only for a unit spin change.

The wave functions of the positron and the neutrino may be chosen to be plane waves at large distances from the nucleons in a box of volume V, so the density of final states can be written as

$$\frac{d\rho_f}{dE_f} = V \frac{4\pi}{h^3} p_e^2 \, dp_e \, V \frac{4\pi}{h^3} p_\nu^2 \frac{dp_\nu}{dE_f}, \tag{6.2.2}$$

where

$$E_f = E_e + E_\nu = 0.94 \text{ MeV}. \tag{6.2.3}$$

It is important to note that statistical weights associated with spin are not at this point assigned to the lepton phase space factors.

After transforming the independent variables from p_e and p_ν to E_e and E_f and taking the derivative with respect to E_f, one has

$$\frac{d\rho_f}{dE_f} = \left(\frac{4\pi}{h^3}\right)^2 V^2 \frac{1}{c^6} \left[E_e^2 - (m_e c^2)^2\right]^{1/2} E_e (E_f - E_e)^2 dE_e. \tag{6.2.4}$$

In constructing eqs. (6.2.2) and (6.2.4), the Coulomb interaction between the escaping positron and the proton in the deuteron has been neglected.

Converting to the dimensionless variable $\epsilon = E_e/m_e c^2$, and integrating over ϵ, one has

$$\rho_f = \left(\frac{4\pi}{h^3}\right)^2 m_e^5 c^4 V^2 F(\epsilon_f), \tag{6.2.5}$$

where

$$F(\epsilon_f) = \int_1^{\epsilon_f} (\epsilon^2 - 1)^{1/2} \epsilon (\epsilon - \epsilon_f)^2 d\epsilon. \tag{6.2.6}$$

Equation (6.2.6) has the solution

$$F(\epsilon_f) = -\frac{1}{4} \eta_f - \frac{1}{12} \eta_f^3 + \frac{1}{30} \eta_f^5 + \frac{1}{4} \epsilon_f \log_e(\eta_f + \epsilon_f), \tag{6.2.7}$$

where

$$\eta_f = \frac{p_f}{m_e c} = \left(\epsilon_f^2 - 1\right)^{1/2}. \tag{6.2.8}$$

For the pp reaction, $\epsilon_f = 0.94$ MeV$/0.51$ MeV $= 1.84$, $\eta_f = 1.54$, and $F_{pp}(1.84) = 0.161$.

On integration, the differential reaction probability, eq. (6.2.1), becomes the total reaction probability

$$\omega_{pp} = \frac{2\pi}{\hbar} g_{weak}^2 \left| \int \psi_d^* \psi_e^* \psi_\nu^* \psi_{pp} \, d\tau_{nuc} \right|^2 \bar{\Sigma}_{pp} \left(\frac{4\pi}{h^3}\right)^2 m_e^5 c^4 V^2 F_{pp}. \tag{6.2.9}$$

Since the positron and neutrino are represented by plane waves, their wave functions have the same amplitude, $|\psi| = 1/\sqrt{V}$, so, finally,

$$\omega_{pp} = \frac{2\pi}{\hbar} g_{weak}^2 \left| \int \psi_d^* \psi_{pp} \, d\tau_{nuc} \right|^2 \bar{\Sigma}_{pp} \left(\frac{4\pi}{h^3}\right)^2 m_e^5 c^4 F_{pp}. \tag{6.2.10}$$

6.3 An estimate of the weak interaction coupling constant

In order to estimate the weak interaction coupling constant appropriate to the pp reaction, one should choose a reaction the properties of which have been well measured in terrestrial laboratories, which involves a unit change of nuclear spin, and which occurs with such a short lifetime that it is likely that the nuclear wave functions strongly overlap. The reaction

$$^6\text{He} \rightarrow {}^6\text{Li} + e^- + \bar{\nu}_e, \tag{6.3.1}$$

where $\bar{\nu}_e$ is an electron antineutrino, is particularly useful since it is very analogous to the pp reaction itself: (a) two neutrons of opposite spin outside of a filled-shell α-particle core transform into something analogous to a deuteron of spin unity: (b) the overlap between the final deuteron-like wave function and the initial di-neutron wave function might be expected to be near unity. Thus, the spin statistics and the coupling constant are the same as for the pp reaction. The expected strong overlap of nuclear wave functions means that the value of the relevant coupling constant estimated on the assumption of complete overlap may be close to the actual value.

In exact analogy with the treatment of the pp reaction, one may write

$$
\omega_{\mathrm{He}\to\mathrm{Li}} = \frac{1}{t_{\mathrm{He}\to\mathrm{Li}}}
$$

$$
= \frac{2\pi}{\hbar} g_{\mathrm{weak}}^2 \left| \int \psi_{\mathrm{Li}}^* \, \psi_{\mathrm{e}^-}^* \, \psi_{\bar{\nu}_e}^* \, \psi_{\mathrm{He}} \, d\tau_{\mathrm{nuc}} \right|^2 \bar{\Sigma}_{\mathrm{He}\to\mathrm{Li}} \left(\frac{4\pi}{h^3} \right)^2 m_e^5 c^4 \, V^2 \, F_{\mathrm{He}\to\mathrm{Li}},
$$

$$(6.3.2)$$

where ψ_{Li} and ψ_{He} are the wave functions of $^6\mathrm{He}$ and $^6\mathrm{Li}$, respectively, $t_{\mathrm{He}\to\mathrm{Li}}$ is the lifetime of $^6\mathrm{He}$ in the terrestial laboratory, $\bar{\Sigma}_{\mathrm{He}\to\mathrm{Li}}$ is the statistical factor for the reaction, and $F_{\mathrm{He}\to\mathrm{Li}}$ follows from eq. (6.1.11) when experimentally determined values of ϵ_f and η_f are inserted. Assuming perfect overlap between the nuclear wave functions,

$$
\frac{1}{t_{\mathrm{He}\to\mathrm{Li}}} = \frac{2\pi}{\hbar} g_{\mathrm{weak}}^2 \left(\frac{4\pi}{h^3} \right)^2 m_e^5 \, c^4 \, F_{\mathrm{He}\to\mathrm{Li}} \, \Sigma_{\mathrm{He}\to\mathrm{Li}}.
\tag{6.3.3}
$$

Combining eqs. (6.2.10) and (6.3.3), one has

$$
\omega_{\mathrm{pp}} = \frac{1}{t_{\mathrm{He}\to\mathrm{Li}}} \frac{F_{\mathrm{pp}}}{F_{\mathrm{He}\to\mathrm{Li}}} \left| \int \psi_{\mathrm{d}}^* \, \psi_{\mathrm{pp}} \, d\tau_{\mathrm{nuc}} \right|^2 \frac{\bar{\Sigma}_{\mathrm{pp}}}{\bar{\Sigma}_{\mathrm{He}\to\mathrm{Li}}}.
\tag{6.3.4}
$$

The cross section σ_{pp} for the pp reaction is, by definition, the probability ω_{pp} divided by the flux. In the adopted approximation, the flux is simply v/V (one pair of protons in volume V at relative velocity v), so

$$
\sigma_{\mathrm{pp}} = \frac{1}{t_{\mathrm{He}\to\mathrm{Li}}} \frac{F_{\mathrm{pp}}}{F_{\mathrm{He}\to\mathrm{Li}}} \frac{V}{v} \left| \int \psi_{\mathrm{d}}^* \, \psi_{\mathrm{pp}} \, d\tau_{\mathrm{nuc}} \right|^2 \frac{\bar{\Sigma}_{\mathrm{pp}}}{\Sigma_{\mathrm{He}\to\mathrm{Li}}}.
\tag{6.3.5}
$$

Although, in the adopted approach, the coupling constant g_{weak} need not be determined explicitly, it is nevertheless of great interest to know its magnitude. Setting $\hbar = 1.0546 \times 10^{-27}$ erg s, $m_e = 0.91095 \times 10^{-27}$ g, and $c = 2.9979 \times 10^{10}$ cm s^{-1}, one obtains from eq. (6.3.3)

$$
\frac{1}{t_{\mathrm{He}\to\mathrm{Li}}} = 0.5632 \times 10^{94} \left(\mathrm{erg} \, \mathrm{cm}^3 \right)^{-2} \mathrm{s}^{-1} \, g_{\mathrm{weak}}^2 \, F_{\mathrm{He}\to\mathrm{Li}} \, \Sigma_{\mathrm{He}\to\mathrm{Li}},
\tag{6.3.6}
$$

or

$$g_{\text{weak}} = 1.3325 \times 10^{-47} \text{ erg cm}^3 \left(\frac{1 \text{ second}}{F_{\text{He} \to \text{Li}} \, t_{\text{He} \to \text{Li}} \, \Sigma_{\text{He} \to \text{Li}}} \right)^{1/2}. \tag{6.3.7}$$

From experiment, one has that $t_{\text{He} \to \text{Li}} = t_{1/2}/\log_e 2 = 1.164$ s. The energy shared by the electron and antineutrino is 4.01 MeV, so that $F_{\text{He} \to \text{Li}} = F(4.01 \text{ MeV}/0.511 \text{ MeV}) = F(7.85) = 924$, as follows from eq. (6.1.11) with $\epsilon_f = 7.85$ and $\eta_f = 7.78$. The logarithm to the base 10 of $F \, t_{1/2}$ is $\log(F \, t_{1/2}) = 2.87$. This $\log(ft)$ value is among the smallest known, placing the transition in the "super allowed" category. This is often interpreted as meaning that the nuclear matrix element is close to unity.

Since the spin of ^{6}Li is unity and since either neutron can change into a proton, the statistical factor is $\Sigma_{\text{He} \to \text{Li}} = 3 \times 2 = 6$. Inserting these results into eq. (6.3.7), it follows that

$$g_{\text{weak}} \sim 1.66 \times 10^{-49} \text{ erg cm}^3, \tag{6.3.8}$$

Another estimate of the relevant coupling strength based on the properties of neutron decay and the decay $^{14}\text{O} \to {}^{14}\text{N}^*$ is discussed in Section 15.2 of Volume 2. The result is

$$g_{\text{weak}} \sim 1.80 \times 10^{-49} \text{ erg cm}^3, \tag{6.3.9}$$

suggesting that the overlap between the ^{6}He and ^{6}Li wave functions is approximately 92% rather than the 100% used to derive the estimate in eq. (6.3.8). The smallness of these numbers relative to the effective coupling constants for electromagnetic and nuclear reactions is responsible for the term "weak" used to describe interactions which involve leptons.

6.4 The nuclear matrix element

It is instructive to estimate the nuclear overlap integral in eq. (6.3.5) in the approximation that the wave function of the deuteron is that of a particle in a square well nuclear potential of depth V_0 and radius R. Assuming that its ground state is primarily an s-state (no orbital angular momentum), the deuteron wave function is a solution of

$$\nabla^2 \psi_{\text{d}} + \frac{2\mu}{\hbar^2}(V_0 - B_{\text{d}})\psi_{\text{d}} = 0, \text{ for } r < R, \tag{6.4.1}$$

and of

$$\nabla^2 \psi_{\text{d}} - \frac{2\mu}{\hbar^2} B_{\text{d}} \psi_{\text{d}} = 0, \text{ for } r > R, \tag{6.4.2}$$

where $\mu \sim M_{\text{p}}/2$ is the reduced mass of the neutron–proton pair, and B_{d} is the binding energy of the deuteron (2.23 MeV). Inside the potential well,

$$\psi_{\text{d}} = A \frac{\sin k_0 r}{r}, \, r < R, \tag{6.4.3}$$

V_0	$-\gamma/k_0$	δ	$k_0 R$	R	γR	s^2	f^2	$f^2 s^2$	$I_i/(I_i + I_0)$
10	−0.536	0.492	2.06	4.76	1.10	5.10	2.44	12.45	0.376
15	−0.418	0.396	1.97	3.55	0.822	3.73	1.98	7.39	0.241
20	−0.354	0.340	1.91	2.92	0.676	3.58	1.76	6.31	0.174
25	−0.313	0.303	1.87	2.53	0.586	3.28	1.65	5.41	0.134

Table 6.4.1 Properties of the deuteron wave function and nuclear integrals

where

$$k_0^2 = \frac{2\mu}{\hbar^2}(V_0 - B_d), \tag{6.4.4}$$

and, outside the well,

$$\psi_d = B\,\frac{e^{-\gamma r}}{r}, \quad r > R, \tag{6.4.5}$$

where

$$\gamma^2 = \frac{2\mu}{\hbar^2}\,B_d. \tag{6.4.6}$$

Matching the logarithmic derivatives of the wave functions at $r = R$ gives

$$\cot k_0 R = -\frac{\gamma}{k_0} = -\sqrt{\frac{B_d}{V_0 - B_d}}. \tag{6.4.7}$$

It is known that the deuteron possesses only one bound state and that the binding energy per particle is an order of magnitude smaller than the binding energy per particle of typical nuclei. Hence $V_0 \gg B_d$, and

$$k_0 R = \frac{\pi}{2} + \delta, \tag{6.4.8}$$

where δ is small compared with $\pi/2$. Thus,

$$\cot k_0 R = -\tan\delta = -\frac{\gamma}{k_0}. \tag{6.4.9}$$

Values of various characteristics of the solution are given in Table 6.4.1 as a function of the potential V_0. The potential is given in units of MeV and the well radius R in column 5 is given in units of 1 Fermi = 10^{-13} cm.

A simple wave function for the di-proton is more difficult to construct. A first approximation which neglects nuclear forces and the finite size of the proton is given by eqs. (3.5.6) and (3.5.7) with $Z_1 = Z_2 = 1$. A better choice, which takes roughly into account the nuclear potential, makes use of the classical Coulomb barrier penetration probabilty. In the WKB approximation, this probability is

$$P_{\text{Gamow}} = \exp\left(-\int_R^{R_E}\sqrt{\frac{2\mu}{\hbar^2}\left(\frac{Z_1 Z_2 e^2}{r} - E\right)}\,dr\right), \tag{6.4.10}$$

where $R_E = Z_1 Z_2 e^2 / E$ and R is the radius of the square well, is named in honor of George Gamow (1928), who first recognized the significance of the Coulomb barrier for nuclear reactions. In first approximation,

$$P_{\text{Gamow}} \sim \exp\left(-2\pi\eta + \frac{4e}{\hbar}(2Z_1 Z_2 \mu R)^{1/2}\right), \tag{6.4.11}$$

where

$$\eta = \frac{Z_1 Z_2 \, e^2}{\hbar v}. \tag{6.4.12}$$

With this as a guide, one may estimate that

$$\psi_{\text{pp}} \sim \sqrt{2}\,\sqrt{\frac{2\pi\eta\,e^{-2\pi\eta}}{V}}\, s, \tag{6.4.13}$$

where

$$s^2 = \exp\left(\frac{4e}{\hbar}(M_p R)^{1/2}\right) = \exp(0.747\,R^{1/2}). \tag{6.4.14}$$

The factor of $\sqrt{2}$ in eq. (6.4.13) comes from the fact that there are two protons in volume V in a spatially symmetric state. In the furthest expression to the right in eq. (6.4.14), R is in units of 1 Fermi. It will here be assumed that R in this equation is identical with the radius of the deuteron square well potential. Values of s^2 are given in the seventh column of Table 6.4.1.

The constant amplitude wave function given by eqs. (6.4.13) and (6.4.14) contains a physically meaningful dependence on the radius R of the nuclear potential well: the smaller R, the thicker is the Coulomb barrier which the di-proton must conquer in order to enter the well and, therefore, the smaller is the amplitude of its wave function within the well, all other things being equal. However, a proper treatment of the di-proton gives a wave function the amplitude of which, for $r < R$, is smaller than that given by eqs. (6.4.13) and (6.4.14) and, for $r > R$, increases with increasing r, eventually becoming larger than given by these equations. The value of the constant amplitude has been chosen in such a way that, for $V_0 \sim 20\,\text{MeV}$, the resultant pp cross section is close to sophisticated standard estimates of the cross section.

The overlap integral in eq. (6.3.5) becomes

$$I_{\text{nuc}} \equiv \int \psi_d^* \psi_{\text{pp}} \mathrm{d}\tau_{\text{nuc}} = \left(\frac{2\pi\eta\,e^{-2\pi\eta}}{V}\right)^{1/2} \sqrt{2}\,s \int_0^\infty \psi_d\, 4\pi r^2 \mathrm{d}r. \tag{6.4.15}$$

The next task is to evaluate

$$I_d \equiv \int_0^\infty \psi_d\, 4\pi r^2 \mathrm{d}r. \tag{6.4.16}$$

From eqs. (6.4.3) and (6.4.5), it follows that $B/A = \sin(k_0 R)e^{\gamma R}$, so

$$I_d = 4\pi A \left(\int_0^R r \sin(k_0 r)\, dr + \sin(k_0 R)\, e^{\gamma R} \int_R^\infty e^{-\gamma r} r\, dr \right). \qquad (6.4.17)$$

The constant A is determined by normalizing the deuteron wave function:

$$\int_0^\infty \psi_d^2\, 4\pi r^2 dr = 1 = 4\pi A^2 \left(\int_0^R \sin^2(k_0 r)\, dr + \sin^2(k_0 R)\, e^{2\gamma R} \int_R^\infty e^{-2\gamma r} dr \right). \qquad (6.4.18)$$

Using eq. (6.4.18) to eliminate A from eq. (6.4.17) gives

$$I_d = \frac{\sqrt{8\pi}}{\gamma^{3/2}} \frac{(\gamma/k_0)^2 \int_0^{k_0 R} x \sin x\, dx + \sin(k_0 R)\, e^{\gamma R} \int_{\gamma R}^\infty e^{-x} x\, dx}{\left((2\gamma/k_0) \int_0^{k_0 R} \sin^2 x\, dx + \sin^2(k_0 R)\, e^{2\gamma R} \int_{2\gamma R}^\infty e^{-x} dx \right)^{1/2}}. \qquad (6.4.19)$$

After some arithmetic,

$$I_d = \frac{\sqrt{8\pi}}{\gamma^{3/2}} \frac{(1 + \gamma R)\cos\delta + (\gamma/k_0)^2 [(\pi/2 + \delta)\sin\delta + \cos\delta]}{\left[\cos^2\delta + (2\gamma/k_0)(\pi/4 + \delta/2 + \sin\delta\cos\delta) \right]^{1/2}}. \qquad (6.4.20)$$

The eighth column in Table 6.4.1 gives values of f^2 in the definition

$$I_d^2 = \frac{8\pi}{\gamma^3} f^2 = 2.03 \cdot 10^{-36} f^2 \text{ cm}^3. \qquad (6.4.21)$$

Equation (6.4.21), along with the relative insensitivity of f^2 to the choice of V_0, reveals that the size of the pp reaction cross section is inversely proportional to the $(3/2)$ power of the binding energy of the deuteron (see eq. (6.4.7)). Thus, the relative weakness of the neutron–proton binding contributes directly to the strength of the pp cross section.

A remarkable characteristic of the overlap integral is that the major contribution to the integral comes from regions outside the nuclear potential well. From eq. (6.4.17), it follows that

$$\frac{I_i}{I_o} = \frac{I_{d,\, \text{inside}}}{I_{d,\, \text{outside}}} = \frac{\int_0^R \psi_d r^2 dr}{\int_R^\infty \psi_d r^2 dr} \qquad (6.4.22)$$

$$= \left(\frac{\gamma}{k_0} \right)^2 (1 + \gamma R). \qquad (6.4.23)$$

The last column in Table 6.4.1 gives the value of $I_i/I_d = I_i/(I_i + I_o)$ as a function of V_0. One infers from these results that the relative weakness of the neutron–proton binding in the deuteron, which is responsible for the "fluffiness" of the deuteron, contributes in still another way to the fact that the pp reaction cross section is as large as it is.

6.5 A numerical estimate of the cross section

Gathering together all quantities required in eq. (6.3.5), one has

$$\sigma_{pp} = \frac{1}{t_{He \to Li}} \frac{F_{pp}}{F_{He \to Li}} \frac{V}{v} \frac{2\pi\eta \, e^{-2\pi\eta} \, 2 \, s^2}{V} I_d^2 \frac{\bar{\Sigma}_{pp}}{\Sigma_{He \to Li}} \tag{6.5.1}$$

$$= \frac{1}{c t_{He \to Li}} \frac{F_{pp}}{F_{He \to Li}} \frac{\pi \, e^2}{2 \, \hbar c} \frac{M_p c^2}{E} e^{-2\pi\eta} \, 2 \, s^2 I_d^2 \frac{\bar{\Sigma}_{pp}}{\Sigma_{He \to Li}} \tag{6.5.2}$$

$$= \frac{7.26 \times 10^{-19} \, s}{t_{He \to Li}} \frac{F_{pp}}{F_{He \to Li}} \frac{e^{-2\pi\eta}}{E} \, 2 \, s^2 f^2 \frac{\bar{\Sigma}_{pp}}{\Sigma_{He \to Li}} \quad \text{keV barn}. \tag{6.5.3}$$

In these expressions, E is the kinetic energy of the di-proton in center of mass coordinates and, in the third version of the cross section, 1 barn $= 10^{-24}$ cm^2. The choice of 1 barn as the preferred unit for describing the cross section is for convenience in comparing with cross sections for subsequent reactions which are measured experimentally and conventionally quoted in units of 1 barn.

The statistical factor $\bar{\Sigma}_{pp}$ follows from (a) averaging over initial states of the di-proton, (b) taking into account that the deuteron has spin unity, and (c) recognizing that either proton can change into a neutron. Since the di-proton is in a triplet state 3/4 of the time and in a singlet state only 1/4 of the time, the initial state configuration contributes a factor of 1/4. The spin factor is 3, and the fact that either proton can transform into a neutron gives a factor of 2. Thus, $\bar{\Sigma}_{pp} = (1/4) \times 3 \times 2 = 3/2$, as opposed to $\Sigma_{He \to Li} = 6$. Inserting relevant approximations for the various quantities in eq. (6.5.3), including the estimate of the weak coupling constant given by eq. (6.3.8), it follows that

$$\sigma_{pp} = \frac{7.26 \times 10^{-19}}{1.164} \frac{0.16}{924} \frac{e^{-2\pi\eta}}{E} \, 2 \, s^2 f^2 \frac{3/2}{6} \quad \text{keV barn} \tag{6.5.4}$$

$$= 3.43 \times 10^{-22} \left(\frac{s^2 f^2}{6.31} \right) \frac{e^{-2\pi\eta}}{E} \quad \text{keV barn}. \tag{6.5.5}$$

In eq. (6.5.5), 6.31 is the value of $f^2 s^2$ when $V_0 = 20$ MeV.

The coefficient of $e^{-2\pi\eta}/E$ in eq. (6.5.5),

$$S_{pp} = 3.43 \times 10^{-22} \left(\frac{s^2 f^2}{6.31} \right) \quad \text{keV barn}, \tag{6.5.6}$$

is called a "cross section factor". If the estimate of the weak coupling constant given by eq. (6.3.9) is adopted, the result is

$$S_{pp} = 3.72 \times 10^{-22} \left(\frac{s^2 f^2}{6.31} \right) \quad \text{keV barn}. \tag{6.5.7}$$

The fact that cross section factors for most other nuclear reactions of importance in stars are many orders of magnitude larger than S_{pp} highlights the "weakness" of the weak interaction strength compared to the strength of nuclear forces responsible for the other reactions.

The original estimate of σ_{pp} was made by H. A. Bethe and C. L. Critchfield (1938). Successively more detailed and sophisticated treatments have followed. Among the first of these are estimates by E. Frieman and L. Motz (1951) and by E. E. Salpeter (1952). What is written here as $s^2 f^2$ is the quantity estimated by Salpeter as $\Lambda^2 = 6.82$. R. J. Gould and N. Guessoum (1990) estimate $S_{pp} = 4.21 \cdot 10^{-22}$ keV barn. M. Kamionkowski and J. Bahcall (1994) estimate $S_{pp} = 3.89 \cdot 10^{-22}$ keV barn. The choice

$$S_{pp} \sim 4 \times 10^{-22} \text{ keV barn.} \tag{6.5.8}$$

represents a reasonable compromise.

6.6 The pp-reaction rate and proton lifetime

To find the rate at which the pp reaction occurs under conditions in the stellar interior, it is necessary to average the cross section times the velocity over all relative velocities. What is involved has already been anticipated in Section 3.5. It is convenient to think in terms of two different kinds of interacting particle. The reaction rate is

$$R_{pp} = \frac{1}{2} \int \int \sigma_{pp} \, v \, dn_1 \, dn_2, \tag{6.6.1}$$

where

$$dn_1 \, dn_2 = n_1 \, n_2 \, \frac{e^{-p_1^2/2M_1kT} \, 4\pi p_1^2 \, dp_1}{(2\pi M_1 kT)^{3/2}} \, \frac{e^{-p_2^2/2M_1kT} \, 4\pi p_2^2 \, dp_2}{(2\pi M_2 kT)^{3/2}}. \tag{6.6.2}$$

The factor of 1/2 in eq. (6.6.1) comes from the fact that two protons are involved in each interaction, so the integration actually counts each reaction twice.

Choosing coordinates $P = p_1 + p_2$ and $p = \mu v = \mu(p_1/M_1 - p_2/M_2)$, where $\mu = M_1 M_2/(M_1 + M_2)$ and $M = M_1 + M_2$, eq. (6.6.2) becomes

$$dn_1 \, dn_2 = n_1 \, n_2 \, \frac{e^{-P^2/2MkT} \, 4\pi P^2 \, dP}{(2\pi M_1 kT)^{3/2}} \, \frac{e^{-p^2/2\mu kT} \, 4\pi p^2 \, dp}{(2\pi M_2 kT)^{3/2}}. \tag{6.6.3}$$

Since the results must be independent of the center of mass motion, one can integrate over P, obtaining

$$n_2 \, dn_1 = n_1 \, dn_2 = n_1 \, n_2 \frac{e^{-p^2/2\mu kT} \, 4\pi p^2 \, dp}{(2\pi \mu kT)^{3/2}} \tag{6.6.4}$$

as the distribution of particle pairs in center of mass coordinates. Equation (6.6.1) becomes

$$R_{\mathrm{pp}} = \frac{n_{\mathrm{p}}^2}{2} S_{\mathrm{pp}} \int_0^\infty \frac{\mathrm{e}^{-p^2/2\mu kT}\, \mathrm{e}^{-2\pi\eta}\, (v/E)\, 4\pi p^2\, \mathrm{d}p}{(2\pi\mu kT)^{3/2}} \tag{6.6.5}$$

$$= \frac{n_{\mathrm{p}}^2}{2} \frac{S_{\mathrm{pp}}}{(2\pi\mu kT)^{3/2}}\, 8\pi\mu \int_0^\infty \mathrm{e}^{-E/kT}\, \mathrm{e}^{-2\pi\eta}\, \mathrm{d}E \tag{6.6.6}$$

$$= \frac{n_{\mathrm{p}}^2}{2} \frac{4S_{\mathrm{pp}}}{(2\pi\mu kT)^{1/2}} \int_0^\infty \mathrm{e}^{-E/kT}\, \exp\left(-\frac{2\pi e^2}{\hbar}\left(\frac{\mu}{2E}\right)^{1/2}\right) \mathrm{d}\left(\frac{E}{kT}\right). \tag{6.6.7}$$

Setting

$$\lambda = 2\pi \frac{e^2}{\hbar c}\left(\frac{\mu c^2}{2kT}\right)^{1/2} = \frac{75.6465}{T_6^{1/2}} \tag{6.6.8}$$

and

$$x = \frac{E}{kT}, \tag{6.6.9}$$

eq. (6.6.7) becomes

$$R_{\mathrm{pp}} = \frac{n_{\mathrm{p}}^2}{2} \frac{4S_{\mathrm{pp}}}{(2\pi\mu kT)^{1/2}} \int_0^\infty \exp\left(-x - \frac{\lambda}{x^{1/2}}\right) \mathrm{d}x. \tag{6.6.10}$$

The term in the exponent can be expanded about its minimum at $x = x_0$, where

$$x_0 = \left(\frac{\lambda}{2}\right)^{2/3} = \frac{11.267\,80}{T_6^{1/3}}. \tag{6.6.11}$$

Keeping just the first two terms in the expansion and changing the independent variable to $z = \sqrt{3/4x_0}\,(x - x_0)$,

$$R_{\mathrm{pp}} = \frac{n_{\mathrm{p}}^2}{2} \frac{4S_{\mathrm{pp}}}{(2\pi\mu kT)^{1/2}}\, \mathrm{e}^{-3x_0} \sqrt{\frac{4x_0}{3}} \int_{-\sqrt{3x_0/4}}^\infty \mathrm{e}^{-z^2}\, \mathrm{d}z. \tag{6.6.12}$$

Finally, replacing the lower limit on the integration by $-\infty$, it follows that

$$R_{\mathrm{pp}} \sim \frac{n_{\mathrm{p}}^2}{2} \frac{4S_{\mathrm{pp}}}{(2\pi\mu kT)^{1/2}} \sqrt{\frac{4\pi x_0}{3}}\, \mathrm{e}^{-3x_0}. \tag{6.6.13}$$

At the center of the Sun, the error committed by extending the integral to $-\infty$ is approximately 3%.

Table 6.6.1 Lifetime (yr) of a proton against capture by a proton ($\rho X_H = 100 \, \text{g cm}^{-3}$)							
T_6	1	2	4	8	12	16	20	24
τ	33.84	26.86	21.32	16.92	14.78	13.43	12.47	11.73
$\tau^2 e^{-\tau}$	2.30/−12	1.56/−9	2.50/−7	1.28/−5	8.33/−5	2.65/−4	5.97/−4	1.11/−3
t_{pp} (yr)	1.91/17	2.82/14	1.76/12	3.44/10	5.28/9	1.66/9	7.37/8	3.97/8

Equation (6.6.13) can be rewritten as

$$R_{pp} = \frac{n_p^2}{2} \frac{8 S_{pp}}{9\sqrt{3}\pi} \frac{\hbar}{\mu e^2} (3x_0)^2 \, e^{-3x_0} \tag{6.6.14}$$

$$= 4.464\,34 \times 10^{14} \, S_{pp}(\text{cm}^2) \, n_p^2 \, \tau^2 e^{-\tau} \text{cm}^{-3} \, \text{s}^{-1} \tag{6.6.15}$$

$$= 7.152\,14 \times 10^{-19} S_{pp}(\text{KeV barn}) \, n_p^2 \, \tau^2 \, e^{-\tau} \, \text{cm}^{-3} \, \text{s}^{-1}, \tag{6.6.16}$$

where

$$\tau = 3x_0 = \frac{33.8034}{T_6^{1/3}}. \tag{6.6.17}$$

Adopting $S_{pp} = 4.0 \times 10^{-22}$ keV barn, as in eq. (6.5.8),

$$R_{pp} = 2.860\,86 \times 10^{-40} \, n_p^2 \, \tau^2 e^{-\tau} \, \text{cm}^{-3} \, \text{s}^{-1}. \tag{6.6.18}$$

An interesting quantity which is related to the reaction rate is the average lifetime of a proton against capture by another proton. This may be defined as

$$\frac{1}{t_{pp}} = -\frac{1}{n_p} \frac{dn_p}{dt} = 2 \frac{R_{pp}}{n_p} = \frac{\rho X_H \, \tau^2 e^{-\tau}}{4.40 \times 10^7 \, \text{yr}}. \tag{6.6.19}$$

Lifetimes for different temperatures and for $\rho X_H = 100 \, \text{g cm}^{-3}$ are given in Table 6.6.1.

The pp reaction controls the rate at which subsequent hydrogen-burning reactions take place. Adding this knowledge to the fact that the Sun has been shining for about 4.6×10^9 yr, one may infer from the last row of Table 6.6.1 that, over the Sun's lifetime, the average temperature in energy-generating regions may not have been much different from $\sim 12 \times 10^6$ K.

6.7 Other hydrogen-burning reactions – laboratory cross sections and extrapolation to stellar conditions

Shining at its present rate, the Sun would emit a total energy of $\sim 5.6 \times 10^{50}$ erg in 4.6×10^9 yr. Since an energy of $(1/2) \times 1.19 \, \text{MeV} \times 1.6 \times 10^{-6} \, \text{erg MeV}^{-1} = 0.95 \times 10^{-6}$ erg is liberated for every proton consumed, if pp reactions were the only source of nuclear energy, a total number of $\Delta N_H \sim 5.9 \times 10^{56}$ protons would have been consumed over the Sun's lifetime. But, at an abundance by mass of 70%, the total number of protons in the initial Sun is only $N_H \sim 8.4 \times 10^{56}$. Given the strong dependence of the pp reaction rate on temperature, ΔN_H and N_H are clearly inconsistent: there is no way that hydrogen could have been exhausted over 70% of the Sun's mass. The first order solution is, of course, that deuterons interact with protons to form ^{3}He at a rate which, not being mediated by the weak interaction on which the pp reaction must rely, is intrinsically much larger than the pp rate. Thus, once two protons combine to form a deuteron, the deuteron reacts with another proton to form a ^{3}He nucleus with the release of 5.49 MeV, almost five times the energy released in the pp reaction. Further, once ^{3}He is formed, a number of other nuclear reactions can take place which produce ^{4}He as a final product with the release of still more energy.

Unlike the pp reaction, most other reactions of relevance for energy generation in main sequence stars have been studied directly in the laboratory and relatively secure estimates for the associated cross section factors at stellar energies exist. In general, cross section factors for the various reactions may be ordered in terms of the nature of the decay product, with those for reactions in which a particle is emitted being the largest, those for reactions in which a γ ray is emitted being smaller, and those involving leptons being the smallest. And, of course, the larger the product of charges of the interacting particles, the smaller is the penetration probability at any given energy.

Most of the reactions of importance are of a non-resonant character, since resonant levels in light nuclei are fairly widely spaced (typically 100 KeV apart) and the range of interaction energies which are relevant in main sequence stars is much smaller than this. Formally, a reaction cross section can be written as

$$\sigma = \pi \lambdabar^2 \sum_{l=0}^{\infty} (2l+1)\, S'_l(E)\, P_l, \qquad (6.7.1)$$

where $\lambdabar = \hbar/(\mu v)$ is $1/2\pi$ times the deBroglie wavelength in the center of mass coordinate system, $\mu = M_1 M_2 /(M_1 + M_2)$ is the reduced mass, and v is the relative velocity. The quantity P_l is a penetration probability for orbital angular momentum quantum number l, and $S'_l(E)$ is a structure factor. Only those partial waves for which r_l in the expression

$$\mu\, v\, r_l = \hbar \sqrt{l(l+1)} \qquad (6.7.2)$$

is of the order of the range of nuclear forces or less will contribute to the sum in eq. (6.7.1). Equation (6.7.2) can be rewritten as

$$r_l = \left(\frac{\mu c^2}{2E} \right)^{1/2} \frac{\hbar}{\mu c} \sqrt{l(l+1)} \qquad (6.7.3)$$

$$= \left(\frac{19.2\ \text{MeV}}{E} \right)^{1/2} \left(\frac{1}{\bar{\mu}} \right)^{1/2} \sqrt{l(l+1)}\ 10^{-13}\ \text{cm}, \qquad (6.7.4)$$

where $\bar{\mu} = \mu/M_p$.

It is obvious that, at temperatures found in main sequence stars, usually only the $l = 0$ or "s" wave will be of importance. Adopting the penetration factor given by eq. (6.4.11), one can write

$$\sigma(E) = \pi \lambda^2 \, S_0'(E) \, \exp\left(-2\pi\eta + \frac{4e}{\hbar}(2Z_1 Z_2 \mu R)^{1/2}\right) = \pi \lambda^2 \, S_0(E) \, \exp\left(-2\pi\eta\right), \tag{6.7.5}$$

where R is a measure of the size of the compound nucleus formed and is given roughly by

$$R = 1.45 \times 10^{-13} \, N^{1/3} \text{ cm}, \tag{6.7.6}$$

N being the total number of nucleons in the compound nucleus. With this choice of R,

$$\frac{4e}{\hbar}(2Z_1 Z_2 \mu R)^{1/2} = 4\left(2 \, Z_2 Z_2 \, \frac{e^2}{\hbar c} \, \frac{R}{\hbar/M_{\mathrm{H}}c} \, \bar{\mu}\right)^{1/2} = 1.269 \, (Z_1 Z_2 N^{1/3} \bar{\mu})^{1/2}. \tag{6.7.7}$$

In the furthest expression to the right in eq. (6.7.5) another structure factor has been defined by absorbing the quantity

$$P_R = \exp\left(+\frac{4e}{\hbar}(2Z_1 Z_2 \mu R)^{1/2}\right) \tag{6.7.8}$$

into the initially defined structure factor. Thus, $S_0 = S_0' \, P_R$. This has been done in part because P_R is, in effect, a structure dependent factor which does not overtly depend on the interaction energy. More importantly, P_R depends on an uncertain prescription for the nuclear potential and, in any case, laboratory experiments determine S_0 rather than S_0'.

One can define yet another structure factor $S(E)$ by setting

$$\sigma(E) = \pi \lambda^2 \, S_0(E) \, \exp\left(-2\pi\eta\right) = \frac{S(E)}{E} \, \exp\left(-2\pi\eta\right). \tag{6.7.9}$$

Laboratory estimates of $S(E)$ are called cross section factors and are normally given in units of keV barn. Since, for non relativistic velocities, $E = (1/2) \, \mu \, v^2 \sim (1/2) \, (\bar{\mu} \, M_{\mathrm{H}}) \, v^2$, one can write

$$\pi \lambda^2 = \pi \left(\frac{\hbar}{\mu v}\right)^2 = \frac{\pi}{\bar{\mu}} \, \frac{M_{\mathrm{H}} c^2}{2E} \left(\frac{\hbar}{M_{\mathrm{H}} c}\right)^2 = \frac{\pi}{\bar{\mu}} \, \frac{207.5 \text{ keV}}{E} \text{ barn} = \frac{1}{\bar{\mu}} \, \frac{652 \text{ keV}}{E} \text{ barn}, \tag{6.7.10}$$

where E in the last two identities is to be expressed in keV.

Combining eqs. (6.7.9) and (6.7.10), one has that the relationship between the dimensionless structure factor $S_0(E)$ and the center of mass cross section factor $S(E)$ is

$$S_0(E) = \frac{\bar{\mu}}{652} \, S(E) \text{ (keV barn)}. \tag{6.7.11}$$

In laboratory experiments, one type of particle of energy $E_{\mathrm{lab}} = (1/2) \, M_1 \, v^2$ strikes a stationary target of mass M_2. The energy in the center of mass system is related to the energy in the laboratory by

$$E = \frac{1}{2} \, \frac{M_1 \, M_2}{M_1 + M_2} \, v^2 = \frac{M_2}{M_1 + M_2} \, E_{\mathrm{lab}}. \tag{6.7.12}$$

The laboratory cross section factor $S_{\text{lab}}(E_{\text{lab}})$ is defined by

$$S_{\text{lab}}(E_{\text{lab}}) = \sigma_{\text{lab}} E_{\text{lab}} \exp\left(2\pi\eta(E)\right). \tag{6.7.13}$$

The center of mass cross section factor $S(E)$ is related to the laboratory cross section factor by $S(E)/E = S_{\text{lab}}(E_{\text{lab}})/E_{\text{lab}}$, giving

$$S(E) = \frac{E}{E_{\text{lab}}} S_{\text{lab}}(E_{\text{lab}}) = \frac{M_2}{M_1 + M_2} S_{\text{lab}}(E_{\text{lab}}). \tag{6.7.14}$$

In many instances, cross section factors vary slowly enough with energy that an extrapolation from data at the lowest energies achievable in the laboratory to energies of relevance in the stellar interior is relatively straightforward.

An integration over a Maxwell–Boltzmann distribution of relative velocities after the fashion described in Sections 3.5 and 6.6 gives, in the general case, the reaction rate

$$R_{ij} = \frac{n_i n_j}{1 + \delta_{ij}} \frac{8\pi\mu}{(2\pi\mu kT)^{3/2}} \int S_{ij}(E) e^{-2\pi\eta_{ij}} \, e^{-\frac{E}{kT}} \, \mathrm{d}E, \tag{6.7.15}$$

where

$$\eta_{ij} = \frac{Z_i Z_j e^2}{\hbar v}, \tag{6.7.16}$$

δ_{ij} is the Kronecker delta function, $\mu = M_i M_j/(M_i + M_j)$, M_i and Z_i are, respectively, the mass and the atomic number of particles of type i, and $S_{ij}(E)$ is the center of mass cross section factor. Assuming that $S_{ij}(E)$ is a slowly varying function of E, one has

$$R_{ij} \approx \frac{n_i n_j}{1 + \delta_{ij}} \frac{8\pi\mu}{(2\pi\mu kT)^{3/2}} S_{ij}(E_0) \int e^{-2\pi\eta_{ij}} \, e^{-\frac{E}{kT}} \, \mathrm{d}E, \tag{6.7.17}$$

where $S_{ij}(E_0)$ is the extrapolated center of mass cross section factor evaluated at the maximum in the integrand, where $E = E_0$.

Following the same steps given by eqs. (6.6.5)–(6.6.17) and inserting numbers, one obtains

$$R_{ij} \approx \frac{2.62 \times 10^{29}}{1 + \delta_{ij}} \frac{\rho X_i}{A_i} \frac{\rho X_j}{A_j} S_{ij}(E_0)(\text{keV barn}) \frac{\tau_{ij}^2 \exp(-\tau_{ij})}{Z_i Z_j A_{ij}} \text{ reactions cm}^{-3} \text{ s}^{-1}, \tag{6.7.18}$$

where ρ is the mass density in g cm^{-3}, X_i and A_i are, respectively, the abundance by mass and atomic number of particles of type i, E_0 is the energy where the maximum in the integrand (after S_{ij} is extracted) occurs,

$$\tau_{ij} \equiv \frac{v_{ij}}{T_6^{1/3}} = 42.48 \left(\frac{Z_i^2 Z_j^2 A_{ij}}{T_6}\right)^{1/3}, \tag{6.7.19}$$

and $A_{ij} = A_i A_j/(A_i + A_j)$.

The maximum in the integrand of eq. (6.7.17) occurs at an energy given by

$$E_0 = \frac{1}{2}\left(2\pi Z_i Z_j \frac{e^2}{\hbar c}(\mu c^2)^{1/2} kT\right)^{2/3}. \tag{6.7.20}$$

Inserting numerical values for the constants, this becomes

$$= 1.22\left(Z_i^2 Z_j^2 A_{ij} T_6^2\right)^{1/3} \text{ keV}. \tag{6.7.21}$$

The width ΔE_0 of the peak in the integral, as measured by the distance between points which are $1/e$ smaller than the maximum in the integrand, is related to E_0 by

$$\frac{\Delta E_0}{E_0} = \frac{4}{\sqrt{3}}\left(\frac{kT}{E_0}\right)^{1/6} = 0.614\left(\frac{T_6}{Z_i^2 Z_j^2 A_{ij}}\right)^{1/6}. \tag{6.7.22}$$

The peak in the integrand is sometimes called the Gamow peak, although it was R. E. d'Atkinson and F. G. Houtermans (1929) who first folded the Gamow penetration factor with the Maxwell–Boltzmann velocity distribution to produce the peak. It is evident that, the larger the temperature, the larger is the relative width of the peak and that the larger the charges and masses of the interacting particles, the narrower is the relative width of the peak. The approximation represented by eq. (6.7.17) is satisfactory if the center of mass cross section factor does not vary rapidly over an energy of the order of ΔE_0. Otherwise, a numerical integration must be performed to find a suitable average value $\langle S_{ij}(E)\rangle$ to replace $S_{ij}(E_0)$ in eq. (6.7.17).

The lifetime of a particle of type j against a reaction with a particle of type i is

$$t_{ij} = (1 + \delta_{ij})\frac{R_{ij}}{n_j} = 2.29 \times 10^{-6}\frac{A_{ij}A_i Z_i Z_j}{S_{ij}\tau_{ij}^2 \exp(-\tau_{ij})\rho X_i} \text{ s}, \tag{6.7.23}$$

where n_j is the number density of particles of type j. It is useful to speak in terms of lifetimes, since β-decays occur and comparisons can most readily be made by comparing the t_{ij}s with β-decay lifetimes.

6.8 The pp-chain reactions

Since the pp-chain reactions have been discussed exhaustively in the literature (e.g., D. D. Clayton, *Principles of Stellar Evolution and Nucleosynthesis*, 1968), it suffices here to restrict the discussion to a catalogue of the reactions and of some of their properties. In Table 6.8.1, the reactions are listed in the first column, E (in MeV) is that portion of the nuclear energy released in the reaction which remains in the star, E_ν is an estimate of the average energy of an emitted neutrino, S (in keV barn) is the center of mass cross section factor defined by eq. (6.7.9), S_0 (also in keV barn) is defined by eq. (6.7.11), the quantity v_{ij} is given by eq. (6.7.16), and a lifetime parameter t_j is defined by

$$t_j = t_{ij}\tau_{ij}^2 \exp(-\tau_{ij}), \tag{6.8.1}$$

		E	E_ν	S	S_0		
Reaction	Rate	MeV	MeV	keV b	keV b	v_{ij}	t_j
$p(p, e^+\nu_e)d$	R_{11}	1.19	0.25	4/−22	3/−25	33.81	3.3/15 yr
$d(p, \gamma)^3\text{He}$	R_{12}	5.49		2/−4	2/−7	37.11	7.6/−3 yr
$^3\text{He}(^3\text{He}, 2p)^4\text{He}$	R_{33}	12.86		5.2/3	12	122.28	8.0/−9 yr
$^3\text{He}(^4\text{He}, \gamma)^7\text{Be}$	R_{34}	1.59		0.54	1.4/−3	128.27	1.1/−4 yr
$^7\text{Be}(p, \gamma)^8\text{B}$	$R_{1\text{Be}7}$	0.14		0.02	3/−5	102.64	4.1/−4 yr
$^8\text{B}(e^+\nu_e)^8\text{Be}^*$	$R_{8\text{B}\rightarrow^8\text{Be}^*}$	6.59	8.35	——	——	——	1.11 s
$^8\text{Be}^* \rightarrow 2\,^4\text{He}$	$R_{8\text{Be}\rightarrow2\alpha}$	3.13		——	——	——	10^{-16} s
$^7\text{Be}(e^-, \nu_e)^7\text{Li}$	$R_{e^7\text{Be}}$	0.05	0.81	——	——	——	~1 mo
$^7\text{Li}(p,^4\text{He})^4\text{He}$	$R_{17\text{Li}}$	17.35		60	0.08	84.15	1/−7 yr

Table 6.8.1　Characteristics of pp-chain reactions

where t_{ij} is given by eq. (6.7.19) and τ_{ij} is given by eq. (6.7.18). The cross section factors are selections used by the author for many years based on estimates by William A. Fowler and coworkers (e.g., Fowler, Caughlan, and Zimmerman, 1975) and do not differ significantly from more recent estimates made by Angulo *et al.* (1999).

In the case of the ^8B positron-emitting reaction, t_j is the decay lifetime, $t_{1/2}/\log_e 2$, where $t_{1/2}$ is the half life in a terrestial laboratory. The lifetime t_j of atomic ^7Be against electron capture is approximately two months in a terrestial laboratory, but in a star ^7Be, though frequently retaining one or more bound electrons, can also capture an electron from the continuum, typically reducing its lifetime relative to the terrestial value. In both beta-decay cases, the reaction rate is given by the number abundance of the unstable nucleus divided by t_j.

In the two cases involving the emission of a position and an electron neutrino, on average approximately 60% of the lepton kinetic energy liberated in the nuclear reaction leaves the star as neutrino energy. The sum $E + E_\nu$ is $2\,m_e c^2$ larger than the total kinetic energy of the leptons and is known in nuclear physics as the Q-value for the associated nuclear reaction. In the decay from the first excited state to the ground state of ^8Be, the energy of the emitted gamma ray is on average 3.04 MeV. This energy is rapidly converted into thermal energy and, as the ground state decays into two alpha particles, the 0.09 MeV kinetic energy of the alpha particles is also rapidly converted into thermal energy. Altogether, in an average decay, 8.35 MeV leaves the star as neutrino energy and 6.59 MeV remains in the form of thermal energy. In the case of ^7Be electron capture, 89.5% of all captures are to the ground state of ^7Li, and the entire energy release of 0.861 MeV escapes from the star as neutrino energy. The remaining 10.5% of the captures are to the first excited state of ^7Li (excitation energy = 0.478 MeV), and the energy release of 0.383 MeV escapes the star as neutrino energy. Decay to the ground state is by gamma emission, with the 0.478 MeV energy of the gamma ray being converted into heat locally. Thus, the net result of ^7Be electron capture is the loss of $0.895 \times 0.861 + 0.105 \times 0.383 = 0.811$ MeV in the form of neutrino energy, and a total of only $0.105 \times 0.478 = 0.050$ MeV remains in the star as heat.

Table 6.8.2 Lifetimes t_j(sec) $\times \rho_0 X_i$ for particles in the pp chains					
Reaction	j	$T_6 = 8$	$T_6 = 16$	$T_6 = 27$	$T_6 = 36$
$p(p, e^+ \nu_e)d$	p	2.4/20	1.2/19	2.0/18	8.5/17
$d(p, \gamma)^3$He	d	6.3/3	2.2/2	2.9/1	1.1/1
^{3}He$(^3$He$,2p)^4$He	^{3}He	6.8/14	3.6/9	2.0/6	6.0/4
^{3}He$(^4$He$,\gamma)^7$Be	^{3}He	1.7/20	4.9/14	1.0/11	4.9/9
^{7}Be$(p, \gamma)^8$B	^{7}Be	1.7/15	7.3/10	1.5/8	8.0/6
^{7}Be$(e^-, \nu_e)^7$Li	^{7}Be	6.67/8	9.45/8	1.23/9	1.42/9
^{7}Li$(p, ^4$He$)^4$He	^{7}Li	7.6/7	2.0/4	1.3/2	1.2/1

The lifetime (in seconds) of isotope j against a reaction with isotope i can be found as a function of temperature and density by dividing the appropriate entry for $t_j \rho_0 X_i$ in Table 6.8.2 by $\rho_0 X_i$, where X_i is the abundance by mass of isotope i and ρ_0 is the density in g cm^{-3}.

The electron-capture reaction ^{7}Be$(e^-, \nu_e)^7$Li is of particular interest because it plays a dual role in the production of neutrinos that are detected in solar neutrino experiments. The reaction itself produces neutrinos which can be detected, but, because it is the dominant reaction by which ^{7}Be is destroyed in the Sun, it limits the rate at which the ^{7}Be$(p, \gamma)^8$B reaction can proceed, thus influencing the rate at which the very high energy neutrinos from the beta decay of ^{8}B can occur. Since it is only partially ionized, ^{7}Be can capture bound electrons as well as electrons from the continuum.

In the terrestrial laboratory, the ^{7}Be nucleus captures a K-shell electron, releases nuclear energy of 0.05 MeV, and has a half life of $t_{1/2} = 53.29$ days. The rate of the reaction can be related simply to the characteristics of a K-shell electron. Following a procedure similar to that described in Section 6.2, one has that, in the terrestrial laboratory,

$$\omega_{\text{lab}} = \frac{2\pi}{\hbar} 2 \left| \int \psi_{\text{Li}}^* \psi_{\nu_e}^* g_{\text{weak}} \psi_{\text{Be}} \psi_{\text{eK}} \, d\tau_{\text{nuc}} \right|^2 \rho_f$$

$$= \frac{1}{t_{\text{lab}}} = \frac{\log_e(2)}{t_{1/2}} = \frac{0.693}{4.60 \times 10^6 \text{ s}} = \frac{1}{6.64 \times 10^6 \text{ s}}, \qquad (6.8.2)$$

where ψ_{Li}, ψ_{ν_e} ψ_{Be}, and ψ_{eK} are, respectively, the wave functions of ^{7}Li, the neutrino, ^{7}Be, and a K-shell electron, and ρ_f is the density of final states. The factor of 2 in front of the square of the matrix element comes from the fact that there are two K-shell electrons.

Over the region occupied by the nucleus, the electron wave function is approximately

$$\psi_{\text{eK}} = f \frac{1}{\sqrt{\pi a^3}}, \qquad (6.8.3)$$

where a is $a_0/4$, a_0 is the Bohr radius of the electron in the ground state of the hydrogen atom, and f is a number smaller than unity which takes into account that, when there are two electrons in the K shell, shielding reduces the probability density at the center of each

K-shell electron relative to the probability density of one electron in the field of a three times ionized ^{7}Be nucleus. Thus, the capture rate becomes

$$\omega_{\text{lab}} = \frac{2\pi}{\hbar} g_{\text{weak}}^2 M_{\text{nuc}}^2 \frac{2f^2}{\pi a^3} \rho_{\text{f}}, \qquad (6.8.4)$$

where

$$M_{\text{nuc}} = \int \psi_{\text{Li}}^* \psi_{\text{Be}} \, d\tau_{\text{nuc}}. \qquad (6.8.5)$$

In a star, the nuclear wave functions, the neutrino wave function, and the density of final states are the same as in the laboratory. For capture from the continuum (i.e., capture of a free electron), the density of K-shell electrons, $2f^2/\pi a^3$, is replaced by

$$|\psi_{\text{free}}|^2 = \int \left| \frac{2\pi \eta}{\exp(2\pi \eta) - 1} \right| dn_{\text{e}}, \qquad (6.8.6)$$

where $\eta = -4 \, e^2/(\hbar v)$, v is the electron speed, and dn_{e} is the density of electrons with speed between v and $v + dv$. The probability of capturing a free electron is thus

$$\omega_{\text{free}} = \frac{2\pi}{\hbar} g_{\text{weak}}^2 M_{\text{nuc}}^2 |\psi_{\text{free}}|^2 \rho_{\text{f}}. \qquad (6.8.7)$$

Under stellar conditions, $\exp(2\pi \eta) = \exp(-2\pi |\eta|)$ is on average very small and can be neglected relative to 1, so that

$$|\psi_{\text{free}}|^2 \sim \langle |2\pi \eta| \rangle \, n_{\text{e}}, \qquad (6.8.8)$$

where brackets denote an average over the electron velocity distribution. Under non-electron-degenerate conditions,

$$\langle |2\pi \eta| \rangle = 2\pi \frac{4e^2}{\hbar c} \frac{n_{\text{e}}}{(2\pi m_{\text{e}} k T)^{3/2}} \int \frac{c}{v} \exp(-E/kT) \, 4\pi p^2 \, dp \qquad (6.8.9)$$

$$= 16 \frac{e^2}{\hbar c} \sqrt{\frac{\pi}{2} \frac{m_{\text{e}} c^2}{kT}} \, n_{\text{e}} \qquad (6.8.10)$$

$$= \frac{11.3}{T_6^{1/2}} n_{\text{e}} = \frac{6.74 \times 10^{24}}{T_6^{1/2}} \frac{\rho_0}{\mu_{\text{e}}}, \qquad (6.8.11)$$

where ρ_0 is the density in g cm^{-3} and μ_{e} is the number of nucleons per free electron. Equations (6.8.4), (6.8.7), and (6.8.11) give

$$\omega_{\text{free}} = \omega_{\text{lab}} \frac{\pi a^3}{2f^2} \langle |2\pi \eta| \rangle n_{\text{e}} = \omega_{\text{lab}} \frac{0.0219}{f^2} \frac{\rho_0}{\mu_{\text{e}} T_6^{1/2}}. \qquad (6.8.12)$$

In the laboratory, lifetime and capture probability are related by eq. (6.8.2) so that, in a star, the lifetime and capture probability are given by

$$t_{\text{free}} = \frac{1}{\omega_{\text{free}}} = 45.7 \, f^2 \, \frac{\mu_e T_6^{1/2}}{\rho} \, t_{\text{lab}} = 3.03 \times 10^8 \, f^2 \, \mu_e \, \frac{T_6^{1/2}}{\rho_0} \text{ s.} \tag{6.8.13}$$

Under appropriate conditions, ^{7}Be nuclei capture bound electrons. The relative abundances f_1 and f_2 of ^{7}Be nuclei with, respectively, one and two bound electrons are (see the development in Section 4.14),

$$f_1 = \frac{\lambda}{1 + \lambda + 0.25\lambda^2 \exp(-\Delta\chi/kT)} \tag{6.8.14}$$

and

$$f_2 = 0.25 \, \lambda \, \exp \Delta\chi/kT \, f_1, \tag{6.8.15}$$

where

$$\lambda = n_e \left(\frac{h^2}{2\pi m_e kT} \right)^{3/2} \exp(-\chi_1/kT), \tag{6.8.16}$$

$\Delta\chi = \chi_1 - \chi_2$, and $\chi_1 = 216.6$ eV and $\chi_2 = 153.1$ eV are, respectively, the binding energies of the first and second K-shell electrons. The total lifetime t_{tot} against captures from the continuum and from bound states may be approximated by

$$\frac{\omega_{\text{tot}}}{\omega_{\text{free}}} = 1 + f_1 \left| \frac{\psi_1(0)}{\psi_{\text{free}}} \right|^2 + 2f_2 \left| \frac{\psi_2(0)}{\psi_{\text{free}}} \right|^2, \tag{6.8.17}$$

where

$$\psi_1(0) = \frac{1}{\sqrt{\pi a^3}} \tag{6.8.18}$$

and

$$\psi_2(0) = f \, \psi_1(0). \tag{6.8.19}$$

With these prescriptions and $f^2 = 0.87$,

$$\frac{\omega_{\text{tot}}}{\omega_{\text{free}}} = 1 + \left(\frac{5.07}{T_6} \right) \exp \left(\frac{2.515}{T_6} \right) \frac{1 + 0.435 \, \lambda \, \exp(-0.735 \, /T_6)}{1 + \lambda + 0.25 \, \lambda^2 \exp(-0.735 \, /T_6)}. \tag{6.8.20}$$

When electron screening is taken into account, application of a variational principle to find the electron wave function leads to the approximation (Iben, Kalata, & Schwartz, 1967)

$$\frac{\omega_{\text{tot}}}{\omega_{\text{free}}} = 1 + \left(\frac{5.07}{T_6} \right) C_R^2 \frac{1 + 0.435 \, \lambda_R \, \exp(-0.735 \, \sigma_R/T_6)}{1 + \lambda_R + 0.25 \, \lambda_R^2 \, \exp(-0.735 \, \sigma_R/T_6)}, \tag{6.8.21}$$

where

$$\sigma_R = \frac{E_R}{\chi} = \frac{a}{a_R} \frac{1 - a_R/2R}{(1 + a_R/2R)^3} \tag{6.8.22}$$

is the ratio of the binding energy in the screening approximation to the binding energy in the absence of screening for an atom with one K-shell electron,

$$C_R^2 = \left| \frac{\psi_R(0)}{\psi_1(0)} \right|^2 = \left(\frac{a}{a_R} \right)^3 \tag{6.8.23}$$

is the square of the amplitude of the electron wave function at the origin relative to the square of the amplitude in the absence of screening,

$$\lambda_R = 0.246 \, \frac{\rho}{\mu_e \, T_6^{3/2}} \, \exp\left(\frac{2.515 \, \sigma_R}{T_6} \right), \tag{6.8.24}$$

and $\alpha_R/2R$ is a solution of the cubic equation

$$\frac{a_R}{2R} \left(1 + 3 \frac{a_R}{2R} \right) \div \left(1 + \frac{a_R}{2R} \right)^3 = \frac{a}{2R}. \tag{6.8.25}$$

From eq. (6.8.22), it is evident that a bound state exists only if $\alpha_R/2R < 1$ and eq. (6.8.25) shows that, when $\alpha_R = 2R$, $R = a$; hence, when $R = a$, $\alpha_R = 2a$. For $R = \infty$, $\alpha_R = a$. Thus, for bound states, $a \leq \alpha_R < 2a$. Using the Newton–Raphson method and a starting estimate of $\alpha_R = a$, at most four iterations are sufficient to solve eq. (6.8.25) for $\alpha_R/2R$ to an accuracy of one part in a million. In rare instances where one obtains $R < a$, the solution must be discarded.

In a typical solar model, the ratio of the total electron-capture probability to the probabilty of capture from the continuum, as given by eqs. (6.8.21)–(6.8.25), varies from $\sim$1.15 at the solar center, where $\rho \sim 150$ g cm^{-3}, $T_6 \sim 15.5$, and $X \sim 0.35$, to $\sim$1.4 at a point where the mass fraction is $\sim$0.3, $\rho \sim 40$ g cm^{-3}, $T_6 \sim 10$, and $X \sim 0.71$.

6.9 Equilibrium abundances and energy-generation rates for pp-chain reactions

Over much of the energy-generating region of a star, many of the isotopes in the pp chains are in equilibrium with respect to creation and destruction mechanisms. Referring to ^{1}H, ^{2}H, ^{3}He, and ^{4}He as particles 1, 2, 3, and 4, respectively, and defining n_j as the number density of particles of type j, the full set of transformation-rate equations is:

$$\frac{dn_1}{dt} = -2R_{11} - R_{12} + 2R_{33} - R_{1^7\text{Li}} - R_{1^7\text{Be}}, \tag{6.9.1}$$

$$\frac{dn_2}{dt} = R_{11} - R_{12}, \tag{6.9.2}$$

$$\frac{dn_3}{dt} = R_{12} - 2R_{33} - R_{34}, \tag{6.9.3}$$

$$\frac{dn_4}{dt} = R_{33} - R_{34} + 2R_{1^7Li} + 2R_{8Be\ decay}, \tag{6.9.4}$$

$$\frac{dn_{7Be}}{dt} = -(R_{8Be} + R_{1^7Be} + R_{34}), \tag{6.9.5}$$

$$\frac{dn_{7Li}}{dt} = R_{e^7Be} - R_{1^7Li}, \tag{6.9.6}$$

$$\frac{dn_{8B}}{dt} = R_{1^7Be} - R_{8B\ decay}, \tag{6.9.7}$$

and

$$\frac{dn_{8Be}}{dt} = R_{8B\ decay} - R_{8Be\ decay}. \tag{6.9.8}$$

In these equations, a quantity R_{ij} gives the rate at which particles of type i and j react in units of cm^{-3} s^{-1}. On the left hand sides of the equations, the number densities are given by $n_i = (\rho/M_H)\,(X_i/A_i)$ and, as shown in Section 8.8, the density ρ is to be held constant in taking time derivatives. That is, dn_i/dt is to be interpreted as $(\rho/M_H\,A_i)(dX_i/dt)$.

By comparing lifetimes in Table 6.8.2, it is evident that, over most of the stellar interior, it is a good approximation to assume that

$$\frac{dn_{7Be}}{dt} = \frac{dn_{7Li}}{dt} = \frac{dn_{8B}}{dt} = \frac{dn_{8Be}}{dt} = \frac{dn_2}{dt} = 0. \tag{6.9.9}$$

Then, the transformation equations reduce to

$$\frac{dn_1}{dt} = -3R_{11} + 2R_{33} - R_{34}, \tag{6.9.10}$$

$$\frac{dn_3}{dt} = R_{11} - 2R_{33} - R_{34}, \quad \text{and} \tag{6.9.11}$$

$$\frac{dn_4}{dt} = R_{33} + R_{34}. \tag{6.9.12}$$

If ^{3}He is in equilibrium, then, from eq. (6.9.11),

$$R_{11} = 2R_{33} + R_{34}, \tag{6.9.13}$$

and, using this result in eqs. (6.9.10) and (6.9.11),

$$\frac{dn_1}{dt} = -2(R_{11} + R_{34}) = -4(R_{33} + R_{34}) = -4(R_{11} - R_{33}) \tag{6.9.14}$$

and

$$\frac{dn_4}{dt} = R_{33} + R_{34} = -\frac{1}{4}\frac{dn_1}{dt}. \tag{6.9.15}$$

By setting

$$R_{ij} = \bar{R}_{ij} X_i X_j, \tag{6.9.16}$$

eqs. (6.9.13)–(6.9.15) can be solved for X_3, the abundance by mass of ^{3}He, as a function of the abundances by mass of hydrogen (X_1) and ^{4}He (X_4). The solution is

$$X_3 = \frac{\bar{R}_{34}}{\bar{R}_{33}} X_4 J, \tag{6.9.17}$$

where

$$J = \frac{1}{4} \left(\sqrt{1 + 8K} - 1 \right), \tag{6.9.18}$$

and

$$K = \frac{\bar{R}_{11} \bar{R}_{33}}{\bar{R}_{34}^2} \left(\frac{X_1}{X_4} \right)^2. \tag{6.9.19}$$

Equations (6.9.18) and (6.9.19) may be manipulated to give

$$\frac{J}{K} = \Lambda \left(\sqrt{1 + \left(\frac{\Lambda}{2} \right)^2} - \frac{\Lambda}{2} \right), \tag{6.9.20}$$

where

$$\Lambda = \frac{1}{\sqrt{2K}}. \tag{6.9.21}$$

This result proves useful when ^{3}He is in equilibrium.

The next task is to construct energy-generation rates ϵ_{ij} in units of erg cm^{-3} s^{-1}. If ^{3}He is not in equilibrium, the best one can do is

$$\epsilon_{pp} = \epsilon_{11} R_{11} + \epsilon_{33} R_{33} + \left(\frac{\epsilon_{e7} + \gamma \epsilon_{17}}{1 + \gamma} \right) R_{34}, \tag{6.9.22}$$

where

$$\gamma = \frac{R_{17}}{R_{e7}}, \tag{6.9.23}$$

$$\epsilon_{11} = (1.19 + 5.49) \text{ MeV} = 6.68 \text{ MeV} = 10.70 \times 10^{-6} \text{ erg}, \tag{6.9.24}$$

$$\epsilon_{33} = 12.85 \text{ MeV} = 20.59 \times 10^{-6} \text{ erg}, \tag{6.9.25}$$

$$\epsilon_{e7} = (0.05 + 17.34 + 1.58) \text{ MeV} = 18.97 \text{ MeV} = 30.39 \times 10^{-6} \text{ erg}, \tag{6.9.26}$$

and

$$\epsilon_{17} = (0.14 + 7.7 + 3.0 + 1.58) \text{ MeV} = 12.42 \text{ MeV} = 19.90 \times 10^{-6} \text{ erg}. \tag{6.9.27}$$

If ^{3}He is in equilibrium,

$$\epsilon_{pp} = \epsilon'_{33} R_{33} + \left(\frac{\epsilon'_{e7} + \epsilon'_{17} \gamma}{1 + \gamma} \right) R_{34}, \tag{6.9.28}$$

where

$$\epsilon'_{33} = 26.21 \text{ MeV} = 41.99 \times 10^{-6} \text{ erg}, \tag{6.9.29}$$

$$\epsilon'_{e7} = 25.65 \text{ MeV} = 41.09 \times 10^{-6} \text{ erg}, \tag{6.9.30}$$

$$\epsilon'_{17} = 19.10 \text{ MeV} = 30.60 \times 10^{-6} \text{ erg}. \tag{6.9.31}$$

A final, and more useful, form is

$$\epsilon_{pp} = \epsilon'_{11} R_{11} \left(1 + \frac{J}{K} \frac{0.9576 + 0.4575 \, \gamma}{1 + \gamma} \right), \tag{6.9.32}$$

where

$$\epsilon'_{11} = 13.105 \text{ MeV} = 20.99 \times 10^{-6} \text{ erg}, \tag{6.9.33}$$

and J/K is given by eqs. (6.9.18)–(6.9.21).

Explicit expressions for the reaction rates in the various equilibrium equations are

$$R_{11} = X_1^2 \frac{\rho_0^2}{T_6^{2/3}} \exp\left(25.51 - \frac{33.81}{T_6^{1/3}} \right), \tag{6.9.34}$$

$$R_{33} = X_3^2 \frac{\rho_0^2}{T_6^{2/3}} \exp\left(81.23 - \frac{122.28}{T_6^{1/3}} \right), \tag{6.9.35}$$

and

$$R_{34} = X_3 X_4 \frac{\rho_0^2}{T_6^{2/3}} \exp\left(72.42 - \frac{128.27}{T_6^{1/3}} \right). \tag{6.9.36}$$

Using these in eq. (6.9.19) to find K and inserting the result in eq. (6.9.21) gives

$$\Lambda = \frac{X_4}{X_1} \exp\left(19.05 - \frac{50.23}{T_6^{1/3}} \right), \tag{6.9.37}$$

and this result may be used in eq. (6.9.20) to find J/K required in eq. (6.9.32). Finally, when ^{3}He is in equilibrium, the pp-chain energy-generation rate in erg g^{-1} s^{-1} is

$$\epsilon_{pp}(\text{erg g}^{-1} \text{ s}^{-1}) = X_1^2 \frac{\rho_0}{T_6^{2/3}} \exp\left(14.74 - \frac{33.81}{T_6^{1/3}} \right) \left(1 + \frac{J}{K} \frac{0.9576 + 0.4575 \gamma}{1 + \gamma} \right). \tag{6.9.38}$$

The first terms in the exponentials in eqs. (6.9.34)–(6.9.37) are the consequences of specific choices for cross section factors: $S_{11}(0) = 4 \times 10^{-22}$ keV barn, $S_{33}(0) = 5200$ keV barn, and $S_{34}(0) = 0.54$ keV barn. It is to be emphasized that these choices are highly subjective extrapolations to temperatures significantly below temperatures at which unambiguous estimates can be made solely from experimental data.

6.10 The CN-cycle reactions

In the Sun and in other Population I stars, at temperatures less than $\sim 15 \times 10^6$ K, reactions between protons and isotopes of elements in the CNO group occur at rates smaller even than the pp reaction. This is due to the fact that the Coulomb barrier between protons and CNO nuclei is so much larger than the Coulomb barrier between protons. However, precisely because the Coulomb barriers are larger, CNO reaction rates increase with temperature much faster than does the pp-reaction rate, with the consequence that, as temperatures are increased only modestly above $\sim 16 \times 10^6$ K, CNO reactions rapidly become dominant in converting hydrogen into helium. The main reactions were identified by C. F. von Weizsäcker (1938) and Hans A. Bethe (1939). They are:

$$^{12}C(p, \gamma)^{13}N(e^+\nu_e)^{13}C(p, \gamma)^{14}N(p, \gamma)^{15}O(e^+\nu_e)^{15}N(p, \alpha)^{12}C. \qquad (6.10.1)$$

This sequence of reactions is known as the CN-cycle and it conserves the total abundance of CNO isotopes. Energies liberated, center of mass cross section factors, and several other characteristics of these reactions are given in the first six rows of Table 6.10.1. The meanings of all entries are the same as for entries in Table 6.8.1.

In the case of the positron-emitting reactions, the maximum energy of the electron neutrino is given by $E_\nu^{max} = Q - 2m_ec^2$, where Q is the conventional measure of the nuclear energy emitted in the reaction. The average kinetic energy of the neutrino is $E_\nu = f E_\nu^{max}$, where $f \sim 0.6$, and the amount of energy that remains in the star in the form of heat is $E = Q - E_\nu$.

The lifetime (in seconds) of an isotope j against proton capture can be found as a function of temperature and density by dividing the appropriate entry in Table 6.10.2 by $\rho_0 X_H$. Table 6.10.2 shows that, during the main sequence phase, the lifetimes of ^{12}C and ^{13}C nuclei are shorter than the lifetime of an ^{14}N nucleus by several orders of magnitude and

Reaction	E MeV	E_ν MeV	S keV b	S_0 keV b	ν_{ij}	t_j yr
$^{12}C(p, \gamma)^{13}N$	1.944		1.3	1.8/−3	136.6	9.5/−6
$^{13}N(e^+\nu_e)^{13}C$	1.50	0.72	—	—	—	14.4 min
$^{13}C(p, \gamma)^{14}N$	7.550		6.0	8.5/−3	136.8	2.1/−6
$^{14}N(p, \gamma)^{15}O$	7.293		3.0	4.3/−3	151.9	5.0/−6
$^{15}O(e^+\nu_e)^{15}N$	1.72	1.04	—	—	—	2.94 min
$^{15}N(p, {}^4He)^{12}C$	4.965		7.2/4	104	152.1	2.1/−10
$^{15}N(p, \gamma)^{16}O$	12.126		32	0.046	152.1	4.7/−7
$^{16}O(p, \gamma)^{17}F$	0.601		5.0	7.2/−3	166.5	3.4/−6
$^{17}F(e^+\nu_e)^{17}O$	1.72	1.04	—	—	—	1.55 min
$^{17}O(p, {}^4He)^{14}N$	1.193		4000	5.8	166.7	4.3/−9

Table 6.10.1 Characteristics of reactions involving CNO isotopes

Table 6.10.2 Lifetimes t_j(s) times ρX_H for isotopes of C, N, and O

Reaction	j	$T_6 = 8$	$T_6 = 16$	$T_6 = 27$	$T_6 = 36$
$^{12}C(p,\gamma)^{13}N$	^{12}C	9.3/20	1.1/15	2.7/11	5.1/9
$^{13}N(e^+\nu_e)^{13}C$	^{13}N	14 min			
$^{13}C(p,\gamma)^{14}N$	^{13}C	2.4/20	2.8/14	6.6/10	1.2/9
$^{14}N(p,\gamma)^{15}O$	^{14}N	8.4/23	2.1/17	1.9/13	2.3/11
$^{15}O(e^+\nu_e)^{15}N$	^{15}O	3 min			
$^{15}N(p,^4He)^{12}C$	^{15}N	3.9/19	9.5/12	8.6/8	1.04/7
$^{15}N(p,\gamma)^{16}O$	^{15}N	8.8/22	2.1/16	1.9/12	2.3/10
$^{16}O(p,\gamma)^{17}F$	^{16}O	7.2/26	4.0/19	1.4/15	1.1/13
$^{17}F(e^+\nu_e)^{17}O$	^{17}F	1.6 min			
$^{17}O(p,^4He)^{14}N$	^{17}O	9.9/23	5.3/16	1.9/12	1.4/10

that the lifetime of ^{15}N against proton capture with the production of ^{12}C and an alpha particle is shorter by roughly four orders of magnitude. The lifetimes of ^{13}N and ^{15}O nuclei against positron decay are extremely short compared with the lifetimes of the other isotopes. This means that ^{14}N becomes by far the most abundant isotope in the CN cycle. It will become evident in later chapters that, in hydrogen-burning shells of luminous red giants, the lifetimes against proton capture can become shorter than positron decay lifetimes, leading to abundance patterns different from those arising during hydrogen burning in main sequence stars. The same is true in the accreting component of a binary star system as it is experiencing a nova explosion.

There is a minor leak out of the CN-cycle due to the fact that, once every $\sim$2200 times that an ^{15}N nucleus reacts with a proton, the resulting compound nucleus decays into an ^{16}O nucleus with the release of a gamma ray rather than into a ^{12}C nucleus and an α particle. However, the ^{16}O nucleus is ultimately converted into an ^{14}N nucleus by the reactions

$$^{16}O(p,\gamma)^{17}F(e^+\nu_e)^{17}O(p,\alpha)^{14}N. \tag{6.10.2}$$

Characteristics of these reactions are given in the last four rows of Table 6.10.1.

In years past, the existence of the leakage channel has motivated the use of the term CNO bi-cycle to describe the entire set of reactions cycling through ^{14}N. Actually, as hydrogen burning progresses, the original store of ^{16}O is converted slowly into ^{14}N until it is reduced to a value given by

$$\frac{n\left(^{16}O\right)}{n\left(^{14}N\right)} = \frac{\tau\left(^{16}O \to {}^{17}F\right)}{\tau\left(^{14}N \to {}^{15}O\right)} \sim \frac{1}{2200} \tag{6.10.3}$$

and, thereafter, for all practical purposes, one can forget about the branching which occurs at ^{15}N. It is therefore much more accurate to continue to describe the process as the CN-cycle process, recognizing that the abundance of ^{14}N, by far the most abundant isotope in the cycle, is at first equal to the initial abundance of ^{14}N, is then enhanced by the conversion of most of the original ^{12}C into ^{14}N, and finally is enhanced by the conversion of most of the original ^{16}O into ^{14}N.

When the isotopes in the CN cycle are at abundances such that destruction and creation rates are identical, the rates of change of hydrogen and helium are given, respectively, by

$$\frac{dn_1}{dt} = -4R_{114} \tag{6.10.4}$$

and

$$\frac{dn_4}{dt} = R_{114}, \tag{6.10.5}$$

where

$$R_{114} = X_N X_H \frac{\rho_0^2}{T_6^{2/3}} \exp\left(74.414 - \frac{152.299}{T_6^{1/3}}\right) \tag{6.10.6}$$

and X_N is the abundance by mass of ^{14}N in equilibrium. In good approximation, this means that

$$\frac{X_N}{14} \sim \frac{X_{C,0}}{12} + \frac{X_{N,0}}{14} + \frac{X_{O,0}}{16}, \tag{6.10.7}$$

where $X_{C,0}$, $X_{N,0}$, $X_{O,0}$ are, respectively, the abundances by mass of ^{12}C, ^{14}N, and ^{16}O with which the star is born.

The rate at which energy produced by CN-cycle reactions per unit volume and unit time remains in the star as heat is

$$\epsilon_{CN} = (1.95 + 1.50 + 7.54 + 7.35 + 1.73 + 4.96) \text{ MeV } R_{114}$$

$$= 25.03 \text{ MeV } R_{114} = 40.10 \times 10^{-6} \text{ erg } R_{114}. \tag{6.10.8}$$

Dividing eq. (6.10.8) by ρ_0 and using eq (6.10.6) for R_{114} gives the energy-generation rate in erg g^{-1} s^{-1}:

$$\epsilon_{CN}(\text{erg g}^{-1} \text{ s}^{-1}) = X_N X_H \frac{\rho_0}{T_6^{2/3}} \exp\left(64.232 - \frac{152.299}{T_6^{1/3}}\right). \tag{6.10.9}$$

The rate at which neutrinos carry off energy in the CN cycle is given by

$$\epsilon_\nu = (0.72 + 1.04 + 1.04) \text{ MeV } R_{114} = 2.80 \text{ MeV } R_{114} = 0.112 \, \epsilon_{CN}. \tag{6.10.10}$$

Throughout the development in this chapter, cross section factors have been treated as if they were constant and precisely determined. This is incorrect on both counts. For every cross section determined in the laboratory there is an error bar which increases with decreasing energy. For every theoretical fit to the data and for every theoretical extrapolation to energies relevant in stars there is an inevitable uncertainty. Finally, because cross section factors vary with energy, this variation should be taken into account if the stellar context warrants it. Then, the analytic approximation to $\langle \sigma \, v \rangle$ must be replaced by the results of a numerical integration (to which analytical fits may be made). A compilation by Angulo *et al.* (1999) describes the cross-section data available at the turn of the twentieth century, presents results of numerical determinations of the $\langle \sigma \, v \rangle$ integral, and provides analytical fits to the results.

6.11 The effect of electrostatic screening on nuclear reaction rates

As described in Section 4.17, electrostatic interactions between ions and electrons produce an electrical potential well about every ion which, in effect, reduces the positive electrical charge which that ion presents to another, approaching, ion. This has the consequence that the Coulomb barrier between two interacting nuclei is decreased, enhancing the probability of penetration to nuclear dimensions. Similarly, if a nucleus retains one or more bound electrons, the effective charge which that nucleus presents to another, approaching, nucleus is reduced. Both phenomena, which may be described as electrostatic screening effects, enhance the probability of a nuclear reaction between two positively charged nuclei. The treatment of screening by free electrons was pioneered by Edwin E. Salpeter (1954). The effect of bound electrons is discussed by Iben (1969).

In stars massive enough to evolve beyond the main sequence in a Hubble time, screening of nuclear charges by the ion plasma has a smaller effect on the rate of hydrogen-burning reactions than does the uncertainty in measured and theoretically extrapolated reaction cross sections. In such stars, the reaction rates presented in previous sections are appropriate approximations. However, there are situations, such as in very low mass main sequence stars, in which screening cannot be neglected.

Nuclear reaction rates are determined by the physics of interactions at particle separations of the order of nuclear dimensions (a few $\times 10^{-13}$ cm), and because these dimensions are quite small compared with typical electrostatic screening radii, which are of the order of interparticle separations $\left(\sim 10^{-8} \text{ cm}/\rho_0^{1/3} \right)$, the electrostatic screening potential is essentially constant over the interaction volume. Thus, the basic effect of screening is to deepen the effective nuclear potential well by the value of the screening potential at zero separation. In the Debye–Hückel weak screening approximation, this potential is described by eq. (4.17.19), with the Debye radius given by eq. (4.17.12).

From the discussion centered on eqs. (6.4.10)–(6.4.12), the cross section for a non-resonant nuclear reaction in the absence of screening can be written as

$$\sigma_{ij}(E) = \frac{S'_{ij}(E)}{E} \, P(E), \tag{6.11.1}$$

where $P(E)$ is a penetration factor defined by eq. (6.4.10) and evaluated in eq. (6.4.11). The cross section factor S'_{ij} in eq. (6.11.1) is related to the cross section factor S_{ij} in eq. (6.4.10) by

$$S'_{ij} = S_{ij} \, \exp\left(\frac{4e}{\hbar} (2Z_i Z_j \mu R)^{1/2} \right). \tag{6.11.2}$$

With screening, the penetration factor becomes

$$P_{\text{scr}}(E) = \exp\left(-\int_R^{R_E} \sqrt{\frac{2\mu}{\hbar^2}\left(\frac{e^2}{r} - U_{ij} - E\right)} dr\right)$$

$$= \exp\left(-\int_R^{R_E} \sqrt{\frac{2\mu}{\hbar^2}\left(\frac{e^2}{r} - (E - U_{ij})\right)} dr\right), \qquad (6.11.3)$$

where, under non electron-degenerate conditions,

$$U_{ij} = -\frac{Z_i Z_j e^2}{R_D}, \qquad (6.11.4)$$

and R_D is the Debye radius. It is evident that, with screening, the penetration factor is effectively that of a particle with kinetic energy $E' = E - U_{ij} = E + |U_{ij}|$.

Noting the relationship between eqs. (6.4.10) and (6.4.11), one has that

$$P_{\text{scr}}(E) \sim \exp\left(-2\pi\eta' + \frac{4e}{\hbar}(2Z_i Z_j \mu R)^{1/2}\right), \qquad (6.11.5)$$

where

$$\eta' = \frac{Z_i Z_j e^2}{\hbar v'} = \eta\frac{v}{v'}, \qquad (6.11.6)$$

with v' and v related by

$$\frac{1}{2}\mu(v')^2 = E' = E + |U_{ij}| = \frac{1}{2}\mu v^2 + |U_{ij}|, \qquad (6.11.7)$$

or

$$\frac{v}{v'} = \left(1 + \frac{|U_{ij}|}{E}\right)^{-1/2}. \qquad (6.11.8)$$

One can therefore write

$$2\pi\eta' = 2\pi\eta - 2\pi\eta\left[1 - \left(1 + \frac{|U_{ij}|}{E}\right)^{-1/2}\right]. \qquad (6.11.9)$$

For values of E such that $|U_{ij}|/E \ll 1$,

$$2\pi\eta' = 2\pi\eta - \pi\eta\frac{|U_{ij}|}{E} + \cdots \qquad (6.11.10)$$

and one can write

$$P_{\text{scr}}(E) \approx P(E)\,\exp\left(\pi\eta\frac{|U_{ij}|}{E}\right). \qquad (6.11.11)$$

Using eq. (6.7.15) for η_0 and eq. (6.7.20) for E_0, one has that, at the maximum in the integrand defined in eq. (6.7.17),

$$\pi \eta_0 \frac{|U_{ij}|}{E_0} = \pi \frac{Z_i Z_j e^2}{\hbar v_0} \frac{|U_{ij}|}{E_0} = \pi \frac{Z_i Z_j e^2}{\hbar (2E_0/\mu)^{1/2} E_0} |U_{ij}| = \frac{1}{kT} |U_{ij}|. \tag{6.11.12}$$

Thus, a reasonable approximation to the result of incorporating screening is

$$\langle \sigma \, v \rangle_{\text{screening}} = \exp \left(\frac{|U_{ij}|}{kT} \right) \langle \sigma \, v \rangle_{\text{no screening}}. \tag{6.11.13}$$

Estimates for $|U_{ij}|/kT$ follow from the explicit expressions for the Debye length given in Section 4.17. When electrons are not degenerate, eq. (4.17.12) in conjunction with eq. (6.11.4) gives

$$-\frac{U_{12}}{kT} = \frac{Z_1 Z_2 e^2}{kT} \frac{1}{R_{\mathrm{D}}} = \frac{Z_1 Z_2 e^2}{kT} \sqrt{\frac{4\pi}{kT} \sum_k n_{0k} (Z_k e)^2} \tag{6.11.14}$$

$$= Z_1 Z_2 \left(\frac{e^2}{kT} \right)^{3/2} \sqrt{4\pi \frac{\rho_0}{M_{\mathrm{H}}} \sum_k Y_k Z_k^2}, \tag{6.11.15}$$

where the summations extend over all ions of charge Z_k and abundance by number Y_k and over all electrons. When the gas is fully ionized, the summation in eq. (6.11.15) is given by

$$\zeta = \sum_{k=\text{all ions}} Y_k \left(Z_k + Z_k^2 \right), \tag{6.11.16}$$

and, numerically,

$$-\frac{U}{kT} = 0.188 \, Z_1 \, Z_2 \left(\zeta \, \frac{\rho_0}{T^3} \right)^{1/2}. \tag{6.11.17}$$

When electrons are partially degenerate, eqs. (4.17.45) and (6.11.4) give

$$-\frac{U}{kT} = \frac{Z_1 Z_2 e^2}{kT} \sqrt{\frac{4\pi e^2}{kT} \left[\sum_j^{\text{ions only}} n_{0j} Z_j^2 + \left| \frac{dn_e}{d\alpha} \right| \right]}, \tag{6.11.18}$$

where $(1/n_e)dn_e/d\alpha$ is given by eqs. (4.17.51)–(4.17.60) for varying degrees of degeneracy. Equation (6.11.17) still holds with ζ replaced by

$$\zeta' = \sum_{j=\text{all ions}} Y_j Z_j \left(Z_j + \left| \frac{1}{n_e} \frac{dn_e}{d\alpha} \right| \right), \tag{6.11.19}$$

The development thus far is called the weak screening approximation; it is strictly applicable only if $\left| (U/kT_{\text{max}}) \right| \ll 1$, where T_{max} is the temperature at which the maximum in the integrand in eq. (6.7.17). It is not reasonable when $|(U/kT_{max})| > 1$. In main sequence stars of mass comparable to or larger than $M_\odot$, weak screening prevails and enhancements of CN-cycle reaction rates vary from factors of ~ 2 at the center of solar mass models to only a few percent at centers of more massive models (see, e.g., Iben, 1969).

When electrons are extremely degenerate, another approach, also due to Salpeter, is customary. Consider two nuclei which are in the process of approaching one another in a sea of nearly constant density electrons. About each nucleus define a radius R_i such that

$$Z_i = n_e \frac{4\pi}{3} R_i^3 = \frac{\rho}{\mu_e M_H} \frac{4\pi}{3} R_i^3, \tag{6.11.20}$$

where n_e is the average electron density and μ_e is the electron molecular weight (number of nucleons per electron). Within the spherical volume of radius R_i, the total electrical charge is zero. The electrostatic energy of the volume is

$$E_i = \int_0^{R_i} \frac{e^2 N_e(r)}{r} \, dN_e(r) - \int_0^{R_i} \frac{Z_i e^2 N_e(r)}{r} \, dn_e(r), \tag{6.11.21}$$

where

$$N_e(r) = n_e \frac{4\pi}{3} r^3. \tag{6.11.22}$$

Integration gives

$$E_i = \frac{3}{5} \frac{e^2 Z_i^2}{R_i} - \frac{3}{2} \frac{e^2 Z_i^2}{R_i} = -\frac{9}{10} \frac{e^2 Z_i^2}{R_i}, \tag{6.11.23}$$

When the two nuclei have approached within nuclear dimensions, the electrostatic energy of the system consisting of the two positively charged nuclei at separation r_{12} and a surrounding hypothetical electron cloud of radius R_{12} is

$$E_{ij} = e^2 \frac{Z_1 Z_2}{r_{12}} - \frac{9}{10} \frac{e^2 (Z_1 + Z_2)^2}{R_{12}}, \tag{6.11.24}$$

where $r_{12} \ll R_{12}$. Hence, the effective Coulomb potential between the two nuclei is

$$E_{Coulomb} = e^2 \frac{Z_1 Z_2}{r_{12}} + U, \tag{6.11.25}$$

where

$$U = -\frac{9}{10} e^2 \left[\frac{(Z_1 + Z_2)^2}{R_{12}} - \frac{Z_1^2}{R_1} - \frac{Z_2^2}{R_2} \right]. \tag{6.11.26}$$

Making use of eq. (6.11.20), one obtains

$$\frac{U}{kT} = -\frac{9}{10} \frac{e^2}{kT} \left(\frac{4\pi}{3} \frac{\rho}{\mu_e M_H} \right)^{1/3} \left[(Z_1 + Z_2)^{5/3} - Z_1^{5/3} - Z_2^{5/3} \right], \tag{6.11.27}$$

where, as usual,

$$\frac{1}{\mu_e} = \sum_j^{\text{all ions}} Y_j Z_j.$$
(6.11.28)

Numerically,

$$-\frac{U}{kT} = 0.205 \left(\frac{\rho_0}{\mu_e T_6^3}\right)^{1/3} \left[(Z_1 + Z_2)^{5/3} - Z_1^{5/3} - Z_2^{5/3}\right].$$
(6.11.29)

6.12 Polytropic models for zero age main sequence stars

With explicit energy-generation rates available, and guided by the results of realistic stellar models in the literature, trends relating radius, mass, and luminosity of observed main sequence stars can be used in conjunction with polytropic models to estimate conditions in the deep interiors of zero age main sequence stars.

As a first exercise, consider a model of homogeneous composition given by $(X, Y, Z) = (0.71, 0.27, 0.02)$ and of global characteristics the same as those of the present Sun, namely, $M = M_\odot$, $L = L_\odot$, and $R = R_\odot$. Adopting a mean density of $\bar{\rho} = 1.41$ g cm^{-3}, entries in the seventh column of Table 5.2.3 can be used to obtain the central densities of polytropes of different indices. Results for indices $N = 2.5$, 3, 3.25, and 3.5 are shown in column 2 of Table 6.12.1. Equations (5.3.11)–(5.3.13) (see the entries for $f(N)$ in the third column of Table 5.3.1) show that, when pressure is given by eq. (5.3.1), $\beta_c \sim 1.000$. Equation (5.3.19), in conjunction with the entries in the first three columns of Table 5.2.3 give the central temperatures shown in the third column of Table 6.12.1.

From eq. (5.3.35), one has that the average energy-generation rate of a polytropic model is related to the energy-generation rates at the center of the model by

$$\epsilon_{\text{average}} = \frac{L(R)}{M(R)} = \Gamma(N) \left[\frac{\epsilon_{\text{CN,c}}}{\alpha_{\text{CN}}^{3/2}} \left(1 + \frac{3N - 5}{16\,\alpha_{\text{CN}}}\right) + \frac{\epsilon_{\text{pp,c}}}{\alpha_{\text{pp}}^{3/2}} \left(1 + \frac{3N - 5}{16\,\alpha_{\text{pp}}}\right)\right],$$
(6.12.1)

Table 6.12.1 Characteristics of polytropic models for mass $M = 1.0\,M_\odot$, radius $R = R_\odot$, and molecular weight $\mu = 0.61$

N	ρ_c g cm^{-3}	T_c 10^6 K	$\frac{J}{K}$	s_{pp}	ϵ_{pp} erg g^{-1} s^{-1}	$\epsilon_{\text{average}}$ erg g^{-1} s^{-1}
2.5	33.00	9.871	0.0035	4.61	1.31	0.144
3.0	76.40	12.06	0.0157	4.36	7.4	0.824
3.25	124.3	13.66	0.0380	4.29	20.72	2.27
3.5	215.6	15.82	0.1203	4.41	67.7	7.15

where $\Gamma(N)$ is given in the sixth column of Table 5.3.1. The αs in eq. (6.12.1) are given by eq. (5.3.32). If electron screening is neglected, $k = 0$, and

$$\alpha_{\text{CN,c}} = \frac{2N + s_{\text{CN}}}{6} \text{ and } \alpha_{\text{pp,c}} = \frac{2N + s_{\text{pp}}}{6}. \tag{6.12.2}$$

The values of s in eq. (6.12.2) are logarithmic temperature gradients of energy-generation rates at the model center.

At the central temperatures shown in Table 6.12.1, ^{7}Be is destroyed primarily by electron captures rather than by proton captures, so, in good approximation, γ in eq. (6.9.38) may be set equal to zero. Then,

$$\epsilon_{\text{pp}}(\text{erg g}^{-1}\text{s}^{-1}) \approx (0.71)^2 \frac{\rho_0}{T_6^{2/3}} \exp\left(14.74 - \frac{33.81}{T_6^{1/3}}\right)\left(1 + 0.9576\frac{J}{K}\right)$$

$$= \frac{\rho_0}{T_6^{2/3}} \exp\left(14.06 - \frac{33.81}{T_6^{1/3}}\right)\left(1 + 0.9576\frac{J}{K}\right), \tag{6.12.3}$$

where J/K is given by eq. (6.9.20), with

$$\Lambda = \exp\left(17.73 - \frac{50.23}{T_6^{1/3}}\right), \tag{6.12.4}$$

as follows from eq. (6.9.37) when $X_4 = 0.27$ and $X_1 = 0.71$. Further,

$$s_{\text{pp}} = \frac{\text{d} \log \epsilon_{\text{pp}}}{\text{d} \log T} = -\frac{2}{3} + \frac{11.27}{T_6^{1/3}} + \frac{0.9576\,(J/K)}{1 + 0.9576\,(J/K)}\frac{\text{d} \log\,(J/K)}{\text{d} \log T}. \tag{6.12.5}$$

To the extent that $J/K \ll 1$,

$$\frac{\text{d} \log\,(J/K)}{\text{d} \log T} = \frac{\text{d} \log \Lambda}{\text{d} \log K} = \frac{1}{3}\frac{50.23}{T_6^{1/3}} = \frac{16.74}{T_6^{1/3}}. \tag{6.12.6}$$

Thus,

$$s_{\text{pp}} \sim -\frac{2}{3} + \frac{11.27}{T_6^{1/3}}\left(1.0 + 1.485\frac{0.9576\,(J/K)}{1 + 0.9576\,(J/K)}\right). \tag{6.12.7}$$

With $X_N = 0.01$ and $X_H = 0.71$, eq. (6.10.9) gives

$$\epsilon_{\text{CN}}(\text{erg g}^{-1}\text{s}^{-1}) = \frac{\rho_0}{T_6^{2/3}} \exp\left(59.342 - \frac{152.3}{T_6^{1/3}}\right), \tag{6.12.8}$$

from which it follows that

$$s_{\text{CN}} = -\frac{2}{3} + \frac{50.766}{T_6^{1/3}}. \tag{6.12.9}$$

Table 6.12.2 Characteristics of polytropic models for mass $M = 9.0\,M_\odot$, radius $R = 3.5\,R_\odot$ and molecular weight $\mu = 0.61$

N	ρ_c g cm^{-3}	β_c	$T_{c,6}$ 10^6 K	s_{CN}	ϵ_{CN} erg g^{-1} s^{-1}	$\epsilon_{average}$ erg g^{-1} s^{-1}
2.5	6.928	0.9624	24.44	16.83	787.0	24.0
3.0	16.04	0.9694	30.06	15.66	52260	1790
3.25	26.09	0.9740	34.20	14.97	6.156/5	3.22/4
3.5	45.25	0.9749	39.66	14.22	9.251/6	3.16/5

Values of J/K, s_{pp}, and ϵ_{pp} at model centers are given, respectively, in columns 4–6 of Table 6.12.1. At the temperatures given in the table, the CN-cycle reactions are of secondary importance to the pp-chain reactions with regard to their contribution to the energy-generation rate. For example, at the highest temperature given, $\epsilon_{CN} = 7.08$ erg g^{-1} s^{-1}, only 13% of ϵ_{pp}. At the next highest temperature, $\epsilon_{CN} = 0.275$ erg g^{-1} s^{-1}, only 1.3% of ϵ_{pp}.

In column 7 of Table 6.12.1 are the average energy-generation rates of the models when only energy generation by pp-chain reactions is taken into account. The appropriate value for N is obtained by insisting that $\epsilon_{average} = L/M = L_\odot/M_\odot = 1.939$ erg g^{-1} s^{-1}. Graphical interpolation shows that $N \sim 3.22$. Then, $\rho_c \sim 117$ g cm^{-3}, $T_c \sim 13.4$ K, and $\epsilon_c \sim 17.6$ erg g^{-1} s^{-1}. From eq. (5.3.38), one has that the relative effective mass of the energy-generating region in the concordant model is

$$\frac{\Delta M_{eff}}{M} = 2\,\frac{L}{\epsilon_c M} \sim 0.117. \tag{6.12.10}$$

Since the Sun has evolved considerably in its 4.6 billion year lifetime, it is not strictly correct to compare with the Sun's current global characteristics. Evolutionary models of a solar mass star suggest that the Sun has increased in brightness by roughly twenty five percent over its lifetime (e.g., Iben, 1967). Choosing a model luminosity $L = 0.75\,L_\odot$, but keeping model mass and radius the same, concordance between $\epsilon_{average}$ and $L/M = 0.75L_\odot/M_\odot = 1.454$ erg g^{-1} s^{-1} is achieved when $N \sim 3.15$. Then, $\rho_c \sim 102$ g cm^{-3}, $T_c \sim 13.0$ K, and $\epsilon_c \sim 13.7$ erg g^{-1} s^{-1}.

As a second example, consider a mass M $= 9\,M_\odot$ and assume that $L \sim 4500\,L_\odot$ and $R = 3.5\,R_\odot$, as given by a model with characteristics similar to those of a zero age main sequence population I star (Iben, 1966). Values of polytropic index N, of density β, and T_6 at the center are given, respectively, in columns 1–4 of Table 6.12.2.

In column 5 of Table 6.12.2 is the value of s_{CN} at the center given by eq. (16.12.9) and in column 6 is the value of ϵ_{CN} defined by eq. (6.12.8). In column 7 is the average energy-generation rate given by eq. (6.12.1) when only energy generation by CN-cycle reactions is taken into account. Interpolating logarithmically, one finds that concordance between $\epsilon_{average}$ and $L/M = (4500\,L_\odot/9\,M_\odot) = 969$ erg g^{-1} s^{-1} occurs when $N \sim 2.92$.

Table 6.12.3 Characteristics of polytropic models for mass $M = 3.0\,M_\odot$, radius $R = 1.75\,R_\odot$, and molecular weight $\mu = 0.61$

N	ρ_c g cm^{-3}	T_c 10^6 K	J/K	ϵ_{pp} erg g^{-1}s^{-1}	s_{pp}	ϵ_{CN} erg g^{-1}s^{-1}	s_{CN}	$\epsilon_{average}$ erg g^{-1}s^{-1}
2.5	18.475	16.844	0.1434	7.15	6.30	2.39	19.1	0.576
3.0	42.768	20.586	0.4193	38.0	7.67	247	17.9	9.45
3.25	69.579	23.330	0.6691	103	6.71	3585	17.1	119
3.5	120.67	27.037	0.8913	309	4.06	72690	16.2	2460

Then, $\rho_c \sim 14.2$ g cm^{-3}, $T_{c,6} \sim 29.0$, and $\epsilon_c \sim 28800$ erg g^{-1} s^{-1}. The relative effective mass of the energy-generating region in the concordant polytrope, as given by eq. (5.3.38), is

$$\frac{\Delta M_{\rm eff}}{M} = 2\,\frac{L}{\epsilon_c M} = 0.067, \qquad (6.12.11)$$

approximately half of the relative effective energy-generating region in the 1 $M_\odot$ polytropic main sequence model.

As a third example, consider a mass $M = 3\,M_\odot$, and, for comparison with a realistic zero age main sequence model, adopt $R = 1.75\,R_\odot$ and $L = 94\,L_\odot$ (Iben, 1965). In Table 6.12.3, N, central density, and central temperature are given in columns 1–3, respectively. The temperatures and densities span regions where the CN cycle and the pp chains are competitive. This time, assuming that proton capture on ^{7}Be is much faster than electron capture, it is appropriate to set $\gamma = \infty$ in eq. (6.9.38), giving for the pp energy-generation rate

$$\epsilon_{pp}(\text{erg g}^{-1}\text{s}^{-1}) = \frac{\rho_0}{T_6^{2/3}}\exp\left(14.06 - \frac{33.81}{T_6^{1/3}}\right)\left(1 + 0.4575\frac{J}{K}\right). \qquad (6.12.12)$$

The logarithmic temperature derivative of this rate is

$$s_{pp} = \frac{d\log\epsilon_{pp}}{d\log T} = -\frac{2}{3} + \frac{11.27}{T_6^{1/3}} + \frac{0.4575(J/K)}{1 + 0.4575\,(J/K)}\frac{d\log\,(J/K)}{d\log T}. \qquad (6.12.13)$$

Equations (6.12.8) and (6.12.9) are retained for the CN-cycle energy-generation rate and its logarithmic temperature derivative, respectively.

The results for energy-generation rates and logarithmic temperature derivatives are given in columns 4–8 in Table 6.12.3. The logarithmic temperature derivative of J/K required to construct s_{pp} as per eq. (6.12.9) has been obtained numerically from the J/K and T_c versus N data in the table. The average energy-generation rate is given in the ninth column of the table. Interpolating logarithmically, one finds that the adopted average energy of $L/M = (94\,L_\odot/3\,M_\odot) = 60.75$ erg g^{-1} s^{-1} corresponds to $\epsilon_{average}$ for a polytrope of index $N \sim 3.18$. Then, $\rho_c \sim 61$ g cm^{-3}, $T_{c,6} \sim 22.5$, and $L_{pp}/(L_{CN} + L_{pp}) \sim 0.089$.

As a fourth example, consider a model of mass $M = (1/9)\,M_\odot$. Noting that the mean value of ρ/T^3 is proportional to $1/M^2$, one may anticipate that the opacities in such a low mass star are large enough that energy flow is primarily by convection and that the temperature gradient is essentially the adiabatic temperature gradient. Therefore, rather than examine luminosity as a function of polytropic index, it makes sense to explore the luminosity of the adiabatic models discussed in Section 5.6. Among these models, as described in Table 5.6.1, a maximum central temperature of $\sim 5 \times 10^6$ K occurs when the central density reaches ~ 750 g cm^{-3} for a model radius of $\sim 0.11\,R_\odot$. From the development in Sections 6.7 and 6.9, when electron screening is neglected, the lifetime t_{33} of a ^{3}He nucleus against capture by another ^{3}He nucleus is

$$\frac{1}{t_{33}} = \tau_{33}^2 \, \exp\left(18.640 - \tau_{33}\right) \rho \, X_3. \tag{6.12.14}$$

where

$$\tau_{33} = \frac{122.737}{T_6^{1/3}} \tag{6.12.15}$$

and X_3 is the abundance by mass of ^{3}He. At $T = 5 \times 10^6$ K and $\rho = 750$ g cm^{-3},

$$t_{33} \sim \frac{7.348 \times 10^{11}}{\rho X_3} \text{ yr.} \tag{6.12.16}$$

Under these conditions, ^{3}He effectively does not react with ^{4}He, and the equilibrium abundance of ^{3}He is given by

$$X_3^{\rm eq} = \frac{X}{\sqrt{2}} \exp\left(-27.6871 + \frac{44.467}{T_6^{1/3}}\right), \tag{6.12.17}$$

or $X_3^{\rm eq} \sim 0.0933$ at $T = 5 \times 10^6$ K and $\rho = 750$ g cm^{-3}. Thus, even when ^{3}He reaches its equilibrium abundance, its lifetime against destruction is of the order of 10^{10} yr. Since the abundance by mass of ^{3}He created in the big bang is only of the order of $X_3 \sim 2 \times 10^{-5}$, it is clear that, initially, $t_{33} \gg 10^{10}$ yr and one is justified in neglecting the destruction of ^{3}He in a zero age main sequence model. Accordingly, apart from a screening correction, the appropriate energy-generation rate is

$$\epsilon_{\rm pp}(\text{erg g}^{-1}\text{ s}^{-1}) = \epsilon_{11} \frac{R_{11}}{\rho_0} = X^2 \frac{\rho_0}{T_6^{2/3}} \exp\left(13.354 - \frac{33.803}{T_6^{1/3}}\right). \tag{6.12.18}$$

Given the high densities and low temperatures in the adiabatic models, it is reasonable to adopt the strong screening correction described by eqs. (6.11.13) and (6.11.29). Altogether, with $X = 0.71$, it follows that

$$\epsilon_{\rm pp}(\text{erg g}^{-1}\text{ s}^{-1}) = \frac{\rho_0}{T_6^{2/3}} \exp\left(12.669 + 0.2286 \frac{\rho_0^{1/3}}{T_6} - \frac{33.803}{T_6^{1/3}}\right). \tag{6.12.19}$$

Table 6.12.4 Characteristics of adiabatic models ($N = 3/2$ polytropes) for mass $M = \frac{1}{9} M_\odot$, molecular weight $\mu = 0.61$, and luminosity determined by the conversion of protons into ^{3}He

δ	ρ_c g cm^{-3}	T_c $10^6 K$	ϵ_c erg g^{-1} s^{-1}	k	s	$\frac{\epsilon_{average}}{\epsilon_c}$	log $L/L_\odot$	log T_e
20	183.3	4.037	0.03002	0.1072	5.937	0.08563	−3.544	3.258
30	326.0	4.523	0.07115	0.1160	5.799	0.08740	−3.158	3.396
40	473.3	4.678	0.13116	0.1270	5.690	0.08876	−2.888	3.490
60	759.1	4.891	0.28807	0.1421	5.545	0.09059	−2.538	3.6123
80	988.6	4.815	0.35775	0.1577	5.533	0.09042	−2.444	3.655
100	1101	4.458	0.26387	0.1765	5.650	0.08825	−2.587	3.627
200	1786	3.878	0.21270	0.2384	5.790	0.08494	−2.697	3.634
500	2466	2.610	0.02936	0.3944	6.334	0.07528	−3.609	3.429

Adopting the power law expression of eq. (5.3.29),

$$k = 0.0762 \, \frac{\rho_0^{1/3}}{T_6} \tag{6.12.20}$$

and

$$s = -\frac{2}{3} - 0.2286 \, \frac{\rho_0^{1/3}}{T_6} + \frac{11.268}{T_6^{1/3}}. \tag{6.12.21}$$

From eq. (5.3.37) and Table 5.3.1, one has that, for $N = 3/2$,

$$\epsilon_{average} = \frac{L(R)}{M(R)} = 0.163\,26 \, \epsilon_c \left[\frac{4}{(2+k) + 2s/3} \right]^{3/2} \left[1 - \frac{0.125}{(2+k) + 2s/3} \right], \tag{6.12.22}$$

where $L(R)$ and $M(R)$ are, respectively, the luminosity and mass of the model in solar units and ϵ_c is the energy generation rate at the center of the model.

The entries in Table 6.12.4 have been obtained by using eqs. (6.12.19)–(6.12.22) in conjunction with the model properties in Table 5.6.1. For example, inserting ρ_c and T_c in eq. (6.12.19) gives $\epsilon_c = \epsilon_{pp}(\rho_c, T_c)$ and eq. (6.12.22) with k from eq. (6.12.20) and s from eq. (6.12.21) gives $\epsilon_{average}$. The luminosity L follows from $\epsilon_{average}$ and the adopted model mass. Finally, the surface temperature T_e comes from L and the radius R in Table 5.6.1.

The binary pair L726-8 A & B have masses very close to $M = (1/9) \, M_\odot$. The observations suggest log $L/L_\odot \sim -2.8$ for component A, log $L/L_\odot \sim -3.0$ for component B, and log $T_e \sim 3.42$ for both stars. Comparing with model properties in Table 6.12.4, it would appear that models with δ in the range 30–40, corresponding to central densities in the range 300–500 g cm^{-3} and central temperatures in the range 4.5–4.7×10^6 K give a rough approximation to conditions at the centers of the observed pair.

Bibliography and references

C. Angulo, M. Arnould, M. Rayet, *et al., Nucl. Phys. A*, **656**, 3, 1999.

R. E. d'Atkinson & F. G. Houtermans, *Zeits. für Phys.*, **54**, 656, 1929.

Hans A. Bethe, *Phys. Rev.*, **55**, 434, 1939.

Hans A. Bethe & C. L. Critchfield, *Phys. Rev.*, **54**, 248, 1938.

Donald D. Clayton, *Principles of Stellar Evolution and Nucleosynthesis*, (New York: McGraw Hill), 1968.

Arthur S. Eddington, *The Internal Constitution of the Stars* (Cambridge: Cambridge University Press), 1926.

William A. Fowler, G. R. Caughlan, & B. A. Zimmerman, *ARAA*, **13**, 69, 1975.

E. Frieman & Lloyd Motz, *Phys. Rev.*, **83**, 202, 1951.

George Gamow, *Zeits. für Physik*, **52**, 510, 1928.

R. J. Gould & N. Guessoum, *ApJL*, **359**, L67, 1990.

Icko Iben, Jr., *ApJ*, **142**, 1447, 1965; **143**, 505, 1966; **147**, 624, 1967; **158**, 1033, 1969.

Icko Iben, Jr., Ken Kalata, & Judah Schwartz, *ApJ*, **150**, 1001, 1967.

M. Kamionkowski & John N. Bahcall, *ApJ*, **420**, 884, 1994.

Edwin E. Salpeter, *Phys. Rev.*, **88**, 547, 1952.

Edwin E. Salpeter, *Aust. J. Phys.*, **7**, 373, 1954.

Carl Friedrich von Weizsäcker, *Zeits. für Physik*, **39**, 633, 1938.

7

Photon–matter interaction probabilities, absorption cross sections, and opacity

An understanding of the manner in which matter and radiation interact is crucial for understanding how the structure of a star is influenced by the flow of energy. In this chapter, the physics of three processes whereby photons are absorbed by electrons interacting through the Coulomb potential with heavy ions is examined. The three processes are photo-ionization, inverse bremsstrahlung on free electrons, and transitions between bound atomic levels. Approximations to the cross sections for these processes are derived and the manner in which calculated cross sections are weighted to obtain the opacity under conditions of thermodynamic equilibrium is described and utilized in sample calculations of the opacity. Of primary interest here is not a presentation of definitive results, but rather a conceptual understanding of the basic ingredients of a quantitative calculation of absorption cross sections and of the related opacity.

An excellent monograph which describes the processes rigorously is *The Quantum Theory of Radiation* by Walter Heitler (1954), a pedagogically excellent text of relevance is *Quantum Mechanics* by Leonard I. Schiff (1949), and a delightfully intuitive approach to the calculation of transition probabilities is presented in *Quantum Electrodynamics*, based on lectures by Richard P. Feynman (1962). It would be remiss not to acknowledge the debt which the theory of quantum electrodynamics owes to Michael Faraday (1791–1867) and James Clerk Maxwell (1831–1879), the principle inventors of classical electrodynamics, as described in Maxwell's two volume *Treatise on Electricity and Magnetism* (1873).

In Section 7.1, the Hamiltonian for the interaction between charged particles and radiation is described. In Section 7.2, first order perturbation theory is employed to construct algorithms for calculating the transition probability for the emission or absorption of a photon by an atomic system. The relationship between emission and absorption probabilities is discussed in Section 7.3. In Section 7.4, the cross section for the ejection of the most tightly bound electron in a hydrogen-like ion by the absorption of a photon is calculated in an approximation which treats the ejected electron as a plane wave. This cross section is compared with the cross section obtained when the Coulomb interaction between the ejected electron and the daughter nucleus is explicitly taken into account. Cross sections for the photo-ejection of less tightly bound electrons are reported.

The absorption of a photon by a free electron in the presence of a heavy charged ion is called free–free absorption and, in the approximation that, both prior to and after absorption, the electron can be described by a plane wave, the process is a second order process involving intermediate states. In Section 7.5, second order perturbation theory is utilized to determine an appropriate composite matrix element and, in Section 7.6, summations over a Maxwell–Boltzmann electron distribution and over the charged ion distribution are performed to estimate the free–free cross section. In Section 7.7, comparisons are made with the Kramers semiclassical cross section for free–free absorption and with the cross section obtained when both the incident and exit electron are described by Coulomb-distorted plane waves.

Building on the development in Sections 7.2 and 7.3, spontaneous emission involving transitions between bound atomic states is discussed in Section 7.8. In Section 7.9, the principle of detailed balance is used to derive the phenomenon of stimulated emission and to outline procedures for determining bound–bound absorption cross sections. Also in Section 7.9, Doppler line broadening is described quantitatively and compared with the natural line width due to the relationship between lifetime and energy given by the Heisenberg uncertainty principle.

The Rosseland mean opacity, which provides a measure of the mean free path of an average photon in a medium in local thermodynamic equilibrium, is defined in Section 7.10 and, in Section 7.11, sample calculations of this opacity are carried through for a gaseous mixture of hydrogen, helium, and oxygen. A first calculation explores temperatures and densities such that hydrogen and helium are completely ionized, but oxygen may have up to two K-shell electrons. Bound–bound absorption is neglected and electron scattering is treated in the elementary form given by eqs. (3.4.16) and (3.4.17). In additional calculations, no restriction is placed on the ionization state of the selected elements, and it is shown that, at constant density, as temperature is decreased from high values such that electron scattering is the dominant contributor to the opacity ($\kappa_{Ross} \sim 0.4$ cm^2 g$^{-1}Y_e$, where Y_e is the electron number-abundance parameter), free–free and bound–free absorption causes the opacity to increase, reaching a maximum in excess of 10^4 cm^2 g^{-1} for densities $\rho > 10^{-4}$ g cm^{-3}, and then declines to nominal values as all elements become electrically neutral.

One of the more important demonstrations in Section 7.11 is that free–free absorption plays a crucial role in regions where, even though bound–free absorption coefficients are large at photon frequencies above threshold, thresholds are at energies comparable to the energy at the peak in the photon energy distribution. In the absence of free–free absorption, most of the photon energy flow would be diverted into frequencies below all bound–free absorption thresholds and the Rosseland mean opacity would be of the order of $\kappa_{Ross} \sim N_A \, \sigma_{TH} \sim 0.4$ cm^2 g^{-1}. However, over much of the density–temperature domain in stars, free–free absorption closes the low frequency window, thus permitting bound–free absorption to make a dominant contribution to the Rosseland mean opacity which it could not otherwise have made. Over an extensive range in temperature, the result is a mean opacity larger by as much as four orders of magnitude than what would be obtained in the absence of free–free absorption and larger by as much as two orders of magnitude than what would be obtained in the absence of bound–free absorption. This means that, as long as the cross section for free–free absorption at low frequencies is significantly larger than the electron-scattering cross section, the existence of the free–free absorption process is more important than the precise value of the free–free cross section.

For intermediate to high temperatures such that bound–bound absorption is not important, fairly accurate analytical approximations to results in the literature of detailed calculations of opacity for different composition choices are possible. Examples are discussed in Section 7.12 and, in Section 7.13, characteristics of the analytical approximations are used to determine a criterion for when a core nuclear burning star has a convective core and, if it does, to estimate the size of and mixing time in this core.

In density–temperature regions where bound–bound absorption is important, general analytical approximations to results in the literature are difficult to devise and it becomes a practical necessity to interpolate in tables which provide opacities at discrete points in the density–temperature plane (or in some equivalent plane) for a set of discrete choices for the composition. In Section 7.14, algorithms for interpolating in opacity tables are presented and, in Section 7.15, an example is given of interpolation in opacity tables for temperatures and densities such that bound–bound absorption and photo-ionization from excited states are important.

In the atmosphere of the Sun and of other cool stars at optical depths of the order of unity and smaller, the major source of opacity is absorption by the negative hydrogen ion, H^-. In Section 7.16, the abundance of H^- is examined as a function of density and temperature and as a function of the abundances of elements of low ionization potential. The cross section for photo-ionization of H^- is discussed and a sample calculation of the opacity due to bound–free and free–free absorption on H^- at very low temperatures (3000–5000 K) typical of surface temperatures of red giants and supergiants is performed.

7.1 Photons and the electron–photon interaction Hamiltonian

Both photons and atoms must be treated quantum mechanically. A photon in a box of volume V can be described by a vector potential

$$\mathbf{A} = \mathbf{a} \cos\left(\mathbf{k} \cdot \mathbf{r} - 2\pi \nu_k t\right), \tag{7.1.1}$$

where the photon momentum $\hbar \mathbf{k}$, wavelength λ_k, and frequencies ν_k and ω_k are related by

$$k = |\mathbf{k}| = \frac{2\pi}{\lambda_k} = \frac{2\pi \nu_k}{c} = \frac{\omega_k}{c}, \tag{7.1.2}$$

and $\mathbf{a}$, which specifies both amplitude and polarization, is a vector perpendicular to $\mathbf{k}$. The magnetic field $\mathbf{H}$ and electric field $\mathbf{E}$ are related to $\mathbf{A}$, respectively, by

$$\mathbf{H} = \nabla \times \mathbf{A} = -\mathbf{k} \times \mathbf{a} \sin\left(\mathbf{k} \cdot \mathbf{r} - \omega_k t\right), \tag{7.1.3}$$

and

$$\mathbf{E} = -\frac{1}{c}\frac{\partial \mathbf{A}}{\partial t} = -\frac{2\pi \nu_k}{c}\, \mathbf{a} \sin\left(\mathbf{k} \cdot \mathbf{r} - \omega_k t\right). \tag{7.1.4}$$

The vector potential may be normalized by equating the quantized energy of the photon to the energy in the associated electromagnetic field:

$$h\nu_k = \frac{1}{8\pi}\langle \mathbf{H} \cdot \mathbf{H} + \mathbf{E} \cdot \mathbf{E}\rangle V = \frac{1}{8\pi}\left(k^2 + \frac{\omega_k^2}{c^2}\right)\frac{a^2}{2} V = \frac{\omega_k^2\, V}{8\pi c^2}\, a^2, \tag{7.1.5}$$

where angle brackets denote time averages and $a = |\mathbf{a}|$. Thus,

$$a = |\mathbf{a}| = 2\sqrt{\frac{c^2\hbar}{\nu_k V}} \tag{7.1.6}$$

and

$$\mathbf{A} = \left(\frac{\mathbf{a}}{a}\right) 2\sqrt{\frac{c^2\hbar}{\nu_k V}} \cos\left(\mathbf{k} \cdot \mathbf{r} - \omega_k t\right) \tag{7.1.7}$$

$$= \sqrt{\frac{c^2\hbar}{\nu_k V}} \left(\hat{\mathbf{a}} e^{-\mathrm{i}(\omega_k t - \mathbf{k}\cdot\mathbf{r})} + \hat{\mathbf{a}}^* e^{\mathrm{i}(\omega_k t - \mathbf{k}\cdot\mathbf{r})}\right), \tag{7.1.8}$$

where

$$\hat{\mathbf{a}} = \frac{\mathbf{a}}{a} \tag{7.1.9}$$

is a unit vector that describes the polarization of the photon.

In quantum electrodynamics, the vector potential for the photon is considered to be an operator and may be written in a form analogous to eq. (7.1.8) as

$$\mathbf{A}_{k\lambda} = \sqrt{\frac{c^2\hbar}{\nu_k V}} \left(a_{k\lambda} e^{-\mathrm{i}(\omega_k t - \mathbf{k}\cdot\mathbf{r})} + a_{k\lambda}^\dagger e^{\mathrm{i}(\omega_k t - \mathbf{k}\cdot\mathbf{r})}\right) \hat{\epsilon}_{k\lambda}, \tag{7.1.10}$$

where $a_{k\lambda}$ is a destruction operator, $a_{k\lambda}^\dagger$ is a creation operator, and $\hat{\epsilon}_{k\lambda}$ is a unit polarization vector that is in a plane perpendicular to $\mathbf{k}$. The Hamiltonian for the radiation field is written as

$$H_{\mathrm{rad}} = \sum h\nu_k\, a_{k\lambda}^\dagger a_{k\lambda}. \tag{7.1.11}$$

After normalizing to do away with the zero-point energy, $h\nu_k/2$,

$$H_{\mathrm{rad}} |n_{k\lambda}\rangle = h\nu_k\, n_{k\lambda} |n_{k\lambda}\rangle, \tag{7.1.12}$$

where $|n_{k\lambda}\rangle$ is a photon state vector and $n_{k\lambda}$ is the number of photons in this state. In a bath of radiation and matter in statistical equilibrium, matrix elements for $a_{k\lambda}$ and $a_{k\lambda}^\dagger$ are

$$\langle n_{k\lambda} - 1 | a_{k\lambda} | n_{k\lambda}\rangle = \sqrt{n_{k\lambda}} \tag{7.1.13}$$

and

$$\langle n_{k\lambda} + 1 | a_{k\lambda}^\dagger | n_{k\lambda}\rangle = \sqrt{n_{k\lambda} + 1}. \tag{7.1.14}$$

A preliminary approximation to the Hamiltonian describing the interaction between the electromagnetic field and charged particles is derivable from elementary considerations. Consider a closed circular current loop with a magnetic dipole moment $\vec{\mu}$ given by

$$\vec{\mu} = \frac{i\mathbf{S}}{c}, \tag{7.1.15}$$

where i is the current, $\mathbf{S}$ is a vector perpendicular to the plane of the loop, and $|\mathbf{S}|$ is the area of the loop. In a constant magnetic field of strength $\mathbf{H}$, the interaction energy between the magnetic field and the loop is

$$H' = -\vec{\mu} \cdot \mathbf{H} = -\frac{i\mathbf{S}}{c} \cdot \mathbf{H} = -\frac{i}{c} \int \mathbf{H} \cdot d\mathbf{S} = -\frac{i}{c} \int \nabla \times \mathbf{A} \cdot d\mathbf{S}, = -\frac{i}{c} \oint \mathbf{A} \cdot d\mathbf{s}.$$

(7.1.16)

In these equations, $d\mathbf{S}$ is an element of area and the integration is carried out over any two-dimensional surface which is bounded by the loop, $d\mathbf{s}$ is a linear distance vector that is tangent to the loop at any point and the integral is carried once around the loop.

Writing

$$i d\mathbf{s} = \mathbf{j} d\tau,$$

(7.1.17)

where $\mathbf{j}$ is a current density and $d\tau$ is a volume element within the material of the loop, in a penultimate step the current loop is replaced by an electron of charge $-e$, mass m_e and momentum $\mathbf{p}$. Then

$$\mathbf{j} = -e \frac{\mathbf{p}}{m_e} \delta(\mathbf{r} - \mathbf{r}_e),$$

(7.1.18)

where $\mathbf{r}_e$ is the location of the electron and $\delta(\mathbf{r} - \mathbf{r}_e)$ is a three dimensional Dirac delta function (∞ when $\mathbf{r} = \mathbf{r}_e$, and 0 otherwise, and $\int_\tau \delta(\mathbf{r} - \mathbf{r}_e) \, d\tau = 1$). Thus, eq. (7.1.16) becomes

$$H' = \frac{e}{m_e c} \mathbf{p} \cdot \mathbf{A}.$$

(7.1.19)

In classical electrodynamics, the Hamiltonian for the interaction between an electron of charge $-e$ and the electromagnetic field is a generalization of this result:

$$H^{\text{int}} = \frac{1}{2m_e} \left(\mathbf{p} + \frac{e}{c}\mathbf{A} \right) \cdot \left(\mathbf{p} + \frac{e}{c}\mathbf{A} \right) - \frac{\mathbf{p} \cdot \mathbf{p}}{2m_e}$$

(7.1.20)

$$= \frac{e}{2m_e c} (\mathbf{p} \cdot \mathbf{A} + \mathbf{A} \cdot \mathbf{p}) + \frac{e^2}{2m_e c^2} \mathbf{A} \cdot \mathbf{A}.$$

(7.1.21)

The first term on the right side of eq. (7.1.21) is a variant of eq. (7.1.19). In quantum electrodynamics, $\mathbf{A}$ is replaced by $\mathbf{A}_{\text{op}}$, which is an operator given by a superposition of the quantities $\mathbf{A}_{k\lambda}$ described by eq. (7.1.10), and $\mathbf{p}$ is replaced by

$$\mathbf{p}_{\text{op}} = \frac{\hbar}{i} \nabla,$$

(7.1.22)

where $i = \sqrt{-1}$. In many applications, the term proportional to $\mathbf{A} \cdot \mathbf{A}$ in eq. (7.1.21) may be neglected, and the quantum-mechanical interaction Hamiltonian may be approximated by

$$H^{\text{int}} = \frac{e}{2m_e c} \left(\mathbf{p}_{\text{op}} \cdot \mathbf{A}_{\text{op}} + \mathbf{A}_{\text{op}} \cdot \mathbf{p}_{\text{op}} \right).$$

(7.1.23)

7.2 First order perturbation theory and the golden rule for a radiative transition between two matter eigenstates

The evolution in time of an atomic system undergoing a radiative transition can be explored with perturbation theory. Let the initial and final states of the matter system be described by the wave functions

$$\psi_i = u_i(\mathbf{r}) \, \exp\left(-i \frac{E_i}{\hbar} t\right) \tag{7.2.1}$$

and

$$\psi_f = u_f(\mathbf{r}) \, \exp\left(-i \frac{E_f}{\hbar} t\right), \tag{7.2.2}$$

where subscripts i and j refer to the initial and final states of the system and the coefficient i in the exponential functions is $i = \sqrt{-1}$. Assume that, during the transition, the wave function for the matter system can be described approximately by

$$\psi = \psi_i + a_f(t) \, \psi_f, \tag{7.2.3}$$

where $a_f(t)$ is a time-dependent amplitude which is initially zero. The development of the system is a solution of the equation

$$-\frac{\hbar}{i} \frac{\partial \psi}{\partial t} = (H_0 + H^{\text{int}}) \, \psi, \tag{7.2.4}$$

the left hand side of which gives

$$-\frac{\hbar}{i} \frac{\partial \psi}{\partial t} = E_i \, \psi_i + a_f \, E_f \, \psi_f - \frac{\hbar}{i} \frac{da_f}{dt} \, \psi_f, \tag{7.2.5}$$

and the right hand side of which gives

$$E_i \, \psi_i + a_f \, E_f \, \psi_f + H^{\text{int}} \, \psi_i + a_f \, H^{\text{int}} \, \psi_f. \tag{7.2.6}$$

Equating eqs. (7.2.5) and (7.2.6), multiplying both sides by ψ_f^*, and integrating over spatial coordinates gives

$$-\frac{\hbar}{i} \frac{da_f}{dt} = H_{fi}^{\text{int}} + a_f \, H_{ff}^{\text{int}}, \tag{7.2.7}$$

where

$$H_{fi}^{\text{int}} = \int \psi_f^*(\mathbf{r}, t) H^{\text{int}} \psi_i(\mathbf{r}, t) d\tau. \tag{7.2.8}$$

Neglecting the second term on the right hand side of eq. (7.2.7), one has finally that

$$\frac{da_f}{dt} = -\frac{i}{\hbar} \, H_{fi}^{\text{int}}. \tag{7.2.9}$$

As an example, consider an atomic transition from an excited state to the ground state, with the emission of a photon. Using eqs. (7.1.23) and (7.1.22) and the second term in eq. (7.1.10),

$$
H_{\mathrm{fi}}^{\mathrm{int}} = \int \psi_{\mathrm{f}}^* \sqrt{\frac{c^2\hbar}{v_k V}} \frac{e}{2m_e c} \left[\left(e^{\mathrm{i}(\omega_k t - \mathbf{k}\cdot\mathbf{r})} \hat{\epsilon}_{k\lambda} \right) \cdot \frac{\hbar}{\mathrm{i}} \nabla + \frac{\hbar}{\mathrm{i}} \nabla \cdot \left(e^{\mathrm{i}(\omega_k t - \mathbf{k}\cdot\mathbf{r})} \hat{\epsilon}_{k\lambda} \right) \right] \psi_{\mathrm{i}} \mathrm{d}\tau.
$$
(7.2.10)

Noting that $\hat{\epsilon}_{k\lambda} \cdot \mathbf{k} = 0$, the second term in square brackets gives the same result as the first and, using eqs. (7.2.1) and (7.2.2), one can write

$$
H_{\mathrm{fi}}^{\mathrm{int}} = \sqrt{\frac{c^2\hbar}{v_k V}} \frac{e}{m_e c} \left(\hat{\epsilon}_{k\lambda} \cdot \mathbf{M}_{\mathrm{fi}} \right) e^{\mathrm{i}(\omega_k - \omega_{\mathrm{if}})t} = \left| H_{\mathrm{fi}}^{\mathrm{int}} \right| e^{\mathrm{i}(\omega_k - \omega_{\mathrm{if}})t},
$$
(7.2.11)

where

$$
\mathbf{M}_{\mathrm{fi}} = \int u_{\mathrm{f}}^*(\mathbf{r}) \, e^{-\mathrm{i}\mathbf{k}\cdot\mathbf{r}} \mathbf{p} \, u_{\mathrm{i}}(\mathbf{r}) \mathrm{d}\tau
$$
(7.2.12)

and

$$
\omega_{\mathrm{if}} = \frac{E_{\mathrm{i}} - E_{\mathrm{f}}}{\hbar}.
$$
(7.2.13)

Inserting eq. (7.2.11) in eq. (7.2.9) and integrating over time, one obtains

$$
a_{\mathrm{f}} = -\frac{\mathrm{i}}{\hbar} \left| H_{\mathrm{fi}}^{\mathrm{int}} \right| \frac{e^{\mathrm{i}(\omega_k - \omega_{\mathrm{if}})t} - 1}{\mathrm{i}(\omega_k - \omega_{\mathrm{if}})},
$$
(7.2.14)

or

$$
|a_{\mathrm{f}}|^2 = \frac{1}{\hbar^2} \left| H_{\mathrm{fi}}^{\mathrm{int}} \right|^2 \left| \frac{e^{\mathrm{i}(\omega_k - \omega_{\mathrm{if}})t/2} - e^{-\mathrm{i}(\omega_k - \omega_{\mathrm{if}})t/2}}{\mathrm{i}(\omega_k - \omega_{\mathrm{if}})} \right|^2
$$

$$
= \frac{1}{\hbar^2} \left| H_{\mathrm{fi}}^{\mathrm{int}} \right|^2 \frac{4\sin^2[(\omega_k - \omega_{\mathrm{if}})t/2]}{(\omega_k - \omega_{\mathrm{if}})^2}.
$$
(7.2.15)

One might be tempted to interpret the quantity

$$
P_{\mathrm{f}\leftarrow\mathrm{i}} = |a_{\mathrm{f}}|^2 / t
$$
(7.2.16)

as the probability per unit time that a transition from state i to state f has occurred, but the quantity has the unsatisfactory property of being a function of time. This suggests that some important physics has been neglected. In fact, from the Heisenberg uncertainty principle it follows that there is an indeterminacy in the energy of the excited state of the atomic system and in the energy of the emitted photon; the problem lies with assuming, as in eq. (7.2.3), that the transition is between sharp energy states.

To rectify the situation, eq. (7.2.3) may be modified to read

$$
\psi = \psi_{\mathrm{i}} + \sum_k a_{k,\mathrm{f}}(t) \, \psi_{\mathrm{f}},
$$
(7.2.17)

where $a_{k,f}$ is the amplitude of the final state arrived at by the emission of a photon of energy $\hbar\omega_k$ and the sum is over photon states characterized by a range of energies corresponding to the energy width ΔE_i which can be assigned to the ith atomic level once a time-independent decay probability has been found. Then, the probability per unit time of a transition becomes

$$P_{f\leftarrow i} = \sum_k |a_{k,f}|^2 / t. \tag{7.2.18}$$

Standard practice is to replace the sum by an integral, making use of the fact that the number of photon states with frequencies in the interval $d\omega_k$, with directions in the solid angle $d\Omega_k$, and with positions in the interval dV is

$$\frac{d^3\mathbf{p}_k\, dV}{h^3} = \frac{(h\nu_k/c)^2\, d(h\nu_k/c)\, d\Omega_k\, dV}{h^3} = \frac{\nu_k^2\, d\nu_k}{c^3}\, d\Omega_k\, dV$$

$$= \frac{\nu_k^2\, d\omega_k}{2\pi c^3}\, d\Omega_k\, dV. \tag{7.2.19}$$

Thus, the differential probability per unit time

$$d^3 P_{f\leftarrow i} = \frac{1}{\hbar^2} \left| H_{fi}^{int} \right|^2 \frac{1}{t} \frac{4\sin^2\left[(\omega_k - \omega_{if})t/2\right]}{[\omega_k - \omega_{if}]^2} \frac{\nu_k^2\, d\omega_k}{c^3\, 2\pi}\, d\Omega_k\, dV \tag{7.2.20}$$

is obtained. Performing integrations over ω_k gives

$$\frac{d^2 P_{f\leftarrow i}}{d\Omega_k\, dV} = \frac{1}{\hbar^2} \frac{1}{c^3} \frac{1}{\pi} \int_{-\Delta\omega t/2}^{\Delta\omega t/2} \left| H_{fi}^{int} \right|^2 \nu_k^2 \frac{\sin^2\left[(\omega_k - \omega_{if})t/2\right]}{[(\omega_k - \omega_{if})t/2]^2}\, d\left[(\omega_k - \omega_{if})t/2\right], \tag{7.2.21}$$

where $\Delta\omega$ is a measure of the frequency width of the excited state.

The function

$$\Theta(\theta) = \frac{1}{\pi} \frac{\sin^2\theta}{\theta^2} \tag{7.2.22}$$

has the value $1/\pi$ when $\theta = 0$ and drops to its first zeros at $\theta = \pm\pi$. Integration from $-\infty$ to $+\infty$ gives

$$\int_{-\infty}^{\infty} \Theta(\theta)\, d\theta = 1, \tag{7.2.23}$$

with over ninety-five percent of this value coming from the region $-\pi \leq \theta \leq \pi$. Note that the area of a triangle of height $1/\pi$ and width 2π is $(1/2)\,(1/\pi)\,2\pi = 1$. The effective width of the function Θ translates into a relationship between an energy width and the duration t of the radiative transition,

$$\Delta\omega\, t \sim 2\pi \rightarrow \Delta E = \hbar\, \Delta\omega \sim \frac{h}{t}, \tag{7.2.24}$$

which is consistent with the Heisenberg uncertainty principle.

Returning to the integral in eq. (7.2.21), if one chooses a value for $\Delta\omega$ such that $\Delta\omega/\omega_{if} \ll 1$ but also such that $\Delta\omega > 1/t$, one may take the factor $|H_{fi}^{int}|^2 \, v_k^2$ outside of the integral and use eq. (7.2.23) to obtain

$$d^2 P_{f\leftarrow i} = \frac{1}{\hbar^2} \frac{1}{c^3} \left| H_{fi}^{int} \right|^2 v_k^2 \, d\Omega_k \, dV = \frac{2\pi}{\hbar} \left| H_{fi}^{int} \right|^2 \frac{v_k^2}{hc^3} \, d\Omega_k \, dV, \qquad (7.2.25)$$

or

$$d^2 P_{f\leftarrow i} = \frac{2\pi}{\hbar} \left| H_{fi} \right|^2 \, d^2\rho_f, \qquad (7.2.26)$$

where

$$d^2\rho_f = \frac{d}{dE_f} \frac{v_k^2 \, dv_k \, d\Omega_k \, dV}{c^3} = \frac{d(v_k^2 \, dv_k)}{d(hv_k)} \frac{d\Omega_k \, dV}{c^3} = \frac{v_k^2}{hc^3} \, d\Omega_k \, dV \qquad (7.2.27)$$

is the differential density of final states. Equation (7.2.26) is sometimes called Fermi's golden rule.

Further insight may be gained by taking the decay of the initial state explicitly into account. First, replace eq. (7.2.17) by

$$\psi = \psi_i \, e^{-t/2\tau_i} + \sum_k a_{k,f}(t) \, \psi_f, \qquad (7.2.28)$$

where τ_i is the lifetime of the initial state defined tentatively by

$$\tau_i = \frac{1}{P_{f\leftarrow i}}. \qquad (7.2.29)$$

Then, from eqs. (7.2.8) and (7.2.9), one obtains

$$a_{k,f} = -\frac{i}{\hbar} \left| H_{fi}^{int} \right| \frac{e^{i(\omega_k - \omega_{if})t} \, e^{-t/2\tau_i} - 1}{i(\omega_k - \omega_{if}) - 1/2\tau_i} \qquad (7.2.30)$$

and

$$|a_{k,f}|^2 = \frac{1}{\hbar^2} \left| H_{fi}^{int} \right|^2 \frac{D(\omega_k - \omega_{if})}{(\omega_k - \omega_{if})^2 + 1/(2\tau_i)^2}, \qquad (7.2.31)$$

where

$$D(\omega_k - \omega_{if}) = 1 - e^{-t/2\tau_i} \left(e^{i(\omega_k - \omega_{if})t} + e^{-i(\omega_k - \omega_{if})t} \right) + e^{-t/\tau_i}$$

$$= 1 - 2 \, e^{-t/2\tau_i} \cos\left[(\omega_k - \omega_{if})t \right] + e^{-t/\tau_i}. \qquad (7.2.32)$$

Performing the sum (integral) over final photon frequencies, one obtains

$$\sum_k |a_{k,f}|^2 = \frac{1}{\hbar^2} \int_V \int_\Omega \int_v \left| H_{fi}^{int} \right|^2 \frac{D(\omega_k - \omega_{if})}{(\omega_k - \omega_{if})^2 + 1/(2\tau_i)^2} \frac{v_k^2}{c^3} \, dv_k \, d\Omega_k \, dV. \qquad (7.2.33)$$

Changing the integration variable from ν_k to $\omega_k - \omega_{if}$, and rearranging, one has

$$\sum_k |a_{k,\mathrm{f}}|^2 = \frac{\tau_i}{\hbar^2} \int_V \int_\Omega \int_\omega \left|H_{\mathrm{fi}}^{\mathrm{int}}\right|^2 \frac{1}{\pi}$$
$$\times \left[\frac{\nu_k^2}{c^3} D(\omega_k - \omega_{if}) \frac{(1/2\tau_i)\,\mathrm{d}(\omega_k - \omega_{if})}{(\omega_k - \omega_{if})^2 + 1/(2\tau_i)^2}\right] \mathrm{d}\Omega_k\,\mathrm{d}V. \qquad (7.2.34)$$

Then, using the fact that

$$\frac{1}{\pi} \int_{-\infty}^\infty \frac{x_0\,\mathrm{d}x}{x^2 + x_0^2} = 1, \qquad (7.2.35)$$

and evaluating $|H_{\mathrm{fi}}^{\mathrm{int}}|^2\,\nu_k^2\,D(\omega_k - \omega_{if})$ at the origin, it follows that

$$\sum_k |a_{k,\mathrm{f}}|^2 = \tau_i \frac{1}{\hbar^2} \int_V \int_\Omega \left|H_{\mathrm{fi}}^{\mathrm{int}}\right|^2 \frac{\nu_k^2}{c^3}\,\mathrm{d}\Omega_k\,\mathrm{d}V \left(1 - \mathrm{e}^{-t/2\tau_i}\right)^2. \qquad (7.2.36)$$

But, from eqs. (7.2.26) and (7.2.29), the double integral between τ_i and the quantity in parentheses squared in eq. (7.2.36) is just $1/\tau_i$, so

$$\sum_k |a_{k,\mathrm{f}}|^2 = \left(1 - \mathrm{e}^{-t/2\tau_i}\right)^2, \qquad (7.2.37)$$

demonstrating that, effectively,

$$\psi = \psi_i\,\mathrm{e}^{-t/2\tau_i} + \psi_\mathrm{f}\left(1 - \mathrm{e}^{-t/2\tau_i}\right), \qquad (7.2.38)$$

thus reinforcing the interpretation in eq. (7.2.29) of the lifetime τ_i as the reciprocal of $P_{\mathrm{f}\leftarrow\mathrm{i}}$.

7.3 The relationship between emission and absorption probabilities

Using the relationship between $(\hat{\epsilon}_{k\lambda} \cdot \mathbf{M}_{fi})$ and H_{fi} given in eq. (7.2.11) and integrating over volume, eq. (7.2.25) becomes

$$\frac{\mathrm{d}P_{\mathrm{f}\leftarrow\mathrm{i}}}{\mathrm{d}\Omega_k} = \frac{1}{\hbar}\frac{1}{c^3}\left(\frac{e}{m_\mathrm{e}}\right)^2 \left(\hat{\epsilon}_{k\lambda} \cdot \mathbf{M}_{\mathrm{fi}}\right)^2 \nu_k. \qquad (7.3.1)$$

In the dipole approximation, the exponential $\mathrm{e}^{-\mathrm{i}\mathbf{k}\cdot\mathbf{r}}$ in the matrix element defined by eq. (7.2.12) is approximated by unity and (as is developed in Section 7.6), the matrix element can be written as

$$\mathbf{M}_{\mathrm{fi}} = \mathbf{p}_{\mathrm{fi}} = m_\mathrm{e}\,\omega_{\mathrm{fi}}\,\mathbf{r}_{\mathrm{fi}}, \qquad (7.3.2)$$

where $\mathbf{r}_{\mathrm{fi}}$ is the matrix element of the radius vector. Then,

$$\frac{\mathrm{d}P_{\mathrm{f}\leftarrow\mathrm{i}}}{\mathrm{d}\Omega_k} = \frac{e^2}{\hbar c}\frac{1}{c^2}\left(\hat{\epsilon}_{k\lambda} \cdot \mathbf{r}_{\mathrm{fi}}\right)^2 \omega_k^2\,\nu_k. \qquad (7.3.3)$$

Continuing,

$$P_{f \leftarrow i} = \frac{e^2}{\hbar c} \frac{1}{c^2} \frac{\omega_k^3}{2\pi} \sum_{k\lambda=1}^{2} \int \left(\hat{\epsilon}_{k\lambda} \cdot \mathbf{r}_{fi}\right)^2 d\Omega_k, \qquad (7.3.4)$$

where the sum is over two states of polarization. As shown in Section 7.6,

$$\sum_{k\lambda=1}^{2} \int \left(\hat{\epsilon}_{k\lambda} \cdot \mathbf{r}_{fi}\right)^2 d\Omega_k = r_{if}^2 \, 2\pi \int_{-1}^{1} \sin^2 \theta \, d(\cos\theta) = \frac{8\pi}{3} r_{if}^2, \qquad (7.3.5)$$

so the probability per unit time of emission (in the dipole approximation) is

$$P_{f \leftarrow i}^{\text{emission}} = \frac{e^2}{\hbar c} \frac{1}{c^2} \frac{\omega_k^3}{2\pi} \frac{8\pi}{3} r_{if}^2 = \frac{e^2}{\hbar c} \frac{\omega_k^3}{c^2} r_{fi}^2 \frac{4}{3}. \qquad (7.3.6)$$

The calculation of an absorption probability develops in exactly the same way as the calculation of an emission probability. The matrix element relevant for absorption of a photon and a transition of the atomic system from state f to state i is

$$H_{if}^{\text{int}} = \int \psi_i^* \sqrt{\frac{c^2 \hbar}{v_k V}} \frac{e}{m_e c} e^{-i(\omega_k t - \mathbf{k} \cdot \mathbf{r})} \hat{\epsilon}_{k\lambda} \cdot \frac{\hbar}{i} \nabla \psi_f d\tau, \qquad (7.3.7)$$

which can be written as

$$H_{if}^{\text{int}} = \sqrt{\frac{c^2 \hbar}{v_k V}} \frac{e}{m_e c} \left(\hat{\epsilon}_{k\lambda} \cdot \mathbf{M}_{if}\right) e^{i(-\omega_k - \omega_{fi})t} = \left|H_{if}^{\text{int}}\right| e^{-i(\omega_k - \omega_{if})t}, \qquad (7.3.8)$$

where ω_{if} is given by eq. (7.2.13) and

$$\mathbf{M}_{if} = \int u_i^*(\mathbf{r}) \, e^{-i\mathbf{k} \cdot \mathbf{r}} \mathbf{p} \, u_f(\mathbf{r}) d\tau. \qquad (7.3.9)$$

Integration by parts gives $|M_{if}| = |M_{fi}|$ and so $|H_{if}^{\text{int}}| = |H_{fi}^{\text{int}}|$. Proceeding with perturbation theory as before, performing an integration of the perturbation over time and summing over a range of incident frequencies to take into account the finite energy width of state i, one arrives again at Fermi's golden rule, eq. (7.2.26). Thus, the probability per unit time for the absorption of a single photon and a transition of an isolated atomic system from state f to state i is exactly equal to the probability per unit time for the emission of a photon and a transition of the atomic system from state i to state f, provided one is clever enough to arrange for the incident photon beam to have the same characteristics as the photon emitted in the transition i to f, namely, traveling as a wave packet of energy width comparable to the width of the excited atomic state and impinging on the atom from all directions with equal probability. This is possible in a system of many atoms and photons in thermodynamic equilibrium, but it does not address the situation in which the photon is initially traveling in a given direction and the atom is oriented randomly with respect to the polarization vector of the photon.

To deal with this latter situation, the normalization of the electrodynamic vector potential may be modified. Since the excited state is characterized by a spread in energies, the incoming photon may be described by an intensity $I(\omega)$ which has the property that, when integrated over all frequencies, $I(\omega)\,d\omega$ is equal to the time average flux of energy given by the Poynting vector

$$\mathbf{S} = \frac{c}{4\pi}\,\mathbf{E}\times\mathbf{H} = \frac{c}{4\pi}\left(\frac{\omega}{c}\right)^2 A^2\,\hat{\mathbf{k}}, \tag{7.3.10}$$

where $\hat{\mathbf{k}}$ is a unit vector in the k direction. For the vector potential one may write

$$\mathbf{A} = \left(\frac{\mathbf{a}}{a}\right)N\,2\,\cos(\mathbf{k}\cdot\mathbf{r}-\omega_k t) \tag{7.3.11}$$

and, taking a time average of A^2, N may be chosen in such a way that

$$\frac{c}{4\pi}\left(\frac{\omega_k}{c}\right)^2 2N^2 = I(\omega_k)\,d\omega_k, \tag{7.3.12}$$

or

$$N^2 = \left(\frac{2\pi c}{\omega_k^2}\right)I(\omega_k)\,d\omega_k. \tag{7.3.13}$$

Replacing $\left(c^2\hbar/\nu_k V\right)$ by N^2 in eq. (7.3.8), one has

$$H_{\mathrm{if}}^{\mathrm{int}} = \left(\frac{2\pi c}{\omega_k^2}\right)I(\omega_k)\,d\omega_k\,\frac{e}{m_e c}\left(\hat{\epsilon}_{k\lambda}\cdot\mathbf{M}_{\mathrm{if}}\right)e^{-\mathrm{i}(\omega_k-\omega_{\mathrm{if}})t}. \tag{7.3.14}$$

Choosing again a total wave function of the form given by eq. (7.2.28), but with the subscripts i and f interchanged, one obtains

$$\sum_k |a_{k,\mathrm{i}}|^2/t = \frac{1}{\hbar^2}\int_\omega\left(\frac{2\pi c}{\omega_k^2}\right)I(\omega_k)\,d\omega_k\left(\frac{e}{m_e c}\right)^2\left(\hat{\epsilon}_{k\lambda}\cdot\mathbf{M}_{\mathrm{if}}\right)^2\frac{4\sin^2[(\omega_k-\omega_{\mathrm{if}})t/2]}{(\omega_k-\omega_{\mathrm{if}})^2}\frac{1}{t} \tag{7.3.15}$$

$$= \frac{4\pi c}{\hbar^2}\left(\frac{e}{m_e c}\right)^2\left(\hat{\epsilon}_{k\lambda}\cdot\mathbf{M}_{\mathrm{if}}\right)^2\int_{-\Delta\omega\,t/2}^{\Delta\omega\,t/2}\frac{I(\omega_k)}{\omega_k^2}$$

$$\times\sin^2\frac{[(\omega_k-\omega_{\mathrm{if}})t/2]}{[(\omega_k-\omega_{\mathrm{if}})t/2]^2}\,d[(\omega_k-\omega_{\mathrm{if}})t/2] \tag{7.3.16}$$

$$= \frac{2\pi c}{\hbar^2}\left(\frac{e}{m_e c}\right)^2\left(\hat{\epsilon}_{k\lambda}\cdot\mathbf{M}_{\mathrm{if}}\right)^2\frac{I(\omega_k)}{\omega_k^2}\,\pi. \tag{7.3.17}$$

Then, using eq. (7.3.2),

$$P_{\mathrm{f}\to\mathrm{i}} = 4\pi^2\,\frac{e^2}{\hbar c}\left(\hat{\epsilon}_{k\lambda}\cdot\mathbf{r}_{\mathrm{if}}\right)^2\frac{\omega_k I(\omega_k)}{\hbar\omega_k}. \tag{7.3.18}$$

The evaluation of $\left(\hat{\epsilon}_{k\lambda}\cdot\mathbf{r}_{\mathrm{if}}\right)^2$ in the case of the absorption of a unidirectional photon of a given polarization is different from the evaluation in the case of emission from an isolated atom. In the latter case, which is hereinafter called spontaneous emission, one does a *sum* over the polarization states and possible directions of the emitted photon, whereas, in the case of absorption from an incident beam, one does an *average* over all possible initial orientations of the atom. Thus,

$$\left(\hat{\epsilon}_{k\lambda} \cdot \mathbf{r}_{if}\right)^2 = r_{if}^2 \frac{1}{4\pi} \int \cos^2 \theta \, d\Omega = \frac{1}{3} r_{if}^2, \tag{7.3.19}$$

giving, finally,

$$P_{f \to i}^{absorption} = \frac{4\pi^2}{3} \frac{e^2}{\hbar c} r_{if}^2 \frac{\omega_k I(\omega_k)}{\hbar \omega_k} \tag{7.3.20}$$

as the probability per unit time of absorption of a photon from a beam of intensity $I(\omega_k)$ per unit frequency in a transition between bound atomic states.

In order to compare a bound–bound absorption probability with probabilities for other kinds of absorption processes, it is convenient to express the absorption probabilities in terms of cross sections. As shown in Section 7.9 (eq. (7.9.21)), when line broadening is neglected, the cross section for absorption $\sigma_{f \to i}(\nu)$ is related to the probability of spontaneous emission $P_{f \leftarrow i}^{spontaneous} = 1/t_{fi}$ by

$$\sigma_{f \to i}(\nu) = \frac{1}{8\pi} \frac{c^2}{\nu^2} P_{f \leftarrow i}^{spontaneous} \delta(\nu - \nu_{fi}), \tag{7.3.21}$$

where $\delta(\nu - \nu_{fi})$ is a delta function or an approximation to a delta function which has the dimension of $1/\nu$, and $P_{f \leftarrow i}^{spontaneous}$ is the probability of emission from state i to state f in the absence of an ambient radiation field. Sources of line broading such as the Doppler effect and pressure or collisional broadening can be taken into account by choosing the function $\delta(\nu - \nu_{fi})$ appropriately, as described, e.g., by R. G. Breene, Jr. (1957) and P. W. Anderson (1949).

Inserting $P_{f \leftarrow i}$ from eq. (7.3.6) into eq. (7.3.21), one has

$$\sigma_{f \to i}(\nu) = \frac{4\pi^2}{3} \frac{e^2}{\hbar c} r_{if}^2 \nu \, \delta(\nu - \nu_{fi}). \tag{7.3.22}$$

To obtain the rate of absorption from a photon beam, the cross section is multiplied by the flux of photons and an integration over photon frequency is performed. Thus, in the current example,

$$P_{f \to i}^{absorption} = \int_\omega \frac{I(\omega)d\omega}{\hbar \omega} \sigma(\omega) = \int_\omega \frac{I(\omega)d\omega}{\hbar \omega} \frac{4\pi^2}{3} \frac{e^2}{\hbar c} r_{if}^2 \nu \, \delta(\nu - \nu_{fi})$$

$$= \frac{4\pi^2}{3} \frac{e^2}{\hbar c} r_{if}^2 \frac{\omega \, I(\omega)}{\hbar \omega}, \tag{7.3.23}$$

in agreement with eq. (7.3.20). The reason for using the word spontaneous in the superscript for the emission probability in eq. (7.3.21) is that, as shown in the next paragraph, emission can also be induced, or stimulated, by the presence of an incident photon beam.

A change in sign of the exponent in eq. (7.3.8) corresponds to a change from absorption to emission. Repeating the development beginning with eq. (7.3.10), it is evident that the probability of the emission of a photon due to the existence of an incident beam is

$$P_{i \to f}^{stimulated} = \frac{4\pi^2}{3} \frac{e^2}{\hbar c} r_{if}^2 \frac{\omega_k I(\omega_k)}{\hbar \omega_k}, \tag{7.3.24}$$

which is precisely the same as the probability of absorption from the beam, as given by eq. (7.3.20). In other words, for a given beam intensity, the probability that an atom in a given state of energy E_f makes a transition to a state of energy E_i by absorbing a photon of energy $h\nu = E_i - E_f$ is exactly equal to the probability that an atom in the state of energy E_i makes a transition to a state of energy E_f by emitting a photon of energy $h\nu$ due to the presence of the beam.

If the beam impinges on a target of thickness dx which consists of a collection of atoms in thermodynamic equilibrium at temperature T, with a number density n_f of atoms in the state f, on leaving the target, the intensity of the beam will have been diminished by

$$dI = -I \frac{4\pi^2}{3} \frac{e^2}{\hbar c} r_{if}^2 (n_f - n_i) \, dx = -I \frac{4\pi^2}{3} \frac{e^2}{\hbar c} r_{if}^2 n_f \left(1 - e^{-h\nu/kT}\right) dx, \quad (7.3.25)$$

where n_f is the number density of atoms in the state f, and the fact that $n_i/n_f = e^{-h\nu/kT}$ has been used. Equation (7.3.25) expresses one of the key operations one must perform in calculating an opacity, namely, the multiplication of absorption cross sections by the factor $(1 - e^{-h\nu/kT})$. A more complete discussion of stimulated emission and its role in the calculation of an opacity is given in Section 7.9.

In Section 7.4, the cross section for bound–free absorption is estimated, in Section 7.5, the second-order matrix element for a free–free transition is constructed, and in Section 7.6, the cross section for free–free absorption is estimated. Dividing a transition probability by the particle flux gives the associated cross section. Beginning with

$$\text{energy flux} = \frac{c}{4\pi} |\mathbf{E} \times \mathbf{H}| = \frac{c}{4\pi} \frac{\omega}{c} k \frac{1}{2} a^2 = \frac{1}{8\pi} \frac{\omega^2}{c} a^2, \quad (7.3.26)$$

and using the normalization of the vector potential as in eq. (7.1.5), one has

$$\text{particle flux} = \frac{\text{energy flux}}{h\nu} = \frac{1}{8\pi} \frac{\omega^2}{c} a^2 \div \frac{1}{8\pi} \frac{\omega^2}{c^2} a^2 V = \frac{c}{V}. \quad (7.3.27)$$

The probability that an atomic system undergoes a transition from state ψ_i to state ψ_f with either the emission or absorption of a single photon is given by an average over initial states and a sum or integral over final states of the differential probability

$$d^2 P_{i \rightarrow f} = \frac{2\pi}{\hbar} |M|^2 \, d^2\rho_f, \quad (7.3.28)$$

where, in the case of absorption,

$$M = \int \psi_f^* \langle n_{k\lambda} - 1 | H^{\text{int}} | n_{k\lambda} \rangle \psi_i d\tau = \sqrt{n_{k\lambda}} \, M_{fi}^+, \quad (7.3.29)$$

and, in the case of emission,

$$M = \int \psi_f^* \langle n_{k\lambda} + 1 | H^{\text{int}} | n_{k\lambda} \rangle \psi_i d\tau = \sqrt{n_{k\lambda} + 1} \, M_{fi}^-, \quad (7.3.30)$$

where

$$M_{\mathrm{fi}}^{\pm} = \int \psi_{\mathrm{f}}^* \sqrt{\frac{c^2\hbar}{\nu_k V}} \frac{e}{2m_e c} \left((e^{\pm i\mathbf{k}\cdot\mathbf{r}}\hat{\epsilon}_{k\lambda}) \cdot \frac{\hbar}{\mathrm{i}}\nabla + \frac{\hbar}{\mathrm{i}}\nabla \cdot (e^{\pm i\mathbf{k}\cdot\mathbf{r}}\hat{\epsilon}_{k\lambda}) \right) \psi_{\mathrm{i}} \, d\tau, \qquad (7.3.31)$$

and the coefficients of M_{fi}^+ and M_{fi}^- follow from eqs. (7.1.13) and (7.1.14).

Averaging over all initial photon polarizations and summing over all final states gives the total probability $P_{\mathrm{i}\to\mathrm{f}}(\nu)$ for the absorption of a photon of frequency ν by the particular process under consideration, and division by the flux of one photon in a box, as described by eq. (7.3.27), gives the cross section for that process. Thus,

$$\sigma_{\mathrm{i}\to\mathrm{f}}(\nu) = P_{\mathrm{i}\to\mathrm{f}}(\nu) \frac{V}{c}. \qquad (7.3.32)$$

7.4 The cross section for bound–free (photoelectric) absorption from the K shell

In this section, the probability that an ion absorbs a photon and ejects an electron is estimated on the assumption that the electron can be described by a plane wave. Attention is focussed on the case of an ion with one electron in its ground state. The resultant cross section is zero at the threshold photon energy, but rises rapidly to approach the Born approximation cross section which results when photon energy is large compared with the ionization potential of the ion. When the electron is described by a Coulomb-distorted plane wave, the resultant cross section is finite at threshold and rises slowly with photon energy, remaining smaller than the Born approximation cross section by a factor of 2–1.5.

The initial state consists of one ion in a box of size V and one linearly polarized photon characterized by momentum $\hbar\mathbf{k}$ and polarization $\hat{\epsilon}_{k\lambda}$. The wave function describing the bound electron is

$$u_{\mathrm{i}} = \frac{1}{\sqrt{\pi a^3}} \exp\left(-\frac{r}{a}\right), \qquad (7.4.1)$$

where $a = a_0/Z$ and Z is the nuclear charge. The final state consists of a nucleus and a free electron. By ignoring the Coulomb interaction between the electron and the nucleus, the wave function of the free electron can be approximated by a plane wave:

$$u_{\mathrm{f}} = \frac{1}{\sqrt{V}} \exp\left(\mathrm{i}\,\mathbf{p}\cdot\mathbf{r}/\hbar\right). \qquad (7.4.2)$$

The differential probability of the reaction taking place, with the electron being ejected into the solid angle $d\Omega$, has the form of eq. (7.2.26), with the matrix element between initial and final states given by

$$H_{\mathrm{f}\leftarrow\mathrm{i}}^{\mathrm{int}} = \int u_{\mathrm{f}}^* \langle 0_{k\lambda} | H^{\mathrm{int}} | 1_{k\lambda} \rangle u_{\mathrm{i}} d\tau \qquad (7.4.3)$$

$$= \int u_{\mathrm{f}}^* \sqrt{\frac{c^2\hbar}{\nu_k V}} \frac{e}{2m_e c} \left((e^{\mathrm{i}\mathbf{k}\cdot\mathbf{r}}\hat{\epsilon}_{k\lambda}) \cdot \frac{\hbar}{\mathrm{i}}\nabla + \frac{\hbar}{\mathrm{i}}\nabla \cdot (e^{\mathrm{i}\mathbf{k}\cdot\mathbf{r}}\hat{\epsilon}_{k\lambda}) \right) u_{\mathrm{i}} d\tau. \qquad (7.4.4)$$

The density of final states, after integration over volume, is given by

$$d\rho_f = 2\frac{Vp^2\,dp\,d\Omega}{h^3}\Big/dE = 2\frac{V}{h^3}m_e p\,d\Omega, \qquad (7.4.5)$$

where the factor of 2 comes from the two possible polarization states of the ejected electron. Since $\hat{\epsilon}_{k\lambda}\cdot\mathbf{k}=0$, the second term in the integrand of eq. (7.4.4) gives the same result as the first. Integration by parts, taking into account the fact that u_i vanishes at $r=\infty$, gives

$$H^{int}_{f\leftarrow i} = -\sqrt{\frac{c^2\hbar}{v_k V}}\frac{e}{m_e c}\frac{\hbar}{i}\int u_i e^{i\mathbf{k}\cdot\mathbf{r}}\hat{\epsilon}_{k\lambda}\cdot\nabla u_f^*d\tau. \qquad (7.4.6)$$

Using eqs. (7.1.29), (7.4.1), and (7.4.2),

$$H^{int}_{f\leftarrow i} = -\sqrt{\frac{c^2\hbar}{v_k V}}\frac{e}{m_e c}\frac{\hbar}{i}\int\left(\frac{1}{\sqrt{\pi a^3}}e^{-r/a}e^{i\mathbf{k}\cdot\mathbf{r}}\hat{\epsilon}_{k\lambda}\cdot\nabla\frac{1}{\sqrt{V}}e^{-i\mathbf{p}\cdot\mathbf{r}/\hbar}\right)d\tau \qquad (7.4.7)$$

$$= \sqrt{\frac{c^2\hbar}{v_k V}}\frac{e}{m_e c}\frac{1}{\sqrt{V}}\frac{1}{\sqrt{\pi a^3}}(\hat{\epsilon}_{k\lambda}\cdot\mathbf{p})\int e^{-i\mathbf{p}\cdot\mathbf{r}/\hbar}e^{i\mathbf{k}\cdot\mathbf{r}}e^{-r/a}d\tau. \qquad (7.4.8)$$

Defining a vector $\Delta\mathbf{k}$ by

$$\Delta\mathbf{k} = \mathbf{k} - \frac{\mathbf{p}}{\hbar}, \qquad (7.4.9)$$

and writing $\Delta\mathbf{k}\cdot\mathbf{r} = \Delta k\,r\,\cos\theta$, where θ is the angle between $\mathbf{k}$ and $\mathbf{r}$, one has

$$\int e^{-i\mathbf{p}\cdot\mathbf{r}/\hbar}e^{i\mathbf{k}\cdot\mathbf{r}}e^{-r/a}d\tau = \int_0^\infty\int_0^\pi\int_0^{2\pi} e^{i\Delta k\,r\cos\theta}e^{-r/a}r^2 dr\,\sin\theta\,d\theta\,d\phi \qquad (7.4.10)$$

$$= 2\pi\int_0^\infty\int_{-1}^1 e^{i\Delta k\,r\mu}e^{-r/a}r^2 dr\,d\mu$$

$$= \frac{2\pi}{i\Delta k}\int_0^\infty\left(e^{i\Delta k\,r}-e^{-i\Delta k\,r}\right)e^{-r/a}r\,dr$$

$$= \frac{8\pi}{a(\Delta k^2+1/a^2)^2}. \qquad (7.4.11)$$

From eq. (7.4.9),

$$\Delta k^2 = k^2 + \frac{p^2}{\hbar^2} - \frac{2kp\cos\theta_{kp}}{\hbar}, \qquad (7.4.12)$$

where θ_{kp} is the angle between the momentum of the ejected electron and the momentum of the incident photon. From conservation of energy

$$h\nu = c\hbar k = \frac{p^2}{2m_e} + \frac{Ze^2}{2a} = \frac{\hbar^2}{2m_e}\left(\frac{p^2}{\hbar^2}+\frac{1}{a^2}\right). \qquad (7.4.13)$$

Eliminating $p^2/\hbar^2$ from eqs. (7.4.12) and (7.4.13) gives

$$\Delta k^2 + \frac{1}{a^2} = \frac{2m_e c k}{\hbar}\left(1 + \frac{h\nu}{2m_e c^2} - \frac{v}{c}\cos\theta_{kp}\right),\tag{7.4.14}$$

where $v/c = \sqrt{2E/m_e c^2}$ is the speed of the electron relative to the speed of light and $h\nu = c\hbar k$ is the photon energy.

Neglecting $h\nu/2m_e c^2$ and v/c compared with 1, the probability per unit time of a reaction taking place is

$$d\omega_{f\leftarrow i} = \frac{2\pi}{\hbar}\,|H_{fi}|^2\,d\rho_f$$

$$= \frac{2\pi}{\hbar}\frac{c^2\hbar}{v_k}\left(\frac{e}{m_e c}\right)^2\frac{2\pi}{a^5}\left(\frac{\hbar}{m_e c k}\right)^4\frac{1}{V^2}(\hat{\epsilon}_{k\lambda}\cdot\mathbf{p})^2\,2\,\frac{V}{h^3}m_e p\,d\Omega.\tag{7.4.15}$$

Integration of $(\hat{\epsilon}_{k\lambda}\cdot\mathbf{p})^2$ over all solid angles yields $(4\pi/3)p^2$, so

$$\omega_{f\leftarrow i}(\nu) = \frac{32}{3}\pi^3\frac{c}{v_k}\left(\frac{e}{m_e c}\right)^2\frac{1}{a^5}\left(\frac{\hbar}{m_e c k}\right)^4\frac{m_e}{h^3}\,p^3\frac{c}{V}.\tag{7.4.16}$$

Division by the incident photon flux c/V (eq. (7.3.27)) gives the cross section for the process, as expressed by eq. (7.3.32). In short, the cross section $\sigma(\nu)$ for the process is the coefficient of c/V in eq. (7.4.16).

The physics involved in producing the cross section can be clarified by making use of relationships between fundamental constants and between various physical quantities. In particular, it is helpful to make the replacements $ck = 2\pi\nu$ and $p = \sqrt{2m_e E}$ and to use the fact that the ionization energy I_K, the charge Z, and the atomic radius a are related by

$$I_K = \frac{Ze^2}{2a} = \frac{\hbar^2}{2m_e a^2} = Z^2\frac{e^2}{2a_0},\tag{7.4.17}$$

so that

$$\frac{1}{a} = \frac{2I_K}{Ze^2} = Z\frac{m_e e^2}{\hbar^2} = \frac{Z}{a_0},\tag{7.4.18}$$

where a_0 is the Bohr radius of the ground state of the hydrogen atom. Using these relationships, one may write the cross section for photo-ejection of a K-shell electron as

$$\sigma_K(\nu) = \frac{32}{3}\,\pi^3\,\frac{ch}{h\nu}\left(\frac{e^2}{m_e c^2}\frac{c}{e}\right)^2\left(\frac{2I_K}{Ze^2}\right)^{7/2}\left(Z\frac{m_e e^2}{\hbar^2}\right)^{3/2}$$

$$\times\left(\frac{\hbar}{m_e}\frac{1}{2\pi\nu}\right)^4\frac{m_e}{h^3}\,(2m_e)^{3/2}\,E^{3/2}.\tag{7.4.19}$$

Collecting powers of various quantities and rearranging, this becomes

$$\sigma_K(\nu) = \frac{8\pi}{3}\left(\frac{e^2}{m_e c^2}\right)^2\left(\frac{ch}{e^2}\right)^3\frac{32}{Z^2}\left(\frac{I_K}{h\nu}\right)^{7/2}\left(\frac{E}{h\nu}\right)^{3/2}.\tag{7.4.20}$$

Further insight comes by using the fact that the classical radius of the electron, r_0, the Compton wavelength of the electron, λ_C divided by 2π $(\lambda_C = \lambda_C/2\pi)$, and a_0 are related to one another by powers of the fine structure constant,

$$\alpha = \frac{e^2}{\hbar c} = \frac{1}{137.036},$$
(7.4.21)

according to

$$r_0 = \frac{e^2}{m_e c^2} = \frac{e^2}{\hbar c}\lambda_C = \left(\frac{e^2}{\hbar c}\right)^2 a_0 = 2.817\,94 \times 10^{-13}\ \text{cm}.$$
(7.4.22)

Thus, for example, one can write

$$\sigma_K(\nu) = \frac{8\pi}{3} r_0^2 \frac{1}{\alpha^3} \frac{32}{Z^2} \left(\frac{I_K}{h\nu}\right)^{7/2} \left(\frac{E}{h\nu}\right)^{3/2},$$
(7.4.23)

$$\sigma_K(\nu) = \frac{8\pi}{3} \lambda_C a_0 \frac{32}{Z^2} \left(\frac{I_K}{h\nu}\right)^{7/2} \left(\frac{E}{h\nu}\right)^{3/2},$$
(7.4.24)

and

$$\sigma_K(\nu) = \frac{8\pi}{3} a_0^2 \alpha \frac{32}{Z^2} \left(\frac{I_K}{h\nu}\right)^{7/2} \left(\frac{E}{h\nu}\right)^{3/2}.$$
(7.4.25)

Noting that

$$\sigma_{Th} = \frac{8\pi}{3} r_0^2$$
(7.4.26)

is the Thomson, or electron-scattering cross section (see eq. (3.3.5)), and that the relationship between the energy of the ejected electron, the photon energy, and the threshold energy is

$$E = h\nu - I_K,$$
(7.4.27)

leads to another insightful conformation:

$$\sigma_K(\nu) = \sigma_{Th} \frac{1}{\alpha^3} \frac{32}{Z^2} \left(\frac{I_K}{h\nu}\right)^{7/2} \left(\frac{h\nu - I_K}{h\nu}\right)^{3/2}.$$
(7.4.28)

Finally, utilizing the fact that the first Bohr radius of a single electron about a nucleus of charge Z is $a_K = a_0/Z$, eq. (7.4.25) may be written as

$$\sigma_K(\nu) = \left[256 \frac{e^2}{\hbar c} \frac{\pi}{3}\right] a_K^2 \left(\frac{I_K}{h\nu}\right)^{7/2} \left(\frac{E}{h\nu}\right)^{3/2}$$

$$= 1.956\,29\, a_K^2 \left(\frac{I_K}{h\nu}\right)^{7/2} \left(\frac{E}{h\nu}\right)^{3/2}.$$
(7.4.29)

As the threshold is approached from above, the cross section goes to zero as the cube of the momentum of the ejected electron. This behavior is a consequence of the phase space factor, eq. (7.4.5), and of the presence in the matrix element, eqs. (7.4.6)–(7.4.8), of the gradient of the plane wave used to represent the electron. Differentiation of eq. (7.4.28) with respect to $h\nu$ shows that the cross section reaches a maximum at $h\nu/I_K = 10/7$ and that this maximum is

$$\sigma_K^{max} = \sigma_{Th} \frac{1}{\alpha^3} \frac{32}{Z^2} \left(\frac{7}{10}\right)^{7/2} \left(1 - \frac{7}{10}\right)^{3/2}$$

$$= \sigma_{Th} \frac{1}{\alpha^3} \frac{32}{Z^2} 0.047\,15 = 0.092\,239\, a_K^2. \tag{7.4.30}$$

The traditional Born approximation follows when $h\nu \gg I_K$ and $E/h\nu \approx 1$, producing

$$\sigma_K^{Born}(\nu) = \sigma_{Th} \frac{1}{\alpha^3} \frac{32}{Z^2} \left(\frac{I_K}{h\nu}\right)^{7/2} = 1.956\,29\, a_K^2 \left(\frac{I_K}{h\nu}\right)^{7/2}. \tag{7.4.31}$$

At threshold, $\sigma_K^{Born} \sim 21\, \sigma_K^{max}$ and, at $h\nu/I_K = 10/7$, $\sigma_K^{Born} \sim 6\, \sigma_K^{max}$.

When the Coulomb interaction between the ejected electron and the daughter ion is taken into account, the cross section for the ejection of a K-shell electron is (M. Stobbe, 1930)

$$\sigma_K(\nu) = \sigma_K^{Born}(\nu)\, 2\pi \left(\frac{I_K}{h\nu}\right)^{1/2} \frac{\exp\left(-4\,\eta\,\cot^{-1}\eta\right)}{1 - \exp\left(-2\pi\,\eta\right)}, \tag{7.4.32}$$

where

$$\eta = Z_K \frac{e^2}{\hbar\nu} = Z_K \frac{e^2}{\hbar c} \left(\frac{m_e c^2/2}{E}\right)^{1/2} = Z_K \frac{e^2}{\hbar c} \left(\frac{m_e c^2}{2I_K}\right)^{1/2} \left(\frac{I_K}{h\nu - I_K}\right)^{1/2}$$

$$= \left(\frac{I_K}{h\nu - I_K}\right)^{1/2}. \tag{7.4.33}$$

Near enough to the absorption edge that η is large compared with unity,

$$\cot^{-1}(\eta) = \frac{1}{\eta} - \frac{1}{3\eta^3} + \frac{1}{5\eta^5} - \cdots, \tag{7.4.34}$$

so, exactly at the absorption edge, where $\eta = \infty$,

$$\sigma_K(\nu_{th} = I_K/h) = 2\pi\, e^{-4}\, \sigma_K^{Born}(\nu_{th} = I_K/h) = 0.115\,08\, \sigma_K^{Born}(\nu_{th} = I_K/h), \tag{7.4.35}$$

which is 2.4 times larger than the maximum cross section when the electron is described by an undistorted plane wave (eq. (7.4.30)). Using eq. (7.4.31), one has

$$\sigma_K(\nu_{th} = I_K/h) = 0.225\,13\, a_K^2 = \frac{0.630\,43}{Z_K^2} \times 10^{-17}\, \text{cm}^2. \tag{7.4.36}$$

The factors by which the plane wave and Coulomb-distorted plane wave solutions differ from the Born approximation solution are shown as functions of $h\nu/I_K$ in Fig. 7.4.1. Except very near threshold, the plane wave factor is approximately twice the distorted plane-wave factor.

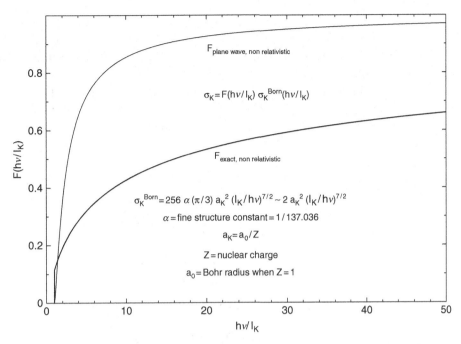

Fig. 7.4.1 Multiplicative factor $F(h\nu/I_K)$ for K–shell photoionization cross-section where I_K = ionization energy and $h\nu$ = photon energy

If the ion has two electrons in the K shell, the cross section for absorption can be estimated in zeroth approximation by replacing, in the equation for absorption by an ion with a single K-shell electron, the charge Z of the bare nucleus by an effective charge Z_{eff} which is consistent with the ionization potential I_K' when two K-shell electrons are present. Thus,

$$I_K' = \frac{1}{2}\frac{e^2}{a_0} Z_{\text{eff}}^2. \tag{7.4.37}$$

For example, for helium, with a first ionization potential of 24.46 eV, one might guess

$$Z_{\text{eff}} = \left(\frac{24.46\,\text{eV}}{13.527\,\text{eV}}\right)^{1/2} = 1.345 = 2 - 0.655. \tag{7.4.38}$$

Because either electron can be ejected, the resultant cross section is to be multiplied by 2.

If the ion has two electrons in the K shell (each here designated as a $1s$ electron) and two electrons in the lowest energy state of the L shell (each here designated as a $2s$ electron), the cross section for the ejection of a $2s$ electron is (Stobbe, 1930; see also Hall, 1936)

$$\sigma_{2\times 2s}(\nu) = \sigma_{\text{Th}}\left(\frac{\hbar c}{e^2}\right)^3 \left(\frac{32}{Z_{2\times 2s}}\right)^2 2\pi \left(\frac{I_{2\times 2s}}{h\nu}\right)^4 \left(2 + 6\frac{I_{2\times 2s}}{h\nu}\right)$$

$$\times \frac{\exp(-8\,\eta\,\cot^{-1}\eta)}{1 - \exp(-4\pi\eta)}, \tag{7.4.39}$$

where $I_{2\times 2s}$ is the ionization potential, $Z_{2\times 2s}$ is an effective charge consistent with the ionization potential, and η is given by eq. (7.4.33) with I_K replaced by $I_{2\times 2s}$. Equation (7.4.39) has been obtained by using a non-relativistic Coulomb-distorted plane wave for the electron.

If there is just one $2s$ electron, an approximate cross section follows by replacing $I_{2\times 2s}$ and $Z_{2\times 2s}$ in eq. (7.4.39) with I_{2s} and Z_{2s}, respectively, and dividing the result by 2. Defining the Bohr radius of a single $2s$ electron by

$$a_{L_1} = 2^2 \frac{a_0}{Z_{2s}}, \tag{7.4.40}$$

one can write

$$\sigma_{2s}(\nu) = \frac{8\pi}{3} a_{L_1}^2 \left(64 \frac{e^2}{\hbar c}\right) \left(\frac{I_{2s}}{h\nu}\right)^4 2\pi \left(1 + 3 \frac{I_{2s}}{h\nu}\right)$$

$$\times \frac{\exp\left(-8\,\eta\,\cot^{-1}\eta\right)}{1 - \exp\left(-4\pi\eta\right)}, \tag{7.4.41}$$

where η is given by eq. (7.4.33) with I_K replaced by I_{2s}. At threshold,

$$\sigma_{2s}(\nu_{\text{th}}) = \frac{8\pi}{3} a_{2s}^2 \left(64 \frac{e^2}{\hbar c}\right) 8\pi\,e^{-8}$$

$$= 0.032\,9874\,a_{2s}^2 = \frac{1.477\,98}{Z_{2s}^2} \times 10^{-17}\ \text{cm}^2. \tag{7.4.42}$$

Comparing with eq. (7.4.36), one has that the ratio of cross sections at the respective thresholds is

$$\frac{\sigma_{2s}(\nu_{\text{th}} = I_{2s}/h)}{\sigma_{1s}(\nu_{\text{th}} = I_{1s}/h)} = 0.146\,43 \left(\frac{a_{2s}}{a_{1s}}\right)^2 = 2.3429 \left(\frac{Z_{1s}}{Z_{2s}}\right)^2. \tag{7.4.43}$$

Additional L-shell electrons are in $2p$ states. If there are six $2p$ electrons, the cross section for the ejection of a $2p$ electrons obtained by Hall (1936) and others is

$$\sigma_{6\times 2p}(\nu) = \sigma_{\text{Th}} \left(\frac{\hbar c}{e^2}\right)^3 \left(\frac{32}{Z_{6\times 2p}}\right)^2 2\pi \left(\frac{I_{6\times 2p}}{h\nu}\right)^5 \left(6 + 16\frac{I_{6\times 2p}}{h\nu}\right)$$

$$\times \frac{\exp\left(-8\,\eta\,\cot^{-1}\eta\right)}{1 - \exp\left(-4\pi\eta\right)}, \tag{7.4.44}$$

where $I_{6\times 2p}$ and $Z_{6\times 2p}$ are, respectively, the ionization potential and the effective charge for any ejected $2p$ electron. Algorithms may be constructed for dealing with situations in which there are fewer than six $2p$ electrons. If, for example, there is just one $2p$ electron, an approximate cross section may be obtained by replacing $I_{6\times 2p}$ and $Z_{6\times 2p}$ in eq. (7.4.44) with I_{2p} and Z_{2p}, respectively, and dividing the result by 6. In another example, if the ion is neutral oxygen with four $2p$ electrons, the total cross section for the ejection of a $2p$ electron may be obtained by replacing $I_{6\times 2p}$ and $Z_{6\times 2p}$ in eq. (7.4.44) with $I_{4\times 2p}$ and $Z_{4\times 2p}$, respectively, and multiplying the result by $4/6$.

The alternative to exercising a certain degree of inventiveness in estimating cross sections, making use of experimental data when available, is a lengthy numerical calculation based on a complicated theoretical formulation (see, e.g., J. J. Boyle & M. D. Kutzner, 1998).

7.4.1 On the determination of the associated opacity coefficient

To find the contribution to the opacity of bound–free absorption by a given type of ion in a particular stage of ionization, one must first solve the relevant Saha equations to find the number abundance of ions in that stage and then multiply by the appropriate cross section. For example, the contribution to the opacity coefficient of ^{16}O nuclei with one bound electron is given by

$$\kappa_{K,O} = \sigma_{K,O}\, S(h\nu - \chi_K)\, N_A\, Y_O^{7+} \sim \sigma_{K,O}\, S(h\nu - \chi_K)\, \frac{X_O^{7+}}{16 M_H}, \qquad (7.4.45)$$

where Y_O^{7+} is the appropriate number abundance parameter, N_A is Avogadro's number, $\sigma_{K,O}$ is given by eqs. (7.4.31)–(7.4.33) with $Z = 8$, $S(h\nu - \chi_{K,O})$ is a step function (0 when $h\nu < \chi_{K,O}$ and 1 when $h\nu > \chi_{K,O}$). In the second formulation, X_O^{7+} is the abundance by mass of oxygen nuclei that have just one K-shell electron.

By itself, the contribution described by eq. (7.4.45) is useless. To it must be added the contributions of all other active ionization stages for oxygen as well as those for all other species of nuclei present. Once these contributions have been estimated and contributions from all other sources of absorption and scattering have been added to the opacity coefficient, a Rosseland mean opacity must be determined (see Sections 7.10–7.11).

Although the cross sections for photoelectric absorption near absorption edges are huge compared with the cross section for electron scattering, the contribution of photo-ejection processes to the Rosseland mean opacity is tempered in many ways. For example, (1) in realistic hydrogen- and helium-rich mixtures at high temperatures, the number abundance of any element heavier than helium is typically at least a thousand times smaller than the number abundance of electrons; (2) when the peak in the photon distribution with respect to energy is near the ionization energy, the fraction of heavy atoms with an electron having this ionization energy is small; and (3) the cross section for absorption falls off rapidly with increasing $h\nu/kT$. Results of explicit calculations which demonstrate these points are described in Section 7.11.

7.5 The matrix element for
free–free (inverse bremsstrahlung) absorption

The process whereby an unbound electron in the vicinity of a charged heavy ion is propelled into a higher energy state by the absorption of a photon is the exact inverse of the process by which an unbound electron emits a photon due to acceleration and deceleration in the field of a charged ion. The designation bremsstrahlung, or braking radiation, describes one aspect of the second process and the designation free–free describes an aspect of both processes.

If the electron wave functions are taken to be exact solutions of the Schrödinger equation for an electron in the Coulomb field of a stationary charged particle, the cross section for the absorption of a photon by a single electron in the field of an isolated heavy charged particle can be solved exactly using first order perturbation theory (Sommerfeld, 1931; Maue, 1932). However, the solution involves hypergeometric functions and the multitude of indices interferes with an understanding of the physics. In an effort to obtain insight into the physics and to identify transparently the dependence of the cross section on fundamental constants and on photon frequency, a more elementary approach using less elaborate mathematical functions is adopted in this section and the next.

In the approximation that the electron wave functions are undistorted plane waves described by eq. (7.4.2), an estimate of the probability for the absorption of a single photon by a single electron of a given velocity in the field of a single heavy ion involves transitions through intermediate states. In this section, second order perturbation theory is used to determine the effective matrix element which characterizes the process. In the next section, this matrix element is used in a calculation of the total cross section for free–free transitions; the calculation involves an average over a Maxwell–Boltzmann probability distribution of initial electron velocities, multiplication by the volume density of electrons to obtain a cross section per ion relevant to the environment in the stellar interior, and summation over all ions to yield the total cross section for the absorption of a single photon by the free–free process.

Adopting eq. (7.1.23) as the perturbation Hamiltonian, the pertinent matrix element for a transition of an electron of momentum $\mathbf{p}_i$ to an electron of momentum $\mathbf{p}_f$ with the absorption of a photon of energy $\hbar\mathbf{k}$ is

$$H_{\mathrm{fi}} = \int u_{\mathrm{f}}^* \langle 0_{k\lambda} | H^{\mathrm{int}} | 1_{k\lambda} \rangle u_i \mathrm{d}\tau = \int u_{\mathrm{f}}^* \sqrt{\frac{c^2 \hbar}{v_k V}} \frac{e}{m_e c} \frac{\hbar}{i} e^{i\mathbf{k}\cdot\mathbf{r}} \hat{\epsilon}_{k\lambda} \cdot \nabla u_i \mathrm{d}\tau. \tag{7.5.1}$$

When both electron wave functions are chosen to be of the form given by eq. (7.4.2), H_{fi} vanishes unless

$$\mathbf{p}_f = \mathbf{p}_i + \hbar\mathbf{k}. \tag{7.5.2}$$

Using the facts that $\hbar k = h\nu/c$, $v_i = p_i/m_e$, and $\cos\theta_{ik} = \hat{\mathbf{p}}_i \cdot \hat{\mathbf{k}}$, it follows from eq. (7.5.2) that

$$\frac{p_f^2}{2m_e} = \frac{p_i^2}{2m_e} + 2\,\mathbf{p}_i \cdot \hbar\mathbf{k} + (\hbar k)^2$$

$$= \frac{p_i^2}{2m_e} + h\nu \left(\frac{v_i}{c} \cos\theta_{ik} + \frac{h\nu}{2m_e c^2} \right). \tag{7.5.3}$$

However, if v_i/c and $h\nu/2m_e c^2$ are small compared with unity, it follows that

$$\frac{p_f^2}{2m_e} \sim \frac{p_i^2}{2m_e} = E - h\nu, \tag{7.5.4}$$

where

$$E = \frac{p_i^2}{2m_e} + h\nu \tag{7.5.5}$$

is the initial energy of the system. Thus, energy and momentum are not simultaneously conserved. Insisting on conservation of energy leads to a zero matrix element for the radiative transition, demonstrating that an isolated electron can neither radiate nor absorb a photon.

The problem is solved by recognizing that, since the static Coulomb interaction between the electrons and the central positive charge has not been taken into account in the choice of electron wave functions, the appropriate perturbation Hamiltonian is not eq. (7.1.23), but rather

$$H^{\text{perturbation}} = H^{\text{int}} + H^{\text{Coulomb}}$$

$$= \frac{e}{2m_e c} \left(\mathbf{p}_{\text{op}} \cdot \mathbf{A}_{\text{op}} + \mathbf{A}_{\text{op}} \cdot \mathbf{p}_{\text{op}} \right) - \frac{Ze^2}{r} \exp\left(-\frac{r}{R_{\text{sc}}}\right), \tag{7.5.6}$$

where the first term expresses the interaction between the electron and the electrodynamic field and the second term expresses the interaction of the electron with the static Coulomb field of the positive ion. In the second term of eq. (7.5.6), Ze is the effective charge of the positive ion and R_{sc} is a screening radius, which could be taken as the Debye radius R_D discussed in Section 4.17 (see eqs. (4.17.16) and (4.17.12)).

The transition between initial and final states is envisioned as a two-stage process in which the electron is scattered by the Coulomb field of the heavy ion either before or after the absorption of the photon. The intermediate states are *virtual* and energy is momentarily not conserved. However, the final electron energy equals the energy of the initial electron plus the energy of the absorbed photon.

Equation (7.2.26) is still valid, but the matrix element has the form

$$M = \sum_j \frac{H_{fj} H_{ji}}{E_f - E_j}, \tag{7.5.7}$$

where H_{fj} and H_{ji} are matrix elements of the perturbation Hamiltonian, eq. (7.5.6). The subscripts f, j, and i in eq. (7.5.7) refer to final, intermediate, and initial states, respectively. Equation (7.5.7) is a standard result of second order perturbation theory. The derivation begins by writing the Schrödinger equation as

$$-\frac{\hbar}{i} \frac{\partial \psi(\vec{r}, t)}{\partial t} = (H + \lambda H') \, \psi(\vec{r}, t), \tag{7.5.8}$$

where H is the unperturbed Hamiltonian, H' is the perturbation Hamiltonian, λ is a parameter used for book-keeping purposes, and ψ is a wave function assumed to be of the form

$$\psi(\vec{r}, t) = \sum_j C_j(t) \, u_j(\vec{r}) \, e^{-(i/\hbar)E_j t}, \tag{7.5.9}$$

where $u_j(\vec{r})\, e^{-(i/\hbar)E_j t}$ is one of a complete set of solutions of eq. (7.5.8) when $\lambda = 0$. Setting ψ into the Schrödinger equation, multiplying by $u_k^*(\vec{r})\, e^{(i/\hbar)E_k t}$, and integrating over all space produces

$$\frac{\partial}{\partial t} C_k = -\frac{i}{\hbar} \lambda \sum_j C_j(t)\, H'_{kj}(t)\, e^{i\omega_{kj} t}, \tag{7.5.10}$$

where

$$\hbar\, \omega_{kj} = E_k - E_j \tag{7.5.11}$$

contains information supplied by the unperturbed Hamiltonian and

$$H'_{kj} = \int u_k^*(\vec{r})\, H'\, u_j(\vec{r})\, d\tau \tag{7.5.12}$$

contains information supplied by the perturbation term in the Hamiltonian. Inserting

$$C_j(t) = C_j^{(0)} + \lambda\, C_j^{(1)} + \lambda^2\, C_j^{(2)} + \cdots \tag{7.5.13}$$

into the expression for $\partial C_k / \partial t$ gives a series of terms in powers of λ. Equating to zero the coefficient of each power of λ produces

$$\frac{\partial}{\partial t} C_k^{(s+1)} = -\frac{i}{\hbar} \sum_j C_j^{(s)}(t)\, H'_{kj}(t)\, e^{i\omega_{kj} t}, \tag{7.5.14}$$

where s is an integer. Finally, assuming that

$$C_j^{(0)}(0) = \delta_{jm}, \tag{7.5.15}$$

one obtains

$$C_k^{(1)}(t_A) = -\frac{i}{\hbar} \int_0^{t_A} H'_{km}(t)\, e^{i\omega_{km} t}\, dt = -\frac{1}{\hbar} H'_{km} \frac{e^{i\omega_{km} t_A} - 1}{\omega_{km}} \tag{7.5.16}$$

and

$$C_k^{(2)}(t_B) = -\frac{i}{\hbar} \sum_n \int_0^{t_B} \left(-\frac{1}{\hbar} H'_{nm} \frac{e^{i\omega_{nm} t_A} - 1}{\omega_{nm}} \right) H'_{kn}\, e^{i\omega_{kn} t_A}\, dt_A \tag{7.5.17}$$

$$= \frac{i}{\hbar^2} \sum_n \int_0^{t_B} \left(\frac{H'_{kn} H'_{nm}}{\omega_{nm}} \right) \left[e^{i\omega_{km} t_A} - e^{i\omega_{kn} t_A} \right] dt_A \tag{7.5.18}$$

$$= \frac{1}{\hbar^2} \sum_n \left(\frac{H'_{kn} H'_{nm}}{\omega_{nm}} \right) \left[\frac{e^{i\omega_{km} t_B} - 1}{\omega_{km}} - \frac{e^{i\omega_{kn} t_B} - 1}{\omega_{kn}} \right]. \tag{7.5.19}$$

In the case at hand, only the first term in square brackets of the last expression is important (see, e.g., Schiff, 1949, Chapter VII) and the development that leads to eqs. (7.2.25) and (7.2.26) can be used, with the only difference being that the first order matrix element is replaced by the sum of the terms in parentheses, i.e., by eq. (7.5.7).

It is clear that, in order for the matrix element given by eq. (7.5.7) to be finite, energy conservation must be violated by transitions into intermediate states. That is, $E_j \neq E$, where

$$E = E_i = h\nu + \frac{|\mathbf{p}_i|^2}{2m_e} = E_f = \frac{|\mathbf{p}_f|^2}{2m_e}. \tag{7.5.20}$$

Here $\mathbf{p}_i$ and $\mathbf{p}_f$ are, respectively, the momenta of the initial and final electron, E_i is the sum of the energies of the initial electron and photon, and E_f is the energy of the final electron. There are actually only two contributions to the sum in eq. (7.5.7). In one case, the electron first is scattered by the Coulomb field and then absorbs a photon. In the other case, the electron first absorbs a photon and then is scattered by the Coulomb field.

In the first case,

$$H_{1i} = \int u_1^* \left(-\frac{Ze^2}{r} \exp(-r/R_{sc}) \right) u_i d\tau, \tag{7.5.21}$$

and

$$H_{f1} = \int u_f^* \langle 0_{k\lambda} | H_{int} | 1_{k\lambda} \rangle u_1 d\tau = \int u_f^* \sqrt{\frac{c^2 \hbar}{\nu_k V}} \frac{e}{m_e c} \frac{\hbar}{i} e^{i\mathbf{k}\cdot\mathbf{r}} \hat{\epsilon}_{k\lambda} \cdot \nabla u_1 d\tau. \tag{7.5.22}$$

With both electron wave functions of the form (7.2.2), H_{f1} vanishes unless

$$\mathbf{p}_1 = \mathbf{p}_f - \hbar\mathbf{k}, \tag{7.5.23}$$

and, since $\hat{\epsilon}_{k\lambda} \cdot \mathbf{k} = 0$,

$$H_{f1} = \sqrt{\frac{c^2 \hbar}{\nu_k V}} \frac{e}{m_e c} \hat{\epsilon}_{k\lambda} \cdot \mathbf{p}_f. \tag{7.5.24}$$

Using the facts that $\hbar k = h\nu/c$, $\nu_f = p_f/m_e$, and $\cos\theta_{fk} = \hat{\mathbf{p}}_f \cdot \hat{\mathbf{k}}$, it follows from eq. (7.5.23) that

$$\frac{p_1^2}{2m_e} = \frac{p_f^2}{2m_e} - 2\,\mathbf{p}_f \cdot \hbar\mathbf{k} + (\hbar k)^2$$

$$= \frac{p_f^2}{2m_e} - h\nu \left(\frac{\nu_f}{c} \cos\theta_{fk} - \frac{h\nu}{2m_e c^2} \right)$$

$$\sim \frac{p_f^2}{2m_e} = E = \frac{p_i^2}{2m_e} + h\nu, \tag{7.5.25}$$

where the last approximate equality follows when ν_f/c and $h\nu/2m_e c^2$ are small compared with unity. Thus, the energy of the intermediate state is

$$E_1 = \frac{p_1^2}{2m_e} + h\nu \sim E + h\nu. \tag{7.5.26}$$

The Coulomb scattering matrix element is

$$H_{1i} = \frac{1}{V} \int \exp(-i\,\mathbf{p}_1 \cdot \mathbf{r}/\hbar) \left(-\frac{Ze^2}{r} \exp(-r/R_{sc}) \right) \exp(i\,\mathbf{p}_i \cdot \mathbf{r}/\hbar) d\tau. \tag{7.5.27}$$

Integration of the sort employed in the development of eqs. (7.4.7)–(7.4.11) (noting that there is one fewer power of r in the initial integral) gives

$$H_{1i} = -\frac{4\pi Z e^2}{V} \frac{\lambda_{i1}^2}{1 + (\lambda_{i1}/R_{sc})^2},$$ (7.5.28)

where

$$\lambda_{i1} = \frac{\hbar}{|\mathbf{p}_i - \mathbf{p}_1|}.$$ (7.5.29)

When the photon is absorbed first, the radiative matrix element vanishes unless

$$\mathbf{p}_2 = \mathbf{p}_i + \hbar\mathbf{k},$$ (7.5.30)

so that

$$H_{2i} = \sqrt{\frac{c^2\hbar}{\nu_k V}} \frac{e}{m_e c} \hat{\epsilon}_{k\lambda} \cdot \mathbf{p}_i.$$ (7.5.31)

It follows from eq. (7.5.30) that the energy of the intermediate state is

$$E_2 = \frac{p_2^2}{2m_e} = \frac{p_i^2}{2m_e} + h\nu \left(\frac{\nu_i}{c} \cos\theta_{ik} + \frac{h\nu}{2m_e c^2} \right)$$

$$\sim \frac{p_i^2}{2m_e} = E - h\nu.$$ (7.5.32)

Here, $\nu_i = p_i/m_e$, $\cos\theta_{ik} = \hat{\mathbf{p}}_i \cdot \hat{\mathbf{k}}$, and again it has been supposed that ν_i/c and $h\nu/2m_e c^2$ are small compared with unity.

Coulomb scattering of the intermediate state electron produces the matrix element

$$H_{f2} = -\frac{4\pi Z e^2}{V} \frac{\lambda_{2f}^2}{1 + (\lambda_{2f}/R_{sc})^2},$$ (7.5.33)

where

$$\lambda_{2f} = \frac{\hbar}{|\mathbf{p}_2 - \mathbf{p}_f|}.$$ (7.5.34)

Although the Coulomb-scattering matrix elements H_{1i} (eq. [7.5.28]) and H_{f2} (eq. (7.5.33)) differ formally from one another, they are essentially the same quantity. From eqs. (7.5.30) and (7.5.23),

$$\Delta p = |\Delta\mathbf{p}| = |\mathbf{p}_i - \mathbf{p}_1| = |\mathbf{p}_f - \mathbf{p}_2| = |\mathbf{p}_f - \mathbf{p}_i - \hbar\mathbf{k}|,$$ (7.5.35)

and this momentum change is balanced by the momentum change of the heavy ion. From eq. (7.5.20),

$$p_f - p_i = \frac{2m_e c}{p_f + p_i} \frac{h\nu}{c} = \frac{2m_e c}{p_f + p_i} \hbar k,$$ (7.5.36)

which means that, for typical values of $\mathbf{p}_i$ and $\mathbf{p}_f$, one can neglect $-\hbar\mathbf{k}$ relative to $\mathbf{p}_f - \mathbf{p}_i$ in the last equality in eq. (7.5.35). Thus, to a very good approximation, $\lambda_{2f} = \lambda_{i1}$, and

$$H_{f2} = H_{1i} \sim -\frac{4\pi Ze^2}{V} \frac{\lambda_{if}^2}{1 + (\lambda_{if}/R_{sc})^2}, \tag{7.5.37}$$

where

$$\lambda_{if} = \frac{\hbar}{|\mathbf{p}_f - \mathbf{p}_i|}. \tag{7.5.38}$$

As an aside, eq. (7.5.36), which states that $p_f - p_i \gg \hbar k$, demonstrates that the heavy ion is primarily responsible for the momentum change of the electron. Equation (7.5.20) shows that the photon supplies the energy for the process.

Making use of eqs. (7.5.24), (7.5.26), (7.5.31), and (7.5.37), the effective matrix element for the free–free transition probability, eq. (7.5.7), becomes

$$M = \frac{1}{h\nu} \frac{e}{m_e c} \sqrt{\frac{c^2\hbar}{v_k V}} \, (\mathbf{p}_f - \mathbf{p}_i) \cdot \hat{\epsilon}_{k\lambda} \frac{4\pi Ze^2}{V} \frac{\lambda_{if}^2}{1 + (\lambda_{if}/R_{sc})^2}. \tag{7.5.39}$$

Setting H_{fi} in eq. (7.2.26) equal to M gives the differential probability for the absorption of a single photon of given direction and polarization by a single electron of a given velocity in the field of a single heavy ion.

7.6 The cross section for free–free absorption

After integration over volume, the differential phase space factor $d^2\rho_f$ needed in eq. (7.2.26) for the final electron is given by eq. (7.4.5), except that the factor of 2 is omitted since the spin of the final electron is constrained to be opposite to that of the initial electron. Inserting M from eq. (7.5.39) and the phase space factor (after integration over volume) into eq. (7.2.26), the probability per unit time for an average electron, after absorption of a photon, to be found in the differential solid angle $d\Omega_f$ is

$$d\omega_{i\to f} = \frac{2\pi}{\hbar} \frac{1}{(h\nu)^2} \left(\frac{e}{m_e c}\right)^2 \frac{c^2\hbar}{v_k V} \left(\frac{4\pi Ze^2}{V}\right)^2 \lambda_{sc}^4 \left[(\mathbf{p}_f - \mathbf{p}_i) \cdot \hat{\epsilon}_{k\lambda}\right]^2 \frac{V m_e p_f}{h^3} \, d\Omega_f, \tag{7.6.1}$$

where

$$\lambda_{sc}^2 = \frac{\lambda_{if}^2}{1 + (\lambda_{if}/R_{sc})^2}. \tag{7.6.2}$$

Only two factors in eq. (7.6.1), the quantity λ_{sc}^4 and the factor in square brackets, depend on the solid angle $d\Omega_f$. Isolating these factors, and assuming temporarily that $R_{sc} = \infty$, the relevant integral over solid angle is $\hbar^4$ times the quantity

$$I = \int \left(\frac{\lambda_{if}}{\hbar}\right)^4 \left[(\mathbf{p}_f - \mathbf{p}_i) \cdot \hat{\epsilon}_{k\lambda}\right]^2 d\Omega_f = \int \frac{\left[(\mathbf{p}_f - \mathbf{p}_i) \cdot \hat{\epsilon}_{k\lambda}\right]^2}{|\mathbf{p}_f - \mathbf{p}_i|^4} \, d\Omega_f. \tag{7.6.3}$$

Choosing the z axis in the direction $\mathbf{p}_i$, let $\theta_{\Delta p}$ be the angle which $\mathbf{p}_f - \mathbf{p}_i$ makes with the z axis, and let $\phi_{\Delta p}$ be the angle which the projection of $\mathbf{p}_f - \mathbf{p}_i$ onto the x-y plane makes with the x axis. Similarly, define $\theta_{k\lambda}$ and $\phi_{k\lambda}$ as the corresponding angles for the vector $\hat{\epsilon}_{k\lambda}$, and θ_{if} and ϕ_{if} as the corresponding angles for the vector $\mathbf{p}_f$. With these choices, one has

$$(\mathbf{p}_f - \mathbf{p}_i) \cdot \hat{\epsilon}_{k\lambda} = |\mathbf{p}_f - \mathbf{p}_i|[\cos\theta_{\Delta p}\cos\theta_{k\lambda} + \sin\theta_{\Delta p}\sin\theta_{k\lambda}\cos(\phi_{\Delta p} - \phi_{k\lambda})]. \quad (7.6.4)$$

Using eq. (7.6.4) in eq. (7.6.3), setting $d\Omega_f = d\phi_{if}\sin\theta_{if}d\theta_{if}$, noting that $\phi_{if} = \phi_{\Delta p}$, and integrating over the angle ϕ_{if} gives

$$I = \int_0^\pi \frac{2\pi\cos^2\theta_{\Delta p}\cos^2\theta_{k\lambda} + \pi\sin^2\theta_{\Delta p}\sin^2\theta_{k\lambda}}{\left(p_f^2 + p_i^2 - 2p_f p_i\cos\theta_{if}\right)}\sin\theta_{if}d\theta_{if}. \quad (7.6.5)$$

One could integrate over θ_{if} at once, but the result is a somewhat cumbersome function of $\theta_{k\lambda}$. It is much simpler to first average over all directions of the photon polarization vector $\hat{\epsilon}_{k\lambda}$, an average which must eventually be performed in any case. Since $\int\cos^2\theta_{k\lambda}d\Omega_{k\lambda} = 4\pi/3$ and $\int\sin^2\theta_{k\lambda}d\Omega_{k\lambda} = 8\pi/3$, this averaging procedure gives

$$\bar{I} = \int I\frac{d\Omega_{k\lambda}}{4\pi} = \frac{2\pi}{3}\int_{-1}^1 \frac{d\cos\theta_{if}}{\left(p_f^2 + p_i^2 - 2p_f p_i\cos\theta_{if}\right)} \quad (7.6.6)$$

$$= \frac{2\pi}{3}\frac{1}{p_i p_f}\log_e\left(\frac{p_f + p_i}{p_f - p_i}\right) = \frac{2\pi}{3}\frac{1}{p_i p_f}\log_e\left(\frac{(p_f + p_i)^2}{2m_e h\nu}\right). \quad (7.6.7)$$

In the term on the rightmost side of eq. (7.6.7), use has been made of eq. (7.5.36).

When the screening radius R_{sc} is included,

$$\bar{I} = \frac{2\pi}{3}\frac{1}{2p_i p_f}L(p_i, p_f, R_{sc}), \quad (7.6.8)$$

where

$$L(p_i, p_f, R_{sc}) = \log_e\left(\frac{(p_f + p_i)^2 + (\hbar/R_{sc})^2}{(p_f - p_i)^2 + (\hbar/R_{sc})^2}\right). \quad (7.6.9)$$

In the Debye approximation $R_{sc} = R_D$ (see eq. (4.17.12)).

Gathering all terms together and rearranging, the differential probability of absorption given by eq. (7.6.1), when averaged over the polarization directions of the absorbed photon and summed (integrated) over all directions of motion of the emitted electron, becomes

$$d\omega'_{i\to f} = \pi\frac{8\pi}{3}\left(\frac{e^2}{m_e c^2}\right)^2 Z^2\left(\frac{\hbar c}{e^2}\right)^2\left(\frac{e^2/a_0}{h\nu}\right)^3 a_0^3\frac{c}{V^2}\frac{m_e c}{p_i}L(p_i, p_f, R_{sc}).$$

$$(7.6.10)$$

Remembering that the Thomson electron-scattering cross section is given by

$$\sigma_{Th} = \frac{8\pi}{3} \left(\frac{e^2}{m_e c^2} \right)^2,$$
(7.6.11)

one may also write

$$d\omega'_{i \to f} = \sigma_{Th} \, \pi \, Z^2 \left(\frac{\hbar c}{e^2} \right)^2 \left(\frac{e^2/a_0}{h\nu} \right)^3 a_0^3 \frac{c}{V^2} \frac{m_e c}{p_i} L(p_i, p_f, R_{sc}).$$
(7.6.12)

The quantity $d\omega'_{i \to f}$ described by eqs. (7.6.10) and (7.6.12) is the probability per unit time that *one* electron of momentum p_i absorbs a photon with the help of *one* ion. But, in the volume V, the number of electrons that have momentum in the range p_i to $p_i + dp_i$ is $V \, dn_e$, where, for non-relativistic electrons (see eqs. (4.3.6) and (4.3.14)),

$$dn_e = \frac{n_e}{(2\pi m_e kT)^{3/2}} \exp\left(-\frac{p_i^2}{2 m_e kT} \right) 4\pi p_i^2 dp_i$$
(7.6.13)

$$= \frac{4\pi n_e}{(2\pi m_e kT)^{3/2}} \exp\left(-\frac{E_i}{kT} \right) m_e \, p_i \, dE_i.$$
(7.6.14)

Using the relationship between a cross section and a transition probability given by eq. (7.3.32) and integrating over the initial electron momentum gives

$$\sigma_Z(\nu) = \int \left(d\omega'_{i \to f} \frac{V}{c} \right) (V \, dn_e)$$
(7.6.15)

as the differential cross section for the absorption of *one* photon by *all of the* electrons in the vicinity of *one* ion. Using eqs. (7.6.12) and (7.6.14) in (7.6.15) gives

$$\sigma_Z(\nu) = \sigma_{Th} \, \pi \, Z^2 \left(\frac{\hbar c}{e^2} \right)^2 \left(\frac{e^2/a_0}{h\nu} \right)^3 (n_e \, a_0^3) \left(\frac{2}{\pi} \frac{m_e c^2}{kT} \right)^{1/2} L_{AV}(\nu, T, \rho),$$
(7.6.16)

where

$$L_{AV}(\nu, T, \rho) = \int_0^\infty L(p_i, p_f, R_{sc}) \exp\left(-\frac{E_i}{kT} \right) d\left(\frac{E_i}{kT} \right).$$
(7.6.17)

Since $p_f = \sqrt{p_i^2 + 2 m_e h\nu}$ (see eq. (7.5.20)) and $E_f = E_i + h\nu$, the integral over energy involving $L(p_i, p_f, R_{sc})$ can be written as a function of the photon frequency, the temperature, and, through R_{sc}, also of the density. If screening is neglected, L_{AV} is a function only of $h\nu/kT$. Writing eq. (7.5.36) as

$$p_f - p_i = \frac{2 m_e}{p_f + p_i} h\nu,$$
(7.6.18)

and inserting this into eq. (7.6.9), along with $R_{sc} = \infty$, one has

$$L(p_i, p_f, R_{sc}) \to L(p_i, p_f) = 2 \log_e \left(\frac{(p_f + p_i)^2}{2m_e kT} \frac{kT}{h\nu} \right) \tag{7.6.19}$$

$$= 2 \log_e \left(\frac{kT}{h\nu} \right) + 2 \log_e \left(\frac{(p_f + p_i)^2}{2m_e kT} \right). \tag{7.6.20}$$

Since

$$p_f + p_i = \sqrt{p_i^2 + 2m_e h\nu} + \sqrt{p_i^2} = \sqrt{2m_e} \left\{ \sqrt{E_i + h\nu} + \sqrt{E_i} \right\}, \tag{7.6.21}$$

it follows that

$$\frac{(p_f + p_i)^2}{2m_e kT} = \left[\sqrt{\frac{E_i}{kT} + \frac{h\nu}{kT}} + \sqrt{\frac{E_i}{kT}} \right]^2. \tag{7.6.22}$$

Setting

$$y = \frac{E_i}{kT} \tag{7.6.23}$$

and using eq. (7.6.22) and (7.6.19) in eq. (7.6.17) gives

$$L_{AV} \to \Lambda_{ff} \left(\frac{h\nu}{kT} \right), \tag{7.6.24}$$

where

$$\Lambda_{ff} \left(\frac{h\nu}{kT} \right) = 2 \int_0^\infty e^{-y} \, dy \, \log_e \left\{ \left(\frac{kT}{h\nu} \right) \left[\left(y + \frac{h\nu}{kT} \right)^{1/2} + y^{1/2} \right]^2 \right\} \tag{7.6.25}$$

$$= 2 \log_e \left(\frac{kT}{h\nu} \right) + 4 \int_0^\infty e^{-y} \, dy \, \log_e \left[\left(y + \frac{h\nu}{kT} \right)^{1/2} + y^{1/2} \right]. \tag{7.6.26}$$

The quantity $\Lambda_{ff}(h\nu/kT)$ is a well behaved function of $h\nu/kT$. Results of a numerical integration are presented in Fig. 7.6.1 where $\log_{10} \Lambda_{ff}$ is shown as a function of $\log_{10} x$, with x being defined by

$$x = \frac{h\nu}{kT}. \tag{7.6.27}$$

It is clear that, for $x \geq 10$, Λ_{ff} is essentially a linear function of $\log x$. This fact can be demonstrated analytically. From eq. (7.6.25), Λ_{ff} can be written as

$$\Lambda_{ff}(x) = 4 \int_0^\infty e^{-y} \, dy \, \log_e \left[\left(1 + \frac{y}{x} \right)^{1/2} + \left(\frac{y}{x} \right)^{1/2} \right]. \tag{7.6.28}$$

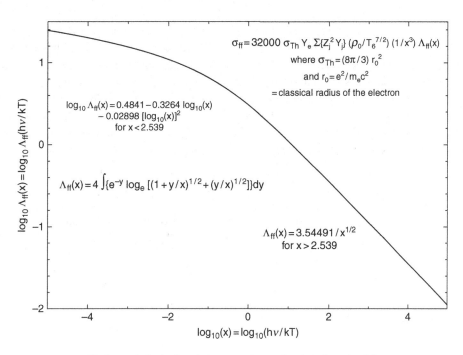

Fig. 7.6.1 The factor $\Lambda_{\rm ff}$ in the free–free cross section as a function of $x = h\nu/kT$

Thanks to the presence of the term e^{-y} in the integrand of eq. (7.6.28), it is possible to expand the term in square brackets and its logarithm in powers of y/x, keeping just the lowest order term in each expansion. Thus,

$$\Lambda_{\rm ff}(x) = 4 \int_0^\infty e^{-y}\, dy\, \log_e\left[1 + \left(\frac{y}{x}\right)^{1/2} + \frac{1}{2}\frac{y}{x} - \frac{1}{8}\left(\frac{y}{x}\right)^2 + \cdots \right] \tag{7.6.29}$$

$$= 4 \int_0^\infty e^{-y}\, dy\, \left[\left(\frac{y}{x}\right)^{1/2} - \frac{1}{2}\frac{y}{x} + \cdots \right] = 4 \int_0^\infty e^{-y}\, dy\, \left(\frac{y}{x}\right)^{1/2} + \cdots . \tag{7.6.30}$$

As $x \to \infty$,

$$\Lambda_{\rm ff}(x) \to \frac{4}{x^{1/2}} \int_0^\infty e^{-y}\, y^{1/2} dy = \frac{4}{x^{1/2}}\frac{\sqrt{\pi}}{2} = \frac{2\sqrt{\pi}}{\sqrt{x}}, \tag{7.6.31}$$

or

$$\Lambda_{\rm ff}(x) = \frac{3.544\,91}{\sqrt{x}}. \tag{7.6.32}$$

A quadratic fit to the numerical values of $\Lambda_{\rm ff}$ in Fig. 7.6.1 at $\log x = 0.0, -2.5$, and -5.0 yields

$$\log_{10}\Lambda_{\rm ff}(x) = 0.4841 - 0.3264\, \log_{10} x - 0.02898\, (\log_{10} x)^2. \tag{7.6.33}$$

Equations (7.6.33) and (7.6.32) match at $\log_{10} x = 0.4046$ ($x = 2.539$) and, there, $\log_{10} \Lambda_{ff} \sim 0.3473$ ($\Lambda_{ff} = 2.2248$). The quadratic in eq. (7.6.33) has a maximum at $\log x = -5.6325$ where $\Lambda_{ff} \sim 10^{1.4033} = 25.31$ and, although it has no practical consequences, Λ_{ff} may be set equal to this maximum value for $\log_{10} x < -5.6325$.

The penultimate step in estimating the contribution of free–free absorption to the opacity begins with the multiplication of the cross section σ_{Z_j} for each type of heavy ion j by Y_j, the number abundance parameter for that ion type, followed by a summation over all ion types to obtain an effective cross section

$$\sigma_{ff}(\nu) = \sum_j Y_j \, \sigma_{Z_j}(\nu) = \sum_j \frac{X_j}{A_j} \sigma_{Z_j}(\nu). \tag{7.6.34}$$

Inserting the cross section for each Z_j from eq. (7.6.16) into eq. (7.6.34) and making the replacement $n_e = (\rho/M_H) Y_e \sim N_A \rho Y_e$, where Y_e is the electron number-abundance parameter, one has

$$\sigma_{ff}(\nu) = \sigma_{Th} \, \pi \left(\frac{\hbar c}{e^2}\right)^2 \left(\frac{\rho \, Y_e}{M_H} a_0^3\right) \left(\frac{e^2/a_0}{kT}\right)^3 \left(\frac{2m_e c^2}{\pi kT}\right)^{1/2} \left(\frac{kT}{h\nu}\right)^3 L_{AV} \sum_j \frac{Z_j^2 X_j}{A_j} \tag{7.6.35}$$

$$= (\sigma_{Th} \rho Y_e) \left[\pi \left(\frac{\hbar c}{e^2}\right)^2 \left(\frac{a_0^3}{M_H}\right) \left(\frac{e^2/a_0}{kT}\right)^3 \left(\frac{2m_e c^2}{\pi kT}\right)^{1/2}\right] \left(\frac{kT}{h\nu}\right)^3 L_{AV} \sum_j Z_j^2 Y_j. \tag{7.6.36}$$

Adopting values for fundamental physical constants determined from 1998 CODATA least squares adjustments (Mohr & Taylor, 2002), the quantity in square brackets in eq. (7.6.36) becomes

$$\pi \, (137.036)^2 \frac{(0.529\,177)^3}{1.672\,62} \left(\frac{27.2114}{86.1734 \, T_6}\right)^2 \left(\frac{2}{\pi} \frac{0.510\,999 \times 10^6}{86.1734 \, T_6}\right)^{1/2}$$

$$= \frac{32\,021.6}{T_6^{7/2}}. \tag{7.6.37}$$

Multiplying $\sigma_{ff}(\nu)$ by Avogadro's number and using the fact that

$$\kappa_{Th} = N_A \, \sigma_{Th} = 0.40 \; \text{cm}^2 \, \text{g}^{-1} \tag{7.6.38}$$

yields the free–free contribution to the absorption coefficient:

$$\kappa_{ff}(\nu) = N_A \, \sigma_{ff}(\nu) = (\kappa_{Th} \, Y_e) \, 32\,021.6 \, \frac{\rho_0}{T_6^{7/2}} \left(\frac{kT}{h\nu}\right)^3 L_{AV} \sum_j Z_j^2 \, Y_j$$

$$= 12\,808.6 \, \frac{\rho_0}{T_6^{7/2}} \left(Y_e \sum_j Z_j^2 \, Y_j\right) \left[\left(\frac{kT}{h\nu}\right)^3 L_{AV}\right] \text{cm}^2 \, \text{g}^{-1}. \tag{7.6.39}$$

As described in Section 7.10, to find the opacity associated with an absorption process requires multiplying the absorption coefficient by a factor which takes stimulated emission into account and then performing an integration which involves a weighting function $W(x)$ given by eqs. (7.10.13) with $x = h\nu/kT$. The factor which takes stimulated emission into account is

$$1 - e^{-x} \tag{7.6.40}$$

and the integral is given by eq. (7.10.16), the result being called a Rosseland mean opacity.

In the present instance, the relevant average is

$$\left\langle \frac{x^3}{\Lambda_{\rm ff}\,(1 - e^{-x})} \right\rangle = \int_0^\infty \frac{x^3}{\Lambda_{\rm ff}\,(1 - e^{-x})}\,W(x)\,dx \div \int_0^\infty W(x)\,dx. \tag{7.6.41}$$

Using the values of $\Lambda_{\rm ff}$ given by eqs. (7.6.32) and (7.6.33), integration gives

$$\left\langle \frac{x^3}{\Lambda_{\rm ff}\,(1 - e^{-x})} \right\rangle = 153.78, \tag{7.6.42}$$

with the stimulated emission factor contributing only 0.3% to the result. Thus, in the absence of any scattering or absorption process other than free–free absorption, the Rosseland mean opacity is of the order of

$$\kappa_{\rm ff} \sim \frac{12\,808.6}{153.78}\,\frac{\rho_0}{T_6^{7/2}}\,\left(Y_{\rm e}\sum_j Z_j^2 Y_j\right)\,{\rm cm^2\,g^{-1}} \tag{7.6.43}$$

$$= 83.292\,\frac{\rho_0}{T_6^{7/2}}\,\left(Y_{\rm e}\sum_j Z_j^2 Y_j\right)\,{\rm cm^2\,g^{-1}}. \tag{7.6.44}$$

In regions where free–free absorption plays an important role, the mean free path of a typical photon can be of the order of

$$l_{\rm ff} = \frac{1}{\rho\,\kappa_{\rm ff}} \sim 10^{-2}\,\frac{T_6^{7/2}}{\rho_0^2}\,{\rm cm}. \tag{7.6.45}$$

This is usually much larger than separations between ions, thus justifying the integration over the electron density described by eq. (7.6.15). For example, when $\rho = 1\,{\rm g\,cm^{-3}}$ and $T_6 = 1$, $l_{\rm ff} \sim 10^{-2}$ cm, $V \sim 10^{-6}\,{\rm cm^2}$, and $n_{\rm e}V \sim 10^{19}$.

7.7 The Kramers semiclassical approximation and Gaunt factors for free–free absorption

The discussion in this section has been introduced because Gaunt factors appear ubiquitously in the literature, in spite of the fact that their inclusion is entirely superfluous when the calculations are done quantum-mechanically. The reader is advised, on the first reading of this book, to skip to the next section and then return to this section before reading Section 7.11 where Gaunt factors are made use of.

The Rosseland mean of the free–free opacity approximated by eq. (7.6.44) is very similar to a standard estimate based on the absorption coefficient derived prior to the development of quantum electrodynamics. In this semiclassical approximation (Kramers, 1923), the cross section for the absorption of a photon of frequency ν by electrons in the velocity range $v \to v + dv$ is

$$d\sigma_{Kramers}(\nu, v) = \frac{4\pi}{3\sqrt{3}} Z^2 \frac{e^6}{h\,c\,m_e^2} \frac{1}{\nu^3} \frac{1}{v} \, dn_e(v), \qquad (7.7.1)$$

where $dn_e(v)$ is the number density of electrons in the stated range. Integration over a Maxwell–Boltzmann distribution produces the total cross section

$$\sigma_{Kramers}(\nu) = \frac{4\pi}{3\sqrt{3}} Z^2 \frac{e^6}{h\,c\,m_e^2} \frac{1}{\nu^3} \left(\frac{2}{\pi} \frac{m_e c^2}{kT} \right)^{1/2}. \qquad (7.7.2)$$

By manipulating fundamental constants, eqs. (7.7.1) and (7.7.2) may be converted into forms which can be compared with the cross section given in the quantum-mechanical plane wave approximation, eq. (7.6.16). Thus,

$$d\sigma_{Kramers}(\nu, v) = \frac{4\pi}{3\sqrt{3}} Z^2 \frac{e^6}{h\,c\,m_e^2} \frac{1}{\nu^3} \frac{h^3}{h^3} \left(\frac{e^2}{m_e c^2} \right)^2 \frac{m_e^2 c^4}{e^4} \left(\frac{\hbar c}{e^2} \right)^2 \frac{e^4}{\hbar^2 c^2} \frac{1}{v} \, dn_e(v)$$

$$= \frac{4\pi}{3\sqrt{3}} Z^2 \left(\frac{e^2}{h\nu} \right)^3 \left(\frac{e^2}{m_e c^2} \right)^2 \left(\frac{\hbar c}{e^2} \right)^2 \frac{h^3}{h\,c\,m_e^2} \frac{m_e^2 c^4}{e^4} \frac{e^4}{\hbar^2 c^2} \frac{1}{v} \, dn_e(v)$$

$$= \frac{4\pi}{3\sqrt{3}} Z^2 \left(\frac{e^2}{m_e c^2} \right)^2 \left(\frac{e^2}{h\nu} \right)^3 \left(\frac{\hbar c}{e^2} \right)^2 \frac{h^2}{\hbar^2} \frac{c}{v} \, dn_e(v)$$

$$= \frac{16\pi^3}{3\sqrt{3}} Z^2 \left(\frac{e^2}{m_e c^2} \right)^2 \left(\frac{e^2}{h\nu} \right)^3 \left(\frac{\hbar c}{e^2} \right)^2 \frac{c}{v} \, dn_e(v)$$

$$= \frac{\pi}{\sqrt{3}} \sigma_{Th} \, \pi \, Z^2 \left(\frac{\hbar c}{e^2} \right)^2 \left(\frac{e^2}{h\nu} \right)^3 \frac{c}{v} \, dn_e(v)$$

$$= \frac{\pi}{\sqrt{3}} \sigma_{Th} \, \pi \, Z^2 \left(\frac{\hbar c}{e^2} \right)^2 \left(\frac{e^2/a_0}{h\nu} \right)^3 a_0^3 \frac{c}{v} \, dn_e(v). \qquad (7.7.3)$$

In the next to last step, the Thomson cross section given by eq. (7.6.11) has been introduced. Integration over a Maxwell–Boltzmann distribution for the electrons produces

$$\sigma_{Kramers}(\nu) = \frac{\pi}{\sqrt{3}} \sigma_{Th} \, \pi \, Z^2 \left(\frac{\hbar c}{e^2} \right)^2 \left(\frac{e^2/a_0}{h\nu} \right)^3 \left(n_e a_0^3 \right) \left(\frac{2}{\pi} \frac{m_e c^2}{kT} \right)^{1/2}. \qquad (7.7.4)$$

Defining the ratio of the total quantum-mechanical cross section in the plane wave approximation to the total semiclassical cross section as

$$\bar{g}_{\text{ff}}(\nu) = \frac{\sigma_Z(\nu)}{\sigma_{\text{Kramers}}(\nu)}, \tag{7.7.5}$$

comparison of eq. (7.7.5) with eq. (7.6.16) shows that

$$\bar{g}_{\text{ff}}(\nu) = \frac{\sqrt{3}}{2\pi} L_{\text{AV}}(\nu, T, \rho) \tag{7.7.6}$$

where $L_{\text{AV}}(\nu, T, \rho)$ is given, in the general case, by eqs. (7.6.17) and (7.6.9). When screening is neglected, $L_{\text{AV}}(\nu, T, \rho)$ is given by the function Λ_{ff} defined by eqs. (7.6.25)–(7.6.28) and approximated by eqs. (7.6.32) and (7.3.33). The quantity $\bar{g}_{\text{ff}}(\nu)$ is called a Gaunt factor (after J. A. Gaunt, 1930).

Beginning with eqs. (7.6.16) and (7.6.17) and reversing the steps which lead from eq. (7.7.1) to eq. (7.7.3), one obtains

$$\sigma_Z(\nu) = \frac{2}{3} Z^2 \frac{e^6}{h (m_e c)^2} \frac{1}{\nu^3} \int \frac{m_e c}{p_i} L(p_i, p_f, R_{\text{sc}}) dn_e, \tag{7.7.7}$$

or, inserting $L(p_i, p_f, R_{\text{sc}})$ from eq. (7.6.9),

$$\sigma_Z(\nu) = \frac{2}{3} Z^2 \frac{e^6}{h (m_e c)^2} \frac{1}{\nu^3} \int \frac{m_e c}{p_i} \log_e \left(\frac{(p_f + p_i)^2 + (\hbar/R_{\text{sc}})^2}{(p_f - p_i)^2 + (\hbar/R_{\text{sc}})^2} \right) dn_e. \tag{7.7.8}$$

Comparing with eq. (7.7.1), one may identify the quantity

$$g_{\text{ff}}(\nu, p_i) = \frac{\sqrt{3}}{2\pi} \log_e \left(\frac{(p_f + p_i)^2 + (\hbar/R_{\text{sc}})^2}{(p_f - p_i)^2 + (\hbar/R_{\text{sc}})^2} \right) \tag{7.7.9}$$

as the factor by which the semiclassical differential cross section per electron must be multiplied to obtain the quantum-mechanical differential cross section in the plane wave approximation when screening is taken into account. When screening is neglected,

$$g_{\text{ff}}(\nu, p_i) = \frac{\sqrt{3}}{\pi} \log_e \left(\frac{p_f + p_i}{p_f - p_i} \right). \tag{7.7.10}$$

The factors $g_{\text{ff}}(\nu, p_i)$ are also known as Gaunt factors. The two types of Gaunt factor are related by

$$\bar{g}_{\text{ff}}(\nu) = \int g_{\text{ff}}(\nu, p_i) \frac{1}{p_i} dn_e \div \int \frac{1}{p_i} dn_e. \tag{7.7.11}$$

Comparing eqs. (7.7.10) and (7.7.11), it is evident that screening has the effect of reducing the Gaunt factors.

In the literature, the free–free opacity coefficient is sometimes written in the form

$$\kappa_{ff}(\nu) = N_A \, \bar{g}_{ff} \, \sigma_{Kramers}(\nu). \tag{7.7.12}$$

Choosing the numerical coefficient from eqs. (7.7.4), (7.6.35), and (7.6.36),

$$\kappa_{ff}(\nu) \sim 23\,232 \, \bar{g}_{ff} \, \frac{\rho_0}{T_6^{7/2}} \left(Y_e \sum_j Z_j^2 \, Y_j \right) \left(\frac{kT}{h\nu} \right)^3 \; cm^2 \; g^{-1}. \tag{7.7.13}$$

Thus, the absorption coefficient is taken to be proportional to

$$\left(\frac{kT}{h\nu} \right)^3 = \left(\frac{1}{x} \right)^3. \tag{7.7.14}$$

The next order of business is to calculate the Rosseland mean opacity described in Section 7.10 and estimated in Section 7.6, eqs. (7.6.39)–(7.6.44). Since

$$\int_0^\infty W(x) \, \frac{x^3}{1 - e^{-x}} \, dx \div \int_0^\infty W(x) \, dx = 195.6, \tag{7.7.15}$$

$$\kappa_{ff} \sim 119 \, \bar{g}_{ff} \frac{\rho_0}{T_6^{7/2}} \left(Y_e \sum_j Z_j^2 Y_j \right) \; cm^2 \; g^{-1}. \tag{7.7.16}$$

It is something of an accident that, with the assumption that $\bar{g}_{ff} \sim 1$, the estimate of the Rosseland mean of the free–free opacity given by eq. (7.7.16) is so close to the estimate given by eq. (7.6.44), which has been obtained in a more complicated way.

Taking the Coulomb interaction between electrons and ions into account in constructing a more rigorously correct free–free cross section is not a trivial matter. Russell M. Kulsrud (1954) argues that, due to the perturbative effects of surrounding ions, the normalization of an electron wave function at high densities differs considerably from the normalization of the wave function in the absence of surrounding ions, and that the Gaunt factor given by eq. (7.7.10) is a good first approximation. In the case of bound–free absorption, there is no question but that the ejected electron is better described by a Coulomb-distorted plane wave centered on the contributing ion than by a simple plane wave distinguished only by its direction of motion. In the case of free–free absorption, the situation is not as clear cut. For example, in hydrogen-rich matter at $\rho = 100$ g cm^{-3} and $T_6 = 1$, the deBroglie wavelength of an average electron with energy $\sim (3/2)kT$ is $\sim 10^{-8}$ cm, about six times larger than the average separation between protons. Under these conditions, it is clear that there are no bound states for electrons about protons, and it is reasonable to suppose that a plane wave is a better description of a typical free electron than is a Coulomb-distorted plane wave centered on a single proton.

Nevertheless, it is standard to assume that, at densities and temperatures such the deBroglie wavelength of a typical electron is small compared with the internuclear spacing, Coulomb-distorted plane waves are the appropriate choice. An idea of the quantitative effect follows from consideration of the amplitude at the origin of a Coulomb-distorted plane wave about an isolated charge center. The square of this amplitude is described by the integrand of eq. (6.8.6), where η is defined by the first identity in eq. (7.4.33). Thus, the cross section is expected to depend on the parameter

$$\eta = Z \frac{e^2}{\hbar v} = Z \frac{e^2}{\hbar c} \left(\frac{m_e c^2}{kT} \right)^{1/2} \left(\frac{E}{kT} \right)^{-1/2} = 0.3973 \frac{Z}{\sqrt{T_6}} \frac{1}{\sqrt{y}}, \qquad (7.7.17)$$

where y is given by eq. (7.6.23), and an integration over the initial electron probability distribution will produce a result which is a function of $Z/\sqrt{T}$ as well as of $x = h\nu/kT$.

Naively, one might expect the exact cross section *in this approximation* to be the cross section given by using simple undistorted plane waves for the electrons participating in the process multiplied by the squares of the amplitudes at the coordinate origin of the Coulomb-distorted plane waves for both electrons. However, the perturbation Hamiltonian involves a spatial derivative and, as shown by A. Sommerfeld (1931), this fact introduces an additional factor of $\exp(-2\pi \eta_f)$, where η_f is the value of η for the final electron. Thus, the exact solution for the Gaunt factor in this approximation is the one given by eq. (7.7.10) multiplied by

$$\frac{2\pi \eta_i}{1 - \exp(-2\pi \eta_i)} \exp(-2\pi \eta_f) \frac{2\pi \eta_f}{1 - \exp(-2\pi \eta_f)} = \frac{2\pi \eta_i}{1 - \exp(-2\pi \eta_i)} \frac{2\pi \eta_f}{\exp(2\pi \eta_f) - 1}, \qquad (7.7.18)$$

where η_i is the value of η for the initial electron.

Using the solution obtained by Sommerfeld and cast in a convenient form by L. C. Biedenharn (1956), W. J. Karzas and R. Latter (1961) construct Gaunt factors averaged over a Maxwell–Boltzmann electron distribution for a wide range of physical conditions and express them as functions of the quantities $x = h\nu/kT$ and Z^2 times the ratio of the Rydberg energy unit to kT:

$$\gamma^2 = Z^2 \frac{e^2/2a_0}{kT} = Z^2 \frac{13.602 \text{ eV}}{86.17 \text{ eV } T_6} = \frac{0.1578 Z^2}{T_6}. \qquad (7.7.19)$$

It is evident that γ^2 is simply η^2 as given by eq. (7.7.17) when $y = E/kT = 1$.

For values of γ^2 small enough that Coulomb distortions might be expected to be small, the Karzas and Latter results for velocity-averaged Gaunt factors are similar to those found when the electron wave functions are taken as undistorted plane waves. Placing eqs. (7.6.25)–(7.6.28) into (7.7.6), the latter Gaunt factors are

$$\bar{g}_{ff}(x) = \frac{\sqrt{3}}{2\pi} \frac{3.5449}{\sqrt{x}} = \frac{0.9772}{\sqrt{x}} \qquad (7.7.20)$$

for $x > 2.54$ and

$$\log_{10} \bar{g}_{ff}(x) = -0.075\,53 - 0.3264 \, \log_{10} x - 0.028\,98 \, (\log_{10} x)^2 \qquad (7.7.21)$$

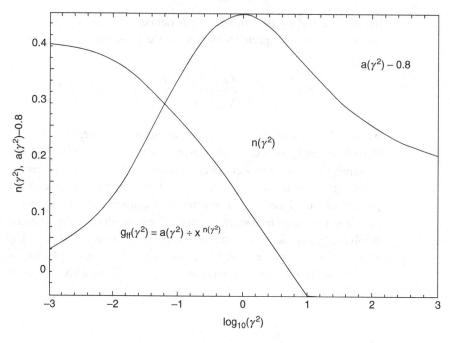

A fit to the Karzas–Latter free–free Gaunt factor when $x = h\nu/kT > 1$

for $x \leq 2.54$. For example, for $x > 2.5$, $\bar{g}_{ff}$ given by eq. (7.7.19) differs from the Karzas &
Latter results for $\gamma^2 = 10^{-3}$ by 10% at most. For $x < 2.5$, $\bar{g}_{ff}$ given by eq. (7.7.20) differs
from the Karzas & Latter results for $\gamma^2 = 10^{-3}$ by 5% at most.

The Karzas and Latter results for $x \geq 1$ can be approximated by the function

$$\bar{g}_{ff}(x, \gamma^2) = \frac{a(\gamma^2)}{x^{n(\gamma^2)}}, \qquad (7.7.22)$$

where $a(\gamma^2)$ and $n(\gamma^2)$ are shown in Fig. 7.7.1. For $10^{-3} \leq \gamma^2 \leq 10$, $a(\gamma^2)$ and $n(\gamma^2)$
are fitted tolerably well by

$$a(\gamma^2) = 1.09 + 0.21 \, \sin\left[\frac{\pi}{3}\left(\log_{10}\gamma^2 + \frac{3}{2}\right)\right] \qquad (7.7.23)$$

and

$$n(\gamma^2) = 0.181\,625 - 0.158 \, \log_{10}\gamma^2 - 0.023\,625\left(\log_{10}\gamma^2\right)^2. \qquad (7.7.24)$$

For $\gamma^2 > 10^3$, there are no data points, but, for $1 \leq \log_{10}\gamma^2 \leq 3.2$,

$$a(\gamma^2) = 1.353 - 0.180 \, \log_{10}\gamma^2 + 0.0265\left(\log_{10}\gamma^2\right)^2 \qquad (7.7.25)$$

and

$$n(\gamma^2) = 0 \qquad (7.7.26)$$

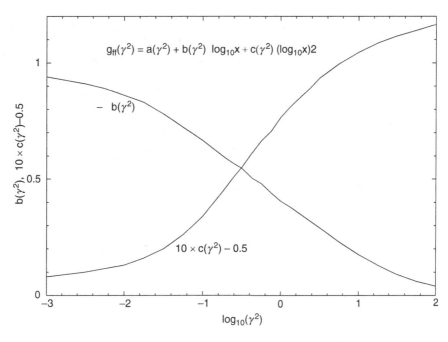

$$g_{ff}(\gamma^2) = a(\gamma^2) + b(\gamma^2)\ \log_{10}x + c(\gamma^2)\ (\log_{10}x)2$$

Fig. 7.7.2   A fit to the Karzas–Latter free–free Gaunt factor when $x = h\nu/kT > 1$

are reasonable approximations. For $\log \gamma^2 > 3.2$, $a(\gamma^2) = 1.048$ and $n(\gamma^2) = 0$ seem plausible extrapolations.

For $x = h\nu/kT \leq 1$, the Karzas and Latter results for $-3 \leq \log_{10} \gamma^2 \leq 3$ can be fitted with quadratics of the form

$$\bar{g}_{ff} = a + b \log_{10} u + c \left(\log_{10} u\right)^2, \tag{7.7.27}$$

where $a = \bar{g}_{ff}(u = 1)$ and b and c are as shown in Fig. 7.7.2. Rough approximations to c and b are, respectively,

$$c = 0.1125 + 0.0575 \frac{2}{\pi} \tan^{-1}\left[1.77(\log_{10} \gamma^2 + 0.3)\right] \tag{7.7.28}$$

and $b = -0.95$ for $\log_{10} \gamma^2 < -2.3$, $b = 0.0$ for $\log_{10} \gamma^2 > 1.8$, and

$$b = -0.95 + 0.2317(\log_{10} \gamma^2 + 2.3) \tag{7.7.29}$$

for $-2.3 \leq \log_{10} \gamma^2 \leq 1.8$.

7.8 Spontaneous emission between bound atomic states

In order to explore what is involved in estimating a cross section for the absorption of a photon which excites an atomic system from a given energy state to a higher energy state, it is convenient to first examine the lifetime of the atomic system against a transition from the higher energy state to the lower energy state with the emission of a photon. A detailed balance argument can then be used to derive the cross section for absorption (Section 7.9).

Consider a transition from atomic state u_j to atomic state u_i with the emission of a photon of energy

$$h\nu_{ij} = E_j - E_i, \tag{7.8.1}$$

where E_j and $E_i(<E_j)$ are the energies of the two atomic states. The formalism of Sections 7.2 and 7.3 gives the differential transition probability per unit time as

$$d\omega_{ij} = \frac{2\pi}{\hbar} \left| \int u_i^* \langle 1_{k\lambda} \left| H^{\text{int}} \right| 0_{k\lambda} \rangle u_j d\tau \right|^2 d\rho_i = \frac{2\pi}{\hbar} \frac{c^2\hbar}{\nu_k V} \left(\frac{e}{m_e c} \right)^2 |M_{ij}|^2 d\rho_i, \tag{7.8.2}$$

where

$$M_{ij} = \int u_i^* e^{-i\mathbf{k}\cdot\mathbf{r}} \hat{\epsilon}_{k\lambda} \cdot \mathbf{p}_{\text{op}} u_j d\tau \tag{7.8.3}$$

and

$$d\rho_i = \frac{1}{c^3} V d\Omega_\nu \nu^2 \frac{d\nu}{dE_\nu}. \tag{7.8.4}$$

In the hydrogenic approximation, the wave functions for an outermost electron are solutions of the equation

$$H_{\text{op}} u = \left(-\frac{\hbar^2}{2m_e} \nabla^2 - \frac{Ze^2}{r} \right) u = Eu, \tag{7.8.5}$$

where Ze is the effective charge of the nucleus (bare or shielded by more tightly bound electrons). The probability function $|u_i|^2$ is large over a region of dimensions

$$a_i = \frac{\hbar}{\sqrt{2m_e E_i}} = 0.53 \cdot 10^{-8} \sqrt{\frac{13.6\,\text{eV}}{E_i}} \text{ cm}, \tag{7.8.6}$$

which is small compared with the wavelength of the emitted photon,

$$\lambda_{ij} = \frac{hc}{E_j - E_i} = 1.46 \cdot 10^{-5} \frac{13.6\,\text{eV}}{E_j - E_i} \text{ cm}. \tag{7.8.7}$$

Thus, where the electron probability density is large, the quantity $\mathbf{k} \cdot \mathbf{r}$ in $e^{-i\mathbf{k}\cdot\mathbf{r}}$ is of order

$$ka \sim \frac{2\pi}{\lambda_{ij}} a_i = \frac{E_j - E_i}{\sqrt{2m_e c^2 E_i}} < \sqrt{\frac{-E_i}{2m_e c^2}} \ll 1, \tag{7.8.8}$$

and, in first approximation, one can set $e^{-i\mathbf{k}\cdot\mathbf{r}} \sim 1$ in eq. (7.8.3).

Using the definitions of the linear momentum and energy operators given by eqs. (7.1.22) and (7.8.5), respectively, one may establish that

$$\mathbf{p}_{\text{op}} \phi = i \frac{m_e}{\hbar} \left(H_{\text{op}} \mathbf{r} - \mathbf{r} H_{\text{op}} \right) \phi, \tag{7.8.9}$$

where ϕ is an arbitrary scalar wave function. The proof is most easily accomplished by taking the required derivatives in a Cartesian coordinate system. Replacing $\mathbf{p}_{\text{op}}$ in eq. (7.8.3) by its equivalent commutation operator given by eq. (7.8.9) and setting $e^{-i\mathbf{k}\cdot\mathbf{r}} = 1$, two integrations by parts yield

$$M_{ij} = i\,\frac{m_e}{\hbar}\left(\int (H_{\text{op}}u_i)^*\mathbf{r}\cdot\hat{\epsilon}_{k\lambda}u_j d\tau - \int u_i^*\mathbf{r}\cdot\hat{\epsilon}_{k\lambda}H_{\text{op}}u_j d\tau\right) \tag{7.8.10}$$

$$= i\,\frac{m_e}{\hbar}(E_i - E_j)\int u_i^*\hat{\epsilon}_{k\lambda}\cdot\mathbf{r}u_j d\tau \tag{7.8.11}$$

$$= -i\,m_e\,2\pi\nu_{ij}\,\hat{\epsilon}_{k\lambda}\cdot\mathbf{r}_{ij}, \tag{7.8.12}$$

where, in eq. (7.8.12),

$$\mathbf{r}_{ij} = \int u_i^*\mathbf{r}u_j d\tau \tag{7.8.13}$$

is called a dipole moment.

Inserting this result into eq. (7.8.2) gives

$$d\omega_{ij} = \frac{2\pi}{\hbar}\frac{c^2\hbar}{\nu_k V}\left(\frac{e}{m_e c}\right)^2 (m_e\,2\pi\nu_{ij})^2\,(\hat{\epsilon}_{k\lambda}\cdot\mathbf{r}_{ij})^2\,V\frac{\nu^2 d\Omega_\nu}{c^3 h} \tag{7.8.14}$$

$$= \frac{1}{h}\frac{e^2}{c^3}(2\pi\nu_{ij})^3(\hat{\epsilon}_{k\lambda}\cdot\mathbf{r}_{ij})^2 d\Omega_\nu. \tag{7.8.15}$$

The two unit polarization vectors $\hat{\epsilon}_{k\lambda 1}$, $\hat{\epsilon}_{k\lambda 2}$ and the unit directional vector $\hat{\mathbf{k}}$ of the photon form a Cartesian coordinate system. Taking the z direction as that of the $\mathbf{k}$ vector and the x and y directions as those of $\hat{\epsilon}_{k\lambda 1}$ and $\hat{\epsilon}_{k\lambda 2}$, respectively, one has $(\hat{\epsilon}_{k\lambda 1}\cdot\mathbf{r})^2 = r^2\sin^2\theta\cos^2\phi$, $(\hat{\epsilon}_{k\lambda 2}\cdot\mathbf{r})^2 = r^2\sin^2\theta\sin^2\phi$, giving $(\hat{\epsilon}_{k\lambda 1}\cdot\mathbf{r})^2 + (\hat{\epsilon}_{k\lambda 1}\cdot\mathbf{r})^2 = r^2\sin^2\theta$. The average of $\sin^2\theta$ over the surface of a sphere is 2/3, so a sum over the two polarization states and an average over $d\Omega_\nu$ yields the total decay probability per unit time:

$$\omega_{j\to i} = \frac{1}{t_{ij}} = \frac{1}{h}\frac{8\pi}{3}\frac{e^2}{c^3}(2\pi\nu_{ij})^3 r_{ij}^2 = \frac{4}{3}\frac{e^2}{c^3}(2\pi\nu_{ij})^3\frac{r_{ij}^2}{\hbar}, \tag{7.8.16}$$

where t_{ij} is defined as the lifetime of state j against transitions into state i and $r_{ij} = |\mathbf{r}_{ij}|$.

Using eq. (7.8.1) and the fact that wavelength $\lambda_{ij} = c/\nu_{ij}$, one can write

$$\omega_{j\to i} = \frac{1}{t_{ij}} = \frac{4}{3}\frac{e^2}{c^3}(2\pi\nu_{ij})^3\frac{r_{ij}^2}{\hbar} = \frac{4}{3}(2\pi)^2\frac{e^2}{\hbar c}\left(\frac{r_{ij}}{\lambda_{ij}}\right)^2\frac{E_j - E_i}{\hbar} \tag{7.8.17}$$

or

$$t_{ij} = 2.61\left(\frac{\lambda_{ij}}{r_{ij}}\right)^2\frac{13.527\,\text{eV}}{E_j - E_i}\,t_{\text{atomic}}, \tag{7.8.18}$$

where

$$t_{\text{atomic}} = \hbar/(m_e e^4/\hbar^2) = 0.482\times10^{-16}\,\text{s} \tag{7.8.19}$$

is Planck's constant divided by twice the binding energy of the ground state of the hydrogen atom. For a transition from the first excited state to the ground state of a hydrogenic atom, $r_{ij} = 0.745a$, where a is the Bohr radius for the ground state (see immediately below). Thus, for hydrogen, $t_{21} \sim 3.7 \times 10^{-11}$ s.

It is instructive to calculate the matrix element r_{ij} for the concrete case of a hydrogen-like ion in which the electron makes a transition from the first excited state to the ground state. The relevant solutions of eq. (7.8.5) are

$$u_i = u_{100} = \frac{1}{\sqrt{\pi a^3}} \exp\left(-\frac{r}{a}\right), \tag{7.8.20}$$

$$u_j = u_{200} = \frac{1}{\sqrt{32\pi a^3}} \left(2 - \frac{r}{a}\right) \exp\left(-\frac{r}{2a}\right), \tag{7.8.21}$$

$$u_j = u_{210} = \frac{1}{\sqrt{32\pi a^3}} \frac{z}{a} \exp\left(-\frac{r}{2a}\right), \text{ or} \tag{7.8.22}$$

$$u_j = u_{21\pm1} = \frac{1}{\sqrt{64\pi a^3}} \frac{(x \pm iy)}{a} \exp\left(-\frac{r}{2a}\right), \tag{7.8.23}$$

where $a = a_0/Z = (\hbar^2/m_e e^4)/Z$. The subscripts of u in the middle of each equation set give, from left to right, the principle quantum number n, the angular momentum quantum number l, and the azimuthal quantum number m. Energies of the n levels are degenerate with respect to l and m and are given, as in the Bohr theory, by $E_n = -m_e e^4 Z^2/(2\hbar^2 n^2)$. States described by eqs. (7.8.20) and (7.8.21) are s states with zero angular momentum, and states described by eqs. (7.8.22) and (7.8.23) are p states which have angular momentum $\sqrt{l(l+1)}\hbar = \sqrt{2}\hbar$, with the projection of angular momentum on the z axis being positive in the state with azimuthal quantum number $m = +1$.

Because the ground state and the excited s state have even parity and $\mathbf{r}$ has odd parity, it is clear that $r_{ij} = r_{(200)(100)} = 0$. If one chooses the direction of the photon to be along the positive z axis, $\hat{\epsilon}_{k\lambda}$ lies in the xy plane; since xz and yz have odd parity, $r_{ij} = r_{(210)(100)} = 0$. This leaves only transitions from the two circularly polarized states $u_{21\pm1}$ to the ground state to consider. One has

$$\mathbf{r}_{ij} = \frac{1}{8\pi a^3} \int \exp\left(-\frac{r}{a}\right) \left(\frac{x^2}{a}\hat{x} \pm i\frac{y^2}{a}\hat{y}\right) \exp\left(-\frac{r}{2a}\right) d\tau \tag{7.8.24}$$

$$= \frac{1}{8\pi a^4} (\hat{x} \pm i\hat{y}) \frac{1}{3} \int_0^\infty \int_{-1}^1 \int_0^{2\pi} \exp\left(-\frac{3r}{2a}\right) r^2 r^2 dr d\mu d\phi \tag{7.8.25}$$

$$= (\hat{x} \pm i\hat{y}) a \frac{2^7}{3^5} = 0.527 a (\hat{x} \pm i\hat{y}), \tag{7.8.26}$$

where the subscripts i and j are given by $i = (21 \pm 1)$ and $j = (100)$. Equation (7.8.24) follows from (7.8.23) because the integral of the square of each of the Cartesian coordinates is the same and equal to one third of the integral of r^2. Finally,

$$r_{ij} = r_{(21\pm1)(100)} = |\mathbf{r}_{ij}| = \sqrt{2}\,\frac{2^7}{3^5}\,a = 0.745\,a. \tag{7.8.27}$$

As shown in Section 7.2, because the amplitude of the initial wave function decays exponentially according to $e^{-t/2t_{ij}}$, the emitted photon has an angular frequency which differs by $d\omega$ from the primary frequency $\omega_{ij} = (E_i - E_j)/\hbar$ with a probability per unit frequency given by (see eq. (7.2.34))

$$\frac{dP_{ij}(\omega)}{d\omega} = \frac{1}{\pi}\,\frac{(1/2t_{ij})}{(\omega_k - \omega_{ij})^2 + 1/(2t_{ij})^2}. \tag{7.8.28}$$

More highly excited states can be assigned a width

$$\Delta E_j = \sum_{i<j}\frac{\hbar}{t_{ij}}, \tag{7.8.29}$$

where the summation extends over all spontaneous transitions accessible to state i. The differential probability that the emitted photon has a frequency other than the primary frequency ν_{ij} is given by

$$\frac{dP_{ij}(\nu)}{d\nu} = \frac{1}{\pi}\,\frac{\gamma_{ij}/4\pi}{(\nu - \nu_{ij})^2 + (\gamma_{ij}/4\pi)^2}, \tag{7.8.30}$$

where

$$\gamma_{ij} = \frac{\Delta E_i + \Delta E_j}{\hbar}. \tag{7.8.31}$$

The quantity γ_{ij} is called the natural breadth of the line. The integral of $dP_{ij}(\nu)/d\nu$ over $d\nu$ from $-\infty$ to $+\infty$ is unity.

7.9 Detailed balance, stimulated emission, bound–bound cross sections, and line broadening

Since matter and radiation in the stellar interior are, to an extremely good approximation, in local thermodynamic equilibrium, every microscopic absorption process is balanced by an emission process which is just the inverse of the absorption process.

The rate per unit volume at which a given absorption process occurs is

$$W_{i\to j} = \left(\frac{2\pi}{\hbar}|M_{ji}|^2\rho_{ja}\bar{n}_{k\lambda}\right)dg_{ik\lambda}n_{ia}, \tag{7.9.1}$$

where M_{ji} is the matrix element for the process, ρ_{ja} is the final density of states for the atomic system after absorption,

$$dg_{ik\lambda} = \frac{\nu_i^2\,d\nu_i\,d\Omega_{ik\lambda}}{c^3} \tag{7.9.2}$$

is the volume density of states of photons of the type absorbed, $\bar{n}_{k\lambda}$ is the occupation number for the photon state $k\lambda$, and n_{ia} is the density of atomic absorbers of type i.

The rate at which the inverse emission process occurs is

$$W_{j \to i} = \left(\frac{2\pi}{\hbar} |M_{ij}|^2 \rho_{ik\lambda} (1 + \bar{n}_{k\lambda}) \right) n_{j\mathrm{a}} \mathrm{d}g_{j\mathrm{a}}, \tag{7.9.3}$$

where

$$\rho_{ik\lambda} = \frac{\mathrm{d}g_{ik\lambda}}{\mathrm{d}(h\nu_i)} \tag{7.9.4}$$

is the density of states for the emitted photons, $n_{j\mathrm{a}}$ is the density of emitters, and

$$\mathrm{d}g_{j\mathrm{a}} = \rho_{j\mathrm{a}} \mathrm{d}E_j = \rho_{j\mathrm{a}} \mathrm{d}(h\nu) \tag{7.9.5}$$

ensures that the number of initial matter states involved in the emission process is equal to the number of final matter states appearing in the absorption process. Because of the symmetry of the Hamiltonian, $|M_{ij}| = |M_{ji}|$.

From eqs. (7.9.4) and (7.9.5), it follows that

$$\rho_{j\mathrm{a}} \mathrm{d}g_{ik\lambda} = \rho_{ik\lambda} \mathrm{d}g_{j\mathrm{a}}. \tag{7.9.6}$$

Hence, the requirement that

$$W_{i \to j} = W_{j \to i} \tag{7.9.7}$$

means that

$$\bar{n}_{k\lambda} n_{i\mathrm{a}} = (1 + \bar{n}_{k\lambda}) n_{j\mathrm{a}}. \tag{7.9.8}$$

But, if the system is in statistical equilibrium and all states are distinguishable, the complication of invoking statistical weights is bypassed and

$$\frac{n_{j\mathrm{a}}}{n_{i\mathrm{a}}} = \exp \left(-\frac{h\nu}{kT} \right). \tag{7.9.9}$$

Combining eqs. (7.9.8) and (7.9.9), one has

$$\bar{n}_{k\lambda} = \frac{1}{\exp(h\nu/kT) - 1}, \tag{7.9.10}$$

demonstrating that $\bar{n}_{k\lambda}$ is indeed the occupation number, or the number of photons per unit cell of phase space (see eq. (4.11.3) when $\alpha = 0$). Turning the argument around, one could just as well have begun with eqs. (7.9.9) and (7.9.10) and derived eq. (7.9.6) from eq. (7.9.7).

The term in eq. (7.9.3) which is proportional to the occupation number $\bar{n}_{k\lambda}$ demonstrates that some of photons absorbed from the radiation field by the process that gives rise to eq. (7.9.1) are replaced by photons having exactly the same properties as those of the absorbed photons. This is the phenomenon of stimulated emission. The process that gives rise to the term in eq. (7.9.3) that is independent of the number of photons present is referred to as spontaneous emission. The two emission processes occur at the rates

$$W_{j \to i}^{\mathrm{stimulated}} = \frac{2\pi}{\hbar} |M_{ij}|^2 \rho_{ik\lambda} \bar{n}_{k\lambda} n_{j\mathrm{a}} \mathrm{d}g_{j\mathrm{a}} \tag{7.9.11}$$

and

$$W_{j\to i}^{\text{spontaneous}} = \frac{2\pi}{\hbar}|M_{ij}|^2 \rho_{ik\lambda} n_{ja}\text{dg}_{ja}, \tag{7.9.12}$$

respectively. The net or "true" rate of absorption is

$$W_{i\to j}^{\text{absorption}} = W_{i\to j} - W_{j\to i}^{\text{stimulated}}. \tag{7.9.13}$$

Making use of eqs. (7.9.6) and (7.9.9), it follows from eqs. (7.9.1) and (7.9.11) that

$$W_{i\to j}^{\text{absorption}} = \frac{2\pi}{\hbar}|M_{ji}|^2 \rho_{ja}\,(\bar{n}_{k\lambda}\text{dg}_{ik\lambda})\,n_{ia}\left[1 - \exp\left(-\frac{h\nu}{kT}\right)\right]. \tag{7.9.14}$$

Detailed balance is still preserved since, if eq. (7.9.7) is satisfied, it is also true that

$$W_{i\to j}^{\text{absorption}} = W_{j\to i}^{\text{spontaneous}}. \tag{7.9.15}$$

Equation (7.9.14) demonstrates that, when forming an absorption coefficient to use in constructing an opacity, the cross section for a bound–bound absorption process must be multiplied by the factor

$$f_{\text{stimulated}} = \frac{\sigma_{\text{abs}}(\text{corrected})}{\sigma_{\text{abs}}(\text{uncorrected})} = 1 - \exp\left(-\frac{h\nu}{kT}\right). \tag{7.9.16}$$

The detailed balance arguments presented here apply just as well to the bound–free and free–free absorption processes, so the cross sections for these processes must also be multiplied by the factor $f_{\text{stimulated}}$. On the other hand, electron scattering is not an absorption process, so no correction is to be made in transforming from a cross section to an opacity.

The rate of spontaneous emission may be related to the cross section for the inverse absorption process. Using eqs. (7.9.6) and (7.9.2) in eq. (7.9.12), one has

$$W_{j\to i}^{\text{spontaneous}} = \frac{2\pi}{\hbar}|M_{ji}|^2 \rho_{ik\lambda} n_{ja}\text{dg}_{ia} \tag{7.9.17}$$

$$= \frac{2\pi}{\hbar}|M_{ij}|^2 \rho_{ja}\text{dg}_{ik\lambda} n_{ja} \tag{7.9.18}$$

$$= \left(\frac{2\pi}{\hbar}|M_{ij}|^2 \rho_{ja}\right)\left(\frac{2V\nu^2 \text{d}\nu \text{d}\Omega_{k\lambda}}{c^3}\right)n_{ja}, \tag{7.9.19}$$

where the 2 in the density of states factor takes into account the two polarization directions for the photon. Integrating over $\text{d}\Omega_{k\lambda}$ and using eq. (7.3.32), it follows that

$$W_{j\to i}^{\text{spontaneous}} = \frac{n_{ja}}{t_{ij}} = \left(\frac{c}{V}\sigma_{i\to j}\right)8\pi \frac{V\nu^2 \text{d}\nu}{c^3}n_{ja}, \tag{7.9.20}$$

where $t_{ij} = t_{j\to i}$ is the lifetime against spontaneous emission. Rearranging, one has

$$\frac{1}{t_{ij}} = 8\pi\sigma_{i\to j}(\nu)\frac{\nu^2}{c^2}\text{d}\nu \tag{7.9.21}$$

or

$$\sigma_{i \to j}(\nu) = \frac{1}{8\pi} \frac{c^2}{\nu^2} \frac{1}{t_{ij}} \delta(\nu - \nu_{ij}). \tag{7.9.22}$$

The appearance of the delta function is a consequence of assuming that all atomic states have sharply defined energies. Taking energy widths into account (Section 7.8), one obtains the more general result that

$$\sigma_{i \to j}(\nu) = \frac{1}{8\pi} \frac{c^2}{\nu^2} \frac{1}{t_{ij}} \frac{dP_{ij}(\nu)}{d\nu}, \tag{7.9.23}$$

where $dP_{ij}(\nu)/d\nu$ is defined by eqs. (7.8.30), (7.8.31), and (7.8.29). Using t_{ij} given by eq. (7.8.16), it follows that

$$\sigma_{i \to j}(\nu) = \frac{4\pi^2}{3} \left(\frac{e^2}{c\hbar} \right) \left(\frac{\nu_{ij}}{\nu} \right)^2 \left(\nu_{ij} \frac{dP_{ij}}{d\nu} \right) r_{ij}^2. \tag{7.9.24}$$

Further insight into eqs. (7.9.22) and (7.9.23) can be obtained by using Einstein's argument involving A and B coefficients (Einstein, 1917). These coefficients are related to the previously defined Ws by

$$W_{j \to i}^{\text{spontaneous}} \to n_{ja} A_{ij}, \tag{7.9.25}$$

$$W_{j \to i}^{\text{stimulated}} \to n_{ja} B_{ij} u_\nu, \tag{7.9.26}$$

and

$$W_{i \to j}^{\text{absorption}} \to n_{ia} B_{ji} u_\nu, \tag{7.9.27}$$

where u_ν is the energy density in the radiation field in thermal equilibrium. The detailed balance equation equivalent of eq. (7.9.7) is

$$n_{ja}(A_{ij} + B_{ij} u_\nu) = n_{ia} B_{ji} u_\nu, \tag{7.9.28}$$

where rates on both sides are in units of reactions per unit volume and per unit of time.
From eqs. (7.9.28) and (7.9.9), it follows that

$$u_\nu = \frac{A_{ij}/B_{ji}}{\exp(h\nu/kT) - B_{ij}/B_{ji}}. \tag{7.9.29}$$

But, Bose–Einstein statistics of the radiation field gives (see eq. (4.11.10))

$$u_\nu = \frac{8\pi}{c^3} \frac{h\nu \, \nu^2}{\exp(h\nu/kT) - 1}. \tag{7.9.30}$$

Equations (7.9.29) and (7.9.30) are compatible at all temperatures only if

$$B_{ij} = B_{ji} \tag{7.9.31}$$

and

$$B_{ji} = \frac{1}{8\pi} A_{ij} \frac{c^3}{v^2} \frac{1}{hv}.$$ (7.9.32)

The rate at which energy is absorbed per unit volume is

$$\dot{E}(v) = (hv)u_v B_{ji} n_{ia} = u_v \frac{c}{8\pi} A_{ij} \frac{c^2}{v^2} n_{ia}.$$ (7.9.33)

Using the concept of a cross section, the rate of absorption between bound states is also given by

$$\dot{E}(v) = n_{ia} \int_0^{4\pi} \left(\frac{c}{4\pi}(u_v dv) \right) \sigma_{i\to j}(v) d\Omega = u_v c(\sigma_{i\to j}(v) dv) n_{ia}.$$ (7.9.34)

Comparing eqs. (7.9.34) and (7.9.33) yields

$$\sigma_{i\to j}(v') = \frac{1}{8\pi} \frac{c^2}{v^2} A_{ij} \delta(v' - v).$$ (7.9.35)

Comparing with eq. (7.9.22), it follows that

$$A_{ij} = \frac{1}{t_{ij}},$$ (7.9.36)

consistent with the definition implicit in eq. (7.9.25). Taking natural line breadths into account leads then again to eq. (7.9.23).

Having arrived at the relationship between a bound–bound cross section and an emission probability in two ways, a closer examination of the resultant cross section as described by eq. (7.9.23) is in order. Introducing the Thomson scattering cross section ($\sigma_{Th} = \sigma_{es}$ from eq. (3.4.16)) and rearranging fundamental constants, eq. (7.9.24) can be written as

$$\sigma_{i\to j}(v) = \sigma_{Th} \frac{\pi}{2} \frac{e^2}{\hbar c} \left(\frac{m_e c^2}{e^2/a_0} \right)^2 \left(\frac{r_{ij}}{a_0} \right)^2 \left(\frac{v_{ij}}{v} \right)^2 \left(v_{ij} \frac{dP_{ij}}{dv} \right)$$ (7.9.37)

$$= \sigma_{Th} \, 6.45 \times 10^7 \left(\frac{r_{ij}}{a_0} \right)^2 \left(v_{ij} \frac{dP_{ij}}{dv} \right).$$ (7.9.38)

Adopting dP_{ij}/dv from eqs. (7.8.30) and (7.8.31), one has that, at line center ($v = v_{ij}$),

$$v_{ij} \frac{dP_{ij}}{dv} = \frac{4v_{ij}}{\gamma_{ij}} = \frac{2}{\pi} \frac{hv_{ij}}{\Delta E_i + \Delta E_j},$$ (7.9.39)

a quantity which can be very large, given the small energy widths of bound states relative to typical emitted photon energies. Thus, as in the case of the free–free and bound–free cross sections, the cross section for a bound–bound transition can be huge compared with σ_{Th}. What can prevent bound–bound absorption from being of overwhelming importance whenever bound states exist (and are well populated) is a small width of the absorption line.

Under stellar conditions, due to collisions between and Doppler motions of absorbing ions, the effective breadth of an absorption line is usually much larger than the natural line breadth. The influence of Doppler motions may be readily estimated. Consider an ion moving in the radial direction with velocity v_r and a photon of frequency ν moving radially outward. The ion sees a frequency

$$\nu_{\text{Doppler}} = \nu \sqrt{\frac{1 - v_r/c}{1 + v_r/c}} \sim \nu(1 - v_r/c), \tag{7.9.40}$$

and, consequently, the value of ν in, e.g., eq. (7.9.22) must be replaced by this modified frequency. Thus, the cross section for absorption vanishes unless

$$\frac{v_r}{c} = \frac{\nu^2 - \nu_{ij}^2}{\nu^2 + \nu_{ij}^2} = (\nu - \nu_{ij})\frac{\nu + \nu_{ij}}{\nu^2 + \nu_{ij}^2} \sim \frac{\nu - \nu_{ij}}{\nu_{ij}}, \tag{7.9.41}$$

where the rightmost near equality follows from the fact that, when ion velocities are non-relativistic, $v_r/c \ll 1$ and therefore $\nu \sim \nu_{ij}$.

The next step is to average over the distribution of ions with regard to the radial velocity coordinate $v_r = (p_r/M)$, where p_r and M are, respectively, the radial momentum and the mass of the ion. This distribution is

$$\frac{\mathrm{d}n_a(v_r)}{\mathrm{d}v_r} = n_a \sqrt{\frac{M}{2\pi kT}} \exp\left(-\frac{Mv_r^2}{2kT}\right), \tag{7.9.42}$$

where $n_a = \int_{-\infty}^{\infty} (\mathrm{d}n_a(v_r)/\mathrm{d}v_r)\mathrm{d}v_r$. Adopting $\mathrm{d}P_{ij}/\mathrm{d}\nu = \delta(\nu - \nu_{ij})$, with ν replaced by ν_{Doppler} from eq. (7.9.40), one obtains

$$\sigma_{i \to j}(\nu) = \frac{1}{n_a} \int_{-\infty}^{\infty} \frac{\mathrm{d}n_a}{\mathrm{d}v_r} \mathrm{d}\left(\nu\frac{v_r}{c}\right) \frac{c}{\nu} \frac{1}{8\pi} \frac{c^2}{\nu_{ij}^2} \frac{1}{t_{ij}} \delta\left((\nu - \nu_{ij}) - \nu\frac{v_r}{c}\right). \tag{7.9.43}$$

Finally,

$$\sigma_{i \to j}(\nu) \sim \frac{1}{8\pi} \frac{c^3}{\nu_{ij}^3} \frac{1}{t_{ij}} \sqrt{\frac{M}{2\pi kT}} \exp\left(-\frac{Mc^2}{2kT}\left(\frac{\nu - \nu_{ij}}{\nu_{ij}}\right)^2\right) \tag{7.9.44}$$

$$= \frac{1}{8\pi} \frac{c^2}{\nu^2} \frac{1}{t_{ij}} \sqrt{\frac{\alpha}{\pi}} \exp\left(-\alpha(\nu - \nu_{ij})^2\right), \tag{7.9.45}$$

where

$$\alpha = \frac{Mc^2}{2kT} \frac{1}{\nu_{ij}^2}. \tag{7.9.46}$$

Since the integral over ν of

$$\frac{\mathrm{d}P'_{ij}}{\mathrm{d}\nu} = \sqrt{\frac{\alpha}{\pi}} \exp\left(-\alpha(\nu - \nu_{ij})^2\right) \tag{7.9.47}$$

is effectively unity, the total absorptive power is maintained, but it is spread out over a different interval than in the case of natural broadening. At half maximum amplitude,

$$\left|\frac{\nu - \nu_{ij}}{\nu_{ij}}\right|_{\text{Doppler}} = \sqrt{\frac{2kT \log_e 2}{Mc^2}} \sim 10^{-3}\sqrt{\frac{T_7}{A}}, \tag{7.9.48}$$

where T_7 is temperature in units of 10^7 K and A is the atomic number.

For the naturally broadened line, half maximum amplitude occurs when

$$\left|\frac{\nu - \nu_{ij}}{\nu_{ij}}\right|_{\text{natural}} = \frac{1}{4\pi}\frac{\gamma_{ij}}{\nu_{ij}}. \qquad (7.9.49)$$

Since $\gamma_{ij} \sim t_{ij}^{-1}$,

$$\left|\frac{\nu - \nu_{ij}}{\nu_{ij}}\right|_{\text{natural}} \sim \frac{1}{4\pi}\frac{1}{\nu_{ij}t_{ij}} = \frac{1}{4\pi}\frac{1}{\nu_{ij}}\frac{4}{3}\frac{e^2}{c^3}(2\pi\nu_{ij})^3\frac{r_{ij}^2}{\hbar} = \frac{e^2}{\hbar c}\left(\frac{2\pi r_{ij}}{\lambda_{ij}}\right)^2. \qquad (7.9.50)$$

For the $2 \to 1$ transition in hydrogen, $|\Delta\nu/\nu|_{\text{natural}} \sim 2.7 \times 10^{-7}$, compared with $|\Delta\nu/\nu|_{\text{Doppler}} \sim 3.2 \times 10^{-5}$ at $T = 10\,000$ K. Since the naturally broadened absorption line is typically much narrower than the Doppler broadened line, essentially the same result can be obtained by using $\mathrm{d}P_{ij}/\mathrm{d}\nu$ given by eq. (7.8.30) as the starting point for finding the Doppler shape of a line.

By interrupting the absorption process, collisions between the absorbing atom and other atoms act as an additional source of broadening. In the first approximation, this source of broadening may be taken into account by adding $\hbar$ times the collision rate to the natural line breadth. See, for example, the discussions by R. G. Breene, Jr. (1957) and by P. W. Anderson (1949). Because it spreads the opacity contribution of each line over a larger range in frequency than would otherwise have been the case, line broadening has a major effect on the contribution of bound–bound transitions to the opacity.

7.10 The Rosseland mean opacity

Svein Rosseland (1924) demonstrated how the frequency-dependent photon–matter absorption coefficient should be folded with the distribution of photons with respect to frequency to obtain l_{ph}, the mean free path of a typical photon at densities and temperatures in stellar interiors where this mean free path is small compared with a temperature scale height. The reciprocal of the product $l_{\text{ph}}\rho$, where ρ is the matter density, is called the Rosseland mean opacity.

The differential momentum balance equation for photons (see eq. (3.4.4)) may be written as

$$\mathrm{d}(P_\nu\,\mathrm{d}\nu) = -\rho\,\kappa_\nu\,\frac{\mathcal{F}_\nu\,\mathrm{d}\nu}{c}\,\mathrm{d}r, \qquad (7.10.1)$$

where $P_\nu\,\mathrm{d}\nu$ is the contribution to the radiation pressure of photons in the frequency range ν to $\nu + \mathrm{d}\nu$, $\mathcal{F}_\nu\,\mathrm{d}\nu$ is the net flux of energy due to photons in this frequency interval, and κ_ν is the total, frequency-dependent absorption coefficient.

In regions where local thermodynamic equilibrium can be assumed, one has from eqs. (4.11.22) and (4.11.10) that

$$P_\nu = \frac{1}{3}U_\nu = \frac{1}{3}\frac{8\pi}{c^3}\frac{(h\nu)\,\nu^2}{\exp(h\nu/kT) - 1} \qquad (7.10.2)$$

and that the total radiation pressure is

$$\int_0^\infty P_\nu \, d\nu = P_{\text{rad}} = \frac{1}{3} a T^4,$$

(7.10.3)

where a is given by eq. (4.11.18). Conservation of energy dictates that

$$\int_0^\infty \mathcal{F}_\nu \, d\nu = F_{\text{rad}} = \frac{L_{\text{rad}}(r)}{4\pi r^2},$$

(7.10.4)

where $F_{\text{rad}} = L_{\text{rad}}(r)/4\pi r^2$ is the total flux of energy by radiation through a spherical shell of radius r.

Since P_ν is a function of the temperature only, eq. (7.10.1) can be rewritten as

$$\mathcal{F}_\nu \, d\nu = -\frac{c}{\rho \kappa_\nu} \frac{d}{dr}(P_\nu \, d\nu) = -\frac{c}{\rho \kappa_\nu} \left(\frac{dP_\nu}{dT} d\nu \right) \frac{dT}{dr}.$$

(7.10.5)

Integration over frequency gives

$$\int_0^\infty \mathcal{F}_\nu \, d\nu = \frac{L_{\text{rad}}}{4\pi r^2} = -\frac{c}{\rho} \left(\int_0^\infty \frac{1}{\kappa_\nu} \frac{dP_\nu}{dT} d\nu \right) \frac{dT}{dr}.$$

(7.10.6)

Defining a Rosseland mean opacity κ_{Ross} by

$$\frac{1}{\kappa_{\text{Ross}}} = \int \frac{1}{\kappa_\nu} \frac{dP_\nu}{dT} d\nu \div \int \frac{dP_\nu}{dT} d\nu$$

(7.10.7)

and noting from eq. (7.10.3) that

$$\int \frac{dP_\nu}{dT} d\nu = \frac{4}{3} a T^3,$$

(7.10.8)

eq. (7.10.6) may be rewritten as

$$\mathcal{F}_{\text{rad}} = \frac{L_{\text{rad}}}{4\pi r^2} = -\frac{4ac}{3} \frac{1}{\kappa_{\text{Ross}}} \frac{T^3}{\rho} \frac{dT}{dr}.$$

(7.10.9)

Comparing eq. (7.10.9) with eq. (3.4.3) and recalling the definition of the average photon mean free path l_{ph} given by eqs. (3.4.1) and (3.4.2), it is evident that

$$l_{\text{ph}} = \frac{1}{\kappa_{\text{Ross}} \rho}.$$

(7.10.10)

Differentiation of eq. (7.10.2) with respect to temperature results in

$$\frac{dP_\nu}{dT} = \frac{8\pi}{3c^3} \left(\frac{kT}{h} \right)^2 k \, W(x),$$

(7.10.11)

where

$$x = \frac{h\nu}{kT}$$

(7.10.12)

and

$$W(x) = \frac{x^4 e^x}{(e^x - 1)^2}. \tag{7.10.13}$$

The quantity $W(x)$ is called a weighting function.

Using in eq. (7.10.11) the fact that

$$dv = \frac{kT}{h} \, dx \tag{7.10.14}$$

yields

$$\frac{dP_v}{dT} \, dv = \frac{8\pi}{3} \left(\frac{kT}{ch} \right)^3 k \, W(x) \, dx, \tag{7.10.15}$$

so eq. (7.10.7) for the Rosseland mean opacity can also be written as

$$\frac{1}{\kappa_{\text{Ross}}} = \int_0^\infty \frac{W(x)}{\kappa_v} \, dx \div \int_0^\infty W(x) \, dx, \tag{7.10.16}$$

where κ_v is to be expressed as a function of x and T.

Using in eq. (7.10.8) the relationship between a and fundamental constants given by eq. (4.11.18), one has that

$$\int_0^\infty \frac{dP_v}{dT} \, dv = \frac{4}{3} \frac{8\pi^5}{15} \left(\frac{kT}{ch} \right)^3 k. \tag{7.10.17}$$

Integrating both sides of eq. (7.10.15) and using eq. (7.10.17) gives

$$\int_0^\infty W(x) \, dx = \frac{4\pi^4}{15} = 25.975\,757\,61. \tag{7.10.18}$$

Insertion into eq. (7.10.15) yields still another form for the Rosseland mean opacity:

$$\frac{1}{\kappa_{\text{Ross}}} = \frac{15}{4\pi^4} \int_0^\infty \frac{W(x)}{\kappa_v} \, dx. \tag{7.10.19}$$

When evaluating the Rosseland mean opacity, it must be remembered that, in order to take stimulated emission into account, all true absorption cross sections and corresponding absorption coefficients are to be multiplied by $(1 - e^{-x})$. That is, the arguments presented explicitly in this section for the case of bound–bound absorption carry through also for bound–free and free–free absorption, as already acknowledged in Sections 7.6 and 7.7. On the other hand, the electron scattering contribution is not affected by the phenomenon of stimulated emission and its contribution to κ_v is not diminished by the factor $(1 - e^{-x})$.

It is worth emphasizing how the frequency distribution characterizing the net flux of photons differs from the frequency distribution of photons in local thermodynamic equilibrium. From eqs. (7.10.5)–(7.10.7), one has that

$$\mathcal{F}_v \, dv \div \frac{L_{\text{rad}}}{4\pi r^2} = \frac{1}{\kappa_v} \frac{dP_v}{dT} \, dv \div \int_0^\infty \frac{1}{\kappa_v} \frac{dP_v}{dT} \, dv$$

$$= \frac{\kappa_{\text{Ross}}}{\kappa_v} \frac{dP_v}{dT} \, dv \div \int_0^\infty \frac{dP_v}{dT} \, dv. \tag{7.10.20}$$

Using the fact that $(dP_\nu/dT)\, d\nu \propto W(x)\, dx$ (see eqs. (7.10.10) and (7.10.13)), it follows that

$$\mathcal{F}_\nu\, d\nu = \frac{L_{\text{rad}}}{4\pi r^2} \left[\frac{\kappa_{\text{Ross}}}{\kappa_{\text{es}} + (1 - e^{-x})\, \kappa_{\text{abs}}(\nu)} \right] \frac{W(x)\, dx}{\int_0^\infty W(x)\, dx}, \tag{7.10.21}$$

where the opacity coefficient has beeen broken into two parts, the first having to do with electron scattering and the second having to do with actual absorption. Equation (7.10.21) contains two important lessons. These lessons are of course already contained in the starting equation, eq. (7.10.5), but in a more disguised form. The first lesson is told by the factor in square brackets in eq. (7.10.21) which states that the transmitted flux is amplified at frequencies where the opacity coefficient is small and diminished at frequencies where this coefficient is large. Thus, the transmitted flux is channeled out of frequency ranges where the opacity is large into frequency ranges where the opacity is small.

The second lesson is told by the third factor in eq. (7.10.21) which, using eqs. (7.10.13) and (7.10.18), may be written as

$$\frac{W(x)\, dx}{\int_0^\infty W(x)\, dx} = \frac{15}{4\pi^4} \frac{x^4\, e^x}{(e^x - 1)^2}\, dx \tag{7.10.22}$$

and compared with the local thermodynamic equilibrium distribution that is given by eq. (4.11.10) which becomes, when rewritten in terms of the variable x,

$$\frac{U_\nu\, d\nu}{\int_0^\infty U_\nu\, d\nu} = \frac{15}{\pi^4} \frac{x^3\, e^x}{(e^x - 1)}\, dx. \tag{7.10.23}$$

Thus, the transmitted flux is further distorted from the local thermodynamic equilibrium distribution in the direction of a distribution which is the temperature derivative of the local thermodynamic equilibrium distribution. In optically thick regions of the stellar interior, where the net outward (or inward) flux is typically very, very small compared with the isotropic flux, the departure of the distribution of the net flux from the local thermodynamic equilibrium distribution has essentially no impact on the local equilibrium distribution. This is not the case in the optically thin layers above the photosphere.

7.11 Sample calculations of the Rosseland mean opacity

The calculation of a Rosseland mean opacity can be a long and tedious process. On the other hand, the process can be a very illuminating and therefore very rewarding exercise, demonstrating the interplay between the ionization state of matter, the dependence on photon frequency of various radiation–matter cross sections, and the distribution of photons with respect to frequency.

The system chosen for analysis in this section is a very simple one, consisting of hydrogen, helium, and oxygen. Abundances by mass of the three elements are chosen as $X = 0.74$, $Y = 0.25$, and $Z = 0.01$, and the corresponding number abundance parameters are $Y_{\text{H}} = 0.74$, $Y_{\text{He}} = 0.0625$, and $Y_{\text{O}} = 0.000\,625$. Only ground states are taken into account in solving the Saha equations for ion abundances. Bound–free absorption from excited levels and bound–bound absorption are neglected; for some densities and temperatures such absorption may make the most important contribution to the Rosseland mean opacity, but pedagogy demands that some limitation be placed on the processes considered.

In a first exercise, temperatures and densities are chosen in such a way that hydrogen and helium are completely ionized and oxygen atoms may have up to two K-shell electrons. How the relative contributions to the Rosseland mean opacity of three different sources of opacity – electron scattering, free–free absorption, and bound–free absorption – vary with temperature and density is examined in a series of examples. In a second exploration, the restriction on ionization state is removed and the temperature–density domain is extended into regions where the chosen elements exist primarily as neutral atoms, showing how, at a given density, the opacity rises from low values at high temperatures, reaches a maximum at temperatures such that hydrogen nuclei begin to acquire electrons, and then declines to very small values at temperatures such that the abundance of free electrons is small compared with the abundance of electrically neutral atoms.

7.11.1 Hydrogen and helium completely ionized, oxygen with zero to two bound electrons

The statistics for oxygen with only two bound states is the same as for helium, so one can use equations from Section 4.15 to solve for the number abundances of the atoms in the three ionization stages considered. For clarity in the present context, the notation in Section 4.15 is inverted so that Y_2, Y_1, and Y_0 denote, respectively, the number abundance parameters for oxygen with two K-shell electrons electrons, one K-shell electron, and zero bound electrons. The ionization potential for atoms with one bound electron is taken to be $\chi_1 = 8^2 \times 13.527$ eV $= 865.7$ eV. If shielding were complete, the ionization energy for oxygen atoms with two bound electrons would be $\chi_2 \sim 7^2 \times 13.527$ eV $= 662.8$ eV. If the shielding of an electron in an oxygen atom with two K-shell electrons were the same as that of an electron in a neutral helium atom (see eqs. (7.4.37) and (7.4.38)), the effective charge seen by either electron would be $Z_{\text{eff}} \sim 8 - 0.655 = 7.345$, and the ionization energy would be

$$\chi_2 = \left(\frac{7.345}{8}\right)^2 \chi_1 = 0.8430 \, \chi_1 = 729.8 \text{ eV}. \tag{7.11.1}$$

The actual ionization threshold differs from this estimate by only ~ 0.5 eV.

The relevant Saha equations are

$$\frac{Y_1 Y_e}{Y_2} = 2 \frac{g_1}{g_2} \frac{1}{N_A \rho} \left(\frac{2\pi m_e kT}{h^2}\right)^{3/2} e^{-\chi_2/kT} = 2 \frac{2}{1} \times \frac{4.01 \, T_6^{3/2}}{\rho} \exp(-\chi_2/kT) \tag{7.11.2}$$

and

$$\frac{Y_0 Y_e}{Y_1} = 2 \frac{g_0}{g_1} \frac{1}{N_A \rho} \left(\frac{2\pi m_e kT}{h^2}\right)^{3/2} e^{-\chi_1/kT} = 2 \frac{1}{2} \times \frac{4.01 \, T_6^{3/2}}{\rho} \exp(-\chi_1/kT). \tag{7.11.3}$$

In light of the discussion in Section 4.13, only the first term in the partition function for each ionization state has been included. In first approximation,

$$Y_e = \frac{1 + X}{2},$$ (7.11.4)

where the small effect on the electron abundance of changes in the ionization state of oxygen has been neglected. Since

$$Y_0 + Y_1 + Y_2 = Y_O = \frac{X_O}{16},$$ (7.11.5)

one has, using eqs. (7.11.2) and (7.11.3), that

$$Y_1 \left[\beta e^{-\chi_1/kT} + 1 + \frac{1}{4\beta e^{-\chi_2/kT}} \right] = Y_O = \frac{X_O}{16},$$ (7.11.6)

where

$$\beta = \frac{4.01 T_6^{3/2}}{\rho_0 Y_e}.$$ (7.11.7)

For example, at a temperature $T_6 = 5$ and density $\rho = 1$ g cm^{-3} ($\rho_0 = 1$), about 87.1% of all oxygen atoms are completely ionized, 12.6% have one electron, and 0.3% have two electrons.

The cross section for ejecting an electron from an ion with one K-shell electron, which is given by eq. (7.4.32), supplemented by eqs. (7.4.31) and (7.4.33), may be written as

$$\sigma_1(\nu) = \sigma_{Th} \frac{5.1741 \times 10^8}{Z_1^2} \left(\frac{\chi_1}{kT} \right)^4 \left(\frac{kT}{h\nu} \right)^4 f_K(\eta_1) S_{1\nu},$$ (7.11.8)

where σ_{Th} is the Thomson cross section given by eq. (7.4.26), Z_1 is the charge of the nucleus ($Z_1 = 8$ in the case at hand),

$$f_K(\eta) = \frac{\exp[-4\eta \cot^{-1}(\eta)]}{1 - \exp(-2\pi\eta)},$$ (7.11.9)

$$\eta_1 = \left(\frac{h\nu}{\chi_1} - 1 \right)^{-1/2} = \left(\frac{\chi_1}{kT} \right)^{1/2} \left(\frac{h\nu}{kT} - \frac{\chi_1}{kT} \right)^{-1/2},$$ (7.11.10)

and $S_{1\nu}$ is a step function: $S_{1\nu} = 0$ if $h\nu < \chi_1$ and $S_{1\nu} = 1$ otherwise.

The cross section for ejecting an electron from an oxygen nucleus with two K-shell electrons follows from eqs. (7.11.8)–(7.11.10) by (1) replacing the subscript 1 with 2, (2) setting $S_{2\nu} = 0$ if $h\nu < \chi_2$ and $S_{2\nu} = 1$ otherwise, (3) setting $Z_2 = Z - s$, where s takes the interaction between the two electrons into account (adopting the argument that leads to eq. (7.11.1), $Z_2 = 7.345$), and (4) multiplying by 2 to account for the fact that either electron may be ejected.

Setting

$$x = \frac{h\nu}{kT},$$ (7.11.11)

the contribution of bound–free absorption to the opacity coefficient may be written as

$$\kappa_{bf}(\nu) = \frac{N_A \sigma_{Th}}{x^3} \, \Gamma_{bf}(x), \tag{7.11.12}$$

where

$$\Gamma_{bf}(x) = \frac{8.0845 \times 10^6}{x} \left[Y_1 \, x_{th,1}^4 \, f_K(\eta_1) \, S_{1\nu} + 2 \, Y_2 \left(\frac{8}{7.345} \right)^2 x_{th,2}^4 \, f_K(\eta_2) \, S_{2\nu} \right] Y_O, \tag{7.11.13}$$

$$x_{th,i} = \frac{\chi_i}{kT}, \tag{7.11.14}$$

and

$$\eta_i = \left(\frac{x_{th,i}}{x - x_{th,i}} \right)^{1/2}. \tag{7.11.15}$$

Since $S_{1\nu}$ and $S_{2\nu}$ are non-zero only when $x > x_{th}$, $\Gamma_{bf}(x)$ is finite for all x.

From eqs. (7.6.35)–(7.6.36), the cross section for free–free absorption when electrons are treated as undistorted plane waves is

$$\sigma_{ff}(\nu) = \sigma_{Th} \, Y_e \, 32021.6 \, \frac{\rho_0}{T_6^{7/2}} \left(\frac{kT}{h\nu} \right)^3 L_{AV} \sum_j \frac{Z_j^2 X_j}{A_j}, \tag{7.11.16}$$

where L_{AV} in its most general form, when electron screening is taken into account, is given by eqs. (7.6.17) and (7.6.9) and, when screening is neglected, by Λ_{ff}, eqs. (7.6.25)–(7.6.26). Here the approximations given by eqs. (7.6.32) and (7.6.33) and graphed in Fig. (7.6.1) are adopted. Using eq. (7.11.11) and doing the sum in the current context,

$$\sigma_{ff}(\nu) = \sigma_{Th} \, Y_e \, 32\,000 \, \frac{\rho_0}{T_6^{7/2}} \frac{1}{x^3} \Lambda_{ff}(x) \, [\, X + Y + 64 \, Y_0 + 49 \, Y_1 + 36 \, Y_2 \,] \tag{7.11.17}$$

$$= \sigma_{Th} \, Y_e \, 3.20 \times 10^5 \, \frac{\rho_0}{T_6^{7/2}} \, [\, 0.99 + 64 \, Y_0 + 49 \, Y_1 + 36 \, Y_2 \,] \, \frac{1}{x^3} \, \Lambda_{ff}(x). \tag{7.11.18}$$

In analogy with eqs. (7.11.12) and (7.11.13), the contribution of free–free absorption to the opacity coefficient may be written as

$$\kappa_{ff}(\nu) = \frac{N_A \sigma_{Th}}{x^3} \, \Gamma_{ff}(x), \tag{7.11.19}$$

where

$$\Gamma_{ff}(x) = Y_e \, 3.20 \times 10^5 \, \frac{\rho_0}{T_6^{7/2}} \, [\, 0.99 + 64 \, Y_0 + 49 \, Y_1 + 36 \, Y_2 \,] \, \Lambda_{ff}(x). \tag{7.11.20}$$

Finally, the electron-scattering contribution to the opacity coefficient may be written as

$$\kappa_{es}(\nu) = \frac{N_A \sigma_{Th}}{x^3} \, \Gamma_{es}(x), \tag{7.11.21}$$

where

$$\Gamma_{es}(x) = Y_e \, x^3. \tag{7.11.22}$$

From the development in Section 7.10, one can write the Rosseland mean opacity for the system as

$$\frac{1}{\kappa_{Ross}} = \int_0^\infty \bar{G}(x) \, dx \div \int_0^\infty W(x) \, dx, \tag{7.11.23}$$

where

$$W(x) = \frac{e^x x^4 dx}{(e^x - 1)^2} \tag{7.11.24}$$

and

$$\bar{G}(x) = \frac{W(x)}{\kappa_{es} + [\kappa_{ff}(\nu) + \kappa_{bf}(\nu)] \, (1 - e^{-x})}. \tag{7.11.25}$$

The numerical value for $\int_0^\infty W(x) \, dx$ given by eq. (7.10.18) may be used to ensure that the integration step is chosen small enough. For integrations performed here, 10 000 equally spaced steps over the range $x = 0 \rightarrow 30$ proves adequate. In the construction of the figures that illustrate numerical results, only one out of twenty points has been used. This accounts for the fact that, near absorption edges, the functions $G(x)$ in the figures exhibit steep gradients rather than discontinuities.

Using eqs. (7.11.12), (7.11.19), and (7.11.21) in eq. (7.11.25), eq. (7.11.23) may be written as

$$\kappa_{Ross} = N_A \sigma_{Th} \frac{I_1}{I_2} = 0.4 \text{ cm}^2 \text{ g}^{-1} \frac{I_1}{I_2}, \tag{7.11.26}$$

where

$$I_1 = \int_0^\infty W(x) \, dx, \tag{7.11.27}$$

$$I_2 = \int_0^\infty G(x) \, dx, \tag{7.11.28}$$

and

$$G(x) = \frac{x^3 \, W(x)}{\Gamma_{es}(x) + [\Gamma_{ff}(x) + \Gamma_{bf}(x)] \, (1 - e^{-x})}. \tag{7.11.29}$$

7.11.2 Anatomy of an opacity when density $= 1$ g cm^{-3} and temperature varies from 10^7 to 10^6 K

Results of a calculation when $\rho = 1$ g cm^{-3} and $T = 10^7$ K are shown in Fig. 7.11.1. The solid curve describes the weighting function $W(x)$ and the dashed curve describes $G_{es+ff+bf}(x)$, the weighting function divided by a measure of the total absorption coefficient as given by eq. (7.11.30). The dotted curve describes $G_{es+bf}(x)$, as given by eq. (7.11.30) with $\Gamma_{ff}(x) = 0$, and the dash-dot curve describes $G_{es+ff}(x)$, as given by

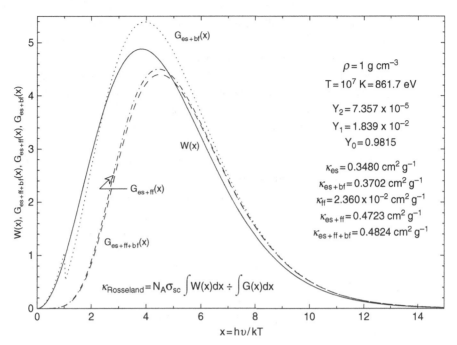

Fig. 7.11.1 Anatomy of an opacity for composition $X_H = 0.74$, $X_{He} = 0.25$, $X_O = 0.01$, density $= 1\,g\,cm^{-3}$ and temperature $= 10^7\,K$

eq. (7.11.30) with $\Gamma_{bf}(x) = 0$. The legend on the right hand side of the figure gives (1) the thermodynamic conditions, (2) the abundances Y_i of oxygen ions in units of the total abundance of oxygen, and (3) the Rosseland mean opacities associated with the various choices for the function $G(x)$. The Rosseland mean opacity is defined at the bottom of the figure.

From eqs. (7.11.29) and (7.11.22), if electron scattering were the only contributor to the opacity,

$$G(x) = W(x)\,\frac{\sigma_{es}}{\sigma_{sc}} = W(x)\,\frac{1}{Y_e}, \qquad (7.11.30)$$

where

$$\sigma_{sc} = Y_e\sigma_{es}, \qquad (7.11.31)$$

and every point on the associated curve $G(x)$ would be larger than the corresponding point on the curve $W(x)$ by the factor $1/Y_e = 1.1494$. The Rosseland mean opacity would be

$$\kappa_{Ross} = 0.4\,cm^2\,g^{-1}\,\frac{1}{1.1494} = 0.348\,cm^2\,g^{-1} \qquad (7.11.32)$$

rather than the value

$$\kappa_{Ross} = 0.4824\,cm^2\,g^{-1} \qquad (7.11.32a)$$

actually found when all considered sources of opacity are included. Most of the difference is due to free–free absorption rather than to bound–free absorption.

How this comes about is elucidated by the curves $G_{es+bf}(x)$ and $G_{es+ff}(x)$ in Fig. 7.11.1. When free–free absorption is omitted, for values of x smaller than $x_{th1} \sim 1$, which marks the absorption edge for oxygen with one bound electron, electron scattering makes the entire contribution to $G_{es+bf}(x)$. At the absorption edge, there is a decrease in $G_{es+bf}(x)$ due to bound–free absorption, but it is small because of the small abundance of oxygen ions with one bound electron ($Y_1 = 0.0184\, Y_O$). Due to the even smaller abundance of oxygen ions with two bound electrons ($Y_2 = 7.36 \times 10^{-5}\, Y_O$), there is no detectable effect of bound–free absorption from such ions. The net result is that bound–free absorption contributes only a fraction of a percent to the Rosseland mean κ_{es+bf}.

On the other hand, when electron scattering and free–free absorption are taken as the only opacity sources, the associated Rosseland mean κ_{es+ff} is 36% larger than the pure electron-scattering opacity. Comparison of the curves $G_{es+ff}(x)$ and $G_{es+bf}(x)$ shows that the reason for this is that the free–free cross section is much larger than the electron-scattering cross section for small $x < 1.5$ and is of comparable size for a much larger range in x. A remarkable feature of the free–free contribution is that, if free–free absorption were the only source of opacity, the Rosseland mean κ_{ff} would be 15 times smaller than κ_{es}, showing dramatically that the Rosseland mean of the free–free opacity is sometimes not a very informative measure of the importance of free–free absorption. The approximation described by eq. (7.6.43) gives $\kappa_{ff} \sim 0.026\ \mathrm{cm^2\ g^{-1}}$, compared with $\kappa_{ff} \sim 0.0236\ \mathrm{cm^2\ g^{-1}}$ given by the detailed calculation, results of which are reported in Fig. 7.11.1.

In Fig. 7.11.2 are shown the results of integrations when $\rho = 1\ \mathrm{g\ cm^{-3}}$ and $T = 5 \times 10^6$ K. This time, the curves for $G_{es+ff+bf}(x)$ and $G_{es+bf}(x)$ are normalized so that the maximum in each curve is at the same height as the height of the weighting function $W(x)$ at its maximum. Both bound–free transitions are well represented along the curve $G_{es+bf}(x)$, even though only $\sim$0.33% of the oxygen ions have two electrons. Once again, the free–free contribution to the opacity plays a role greater than expected from the size of its Rosseland mean: $\kappa_{ff} = 0.267\ \mathrm{cm^2\ g^{-1}}$.

Even more startling are the results for $\rho = 1\ \mathrm{g\ cm^{-3}}$ and $T = 3 \times 10^6$ K shown in Fig. 7.11.3. Both bound–free transitions contribute substantially to the total Rosseland mean, but do so primarily because of the action of free–free absorption. When free–free absorption is omitted, strong bound–free absorption cuts off photon transmission at frequencies above the active absorption edges and opens a "window" which allows photons at frequencies below those at the absorption edges to be transmitted more easily. That is, photon transmission concentrates into the frequency range below the thresholds for bound–free absorption.

The major contribution to $G_{es+bf}(x)$ is due to electron scattering within the window and is given by the area of the nearly triangular region defined by the dotted curve in Fig. 7.11.3 for values of x below the threshold x_{th2} for ionization of oxygen ions with two electrons. By inspection, this area is roughly 0.25 times the area under the weighting curve and is about four times larger than the area under the remainder of the dotted curve for which bound–free absorption is overwhelmingly responsible. The Rosseland mean is therefore

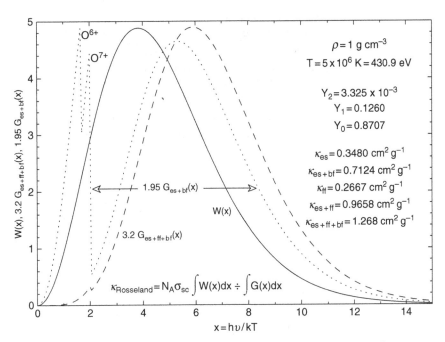

Fig. 7.11.2   Anatomy of an opacity for composition $X_H = 0.74$, $X_{He} = 0.25$, $X_0 = 0.01$, density $= 1\,g\,cm^{-3}$ and temperature $= 5 \times 10^6\,K$

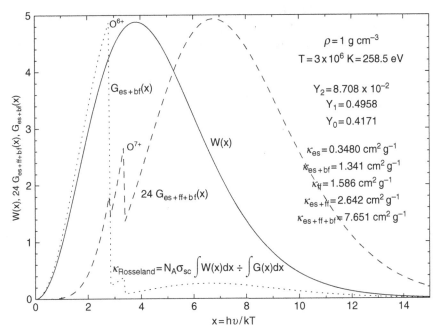

Fig. 7.11.3   Anatomy of an opacity for composition $X_H = 0.74$, $X_{He} = 0.25$, $X_0 = 0.01$, density $= 1\,g\,cm^{-3}$ and temperature $= 3 \times 10^6\,K$

roughly $\kappa_{es+bf} \sim 0.4/(0.25 \times 1.25) = 3.2 \times 0.4 = 1.28 \text{ cm}^2 \text{ g}^{-1}$, close to the more exact value $\kappa_{es+bf} = 1.34 \text{ cm}^2 \text{ g}^{-1}$ given in the legend at the right hand side of the figure.

The argument is worth repeating in a slightly different way. Consider an opacity $\kappa = \kappa_A + \kappa_B$ such that, for $x < x_{th}$, κ_A is a constant $\neq 0$ and $\kappa_B = 0$, and that, for $x > x_{th}$, κ_B is a constant $\gg \kappa_A$. Then

$$\int_0^\infty \frac{W(x)\,dx}{\kappa_A + \kappa_B} = \int_0^{x_{th}} \frac{W(x)\,dx}{\kappa_A} + \int_{x_{th}}^\infty \frac{W(x)\,dx}{\kappa_B} \qquad (7.11.33)$$

and

$$\frac{I_A + I_B}{\kappa_{Ross}} = \frac{I_A}{\kappa_A} + \frac{I_B}{\kappa_B}, \qquad (7.11.34)$$

where $I_A = \int_0^{x_{th}} W(x)\,dx$ and $I_B = \int_{x_{th}}^\infty W(x)\,dx$. If, also, $I_A/\kappa_A \gg \frac{I_B}{\kappa_B}$, then

$$\kappa_{Ross} = \left(1 + \frac{I_B}{I_A}\right)\kappa_A. \qquad (7.11.35)$$

In the case at hand, $I_B \sim 3I_A$ and $\kappa_A = 0.348 \text{ cm}^2 \text{ g}^{-1}$, which gives $\kappa_{Ross} \sim 1.39 \text{ cm}^2 \text{ g}^{-1}$.

The introduction of free–free absorption closes the low frequency window and photon transmission is focussed into the frequency range above the bound–free absorption edges where bound–free absorption can make its presence felt and provide the major contribution to $G_{es+ff+bf}(x)$. The net result is that $\kappa_{es+ff+bf}$ ($7.65 \text{ cm}^2 \text{ g}^{-1}$) is almost six times larger than when free–free absorption is ignored. Thus, free–free absorption at frequencies below absorption edges permits bound–free absorption to play a role which it would otherwise have been denied.

This is a wonderfully delightful result, but not as easily modeled as when free–free absorption is omitted. One could suppose that

$$\int_0^\infty \frac{W(x)\,dx}{\kappa_A + \kappa_C + \kappa_B} = \int_0^{x_{th}} \frac{W(x)\,dx}{\kappa_A + \kappa_C} + \int_{x_{th}}^\infty \frac{W(x)\,dx}{\kappa_A + \kappa_C' + \kappa_B}, \qquad (7.11.36)$$

$$\sim \int_0^{x_{th}} \frac{W(x)\,dx}{\kappa_C} + \int_{x_{th}}^\infty \frac{W(x)\,dx}{\kappa_C + \kappa_B}, \qquad (7.11.37)$$

where it has been assumed that, for $x < x_{th}$, $\kappa_C \gg \kappa_A$ and, for $x > x_{th}$, $\kappa_B \gg \kappa_A$. One can then estimate that

$$\frac{I_A + I_B}{\kappa_{Ross}} = \frac{I_A}{\kappa_C} + \frac{I_B}{\kappa_C' + \kappa_B}, \qquad (7.11.38)$$

or

$$\kappa_{Ross} = \left(\kappa_C' + \kappa_B\right) \div \left(1 + \frac{I_A}{I_A + I_B} \frac{\kappa_C' + \kappa_B}{\kappa_C}\right), \qquad (7.11.39)$$

where κ_C is an average over the range $0 < x < x_{\rm th}$ and κ'_C and κ_B are appropriate averages for $x > x_{\rm th}$. Supposing further that $\kappa'_C \ll \kappa_B$,

$$\kappa_{\rm Ross} = \kappa_B \div \left(1 + \frac{I_A}{I_A + I_B}\frac{\kappa_B}{\kappa_C}\right). \tag{7.11.40}$$

Supposing still further that $\kappa_C \ll \kappa_B$,

$$\kappa_{\rm Ross} = \left(1 + \frac{I_B}{I_A}\right)\kappa_C, \tag{7.11.41}$$

which is the same as eq. (7.11.35) except that $\kappa_C \gg \kappa_A$ replaces κ_A. Whichever of the last two approximations best fits the case at hand, the contribution of electron scattering to the opacity has been marginalized.

Results of integrations for $\rho = 1$ g cm^{-3} and $T = 2 \times 10^6$ K, $T = 1.5 \times 10^6$ K, and $T = 10^6$ K are shown, respectively, in Figs. 7.11.4–7.11.6. These results continue and enlarge the lessons from Figs. (7.11.1)–(7.11.3). As temperature decreases, the thresholds for absorption by oxygen ions with one and two electrons shift to larger values of $x_{\rm th1}$ and $x_{\rm th2}$, respectively, and the window for the transmission of low energy photons, which is open when free–free absorption is neglected, widens to the extent that the effect of bound–free absorption becomes barely discernable (the ratio $\kappa_{\rm es+bf}/\kappa_{\rm es}$ approaches unity). Nevertheless, when free–free absorption is included, the ratio $\kappa_{\rm es+ff+bf}/\kappa_{\rm es+ff}$ remains significantly larger than unity, demonstrating again that, by closing the transmission window

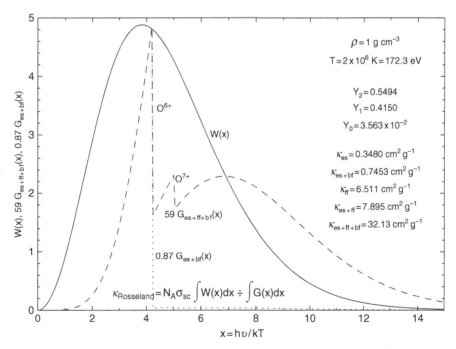

Fig. 7.11.4 Anatomy of an opacity for composition $X_H = 0.74$, $X_{He} = 0.25$, $X_O = 0.01$, density $= 1$ g cm^{-3} and temperature $= 2 \times 10^6$ K

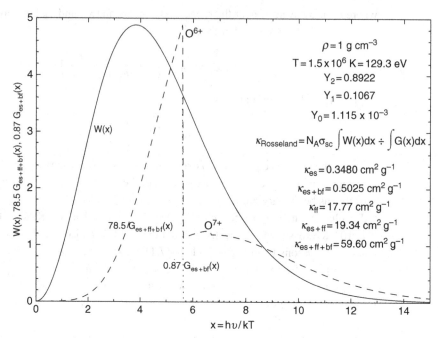

Fig. 7.11.5 Anatomy of an opacity for composition $X_H = 0.74$, $X_{He} = 0.25$, $X_0 = 0.01$, density $= 1\,\mathrm{g\,cm^{-3}}$ and temperature $= 1.5 \times 10^6\,\mathrm{K}$

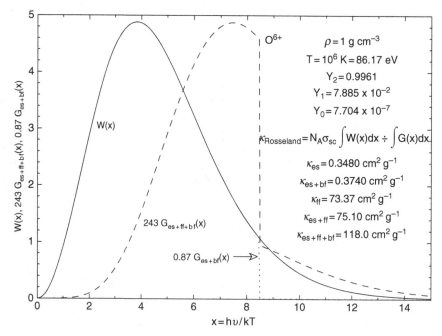

Fig. 7.11.6 Anatomy of an opacity for composition $X_H = 0.74$, $X_{He} = 0.25$, $X_0 = 0.01$, density $= 1\,\mathrm{g\,cm^{-3}}$ and temperature $= 10^6\,\mathrm{K}$

below bound–free absorption thresholds, free–free absorption enables bound–free absorption to contribute subtantially to the Rosseland mean opacity.

The dramatic enhancement of the bound–free contribution to the Rosseland mean opacity which free–free absorption can bring about does not prevail throughout the-temperature–density domain. Comparing entries for $\kappa_{es+ff+bf}$ and κ_{es+ff} in Figs. 7.11.1–7.11.6, it is evident that, at fixed density, as temperature decreases, the ratio $\kappa_{es+ff+bf}/\kappa_{es+ff}$ rises to a maximum and then decreases. This behavior can be understood as a consequence of the fact that, as temperature decreases, the thresholds for bound–free absorption shift from being far to the left of the peak in the weighting function $W(x)$ to being far to the right of the peak.

7.11.3 Arbitrary states of ionization for hydrogen, helium, and oxygen

Exploration of regions of lower temperature and lower density requires that the restriction on allowed ionization states be removed. Reverting to the notation of Section 4.15, the relationship between the abundances of two adjacent states of ionization of a given element may be written as

$$Y_{i,n} = \frac{Z_{i,n}}{Y_e}\, Y_{i,n-1}, \tag{7.11.42}$$

where i designates the element, n designates the degree of ionization, and the quantity

$$Z_{i,n} = 2\, \frac{g_{i,n}}{g_{i,n-1}}\, \frac{4.01\, T_6^{3/2}}{\rho_0}\, e^{-\chi_{i,n-1}/kT} \tag{7.11.43}$$

includes only the first term in the appropriate partition functions. The designation of the partition function by the letter Z comes from the German words for this function: Zustand Summe.

For hydrogen, conservation of nuclei gives

$$Y_e\, Y_H = (Y_e + Z_{H,1})\, Y_{H,0}, \tag{7.11.44}$$

and, for helium, conservation of nuclei gives

$$Y_e^2\, Y_{He} = (Y_e^2 + Y_e\, Z_{He,1} + Z_{He,2}\, Z_{He,1})\, Y_{He,0}, \tag{7.11.45}$$

where Y_H and Y_{He} are proportional, respectively, to the total abundances of hydrogen and helium nuclei. Given an estimate of Y_e, eqs. (7.11.44) and (7.11.45) provide estimates of $Y_{H,0}$ and $Y_{He,0}$ and eq. (7.11.42) can be used to determine the abundances of ionized species. Explicitly,

$$Y_{H,1} = Y_H\, \frac{Y_e}{Y_e + Z_{H,1}}, \tag{7.11.46}$$

$$Y_{He,1} = Y_{He}\, \frac{Y_e\, Z_{He,1}}{Y_e^2 + Y_e\, Z_{He,1} + Z_{He,2}\, Z_{He,1}}, \tag{7.11.47}$$

and

$$Y_{He,2} = Y_{He} \frac{Z_{He,2} Z_{He,1}}{Y_e^2 + Y_e Z_{He,1} + Z_{He,2} Z_{He,1}}. \qquad (7.11.48)$$

Were it not for the presence of oxygen, a solution would consist of (1) adopting an initial estimate of Y_e, (2) using eqs. (7.11.43)–(7.11.48) to obtain first estimates of $Y_{H,0}$, $Y_{H,1}$, $Y_{He,0}$, $Y_{He,1}$, and $Y_{He,2}$, and (3) using the conservation of charge,

$$Y_e = Y_{H,1} + Y_{He,1} + 2 Y_{He,2}, \qquad (7.11.49)$$

to obtain a next estimate of Y_e which could, in turn, be used to begin another iteration.

To include oxygen, eq. (7.11.42) and the conservation of nuclei may be invoked successively to obtain

$$Y_e^8 Y_O = \sum_{j=0}^{8} Y_e^{8-j} P_j Y_{O,0}, \qquad (7.11.50)$$

where $P_0 = 1$ and

$$P_{j>0} = Z_{O,j} P_{j-1}. \qquad (7.11.51)$$

Equation (7.11.50) also follows by induction from eq. (7.11.45). Values of the first six ionization potentials for oxygen, as given by the *Handbook of Chemistry and Physics* (Hodgman, Weast, & Selby, 1957), are $\chi_1 = 13.55$ eV, $\chi_2 = 34.93$ eV, $\chi_3 = 54.87$ eV, $\chi_4 = 76.99$ eV, $\chi_5 = 113.$ eV, and $\chi_6 = 137.5$ eV. Inserting a first estimate of Y_e into eq. (7.11.50) provides a first estimate of $Y_{O,0}$ and eq. (7.11.42) can be used eight times to yield estimates of $Y_{1,0}$ through $Y_{8,0}$.

Conservation of charge for the entire assembly of hydrogen, helium, and oxygen requires that

$$Y_e = Y_{H,1} + Y_{He,1} + 2 Y_{He,2} + \sum_{j=1}^{8} j Y_{O,j}. \qquad (7.11.52)$$

In the following, an oxygen atom which has been ionized n times and therefore has $8 - n$ bound electrons is described by O^{n+}. For O^{5+}, the outer electron is in a $2s$ state and the cross section for bound–free absorption is taken to be (see eq. (7.4.41))

$$\sigma_{5+}(\nu) = \sigma_{Th} \frac{1.6557 \times 10^{10}}{Z_{5+}^2} \left(\frac{\chi_5}{kT}\right)^4 \left(\frac{kT}{h\nu}\right)^4 \left(1 + 3 \frac{\chi_5}{h\nu}\right) f_L(\eta_{5+}) S_{5+\nu}, \qquad (7.11.53)$$

where Z_{5+} is the effective charge of the oxygen ion with three bound electrons,

$$f_L(\eta) = \frac{\exp[-8\eta \cot^{-1}(\eta/2)]}{1 - \exp(-2\pi\eta)}, \qquad (7.11.54)$$

and η_{5+} is given by eq. (7.11.10) with Z_1 replaced by Z_{5+}. For O^{4+}, the cross section is as described by eqs. (7.11.53) and (7.11.54), except that quantities labeled with 5+ are replaced by quantities labeled with 4+ and the entire expression is multiplied by 2 to take into account that there are two electrons in the $2s$ shell (see eq. (7.4.39)).

Similar expressions hold for the ionic species O^{3+} to O^{0+} in which the outermost electrons are in $2p$ states (see the discussion in Section 7.4 and eq. (7.4.44)). For example,

$$\sigma_{3+}(\nu) = \sigma_{Th} \frac{1.6557 \times 10^{10}}{Z_{3+}^2} \left(\frac{\chi_3}{kT}\right)^5 \left(\frac{kT}{h\nu}\right)^4 \left(1 + \frac{8}{3}\frac{\chi_3}{h\nu}\right) f_L(\eta_{3+}) S_{3+\nu}. \quad (7.11.55)$$

7.11.4   Number abundances and opacities as functions of temperature when density $= 0.01\,\mathrm{g\,cm^{-3}}$

Number abundance parameters for most ion species considered and the total Rosseland mean opacity (excluding bound–bound transitions and photoionization from excited states) are shown on a logarithmic scale in Fig. 7.11.7 as functions of the temperature for a density $\rho = 10^{-2}\,\mathrm{g\,cm^{-3}}$. The number abundance parameters for oxygen ions have been multiplied by 100. At the highest temperatures shown, a determination of which ions dominate in contributing through bound–free absorption to the overall Rosseland mean requires, in general, an analysis of the sort shown in Figs. 7.11.1–7.11.6. However, one may anticipate that, at temperatures such that the abundances of singly ionized and neutral helium exceed $\sim 10^{-3}$, these ions will contribute more than oxygen ions. And, of course, at temperatures such that helium exists primarily as a neutral atom, bound–free absorption from neutral hydrogen plays the dominant role.

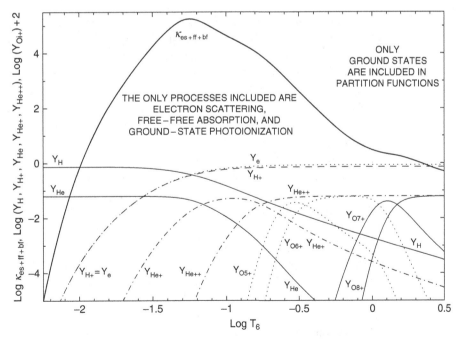

Fig. 7.11.7   Number abundances and Rosseland mean opacity versus temperature when density $= 10^{-2}\,\mathrm{g\,cm^{-3}}$ and $X = 0.74$, $Y = 0.25, Z_0 = 0.01$

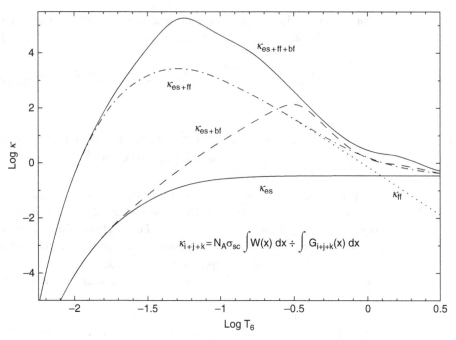

Fig. 7.11.8 Rosseland means of partial absorption coefficients versus temperature when density $= 10^{-2}$ g cm^{-3} and $X = 0.74, Y = 0.25, Z_0 = 0.01$

Variations with temperature of Rosseland means of various partial opacities are shown in Fig. 7.11.8. Note that, at temperatures smaller than $\sim$300 000 K, $\kappa_{\rm ff}$ and $\kappa_{\rm es+ff}$ are indistinguishable. This is because, at frequencies where the weighting function is large, the free–free absorption cross section is much larger than the electron-scattering cross section. Beginning at $T \sim$ 100 000 K, the number abundance of free electrons declines ever more rapidly with decreasing temperature. At temperatures smaller than $\sim$20 000 K, $\kappa_{\rm es}$ and $\kappa_{\rm es+bf}$ are indistinguishable, indicating that the thresholds for bound–free absorption are at energies much larger than the photon energy at the maximum in the weighting function. For temperatures smaller than $\sim$13 000 K, $\kappa_{\rm ff}$, $\kappa_{\rm es+ff}$, and $\kappa_{\rm es+ff+bf}$ are indistinguishable, indicating that free–free absorption is the dominant contributor to the Rosseland mean opacity. Responsible for the decrease with decreasing temperature of all opacity measures at temperatures less than $\sim$40 000 K is the decrease in the number abundance of free electrons.

The most striking lesson told by Fig. 7.11.8 is that the interplay between free–free and bound–free absorption revealed by an analysis of Figs. 7.11.1–7.11.6 is very widespread. That is, over most of the temperature domain considered, bound–free absorption is the major contributor to the overall Rosseland mean opacity, but it is the major contributor only because free–free absorption dominates electron scattering at low photon frequencies.

The anatomy of the opacity at $\rho = 0.01$ g cm^{-3} and at nine selected temperatures between 1.5×10^6 K and 10^4 K is shown in Figs. 7.11.9–7.11.17. In these figures, $Y_{{\rm O}n+}$ represents the absolute number-abundance parameter of the oxygen ion stripped of n

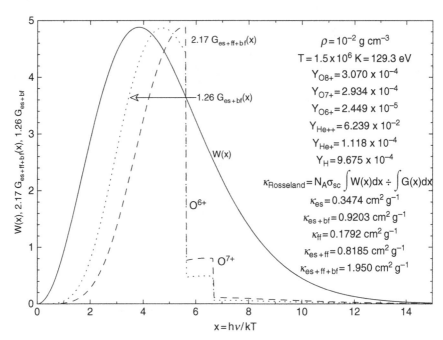

Fig. 7.11.9 Anatomy of an opacity for composition $X_H = 0.74$, $X_{He} = 0.25$, $X_O = 0.01$ when density $= 10^{-2}$ g cm^{-3} and temperature $= 1.5 \times 10^6$ K

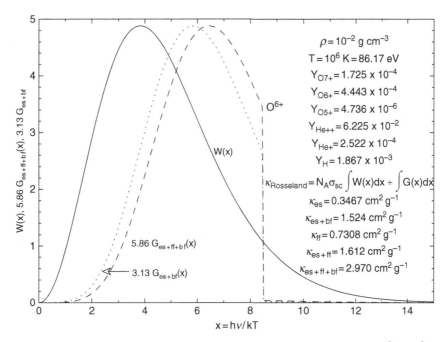

Fig. 7.11.10 Anatomy of an opacity for composition $X_H = 0.74$, $X_{He} = 0.25$, $X_O = 0.01$ when density $= 10^{-2}$ g cm^{-3} and temperature $= 10^6$ K

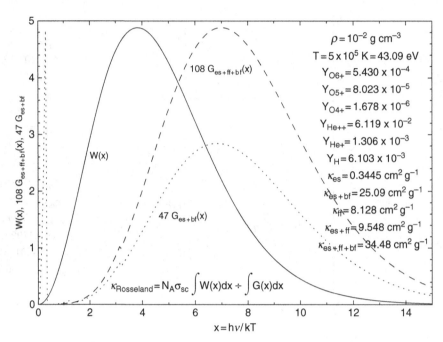

Fig. 7.11.11   Anatomy of an opacity for composition $X_H = 0.74$, $X_{He} = 0.25$, $X_0 = 0.01$ when density $= 10^{-2}$ g cm^{-3} and temperature $= 5 \times 10^5$ K

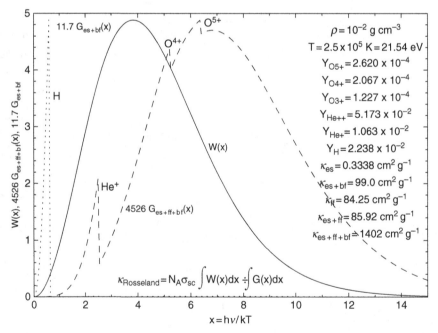

Fig. 7.11.12   Anatomy of an opacity for composition $X_H = 0.74$, $X_{He} = 0.25$, $X_0 = 0.01$ when density $= 10^{-2}$ g cm^{-3} and temperature $= 2.5 \times 10^5$ K

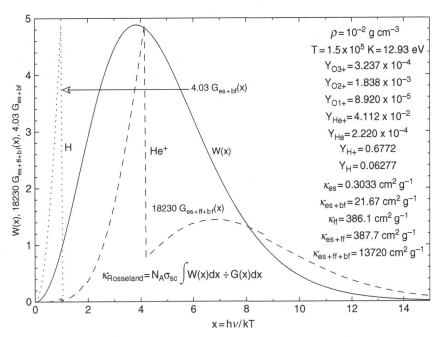

Fig. 7.11.13 Anatomy of an opacity for composition $X_H = 0.74$, $X_{He} = 0.25$, $X_O = 0.01$ when density $= 10^{-2}$ g cm^{-3} and temperature $= 1.5 \times 10^5$ K

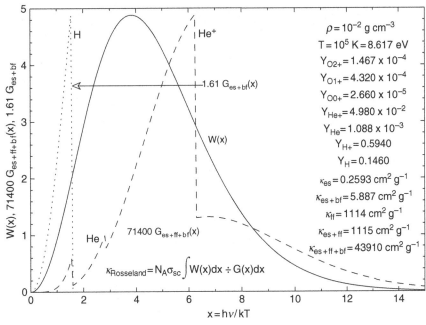

Fig. 7.11.14 Anatomy of an opacity for composition $X_H = 0.74$, $X_{He} = 0.25$, $X_O = 0.01$ when density $= 10^{-2}$ g cm^{-3} and temperature $= 10^5$ K

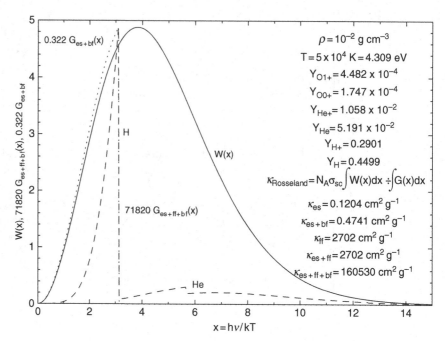

Anatomy of an opacity for composition $X_H = 0.74$, $X_{He} = 0.25$, $X_0 = 0.01$ when density $= 10^{-2}$ g cm^{-3} and temperature $= 5 \times 10^4$ K

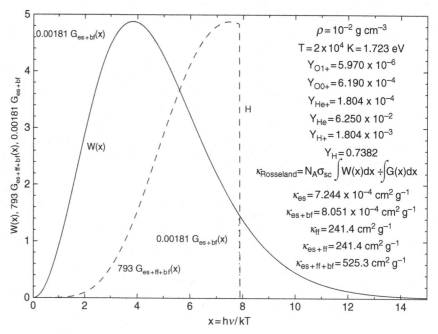

Anatomy of an opacity for composition $X_H = 0.74$, $X_{He} = 0.25$, $X_0 = 0.01$ when density $= 10^{-2}$ g cm^{-3} and temperature $= 2 \times 10^4$ K

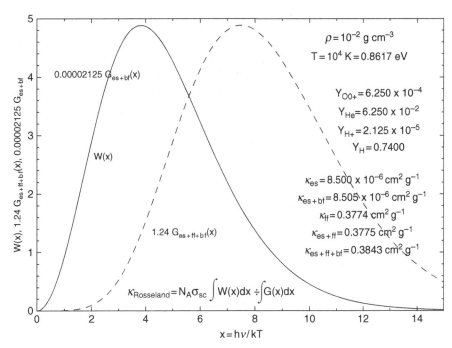

Fig. 7.11.17 Anatomy of an opacity for composition $X_H = 0.74$, $X_{He} = 0.25$, $X_O = 0.01$ when density $= 10^{-2}$ g cm^{-3} and temperature $= 10^4$ K

electrons, and not the abundance relative to the total oxygen abundance as in Figs. 7.11.1–7.11.6. Along the curve for $\kappa_{es+ff+bf}$ in Fig. 7.11.7, the logarithmic slope is smaller in the temperature range $\sim$1.5–1.0 $\times$ 10^6 K than on either side of this range. As revealed by Figs. 7.11.9 and 7.11.10, bound–free absorption from O^{7+} and O^{6+}, aided and abetted by free–free absorption *and by electron scattering* is responsible for this plateau. As indicated in the legend in Fig. 7.11.9, at $T = 1.5 \times 10^6$ K, the number abundances of O^{7+} and O^{8+} are comparable at $\sim$3 $\times$ 10^{-4} and the number abundance of O^{6+} is an order of magnitude smaller. The legend in Fig. 7.11.10 indicates that, at $T = 10^6$ K, the dominant oxygen ions are O^{7+} and O^{6+} at number abundances of $\sim$4.4 $\times$ 10^{-4} and $\sim$1.7 $\times$ 10^{-4}, respectively; the number abundance of O^{8+} is two orders of magnitude smaller than the sum of the abundances of the dominant isotopes. The curves for $G_{es+bf}(x)$ and $G_{es+ff+bf}(x)$ in Fig. 7.11.9 demonstrate that, at $T = 1.5 \times 10^6$ K, bound–free absorption, primarily from O^{7+} but also from O^{6+}, contributes significantly to the overall Rosseland mean opacity. These same curves in Fig. 7.11.10 show that, at $T = 10^6$ K, bound–free absorption from O^{6+} is the major bound–free contributor to the overall Rosseland mean. Even though O^{7+} is $\sim$2.5 times more abundant than O^{6+}, its contribution to the opacity is so far into the outer wing of the weighting function that this contribution is not detectable.

It is interesting to compare opacities and ionic abundances at the same temperature but at different densities. For example, at $T_6 = 1.5$ (compare Figs. 7.11.5 and 7.11.9), the overall Rosseland mean opacity is $\sim$30 times larger at 1 g cm^{-3} than it is at 0.01 g cm^{-3}. Some of the difference may be attributed to the fact that the average number of oxygen

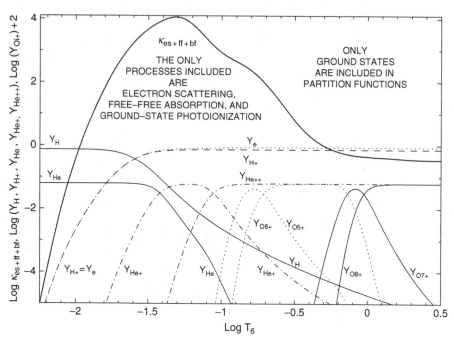

Fig. 7.11.18 Number abundances and Rosseland mean opacity versus temperature when density $= 10^{-4}$ g cm^{-3} and $X = 0.74$, $Y = 0.25, Z_0 = 0.01$

nuclei with bound electrons is larger at the larger density: at the higher density, 90% of the oxygen nuclei bind two electrons, and 10% bind one electron. At the lower density, 49% of the oxygen nuclei bind no electrons and 47% bind one. However, the major factor contributing to the difference in opacity is the factor of 100 difference in the density of free electrons. This difference in free electron density is reflected both in the direct contribution of the free–free absorption coefficient to the Rosseland mean opacity and in the indirect contribution of free–free absorption in enabling the bound–free absorption coefficient to express itself fully in the Rosseland mean. The same story is told at $T_6 = 1$ by a comparison between Figs. 7.11.6 and 7.11.10.

As is evident from Figs. 7.11.11–7.11.18, bound–free absorption by oxygen ions is of negligible importance at temperature $T_6 \leq 0.5$ when $\rho = 0.01$ g cm^{-3}. However, as detailed in Figs. 7.11.11–7.11.14, with decreasing temperature, bound–free absorption from singly ionized and neutral helium plays an increasingly important role for temperatures in the range $0.1 < T_6 < 0.3$. It is also evident that, although the number abundance of neutral hydrogen becomes comparable with and then much larger than the number abundance of helium in this temperature range, the contribution to bound–free absorption by neutral hydrogen at higher temperatures in the range is limited because it occurs at frequencies in the low-frequency tail of the weighting function and does not have the same impact on the Rosseland mean as does bound–free absorption from neutral and singly ionized helium. This characterization becomes less and less evident at the lowest temperatures in the range.

At temperatures only slightly smaller than 100 000 K, thanks to the assistance of free–free absorption, bound–free absorption by neutral hydrogen becomes the dominant contributor to the Rosseland mean opacity and is responsible for the peak along the $\kappa_{es+ff+bf}$ opacity profile in Fig. 7.11.8. An example of the anatomy of the opacity near the peak is given by Fig. 7.11.15. In this instance, the assistance of free–free absorption in enabling bound–free absorption to contribute is truly massive. But, as shown in Figs. 7.11.12–7.11.14, this is true also along the bump in $\kappa_{es+ff+bf}$ centered at $T \sim 160\,000$ K, where bound–free absorption by helium assisted by free–free absorption is the dominant contributor to the Rosseland mean opacity.

After the peak in $\kappa_{es+ff+bf}$, the steady decline with decreasing temperature in $\kappa_{es+ff+bf}$ is due to the decline in the free electron density. Also contributing to the decline is the fact that the absorption edge for photo-ionization of hydrogen moves through the maximum and then far into the high frequency tail of the weighting function. That both factors are at work is made abundantly clear in Figs. 7.11.16 and 7.11.17.

7.11.5 Number abundances and opacities as functions of temperature when density $= 10^{-4}$ g cm^{-3}

Fig. 7.11.18 shows number abundances and the total Rosseland mean opacity as functions of temperature and Fig. 7.11.19 shows Rosseland means of various combinations of absorption coefficients as functions of temperature. Figs. 7.11.20–7.11.27 describe the

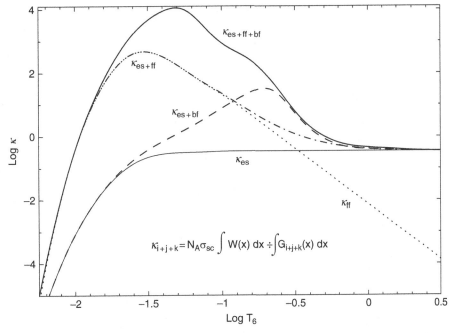

Fig. 7.11.19 Rosseland means of partial absorption coefficients versus temperature when density $= 10^{-4}$ g cm^{-3} and $X = 0.74, Y = 0.25, Z_0 = 0.01$

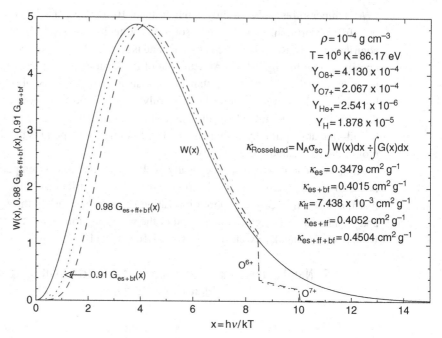

Fig. 7.11.20 Anatomy of an opacity for composition $X_H = 0.74$, $X_{He} = 0.25$, $X_0 = 0.01$ when density $= 10^{-4}$ g cm^{-3} and temperature $= 10^6$ K

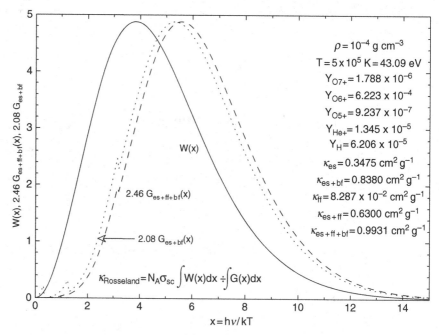

Fig. 7.11.21 Anatomy of an opacity for composition $X_H = 0.74$, $X_{He} = 0.25$, $X_0 = 0.01$ when density $= 10^{-4}$ g cm^{-3} and temperature $= 500,000$ K

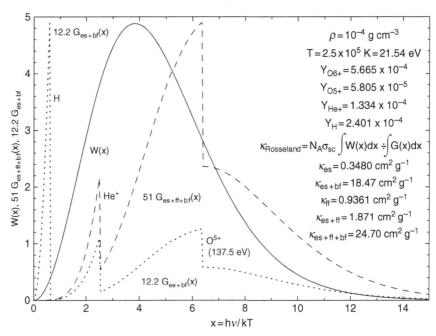

Fig. 7.11.22 Anatomy of an opacity for composition $X_H = 0.74$, $X_{He} = 0.25$, $X_O = 0.01$ when density $= 10^{-4}$ g cm^{-3} and temperature $= 250,000$ K

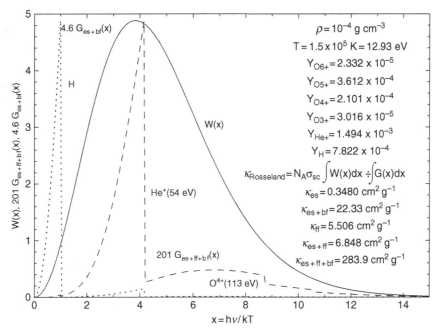

Fig. 7.11.23 Anatomy of an opacity for composition $X_H = 0.74$, $X_{He} = 0.25$, $X_O = 0.01$ when density $= 10^{-4}$ g cm^{-3} and temperature $= 150,000$ K

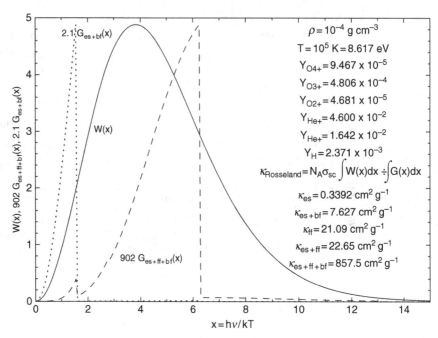

Fig. 7.11.24 Anatomy of an opacity for composition $X_H = 0.74$, $X_{He} = 0.25$, $X_0 = 0.01$ when density $= 10^{-4}$ g cm^{-3} and temperature $= 100,000$ K

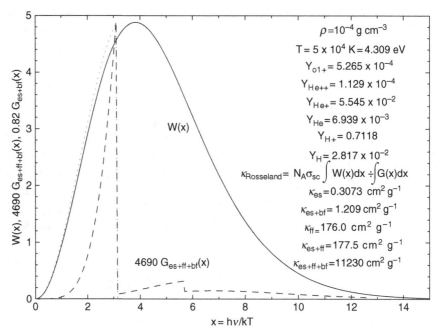

Fig. 7.11.25 Anatomy of an opacity for composition $X_H = 0.74$, $X_{He} = 0.25$, $X_0 = 0.01$ when density $= 10^{-4}$ g cm^{-3} and temperature $= 50,000$ K

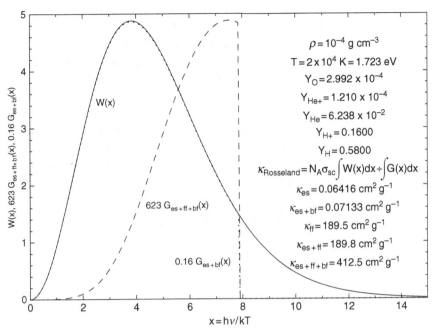

Fig. 7.11.26 Anatomy of an opacity for composition $X_H = 0.74$, $X_{He} = 0.25$, $X_O = 0.01$ when density $= 10^{-4}$ g cm^{-3} and temperature $= 20,000$ K

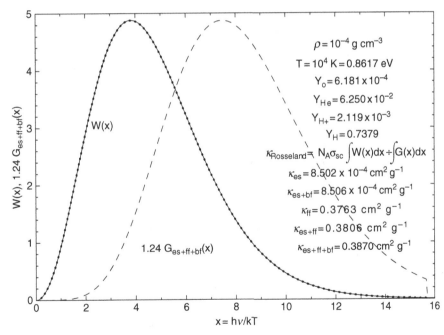

Fig. 7.11.27 Anatomy of an opacity for composition $X_H = 0.74$, $X_{He} = 0.25$, $X_O = 0.01$ when density $= 10^{-4}$ g cm^{-3} and temperature $= 10,000$ K

anatomy of the opacity at eight different temperatures from $T = 10^6$ K to $T = 10^4$ K. Comparing with the corresponding figures for the case $\rho = 10^{-2}$ g cm^{-3}, it is evident that the primary reason for the smaller opacities in the lower density case is the fact that the free–free absorption coefficient is essentially linearly proportional to the density. The fact that the maximum overall mean Rosseland opacity is only about one and not two orders of magnitude smaller in the lower density case is a consequence of the fact that free–free absorption plays a supporting role in enhancing the contribution of bound–free absorption. The anatomy of the opacity at $\rho = 10^{-4}$ g cm^{-3} and $T = 250\,000$ described in Fig. 7.11.22 is particularly interesting. It shows that, under appropriate conditions, bound–free absorption from an oxygen ion with an outer $2s$ electron, in this case O^{5+}, can compete with bound–free absorption from He$^+$ in contributing to the Rosseland mean opacity.

7.11.6 Number abundances and opacities as functions of temperature when density $= 10^{-6}$ g cm^{-3}

To provide further perspective, number-abundance parameters and the total Rosseland mean opacity are displayed in Fig. 7.11.28 as functions of temperature for a density of $\rho = 10^{-6}$ g cm^{-3}. Rosseland means of various combinations of absorption coefficients at the same density are displayed as functions of temperature in Fig. 7.11.29. Comparison with the same quantities at $\rho = 10^{-4}$ g cm^{-3} in Figs. 7.11.18 and 7.11.19 and at

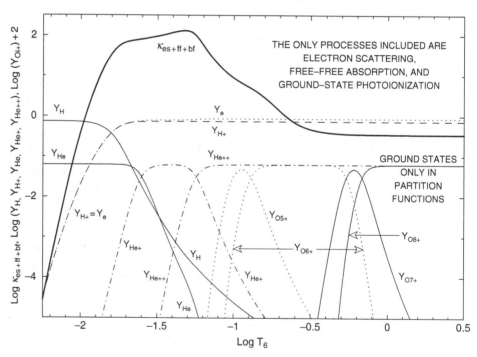

Fig. 7.11.28 Number abundances and Rosseland mean opacity versus temperature when density $= 10^{-6}$ g cm^{-3} and $X = 0.74$, $Y = 0.25$, $Z_0 = 0.01$

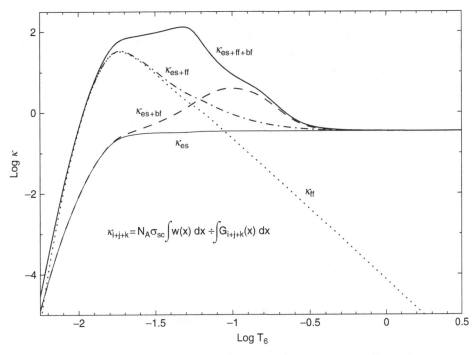

Rosseland means of partial absorption coefficients versus temperature when density $= 10^{-6}$ g cm^{-3} and $X = 0.74, Y = 0.25, Z_0 = 0.01$

$\rho = 10^{-2}$ g cm^{-3} in Figs. 7.11.7 and 7.11.8 reinforce the trends established already by the results at the two larger densities. Again, the major story is the supporting role of the free–free absorption coeffient in enabling bound–free opacity to express itself in the overall Rosseland mean opacity.

7.11.7 The effect on the Rosseland mean opacity of using Coulomb-distorted plane waves for electrons to obtain the free–free absorption coefficient

Given the importance of free–free absorption in enabling bound–free absorption to contribute to the Rosseland mean opacity, it is of relevance to determine whether the uncertainty in the free–free absorption coefficient associated with the choice of free electron wave functions is of concern. The concrete opacities thus far described have made use of a free–free absorption coefficient calculated by assuming that the electrons are undistorted plane waves and that electron screening, which would reduce the free–free absorption coefficient, may be neglected. An upper bound to the free–free absorption coefficient is obtained by neglecting electron screening and assuming that the electrons are plane waves which are distorted by the Coulomb field of an isolated ion, as described in Section 7.6. Schematically, the free–free absorption coefficient may be written as

$$\kappa_{ff} = \frac{N_A\,\sigma_{Th}}{x^{3.5}}\,\Gamma_{ff}(x) = N_A\,\sigma_{Kramers}\,\bar{g}_{ff}, \qquad (7.11.56)$$

where, as described in Section 7.7, σ_{Kramers} is the cross section in a semiclassical treatment of absorption coefficients and $\bar{g}_{\text{ff}}$ is a Gaunt factor relating the semiclassical cross section to the full quantum-mechanical cross section. For each ion type, there is a different Gaunt factor $\bar{g}_{\text{ff,i}}$ and, for each ion type,

$$\sigma_{\text{Kramers,i}} = \frac{\pi}{\sqrt{3}} \, \sigma_{\text{Th}} \, \pi \, Z_i^2 \left(\frac{\hbar c}{e^2}\right)^2 \left(\frac{e^2/2a_0}{h\nu}\right)^3 n_e \, a_0^2 \left(\frac{2}{\pi} \frac{m_e c^2}{kT}\right)^{1/2} \tag{7.11.57}$$

$$= \frac{\pi}{\sqrt{3}} \, \sigma_{\text{Th}} \, \rho \, Y_e \, \frac{32021.6}{T_6^{7/2}} \frac{1}{x^3} \, Z_i^2, \tag{7.11.58}$$

and

$$\kappa_{\text{ff,i}} = (N_A \, \sigma_{\text{Th}} \, \rho \, Y_e) \frac{58080.8}{T_6^{7/2}} \frac{1}{x^3} \left(Z_i^2 \, \bar{g}_{\text{ff,i}} \, Y_i\right). \tag{7.11.59}$$

The Rosseland means of the free–free opacity coefficient obtained by Karzas and Latter using Coulomb-distorted plane waves and approximated by the analytical expressions in Section 7.3 are displayed as functions of temperature in Fig. 7.11.30 for three different densities. The kink at $T \sim 90\,000$ K along the curve for $\rho = 10^{-4}$ g cm^{-3} and the kink at $T \sim 50\,000$ K along the curve for $\rho = 10^{-6}$ g cm^{-3} are possibly consequences of shortcomings in the analytic approximations.

The dotted curves in Fig. 7.11.30 give the Rosseland means of the free–free opacity coefficient obtained with the use of undistorted plane waves for the electrons. The fact that

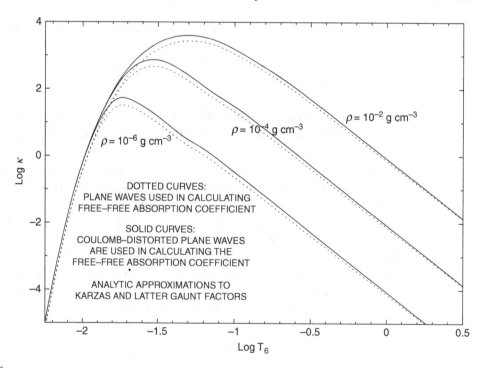

Fig. 7.11.30

Dependence of the free–free Rosseland mean opacity on the choice of electron wave functions

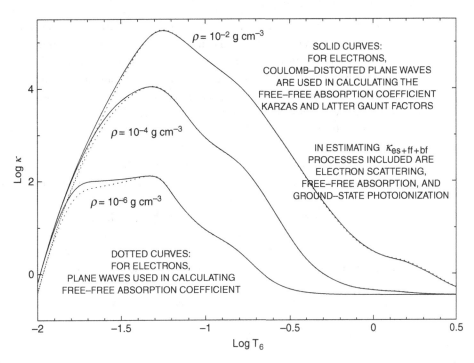

Fig. 7.11.31 Dependence of the Rosseland mean opacity on the choice of electron wave functions used in estimating the free–free absorption coefficient

there are also kinks along the dotted curves for $\rho = 10^{-4}$ g cm^{-3} and $\rho = 10^{-6}$ g cm^{-3} at approximately the same temperatures as along the corresponding solid curves suggest that physics rather than shortcomings in the analytical approximations is responsible for the kinks.

It is significant that, although the Rosseland mean of the free–free opacity is detectably larger when Coulomb-distorted plane waves are employed, the maximum difference is nowhere as large as a factor of 2.

In Fig. 7.11.31, the Rosseland means of the overall opacity obtained with the two approximations for the free–free absorption coefficient are compared, the solid curves being for the case when Coulomb-distorted electron plane waves are adopted, and the dotted curves being for the case when simple plane waves are employed. It is evident that, at temperatures higher than the temperatures at the peaks in the curves, the precise choice of the free–free absorption coefficient is of absolutely no consequence, demonstrating once again the catalytic role of free–free absorption when there are potentially potent sources of bound–free absorption present. At temperatures lower than at the peaks in the curves, the differences between pairs of opacity at a given density are modest, the largest difference being of the order of 40% along the curves for $\rho = 10^{-6}$ g cm^{-3} at $T \sim 20\,000$ K.

7.11.8 Concluding comments

The examples presented in this section offer a glimpse into the complexity involved in estimations of Rosseland mean opacities for conditions of relevance in real stars. It is clear from these examples that, even for the simple abundance mix chosen for illustration, at some densities and temperatures the inclusion of bound–free absorption from excited states and of absorption involving transitions between bound levels can increase the calculated opacity by an order of magnitude or more. This means, in turn, that the calculation of the ionization and excitation states of ions cannot be confined, as in the examples, to a consideration of just the first term in partition functions. Further, the inclusion of heavy elements other than oxygen has the effect of enhancing the opacity over a wider range of temperatures than shown in the examples. Finally, the effects of electron screening and electron degeneracy must be properly taken into account.

Although considerable effort over the years has been expended on producing realistic opacities for realistic conditions in stars, future generations of astrophysicists have the opportunity to contribute substantially to further development in the field. One aspect of opacities which deserves much more attention than it has received in the past is the construction of analytic expressions for opacity which are functions not only of temperature and density for any given choice of element abundances, but also of the elemental number abundances which change as evolution progresses. The preparation and use of opacity tables for every change encountered is exceedingly cumbersome. Furthermore, continuous derivatives with respect to all relevant variables are absolutely essential for achieving convergence in evolutionary calculations and analytical functions are the best way to ensure continuous derivatives.

7.12 Analytical approximations to results of opacity calculations for intermediate to high temperatures

In previous sections, the focus has been on the basic concepts involved in determining the major contributions to the opacity, leaving out details which a rigorous treatment requires. The objective has been to give insight into the physical processes responsible for the opacity rather than to provide quantitatively satisfactory estimates. It has hopefully become apparent that the determination of accurate opacities is basically a numerical exercise and that the results of this exercise are not easily summarized in closed analytical form.

Since the middle of the twentieth century, both the speed and the memory of electronic computers has accelerated exponentially with time and the sophistication and extent of opacity calculations has increased apace. The results of these calculations appear typically in the form of tables describing, for various choices of element abundances, the opacity as a function of density and temperature.

Examples of the necessary calculations which make use of theoretical information and computer technology available at the time are: G. Keller & R. E. Meyerott (1955), who include electron-scattering at low photon energies, free–free absorption, and bound–free absorption to estimate the opacity at high temperatures ($T_6 = 0.1$–20.0) for hydrogen- and helium-rich compositions at high metallicity; Arthur N. Cox & John N. Stewart (1970), who include bound–bound aborption for selected chemical species to obtain opacity estimates for a larger range of composition and thermodynamic variables; and Forrest J. Rogers & Carlos A. Iglesias (1992; see also Iglesias & Rogers, 1996) who make use of an extensive body of laboratory and theoretical data on atomic structure relevant to bound–bound absorption to produce extensive tables of opacity for hydrogen–helium mixtures and a wide variety of metallicities.

Tabular results are indispensible, but, in calculations of stellar structure, smooth derivatives of all relevant functions are essential for achieving convergence. For this reason, much of the knowledge gained about the evolution of stars over the second half of the twentieth century is the consequence of evolutionary calculations which have relied on analytical approximations to tabular opacities.

7.12.1 Keller–Meyerott opacities

Keller–Meyerott opacity estimates, based on principles elaborated by Bengt Strömgren (1932) and developed by Philip M. Morse (1940) have long been superceded, but it is useful to retain them as a reference point. They can be fitted by an expression of the form (Iben & Ehrman, 1962)

$$\kappa = \kappa_e + \bar{\kappa}, \tag{7.12.1}$$

where

$$\kappa_e = 0.19(1 + X) \tag{7.12.2}$$

and

$$\bar{\kappa} = 10^A \, \frac{\rho_e^\lambda}{T_6^B}. \tag{7.12.3}$$

In eq. (7.12.3),

$$\rho_e = \rho \, \frac{1 + X}{2} \tag{7.12.4}$$

is the electron "density". The coefficient of $(1 + X)$ in eq. (7.12.2) (0.19 instead of 0.20) comes from an error in the literature extant at the time which held that the electron-scattering contribution to the opacity is diminished by stimulated emission. In any case, the adopted form for the analytical approximation is a purely mathematical construct, justified solely by the fact that it permits tolerable fits to numerical results with a minimum of parameters.

The powers A, B, and λ are given, respectively, by

$$A = 2.0915 + 22.35Z \quad \text{when } \log T_6 > 0.1, \tag{7.12.5}$$

$$A = (2.3445 - 0.1392X) + (19.05 + 0.79X)Z \quad \text{when } \log T_6 < 0.1, \tag{7.12.6}$$

$$B = 3.065 + 8.5Z \quad \text{when } \log T_6 > 0.1, \tag{7.12.7}$$

$$B = (5.595 - 1.392X) - (24.5 - 7.9X)Z \quad \text{when } \log T_6 < 0.1, \tag{7.12.8}$$

and

$$\lambda = \max([0.62], [0.62 + 0.425(0.7 - \log T_6)]) \quad \text{when } \log T_6 > 0.3, \tag{7.12.9}$$

$$\text{and } \lambda = \min([1], [0.79 + 0.1573(0.3 - \log T_6)]) \quad \text{when } \log T_6 < 0.3. \tag{7.12.10}$$

For temperatures in the range $0.1 < T_6 < 23$ and for over nine orders of magnitude in density, the analytical fit provides an excellent match to the tabular values obtained by Keller and Meyerott (see Fig. 1 in Iben & Ehrman, 1962). From the development in Section 7.3, the zeroth order approximation to the free–free contribution to opacity is proportional to $\rho/T^{3.5}$. At temperatures such that electrons are scattered isotropically from electrons and at sufficiently low densities, the Rosseland mean opacity as calculated by Keller and Meyerott is a constant (only electron scattering contributes) and at high enough densities the mean opacity increases linearly with the density (only free–free absorption is important). The fact that the non-constant term in the analytical approximation depends on a power of the density which can be substantially different from unity and depends on an inverse power of the temperature different from 3.5 is simply a mathematical quirk that disguises the underlying physics.

7.12.2 Metal-free opacities at high temperatures

It is of interest to compare opacities contructed at two different epochs for temperatures and densities such that hydrogen and helium are completely ionized. The comparison is particularly valuable when no elements heavier than helium are present, i.e., when $Z = 0$. Under such conditions bound–free and bound–bound absorption do not occur and the comparison is a check on whether electron-scattering and free–free absorption have been treated similarly.

For $Z = 0$ and high temperatures ($T_6 > 5$), the fit to Keller–Meyerott opacities gives

$$\kappa = \kappa_{KM} = \kappa_e + 10^{1.90} (1 + X)^{0.62} \frac{\rho^{0.62}}{T_6^{3.07}} \tag{7.12.11}$$

$$\sim \kappa_e + 10^{1.90 + 0.27X} \frac{\rho^{0.62}}{T_6^{3.07}}. \tag{7.12.12}$$

The coefficient of X in eq. (7.12.12) comes from the fact that $10^{0.27X} = e^{0.62X} = 1 + 0.62X + \cdots \sim (1 + X)^{0.62}$.

At very high temperatures, a typical photon is scattered by an electron preferentially in the forward direction, with the degree of forward scattering increasing with the energy of the photon. The consequence is that κ_e decreases with increasing temperature. Using the quantum-mechanical cross section obtained by O. Klein and Y. Nishina (Klein & Nishina, 1929; Nishina, 1929), Cox & Stewart (1970) include the forward scattering effect in their opacity calculations. A fit to Cox–Stewart metal-poor ($Z = 0.0001$) hydrogen–helium opacities at densities sufficiently small that no other sources of opacity are important is the hyperbola:

$$\kappa_e = \left[0.2 - D - \sqrt{(D^2 + (0.02)^2)} \right] (1 + X), \tag{7.12.13}$$

where

$$D = 0.05 \, (\log T_6 - 1.7). \tag{7.12.14}$$

In the first published version of eq. (7.12.13) (Iben, 1975), $(0.02)^2$ is misprinted as 0.004. Equation (7.12.13) fits the data to better than 2% out to temperatures at least as high as $T_6 = 10^3$, where $\kappa_e = 0.0443(1 + X)$.

Subtracting κ_e given by eq. (7.12.13) from Cox–Stewart opacities for $Z = 0.0001$ and constructing linear fits in $\log \rho$ and $\log T_6$ yields an opacity expression of the form given by eq. (7.12.1), where κ_e is given by eqs. (7.12.13) and (7.12.14) and $\bar{\kappa}$ may be written as (Iben, 1975)

$$\bar{\kappa} = \left(\frac{\rho}{\bar{\rho}} \right)^{\lambda}, \tag{7.12.15}$$

where

$$\bar{\rho} = \min \left(\bar{\rho}_1, \bar{\rho}_2 \right), \tag{7.12.16}$$

$$\log \bar{\rho}_1 = -2.3 + 4.833 \, \log T_6, \tag{7.12.17}$$

and

$$\log \bar{\rho}_2 = -(1.86 + 0.3125 \, X) + 3.86 \, \log T_6. \tag{7.12.18}$$

The power λ is given by

$$\log \lambda = \max \left[\left(-0.174 - \Delta + \sqrt{\Delta^2 + (0.053)^2} \right), (\log 0.67) \right], \tag{7.12.19}$$

where

$$\Delta = 0.15 \, (0.13 + 0.212 \, X \, \log T_6). \tag{7.12.20}$$

The first member of the max function in eq. (7.12.19) was inadvertently omitted in the published version of the fit.

The free–free contribution to the opacity is unity along the curves relating the $\bar{\rho}$s to T_6; the intersection of the two curves occurs at

$$T_6 = 10^{(0.45 - 0.32 \, X)}. \tag{7.12.21}$$

At high temperatures ($T_6 \geq 3$),

$$\kappa = \kappa_{CS} = \kappa_e + 10^{1.25 + 0.21 \, X} \frac{\rho^{0.67}}{T_6^{2.59}}. \tag{7.12.22}$$

At temperatures $T_6 < 20$, the opacity given by eq. (7.12.22) is noticeably smaller than the opacity given by eq. (7.12.12). The Keller–Meyerott opacities on which eq. (7.12.12) is based are all metal rich and include bound–free absorption; the extrapolation to $Z = 0$ is therefore not supported by adequate data and this may account for most of the discrepancy between the two fits.

Proceeding in the same way with the entries in the Rogers–Iglesias (1992) opacity tables for $Z = 0$ hydrogen–helium mixtures (Tables 10 ($Z = 0$, $X = 0.7$), 11 ($Z = 0$, $X = 0.35$), and 12 ($Z = 0$, $X = 0$)), it transpires that

$$\log \bar{\rho}_1 = -2.27 + 5.86 \log T_6, \tag{7.12.23}$$

$$\log \bar{\rho}_2 = -(2.00 + 0.35\, X) + (4.00 + 0.16\, X) \log T_6, \tag{7.12.24}$$

and

$$\lambda = \max\left[\left(\frac{10^{-(0.106+0.0876X)}}{T_6^{(0.147+0.052X)}}\right),\ 0.675\right]. \tag{7.12.25}$$

Again, the free–free contribution to the opacity is unity along the curves relating the $\bar{\rho}$s to T_6, and the intersection of the two curves occurs at

$$\log T_6 = \frac{0.27 - 0.35X}{1.36 - 0.16X}. \tag{7.12.26}$$

At high temperatures ($T_6 \geq 1.6$),

$$\kappa = \kappa_{\mathrm{RI}} = \kappa_{\mathrm{e}} + 10^{1.35+0.24X}\, \frac{\rho^{0.675}}{T_6^{2.70+0.11X}}. \tag{7.12.27}$$

Equation (7.12.27) fits the Rogers–Iglesias tabular values to better than 20% over the entire range of tabulated values for $T_6 > 1$. For most of the region $0.2 < T_6 < 1$, the fit with $\bar{\rho} = \bar{\rho}_1$ given by eq. (7.12.23) and the appropriate value of λ from eq. (7.12.25) is also better than 20%. It is encouraging that, over their regions of validity, eqs. (7.12.27) and (7.12.22) give similar results.

7.12.3   Cox–Stewart opacities at intermediate to high temperatures for mixtures of hydrogen and helium when $Z \leq 0.02$

Cox–Stewart calculations include bound–bound absorption and are thus applicable at temperatures and densities such that heavy elements are only partially ionized. A fit to the opacities constructed for the compositions $(X, Y, Z) = (0.8, 0.18, 0.02)$, $(0.0, 0.98, 0.02)$, $(0.8, 0.1999, 0.0001)$, and $(0.0, 0.9999, 0.0001)$ is given by eq. (7.12.1), where κ_{e} is given by eqs. (7.12.13) and (7.12.14) and $\bar{\kappa}$ is given by (Iben, 1975)

$$\log \bar{\kappa}(Z) = \lambda\, \log\left(\frac{\rho}{\bar{\rho}}\right) + \left(\frac{Z}{0.02}\right)\, A_1\, \exp\left[-2.76 \log\left(\frac{\rho}{\rho_0}\right)^2 \frac{1}{W^2}\right], \tag{7.12.28}$$

where λ is given by eqs. (17.12.19) and (7.12.20), $\bar{\rho}$ is given by eqs. (7.12.16)–(7.12.18), $\log \rho_0$ is given by

$$\log \rho_0 = -(1.68 + 0.35\ X) + 1.8\ \log T_6, \tag{7.12.29}$$

when $T_6 > 1$, and by

$$\log \rho_0 = -(1.68 + 0.35\ X) + (3.42 - 0.52\ X)\ \log T_6 \tag{7.12.30}$$

when $T_6 < 1$. Further,

$$A_1 = 1.22\ \exp\left[-(1.74 - 0.755\ X)\ (\log T_6 - 0.22 + 0.1375\ X)^2\right] \tag{7.12.31}$$

and

$$W = 4.05\ \exp\left[-(0.306 - 0.04125\ X)\ (\log T_6 - 0.18 + 0.1625\ X)^2\right]. \tag{7.12.32}$$

Although it has been shown by Rogers and Iglesias that the calculations by Cox and his collaborators considerably underestimate the contribution of bound–bound transitions in heavy elements such as iron (in some instances by factors up to three for metal-rich mixtures at temperatures of a few hundred thousand degrees), the analytic fits have the virtue of demonstrating the flavor of bound–free and bound–bound contributions more transparently than do tabular data. They also prevent the appearance of discontinuous derivatives which can create mischief during stellar evolution calculations. Finally, the huge opacities which can arise when absoption by bound–free and bound–bound transitions are dominant contributors typically lead to such large radiative gradients that convection dominates energy flow, making it unnecessary to know the precise value of the opacity.

7.12.4 Cox–Stewart opacities at high temperatures for mixtures of helium, carbon, and oxygen

In regions of high density and temperature where hydrogen has been exhausted, useful estimates of the opacity are those of Cox & Stewart (1970). A fit to their estimates for the compositions $(X, Y, X_{12}, X_{16}) = (0, 0.9999, 0, 0)$, $(0, 0.5, 0.5, 0)$, $(0, 0, 1, 0)$, $(0, 0, 0.5, 0.5)$, and $(0, 0, 0, 1)$, where X_{12} and X_{16} are the abundances by mass of ^{12}C and ^{16}O, respectively, is

$$\kappa = \kappa_e + 10^{1.25 + 0.488\sqrt{\mu} + 0.092\mu}\left[1 + C\left(1 + \frac{C}{24.55}\right)\right]\frac{\rho^{0.67}}{T_6^{2.59 + 0.169\sqrt{\mu} + 0.012\mu}}, \tag{7.12.33}$$

where

$$\mu = \sum_i \frac{Z_i^2 X_i}{A_i} - 1 \tag{7.12.34}$$

and

$$C = \left(\frac{2.019 \times 10^{-4} \, \rho}{T_6^{1.7}} \right)^{2.425} . \tag{7.12.35}$$

In eq. (7.12.34), the sum can, in principle, be extended to include all isotopes heavier than hydrogen. The factor in square brackets in eq. (7.12.33) which involves the parameter C takes the plasma frequency cutoff roughly into account. This cutoff becomes important in the cores of AGB stars where conduction is the primary mode of energy transfer; it has the effect of increasing the radiative opacity, thereby reinforcing the role of conduction as the primary mode of energy transfer.

To understand how the plasma cutoff comes about, one may begin with Maxwell's equations:

$$\nabla \times \mathbf{H} = \frac{4\pi}{c} \mathbf{j} + \frac{1}{c} \frac{\partial \mathbf{E}}{\partial t}, \quad \nabla \times \mathbf{E} = -\frac{1}{c} \frac{\partial \mathbf{H}}{\partial t},$$

$$\nabla \cdot \mathbf{E} = 4\pi \rho_e, \quad \text{and} \quad \nabla \cdot \mathbf{H} = 0, \tag{7.12.36}$$

where ρ_e is the electric charge density. From these equations,

$$\nabla \times (\nabla \times \mathbf{H}) = \frac{4\pi}{c} (\nabla \times \mathbf{j}) + \frac{1}{c} \frac{\partial}{\partial t} (\nabla \times \mathbf{E}) = \frac{4\pi}{c} (\nabla \times \mathbf{j}) - \frac{1}{c^2} \frac{\partial^2 \mathbf{H}}{\partial t^2} \tag{7.12.37}$$

and

$$\nabla \times (\nabla \times \mathbf{H}) = \nabla (\nabla \cdot \mathbf{H}) - \nabla^2 \mathbf{H} = -\nabla^2 \mathbf{H}, \tag{7.12.38}$$

giving

$$\frac{1}{c^2} \frac{\partial^2 \mathbf{H}}{\partial t^2} - \nabla^2 \mathbf{H} = \frac{4\pi}{c} (\nabla \times \mathbf{j}). \tag{7.12.39}$$

Next, send a plane wave described by

$$\mathbf{E} = \mathbf{E}_0 \, e^{i(\omega t - kx)}, \tag{7.12.40}$$

where $\mathbf{E}_0$ is in the yz plane, through a medium with a free electron density of n_e. Assuming that each electron oscillates according to the equation

$$m_e \frac{d^2 \mathbf{y}}{dt^2} = -e\mathbf{E} \tag{7.12.41}$$

and integrating, one has

$$\frac{d\mathbf{y}}{dt} = -\frac{e\mathbf{E}}{i\omega \, m_e}, \tag{7.12.42}$$

or

$$\mathbf{j} = \frac{e^2 \mathbf{E}\, n_e}{i\omega\, m_e}. \tag{7.12.43}$$

Combining eqs. (7.12.43) and (7.12.39) gives, when n_e is constant,

$$\frac{1}{c^2} \left(\frac{\partial^2 \mathbf{H}}{\partial t^2} + \frac{\omega_p^2}{i\omega} \frac{\partial \mathbf{H}}{\partial t} \right) = \nabla^2 \mathbf{H}, \tag{7.12.44}$$

where

$$\omega_p = \sqrt{\frac{4\pi e^2 n_e}{m_e}} \tag{7.12.45}$$

is the plasma frequency. Inserting the choice

$$\mathbf{H} = \mathbf{H}_0\, e^{i(\omega t - k\mathbf{x})} \tag{7.12.46}$$

into eq. (7.12.44) gives

$$k^2 = \left(\frac{2\pi}{\lambda} \right)^2 = \frac{1}{c^2} \left(\omega^2 - \omega_p^2 \right). \tag{7.12.47}$$

When $\omega < \omega_p$, write $\gamma^2 = \omega_p^2 - \omega^2$, giving

$$\mathbf{H} = \mathbf{H}_0\, e^{i\omega t}\, e^{\pm|\gamma|\, x}. \tag{7.12.48}$$

The physically meaningful choice of sign is the one which implies an amplitude which decays rather than grows with increasing distance. Thus, electromagnetic waves of frequency smaller than the plasma frequency cannot propagate and therefore make no contribution to the inverse of the Rosseland mean opacity, resulting in a mean opacity larger than would otherwise be the case. An idea of when the plasma cutoff plays a significant role comes from a comparison of the energy of a photon at the plasma frequency ($\hbar\omega_p$) with the energy of an average photon ($\sim 2.70\, kT$). When electrons are not relativistic, the electrical force on an electron is given by eq. (7.12.41), and

$$\frac{\hbar\omega_p}{2.7\, kT} = \hbar \sqrt{\frac{4\pi e^2 n_e}{m_e}} \frac{1}{2.7\, kT} = \frac{123}{T_6} \sqrt{\frac{\rho_6}{\mu_e}}. \tag{7.12.49}$$

As shown in Chapter 15 of Volume 2 (Section 15.7, eq. (15.7.52)), when electrons are degenerate,

$$\omega_p = \sqrt{\frac{4\pi e^2 n_e}{E_F/c^2}}, \tag{7.12.50}$$

where $E_F = \sqrt{(m_e c^2)^2 + (c p_F)^2}$ is the electron Fermi energy. Thus, when electrons are degenerate and relativistic

$$\frac{\hbar \omega_p}{2.7 \, kT} = \frac{123}{T_6} \sqrt{\frac{\rho_6}{\mu_e}} \, f, \tag{7.12.51}$$

where

$$f = \frac{m_e c^2}{E_F} \tag{7.12.52}$$

can be considerably smaller than 1.

7.13 Convective cores in stars burning nuclear fuel at the center

The availability of analytical approximations to the opacity at high temperatures and densities permits one to address the question of whether or not a star which is burning a nuclear fuel at the center possesses a convective core and, if it does, to estimate the mass of the convective core relative to the total mass of the star and relative to the mass of the region over which the fuel is being burned. The analysis is based on a comparison between the radiative and adiabatic gradients in central regions. An estimate of the radiative gradient requires an equation of state, a nuclear energy-generation rate, and an estimate of the opacity.

The nature of composition changes in a core nuclear burning star and the lifetime of such a star are influenced by the rate at which particle transport occurs in the region in which burning takes place. If burning takes place in a convective core, and the mixing rate in this core is large enough (see the discussion in Section 8.9) the local rate at which the abundance of the fuel declines with time is independent of position in the core, and the total amount of fuel in the star which is consumed before the fuel is exhausted at the center is larger than if energy transport in the burning region is only by radiation and conduction. This means that the lifetime of the central burning phase is larger if burning takes place in a convective core than if the only mixing occurring in the burning region is by ordinary diffusion. Furthermore, when the fuel is exhausted at the center of a convective core, it is exhausted almost simultaneously over the rest of the core, and this leads to changes in observable characteristics (luminosity versus surface temperature) that differ from those changes experienced by stars which do not have convective cores while burning fuel at the center.

It is possible to determine with the help of polytropic models and analytical approximations to nuclear energy-generation rates and opacities that an initial main sequence star which relies on CN-cycle reactions as its primary source of nuclear energy has a convective core and that the ratio of convective core mass to the total stellar mass increases with increasing stellar mass. On the other hand, without appealing to more sophisticated models, it is not possible to determine definitively whether or not a main sequence star which relies on the pp chains as its primary source of energy has a convective core.

Exploration of the question of convective cores has a long history, beginning with T. G. Cowling (1934), who examined the conditions necessary for the existence of convective cores in massive main sequence stars even before von Weizsäcker and Bethe identified the CN-cycle reactions as the source of nuclear energy in such stars. Peter Naur and Donald E. Osterbrock (1953) and Roger Tayler (1950) introduced a criterion bearing on the size of convective cores which might possibly occur in solar mass main sequence stars which rely on the pp chains. Interesting discussions of the NOT (Naur, Osterbrock, Tayler) criterion are given by Roger Tayler (1967), Lloyd Motz (*Astrophysics and Stellar Structure*, 1970), and Ben Dorman and Robert T. Rood (1993). Although the NOT criterion cannot predict whether or not a convective core exists, it can make it plausible that, in stars which do not have a deep convective envelope and in which the burning rate in central regions is highly temperature sensitive, a convective region which begins at the center does not extend to the surface.

7.13.1 The criterion for convection at the center and evidence for the composite nature of models relying on CN-cycle energy generation

Energy flow by turbulent convection is expected when the radiative gradient V_{rad} defined by eq. (3.4.28) is larger than the adiabatic gradient V_{ad} defined by eq. (3.4.29). The radiative gradient may be written as

$$V_{\mathrm{rad}} = \frac{3}{4ac} \frac{\kappa P}{T^4} \frac{L(r)}{4\pi G M(r)} = \frac{3}{64\pi\sigma G} \frac{\kappa P}{T^4} \frac{L(r)}{M(r)} = 3.945 \times 10^9 \frac{\kappa P}{T^4} \frac{L(r)}{M(r)}, \quad (7.13.1)$$

where the luminosity $L(r)$ and mass $M(r)$ are functions of the distance r from the stellar center. All quantities to the right of the last equal sign in eq. (7.13.1) are in cgs units. For a completely ionized, weakly electron-degenerate, non-relativistic gas (Section 4.5) supplemented by radiation,

$$\frac{P}{T^4} = \frac{k}{\mu M_{\mathrm{H}}} \frac{\rho}{T^3} \left[1 + \frac{\mu}{\mu_e} \left(0.0219 \frac{\rho}{\mu_e T_6^{3/2}} + \cdots\right)\right] + \frac{1}{3} a$$

$$\sim 0.826 \times 10^{-10} \frac{\rho}{\mu T_6^3} \left[1 + \frac{\mu}{\mu_e} \left(0.0219 \frac{\rho}{\mu_e T_6^{3/2}}\right)\right] + 2.52 \times 10^{-15}, \quad (7.13.2)$$

where μ is the total molecular weight and μ_e is the electron molecular weight. Electron degeneracy is responsible for the term in parentheses on the right hand sides of these equations and radiation pressure is responsible for the last term in the equations. Inserting eq. (7.13.2) into eq. (7.13.1) gives

$$V_{\mathrm{rad}} \sim 0.326 \kappa \left\{\frac{\rho}{\mu T_6^3} \left[1 + \frac{\mu}{\mu_e} \left(0.0219 \frac{\rho}{\mu_e T_6^{3/2}}\right)\right] + 3.05 \times 10^{-5}\right\} \frac{L(r)}{M(r)}. \quad (7.13.3)$$

At the stellar center, as $r \to 0$, $L(r)/M(r) \to \epsilon(0) = \epsilon_c$, so

$$
V_{rad} \sim 0.326 \, \kappa \left\{ \left(\frac{\rho}{\mu T_6^3} \right)_c \left[1 + \frac{\mu}{\mu_e} \left(0.0219 \, \frac{\rho}{\mu_e T_6^{3/2}} \right) \right]_c + 3.05 \times 10^{-5} \right\} \epsilon_c,
$$

$$(7.13.4)$$

where the subscript c denotes evaluation at the center.

The adiabatic gradient in the nearly perfect gas plus radiation mixture can vary over the range

$$
V_{ad} = \left(\frac{d \log P}{d \log T} \right)_{ad} = \frac{2}{5} \to \frac{1}{4}, \tag{7.13.5}
$$

the larger value obtaining when radiation pressure is unimportant, and the smaller obtaining when radiation is the dominant contributor to pressure. For the present purpose, it is a good enough approximation to adopt $V_{ad} \sim 0.4 = $ constant. If a finite convective core exists, $V_{rad} > V_{ad}$ at the center of the model and V_{rad} must decrease outward until $V_{rad} < V_{ad}$ before the model surface is reached. That is, $V_{rad} > V_{ad}$ everywhere in the core, and, at the outer edge of the core, the derivative with respect to position of V_{rad}/V_{ad} must be negative.

The polytropic models described in Section 6.4 provide estimates of density and temperature at the stellar center for homogeneous main sequence stars of a typical Population I composition of $X = 0.71$, $Y = 0.27$, and $Z = 0.02$, permitting estimates of all of the quantities required in eq. (7.13.4). Adopting the approximation to Keller–Meyerott opacities given by eqs. (7.12.1)–(7.12.10) (with the exception that the factor 0.19 in eq. (7.12.2) is replaced by 0.2), for relevant densities and temperatures one has

$$
\kappa \sim 0.34 + 313 \, \frac{\rho_0^{0.62}}{T_6^{3.23}}. \tag{7.13.6}
$$

The $9 \, M_\odot$, $L = 6500 \, L_\odot$ model discussed in Section 6.4 is characterized by a polytropic index $N \sim 3.07$, central density $\rho_c \sim 18.3$ gm cm^{-3}, central temperature $(T_6)_c \sim 30.5$, and an energy-generation rate at the center of $\epsilon_c \sim 29\,500$ erg s^{-1}. From eq. (7.13.6), it is evident that by far the major contribution to the opacity at the model center is electron scattering and that the total opacity is $\kappa \sim 0.37$ cm^2 g^{-1}. Equation (7.13.4) gives $(V_{rad})_c \sim 0.326 \times 0.37 \times 0.001\,09 \times 29\,500 = 4.84 \sim 12 \, V_{ad}$, indicating that the model is strongly convective at the center. A similar exercise with the $3 \, M_\odot$ model in Section 6.4 characterized by $N = 3.25$ gives $(V_{rad})_c \sim 0.326 \times 0.51 \times 0.0090 \times 1670 = 2.26 \sim 6.2 \, V_{ad}$, indicating that it, too, is convective at the center. The fact that, at the centers of both models, V_{rad} is several times larger than V_{ad} suggests that the mass M_{CC} of the convective core may be a considerable fraction of the total mass M of each model. The fact that $V_{rad} - V_{ad}$ at the center increases with increasing M, suggests that, at least in models which rely on CN-cycle reactions for nuclear energy generation, M_{CC} increases with increasing M.

At the center of the solar mass model in Section 6.4 characterized by $L = L_\odot$, the temperature, density, and energy-generation rates are, respectively, $(T_6)_c \sim 13.5$, $\rho_c \sim 119$ g cm^{-3}, and $\epsilon_c \sim 17.8$ erg g^{-1} s^{-1}. Equation (7.13.6) states that the free–free contribution to the opacity at the model center is about four times larger than the contribution of electron scattering, giving a total opacity at the center of the model of $\kappa_c \sim 1.7$ cm^2 g^{-1}.

Equation (7.13.4) then gives $(V_{rad})_c \sim 0.326 \times 1.7 \times 17.8 \times 0.081 = 0.80$. Realistic evolutionary calculations show that the Sun's luminosity has increased by about one quarter to a third of its initial value over its history (Chapter 10). Thus, a more appropriate polytropic model of a solar mass zero age main sequence star has a luminosity of $L \sim 0.75\, L_\odot$. For the less luminous model, the entries for ϵ_c in Table 6.4.1 are to be reduced by $\sim 25\%$, leading to an interpolated best fit model with characteristics $N \sim 3.16$, $(T_6)_c \sim 13.0$, $\rho_c \sim 105$ g cm^{-3}, and $\epsilon_c \sim 13.2$ erg gm^{-1} s^{-1}. With these characteristics, eq. (7.13.6) gives $\kappa_c \sim 1.75$ cm^2 g^{-1} and eq. (7.13.4) produces $(V_{rad})_c \sim 0.326 \times (1.75 \times 13.2 \times 0.080) = 0.60$. Both estimates of $(V_{rad})_c$ are larger than V_{ad}, suggesting that a real zero age main sequence star of solar mass and composition might possess a small convective core. This inference is not corroborated by more realistic models constructed with the full set of stellar structure equations in which the relationship between pressure and density is not fixed by the artificial condition that $P \propto \rho^{1+\frac{1}{N}}$ but is found from an equation of state which specifies pressure as a function of both temperature and density.

The indication that, for intermediate mass main sequence stars, $(V_{rad})_c$ increases with stellar mass can be understood qualitatively without appealing to properties of detailed models. Adopting estimates of central temperature and density given in Chapter 3 by considerations involving elementary physics and observed relationships between stellar mass, radius, and luminosity for main sequence stars, it follows that, for intermediate mass main sequence stars, the opacity at the center is supplied primarily by electron scattering and can be viewed as effectively constant. Dimensional considerations give $P/T^4 \propto \rho/T^3 \propto (M/R^3)/(M/R)^{-3} \propto 1/M^2$. For CN-cycle burning stars, the discussion in Section 5.3 reveals that, for polytropes of index $1 < N < 4.5$, ϵ_c/ϵ_{Av} is nearly constant, varying from ~ 34 to ~ 29. From the observations, one has that, roughly, $\epsilon_{Av} = L/M \propto M^4/M = M^3$. Inserting these relationships into eq. (7.13.4) and neglecting the contribution of radiation pressure (the last term in curly brackets in eq. (7.13.4)) and electron degeneracy (the term in parentheses in eq. (7.13.4)), one has

$$(V_{rad})_c \propto \frac{1}{M^2}\, M^3 = M. \tag{7.13.7}$$

In addition to supporting the inference based on polytropic models that M_{CC} increases with increasing M, this result suggests also that M_{CC}/M may be large in all stars shining because of CN-cycle reactions occurring in central regions.

To summarize the arguments thus far, one may state that general considerations involving (1) a mass–luminosity relationship based on observations, (2) a temperature dependence of nuclear reactions based on a combination of experiment, nuclear reaction theory, and equilibrium thermodynamics, (3) a temperature and density dependence of opacity based on theoretical quantum-mechanical considerations, and (4) relationships between conditions at the center of a star of homogeneous composition and observed global properties of stars revealed by an examination of polytropic models show that intermediate mass main sequence stars which rely on CN-cycle reactions for energy generation possess convective cores of mass M_{CC} which increases with increasing stellar mass M. Exactly how M_{CC}/M depends on M requires the construction of realistic models.

It is clear that realistic models cannot be structures in which the local polytropic index, $N_{loc} = \left(d \log P / d \log \rho - 1\right)^{-1}$, is independent of position in the model. To a very good approximation, in a convective region in the deep interior of a stellar model (see the discussion in Section 3.4 related to eq. (3.4.52)), $V \sim V_{ad}$. To the extent that the perfect gas relationship is a good approximation, $V_{ad} = \left(d \log T / d \log P\right)_{ad} = 2/5$ and $d \log P / d \log \rho \sim \left(d \log P / d \log \rho\right)_{ad} = 5/3$. Hence, in the convective core of an intermediate mass main sequence star, $N_{loc} \sim 1.5$, and the core may be reasonably well represented by the inner portion of a polytrope of index $N \sim 1.5$. On the other hand, the discussion in Section 6.4 suggests that the global properties of real main sequence stars of intermediate mass can be understood in terms of polytropic models of a single index $N \gtrsim 3$. One infers that a more appropriate analytic approximation to intermediate mass main sequence stars is a composite polytrope consisting of an inner $N = 1.5$ polytopic segment and an outer $N \gtrsim 3$ polytropic segment. The mathematics involved in constructing such a composite, as addressed in Chapter IV, Section 28 of Chandrasekhar's book *Stellar Structure* (1939), is sufficiently convoluted that such a construction is not recommended.

7.13.2   Estimates of the size of a convective core in CN-cycle-burning main sequence stars

Insight into what determines the size of a convective core can be provided by making use of the spatial derivative of the ratio of V_{rad} to V_{ad} near the model center. Assuming that the criterion for convective flow is fulfilled at the center, the idea is that, if the convective region extending outward from the center does not reach the surface, the radiative gradient must decrease outward and fall below the adiabatic gradient at a mass smaller than the model mass. From eq. (7.13.1), it follows that, for a change $d \log r > 0$,

$$(d \log V_{rad})_{r>0} = \left[d \log \kappa + d \log P - 4 d \log T + d \log \frac{L(r)}{M(r)} \right]_{r>0} < 0. \qquad (7.13.8)$$

Assuming again that the equation of state may be approximated by a perfect gas, the temperature T and the dimensionless distance parameter z near the center of a polytrope are related by $T \propto u \overset{\sim}{=} \exp\left(-z^2/6\right)$, independent of polytropic index. Using this fact and the approximation to the luminosity–mass ratio given by eq. (5.4.6), one has that, near the center,

$$\frac{d \log \left[L(r)/M(r) \right]}{d \log T} \sim \frac{3}{5} \left[N\left(1 + k\right) + s \right]. \qquad (7.13.9)$$

It is convenient to approximate the opacity by the expression

$$\kappa = \kappa_e + \kappa_c \left(\frac{\rho}{\rho_c} \right)^{\lambda} \left(\frac{T}{T_c} \right)^{-\gamma}, \qquad (7.13.10)$$

where

$$\kappa_e = 0.2 \, (1 + X) \, \mathrm{cm}^2 \, \mathrm{g}^{-1}, \qquad (7.13.11)$$

λ and γ are constants which depend on the adopted composition, and κ_c is a function of the central density ρ_c and temperature T_c, but not of ρ and T. Using $\rho \propto T^N$, differentiation gives

$$\frac{d \log \kappa}{d \log T} = \left(1 - \frac{\kappa_e}{\kappa}\right)(N\lambda - \gamma). \qquad (7.13.12)$$

Setting

$$\left(\frac{d \log P}{d \log T}\right) = \frac{1}{V} \qquad (7.13.13)$$

and gathering together all relevant terms, the criterion for a radiative gradient which decreases outwards is

$$\frac{d \log V_{\mathrm{rad}}}{d \log T} = \left(1 - \frac{\kappa_e}{\kappa}\right)(N\lambda - \gamma) + \frac{1}{V} - 4 + \frac{3}{5}[N(1+k) + s] > 0. \qquad (7.13.14)$$

In a convective core, in the approximation that $V = V_{\mathrm{ad}} = 0.4$ and $N = 1.5$, eq. (7.13.14) becomes

$$\frac{d \log V_{\mathrm{rad}}}{d \log T} = \left(1 - \frac{\kappa_e}{\kappa}\right)\left(\frac{3}{2}\lambda - \gamma\right) - \frac{3}{2} + \frac{3}{5}\left[\frac{3}{2}(1+k) + s\right] > 0. \qquad (7.13.15)$$

The inequality in eq. (7.13.15) can be rewritten as

$$s > 1 + \frac{5}{3}\left(\gamma - \frac{3}{2}\lambda\right)\left(1 - \frac{\kappa_e}{\kappa}\right) - \frac{3}{2}k, \qquad (7.13.16)$$

which is a generalized form of the NOT criterion (see Iben & Ehrman, 1962). With $k = 0$ (electron screening neglected), $\gamma = 3.23$, and $\lambda = 0.62$, as given by eq. (7.13.6), one has

$$s > 1 + 3.83\left(1 - \frac{\kappa_e}{\kappa}\right). \qquad (7.13.17)$$

It is evident that, in models relying on CN-cycle reactions for energy generation, since $s \gtrsim 14$, the NOT criterion is easily met. At the centers of the best fit 1 $M_\odot$ models of Section 6.4, $\kappa_e/\kappa \sim 1/5$; with $s \sim 4.35$, the NOT criterion is only marginally satisfied.

Satisfying the NOT criterion says absolutely nothing about whether or not a convective core exists. The necessary and sufficient condition for the occurrence of a convective region extending outward from the center of a model is that $V_{\mathrm{rad}} > V_{\mathrm{ad}}$ at the center. On the other hand, if $V_{\mathrm{rad}} > V_{\mathrm{ad}}$ at the center, eq. (7.13.15) can be used to make a guess as to how far outward the convective region extends. In the deep interior of a main sequence model, the gradient of V_{ad} is typically small, and a rough estimate of the drop in temperature through the convective region is given by

$$(\Delta \log T)_{CC} = \log \left(\frac{T_{edge}}{T_c} \right) \sim \frac{\log \left[V_{ad}/(V_{rad})_c \right]}{(d \log V_{rad}/d \log T)_c}, \tag{7.13.18}$$

where T_{edge} is the temperature at the edge of the convective region and the derivative in the denominator on the right hand side may be estimated from eq. (7.13.15). Setting $k = 0$, eq. (7.13.15) becomes

$$\frac{d \log V_{rad}}{d \log T} \sim \frac{3}{5} (s - 1) - \left(1 - \frac{\kappa_e}{\kappa} \right) \left(\gamma - \frac{3}{2} \lambda \right)$$

$$\sim \frac{3}{5} (s - 1) - 2.3 \left(1 - \frac{\kappa_e}{\kappa} \right), \tag{7.13.19}$$

where the second line follows from choosing γ and λ from eq. (7.13.6).

At the centers of the the 3 $M_\odot$ and 9 $M_\odot$ models discussed in Section 6.4, the quantity $1 - \kappa_e/\kappa$ is small and eq. (7.13.19) becomes approximately

$$\frac{d \log V_{rad}}{d \log T} \sim \frac{3}{5} (s - 1) \sim 9. \tag{7.13.20}$$

For the 9 $M_\odot$ model, one infers from eqs. (7.13.18) and (7.13.20) that $\log \left(T_{edge}/T_c \right) \sim \log(0.4/4.84)/9 \sim -0.12$, or $u_{edge} = T_{edge}/T_c \sim 0.76$. Similarly, for the 3 $M_\odot$ model, one infers that $\log T_{edge}/T_c \sim \log(0.4/2.49)/9 \sim -0.088$, or $u_{edge} = T_{edge}/T_c \sim 0.82$. Since the effective polytropic index outside of the core ($N \sim 3$, perhaps) is different from the effective index in the core ($N = 1.5$), the change in temperature in the convective core cannot be translated into a firm statement about the mass of the core without constructing a composite polytropic model. On the other hand, the estimated drop in temperature through the core in both the 9 $M_\odot$ and 3 $M_\odot$ cases is only of the order of 20% and this is an indication that the mass of the convective region extending outward from the center may be small compared with the mass of the model. In Tables (5.2.1) and (5.2.2), the quantity $\left(z^2 \, du/dz \right)/\left(z^2 \, du/dz \right)_s$ for values of $u = 0.76$–0.82 suggest that, for both models, the mass of the convective core may be of the order of twenty percent or so of the mass of the model.

These estimates of convective core mass rely on a rate of change of V_{rad} with respect to $\log T$ (eqs. (7.13.15) and (7.13.18)) based on a luminosity to mass ratio (eq. (5.4.6)) which is valid over a domain small compared with the estimated core mass. A more respectable estimate might be obtained by adopting the luminosity to mass ratio given by eqs. (5.4.24) and (5.4.7). On the other hand, the central densities and temperatures estimated in Section 6.4 have been obtained by using polytropic models of a single index $N \gtrsim 3$, and are therefore not appropriate as starting points for a more careful estimate of M_{CC}. In particular, since index $N = 3$ polytropes are much more centrally condensed than $N = 1.5$ polytropes, the central densities estimated in Section 6.4 are larger than would be found in a composite model with a core characterized by $N = 1.5$.

7.13.3   Convective regions in realistic models of zero age main sequence models

Comparing properties of single-index polytropes with observed properties of zero age main sequence stars and white dwarfs is both instructive and enjoyable. Constructing composite-index polytropes and comparing with the observations could also be an instructive and entertaining exercise. However, given the fact that all of the ingredients for the construction of realistic models, including relatively well developed input physics, are available, and given the fact that modern day computers have the speed and memory to make the calculation of realistic models almost trivial, it does not seem appropriate to pursue in this book the construction of composite-index polytropes.

As an aside, it is worth remarking that the continued rapid growth in the speed and memory of accessible computational tools helps account for the fact that, since the middle of the twentieth century, essentially all of the progress in understanding stellar structure and evolution has involved the construction of realistic numerical models which make use of the currently most up-to-date input physics. Some of the early progress achieved by this approach is described by Bengt Strömgren in Chapter 4 (pages 269–296) of a book *Stellar Structure* (1965). Classic examples of early techniques employed and the attendant results are given by Martin Schwarzschild (*Structure and Evolution of the Stars*, 1958) and by Chuchiro Hayashi, R. Hoshi, and Daiichiro Sugimoto (1962).

Properties relevant to the occurrence of convective regions in zero age main sequence stars, as provided by models considered to be realistic in the mid-sixties of the twentieth century, are shown in Table 7.13.1. The manner in which the models have been selected from a set of evolutionary models is described in Section 5.5 and additional properties of the models are shown in Table 5.5.1.

Most of the information in Table 7.13.1 has been taken from a review article by Iben (1967) or from source material used in the review article. Entries in the first five columns of the table follow directly from the models described in the review, with the average

Table 7.13.1 Convective-core related properties of initial main sequence models based on evolutionary calculations (Iben, 1967)

$M/M_\odot$	$L/L_\odot$	ϵ_{Av}	ρ_c	T_c	ϵ_c	ρ_c/T_c^3	κ_c	$(V_{rad})_c$	M_{CC}/M	ϵ_{Av}/ϵ_c
15	20900	2700	6.17	34.4	82100	1.52/−4	0.34	2.55	0.40	0.033
9	4470	963	10.5	31.0	29200	3.52/−4	0.35	2.03	0.31	0.033
5	631	245	21.4	27.3	8070	1.05/−3	0.37	1.41	0.23	0.030
3	93.3	60.3	40.4	24.1	2030	2.89/−3	0.43	1.36	0.18	0.030
2.25	30.2	26.0	58.8	22.2	748	5.37/−3	0.50	1.08	0.15	0.035
1.5	5.50	7.11	87.7	18.8	144	1.32/−2	0.71	0.72	0.06	0.049
1.25	2.37	3.68	93.5	16.6	40.7	2.04/−2	0.92	0.41	0.009	0.09
1	0.75	1.45	90.0	13.9	12.1	3.35/−2	1.36	0.29	0.0	0.12
0.5	0.05	0.19	73.6	9.01	1.58	1.01/−1	4.05	0.36	0.0	0.12
0.25	0.01	0.078	155	7.46	0.65	3.73/−1	11.2	1.56	1.0	0.12

energy generation rate ϵ_{Av} being in units of erg g^{-1} s^{-1}, central density ρ_c in units of g cm^{-3}, and central temperature in units of 10^6 K. For models of mass $\geq 2.25\ M_\odot$, the rate of nuclear energy generation at the center, ϵ_c, which is in units of erg g^{-1} s^{-1}, has been found by using the listed values of central density and temperature in energy-generation rates given in Chapter 6. For the 1.25 $M_\odot$ and 1.5 $M_\odot$ models, the values of ϵ_c have been estimated by using the luminosity–mass relationships provided in the source material used in the review article The estimates of ϵ_c for $M \leq 1\ M_\odot$ have been obtained by assuming that $\epsilon_{Av}/\epsilon_c = 0.12$. The opacity at the center, in units of cm^2 g^{-1}, has been calculated using eq. (7.13.6) and the radiative gradient at the center has been calculated using eq. (7.13.4).

The quantity M_{CC}/M in the tenth column of the table is the ratio of the mass of the convective core to the total model mass given by the evolutionary calculations. The ratio ϵ_{Av}/ϵ_c in the last column of the table is an indication of the nature of the dominant nuclear burning sources in the models, with values $\gtrsim 0.033$ indicating that CN-cycle energy generation dominates and values ~ 0.12 indicating that pp-chain energy generation dominates. Intermediate values of the ratio indicate that both modes of energy generation are important. The variation of the convective core mass with model mass in static main sequence models of homogeneous composition is described in Fig. 4 of I. Iben, Jr. & John Ehrman (1962). For solar-like metallicity, the stellar mass at which the transition from radiative cores to convective cores occurs is $M \sim 1.1\ M_\odot$, and, for models of mass $M \simeq 2.5\ M_\odot$ and larger, essentially all nuclear energy production is confined to the convective core.

The solar mass model and the 0.5 $M_\odot$ model do not have convective cores. But, thanks to the increase in opacity with decreasing temperature in outer layers where bound–free absorption becomes important, both models possess convective envelopes, with the mass of the envelope relative to model mass being quite large in the 0.5 $M_\odot$ model. In the 0.25 $M_\odot$ model, the convective envelope extends all the way to the center. For this model, the development in Section 5.6 is relevant.

It is interesting to explore the variation of V_{rad} near the outer edge of the convective core of one of the models. Combining eqs. (7.13.2) and (5.4.24), and using $T/T_c = u$ and $\rho/\rho_c = u^{1.5}$, one has

$$V_{rad} \stackrel{\sim}{=} 0.326 \left[0.34 + 313\, \frac{\rho_c^{0.62}}{T_c^{3.23}}\, \frac{u^{0.62 \cdot 1.5}}{u^{3.23}} \right]$$

$$\times \left[\frac{\rho_c}{\mu T_{c6}}\, \frac{u^{3/2}}{u^3} \left(1 + \frac{\mu}{\mu_e} 0.0219 \frac{\rho_c}{\mu_e T_{c6}^{3/2}} \right) + 3.05 \times 10^{-5} \right] \times \left[\epsilon_c \frac{J_L(\alpha, z)}{-z^2 (du/dz)} \right],$$

$$(7.13.21)$$

where, as defined by eq. (5.4.38) when $N = 1.5$,

$$J_L(\alpha, z) = J_2(\alpha, z) - \left(\frac{\alpha}{120} \right) J_6(\alpha, z), \qquad (7.13.22)$$

J_2 and J_6 are given, respectively, by eqs. (5.4.19) and (5.4.20) in conjunction with eqs. (5.4.17) and (5.4.18), α is given by eq. (5.3.32) with $k = 0$ and $N = 1.5$, and s is to be determined.

Using the temperature at the center of the 3 $M_\odot$ model given in Table 7.13.1, eq. (6.4.7) gives $s \sim 16.9$, so $\alpha \sim 3.32$ and

$$V_{\text{rad}} \stackrel{\sim}{=} 0.326 \left(0.34 + \frac{0.107}{u^{2.3}} \right) \left(\frac{0.004\,73}{u^{1.5}} + 0.000\,0305 \right)$$

$$\times \left(2030 \, \frac{J_2(3.32, z) - 0.0277 \, J_6(3.32, z)}{-z^2 \, (\mathrm{d}u/\mathrm{d}z)} \right), \quad (7.13.23)$$

or

$$V_{\text{rad}} \stackrel{\sim}{=} 1.064 \left(1 + \frac{0.315}{u^{2.3}} \right) \left(\frac{1}{u^{1.5}} + 0.006\,45 \right) \left(\frac{J_2(3.32, z) - 0.0277 \, J_6(3.32, z)}{-z^2 \, (\mathrm{d}u/\mathrm{d}z)} \right). \quad (7.13.24)$$

When $z = 1$, Table 5.2.1 gives $u \sim 0.845$ and $z^2 \, (\mathrm{d}u/\mathrm{d}z) = -0.287$. Equations (5.4.17)–(5.4.20) give $J_0(3.32, 1) \sim 0.481$, $J_2(3.32, 1) \sim 0.0668$, and $J_6(3.32, 1) \sim 0.0132$. The numerator in the coefficient of ϵ_c in eq. (7.13.24) is $J_L(3.32, 1) = 0.0664$. Altogether, one has

$$V_{\text{rad}} \stackrel{\sim}{=} 1.064 \cdot 1.464 \cdot 1.294 \cdot 0.231 \sim 0.47. \quad (7.13.25)$$

When $z = 1.1$, rather than interpolate in Table 5.2.1, one may use eqs. (5.3.25) and (5.4.7) to estimate, respectively, $u \sim \exp\left(-z^2/6\right) \left(1 - z^4/720\right) \sim 0.816$ and $z^2 \, (\mathrm{d}u/\mathrm{d}z) = -(z^3/3) \exp(-z^2/6)[1 + (z^2/60) \, (1 - z^2/12)] \sim -0.369$. Further, $J_0(3.32, 1) \sim 0.484$, $J_2(3.32, 1) \sim 0.0699$, and $J_6(3.32, 1) \sim 0.0167$, so $J_L(3.32, 1.1) = 0.0694$ and

$$V_{\text{rad}} \stackrel{\sim}{=} 1.064 \cdot 1.500 \cdot 1.356 \cdot 0.188 \sim 0.41. \quad (7.13.26)$$

When $z = 1.2$, Table 5.2.1 gives $u = 0.784$ and $z^2 \, (\mathrm{d}u/\mathrm{d}z) = -0.465$. Since $J_0(3.32, 1.2) \sim 0.485$, $J_2(3.32, 1.2) \sim 0.0715$, $J_6(3.32, 1) \sim 0.0195$, and $J_L(3.32, 1.2) = 0.0710$ and

$$V_{\text{rad}} \stackrel{\sim}{=} 1.064 \cdot 1.550 \cdot 1.446 \cdot 0.153 \sim 0.36. \quad (7.13.27)$$

Thus, the transition from convective to radiative flow occurs at $z \sim 1.12$ and, interpolating in both Tables 5.2.1 and 5.2.2, one has that $(z^2 \, \mathrm{d}u/\mathrm{d}z)/(z^2 \, \mathrm{d}u/\mathrm{d}z)_s \sim 0.18$ in agreement with the evolutionary model estimate of $M_{\text{CC}}/M = 0.18$. From eq. (5.5.48) one has that the limiting value of $J_L(3.32, z > 2) = 0.0726$, indicating that over 96% of the luminosity of the model is generated in the convective core.

It is instructive to examine the variations in the factors that make up the products in eqs. (7.13.25)–(7.13.27). The first factor in each equation is the same, being a function of conditions at the model center. The next three factors are proportional, respectively, to the three terms in square brackets in eq. (7.13.21), and the three terms in parentheses in eq. (7.13.23) are proportional to, respectively, the opacity κ, P/T^4, and $L(r)/M(r) \propto L(r)/4\pi r^2 \div GM(r)/r^2$, the latter quantity being the ratio of the energy flux to the gravitational acceleration. In the three eqs. (7.13.25)–(7.13.27), the variation in the opacity related factor is by about 3% from one equation to the next, the variation in the P/T^4 related factor is by $(6\pm1)\%$, and the variation in the ratio of energy flux to gravitational acceleration is by 23%. Thus, it is the decrease in the ratio of the energy flux to the gravitational acceleration that is most responsible for the finite size of the convective core.

Assuming that, once outside of the convective core, the effective local polytropic index switches to something like $N \sim 3$, the P/T^4 related factor becomes approximately constant. As long as the analytical opacity adopted remains a reasonable approximation, the second term in the opacity related factor becomes inversely proportional to a smaller power of u. On the other hand, having almost reached its final value, the luminosity remains nearly constant but the mass continues to increase steadily. Thus, proceeding outward beyond the edge of the convective core, one may expect V_{rad} to decrease for some distance because of the continued decline in the energy flux relative to the gravitational acceleration. At some point, however, as $M(r)$ begins to approach the model mass M, the change in $L(r)/M(r)$ ceases to play a decisive role and the opacity factor becomes the controlling factor in determining V_{rad}, which may in general be expected to increase outward and perhaps exceed V_{ad}, especially if a region is encountered in which bound–free and/or bound–bound absorption becomes important. This does not occur in the 3 $M_\odot$ model main sequence model, but does occur in models significantly less massive in which extensive zones of partial ionization exist below the surface. In such regions, not only can κ be much larger than given by eq. (7.13.6) but V_{ad} can be substantially smaller than 0.4.

A rough estimate of the time scale τ_{mix} for complete mixing in a convective core in which the mixing length is comparable to the core size is, as discussed in Section 8.8,

$$\tau_{mix} \sim \frac{l_{core}^2}{l_{mix}\,\bar{v}_{conv}} \sim \frac{l_{core}}{\bar{v}_{conv}}, \tag{7.13.28}$$

where l_{core} is the radial dimension of the core and $\bar{v}_{conv}$ is an average turbulent speed in the core. Approximating $\left(d\log P/d\log \rho\right)_{ad} \sim (5/3)$ in eq. (3.4.41), and using eq. (3.4.52) to eliminate $(V - V_{ad})$ from the same equation, one obtains

$$v_{conv} \sim 0.540 \times 10^{-4}\, \frac{(L(r)/L_\odot)^{1/3}\,(R_\odot/r)^{2/3}}{\rho_0^{1/3}\,(T_6/\mu\,)^{1/2}} \left(1 - \frac{V_{ad}}{V_{rad}}\right)^{1/3} \left(\frac{l_{mix}}{H_P}\right)^{1/3} v_{sound}. \tag{7.13.29}$$

The sound speed is given approximately by

$$v_{sound} = \left(\frac{dP}{d\rho}\right)_{ad} \sim \left(\frac{5}{3}\frac{P}{\rho}\right)^{1/2} \sim \left(\frac{5}{3}\frac{kT}{\mu M_H}\right)^{1/2} \sim 1.17 \times 10^7 \left(\frac{T_6}{\mu}\right)^{1/2}\ \text{cm s}^{-1}. \tag{7.13.30}$$

From eqs. (7.13.29) and (7.13.30), a crude estimate of an average turbulent speed in the convective core of the 3 $M_\odot$ model is

$$\bar{v}_{\text{conv}} \sim 0.54 \times 10^{-4} \frac{(93/2)^{1/3} (1/0.26)^{2/3}}{33^{1/3} (21/0.61)^{1/2}} (1 - 0.5)^{1/3}$$

$$\times 1.17 \times 10^7 (21/0.61)^{1/2} \text{ cm s}^{-1}$$

$$\sim 2 \times 10^{-5} \times 6.88 \times 10^7 \text{ cm s}^{-1}. \sim 1.4 \times 10^3 \text{ cm s}^{-1}. \tag{7.13.31}$$

Thus, in the convective core of the 3 $M_\odot$ model, the mixing time is estimated to be of the order of

$$\tau_{\text{mix}} \sim \frac{0.35 R_\odot}{1.4 \times 10^3 \text{ cm s}^{-1}} = 1.74 \times 10^7 \text{ s} \sim 6.6 \text{ months}. \tag{7.13.32}$$

This time is quite short compared with the time scale for destroying the stable isotopes in the CN cycle by proton-capture reactions, but it is long compared with the lifetimes of the beta-unstable isotopes in the cycle (see Section 6.2). Thus, the abundances of isotopes such as ^{12}C, ^{13}C, ^{14}N, and ^{15}N are constant as functions of position in the core, but isotopes such as ^{13}N and ^{15}O are everywhere in local equilibrium with respect to creation and destruction mechanisms.

7.14 Algorithms for interpolation in opacity tables

Ideally, it would be desirable to be able to approximate opacities over the entire range of conditions of relevance in stellar interiors by means of analytical functions of the type described in Section 7.12. Although this is in principle possible, the number of parameters necessary for an accurate approximation to modern day opacities is such as to make the search for self-contained analytical approximations less attractive than implementing routines for smooth interpolation among tabular values. The downside of relying solely on interpolation in tables is that instructive analyses of the sort conducted in Section 7.13 are not possible.

The easiest kinds of table to deal with are rectangular, with a dependent variable $z(x, y)$ being presented as a set of entries z_{ij} in a matrix in which rows are specified by specific values of the independent variable x_i and columns are specified by specific values of the independent variable y_j. As long as $x_1 < x < x_I$, where x_1 and x_I designate the first and last rows, respectively, and $y_1 < y < y_J$, where y_1 and y_J designate the first and last columns, respectively, any of a number of interpolation schemes is possible. In general, the simpler the interpolation scheme one can employ and yet achieve convergence in an evolutionary calculation which relies on the availability of smoothly varying opacities and smooth opacity derivatives, the better.

7.14.1 Linear interpolation

The simplest interpolation scheme by far is linear and implementation is trivial. Consider the point (x, y) bounded by the points, in counterclockwise order, (x_1, y_1), (x_2, y_1),

(x_2, y_2), and (x_1, y_2). Along the boundaries of the square grid defined by these four points, values of z defined by linear interpolation are

$$z(x, y_1) = z(x_1, y_1) + \frac{z(x_2, y_1) - z(x_1, y_1)}{x_2 - x_1} (x - x_1), \quad (7.14.1)$$

$$z(x, y_2) = z(x_1, y_2) + \frac{z(x_2, y_2) - z(x_1, y_2)}{x_2 - x_1} (x - x_1), \quad (7.14.2)$$

$$z(x_1, y) = z(x_1, y_1) + \frac{z(x_1, y_2) - z(x_1, y_1)}{y_2 - y_1} (y - y_1), \quad (7.14.3)$$

and

$$z(x_2, y) = z(x_2, y_1) + \frac{z(x_2, y_2) - z(x_2, y_1)}{y_2 - y_1} (y - y_1). \quad (7.14.4)$$

Linear interpolation with eqs. (7.14.1) and (7.14.2) gives, at an arbitrary point in the grid,

$$z_a(x, y) = z(x, y_1) + \frac{z(x, y_2) - z(x, y_1)}{y_2 - y_1} (y - y_1), \quad (7.14.5)$$

while linear interpolation with eqs. (7.14.3) and (7.14.4) gives, at the same point,

$$z_b(x, y) = z(x_1, y) + \frac{z(x_2, y) - z(x_1, y)}{x_2 - x_1} (x - x_1). \quad (7.14.6)$$

Using eqs. (7.14.1)–(7.14.4) in eqs. (7.14.5) and (7.14.6) shows that

$$z_a(x, y) = z_b(x, y)$$

$$= z(x_1, y_1) + (z(x_2, y_1) - z(x_1, y_1)) \frac{x - x_1}{x_2 - x_1}$$

$$+ (z(x_1, y_2) - z(x_1, y_1)) \frac{y - y_1}{y_2 - y_1}$$

$$+ (z(x_2, y_2) - z(x_1, y_2) - z(x_2, y_1) + z(x_1, y_1)) \frac{x - x_1}{x_2 - x_1} \frac{y - y_1}{y_2 - y_1}. \quad (7.14.7)$$

Derivatives are

$$\left(\frac{\partial z}{\partial x} \right)_y = (z(x_2, y_1) - z(x_1, y_1)) \frac{1}{x_2 - x_1}$$

$$+ (z(x_2, y_2) - z(x_1, y_2) - z(x_2, y_1) + z(x_1, y_1)) \frac{1}{x_2 - x_1} \frac{y - y_1}{y_2 - y_1}, \quad (7.14.8)$$

and

$$\left(\frac{\partial z}{\partial y}\right)_x = (z(x_1, y_2) - z(x_1, y_1)) \; \frac{1}{x_2 - x_1}$$

$$+ (z(x_2, y_2) - z(x_1, y_2) - z(x_2, y_1) + z(x_1, y_1)) \; \frac{1}{y_2 - y_1} \frac{x - x_1}{x_2 - x_1}. \quad (7.14.9)$$

A major advantage of linear interpolation is that, within the confines of a four point grid, the interpolated function and its derivatives are unique. Major disadvantages are that any information related to curvature in the surface defined by the exact function $z(x, y)$ is absent in the interpolated function and, in general, derivatives at grid boundaries are discontinuous.

7.14.2 Quadratic interpolation

The next simplest interpolation scheme is quadratic and requires the availability of nine grid points (x_i, y_j), where, say, $i = 1, 3$ and $j = 1, 3$. At grid points along a column specified by y_j, the function z has the values $z(x_1, y_j), z(x_2, y_j)$, and $z(x_3, y_j)$, and the quadratic that passes through the three grid-point values of x has the general form

$$z(x, y_j) = c_{1j} + c_{2j} \, x + c_{3j} \, x^2. \quad (7.14.10)$$

The coefficients c_{mj} are obtained by solving the linear equations

$$z(x_1, y_j) = c_{1j} + c_{2j} \, x_1 + c_{3j} \, x_1^2,$$

$$z(x_2, y_j) = c_{1j} + c_{2j} \, x_2 + c_{3j} \, x_2^2, \quad \text{and}$$

$$z(x_3, y_j) = c_{1j} + c_{2j} \, x_3 + c_{3j} \, x_3^2. \quad (7.14.11)$$

Subtracting successive pairs of expressions in eq. (7.14.11) gives

$$z(x_1, y_j) - z(x_2, y_j) = c_{2j} \, (x_1 - x_2) + c_{3j} \left(x_1^2 - x_2^2\right), \quad \text{and}$$

$$z(x_2, y_j) - z(x_3, y_j) = c_{2j} \, (x_2 - x_3) + c_{3j} \left(x_2^2 - x_3^2\right), \quad (7.14.12)$$

which may be rearranged to give

$$c_{2j} + c_{3j} \, (x_1 + x_2) = \frac{z(x_1, y_j) - z(x_2, y_j)}{x_1 - x_2}, \quad \text{and}$$

$$c_{2j} + c_{3j} \, (x_2 + x_3) = \frac{z(x_2, y_j) - z(x_3, y_j)}{x_2 - x_3}. \quad (7.14.13)$$

Subtracting the two expressions in eq. (7.14.13) gives the coefficient c_{3j} in terms of known quantities:

$$c_{3j} = \frac{1}{x_1 - x_3} \left(\frac{z(x_1, y_j) - z(x_2, y_j)}{x_1 - x_2} - \frac{z(x_2, y_j) - z(x_3, y_j)}{x_2 - x_3}\right). \quad (7.14.14)$$

From the first expression in eq. (7.14.13), one has

$$c_{2j} = -c_{3j}(x_1 + x_2) + \frac{z(x_1, y_j) - z(x_2, y_j)}{x_1 - x_2}. \tag{7.14.15}$$

Using eq. (7.14.14) in eq. (7.14.15) and rearranging yields

$$c_{2j} = \frac{1}{x_1 - x_3}\left(-(x_2 + x_3)\frac{z(x_1, y_j) - z(x_2, y_j)}{x_1 - x_2} + (x_1 + x_2)\frac{z(x_2, y_j) - z(x_3, y_j)}{x_2 - x_3}\right). \tag{7.14.16}$$

The first expression in eq. (7.14.11) gives

$$c_{1j} = z(x_1, y_j) - c_{2j}\,x_1 - c_{3j}\,x_1^2, \tag{7.14.17}$$

which becomes, on using eqs. (7.14.14) and (7.14.16) and rearranging,

$$c_{1j} = \frac{z(x_2, y_j)\,x_1 - z(x_1, y_j)\,x_2}{x_1 - x_2}$$
$$+ \frac{x_1 x_2}{x_1 - x_3}\left(\frac{z(x_1, y_j) - z(x_2, y_j)}{x_1 - x_2} - \frac{z(x_2, y_j) - z(x_3, y_j)}{x_2 - x_3}\right). \tag{7.14.18}$$

Although eqs. (7.14.14), (7.14.15), and (7.14.17) in conjunction with eq. (7.14.10) totally specify the functions $z(x, y_j)$, it is interesting to use eqs. (7.14.16) and (7.14.18) as well as eq. (7.14.14) in eq. (7.14.10) to show explicitly how these functions of x depend on values of the functions at grid points. The result is

$$z(x, y_j) = \frac{z(x_2, y_j)(x - x_1) - z(x_1, y_j)(x - x_2)}{x_2 - x_1}$$
$$+ \frac{(x - x_1)(x - x_2)}{x_3 - x_1}\left(\frac{z(x_3, y_j) - z(x_2, y_j)}{x_3 - x_2} - \frac{z(x_2, y_j) - z(x_1, y_j)}{x_2 - x_1}\right). \tag{7.14.19}$$

The same manipulations along each row produce quadratics of the form

$$z(x_i, y) = r_{i1} + r_{i2}\,y + r_{i3}\,y^2, \tag{7.14.20}$$

with coefficients

$$r_{i3} = \frac{1}{y_1 - y_3}\left(\frac{z(x_i, y_1) - z(x_i, y_2)}{y_1 - y_2} - \frac{z(x_i, y_2) - z(x_i, y_3)}{y_2 - y_3}\right), \tag{7.14.21}$$

$$r_{i2} = -r_{i3}(y_1 + y_2) + \frac{z(x_i, y_1) - z(x_i, y_2)}{y_1 - y_2}, \tag{7.14.22}$$

and

$$r_{i1} = z(x_i, y_1) - r_{i2}\,y_1 - r_{i3}\,y_1^2. \tag{7.14.23}$$

The next task is to estimate $z(x, y)$ for an arbitrary position (x, y) within the nine-point grid such that $x_1 \leq x \leq x_3$ and $y_1 \leq y \leq y_3$. One approach is to construct a quadratic in the y direction through the points (z_{xj}, y_j), where

$$z_{xj} = c_{1j} + c_{2j}\, x + c_{3j}\, x^2, \tag{7.14.24}$$

$j = 1, 3$, and the coefficients c_{mj} are given by eqs. (7.14.14), (7.14.15), and (7.14.17). The result is

$$z_a(x, y) = d_{x1} + d_{x2}\, y + d_{x3}\, y^2, \tag{7.14.25}$$

where the quantities d_{xj} are constrained by

$$z_{xj} = d_{x1} + d_{x2}\, y_j + d_{x3}\, y_j^2, \tag{7.14.26}$$

and $j = 1, 3$. Repeating earlier steps, one obtains

$$d_{x3} = \frac{1}{y_1 - y_3} \left(\frac{z_{x2} - z_{x1}}{y_2 - y_1} - \frac{z_{x3} - z_{x2}}{y_3 - y_2} \right), \tag{7.14.27}$$

$$d_{x2} = -d_{x3}\, (y_1 + y_2) + \frac{z_{x2} - z_{x1}}{y_2 - y_1}, \tag{7.14.28}$$

and

$$d_{x1} = z_{x1} - d_{x2}\, y_1 - d_{x3}\, y_1^2. \tag{7.14.29}$$

Using eqs. (7.14.27)–(7.14.29) in eq. (7.14.25), one obtains

$$
\begin{aligned}
z_a(x, y) = z_{x1} + \frac{z_{x2} - z_{x1}}{y_2 - y_1}\, (y - y_1) \\
+ \frac{1}{y_1 - y_3} \left(\frac{z_{x2} - z_{x1}}{y_2 - y_1} - \frac{z_{x3} - z_{x2}}{y_3 - y_2} \right) (y - y_1)\, (y - y_2),
\end{aligned} \tag{7.14.30}
$$

and it is to be remembered that the variable x enters this expression through the quantities z_{xj} defined by eq. (7.14.24) and eqs. (7.14.14), (7.14.15), and (7.14.17).

An alternative approach is to construct a quadratic in the x direction through the points (z_{iy}, x_i), where

$$z_{iy} = r_{i1} + r_{i2}\, x + r_{i3}\, x^2, \tag{7.14.31}$$

$i = 1, 3$, and the coefficients r_{im} are given by eqs. (7.14.21)–(7.14.23). The result is

$$z_b(x, y) = e_{1y} + e_{2y}\, y + e_{3y}\, y^2, \tag{7.14.32}$$

where the quantities e_{iy} are constrained by

$$z_{iy} = e_{1y} + e_{2y}\, x_i + e_{3y}\, x_i^2, \tag{7.14.33}$$

and $i = 1, 3$. Proceeding as before,

$$e_{3y} = \frac{1}{x_1 - x_3} \left(\frac{z_{2y} - z_{1y}}{x_2 - x_1} - \frac{z_{3y} - z_{2y}}{x_3 - x_2} \right), \tag{7.14.34}$$

$$e_{2y} = -e_{3y}(x_1 + x_2) + \frac{z_{2y} - z_{1y}}{x_2 - x_1}, \tag{7.14.35}$$

and

$$e_{1y} = z_{1y} - e_{2y} x_1 - e_{3y} x_1^2. \tag{7.14.36}$$

Using eqs. (7.14.34)–(7.14.36) in eq. (7.14.32) gives

$$z_b(x, y) = z_{1y} + \frac{z_{2y} - z_{1y}}{x_2 - x_1}(x - x_1)$$
$$+ \frac{1}{x_1 - x_3} \left(\frac{z_{2y} - z_{1y}}{x_2 - x_1} - \frac{z_{3y} - z_{2y}}{x_3 - x_2} \right)(x - x_1)(x - x_2), \tag{7.14.37}$$

where the variable y enters the expression through the quantities z_{iy} defined by eq. (7.14.31) and eqs. (7.14.21)–(7.14.23).

A major advantage of quadratic interpolation is that it takes curvature into account. Disadvantages are that, in general, $z_a(x, y) \neq z_b(x, y)$, and the derivatives of the interpolated zs are also not unique. In addition, derivatives are discontinuous at the boundaries of the nine-point grid. As a compromise, one may approximate

$$z(x, y) = \frac{1}{2}(z_a(x, y) + z_b(x, y)), \tag{7.14.38}$$

and, as derivatives, choose

$$\frac{\partial z(x, y)}{\partial y} \sim \frac{\partial z_a(x, y)}{\partial y} = \frac{z_{x2} - z_{x1}}{y_2 - y_1} + \frac{1}{y_1 - y_3} \left(\frac{z_{x2} - z_{x1}}{y_2 - y_1} - \frac{z_{x3} - z_{x2}}{y_3 - y_2} \right)(2y - y_2 - y_1) \tag{7.14.39}$$

and

$$\frac{\partial z(x, y)}{\partial x} \sim \frac{\partial z_b(x, y)}{\partial x} = \frac{z_{2y} - z_{1y}}{x_2 - x_1}$$
$$+ \frac{1}{x_1 - x_3} \left(\frac{z_{2y} - z_{1y}}{x_2 - x_1} - \frac{z_{3y} - z_{2y}}{x_3 - x_2} \right)(2x - x_2 - x_1). \tag{7.14.40}$$

7.14.3 Cubic spline interpolation

A more sophisticated interpolation scheme produces a function $z(x, y)$ which is continuous and has continuous first derivatives everywhere along each column and along each row. Consider a function $y(x)$, values $y_i = y(x_i)$ for which are known at a discrete set of points

x_i, where $i = 1, N$. Linear interpolation between each pair of points along the x axis gives a set of $N - 1$ equations,

$$y(x) = y_i + (x - x_i)\,\frac{y_{i+1} - y_i}{x_{i+1} - x_i}, \quad x_i < x < x_{i+1}, \tag{7.14.41}$$

which can also be written as

$$y(x) = \frac{y_i\,(x_{i+1} - x) + y_{i+1}\,(x - x_i)}{x_{i+1} - x_i}. \tag{7.14.42}$$

In eq. (7.14.42), the coefficient of y_i vanishes at $x = x_{i+1}$ and the coefficient of y_{i+1} vanishes at $x = x_i$. In general, unless $y(x)$ is linear, both the first and second derivatives with respect to x of the piecewise linear function expressed by eq. (7.14.41) are discontinuous at grid points. The idea of the cubic spline is to add to eq. (7.14.42) two cubic functions $C(x)$ and $D(x)$ which vanish at the points x_i and x_{i+1} and which have derivatives such that the first and second derivatives of the complete function are continuous everywhere between x_1 and x_N. The equations which these requirements produce connect quantities between adjacent intervals. Once a boundary condition is prescribed at one end point, e.g., $(\mathrm{d}^2 y(x)/\mathrm{d}x^2)_{x=x_1} = 0$, the connections between quantities in adjacent intervals can be constructed by moving from one interval to the next until the other end point is reached. After applying a boundary condition at this second end point, all of the intermediate quantities necessary for finding the coefficients of x in the cubic equations $C(x)$ and $D(x)$ are determined, and one may proceed backward through the set of intervals, finding the coefficients.

Thus, in each interval i to $i + 1$, setting

$$y(x) = y_i + (x - x_i)\,\frac{y_{i+1} - y_i}{x_{i+1} - x_i} + C(x) + D(x) \tag{7.14.43}$$

gives

$$y'(x) \equiv \frac{\mathrm{d}y(x)}{\mathrm{d}x} = \frac{y_{i+1} - y_i}{x_{i+1} - x_i} + \frac{\mathrm{d}C(x)}{\mathrm{d}x} + \frac{\mathrm{d}D(x)}{\mathrm{d}x} \tag{7.14.44}$$

and

$$y''(x) \equiv \frac{\mathrm{d}^2 y(x)}{\mathrm{d}x^2} = \frac{\mathrm{d}^2 C(x)}{\mathrm{d}x^2} + \frac{\mathrm{d}^2 D(x)}{\mathrm{d}x^2}. \tag{7.14.45}$$

Since, by assumption, $C(x)$ and $D(x)$ are cubic in x, one has that

$$\frac{\mathrm{d}^2 C(x)}{\mathrm{d}x^2} = \alpha\,x + \beta \tag{7.14.46}$$

and

$$\frac{\mathrm{d}^2 D(x)}{\mathrm{d}x^2} = \gamma\,x + \delta, \tag{7.14.47}$$

where α, β, γ, and δ are constants. The standard practice is to choose the constants in such a way that the second derivative of $C(x)$ vanishes at $x = x_{i+1}$ and the second derivative of $D(x)$ vanishes at $x = x_i$. Setting

$$y_i'' = \left(\frac{d^2 C(x)}{dx^2} \right)_{x=x_i} = \alpha\, x_i + \beta \tag{7.14.48}$$

and

$$y_{i+1}'' = \left(\frac{d^2 D(x)}{dx^2} \right)_{x=x_{i+1}} = \gamma\, x_{i+1} + \delta, \tag{7.14.49}$$

and applying the conditions

$$\left(\frac{d^2 C(x)}{dx^2} \right)_{x=x_{i+1}} = \alpha\, x_{i+1} + \beta = 0 \tag{7.14.50}$$

and

$$\left(\frac{d^2 D(x)}{dx^2} \right)_{x=x_i} = \gamma\, x_i + \delta = 0, \tag{7.14.51}$$

one has four equations in the four unknowns α, β, γ, and δ. Solving, one obtains

$$\frac{d^2 C(x)}{dx^2} = \frac{x_{i+1} - x}{x_{i+1} - x_i}\, y_i'' \tag{7.14.52}$$

and

$$\frac{d^2 D(x)}{dx^2} = \frac{x - x_i}{x_{i+1} - x_i}\, y_{i+1}''. \tag{7.14.53}$$

Two integrations of eq. (7.14.52) yield

$$C(x) = \left(x_{i+1} \frac{x^2}{2} - \frac{x^3}{6} + a\, x + b \right) \frac{y_i''}{x_{i+1} - x_i}, \tag{7.14.54}$$

where a and b are constants which are determined by setting $C(x_{i+1}) = C(x_i) = 0$. The result, after rearrangement, is

$$C(x) = -\frac{1}{6} \frac{y_i''}{x_{i+1} - x_i} (x - x_{i+1})(x - x_i) \left\{ x - \left[x_{i+1} + (x_{i+1} - x_i) \right] \right\}. \tag{7.14.55}$$

In a similar fashion, one finds that

$$D(x) = +\frac{1}{6} \frac{y_{i+1}''}{x_{i+1} - x_i} (x - x_i)(x - x_{i+1}) \left\{ x - \left[x_i - (x_{i+1} - x_i) \right] \right\}. \tag{7.14.56}$$

Note that $x = x_i$ and $x = x_{i+1}$ are roots of both $C(x)$ and $D(x)$. The third root of $C(x)$ is a distance $x_{i+1} - x_i$ to the right of x_{i+1} and the third root of $D(x)$ is the same distance to the left of x_i.

The second derivatives y_i'' are fixed by insisting that the first derivative of the interpolation formula be continuous across the boundaries between adjacent intervals. Inserting derivatives of eqs. (7.14.55) and (7.14.56) into eq. (7.14.44), it follows that, just to the left of point x_i,

$$\left(\frac{dy(x)}{dx} \right)_- = \frac{y_i - y_{i-1}}{x_i - x_{i-1}} + \frac{1}{6} (x_i - x_{i-1}) \left(y_{i-1}'' + 2 y_i'' \right). \qquad (7.14.57)$$

Just to the right of point x_i,

$$\left(\frac{dy(x)}{dx} \right)_+ = \frac{y_{i+1} - y_i}{x_{i+1} - x_i} - \frac{1}{6} (x_{i+1} - x_i) \left(2 y_i'' + y_{i+1}'' \right). \qquad (7.14.58)$$

Equating the two derivatives yields $N - 2$ equations,

$$y_{i-1}'' \frac{x_i - x_{i-1}}{6} + y_i'' \frac{x_{i+1} - x_{i-1}}{3} + y_{i+1}'' \frac{x_{i+1} - x_i}{6} = \frac{y_{i+1} - y_i}{x_{i+1} - x_i} - \frac{y_i - y_{i-1}}{x_i - x_{i-1}}, \qquad (7.14.59)$$

which hold for $i = 2, N - 1$.

Two additional equations are necessary for finding all N second derivatives. At the left-hand boundary, eq. (7.14.58) with $i = 1$ gives

$$\left(\frac{dy(x)}{dx} \right)_{x=x_1} = \frac{y_2 - y_1}{x_2 - x_1} - \frac{1}{6} (x_2 - x_1) \left(2 y_1'' + y_2'' \right), \qquad (7.14.60)$$

and a choice must be made as to the value of one or both of $(dy/dx)_{x=x_1}$ and y_1''. At the right-hand boundary, eq. (7.14.57) with $i = N$ gives

$$\left(\frac{dy(x)}{dx} \right)_{x=x_N} = \frac{y_N - y_{N-1}}{x_N - x_{N-1}} + \frac{1}{6} (x_N - x_{N-1}) \left(y_{N-1}'' + 2 y_N'' \right), \qquad (7.14.61)$$

and a choice must be made as to the value of one or both of $(dy/dx)_{x=x_N}$ and y_N''.

For all $i = 2, N - 1$, define

$$M_i = \frac{y_{i+1} - y_i}{x_{i+1} - x_i} - \frac{y_i - y_{i-1}}{x_i - x_{i-1}}, \qquad (7.14.62)$$

and rewrite eq. (7.14.59) as

$$y_{i-1}'' J_i + y_i'' K_i + y_{i+1}'' L_i = M_i, \qquad (7.14.63)$$

where

$$J_i = \frac{x_i - x_{i-1}}{6},$$

(7.14.64)

$$K_i = \frac{x_{i+1} - x_{i-1}}{3},$$

(7.14.65)

and

$$L_i = \frac{x_{i+1} - x_i}{6}.$$

(7.14.66)

In the first interval, define

$$M_1 = -\left(\frac{dy}{dx}\right)_{x=x_1} + \frac{y_2 - y_1}{x_2 - x_1},$$

(7.14.67)

and rewrite eq. (7.14.60) as

$$y_1'' K_1 + y_2'' L_1 = M_1,$$

(7.14.68)

where

$$K_1 = \frac{1}{3}(x_2 - x_1)$$

(7.14.69)

and

$$L_1 = \frac{1}{6}(x_2 - x_1).$$

(7.14.70)

If, as the left-hand boundary condition, it has been decided to assign a value to $(dy(x)/dx)_{x=x_1}$ (or, equivalently, to M_1 through eq. (7.14.67)), the second derivative y_1'' is given by

$$y_1'' = S_1 - V_1 y_2'',$$

(7.14.71)

where

$$S_1 = \frac{M_1}{K_1}$$

(7.14.72)

and

$$V_1 = \frac{L_1}{K_1}.$$

(7.14.73)

Equations (7.14.71) and (7.14.63) (with $i = 2$) give

$$y_2'' = S_2 - V_2 y_3'',$$

(7.14.74)

where

$$S_2 = \frac{M_2 - S_1 J_2}{K_2 - V_1 J_2}$$

(7.14.75)

and

$$V_2 = \frac{L_2}{K_2 - V_1 \, J_2}. \tag{7.14.76}$$

If, instead, a choice is made as to the value of y_1'', this choice can be inserted directly into eq. (7.14.63) for $i = 2$ and solved for the relationship between y_2'' and y_3''. The so-called natural spline follows from the choice $y_1'' = 0$. In this case, y_2'' is still given by eq. (7.14.74), but

$$S_2 = \frac{M_2}{K_2} \tag{7.14.77}$$

and

$$V_2 = \frac{L_2}{K_2}. \tag{7.14.78}$$

By induction from eqs. (7.14.74)–(7.14.76), one has, for all subsequent intervals up to and including $i = N - 1$, that

$$y_i'' = S_i - V_i \, y_{i+1}'', \tag{7.14.79}$$

where

$$S_i = \frac{M_i - S_{i-1} \, J_i}{K_i - V_{i-1} \, J_i} \tag{7.14.80}$$

and

$$V_i = \frac{L_i}{K_i - V_{i-1} \, J_i}. \tag{7.14.81}$$

Equations (7.14.79)–(7.14.81) are valid for all $i = 3, N - 1$. In particular, for $i = N - 1$,

$$y_{N-1}'' = S_{N-1} - V_{N-1} \, y_N''. \tag{7.14.82}$$

At the right-hand boundary, set

$$M_N = \left(\frac{dy}{dx}\right)_{x=x_N} - \frac{y_N - y_{N-1}}{x_N - x_{N-1}}, \tag{7.14.83}$$

and rewrite eq. (7.14.61) as

$$y_{N-1}'' \, J_N + y_N'' \, K_N = M_N, \tag{7.14.84}$$

where

$$J_N = \frac{1}{6} \, (x_N - x_{N-1}) \tag{7.14.85}$$

and

$$K_N = \frac{1}{3} \, (x_N - x_{N-1}). \tag{7.14.86}$$

The solution of eq. (7.14.84),

$$y''_{N-1} = \frac{M_N - K_N \, y''_N}{J_N}, \tag{7.14.87}$$

complements the solution in the first interval (eqs. (7.14.66)–(7.14.68)):

$$y''_2 = \frac{M_1 - K_1 \, y''_1}{L_1}, \tag{7.14.88}$$

which can also be written as

$$y''_1 = \frac{M_1 - L_1 \, y''_2}{K_1}. \tag{7.14.89}$$

Equations (7.14.82) and (7.14.87) yield a connection between y''_N and M_N:

$$y''_N \left(V_{N-1} - \frac{K_N}{J_N} \right) = S_{N-1} - \frac{M_N}{J_N}. \tag{7.14.90}$$

Choices must now be made for M_N and y''_N. For the natural spline, set $y''_N = 0$ and eq. (7.14.90) gives $M_N = J_N S_{N-1}$ and eq. (7.14.82) gives $y''_{N-1} = S_{N-1}$. Equation (7.14.79) may now be applied until all of the y''_i up to and including y''_2 have been evaluated; application of eq. (7.14.87) completes the determination of all required coefficients. If the spline is natural, eq. (7.14.89) gives M_1; otherwise, it determines y''_1.

Since an infinite number of other choices may be made, the user of a table must experiment to find the choice which provides the best approximation for that table. This inevitably involves a graphical exploration and a possible adjustment of opacity entries near table boundaries to eliminate non-physical oscillations which may have been produced by the spline fits. It may also be the case that table boundaries are a consequence of the fact that the physics required for doing the opacity calculations near and beyond the boundaries is sufficiently unclear to make the results of calculations unreliable. If this is so, it may be important to adjust the entries near table boundaries and/or to extend the tables by eye, using the systematics available in the main body of the table and a certain amount of common sense.

7.14.4   Bicubic spline interpolation

The most ambitious scheme for interpolation in a table is the bicubic spline which produces a smooth opacity surface that passes through the tabulated opacity values at all grid points and has the property that first derivatives with respect to the temperature and density variables are continuous everywhere on the surface.

Let the opacity variable be called $z(x, y)$, where x and y are, for example, the temperature and density variables, respectively. Normally, grid points in the x–y plane lie along straight lines in the x and y directions. Through each straight line, a cubic spline can be constructed to provide continuous first derivatives dz/dx and dz/dy at every point along the lines, and, in particular, at grid points. Cross derivatives $\frac{d^2 z}{dx \, dy}$ can also be assigned, but this involves a somewhat cumberson numerical procedure the description of which will be delayed to the end.

In any given rectangle formed by four adjacent grid points bounded by two values of x and two values of y, label the grid points in a counterclockwise fashion 1 through 4 so that $z_1 = z(x_1, y_1)$, $z_2 = z(x_2, y_1)$, $z_3 = z(x_2, y_2)$, and $z_4 = z(x_1, y_2)$, where x_1 and $x_2 > x_1$ are the two boundary values of x and y_1 and $y_2 > y_1$ are the two boundary values of y. Next, set

$$z(x, y) = \sum_{i=1}^{4} \sum_{j=1}^{4} c_{ij} \, u^{i-1} \, v^{j-1}, \tag{7.14.91}$$

where

$$u = \frac{x - x_1}{x_2 - x_1} = \frac{x - x_1}{\Delta x} \tag{7.14.92}$$

and

$$v = \frac{y - y_1}{y_2 - y_1} = \frac{y - y_1}{\Delta y}. \tag{7.14.93}$$

The coefficients c_{ij} are to be chosen in such a way as to reproduce the tabular values of z and its first derivatives at grid points and to fit the assigned values of cross derivatives.

Note that, at the four grid points, (1) $u = 0$, $v = 0$, (2) $u = 1$, $v = 0$, (3) $u = 1$, $v = 1$, and (4) $u = 0$, $v = 1$. Using the four values of z at the grid points provides four conditions on the c_{ij}s:

$$z_1 = c_{11}, \tag{7.14.94}$$

$$z_2 = c_{11} + c_{21} + c_{31} + c_{41}, \tag{7.14.95}$$

$$z_3 = (c_{11} + c_{12} + c_{13} + c_{14}) + (c_{21} + c_{22} + c_{23} + c_{24})$$
$$+ (c_{31} + c_{32} + c_{33} + c_{34}) + (c_{41} + c_{42} + c_{43} + c_{44}), \tag{7.14.96}$$

and

$$z_4 = c_{11} + c_{12} + c_{13} + c_{14}. \tag{7.14.97}$$

Fitting first derivatives gives eight additional conditions:

$$\Delta x \left(\frac{dz}{dx} \right)_1 = c_{21}, \tag{7.14.98}$$

$$\Delta x \left(\frac{dz}{dx} \right)_2 = (c_{21} + 2\,c_{31} + 3\,c_{41}), \tag{7.14.99}$$

$$\Delta x \left(\frac{dz}{dx} \right)_3 = (c_{21} + c_{22} + c_{23} + c_{24})$$
$$+ 2\,(c_{31} + c_{32} + c_{33} + c_{34}) + 3\,(c_{41} + c_{42} + c_{43} + c_{44}), \tag{7.14.100}$$

$$\Delta x \left(\frac{dz}{dx}\right)_4 = (c_{21} + c_{22} + c_{23} + c_{24}),\tag{7.14.101}$$

$$\Delta y \left(\frac{dz}{dy}\right)_1 = c_{12},\tag{7.14.102}$$

$$\Delta y \left(\frac{dz}{dy}\right)_2 = (c_{12} + c_{22} + c_{32} + c_{42}),\tag{7.14.103}$$

$$\Delta y \left(\frac{dz}{dy}\right)_3 = (c_{12} + c_{22} + c_{32} + c_{42})$$

$$+ 2\,(c_{13} + c_{23} + c_{33} + c_{43}) + 3\,(c_{14} + c_{24} + c_{34} + c_{44}),\tag{7.14.104}$$

and

$$\Delta y \left(\frac{dz}{dy}\right)_4 = (c_{12} + 2\,c_{13} + 3\,c_{14}).\tag{7.14.105}$$

Four more conditions are necessary. One might consider using the second derivatives d^2z/dx^2 and d^2z/dy^2, but this provides eight conditions, overdetermining the system of equations. The standard procedure is to fit to the four cross derivatives. Thus,

$$\Delta x \Delta y \left(\frac{d^2z}{dy\,dx}\right)_1 = c_{22},\tag{7.14.106}$$

$$\Delta x \Delta y \left(\frac{d^2z}{dy\,dx}\right)_2 = c_{22} + 2\,c_{32} + 3\,c_{42},\tag{7.14.107}$$

$$\Delta x \Delta y \left(\frac{d^2z}{dy\,dx}\right)_3 = (c_{22} + 2\,c_{23} + 3\,c_{24})$$

$$+ 2(c_{32} + 2\,c_{33} + 3\,c_{34}) + 3\,(c_{42} + 2\,c_{43} + 3\,c_{44}),\tag{7.14.108}$$

and

$$\Delta x \Delta y \left(\frac{d^2z}{dy\,dx}\right)_4 = c_{22} + 2\,c_{23} + 3\,c_{24}.\tag{7.14.109}$$

A solution of eqs. (7.14.94)–(7.14.109) produces twelve relatively simple coefficients and four more complicated ones. The simple ones are:

$$c_{11} = z_1,\tag{7.14.110}$$

$$c_{21} = \Delta x \left(\frac{dz}{dx}\right)_1,\tag{7.14.111}$$

$$c_{12} = \Delta y \left(\frac{dz}{dy}\right)_1,\tag{7.14.112}$$

$$c_{22} = \Delta y \, \Delta x \left(\frac{d^2 z}{dy dx} \right)_1, \tag{7.14.113}$$

$$c_{41} = 2 \, (z_1 - z_2) + \Delta x \left(\left(\frac{dz}{dx} \right)_1 + \left(\frac{dz}{dx} \right)_2 \right), \tag{7.14.114}$$

$$c_{31} = 3 \, (z_2 - z_1) - \Delta x \left(2 \left(\frac{dz}{dx} \right)_1 + \left(\frac{dz}{dx} \right)_2 \right), \tag{7.14.115}$$

$$c_{14} = 2 \, (z_1 - z_4) + \Delta y \left(\left(\frac{dz}{dy} \right)_1 + \left(\frac{dz}{dy} \right)_4 \right), \tag{7.14.116}$$

$$c_{13} = 3 \, (z_4 - z_1) - \Delta y \left(2 \left(\frac{dz}{dy} \right)_1 + \left(\frac{dz}{dy} \right)_4 \right), \tag{7.14.117}$$

$$c_{24} = 2 \, \Delta x \left(\left(\frac{dz}{dx} \right)_1 - \left(\frac{dz}{dx} \right)_4 \right) + \Delta y \, \Delta x \left(\left(\frac{d^2 z}{dy dx} \right)_1 + \left(\frac{d^2 z}{dy dx} \right)_4 \right), \tag{7.14.118}$$

$$c_{23} = -3 \, \Delta x \left(\left(\frac{dz}{dx} \right)_1 - \left(\frac{dz}{dx} \right)_4 \right) - \Delta y \, \Delta x \left(2 \left(\frac{d^2 z}{dy dx} \right)_1 + \left(\frac{d^2 z}{dy dx} \right)_4 \right), \tag{7.14.119}$$

$$c_{42} = 2 \, \Delta y \left(\left(\frac{dz}{dy} \right)_1 - \left(\frac{dz}{dy} \right)_2 \right) + \Delta y \, \Delta x \left(\left(\frac{d^2 z}{dy dx} \right)_1 + \left(\frac{d^2 z}{dy dx} \right)_2 \right), \tag{7.14.120}$$

and

$$c_{32} = -3 \, \Delta y \left(\left(\frac{dz}{dy} \right)_1 - \left(\frac{dz}{dy} \right)_2 \right) - \Delta y \, \Delta x \left(2 \left(\frac{d^2 z}{dy dx} \right)_1 + \left(\frac{d^2 z}{dy dx} \right)_2 \right), \tag{7.14.121}$$

The final four coefficients are

$$c_{44} = 4 \, \alpha - 2 \, \beta - 2 \, \gamma + \delta, \tag{7.14.122}$$

$$c_{34} = -6 \, \alpha + 2 \, \beta + 3 \, \gamma - \delta, \tag{7.14.123}$$

$$c_{43} = -6 \, \alpha + 3 \, \beta + 2 \, \gamma - \delta, \tag{7.14.124}$$

and

$$c_{33} = 9 \, \alpha - 3 \, \beta - 3 \, \gamma + \delta, \tag{7.14.125}$$

where

$$\alpha = z_3 - (c_{11} + c_{12} + c_{13} + c_{14}) - (c_{21} + c_{22} + c_{23} + c_{24}) - (c_{31} + c_{32} + c_{41} + c_{42}), \tag{7.14.126}$$

$$\beta = \Delta x \left(\frac{dz}{dx}\right)_3 - (c_{21} + c_{22} + c_{23} + c_{24}) - 2\,(c_{31} + c_{32}) - 3\,(c_{41} + c_{42}),$$

$$(7.14.127)$$

$$\gamma = \Delta y \left(\frac{dz}{dy}\right)_3 (c_{12} + c_{22} + c_{32} + c_{42}) - 2\,(c_{13} + c_{23}) - 3\,(c_{14} + c_{24}), \quad (7.14.128)$$

and

$$\delta = \Delta y\, \Delta x \left(\frac{d^2 z}{dx\,dy}\right)_3 - (c_{22} + 2\,c_{23} + 3\,c_{24}) - 2\,c_{32} - 3\,c_{42}. \qquad (7.14.129)$$

The matter of estimating cross derivatives is the next item of business. One approach is to make use of spline-derived first derivatives at points in a secondary grid displaced from the main grid. For clarity, colors are ascribed to the various spline curves of which use is to be made. In the main grid, let the one-dimensional splines which pass through opacity points anchored along straight lines parallel to the y axis be blue and those which pass through opacity points anchored along straight lines parallel to the x axis be green. Along a blue curve corresponding to a fixed value of x, e.g., $x = x_k$, define secondary grid points at values of $y'_l = y_{l-1} + \frac{1}{2}\,(y_l - y_{l-1})$, for all values of l; do this for all values of k. Then, from the blue spline solutions, find the opacity variable at each new grid point $z\,(x_k, y'_l)$ for all values of k and l. Now, for each value of y'_l, pass a cubic spline through the new opacity values and color the resulting curve orange. Next, along each green curve corresponding to a fixed value of y, e.g., $y = y_j$, define secondary grid points at values of $x'_m = x_{m-1} + \frac{1}{2}\,(x_m - x_{m-1})$, for all values of m; do this for all values of j. Then, from the green spline solutions, find the opacity variable at each new grid point $z\,(x'_m, y_j)$ for all values of j and m. Now, for each value of x'_m, pass a cubic spline through the new opacity values and color the resulting curve pink. The secondary grid is now complete, with values of z, dz/dx, and dz/dy defined at all points in this grid.

Next, consider a point (x_j, y_i) in the primary grid. Call the points in the secondary grid along the blue line which passes through this point points c and d, respectively, so that $y_d = y_i + \frac{1}{2}\,(y_{i+1} - y_i)$ and $y_c = y_i - \frac{1}{2}\,(y_i - y_{i-1})$. Points c and d lie at the intersection of the blue curve with orange curves. Call the points in the secondary grid along the green line which passes through the point (x_j, y_i) points a and b, respectively, with $y_b = y_i + \frac{1}{2}\,(y_{i+1} - y_i)$ and $y_a = y_i - \frac{1}{2}\,(y_i - y_{i-1})$. Points a and b lie at intersections of the green curve with pink curves.

Defining two approximations to a cross derivative by

$$\frac{d}{dx}\left(\frac{dz}{dy}\right) = \left[\left(\frac{dz}{dy}\right)_b - \left(\frac{dz}{dy}\right)_a\right] \frac{1}{x_b - x_a} \qquad (7.14.130)$$

and

$$\frac{d}{dy}\left(\frac{dz}{dx}\right) = \left[\left(\frac{dz}{dx}\right)_d - \left(\frac{dz}{dx}\right)_c\right] \frac{1}{y_d - y_c}, \qquad (7.14.131)$$

at point (x_j, y_i) the average

$$\frac{d^2z}{dx\,dy} = \frac{1}{2}\left[\frac{d}{dx}\left(\frac{dz}{dy}\right) + \frac{d}{dy}\left(\frac{dz}{dx}\right)\right] \tag{7.14.132}$$

may be adopted as a compromise cross derivative. This procedure can be used everywhere in the table except near table boundaries, where another scheme must be invented.

Clearly, a lot of work is required to implement the bicubic scheme. A lot of work and a lot of memory. It would be foolish to go through the outlined procedure every time an opacity is needed. The sensible thing to do is to store, for every rectangle in the primary grid, the 16 quantities defined by eqs. (7.14.110)–(7.14.129). Of course, since composition changes, coefficients for tables for other compositions must be stored and algorithms devised for interpolating among these coefficients.

7.15 Interpolation in opacity tables: a concrete example

Opacity tables produced by Forrest Rogers & Carlos Iglesias for a wide variety of compositions (1992 and subsequent publications) are conveniently available at the website www-phys.llnl.gov/Research/OPAL/index.html. Also available on this website is the Fortran source code XZTRIN21.f containing routines which incorporate algorithms prepared by Mike J. Seaton (1993) for fitting and smoothing opacity data.

In the OPAL tables, the independent variable labeling rows is $x = \log(T_6)$ and the independent variable labeling columns is $y = \log R$, where

$$R = \frac{\rho(\text{g cm}^{-3})}{T_6^3}. \tag{7.15.1}$$

Table entries are $\log(\kappa)$, the logarithm of the radiative opacity. In general, $\log(T_6)$ varies from 3.75 to 8.70 in steps of $\Delta \log(T_6) = 0.05$, while $\log(R)$ varies from -8.0 to $+1.0$ in steps of $\Delta \log(R) = 0.5$. For the most part, the tables are bimodally piecewise rectangular, and can be described as rectangular with rectangular cutouts at higher temperatures.

Composition choices include the hydrogen abundance by mass $X = 0.0, 0.03, 0.10, 0.35,$ and 0.70 and the heavy element abundance by mass $Z = 0.03, 0.02, 0.01, 0.004, 0.001,$ 0.0003, 0.0001, and 0.0. Relative abundances among heavy elements comprising Z are taken from N. Grevesse & A. Noels (in *Origin and Evolution of the Elements*, 1993). In addition, eight different combinations for the abundances of carbon and oxygen, over and above their abundances in Z are represented. Thus, opacities are available for all stages of hydrogen and helium burning.

At very low temperatures such that molecules, grains, and H^- contribute to the opacity, the OPAL tables are supplemented by results of calculations by R. Alexander & J. W. Ferguson (1994) and by Ferguson *et al.* (2005). Relevant tables are rectangular, with $\log(T) = 2.85$ to 4.10 in steps of $\Delta \log(T) = 0.05$ and $\log(R) = -7.0$ to 1.0 in steps

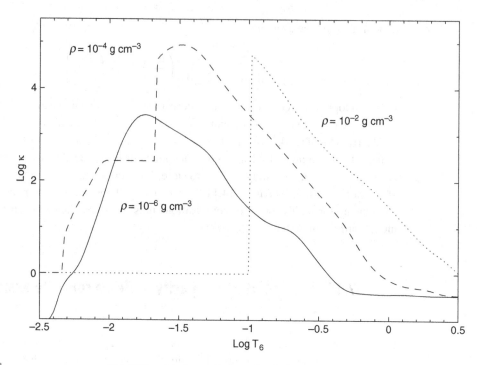

Fig. 7.15.1   Fit to OPAL and Anderson–Ferguson data for $Z = 0.02, X = 0.74$

of $\Delta \log(R) = 0.5$. Compositions represented are $Z = 0.01$ and 0.02 and $X = 0.35, 0.5$, 0.7, and 0.8.

These tables have been used in conjunction with the interpolation routines from the OPAL website to construct Figs. 7.15.1 and 7.15.2. The two major discontinuities in the $\log(\kappa)$ versus $\log(T_6)$ curves in Fig. 7.15.1 for $\rho = 10^{-2}$ g cm^{-3} at $\log(T_6) = -1$ and for $\rho = 10^{-4}$ g cm^{-3} at $\log(T_6) \sim -1.7$ are consequences of the fact that the OPAL tables do not extend beyond $\log(R) = 1$. The discontinuity at $\log(T_6) \sim -2.3$ along the $\rho = 10^{-4}$ g cm^{-3} curve is a consequence of the fact that the Alexander–Ferguson tables do not extend beyond $\log(R) = 1$. The segment of the curve for $\rho = 10^{-4}$ g cm^{-3} between $\log(T_6) = -2.1$ to -2.3 is solely from the Anderson–Ferguson tables. For temperatures in the range $8\,000$ K to $10\,000$ K, the logarithm of the opacity is a sinusoidal fit between the two sets of opacity tables.

The pedagogical content of Fig. 7.15.1 would be greater had tabular values been available for larger values of R. This is evident from a comparison of Fig. 7.15.1 with Fig. 7.11.31 where bare-bones opacities for the three different choices of density are presented for the entire range of temperatures considered. As it is, one must be content to note that, over a wide range of temperatures about the peaks in the opacity for each density, the inclusion of the whole range of possible opacity sources for a population I composition increases the calculated opacity by at least an order of magnitude over what is estimated when only free–free absorption and bound–free absorption from the ground state of hydrogen and the ground states of helium and oxygen isotopes is considered.

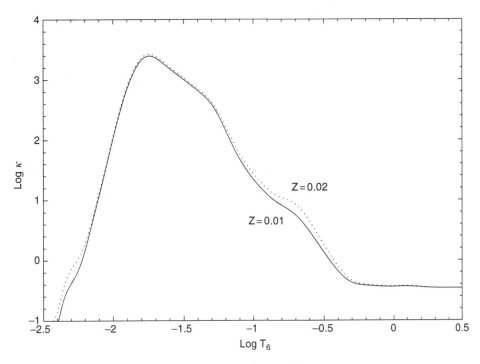

Fig. 7.15.2

Fit to OPAL and Alexander–Ferguson data for $\rho = 10^{-6}$ g cm^{-3}, $X = 0.74$, $Z = 0.01$ and $Z = 0.02$

Fortunately, for the lowest density considered, a comparison of opacities for different metallicities is possible over the entire temperature range considered. As demonstrated in Fig. 7.15.2, a factor of two difference in metallicity produces at most only a ~60 % effect, and does so only over a very narrow range of temperatures around 200 000 K and over another narrow range around 5000 K. The enhancement at temperatures around 200 000 K is primarily due to transitions between bound levels of iron-peak elements and the enhancement at temperatures around 5000 K is due to the facts that (1) metals of ionization potential smaller than the ionization potential of hydrogen become a more important source of electrons than neutral hydrogen and (2) absorption by the negative hydrogen ion H^{-} becomes the major contributor to the opacity.

The opacity for the $\rho = 10^{-6}$ g cm^{-3} case in Figs. 7.15.1 and the opacity for the bare-bones case in Fig. 7.11.28 are contrasted in Fig. 7.15.3. For both cases, the abundances of hydrogen and helium are the same, but, in the bare-bones case, the only heavy element is oxygen and the only processes included are electron scattering, free–free absorption, and bound–free absorption from ground states. For guidance, locations are shown where abundances of related ion types are equal.

It is evident that, at temperatures larger than 600 000 K, where both the complete opacity and the bare-bones opacity are essentially identical, electron scattering is the primary source of opacity. At temperatures in the range 50 000 K to 500 000 K, while bound–bound absorption and photo-ionization from excited states of iron peak elements and CNO elements make substantial contributions (factors of 2 to 5 enhancements over the bare-bones

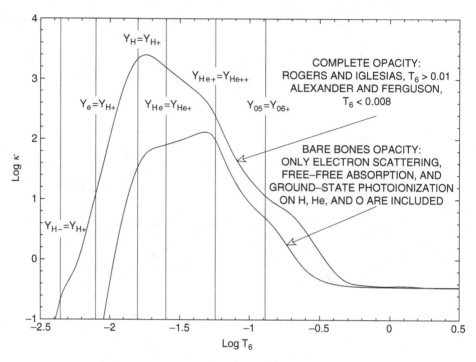

Fig. 7.15.3 Complete opacity compared with bare–bones opacity when $\rho = 10^{-6}\,\mathrm{g\,cm^{-3}}, X = 0.74$, and $Z = 0.01$

approximation), it is nevertheless true that photo-ionization from ground states of He^+, O^{4+}, and O^{5+} and free–free absorption from H^+, He^+, and He^{++} make substantial contributions to the opacity (one to two orders of magnitude enhancements over a pure electron scattering opacity). In the temperature range 8000 K to 50 000 K, absorption between excited states and photo-ionization from excited states of hydrogen clearly make huge contributions to the opacity.

In the bare-bones opacity case, for temperatures less than 16 000 K, the rapid decrease with decreasing temperature in the abundance of free electrons is directly responsible for the decrease with temperature in the estimated opacity. A similar decrease occurs in the complete opacity case, except that, thanks to the greater richness of opacity sources, the decrease begins at a value two orders of magnitude larger than in the bare-bones opacity case. The decrease is also considerably less precipitous in the complete opacity case than in the bare-bones case. In the bare-bones case, the opacity decreases by $\Delta \log(\kappa) = 6.1$ as temperature declines from $\log(T_6) = -1.8$ to $\log(T_6) = -2.25$ (see Fig. 7.11.28), whereas, in the complete case, the opacity decreases by only $\Delta \log(\kappa) = 3.5$ as temperature declines from $\log(T_6) = -1.8$ to $\log(T_6) = -2.25$ (see Fig. 7.15.3). The less precipitous decline in the complete opacity case is due to the fact that electrons are contributed by elements with first ionization potentials smaller than that of hydrogen, so that the free–free contribution to opacity drops less rapidly with temperature than would otherwise be the case, and to the fact that neutral hydrogen can actually bind an additional electron and act as a primary contributor to the opacity through bound–free and free–free transitions.

7.16 Absorption by the negative hydrogen ion

At temperatures in the range $T = 3000$–6000 K and at densities prevailing in the photospheric layers of main sequence stars of spectral type similar to and later than that of the Sun, as well as in the photospheres of T-Tauri stars, red giants, and AGB stars, the dominant source of opacity is the H^- ion.

Applying the variational principle established by H. Ritz (1909), Egil A. Hylleraas (1929) determined accurately the first ionization energy of the helium atom. Applying the same technique to the H^- ion, Hans A. Bethe (1929) and Hylleraas (1930) found, respectively, ionization energies for H^- of 0.74 ± 0.04 eV and 0.700 ± 0.015 eV.

Rupert Wildt (1939) used the Hylleraas estimate of 0.7 eV to show that, in the Sun, at optical depths less than unity, the abundance of H^- exceeds the abundance of neutral H atoms in the first excited state (at 10.16 eV) by roughly two orders of magnitude. The extra electrons for the H^- ions originate primarily from elements of ionization energy less than that of neutral hydrogen. Wildt used element abundances estimated by Russell (see Section 2.1).

A calculation by H. S. W. Massey and R. A. Smith (1936) showed that, when plotted against wavelength, the cross section for the photo-ionization of H^- rises slowly from zero at the photo-ionization threshold near $16\,000$ Å, reaching a maximum near 8000 Å; additional estimates suggested a cross section of the order of 2×10^{-17} cm^2 at near infrared wavelengths and over the entire range of visible wavelengths. Building on these facts, and including estimates of the cross section for free–free absorption by the H^- ion (Pannekoek), Wildt (1939) demonstrated that the decrease with increasing spectral type in the magnitude of the Balmer jump can be accounted for by absorption by H^-. This result removed the need to invoke abundances of metals relative to hydrogen which are orders of magnitude larger than determined by Russell from an analysis of solar spectral line strengths.

The Saha equation for H^-, which has only one bound state, is

$$\frac{n_H\, n_e}{n_{H^-}} = \frac{2g_H}{g_{H^-}} \left(\frac{2\pi\, m_e kT}{h^2} \right)^{3/2} e^{-E_{H^-}/kT}, \tag{7.16.1}$$

where $E_{H^-} = 0.754$ eV. The electron in the neutral hydrogen atom can be in either of two spin 1/2 states, so $g_H = 2$, whereas, for H^-, the spins of the two electrons are opposed, so $g_{H^-} = 1$. For an element i of first ionization potential E_i, the Saha equation is, in first approximation,

$$\frac{n_{i^+}\, n_e}{n_i} = \frac{2g_{i^+}}{g_i} \left(\frac{2\pi\, m_e kT}{h^2} \right)^{3/2} e^{-E_i/kT}. \tag{7.16.2}$$

Ionization potentials and number abundances relative to hydrogen in the solar atmosphere of several elements with low ionization potentials are listed in Table 7.16.1. Number abundances relative to hydrogen are taken from Asplund, Grevesse, & Sauval (2005, see

Table 7.16.1 Atoms with low ionization potentials

Atom	H^-	Li	Na	Mg	Al	Si	K	Ca	Fe
Abundance	—	1/−11	1.5/−6	3.4/−5	2.3/−6	3.2/−5	1.2/−7	2/−6	2.8/−5
IP (eV)	0.754	5.36	5.12	7.61	5.96	8.12	4.32	6.09	7.83
IP (eV)	0.754	5.39	5.14	7.65	5.99	8.15	4.34	6.11	7.90
$2g_{i+}/g_i$	4	1	1	4	1	4	1	4	1

Section 2.1 and Table 2.1.1) and the two sets of ionization potentials (IPs) are taken from versions of the *Handbook of Chemistry and Physics* separated by four temporal decades.

Defining number abundance parameters Y_j by

$$n_j = N_A \rho Y_j, \tag{7.16.3}$$

where N_A is Avogadro's number and j designates a particle type, one has for the element hydrogen and free electrons

$$\frac{Y_H Y_e}{Y_{H^-}} = Z_{H^-}(\rho, T), \tag{7.16.4}$$

where

$$Z_{H^-}(\rho, T) = \frac{4}{N_A \rho} \left(\frac{2\pi m_e k T}{h^2} \right)^{3/2} e^{-E_{H^-}/kT}, \tag{7.16.5}$$

and

$$\frac{Y_{H^+} Y_e}{Y_H} = Z_H(\rho, T), \tag{7.16.6}$$

where

$$Z_H(\rho, T) = \frac{1}{N_A \rho} \left(\frac{2\pi m_e k T}{h^2} \right)^{3/2} e^{-E_H/kT}. \tag{7.16.7}$$

Similarly, for elements of low ionization potential,

$$\frac{Y_{i+} Y_e}{Y_i} = Z_i(\rho, T), \tag{7.16.8}$$

where

$$Z_i(\rho, T) = \frac{2g_{i+}}{g_i} \frac{1}{N_A \rho} \left(\frac{2\pi m_e k T}{h^2} \right)^{3/2} e^{-E_i/kT}. \tag{7.16.9}$$

Conservation of particles dictates that

$$Y_{i+} + Y_i = Y_{it}, \tag{7.16.10}$$

where Y_{it} is proportional to the total abundance of the ith species of atom, regardless of ionization state, and

$$Y_{H^+} + Y_H + Y_{H^-} = Y_{Ht}, \tag{7.16.11}$$

where Y_{Ht} is proportional to the total abundance of hydrogen atoms and bare protons.

Conservation of charge dictates that

$$Y_e + Y_{H^-} = \sum_i Y_{i+} + Y_{H^+}. \tag{7.16.12}$$

Equations (7.16.10)–(7.16.12) may be used in conjunction with eqs. (7.16.4)–(7.16.9) to find that

$$Y_e^2 \left(1 + \frac{Y_H}{Z_{H^-}}\right) = \sum Y_i \, Z_i + Y_H \, Z_H, \tag{7.16.13}$$

$$Y_H = Y_{Ht} \div \left(1 + \frac{Z_H}{Y_e} + \frac{Y_e}{Z_{H^-}}\right), \tag{7.16.14}$$

and

$$Y_i = Y_{it} \div \left(1 + \frac{Z_i}{Y_e}\right). \tag{7.16.15}$$

Before examining detailed solutions of these equations, suppose, for illustrative purposes, that there is only one electron donor element characterized by $i = m$ and that the temperature and density are such that this donor is only slightly ionized. Then, $n_e \sim n_{m+}$, $n_m \sim n_{mt} \gg n_{m+}$, and

$$n_{H^-} \sim n_{Ht} \, n_{mt}^{1/2} \left(\frac{2g_{m+}}{g_m}\right)^{1/2} \frac{1}{4} \left(\frac{h^2}{2\pi m_e kT}\right)^{3/4} e^{(E_{H^-} - E_m/2)/kT}, \tag{7.16.16}$$

or

$$Y_{H^-} = Y_{Ht} \, Y_{mt}^{1/2} \, (N_A \rho)^{1/2} \left(\frac{2g_{m+}}{g_m}\right)^{1/2} \frac{1}{4} \left(\frac{h^2}{2\pi m_e kT}\right)^{3/4} e^{-(E_m/2 - E_{H^-})/kT} \tag{7.16.17}$$

$$\propto Y_{Ht} \, Y_{mt}^{1/2} \, \rho^{1/2} \frac{1}{T^{3/4}} \, e^{-(E_m/2 - E_{H^-})/kT}. \tag{7.16.18}$$

At the other extreme, when the ion donor is almost completely ionized, to the extent that $n_e \sim n_{mt}$,

$$Y_{H^-} \sim Y_{Ht} \, Y_{mt} \frac{N_A \rho}{4} \left(\frac{h^2}{2\pi m_e kT}\right)^{3/2} e^{+E_{H^-}/kT}. \tag{7.16.19}$$

In words, at temperatures such that the donor begins to be ionized, the abundance of H^- ions increases as the square root of the density and, when the donor is completely ionized,

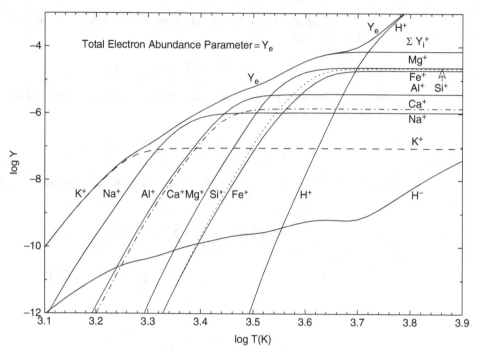

Fig. 7.16.1 Number abundance parameters for H^-, H^+, K^+, Na^+, Al^+, Ca^+, Mg^+ Si^+, Fe^+, and free electrons versus temperature (K) when $\rho = 10^{-8}$ g cm^{-3}

the abundance of H^- ions increases linearly with the density. One may infer that, in the general case, with several donor species present and at temperatures such that hydrogen is not appreciably ionized, the H^- abundance varies with a power of the density between 0.5 and 1. Differentiating eq. (7.16.18) logarithmically with respect to temperature, one has

$$\frac{d \log Y_{H^-}}{d \log T} = -\frac{3}{4} + \frac{E_m/2 - E_{H^-}}{kT} = -0.75 + \frac{E_m/2 - 0.754}{0.865 T_4}, \qquad (7.16.20)$$

which, for potassium, is $d \log Y_{H^-}/d \log T \sim -0.75 + 1.64/T_4 \sim 9.6$ when $T = 1585$ K. Differentiating eq. (7.16.19) logarithmically with respect to temperature, one has

$$\frac{d \log Y_{H^-}}{d \log T} = -\frac{3}{2} - \frac{E_{H^-}}{kT} = -1.5 - \frac{8.67}{T_4}, \qquad (7.16.21)$$

which is always negative.

When many elements with a variety of first ionization potentials are taken into account, the inferences with respect to the density dependence of Y_{H^-} are borne out but the inferences with regard to temperature dependences are found to be quite misleading. Some results of calculations with the electron donors listed in Table 7.16.1, taking $Y_{Ht} = 0.7$, are shown in Figs. 7.16.1–7.16.4. Number abundances of the singly ionized donors and of H^+ and H^- are shown as functions of temperature in Figs. 7.16.1 and 7.16.2 for densities of 10^{-8} g cm^{-3} and 10^{-6} g cm^{-3}, respectively. The number abundance of free electrons

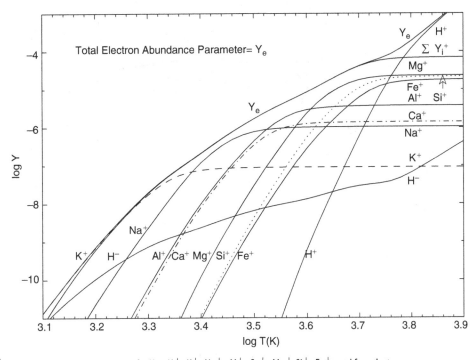

Fig. 7.16.2 Number abundance parameters for H^-, H^+, K^+, Na^+, Al^+, Ca^+, Mg^+ Si^+, Fe^+, and free electrons versus temperature (K) when $\rho = 10^{-6}$ g cm^{-3}

and the total number abundance of singly ionized elements of low ionization potential are shown by curves labeled Y_e and $\sum Y_i^+$, respectively.

Comparing the curves for Y_{H^-} in the two figures, it is evident that, at the lowest temperatures ($T \leq 1800$ K), where potassium is the only significant electron donor, and at temperatures ($T \geq 6500$ K) such that the elements of low ionization potential are all singly ionized and neutral hydrogen is the primary electron donor, the H^- abundance increases approximately as $\sqrt{\rho}$, in agreement with the expectation provided by eq. (7.16.18). At intermediate temperatures, the exponent n in the relationship $Y_{H^-} \propto \rho^n$ increases, reaching a maximum of $n \sim 0.8$ at $T \sim 5000$ K, where magnesium, silicon, and iron are the major electron donors, and then decreases with decreasing temperature.

In the $\rho = 10^{-8}$ g cm^{-3} case, at $T \sim 5250$ K, elements of low ionization potential contribute the same number of free electrons (curve labeled $\sum Y_i^+$) as does hydrogen (curve labeled H^+) and, as temperature decreases below this, the elements of low ionization potential rapidly become the dominant contributors to Y_e. In the $\rho = 10^{-6}$ g cm^{-3} case, the temperature at which $Y_H = \sum Y_i$ is $T \sim 6025$ K. The temperatures where $Y_{H^-} = Y_{H^+}$ and $Y_e = Y_{H^+}$ when $\rho = 10^{-6}$ g cm^{-3} are indicated in Fig. 7.15.3. To the left of a point roughly midway between the two indicated temperature locations, the opacity is dominated by H^- absorption, with the electrons in the negative hydrogen ion being donated primarily by elements of low ionization potential.

Over the entire range of temperatures between 2000 K and 5000 K in Figs. 7.16.1 and 7.16.2, the temperature dependence of Y_{H^-} is significantly less steep than outside of

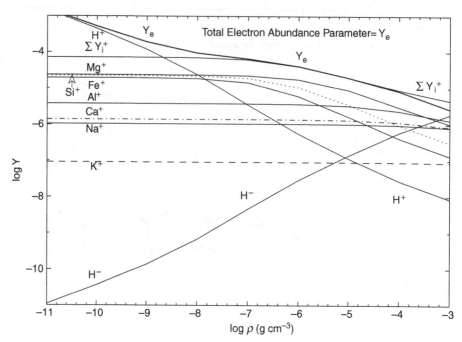

Fig. 7.16.3   Number abundance parameters for H^-, H^+, K^+, Na^+, Al^+, Ca^+, Mg^+ Si^+, Fe^+, and free electrons versus density $(g\ cm^{-3})$ when T $=$ 5000 K

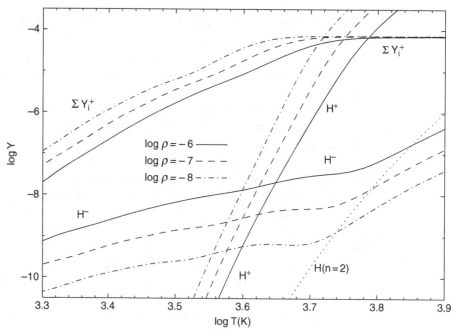

Fig. 7.16.4   Abundance parameters for H^-, $H(n=2)$, H^+, and electron donors of low ionization potential as functions of temperature (K) and density $(g\ cm^{-3})$

this range. Below this range, potassium is the principal electron donor and, as deduced from eq. (7.16.20), at $T = 1500$ K, $Y_{H^-} \overset{\sim}{\propto} T^{10}$. Above this range, where hydrogen is the primary donor, as deduced from eq. (7.16.20) when $T = 6000$ K, $Y_{H^-} \overset{\sim}{\propto} T^{11}$. In the intermediate range of temperatures, the temperature exponent is, on average, of the order of 3–4. In the $\rho = 10^{-8}$ g cm^{-3} case, the exponent actually becomes negative over a narrow range of temperatures $T \sim 4400$–5000 K, where essentially all elements of low ionization potential are ionized and the number abundance of free electrons supplied by hydrogen is still smaller than the number abundance of electrons supplied by the elements of low ionization potential. This is at least in qualitative accord with the prediction of eq. (7.16.21). In the $\rho = 10^{-6}$ g cm^{-3} case, the temperature exponent remains near zero over temperatures in the range $T \sim 4900$–5800 K.

The density dependences of degrees of ionization are shown in Fig. 7.16.3 at $T = 5000$ K. At both the lowest densities ($\log \rho \le -9$), when all of the elements of low ionization potential are singly ionized and neutral hydrogen is the major electron donor, and at the highest densities ($\log \rho \ge -4$), when magnesium and aluminum are the major electron donors, the exponent n in the relationship $Y_{H^-} \propto \rho^n$ is $n \sim 0.5$. At intermediate densities ($-8 \le \log \rho \le -6$), where magnesium, silicon, and iron are the major electron donors, $n \sim 0.85$.

Abundances of H$^-$ and of the ionized species are shown in Fig. 7.16.4 as functions of temperature for a range of densities that encompass these typically encountered in stellar photospheres at optical depths of the order of unity. It is interesting that, for stars with surface temperatures similar to or less than that of the Sun, the H$^-$ abundance is many orders of magnitude smaller than $\sum Y_i^+$, the sum of the abundances of singly ionized elements of low ionization potential. The reason for this is, of course, that the ionization potential of H$^-$ is so much smaller than the ionization potentials of the other elements. Also shown in Fig. 7.16.4 is the number abundance parameter for neutral hydrogen atoms in the first excited state. This parameter is given by

$$Y_H(n = 2) = Y_H(n = 1) \, 2^2 \, \exp\left(-\chi_H \left(1 - \frac{1}{2^2}\right)\right)$$

$$= Y_H(n = 1) \, \exp\left(1.386 - \frac{10.145 \text{eV}}{kT}\right)$$

$$= Y_H(n = 1) \, \exp\left(1.386 - 11.773 \, \frac{10\,000 \text{ K}}{T}\right). \qquad (7.16.22)$$

Since the cross section for absorption by hydrogen atoms in the first excited state is comparable to the photo-ionization cross section for H$^-$, it is evident that only for temperatures larger than $\sim$5500 K do the two sources of opacity compete, a point stressed by Wildt in his 1939 discussion.

For temperatures in the range 2000 K $< T <$ 5000 K, a very rough compromise approximation to the number abundance of the H$^-$ ion consistent with the information in all four figures, Figs. 7.16.1–7.16.4, is the relationship

$$Y_{H^-} \sim 2.7 \times 10^{-9} \left(\frac{T}{4000 \text{ K}} \right)^{3.26} \left(\frac{\rho}{10^{-7} \text{ g cm}^{-3}} \right)^{0.79}. \qquad (7.16.23)$$

On either side of the region in which the compromise relationship is applicable, the temperature exponent can be of the order of three or more times larger than 3.26 and the density exponent is of the order of 0.5 rather than 0.8.

The starting point for an estimate of the cross section for the photo-ionization of H^- is the matrix element given by eq. (7.4.6). The term $e^{i\mathbf{k}\cdot\mathbf{r}}$ describing the spatial dependence of the photon wave function is approximated by unity (over the atomic volume) and, assuming that the photon moves in a plane perpendicular to the z axis, so that the polarization vector $\hat{\epsilon}_{k\lambda}$ is in the z direction, the matrix element becomes

$$H_{f \leftarrow i}^{\text{int}} = -\sqrt{\frac{c^2 \hbar}{\nu_k V}} \frac{e}{m_e c} \frac{\hbar}{i} \left(\int u_i \left(\frac{\text{d}}{\text{d}z_1} + \frac{\text{d}}{\text{d}z_2} \right) u_f^* \, \text{d}\tau \right) \cos\theta, \qquad (7.16.24)$$

where the fact that either of the two electrons in H^- can be ejected is taken into account and θ is the angle between the direction of the outgoing electron and the direction of the incident photon. The differential probability for the reaction taking place, with a weakly bound electron being ejected into the solid angle $\text{d}\Omega$, is

$$\text{d}\omega_{f \leftarrow i} = \frac{2\pi}{\hbar} |H_{fi}|^2 \, \text{d}\rho_f, \qquad (7.16.25)$$

where

$$\text{d}\rho_f = \frac{V}{h^3} m_e \, p \, \text{d}\Omega. \qquad (7.16.26)$$

Inserting eqs. (7.16.24) and (7.16.26) into (7.16.25) gives

$$\text{d}\omega_{f \leftarrow i}(\nu) = \frac{2\pi}{\hbar} \left[\frac{c^2 \hbar}{\nu V} \left(\frac{e}{m_e c} \right)^2 \hbar^2 M^2 \right] \cos^2\theta \frac{V}{h^3} m_e \, p \, \text{d}\Omega, \qquad (7.16.27)$$

where

$$M = \int u_i \left(\frac{\text{d}}{\text{d}z_1} + \frac{\text{d}}{\text{d}z_2} \right) u_f^* \, \text{d}\tau. \qquad (7.16.28)$$

The total cross section σ_{bf} for photo-ionization is obtained by integrating over the solid angle and dividing by the incident photon flux. Thus,

$$\sigma_{bf}(\nu) = \int_{\Omega} \text{d}\omega_{f \leftarrow i} \div \frac{c}{V} = \frac{1}{2\pi} \frac{1}{h\nu} \frac{\nu}{c} e^2 V M^2 \int \cos^2\theta \, \text{d}\Omega \qquad (7.16.29)$$

$$= \frac{2}{3} \frac{1}{h\nu} \frac{\nu}{c} \frac{e^2}{a_0} a_0^2 \left[\frac{V}{a_0} M^2 \right] \qquad (7.16.30)$$

$$= \frac{2}{3} a_0^2 \frac{e^2/a_0}{h\nu} \frac{\nu}{c} \left[\frac{V}{a_0} M^2 \right]. \qquad (7.16.31)$$

The wave function u_f has a component which is an unbound state and yet must satisfy $\int |u_f|^2 d\tau = 1$. Hence, $u_f \propto 1/V^{1/2}$. Similarly, since the bound state must satisfy $\int |u_i|^2 d\tau = 1$, but is finite only over atomic dimensions, $u_i \propto 1/a_0^{3/2}$. Further, $4\pi r^2 dr \propto a_0^3$ and $d/dz \propto 1/a_0$. Thus, M in eq. (7.16.28) has the dimensions of $a_0^{1/2}/V^{1/2}$ and the quantity

$$\bar{M}^2 = \frac{V}{a_0} M^2 \tag{7.16.32}$$

is dimensionless.

In atomic units, energies are expressed in units of twice the ionization energy of hydrogen, $e^2/a_0 = 27.211$ eV, radial distances r are expressed in units of the Bohr radius $a_0 = 0.529\,177 \times 10^{-8}$ cm. The wave number k for the ejected electron is defined by

$$\frac{1}{2} m_e v^2 = \frac{1}{2} m_e c^2 \left(\frac{v}{c}\right)^2 = \frac{1}{2} \frac{e^2}{a_0} k^2, \tag{7.16.33}$$

so that

$$\frac{v}{c} = k \left(\frac{e^2/a_0}{m_e c^2}\right)^{1/2}$$

$$= 7.2973 \times 10^{-3} k. \tag{7.16.34}$$

Conservation of energy gives

$$\frac{h\nu}{e^2/a_0} = \frac{E_{\mathrm{H}^-}}{e^2/a_0} + \frac{1}{2} k^2 = I + \frac{1}{2} k^2, \tag{7.16.35}$$

where

$$I = \frac{E_{\mathrm{H}^-}}{e^2/a_0} = \frac{0.754\ \mathrm{eV}}{27.211\ \mathrm{eV}} = 0.0277 \tag{7.16.36}$$

is the ionization energy in atomic units.

In these units, the photo-ionization cross section is

$$\sigma_{\mathrm{bf}} = \frac{4}{3} \sqrt{\frac{e^2/a_0}{m_e c^2}}\, a_0^2\, \frac{k}{2I + k^2}\, \bar{M}^2 = 2.7246 \times 10^{-19}\ \mathrm{cm}^2\, \frac{k}{2I + k^2}\, \bar{M}^2, \tag{7.16.37}$$

where $\bar{M}$ is given by eq. (7.16.32). The numerical factor in eq. (7.16.37) is equivalent to the value of 2.725 quoted in the literature.

That the matrix element is large compared with unity could be anticipated from the earliest estimates of the bound electron wave function. For example, in a third approximation, Bethe (1929) chooses as a trial function

$$\phi(r_1, r_2) = e^{-(r_1+r_2)/2} \left(1 + \alpha\, |r_2 - r_1| + \beta\, (r_2 - r_1)^2\right), \tag{7.16.38}$$

finding $\alpha = 0.2$ and $\beta = 0.05$ and an electron density distribution in one dimension described by

$$\rho_e = 4\pi r^2 \phi^2(r) = e^{-1.535r}$$

$$\times \left(1.69r + 1.82r^2 - 0.536r^3 + 0.322r^4 - 0.0273r^5 + 0.0096r^6\right)$$

$$+ e^{-3.07r}\left(-1.69r + 0.52r^2\right), \tag{7.16.39}$$

which has a maximum at $r = 1.13$ and gives a mean radius of

$$r_{\text{mean}} = \frac{\int \rho_e(r) r \, dr}{\int \rho_e(r) \, dr} = 2.43 \tag{7.16.40}$$

in units of a_0.

From a more naive point of view, if one overlooks the fact that the wave function of the H^- atom is symmetric in the spatial coordinates of two electrons and thinks in terms of an electrically neutral core and a valence electron which is very loosely bound, one might guess that the peak in the density distribution of the valence electron is of the order of r_{valence} given by

$$\frac{1}{2}\frac{e^2}{r_{\text{valence}}} \sim \frac{1}{2}\frac{e^2}{a_0}\frac{0.754}{13.6}, \tag{7.16.41}$$

or $r_{\text{valence}} \sim 18a_0$, and that the dimensionless matrix element squared could be of the order of 10^2. In fact, detailed estimates of the dimensionless matrix element by Chandrasekhar and his associates (Chandrasekhar, 1945; Chandrasekhar and F. H. Breen, 1946; Chandrasekhar, 1958) show this to be the case. Experimental determinations of the cross section by Lewis M. Branscomb and Stephan J. Smith (1955) and by S. J. Smith and D. S. Burch (1959) confirm this guess. In a definitive work, S. Geltman (1962) demonstrates that a cross section estimated with the velocity matrix element formulation (the one described here) is more consistent with the experimental results than those estimated with other formulations such as the dipole moment formulation.

The cross section which best fits the data is described in Geltman's Table 3 by 34 values at intervals of 500 Å. These values have been used to construct Fig. 7.16.5. To obtain the cross section as a function of photon energy, a cubic spline in the energy coordinate has been fitted to the Table 3 values between $\lambda = 500$ Å and the threshold at $\lambda = 16\,419$ Å; for extrapolation to energies larger than the energy corresponding to 500 Å, an exponential has been fitted to the two points at 500 Å and 1000 Å. The result is plotted in Fig. 7.16.6 at intervals of 0.1 eV. It is interesting that the maximum in the cross section is not very different from the cross section given at threshold by the Born approximation (eq. (7.4.31)): 5.47×10^{-17} cm^2.

Combining the information in Figs. (7.16.4), (7.16.5), and (7.16.6), one might guess that the opacity in regions where bound–free absorption from H^- dominates is of the order of

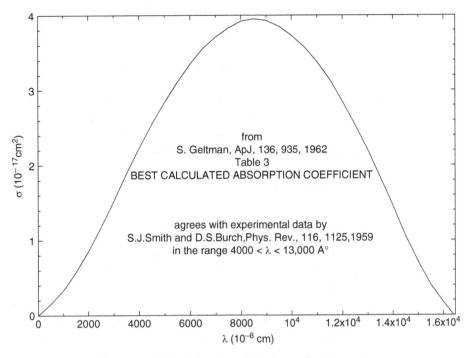

from
S. Geltman, ApJ, 136, 935, 1962
Table 3
BEST CALCULATED ABSORPTION COEFFICIENT

agrees with experimental data by
S.J.Smith and D.S.Burch,Phys. Rev., 116, 1125,1959
in the range 4000 < λ < 13,000 A°

Fig. 7.16.5 Cross section for bound–free absorption by the negative hydrogen ion

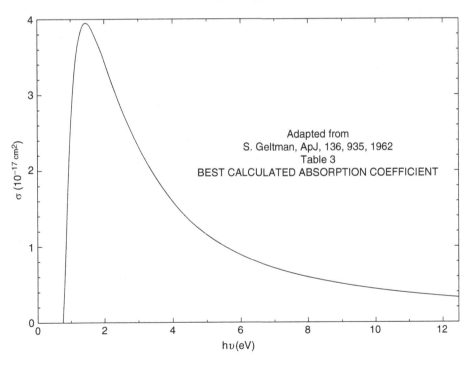

Adapted from
S. Geltman, ApJ, 136, 935, 1962
Table 3
BEST CALCULATED ABSORPTION COEFFICIENT

Fig. 7.16.6 Cross section for bound–free absorption by the negative hydrogen ion

$$\kappa_{\rm bf} \sim N_A\, 3 \times 10^{-17}\ {\rm cm}^2 \times 10^{-9}\ \left(\frac{Y_{\rm H^-}}{10^{-9}}\right)$$

$$= 1.8 \times 10^{-2}\ {\rm cm}^2\ {\rm g}^{-1}\ \left(\frac{Y_{\rm H^-}}{10^{-9}}\right) \sim (0.02 - 2.0)\ {\rm cm}^2\ {\rm g}^{-1}, \qquad (7.16.42)$$

the pair of numerical values given at the far right being for $Y_{\rm H^-} = 10^{-9}$–10^{-7}. However, even in regions where bound–free absorption from H^- provides the major contribution to the opacity for photon energies greater than the threshold energy of 0.75 eV, the construction of a Rosseland mean opacity as given by eq. (7.10.7) requires an estimate of the contribution of other sources of opacity at photon energies below threshold. That is, in evaluating a Rosseland mean cross section,

$$\frac{1}{\sigma_{\rm Ross}} = \int \frac{1}{\sigma(\nu)[1 - \exp(-h\nu/kT)]}\,\frac{dP_\nu}{dT}\,d\nu \div \int \frac{dP_\nu}{dT}\,d\nu, \qquad (7.16.43)$$

if the cross section in the region below the threshold in Fig. 7.16.6 is neglected, the calculated opacity could be considerably smaller than naively expected.

The number abundance of free electrons provided by atoms of low ionization potential is larger than the number abundance of H^- ions by two to five orders of magnitude over the range of densities and temperatures illustrated in Fig. 7.16.4. Even so, because the electron scattering cross section is over seven orders of magnitude smaller than the typical H^- bound–free cross section, the contribution to opacity by electron scattering is several orders of magnitude smaller. That is, since, as follows from eqs. (3.4.16) and (3.4.17),

$$\kappa_{\rm e} = N_A \sigma_{\rm es} Y_{\rm e} = N_A\, 0.665 \times 10^{-24}{\rm cm}^2\ Y_{\rm e} = 4 \times 10^{-6}{\rm cm}^2\ {\rm g}^{-1}\ \left(\frac{Y_{\rm e}}{10^{-5}}\right), \qquad (7.16.44)$$

the number abundance of free electrons would have to be of the order of 10^7 or larger than the number abundance of H^- ions for the electron-scattering contribution to opacity to compete with the H^- bound–free contribution. From Fig. 7.16.4 it is evident that $Y_{\rm e}/Y_{\rm H^-} < 10^5$ for temperatures $T < 5000$ K and $\rho > 10^{-8}$ g cm^{-3}.

Estimates of the cross sections for free–free absorption from electron donors of low ionization potential follow from the development in Section 7.6. In rough approximation, since $\sum_j Y_j^+ \sim Y_{\rm e}$,

$$\sigma_{\rm ff}(\nu)\, Y_{\rm e} = \sigma_{\rm es}\, Y_{\rm e}^2\, 1.92 \times 10^4\, x^{3+\delta}\,\frac{\rho_0}{T_6^{3.5}}, \qquad (7.16.45)$$

where $x = h\nu/kT$, $\delta = 0.5$ for $x \geq 1$, and $\delta = 0.25$ for $x < 1$. This translates into

$$\kappa_{\rm ff}(\nu) = 0.4\ {\rm cm}^2\ {\rm g}^{-1} 10^{-10}\ \left(\frac{Y_{\rm e}}{10^{-5}}\right)^2\, 1.92 \times 10^4\, x^{3+\delta} 10^{-8}$$

$$\times \left(\frac{\rho_0}{10^{-8}}\right)\, 2.5 \times 10^8\, \left(\frac{4000\ {\rm K}}{T}\right)^{3.5}$$

$$= 1.92 \times 10^{-6}\ {\rm cm}^2\ {\rm g}^{-1}\ \left(\frac{Y_{\rm e}}{10^{-5}}\right)^2 \left(\frac{\rho_0}{10^{-8}}\right) \left(\frac{4000\ {\rm K}}{T}\right)^{3.5} x^{3+\delta}. \qquad (7.16.46)$$

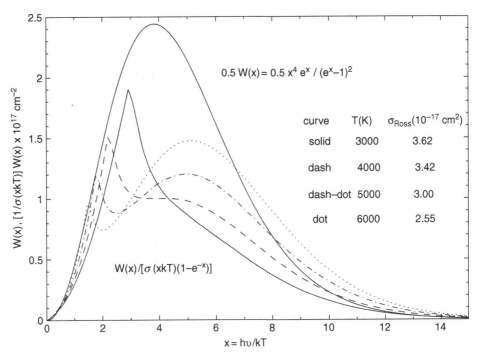

Fig. 7.16.7 Sample calculation of the Rosseland mean cross section for H^- at $\rho = 10^{-6}$ g cm^{-3} and T = 3000K, 4000K, 5000K, and 6000K

At the threshold for bound–free absorption from H^-, $x \sim 3$–1.5 for $T = 3000$–6000 K and $x^{3+\delta} \sim 0.028$–0.27.

Free–free absorption mediated by H^- ions also occurs, but because the electron density distribution extends to much larger radii than is the case in positive ions, the wave functions of incident and exit electrons deviate considerably from plane wave functions in the region over which overlap integrals are calculated to determine the appropriate matrix elements. Adoption of the expression given by eq. (7.16.46) for free–free transitions (with one of the two Y_es replaced by Y_{H^-}) therefore leads to a lower limit on the H^- Rosseland mean opacity. The weighting function $W(x)$ and the weighting function divided by the cross section corrected for stimulated emission are graphed in Fig. 7.16.7 for a density of $\rho = 10^{-6}$ g cm^{-3} and for four different temperatures. The resulting lower limit on the Rosseland mean H^- opacity and its ingredients are listed in Table 7.16.2. Values of the Rosseland mean cross section $(\sigma_{H^-})_{Ross}$ are taken from Fig. 7.16.6, values of Y_{H^-} are read from Fig. 7.16.2, and

$$(\kappa_{H^-})_{Ross} = N_A\,(\sigma_{H^-})_{Ross}\,Y_{H^-}. \tag{7.16.47}$$

The major lessons from Table 7.16.2 are that the Rosseland mean cross section for H^- decreases with increasing temperature and that the Rosseland mean opacity increases with increasing temperature because the number abundance of H^- ions increases with increasing temperature faster than the mean cross section decreases.

T (K)	$(\sigma_{\bar{H}})_{Ross}$ (10^{-17} cm^2)	Y_{H^-} (10^{-8})	$(\kappa_{H^-})_{Ross}$ (cm^2 g^{-1})
	Table 7.16.2 The Rosseland mean H$^-$ opacity and its ingredients when $\rho = 10^{-6}$ g cm^{-3}		
3000	3.62	0.541	0.118
4000	3.42	1.36	0.280
5000	3.00	2.93	0.529
6000	2.55	4.89	0.751

A straight line logarithmic fit to the entries in the last three rows of Table 7.16.2 gives

$$(\sigma_{H^-})_{Ross} \sim 3.44 \left(\frac{4000 \text{ K}}{T}\right)^{0.647} 10^{-17} \text{cm}^2. \tag{7.16.48}$$

Combining with eq. (7.16.23), one obtains

$$(\kappa_{H^-})_{Ross} \sim N_A \, (\sigma_{H^-})_{Ross} \, Y_{H^-} \sim 0.0559 \left(\frac{T}{4000 \text{ K}}\right)^{2.61} \left(\frac{\rho}{10^{-7}\text{g cm}^{-3}}\right)^{0.79} \text{cm}^2 \text{ g}^{-1}. \tag{7.16.49}$$

This last expression is an approximation designed for the temperature range T = 3000–6000 K. At the low end of the range, it is an overestimate because of an overestimate in the H$^-$ Rosseland mean cross section as approximated by eq. (7.16.48). At the upper end of the range, it is an underestimate because bound–free absorption by hydrogen atoms in the first excited state and by hydrogen molecules and free–free absorption by free protons have not been taken into account. One way of taking these processes into account is to increase the exponent of the temperature in eq. (7.16.49).

Typical estimates of the temperature exponent quoted in textbooks are of the order of 5, but estimates of the opacity tend to be smaller than the lower limits presented here. For example, calculations by Dilan Ezer and A. G. W. Cameron (1963) using Los Alamos opacity programs and quoted in Chiu's *Stellar Physics* (1968) can be summarized, for the range of temperatures and densities examined here, by

$$\kappa \sim 0.0126 \left(\frac{T}{4000 \text{ K}}\right)^{5.1} \left(\frac{\rho}{10^{-7} \text{ g cm}^{-3}}\right)^{0.7} \text{cm}^2 \text{ g}^{-1}. \tag{7.16.50}$$

Even though the number abundances adopted by Ezer and Cameron (1963) do not differ by more than a factor of ~2 (in either direction) from those adopted here, the coefficient preceeding the temperature and density dependent factors in eq. (7.16.50) is more than four times smaller than the coefficient in eq. (7.16.49).

A final word on nomenclature is appropriate. In astrophysical literature, a cross section is sometimes called an absorption coefficient and designated by κ, thus differing by name as well as by symbol from the word cross section and the symbol σ used in physics. Since κ is also used in astrophysics to denote an opacity which is related to a cross section by

$\kappa\rho = n\sigma$, where n is the number of absorbers per unit volume, it is important to ascertain which useage is being invoked.

Bibliography and references

R. Alexander & J. W. Ferguson, *ApJ*, **437**, 879, 1994.

P. W. Anderson, *Phys. Rev.*, **76**, 647, 1949.

M. Asplund, N. Grevesse, & A. J. Sauval, in *Cosmic Abundances as Records of Stellar Evolution and Nucleosynthesis*, eds. F. N. Bash & T. G. Barnes, ASP Conf. Ser., Vol. **30**, 2005.

Hans A. Bethe, *Zeits. für Physik*, **57**, 815, 1929.

L. C. Biedenharn, *Phys. Rev.*, **102**, 262, 1956.

J. J. Boyle & M. D. Kutzner, in *Many Body Atomic Physics*, eds. J. J. Boyle & M. S. Pinzola, Cambridge: Cambridge University Press, 1998.

Lewis M. Branscomb & Stephan J. Smith, *Phys. Rev.*, **98**, 1028, 1955.

R. G. Breene, Jr., *Rev. Mod. Phys.*, **29**, 94, 1957.

Subrahmanyan Chandrasekhar, *The Study of Stellar Structure*, (Chicago: University of Chicago Press), 1939.

Subrahmanyan Chandrasekhar, *ApJ*, **102**, 223, 395, 1945

Subrahmanyan Chandrasekhar, *ApJ*, **128**, 114, 1958.

S. Chandrasekhar & F. H. Breen, *ApJ*, **104**, 430, 1946.

Hong-Yee Chiu, *Stellar Physics*, (Waltham, Mass.: Blaisdell), 1968.

Arthur N. Cox & John N. Stewart, *ApJS*, **19**, 243, 261, 1970.

Thomas George Cowling, *MNRAS*, **94**, 768, 1934.

Ben Dorman & Robert T. Rood, *ApJ*, **409**, 387, 1993.

Albert Einstein, *Phys. Zeits.*, **18**, 121, 1917.

Dilan Ezer and Alastair G. W. Cameron, *Icarus 1*, No. **5–6**, 1963.

J. W. Ferguson, *ApJ*, **437**, 879, 1994.

J. W. Ferguson, D. R. Alexander, F. Allard, *et al., ApJ*, **623**, 585, 2005.

Richard P. Feynman, *Quantum Electrodynamics* (New York: Benjamin), 1962.

John Arthur Gaunt, *Proc. Roy. Soc. (A) London*, **126**, 654, 1930.

S. Geltman, *ApJ*, **136**, 935, 1962.

N. Grevesse & A. Noels, in *Origin and Evolution of the Elements*, eds. N. Pranzo, E. Vangioni-Flam, & M. Casse (Cambridge: Cambridge University Press), 1993.

Harvey Hall, *Rev. Mod. Phys.*, **8**, 358, 1936.

Chuchiro Hayashi, R. Hoshi, & Daíchiro Sugimoto, *Prog. Theoret. Phys.*, Supp. No. **22**, 1962.

Walter Heitler, *The Quantum Theory of Radiation* (Oxford: Clarendon Press), 1954.

Charles D. Hodgman, Robert C. Weast, & Samuel M. Selby, eds., *Handbook of Chemistry and Physics*, 39th edition, (Cleveland, Ohio: Chemical Rubber Publishing Co.), 1957.

Egil A. Hylleraas, *Zeits. für Physik*, **54**, 347, 1929; 60, 624, 1930.

Icko Iben, Jr., *ARAA*, **5**, 571, 1967.

Icko Iben,Jr., *ApJ*, **196**, 525, 1975.

Icko Iben, Jr. & John Ehrman, *ApJ*, **135**, 770, 1962.

Carlos A. Iglesias & Forrest J. Rogers, *ApJ*, **412**, 752, 1993; **464**, 943, 1996.

J. David Jackson, *Classical Electrodynamics* (New York: Wiley), third edition, 1999

William J. Karzas & R. Latter, *ApJS*, **6**, 167, 1961.

G. Keller & R. E. Meyerott, *ApJ*, **122**, 32, 1955.

O. Klein & Y. Nishina, *Zeits. für Phys.*, **52**, 853, 1929.

Hendrik Anthony Kramers, *Phil. Mag.*, **436**, 836, 1923.

Russell M. Kulsrud, *ApJ*, **119**, 386, 1954.

H. S. W. Massey & R. A. Smith, *Proc. Roy. Soc., A.*, **155**, 472, 1936.

A. W. Maue, *Ann. Phys.*, **13**, 161, 1932.

James Clerk Maxwell, *Treatise on Electricity and Magnetism* (Oxford: Clarendon Press), 1873.

Peter J. Mohr and Barry N. Taylor, *Physics Today*, August, BG5, 2002.

Philip M. Morse, *ApJ*, **92**, 27, 1940.

Lloyd Motz, *Astrophysics and Stellar Structure* (Waltham, Mass.: Ginn), 1970.

Peter Naur and Donald E. Osterbrock, *ApJ*, **117**, 306, 1953.

Y. Nishina, *Zeits. für Phys.*, **52**, 869, 1929.

T. Ohmura and H. Ohmura, *ApJ*, **131**, 8, 1960.

Pannekoek, Pub. Astr. Inst. U. Amsterdam, No. 4 (cited in this way with no date given by Rupert Wildt, *ApJ*, **90**, 611, 1939).

H. Ritz, *J. Reine Angewante Math.*, **135**, 1, 1909.

Forrest J. Rogers & Carlos A. Iglesias, *ApJS*, **79**, 507, 1992; **401**, 361, 1992.

P. W. Svein Rosseland, *MNRAS*, **84**, 525, 1924.

Martin Schwarzschild, *Structure and Evolution of the Stars*, (Princeton: Princeton University Press), 1958.

Mike J. Seaton, *MNRAS*, **265**, L25, 1993.

Leonard I. Schiff, *Quantum Mechanics* (New York: McGraw-Hill), 1949.

S. J. Smith & D. S. Burch, *Phys. Rev.*, **116**, 248, 1959.

Arnold Sommerfeld, *Ann. Phys.*, **11**, 257, 1931.

M. Stobbe, *Ann. der Phys.*, **7**, 661, 1930.

Bengt Strömgren, *Zeits. für Phys.*, **4**, 118, 1932.

Bengt Strömgren, Chapter 4 (pages 269–296) in *Stellar Structure*, eds. L. H. Aller & D. B. McLaughlin (Chicago: University of Chicago Press), 1965.

Roger Tayler, PhD Thesis, University of Cambridge, 1950.

Roger Tayler, *MNRAS*, **135**, 225, 1967.

Rupert Wildt, *ApJ*, **90**, 611, 1939.

8 Equations of stellar evolution and methods of solution

The discipline of stellar structure asks: given the opacity and the energy-generation rate as functions of composition, density, and temperature, and given the composition as a function of mass, what is the model structure in the static approximation? The discipline of stellar evolution asks: how, due to a combination of nuclear transformations and mixing processes, does the distribution of composition variables in a model star change with time, and how does the structure respond to these changes and to the loss of energy in the form of photons from the surface and neutrinos from the interior by the conversion of gravitational potential energy into heat and work and by the conversion of heat and work into gravitational potential energy. For a wide variety of situations, it is possible to explore evolution in the quasistatic approximation, which follows when bulk acceleration in an equation relating pressure-gradient and gravitational forces is neglected and the contribution to the internal energy of the kinetic energy of bulk motions is neglected. Nevertheless, meaningful estimates of bulk velocities follow as a consequence of changes in gravothermal characteristics required by the conservation of energy.

In order to reveal the full character of the quasistatic approximation, structure equations are derived in Section 8.1 without assuming spherical symmetry or placing restrictions on the acceleration. By invoking the conservation of mass, linear momentum, and energy, it is shown how the work done by gravity is translated by pressure-gradient forces into a primary component of the local gravothermal energy-generation rate ϵ_{grav} and how, in regions where particles are being created and destroyed, another component of ϵ_{grav} depends on the rates of creation and destruction of particles. An important theorem is derived which shows that, although the local rate at which gravity does work differs from the local rate at which pressure-gradient forces do work, the global rate at which gravity does work is identical with the global rate at which pressure-gradient forces do work.

The gravothermal energy source plays the extremely important role of mediating the transition between essentially static phases of evolution during which the structure of the star is stabilized not only by a balance between pressure-gradient and gravitational forces, but also by a balance between the rates of nuclear and gravothermal energy generation in the interior and the rate of energy outflow from the surface. In Section 8.2, some of the subtleties involved in interpreting the contribution of entropy changes to ϵ_{grav} are explored and the concept of a creation–destruction potential is elaborated. Examples involving beta-decay reactions in regions where electron degeneracy plays a role are given.

In Section 8.3, the complete set of equations in the quasistatic approximation for spherically symmetric model interiors is displayed. The theorem that the global rate at which gravity does work is identical to the global rate at which pressure-gradient forces do work is rederived, and a theorem which equates the rate of change of the net binding energy of a star with the global gravothermal energy-generation rate is derived.

As the stellar surface is approached, the radiation field changes from being essentially isotropic to being completely outwardly directed. Section 8.4 describes how, in an elementary but elegant manner (see Arthur S. Eddington in *The Internal Constitution of the Stars*, 1926), this change may be utilized to construct a surface boundary condition that is entirely adequate in the context of models of stellar evolution.

An evolutionary calculation requires an initial model. Often, this model can be a static structure obtained by setting $\epsilon_{grav} = 0$. However, solutions of even the simplest form of the equations of stellar structure are not entirely trivial. For example, straightforward integrations from the surface inward diverge, as do straightforward integrations from the center outward. One way to circumvent these divergences is given by the classical fitting technique which consists of both inward and outward integrations and fitting at some intermediate point in the interior, a technique described in the first part of Section 8.5. The concepts involved and techniques used in the construction of numerical integration algorithms are described in Section 8.6.

Evolutionary models are most easily constructed by employing an implicit relaxation technique which begins by dividing the model star into a large number of shells and replacing the differential equations by difference equations for each shell. As shown in Section 8.7, by perturbing the initial choice of structure variables in each shell and using the difference equations relating variables in adjacent shells, a series of equations relating the perturbations can be obtained. For small enough time steps, the equations can be linearized and solved for the perturbations.

Nuclear transformations are the main drivers of stellar evolution prior to the end stages of white dwarf, neutron star, or black hole, and they produce, with the help of various mixing mechanisms, the changing abundance distributions seen in the spectra of real stars at different stages in their evolution. In Section 8.8, nuclear transformation equations for an arbitrary choice of isotopes are obtained and it is shown that an implicit solution can be constructed by successive iterations involving the solution of a set of equations which are linear in abundance increments. In Section 8.9, it is shown how these linear equations may be solved by Gaussian elimination with pivoting or by the use of matrix decomposition.

In Section 8.10, issues which must be addressed in estimating composition changes in a convective zone are discussed. The turbulent convective mixing time (or turnover time) dt_{mix} in the zone must be estimated and compared with the average time scale $dt_{eq,i}$ for each isotope i to achieve a local equilibrium abundance with respect to nuclear creation and destruction mechanisms. It must also be compared with dt_{evo}, the time step which evolutionary considerations suggest should be taken between adjacent models. If $dt_{mix} < dt_{evo} \ll dt_{eq,i}$, it may be assumed that the ith isotope is completely mixed in the convective zone. If $dt_{evo} < dt_{mix} < dt_{eq,i}$, partial or diffusive mixing is in order and an algorithm for calculating the extent of mixing is presented. Otherwise, the abundance of isotope i is taken to be the local equilibrium abundance.

Composition changes due to particle diffusion occur in radiative zones, even in the absence of nuclear transformations. Gravity-induced diffusion leads to a tendency for element species to separate, with species composed of more massive particles migrating inward in mass and those composed of lighter particles migrating outward in mass. Diffusion of any species also occurs if there is a gradient in the number abundance of that species. Both types of diffusion continue to operate even when local equilibrium with respect to diffusion has been established and the abundance distribution has been stabilized. That is, even under equilibrium conditions, there is a flow of particle species in both directions, the downward flow being maintained by acceleration due to gravity and the upward flow being maintained by the gradients in number abundances. The role of particle diffusion in stellar evolution is outlined in Section 8.11, but a discussion of the physics of particle diffusion and methods of solution is deferred to Chapter 12 in Volume 2. Although real stars rotate, and mixing currents driven by differential rotation work to offset gravitational settling, no attempt is made in this book to treat rotationally induced diffusion.

In Section 8.12, issues related to spatial zoning and choice of time step are addressed. Criteria are established for ensuring that solutions of the difference equations used in describing stellar models are reasonably close approximations to the solutions of the differential equations they replace, that spatial zoning is sufficiently fine for profiles of abundances and other physical characteristics of interest to be properly resolved, and that time steps are such that convergence is achieved after an optimal number of iterations. Algorithms for rezoning are presented.

For over two-thirds of a century, the rate of development of our understanding of stellar evolution has been coupled with the rate at which computers have evolved with respect to speed and storage capacity. Reflections on this phenomenon are offered in Section 8.13.

8.1 Consequences of the conservation of mass, momentum, and energy

In this section, three of the equations of stellar structure are derived without making any restrictions regarding symmetry, but assuming that viscosity and shear forces can be neglected. Consider a hypothetical bubble encased in an invisible membrane through which no matter is allowed to pass but which can be distorted into any shape in response to pressure forces acting across the membrane. The internal energy of matter and radiation inside the bubble can change in response to the work done by pressure forces across the membrane and in response to a flow of energy across the membrane. And, of course, gravity exerts a force on all of the matter within the bubble.

The mass M, linear momentum $\mathbf{P}_m$, and energy content E of the bubble may be written, respectively, as

$$M = \int \rho \, d\tau, \tag{8.1.1}$$

$$\mathbf{P}_m = \int \rho \mathbf{v} \, d\tau, \tag{8.1.2}$$

and

$$E = \int \rho \left(U + \frac{1}{2} v^2 \right) \, d\tau, \tag{8.1.3}$$

where ρ is the mass density, $\mathbf{v}$ is the macroscopic velocity, $d\tau$ is a volume element, U is the internal energy per gram (including microscopic thermal motions, electromagnetic energy, etc.), and $v^2/2$ is the kinetic energy per gram in macroscopic motions. The integrations extend over the matter in the bubble.

Conservation of mass, linear momentum, and energy give, respectively,

$$\frac{dM}{dt} = 0, \tag{8.1.4}$$

$$\frac{d\mathbf{P}_m}{dt} = -\int P \, d\mathbf{S} + \int \mathbf{g}\rho \, d\tau, \tag{8.1.5}$$

and

$$\frac{dE}{dt} = \int (\epsilon_{\mathrm{nuc}} - \epsilon_\nu + \mathbf{g} \cdot \mathbf{v})\rho \, d\tau - \int \mathbf{F}_{\mathrm{tot}} \cdot d\mathbf{S} - \int P\mathbf{v} \cdot d\mathbf{S}. \tag{8.1.6}$$

Here, $\mathbf{g}$ is the acceleration due to gravity, $d\mathbf{S}$ is an area element pointing outward in a direction perpendicular to the surface, and the integrals containing $d\mathbf{S}$ are over the surface of the bubble. Further, P is the pressure, assumed to be isotropic, ϵ_{nuc} is the rate at which nuclear energy is liberated, and ϵ_ν is the rate at which nuclear energy is converted into the kinetic energy of neutrinos. Energy flow by radiation, conduction, and convection have been lumped together in the flux $\mathbf{F}_{\mathrm{tot}}$. The minus sign before ϵ_ν comes from the assumption that neutrinos do not interact with matter in the star, but escape with their initial energy undiminished. This assumption must be modified to follow the formation of neutron stars in supernova explosions.

It is worth noting that, due to the fact that nuclear fusion reactions convert rest mass energy into photon energy which ultimately leaks out of the star, mass conservation (eq. (8.1.4)) is only approximately satisfied; for example, during hydrogen burning, roughly one per cent of the initial rest mass energy in regions where burning has gone to completion is lost from the star.

Be that as it may, for short periods of time, eq. (8.1.4) is a quite good enough approximation. Differentiation of eq. (8.1.1) gives

$$\frac{dM}{dt} = \int \frac{\partial \rho}{\partial t} \, d\tau + \int \rho \, \mathbf{v} \cdot d\mathbf{S} = 0. \tag{8.1.7}$$

The integral over the surface of the bubble in eq. (8.1.7) can be converted into an integral over volume, so that

$$\frac{dM}{dt} = \int \left(\frac{\partial \rho}{\partial t} + \nabla \cdot (\rho\mathbf{v}) \right) \, d\tau. \tag{8.1.8}$$

The only way for this to vanish for an arbitrary bubble is for the integrand to vanish. Thus,

$$\frac{\partial \rho}{\partial t} + \nabla \cdot (\rho \mathbf{v}) = \frac{d\rho}{dt} + \rho \, (\nabla \cdot \mathbf{v}) = 0, \tag{8.1.9}$$

where

$$\frac{d\rho}{dt} = \frac{\partial \rho}{\partial t} + \mathbf{v} \cdot \nabla \rho \tag{8.1.10}$$

is the time derivative of the density in the frame of reference moving with velocity $\mathbf{v}$.

Differentiation of eq. (8.1.2) gives

$$\frac{d\mathbf{P}_m}{dt} = \int \frac{\partial (\rho \mathbf{v})}{\partial t} \, d\tau + \int (\rho \mathbf{v}) \, \mathbf{v} \cdot d\mathbf{S}$$

$$= \int \left(\frac{\partial (\rho \mathbf{v})}{\partial t} + \nabla \cdot (\rho \mathbf{v} \mathbf{v}) \right) d\tau$$

$$= \int \left(\rho \frac{\partial \mathbf{v}}{\partial t} + \mathbf{v} \frac{\partial \rho}{\partial t} + \mathbf{v} \nabla \cdot (\rho \mathbf{v}) + (\rho \mathbf{v}) \cdot \nabla \mathbf{v} \right) d\tau$$

$$= \int \left[\mathbf{v} \left(\frac{\partial \rho}{\partial t} + \nabla \cdot (\rho \mathbf{v}) \right) + \rho \left(\frac{\partial \mathbf{v}}{\partial t} + \mathbf{v} \cdot \nabla \mathbf{v} \right) \right] d\tau. \tag{8.1.11}$$

By virtue of eq. (8.1.9), the first term in parentheses in the last form of eq. (8.1.11) vanishes. Replacing ρ by $\mathbf{v}$ in eq. (8.1.10) to define the derivative $d\mathbf{v}/dt$ in the frame of reference moving with velocity $\mathbf{v}$, eq. (8.1.11) becomes

$$\frac{d\mathbf{P}_m}{dt} = \int \rho \left(\frac{\partial \mathbf{v}}{\partial t} + \mathbf{v} \cdot \nabla \mathbf{v} \right) d\tau = \int \rho \frac{d\mathbf{v}}{dt} \, d\tau. \tag{8.1.12}$$

Coupling this with eq. (8.1.5), one has that

$$\int \rho \frac{d\mathbf{v}}{dt} \, d\tau = - \int P \, d\mathbf{S} + \int \mathbf{g} \, \rho \, d\tau = \int (-\nabla P + \mathbf{g} \, \rho) \, d\tau, \tag{8.1.13}$$

which has the general solution

$$\rho \frac{d\mathbf{v}}{dt} = -\nabla P + \mathbf{g} \, \rho. \tag{8.1.14}$$

Differentiation of eq. (8.1.3) gives

$$\frac{dE}{dt} = \int \frac{\partial}{\partial t} \left[\rho \left(U + \frac{1}{2} v^2 \right) \right] d\tau + \int \rho \left(U + \frac{1}{2} v^2 \right) \mathbf{v} \cdot d\mathbf{S} \tag{8.1.15}$$

$$= \int \left(\frac{\partial}{\partial t} \left[\rho \left(U + \frac{1}{2} v^2 \right) \right] + \nabla \cdot \left[\rho \mathbf{v} \left(U + \frac{1}{2} v^2 \right) \right] \right) d\tau \tag{8.1.16}$$

$$= \int \left[\rho \frac{\partial}{\partial t} \left(U + \frac{1}{2} v^2 \right) + \left(U + \frac{1}{2} v^2 \right) \frac{\partial \rho}{\partial t} \right] d\tau$$

$$+ \int \left[\rho \, \mathbf{v} \cdot \nabla \left(U + \frac{1}{2} v^2 \right) + \left(U + \frac{1}{2} v^2 \right) \nabla \cdot (\rho \mathbf{v}) \right] d\tau. \tag{8.1.17}$$

Using the fact that for any scalar quantity q,

$$\frac{dq}{dt} = \frac{\partial q}{\partial t} + \mathbf{v} \cdot \nabla q, \tag{8.1.18}$$

it follows that

$$\frac{dE}{dt} = \int \left(\rho \frac{d}{dt} \left[U + \frac{1}{2} v^2 \right] + \left[U + \frac{1}{2} v^2 \right] \left[\frac{\partial \rho}{\partial t} + \nabla \cdot (\rho \mathbf{v}) \right] \right) d\tau. \tag{8.1.19}$$

Employing eq. (8.1.9) again, one has finally that

$$\frac{dE}{dt} = \int \rho \frac{d}{dt} \left(U + \frac{1}{2} v^2 \right) d\tau. \tag{8.1.20}$$

Rewriting eq. (8.1.6) as

$$\frac{dE}{dt} = \int \left[\rho(\epsilon_{nuc} - \epsilon_\nu + \mathbf{g} \cdot \mathbf{v}) - \nabla \cdot (\mathbf{F}_{tot} + P\mathbf{v}) \right] d\tau \tag{8.1.21}$$

$$= \int \left[\rho(\epsilon_{nuc} - \epsilon_\nu) + (\rho \mathbf{g} - \nabla P) \cdot \mathbf{v} - \nabla \cdot \mathbf{F}_{tot} - P\nabla \cdot \mathbf{v} \right] d\tau, \tag{8.1.22}$$

and using eqs. (8.1.9) and (8.1.14), it follows that

$$\frac{dE}{dt} = \int \left[\rho(\epsilon_{nuc} - \epsilon_\nu) - \nabla \cdot \mathbf{F}_{tot} + \rho \frac{d}{dt} \left(\frac{1}{2} v^2 \right) + \frac{P}{\rho} \frac{d\rho}{dt} \right] d\tau. \tag{8.1.23}$$

Finally, equating the integrands in eqs. (8.1.20) and (8.1.23), one has

$$\nabla \cdot \mathbf{F}_{tot} = \rho(\epsilon_{nuc} - \epsilon_\nu + \epsilon_{grav}), \tag{8.1.24}$$

where

$$\epsilon_{grav} = -\frac{dU}{dt} + \frac{P}{\rho^2} \frac{d\rho}{dt}. \tag{8.1.25}$$

The quantity dU/dt is the rate at which the internal energy per unit mass increases, (P/ρ^2) $(d\rho/dt)$ is the rate at which work is being done per unit mass to compress matter, and ϵ_{grav} is the "gravothermal" energy-generation rate, which can be either positive or negative. For example, in a region which is both heating and expanding, energy is absorbed locally from the ambient flow of energy and, therefore, $\epsilon_{grav} < 0$.

In contrast to ϵ_{nuc} and ϵ_ν, ϵ_{grav} is not an explicit function of variables in the model being constructed at time t. That is, dU/dt and $d\rho/dt$ require information about two models, one at time t and one at time $t - dt$. This feature, in addition to the change in composition wrought by nuclear transformations and diffusive mixing mechanisms, contributes to the distinction between stellar "evolution" and stellar "structure". In principle, the simplest way to approximate the gravothermal energy-generation rate is to set

$$\epsilon_{grav} \approx -\frac{\delta U}{\delta t} + \frac{P}{\rho^2} \frac{\delta \rho}{\delta t}, \tag{8.1.26}$$

where δU and $\delta \rho$ are, respectively, the change in the internal energy per unit mass and the change in the density between the model being constructed at time t and its predecessor at time $t - \delta t$. In this formulation, the changes in particle number abundances due to nuclear transformations and mixing mechanisms are automatically taken into account.

On the other hand, it is instructive to understand explicitly how local number-abundance changes contribute to ϵ_{grav}. Considering the internal energy U to be a function of any two thermodynamic variables A and B and of the abundances by number Y_i of all particle species, one may formally break dU/dt into two parts:

$$\frac{dU}{dt} = \left(\frac{dU}{dt}\right)_Y + \sum_i \left(\frac{\partial U}{\partial Y_i}\right)_{A,B} \frac{dY_i}{dt}, \tag{8.1.27}$$

where the first term on the right hand side of the equation is the rate at which the internal energy per unit mass changes when number abundances of nuclei and free electrons are held constant and the second term is the rate at which the internal energy per unit mass changes due to the creation and destruction of particles via nuclear reactions and/or mixing processes when the thermodynamic variables are held constant.

Choosing ρ and T as the independent thermodynamic variables, one may set

$$-\epsilon_{\mathrm{th}} = \left(\frac{dU}{dt}\right)_Y = \left(\frac{\partial U}{\partial \rho}\right)_{T,Y} \frac{d\rho}{dt} + \left(\frac{\partial U}{\partial T}\right)_{\rho,Y} \frac{dT}{dt}, \tag{8.1.28}$$

where ϵ_{th} is the thermal energy-generation rate in the absence of changes in particle number abundances. In the second term of eq. (8.1.27), differentiation is with respect to particle number abundance parameters Y_i, with ρ and T kept constant, and the sum is over all nuclear species and the free electron abundance. Using eqs. (8.1.26)–(8.1.28), one may write

$$\epsilon_{\mathrm{grav}} = \epsilon_{\mathrm{gth}} + \epsilon_{\mathrm{cdth}}, \tag{8.1.29}$$

where

$$\epsilon_{\mathrm{gth}} = \epsilon_{\mathrm{th}} + \frac{P}{\rho^2} \frac{d\rho}{dt} = -\left(\frac{dU}{dt}\right)_Y + \frac{P}{\rho^2} \frac{d\rho}{dt} \tag{8.1.30}$$

$$= -\left(\frac{\partial U}{\partial T}\right)_{\rho,Y} \frac{dT}{dt} + \left[\frac{P}{\rho^2} - \left(\frac{\partial U}{\partial \rho}\right)_{T,Y}\right] \frac{d\rho}{dt} \tag{8.1.31}$$

and

$$\epsilon_{\mathrm{cdth}} = -\sum_i \left(\frac{\partial U}{\partial Y_i}\right)_{\rho,T} \frac{dY_i}{dt}. \tag{8.1.32}$$

In this formulation, ϵ_{th} and ϵ_{gth} always require information from models at time t and time $t - dt$ to form $d\rho/dt \sim \delta\rho/\delta t$ and $dT/dt \sim \delta T/\delta t$, whereas, in the absence of mixing currents, the quantities dY_i/dt and therefore ϵ_{cdth} are explicit functions of variables at time t, making the approximations $dY_i/dt \sim \delta Y_i/\delta t$ unnecessary.

For nuclei, the dimensionless number abundance parameter Y_i can be defined as

$$Y_i = \frac{1}{N_{\mathrm{A}}} \frac{n_i}{\rho}, \tag{8.1.33}$$

where n_i is the abundance per unit volume of nuclei of type i, ρ is the density, and $N_A = 6.023 \times 10^{23} \mathrm{g}^{-1}$ is related to Avogadro's number (6.023×10^{23} particles per mole). For electrons,

$$Y_e = \frac{1}{N_A} \frac{n_e}{\rho} = \sum_i Z_i \, Y_i, \tag{8.1.34}$$

where Z_i is the net positive charge of the ith nucleus or ion.

Defining a "creation–destruction" potential as

$$\bar{\mu}_i = \frac{1}{N_A} \left(\frac{\partial U}{\partial Y_i} \right)_{\rho,T}, \tag{8.1.35}$$

eq. (8.1.32) becomes

$$\epsilon_{\text{cdth}} = -N_A \sum_i \bar{\mu}_i \, \frac{dY_i}{dt}. \tag{8.1.36}$$

The quantity $\bar{\mu}_i$ has the dimensions of energy and is always positive. The fact that it appears in eq. (8.1.36) preceded by a minus sign can be understood by considering a set of N neutral perfect-gas particles initially in thermal equilibrium at temperature T in a box of fixed size with insulating walls. Add $\delta N \ll N$ particles with zero kinetic energy. After thermal equilibrium is again achieved, the temperature of the ensemble has dropped because the original store of thermal energy is shared by more than the original N particles. Mathematically,

$$N \, \delta \left(\frac{3}{2} kT \right) = - \left(\frac{3}{2} kT \right) \delta N = -\bar{\mu} \, \delta N. \tag{8.1.37}$$

That is, the addition of particles with zero kinetic energy acts as an energy drain for the particles already in the box.

It is instructive to explore the structure of ϵ_{grav} in a nuclear burning region. For simplicity, assume that all electrons are free and that all particles obey the perfect gas law. Under these circumstances,

$$U = N_A \frac{3}{2} kT \sum_i Y_i \tag{8.1.38}$$

and

$$P = N_A kT \rho \sum_i Y_i, \tag{8.1.39}$$

where the number abundance parameters for nuclei and electrons are given, respectively, by eqs. (8.1.33) and (8.1.34). The gravothermal energy-generation rate is

$$\epsilon_{\text{grav}} = -\frac{d}{dt} \left(N_A \frac{3}{2} kT \sum_i Y_i \right) - N_A kT \rho \sum_i Y_i \frac{d}{dt} \left(\frac{1}{\rho} \right) \tag{8.1.40}$$

$$\sim - N_A kT \sum_i Y_i \frac{1}{\delta t} \left(\frac{3}{2} \frac{\delta T}{T} - \frac{\delta \rho}{\rho} \right) - N_A \frac{3}{2} kT \sum_i \frac{dY_i}{dt}, \tag{8.1.41}$$

Thus,

$$\epsilon_{\text{gth}} = - N_A \, kT \sum Y_i \frac{1}{\delta t} \left(\frac{3}{2} \frac{\delta T}{T} - \frac{\delta \rho}{\rho} \right) \tag{8.1.42}$$

and, in the absence of mixing currents,

$$\epsilon_{\text{cdth}} = -N_A \frac{3}{2} kT \sum_i \frac{dY_i}{dt}. \tag{8.1.43}$$

It follows from eqs. (8.1.43) and (8.1.36) that

$$\bar{\mu}_i = \bar{\mu}_e = \frac{3}{2} kT, \tag{8.1.44}$$

as expected.

Differentiation of eq. (8.1.39) gives that, over a time step of size δt, changes in thermodynamic and number-abundance variables are related by

$$\frac{\delta P}{P} = \frac{\delta T}{T} + \frac{\delta \rho}{\rho} + \frac{\sum_i \delta Y_i}{\sum_i Y_i}. \tag{8.1.45}$$

Replacing $\delta \rho / \rho$ in eq. (8.1.42) by its equivalent from eq. (8.1.45), with $\delta Y_i / \delta t$ replaced by dY_i / dt, one obtains

$$\epsilon_{\text{gth}} = - N_A \, kT \sum_i Y_i \frac{1}{\delta t} \left(\frac{5}{2} \frac{\delta T}{T} - \frac{\delta P}{P} \right) - N_A \, kT \sum_i \frac{dY_i}{dt}, \tag{8.1.46}$$

demonstrating that the separation of ϵ_{grav} into two parts, one of which is identified with particle creation and destruction, is not a unique separation, but depends on which pair of thermodynamic variables is chosen to be independent. This is important to keep in mind if variables other than density and temperature are chosen in constructing a stellar evolution program and it is a fact worth belaboring. In the expressions

$$\epsilon_{\text{grav}} = -\frac{P}{\rho} \frac{1}{\delta t} \left[\frac{3}{2} \frac{\delta T}{T} - \frac{\delta \rho}{\rho} \right] - \frac{3}{2} \frac{P}{\rho} \frac{1}{\delta t} \frac{\sum_i \delta Y_i}{\sum_i Y_i} \tag{8.1.47}$$

and

$$\epsilon_{\text{grav}} = -\frac{P}{\rho} \frac{1}{\delta t} \left[\frac{5}{2} \frac{\delta T}{T} - \frac{\delta P}{P} \right] - \frac{5}{2} \frac{P}{\rho} \frac{1}{\delta t} \frac{\sum_i \delta Y_i}{\sum_i Y_i}, \tag{8.1.48}$$

the first terms on the right hand sides of the expressions are identical when $\sum_i \delta Y_i = 0$, but are different when $\sum_i \delta Y_i \neq 0$. That is,

$$\epsilon_{\text{grav}} \neq \epsilon_{\text{grav}} \left(\text{when} \sum_i \delta Y_i = 0 \right) + \text{a unique } \epsilon_{\text{cdth}}. \tag{8.1.49}$$

The reason for this is that, in a given mass shell, both pressure and density are functions of the composition.

Returning to ρ and T as the independent variables of choice, eq. (8.1.42) may be recast as

$$\epsilon_{\text{gth}} = -\left(N_A \sum_i Y_i\right) kT \frac{\mathrm{d}}{\mathrm{d}t} \left(\log_e \frac{T^{3/2}}{\rho}\right). \tag{8.1.50}$$

By setting $V = 1/\rho$ in eq. (8.1.30), the rate of gravothermal energy generation due to the action of gravity and thermodynamic changes in internal energy can be written as

$$\epsilon_{\text{gth}} = -\left[\left(\frac{\mathrm{d}U}{\mathrm{d}t}\right)_Y + P \frac{\mathrm{d}V}{\mathrm{d}t}\right] = -T \left(\frac{\mathrm{d}S}{\mathrm{d}t}\right)_Y, \tag{8.1.51}$$

where S is the entropy as defined in classical thermodynamics. Equations (8.1.50) and (8.1.51) then give the familiar result that, for a perfect gas, the entropy per gram is

$$S = \left(N_A \sum_i Y_i\right) k \left(\log_e \frac{T^{3/2}}{\rho}\right) + \text{constant.} \tag{8.1.52}$$

To summarize the development in this section, the consequences of energy conservation can be written as

$$\nabla \cdot \mathbf{F}_{\text{tot}} = \rho \left(\epsilon_{\text{nuc}} - \epsilon_\nu + \epsilon_{\text{gth}} + \epsilon_{\text{cdth}}\right), \tag{8.1.53}$$

where ϵ_{nuc}, ϵ_ν, ϵ_{gth}, and ϵ_{cdth} are, respectively, the rate of liberation of nuclear energy, the rate of energy loss by neutrinos, the rate of change in internal energy plus the rate of work done by pressure-gradient forces due to the action of gravitational forces, and the rate of internal energy changes due to particle creation and annihilation. The gravothermal energy-generation can be written as

$$\epsilon_{\text{grav}} = \epsilon_{\text{gth}} + \epsilon_{\text{cdth}} = -\left[T \left(\frac{\mathrm{d}S}{\mathrm{d}t}\right)_Y + N_A \sum_i \bar{\mu}_i \frac{\mathrm{d}Y_i}{\mathrm{d}t}\right], \tag{8.1.54}$$

where S is the classical entropy and $\bar{\mu}_i$ is a creation–destruction potential given by eq. (8.1.35).

In regions in a star where beta-unstable nuclei are formed, the emission of electrons has exactly the same consequences as described by eq. (8.1.54), even though the electrons appear with a finite kinetic energy determined by the physics of the decay reaction. The initial kinetic energy of the emitted electron, after accounting for neutrino losses, is traditionally included in the term $\epsilon_{\text{nuc}} - \epsilon_\nu$; the fact that the electron comes rapidly into thermal equilibrium is taken into account by including $\epsilon_{\text{cdth}} < 0$ in the energy budget. When positrons are emitted, they annihilate with electrons, so, for positrons, creation is exactly balanced by destruction. The rest mass energy liberated in the annihilation process is normally taken into account in ϵ_{nuc}. The kinetic energies of annihilated electrons, which contribute to the energy of the gamma rays emitted in the annihilation process, are taken into account in $\epsilon_{\text{cdth}} > 0$.

It is to be emphasized again that, under all circumstances, ϵ_{gth} is a time-dependent quantity which depends on the structure of two adjacent models. In the absence of mixing currents, ϵ_{cdth} is a local density-, temperature-, and composition-dependent quantity. Triggered both by composition changes (due to nuclear reactions and mixing mechanisms such as diffusion) and by energy transfer and loss mechanisms, the dynamical action of gravity and changes in internal energy are responsible for ϵ_{gth}.

Although the gravitational field does work in contributing to ϵ_{gth}, the gravitational acceleration $\mathbf{g}$ (or, more importantly, Newton's gravitational constant) does not explicitly enter the expression for ϵ_{gth}. This is because gravitational energy is not locally defined. The energy which is formally produced at the rate $\mathbf{g} \cdot \mathbf{v}$ by the action of gravity at a given point actually appears as compressional work over an extended region displaced from this point. Multiplying eq. (8.1.14) by $\mathbf{v}\cdot$ and rearranging, one has that the rate at which gravity does work locally is

$$\mathbf{g} \cdot \mathbf{v} = \mathbf{v} \cdot \frac{d\mathbf{v}}{dt} + \frac{1}{\rho} \mathbf{v} \cdot \nabla P \tag{8.1.55}$$

or

$$\mathbf{g} \cdot \mathbf{v} = \frac{d}{dt}\left(\frac{1}{2} v^2\right) + \frac{1}{\rho} \mathbf{v} \cdot \nabla P. \tag{8.1.56}$$

One may interpret eq. (8.1.56) to mean that part of the work done by gravity is converted into the kinetic energy of bulk motion at the point where the gravitational work is being done. However, the second term in eq. (8.1.56) involving the gradient of pressure is not the rate at which energy is being delivered locally by compression.

Multiplying eq. (8.1.56) by ρ and using the vector identity

$$\nabla \cdot (P\mathbf{v}) = \mathbf{v} \cdot \nabla P + P \nabla \cdot \mathbf{v} \tag{8.1.57}$$

to replace $\mathbf{v} \cdot \nabla P$ in the equation results in

$$\rho \, \mathbf{g} \cdot \mathbf{v} = \rho \frac{d}{dt}\left(\frac{1}{2} v^2\right) + \nabla \cdot (P\mathbf{v}) - P \nabla \cdot \mathbf{v}. \tag{8.1.58}$$

Replacing $\nabla \cdot \mathbf{v}$ in eq. (8.1.58) by its equivalent $-(1/\rho)\, d\rho/dt$, as given by eq. (8.1.9), produces

$$\rho \, \mathbf{g} \cdot \mathbf{v} = \rho \frac{d}{dt}\left(\frac{1}{2} v^2\right) + \nabla \cdot (P\mathbf{v}) + \frac{P}{\rho} \frac{d\rho}{dt}. \tag{8.1.59}$$

Multiplying eq. (8.1.59) by a volume element $d\tau$ and integrating over the space occupied by the model star gives

$$\int \mathbf{g} \cdot \mathbf{v} \, \rho \, d\tau = \frac{d}{dt} \int \left(\frac{1}{2} v^2\right) \rho \, d\tau + \int \nabla \cdot (P\mathbf{v}) \, d\tau + \int \frac{P}{\rho^2} \frac{d\rho}{dt} \rho \, d\tau. \tag{8.1.60}$$

The integral over volume in the second term on the right hand side of eq. (8.1.60) may be converted into an integral over the surface of the model star:

$$\int \nabla \cdot (P\mathbf{v}) \, d\tau = \int (P\mathbf{v}) \cdot d\mathbf{S}. \tag{8.1.61}$$

Defining a mass element by $dM = \rho \, d\tau$, one has finally that

$$\int \mathbf{g} \cdot \mathbf{v} \, dM = \frac{d}{dt} \int \left(\frac{1}{2} v^2 \right) dM + \int P\mathbf{v} \cdot d\mathbf{S} + \int \frac{P}{\rho^2} \frac{d\rho}{dt} \, dM, \tag{8.1.62}$$

or

$$\dot{E}_{\text{work}} = \dot{E}_{\text{kinetic}} + \int P\mathbf{v} \cdot d\mathbf{S} + \dot{E}_{\text{compression}}, \tag{8.1.63}$$

where

$$\dot{E}_{\text{work}} = \int (\mathbf{g} \cdot \mathbf{v}) \, dM, \tag{8.1.64}$$

is the global rate at which gravitational forces do work,

$$\dot{E}_{\text{compression}} = \int \frac{P}{\rho^2} \frac{d\rho}{dt} \, dM \tag{8.1.65}$$

is the global rate at which compressional forces do work, and

$$\dot{E}_{\text{kinetic}} = \frac{d}{dt} \int \left(\frac{1}{2} v^2 \right) dM \tag{8.1.66}$$

is the global rate at which the kinetic energy associated with bulk motion increases.

Although it is not transparently obvious, the first two terms on the right hand sides of eqs. (8.1.62) and (8.1.63) in general turn out to be quite small compared with the other two terms and it transpires that, in very good approximation,

$$\dot{E}_{\text{work}} = \dot{E}_{\text{compression}}. \tag{8.1.67}$$

At any given position, the integrands in eqs. (8.1.64) and (8.1.65) are typically quite different in magnitude. For example, the quantity $(P/\rho^2) \, d\rho/dt$ is usually finite at the center, whereas, $\mathbf{g} \cdot \mathbf{v}$ must vanish at the center. The fact that the two rates are generally quite different locally but are globally the same demonstrates that the work done by compression at a given location is a consequence of work done by gravity in other parts of the star and communicated to the given location by pressure-gradient forces.

8.2 Examples of the creation–destruction potential for ions and electrons

The creation-destruction potential $\bar{\mu}_i$ defined by eq. (8.1.35) is not to be confused with the chemical potential defined in thermodynamics. As an example, consider a box of size V containing N neutral particles at temperature T and pressure P. With $E = \frac{3}{2} NkT$ and $PV = NkT$, it follows that

$$dQ = dE + PdV = NkT \, d\left(\log_e\left((NT)^{3/2} V\right)\right)$$

$$= T \, d\left(Nk \, \log_e\left((NT)^{3/2} V\right)\right) - kT \, \log_e\left((NT)^{3/2} V\right) \, dN$$

$$= T \, dS + \mu \, dN, \tag{8.2.1}$$

where

$$S = Nk \, \log_e\left((NT)^{3/2} V\right) \tag{8.2.2}$$

is defined as the entropy and

$$\mu = -kT \, \log_e\left((NT)^{3/2} V\right) \tag{8.2.3}$$

is defined, in accordance with thermodynamical tradition, as the chemical potential of the system. On the other hand, one can also write

$$dQ = NkT \, d\log_e\left(T^{3/2}V\right) + \frac{3}{2} kT \, dN$$

$$= T \, d\left(Nk \, \log_e\left(T^{3/2}V\right)\right)_{N=\text{const}} + \frac{3}{2} kT \, dN$$

$$= T \, d\bar{S}_{N=\text{const}} + \bar{\mu} \, dN$$

$$= T \, N \, d\bar{s} + \bar{\mu} \, dN, \tag{8.2.4}$$

where

$$\bar{S} = N \, k \, \log_e\left(T^{3/2}V\right), \tag{8.2.5}$$

$$\bar{s} = k \, \log_e\left(T^{3/2}V\right), \tag{8.2.6}$$

and

$$\bar{\mu} = +\frac{3}{2} kT, \tag{8.2.7}$$

a result already anticipated in Section 4.14 (see eqs. (4.14.16)–(4.14.23)).

Because it is not explicitly dependent on an abundance, the quantity $\bar{s}$ may be called a "specific" entropy, even though the quantity $\bar{S} = N \, \bar{s}$ is not the entropy in the sense of classical thermodynamics. From the computational point of view, the formulation involving $\bar{s}$ and $\bar{\mu}$ is more convenient than the one involving the entropy and the chemical potential. From the physical point of view in the present context, it is also more enlightening. In order to distinguish it from the traditional chemical potential μ used in chemical thermodynamics, $\bar{\mu}$ is in this book called a creation–destruction potential.

One can apply the same logic to different kinds of particles in a stellar environment. In this environment, the number abundance parameter Y_i for the ith type of particle can be defined by (see also eq. (8.1.33))

$$n_i = \frac{\rho}{M_H} Y_i, \tag{8.2.8}$$

where n_i is the number per unit volume of the ith type of particle and $M_H \sim 1/N_A$ is the mass of a hydrogen atom.

Consider first non-degenerate particles. For the ith type of particle, the energy per gram is

$$U_i = \frac{3}{2} \frac{k}{M_H} (Y_i\, T),$$ (8.2.9)

and the partial pressure is

$$P_i = \frac{k}{M_H} \rho\, (Y_i\, T).$$ (8.2.10)

To the extent that ρV is constant, the first law of thermodynamics gives

$$dQ_i = dU_i - \frac{P_i}{\rho^2}\, d\rho = \frac{kT}{M_H} Y_i\, d \log_e \left(\frac{(Y_i\, T)^{3/2}}{\rho} \right)$$ (8.2.11)

$$= T\, dS_i + \mu_i\, dY_i,$$ (8.2.12)

where dQ_i is the change in heat per gram for the ith type of particle,

$$S_i = \frac{1}{M_H} Y_i\, k\, \log_e \left(\frac{(Y_i\, T)^{3/2}}{\rho} \right),$$ (8.2.13)

and

$$\mu_i = -\frac{1}{M_H} T\, k\, \log_e \left(\frac{(Y_i\, T)^{3/2}}{\rho} \right).$$ (8.2.14)

One can also write

$$dQ_i = \frac{1}{M_H} (Y_i\, T\, d\bar{s}_i + \bar{\mu}_i\, dY_i),$$ (8.2.15)

where

$$\bar{s}_i = k\, \log_e \left(\frac{T^{3/2}}{\rho} \right)$$ (8.2.16)

and

$$\bar{\mu}_i = +\frac{3}{2}\, kT.$$ (8.2.17)

In the interiors of almost all stars, except for neutron stars and the coldest of white dwarfs, expression (8.2.17) adequately describes the creation–destruction potential for atoms, ions, and bare nuclei.

For free electrons, the form of the creation–destruction potential $\bar{\mu}_e$ depends on the degree of degeneracy. Using the definition of Y_e given by eq. (8.1.34), for non-degenerate electrons one has

$$U_e = \frac{1}{\rho} n_e \frac{3}{2} kT = N_A \frac{3}{2} kT\, Y_e$$ (8.2.18)

and the contribution of free electrons to ϵ_{cdth} becomes

$$\epsilon_{\text{cdth},e} = -N_A \, \frac{3}{2} \, kT \, \frac{dY_e}{dt}. \tag{8.2.19}$$

Thus, again, as in eq. (8.2.7),

$$\bar{\mu}_e = \frac{3}{2} \, kT. \tag{8.2.20}$$

In the case of partially degenerate electrons for which the parameter ϵ_F' is negative, eqs. (4.5.5), (4.5.6), and (4.5.18) give that the number of electrons per unit volume is

$$n_e = N_A \, \frac{\rho}{\mu_e} = N_A \, \rho \, Y_e = \frac{2}{\lambda_e^3} \, \Sigma_{3/2}(\alpha), \tag{8.2.21}$$

and eqs. (4.5.31), (4.5.17), and (4.5.18) give that the kinetic energy of electrons per unit mass is

$$U_e = \frac{E_e'}{\rho} = \frac{3}{2} \, \frac{n_e}{\rho} \, kT \, \frac{\Sigma_{5/2}(\alpha)}{\Sigma_{3/2}(\alpha)} = N_A \, \frac{3}{2} \, kT \, \frac{\Sigma_{5/2}(\alpha)}{\Sigma_{3/2}(\alpha)} \, Y_e. \tag{8.2.22}$$

In these equations, α is a degeneracy parameter and the Σs have the property that

$$\frac{1}{\Sigma_k(\alpha)} \, \frac{\partial \Sigma_k(\alpha)}{\partial \alpha} = -\Sigma_{k-1}(\alpha). \tag{8.2.23}$$

It follows that

$$\left(\frac{\partial U_e}{\partial n_e} \right)_{\rho,T} = \frac{E_e}{n_e} \, \frac{\Sigma_{3/2}^2}{\Sigma_{5/2}\Sigma_{1/2}}, \tag{8.2.24}$$

and, using eq. (8.2.22),

$$\left(\frac{\partial U_e}{\partial Y_e} \right)_{\rho,T} = \left(\frac{\partial U_e}{\partial n_e} \right)_T \left(\frac{\partial n_e}{\partial Y_e} \right)_\rho = N_A \, \frac{3}{2} \, kT \, \frac{\Sigma_{3/2}}{\Sigma_{1/2}}, \tag{8.2.25}$$

so that

$$\bar{\mu}_e = \frac{3}{2} \, kT \, \frac{\Sigma_{3/2}}{\Sigma_{1/2}}. \tag{8.2.26}$$

When the degeneracy parameter vanishes ($\alpha = 0.0$), $\Sigma_{1/2} \to S_{1/2} = 0.582\,549$, $\Sigma_{3/2} \to S_{3/2} = 0.765\,102$, and

$$\bar{\mu}_e \to \frac{3}{2} \, kT \, \frac{S_{3/2}}{S_{1/2}} = 1.313\,370 \, \frac{3}{2} \, kT. \tag{8.2.27}$$

For conditions of intermediate, non-relativistic, degeneracy, one may differentiate eq. (4.7.6) with respect to Y_e and use the fact that $U_e = (3/2)\,(P_e/\rho)$ to obtain

$$\bar{\mu}_e = \frac{3}{2}\,kT\,(1 + 0.043\,7452\,\delta\,(1 - 0.003\,465\,\delta\,(1 - 0.005\,220\,13\,\delta))), \tag{8.2.28}$$

where δ is given by eq. (4.5.10). This approximation is applicable for values of δ up to $\delta \sim 64$ (equivalent to $\epsilon'_F/kT \sim 4.6$, where $\epsilon'_F = \epsilon_F - 1$ is the electron Fermi kinetic energy).

For higher degrees of non-relativistic electron degeneracy and very low temperatures,

$$U_e = \frac{3}{5}\,\epsilon'_F(0)\frac{n_e}{\rho} = \frac{3}{5}\,\epsilon'_F(0)N_A\,Y_e \tag{8.2.29}$$

and

$$\epsilon'_F(0) = \frac{p_F^2(0)}{2m_e} = \frac{1}{2m_e}\left(\frac{3}{8\pi}\,h^3\,N_A\,\rho\right)^{2/3} Y_e^{2/3}, \tag{8.2.30}$$

where $\epsilon'_F(0) = p_F^2(0)/2m_e$ is the electron Fermi kinetic energy at zero temperature. Differentiation gives

$$\epsilon_{\mathrm{cdth}} = -\left(\frac{\partial U_e}{\partial Y_e}\right)_{\rho,T} = -\frac{3}{5}\,N_A\,\left(\epsilon'_F(0) + \left(\frac{\partial \epsilon'_F(0)}{\partial Y_e}\right)_{\rho,T} Y_e\right)\frac{dY_e}{dt}$$

$$= -N_A\,\epsilon'_F(0)\,\frac{dY_e}{dt}. \tag{8.2.31}$$

Thus,

$$\bar{\mu}_e = \epsilon'_F(0). \tag{8.2.32}$$

For higher temperatures, use of eqs. (4.6.27), (4.6.41), and (4.6.43) yields

$$\bar{\mu}_e = \epsilon'_F(0)\left(1 + \frac{\pi^2}{12}\left(\frac{kT}{\epsilon'_F(0)}\right)^2 + \frac{3\pi^4}{80}\left(\frac{kT}{\epsilon'_F(0)}\right)^4\right). \tag{8.2.33}$$

For extremely relativistically degenerate electrons at low temperatures, $\epsilon'_F(0) \propto p_F \propto (\rho\,Y_e)^{1/3}$ and $U_e = (3/4)\,\epsilon'_F(0)\,N_A\,Y_e$, so $\bar{\mu}_e$ is given by eq. (8.2.32), just as in the case of non-relativistically degenerate electrons.

The energy budget is particularly easy to understand when electrons are very degenerate. In the case of electron decay, the emitted electron has nowhere to go in reaching thermodynamic equilibrium except to the top of the Fermi sea, where it finally resides with a kinetic energy ϵ'_F to which already present electrons contribute; thus the average kinetic energy of the initially present electrons is (very slightly) reduced. In the case of positron emission, regardless of which electron is chosen by the positron for annihilation, the final result after thermodynamic equilibrium is re-established is one fewer electron at the top of the Fermi sea; the total internal kinetic energy of the remaining electrons has been increased by ϵ'_F, thus (very slightly) increasing the average kinetic energy of these remaining electrons.

A general, temperature-dependant expression for relativistically degenerate electrons follows from the development in Section 4.9. From eq. (4.9.24) and the fact that U_e, the electron kinetic energy per unit mass, is related to E_e, the total energy per unit volume, by

$$U_e = \frac{E_e}{\rho} - N_A\, m_e c^2\, Y_e, \tag{8.2.34}$$

one has that

$$\frac{U_e}{m_e c^2} = \frac{8\pi}{\lambda_C^3} \frac{1}{\rho}(I_E + I_{ET}) - N_A\, Y_e, \tag{8.2.35}$$

where I_E and I_{ET} are defined, respectively by eqs. (4.9.29) and (4.9.32). In the latter two equations, the Fermi energy $\epsilon_F = \sqrt{1 + p_F^2}$ includes the electron rest mass energy and is in units of $m_e c^2$ and p_F is in units of $m_e c$. Differentiation with respect to the electron abundance by number gives

$$\bar\mu_e = \frac{1}{N_A}\left(\frac{\partial U_e}{\partial Y_e}\right)_{\rho,T} = \frac{8\pi}{\lambda_C^3}\frac{1}{N_A\,\rho}m_e c^2\left[\left(\frac{\partial I_E}{\partial Y_e}\right)_{\rho,T} + \left(\frac{\partial I_{ET}}{\partial Y_e}\right)_{\rho,T}\right] - m_e c^2. \tag{8.2.36}$$

Derivatives of I_E and I_{ET} with respect to Y_e at constant temperature and density can be obtained by differentiating successively with respect to p_F, $p_F(0)$, and then Y_e. Thus, for example,

$$\left(\frac{\partial I_E}{\partial Y_e}\right)_{\rho,T} = \left(\frac{\partial p_F(0)}{\partial Y_e}\right)_{\rho,T}\left(\frac{\partial p_F}{\partial p_F(0)}\right)_{\rho,T}\left(\frac{\partial I_E}{\partial p_F}\right)_{\rho,T}. \tag{8.2.37}$$

Using $n_e = N_A\,\rho\,Y_e$ in eq. (4.8.7), it follows that

$$p_F(0) = \lambda_C\left(\frac{3}{8\pi}N_A\,\rho\,Y_e\right)^{1/3}, \tag{8.2.38}$$

so that

$$\left(\frac{\partial p_F(0)}{\partial Y_e}\right)_{\rho,T} = \frac{1}{3}\frac{p_F(0)}{Y_e}. \tag{8.2.39}$$

From eq. (4.9.11) for $R = p_F(0)/p_F$, it follows that

$$\left(\frac{\partial p_F}{\partial p_F(0)}\right)_{\rho,T} = \left(\frac{p_F(0)}{p_F}\right)^2\frac{1}{J} = \frac{R^2}{J}, \tag{8.2.40}$$

where

$$J = 1 - \frac{\pi^2}{6} (1 - 2 p_F^2) \left(\frac{kT}{p_F^2}\right)^2 - \frac{7\pi^4}{24} \left(\frac{kT}{p_F^2}\right)^4 - \cdots .$$ (8.2.41)

Coupling eqs. (8.2.36)–(8.2.41) with eq. (4.8.6) gives

$$\frac{\bar{\mu}_e}{m_e c^2} = \frac{1}{J} \frac{1}{p_F^2} \left[\left(\frac{\partial I_E}{\partial p_F}\right)_{\rho,T} + \left(\frac{\partial I_{ET}}{\partial p_F}\right)_{\rho,T} \right] - 1.$$ (8.2.42)

Differentiation of eq. (4.9.29) yields

$$\left(\frac{\partial I_E}{\partial p_F}\right)_{\rho,T} = \frac{d I_E}{d p_F} = \epsilon_F \, p_F^2,$$ (8.2.43)

and differentiation of eq. (4.9.32) yields

$$\left(\frac{\partial I_{ET}}{\partial p_F}\right)_{\rho,T} = -\frac{\pi^2}{6} \left(\frac{kT}{p_F^2}\right)^2 \frac{p_F^2}{\epsilon_F} \left[\left(1 - 3p_F^2 - 6p_F^4\right) + \frac{7\pi^2}{20} \left(\frac{kT}{p_F^2}\right)^2 \left(5 + p_F^2\right) + \cdots \right].$$ (8.2.44)

Inserting eqs. (8.2.43) and (8.2.44) into eq. (8.2.42) yields

$$\frac{\bar{\mu}_e}{m_e c^2} = \left(\frac{\epsilon_F}{J} - 1\right) - \frac{\pi^2}{6} \left(\frac{kT}{p_F^2}\right)^2 \frac{1}{\epsilon_F} \frac{1}{J}$$

$$\times \left[\left(1 - 3p_F^2 - 6p_F^4\right) + \frac{7\pi^2}{4} \left(1 + \frac{p_F^2}{5}\right) \left(\frac{kT}{p_F^2}\right)^2 + \cdots \right].$$ (8.2.45)

Expanding J^{-1} from eq. (8.2.41) in powers of $\left(\frac{kT}{p_F^2}\right)$,

$$\frac{1}{J} = 1 + \frac{\pi^2}{6} \left(1 - 2 p_F^2\right) \left(\frac{kT}{p_F^2}\right)^2 + \frac{\pi^4}{9} \left[\frac{23}{8} - p_F^2 + p_F^4\right] \left(\frac{kT}{p_F^2}\right)^4 + \cdots ,$$ (8.2.46)

and then inserting in eq. (8.2.45), the final result for the creation–destruction potential is

$$\frac{\bar{\mu}_e}{m_e c^2} = (\epsilon_F - 1) + \frac{\pi^2}{3} \frac{p_F^2}{\epsilon_F} \left(1 + 2 p_F^2\right) \left(\frac{kT}{p_F^2}\right)^2 + \frac{\pi^4}{9} \frac{p_F^2}{\epsilon_F} \left[\frac{13}{5} - 2p_F^4\right] \left(\frac{kT}{p_F^2}\right)^4 + \cdots .$$ (8.2.47)

Once again, it transpires that, at $T = 0$, $\bar{\mu}_e$ equals the electron Fermi kinetic energy. In the limit $\epsilon_F' = \epsilon_F - 1 \ll 1$, the first two terms of eq. (8.2.47) are equivalent to the first two terms of eq. (8.2.33). Because of the implicit nature of the relationship between p_F and $p_F(0)$ given by eq. (4.9.11), establishing explicitly the identity of the third term is more difficult.

8.3 The quasistatic equations of stellar structure in spherical symmetry

A zeroth order approximation to the relationship between the pressure gradient and the force of gravity follows when it is assumed that the term $d\mathbf{v}/dt$ in eq. (8.1.14) may be neglected. This is the so-called quasistatic approximation. In making this approximation, the possibility of pulsations is ignored and the effects of rotation and of other dynamical motions are neglected. Assuming spherical symmetry, the equations of stellar structure reduce to four equations in one dimension. If one chooses the radial coordinate r as the independent variable, differentiation of eq. (8.1.1) gives

$$\frac{dM(r)}{dr} = 4\pi r^2 \rho(r), \tag{8.3.1}$$

which expresses the conservation of mass.

The conservation of linear momentum, eq. (8.1.14) with $d\mathbf{v}/dt = 0$, becomes

$$\frac{dP(r)}{dr} = -\frac{GM(r)}{r^2}\,\rho(r). \tag{8.3.2}$$

By setting

$$\mathbf{F}_{\text{tot}} = \frac{L(r)}{4\pi r^2}\,\hat{\mathbf{r}} \tag{8.3.3}$$

in eqs. (8.1.24) and (8.1.54), which express the conservation of energy, one has that

$$\left(\frac{2}{r} + \frac{d}{dr}\right)\frac{L(r)}{4\pi r^2} = \frac{1}{4\pi r^2}\frac{dL(r)}{dr} = \rho(r)\,\epsilon_{\text{tot}}, \tag{8.3.4}$$

where

$$\epsilon_{\text{tot}} = \epsilon_{\text{nuc}} - \epsilon_\nu + \epsilon_{\text{grav}} = \epsilon_{\text{nuc}} - \epsilon_\nu + \epsilon_{\text{gth}} + \epsilon_{\text{cdth}} \tag{8.3.5}$$

and all of the ϵs are evaluated at the point r. Rearranging, one can also write

$$\frac{dL(r)}{dr} = 4\pi r^2 \rho(r)\,\epsilon_{\text{tot}}. \tag{8.3.6}$$

The final equation of stellar structure in the approximation of spherical symmetry has been discussed at length in Section 3.4 and reduces to

$$\frac{dT}{dr} = \frac{dT}{dP}\frac{dP}{dr} = -\frac{T}{P}\,V\frac{GM(r)}{r^2}\,\rho(r), \tag{8.3.7}$$

where the logarithmic derivative

$$V = \frac{d \log T}{d \log P} \tag{8.3.8}$$

is either (a) $V_{\text{rad+cond}}$, as given by eq. (3.4.64) if energy flow is entirely by radiation and conduction, (b) the adiabatic derivative V_{ad}, if matter is in a convective zone ($V_{\text{ad}} < V_{\text{rad}}$) where convective speeds are small, or (c) a value intermediate between V_{ad} and $V_{\text{rad+cond}}$ (given crudely by algorithms in Section 3.4) in convective zones where convective speeds are large.

Choosing $M = M(r)$ as the independent variable instead of r, the equations of stellar structure are

$$\frac{dr}{dM} = \frac{1}{4\pi r^2 \rho}, \tag{8.3.9}$$

$$\frac{dP}{dM} = -\frac{GM}{4\pi r^4}, \tag{8.3.10}$$

$$\frac{dL}{dM} = \epsilon_{\text{tot}}, \tag{8.3.11}$$

and

$$\frac{dT}{dM} = -\frac{T}{P} V \frac{GM}{4\pi r^4}. \tag{8.3.12}$$

In Sections 8.5 and 8.7, both forms of the equations are used, (eqs. (8.3.1), (8.3.2), (8.3.6), and (8.37) when r is the independent variable and eqs. (8.3.9)–(8.3.12) when M is the independent variable), as convenience dictates.

As discussed near the end of Section 8.1, the global rate at which gravity does work is very close to the global rate at which pressure-gradient forces do compressional work. Since, locally, the two rates are quite different, this result is not trivially obvious and it is worth re-examining the argument again in the spherically symmetric quasistatic approximation. Multiplying both sides of eq. (8.1.57) by $dM = 4\pi r^2 \rho \, dr$ and integrating, one has that

$$\int_0^{M(R)} (\mathbf{g} \cdot \mathbf{v}) \, dM = \int_0^{M(R)} \frac{d}{dt} \left(\frac{1}{2} v^2 \right) dM + \int_0^R \frac{1}{\rho} v \frac{dP}{dr} 4\pi r^2 \rho \, dr. \tag{8.3.13}$$

Taking the time derivative outside of the first integral on the right hand side of this equation and simplifying the last integral produces

$$\int_0^{M(R)} (\mathbf{g} \cdot \mathbf{v}) \, dM = \frac{d}{dt} \int_0^{M(R)} \left(\frac{1}{2} v^2 \right) dM + \int_{P(0)}^{P(R)} 4\pi r^2 v \, dP. \tag{8.3.14}$$

Integration by parts of the last integral in eq. (8.3.14) gives

$$\int_{P(0)}^{P(R)} 4\pi r^2 v \, dP = \left(4\pi r^2 v(r) \, P(r) \right) \Big|_0^R - \int 4\pi P \, d(r^2 v). \tag{8.3.15}$$

Since the pressure $P(R)$ at the formal surface of a stellar model does not vanish ($P(R) \sim P_{\text{rad}}(R) \propto T_{\text{e}}^4 \neq 0$), the first term on the right hand side of eq. (8.3.15) does not formally vanish. That is,

$$\left(4\pi r^2 v(r) \, P(r)\right)_0^R = 4\pi R^2 v(R) \, P(R) - 4\pi 0^2 v(0) \, P(0)$$

$$= 4\pi R^2 v(R) \, P(R) \neq 0. \tag{8.3.16}$$

In the spherical approximation, the continuity equation, eq. (8.1.9), may be written as

$$\frac{d\rho}{dt} = -\rho \, \nabla \cdot \mathbf{v} = -\rho \, \frac{1}{r^2} \frac{d}{dr}(r^2 v), \tag{8.3.17}$$

so that

$$d(r^2 v) = -\frac{1}{\rho^2} \frac{d\rho}{dt} \, \rho \, r^2 dr. \tag{8.3.18}$$

Using eqs. (8.3.16) and (8.3.18) in eq. (8.3.15) gives

$$\int_{P(0)}^{P(R)} 4\pi r^2 v \, dP = 4\pi R^2 \, v(R) P(R) + \int_0^{M(R)} \frac{P}{\rho^2} \frac{d\rho}{dt} \, dM. \tag{8.3.19}$$

Equation (8.3.14) may now be written as

$$\int_0^{M(R)} (\mathbf{g} \cdot \mathbf{v}) \, dM = \frac{d}{dt} \int_0^{M(R)} \left(\frac{1}{2} v^2\right) dM + 4\pi R^2 \, v(R) \, P(R) + \int_0^{M(R)} \frac{P}{\rho^2} \frac{d\rho}{dt} \, dM, \tag{8.3.20}$$

or, using the definitions given by eqs. (8.1.64)–(8.1.66), as

$$\dot{E}_{\text{work}} = \dot{E}_{\text{kinetic}} + 4\pi R^2 \, v(R) \, P(R) + \dot{E}_{\text{compression}}. \tag{8.3.21}$$

The only difference between eqs. (8.3.21) and (8.1.63) is that the surface term $\int P\mathbf{v} \cdot d\mathbf{S}$ in eq. (8.1.63) is approximated explicitly in eq. (8.3.21) by $4\pi R^2 \, v(R) \, P(R)$.

One might anticipate that, since $dv/dt = 0$ in the quasistatic approximation, $\dot{E}_{\text{kinetic}}$ would be formally zero. However, a numerical solution actually produces a realistic velocity distribution, so E_{kinetic} does not vanish and one must resort to numerical comparisons in detailed models to discover that E_{kinetic} is in general very, very small compared with the microscopic internal energy E_{internal} and compared with $\dot{E}_{\text{compression}}$. At the photosphere of a realistic model,

$$P(R) \sim P_{\text{rad}}(R) \sim \frac{1}{3} a \, T_{\text{e}}^4 = \frac{1}{3} \frac{4\sigma}{c} \, T_{\text{e}}^4 = \frac{4}{3} \frac{1}{c} \frac{L}{4\pi R^2}, \tag{8.3.22}$$

so that the surface term in eq. (8.3.21) is

$$4\pi R^2 v(R) \, P(R) \sim \frac{4}{3} \frac{v_{\text{surface}}}{c} \, L. \tag{8.3.23}$$

In general, $v_{surface}/c \ll 1$ so, to an exceedingly good approximation,

$$\dot{E}_{work} = \dot{E}_{compression}, \tag{8.3.24}$$

in agreement with eq. (8.1.67).

In spherical symmetry, the scalar product of the gravitational acceleration and the velocity is

$$\mathbf{g} \cdot \mathbf{v} = -\frac{GM(r)}{r^2} \frac{dr}{dt} = \frac{d}{dt} \frac{GM(r)}{r}, \tag{8.3.25}$$

where the second identity in eq. (8.3.25) follows because the value of $M(r)$ to which r is attached does not change with time. Multiplying eq. (8.3.25) by dM and integrating over the whole star, one has that

$$\dot{E}_{work} = \int (\mathbf{g} \cdot \mathbf{v}) \, dM = \frac{d}{dt} \int \frac{GM \, dM}{r} = -\frac{d\Omega}{dt}, \tag{8.3.26}$$

where $-\Omega > 0$ is the gravitational binding energy introduced in Chapter 3. Defining the total internal energy of the star as

$$E_{internal} = \int U \, dM, \tag{8.3.27}$$

one has that

$$E_{bind} = -\Omega - E_{internal} \tag{8.3.28}$$

is the energy required to convert the star into a gas at zero density and zero temperature. Taking the time derivative of eq. (8.3.28) and using eqs. (8.3.26) and (8.3.24), one has that

$$\dot{E}_{bind} = -\dot{\Omega} - \dot{E}_{internal} = \dot{E}_{work} - \dot{E}_{internal}$$

$$= \dot{E}_{compression} - \dot{E}_{internal}. \tag{8.3.29}$$

On the other hand, eqs. (8.1.25), (8.3.5), and (8.3.11) show that compressional work and changes in internal energy contribute to the stellar luminosity at the rate

$$\dot{E}_{gravothermal} = \int \epsilon_{grav} \, dM = \int \left[\frac{P}{\rho^2} \frac{d\rho}{dt} - \frac{dU}{dt} \right] dM$$

$$= \dot{E}_{compression} - \dot{E}_{internal}. \tag{8.3.30}$$

Comparing eqs. (8.3.30) and (8.3.29), one has that

$$\dot{E}_{bind} = \dot{E}_{gravothermal}. \tag{8.3.31}$$

One might argue that the relationship expressed by eq. (8.3.31) is intuitively obvious. Indeed, during pre-main sequence phases when nuclear reactions and neutrino losses are not important and the release of gravitational potential energy is the primary source of stellar luminosity, the equation is self-evidently fulfilled: it is a simple consequence of energy conservation expressed by

$$L_{surface} = \dot{E}_{gravothermal} = \dot{E}_{bind}. \tag{8.3.32}$$

In this case, one may begin with eq. (8.3.32) and, reversing the arguments which led from eq. (8.3.24) to (8.3.31), arrive at eq. (8.3.24) as a derived theorem.

However, when nuclear reactions are the major source of stellar luminosity, when energy released by nuclear reactions can contribute locally to the increase in internal energy and when, outside of regions of nuclear energy production (through absorption from the ambient flow of energy to which nuclear sources have made the dominant contribution), nuclear energy contributes to the work of expansion, it is a miracle that the theorems relating rates of change of global gravothermal energy components remain completely oblivious to the undeniable influence of nuclear energy release on local gravothermal characteristics everywhere in the star.

For these reasons, in discussions of stellar models in subsequent chapters, pains have been taken to verify numerically that, even when nuclear reactions are the dominant contributor to the stellar luminosity, the theorems expressed by eqs. (8.3.24) and (8.3.31) are, to the accuracy of the calculations, exact.

8.4 The photospheric boundary condition

In the deep interior of a star, the radiation field is essentially isotropic, with the intensity of radiation depending only slightly on direction. Near the surface, however, the intensity in forward directions increases relative to that in backward directions until there is no backward flow of radiation. In this section, Eddington's elegant zeroth order approximation to photospheric structure is presented.

Consider a cylinder of unit cross section and length ds which makes an angle θ with respect to the radial direction r. The change in the intensity of radiation $I(\theta)$ through the cylinder is given by

$$dI(\theta) = -I(\theta)\kappa\rho ds + j\rho ds, \tag{8.4.1}$$

where j is the emissivity. Using $dr = \cos\theta ds$, $dx = -dr$, and defining

$$\tau = \int_0^x \kappa\rho dx \tag{8.4.2}$$

as the "optical depth" at a distance x below the surface, eq. (8.4.1) can be written as

$$\cos\theta \frac{dI(\theta)}{d\tau} = I(\theta) - \frac{j}{\kappa}. \tag{8.4.3}$$

Defining moments by

$$J = \frac{1}{4\pi}\int I(\theta)d\Omega, \tag{8.4.4}$$

$$H = \frac{1}{4\pi}\int I(\theta)\cos\theta d\Omega, \tag{8.4.5}$$

and

$$K = \frac{1}{4\pi} \int I(\theta) \cos^2 \theta \, d\Omega,$$ (8.4.6)

where $d\Omega = \sin \theta \, d\phi \, d\theta$ is an increment in solid angle, integration of eq. (8.4.2) over $d\Omega$ gives

$$\frac{dH}{d\tau} = J - \frac{j}{\kappa}$$ (8.4.7)

and

$$\frac{dK}{d\tau} = H.$$ (8.4.8)

The moments have the following significance:

$$\frac{J}{c} = aT^4$$ (8.4.9)

is the energy per unit volume in the radiation field,

$$H = \frac{L}{4\pi R^2} = \sigma T_e^4$$ (8.4.10)

is the flux of energy in the x direction, and

$$\frac{K}{c} = \frac{1}{3} aT^4$$ (8.4.11)

is the radiation pressure.

The simplifying trick is to set $I(\theta) = I_1$ for $\theta < \pi/2$ and $I(\theta) = I_2$ for $\theta > \pi/2$, where I_1 and I_2 are angle independent but vary with depth below the surface in such a way that $I_2 = 0$ at the surface. With this prescription, the moments become

$$J = \frac{1}{2}(I_1 + I_2),$$ (8.4.12)

$$H = \frac{1}{4}(I_1 - I_2),$$ (8.4.13)

and

$$K = \frac{1}{6}(I_1 + I_2) = \frac{1}{3}J.$$ (8.4.14)

Since H is constant, eqs. (8.4.8) and (8.4.14) give

$$J = 3H\tau + \text{constant}.$$ (8.4.15)

At the surface, $I_2 = 0$ and $\tau = 0$, so eqs. (8.4.12) and (8.4.13) give

$$J = 2H = \text{constant}.$$ (8.4.16)

Altogether, then,

$$J = caT^4 = 2H\left(1 + \frac{3}{2}\tau\right) \tag{8.4.17}$$

or

$$T^4 = \frac{1}{2}T_e^4\left(1 + \frac{3}{2}\tau\right). \tag{8.4.18}$$

In other words, T_e, the "effective" or "surface" temperature of a star is the temperature of the radiation field at optical depth $\tau = 2/3$.

8.5 The classical fitting technique for model construction

The equations given in Section 8.3 may be solved in many different ways. For models of homogeneous composition, a straightforward algorithm for obtaining a solution is to (1) choose two physical variables at the center as parameters to be varied, and make a first outward integration to some convenient fitting point (e.g., half the mass of the chosen model), (2) choose two variables at the surface as parameters to be varied and make a first integration inwards to the fitting point, (3) perform additional integrations after varying central and surface parameters, one at a time, to obtain numerical derivatives with respect to the parameters of the discrepancies in physical variables (P, T, L, R) at the fitting point, (4) adjust the central and surface parameters in such a way as to formally reduce the discrepancies in physical variables at the fitting point, and finally (5) begin the process (1)–(4) again with the new guesses for central and surface parameters.

8.5.1 Development close to the model center and near the surface

At the center, where $r = 0$,

$$M(0) = L(0) = 0. \tag{8.5.1}$$

Central pressure and temperature,

$$P_c = P(0) \quad \text{and} \quad T_c = T(0), \tag{8.5.2}$$

may be chosen as the parameters to be varied at the center, although any other pair of physical variables would do as well.

Very close to the origin, one may neglect the variation of density with distance from the center and, by integrating the stellar structure equations, the following approximations to the radial dependence of physical variables emerge:

$$M(r) = \frac{4\pi}{3}r^3\rho_c, \tag{8.5.3}$$

$$P(r) = P_c - \frac{2\pi}{3} G \rho_c^2 r^2 = P_c - \frac{1}{2} \frac{GM(r)}{r} \rho_c, \tag{8.5.4}$$

$$T(r) = T_c V_c \left(1 - \frac{2\pi}{3} G \frac{\rho_c^2}{P_c} r^2\right) = T_c - \frac{T_c V_c}{P_c} \frac{1}{2} \frac{GM(r)}{r}, \tag{8.5.5}$$

and

$$L(r) = M(r) \, \epsilon_c = \frac{4\pi}{3} \rho_c \, \epsilon(\rho_c, T_c) \, r^3. \tag{8.5.6}$$

In these equations, the subscript c denotes evaluation at the center.

Integration may be continued outward with, say, a fourth order Runge–Kutta scheme, as discussed in Section 8.6. If continued indefinitely, the solution does not, in general, satisfy the surface boundary condition (Section 8.4). For example, the temperature can approach an asymptotic value far in excess of the value given by eq. (8.4.18).

At the "surface", model mass M is specified and, for example, L and R (defining $T_e^4 = L/4\pi\sigma R^2$) can be chosen as surface parameters. The initial temperature may be chosen as $T = T_e/2^{1/4}$ (see eq. (8.4.18)) and the density may be chosen in such a way that gas pressure is small compared with the radiation pressure. Integration of eq. (8.4.2) along with eqs. (8.3.1)–(8.3.2) is continued until $\tau = 2/3$. The radius at this point may be defined as the radius of the model. The calculation of temperature during further inward integration makes use of eq. (8.3.8) instead of eqs. (8.4.2) and (8.4.18). As in the case of integrations outward from the center, the physical quantities obtained by inward integration eventually diverge from a realistic solution. For example, it can occur that $M(r) \to 0$ at a finite value of r.

8.5.2 Matching results of inward and outward integrations

The divergences may be circumvented by comparing inward and outward solutions at some intermediate point in the model and then varying P_c, T_c, L, and R until, at the matching point, all physical variables determined by inward integrations correspond as closely as desired to the physical variables determined by outward integrations.

A simple algorithm for approaching an acceptable solution follows if one assumes that the discrepancies at the fitting point are linear functions of P_c, T_c, L, and R. The assumption of linearity is not in general true, but the direction in which the central and surface variables should be altered (increased or decreased) to reduce fitting discrepancies are normally correctly given and, as discrepancies are reduced, the assumption of linearity comes ever closer to being fulfilled. The procedure begins with a choice of fitting point, say $M(r) = M/2$, and five sets of integrations. The first set produces the discrepancies

$$P_1 - P_2 = Q_P(P_c, T_c, L, R), \tag{8.5.7}$$

$$T_1 - T_2 = Q_T(P_c, T_c, L, R), \tag{8.5.8}$$

$$L_1 - L_2 = Q_L(P_c, T_c, L, R), \tag{8.5.9}$$

and

$$R_1 - R_2 = Q_R(P_c, T_c, L, R). \tag{8.5.10}$$

Here the quantities labeled with the subscript 1 are, respectively, the pressure, temperature, luminosity, and radius at the fitting point given by the outward integration beginning with P_c and T_c. Quantities labeled with the subscript 2 are these same variables at the fitting point given by inward integrations beginning with L and R.

Next, each of the boundary variables is altered to find new discrepancies which may be used to define derivatives. For example,

$$\frac{Q_P(P_c + \Delta P_c, T_c, L, R) - Q_P(P_c, T_c, L, R)}{\Delta P_c} \equiv \frac{\partial Q_P}{\partial P_c}. \tag{8.5.11}$$

After all five integrations have been completed, and with all of the derivatives in hand, one may set up and solve the equations

$$Q_P + \frac{\partial Q_P}{\partial P_c} \delta P_c + \frac{\partial Q_P}{\partial T_c} \delta T_c + \frac{\partial Q_P}{\partial L} \delta L + \frac{\partial Q_P}{\partial R} \delta R = 0,$$

$$Q_T + \frac{\partial Q_T}{\partial P_c} \delta P_c + \frac{\partial Q_T}{\partial T_c} \delta T_c + \frac{\partial Q_T}{\partial L} \delta L + \frac{\partial Q_T}{\partial R} \delta R = 0,$$

$$Q_L + \frac{\partial Q_L}{\partial P_c} \delta P_c + \frac{\partial Q_L}{\partial T_c} \delta T_c + \frac{\partial Q_L}{\partial L} \delta L + \frac{\partial Q_L}{\partial R} \delta R = 0,$$

$$Q_R + \frac{\partial Q_R}{\partial P_c} \delta P_c + \frac{\partial Q_R}{\partial T_c} \delta T_c + \frac{\partial Q_R}{\partial L} \delta L + \frac{\partial Q_R}{\partial R} \delta R = 0 \tag{8.5.12}$$

for the increments δP_c, δT_c, δL, and δR. The increments may then be added to the first choices of central and surface parameters and the process of finding new appropriate derivatives and new increments may be begun afresh. The procedure is to be continued until the discrepancies at the fitting point become satisfactorily smaller than the average values of the associated physical variables there, with, e.g.,

$$\left| \frac{2 Q_P}{P_1 + P_2} \right| < 10^{-5}. \tag{8.5.13}$$

The algorithm just described is actually an oversimplification of the procedure which should be employed in practice. Since structure variables not only have different dimensions but also vary over widely different ranges, the Q_js should be defined in terms of logarithms of ratios of quantities at the fitting point, with, e.g.,

$$Q_P = \log \frac{P_1}{P_2}. \tag{8.5.14}$$

Derivatives of the Q_js should be with respect to changes in the logarithms of the four boundary variables, and, instead of solving eqs. (8.5.12) for increments in δP_c, etc., one should solve

$$
\begin{pmatrix}
\frac{\partial Q_P}{\partial \log P_c} & \frac{\partial Q_P}{\partial \log T_c} & \frac{\partial Q_P}{\partial \log L} & \frac{\partial Q_P}{\partial \log R} \\[2ex]
\frac{\partial Q_T}{\partial \log P_c} & \frac{\partial Q_T}{\partial \log T_c} & \frac{\partial Q_T}{\partial \log L} & \frac{\partial Q_T}{\partial \log R} \\[2ex]
\frac{\partial Q_L}{\partial \log P_c} & \frac{\partial Q_L}{\partial \log T_c} & \frac{\partial Q_L}{\partial \log L} & \frac{\partial Q_L}{\partial \log R} \\[2ex]
\frac{\partial Q_R}{\partial \log P_c} & \frac{\partial Q_R}{\partial \log T_c} & \frac{\partial Q_R}{\partial \log L} & \frac{\partial Q_R}{\partial \log R}
\end{pmatrix}
\begin{pmatrix}
\frac{\delta P_c}{P_c} \\[2ex]
\frac{\delta T_c}{T_c} \\[2ex]
\frac{\delta L}{L} \\[2ex]
\frac{\delta R}{R}
\end{pmatrix}
= -
\begin{pmatrix}
Q_P \\[2ex]
Q_T \\[2ex]
Q_L \\[2ex]
Q_R
\end{pmatrix}
\tag{8.5.15}
$$

for the logarithmic increments $\delta P_c/P_c$, $\delta T_c/T_c$, $\delta L/L$, and $\delta R/R$. The solution of these equations is given by Cramer's rule, with, e.g.,

$$
\frac{\delta P_c}{P_c} =
\begin{vmatrix}
-Q_P & \frac{\partial Q_P}{\partial \log T_c} & \frac{\partial Q_P}{\partial \log L} & \frac{\partial Q_P}{\partial \log R} \\[2ex]
-Q_T & \frac{\partial Q_T}{\partial \log T_c} & \frac{\partial Q_T}{\partial \log L} & \frac{\partial Q_T}{\partial \log R} \\[2ex]
-Q_L & \frac{\partial Q_L}{\partial \log T_c} & \frac{\partial Q_L}{\partial \log L} & \frac{\partial Q_L}{\partial \log R} \\[2ex]
-Q_R & \frac{\partial Q_R}{\partial \log T_c} & \frac{\partial Q_R}{\partial \log L} & \frac{\partial Q_R}{\partial \log R}
\end{vmatrix}
\div
\begin{vmatrix}
\frac{\partial Q_P}{\partial \log P_c} & \frac{\partial Q_P}{\partial \log T_c} & \frac{\partial Q_P}{\partial \log L} & \frac{\partial Q_P}{\partial \log R} \\[2ex]
\frac{\partial Q_T}{\partial \log P_c} & \frac{\partial Q_T}{\partial \log T_c} & \frac{\partial Q_T}{\partial \log L} & \frac{\partial Q_T}{\partial \log R} \\[2ex]
\frac{\partial Q_L}{\partial \log P_c} & \frac{\partial Q_L}{\partial \log T_c} & \frac{\partial Q_L}{\partial \log L} & \frac{\partial Q_L}{\partial \log R} \\[2ex]
\frac{\partial Q_R}{\partial \log P_c} & \frac{\partial Q_R}{\partial \log T_c} & \frac{\partial Q_R}{\partial \log L} & \frac{\partial Q_R}{\partial \log R}
\end{vmatrix}.
\tag{8.5.16}
$$

Vertical bars indicate that determinants of the matrices are to be taken, and it is not normally necessary to invoke pivoting (see Section 8.8) to evaluate the determinants.

Starting at a point near the center where eqs. (8.5.3)–(8.5.6) are acceptable approximations, numerical integrations may be continued outward with a fourth order Runge–Kutta algorithm. Beginning at the optical depth $\tau = 2/3$, integrations may be continued inward with the same algorithms.

8.6 On the construction of integration algorithms

As an introduction to the concepts involved in constructing a fourth order Runge–Kutta algorithm, consider solutions to the simple first order differential equation

$$
\frac{dy}{dx} = f(x).
\tag{8.6.1}
$$

Assuming that $f(x)$ is analytic, one can write

$$
\left(\frac{dy}{dx} \right)_{x=x_0+h} = f(x_0 + h) = f(x_0) + \left(\frac{df}{dx} \right)_0 h
$$

$$
+ \frac{1}{2!} \left(\frac{d^2 f}{dx^2} \right)_0 h^2 + \frac{1}{3!} \left(\frac{d^3 f}{dx^3} \right)_0 h^3 + \cdots ,
\tag{8.6.2}
$$

where the subscript 0 implies evaluation at $x = x_0$. Integration gives

$$y(x_0 + h) = y(x_0) + f(x_0)\, h + \left(\frac{df}{dx}\right)_0 \frac{h^2}{2} + \frac{1}{2!}\left(\frac{d^2 f}{dx^2}\right)_0 \frac{h^3}{3}$$

$$+ \frac{1}{3!}\left(\frac{d^3 f}{dx^3}\right)_0 \frac{h^4}{4} + \frac{1}{4!}\left(\frac{d^4 f}{dx^4}\right)_0 \frac{h^5}{5} + \cdots . \tag{8.6.3}$$

Noting that

$$\left(\frac{dy}{dx}\right)_{x=x_0+h/2} = f(x_0) + \left(\frac{df}{dx}\right)_0 \left(\frac{h}{2}\right) + \frac{1}{2!}\left(\frac{d^2 f}{dx^2}\right)_0 \left(\frac{h}{2}\right)^2$$

$$+ \frac{1}{3!}\left(\frac{d^3 f}{dx^3}\right)_0 \left(\frac{h}{2}\right)^3 + \cdots , \tag{8.6.4}$$

it follows that

$$y(x_0 + h) = y(x_0) + h \left(\frac{dy}{dx}\right)_{x=x_0+h/2} + O(h^3), \tag{8.6.5}$$

where

$$O(h^3) = \left(\frac{1}{3!} - \frac{1}{2!2^2}\right)\left(\frac{d^2 f}{dx^2}\right)_0 h^3 + \cdots = \frac{1}{24}\left(\frac{d^2 f}{dx^2}\right)_0 h^3 + \cdots . \tag{8.6.6}$$

The approximation

$$y^{\text{approx}}(x_0 + h) = y(x_0) + h \left(\frac{dy}{dx}\right)_{x=x_0+h/2} \tag{8.6.7}$$

is said to be good to the second order in step size, meaning that, if the series given by eq. (8.6.2) converges, the error in the approximation may be of the order of the first term in $O(h^3)$. In practice, the accuracy of the solution is best ascertained by experimenting with different step sizes chosen in such a way that h is small compared with the size of the region over which the dependent variable y changes appreciably. When y and its derivative are finite, one can define a length L by

$$\left|\frac{1}{y}\frac{dy}{dx}\right| L = 1 \tag{8.6.8}$$

and insist that $h \ll L$. One may set $h = \epsilon L$, choose an arbitrary initial value for ϵ, and either decrease ϵ until a further decrease does not result in a change in the end result or increase ϵ until the solution begins to change, settling on an optimum value of ϵ with regard to accuracy and execution time.

An approximation better than that given by eq. (8.6.7) may be found if one assumes that

$$y(x_0 + h) = y(x_0) + h \left[a \left(\frac{dy}{dx} \right)_{x=x_0} + b \left(\frac{dy}{dx} \right)_{x=x_0+h/2} + c \left(\frac{dy}{dx} \right)_{x=x_0+h} \right] + \cdots$$

(8.6.9)

$$= y(x_0) + a \, hf(x_0)$$

$$+ b \, h \left[f(x_0) + \left(\frac{df}{dx} \right)_0 \frac{h}{2} + \frac{1}{2!} \left(\frac{d^2 f}{dx^2} \right)_0 \left(\frac{h}{2} \right)^2 \right.$$

$$\left. + \frac{1}{3!} \left(\frac{d^3 f}{dx^3} \right)_0 \left(\frac{h}{2} \right)^3 + \frac{1}{4!} \left(\frac{d^4 f}{dx^4} \right)_0 \left(\frac{h}{2} \right)^4 \right]$$

$$+ c \, h \left[f(x_0) + \left(\frac{df}{dx} \right)_0 h + \frac{1}{2!} \left(\frac{d^2 f}{dx^2} \right)_0 h^2 \right.$$

$$\left. + \frac{1}{3!} \left(\frac{d^3 f}{dx^3} \right)_0 h^3 + \frac{1}{4!} \left(\frac{d^4 f}{dx^4} \right)_0 h^4 \right] + \cdots .$$

(8.6.10)

Comparing the coefficient of a given power of h in eq. (8.6.10) with the coefficient of the same power of h in the exact solution, eq. (8.6.3), one finds that

$$a + b + c = 1,$$
$$\tfrac{1}{2} b + c = \tfrac{1}{2}, \text{ and}$$
$$\tfrac{1}{4} b + c = \tfrac{1}{3},$$

(8.6.11)

which means that

$$b = \tfrac{2}{3}, \text{ and}$$
$$c = a = \tfrac{1}{6}.$$

(8.6.12)

Thus, the approximate solution, eq. (8.6.9), becomes

$$y^{\text{approx}}(x_0 + h) = y(x_0) + h \left[\frac{1}{6} \left(\frac{dy}{dx} \right)_{x=x_0} + \frac{2}{3} \left(\frac{dy}{dx} \right)_{x=x_0+h/2} + \frac{1}{6} \left(\frac{dy}{dx} \right)_{x=x_0+h} \right].$$

(8.6.13)

Comparing with the exact solution, eq. (8.6.3), one has that

$$y^{\text{approx}}(x_0 + h) = y^{\text{exact(6)}}(x_0 + h) + \left(\frac{5}{24} - \frac{1}{5} \right) \frac{1}{4!} \left(\frac{d^4 f}{dx^4} \right)_0 h^5 + \cdots , \qquad (8.6.14)$$

where $y^{\text{exact(6)}}(x_0 + h)$ is given by the first six terms in eq. (8.6.3). The second term on the right hand side of eq. (8.6.14) is 24 times smaller than the term proportional to h^5 in the Taylor expansion of the exact solution. Thus, a solution good to the fourth (and almost to the fifth) order in step size has been achieved very simply. Equation (8.6.13) is known as Simpson's rule.

Next, consider the differential equation

$$\frac{dy}{dx} = f(x, y), \qquad (8.6.15)$$

which, with the derivative of the dependent variable being an explicit function of the dependent variable as well as of the independent variable, is a simple prototype of any one of the stellar structure equations. If an expansion of the sort

$$f(x, y) = f(x_0 + h, y_0 + k) = f(x_0, y_0) + \left(\frac{\partial f}{\partial x} \right)_0 h + \left(\frac{\partial f}{\partial y} \right)_0 k$$

$$+ \frac{1}{2} \left[\left(\frac{\partial^2 f}{\partial x^2} \right)_0 h^2 + 2 \left(\frac{\partial^2 f}{\partial x \partial y} \right)_0 hk + \left(\frac{\partial^2 f}{\partial y^2} \right)_0 k^2 \right] + \cdots \qquad (8.6.16)$$

is possible, it must also be true that

$$f(x, y) = f(x_0 + h, y_0 + k) = f(x_0, y_0) + \left(\frac{df}{dx} \right)_0 h + \frac{1}{2} \left(\frac{d^2 f}{dx^2} \right)_0 h^2 + \cdots . \qquad (8.6.17)$$

Here, the total first derivative of f with respect to x is related to partial derivatives with respect to x and y by

$$\frac{df}{dx} = \frac{d^2 y}{dx^2} = \frac{\partial f}{\partial x} + \frac{\partial f}{\partial y} \frac{dy}{dx} = \frac{\partial f}{\partial x} + \frac{\partial f}{\partial y} f \qquad (8.6.18)$$

and the second total derivative with respect to x is a special case of the general relationship

$$\frac{dg}{dx} = \frac{\partial g}{\partial x} + \frac{\partial g}{\partial y} \frac{dy}{dx} = \frac{\partial g}{\partial x} + \frac{\partial g}{\partial y} f, \qquad (8.6.19)$$

where $g(x, y)$ is any function of $f(x, y)$ or of its derivatives. Using eq. (8.6.18) in eq. (8.6.17), one has that

$$f(x, y) = f(x_0, y_0) + \left[\left(\frac{\partial f}{\partial x} \right)_0 + \left(\frac{\partial f}{\partial y} \right)_0 f(x_0, y_0) \right] h + \cdots . \qquad (8.6.20)$$

Comparing with eq. (8.6.16), it follows that

$$k = f(x_0, y_0) h. \qquad (8.6.21)$$

Using eq. (8.6.17) in eq. (8.6.15) and integrating yields the exact solution

$$y(x_0 + h, y_0 + k) = y_0 + f(x_0, y_0) h + \left(\frac{df}{dx}\right)_0 \frac{h^2}{2} + \frac{1}{2}\left(\frac{d^2 f}{dx^2}\right)_0 \frac{h^3}{3} + \cdots .$$

$$(8.6.22)$$

By analogy with eq. (8.6.7), one may adopt

$$y^{approx}(x_0 + h, y_0 + k) = y_0 + f(x_0 + h/2, y_0 + k/2) h \qquad (8.6.23)$$

as a first approximate solution. The first few terms in a Taylor expansion of eq. (8.6.23) are

$$y^{approx}(x_0 + h, y_0 + k) = y_0 + f(x_0, y_0) h + \left(\frac{\partial f}{\partial x}\right)_0 \frac{h^2}{2} + \left(\frac{\partial f}{\partial y}\right)_0 \frac{hk}{2} + \cdots .$$

$$(8.6.24)$$

Using in this equation first eq. (8.6.21) and then eq. (8.6.18), it follows that

$$y^{approx}(x_0 + h, y_0 + k) = y_0 + f(x_0, y_0) h + \left[\left(\frac{\partial f}{\partial x}\right)_0 + f_0 \left(\frac{\partial f}{\partial y}\right)_0\right] \frac{h^2}{2} + \cdots$$

$$= y_0 + f(x_0, y_0) h + \left(\frac{df}{dx}\right)_0 \frac{h^2}{2} + \cdots , \qquad (8.6.25)$$

which, up to and including the term proportional to h^2, is identical with the exact solution given by eq. (8.6.22). Thus, the approximation given by eq. (8.6.23) is good to at least the second order in step size.

The approximation is far from unique. An entire family of algorithms good to the second order in step size can be constructed. Suppose that

$$y^{approx} = y_0 + a k_1 + b k_2, \qquad (8.6.26)$$

where

$$k_1 = hf(x_0, y_0) = hf_0 = h \left(\frac{dy}{dx}\right)_0 \qquad (8.6.27)$$

and

$$k_2 = hf(x_0 + \alpha h, y_0 + \beta k_1). \qquad (8.6.28)$$

The coefficients a, b, α, and β are parameters to be determined by comparing the approximate solution with the exact solution. Expanding k_2 in a Taylor series, one has

$$\frac{1}{h}k_2 = f(x_0, y_0) + \left(\frac{\partial f}{\partial x}\right)_0 \alpha h + f_0 \left(\frac{\partial f}{\partial y}\right)_0 \beta k_1$$

$$+ \frac{1}{2!}\left[\left(\frac{\partial^2 f}{\partial x^2}\right)_0 (\alpha h)^2 + f_0 \left(\frac{\partial^2 f}{\partial x \partial y}\right)_0 \alpha\beta h k_1 + f_0^2 \left(\frac{\partial^2 f}{\partial y^2}\right)_0 \beta^2 k_1^2\right] + \cdots .$$

$$(8.6.29)$$

Inserting k_1 from eq. (8.6.27) into eq. (8.6.29) and multiplying by h, one obtains

$$
k_2 = h\, f_0 + h^2 \left[\alpha \left(\frac{\partial f}{\partial x} \right)_0 + \beta f_0 \left(\frac{\partial f}{\partial y} \right)_0 \right]
$$
$$
+ \frac{1}{2!} h^3 \left[\left(\frac{\partial^2 f}{\partial x^2} \right)_0 \alpha^2 + 2\, f_0 \left(\frac{\partial^2 f}{\partial x \partial y} \right)_0 \alpha\beta + f_0^2 \left(\frac{\partial^2 f}{\partial y^2} \right)_0 \beta^2 \right] + \cdots .
$$
(8.6.30)

Inserting k_1 given by eq. (8.6.27) and k_2 given by eq. (8.6.30) into eq. (8.6.26) produces

$$
y^{\text{approx}} = y_0 + (a+b)\, h\, f_0 + b\, h^2 \left[\alpha \left(\frac{\partial f}{\partial x} \right)_0 + \beta f_0 \left(\frac{\partial f}{\partial y} \right)_0 \right]
$$
$$
+ b \frac{h^3}{2} \left[\alpha^2 \left(\frac{\partial^2 f}{\partial x^2} \right)_0 + 2\alpha\beta f_0 \left(\frac{\partial^2 f}{\partial x \partial y} \right)_0 + \beta^2 f_0^2 \left(\frac{\partial^2 f}{\partial x^2} \right)_0 \right] + \cdots .
$$
(8.6.31)

The exact solution, eq. (8.6.22), may be rewritten as

$$
y^{\text{exact}} = y_0 + h\, f_0 + \frac{h^2}{2} \left[\left(\frac{\partial f}{\partial x} \right)_0 + f_0 \left(\frac{\partial f}{\partial y} \right)_0 \right] + \frac{h^3}{6} \left(\frac{d^3 y}{dx^3} \right)_0 + \cdots , \qquad (8.6.32)
$$

where eq. (8.6.18) has been used to recast the term proportional to h^2. Comparing terms proportional to the same power of h, it is evident that, in the general case, the approximate solution and the exact solution are identical out to the second order in h only if

$$
a + b = 1,
$$
$$
\alpha = \beta, \text{ and} \qquad (8.6.33)
$$
$$
b\alpha = \tfrac{1}{2}.
$$

The fact that there are only three relationships connecting the four parameters a, b, α, and β means that an additional criterion may be imposed by the user. This criterion is a matter of taste which could be based, for example, on considerations of symmetry, economy of execution, or choice of where derivatives are to be estimated over the step interval h. Choosing $\alpha = \beta = 1/2$, it follows that $b = 1$, $a = 0$, and

$$
y^{\text{approx}} = y_0 + k_2 = y_0 + h\, f(x_0 + h/2, y_0 + k_1/2), \qquad (8.6.34)
$$

which reproduces eq. (8.6.23). Choosing $\alpha = \beta = 1$, it follows that $a = b = 1/2$ and

$$
y^{\text{approx}} = y_0 + \frac{1}{2}(k_1 + k_2) = y_0 + \frac{h}{2}[f(x_0, y_0) + f(x_0 + h, y_0 + k_1)]. \qquad (8.6.35)
$$

Combining the two solutions in the form

$$y^{\text{approx}} = y_0 + \frac{h}{4} \left[f(x_0, y_0) + 2 f(x_0 + h/2, y_0 + k_1/2) + f(x_0 + h, y_0 + k_1) \right]$$

(8.6.36)

produces a result which incorporates information from three equally spaced points covering the interval h.

Only in special circumstances can the approximate solution be made to match the exact solution to third order in step size. To prepare for a comparison of third order terms, combine eqs. (8.6.18) and (8.6.19) to obtain

$$\frac{d^3 y}{dx^3} = \frac{\partial^2 f}{\partial x^2} + 2 f \frac{\partial^2 f}{\partial x \partial y} + f^2 \frac{\partial^2 f}{\partial y^2} + \frac{\partial f}{\partial y} \frac{d^2 y}{dx^2}.$$

(8.6.37)

In the term proportional to h^3 in eq. (8.6.31), set $\alpha = \beta$ and use eq. (8.6.37) to replace the three second order partial derivatives. Equating the terms proportional to h^3 in eqs. (8.6.31) and (8.6.32) yields

$$\frac{1}{2} b \alpha^2 \left[\left(\frac{d^3 y}{dx^3} \right)_0 - \left(\frac{d^2 y}{dx^2} \right)_0 \left(\frac{\partial f}{\partial y} \right)_0 \right] = \frac{1}{6} \left(\frac{d^3 y}{dx^3} \right)_0.$$

(8.6.38)

It is evident that there is no generally applicable solution unless

$$\left(\frac{\partial f}{\partial y} \right)_0 \left(\frac{d^2 y}{dx^2} \right)_0 \ll \left(\frac{d^3 y}{dx^3} \right)_0.$$

(8.6.39)

In this special circumstance, $b \alpha^2 = 1/3$, so that $\alpha = 2/3$, $b = 3/4$, $a = 1/4$, and

$$y^{\text{approx}} = y_0 + \frac{1}{4} (k_1 + 3 k_2) = y_0 + \frac{h}{4} \left[f(x_0, y_0) + 3 f(x_0 + 2h/3, y_0 + 2k_1/3) \right],$$

a solution much less symmetric than that given by eq. (8.6.36).

An approximation good in all circumstances to the third order in h can be found by adding another term to eq. (8.6.26). For example, one can adopt

$$y^{\text{approx}} = y_0 + a k_1 + b k_2 + c k_3,$$

(8.6.40)

where

$$k_2 = hf(x_0 + \alpha h, y_0 + \alpha k_1)$$

(8.6.41)

and

$$k_3 = hf(x_0 + \alpha_1 h, y_0 + \beta_1 k_1 + \gamma_1 k_2).$$

(8.6.42)

Relationships between a, b, c, α, α_1, β_1, and γ_1 follow from comparisons between coefficients of equal powers of h in the Taylor expansions of the exact and approximate solutions. The parameter k_1 is given by eq. (8.6.37) and the first four terms in the Taylor expansion of k_2 are

$$
k_2 = h\, f_0 + h^2\, \alpha \left[\left(\frac{\partial f}{\partial x} \right)_0 + f_0 \left(\frac{\partial f}{\partial y} \right)_0 \right]
$$

$$
+ \frac{h^3}{2} \alpha^2 \left[\left(\frac{\partial^2 f}{\partial x^2} \right)_0 + 2 f_0 \left(\frac{\partial^2 f}{\partial x \partial y} \right)_0 + f_0^2 \left(\frac{\partial^2 f}{\partial x^2} \right)_0 \right]
$$

$$
+ \frac{\alpha^3}{3!} h^4 \left[\left(\frac{\partial^3 f}{\partial x^3} \right)_0 + 3 f_0 \left(\frac{\partial^3 f}{\partial x^2 \partial y} \right)_0 + 3 f_0^2 \left(\frac{\partial^3 f}{\partial x \partial y^2} \right)_0 + f_0^3 \left(\frac{\partial^3 f}{\partial y^3} \right)_0 \right] + \cdots .
$$

$$(8.6.43)$$

The first three terms in eq. (8.6.43) have been obtained by setting $\beta = \alpha$ in eq. (8.6.30) and the fourth term has been obtained by using a procedure which makes use of the fact that, for any analytic function $f(x, y)$,

$$
f(x + p, y + q) = f(x, y) + \sum_{n=1}^{\infty} \frac{1}{n!} D^n f,
\tag{8.6.44}
$$

where p and q are constants,

$$
D^1 f = p \frac{\partial f}{\partial x} + q \frac{\partial f}{\partial y},
\tag{8.6.45}
$$

$$
D^2 f = \left(p \frac{\partial}{\partial x} + q \frac{\partial}{\partial y} \right) \left(p \frac{\partial f}{\partial x} + q \frac{\partial f}{\partial y} \right) = p^2 \frac{\partial^2 f}{\partial x^2} + 2 p q \frac{\partial^2 f}{\partial x \partial y} + q^2 \frac{\partial^2 f}{\partial y^2},
\tag{8.6.46}
$$

$$
D^3 f = p^3 \frac{\partial^3 f}{\partial x^3} + 3 p^2 q \frac{\partial^3 f}{\partial x^2 \partial y} + 3 p q^2 \frac{\partial^3 f}{\partial x \partial y^2} + q^3 \frac{\partial^3 f}{\partial y^3},
\tag{8.6.47}
$$

and so forth. Setting $p = \alpha\, h$ and $q = \alpha\, k_1 = \alpha\, h\, f_0$ into eq. (8.6.62) and multiplying by $h/3!$ produces the fourth term on the right hand side of eq. (8.6.43).

To construct the Taylor series for k_3, set $p = \alpha_1 h$, and

$$
q = \beta_1\, k_1 + \gamma_1\, k_2
$$

$$
= (\beta_1 + \gamma_1)\, h f_0 + \gamma_1\, \alpha h^2 \left[\left(\frac{\partial f}{\partial x} \right)_0 + f_0 \left(\frac{\partial f}{\partial y} \right)_0 \right]
$$

$$
+ \gamma_1 \frac{\alpha^2}{2} h^3 \left[\left(\frac{\partial^2 f}{\partial x^2} \right)_0 + 2 f_0 \left(\frac{\partial^2 f}{\partial x \partial y} \right)_0 + f_0^2 \left(\frac{\partial^2 f}{\partial x^2} \right)_0 \right] + \cdots .
\tag{8.6.48}
$$

Applying eqs. (8.6.45)–(8.6.47) and multiplying by h yields

$$
D^1 k_3 = \left[\alpha_1 \left(\frac{\partial f}{\partial x} \right)_0 + (\beta_1 + \gamma_1) f_0 \left(\frac{\partial f}{\partial y} \right)_0 \right] h^2 + \gamma_1 \alpha \left(\frac{\partial f}{\partial y} \right)_0 \left[\left(\frac{\partial f}{\partial x} \right)_0 + f_0 \left(\frac{\partial f}{\partial y} \right)_0 \right] h^3
$$

$$
+ \gamma_1 \frac{\alpha^2}{2} \left(\frac{\partial f}{\partial y} \right)_0 \left[\left(\frac{\partial^2 f}{\partial x^2} \right)_0 + 2 f_0 \left(\frac{\partial^2 f}{\partial x \partial y} \right)_0 + f_0^2 \left(\frac{\partial^2 f}{\partial y^2} \right)_0 \right] h^4 + \cdots ,
$$

$$(8.6.49)$$

$$D^2 k_3 = \left[\alpha_1^2 \left(\frac{\partial^2 f}{\partial x^2} \right)_0 + 2\alpha_1 (\beta_1 + \gamma_1) \left(\frac{\partial^2 f}{\partial x \partial y} \right)_0 f_0 + (\beta_1 + \gamma_1)^2 \left(\frac{\partial^2 f}{\partial y^2} \right)_0 f_0^2 \right] h^3$$

$$+ \left[2\alpha_1 \left(\frac{\partial^2 f}{\partial x \partial y} \right)_0 + (\beta_1 + \gamma_1) f_0 \left(\frac{\partial^2 f}{\partial y^2} \right)_0 \right]$$

$$\times \gamma_1 \alpha \left[\left(\frac{\partial f}{\partial x} \right)_0 + f_0 \left(\frac{\partial f}{\partial y} \right)_0 \right] h^4 + \cdots , \tag{8.6.50}$$

and

$$D^3 k_3 = \alpha_1^3 \left(\frac{\partial^3 f}{\partial x^3} \right)_0 h^4 + \cdots . \tag{8.6.51}$$

Altogether,

$$k_3 = h f_0 + D^1 k_3 + \frac{1}{2!} D^2 k_3 + \frac{1}{3!} D^3 k_3 + \cdots . \tag{8.6.52}$$

Inserting eqs. (8.6.52), (8.6.43), and (8.6.27) into eq. (8.6.40) and equating the coefficient of h in the resulting equation with the coefficient of h in the exact solution, eq. (8.6.22), gives

$$a + b + c = 1. \tag{8.6.53}$$

Equating coefficients of h^2 in the approximate and exact solutions gives

$$b\alpha \left[\left(\frac{\partial f}{\partial x} \right)_0 + f_0 \left(\frac{\partial f}{\partial y} \right)_0 \right] + c \left[\alpha_1 \left(\frac{\partial f}{\partial x} \right)_0 + (\beta_1 + \gamma_1) f_0 \left(\frac{\partial f}{\partial y} \right)_0 \right]$$

$$= \frac{1}{2} \left[\left(\frac{\partial f}{\partial x} \right)_0 + f_0 \left(\frac{\partial f}{\partial y} \right)_0 \right], \tag{8.6.54}$$

which is satisfied in general only if

$$\alpha_1 = \beta_1 + \gamma_1, \quad \text{and}$$

$$b\,\alpha + c\,\alpha_1 = \frac{1}{2}. \tag{8.6.55}$$

Equating coefficients of h^3 gives

$$b \frac{\alpha^2}{2} \left[\left(\frac{\partial^2 f}{\partial x^2} \right)_0 + 2 f_0 \left(\frac{\partial^2 f}{\partial x \partial y} \right)_0 + f_0^2 \left(\frac{\partial^2 f}{\partial y^2} \right)_0 \right]$$

$$+ c\, \gamma_1 \alpha \left(\frac{\partial f}{\partial y} \right)_0 \left[\left(\frac{\partial f}{\partial x} \right)_0 + f_0 \left(\frac{\partial f}{\partial y} \right)_0 \right]$$

$$+ c \frac{1}{2} \left[\alpha_1^2 \left(\frac{\partial^2 f}{\partial x^2} \right)_0 + 2\alpha_1(\beta_1 + \gamma_1) f_0 \left(\frac{\partial^2 f}{\partial x \partial y} \right)_0 \right.$$

$$\left. + (\beta_1 + \gamma_1)^2 f_0^2 \left(\frac{\partial^2 f}{\partial y^2} \right)_0 \right] = \frac{1}{6} \left(\frac{d^3 y}{dx^3} \right)_0, \tag{8.6.56}$$

which becomes, on using eqs. (8.6.55) and (8.6.36),

$$b \frac{\alpha^2}{2} \left[\left(\frac{d^3 y}{dx^3} \right) - \left(\frac{\partial f}{\partial y} \right)_0 \left(\frac{d^2 y}{dx^2} \right)_0 \right] + c \gamma_1 \alpha \left(\frac{\partial f}{\partial y} \right)_0 \left(\frac{d^2 y}{dx^2} \right)_0$$

$$+ c \frac{\alpha_1^2}{2} \left[\left(\frac{d^3 y}{dx^3} \right)_0 - \left(\frac{\partial f}{\partial y} \right)_0 \left(\frac{d^2 y}{dx^2} \right)_0 \right] = \frac{1}{6} \left(\frac{d^3 y}{dx^3} \right)_0. \tag{8.6.57}$$

For eq. (8.6.57) to be satisfied in all cases, it is necessary that

$$b \, \alpha^2 + c \, \alpha_1^2 = \frac{1}{3} \text{ and}$$

$$c \, \gamma_1 \, \alpha = \frac{1}{6}. \tag{8.6.58}$$

Equations (8.6.53), (8.6.55), and (8.6.58) provide five relationships between the seven parameters a, b, c, α, α_1, β_1, and γ_1. The first relationship in eq. (8.6.55) may be used to rewrite k_3 as

$$k_3 = hf(x_0 + \alpha_1 h, \ y_0 + [\alpha_1 - \gamma_1] k_1 + \gamma_1 k_2), \tag{8.6.59}$$

which makes it clear that the sum of the coefficients of the ks in the y argument of $f(x, y)$ is equal to the coefficient of h in the x argument.

The fact that the number of parameters exceeds the number of constraints on the parameters means that, once again, user taste enters the picture. An entire class of solutions follows from the assumption that

$$a = c = \frac{1 - b}{2}. \tag{8.6.60}$$

By inserting into the first relationship in eq. (8.6.58) the quantity α_1 given by the second relationship in eq. (8.6.55) and then using eq. (8.6.60) to replace c in terms of b, one obtains

$$b \, \alpha^2 + \frac{2}{1 - b} \left(\frac{1}{2} - b\alpha \right)^2 = \frac{1}{3}, \text{ or}$$

$$\alpha^2 - \frac{2}{1 + b} \alpha + \frac{1}{6} \frac{1 + 2b}{b(1 + b)} = 0. \tag{8.6.61}$$

The solution of this quadratic equation can be written as

$$\alpha = \frac{1}{1 + b} \left(1 \pm \sqrt{\frac{1}{3b}(1 - b)\left(b - \frac{1}{2} \right)} \right), \tag{8.6.62}$$

which shows that, if a real solution is desired, b is constrained to be in the range

$$\frac{1}{2} \le b \le 1.$$
(8.6.63)

Using Simpson's rule (eq. (8.6.13)) as a guide, one might choose

$$c = a = \frac{1}{6}, \text{ and}$$

$$b = \frac{2}{3}.$$
(8.6.64)

Then, from eq. (8.6.62) one has that

$$\alpha = \frac{6}{10} \pm \frac{1}{10}.$$
(8.6.65)

Choosing $\alpha = 7/10$ produces an unattractive result that lacks symmetry and does not resemble Simpson's rule. Choosing $\alpha = 1/2$, it follows from eqs. (8.6.55) and (8.6.58) that $\alpha_1 = 1$, $\gamma_1 = 2$, and $\beta_1 = -1$. Thus,

$$k_2 = hf(x_0 + h/2, y_0 + k_1/2),$$

$$k_3 = hf(x_0 + h, y_0 - k_1 + 2k_2),$$
(8.6.66)

and

$$y^{\text{approx}} = y_0 + \frac{h}{6} \left[f_0 + 4 f(x_0 + h/2, y_0 + k_1/2) + f(x_0 + h, y_0 - k_1 + 2k_2) \right].$$
(8.6.67)

To the extent that $k_2 \sim k_1$, eq. (8.6.67) resembles Simpson's rule.

In another example, choose $b = 1/2$, so that $a = c = 1/4$. Then, $\alpha = \alpha_1 = 2/3$, $\gamma_1 = 1$, $\beta_1 = \alpha_1 - \gamma_1 = -1/3$,

$$k_2 = hf(x_0 + 2h/3, y_0 + 2k_1/3),$$

$$k_3 = hf(x_0 + 2h/3, y_0 - k_1/3 + k_2),$$
(8.6.68)

and

$$y^{\text{approx}} = y_0 + \frac{h}{4} \left(f_0 + 2 f(x_0 + 2h/3, y_0 + 2k_1/3) + f(x_0 + 2h/3, y_0 - k_1/3 + k_2) \right).$$
(8.6.69)

Given that it makes use of a derivative at an asymmetric position in the interval h, this solution is not particularly attractive.

Consider two additional examples: (1) if $b = 6a = 6c = 3/4$, then

$$\alpha = \frac{4}{7} \left(1 \pm \frac{1}{4}\sqrt{\frac{7}{12}} \right), \alpha_1 = \frac{4}{7} \left(1 \mp \frac{3}{2}\sqrt{\frac{7}{12}} \right), \text{ and } \gamma_1 = \frac{7}{3} \div \left(1 \pm \frac{1}{4}\sqrt{\frac{7}{12}} \right);$$
(8.6.70)

(2) if $b = 3a = 3c = 3/5$, then

$$\alpha = \frac{5}{8}\left(1 \pm \frac{1}{3}\sqrt{\frac{1}{5}}\right), \alpha_1 = \frac{5}{8}\left(1 \mp \sqrt{\frac{1}{5}}\right), \text{ and } \gamma_1 = \frac{4}{3} \div \left(1 \pm \frac{1}{3}\sqrt{\frac{1}{5}}\right). \quad (8.6.71)$$

In both cases, the expressions for y^{approx} are quite untidy.

Another approach to finding solutions begins with a choice of relationships between α, α_1, and γ_1. For example, setting $\alpha_1 = \alpha$ and $\beta_1 = \alpha_1 - \gamma_1 = 0$ leads to $\gamma_1 = \alpha_1 = \alpha = 2/3$, $c = b = 3/8$, $a = 2/8$, and

$$y^{\text{approx}} = y_0 + \frac{h}{8}\left(2f_0 + 3f(x_0 + 2h/3, y_0 + 2k_1/3) + 3f(x_0 + 2h/3, y_0 + 2k_2/3)\right). \quad (8.6.72)$$

One could continue to explore additional options indefinitely, but there do not appear to be any solutions obviously more attractive than given by eqs. (8.6.67) and (8.6.66). A comparison of the term proportional to h^4 in an approximate third order solution with that in the exact solution reveals the same difficulty encountered in the case of second order approximations: to wit, in the general case, an evaluation of the error in the fourth order term is not possible. Thus, if a solution good to the fourth order in step size is desired, the best approach is to add to eq. (8.6.40) yet another term, $d\, k_4$, where

$$k_4 = hf(x_0 + \alpha_2\, h, y_0 + \beta_2\, k_1 + \gamma_2\, k_2 + \delta_2\, k_3), \quad (8.6.73)$$

and to find, following the procedures used to obtain third order solutions, relationships involving the new parameters d, α_2, β_2, and δ_2, in addition to the parameters in the third order solutions.

The earliest investigation into an algorithm good to the fourth order in step size is that of the spectroscopist Carl David Runge (1895) and the first systematic study of broad classes of fourth order solutions is that of the aerodynamicist M. Wilhelm Kutta (1901). Informative discussions are given by E. L. Ince (1926) and by F. Ceschino and J. Kuntzmann (1963; translation by D. Boyanovitch 1966).

The Runge algorithm is

$$y^{\text{approx}}(x_0 + h, y_0 + k) = y_0 + \frac{1}{6}(k_1 + 2k_2 + 2k_3 + k_4), \quad (8.6.74)$$

where

$$k_1 = h\, f(x_0, y_0),$$

$$k_2 = h\, f(x_0 + h/2, y_0 + k_1/2),$$

$$k_3 = h\, f(x_0 + h/2, y_0 + k_2/2), \text{ and}$$

$$k_4 = h\, f(x_0 + h, y_0 + k_3), \quad (8.6.75)$$

and the most frequently cited Kutta algorithm is

$$y^{\text{approx}}(x_0 + h, y_0 + k) = y_0 + \frac{1}{8}(k_1 + 3k_2 + 3k_3 + k_4), \quad (8.6.76)$$

where

$$k_1 = h\, f(x_0, y_0),$$

$$k_2 = h\, f(x_0 + h/3, y_0 + k_1/3),$$

$$k_3 = h\, f(x_0 + 2h/3, y_0 - k_1/3 + k_2), \text{ and}$$

$$k_4 = h\, f(x_0 + h, y_0 + k_1 - k_2 + k_3). \tag{8.6.77}$$

Note that in both algorithms there is a high degree of symmetry ($a = d$ and $b = c$) and that derivatives are sampled at evenly spaced intervals over the interval h.

Apart from the fact that, in both cases, $k_1 = h\, f(x_0, y_0)$, the two algorithms have in common built-in relationships between coefficients of the k_is in the y_i arguments of $f(x_i, y_i)$. In both cases, one can write

$$k_2 = h\, f(x_0 + m\, h, y_0 + m\, k_1),$$

$$k_3 = h\, f(x_0 + nh, y_0 + [\, n - r\,]\, k_1 + r\, k_2), \text{ and}$$

$$k_4 = h\, f(x_0 + h, y_0 + [\, 1 - s - t\,]\, k_1 + s\, k_2 + t\, k_3). \tag{8.6.78}$$

Thus, not only does the sum of the coefficients of the k_is in y^{approx} equal unity but, in each $k_i = f(x_i, y_i)$, the sum of the coefficients of the ks in y_i is equal to the coefficient of h in x_i. These features are true of all of the algorithms devised by Kutta. Capitalizing on this commonality and extending the investigation to minimize the number of storage registers required in a computer calculation and to minimize the error in the term in the Taylor expansion proportional to h^5, S. Gill (1951) obtains

$$y^{\text{approx}}(x_0 + h, y_0 + k) = y_0 + \frac{1}{6}(k_1 + k_4) + \frac{1}{3}\left[\left(1 - \frac{1}{\sqrt{2}}\right)k_2 + \left(1 + \frac{1}{\sqrt{2}}\right)k_3\right],$$
$$\tag{8.6.79}$$

where

$$k_1 = h\, f(x_0, y_0),$$

$$k_2 = h\, f(x_0 + h/2, y_0 + k_1/2),$$

$$k_3 = h\, f(x_0 + h/2, y_0 + (1/\sqrt{2} - 1/2)\, k_1 + (1 - 1/\sqrt{2})\, k_2), \text{ and}$$

$$k_4 = h\, f(x_0 + h, y_0 - k_2/\sqrt{2} + (1 + 1/\sqrt{2})\, k_3). \tag{8.6.80}$$

The results of the three fourth order algorithms described here are summarized in Table 8.6.1, along with two of Kutta's general results defined as functions of the parameters t (Case 5) and m (Case 1). Setting $t = 1$ in the Case 5 solution gives the Runge algorithm and setting $m = 1/3$ in the Case 1 solution gives what common usage calls the Kutta algorithm. Setting $t = 1 + 1/\sqrt{2}$ in the Case 5 solution gives the algorithm recommended by Gill.

Table 8.6.1 Parameters in several fourth order Runge–Kutta algorithms

quantity	m	n	r	s	t	$a = d$	b	c
Case 5	$\frac{1}{2}$	$\frac{1}{2}$	$\frac{1}{2t}$	$1-t$	t	$\frac{1}{6}$	$\frac{2-t}{3}$	$\frac{t}{3}$
Runge	$\frac{1}{2}$	$\frac{1}{2}$	$\frac{1}{2}$	0	1	$\frac{1}{6}$	$\frac{1}{3}$	$\frac{1}{3}$
Case 1	m	$l = m-1$	$\frac{l}{2m}$	$\frac{l(l+1)/2m}{6ml-1}$	$\frac{m}{6ml-1}$	$\frac{6ml-1}{12ml}$	$\frac{1}{12ml}$	$\frac{1}{12ml}$
Kutta	$\frac{1}{3}$	$\frac{2}{3}$	1	-1	1	$\frac{1}{8}$	$\frac{3}{8}$	$\frac{3}{8}$
Gill	$\frac{1}{2}$	$\frac{1}{2}$	$1-\frac{1}{\sqrt{2}}$	$-\frac{1}{\sqrt{2}}$	$1+\frac{1}{\sqrt{2}}$	$\frac{1}{6}$	$\frac{\sqrt{2}-1}{3\sqrt{2}}$	$\frac{\sqrt{2}+1}{3\sqrt{2}}$

There are eight relationships between coefficients in fourth order solutions. The four that bear the closest resemblance to relationships between coefficients in the third order solutions are

$$(a + b + c) + d = 1,$$

$$(b\, m + c\, n) + d = \frac{1}{2},$$

$$(b\, m^2 + c\, n^2) + d = \frac{1}{3}, \text{ and}$$

$$(c\, m\, r) + d\,[\,n\, t + m\, s\,] = \frac{1}{6}. \tag{8.6.81}$$

Noting that $m = \alpha$, $n = \alpha_1$, and $r = \gamma_1$, and setting $d = 0$ in these relationships, one recovers four of the five relationships between coefficients in third order solutions, as given by eqs. (8.6.53), (8.6.55), and (8.6.58). The relationship $\alpha_1 = \beta_1 + \gamma_1$ in eq. (8.6.55) is equivalent to the relationship $\beta_1 = \alpha_1 - \gamma_1 = n - r$ which appears as the coefficient of k_1 in the expression for k_3 in eq. (8.6.59). The final four relationships between coefficients in fourth order solutions are

$$(b\, m^3 + c\, n^3) + d = \frac{1}{4},$$

$$(c\, m\, n\, r) + d\,[\,n\, t + m\, s\,] = \frac{1}{8},$$

$$(c\, m^2\, r) + d\,[\,n^2\, t + m^2\, s\,] = \frac{1}{12}, \text{ and}$$

$$d\, m\, r\, t = \frac{1}{24}. \tag{8.6.82}$$

Taken together, eqs. (8.6.81) and (8.6.82) comprise eight relationships between nine coefficients. The universal choice for a ninth relationship is $a = d$. Another often imposed symmetry is the choice $b = c$.

In the modern literature, all fourth order algorithms tend to be referred to as Runge–Kutta algorithms, the distinction between algorithms being ignored. The Gill algorithm was designed to minimize the computer storage required for solving systems of simultaneous linear equations and to minimize the discrepancy between the approximate solution and the exact solution in the term proportional to h^5 in the Taylor expansions of the two solutions. E. K. Blum (1962) shows that, by a suitable arrangement of steps taken in the computation, the storage requirements for the Runge algorithm can be made as small as the storage requirements for the Gill algorithm, but does not address the question of accuracy to the fifth order in h.

It is obvious that, if integrations are performed with step size small enough, all methods must give the same end results. To achieve the desired accuracy in any given application, the user is free to repeat the calculation with several choices of the criteria for determining the interval h (see the discussion centered on eq. (8.6.8)).

8.6.1 Application to stellar structure

To apply a Runge–Kutta algorithm to obtain solutions of the equations of stellar structure, the equations may be written as

$$\frac{dy_i}{dx} = f_i(x, y_1, y_2, y_3, y_4), \tag{8.6.83}$$

where, when $x = r$ has been chosen as the independent variable, $i = 1, 4$ denotes one of the four equations, eqs. (8.3.1), (8.3.2), (8.3.6), and (8.3.7), with $y_1 = M$, $y_2 = P$, $y_3 = T$, and $y_4 = L$ being the dependent variables. For each step in r of size h, the quantities k_j, $j = 1, 4$ may be found for each of the four dependent variables and the values of the variables at the end of the step may be obtained by applying one of the Runge–Kutta algorithms. The choice of independent variable is, of course, a matter of taste, and, for some, from the point of view of preparing a model for use in an evolution program which uses the relaxation technique, mass may be preferable to radius as the independent variable. Then, $i = 1, 4$ denotes one of the four equations, eqs. (8.3.9)–(8.3.12), with $y_1 = r$, $y_2 = P$, $y_3 = L$, and $y_4 = T$ being the dependent variables.

In practice, it is convenient to set $y_0 = x$ and $f_0 = 1$, and to solve the equations

$$\frac{dy_i}{dx} = f_i(y_0, y_1, y_2, y_3, y_4), \tag{8.6.84}$$

with $i = 0, 4$. Step size may be adjusted by requiring that, at the beginning of each step,

$$\frac{dy_i}{dx} h \leq \text{dymax}, \tag{8.6.85}$$

where dymax can be adjusted with experience. Typically, dymax ~ 0.05–0.10 is adequate.

8.7 The relaxation technique for model construction

Having constructed a model in which the most abundant composition variables are independent of position (e.g., a zero age main sequence model), one could in principle continue to use the classical fitting technique to follow the evolution of the model due to changes in interior composition caused by nuclear transformations. However, a much more powerful technique exists. It begins by replacing the differential equations of stellar structure with difference equations relating physical quantities defined at a finite set of mesh points and/or defined at positions between mesh points. Since physical quantities associated with any given mesh point are related to physical quantities associated with adjacent mesh points, the process of finding a solution requires making simultaneous adjustments in the physical variables associated with all mesh points in an effort to satisfy simultaneously the difference equations connecting all adjacent mesh points. In the process of arriving at a solution, the equations are linearized in the perturbations, and, by matrix inversion, a first approximation to the perturbations is obtained and all physical variables are updated by adding the perturbations to the initial choice of variables. The process is repeated as many times as required to satisfy the difference equations to the desired accuracy. The fact that several iterations are required and that, in general, the magnitudes of the perturbations decrease with each iteration gives rise to the term relaxation to describe the process of achieving convergence toward a solution.

The first task is to break a pre-existing initial model into zones, assigning a mass value M_i to every zone boundary. In principle, the number of zones chosen should be large enough that the difference equations provide a good approximation to the differential equations on which they are patterned. However, the larger the number of zones, the larger is the number of variables that must be stored in memory, the larger is the potential for roundoff error to inhibit progress toward convergence, and the larger is the time required to obtain a solution. In practice, a zoning algorithm which ensures that no physical variable changes by more than, say, 5–10% across a zone provides an adequate approximation to the differential equations.

The next task is to decide where to define physical variables other than mass, whether at zone boundaries or at some position between zone boundaries. Given the fact that density is a mass divided by a volume, it makes sense to define density ρ in the "middle" or "center" of a mass zone and to define radius R at zone boundaries. Since pressure P is related to ρ and temperature T by an equation of state, it makes further sense to define both T and P in the middle of a zone where ρ is defined. For similar reasons, the composition and the rate of energy generation are best defined at zone centers and luminosity L is best defined at zone boundaries.

Assume that a first guess exists for all variables P_i, T_i, L_i, and R_i for $i = 1, N$. This first guess could come from a model constructed with the classical fitting technique described in Section 8.5, or it could come from the last of a series of time-dependent solutions obtained by the relaxation technique here being described. In any case, finding a zeroth order solution is a straightforward process.

The first, or central zone, is taken to be a sphere of mass M_2 and radius R_2. The center is assigned the number 1. The mean density of the central sphere is

$$\rho_2 = \frac{3}{4\pi} \frac{M_2}{R_2^3}. \tag{8.7.1}$$

With this value of density and a guess T_2 for the mean temperature, the equation of state gives pressure $P = P(\rho_2, T_2)$. In general, $P(\rho_2, T_2) \neq P_2$, and the quantity

$$B(1) = P(\rho_2, T_2) - P_2 \tag{8.7.2}$$

is not initially equal to zero. In the process of finding a solution, adjustments in ρ_2, T_2, and P_2 are made in such a way that, as iterations progress, $B(1)$ eventually tends toward zero.

Assuming that the density throughout the sphere is nearly constant and equal to the mean density, mass conservation gives

$$M(r) = \frac{4\pi}{3} r^3 \rho_2, \tag{8.7.3}$$

The balance between pressure-gradient forces and gravitational forces may be expressed by

$$\frac{\partial P}{\partial r} = -\frac{GM(r)}{r^2} \rho_2 = -G \frac{(4\pi/3) r^3 \rho_2}{r^2} \rho_2, \tag{8.7.4}$$

which may be integrated with respect to r to relate the pressure P_1 at the center of the sphere to the pressure $P(r)$ at any other point r within the sphere:

$$P(r) = P_1 - G \frac{2\pi}{3} \rho_2^2 r^2 = P_1 - \frac{1}{2} \frac{GM(r)}{r} \rho_2. \tag{8.7.5}$$

The radius $R_{3/2}$ at the point midway in mass within the central shell is related to the radius R_2 of the sphere approximately by

$$\frac{4\pi}{3} \rho_2 R_{3/2}^3 = \frac{M_2}{2} = \frac{1}{2} \frac{4\pi}{3} \rho_2 R_2^3, \tag{8.7.6}$$

which gives

$$R_{3/2} = \frac{1}{2^{1/3}} R_2. \tag{8.7.7}$$

In the approximation that $P(R_{3/2}) = P_2$, eqs. (8.7.5) and (8.7.7) give

$$P_1 = P_2 + \frac{1}{2^{5/3}} \frac{GM_2}{R_2} \rho_2, \tag{8.7.8}$$

which, using eq. (8.7.1), may also be written as

$$P_1 = P_2 + \frac{3}{4\pi} \frac{1}{2^{5/3}} \frac{GM_2^2}{R_2^4}. \tag{8.7.9}$$

In general, the quantity

$$B(2) = (P_1 - P_2) - \frac{1}{2^{5/3}} \frac{GM_2}{R_2} \rho_2 = (P_1 - P_2) - \frac{3}{4\pi} \frac{1}{2^{5/3}} \frac{GM_2^2}{R_2^4} \qquad (8.7.10)$$

is initially non-zero, but tends toward zero as the solution progresses.

The same procedures may be used to estimate T_2 at the point $M(R_{3/2}) = M_2/2$ relative to the temperature T_1 at the center. For example, if energy transport is entirely by radiation, T_1 and T_2 in a converged model are related by

$$T_1 - T_2 = K_2, \qquad (8.7.11)$$

where

$$K_2 = \frac{1}{2^{5/3}} \frac{3}{4ac} \left(\frac{\kappa\rho}{T^3}\right)_2 \frac{L_2}{4\pi R_2}. \qquad (8.7.12)$$

Using the definition of $V_{\rm rad}$ expressed by eq. (3.4.28), this can also be written as

$$T_1 - T_2 = \frac{1}{2^{5/3}} \left(\frac{T}{P} V_{\rm rad}\right)_2 \frac{GM_2}{R_2} \rho_2 \qquad (8.7.13)$$

or, making use of eq. (8.7.8),

$$T_1 - T_2 = \left(\frac{T}{P} V_{\rm rad}\right)_2 (P_1 - P_2). \qquad (8.7.14)$$

The quantity

$$B(3) = (T_1 - T_2) - K_2, \qquad (8.7.15)$$

while not expected to be zero in an initial model, tends toward zero as the solution progresses.

If energy flow is primarily by convection and one can assume that the temperature gradient is close to the adiabatic gradient,

$$T_1 - T_2 = \frac{1}{2^{5/3}} \left(\frac{T}{P} V_{\rm ad}\right)_2 \frac{GM_2}{R_2} \rho_2, \qquad (8.7.16)$$

or, making use of eq. (8.7.8),

$$T_1 - T_2 = \left(\frac{T}{P} V_{\rm ad}\right)_2 (P_1 - P_2). \qquad (8.7.17)$$

If matter is unstable against convection, but the temperature gradient is superadiabatic, $V_{\rm ad}$ in eq. (8.7.16) is replaced by V as given, for example, by solutions of eqs. (3.4.49) and (3.4.46). Thus, in the general situation, one can write

$$B(3) = (T_1 - T_2) - \left(\frac{T}{P} V\right)_2 (P_1 - P_2). \qquad (8.7.18)$$

Conservation of energy supplies the final quantity to be minimized:

$$B(4) = L_2 - \epsilon(\rho_2, T_2) M_2. \qquad (8.7.19)$$

The equations (8.7.2), (8.7.10), (8.7.15) or (8.7.18), and (8.7.19) can be differentiated with respect to the seven variables P_1, T_1, P_2, T_2, L_2, R_2, and ρ_2, with each derivative being multiplied by the differential of the variable with respect to which the derivative has been taken. The aim is to find increments δQ_j in all of the variables Q_j such that

$$B(k) + \sum_{j=1}^{7} \left(Q_j \frac{\partial B(k)}{\partial Q_j} \right) \frac{\delta Q_j}{Q_j} = 0. \tag{8.7.20}$$

In the resulting equations, the differential $\delta \rho_2$ may be eliminated by using eq. (8.7.1), which gives

$$\frac{\delta \rho_2}{\rho_2} = -3 \frac{\delta R_2}{R_2}. \tag{8.7.21}$$

At the end of the construction process, one has four equations in the six unknowns $\delta P_1/P_1$, $\delta T_1/T_1$, $\delta P_2/P_2$, $\delta T_2/T_2$, $\delta L_2/L_2$, and $\delta R_2/R_2$.

The first equation is

$$B(1) - 3 P(\rho_2, T_2) \left(\frac{\partial \log P}{\partial \log \rho} \right)_2 \frac{\delta R_2}{R_2} + P(\rho_2, T_2) \left(\frac{\partial \log P}{\partial \log T} \right)_2 \frac{\delta T_2}{T_2} - P_2 \frac{\delta P_2}{P_2} = 0, \tag{8.7.22}$$

and the second is

$$B(2) + P_1 \frac{\delta P_1}{P_1} - P_2 \frac{\delta P_2}{P_2} + \frac{4}{2^{5/3}} \frac{GM_2}{R_2} \rho_2 \frac{\delta R_2}{R_2} = 0. \tag{8.7.23}$$

When energy flow is by radiation, the third equation may be written as

$$B(3) + T_1 \frac{\delta T_1}{T_1} - \left(T_2 + K_2 \left[\left(\frac{\partial \log \kappa}{\partial \log T} \right)_2 - 3 \right] \right) \frac{\delta T_2}{T_2} - K_2 \frac{\delta L_2}{L_2}$$
$$+ K_2 \left[3 \left(\frac{\partial \log \kappa}{\partial \log \rho} \right)_2 + 4 \right] \frac{\delta R_2}{R_2} = 0, \tag{8.7.24}$$

where K_2 is given by eq. (8.7.12). When energy flow is primarily by convection, so that $V \sim V_{\text{ad}} > 0$,

$$B(3) - U_2 P_1 \frac{\delta P_1}{P_1} + T_1 \frac{\delta T_1}{T_1} + U_2 P_1 \frac{\delta P_2}{P_2}$$
$$- \left(T_2 + U_2 (P_1 - P_2) \left[1 + \left(\frac{\partial \log V_{\text{ad}}}{\partial \log T} \right)_2 \right] \right) \frac{\delta T_2}{T_2}$$
$$+ 3 U_2 (P_1 - P_2) \left(\frac{\partial \log V_{\text{ad}}}{\partial \log \rho} \right)_2 \frac{\delta R_2}{R_2} = 0, \tag{8.7.25}$$

where

$$U_2 = \left(\frac{T}{P} \, V_{\text{ad}} \right)_2 . \tag{8.7.26}$$

Differentiation of eq. (8.7.19) and use of eq. (8.7.21) produces the fourth equation:

$$B(4) + L_2 \frac{\delta L_2}{L_2} - M_2 \left[\left(\frac{\partial \epsilon}{\partial \log_e T} \right)_2 \frac{\delta T_2}{T_2} - 3 \left(\frac{\partial \epsilon}{\partial \log_e \rho} \right)_2 \frac{\delta R_2}{R_2} \right] = 0. \tag{8.7.27}$$

The fact that ϵ can be negative as well as positive has dictated the form of the derivatives in parentheses.

In practice, derivatives of ϵ, κ, V, and $P(\rho, T)$ in the four equations must, in general, be found by numerical differentiation. The equations have been written in terms of the ratios $\delta Q_j / Q_j$, given the potential that these ratios will become vanishingly small as the solution progresses. One may also expect that, if the initial model has been chosen with care, the ratios are also initially small compared with unity.

As they stand, eqs. (8.7.22)–(8.7.27) are a diverse lot, of different dimensions and of different magnitudes. Yet, in finding solutions, the equations are coupled to one another and to additional equations of the same kind in all of the other mass shells in the model. To put them on a more equal numerical footing, each of the equations may be renormalized in such a way that, once convergence has been achieved, each of the two balancing terms in the quantities $B(k)$, $k = 1, 4$ is close to unity in absolute value. For example, dividing $B(1)$ in eq. (8.7.2) by the first term in the equation gives

$$B'(1) = \frac{B(1)}{P(\rho_2, T_2)} = 1 - \frac{P_2}{P(\rho_2, T_2)} \tag{8.7.28}$$

and the renormalized version of eq. (8.7.22) is achieved by dividing all terms in the equation by $P(\rho_2, T_2)$. Similarly, as inspection of eq. (8.7.10) shows, renormalization of eq. (8.7.23) can be achieved by dividing all terms in the equation by either $(P_1 - P_2)$ or by the second term on the right hand side of eq. (8.7.10). From eq. (8.7.15), it is evident that renormalization of that equation and of eq. (8.7.24) can be achieved by dividing all terms by K_2. Renormalization of eqs. (8.7.17) and (8.7.25) is achieved by dividing by $U_2 (P_1 - P_2)$. The equations for energy conservation, eqs. (8.7.19) and (8.7.27), present a potential problem because ϵ can be either positive or negative and therefore possibly zero. However, the duration of an $\epsilon \sim 0$ phase is normally small compared with the value of an evolutionary time step. In the general case, the best approach is to find the absolute value of each of the two terms in eq. (8.7.19) and to choose the divisor of eq. (8.7.27) as the maximum of the two absolute values. As evolution progresses, one should monitor the behavior of $\epsilon(\rho_2, T_2)$ and be prepared to treat as a special case the specific model for which $\epsilon(\rho_2, T_2)$ passes through zero.

It is useful to define for every zone in the model a four by eight matrix $C(k, m)$, where k (=1,4) designates one of the four equations to be satisfied in each zone and m (=1,8) designates one of the eight variables $\delta P_{i-1}/P_{i-1}$, $\delta T_{i-1}/T_{i-1}$, $\delta L_{i-1}/L_{i-1}$, $\delta R_{i-1}/R_{i-1}$,

$\delta P_i/P_i$, $\delta T_i/T_i$, $\delta L_i/L_i$, and $\delta R_i/R_i$ which are associated with the ith zone. That is, following the pattern given by eq. (8.7.20), one may write

$$B'(k) + \sum_{m=1}^{8} C(k, m) \frac{\delta Q(m)}{Q(m)} = 0, \ k = 1, 4. \tag{8.7.28a}$$

In the first ($i = 2$) zone, since $L_1 = R_1 = 0$, $C(k, 3) = C(k, 4) = 0$ for all k. From eqs. (8.7.22) and (8.7.28),

$$B'(1) + C(1, 5) \frac{\delta P_2}{P_2} + C(1, 6) \frac{\delta T_2}{T_2} + C(1, 8) \frac{\delta R_2}{R_2} = 0, \tag{8.7.29}$$

where

$$C(1, 5) = -\frac{P_2}{P(\rho_2, T_2)}, \ C(1, 6) = \left(\frac{\partial \log P}{\partial \log T}\right)_2, \ \text{and } C(1, 8) = -3 \left(\frac{\partial \log P}{\partial \log \rho}\right)_2. \tag{8.7.30}$$

Renormalizing eq. (8.7.10) as

$$B'(2) = B(2) \div \left(\frac{1}{2^{5/3}} \frac{GM_2}{R_2} \rho_2\right) = (P_1 - P_2) \, 2^{5/3} \frac{R_2}{GM_2} \frac{1}{\rho_2} - 1, \tag{8.7.31}$$

one has from from eq. (8.7.23) that

$$B'(2) + C(2, 1) \frac{\delta P_1}{P_1} + C(2, 5) \frac{\delta P_2}{P_2} + C(2, 8) \frac{\delta R_2}{R_2} = 0, \tag{8.7.32}$$

where

$$C(2, 1) = 2^{5/3} \frac{R_2}{GM_2} \frac{1}{\rho_2} P_1, \ C(2, 5) = -\frac{P_2}{P_1} C(2, 1), \ \text{and } C(2, 8) = 4. \tag{8.7.33}$$

If energy flow is by radiation, renormalization is achieved by dividing all quantities in eqs. (8.7.15) and (8.7.24) by K_2 defined by eq. (8.7.12), resulting in

$$B'(3) = \frac{T_1 - T_2}{K_2} - 1 \tag{8.7.33a}$$

and

$$B'(3) + C(3, 2) \frac{\delta T_1}{T_1} + C(3, 6) \frac{\delta T_2}{T_2} + C(3, 7) \frac{\delta L_2}{L_2} + C(3, 8) \frac{\delta R_2}{R_2} = 0, \tag{8.7.34}$$

where

$$C(3, 2) = \frac{T_1}{K_2}, \ C(3, 6) = 3 - \left(\frac{\partial \log \kappa}{\partial \log T}\right)_2 - \frac{T_2}{K_2}, \ C(3, 7) = -1,$$

$$\text{and } C(3, 8) = 4 + 3 \left(\frac{\partial \log \kappa}{\partial \log \rho}\right)_2. \tag{8.7.35}$$

If energy flow is primarily by convection, one has from eqs. (8.7.17) and (8.7.25) that

$$B'(3) = \frac{T_1 - T_2}{P_1 - P_2} \frac{1}{U_2} - 1, \tag{8.7.36}$$

$$B'(3) + C(3, 1) \frac{\delta P_1}{P_1} + C(3, 2) \frac{\delta T_1}{T_1} + C(3, 6) \frac{\delta T_2}{T_2} + C(3, 8) \frac{\delta R_2}{R_2} = 0, \tag{8.7.37}$$

where

$$C(3, 1) = -\frac{P_1}{P_1 - P_2}, \quad C(3, 2) = \frac{T_1}{U_2 (P_1 - P_2)}, \quad C(3, 5) = -C(3, 1),$$

$$C(3, 6) = -\left[\frac{T_2}{U_2 (P_1 - P_2)} + \left(\frac{\partial \log V_{\mathrm{ad}}}{\partial \log T} \right)_2 \right], \quad \text{and } C(3, 8) = 3 \left(\frac{\partial \log V_{\mathrm{ad}}}{\partial \log \rho} \right)_2. \tag{8.7.38}$$

Finally, one has from eq. (8.7.19) that

$$B'(4) = \frac{L_2}{D_4} - \frac{\epsilon(\rho_2, T_2) M_2}{D_4}, \tag{8.7.39}$$

where

$$D_4 = \max \left(|L_2|, |\epsilon(\rho_2, T_2)| M_2 \right), \tag{8.7.40}$$

and, from eq. (8.7.27),

$$B'(4) + C(4, 6) \frac{\delta T_2}{T_2} + C(4, 7) \frac{\delta L_2}{L_2} + C(4, 8) \frac{\delta R_2}{R_2} = 0, \tag{8.7.41}$$

where

$$C(4, 6) = -\frac{M_2}{D_4} \left(\frac{\partial \epsilon}{\partial \log_e T} \right)_2, \quad C(4, 7) = \frac{L_2}{D_4}, \quad \text{and } C(4, 8) = 3 \frac{M_2}{D_4} \left(\frac{\partial \epsilon}{\partial \log_e \rho} \right)_2. \tag{8.7.42}$$

Construction of the derivatives of ϵ in eqs. (8.7.42) is complicated by the fact that the two components of $\epsilon = \epsilon_{\mathrm{nuc}} + \epsilon_{\mathrm{grav}}$ have different structures. Finding the derivatives of ϵ_{nuc} is straightforward, but the form of ϵ_{grav} which is most useful in stellar evolution calculations itself involves derivatives so that quantities such as $\partial \epsilon_{\mathrm{grav}} / \partial \log_e T$ involve a mixture of first and second derivatives, as will be discussed after eq. (8.7.75).

It proves convenient to choose $\delta P_2 / P_2$ and $\delta T_2 / T_2$ as the independent differentials, and the solution of eqs. (8.7.29), (8.7.32), (8.7.41), and either eq. (8.7.34) or eq. (8.7.37) in, terms of these two differentials may be written as

$$\frac{\delta P_1}{P_1} = \alpha(1, 2) \frac{\delta P_2}{P_2} + \beta(1, 2) \frac{\delta T_2}{T_2} + \gamma(1, 2), \tag{8.7.43}$$

$$\frac{\delta T_1}{T_1} = \alpha(2, 2) \frac{\delta P_2}{P_2} + \beta(2, 2) \frac{\delta T_2}{T_2} + \gamma(2, 2), \tag{8.7.44}$$

$$\frac{\delta L_2}{L_2} = \alpha(3, 2) \frac{\delta P_2}{P_2} + \beta(3, 2) \frac{\delta T_2}{T_2} + \gamma(3, 2), \tag{8.7.45}$$

and

$$\frac{\delta R_2}{R_2} = \alpha(4, 2) \frac{\delta P_2}{P_2} + \beta(4, 2) \frac{\delta T_2}{T_2} + \gamma(4, 2), \tag{8.7.46}$$

where the integer 2 following the comma in all of the αs, βs, and γs signifies that these equations hold for the central, spherical zone with outer boundary at $i = 2$. All of the information from the central sphere necessary for a final solution is contained in the αs, βs, and γs so, having fulfilled its role in the construction of these quantities, the matrix $C(k, m)$ is no longer needed.

Outside of the central sphere, zones are spherical *shells* and the density in the shell of mass dM_i between mass points M_{i-1} and M_i may be defined by

$$\rho_i = \frac{dM_i}{4\pi R_{i-1/2}^2 (R_i - R_{i-1})} = \frac{M_i - M_{i-1}}{4\pi R_{i-1/2}^2 (R_i - R_{i-1})}, \tag{8.7.47}$$

where M_{i-1} and M_i are, respectively, the masses at the inner and outer boundaries of the ith shell, R_{i-1} and R_i are, respectively, the radii at these same boundaries, and, except when $i = 2$,

$$R_{i-1/2} = \frac{1}{2} (R_i + R_{i-1}) \tag{8.7.48}$$

is defined as the radius at the midpoint of the ith shell. When $i = 2$, $R_{3/2}$ is defined by eq. (8.7.7).

It is worth repeating again that the labeling scheme which has been adopted assigns the integer 1 to the very center of the model and the integer 2 to the outer edge of the central sphere. Thus, the inner edge of the *first* spherical shell is also assigned the integer $i = 2$, its outer edge is assigned the integer $i = 3$, and this shell is referred to as the third shell even though it is the first spherical shell outside of the central sphere. Similarly, the shell referenced as the ith shell is defined as the shell with outer boundary labeled i although it is actually the $(i - 2)$th spherical shell outside of the central sphere.

In each spherical shell, four equations can be constructed which relate differentials of the eight variables P_{i-1}, T_{i-1}, L_{i-1}, R_{i-1}, P_i, T_i, L_i, and R_i in such a way as to minimize deviations from equalities based on (1) the equation of state, (2) the balance between gravitational forces and pressure-gradient forces, (3) the relationship between energy flux and temperature gradient, and (4) the conservation of energy. Of the four quantities to be minimized, the three which are not explicitly dependent on the mode of energy transfer are

$$B(1) = P(\rho_i, T_i) - P_i, \tag{8.7.49}$$

$$B(2) = (P_i - P_{i-1}) + \frac{GM_{i-1}}{R_{i-1}^2} \frac{M_i - M_{i-2}}{8\pi R_{i-1}^2}, \tag{8.7.50}$$

and

$$B(4) = (L_i - L_{i-1}) - \epsilon(\rho_i, T_i)(M_i - M_{i-1}). \tag{8.7.51}$$

The rightmost term in eq. (8.7.50) is a consequence of the definition

$$4\pi R_{i-1}^2 (R_{i-1/2} - R_{i-3/2}) \rho_{i-1/2} = \frac{dM_i + dM_{i-1}}{2} = \frac{M_i - M_{i-2}}{2}, \tag{8.7.52}$$

where $\rho_{i-1/2}$ is taken to be the density at the base of the ith spherical shell. Except for $R_{3/2}$, which is given by eq. (8.7.7), the Rs with half integer subscripts are given by eq. (8.7.48).

With regard to energy transport, the choice of an appropriate quantity to minimize is not straightforward. In this book, in regions where transport is by radiation and/or conduction, the radiative gradient is taken to be

$$V_{\text{rad}} = \frac{3}{4ac} \left(\kappa \frac{P}{T^4} \right)_i \frac{L_{i-1}}{4\pi G M_{i-1}}, \tag{8.7.53}$$

where κ is κ_{eff} defined by eq. (3.4.62), and the relevant quantity to be minimized is taken to be

$$B(3) = (T_i - T_{i-1}) + \frac{3}{4ac} \left(\frac{\kappa}{T^3} \right)_i \frac{L_{i-1}}{4\pi R_{i-1}^2} \frac{M_i - M_{i-2}}{8\pi R_{i-1}^2}. \tag{8.7.54}$$

If $V_{\text{ad}} < V_{\text{rad}}$ and it is reasonable to suppose that the superadiabatic gradient is small, eq. (8.7.54) is replaced by

$$B(3) = (T_i - T_{i-1}) - V_{\text{ad}}(\rho_i, T_i) \frac{T_i}{P_i} (P_i - P_{i-1}), \tag{8.7.55}$$

where V_{ad} is evaluated in the middle of the ith shell. In regions where the superadiabatic gradient is large,

$$B(3) = \frac{P_i}{T_i} \frac{T_i - T_{i-1}}{P_i - P_{i-1}} - V_i(V_{\text{rad}}, V_{\text{ad}}, \rho_i, T_i), \tag{8.7.56}$$

where V_i is the logarithmic gradient of temperature relative to pressure as estimated by, say, the mixing length algorithm for convection described in Section 3.4.

Equations (8.7.50) and (8.7.53) are suggested by stability considerations (Sugimoto, 1970; Sugimoto, Nomoto, & Eriguchi, 1981). The balance between pressure forces and gravitational forces and the transport of energy are evaluated across the base of the ith zone and rely on characteristics of the $(i - 1)$th zone as well as on those of the ith zone. This has the effect of producing an additional coupling between shells that contributes to the stability of the solution.

As illustrated by eqs. (8.7.54)–(8.7.56), there is no sacred way of writing any of the $B(m)$s. For example, taking eq. (8.7.56) as standard, multiplication of the right hand side by $T_i (P_i - P_{i-1})/P_i$ and replacing V_i by V_{rad} as given by eq. (8.7.53) reproduces $B(3)$

given by eq. (8.7.54). Similarly, multiplication of the right hand side of eq. (8.7.56) by $(P_i - P_{i-1})(T_i/P_i)$ and replacing V_i by $V_{ad}(\rho_i, T_i)$ reproduces $B(3)$ given by eq. (8.7.55).

A solution of eqs. (8.7.49)–(8.7.50) and eq. (8.7.54) or (8.7.55) consists of finding the values of the variables which cause the right hand side of each equation to vanish. Using eqs. (8.7.47) and (8.7.48), the logarithmic differential $\delta\rho_i/\rho_i$ can be replaced by

$$\frac{\delta\rho_i}{\rho_i} = A_i \frac{\delta R_{i-1}}{R_{i-1}} + B_i \frac{\delta R_i}{R_i}, \tag{8.7.57}$$

where

$$A_i = \frac{R_i R_{i-1}}{R_i^2 - R_{i-1}^2}\left(3\frac{R_{i-1}}{R_i} - 1\right) = \frac{R_{i-1}(3R_{i-1} - R_i)}{R_i^2 - R_{i-1}^2} \tag{8.7.58}$$

and

$$B_i = -\frac{R_i R_{i-1}}{R_i^2 - R_{i-1}^2}\left(3\frac{R_i}{R_{i-1}} - 1\right) = -\frac{R_i(3R_i - R_{i-1})}{R_i^2 - R_{i-1}^2}. \tag{8.7.59}$$

Differentiation of eqs. (8.7.49)–(8.7.52), and use of eqs. (8.7.57)–(8.7.59) yields four equations having the structure of eqs. (8.7.20). To repeat, for emphasis, one may write

$$B(k) + \sum_{m=1}^{8} C(k, m)\frac{\delta Q(m)}{Q(m)} = 0, \; k = 1, 4, \tag{8.7.60}$$

where the $Q(m)$s stand for the eight quantities $P_{i-1}, T_{i-1}, L_{i-1}, R_{i-1}, P_i, T_i, L_i$, and R_i in the order chosen; the $C(k, m)$s are the logarithmic derivatives of the $B(k)$s with respect to the $Q(m)$s, and the integer k goes from 1 to 4 in the order given by mass conservation, pressure balance, energy flow, and energy conservation.

As in the case of the central sphere, each of the k equations described by eq. (8.7.60) can be renormalized in such a way that, once convergence has been achieved, each of the two balancing terms in the quantities $B(k)$ is, in absolute value, close to unity. For example, from eq. (8.7.49), the renormalizing divisor for $B(1)$ can be chosen as $P(\rho_i, T_i)$ and the renormalized version of the equation to be solved for the increments $\delta Q(m)$ when $k = 1$ is achieved by dividing $B(1)$ and all of the $C(1, m)$ terms in eq. (8.7.60) by $P(\rho_i, T_i)$. The renormalized quantities are

$$B'(1) = 1 - \frac{P_i}{P(\rho_i, T_i)}, \tag{8.7.61}$$

$$C(1, 4) = \left(\frac{\partial \log P}{\partial \log \rho}\right)_i \frac{R_{i-1}(3R_{i-1} - R_i)}{R_i^2 - R_{i-1}^2}, \; C(1, 5) = -\frac{P_i}{P(\rho_i, T_i)},$$

$$C(1, 6) = \left(\frac{\partial \log P}{\partial \log T}\right)_i, \; \text{and } C(1, 8) = -\left(\frac{\partial \log P}{\partial \log \rho}\right)_i \frac{R_i(3R_i - R_{i-1})}{R_i^2 - R_{i-1}^2}. \tag{8.7.62}$$

All other $C(1, m)$s are zero.

The equation for pressure balance, eq. (8.7.50), can be made dimensionless by dividing by P_i, giving

$$B'(2) = 1 - \frac{1}{P_i} \left(P_{i-1} - \frac{GM_{i-1}(M_i - M_{i-2})}{8\pi R_{i-1}^4} \right). \tag{8.7.63}$$

Differentiating $B(2)$ as prescribed in eq. (8.7.60) and dividing by P_i gives the coefficients

$$C(2,1) = -\frac{P_{i-1}}{P_i}, \quad C(2,4) = -\frac{GM_{i-1}(M_i - M_{i-2})}{2\pi R_{i-1}^4 P_i}, \quad \text{and } C(2,5) = 1. \tag{8.7.64}$$

As an aside, it is a curiosity that differentiating $B'(2)$ directly gives $C(2,5) = 1 - B'(2)$

When energy flow is by radiation, the choice

$$B'(3) = \frac{1}{T_i}(T_i - T_{i-1}) + K_i = 1 - \left(\frac{T_{i-1}}{T_i} - \frac{K_i}{T_i} \right), \tag{8.7.65}$$

where

$$K_i = \frac{3}{4ac} \left(\frac{\kappa}{T^3} \right)_i \frac{M_i - M_{i-2}}{32\pi^2} \frac{L_{i-1}}{R_{i-1}^4}, \tag{8.7.66}$$

leads to

$$C(3,2) = -\frac{T_{i-1}}{T_i}, \quad C(3,3) = \frac{K_i}{T_i}, \quad C(3,4) = \frac{K_i}{T_i} \left(A_i \left(\frac{\partial \log \kappa}{\partial \log \rho} \right)_i - 4 \right),$$

$$C(3,6) = 1 + \frac{K_i}{T_i} \left(\left(\frac{\partial \log \kappa}{\partial \log T} \right)_i - 3 \right), \quad \text{and } C(3,8) = \frac{K_i}{T_i} B_i \left(\frac{\partial \log \kappa}{\partial \log \rho} \right)_i, \tag{8.7.67}$$

where A_i and B_i are given by eqs. (8.7.58) and (8.7.59), respectively.

When energy transfer is primarily by convection, the choice of an appropriate transfer equation is complicated by the fact that, whereas the temperature, pressure, and adiabatic gradient of temperature with respect to pressure are all defined in the interior of a zone, the adiabatic temperature gradient at zone center should actually be compared with the logarithmic gradient of temperature with respect to pressure across the zone. For example, one might choose

$$V_{ad}(\rho_{i-1/2}, T_{i-1/2}) \sim \frac{P_{i-1/2}}{T_{i-1/2}} \frac{T_i - T_{i-1}}{P_i - P_{i-1}}, \tag{8.7.68}$$

or, perhaps,

$$V_{ad}(\rho_i, T_i) \sim \frac{P_i}{T_i} \frac{T_{i+1/2} - T_{i-1/2}}{P_{i+1/2} - P_{i-1/2}}. \tag{8.7.68a}$$

In the program used to construct models for this book, the much simpler approximation

$$V_{ad}(\rho_i, T_i) \sim \frac{P_i}{T_i} \frac{T_i - T_{i-1}}{P_i - P_{i-1}} \tag{8.7.69}$$

has been adopted. Setting

$$B'(3) = \frac{B(3)}{T_i},$$ (8.7.70)

where

$$B(3) = (T_i - T_{i-1}) - V_{\text{ad}}(\rho_i, T_i) \frac{T_i}{P_i} (P_i - P_{i-1}),$$ (8.7.71)

one finds

$$C(3, 1) = V_{\text{ad}}(\rho_i, T_i) \frac{P_{i-1}}{P_i}, \quad C(3, 2) = -\frac{T_{i-1}}{T_i},$$

$$C(3, 4) = V_{\text{ad}}(\rho_i, T_i) \frac{P_i - P_{i-1}}{P_i} \left(\frac{\partial \log V_{\text{ad}}}{\partial \log \rho} \right)_i A_i, \quad C(3, 5) = -C(3, 1),$$

$$C(3, 6) = 1 - V_{\text{ad}}(\rho_i, T_i) \frac{P_i - P_{i-1}}{P_i} \left(1 + \left(\frac{\partial \log V_{\text{ad}}}{\partial \log T} \right)_i \right),$$

$$\text{and } C(3, 8) = V_{\text{ad}}(\rho_i, T_i) \frac{P_i - P_{i-1}}{P_i} \left(\frac{\partial \log V_{\text{ad}}}{\partial \log \rho} \right)_i B_i,$$ (8.7.72)

From the equation for energy conservation, eq. (8.7.51), one can write

$$B'(4) = \frac{B(4)}{D_4},$$ (8.7.73)

where

$$D_4 = \max \left(|(L_i - L_{i-1})|, |\epsilon(\rho_i, T_i) (M_i - M_{i-1})| \right).$$ (8.7.74)

Non-vanishing elements of the matrix $C(4, m)$ are

$$C(4, 3) = \frac{L_i}{D_4}, \quad C(4, 4) = - \left(\frac{\partial \epsilon}{\partial \log_e \rho} \right)_i A_i \frac{M_i - M_{i-1}}{D_4},$$

$$C(4, 6) = - \left(\frac{\partial \epsilon}{\partial \log_e T} \right)_i \frac{M_i - M_{i-1}}{D_4}, \quad C(4, 7) = -\frac{L_{i-1}}{D_4},$$

$$\text{and } C(4, 8) = - \left(\frac{\partial \epsilon}{\partial \log_e \rho} \right)_i B_i \frac{M_i - M_{i-1}}{D_4}.$$ (8.7.75)

A discussion of derivatives of ϵ_{grav}, the gravothermal component of ϵ, is appropriate at this point. As follows from eq. (8.1.26), the gravothermal energy-generation rate in the ith shell bounded by the mass points M_{i-1} and M_i can be approximated by

$$\epsilon_{\text{grav},i} \approx -\frac{\delta U_i}{\delta t} + \frac{P_i(t)}{\rho_i^2(t)} \frac{\delta \rho_i(t)}{\delta t},$$ (8.7.76)

where $U_i(t)$, $P_i(t)$, and $\rho_i(t)$ are, respectively, the total internal energy per unit mass of matter and radiation, the total pressure, and the matter density at the center of the ith shell of the model currently being constructed at time t, δt is the magnitude of the time step chosen,

$$\delta U_i(t) = U_i(t) - U_i(t - \delta t), \tag{8.7.77}$$

and

$$\delta \rho_i(t) = \rho_i(t) - \rho_i(t - \delta t). \tag{8.7.78}$$

Because internal energies at times t and $t - \delta t$ are often quite large compared with the difference $|\delta U| = |U(t) - U(t - \delta t)|$, roundoff error in the first term on the right hand side of eq. (8.7.76) can be quite large and slow progress toward convergence. Suppose, for example, that only six digit accuracy is possible and that, in a time step, U increases from $U_1 = x.xxxxx$ to $U_2 = x.xxxyy$. Then, $\Delta U = U_2 - U_1 = (y.y - x.x) \times 10^{-4}$, accurate to only two digits. If an analytic expression for U is available, the problem can be minimized by adopting the formulation given by eqs. (8.7.29)–(8.7.32). Then,

$$\epsilon_{\mathrm{grav},i} = \epsilon_{\mathrm{th},i} + \epsilon_{\mathrm{cdth},i}$$

$$\approx -\left(\frac{\partial U_i}{\partial T}\right)_{\rho,Y} \frac{\delta T_i(t)}{\delta t} + \left[\frac{P_i(t)}{\rho_i^2(t)} - \left(\frac{\partial U_i}{\partial \rho}\right)_{T,Y}\right] \frac{\delta \rho_i(t)}{\delta t} - \sum_j \left(\frac{\partial U_i}{\partial Y_j}\right)_{\rho,T} \frac{\delta Y_{ji}}{\delta t}. \tag{8.7.79}$$

The quantities $\left(\partial U_i/\partial Q\right)_{A,B}$ in eq. (8.7.79) are all evaluated at time t at the center of the ith shell and Y_{ji} is the number abundance parameter of the jth type of particle at the center of the ith shell.

In a radiative zone, the third term in eq. (8.7.79) may be written as

$$\epsilon_{\mathrm{cdth},i} = -N_{\mathrm{A}} \sum_j \bar{\mu}_{ji}(t) \frac{\mathrm{d} Y_{ji}(t)}{\mathrm{d} t}, \tag{8.7.80}$$

where $\bar{\mu}_{ji}(t)$, the local creation–destruction potential at time t, and $\mathrm{d}Y_{ji}/\mathrm{d}t$, the local rate of change in Y_{ji}, are both functions of local conditions. In a convective zone, the contribution to $\epsilon_{\mathrm{cdth},i}$ by an isotope which is fully mixed is

$$-N_{\mathrm{A}}\, \bar{\mu}_{ji} \frac{\delta Y_{ji}}{\delta t}, \tag{8.7.81}$$

where $\bar{\mu}_{ji}$ is a locally determined quantity, but δY_{ji} is everywhere the same in the convective zone. In either case, finding derivatives of ϵ_{cdth} is as straightforward as finding derivatives of ϵ_{nuc}.

Altogether, derivatives of ϵ_{grav} with respect to the logarithms of temperature and density are

$$\left(\frac{\partial \epsilon_{\mathrm{grav}}}{\partial \log_e \rho}\right)_i = -\left(\rho \frac{\partial^2 U}{\partial \rho\, \partial T}\right)_i \frac{\delta T_i}{\delta t} + \left(\frac{1}{\rho} \frac{\partial P}{\partial \rho} - \frac{2P}{\rho^2} - \rho \frac{\partial^2 U}{\partial \rho^2}\right)_i \frac{\delta \rho_i}{\delta t}$$

$$+ \left(\frac{P}{\rho} - \rho \frac{\partial U}{\partial \rho}\right)_i \frac{1}{\delta t} + \left(\frac{\partial \epsilon_{\mathrm{cdth}}}{\partial \log_e \rho}\right)_i. \tag{8.7.82}$$

and

$$\left(\frac{\partial \epsilon_{\text{grav}}}{\partial \log_e T}\right)_i = -\left(T \frac{\partial^2 U}{\partial^2 T}\right)_i \frac{\delta T_i}{\delta t} + \left(T \frac{\partial U}{\partial T}\right)_i \frac{1}{\delta t}$$

$$+ \left(\frac{T}{\rho^2} \frac{\partial P}{\partial T} - T \frac{\partial^2 U}{\partial T \partial \rho}\right)_i \frac{\delta \rho_i}{\delta t} + \left(\frac{\partial \epsilon_{\text{cdth}}}{\partial \log_e T}\right)_i. \tag{8.7.83}$$

If an analytic aproximation to U is not available, so that derivatives of U must be numerical, the accuracy achieved using eq. (8.7.79) is the same as that achieved using eq. (8.7.76).

Using eqs. (8.7.61)–(8.7.75) in the renormalized version of eq. (8.7.60), one can solve for any four of the eight differentials $\delta P_{i-1}/P_{i-1}$, $\delta T_{i-1}/T_{i-1}$, $\delta L_{i-1}/L_{i-1}$, $\delta R_{i-1}/R_{i-1}$, $\delta P_i/P_i$, $\delta T_i/T_i$, $\delta L_i/L_i$, and $\delta R_i/R_i$ in terms of the other four. However, making use of the relationships between differentials established by the equations constructed for shell $i-1$, it is possible to reduce the number of unknown differentials in shell i to two. For example, setting $i=3$ and making use of eqs. (8.7.45) and (8.7.46), one can eliminate the differentials $\delta L_2/L_2$ and $\delta R_2/R_2$ and solve the resulting equations to obtain

$$\frac{\delta P_2}{P_2} = \alpha(1, 3) \frac{\delta P_3}{P_3} + \beta(1, 3) \frac{\delta T_3}{T_3} + \gamma(1, 3), \tag{8.7.84}$$

$$\frac{\delta T_2}{T_2} = \alpha(2, 3) \frac{\delta P_3}{P_3} + \beta(2, 3) \frac{\delta T_3}{T_3} + \gamma(2, 3), \tag{8.7.85}$$

$$\frac{\delta L_3}{L_3} = \alpha(3, 3) \frac{\delta P_3}{P_3} + \beta(3, 3) \frac{\delta T_3}{T_3} + \gamma(3, 3), \tag{8.7.86}$$

and

$$\frac{\delta R_3}{R_3} = \alpha(4, 3) \frac{\delta P_3}{P_3} + \beta(4, 3) \frac{\delta T_3}{T_3} + \gamma(4, 3). \tag{8.7.87}$$

Having served its purpose, the matrix $C(k, m)$ for the $i=3$ shell is no longer needed.

Continuing in this way, one has for the shell bounded by the $(i-1)$th and ith grid points:

$$\frac{\delta P_{i-1}}{P_{i-1}} = \alpha(1, i) \frac{\delta P_i}{P_i} + \beta(1, i) \frac{\delta T_i}{T_i} + \gamma(1, i), \tag{8.7.88}$$

$$\frac{\delta T_{i-1}}{T_{i-1}} = \alpha(2, i) \frac{\delta P_i}{P_i} + \beta(2, i) \frac{\delta T_i}{T_i} + \gamma(2, i), \tag{8.7.89}$$

$$\frac{\delta L_i}{L_i} = \alpha(3, i) \frac{\delta P_i}{P_i} + \beta(3, i) \frac{\delta T_i}{T_i} + \gamma(3, i), \tag{8.7.90}$$

and

$$\frac{\delta R_i}{R_i} = \alpha(4, i) \frac{\delta P_i}{P_i} + \beta(4, i) \frac{\delta T_i}{T_i} + \gamma(4, i). \tag{8.7.91}$$

For the last interior shell, one has

$$\frac{\delta P_{N-1}}{P_{N-1}} = \alpha(1, N) \frac{\delta P_N}{P_N} + \beta(1, N) \frac{\delta T_N}{T_N} + \gamma(1, N), \tag{8.7.92}$$

$$\frac{\delta T_{N-1}}{T_{N-1}} = \alpha(2, N) \frac{\delta P_N}{P_N} + \beta(2, N) \frac{\delta T_N}{T_N} + \gamma(2, N), \tag{8.7.93}$$

$$\frac{\delta L_N}{L_N} = \alpha(3, N) \frac{\delta P_N}{P_N} + \beta(3, N) \frac{\delta T_N}{T_N} + \gamma(3, N), \tag{8.7.94}$$

and

$$\frac{\delta R_N}{R_N} = \alpha(4, N) \frac{\delta P_N}{P_N} + \beta(4, N) \frac{\delta T_N}{T_N} + \gamma(4, N). \tag{8.7.95}$$

The next task is to compare structure variables at the outer edge of the model interior with the same variables at the base of a static envelope. Pressure P_E and temperature T_E at the outer edge of the interior (remember that P and T are defined at shell centers) may be defined by

$$P_E = P_N + (P_N - P_{N-1})\frac{M_N - M_{N-1}}{M_N - M_{N-2}}, \tag{8.7.96}$$

and

$$T_E = T_N + (T_N - T_{N-1})\frac{M_N - M_{N-1}}{M_N - M_{N-2}}. \tag{8.7.97}$$

Let P_B, T_B, L_B, and R_B be the values of the variables at the base of a static envelope of mass $M_{\rm env} = M_{\rm model} - M_N$, luminosity L_S, and radius R_S. Initially, the quantities $B(k)$ defined by

$$\frac{P_B}{P_N} - 1 = B(1), \quad \frac{T_B}{T_N} - 1 = B(2), \quad \frac{L_B}{L_N} - 1 = B(3), \quad \text{and} \quad \frac{R_B}{R_N} - 1 = B(4) \tag{8.7.98}$$

are sensibly different from zero. Two additional static envelopes are sufficient to construct derivatives of P_B, T_B, L_B, and R_B with respect to L_S and R_S. In an attempt to match interior and envelope solutions at the interface between the two solutions, one may set

$$P_E \left(1 + \frac{\delta P_E}{P_E}\right) = P_B \left(1 + \frac{\partial \log P_B}{\partial \log L_S} \frac{\delta L_S}{L_S} + \frac{\partial \log P_B}{\partial \log R_S} \frac{\delta R_S}{R_S}\right), \tag{8.7.99}$$

$$T_E \left(1 + \frac{\delta T_E}{T_E}\right) = T_B \left(1 + \frac{\partial \log T_B}{\partial \log L_S} \frac{\delta L_S}{L_S} + \frac{\partial \log T_B}{\partial \log R_S} \frac{\delta R_S}{R_S}\right), \tag{8.7.100}$$

$$L_N \left(1 + \frac{\delta L_N}{L_N}\right) = L_B \left(1 + \frac{\partial \log L_B}{\partial \log L_S} \frac{\delta L_S}{L_S} + \frac{\partial \log L_B}{\partial \log R_S} \frac{\delta R_S}{R_S}\right) = L_S \left(1 + \frac{\delta L_S}{L_S}\right), \tag{8.7.101}$$

and

$$R_N \left(1 + \frac{\delta R_N}{R_N}\right) = R_B \left(1 + \frac{\partial \log R_B}{\partial \log L_S} \frac{\delta L_S}{L_S} + \frac{\partial \log R_B}{\partial \log R_S} \frac{\delta R_S}{R_S}\right). \tag{8.7.102}$$

The rightmost identity in eq. (8.7.100) follows from the fact that, in a static envelope, $L_B = L_S$.

Equations (8.7.94) and (8.7.95) may be used in conjunction with eqs. (8.7.90) and (8.7.91) to replace the left hand sides of eqs. (8.7.99) and (8.7.100) with terms involving just the differentials $\delta P_N/P_N$ and $\delta T_N/T_N$. Equations (8.7.88) and (8.7.89) may be used to replace the left hand sides of eqs. (8.7.96) and (8.7.97) with terms also involving just the differentials $\delta P_N/P_N$ and $\delta T_N/T_N$. The net result is a set of four equations

$$e(1, 1)\frac{\delta P_N}{P_N} + e(1, 2)\frac{\delta T_N}{T_N} + e(1, 3)\frac{\delta L_S}{L_S} + e(1, 4)\frac{\delta R_S}{R_S} = d(1), \tag{8.7.103}$$

$$e(2, 1)\frac{\delta P_N}{P_N} + e(2, 2)\frac{\delta T_N}{T_N} + e(2, 3)\frac{\delta L_S}{L_S} + e(2, 4)\frac{\delta R_S}{R_S} = d(2), \tag{8.7.104}$$

$$e(3, 1)\frac{\delta P_N}{P_N} + e(3, 2)\frac{\delta T_N}{T_N} + e(3, 3)\frac{\delta L_S}{L_S} + e(3, 4)\frac{\delta R_S}{R_S} = d(3), \tag{8.7.105}$$

and

$$e(4, 1)\frac{\delta P_N}{P_N} + e(4, 2)\frac{\delta T_N}{T_N} + e(4, 3)\frac{\delta L_S}{L_S} + e(4, 4)\frac{\delta R_S}{R_S} = d(4), \tag{8.7.106}$$

which can be solved for the four differentials $\delta L_S/L_S$, $\delta R_S/R_S$, $\delta P_N/P_N$, and $\delta T_N/T_N$.

Using eqs. (8.7.92)–(8.7.95), (8.7.88)–(8.7.91), and (8.7.43)–(8.7.46), one may now work back through the interior, finding differentials in all other shells. With the differentials at all grid points in hand, all of the structure variables are updated by setting $Q(m) \rightarrow Q'(m) = Q(m)\left(1 + \delta Q(m)/Q(m)\right)$ and the quantities $B(k)$ described by eqs. (8.7.2), (8.7.9), (8.7.14), (8.7.15), (8.7.35)–(8.7.38), and (8.7.60) are recalculated. In general, the $B(k)$ are still non-zero, and it is necessary to repeat the exercise of constructing the α, β, and γ matrices for interior shells and the $e(m, k)$ matrices at the interface between interior and envelope solutions until, in each shell and at the interface, the $B(k)$s are sufficiently small for one to declare that convergence has been achieved.

Each $B(k)$ in, e.g., eqs. (8.7.35)–(8.7.38), is composed of two terms, $A_1(k)$ and $A_2(k)$, and one must examine in every shell the quantities

$$S(k) = \frac{|B(k)|}{\min(|A_1(k)|, |A_2(k)|)}, \tag{8.7.107}$$

where $\min(|A_1(k)|, |A_2(k)|)$ is the smaller of the absolute values of $A_1(k)$ and $A_2(k)$. Because of the inevitability of roundoff error in a numerical calculation, it is impossible to force the $S(k)$ to vanish, but it is frequently possible to converge to the point where the maximum value of $S(k)$ in the entire model satisfies

$$S(k)_{max} < 10^{-5}. \tag{8.7.108}$$

Not all quantities necessary for constructing a model may be present in the choice of an initial trial model. Because the gravothermal energy source is explicitly time dependent (see the discussion in Section 8.1 and particularly eq. (8.1.25)) and because nuclear transformations continuously change the distribution of composition variables in a star, one cannot, in principle, construct a realistic quasistatic model at any given time without making use of a precursor model of some sort. In practice, the initial precursor model can be relatively crude, with, e.g., gravothermal energy generation neglected; in a relatively few time steps, models constructed with the relaxation technique lose memory of the inadequacies of the initial model on a time scale short compared with nuclear burning time scales.

Once a converged model at time t has been achieved, structure and composition variables in the next model at time $t + \delta t$ must be estimated. One approach is to estimate changes in composition variables explicitly using derivatives based on conditions in the model at time t and to estimate structure variables at time $t + \delta t$ by extrapolating from the structure variables at times t and $t - \delta t$. Almost fifty years of experience suggests that, surprisingly, the best choices for initial estimates of structure variables at time $t + \delta t$ are the structure variables at time t even though, in regions where gravothermal energy-generation rates are important, the energy balance equation is maximally violated during the first iteration. With this prescription, first estimates for composition variables at time $t + \delta t$ are explicit. During subsequent iterations, the best results are achieved by calculating changes in composition variables implicitly, using both structure and composition variables at time $t + \delta t$ in the model under construction.

8.8 Composition changes in radiative regions due to nuclear transformations

Except during the early pre-main–sequence phase of gravitational contraction and, of course, during the final stage of evolution after all nuclear energy sources have been exhausted, the main driver of stellar evolution is nuclear transformations. These transformations could, in principle, be incorporated directly into the relaxation solution. In practice, it is usually more convenient and practical to treat them separately.

Proceeding as in Section 8.1, let

$$N_i = \int n_i d\tau = \int \rho N_A Y_i d\tau \tag{8.8.1}$$

be the number of nuclei of type i in a co-moving bubble of gas. In the nuclear astrophysics literature, the number abundance parameter Y_i is typically defined by eq. (8.1.33) rather than by eq. (8.2.8). That is,

$$Y_i = \frac{1}{N_A} \frac{n_i}{\rho}, \tag{8.8.2}$$

where n_i is the abundance of particles of type i per cm^3, ρ is the density, and $N_A = 6.2030 \times 10^{23}$ particles per mole is Avogadro's number.

Differentiating eq. (8.8.1) with respect to time, one has

$$\frac{dN_i}{dt} = \int \frac{\partial n_i}{\partial t} d\tau + \int n_i \mathbf{v}_i \cdot d\mathbf{S} = \int \left(\frac{\partial n_i}{\partial t} + \nabla \cdot (n_i \mathbf{v}_i) \right) d\tau, \tag{8.8.3}$$

where $\mathbf{v}_i$ is the mean speed of the particles i at any given location in the bubble. Assuming that diffusion can be neglected, $\mathbf{v}_i = \mathbf{v}$, where $\mathbf{v}$ is the bulk speed. Using eq. (8.1.9), it follows that

$$\frac{\partial n_i}{\partial t} + \nabla \cdot (n_i \mathbf{v}) = \frac{dn_i}{dt} + n_i \nabla \cdot \mathbf{v} = \frac{dn_i}{dt} - \frac{n_i}{\rho} \frac{d\rho}{dt}$$

$$= \frac{d}{dt}(N_A \rho Y_i) - N_A Y_i \frac{d\rho}{dt} = N_A \rho \frac{dY_i}{dt}. \tag{8.8.4}$$

Thus,

$$\frac{dN_i}{dt} = \int N_A \rho \frac{dY_i}{dt} d\tau. \tag{8.8.5}$$

The rate at which nuclei of type i are destroyed by binary reactions with other nuclei (of type j) in the bubble may be written as

$$\left(\frac{dN_i}{dt} \right)_- = \int n_i \sum_j n_j \langle \sigma_{ij} v \rangle \, d\tau, \tag{8.8.6}$$

where σ_{ij} is a cross section, v is a relative speed, and angle brackets denote an average over the thermal distribution.

The rate at which nuclei of type i are created by binary reactions with other nuclei may be written as

$$\left(\frac{dN_i}{dt} \right)_+ = \int \sum_k n_k \sum_j n_j \langle \sigma_{kj} v \rangle \, d\tau, \tag{8.8.7}$$

where the sums are over the relevant nuclear types.

Ignoring, for the moment, gamma capture, electron capture, and electron and positron decay, one can set

$$\frac{dN_i}{dt} = \left(\frac{dN_i}{dt} \right)_+ - \left(\frac{dN_i}{dt} \right)_-, \tag{8.8.8}$$

and using eq. (8.8.2) in combination with eqs. (8.8.5)–(8.8.7), one has

$$\frac{dY_i}{dt} = N_A \rho \left(\sum_k \sum_l Y_k Y_l \langle \sigma_{kl} v \rangle - Y_i \sum_j Y_j \langle \sigma_{ij} v \rangle \right). \tag{8.8.9}$$

$$
\begin{pmatrix}
 & & & & \begin{matrix} in\alpha \\ nzp2, nnp1 \end{matrix} & \begin{matrix} i\gamma\alpha \\ nzp2, nnp2 \end{matrix} \\[2ex]
 & \begin{matrix} inp \\ nzp1, nnm1 \end{matrix} & \begin{matrix} i\gamma p \\ nzp1, nn \end{matrix} & & \begin{matrix} ip\alpha \\ nzp1, nnp2 \end{matrix} \\[2ex]
 & \begin{matrix} in\gamma \\ nz, nnm1 \end{matrix} & \begin{matrix} i \\ nz, nn \end{matrix} & \begin{matrix} i\gamma n \\ nz, nnp1 \end{matrix} \\[2ex]
\begin{matrix} i\alpha p \\ nzm1, nnm2 \end{matrix} & & \begin{matrix} ip\gamma \\ nzm1, nn \end{matrix} & \begin{matrix} ipn \\ nzm1, nnp1 \end{matrix} \\[2ex]
\begin{matrix} i\alpha\gamma \\ nzm2, nnm2 \end{matrix} & \begin{matrix} i\alpha n \\ nzm2, nnm1 \end{matrix}
\end{pmatrix}
\begin{pmatrix} \uparrow \\ \text{proton} \\ \text{number} \end{pmatrix}
$$

(neutron   number   →   )

Fig. 8.8.1   Matrix Building Block

As an example, consider the transformations which involve reactions of isotopes of various kinds with protons, neutrons, alpha particles, gamma rays, and electrons. As illustrated in the "matrix building block" of Fig. 8.8.1, there are potentially as many as twelve isotopes that are linked to any given isotope by reactions involving a proton (p), alpha particle (α), neutron (n), gamma ray (γ), electron (e^-), or positron (e^+). The isotope labeled i at the center of the block is made up of nz protons and nn neutrons. An (α, γ) reaction on this isotope produces the isotope with $nzp2 = nz + 2$ protons and $nnp2 = nn + 2$ neutrons at the upper right of the figure. By labeling the product the $i\gamma\alpha$th isotope, one recognizes that a (γ, α) reaction on it transforms it back into the ith isotope. In a similar way, a (p, n) reaction transforms the ith isotope into the inpth isotope with $nzp1 = nz + 1$ protons and $nnm1 = nn - 1$ neutrons and it, in turn, is transformed back into the ith isotope by an (n, p) reaction.

Taking into account all of the potential reactions, the time rate at which the abundance Y_i of the ith isotope decreases (due to reactions involving protons, neutrons, alpha particles, gamma rays, and electrons) can be written as

$$
\begin{aligned}
(\dot{b}_i)_- = \; & Y_i \, Y_1 \left(R_{p\gamma,i} + R_{p\alpha,i} + R_{pn,i} \right) \\
& + Y_i \, Y_2 \left(R_{n\gamma,i} + R_{np,i} + R_{n\alpha,i} \right) \\
& + Y_i \, Y_3 \left(R_{\alpha\gamma,i} + R_{\alpha p,i} + R_{\alpha n,i} \right) \\
& + Y_i \left(R_{\gamma p,i} + R_{\gamma\alpha,i} + R_{\gamma n,i} + R_{ec,i} + R_{ed,i} \right).
\end{aligned}
\tag{8.8.10}
$$

In eq. (8.8.10) protons, neutrons, and alpha particles are labeled as particles of type 1, 2, and 3, respectively, $R_{p\gamma,i}$ is the rate of the (p, γ) reaction on the ith isotope, $R_{p\alpha,i}$ is the rate of the (p, α) reaction on this same isotope, and so on. The quantity $R_{ed,i}$ is the e^- decay rate of the ith isotope and $R_{ec,i}$ is the electron-capture rate on, or the e^+ decay rate of, this isotope. All rates are in units of number $s^{-1} \, \#^{-1} \#^{-1}$, where the #s are appropriate number abundance parameters. Note that, in the last term of eq. (8.8.10), there is no explicit dependence on a particle abundance other than that of the ith isotope. This is because the photon "abundance" is incorporated in the temperature dependence of the (γ, p), (γ, α), and (γ, n) rates, the free electron abundance (if relevant) is incorporated in the density dependence of the electron-capture rate, and (except under electron-degenerate conditions) the e^- and e^+ decay rates do not depend sensitively on the electron abundance.

In a similar fashion, the time rate at which the abundance of the ith isotope increases due to reactions involving protons, neutrons, alpha particles, gamma rays, and electrons with isotopes other than the ith can be written as

$$
\begin{aligned}
(b_i)_+ = &\ Y_1 \left(Y_{ip\gamma}\, R_{p\gamma,ip\gamma} + Y_{ip\alpha}\, R_{p\alpha,ip\alpha} + Y_{ipn}\, R_{pn,ipn} \right) \\
&+ Y_2 \left(Y_{in\gamma}\, R_{n\gamma,in\gamma} + Y_{in\alpha}\, R_{n\alpha,in\alpha} + Y_{in\alpha}\, R_{n\alpha,in\alpha} \right) \\
&+ Y_3 \left(Y_{i\alpha\gamma}\, R_{\alpha\gamma,i\alpha\gamma} + Y_{i\alpha p}\, R_{\alpha p,i\alpha p} + Y_{i\alpha n}\, R_{\alpha n,i\alpha n} \right) \\
&+ \left(Y_{i\gamma p}\, R_{\gamma p,i\gamma p} + Y_{i\gamma\alpha}\, R_{\gamma\alpha,i\gamma\alpha} + Y_{i\gamma n}\, R_{\gamma n,i\gamma n} \right) \\
&+ \left(R_{\mathrm{ed},ipn} + R_{\mathrm{ec},inp} \right).
\end{aligned}
\tag{8.8.11}
$$

Here, $Y_{ip\gamma}$ is the number abundance parameter for the isotope labeled $ip\gamma$ which experiences a (p, γ) reaction at the rate $R_{p\gamma,ip\gamma}$ to become an isotope of type i, $Y_{ip\alpha}$ is the number abundance parameter of the isotope labled $ip\alpha$ which experiences a (p, α) reaction at the rate $R_{p\alpha,ip\alpha}$ to become an isotope of type i, and so forth.

Altogether,

$$
\frac{\mathrm{d}Y_i}{\mathrm{d}t} = b_i = (b_i)_+ - (b_i)_- , \quad i = 1, n_{\mathrm{e}},
\tag{8.8.12}
$$

where n_{e} is the total number of isotopes in the nuclear burning network. To solve equation (8.8.4), let Y_i now be the abundance of an isotope at time t and let

$$
Y_i' = Y_i + \delta Y_i
\tag{8.8.13}
$$

be its abundance at time $t + \mathrm{d}t$. For small enough $\mathrm{d}t$,

$$
\frac{\delta Y_i}{\delta t} = b_i + \sum_{j=1}^{n_{\mathrm{e}}} \left(\frac{\partial b_i}{\partial Y_j} \right) \delta Y_j
$$

$$
= b_i - \sum_{j=1}^{n_{\mathrm{e}}} a_{ij}\, \delta Y_j,
\tag{8.8.14}
$$

where, on the left hand side of the first inequality, both δY_i and δt are numbers and, on the right hand side, $\partial b_i / \partial Y_j$ is a partial derivative of b_i in eq. (8.8.13) with respect to Y_j:

$$
a_{ij} = - \frac{\partial b_i}{\partial Y_j}.
\tag{8.8.15}
$$

Equation (8.8.15) can also be obtained by inserting $Y_j' = Y_j + \delta Y_j$ for all j in equation (8.8.10) and neglecting all products $\delta Y_j\, \delta Y_k$. Rearranging terms in eq. (8.8.14), one has, for every $i(= 1, n_e)$,

$$\delta Y_i \left(a_{ii} + \frac{1}{\delta t} \right) + \sum_{j=1}^{n_e}{}' a_{ij}\, \delta Y_j = b_i, \tag{8.8.16}$$

where $\sum'$, the sum over j, does not include $j = i$. If only a few isotopes are being followed, eqs. (8.8.16) can be solved for the δY_i by using determinants (Cramer's rule). As a rule, however, it is more convenient to solve the equations by Gaussian elimination (see Section 8.9).

For a small enough time step, the approximation expressed by eq. (8.8.13), with δY_i a solution of eq. (8.8.16), is an adequate approximation. In general, successive iterations are necessary. Define the nth approximation by $Y_i^{(n)}$ and set

$$Y_i^{(n)} = Y_i^{(n-1)} + \delta Y_i^{(n-1)}. \tag{8.8.17}$$

Then $\delta Y_i^{(n)}$ is a solution of

$$\frac{\delta Y_i^{(n)} + Y_i^{(n)} - Y_i^{(0)}}{\delta t} = b_i^{(n)} + \sum_{j=1}^{n_e} \left(\frac{\partial b_i^{(n)}}{\partial Y_j^{(n)}} \right) \delta Y_j^{(n)}$$

$$= b_i^{(n)} - \sum_{j=1}^{n_e} a_{ij}^{(n)}\, \delta Y_j^{(n)}, \tag{8.8.18}$$

where the quantities $b_i^{(n)}$ are given by eqs. (8.8.10)–(8.8.11) evaluated with $Y_i = Y_i^{(n)}$ and the quantities $a_{ij}^{(n)}$ are given by eqs. (8.8.15) evaluated with $Y_i = Y_i^{(n)}$. Rearranging, one has

$$\delta Y_i^{(n)} \left(a_{ii}^{(n)} + \frac{1}{\delta t} \right) + \sum_{j=1}^{n_e}{}' a_{ij}^{(n)}\, \delta Y_j^{(n)} = b_i^{(n)} - \frac{Y_i^{(n)} - Y_i^{(0)}}{\delta t}. \tag{8.8.19}$$

With luck, successive iterations lead to decreasing values of $\delta Y_i^{(n)}$ and may be terminated when

$$\frac{\delta Y_i^{(n)}}{Y_i^{(n)}} < 10^{-m}, \tag{8.8.20}$$

where m is a predetermined quantity of the desired size (say, ≥ 5). Note that, to the extent that eq. (8.8.20) is satisfied when $m = \infty$,

$$Y_i^{(n)} = Y_i^{(0)} + b_i^{(n)} \delta t. \tag{8.8.21}$$

In a model star, the density and temperature also change in a time step, and one may take this into account by first solving for the abundances $Y_i^{(n)}$ at time $t + \delta t$ on the assumption that ρ and t are constant and equal to those at the beginning of the time step, obtaining the estimates $(Y_i)_{\text{start}} = Y_i^{(n)}$. Then, in each iteration k for the solution of the stellar structure equations at time $t + \delta t$, one may again solve eqs. (8.8.19) and (8.8.17), evaluating the b_is and a_{ij}s at the densities and temperatures found at the start of the kth iteration. This time, however, although the value of $Y_i^{(0)}$ on the right hand side of eq. (8.8.19) remains as before, the starting value of $Y_i^{(n)}$ for $n = 0$ is given by $(Y_i)_{\text{start}}$. Thus, the first equation to solve is

$$\delta Y_i^{(0)} \left(a_{ii}^{(0)} + \frac{1}{\delta t} \right) + \sum_{j=1}^{n_e} {}' \, a_{ij}^{(0)} \, \delta Y_j^{(0)} = b_i^{(0)} - \frac{(Y_i)_{\text{start}} - Y_i^{(0)}}{\delta t}, \tag{8.8.22}$$

where the b_is and a_{ij}s are evaluated with the $(Y_i)_{\text{start}}$s, and the solution gives

$$Y_i^{(1)} = (Y_i)_{\text{start}} + \delta Y_i^{(0)}. \tag{8.8.23}$$

In subsequent iterations eqs. (8.8.19) and (8.8.17) are again to be used.

Because the solution algorithm just outlined is implicit, care must be exercised in its use, particularly when there are isotopes in the nuclear burning network which can achieve an equilibrium abundance on a timescale $\tau_{i,\text{eq}}$ which is much smaller than the time step δt appropriate for following structural changes in the model star. One approach is to isolate those isotopes for which $\tau_{i,\text{eq}} \ll \delta t$ and to solve for the equilibrium abundances of these isotopes before calculating the abundance changes of isotopes for which $\tau_{i,\text{eq}} > \delta t$. Equation (8.8.12) can be rewritten as

$$\frac{dY_i}{dt} = (b_i)_+ - d_i \, Y_i$$

$$= d_i \left(Y_{i,\text{eq}} - Y_i \right), \tag{8.8.24}$$

where

$$d_i \, Y_i = (b_i)_- \tag{8.8.25}$$

and

$$Y_{i,\text{eq}} = \frac{(b_i)_+}{d_i}. \tag{8.8.26}$$

Defining lifetimes $\tau_{i,\text{eq}}$ by

$$\tau_{i,\text{eq}} \left| \frac{dY_i}{dt} \right| = \left| Y_{i,\text{eq}} - Y_i \right|, \tag{8.8.27}$$

one has

$$\tau_{i,\text{eq}} = \frac{1}{d_i}. \tag{8.8.28}$$

Equations (8.8.24)–(8.8.28) can be generalized by attaching a superscript (n) to all quantities. Since, by definition, when convergence has been achieved,

$$\frac{dY_i}{dt} = \frac{Y_i^{(n)} - Y_i^{(0)}}{\delta t}, \tag{8.8.29}$$

one has that

$$\tau_{i,eq} \frac{Y_i^{(n)} - Y_i^{(0)}}{\delta t} = \pm \left(Y_{i,eq}^{(n)} - Y_i^{(n)} \right), \tag{8.8.30}$$

or

$$Y_i^{(n)} = Y_{i,eq}^{(n)} \mp \frac{\tau_{i,eq}^{(n)}}{\delta t} \left(Y_i^{(n)} - Y_i^{(0)} \right). \tag{8.8.31}$$

If $Y_i^{(n)} > Y_i^{(0)}$ and $Y_{i,eq}^{(n)} > Y_i^{(n)}$, or if $Y_i^{(n)} < Y_i^{(0)}$ and $Y_{i,eq}^{(n)} < Y_i^{(n)}$, the upper sign is appropriate in eqs. (8.8.30) and (8.8.31). If the two pairs of inequalities are of opposite sign, the lower sign is appropriate. To the extent that $\tau_{i,eq}/\delta t \ll 1$ and $\left| Y_i^{(n)} - Y_i^{(0)} \right| < \min\left(Y_i^{(n)}, Y_{i,eq}^{(n)} \right)$, it is evident that $Y_i^{(n)} = Y_{i,eq}^{(n)}$. One can also write

$$Y_i^{(n)} = Y_{i,eq}^{(n)} + b_i^{(n)} \tau_{i,eq}^{(n)}, \tag{8.8.32}$$

which is to be compared with eq. (8.8.21). From this comparison and from eq. (8.8.30), it is also evident that, when $\tau_{i,eq} \ll \delta t$, one may set

$$Y_i^{(n)} = Y_{i,eq}^{(n)} = \frac{(b_i)_+^{(n)}}{d_i^{(n)}}. \tag{8.8.33}$$

Energy generation rates can be very sensitive to the abundances of isotopes which are nearly in equilibrium and, in constructing a structural model, it is necessary to find the equilibrium abundances. In radiative regions, this is done automatically, as just outlined. In convective regions, it is necessary to construct the matrix obtained from eq. (8.8.15) by restricting attention to just the j_{eq} isotopes which are assumed to be in equilibrium. One could set $\delta t = \infty$ in eq. (8.8.16) and attempt to solve

$$\delta Y_i^{(n)} a_{ii}^{(n)} + \sum_{j=1}^{j_{eq}} {}' a_{ij}^{(n)} \delta Y_j^{(n)} = b_i, \tag{8.8.34}$$

but, in practice, the matrix is sometimes singular, and it is best to include the term $1/\delta t$ in the diagonal matrix elements.

Under some circumstances, it may be desirable to omit from explicit consideration those isotopes which are expected to be in equilibrium most of the time and to calculate nuclear transformations explicitly, being careful to choose the time step δt small enough that the abundance of no isotope retained explicitly changes by a large amount relative to its initial (or final) value. In this case, the equations to solve are

$$Y_i' = Y_i + \frac{dY_i}{dt} \delta t, \tag{8.8.35}$$

and a Simpson's rule algorithm (eq. (8.6.13)) is adequate for solution.

8.9 Solution of linear equations by Gaussian elimination and LU decomposition

Each step of the solution of the implicit abundance-change equations discussed in Section 8.8 requires the solution of a set of linear equations. In the general case, abundances and abundance changes vary over many orders of magnitude and great care must be exercised in obtaining solutions. One tried and true method, named after the mathematician Carl Friedrich Gauss, is Gaussian elimination with pivoting.

The set of n equations

$$\sum_{j=1}^{n} a_{ij} x_j = b_i \, , \, i = 1, n \tag{8.9.1}$$

can be written as

$$a_{11} \, x_1 + a_{12} \, x_2 + \cdots + a_{1n} \, x_n = b_1,$$

$$a_{21} \, x_1 + a_{22} \, x_2 + \cdots + a_{2n} \, x_n = b_2,$$

$$\cdots$$

$$a_{n1} \, x_1 + a_{n2} \, x_2 + \cdots + a_{nn} \, x_n = b_n. \tag{8.9.2}$$

The coefficients a_{ij} are the known elements of a square $n \times n$ matrix $\mathbf{a}$, the b_i are the known elements of a column vector $\mathbf{b}$, and the x_j are the elements of a column vector $\mathbf{x}$ for which a solution is sought. That is, $\mathbf{ax} = \mathbf{b}$.

It is prudent to first normalize the coefficients in each row of eq. (8.9.2) so that, in absolute value, the maximum coefficient is unity. That is, in each row i, find the maximum $|a_{ij}|$ and divide b_i and all coefficients a_{ij} in the row by this maximum, calling the resultant quantities b_i' and a_{ij}'. Next, find the row $i = I$ containing the largest $|a_{i1}'|$ and rearrange the rows in $\mathbf{a}'$ in such a way that row I is the first row in the final matrix $\mathbf{a}''$, and adjust the elements in the vector $\mathbf{b}'$ accordingly to obtain the vector $\mathbf{b}''$. This process is called pivoting and is often crucial for achieving satisfactory accuracy. Note that by rearranging rows rather than columns to insure that the top left element in the matrix is the largest element in the first column of the matrix, the vector $\mathbf{x}$ is unchanged. Finally, define $b_i^{(0)} = b_i''$ for all i and $a_{ij}^{(0)} = a_{ij}''$ for all i and j.

Solving the first row in the equation $\mathbf{a}^{(0)} \mathbf{x} = \mathbf{b}^{(0)}$ for x_1 in terms of the other elements in $\mathbf{x}$, one has

$$x_1 = \frac{b_1^{(0)}}{a_{11}^{(0)}} - \sum_{j=2}^{n} \frac{a_{1j}^{(0)}}{a_{11}^{(0)}} x_j. \tag{8.9.3}$$

Inserting this expression for x_1 into the remaining $n - 1$ rows of eq. (8.9.3) produces

$$a_{22}^{(1)} x_2 + a_{23}^{(1)} x_3 + \cdots + a_{2n}^{(1)} x_n = b_2^{(1)},$$

$$a_{32}^{(1)} x_2 + a_{33}^{(1)} x_3 + \cdots + a_{3n}^{(1)} x_n = b_3^{(1)},$$

$$\cdots$$

$$a_{n2}^{(1)} x_2 + a_{n3}^{(1)} x_3 + \cdots + a_{nn}^{(1)} x_n = b_n^{(1)}, \tag{8.9.4}$$

where

$$a_{2j}^{(1)} = a_{2j}^{(0)} - \frac{a_{21}^{(0)}}{a_{11}^{(0)}} a_{1j}^{(0)}, \quad j \geq 2,$$

$$a_{3j}^{(1)} = a_{3j}^{(0)} - \frac{a_{31}^{(0)}}{a_{11}^{(0)}} a_{1j}^{(0)}, \quad j \geq 2,$$

$$\cdots$$

$$a_{nj}^{(1)} = a_{nj}^{(0)} - \frac{a_{n1}^{(0)}}{a_{11}^{(0)}} a_{1j}^{(0)}, \quad j \geq 2, \tag{8.9.5}$$

and

$$b_i^{(1)} = b_i^{(0)} - \frac{b_1^{(0)}}{a_{11}^{(0)}} a_{i1}^{(0)}, \quad i \geq 2. \tag{8.9.6}$$

In each row i of the $(n - 1) \times (n - 1)$ matrix described by eq. (8.9.4), find the maximum $\left| a_{ij}^{(1)} \right|$ and divide $b_i^{(1)}$ and all coefficients $a_{ij}^{(1)}$ in the row by this maximum. Then, find the row $i = I$ containing the largest value of $\left| a_{i2}^{(1)} \right|$, rearrange the rows in $\mathbf{a}^{(1)}$ so that this row is at the top, and adjust the vector $\mathbf{b}^{(1)}$ accordingly, calling the new matrix elements and vector elements a_{ij}' and b_i', respectively. Next, rename the new set of members in the normalized and pivoted matrix and in the adjusted vector $a_{ij}^{(1)} = a_{ij}'$ and $b_i^{(1)} = b_i'$, respectively. Solve the first row of the revised eq. (8.9.4) for x_2 in terms of the other x_is:

$$x_2 = \frac{b_2^{(1)}}{a_{22}^{(1)}} - \sum_{j=3}^{n} \frac{a_{2j}^{(1)}}{a_{22}^{(1)}} x_j. \tag{8.9.7}$$

Eliminate x_2 from the remaining $n - 2$ equations, resulting in

$$a_{33}^{(2)} x_3 + a_{34}^{(2)} x_3 + \cdots + a_{3n}^{(2)} x_n = b_3^{(2)},$$

$$a_{43}^{(2)} x_3 + a_{44}^{(2)} x_4 + \cdots + a_{4n}^{(2)} x_n = b_4^{(2)},$$

$$\cdots$$

$$a_{n3}^{(2)} x_3 + a_{n4}^{(2)} x_4 + \cdots + a_{nn}^{(2)} x_n = b_n^{(2)}, \tag{8.9.8}$$

where

$$a_{3j}^{(2)} = a_{3j}^{(1)} - \frac{a_{32}^{(1)}}{a_{22}^{(1)}} a_{2j}^{(1)},$$

$$a_{4j}^{(2)} = a_{4j}^{(1)} - \frac{a_{42}^{(1)}}{a_{22}^{(1)}} a_{2j}^{(1)},$$

$$\cdots$$

$$a_{nj}^{(2)} = a_{nj}^{(1)} - \frac{a_{n2}^{(1)}}{a_{22}^{(1)}} a_{2n}^{(1)}, \qquad (8.9.9)$$

and

$$b_i^{(2)} = b_i^{(1)} - \frac{a_{i2}^{(1)}}{a_{22}^{(1)}} b_2^{(1)}. \qquad (8.9.10)$$

Continue in this fashion until obtaining

$$a_{n-1,n-1}^{(n-2)} x_{n-1} + a_{n-1,n}^{(n-2)} x_n = b_{n-1}^{(n-2)} \qquad (8.9.11)$$

and

$$a_{n,n-1}^{(n-2)} x_{n-1} + a_{n,n}^{(n-2)} x_n = b_n^{(n-2)}. \qquad (8.9.12)$$

From this last pair of equations emerge concrete solutions for x_{n-1} and x_n in terms of known quantities:

$$x_{n-1} = \frac{b_{n-1}^{(n-2)} a_{n,n}^{(n-2)} - b_n^{(n-2)} a_{n-1,n}^{(n-2)}}{a_{n-1,n-1}^{(n-2)} a_{n,n}^{(n-2)} - a_{n,n-1}^{(n-2)} a_{n,n}^{(n-2)}} \qquad (8.9.13)$$

and

$$x_n = \frac{b_n^{(n-2)} a_{n-1,n-1}^{(n-2)} - b_{n-1}^{(n-2)} a_{n,n-1}^{(n-2)}}{a_{n-1,n-1}^{(n-2)} a_{n,n}^{(n-2)} - a_{n,n-1}^{(n-2)} a_{n,n}^{(n-2)}}. \qquad (8.9.14)$$

From eqs. (8.9.3) and (8.9.7), it follows by induction that concrete values for all other elements in the vector **x** may be obtained by using the recurrence relationship

$$x_k = \frac{1}{a_{kk}^{(k-1)}} \left(b_k^{(k-1)} - \sum_{j=k+1}^{n} a_{kj}^{(k-1)} x_j \right), \quad k = n - 3, 1. \qquad (8.9.15)$$

The notation used in reaching eqs. (8.9.13)–(8.9.15), although cumbersome, has been adopted to emphasize that, at the end of each step of the Gaussian elimination and pivoting process, the elements in the reduced matrix and vector are usually different from the elements at the beginning of the process. This point being made, it becomes appropriate to simplify the notation. Defining the numerator and denominator in eq. (8.9.14) as b'_n and a'_{nn}, respectively, one has that

$$x_n = \frac{b'_n}{a'_{n,n}}. \tag{8.9.16}$$

Similarly, b'_{n-1} and $a'_{n-1,n-1}$ may be defined in such a way that eq. (8.9.13) reads

$$x_{n-1} = \frac{1}{a'_{n-1,n-1}} \left(b'_{n-1} - a'_{n-1,n} x_n \right). \tag{8.9.17}$$

Finally, defining $b'_k = b_k^{(k-1)}$ and $a'_{kj} = a_{kj}^{(k-1)}$, eqs. (8.9.15)–(8.9.17) can be described by a single equation:

$$x_k = \frac{1}{a'_{kk}} \left(b'_k - \sum_{j=k+1}^{n} a'_{kj} x_j \right), \quad k = n, 1, \tag{8.9.18}$$

with the understanding that the sum over j is zero when $k = n$. It is easy to verify that eq. (8.9.18) is the solution of the matrix equation

$$
\begin{pmatrix}
a'_{11} & a'_{12} & a'_{13} & \cdots & a'_{1,n-2} & a'_{1,n-1} & a'_{1n} \\
0 & a'_{22} & a'_{23} & \cdots & a'_{2,n-2} & a'_{2,n-1} & a'_{2n} \\
0 & 0 & a'_{33} & \cdots & a'_{3,n-2} & a'_{3,n-1} & a'_{3n} \\
\vdots & \vdots & \vdots & \ddots & \vdots & \vdots & \vdots \\
0 & 0 & 0 & \cdots & a'_{n-2,n-2} & a'_{n-2,n-1} & a'_{n-2,n} \\
0 & 0 & 0 & \cdots & 0 & a'_{n-1,n-1} & a'_{n-1,n} \\
0 & 0 & 0 & \cdots & 0 & 0 & a'_{nn}
\end{pmatrix}
\begin{pmatrix}
x_1 \\ x_2 \\ x_3 \\ \vdots \\ x_{n-2} \\ x_{n-1} \\ x_n
\end{pmatrix}
=
\begin{pmatrix}
b'_1 \\ b'_2 \\ b'_3 \\ \vdots \\ b'_{n-2} \\ b'_{n-1} \\ b'_n
\end{pmatrix}. \tag{8.9.19}
$$

The product of the last row of the matrix with the vector $\mathbf{x}$ produces $a'_{nn} x_n = b'_n$, which is just eq. (8.9.16). Equating the product of the next to the last row of the matrix with the vector $\mathbf{x}$ gives eq. (8.9.17), and so on. Thus, the process of Gaussian elimination has converted the initial set of linear equations into a form such that the product of an "upper" triangular matrix and the desired vector equals another vector. The solution of this equation is enjoyably straightforward.

The same is true if the matrix is a "lower" triangular matrix, the only difference being that, instead of finding the last element of the vector $\mathbf{x}$ and solving for other elements by back substitution, one finds instead the first element of the vector and solves for additional elements by forward substitution. That is, if the sought after vector satisfies the matrix equation

$$
\begin{pmatrix}
\alpha_{11} & 0 & \cdots & 0 & 0 \\
\alpha_{21} & \alpha_{22} & \cdots & 0 & 0 \\
\vdots & \vdots & \ddots & \vdots & \vdots \\
\alpha_{n-1,1} & \alpha_{n-1,2} & \cdots & \alpha_{n-1,n-1} & 0 \\
\alpha_{n1} & \alpha_{n2} & \cdots & \alpha_{n,n-1} & \alpha_{nn}
\end{pmatrix}
\begin{pmatrix}
x_1 \\ x_2 \\ \vdots \\ x_{n-1} \\ x_n
\end{pmatrix}
=
\begin{pmatrix}
b_1 \\ b_2 \\ \vdots \\ b_{n-1} \\ b_n
\end{pmatrix}, \tag{8.9.20}
$$

then

$$x_k = \frac{1}{\alpha_{kk}} \left(b_k - \sum_{j=1}^{k-1} a_{kj} \, x_j \right), \quad k = 1, n, \tag{8.9.21}$$

where it is understood that the sum over j is zero when $k = 1$.

With rare exceptions, any square matrix $\mathbf{A}$ can be converted into the product of a lower triangular matrix $\mathbf{L}$ and an upper triangular matrix $\mathbf{U}$, with the consequence that any equation of the form $\mathbf{Ax} = \mathbf{b}$ can be written and solved in a manner which involves the simple forward and backward substitution procedures described by eqs. (8.9.21) and (8.9.18), respectively. That is, one may write

$$\mathbf{Ax} = \mathbf{LUx} = \mathbf{L(Ux)} = \mathbf{Ly} = \mathbf{b}, \tag{8.9.22}$$

where the vector $\mathbf{y}$ is the solution of

$$\mathbf{Ly} = \mathbf{b} \tag{8.9.23}$$

and the vector $\mathbf{x}$ is the solution of

$$\mathbf{Ux} = \mathbf{y}. \tag{8.9.24}$$

Back substitution prescribed by eq. (8.9.21) may be used to find $\mathbf{y}$ and forward substitution prescribed by eq. (8.9.18) may be used to find $\mathbf{x}$.

The process of matrix decomposition appears to have been described first by André-Louis Cholesky prior to his death on the battlefield in 1918, with the description appearing posthumously in the 1922 issue of the *Bulletin Geodésique*. Two other names — Banachiewicz (1938, 1942) and Crout (1941) — are frequently associated with the process. Since it involves fewer operations than ordinary Gaussian elimination, the solution of linear equations by LU decomposition is a popular choice, particulary when the initial matrix is dense.

The basic ideas involved in LU decomposition are easily demonstrated for 2×2 matrices, with generalization to higher order matrices being accomplished readily by induction. Assume that there exist three 2×2 matrices with properties such that

$$\begin{pmatrix} \alpha_{11} & 0 \\ \alpha_{21} & \alpha_{22} \end{pmatrix} \begin{pmatrix} \beta_{11} & \beta_{12} \\ 0 & \beta_{22} \end{pmatrix} = \begin{pmatrix} a_{11} & a_{12} \\ a_{21} & a_{22} \end{pmatrix}, \tag{8.9.25}$$

where the four non-vanishing quantities a_{ij} in the third matrix are known and the six quantities in the two triangular matrices are to be found. Multiplying the matrices on the left hand side of eq. (8.9.25) produces the matrix equation

$$\begin{pmatrix} \alpha_{11}\beta_{11} & \alpha_{11}\beta_{12} \\ \alpha_{21}\beta_{11} & \alpha_{21}\beta_{12} + \alpha_{22}\beta_{22} \end{pmatrix} = \begin{pmatrix} a_{11} & a_{12} \\ a_{21} & a_{22} \end{pmatrix}, \tag{8.9.26}$$

which corresponds to four linear equations with six unknowns. The Crout algorithm begins by setting $\alpha_{11} = \alpha_{22} = 1$, converting eq. (8.9.26) into

$$\begin{pmatrix} \beta_{11} & \beta_{12} \\ \alpha_{21}\beta_{11} & \alpha_{21}\beta_{12} + \beta_{22} \end{pmatrix} = \begin{pmatrix} a_{11} & a_{12} \\ a_{21} & a_{22} \end{pmatrix}. \tag{8.9.27}$$

Thus, the top row of the product matrix has been identified with the top row of the a matrix, thereby allowing two unknowns in the first and second rows of the product matrix to be replaced by knowns:

$$\begin{pmatrix} a_{11} & a_{12} \\ \alpha_{21}a_{11} & \alpha_{21}a_{12} + \beta_{22} \end{pmatrix} = \begin{pmatrix} a_{11} & a_{12} \\ a_{21} & a_{22} \end{pmatrix}. \tag{8.9.28}$$

Equating lower left hand elements in the two matrices then gives α_{21} in terms of knowns and this result can be used to reduce the number of unkowns in the lower right hand element of the product matrix to one, which is then found by equating lower right hand elements in the two matrices. The final result is

$$\alpha = \begin{pmatrix} 1 & 0 \\ a_{21}/a_{11} & 1 \end{pmatrix} \text{ and } \beta = \begin{pmatrix} a_{11} & a_{12} \\ 0 & a_{22} - a_{21}a_{12}/a_{11} \end{pmatrix}. \tag{8.9.29}$$

Adopting the same procedure for 3×3 matrices, i.e., setting all diagonal elements in the L triangular matrix equal to unity ($\alpha_{jj} = 1$ for all $j = 1, 3$), one may go immediately to the equivalent of eq. (8.9.28):

$$\begin{pmatrix} a_{11} & a_{12} & a_{13} \\ \alpha_{21}a_{11} & \alpha_{21}a_{12} + \beta_{22} & \alpha_{21}a_{13} + \beta_{23} \\ \alpha_{31}a_{11} & \alpha_{31}a_{12} + \alpha_{32}\beta_{22} & \alpha_{31}a_{13} + \alpha_{32}\beta_{23} + \beta_{33} \end{pmatrix} = \begin{pmatrix} a_{11} & a_{12} & a_{13} \\ a_{21} & a_{22} & a_{23} \\ a_{31} & a_{32} & a_{33} \end{pmatrix}. \tag{8.9.30}$$

Equating elements in the first columns of the two matrices yields the unknowns α_{21} and α_{31} in terms of knowns and the results may be used in the second column of the left hand matrix to leave one unknown, β_{22}, in the second row and two unknowns, β_{22} and α_{32}, in the third row. Equating the $ij = 22$ elements in the two matrices then yields β_{22} in terms of knowns and this may be used in the $ij = 32$ element on the left hand side. Since α_{31} in this element is already known, equating the $ij = 32$ components of the two matrices then yields α_{32} in terms of knowns. Finally, one proceeds downward through the third columns of the two matrices. In each successive row in the third column of the left hand matrix, all but one of the remaining unknowns has been determined by a previous operation and this unknown is found in terms of knowns by equating the elements in the left and right hand matrices. In this way, one obtains, in the following order,

$$\beta_{11} = a_{11}, \ \alpha_{21} = \frac{a_{21}}{a_{11}}, \ \alpha_{31} = \frac{a_{31}}{a_{11}}, \tag{8.9.31}$$

$$\beta_{12} = a_{12}, \ \beta_{22} = \frac{a_{22}a_{11} - a_{21}a_{12}}{a_{11}}, \ \alpha_{32} = \frac{a_{11}a_{32} - a_{31}a_{12}}{a_{22}a_{11} - a_{21}a_{12}}, \tag{8.9.32}$$

$$\beta_{13} = a_{13}, \ \beta_{23} = \frac{a_{23}a_{11} - a_{21}a_{13}}{a_{11}}, \tag{8.9.33}$$

and

$$\beta_{33} = \frac{a_{33}a_{11} - a_{31}a_{13}}{a_{11}} - \frac{a_{11}a_{32} - a_{31}a_{12}}{a_{22}a_{11} - a_{21}a_{12}} \frac{a_{23}a_{11} - a_{21}a_{13}}{a_{11}}. \tag{8.9.34}$$

For still higher order matrices, exactly the same procedure is followed. Converting to what has become a standard notation, designate elements in the **L** and **U** matrices by l_{ij} and u_{ij}, respectively. The first order of business is to set all of the diagonal elements l_{ii} of the lower triangular matrix equal to unity. Then, starting at the top of the first column of the product matrix **LU**, one progresses downward, finding u_{11} and $(l_{i1}, \; i = 2, n)$, where n is the order of the matrix. In the second column, again starting at the top, one progresses downward, finding u_{12}, u_{22}, and $(l_{i2}, \; i = 3, n)$. In the jth column, one finds $(u_{ij}, \; i = 1, j)$ and $(l_{ij}, \; i = j + 1, n)$. In the $(n - 1)$th column, one finds $(u_{i,n-1}, \; i = 1, n - 1)$ and $l_{n,n-1}$ and, in the last column, one finds $(u_{in}, \; i = 1, n)$.

In this elementary presentation of the process of decomposition, the process of pivoting has been ignored. A description of the decomposition process with pivoting is given in the book *Numerical Recipes in FORTRAN: the Art of Scientific Computing* by William H. Press, Brian P. Flannery, Saul A. Teukolsky, & William T. Vetterling (first edition 1966 and many editions since). The description includes a Fortran code implementing the process. Information about coding in languages other than Fortran is available on the website www.ns.com.

8.10 Composition changes in convective regions

Convective currents appear ubiquitously in evolving stars. They influence structural characteristics and can act as a principle vehicle for carrying into surface layers the products of nuclear transformations occurring in the deep interior. Convective zones in the deep interior of a star are typically associated with a nuclear burning source with a strong temperature dependence. An example is a main sequence star which is massive enough that CN-cycle reactions are the dominant manner in which hydrogen burns near the center. The convective zone typically contains more mass than the region within which energy production is confined, with the result that mixing of unburned matter with burned matter prolongs the main sequence lifetime.

Another example is a thermally pulsing asymptotic giant branch (TPAGB) star which periodically ignites helium in a shell sandwiched between a hot white dwarf core and a hydrogen-rich convective envelope. During a mild helium-shell flesh, convection carries products of helium burning outward to just below the hydrogen-rich region. Still another example is the dredge-up episode following a helium-shell flash. After shell convection has died out, energy stored as heat in the erstwhile shell is converted into an outward flow of radiant energy and the increase in the energy flux at the base of the hydrogen-rich region causes the base of the hydrogen-rich convective envelope to move inward in mass into the region where fresh products of helium burning exist. These products are mixed throughout the convective envelope from which they may be expelled through winds into the interstellar medium. Thermally pulsing evolution is discussed extensively in Chapters 17–19 in Volume 2.

The theory of turbulent flow in convective regions is complex and, in practice, its application in the stellar structure and evolution context has been necessarily rather rudimentary, with the mixing-length treatment of convection discussed in Section 3.4 often being the default option adopted.

8.10.1 Time scales of relevance

From the point of view of compositional changes, an important feature of a convective region is the "turnover" time, the time required for matter to mix thoroughly in the region. In the framework of the mixing-length treatment, a lower limit on this time can be estimated as

$$dt_{mix} > dt_{min} = \int_{x_{base}}^{x_{edge}} \frac{dx}{v_{conv}}, \tag{8.10.1}$$

where the convective speed v_{conv} is given in various guises by eqs. (3.4.32), (3.4.38), and (3.4.41) and the integral extends from the base to the outer edge of the convective region. This is a lower limit since, in the mixing length picture, convective elements move back and forth in either direction with respect to the temperature gradient. Given that convective elements are formed and decay randomly, one might guess that the turnover time is related to the duration of a convective cell, the mixing length, and the dimensions of the convective region in a manner similar to that which describes Brownian motion in the random walk approximation. In this approximation, if the time dt_{step} for a particle to travel a mean distance L_{step} in a straight line at mean speed v_{step} before changing directions is

$$t_{step} = \frac{L_{step}}{v_{step}}, \tag{8.10.2}$$

then the time t_{total} to traverse a distance L_{total} in any specific direction is related the distance traversed in each of a series of randomly oriented directions by

$$t_{total} = \left(\frac{L_{total}}{L_{step}}\right)^2 t_{step} = \frac{L_{total}^2}{v_{step}L_{step}}. \tag{8.10.3}$$

Thus, an estimate of the turnover time in a convective region might be

$$dt_{mix} \sim \frac{(\Delta R_{conv})^2}{v_{conv}l_{mix}} \sim \frac{\Delta R_{conv}}{l_{mix}} dt_{min}, \tag{8.10.4}$$

where ΔR_{conv} is the radial distance across the convective region. While this estimate makes intuitive sense when the ratio of ΔR_{conv} to l_{mix} is large, in many situations encountered during the course of evolution, the exact reverse is true. This highlights another major conceptual shortcoming of the mixing-length treatment.

Two other time scales enter into the determination of how changes in the abundance of a given isotope are to be estimated. These are dt_{eq}, the time for an isotope to come into equilibrium locally in consequence of nuclear transformations (discussed in Section 8.7), and dt_{evo}, the time step chosen in the construction of a set of models. This latter time may be chosen to be some fraction of the time over which significant changes in structural characteristics or in isotopic abundances occur. In practice, it is often dictated by experience as the time step necessary for achieving convergence in a given stage of evolution.

Table 8.10.1 Time scales of relevance in convective zones

Case	Shortest	Middle	Longest	Action
(1)	dt_{eq}	dt_{mix}	dt_{evo}	local equilibrium, no mixing
(2)	dt_{eq}	dt_{evo}	dt_{mix}	local equilibrium, no mixing
(3)	dt_{mix}	dt_{eq}	dt_{evo}	mix uniformly, global equilibrium
(4)	dt_{mix}	dt_{evo}	dt_{eq}	mix uniformly, no equilibrium
(5)	dt_{evo}	dt_{mix}	dt_{eq}	no equilibrium, diffusive mixing
(6)	dt_{evo}	dt_{eq}	dt_{mix}	no mixing, no equilibrium

In every interior shell of a model, each isotope may be assigned an "equilibrium index" according to which of the six permutations of the three time scales applies. The permutations and the action to be taken are shown in Table 8.10.1 In some instances, a given isotope belongs to category 1 in some parts of the convective zone, and to category 3 in other parts. One way of handling this situation is to "take a vote", weighting according to the mass of the shell; the category with the most mass wins and the isotope of the given kind is treated in the same way throughout the convective shell. That is, the isotope is either mixed uniformly throughout the convective zone, or it takes on the local equilibrium value in each mass shell in the convective zone.

When uniform mixing is appropriate for a set of isotopes, all isotopes in the set are mixed at the beginning of a time step. Next, in each mass shell in the convective zone, changes which occur *locally* over a time step dt_{evo} in all isotopic abundances are calculated according to the nuclear transformation algorithms of Section 8.7. Then, each isotope in the to-be-mixed set is again mixed over the convective zone. In a final step, the abundances of all isotopes which are in local equilibrium at the densities and temperatures characterizing the model at the beginning of the time step may be redetermined, although these equilibrium abundances will change in the process of determining the density and temperature distribution at the end of the time step.

8.10.2 Convective diffusion in the mixing length approximation

If, as during nova explosions, the time scale set by the rate of change of structural characteristics is smaller than either the time scale for achieving local equilibrium or the convective mixing time scale, the variation in space of composition variables may be estimated by employing a "convective diffusion" algorithm modeled after the treatment of molecular diffusion in terrestial laboratory experiments. When gradients in the mass density are negligible, experimental results can be expressed as solutions of Fricke's law, which can be written as

$$\mathbf{f}_i = -D_i \, \nabla n_i, \tag{8.10.5}$$

where $\mathbf{f}_i$ is the flux of particles of type i, n_i is the number density of such particles, and D_i is a diffusion coefficient which depends on local conditions, including, in principle, the number abundances of all types of particles present.

In the stellar context, where the density is variable, one may define a "diffusivity" D_i in such a way that the diffusive flux is

$$\mathbf{f}_i = -D_i \, N_A \, \rho \, \nabla Y_i, \tag{8.10.6}$$

where Y_i is the abundance parameter defined by eq. (8.1.33). In this formulation, diffusion occurs in a region of variable density only if there are gradients in relative abundance parameters.

The total flux of particles of type i may be written as

$$n_i \mathbf{v}_i = n_i \mathbf{v} + \mathbf{f}_i, \tag{8.10.7}$$

where $\mathbf{v}$ and $\mathbf{v}_i$ are, respectively, the bulk velocity and the mean velocity of the ith type of particle, and $\mathbf{f}_i$ is the flux of the ith type of particle in a frame of reference moving with the bulk velocity. The total number of particles of type i can be written as

$$N_i = \int n_i d\tau, \tag{8.10.8}$$

and, in analogy with eqs. (8.1.4) and (8.1.7), the equation for particle conservation is

$$\frac{dN_i}{dt} = \int \frac{\partial n_i}{\partial t} \, d\tau + \int n_i \mathbf{v}_i \cdot d\mathbf{S} = \int \frac{\partial n_i}{\partial t} \, d\tau + \int n_i \mathbf{v} \cdot d\mathbf{S} + \int \mathbf{f}_i \cdot d\mathbf{S} = 0. \tag{8.10.9}$$

Using Green's theorem to convert the surface integrals to integrals over volume, eq. (8.10.9) becomes

$$\frac{\partial n_i}{\partial t} + \nabla \cdot (n_i \mathbf{v}) = -\nabla \cdot \mathbf{f}_i. \tag{8.10.10}$$

Replacing n_i in terms of ρ, Y_i, and Avogadro's number, as given by eq. (8.1.33), the left hand side of eq. (8.10.10) becomes

$$\frac{\partial n_i}{\partial t} + \nabla \cdot (n_i \mathbf{v}) = N_A \left[\frac{\partial (\rho Y_i)}{\partial t} + \nabla \cdot (\rho Y_i \mathbf{v}) \right] \tag{8.10.11}$$

$$= N_A \left[Y_i \left(\frac{\partial \rho}{\partial t} + \nabla \cdot (\rho \mathbf{v}) \right) + \rho \left(\frac{\partial Y_i}{\partial t} + \mathbf{v} \cdot \nabla Y_i \right) \right]. \tag{8.10.12}$$

In the expression in square brackets in eq. (8.10.12), the coefficient of Y_i vanishes by virtue of the conservation of mass as expressed by eq. (8.1.9). From the definition given by eq. (8.1.18), it is evident that the coefficient of ρ in the square brackets is the moving time derivative dY_i/dt. Thus,

$$\frac{\partial n_i}{\partial t} + \nabla \cdot (n_i \mathbf{v}) = N_A \, \rho \, \frac{dY_i}{dt}. \tag{8.10.13}$$

Using eqs. (8.10.6) and (8.10.13) in eq. (8.10.10) gives

$$\rho \, \frac{dY_i}{dt} = -\nabla \cdot \mathbf{f}_i = \nabla \cdot (\rho \, D_i \, \nabla Y_i). \tag{8.10.14}$$

Invoking spherical symmetry, one arrives at

$$\rho \, \frac{dY_i}{dt} = \frac{1}{r^2} \, \frac{\partial}{\partial r} \left(r^2 \, \rho \, D_i \, \frac{\partial Y_i}{\partial r} \right). \tag{8.10.15}$$

In molecular diffusion, the diffusivity is the product of a shear velocity and mean free path and differs from one type of particle to the next. In the case of convective diffusion, one might suppose that, since turbulent currents carry along all particles, the diffusive coefficient is the same for every type of particle, so that $D_i = D$, where D is a function of characteristics of the turbulent medium. For example, one might guess that

$$D \sim \lambda' \, v_{\mathrm{conv}} \, l_{\mathrm{mix}} \sim \lambda \, v_{\mathrm{conv}} \, H_{\mathrm{P}}, \tag{8.10.16}$$

where v_{conv} is a measure of the speed of convective motions relative to the bulk velocity, l_{mix} is an estimate of the mixing length, which is commonly taken to be some fraction of the pressure scale height H_{P}, and λ' and λ are parameters which are commonly assumed to be of the order of unity.

In general, compositional changes should be estimated by solving the equation

$$\frac{dY_i}{dt} = \left(\frac{dY_i}{dt} \right)_{\mathrm{turbulent \ diffusion}} + \left(\frac{dY_i}{dt} \right)_{\mathrm{nuclear \ transformations}} \tag{8.10.17}$$

for each isotope. This approach has been used with some success to model abundance changes in the convective envelopes of intermediate mass AGB stars which become super lithium rich (Sackmann, Smith, & Despain, 1974). In this case, the time scale for the convective diffusion of $^7\mathrm{Li}$ and of its electron-capture progenitor $^7\mathrm{Be}$ and the time scales for nuclear transformations which create $^7\mathrm{Be}$ and destroy $^7\mathrm{Li}$ are comparable.

However, in other circumstances, such as when hydrogen is ingested into a region in which convection is driven by helium burning, this approach is not practical and it is more useful to think of the two types of transformation – nuclear transitions and convective mixing – as processes which can be treated separately. In this case, it is necessary to treat eqs. (8.10.15) and (8.10.16) as an empirical algorithm in which the coefficient λ is chosen small enough to prevent ignition on a dynamic time scale of the ingested hydrogen.

To the extent that the product $r^2 \, \rho \, D$ varies slowly with respect to $(\partial Y_i / \partial r)$, one may write the diffusion equation as

$$\frac{dY_i}{dt} \sim D \, \frac{\partial^2 Y_i}{\partial r^2}. \tag{8.10.18}$$

In a region of linear dimensions ΔR, the time scale for diffusion suggested by eq. (8.10.18) is of the order of

$$t_{\mathrm{diff}} \sim \frac{(\Delta R)^2}{D}. \tag{8.10.19}$$

With the choice for D given by eq. (8.10.16), eq. (8.10.17) resembles eq. (8.10.4). This provides some support for invoking the concept of a random walk to describe the motions of turbulent elements in a convective region.

8.10.3 Solution of the convective diffusion equation: conversion to a difference equation and construction of recurrence relationships

In cases when

$$dt_{evo} \ll dt_{mix} \ll dt_{eq}, \tag{8.10.20}$$

the convective diffusion equation may be solved separately from the equations for nuclear transformations. Since $dM(r) = \rho\, 4\pi r^2 dr$, eq. (8.10.15) may be written as

$$\frac{dY_i}{dt} = \frac{d}{dM(r)}\left((4\pi r^2 \rho)^2\, D_i\, \frac{dY_i}{dM(r)}\right). \tag{8.10.21}$$

The next step is to convert the differential equation into a difference equation. Since the form of the equation is the same for all isotopes, the continuous variable Y_i describing the ith isotope may be replaced by a generic, unsubscripted, continuous variable Y. However, in converting the differential equation into a difference equation, Y must be defined at a discrete set of points, so, perversely, a subscript must be re-introduced. Thus, the continuous variable Y is replaced by a set of quantities Y_j, where the subscript j identifies the outer edge of the jth spherical shell (see Section 8.6 for the definition of the jth shell). Calling $Y_j^{(0)}$ the initial value of an arbitrary isotope in the middle of the jth shell and Y_j the value of this isotope at the end of a time step δt, eq. (8.10.21) may be approximated by

$$\frac{Y_j - Y_j^{(0)}}{\delta t} = -\frac{F_j - F_{j-1}}{dM_j}, \tag{8.10.22}$$

where

$$F_j = -\left(4\pi R_j^2 \rho_{j+1/2}\right)^2 D_j\, \frac{Y_{j+1} - Y_j}{(dM_j + dM_{j+1})/2} \tag{8.10.23}$$

is the flux through the outer boundary (the *top*) of the jth shell. In these equations, M_j is the total mass existing below the outer boundary of the jth shell and $dM_j = M_j - M_{j-1}$. Further,

$$\rho_{j+1/2} = \frac{\rho_j dM_{j+1} + \rho_{j+1} dM_j}{dM_{j+1} + dM_j} \tag{8.10.24}$$

is an approximation to the density at the outer boundary of the jth shell, and one may adopt

$$D_j = \lambda\, (H_P)_j\, v_j = \lambda\left(-\frac{1}{P}\frac{dP}{dr}\right)_j^{-1} v_j$$

$$= \lambda\left(\frac{1}{P_{j+1/2}}\frac{GM_j}{R_j^2}\rho_{j+1/2}\right)^{-1} v_j, \tag{8.10.25}$$

where $(H_P)_j$ is the pressure scale height and v_j is the convective velocity at the outer edge of the jth shell.

In preparation for solving the diffusion equation, it is convenient to define three variables,

$$s(1, j) = D_j,$$ (8.10.26)

$$s(2, j) = \frac{\delta t}{dM_j},$$ (8.10.27)

and

$$s(3, j) = \left(4\pi R_j^2 \rho_{j+1/2}\right)^2 \frac{s(1, j)}{(dM_{j+1} + dM_j)/2},$$ (8.10.28)

in terms of which eq. (8.10.23) becomes

$$F_j = -s(3, j)(Y_{j+1} - Y_j)$$ (8.10.29)

and eq. (8.10.22) becomes

$$Y_j - Y_j^{(0)} = s(2, j)[s(3, j)(Y_{j+1} - Y_j) - s(3, j - 1)(Y_j - Y_{j-1})].$$ (8.10.30)

Equation (8.10.30) may be rearranged to read

$$Y_j = \alpha_j Y_j^{(0)} + \beta_j Y_{j+1} + \gamma_j Y_{j-1},$$ (8.10.31)

where

$$\alpha_j = \frac{1}{1 + s(2, j)[s(3, j) + s(3, j - 1)]},$$ (8.10.32)

$$\beta_j = \alpha_j s(2, j) s(3, j),$$ (8.10.33)

and

$$\gamma_j = \alpha_j s(2, j) s(3, j - 1).$$ (8.10.34)

Formally setting

$$Y_{j+1} = a_{j+1} Y_j + b_{j+1}$$ (8.10.35)

in eq. (8.10.31) yields

$$Y_j = a_j Y_{j-1} + b_j,$$ (8.10.36)

where

$$a_j = \frac{\gamma_j}{1 - \beta_j a_{j+1}}$$ (8.10.37)

and

$$b_j = \frac{\alpha_j Y_j^{(0)} + \beta_j b_{j+1}}{1 - \beta_j a_{j+1}}.$$ (8.10.38)

It is evident that if, at the outer edge of the convective zone where, say, $j = K$, a_K and b_K can be determined by the application of a boundary condition, then eqs. (8.10.37) and (8.10.38) may be used to determine a_j and b_j for all $J < j < K$, where J denotes the index of the mass shell at the base of the convective zone. Application of a boundary condition at the lower boundary provides a new abundance Y_J and eq. (8.10.30) may then be used to find new values Y_j for all $J < j \leq K$.

8.10.4 The outer boundary condition and the quantities a_K and b_K

If the formal outer edge of the convective zone is at a grid point $j = K < N$, so that convection does not extend into the static envelope, and if overshoot is neglected, one has that

$$F_K = 0. \tag{8.10.39}$$

From eqs. (8.10.22), (8.10.27), and (8.10.29), it follows that

$$Y_K - Y_K^{(0)} = \frac{\delta t}{\mathrm{d}M_K} F_{K-1} = -s(2, K)\, s(3, K - 1)\, (Y_K - Y_{K-1}), \tag{8.10.40}$$

which may be written as

$$Y_K = \frac{Y_K^{(0)} + s(2, K)\, s(3, K - 1)\, Y_{K-1}}{1 + s(2, K)\, s(3, K - 1)}. \tag{8.10.41}$$

Comparing with eq. (8.10.36), it is apparent that

$$a_K = \frac{s(2, K)\, s(3, K - 1)}{1 + s(2, K)\, s(3, K - 1)} \tag{8.10.42}$$

and

$$b_K = \frac{Y_K^{(0)}}{1 + s(2, K)\, s(3, K - 1)}. \tag{8.10.43}$$

If the outer boundary of the interior convective zone is at the base of the static envelope, it is usually the case that convection extends all the way to the photosphere. Typically, the mass above the photosphere is small compared with the mass of the static envelope. Since the mass of the last interior zone is also small compared with the mass of the static envelope, it is incorrect to assume that $F_N = 0$. Rather than attempting a definition of F_N involving properties of the envelope structure, one can think of the static envelope and the last interior zone as comprising a single zone at the base of which the flux is F_{N-1} and at the surface of which $F_{N+1} = 0$.

For an arbitrary isotope, conservation of particles dictates that

$$\sum_{j=J}^{N-1} \left(Y_j - Y_j^{(0)}\right) \mathrm{d}M_j + \left(Y_N - Y_N^{(0)}\right) \mathrm{d}M_N + \left(Y_{N+1} - Y_{N+1}^{(0)}\right) \mathrm{d}M_{N+1} = 0, \tag{8.10.44}$$

where the $(N+1)$th shell is the static envelope of mass dM_{N+1} and abundance parameter Y_{N+1}. A mean abundance $Y_{N+1/2}$ may be defined by

$$Y_{N+1/2} = \frac{Y_N dM_N + Y_{N+1} dM_{N+1}}{dM_N + dM_{N+1}}. \tag{8.10.45}$$

Using eqs. (8.10.22) and (8.10.45) in eq. (8.10.44) gives

$$-\delta t \sum_{j=J}^{N-1} (F_j - F_{j-1}) + \left(Y_{N+1/2} - Y_{N+1/2}^{(0)}\right)(dM_N + dM_{N+1}) = 0. \tag{8.10.46}$$

Performing the sum in this equation gives

$$-\delta t \ (F_{N-1} - F_{J-1}) + \left(Y_{N+1/2} - Y_{N+1/2}^{(0)}\right)(dM_N + dM_{N+1}) = 0. \tag{8.10.47}$$

Since, by construction, $F_{J-1} = 0$, one has that

$$Y_{N+1/2} - Y_{N+1/2}^{(0)} = \frac{\delta t}{dM_N + dM_{N+1}} F_{N-1}. \tag{8.10.48}$$

If, following the notation of eq. (8.10.27), one defines

$$s(2, N+1/2) = \frac{\delta t}{dM_N + dM_{N+1}}, \tag{8.10.49}$$

then eq. (8.10.48) may be written as

$$Y_{N+1/2} = Y_{N+1/2}^{(0)} - s(2, N+1/2)\, s(3, N-1)\, (Y_N - Y_{N-1}). \tag{8.10.50}$$

In order to complete the set of equations joining the surface to the interior, it is necessary to make a choice as the relationship between Y_N and $Y_{N+1/2}$. Given the rudimentary nature of the treatment adopted here, the choice $Y_N = Y_{N+1/2}$ is as reasonable as any, so

$$Y_N \sim \frac{Y_N^{(0)} + s(2, N+1/2)\, s(3, N-1)\, Y_{N-1}}{1 + s(2, N+1/2)\, s(3, N-1)}. \tag{8.10.51}$$

Comparing with eq. (8.10.36), it follows that

$$a_N = \frac{s(2, N+1/2)\, s(3, N-1)}{1 + s(2, N+1/2)\, s(3, N-1)} \tag{8.10.52}$$

and

$$b_N = \frac{Y_N^{(0)}}{1 + s(2, N+1/2)\, s(3, N-1)}. \tag{8.10.53}$$

In summary, the only difference between the case when $K = N$ and the case when $K < N$ is that, for $K = N$, instead of adopting eq. (8.10.27), one chooses

$$s(2, N) = s(2, N+1/2) = \frac{\delta t}{dM_N + dM_{N+1}}. \tag{8.10.54}$$

This choice makes sense intuitively. Since the mass of the Nth shell is, in general, much smaller than the mass of the static envelope, particles flowing from below into the Nth shell are spread out over a mass much larger than the mass of the Nth shell, and, in any time step, the change in the particle abundance in the Nth shell is much smaller than if convection had terminated in the Nth shell.

Having obtained a_K and b_K at the outer edge of the convective zone, eqs. (8.10.37) and (8.10.38) may be used to find a_j and b_j for all j from $j = K - 1$ to $j = J$, at the base of the convective zone.

8.10.5 The inner boundary condition and determination of new composition variables

The final task is to apply a boundary condition at the base of the convective region to find the new composition Y_J in the innermost shell of the convective region and to then use the recurrence relation, eq. (8.10.36), to find all the other Y_j in the convective zone. Suppose that convection does not extend to the center of the model, so that $J > 2$. If it is assumed that the flux through the outer edge of the radiative shell just below the base of the convective zone vanishes, then

$$F_{J-1} = 0. \tag{8.10.55}$$

Using this equation and eqs. (8.10.22) and (8.10.27), one has that

$$Y_J - Y_J^{(0)} = -\delta t \, \frac{F_J}{\mathrm{d}M_J} = -s(2, J) \, F_J. \tag{8.10.56}$$

Then, using eq. (8.10.29),

$$Y_J - Y_J^{(0)} = s(2, J) \, s(3, J) \, (Y_{J+1} - Y_J). \tag{8.10.57}$$

Finally, combining eqs. (8.10.35) and (8.10.57) produces

$$Y_J = \frac{Y_J^{(0)} + s(2, J) \, s(3, J) \, b_{J+1}}{1 + s(2, J) \, s(3, J) \, (1 - a_{J+1})}. \tag{8.10.58}$$

If the base of the convective zone is at the center, where $j = 1$, then $F_1 = 0$ and eqs. (8.10.22), (8.10.27), and (8.10.29) give

$$Y_2 - Y_2^{(0)} = -(F_2 - F_1) \, \frac{\delta t}{M_2} = -F_2 \, \frac{\delta t}{M_2} = s(2, 2) \, s(3, 2) \, (Y_3 - Y_2). \tag{8.10.59}$$

Combining eqs. (8.10.27) and (8.10.34) results in

$$Y_2 = \frac{Y_2^{(0)} + s(2, 2) \, s(3, 2) \, b_3}{1 + s(2, 2) \, s(3, 2) \, (1 - a_3)}, \tag{8.10.60}$$

which is consistent with eq. (8.10.58).

Alternatively, setting $Y_1 = Y_2$ in eq. (8.10.35) directly produces another estimate for Y_2:

$$Y_2' = \frac{b_2}{1 - a_2}.$$
(8.10.61)

Replacing a_2 and b_2 in eq. (8.10.61) by their equivalents as given by eqs. (8.10.37) and (8.10.38), and then replacing α_2, β_2, and γ_2 in the resulting expression by their equivalents as given by eqs. (8.10.32)–(8.10.34), eq. (8.10.60) is reproduced with Y_2' on the left hand side. Thus, the condition that the composition gradient vanishes in the central zone has an effect which is identical with the effect produced by the condition that the particle flux vanishes at the center.

In summary, if $J = 1$, $Y_1 = Y_2$, where Y_2 is given by eq. (8.10.60) and, if $J \geq 2$, Y_J is given by eq. (8.10.58) and Y_{J-1} is determined by conditions in the $(J - 1)$th shell.

8.10.6 Composition in the static envelope

In practice, it is not essential to consider static surfaces in which abundances vary with position, except to take into account the spatial variations associated with ionization. It is also not practical to change the element abundances assigned to the the static envelope during the course of a calculation. Such changes would necessitate making at least four new static surfaces during the construction of every new model. In many instances, changes in surface composition may be handled by setting the abundances in the envelope at the start of a calculational run equal to the abundances in the last interior shell of the final model obtained in the immediately previous run. If the envelope convective zone extends into the static surface, this fact must be taken into account in determining time changes in the abundances of elements in the entire convective zone.

8.11 Remarks on mixing in radiative zones due to particle diffusion: gravitational settling, abundance-gradient induced diffusion and rotation-induced diffusion

Even when there is locally an overall balance between the bulk pressure-gradient force acting outward and the bulk gravitational force acting inward, if there are no spatial gradients in the ratios of element abundances, the bulk gravitational force acting inward on a species composed of particles which are more massive than the average particle mass is larger than the outward-directed pressure-gradient force generated by this species. In consequence of this disparity, a separation of species must occur. In general, particles of larger atomic mass diffuse inward in the direction of the gravitational field acceleration and particles of smaller atomic mass diffuse outward. The separation process, which can be called gravitational settling or gravity-induced diffusion, continues to separate element species according to mass as long as the bulk pressure-gradient force generated by the ions and associated electrons of an individual species fails to match the bulk gravitational force on that species. As the tendency towards equilibrium with respect to gravity-induced diffusion progresses, the gradient in the relative number abundance associated with each species increases towards an asymptotic value characterized by a scale height which is smaller, the larger the atomic mass of the species.

A gradient in the number abundance of any species leads to a tendency for that species to diffuse in the direction in which the number abundance decreases because, in any plane perpendicular to the abundance gradient, there are fewer particles per unit time reaching the plane from above where the number abundance is smaller than are reaching the plane from below where the number abundance is larger. This process may be called abundance-gradient-induced diffusion. In complete equilibrium, the rate at which particles of a given type flow outward due to a number-abundance gradient matches the rate at which particles of this type flow inward in response to acceleration by the gravitational field.

Prominent examples of gravitational settling at work are white dwarfs which, with few exceptions, exhibit almost monelemental surface abundance characteristics. The exceptions include newly formed white dwarfs of high luminosity and white dwarfs accreting from a companion in a close binary system. Members of the most prominent class of white dwarfs (the DA variety) exhibit essentially pure hydrogen spectra while members of the class of non-DA white dwarfs exhibit nearly pure helium spectra. These examples tell two stories: (1) there are at least two evolutionary paths that lead to white dwarfs, and (2) gravitational settling at the high gravity surfaces of white dwarfs causes heavy elements to sink below the surface on a time scale short compared with the cooling time scales of white dwarfs.

Abundance-gradient-induced diffusion can have still other observational consequences. In radiative regions in the interior where nuclear transformations are active, gradients in relative number abundances arise because nuclear transformations in fuel-rich regions proceed more rapidly with proximity to the center. If the gradient in the number abundance of any species becomes larger in absolute value than is consistent with equilibrium with respect to gravitational settling, that species will diffuse in the direction in which the number abundance decreases. Thus, near the center of main sequence stars with radiative cores, the abundance of helium per unit mass increases inward while that of hydrogen increases outward, and there is a tendency for helium to diffuse outward and for hydrogen to diffuse inward, thus prolonging the time required for the exhaustion of hydrogen near the center. This phenomenon obviously can affect estimates of the ages of globular cluster stars obtained by comparing characteristics of cluster stars with models of evolved stars of low metallicity.

The caution with which diffusion is presented in this book is a consequence of the facts that (1) real stars rotate and an appropriate three-dimensional treatment of the mixing associated with rotation goes considerably beyond the professional competence of the author, and (2) in the case of main sequence stars, the operation of rotation-induced mixing currents could significantly inhibit the action of gravitational settling. In rapidly rotating main sequence stars of spectral type earlier than F, the second caveat is almost certainly the case, and it is probably best not to include diffusion in constructing models of such stars, not only because of the probable importance of rotation-induced mixing, but because the time scale for nuclear evolution can be quite small relative to the time scale for diffusion even in the absence of rotation-induced currents.

Even for solar models, it is not clear that including diffusion is a better approximation than leaving it out. Although the Sun completes a rotation approximately 30 times more slowly than does the Earth, its radius is 100 times larger, so that the velocity of rotation in envelope layers is three times larger than in the Earth's atmosphere. It is a matter of common experience that the Earth's atmosphere is continuously stirred up by winds driven by the operation of many physical mechanisms which include rotation-induced processes, and it is not difficult to imagine that similar and perhaps even more exotic processes could be at work below the base of the Sun's convective envelope to effectively negate the influence of diffusion in building up abundance gradients. Nevertheless, an ability to estimate the maximum effect of diffusion in influencing evolutionary properties is ample justification for describing later, in Chapter 12 of Volume 2, the procedures for estimating diffusion velocities in non-rotating models.

The situation is quite different for the white dwarf phase of evolution. Because of the comparatively huge gravitational forces in white dwarfs, $\sim 10^4$ times larger than in the Sun, coupled with the 10^{10} yr cooling time scales for observable white dwarfs, not to mention the monoelemental character of most observed white dwarf spectra, it is clear that gravitational settling plays an essential role in determining surface abundance characteristics. At the same time, however, the basic structure of a white dwarf is quite well approximated by a model which preserves the composition with which it emerges from the planetary nebula phase and, as shown by comparing model evolution with and without diffusion, the time scale for cooling is not dramatically affected by the inclusion of diffusion.

White dwarfs provide a prime example of diffusion induced by gradients in relative abundances. In the traditional view, the luminosity of a cooling white dwarf is due entirely to the release of internal thermal energy, and comparisons between the luminosity function defined by nearby white dwarfs and theoretical luminosity functions based on models of white dwarfs which do not burn hydrogen have been used to estimate the birthrate of nearby white dwarfs and the age of the galactic disk. However, cooling white dwarfs with hydrogen-rich spectra have descended from the bolometrically luminous central stars of planetary nebulae. Central stars are, in turn, descendents of AGB stars which have ejected most of their hydrogen-rich envelopes. The mass of the hydrogen-rich envelope of the central star at the end of the planetary nebula phase can be estimated from theoretical models.

During the initial white dwarf cooling phase at high luminosity, diffusion of hydrogen inward to helium-rich layers hot enough for hydrogen to burn depletes the mass of the hydrogen-rich layer to the extent that, during later phases, the temperature at the base of the surface layer is too small for hydrogen burning to compete with cooling as a major source of surface luminosity. Were it not for diffusion during the high luminosity phase of cooling, hydrogen burning during evolution at low luminosity could prolong the time for evolution through the lowest luminosity interval in which white dwarfs are observed and cause modifications in estimates of the age of the galactic disk and of the birthrate of nearby white dwarfs obtained by comparing with models which include hydrogen burning but do not include diffusion. See, for example, the discussions in Iben & Tutukov (1984) and Iben & MacDonald (1985, 1986).

8.12 Zoning considerations and choice of time step

As evolution progresses, the gradients of all structure and composition variables change with time, and, in order for the difference equations to satisfactorily approximate the differential equations from which they are derived, it is necessary to rezone from time to time. As a rule of thumb, insisting that no structure or composition variable change by more than, say, 5–10% from shell to shell or from one time step to the next is usually adequate. When a nuclear shell source is present, it is wise also to insist that the rate of nuclear energy generation change by no more than $\sim 10\%$ from shell to shell and that the integrated nuclear energy-generation rate change by no more than $\sim 10\%$ from one time step to the next. To make sure that the profile of the rate of energy generation for a given energy source is resolved, one may insist that the energy-generation rate for that source change from shell to shell by no more than, say, five percent or so of the maximum energy-generation rate for that source in the model.

Because the locations of interfaces between convective and radiative zones change with time as evolution progresses, zoning in regions encompassing these interfaces requires attention. The criteria to be used in determining spatial mesh size and time-step size may require tailoring in special circumstances, but, in general, application of common sense principles is sufficient. For example, in any time step, the change in the mass of a convective zone should be small compared with the mass of the convective zone. Ensuring this requires both control of the size of the time step and suitably fine zoning in mass in the region ahead of an advancing interface or behind a retreating interface. At the same time, to minimize the numerically generated noise that accompanies rezoning, it is important not to coarsen too quickly the mass zoning behind an advancing interface or above a retreating interface. If M_{CS} is the mass of a convective zone, it is reasonable to require of the order of 30 equal-mass shells in a region extending from a mass $0.1 \times M_{CS}$ ahead of the radiative–convective interface to a mass $0.05 \times M_{CS}$ behind the interface, with time step Δt being chosen so that $|(\mathrm{d}M_{CS}/\mathrm{d}t)\,\Delta t| \leq 0.01 \times M_{CS}$.

In contrast with other structure variables, the luminosity variable can be negative as well as positive. During the evolution of most stars that leave the main sequence in less than a Hubble time, sign changes in interior luminosity appear and, near where such changes occur, use of a zoning criterion based on restricting the percentage change in luminosity across mass shells can lead to many more shells than are necessary. A better approach is patterned after the algorithm for zoning near radiative–convective interfaces. Flagging the locations in mass where the ith sign change occurs as $M_{\pm}(i)$, one may require, say, roughly 30 equal-mass shells over the region between $M_{\pm}(i) - \Delta M_{-}(i)$ and $M_{\pm}(i) + \Delta M_{+}(i)$, the larger $\Delta M_{\pm}(i)$ being in the direction in which $M_{\pm}(i)$ is moving. The magnitude of $\Delta M_{\pm}(i)/M_{\pm}(i)$ can vary greatly with the circumstances. In the case of a sign change in the electron-degenerate helium core of a red giant model or in the carbon–oxygen or oxygen–neon core of an AGB model, $\Delta M_{\pm}(i)/M_{\pm}(i) \sim 0.05 - 0.1$ is a reasonable choice. In contrast, values of $\Delta M_{\pm}(i)/M_{\pm}(i) \sim 10^{-5}$ are appropriate near the edges of the negative luminosity region just ahead of the convective zone formed during a helium shell flash in an AGB model star.

In order to avoid the formation of too many zones, adjacent shells should be merged if the variations in structural and composition variables across the enlarged shell still meet the minimum requirements for accuracy that have been established. Number conservation requires that the abundance of the mth isotope in the new shell k be related to the abundances in the old shells i and $i + 1$ by

$$Y(m, k)\, \mathrm{d}M_k = Y(m, i)\, \mathrm{d}M_i + Y(m, i + 1)\, \mathrm{d}M_{i+1}, \qquad (8.12.1)$$

where

$$\mathrm{d}M_k = \mathrm{d}M_i + \mathrm{d}M_{i+1}. \qquad (8.12.2)$$

The best way to add a shell is to simply split a pre-existing shell into two shells of equal mass. If shell i is to be split into two new shells k and $k + 1$, interpolation between abundances in the three adjacent shells $i - 1$, i, and $i + 1$ may be used to obtain approximations to the new abundances $Y(m, k)$ and $Y(m, k + 1)$. For example, linear interpolation gives

$$Y(m, k) = Y(m, i) - \frac{Y(m, i) - Y(m, i - 1)}{\mathrm{d}M_i + \mathrm{d}M_{i-1}} \frac{\mathrm{d}M_i}{2} \qquad (8.12.3)$$

and

$$Y(m, k + 1) = Y(m, i) + \frac{Y(m, i + 1) - Y(m, i)}{\mathrm{d}M_{i+1} + \mathrm{d}M_i} \frac{\mathrm{d}M_i}{2}. \qquad (8.12.4)$$

One way to conserve numbers in this approximation is to set

$$Y(m, k) = N(m)\, Y(m, k) \qquad (8.12.5)$$

and

$$Y(m, k + 1) = N(m)\, Y(m, k + 1), \qquad (8.12.6)$$

where

$$N(m) = \frac{2\, Y(m, i)}{Y(m, k) + Y(m, k + 1)}. \qquad (8.12.7)$$

This procedure preserves numbers, but one has to watch out not to divide by zero in finding $N(m)$. Furthermore, the procedure does not preserve mass in each shell. If one wishes to preserve mass, numbers may be renormalized in such a way that

$$\sum_m A_m\, Y(m, k) = \sum_m A_m\, Y(m, k + 1) = 1, \qquad (8.12.8)$$

where A_m is the atomic mass of the mth isotope.

Rezoning in the mass coordinate leads to artificial "bumps" in the rate of gravothermal energy release, and it is generally a good practice to choose time steps in such a way that rezoning to meet the established accuracy requirements need take place only once every three or so time steps. In some evolutionary phases, such as during the thermal pulse phase of AGB model stars, when the time scale for changes in nuclear burning shells is small compared with the time scale for energy diffusion in the white-dwarf-like core, it is necessary to "freeze" the zoning in the core to prevent spurious oscillations in the thermal and luminosity structure of the core.

One must be particularly careful in limiting the time step based on changes in composition in convective zones where many isotopes are mixed throughout the zone on a time scale short compared with the time scale on which they are altered locally by nuclear transformations, resulting in an overall rate of change in a given composition variable which is small compared to the local rate of change at the highest temperature location in the convective zone. The evolutionary time step should be chosen so that the local rate of change times the evolutionary time step is less than 5–10%. In the case of a composition variable that is decreasing with time, this algorithm must of course be suspended when the abundance of the variable becomes very small.

For a fixed time-step size Δt, the number of steps N required to evolve for a total time T is given by $N = T/\Delta t$ and the total number of iterations is given by $NI = TI/\Delta t$, where I is the average number of iterations required to achieve a converged model. If the time step is too large, I increases with increasing Δt faster than N decreases. On the other hand, if the time step is small enough, the number of iterations required for convergence approaches $I = 1$ but NI increases with decreasing Δt. Thus, for any given phase of evolution, there exists an optimum (range of) time step size(s) Δt_{opt} such that the total number of operations (the total computational time) is a minimum. This optimum can be found with experience.

8.13 On the evolution of the computing environment

It is fascinating to consider the huge difference that has occurred in the way that stellar evolution can be studied since the first appearance of electronic digital computers. Some of the sociology associated with the early development of computers and some of the results due to their use is described in the book *Computers and their Role in the Physical Sciences* edited by Fernbach and Taub (1970).

My own experience in constructing stellar models began in 1960 at Williams College where, following the prescriptions given in Martin Schwarzschild's Book *Stellar Structure and Evolution* (1958), I attempted to make static models of homogeneous composition (zero age main sequence models) using a Frieden hand-crank desk calculator. I abandoned the project after it became clear that many, many years of effort would be required to achieve publishable results. While teaching at the University of Illinois during the summer of 1960, I began a collaboration with John Ehrman to calculate homogeneous main sequence models using the ILLIAC I computer. The computer, which served the entire university community, consisted of many racks of vacuum tubes in a large air conditioned room. Information was entered as punched holes in a paper tape which was fed into the computer on a hand-cranked wheel. Every now and again the computer would stop; restarting was achieved by waving a metal key in front of the glass wall of a rack of vacuum tubes. Results of calculations were returned as punched holes in a paper tape. John would translate the holes obtained for each integration into numbers which I would enter on graph paper until the integration was complete. Interpolation in the graphs resulting from several integrations led to eye-fit guesses for starting values of variables for additional integrations. The process would continue until convergence was deemed to have occurred. Thirty homogeneous models were constructed over a period of several months of computing several hours a day and the results demonstrated, as a function of mass and metallicity, the transition from burning primarily by pp reactions to burning primarily by CN-cycle reactions (Iben & Ehrman, 1962).

My next computational experience occurred during a three year stint (1961–1964) as a senior postdoctoral fellow at Cal Tech where, for the first year, computations continued on an IBM 7094 at the off-campus Jet Propulsion Laboratory (JPL). The computer again consisted of many racks of devices filling a large air conditioned room. This time, instead of vacuum tubes, the devices were early generation transistors. Disks coated with magnetizable metals and vertically suspended magnetic tapes served as information storage devices. Instructions for the machine (the program) and data were entered by using a mechanical punching machine to punch holes into a series of rectangular cards. The assembled cards were placed in one or more boxes and delivered by hand to the computing room and submitted as a job at a scheduling desk. The cards were ultimately fed into the computer by an operator according to algorithms which were never clearly stated. Once entered into the computer as a set of instructions, a job would be activated according to algorithms presumably designed to maximize the computer's efficiency and take into account priorities related to the cost per unit of computing time the programmer was willing to bear. The daytime competition for computer time was such and the computational time required by a stellar structure program was such that, since job submission and result retrieval required the physical presence of the submitter, I was forced, during my first year at Cal Tech, to commute late at night between Cal Tech and JPL.

The situation improved dramatically in 1962 when Cal Tech acquired its own IBM 7094. Nevertheless, many commutes on foot between the punch machine in the Kellog Radiation Laboratory and the Computing Center were necessary both late at night and on weekends. Results of computations came on large sheets of paper as printed text and as machine-produced graphs in formats chosen by the programmer. Additional graphs could be constructed by hand from the content of the printed text. The quality of all graphs was such that the attention of a professional draftsperson was necessary to convert a graph into a publishable illustration. Much of the funding for the computational work came from the Office of Naval Research and the National Aeronautics and Space Administration obtained by William A. Fowler, who had hired me, Dick Sears, Donald Clayton, and John Bahcall to address problems in stellar evolution and nucleosynthesis (SINS = Stellar Interiors and Nucleosynthesis). Additional funding was provided by G. D. McCann, the director of the Computational Center at Cal Tech.

My third computing experience was during eight years (1964–1972) as a professor at MIT. The first three years were consumed in preparing for publication most of the results obtained at Cal Tech. The next four years were devoted to additional calculations of stellar evolution on an IBM 7044 computer at the Laboratory for Nuclear Science at MIT supported by funds from the Center for Space Research at MIT. The procedures for programming, submission of jobs, and protocols for running jobs were very much the same as those at Cal Tech. During my final MIT-employed year, I was on sabbatical at the University of

Colorado and at the University of California at Santa Cruz and most of my computing was done on CDC 6600 and CDC 7600 computers at the National Center for Atmospheric Research (NCAR) in Boulder, Colorado.

Two very interesting interactions with computer directors at MIT and NCAR highlight the obstacles which faced those of us trying to do computer time- and memory-intensive calculations on central computers which attempted to serve an entire academic community. At MIT, I developed friendly relationships with several operators, and these relationships undoubtedly led to favored access to the computer. When the director become aware of this, I was explicitly requested to refrain from continuing the friendly relations. At NCAR, the director insisted that, before being granted computer time, I would have to make a concerted effort to streamline my programs, ignoring the fact that streamlining had been an essential component for many years of my effort to conform requirements of accuracy and generality to the constraints imposed by computers whose memory capacity and instruction-performance times were just barely adequate to do a proper job calculating fairly simple stellar evolution models.

The remainder of my teaching career was spent at the University of Illinois where, among other activities, I embarked on a multi-year program to calculate the thermal pulse evolution of model TPAGB stars. I initially made use of the IBM 360/75 computer run by the Computational Center at Illinois, paying for computer time with funds supplied by the National Science Foundation (NSF) and by the University of Illinois Research Board. The procedures for preparing and submitting programs were initially pretty much the same as at Cal Tech and MIT, but, fairly soon, punched cards were replaced by electronic (Silent 700) data terminals from which both programs and data could be transferred across phone lines to the central computer.

The competition for computer time being as intense as at MIT, I became well acquainted with the director and the operating staff of the Computational Center, and phone calls late at night and on weekends sometimes expedited the running of my programs. However, it became apparent that obtaining publishable results concerning the TPAGB phase of evolution would require far more computer time than in the case of earlier evolutionary phases and it was necessary to cast about for access to supplementary computer facilities. The university administration had an IBM 360/158 computer which did not operate during 8 hours at night or on weekends until I lobbied for access during the unused time. Interestingly, after only a few months of granted access, the administration discovered that it required the exclusive use of the machine 24 hours a day for the entire week. Ultimately, I had to resort to trips to NCAR in Colorado to use the CDC computers there. Some of the frustrations involved in transporting a program and data tailored for use on a computer at a home institution to a computer at a remote installation and translating the program into a new form tailored for use on the remote computer are described in an article beginning on page 109 of the book *Computer Science and Scientific Computing* edited by J. M. Ortega (1976).

Among the results of the calculations of thermal pulses was a demonstration that stars which develop helium exhausted cores of mass as large as $1\,M_\odot$ activate the ^{22}Ne$(\alpha, n)^{25}$Mg neutron source for the production of s-process elements and dredge these elements as well as freshly made carbon into the convective envelope from which they can be expelled into the interstellar medium by the action of a superwind (Iben, 1975, 1976, see also the discussions in Chapters 17–19 in Volume 2 of this book). My feeling at the time was and still is that these results added significantly to mankind's understanding of the origin of elements, including those that play a vital part in the operation of electronic computers (rare-earth elements) and life (carbon). My chief competition for computer time on the Illinois Computational Center computer was a brute force calculation by Kenneth Appel and Wolfgang Haken to prove the four color theorem and a brute force calculation by Samuel S. Wagstaff to prove Poincarè's last theorem. This latter calculation took very little memory but huge amounts of computer time. Arguments about the relative importance of the three competing programs always fell upon deaf ears, but the offer of money from the NSF was sometimes persuasive.

I could continue to relate the struggles experienced over the years in attempting to fit programs with ever more ambitious objectives into computers which always seemed inadequate to the task and fighting for computer time, but enough is enough. Happy to relate, primarily in consequence of the development of ever smaller transistors, the computing environment has been revolutionized relative to that which my anecdotal account describes. I will not attempt to detail the transformation, but only compare the situation in the first years of the twenty-first century with the situation prevailing in my anecdotal account. No more punch-card machines, no more submission of jobs to be run on a computer which serves an entire community of researchers, no more jockeying with other programmers for running priority on a single machine, no more reams of paper output, no more necessity to submit grant proposals to funding agencies, no more commuting back and forth between office and computer center, no more need to make late night phone calls to expedite the execution of a program, and so forth. Thanks to the development of suitable software, publishable graphs can be prepared without the need for buying the service of a draftsperson.

For the past ten years, I have sat in comfort before a personal computer in my den at home and communicated by various means (ethernet, cable, phoneline, etc.) either with a workstation or with a personal computer at my office. The business ends of the computers occupy no more than 1.26 cubic feet and perform all of the required calculations at my convenience and in a timely fashion. There are no more sleepless nights waiting for a job to be run or results to be obtained. All that remains as a source of frustration is the unavoidable necessity of organizing computations in such a way as to achieve convergent, physically believable solutions. Utilizing personal computers, the calculation of stellar evolutionary models in the spherically symmetric and quasistatic approximations is now a straightforward exercise, free from sociological influences. This has enabled me to enjoy ten years of pleasurable extraction and contemplation of the intricate and fascinating changes in structure which occur in stars as they evolve and which I am privileged to share with the readers of this book.

This account is not intended to imply that the struggles experienced by researchers in obtaining results during the early days of electronic digital computers were a waste of time. The earliest results made lasting contributions to mankind's understanding of many physical problems. Just as importantly, the pressure exerted by those eager to obtain results undoubtedly contributed to accelerating the rate of evolution of the computing environment into one which is, from the perspective of a person who continues to explore problems which were difficult but just barely possible to attack in the early days, now very benign.

The lesson for future generations is clear: (1) continue to work on challenging problems the solution of which stretches the facilities available for their solution, and (2) live long enough to be able to enjoy extracting and contemplating solutions to the once-difficult problems using facilities which make the task of finding these solutions pure pleasure.

Bibliography and references

Tadeasz Barachiewiez, *Bull. Ac. Pol., Ser. A.,* **393**, 1938; *AJ*, **50**, 38, 1842.

E. K. Blum, *Math. Comp.*, **16**, 176, 1962.

F. Ceschino & J. Kuntzmann, *Problèmes Differentiels de Conditions Initiales*, (Paris: Dunod), 1963; translation by D. Boyanovitch (Englewood Cliffs, NJ: Prentice Hall), 1966.

André-Louis Colesley, *Bulletin Geodésique*, **1**, 159, 1922; **2**, 67, 1924 (cdt Benoit, posthumous).

P. D. Crout, *Trans. Amer. Inst. Elec. Engineers*, **60**, 1235, 1941.

Arthur S. Eddington, *The Internal Constitution of the Stars* (Cambridge: Cambridge University Press), 1926.

S. Fernbach & A. H. Taub, eds., *Computers and their Role in the Physical Sciences* (New York: Gordon & Breach), 1970.

S. Gill, *Proc. Camb. Phil. Soc.*, **47**, 96, 1951.

Louis George Henyey, J. E. Forbes, & N. L. Gould, *ApJ*, **139**, 306, 1964.

Icko Iben, Jr., *ApJ*, **196**, 525, 549, 1975; **208**, 165, 1976.

Icko Iben, Jr. & John Ehrman, *ApJ*, **135**, 770, 1962.

Icko Iben, Jr. & Alexander V. Tutukov, *ApJ*, **282**, 615, 1984.

Icko Iben, Jr. & Jim MacDonald, *ApJ*, **296**, 540, 1985; **301**, 164, 1986.

E. L. Ince, *Ordinary Differential Equations*, Appendix B, (New York: Dover), 1926.

M. Wilhelm Kutta, *Zeits. für Math. Phys.*, **46**, 435, 1901.

James M. Ortega, ed., *Computer Science and Scientific Computing* (New York: Academic Press), 1976.

William H. Press, Brian P. Flannery, Saul A. Teukolsky, & William T. Vetterling, *Numerical Recipes in FORTRAN: the Art of Scientific Computing*, (Cambridge: Cambridge University Press), 1966.

Robert D. Richtmeyer, *Difference Methods for Initial Value Problems* (New York: Interscience), 1957.

Carl David Runge, *Math. Ann.*, **46**, 167, 1895.

I.-J. Sackmann, R. L. Smith, & K. M. Despain, *ApJ*, **187**, 555, 1974.

A. Schwarzenberg-Czerny, Astron. Astrophys. Suppl. Ser., 110, 405, 1995.

Martin Schwarzschild, *Stellar Structure and Evolution* (Princeton: Princeton University Press), 1958.

Dai'chiro Sugimoto, *ApJ*, **159**, 619, 1970.

Dai'ichiro Sugimoto, Ken'ichi Nomoto, & Y. Eriguchi, *Prog. Theoret. Phys.*, Suppl. No. **7**, 1981.

PART III

PRE-MAIN SEQUENCE, MAIN SEQUENCE, AND SHELL HYDROGEN-BURNING EVOLUTION OF SINGLE STARS

9 Star formation, pre-main sequence evolution, and the zero age main sequence

The saga of pre-stellar evolution begins with the Big Bang and the evolution of large scale structure, continues with the formation of galaxies and giant molecular clouds, and extends finally to the formation of protostellar condensations that collapse into objects recognized as stars which, evolving on a quasistatic time scale, form the main subject of this book.

Nucleosynthesis, beginning with protons and neutrons in the early phases of the expanding Universe, prior to the formation of structure, is responsible for the fact that hydrogen and helium are the most abundant elements in the Universe. The presence in the current Universe of elements such as carbon, nitrogen, oxygen, and iron, coupled with the fact that these elements are not produced in models of the early Universe, is evidence that stars make these elements and inject them into the interstellar medium. Thus, the initial composition and therefore the detailed evolutionary history of stars of a given generation differs from the composition and history of stars of earlier and later generations.

Whatever the physics is that is responsible for their existence, giant molecular clouds are the birthplaces of smaller condensations, or protostellar clouds, which are thought to develop into protostars consisting initially of small, quasistatic cores that accrete from dynamically collapsing envelopes. In Section 9.1, some considerations involved in understanding the properties of giant molecular clouds and early protostar evolution are presented. After almost a century of thought, where a real star appears in the HR diagram as an isolated, quasistatically evolving object has not been determined from first principles.

However, the properties of a class of low mass stars in young clusters known as T-Tauri stars, many of which are cooler and brighter than stars of comparable mass on the main sequence, suggest that the accretion phase terminates prior to the onset of hydrogen burning and that the core of the protostar emerges as an opaque, quasistatically contracting object which relies for its surface luminosity on the release of gravitational potential energy in its interior. This inference follows from theoretical considerations by Chushiro Hayashi who showed that, when the structure of photospheric layers (in which the opacity increases with both temperature and density) is properly taken into account, quasistatic models of large radius are constrained to be in a nearly vertical band in the HR diagram at a surface temperature of the order of 3000–5000 K (Hayashi, 1961). This band extends from the region of red supergiants at large luminosity, through the region of red giants, all the way to the main sequence band at very low luminosity. The most luminous T-Tauri stars lie within the band. At a luminosity comparable to or larger than the luminosity of a main sequence model of the same mass, a hydrogen-rich quasistatic model in the Hayashi band is fully convective from the base of its photosphere to its center and relies for its luminosity on the gravitational potential energy released in its interior.

In Section 9.2, as a first exercise in pursuing stellar evolution with the full panoply of techniques described in Chapter 8, a 1 $M_\odot$ model is constructed in the Hayashi band and evolved to the main sequence. The composition of the model is $(X, Y, Z = 0.74, 0.25, 0.01)$ and temperatures in the initial model are small enough that deuterium has not begun to burn. As it contracts, the model descends to lower luminosities at nearly constant surface temperature. After a brief episode of deuterium burning, during which the rate of contraction is temporarily slowed, the sole source of surface luminosity again becomes the liberation of gravitational potential energy. The model evolves downward in the HR diagram until it develops a radiative core. When the radiative core has grown to encompass most of its mass, the model evolves at nearly constant radius towards the hydrogen-burning main sequence, luminosity and surface temperature increasing all the while. The entire track of the model from any point along the fully convective Hayashi portion of the track, through the phase of radiative core growth, to the onset of hydrogen burning on the main sequence (Hayashi, Hoshi, & Sugimoto, 1962) may be defined as the *canonical* evolutionary track of an isolated 1 $M_\odot$ star during its gravitationally contracting phase.

In Section 9.3, the final transition of the 1 $M_\odot$ model from the phase during which the release of gravitational potential energy is the primary source of surface luminosity to the phase when the primary source of surface luminosity is nuclear energy released by the pp-chain hydrogen-burning reactions is examined. The final model is called a ZAMS (zero age main sequence) model.

Changing the composition of the initial Hayashi-band model of 1 $M_\odot$ described in Section 9.2 to $(X, Y, Z) = (0.71, 0.275, 0.015)$, in Section 9.4 a gravitationally contracting model of mass 5 $M_\odot$ is constructed, beginning with a composition-modified 1 $M_\odot$ model, by artificially increasing, prior to each time step, the mass at every point in the evolving model by a factor of 1.05 until the desired final mass has been achieved. In the process of construction, which takes place at nearly constant radius, interior temperatures become large enough that deuterium is destroyed over the interior and the 5 $M_\odot$ model emerges on the radiative portion of its Hayashi track with all original deuterium having been converted into ^{3}He. The larger temperatures ultimately achieved in the interior of the 5 $M_\odot$ model result in an episode of nuclear burning during which, over a substantial central region of the model, ^{12}C is converted into ^{14}N, producing a short lived C $\rightarrow$ N main sequence phase, following which the model contracts further to become a ZAMS star in which the dominant contributor to surface luminosity is burning by the full set of equilibrium CN-cycle reactions.

In Section 9.5, the evolution to the main sequence of a 25 $M_\odot$ model is described as an example of models in which radiation pressure is a significant fraction of the total pressure. The initial 25 $M_\odot$ model is achieved by increasing by a factor of 25 the mass and radius at every point in the 1 $M_\odot$ initial Hayashi track model described in Section 9.2 and by decreasing the pressure at every point by a factor of $625 = (25)^2$ and increasing the luminosity at every point by the same factor of 625. After evolving the resultant model for several time steps, the net result is a 25 $M_\odot$ model of roughly the same interior temperatures and surface temperature as the initial 1 $M_\odot$ model, but with a large radiative core. The 25 $M_\odot$ model experiences an explicit deuterium-burning episode at low surface temperatures which does not significantly slow its rate of contraction. As it evolves toward the ZAMS state, its contraction rate is substantially reduced during two phases of $C \rightarrow N$ burning, following which it becomes a ZAMS star burning hydrogen by the full set of equilibrium CN-cycle reactions.

9.1 Some concepts relevant to star formation

In our Galaxy, molecular clouds are concentrated in spiral arms where particle densities are larger than between the arms. The giant molecular clouds are not themselves collapsing, but rather act as nurseries within which smaller, more dense molecular cores are formed. The efficiency of star formation is not very high. For example, the mean density of matter locked into stars in young open clusters is comparable to the mean density of gaseous matter in giant molecular clouds, but the total mass in a typical open cluster is several orders of magnitude smaller than the mass of a typical giant molecular cloud. These two facts demonstrate that (1) giant molecular clouds fragment into dense cores, some parts of which evolve into individual stars and (2) most of the matter in the parent cloud is dissipated into space as a star cluster is formed. The radiation from the most massive stars in the emerging cluster ionizes the remaining cloud material and shocks from the first supernovae exploding in the cluster blow away into interstellar space most of the original matter in the parent giant molecular cloud.

In the present context, a single star may be defined as an isolated, optically visible object in quasistatic equilibrium, with a well-defined photosphere and an observed luminosity which is a consequence of energy liberated in its interior, independent of whether the energy source is nuclear energy, internal energy, gravitational potential energy, or a combination of the three.

There is a wealth of knowledge about the physical conditions prevailing in regions of star formation, but a theoretical understanding of how these conditions come about and how they lead to the formation of isolated stars which have well defined photospheres and are in quasistatic equilibrium is far from complete. It is well established that, in our Galaxy and in others for which such information can be derived, mean temperatures in giant molecular clouds are in the neighborhood of 5–20 K and particle densities are of the order of 100–1000 cm^{-3}. The mass of a typical molecular cloud varies from $\sim 100\, M_\odot$ to $\sim 10^5\, M_\odot$ and it appears that magnetic fields of magnitudes in the range of 10–150 micro-Gauss play a major role in supporting the clouds against gravity.

9.1.1 The Jeans criterion

An important step in establishing a dialogue between observational facts and theoretical speculation is a criterion established by James H. Jeans in 1902 (see J. H. Jeans, *Astronomy and Cosmology*, 1961) regarding the stability against gravitational collapse of an initially homogeneous isothermal gas. Assume a very large space filled with a large mass of gas initially of constant density ρ_0, pressure P_0, and temperature T_0. The bulk velocity $\mathbf{v}$ is everywhere zero and the gravitational potential ϕ may also be chosen as zero. When a perturbation is applied, the changes in physical quantities can be described by

$$\rho = \rho_0 + \rho_1, \quad P = P_0 + P_1, \quad \phi = \phi_0 + \phi_1 = \phi_1, \text{ and } \mathbf{v} = \mathbf{v}_0 + \mathbf{v}_1 = \mathbf{v}_1. \tag{9.1.1}$$

By supposing that the temperature remains constant, it is assumed that heat transfer occurs on a time scale much shorter than the time scale for dynamical motions and that the relationship

$$P = \frac{kT_0}{\mu M_{\mathrm{H}}} \rho = v_{\mathrm{s}}^2 \rho, \tag{9.1.2}$$

where v_{s} is the isothermal velocity of sound, may be adopted as an equation of state. The assumption of isothermality, of course, contrasts with the behavior of sound waves in a terrestial gas for which $v_{\mathrm{s}}^2 = (\partial P / \partial \rho)_{\mathrm{adiabatic}}$

The three equations governing the motion are eq. (8.1.9), the equation for mass conservation, eq. (8.1.14), the relationship between bulk acceleration, the pressure gradient, and the gravitational force, and eq. (5.1.9), Poisson's equation connecting the gravitational potential and the density. Inserting the relationships given by eqs. (9.1.1) into the equations of motion, subject to the constraint given by eq. (9.1.2), and retaining only terms linear in the perturbations produces the equations

$$\frac{\partial \rho_1}{\partial t} + \rho_0 \, \nabla \cdot \mathbf{v}_1 = 0, \tag{9.1.3}$$

$$\nabla^2 \phi_1 = 4\pi G \rho_1, \tag{9.1.4}$$

and

$$\frac{\partial \mathbf{v}_1}{\partial t} = \frac{v_{\mathrm{s}}^2}{\rho_0} \, \nabla \rho_1 - \nabla \phi_1. \tag{9.1.5}$$

Inserting into these equations a plane wave solution

$$q_i = q_{i0} \mathrm{e}^{\mathrm{i}(\mathbf{k} \cdot \mathbf{r} - \omega t)}, \tag{9.1.6}$$

where q_i is any one of the three variables ϕ_1, ρ_1, and v_1, produces another three equations,

$$-\mathrm{i}\omega \, \rho_{10} + \mathrm{i}\rho_0 \, \mathbf{v}_{10} \cdot \mathbf{k} = 0, \tag{9.1.7}$$

$$4\pi G \, \rho_{10} + k^2 \phi_{10} = 0, \tag{9.1.8}$$

where $k = |\mathbf{k}|$, and

$$\frac{v_{\mathrm{s}}^2}{\rho_0} \, \mathrm{i}\mathbf{k} \, \rho_{10} - \mathrm{i}\omega \, \mathbf{v}_{10} - \mathrm{i}\mathbf{k} \, \phi_{10} = 0, \tag{9.1.9}$$

from which it follows that $\mathbf{k}$ and $\mathbf{v}$ are in the same direction (longitudinal waves). The wave number k and the wavelength λ of the perturbation are related by $k = 2\pi/\lambda$. The three equations (9.1.7)–(9.1.9) have a solution such that

$$\omega^2 = v_{\mathrm{s}}^2 \, k^2 - 4\pi G \, \rho_0. \tag{9.1.10}$$

If ω^2 is negative, then the general solution is made up of two terms, one of which is proportional to $e^{+\sqrt{|\omega^2|}\,t}$ and is therefore unstable. That is, if the wavelength of the perturbation is larger than λ_{crit}, where

$$v_s^2 \left(\frac{2\pi}{\lambda_{\text{crit}}} \right)^2 = 4\pi G \rho_0, \tag{9.1.11}$$

or

$$\lambda_{\text{crit}} = v_s \sqrt{\frac{\pi}{G\rho_0}} = \left(\frac{kT_0}{\mu M_H} \right)^{1/2} \sqrt{\frac{\pi}{G\rho_0}}$$

$$= 0.626 \times 10^8 \left(\frac{T_0}{\mu \rho_0} \right)^{1/2} \text{cm} = 0.486 \times 10^{20} \left(\frac{T_0}{\mu^2 n_0} \right)^{1/2} \text{cm}, \tag{9.1.12}$$

the perturbation will begin to grow in amplitude. In the final expression of eq. (9.1.12), $n_0 = \rho_0/\mu M_H$ is the number density. The mass of the critically unstable region is of the order of

$$M_{\text{crit}} = \rho_0\, \lambda_{\text{crit}}^3 = \rho_0 \left(\frac{kT_0}{\mu M_H} \right)^{3/2} \left(\frac{\pi}{G\rho_0} \right)^{3/2}$$

$$= 1.22 \times 10^{-10} M_\odot \frac{T_0^{3/2}}{\mu^{3/2} \rho_0^{1/2}} \sim 95 M_\odot \frac{T_0^{3/2}}{\mu^2\, n_0^{1/2}}. \tag{9.1.13}$$

The quantities λ_{crit} and M_{crit} are known, respectively, as the Jeans length and the Jeans mass.

A simple interpretation of the Jeans criterion can be constructed by considering the time scale for the free fall collapse of an isolated sphere of finite mass in which there are, contrary to fact, no pressure forces. This time can be most easily found by using Kepler's third law for a point mass which is initially located at a distance r from the center of the spherical distribution and moves in an orbit of eccentricity $e = 1$ under the influence of a central stationary object of mass $M(r)$ equal to the mass contained in the sphere of radius r. Kepler's third law gives

$$P_{\text{orb}}^2 = (2\pi)^2 \frac{A^3}{GM(r)} = \frac{(2\pi)^2}{G\,(4\pi/3)\rho_0} \frac{A^3}{r^3}, \tag{9.1.14}$$

where P_{orb} is the period for a complete orbit over a distance $4A$ and A is the semimajor axis of the orbit. In the present context, $r = 2A$ and the particle reaches the center in a time

$$t_{\text{ff}} = \frac{P_{\text{orb}}}{2} = \left(\frac{3\pi}{G\rho_0} \right)^{1/2} \left(\frac{A}{r = 2A} \right)^{3/2} = \sqrt{\frac{3\pi}{32 G\rho_0}}. \tag{9.1.15}$$

This time is independent of the initial radius r, so, when pressure forces are neglected, all point masses in the sphere reach the center at the same time. Using eq. (9.1.15) in eq. (9.1.12), it follows that

$$\lambda_{\text{crit}} \sim \sqrt{10}\, v_s\, t_{\text{ff}}. \tag{9.1.16}$$

Thus, if the size of the region which is perturbed is larger than about three times the distance which sound can travel in a free fall time, the region is unstable against collapse.

9.1.2 The roles of magnetic fields and cosmic rays

For pure molecular hydrogen at $T_0 \sim 10$ K and $n_0 \sim 100$ cm^{-3}, the Jeans length is $\lambda_{\mathrm{crit}} \sim 7.7 \times 10^{18}$ cm and the Jeans mass is $M_{\mathrm{crit}} \sim 76 \, M_\odot$, small compared with the masses of typical giant molecular clouds. This suggests that forces other than gas pressure are helping to support these clouds against gravitational collapse. Over the years, turbulent pressure (hydromagnetic waves) and magnetic energy have been invoked as stabilizing agents. Telemachos Ch. Mouschovias (1976a, 1976b) has constructed exact equilibrium states for isothermal, magnetic, self-gravitating clouds which have been used (Mouschovias and Spitzer, 1976) to establish a criterion for stability against collapse. This criterion is

$$\frac{M_{\mathrm{cloud}}}{\Phi_B} < \frac{M_{\mathrm{crit}}}{\Phi_B} = \left(\frac{1}{63G}\right)^{1/2}, \tag{9.1.17}$$

where Φ_B is the flux of a frozen-in field which threads a cloud of mass M_{cloud}. The critical mass can also be written as

$$M_{\mathrm{crit}} = \frac{5}{27} \times 10^5 \frac{(B(\mu\mathrm{Gauss}))^3}{\left(n_{\mathrm{p}}(\mathrm{cm}^{-3})\right)^2}, \tag{9.1.18}$$

where B is the magnetic field strength and n_{p} is the proton density, whether in atomic or molecular form. At $B \sim 30 \, \mu\mathrm{Gauss}$ and $n_{\mathrm{p}} \sim 100$ cm^{-3}, $M_{\mathrm{crit}} \sim 5 \times 10^4 \, M_\odot$. Since this is near the upper end of the distribution of cloud masses, one may infer that the balance between gravity and magnetic forces is responsible for setting the upper bound on the observed distribution.

Since masses of stars are typically orders of magnitude smaller than giant molecular cloud masses, it is clear that other factors must be involved which persuade giant molecular clouds to fragment into smaller pieces, some of which become unstable to gravitational collapse. Mouschovias has argued (1977, 1979) that the two key ingredients of a solution are the existence of one or more ionizing agents (e.g., cosmic rays, radioactivity, ultraviolet radiation) and the mechanism of ambipolar diffusion (Mestel & Spitzer, 1956). In the ambipolar diffusion mechanism, charged particles (free electrons and dust grains in particular) are tied to the magnetic field, but neutral particles are able to diffuse inward under the action of gravity and exert a drag on the charged particles. Regions of high density grow denser; in these regions, shielding against ionizing cosmic rays increases, decreasing the strength of the coupling between magnetic field and charged particles; eventually a central core forms in which both the ratio of gravitational binding energy to thermal energy content and the ratio of mass to magnetic flux exceed critical values. An added bonus achieved during the contraction phase controlled by ambipolar diffusion is that magnetic braking reduces the angular momentum of the contracting core to the extent that gravity can effectively counter "centrifugal" forces which might otherwise prevent evolution into a compact object in which the force associated with a gas pressure gradient balances gravity and in which central temperatures are large enough for energy generation by nuclear burning to be the primary source of luminosity.

Mouschovias and his collaborators have followed in various geometries the evolution of structures beginning with conditions in giant molecular clouds and continuing to the formation of supercritical cores and have followed the subsequent core collapse, which is also driven by ambipolar diffusion, until particle densities reach $n \sim 10^{10}$ cm^{-3} (Fiedler & Mouschovias, 1992; 1993; Ciolek & Mouschovias 1993, 1994, 1995; Basu & Mouschovias 1994, 1995a, 1995b). Following the evolution further, Konstantinos Tassis and Mouschovias (2005a, 2005b) find that an opaque hydrostatic protostellar core is formed which, for a wide range of initial conditions, has properties which are independent of the initial conditions. The rearrangement of magnetic flux is such that magnetic pressure in the protostar does not significantly affect its structure which has become essentially spherically symmetric. Treating the hydrostatic core, which by now has a central density of $\sim 10^{15}$ cm^{-3} and a central temperature $\sim 10^3$ K, as a spherical sink of radius ~ 7 AU, they find that accretion onto this sink is spasmotic (involving the quasiperiodic formation and dissipation of magnetic shocks), with the accretion rate varying from $\dot{M} \sim 4 \times 10^{-7}$ $M_\odot$ yr^{-1} to $\dot{M} \sim 10^{-4}$ $M_\odot$ yr^{-1}. Thus, although the magnetic field in the nearly hydrostatic core of the protostar does not play a direct role in the evolution of the core, the magnetic field outside of the core controls the rate of accretion. The time for the mass of the sink to reach $1 M_\odot$ is $\sim 2.5 \times 10^5$ years, giving an average accretion rate of $\dot{M} \sim 4 \times 10^{-6}$ $M_\odot$ yr^{-1}.

9.1.3 Other studies of collapse and formation of a quasistatic core

That the formation of an opaque, quasistatic central core is a penultimate step in protostar evolution was recognized by, among others, M. V. Penston (1966), Peter Bodenheimer & Alan Sweigert (1968), and Richard B. Larson (1969a, 1969b), who explored the dynamical evolution of spherically symmetric configurations of finite mass, initially of constant density and constant temperature and marginally unstable against collapse in the absence of magnetic fields. In Larson's work, the evolution is followed through a sequence of stages culminating in the formation of a central core in quasistatic equilibrium which accretes, through a standing shock, matter from a nearly free falling envelope.

Larson finds that collapse occurs if the radius of the configuration is larger than R_{crit}, where

$$\frac{kT_0}{\mu M_{\mathrm{H}}} \gtrsim 0.43 \, \frac{G M_{\mathrm{protostar}}}{R_{\mathrm{crit}}} = 0.43 \, G\rho_0 \, \frac{4\pi}{3} \, R_{\mathrm{crit}}^2. \tag{9.1.19}$$

If the coefficient of $G M_{\mathrm{protostar}}/R_{\mathrm{crit}}$ in eq. (9.1.19) is chosen as large as 0.46, the collapse fizzles. If the coefficient is chosen as 0.41, collapse proceeds initially on the free fall time scale. Note that

$$2 \, R_{\mathrm{crit}} \sim 1.56 \left(\frac{kT_0}{M_{\mathrm{H}}} \right)^{1/2} \frac{1}{\sqrt{G\rho_0}} = 0.88 \, \lambda_{\mathrm{crit}}, \tag{9.1.20}$$

where λ_{crit} is given by eq. (9.1.12).

For models of composition $X = 0.626$, $Y = 0.364$, and $Z = 0.01$ at a temperature $T_0 = 10$ K and density $\rho_0 = 1.1 \times 10^{-19}$ g cm^{-3}, hydrogen is in molecular form and metals are presumably locked into grains so $1/\mu \sim (1 + X)/4 = 1/2.46$ and $n_0 = \rho_0/(\mu M_H) = 2.67 \times 10^4$ cm^{-3}. For a 1 $M_\odot$ model, the starting radius given by eq. (9.1.18) is $R = 1.63 \times 10^{17}$ cm $= 2.34 \times 10^6 R_\odot = 1.08 \times 10^4$ AU, and the free fall time scale is $\sim 2 \times 10^5$ yr. Under the adopted initial conditions, the time scale for cooling by grains (Gaustad, 1963) is much shorter than the time scale for heating due to compression, so, during the first part of the collapse, the opacity is taken to be zero. Even though the initial collapse is isothermal (by construction), a density (hence, pressure) gradient is built up which prevents the collapse from being truly free fall. The establishment of pressure gradients is not an artifact of the adopted outer boundary condition (fixed volume and gas velocity equal to zero at the fixed outer edge of the configuration), as subsequent calculations with other boundary conditions (such as a fixed external pressure at the outer boundary and free fall velocity at this boundary) have shown.

Assuming that the opacity is due to grains and is constant at the value $\kappa = 0.15$ cm^2 g^{-1}, Larson finds that, after 1.34×10^5 yr of evolution (two thirds of the initial free fall time), the inner ~ 0.001 $M_\odot$ of the model has become opaque and the assumption of isothermality is replaced by the equation for radiative flow. A nearly hydrostatic core of mass ~ 0.005 $M_\odot$ forms, with a central mass density of $\rho_c \sim 2 \times 10^{-10}$ g cm^{-3} (particle density $n_c \sim 4.85 \times 10^{13}$ cm^{-3}) and central temperature of $T_c \sim 170$ K. The core continues to grow in mass and to heat in consequence of the compression until, when core temperatures reach the dissociation temperature for molecular hydrogen (~ 2000 K), a second core forms and experiences a phase of dynamical collapse. Ultimately, after a rather complicated phase of core expansion followed by contraction, the configuration consists of a core of atomic hydrogen of mass ~ 0.01 $M_\odot$ which is in quasistatic equilibrium and which is accreting from an extended envelope. Matter from the envelope reaches the surface of the core with a velocity essentially equal to the velocity achieved in free fall from infinity to the core radius. Accretion occurs through a shock. All of the energy radiated outward by the shock and from the effective photosphere of the core is absorbed by grains in the envelope and reradiated in the infrared. Thus, during the main accretion phase, which lasts for several times 10^5 years, the model is an infrared protostar. Larson finds that the final configuration lies above the main sequence at a radius of approximately ~ 2 $R_\odot$. Repeating the calculations for more massive configurations, Larson concludes that protostars which become main sequence stars with masses larger than ~ 5–6 $M_\odot$ ignite hydrogen in central regions before the accretion phase has ended, a conclusion reinforced by subsequent studies by others (e.g., Appenzeller & Tscharnuter, 1975).

Several of the features found in the spherical collapse calculations can be understood in terms of similarity solutions (e.g., Shu, 1977a; Hunter, 1977). There exists one relevant solution which is particularly simple to derive. For an isothermal sphere in hydrostatic equilibrium, one can write

$$\frac{\partial P}{\partial r} = v_s^2 \frac{\partial \rho}{\partial r} = -\frac{GM(r)}{r^2} \rho. \tag{9.1.21}$$

Assuming that

$$\rho = \frac{A}{r^m}, \tag{9.1.22}$$

it follows from eq. (9.1.21) that

$$-\frac{A}{\rho}\frac{m}{r^{m+1}} = \frac{G}{v_s^2}\frac{M(r)}{r^2}, \tag{9.1.23}$$

or

$$M(r) = -A\frac{v_s^2}{G}\frac{r^2}{\rho}\frac{m}{r^{m+1}}. \tag{9.1.24}$$

But, since

$$M(r) = \int_0^r 4\pi\rho r^2\,\mathrm{d}r = -4\pi A\,\frac{1}{m-3}\frac{1}{r^{m-3}}, \tag{9.1.25}$$

it is also true that, provided $0 < m < 3$,

$$\rho = \frac{m}{3-m}\frac{1}{4\pi}\frac{v_s^2}{G}\frac{1}{r^2}, \tag{9.1.26}$$

which is consistent with the initial assumption, eq. (9.1.22), only if $m = 2$. Thus,

$$\rho = \frac{1}{2\pi}\frac{v_s^2}{G}\frac{1}{r^2}, \tag{9.1.27}$$

$$A = \frac{1}{2\pi}\frac{v_s^2}{G}, \tag{9.1.28}$$

and

$$M(r) = 2\frac{v_s^2}{G}r. \tag{9.1.29}$$

Shu (1977b) points out that this singular solution forms a boundary to a class of isothermal spheres explored by R. Ebert (1955; 1957) and by W. B. Bonner (1956) that resemble the interior structure of Larson's models after the formation of a hydrostatic core. Isothermal Ebert–Bonner spheres are in hydrostatic equilibrium, subject to an external pressure. Their stability can be classified according to the degree of central concentration $\rho_c/\bar{\rho}$. Those with $\rho_c/\bar{\rho} < 14.3$ are stable and those with $\rho_c/\bar{\rho} > 14.3$ are unstable. The unstable Ebert–Bonner spheres consist of a core in which the density decreases rather slowly outward, an envelope in which the density declines roughly as the inverse square of the radius, and a transition region between core and envelope. With increasing concentration, the size of the core decreases and the size of the envelope increases. In the limit of infinite central concentration, the singular solution, eq. (9.1.27), is approached. It is the unstable Ebert–Bonner spheres that most closely resemble some of Larson's models and it is the inverse square law density distribution predicted for the matter in the infalling dusty envelope that may be susceptible to observational check. Indeed, in a study of the 2.7 mm continuum emission from the dusty envelopes around a handful of real objects, Leslie Looney, Lee G. Mundy, and W. J. Welch (2003) find that $\rho \propto 1/r^2$ rather than, say, $\rho \propto 1/r^{3/2}$.

9.1.4 Further description of the accretion phase

Steven W. Stahler, Frank H. Shu, and Ronald E. Taam (1980a; 1980b; 1981) propose that, once the quasistatic core of atomic hydrogen is formed, the protostar may be broken into three parts which may be treated separately: (1) an opaque, accreting quasistatic core, (2) an accretion shock, and (3) an envelope which, for much of the evolution, is in free fall, and through which matter flows at a constant rate. Stahler *et al.* estimate the appropriate accretion rate by equating the free fall time scale for the envelope at any instant with the time for a sound wave to cross the envelope. Thus, using eq. (9.1.15),

$$t_{\text{crossing}} \sim \frac{R}{v_s} = t_{\text{ff}} \sim \sqrt{\frac{3\pi}{32} \frac{4\pi}{3} \frac{R^3}{M}} = \sqrt{\frac{\pi^2}{8} \frac{R^3}{GM}}, \tag{9.1.30}$$

or

$$R \sim \frac{GM}{v_s^2}, \tag{9.1.31}$$

giving an accretion rate which is roughly

$$\dot{M}_{\text{accrete}} \sim \frac{M}{t_{\text{ff}}} = \frac{M}{R/v_s} \sim \frac{v_s^3}{G}$$

$$\sim \frac{3}{2} \times 10^{-7} \left(\frac{T}{\mu}\right)^{3/2} M_{\odot} \, \text{yr}^{-1}. \tag{9.1.32}$$

Beginning with a hydrostatic core of mass $0.01 \, M_{\odot}$ in which the initial entropy distribution is chosen somewhat arbitrarily, and depositing matter on the core at a constant rate of $\dot{M} = 10^{-5} \, M_{\odot} \, \text{yr}^{-1}$ for 10^5 years, the end result is a quasistatic star on the fully convective portion of a canonical track rather than on the radiative portion of the track. However, by assuming a constant mass flow rate in the envelope, which requires that

$$\dot{M} = -4\pi \, r^2 \, \rho \, u = \text{constant}, \tag{9.1.33}$$

where u is the flow velocity, and assuming

$$u^2 \propto \frac{1}{r}, \tag{9.1.34}$$

as required by free fall, it follows that

$$\rho \propto \frac{1}{r^{3/2}}, \tag{9.1.35}$$

which disagrees with the observations.

In order to be consistent with the distribution of T-Tauri stars in the HR diagram, the final model star which emerges from protostar collapse calculations should be near the low temperature edge of the observed distribution. In a study of $\sim$500 young stars in the Taurus–Auriga complex, Orion, NGC 2264, NGC 7000/IC 5070, and the ρ-Ophiuchi association,

Martin Cohen and Leonard V. Kuhi (1979) show that the locations in the HR diagram of stars less massive than $\sim$2 $M_\odot$ are in much better agreement with the characteristics of models along the convective portion of canonical tracks (see Section 9.1) than with the characteristics predicted by the then extant spherical collapse models, which emerge as isolated stars closer to the main sequence.

The consequences of deuterium burning in non-accreting, quasistatic models were first examined by Peter Bodenheimer (1966a, 1966b). Based on properties of a series of accretion models, Stahler (1988) argues that the photosphere of an accreting protostar during a deuterium-burning phase defines a mass–radius relationship that, in the HR diagram, corresponds with the low temperature envelope of the stellar distribution defined by the Cohen and Kuhi samples, and calls this envelope a birth line. However, deuterium burning releases a total amount of energy which is quite small compared with the potential energy released in the interior during the entire accretion phase (being about 20% in the case of a 1 $M_\odot$ star), so this correspondence may simply be a coincidence. Francesco Palla & Stahler (1991) concentrate on models of mass $\sim$5–6 $M_\odot$ and compare with Herbig Ae and Be stars (Herbig, 1960), arguing that these stars have burned deuterium during the protostar accretion phase (see also Palla & Stahler, 1993). Deuterium burning is an interesting phenomenon, but whether it occurs during the accretion phase or during the following phase of quasistatic evolution may not seriously impact the observed distribution of young stars in the HR diagram.

In summary, extant hydrodynamical explorations provide crucial insights into protostar behavior during the phase of accretion onto a quasistatic core, but numerical descriptions which are fully consistent with physical principles and with the observations have yet to be achieved. Comparison with the locations in the HR diagram of T-Tauri stars suggests that, for population I stars of mass $\gtrsim$2 $M_\odot$, a quasistatic phase of gravitational contraction not significantly affected by accretion or deuterium burning precedes the ZAMS configuration and that, although deuterium burning occurs near the transition from the accretion phase to the subsequent isolated star contraction phase, this proximity is essentially a coincidence.

For more massive stars, the situation is less clear. That is, stars as massive as $\sim$5–6 $M_\odot$ may or may not appear on canonical gravitational contraction tracks before reaching the ZAMS (Kuhi, 1966). The caveat "may" is a consequence of the fact that the lifetime of a massive star on a canonical track is much smaller than the ages of typical young stellar clusters, with the consequence that definitive observational tests are lacking. In this book, it will be assumed that all protostars ultimately make a transition from a dynamical accretion phase to a quasistatic pre-main sequence phase during which the release of gravitational potential energy is the primary source of surface luminosity.

9.2 Pre-main sequence quasistatic evolution of a solar mass population I model with deuterium burning

To construct a quasistatic model in the Hayashi band, one may take advantage of the fact that, since the entropy $S(t)$ is the same everywhere in the fully convective portion of a model, the time rate of change of entropy $dS(t)/dt$ is also independent of position, allowing the gravothermal energy-generation rate to be written as

$$\epsilon_{grav} = -T\,\frac{dS(t)}{dt} = C(t)\,T, \qquad (9.2.1)$$

where $C(t) > 0$ depends only on time. Adopting an energy-generation rate of the form

$$\epsilon_{tot} = \epsilon_{nuc} + C(t)\,T, \qquad (9.2.2)$$

and using the classical fitting technique described in Section 8.5, one can construct a series of static models (of index i) characterized by increasing values of $C_i = C(t_i)$. Equivalently, one can adopt a simple expression for $C(t)$, say, $C(t) = A\,t$, where $A > 0$ is a constant, and use the relaxation technique described in Section 8.7 to evolve models. The first model, with $C_1 = C_1(t_1) = 0$, can be of homogeneous composition and burn hydrogen. For subsequent models, the characteristics of model $(i-1)$ can be used as starting values for constructing model (i). As $C(t)$ is increased, the energy produced by the term linear in T in eq. (9.2.2) forces successive models to become larger and cooler, until eventually nuclear burning becomes negligible. With increasing $C(t)$, energy fluxes become larger; with decreasing temperatures, interior opacities become larger; eventually, models become unstable against convection over all but the very outermost photospheric layers and expression (9.2.1) becomes an exact representation of the rate of gravothermal energy release.

Since ultimately $\epsilon_{nuc} \ll C(t)\,T$, any reasonable approximation to ϵ_{nuc}, such as given by the algorithms described in Section 6.2.3 and 6.2.4, leads to the same final result. Since the final model is convective everywhere except in the neighborhood of the photosphere, the model is insensitive to the details of the opacity chosen for the interior, e.g., the analytic expressions given in Section 7.12. On the other hand, where the final model appears in the HR diagram depends crucially upon the opacity chosen for matter in photospheric layers. As long as this latter opacity has the property that it decreases with decreasing temperature and with decreasing density, as is the case for the electron-scattering contribution to the opacity when the free electron density decreases with decreasing temperature and in the case of the H^- contribution to the opacity (see, e.g., eq. (7.16.48)), the smaller the final surface temperature, the larger is the final luminosity.

Where in the Hayashi band one wishes to terminate the construction of models is a matter of taste. For this book, construction of a $1\,M_\odot$ model has been terminated when the central temperature of the model is $T_c \sim 0.64 \times 10^6$ K and the energy-generation rate by deuterium burning at the center is several orders of magnitude smaller than the rate of gravothermal energy release there. Models with masses in the range 0.5–$15\,M_\odot$ constructed in the fashion outlined here are described by Iben (1965).

9.2.1 Input physics and initial abundances for evolutionary calculations

Once a starting Hayashi-band model has been settled upon, the artificial energy-generation rate can be discarded and more sophisticated nuclear energy-generation rates and opacities can be introduced. In this work, the nuclear reaction rates given by Angulo *et al.* (1999) have been adopted. For opacities at high temperatures, Livermore tables for $Z = 0.01$ and $Z = 0.02$ (Rogers & Iglesias, 1996; Iglesias & Rogers, 1996) have been chosen and, for opacities at low temperatures, the tables of D. R. Alexander & J. W. Ferguson (1994) have been incorporated.

Choices for initial number abundance parameters and a choice for the mixing length parameter depend on the desired objectives. In the present context, the mass abundance parameters X, Y, Z, and the abundances of ^{2}H, ^{3}He, and the lithium isotopes ^{7}Li and ^{6}Li are of particular interest. In an initial exercise, the primary mass abundance parameters have been chosen as $X = 0.74$, $Y = 0.25$, and $Z = 0.01$, and the mixing length to pressure scale height ratio has been chosen as $l_{rmix}/H_{pres} = 0.6$. For a model of 1 $M_\odot$, these choices lead, after 4.6×10^9 yr of evolution, to an evolved main sequence model of luminosity $L = L_\odot$ and radius $R = R_\odot$.

For deuterium and ^{3}He, which are consequences of Big Bang nucleosynthesis and processing in earlier generations of stars, initial number abundances relative to hydrogen have been chosen as $Y(^2\text{H})/Y(^1\text{H}) = 2 \times 10^{-5}$, and $Y(^3\text{He})/Y(^1\text{H}) = 1.7 \times 10^{-5}$. For the lithium isotopes, which are a consequence of Big Bang nucleosynthesis, spallation by cosmic rays on interstellar matter, and processing in stars, initial abundances relative to hydrogen have been chosen as $Y(^6\text{Li})/Y(\text{H}) = 1.10 \times 10^{-10}$ and $Y(^7\text{Li})/Y(\text{H}) = 1.35 \times 10^{-9}$. The lithium abundance choices have no influence on the evolutionary time scale, but their history is of relevance for understanding the nature of mixing in subphotospheric layers of the Sun over its lifetime.

The entire set of initial isotope abundances that has been adopted is shown in Table 9.2.1. In this table, the third column gives the chosen number abundance Y_i' of each isotope relative to hydrogen, with the hydrogen abundance taken as unity, the second column gives $\log_{10} Y_i' + 12$, the fourth column gives the number abundance parameter Y_i defined by

$$Y_i = \frac{Y_i'}{\sum_i Y_i' A_i}, \tag{9.2.3}$$

and the fifth column gives

$$X_i = A_i Y_i, \tag{9.2.4}$$

where A_i is an atomic mass. With these definitions, $\sum_i X_i = 1$. The electron number abundance parameter is given by

$$Y_e = \sum_i Z_i Y_i, \tag{9.2.5}$$

where Z_i is an atomic number, and $X_e = Y_e \, (m_e/M_H)$. In most instances, the abundances of elements heavier than helium are smaller by a factor of 1.22 than abundances quoted by Asplund, Grevesse, and Sauval (see the discussion in Section 9.2) who give $Z = 0.0122$. The abundances of deuterium and of ^{3}He are based on the discussion by Brian D. Fields (1996) of Big Bang yields and observations of abundances in the interstellar medium and on the discussion by C. Chiappini, A. Renda, and F. Matteucci (2002) regarding the

	Table 9.2.1 Abundances in starting model ($Z = 0.01$)			
Isotope	$\log Y_i' + 12$	Y_i'	Y_i	X_i
proton	12	1	0.740	0.740
neutron	$-\infty$	0.0	0.0	0.0
He4	10.93	0.08446	0.0625	0.250
electron			0.867	4.71/$-$4
deuteron	7.30	2.00/$-$5	1.48/$-$5	2.96/$-$5
He3	7.23	1.70/$-$5	1.26/$-$5	3.77/$-$5
Li6	2.04	1.10/$-$10	8.10/$-$11	4.86/$-$10
Li7	3.13	1.35/$-$9	9.99/$-$10	6.99/$-$9
Be7	$-\infty$	0.0	0.0	0.0
Be9	1.11	1.30/$-$11	9.62/$-$12	8.66/$-$11
B8	$-\infty$	0.0	0.0	0.0
B10	1.90	8.20/$-$11	6.07/$-$11	5.94/$-$10
B11	2.52	3.30/$-$10	2.44/$-$10	2.69/$-$9
C12	8.30	2.00/$-$4	1.48/$-$4	1.78/$-$3
C13	6.35	2.23/$-$6	1.65/$-$6	2.14/$-$5
C14	$-\infty$	0.0	0.0	0.0
N13	$-\infty$	0.0	0.0	0.0
N14	7.69	4.95/$-$5	3.66/$-$5	5.13/$-$4
N15	5.26	1.82/$-$7	1.35/$-$7	2.02/$-$6
O15	$-\infty$	0.0	0.0	0.0
O16	8.57	3.75/$-$4	2.78/$-$4	4.44/$-$3
O17	5.16	1.43/$-$7	1.06/$-$7	1.80/$-$6
O18	5.88	7.55/$-$7	5.59/$-$7	1.01/$-$5
F17	$-\infty$	0.0	0.0	0.0
F18	$-\infty$	0.0	0.0	0.0
F19	4.18	1.51/$-$8	1.12/$-$8	2.12/$-$7
Ne20	7.70	5.00/$-$5	3.70/$-$5	7.40/$-$4
Ne21	6.11	1.30/$-$6	9.62/$-$7	2.02/$-$5
Ne22	6.70	5.00/$-$6	3.70/$-$6	8.14/$-$5
Na21	$-\infty$	0.0	0.0	0.0
Na22	$-\infty$	0.0	0.0	0.0
Na23	6.09	1.22/$-$6	8.99/$-$7	2.07/$-$5
Mg23	$-\infty$	0.0	0.0	0.0
Mg24	7.33	2.20/$-$5	1.63/$-$5	3.90/$-$4
Mg25	6.44	2.78/$-$6	2.06/$-$6	5.14/$-$5
Mg26	6.49	3.06/$-$6	2.26/$-$6	5.89/$-$5
Si28	7.42	2.66/$-$5	1.97/$-$5	5.50/$-$4
S32	7.05	1.13/$-$5	8.36/$-$6	2.68/$-$5
Fe56	7.36	2.31/$-$5	1.71/$-$5	9.57/$-$4

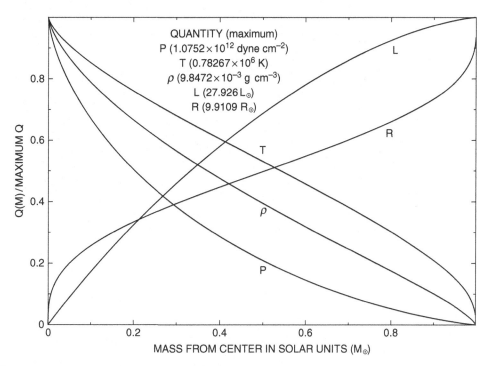

QUANTITY (maximum)
P (1.0752×10^{12} dyne cm^{-2})
T (0.78267×10^6 K)
ρ (9.8472×10^{-3} g cm^{-3})
L (27.926 $L_\odot$)
R (9.9109 $R_\odot$)

Fig. 9.2.1 Structure of a gravitationally contracting model before deuterium burning (M = 1$M_\odot$, Z = 0.01, Y = 0.25)

influence of galactic evolution. The chosen abundances are consistent with the analysis of abundances in the protosolar cloud by J. Geiss and G. Gloeckler (1998).

9.2.2 Structure and gravothermal characteristics of a Hayashi-band model

After 0.9×10^4 yr of evolution with the modified input physics, the model arrives in the HR diagram at the point $\log L/L_\odot = 1.446$, $\log(T_e) = 3.626$, with a radius of $R = 9.91$ $R_\odot$. The structure characteristics of the model are described in Figs. 9.2.1 and 9.2.2. These characteristics have been plotted using the WIP (Work in Progress) graphics package (see J. A. Morgan WIP – an interactive graphics software package, in Shaw, Payne, and Hayes, 1995) which draws straight lines between points. From the smoothness of the curves in Figs. 9.2.1 and 9.2.2, it is clear that the zoning criteria (maximum 10% variation from zone to zone in all structure variables and in the gravothermal energy-generation rate) are more than adequate. The size of the central sphere is determined by the requirement that $(P_1 - P_2)/P_1 < 0.003$, where P_1 and P_2 are, respectively, the pressure at the center and at the edge of the central sphere. Altogether, there are 254 grid points in the model. It is worth emphasizing that P_i, T_i, and ρ_i, which are defined at shell centers, are plotted against $R_{i-1/2}$ or $M_{i-1/2}$, whereas L_i, which is defined at grid points, is plotted against R_i or M_i, both of which are also defined at grid points; plotting P_i, T_i, and ρ_i against R_i or M_i and L_i against $R_{i-1/2}$ or $M_{i-1/2}$ results in curves which are noticeably bumpy. Although the static envelope has a mass of only 0.0005 $M_\odot$ (about half the mass of Jupiter), its radial

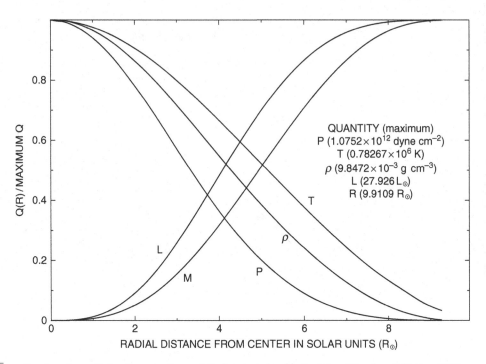

QUANTITY (maximum)
P $(1.0752\times10^{12}$ dyne cm$^{-2})$
T $(0.78267\times10^{6}$ K)
ρ $(9.8472\times10^{-3}$ g cm$^{-3})$
L $(27.926$ L$_{\odot})$
R $(9.9109$ R$_{\odot})$

Fig. 9.2.2   Structure of a gravitationally contracting model before deuterium burning (M $= 1M_{\odot}, Z = 0.01, Y = 0.25$)

extent, $\Delta R \sim 0.075\, R_{\odot}$, is almost 1% of the radius of the model. This accounts for the fact that the profile of the interior temperature in Fig. 9.2.2 terminates at $T \sim 26\,000$ K, compared with the surface temperature of $T_e \sim 4230$ K.

Characteristics of participants in the gravothermal energy-generation process are shown in Figs. 9.2.3 and 9.2.4. The rate at which work is done locally by gravity is given in Fig. 9.2.3 by the curve

$$\epsilon_{\text{work}} = \mathbf{g} \cdot \mathbf{v}. \tag{9.2.6}$$

The components g and v of ϵ_{work} are described in Fig. 9.2.4, where g_i is the gravitational acceleration at grid point i, $v_i = \delta r_i/\delta t$ is the velocity at grid point i, and δr_i is the change in the radius at grid point i in the time step δt which, for this model, is $\delta t = 100$ yr. The rate at which compressional work is done locally is given in Fig. 9.2.3 by the curve labeled

$$\epsilon_{\text{compression}} = \frac{P}{\rho^2} \frac{\delta \rho}{\delta t}, \tag{9.2.7}$$

where $\delta\rho$ is the change in ρ in time step δt. The curves $\epsilon_{\text{compression}}$ and $\mathbf{g} \cdot \mathbf{v}$ bear no obvious relationship to one another, and yet, when the curves are integrated over mass, the results are identical, thus explicitly corroborating the arguments leading to eqs. (8.1.67) and (8.3.24). The fact that $\mathbf{g} \cdot \mathbf{v} = \mathbf{v} \cdot (\rho^{-1}\, \nabla P)$ provides the clue that the work done by

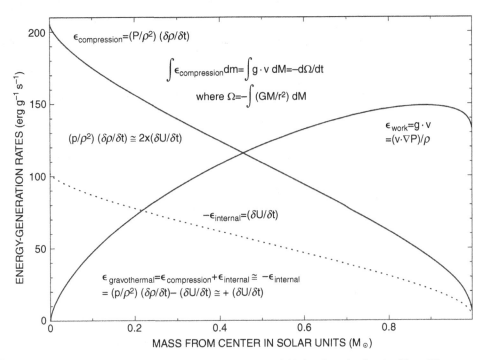

Fig. 9.2.3   Gravothermal energy generation in a gravitationally contracting model before deuterium burning ($M = 1M_\odot$, $Z = 0.01, Y = 0.25$)

gravity at any point is communicated by pressure-gradient forces to appear elsewhere as compressional work.

The rate at which U, the internal energy per gram, increases is given in Fig. 9.2.3 by the dotted curve labeled

$$-\epsilon_{internal} = \frac{\delta U}{\delta t}. \tag{9.2.8}$$

This quantity is indistinguishable from

$$\epsilon_{gravothermal} = \epsilon_{compression} + \epsilon_{internal}. \tag{9.2.9}$$

Thus,

$$\frac{P}{\rho^2}\frac{\delta\rho}{\delta t} - \frac{\delta U}{\delta t} = \frac{\delta U}{\delta t}, \tag{9.2.10}$$

which says that the rate at which gravitational work is manifested locally in the form of compressional work minus the rate at which the internal energy increases locally is equal to the rate at which the internal energy increases locally. In short, the global virial theorem (see Section 3.3) which states that, at any given time, the total gravitational binding energy minus the total internal energy is equal to the total gas internal energy is, in the present instance, also a local theorem relating the time rate of change of the local contribution to the binding energy to the time rate of change of the local internal energy.

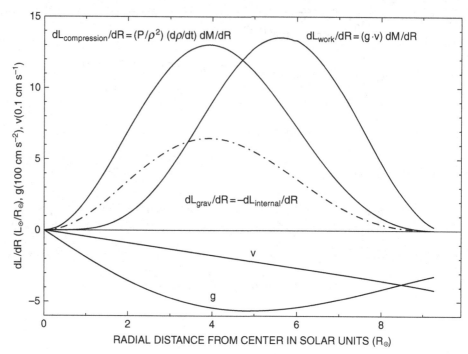

Fig. 9.2.4 Contributions to gravothermal luminosities in a gravitationally contracting model before deuterium burning ($M = 1M_\odot, Z = 0.01, Y = 0.25$)

In the present instance, the perfect gas equation of state is a very good approximation, and so eq. (9.2.10) can be written as

$$\frac{P}{\rho^2}\,\delta\rho - \delta U = N_A\,kT\,\sum_i Y_i\left(\frac{\delta\rho}{\rho} - \frac{3}{2}\frac{\delta T}{T}\right)$$

$$= \delta U = N_A\,kT\,\sum_i Y_i\left(\frac{3}{2}\frac{\delta T}{T}\right), \tag{9.2.11}$$

from which it follows that

$$\frac{\delta\rho}{\rho} = 3\frac{\delta T}{T}. \tag{9.2.12}$$

Thus, in every mass shell i in the model interior,

$$\left(\frac{\rho}{T^3}\right)_i = \text{constant}. \tag{9.2.13}$$

This means that the quantity $\langle\rho/T^3\rangle$, where brackets denote an average over the mass of the model, is also a constant, supporting the reasoning in Chapter 3 which leads to the statement that, independent of time, $\langle\rho/T^3\rangle \propto 1/M^2$ (see eq. (3.1.15)).

If in the equation for entropy changes,

$$-T\frac{\delta S}{\delta t} = \frac{P}{\rho^2}\frac{\delta\rho}{\delta t} - \frac{\delta U}{\delta t}, \tag{9.2.14}$$

one uses the first identity in eq. (9.2.11), integration produces

$$S = -\left(N_A\, k\, \sum_i Y_i\right) \log_e \frac{\rho}{T^{3/2}} + \text{constant}. \tag{9.2.15}$$

Differencing eq. (9.2.15) (with respect to time) and using eq. (9.2.12), it follows that

$$\frac{\delta S}{S} = -\left(\frac{\delta\rho}{\rho} - \frac{3}{2}\frac{\delta T}{T}\right) = -\frac{3}{2}\frac{\delta T}{T} = -\frac{1}{2}\frac{\delta\rho}{\rho}. \tag{9.2.16}$$

Since, at any given time, S in a fully convective region is independent of position, it is also true that logarithmic changes in temperature are the same in every interior shell. The same is true for logarithmic changes in the density. That is, for all i,

$$\frac{\delta T_{i-1}}{T_{i-1}} = \frac{\delta T_i}{T_i} = \frac{\delta T_{i+1}}{T_{i+1}} \tag{9.2.17}$$

and

$$\frac{\delta\rho_{i-1}}{\rho_{i-1}} = \frac{\delta\rho_i}{\rho_i} = \frac{\delta\rho_{i+1}}{\rho_{i+1}}. \tag{9.2.18}$$

By looking at the contribution to energy-generation rates of shells of constant thickness in the radial coordinate, a different perspective which is complementary to the one provided by Fig. 9.2.3 emerges. Multiplication of an energy-generation rate given in units of erg g^{-1} s^{-1} by

$$\frac{dM}{dr} = 4\pi r^2 \rho \tag{9.2.19}$$

produces the differential contribution of a spherical shell to a quantity having the dimensions of luminosity over distance. In Fig. 9.2.4 are plotted the quantities

$$\frac{dL_{\text{compression}}}{dr} = \frac{P}{\rho^2}\frac{\delta\rho}{\delta t}\frac{dM}{dr}, \tag{9.2.20}$$

$$\frac{dL_{\text{internal}}}{dr} = -\frac{\delta U}{\delta t}\frac{dM}{dr}, \tag{9.2.21}$$

$$\frac{dL_{\text{work}}}{dr} = (\mathbf{g} \cdot \mathbf{v})\frac{dM}{dr}, \tag{9.2.22}$$

and

$$\frac{dL_{\text{grav}}}{dr} = \frac{dL_{\text{gravothermal}}}{dr} = \frac{dL_{\text{compression}}}{dr} + \frac{dL_{\text{internal}}}{dr} = -\frac{dL_{\text{internal}}}{dr}. \tag{9.2.23}$$

In the mass-coordinate representation, the distributions $\epsilon_{\text{compression}}$ and $\epsilon_{\text{gravothermal}}$ peak at the center of the model. In contrast, in the radius-coordinate representation, the major contributions to the compressional work rate and the major contributions to the surface luminosity occur in the middle regions of the model. The relationship between the rate at which gravity does work and the rate at which compressional work is being done are more transparently related in the radius-coordinate representation.

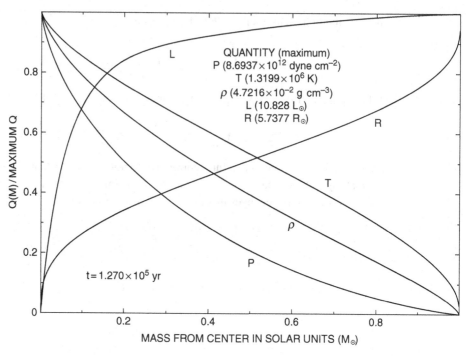

Fig. 9.2.5   Structure of a fully convective deuterium burning model ($M = 1 M_\odot$, $Z = 0.01$, $Y = 0.25$)

9.2.3 The deuterium-burning phase

After about 50 000 yr of further contraction and heating, the rate at which energy is produced by deuterium burning begins to compete with and then exceed the rate of release
of gravothermal energy due to contraction. The rate of release of gravothermal energy
L_{grav} never decreases below ~15% of the total surface luminosity, but, for approximately
140 000 yr, the rate L_{pd} at which nuclear energy is produced by the $p + d \rightarrow {}^3\mathrm{He} + \gamma$ reaction dominates as a contributor to the surface luminosity. When the deuterium abundance
has decreased to $Y_d = 6.80 \times 10^{-6}$, which is slightly less than half of the initial deuterium
abundance, structure characteristics in the model are as shown in Figs. 9.2.5 and 9.2.6.

The feature which best distinguishes the structure of a model in which the dominant
source of power is a centrally concentrated nuclear source from a model in which the dominant source of power is the liberation of gravitational potential energy is the shape of the
luminosity profile. One may characterize the nuclear burning source by adopting as its
midpoint the spherical shell below which 50% of the nuclear energy is produced and defining its thickness by the difference between the two spherical shells below which, respectively, the inner 10% of the nuclear energy is produced and the inner 90% is produced.
For the deuterium-burning model described by Figs. 9.2.5 and 9.2.6, $M_{0.5} = 0.0396\ M_\odot$,
$M_{0.1} = 0.004\,98\ M_\odot$, and $M_{0.9} = 0.141\ M_\odot$, giving a thickness of 0.136 $M_\odot$. At the
midpoint $M_{0.5}$, $P = 7.11 \times 10^{12}$ dyne cm^{-2}, $T_6 = 1.22$, $\rho = 4.19$ g cm^{-3}, $R = 1.08\ R_\odot$,
and $L = 4.55\ L_\odot$.

Table 9.2.2 Characteristics of the convective region in a deuterium-burning model

$\dfrac{M}{M_\odot}$	$\dfrac{R}{R_\odot}$	$\dfrac{T_6}{10^6 \text{ K}}$	$\dfrac{\rho}{\text{g cm}^{-3}}$	$\dfrac{v_{sound}}{\text{cm s}^{-1}}$	V_{ad}	Opacity $\text{cm}^2\,\text{g}^{-1}$	V_{rad}	$V - V_{ad}$	$\dfrac{v_{conv}}{\text{cm s}^{-1}}$
0.0	0.0	1.32	4.72/−2	1.75/7	0.399	29.3	124.	7.64/−7	7.12/3
0.661	3.41	0.555	1.33/−2	1.13/7	0.400	117.	159.	3.76/−6	1.02/4
0.9995	5.430	3.63/−2	1.18/−4	2.47/6	0.235	9.39/4	2.30/6	3.79/−4	3.03/4

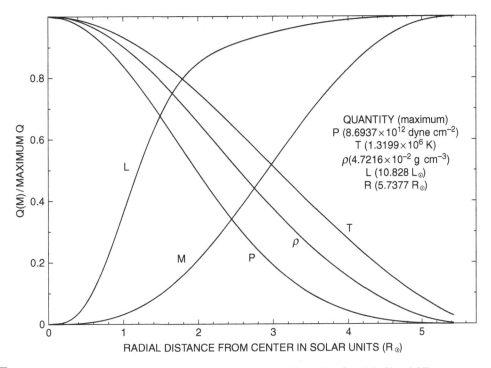

Fig. 9.2.6 Structure of a fully convective deuterium burning model ($M = 1M_\odot$, $Z = 0.01$, $Y = 0.25$)

As discussed in Section 8.10, in a convective zone, the treatment of any isotope with regard to abundance changes requires the consideration of three time scales: the time t_{mix} over which turbulence completely mixes matter in the zone, the time t_{eq} for the isotope to come into local equilibrium with respect to nuclear creation and destruction reactions, and the size t_{evo} of an evolutionary time step. In the case at hand, $t_{evo} = 3.156 \times 10^9$ s = 100 yr, and, as defined by eq. (8.10.1), $t_{mix} = 3.43 \times 10^7$ s ~ 1 yr. A sample of the ingredients for an estimate of the mixing time are given in Table 9.2.2. All quantities in the table have been previously defined. The algorithm for estimating v_{conv} from the listed quantities is given in Section 3.4.

At the center of the model, the local equilibrium abundance of deuterium is $Y_{eq} = 2.47 \times 10^{-9}$ and $t_{eq}/t_{mix} = 2740$. In the outermost zone of the interior of the model, $Y_{eq} = 2.41 \times 10^{-14}$ and $t_{eq}/t_{mix} = 1.04 \times 10^{37}$. Clearly, deuterium is uniformly distributed throughout

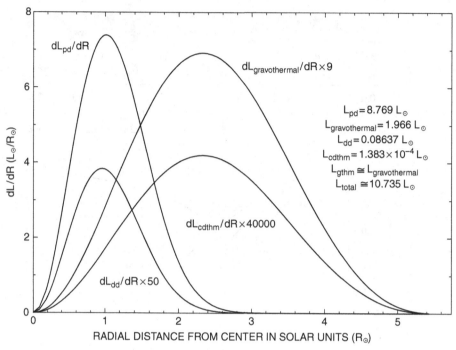

Fig. 9.2.7   Contributions to luminosity in a fully convective deuterium burning model ($M = 1M_\odot, Z = 0.01, Y = 0.25$)

the model and changes everywhere in the model occur at the same rate. In the present instance,

$$\frac{\delta Y_{nuc}}{\delta t} \sim \frac{\delta}{\delta t} \left(Y_p + Y_{deuterium} + Y_{^3He} \right) = -\frac{\delta Y_{deuterium}}{\delta t} \tag{9.2.24}$$

$$= -3.136 \times 10^{-18} \; s^{-1} = -0.990 \times 10^{-8}/100 \; yr. \tag{9.2.25}$$

The global destruction rate is 4.71×10^{39} deuterons s^{-1}. Deuterons are also being created by the pp reaction at the global rate of 1.21×10^{36} deuterons s^{-1}.

The time step between two evolutionary models is controlled by the requirement that, at any point in the models, the maximum change in the deuterium abundance which would occur in the absence of mixing is smaller than some predetermined fraction f_{max} of the deuterium abundance at the beginning of a time step. In the present instance, this fraction has been chosen rather arbitrarily as $f_{max} \sim 0.25$ and, since the maximum local rate of change is at the center, the time step is controlled by the rate of change of deuterium at the center.

In Fig. 9.2.7 are shown the differential contributions to the surface luminosity of two major sources of energy, the $d(p, \gamma)^3He$ nuclear source and the gravothermal source, as well as the contributions of two very minor but interesting sources of energy, the dd reactions $(d(d, n)^3He + d(d, p)^3H + {}^3H(e^- \nu_{\bar{e}})^3He)$ and the creation–destruction component of the gravothermal source. Both nuclear source contributions dL_{pd}/dR and dL_{dd}/dR peak near the model center at $r \sim R_\odot$, but the contribution of the creation–destruction component dL_{cdthm}/dR peaks at a point over twice as far from the model center, at the same place where the total gravothermal contribution dL_{grav}/dR peaks. In fact, apart from the difference in scale, the distributions dL_{cdthm}/dR and dL_{grav}/dR are identical. This is because, although ϵ_{cdthm} is proportional both to the temperature and to the rate of change of the particle abundance, the rate of change of the nuclear particle abundance dY_{nuc}/dt is everywhere the same. Hence, ϵ_{cdthm} is linearly proportional the temperature, just as is the rate of change of the internal energy (due to the fact that dS/dt is independent of position). For the same reason, dL_{cdthm}/dR is less concentrated toward the center than it would be if deuterium was everywhere in local equilibrium. Another way of describing the difference between the nuclear energy-source distributions and the "cdthm" distribution is that the nuclear sources produce energy at rates which are proportional to the rates at which reactions occur locally and are therefore not proportional to the rates at which particle abundances change locally, the latter changes being controlled by mixing currents which maintain equality between abundances at every point in the interior.

In the Big Bang, because neutron and proton abundances are comparable, reactions between deuterons are of paramount importance. In a star composed mostly of protons and alpha particles, reactions between deuterons and their consequences are basically only of pedagogical interest. The reactions $d + d \rightarrow {}^3\text{He} + n$ and $d + d \rightarrow {}^3\text{H} + p$ occur at approximately the same rates. The neutron produced in the first reaction reacts primarily with protons to produce deuterium and a gamma ray. Decay of ${}^3\text{H}$ (tritium) produces ${}^3\text{He}$, an electron, and an antineutrino. Comparing the distributions dL_{dd}/dR and dL_{pd}/dR in Fig. 9.2.7 shows that the entire set of reactions initiated by deuteron–deuteron reactions produces about 1% of the energy produced by $p + d \rightarrow {}^3\text{He} + \gamma$ reactions. The channel that passes through tritium (${}^3\text{H}$) yields electrons to the extent that

$$\frac{\delta Y_e}{\delta t} = +6.213 \times 10^{-20}\ \text{s}^{-1} = +1.961 \times 10^{-10}/100\ \text{yr},\qquad(9.2.26)$$

and thus (compare with eq. (9.2.25)) reduces the creation–destruction contribution to the gravothermal energy-production rate by about 2%.

Further characteristics of gravothermal energy generation in the deuterium-burning model are shown in Figs. 9.2.8 and 9.2.9. Comparing these figures with Figs. 9.2.3 and 9.2.4, it is evident that, at least in models which are fully convective below the photosphere, the nature of the conversion of gravitational potential energy into compressional work and a change in internal energy is independent of whether or not nuclear energy sources are acting. That is, even though nuclear energy provides most of the surface luminosity, an exact balance is maintained between the rate of work done by gravity and the rate at which compressional work is being done, with

$$\frac{dW_{\text{compression}}}{dt} = \int \frac{P}{\rho^2}\frac{\delta\rho}{\delta t}\,dM = \int \mathbf{g}\cdot\mathbf{v}\,dM.\qquad(9.2.27)$$

In the present instance, $dW_{\text{compression}}/dt = 3.932\ L_\odot$ and the rate at which gravothermal energy is liberated locally is essentially equal to the rate $\delta U/\delta t$ at which the internal energy increases locally, just as in the case of the earlier models in which nuclear energy sources are unimportant.

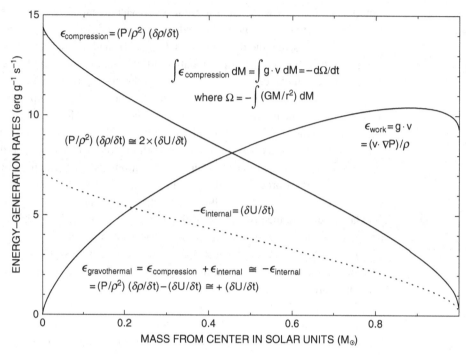

Fig. 9.2.8   Gravothermal energy generation in a fully convective deuterium burning model ($M = 1M_\odot, Z = 0.01, Y = 0.25$)

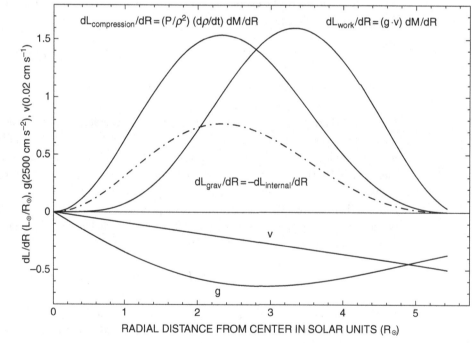

Fig. 9.2.9   Contributions to gravothermal luminosities in a deuterium burning model ($M = 1M_\odot, Z = 0.01, Y = 0.25$)

Thus, the sole role of nuclear energy sources is to supply energy which escapes from the star, reducing the flux of energy which the gravothermal source must supply to help satisfy the surface boundary condition. Nuclear energy contributes neither to the compressional work done nor to the increase in internal energy.

A feeling for the numerical accuracy achieved in the deuterium-burning models comes from an examination of global rates of change. The gravitational binding energy of the model currently under scrutiny is

$$\Omega = -\int \frac{GM}{r^2}\, dM = -5.808\,674\,0371 \times 10^{47}\ \text{erg} \tag{9.2.28}$$

and that of its immediate precursor is $\Omega = -5.808\,194\,6484 \times 10^{47}$ erg. Thus,

$$-\frac{1}{2}\frac{\delta\Omega}{\delta t} = 7.5949 \times 10^{33}\ \text{erg s}^{-1} = 1.9676\,L_\odot, \tag{9.2.29}$$

which differs by only 0.1% from half the rate at which compressional work is being done, as given by eq. (9.2.28). The total internal energy of the model is

$$E_{\text{internal}} = \int U\, dM = 2.904\,521\,0925 \times 10^{47}\ \text{erg}, \tag{9.2.30}$$

and that of its immediate precursor is $E_{\text{internal}} = 2.904\,281\,0444 \times 10^{47}$ erg, so the rate at which the total internal energy changes is

$$\frac{\delta E_{\text{internal}}}{\delta t} = 7.6061 \times 10^{33}\ \text{erg s}^{-1} = 1.9705\,L_\odot, \tag{9.2.31}$$

which is within 0.15% of the rate given by eq. (9.2.29). On the other hand, for both models under examination,

$$\frac{E_{\text{internal}}}{\Omega} = -0.5000, \tag{9.2.32}$$

as predicted by the virial theorem for a perfect gas equation of state.

It is interesting to compare properties of the numerical models with the properties of a polytrope of index $N = 1.5$. For the deuterium-burning model under scrutiny,

$$\Omega = -0.1528\,\frac{GM_\odot^2}{R_\odot}, \tag{9.2.33}$$

whereas, for an index $N = 3/2$ polytrope of the same radius, $R = 5.744\,R_\odot$,

$$\Omega_{N=3/2} = -0.857\,\frac{GM_\odot^2}{R_\odot}\frac{R_\odot}{R} = -0.1494\,\frac{GM_\odot^2}{R_\odot}. \tag{9.2.34}$$

The central concentration of the deuterium-burning model is

$$\frac{\rho_{\text{central}}}{\rho_{\text{mean}}} = 6.3262, \tag{9.2.35}$$

compared with a central concentration for an index $N = 3/2$ polytrope (see Table 5.2.3) of

$$\left(\frac{\rho_{\text{central}}}{\rho_{\text{mean}}}\right)_{N=1.5} = 5.9908. \tag{9.2.36}$$

The differences can be ascribed to the fact that the numerical model is capped by an envelope in which the temperature gradient is superadiabatic in convective regions and becomes radiative in the photosphere, leading to a radius which is larger and a mean density which is smaller than for an index $N = 3/2$ polytrope of the same central density.

In the numerical model, the total kinetic energy of bulk motions for which gravitational acceleration is responsible is

$$E_{\text{kin}} = \int \frac{1}{2} v^2 \, dM = 3.1695 \times 10^{28} \text{ erg} = 1.0912 \times 10^{-19} \, E_{\text{internal}}. \tag{9.2.37}$$

The ratio of E_{kin} to E_{internal} is not a proper measure of the accuracy of the quasistatic approximation. A more appropriate measure is the ratio

$$\int \left(\mathbf{v} \cdot \frac{d\mathbf{v}}{dt}\right) dM \div \int (\mathbf{v} \cdot \mathbf{g}) \, dM$$

$$= \frac{dE_{\text{kin}}}{dt} \div \frac{dW_{\text{grav}}}{dt} = \frac{dE_{\text{kin}}}{dW_{\text{grav}}}, \tag{9.2.38}$$

where

$$\frac{dW_{\text{grav}}}{dt} = \int \epsilon_{\text{work}} \, dM \tag{9.2.39}$$

and ϵ_{work} is given by eq. (9.2.6). From the model and its precursor, one has

$$\frac{dE_{\text{kin}}}{dW_{\text{grav}}} \sim 0.952 \times 10^{-18}, \tag{9.2.40}$$

showing that the quasistatic assumption is an exceedingly good approximation.

The energy in bulk motion due to contraction is completely dwarfed by the turbulent energy due to convection. Adopting an average convective speed of $v_{\text{conv}} \sim 10^4$ cm s^{-1} (see Table 9.2.1), one has

$$E_{\text{turbulence}} = \int \frac{1}{2} v_{\text{conv}}^2 \, dM \sim \frac{1}{2} \left(10^4 \text{ cm s}^{-1}\right)^2 M_\odot$$

$$\sim 10^{41} \text{ erg} \sim 3.4 \times 10^{-7} \, E_{\text{internal}} \gg E_{\text{kin}}. \tag{9.2.41}$$

In summary,

$$E_{\text{thermal motion}} \gg E_{\text{turbulent motion}} \gg E_{\text{bulk motion due to evolution}}. \tag{9.2.42}$$

The time dependences of several model characteristics during the deuterium-burning phase are given in Table 9.2.3, where L_s, $L_{\text{nuc}} = L_{\text{pd}}$, and $L_{\text{grav}} = L_{\text{gravothermal}} = L_g$ are, respectively, the surface luminosity and the contributions to the surface luminosity of deuterium burning and of gravothermal energy generation. The luminosities L_s and L_g and the ratio L_g/L_s are given as functions of time in Fig. 9.2.10.

Table 9.2.3 Time dependences during the deuterium-burning phase

$t\ 10^{12}$ s	$\log \frac{L_s}{L_\odot}$	$\frac{R_s}{R_\odot}$	E_{bind} 10^{47} erg	Y_{deut}	$T_c\ 10^6$ K	ρ_c g cm^{-3}	$\frac{L_{nuc}}{L_\odot}$	$\frac{L_{grav}}{L_\odot}$
1.18350	1.1854	6.980	2.40642	1.469/−5	1.093	2.683/−2	1.271	14.055
1.37286	1.1546	6.703	2.50156	1.456/−5	1.137	3.015/−2	3.015	12.040
1.56222	1.1295	6.486	2.58176	1.436/−5	1.173	3.315/−2	3.499	9.975
1.84626	1.1014	6.252	2.67416	1.388/−5	1.215	3.684/−2	5.626	7.006
2.13030	1.0828	6.103	2.73697	1.317/−5	1.244	3.951/−2	7.454	4.648
2.50902	1.0675	5.984	2.78945	1.200/−5	1.268	4.183/−2	8.895	2.788
2.88774	1.0576	5.909	2.82333	1.070/−5	1.283	4.338/−2	9.404	2.017
3.26646	1.0497	5.850	2.85063	9.372/−6	1.295	4.465/−2	9.423	1.790
3.64518	1.0423	5.794	2.87675	8.065/−6	1.307	4.589/−2	9.207	1.818
4.02390	1.0345	5.738	2.90415	6.803/−6	1.370	4.722/−2	8.863	1.966
4.59198	1.0217	5.644	3.02842	5.038/−6	1.341	4.953/−2	8.190	2.323
5.34942	1.0002	5.493	3.02842	2.983/−6	1.376	5.355/−2	6.950	3.056
6.10686	0.9720	5.301	3.13390	1.372/−6	1.424	5.935/−2	5.123	4.253
6.86430	0.9341	5.053	3.28251	3.729/−7	1.492	6.822/−2	2.623	5.970
7.62174	0.8860	4.758	3.47872	3.114/−8	1.581	8.121/−2	0.480	7.212

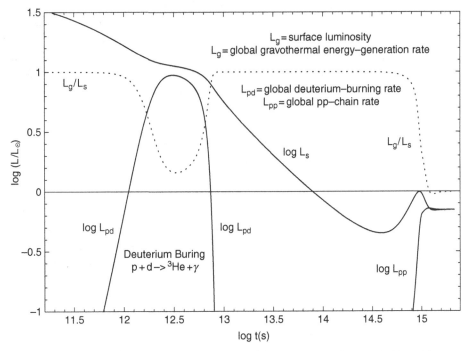

Fig. 9.2.10   Time evolution of global energy-generation rates in a low mass gravitationally contracting model (M = 1M$_\odot$, Z = 0.01, Y = 0.25)

The total time during which $L_{pd} > L_g$ is $\Delta t_1 = 1.39 \times 10^5$ yr. The total energy liberated by deuterium burning is

$$\Delta E_{deuterium} \sim N_A \, Y_{deuterium}^{start} \times 5.493 \text{ MeV} \times 1.6022 \times 10^{-6} \text{ erg MeV}^{-1}$$

$$= 1.562 \times 10^{47} \text{erg}. \tag{9.2.43}$$

Adopting an average mean global burning rate of $\langle L_{pd} \rangle \sim 8 \, L_\odot$, the time interval

$$\Delta t_2 = \frac{\Delta E_{deuterium}}{8 L_\odot} = 5.06 \times 10^{12} \text{ s}$$

$$= 1.60 \times 10^5 \text{ yr} \tag{9.2.44}$$

is a measure of the duration of the deuterium-burning phase.

The character of evolution after the deuterium-burning phase is essentially identical with that prior to this phase. The interior is completely convective and, hence, describable by an index $N = 1.5$ polytrope. The structure and gravothermal energy-generation properties are qualitatively exactly the same as given by Figs. 9.2.1–9.2.4.

9.2.4 Evolution in the HR diagram and characteristics of models along the approximately vertical portion of the track

Evolution in the HR diagram is shown in Fig. 9.2.11. Of over 4000 models constructed, a selection of approximately 90 has been chosen for display. The spacing in time between adjacent circles is indicated by the sizes of the circles. There are four sets of spacings: $\Delta t \, (\text{yr}) = 10^4$, 5×10^4, 2×10^5, and 10^6. The size of the region designated as the deuterium-burning region has been slightly exaggerated to encompass 2×10^5 yr of evolution. In comparing with the distribution of T-Tauri stars in the HR diagram, one expects that the number of stars found in any given luminosity interval is proportional to the time necessary for models to evolve through this interval. Thus, assuming that all low mass stars appear in the Hayashi band before deuterium burning begins, one might expect the number of 1 $M_\odot$ stars in the four 0.75 magnitude intervals $\log L = 1.2$–1.5, $\log L = 0.9$–1.2, $\log L = 0.6$–0.9, and $\log L = 0.3$–0.6 to be in the ratios 0.3, 2, 2, and 5.5. That is, the number–magnitude distribution should begin suddenly, remain flat, and then steepen upwards noticeably with increasing magnitude. Of course, comparison with observations encounters the problem of small number statistics, with the likelihood that no stars or very few stars are to be found in either the highest luminosity or in the flat portion of the distribution, so that no definitive conclusions can be drawn. This certainly appears to be the case in the samples studied by Cohen and Kuhi (1979). In the Taurus-Auriga dark clouds, the Ophiuchus dark clouds, and NGC 2264, the distributions for 1 $M_\odot$ stars begin near the end of or below the deuterium-burning region. In the Orion complex and NGC 7000/IC 5070, there is more positive evidence that several stars may be in the deuterium-burning phase.

Outside of the deuterium-burning region, the time scale for evolution along the nearly vertical portion of the track can be estimated very simply by using the facts that (1) the interior portion of the model is essentially a polytrope of index $N = 1.5$, (2) the surface

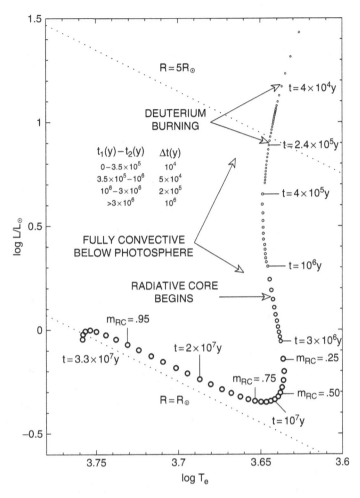

Fig. 9.2.11   Pre main sequence evolution in the HR diagram of a low mass model ($M = 1M_\odot, Z = 0.01, Y = 0.25$)

temperature along the track is nearly constant, and (3) the luminosity is equal to the rate of
change of the binding energy. Taking the binding energy from eq. (5.2.31) in the approximation that radiation pressure is negligible, the first and third facts give

$$L = \frac{dE_{\text{bind}}}{dt} = \frac{d}{dt}\left(\frac{3}{7}\frac{GM^2}{R}\right). \tag{9.2.45}$$

Combining with the approximate relationship

$$\frac{L}{4\pi R^2} = \sigma T_e^4 = \text{constant} \tag{9.2.46}$$

gives

$$dt = \frac{1}{7}\frac{GM^2}{4\pi\sigma T_e^4}\, d\left(\frac{1}{R^3}\right), \tag{9.2.47}$$

or

$$t - t_0 = \frac{1}{7} \frac{GM^2}{4\pi\sigma T_e^4} \left(\frac{1}{R^3} - \frac{1}{R_0^3} \right) \tag{9.2.48}$$

$$= \frac{1}{7} \frac{GM^2}{RL} \left[1 - \left(\frac{L}{L_0} \right)^{3/2} \right] \tag{9.2.49}$$

$$= 1.4064 \times 10^{14} \text{ s } \frac{R_\odot}{R} \frac{L_\odot}{L} \left(\frac{M}{M_\odot} \right)^2 \left[1 - \left(\frac{L}{L_0} \right)^{3/2} \right]$$

$$= 4.456 \times 10^6 \text{ yr } \frac{R_\odot}{R} \frac{L_\odot}{L} \left(\frac{M}{M_\odot} \right)^2 \left[1 - \left(\frac{L}{L_0} \right)^{3/2} \right]. \tag{9.2.50}$$

An example of the utility of this result is that, at $R = \sqrt{5} \, R_\odot$ and $L = 1.7 \, L_\odot$, $t - t_0 = 1.15 \times 10^6$ yr, almost identical with the time given by the detailed models with deuterium burning included.

Any attempt to make more than a qualitative comparison between Hayashi-band tracks and global characteristics of real stars faces the problem that the mean surface temperature along any theoretical track is a function of several parameters, including the mass of the model. In particular, the choice of the parameter l_{mix}/H_P is essentially arbitrary, compounding the shortcoming that the mixing-length treatment is itself at best a crude approximation. Further, for any choice of l_{mix}/H_P, the mean temperature along a track depends not only on the choice of metallicity, but on the detailed abundances of those atoms of low ionization potential which contribute electrons for the formation of H^- ions. Given these facts, it may make more sense in comparing with the number–magnitude distributions of real stars to dispense with the details of numerical models and to use instead an analytical lifetime relationship such as eq. (9.2.48) and to treat $\langle T_e^4 \rangle$ as a parameter determined by the observations.

The near constancy of model surface temperature during the phase when most of the interior is fully convective can be understood qualitatively in terms of a very simplified model in which (1) physical variables in supraphotospheric layers are determined by an opacity which is a rough analytical approximation to the Rosseland mean opacity due to the H^- ion and (2) variables at the photosphere are related to variables in the interior exactly as in an index $N = 1.5$ polytrope.

Inserting eq. (3.4.5) into the equation for radiative transport as given by eq. (3.4.4) and combining with the equation for pressure balance, eq. (3.1.4), the equation for radiative transfer above the photosphere may be approximated by

$$\frac{d}{dP} \left(\frac{1}{3} aT^4 \right) = \frac{dP_{\text{rad}}}{dP} = \frac{L}{4\pi c \, GM} \kappa, \tag{9.2.51}$$

where L and M are constants. Using $(ca/4) \, T_e^4 = L/4\pi R^2$, $g = GM/R^2$, and $T_0^4 = (1/2) \, T_e^4$, eq. (9.2.51) can be written as

$$d \left(\frac{T}{T_0} \right)^4 = \frac{3}{2} \frac{\kappa}{g} dP. \tag{9.2.52}$$

Assuming that the opacity can be written as

$$\kappa = \bar{\kappa}_0 \, \rho^m \, T^n, \tag{9.2.53}$$

where $\bar{\kappa}_0$ is a constant, and adopting the perfect gas equation of state in eq. (9.2.53) gives

$$\kappa = \kappa_0 \, P^m \, T^{n-m}, \tag{9.2.54}$$

where κ_0 is another constant. Inserting eq. (9.2.54) into eq. (9.2.52) and integrating gives

$$T_e^{4+m-n} - T_0^{4+m-n} = \frac{3}{8} \frac{4+m-n}{m+1} \frac{\kappa_0}{g} T_0^4 \left(P_e^{m+1} - P_0^{m+1} \right), \tag{9.2.55}$$

where the subscript e implies evaluation at the base of the photosphere and the subscript 0 implies evaluation at the outer edge of the photosphere. Since $T_e^4 = 2 \, T_0^4$ and $P_0 = 0$, one has that

$$T_e^{n-m} \, P_e^{m+1} = \left(1 - \frac{1}{2^{(4+m-n)/4}} \right) \frac{16}{3} \frac{m+1}{4+m-n} \frac{g}{\kappa_0} \propto \frac{g}{\kappa_0} \propto \frac{M}{\kappa_0 \, R^2}. \tag{9.2.56}$$

Convection extends all the way from the base of the photosphere to the center of the model and, over most of the interior, the relationship between state variables is the same as in an index $N = 3/2$ polytrope, so that $P/\rho^{5/3} = P_c/\rho_c^{5/3} \propto M^{1/3} \, R$. With $\rho \propto P/T$, this means that, everywhere in the deep interior,

$$P \propto \frac{T^{5/2}}{M^{1/2} R^{3/2}}, \tag{9.2.57}$$

with the constant of proportionality being a function of the mean molecular weight. Setting $T = T_e$ and $P = P_e$ in eq. (9.2.57) and combining with eq. (9.2.56) gives

$$T_e^{n+(5+3m)/2} \propto M^{(3+m)/2} \frac{1}{\kappa_0} R^{(3m-1)/2}. \tag{9.2.58}$$

If it is due mostly to the H^- ion, the opacity in photospheric layers is roughly proportional to a power of the density between 0.5 and 1 and to a power of the temperature between 2.6 and 5 (see Section 7.16). Setting $m = 0.5$ and $n = 5$, eq. (9.2.58) becomes

$$T_e \propto \frac{M^{7/33}}{\kappa_0^{4/33}} R^{1/33}, \tag{9.2.59}$$

which states that T_e is quite insensitive to R. Using the relationship $T_e^4 \propto L/R^2$ to replace R in terms of L and T_e in eq. (9.2.59) yields

$$T_e \propto \frac{M^{1/5}}{\kappa_0^{4/35}} L^{1/70}, \tag{9.2.60}$$

which states that the evolutionary track is essentially vertical in the HR diagram, consistent with the results of the detailed computations. The dependences on model mass and on metallicity (κ_0 increases with increasing metallicity) are also qualitatively consistent with the results of model calculations in the literature. As noted in Section 7.16, the analytic relationships constructed by various investigators for the H^- contribution to the opacity cover a wide range. For example, the approximation derived in this book and summarized in eq. (7.16.48) gives $n \sim 2.61$ and $m \sim 0.79$, so that $T_e \propto M^{0.30}/\kappa_0^{0.16} R^{0.11}$ and $T_e \propto M^{0.25}/\kappa_0^{0.13} L^{0.045}$, leading again to the conclusion that the evolutionary track is essentially vertical in the HR diagram.

The numerical relationships presented here are consequence of assuming a particular form for the opacity in photospheric layers on the assumption that H^- absorption makes the main contribution to the opacity in these layers. A similar set of relationships would result even if the contribution of H^- ions is ignored, as one might adopt as a first approximation in constructing contracting models for first generation stars which lack donors of ionization potential smaller than the ionization potential of hydrogen. This is evident from the fact that, at temperatures small enough that the abundance of free electrons increases steeply with increasing temperature, the opacity also increases steeply with increasing temperature, whether or not absorption by H^- ions is important. In fact, in the absence of H^- absorption, the effective value of n in eqs. (9.2.53) and (9.2.54) can be of the order of $n \sim 10$, significantly larger than the values $n \sim 2.6$–5 that characterize the H^- opacity.

The simple analytical model is a popular description of the fact that gravitationally contracting models which are convective everywhere below the photosphere define a Hayashi band which is almost vertical in the HR diagram and the mean temperature of which increases with increasing model mass and with decreasing metallicity. Truth to tell, however, the assumption of a perfect match at the photosphere between a supraphotospheric solution and an interior solution that produces these qualitatively correct conclusions is quantitatively completely unjustified: there is no good reason to assume that, at the photosphere, there is a match between the solution for a completely ionized interior in adiabatic equilibrium and a radiative solution for supraphotospheric layers in which most atoms are not ionized. The match actually occurs over an extended region below the photosphere where, with increasing distance from the photosphere, the temperature gradient is at first highly superadiabatic and then, although the temperature gradient becomes nearly identical with the adiabatic gradient, the adiabatic gradient only gradually approaches the value of $V_{ad} = 0.4$ which is associated with an index 3/2 polytrope that describes a star composed of fully ionized matter in adiabatic equilibrium.

In Figs. 9.2.12 and 9.2.13, variations of several relevant quantities are shown as functions of distance near the surface of a model which, in the HR diagram of Fig. 9.2.11, lies about midway along the nearly vertical portion of the Hayashi track, at $\log(L/L_\odot) \sim 0.5$ and $\log(T_e) \sim 3.65$. The photosphere of the model is located at $\Delta R/R_\odot = 0.01425$, where the zero point for the distance ΔR is essentially arbitrary. The run of state variables and the Rosseland mean opacity over a more extended region beyond the photosphere are shown in Fig. 9.2.14. Degrees of ionization of hydrogen and helium and the adiabatic temperature gradient are shown in Fig. 9.2.15 over an even more extended region.

The quantity c_1 in Fig. 9.2.12 is the degree of ionization for hydrogen and Y_{H+} in Fig. 9.2.13 is the free proton abundance parameter ($Y_{H+} = c_1 X$). The quantities V_{rad}, V_{ad}, and V in Fig. 9.2.12 are, respectively, the radiative, adiabatic, and the actually adopted logarithmic gradients of the temperature with respect to pressure. It is evident from the curves for c_1 and Y_{H+} versus distance that hydrogen is minimally ionized over an extended region below the photosphere. Thus, the assumption that a photospheric solution can be

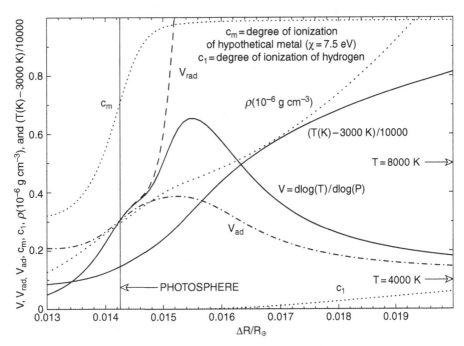

Fig. 9.2.12 Temperature, density, temperature gradients, and degrees of ionization near the surface of a 1 $M_\odot$ model in the Hayashi band (L = 3.16 $L_\odot$, R = 2.99 $R_\odot$)

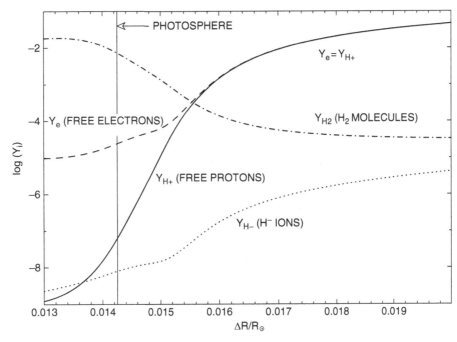

Fig. 9.2.13 Number abundance parameters of relevance near the surface of a 1 $M_\odot$ model in the Hayashi band (L = 3.16 $L_\odot$, R = 2.99 $R_\odot$)

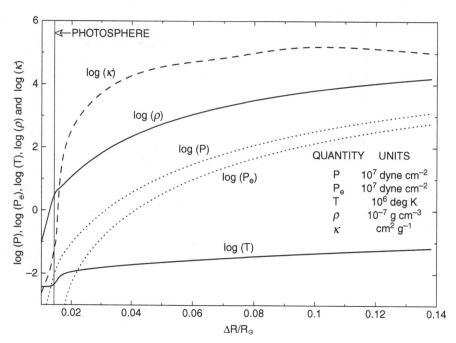

Fig. 9.2.14 State variables (P, T, and ρ), electron pressure P_e, and opacity κ near the surface of a 1 $M_\odot$ model in the Hayashi band ($L = 3.16 L_\odot$, $R = 2.99 R_\odot$)

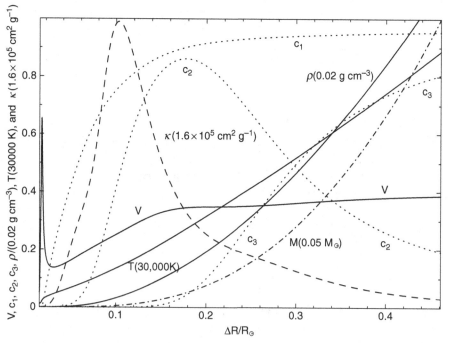

Fig. 9.2.15 T, ρ, V, κ, and degrees of ionization of H and He in the envelope of a 1 $M_\odot$ model in the Hayashi band ($L = 3.16 L_\odot$, $R = 2.99 R_\odot$)

matched to an interior solution where hydrogen is everywhere completely ionized is obviously incorrect. Even more damaging are the facts that, in an extended region below the photosphere, the adopted temperature gradient V is significantly larger than the adiabatic temperature gradient V_{ad}, and that, in regions far enough inward that V and V_{ad} approach one another, V_{ad} differs considerably from the value $V_{ad} \sim 0.4$ which characterizes interior regions where hydrogen and helium are completely ionized. As demonstrated in Fig. 9.2.15, ionization is incomplete and the adiabatic gradient has not reached the value 0.4 even at a depth of 0.05 $M_\odot$ below the photosphere.

The fact that H^- absorption makes the dominant contribution to the opacity in supraphotospheric layers is evident from a quantitative comparison between the Rosseland mean opacity shown in Fig. 9.2.14 and an estimate of the H^- contribution to the opacity deduced from the number abundance parameter Y_{H-} for H^- ions given in Fig. 9.2.13 and the photoabsorption cross section for H^- described in Section 7.16. At the photosphere, $Y_{H-} \sim 10^{-8}$. The cross section for absorption at the photospheric temperature (~ 4500 K) and density ($\sim 3 \times 10^{-7}$ g cm^{-3}) is $\sigma_{bf} \sim 10^{-17}$ cm^2 g^{-1} (see Section 7.16), so the H^- opacity is of the order of $N_A \sigma_{bf} Y_{H-} \sim 0.06$ cm^2 g^{-1}. This is close to the value of the Rosseland mean opacity at the photosphere recorded in Fig. 9.2.14.

The curve labeled c_m in Fig. 9.2.12 describes the degree of ionization of a hypothetical element with an ionization potential of 7.5 eV. Comparing with c_1, the degree of ionization of hydrogen, one might infer that, in supraphotospheric layers, most free electrons are supplied by elements of ionization potential much lower than the ionization potential of hydrogen. This inference is confirmed in Fig. 9.2.13 where the free-electron number abundance parameter Y_e and the free-proton number abundance parameter Y_{H+} are compared. The parameter Y_e, which has been constructed using the mix of elements of low ionization potential given in Table 7.16.1, is over two orders of magnitude larger than Y_{H+} in supraphotospheric layers, and remains larger than Y_{H+} until the temperature exceeds $T \sim 7000$ K near $\Delta R/R_\odot \sim 0.016$.

On the other hand, the dominant contribution to the opacity begins to shift from H^- absorption to free–free absorption mediated by free protons at $T \sim 5000$ K. The inflection in the curve for V_{rad} in Fig. 9.2.12 and the corresponding inflection in the curve for $\log(\kappa)$ in Fig. 9.2.14 near $\Delta R/R_\odot \sim 0.01475$ mark the center of the transition region.

At the low temperatures in photospheric regions, molecular species are expected to be at detectable abundances. The number abundance parameter Y_{H_2} for hydrogen molecules shown in Fig. 9.2.13 bears out this expectation and demonstrates further that, in photospheric layers, the molecular weight is affected at the 1% level by the presence of molecules.

9.2.5 Development of a radiative core and the transition from vertically downward to upward and leftward in the HR diagram

The big picture with regard to evolution in time

Shown in Fig. 9.2.16 as functions of time are the surface radius R_s, the central temperature T_c, the quantity ρ_c/T_c^3, where ρ_c is the central density, and the central concentration

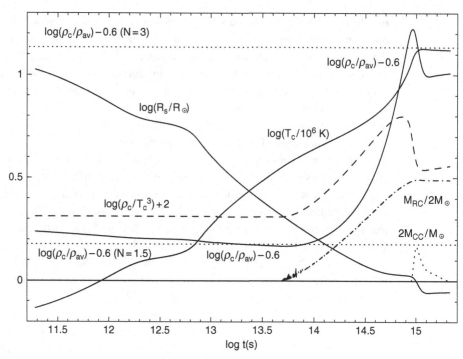

Fig. 9.2.16 Time evolution of the radius, convective regions, and central characteristics of a gravitationally contracting model ($M = 1 M_\odot, Z = 0.01, Y = 0.25$)

ρ_c/ρ_{av}, where $\rho_{av} = (3/4\pi)\,(M/R^3)$ is the mean density. As the model descends in the HR diagram, the increase in interior temperatures leads to a decrease in the interior opacity, with a consequent decrease in the radiative gradient. Ultimately, over a central region of increasing mass, the radiative gradient becomes smaller than the adiabatic gradient and matter becomes stable against convective flow. The mass M_{RC} of the radiative core is described as a function of time by the dash-dot curve in Fig. 9.2.16.

Two technical details are to be noted in connection with the development of the radiative core. The first has to do with the fact that the mass in the subphotospheric convective envelope decreases with time, so that the convective portion of the interior model becomes smaller relative to the mass of the static envelope. During this phase, lithium isotopes burn and, in the case of ^{7}Li, the degree of destruction in subphotospheric regions could be significantly overestimated by neglecting the mixing between matter in interior shells and the static envelope. To minimize this overestimate, the mass of the static envelope has been decreased discontinuously four times prior to the formation of a radiative core, from $0.0005\ M_\odot$ to $0.0001\ M_\odot$. Because of a small discontinuity in molecular weight between the interior and the static envelope (only eight isotopes are considered in the static envelope), the decreases in envelope mass are responsible for the discontinuities that are evident along the evolutionary track in Fig. 9.2.11 between $\log L/L_\odot = 0.7$ and $\log L/L_\odot = 0.3$. The discontinuities could have been avoided by reducing the mass in the static envelope at the beginning of the computation, but then the pedagogical value of demonstrating the consequences of changing the mass of the static envelope during a computation would have been lost.

By taking explicitly into account the fact that effectively no reactions occur in the envelope, the reduction in the rate of change of abundances in a surface convective zone due to the fact that convection extends into the static envelope can be estimated without reducing the mass of the static envelope. On the other hand, potential contributions to the surface luminosity by gravitational energy generation in surface layers can be substantial and, the smaller the mass in the static envelope, the smaller is the error committed by neglecting the energy generation in this envelope.

The second detail has to do with the effect of rezoning in the neighborhood of the boundary between the radiative core and the convective envelope. The algorithm that has been adopted is to request that at least 20 zones be present over a region that extends in mass from $0.9\,M_{\rm RC}$ to $1.1\,M_{\rm RC}$. At the same time, in order to reduce the numerical noise which is introduced when rezoning occurs, rezoning has been permitted for only every third model. The result is that, when the radiative core first forms and while it is small, there are erratic variations in $M_{\rm RC}$. These variations could have been minimized by introducing, prior to the appearance of the radiative core, finer zoning in the inner part of the model and then freezing (i.e., forbidding rezoning in) the finely zoned region until the radiative core was well established. Then, again, doing so would have resulted in another missed pedagogical opportunity.

As the model star makes the transition from being almost entirely convective to being almost entirely radiative, the character of evolution in the HR diagram changes dramatically. During the predominantly convective phase, the model becomes dimmer and smaller with time at nearly constant surface temperature. During the predominantly radiative phase, it becomes brighter with time at nearly constant radius.

The characteristics of evolution during the fully convective phase are a consequence of the action of three factors: (1) the luminosity is equal to the rate of change of the binding energy of the model; (2) the binding energy of the model is to good approximation a constant divided by the radius of the model; and (3) the photospheric opacity imposes the condition of a nearly constant surface flux. The result is that $dR/dt \propto -LR^2$ and $L \propto R^2$. In words, the star must shrink because it shines and it must dim because its surface area decreases.

The factors responsible for the characteristics of evolution during the predominantly radiative phase can also be identified: (1) the binding energy must increase to supply L; (2) the global virial theorem requires that the mean interior temperature must increase at half the rate at which the binding energy increases; (3) larger interior temperatures mean smaller opacities and, hence, an enhanced probability that radiative flow is the primary way in which energy released in the deep interior reaches surface layers; (4) when most of the interior has become radiative, the theorem derived in Chapter 3 to the effect that $L \propto \mu^4\,M^3/\kappa$, where κ is a mean opacity, requires that L must increase as the mean opacity decreases. The final ingredient is that, because the binding energy increases, (5) either the star continues to shrink or its central concentration must increase, or some combination of the two processes must occur. It is evident from the time evolution of R_s, $\rho_c/\rho_{\rm av}$, and $M_{\rm RC}$ shown in Fig. 9.2.16 that both processes occur: the model shrinks and the central condensation of the model increases as the mass of the radiative core grows.

9.2.6 Characteristics of a model in transition from the convective phase to the predominantly radiative phase

Several characteristics of a model of age 1.41496×10^7 yr which lies just beyond the minimum in luminosity along the evolutionary track in Fig. 9.2.11, at a point where $\log(L/L_\odot) = -0.3422$, $\log(T_e) = 3.6542$, and $R = 1.1088\ R_\odot$, are shown in Figs. 9.2.17–9.2.22. In Figs. 9.2.17 and 9.2.18, profiles of structure variables are shown as functions of the mass coordinate and of the radial coordinate, respectively. The transition between radiative and predominantly convective flow occurs at mass $\sim 0.74\ M_\odot$ and radius $\sim 0.66\ R_\odot$.

In predominantly convective models, it is the character of the H^- opacity in the tiny photospheric layer that forces the evolution to be at constant surface temperature, or, equivalently, at a nearly constant surface energy flux. However, as the radiative core grows, the interior opacity and the radiative flux in the interior play increasingly important roles in determining the character of evolution. In Fig. 9.2.19 are shown the logarithmic adiabatic gradient $V_{ad} = \left(d \log T / d \log P \right)_{ad}$ and the logarithmic radiative gradient V_{rad} in the model under consideration. In the radiative portion of the model, the actual logarithmic gradient is $V = V_{rad}$. Over most of the convective portion of the model, $V = V_{ad}$, but in subphotospheric layers, V is somewhere between the two extreme gradients.

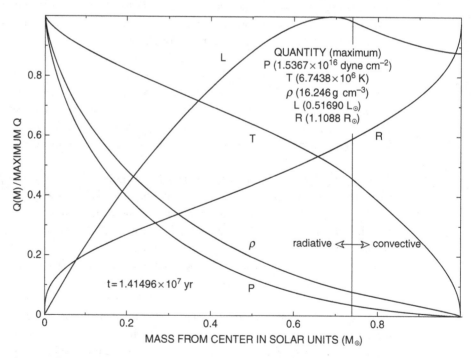

Structure of a gravitationally contracting model with a growing radiative core ($M = 1M_\odot$, $Z = 0.01$, $Y = 0.25$)

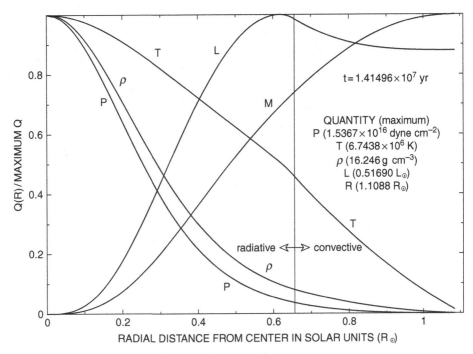

Fig. 9.2.18 Structure of a gravitationally contracting model with a growing radiative core ($M = 1M_\odot, Z = 0.01, Y = 0.25$)

From eq. (3.4.28) and the fact that radiation pressure is related to temperature by $P_{rad} = (1/3)\, aT^4$, it follows that

$$V_{rad} = \left(\frac{d\log T}{d\log P}\right)_{rad} = \frac{3}{4ac}\frac{\kappa P}{T^4}\frac{L}{4\pi GM} = \frac{1}{4}\frac{P}{P_{rad}}\frac{\kappa L}{4\pi cGM}, \qquad (9.2.61)$$

or

$$V_{rad} \propto \left(\frac{P}{P_{rad}}\right)\kappa\left(\frac{L}{M}\right). \qquad (9.2.62)$$

The three variable factors of V_{rad} given by eq. (9.2.62) are displayed in Fig. 9.2.19. It is remarkable that, over much of the radiative interior of the model, the product of the three factors is nearly constant. This near constancy permits one to entertain the concept of an effective polytropic index for the radiative interior.

Defining a local polytropic index $N(r)$ by

$$P(r) = \rho(r)^{1+1/N(r)} \qquad (9.2.63)$$

and adopting the perfect gas equation of state, one has that

$$V = \frac{d\log T}{d\log P} = 1 - \frac{d\log\rho}{d\log P} = 1 - \frac{N(r)}{N(r)+1} = \frac{1}{N(r)+1}, \qquad (9.2.64)$$

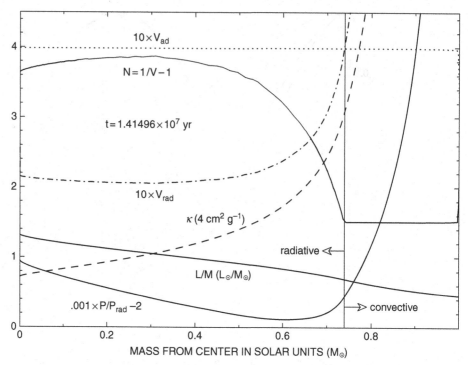

Fig. 9.2.19 Logarithmic temperature–pressure gradients and ingredients of V_{rad} in a gravitationally contracting model with a growing radiative core (M $= 1M_\odot$, Z $= 0.01$, Y $= 0.25$)

or

$$N(r) = \frac{1}{V} - 1. \qquad (9.2.65)$$

For the model under consideration (see Fig. 9.2.19), the mass averaged value of $N(r)$ in the radiative core is $\langle N \rangle \sim 3.4$ and, in the convective envelope, $N = 1.5$. One may think of the model as a composite of two polytropic segments of index 3.4 and 1.5, respectively, combining to form a model of effective polytropic index $N_{eff} \sim 0.74 \times 3.4 + 0.26 \times 1.5 \sim 2.9$.

The composite polytropic nature of the model is evident from other considerations. Inserting the model radius $R/R_\odot = 1.1088$ into the equation for the negative of the gravitational binding energy of a polytrope of index N (eq. (5.2.30)) and equating to $|\Omega| = 3.402\,143\,0597 \times 10^{48}$ erg, the gravitational binding energy of the model, produces $N = 2.00$. The central concentration of the model is $\rho_c/\bar{\rho} = 15.7098$. Comparison with the value of 11.403 for an index $N = 2$ polytrope and the value of 23.407 for an $N = 2.5$ polytrope (see Table 5.2.3) suggests $2 < N < 2.5$. The fact that an estimate of an effective polytropic index is quite sensitive to the manner in which the estimate is made demonstrates that the concept of an effective polytropic index is quite imprecise.

Nevertheless, the fact that the density of points along the evolutionary curve in Fig. 9.2.11 passes through a relative maximum in the neighborhood of the bend in the track where the

Table 9.2.4 Transition times for changes in effective polytropic index at constant radius and constant luminosity

δN	t $(10^6$ yr) $\bar{L}\bar{R}/\bar{M}^2 = 1$	t $(10^6$ yr) $\bar{L}\bar{R}/\bar{M}^2 = 0.6$
0.5	2.228	3.7
1	5.348	8.9
1.5	10.028	16.7

radiative core increases in mass from ~ 0.5 $M_{\odot}$ to ~ 0.75 $M_{\odot}$ can be understood as a consequence of the change in the effective polytropic index of the model. At a given radius, the binding energy of an index $N = 3/2 + \delta N$ polytrope is greater than that of an index $N = 3/2$ polytrope of the same radius by (see eq. (5.2.22) with $\tilde{\beta} = 1$)

$$\delta E_{\text{bind}} = \frac{6}{49} \frac{\delta N}{1 - (2/7)\delta N} \frac{GM^2}{R}. \tag{9.2.66}$$

Assuming that the transition in polytropic index occurs at constant radius $\bar{R}$ and constant luminosity $\bar{L}$, the time for a change δN in polytropic index is

$$\delta t \sim \frac{\delta E_{\text{bind}}}{\bar{L} L_{\odot}} = \frac{6}{49} \frac{\delta N}{1 - (2/7)\delta N} \frac{GM_{\odot}^2}{L_{\odot} R_{\odot}} \frac{\bar{M}^2}{\bar{L}\bar{R}}, \tag{9.2.67}$$

where both $\bar{R}$ and $\bar{L}$ are in solar units. Inserting numbers,

$$\delta t = 12.0565 \times 10^{13} \text{ s } \frac{\delta N}{1 - (2/7)\delta N} \frac{\bar{M}^2}{\bar{L}\bar{R}}$$

$$= 3.820 \times 10^6 \text{ yr } \frac{\delta N}{1 - (2/7)\delta N} \frac{\bar{M}^2}{\bar{L}\bar{R}}. \tag{9.2.68}$$

In Table 9.2.4, the second column gives transition times in units of $\bar{L}\bar{R}/\bar{M}^2 = 1$ and the third column gives transition times for $\bar{M} = 1$, $\bar{L} = 0.5$, and $\bar{R} = 1.2$, appropriate for the model track under scrutiny. Comparison with the density of points along the model track suggests that the choice $\delta N \sim 0.7$ leads to a reasonable approximation of the rate of evolution in the HR diagram during the transition from the phase of evolution at nearly constant surface temperature to the phase of evolution at nearly constant radius.

An interesting feature of the model with a growing radiative core is the fact that (see Figs. 9.2.17 and 9.2.18) the luminosity decreases outward through a region extending from slightly below the base of the convective envelope to the surface, showing that a fraction of the gravitational potential energy which is released and appears as compressional work in the deep interior is carried into the outer portions of the model where it is deposited as heat. The ingredients of this tranfer are evident in Fig. 9.2.20, where various energy-generation rates are given as functions of the mass coordinate. In the outer $\sim 30\%$ of the mass of the model, from slightly below the base of the convective envelope to the photosphere, the rate at which local heating occurs exceeds the rate at which compressional work is being done. The energy needed to make up the deficit is abstracted from the ambient

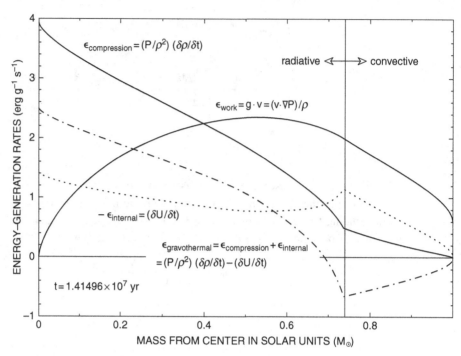

Fig. 9.2.20   Gravothermal energy-generation rates in a gravitationally contracting model with a growing radiative core ($M = 1M_\odot, Z = 0.01, Y = 0.25$)

outward flow of energy. Responsible for the outward flow of energy is the cumulative contribution of compressional work done in deeper layers minus the heat absorbed locally in these layers.

Note that, in contrast with the case during the fully convective phase of evolution (compare Figs. 9.2.3 and 9.2.8), nowhere in the model with a growing radiative core (except by accident at the point where the dash-dot and dot-dot curves in Fig. 9.2.20 cross) does a local virial theorem operate. That is,

$$\frac{P}{\rho^2} \frac{\delta\rho}{\delta t} - \frac{\delta U}{\delta t} \neq \frac{\delta U}{\delta t}.$$

(9.2.69)

On the other hand, the integrals under the dash-dot and dot-dot curves are identical,

$$\int \left[\frac{P}{\rho^2} \frac{\delta\rho}{\delta t} - \frac{\delta U}{\delta t}\right] dM = \int \frac{\delta U}{\delta t} dM,$$

(9.2.70)

which means that

$$\frac{1}{2} \int \frac{P}{\rho^2} \frac{\delta\rho}{\delta t} dM = \int \frac{\delta U}{\delta t} dM.$$

(9.2.71)

Furthermore, as the discussion in Section 8.3 suggests, the total rate at which gravity does work (the integral under the ϵ_{work} curve in Fig. 9.2.20) is equal to the total rate at

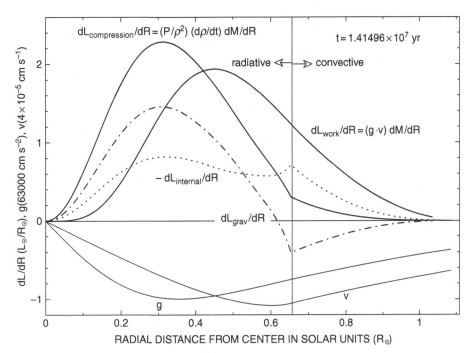

Fig. 9.2.21 Contributions to gravothermal luminosities in a gravitationally contracting model with a growing radiative core ($M = 1M_\odot$, $Z = 0.01$, $Y = 0.25$)

which compressional work is being done, even though the distributions of the two types of work are quite different. That is, the integrals under the two solid curves in Fig. 9.2.20 are identical. Thus,

$$\int \frac{P}{\rho^2} \frac{\delta\rho}{\delta t} \, dM = \int (\mathbf{g} \cdot \mathbf{v}) \, dM = \int \frac{GM}{r^2} \frac{dr}{dt} \, dM = \frac{d}{dt} \left[-\int \frac{GM}{r} \, dM \right] = -\frac{d\Omega}{dt},$$
(9.2.72)

where Ω is the gravitational binding energy. Equations (9.2.71) and (9.2.72) give

$$-\frac{1}{2} \frac{d\Omega}{dt} = \int \frac{\delta U}{\delta t} \, dM = \frac{d}{dt} \int U \, dM = \frac{dE_{\text{internal}}}{dt},$$
(9.2.73)

where E_{internal} is the total internal energy, demonstrating explicitly that the time derivative of the global virial theorem (Section 8.3) holds true.

Contributions to the surface luminosity that are produced in spherical shells of constant radial thickness by the different energy sources are shown as functions of the radial coordinate in Fig. 9.2.21. Although the distributions with respect to radius tell the same story as do the distributions versus mass in Fig. 9.2.20, they do so from a perspective that permits comparison with the distributions in Figs. 9.2.5 and 9.2.9 for completely convective models.

The variation of velocity with radius in Fig. 9.2.21 shows that, in contrast with the completely convective models, the inward velocity in the model with a growing radiative core

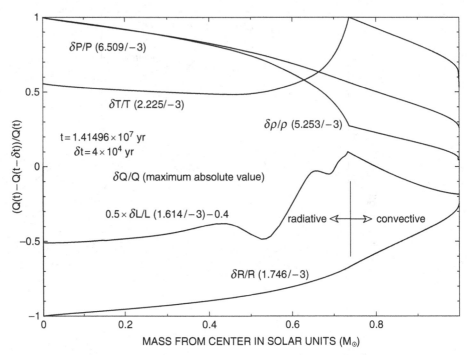

Fig. 9.2.22 Logarithmic increments in structure variables in a gravitationally contracting model with a growing radiative core $(M = 1M_\odot, Z = 0.01, Y = 0.25)$

does not increase everywhere monotonically outwards. It increases outwards through most of the radiative core, but decreases outwards through the convective envelope, with the result that densities in the convective envelope do not increase logarithmically as rapidly as do densities in the radiative core. In short, the model is becoming more centrally condensed. This inference is reinforced by Fig. 9.2.22 which shows the logarithmic increments in structure variables over the time step $\delta t = 4 \times 10^4$ yr separating the current model and its immediate precursor. In parentheses beside each label $\delta Q/Q$ is given the maximum absolute value in the model of the labeled quantity. It is evident from the curve $\delta\rho/\rho$ that the incremental logarithmic increase in density in the region where $\epsilon_{\text{gravothermal}} < 0$ is on average four times smaller than the logarithmic increase in interior regions, further demonstrating that the model is becoming more centrally condensed. Numerical comparison of density and temperature changes verifies that the entropy is increasing throughout the region in which energy is absorbed from the ambient flow ($\epsilon_{\text{gravothermal}} < 0$) and is decreasing elsewhere.

For display purposes, the quantity $\delta L/L$ has been halved and shifted vertically in Fig. 9.2.22; to determine the actual value of the logarithmic increment, the curve is to be mentally raised by 0.4 and the value of an ordinate on the left hand scale is to be multiplied by $2 \times 1.614 \times 10^{-3}$. The bumpiness of the $\delta L/L$ curve shows that the zoning could be improved upon, even though, as it is, the model has 326 zones. On the other hand, no new insight would be achieved by insisting on finer zoning.

9.3 Approach of a solar mass model to the main sequence: the onset and ascendancy of hydrogen burning by pp-chain reactions and properties of a zero-age main sequence model

The alterations in structural and compositional properties which a model star experiences as it approaches and settles onto the hydrogen-burning main sequence are described by an analysis of three models: (1) a model in which, at the center, the rate of energy generation by nuclear burning is approximately one half the rate of gravothermal energy generation but, globally, gravothermal energy generation remains the dominant contributor to the surface luminosity; (2) a model in which, at the center, the rate of energy generation by nuclear reactions is approximately fifty times larger than the rate of gravothermal energy generation but, globally, nuclear energy generation and gravothermal energy generation contribute comparably to the surface luminosity; and (3) a model in which energy generation by nuclear reactions is by far the dominant contributor to the energy generation rate, both locally and globally, but in which very little hydrogen has been converted into helium.

9.3.1 Hydrogen burning begins

The first model is described by Figs. 9.3.1–9.3.6. At an age of $2.374\,95 \times 10^7$ yr, it is approximately 10^7 yr older than the model described by Figs. 9.2.17–9.2.22. Located in the HR diagram at $\log L = -0.1336\,L_\odot$ and $\log T_e = 3.715$, it has a radius of $R = 1.06763$ $R_\odot$. At the center of the model, the rate of energy generation by pp-chain reactions is approximately half the rate of gravothermal energy production and, overall, nuclear energy accounts for $\sim 7.1\%$ of the surface luminosity. In detail, the global rate of nuclear energy generation is $L_{\text{nuc}} = 0.0522\,L_\odot$, the global rate of gravothermal energy generation is $L_{\text{grav}} = 0.6831\,L_\odot$, and the surface luminosity is $L_{\text{surf}} = 0.7353\,L_\odot$.

The gravitational binding energy of the model is $|\Omega| = 4.682\,409\,9024 \times 10^{48}$ erg and the net binding energy and the gravitational binding energy are related by $E_{\text{bind}}/|\Omega| = 0.5014$. From eq. (5.2.20) it follows that the gravitational binding energy of the model is the same as that of an index $N = 2.86$ polytrope of the same radius. The central concentration of the model, at $\rho_c/\bar{\rho} = 44.866$, is closer to that of an index $N = 3$ polytrope ($\rho_c/\bar{\rho} = 54.185$) than to that of an index $N = 2.5$ polytrope ($\rho_c/\bar{\rho} = 23.407$).

Comparison of Fig. 9.3.1 with Fig. 9.2.17 reveals that, in $\sim 10^7$ yr of evolution at nearly constant radius, the mass of the convective envelope has decreased by a factor of ~ 3.5, from $\sim 0.26\,M_\odot$ to $\sim 0.075\,M_\odot$. Comparison of Fig. 9.3.2 with Fig. 9.2.18 shows that the decrease in the volume of the convective envelope has been much less pronounced. The central concentration of the more evolved model is approximately three times larger than that of the less evolved model, as is evident from a comparison of the density profiles in Figs. 9.3.2 and 9.2.18.

Comparing Figs. 9.3.3 and 9.2.19, it is interesting to observe that, even though the three factors in eq. (9.2.62) for the radiative gradient differ in detail between the two models, V_{rad}

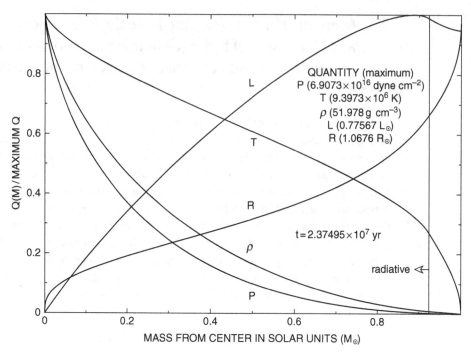

Fig. 9.3.1   Structure of a gravitationally contracting model beginning to burn hydrogen ($M = 1M_\odot$, $Z = 0.01$, $Y = 0.25$)

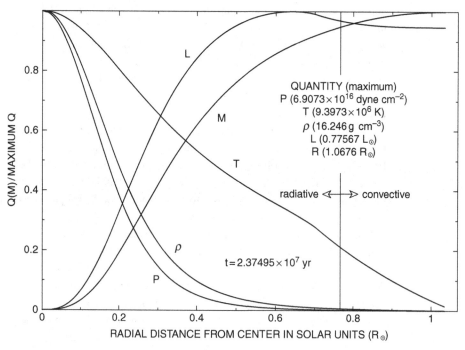

Fig. 9.3.2   Structure of a gravitationally contracting model beginning to burn hydrogen ($M = 1M_\odot$, $Z = 0.01$, $Y = 0.25$)

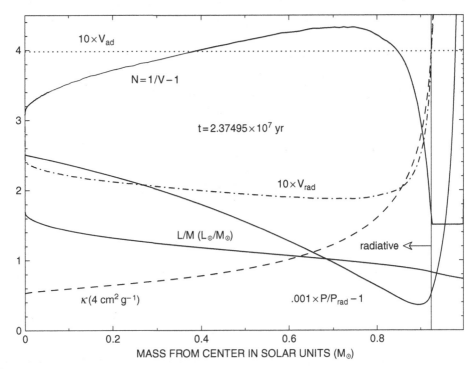

Logarithmic gradients and the ingredients of V_{rad} in a gravitationally contracting model beginning to burn hydrogen ($M = 1M_\odot, Z = 0.01, Y = 0.25$)

is approximately constant (~ 0.2–0.24) over most of the radiative region in both models. Since V_{rad} can also be written as

$$V_{\text{rad}} = \frac{1}{4} \frac{d\log P_{\text{rad}}}{d\log P}, \tag{9.3.1}$$

the statement that $P_{\text{rad}} \propto P$ (or that $T^3 \propto \rho$) is equivalent to the statement that

$$V_{\text{rad}} = \frac{1}{4}. \tag{9.3.2}$$

Thus, it appears that, for whatever reason, in radiative regions, the radiative pressure is roughly proportional to the total pressure, ρ is roughly proportional to T^3, and the local polytropic index is not dramatically different from $N = 3$. The mass-averaged value for $N(r)$ over the radiative region in Fig. 9.2.19 is actually approximately $N = 3.5$ and the mass-averaged value in Fig. 9.3.3 is approximately 3.9.

Components of gravothermal energy-generation rates are shown as functions of mass in Fig. 9.3.4, along with the nuclear energy-generation rate. An interesting feature of the gravothermal rates is that, in contrast with the distribution of rates in the model described by Fig. 9.2.20, over much of the radiative core,

$$\frac{P}{\rho^2} \frac{\delta\rho}{\delta t} - \frac{\delta U}{\delta t} \sim \frac{\delta U}{\delta t},$$

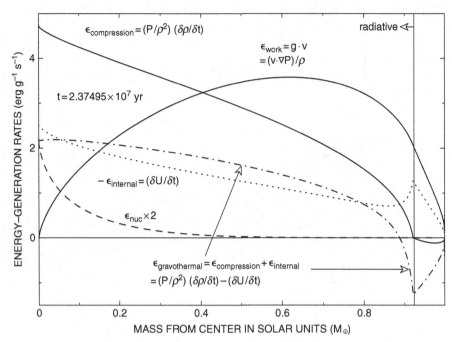

Fig. 9.3.4   Gravothermal energy-generation rates in a gravitationally contracting model beginning to burn hydrogen ($M = 1M_\odot, Z = 0.01, Y = 0.25$)

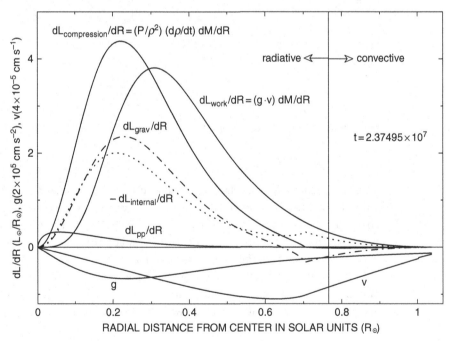

Fig. 9.3.5   Contributions to gravothermal luminosities in a gravitationally contracting model beginning to burn hydrogen ($M = 1M_\odot, Z = 0.01, Y = 0.25$)

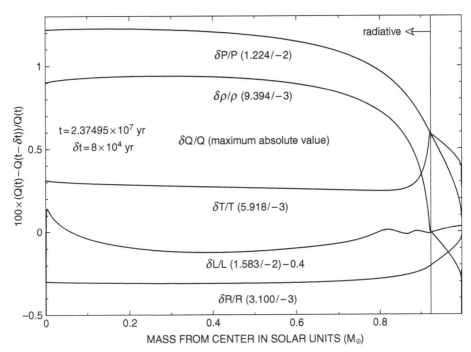

Fig. 9.3.6 Logarithmic increments in structure variables in a gravitationally contracting model beginning to burn hydrogen ($M = 1M_\odot$, $Z = 0.01$, $Y = 0.25$)

or

$$\frac{1}{2} \frac{P}{\rho^2} \frac{\delta\rho}{\delta t} \sim \frac{\delta U}{\delta t},$$

demonstrating the existence of an approximate local virial theorem, with half of the compressional work done locally going into local heating and the other half contributing to the outward flow of energy, just as in the convective Hayashi-track model as described in Fig. 9.2.3 and by eq. (9.2.10).

Comparing luminosity distributions in the radius-coordinate representation (Figs. 9.2.21 and 9.3.5), it is evident that, over the $\sim 10^7$ yr elapsing between the models, the peak in each distribution has shifted toward the center, consistent with the increase in central concentration. Comparing logarithmic increments in structure variables between the two models (Figs. 9.3.6 and 9.2.22), it is evident that, in the more evolved model, all increments are much less dependent on position. In particular, the radius and temperature logarithmic increments are essentially flat over the entire radiative region in the more evolved model. The incipient appearance of nuclear burning near the center of the more advanced model is reflected in the increase in $\delta L/L$ toward the model center.

Abundances of several light isotopes in the model at $t = 2.374\,95 \times 10^7$ yr are shown in Fig. 9.3.7. Comparing with the initial abundances in Table 9.2.1, it is evident that the initial stores of both deuterium and ^{6}Li have been completely destroyed over the entire model, deuterium having been converted into ^{3}He and ^{6}Li having been converted into ^{3}He

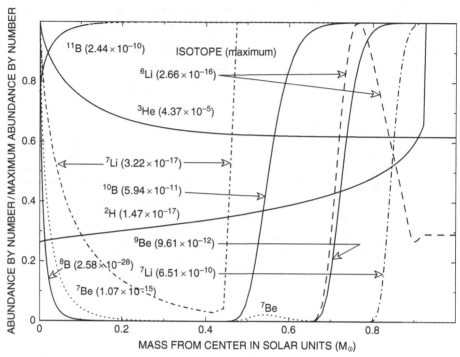

Fig. 9.3.7 Light isotope abundances in a gravitationally contracting model beginning to burn hydrogen (M = 1M☉, Z = 0.01, Y = 0.25)

and α particles. The ^{6}Li now in the model, with a peak abundance at a mass $\sim$0.77 $M_\odot$, is a consequence of (p, α) reactions on ^{9}Be. Although the initial store of ^{7}Li has been destroyed over roughly the inner $\sim$85% of the mass of the model, in the outer $\sim$15% of the mass of the model, the ^{7}Li abundance is only about 30% smaller than it was initially.

The abundance of ^{3}He in the outer half of the model is precisely equal to the sum of the initial abundances of ^{3}He and deuterium. The increase in ^{3}He with increasing proximity to the center is a consequence of nucleosynthesis beginning with the pp reaction, which makes deuterium, and continuing with the (p, γ) reaction on deuterium to make ^{3}He. Thus, the tiny deuterium abundance in the interior is an equilibrium abundance achieved by a balance between creation by the pp reaction and destruction by the $d(p, \gamma)^3$He reaction.

The beta-unstable nuclei ^{7}Be and ^{8}B appearing in central regions are similarly the consequence of nucleosynthesis in the pp chains, ^{7}Be being a result of the capture of ^{3}He nuclei by ^{4}He and ^{8}B being the result of (p, γ) reactions on ^{7}Be. The ^{7}Li in central regions is the result of electron capture on ^{7}Be.

In addition to the outermost region where the initial store of ^{7}Li has escaped substantial nuclear burning and the central region where the abundance of ^{7}Li is entirely a consequence of nucleosynthesis involving helium isotopes and electron capture, there is a third region located between the other two in which the ^{7}Li abundance is the consequence of the reactions ^{10}B$(p, \alpha)^7$Be and ^{7}Be$(e^-, \nu_e)^7$Li. The intermediate region extends from $M \sim 0.43\ M_\odot$ to $\sim$0.767 $M_\odot$. In this middle region, the abundance of ^{7}Li reaches a

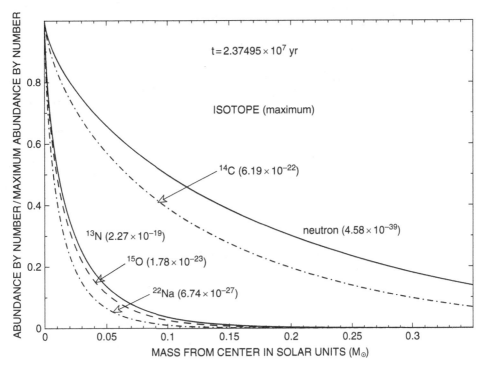

Fig. 9.3.8 Beta–unstable isotope abundances in a gravitationally contracting model beginning to burn hydrogen (M $= 1M_\odot$, Z $= 0.01$, Y $= 0.25$)

maximum which is approximately 120 times larger than the abundance of ^{7}Li at the center, but about 170 000 times smaller than the abundance of ^{7}Li at the surface; it then drops by a factor of about 4 before rising toward the surface value.

Nucleosynthesis extends to the formation of unstable isotopes including ^{13}N and ^{15}O which play roles in the CN cycle. Abundances of these two isotopes and of several others are shown in Fig. 9.3.8. Although the other isotopes are of no importance for evolution at this stage, it is interesting to be aware of their presence. Because of the large temperature dependences of the reactions that produce them, some of them could conceivably be of importance during more advanced stages of evolution.

9.3.2 A model in which nuclear burning and gravitational work contribute comparably to the surface luminosity

After a further $\sim 10^7$ yr of evolution, nuclear burning has taken over as the dominant source of energy liberated in central regions, but gravitational potential energy liberated in the still contracting envelope continues to make a significant contribution to the surface luminosity. Structural characteristics of a model of age $3.334\,95 \times 10^7$ yr are shown in Figs. 9.3.9 and 9.3.10 as functions, respectively, of the mass and radius coordinates. In both representations, the slope of the luminosity profile gives an indication of where each type of energy

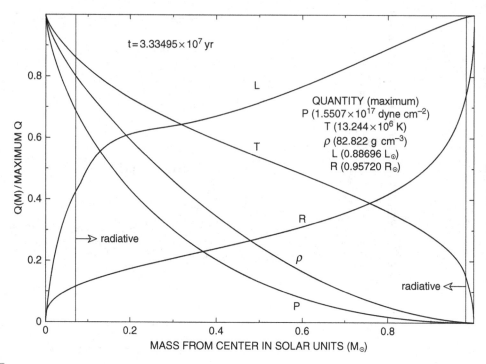

Fig. 9.3.9   Structure of a model settling onto the main sequence (M = 1M$_\odot$, Z = 0.01, Y = 0.25)

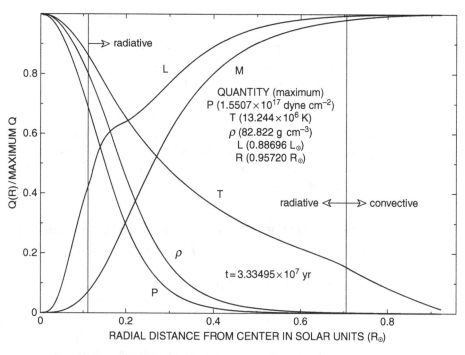

Fig. 9.3.10   Structure of a model settling onto the main sequence (M = 1M$_\odot$, Z = 0.01, Y = 0.25)

source dominates, the region of steeper slope indicating predominantly nuclear energy production and the region of shallower slope indicating predominantly gravothermal energy production.

The model is located in the HR diagram at $\log(L) = -0.0521$ and $\log(T_e) = 3.7586$ and its radius is $R = 0.957\,20\ R_\odot$. Although the mass of the subsurface convective zone has decreased to $0.0186\ M_\odot$, about a factor of 4 smaller than its mass $\sim 10^7$ yr earlier, the volume of the subsurface zone is comparable to its volume in the earlier model (compare Figs. 9.3.10 and 9.3.2).

The appearance of a convective zone in the central region of the model is due to the large fluxes generated by nuclear burning reactions. The mass and radial extent of the central convective zone are, respectively, $M_{CC} = 0.0707\ M_\odot$ and $R_{CC} = 0.1111\ R_\odot$ and the mixing time in the zone is ~ 85 days.

Using the convention for characterizing a nuclear burning shell set in Section 9.2 in the context of deuterium burning, the center of the hydrogen-burning shell (the mass shell below which 50% of the nuclear energy is released) is at the mass coordinate $M_{0.5} = 0.0494\ M_\odot$, or about 70% of the distance in mass from the center to the outer edge of the convective core. The mass shells below which 10% and 90% of the nuclear energy is released are, respectively, $M_{0.1} = 0.004\,98\ M_\odot$ and $M_{0.9} = 0.207\ M_\odot$, giving the burning shell a thickness of $0.202\ M_\odot$. At the shell center, $P = 1.17 \times 10^{17}$ dyne cm^{-2}, $T_6 = 11.85$, $\rho = 70.18$ g cm^{-3}, $R = 0.097\,58\ R_\odot$, and $L = 0.3186\ L_\odot$. The total rate at which nuclear energy is released is $L_{nuc} = 0.6787\ L_\odot$, with somewhat more energy being released in the convective core than in the region outside the core.

In the radiative region between the central and subsurface convective zones, the mass-averaged logarithmic temperature gradient is close to $V_{rad} \sim 0.25$, as may be seen from the curve for V_{rad} versus M in Fig. 9.3.11. Thus, once again, $\rho \propto T^3$ and the model has an average structure similar to that of an $N \sim 3$ polytrope, an inference strengthened by the fact that the mass-averaged value of $N(r)$ in Fig. 9.3.11 is also of the order of $N \sim 3$.

Nuclear and gravothermal energy-generation rates are displayed in Fig. 9.3.12 as functions of the mass coordinate, and contributions to the surface luminosity by various energy sources are shown as functions of the radial coordinate in Fig. 9.3.13. Here, $dL_{process}/dR = \epsilon_{process}\ dM/dR$, where $\epsilon_{process}$ is an energy-generation rate in units of erg g^{-1} s^{-1}. In Fig. 9.3.12, ϵ_{nuc} is 10 times larger than shown on the left hand scale and, in Fig. 9.3.13, dL_{pp}/dR is three times larger than shown on the left hand scale. Integration under the curves gives $L_{pp} = 0.663\ L_\odot$, $L_{gravothermal} = 0.208\ L_\odot$, $L_{work} = 0.418\ L_\odot$, and $L_{internal} = -0.209\ L_\odot$. CN-cycle reactions contribute $L_{CN} = 0.0154\ L_\odot$, and the total rate at which nuclear energy is being generated is $L_{nuc} = 0.679\ L_\odot$.

In the region of nuclear energy production, some of the nuclear energy released is used up in heating matter and the rate at which nuclear sources contribute to the surface luminosity is $L_{nuc}^{out} = L_{nuc} - (L_{grav})^-$, where $(L_{grav})^-$ is the absolute value of $\int dL_{grav}/dR\ dR$ over the region where $dL_{grav}/dR < 0$. Thus, with $(L_{grav})^- \sim 0.078\ L_\odot$, the rate at which energy of nuclear origin actually reaches the surface is $(L_{nuc})^{out} \sim 0.60\ L_\odot$, and the rate at which energy of gravothermal origin reaches the surface is $(L_{grav})^+ \sim 0.29\ L_\odot$, which is the value of $\int dL_{grav}/dR\ dR$ over the region where $dL_{grav}/dR > 0$. Doing the arithmetic in this way, one has that gravothermal energy production contributes to the surface

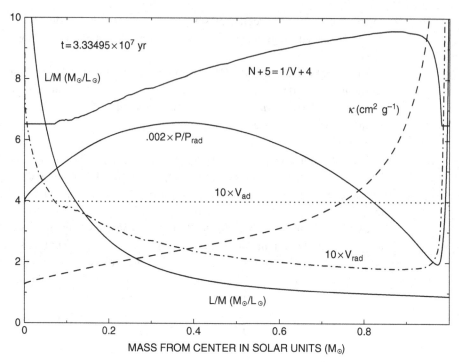

Fig. 9.3.11 Logarithmic gradients and the ingredients of V_{rad} in a model settling onto the main sequence ($M = 1M_{\odot}$, $Z = 0.01$, $Y = 0.25$)

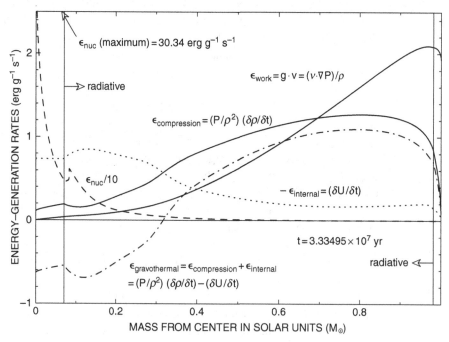

Fig. 9.3.12 Gravothermal energy–generation rates in a model settling onto the main sequence ($M = 1M_{\odot}$, $Z = 0.01$, $Y = 0.25$)

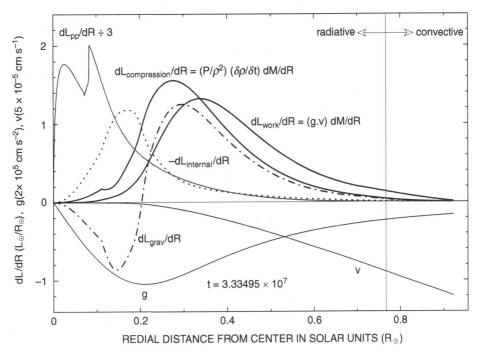

Fig. 9.3.13 Contributions to gravothermal luminosities in a model settling onto the main sequence (M = 1M$_\odot$, Z = 0.01, Y = 0.25)

luminosity at a rate which is about half the rate at which nuclear reactions supply energy to the surface rather than about one-third the rate at which nuclear reactions supply energy, as might be inferred from a comparison between L_{nuc} and $L_{gravothermal}$.

In contrast with the situation during deuterium burning in a completely convective model (see Figs. 9.2.8 and 9.2.9), a local virial theorem does not operate in the model settling onto the main sequence. That is, locally, $\epsilon_{gravothermal} \neq -\epsilon_{internal}$. On the other hand, direct integration of the relevant curves in Figs. 9.3.12 and 9.3.13 verifies that $L_{compression} = L_{work}$ and that the global virial theorem operates: $L_{gravothermal} = L_{compression} + L_{internal} = -L_{internal}$.

The gravitational binding energy of the model is $|\Omega| = 6.032\,735\,3400 \times 10^{48}$ erg and the net binding energy is related to the gravitational binding energy by $E_{bind}/|\Omega| = 0.5013$. Equation (5.2.20) gives that E_{bind} is the same as the gravitational binding energy of an index $N = 3.026$ polytrope of the same radius. At $\rho_c/\bar{\rho} = 51.5216$, the central concentration of the model is nearly the same as that of an index $N = 3$ polytrope ($\rho_c/\bar{\rho} = 54.185$). The energy associated with bulk motion is $E_{kin} = 1.370\,87 \times 10^{23}$ erg, which means that $E_{kin}/E_{internal} = 4.556 \times 10^{-26}$. This is tiny compared with the ratio $E_{kin}/E_{internal} = 1.0912 \times 10^{-19}$ which characterizes the deuterium-burning model discussed in Section 9.2.

The differences between the profiles for logarithmic increments in structure variables characterizing the model settling onto the main sequence, Fig. 9.3.14, and those characterizing the model just beginning to ignite nuclear fuel at the center, Fig. 9.3.6, are instructive.

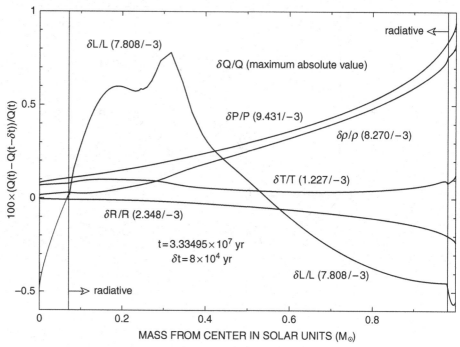

Fig. 9.3.14   Logarithmic increments in structure variables in a model settling onto the main sequence ($M = 1M_\odot$, $Z = 0.01$, $Y = 0.25$)

From the curves for $\delta R/R$ and for $\delta \rho/\rho$, it is clear that the ascendancy of nuclear burning in the central regions of the model leads to a pronounced decrease in the rate of contraction over much of the interior.

The velocity profile in the model settling onto the main sequence (Fig. 9.3.13) makes an interesting contrast with the velocity profile in the model just beginning to burn hydrogen in central regions (Fig. 9.3.5). In the model just beginning to burn hydrogen, the inward velocity increases rapidly outward from the center but then decreases sharply as the surface is approached, showing that the contraction rate is comparatively large near the center and comparatively small near the surface. In the model settling onto the main sequence, the inward velocity is nearly zero over a relatively large region extending outward from the center but it then increases steeply toward the surface, showing that the contraction rate is small near the center, consistent with the fact that nuclear burning is the primary source of energy, and it is comparatively large near the surface, consistent with the fact that gravitational energy production dominates in the envelope of the model.

The existence of two relative maxima in the curve for ϵ_{nuc} in Fig. 9.3.12 and the double peaked structure of dL_{pp}/dR evident in Fig. 9.3.13 are related to the abundance profile of ^{3}He which, as shown in Fig. 9.3.15, is uniform in the convective core but is variable in radiative regions. Outside of the convective core, the abundance of ^{3}He is everywhere determined by a local equilibrium between creation and destruction mechanisms, the rates of which depend on the density and temperature. However, at any point in the convective

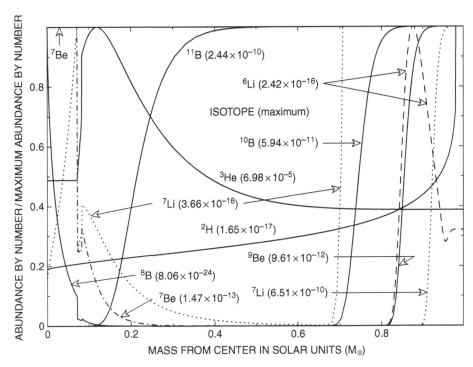

Fig. 9.3.15   Light isotope abundances in a model settling onto the main sequence (M = 1$M_\odot$, Z = 0.01, Y = 0.25)

core, the rate at which ^{3}He is either created or destroyed is large compared with the mixing time in the core, so that global rather than local equilibrium is maintained.

From the distribution of the ^{7}Li number abundance parameter near the model center, shown in Fig. 9.3.15, one may infer that ^{7}Li is in local equilibrium with respect to creation and destruction mechanisms not only outside the convective core but inside the core as well. On the other hand, the abundance of ^{7}Be is independent of position in the convective core, which might at first appear surprising, given that the lifetime of a ^{7}Be nucleus in a terrestial laboratory (∼77 days) is shorter than the estimated mixing time in the core (∼85 days). In the stellar context, however, only a small fraction of ^{7}Be isotopes bind K-shell electrons and most electron-capture events involve free electrons. From eq. (6.2.32) with the choice $f = 0.87$ one has that, at the center of the hydrogen-burning shell, which is inside the convective core, the lifetime of a ^{7}Be nucleus against capture of a free electron is of the order of 150 days, compared with the estimated convective mixing time of ∼85 days, and this accounts for the fact that ^{7}Be is uniformly distributed in the convective core. Given that the abundance of ^{7}Be is constant in the convective core, the profile of the ^{7}Li abundance parameter in the core reflects exactly the temperature and density dependences of the ^{7}Li$(p, \alpha)^3$He reaction.

Abundances of several stable isotopes which play a role in the CN cycle and of several beta-unstable isotopes, two of which also play roles in the CN cycle, are shown in Fig. 9.3.16. Comparing with isotope abundances in Fig. 9.3.8, it is striking (but not surprising) that only modest increases in density and temperature lead to increases in the

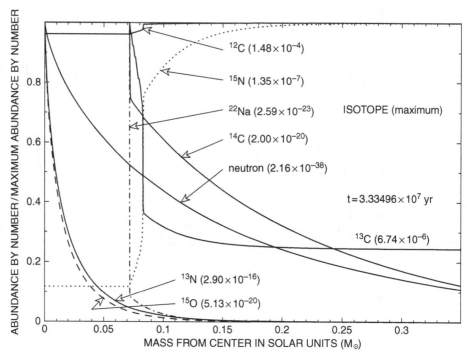

Fig. 9.3.16 Abundances of ^{12}C, ^{13}C, ^{15}N, and of several beta–unstable isotopes in a model settling onto the main sequence ($M = 1M_\odot, Z = 0.01, Y = 0.25$)

abundances of beta-unstable nuclei by orders of magnitude. For example, at the center, where temperature increases from $T_6 \sim 9.4$ to $T_6 \sim 13.2$ and density increases from $\rho \sim 52$ g cm^{-3} to $\rho \sim 83$ g cm^{-3}, $Y(^{15}$O) increases by a factor of ~ 3000 from 1.78×10^{-23} to 5.13×10^{-20}. In the convective core, 5% of the initial ^{12}C has been converted into ^{13}C. Since it has not changed noticeably from its initial value, the abundance of ^{14}N is not shown.

The differential contributions to the luminosity of four reactions in the pp chains are shown in the radial coordinate representation in Fig. 9.3.17. In the legend on the right hand side of the figure, the maximum differential contribution for every reaction in the pp chains is given in parentheses beside the description of the reaction. Keep in mind that the numbers in parentheses refer to energy which appears locally and then diffuses outward by radiation or convection and that to translate these numbers into reaction rates one must divide by the total energy liberated minus the average energy lost via neutrinos (see E in Table 6.2.1). Thus, for example, the maximum rate at which ^{7}Be captures electrons relative to the maximum rate at which it captures protons is $(2.069 \times 10^{-4}/0.05) \, (0.14/1.627 \times 10^{-7}) = 3.561 \times 10^3$.

Integration under the profiles in Fig. 9.3.17 shows that the ^{3}He(^{3}He,$2p$)^{4}He and the ^{2}H(p, γ)^{3}He reactions contribute comparably to the global luminosity and that, together, they produce most of the nuclear energy released. About three quarters of the energy released by the ^{3}He(^{3}He,$2p$)^{4}He reaction is produced in the convective core, whereas energy production by the ^{2}H(p, γ)^{3}He reaction is nearly equally divided between the

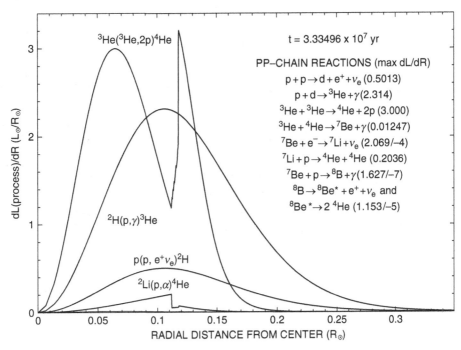

Fig. 9.3.17 pp–chain contributions to luminosity in a model settling onto the main sequence ($M = 1M_\odot$, $Z = 0.01$, $Y = 0.25$)

convective core and outside the convective core. Most of the energy produced by the $^7\mathrm{Li}(p, \alpha)^3\mathrm{He}$ reactions is released in the convective core.

9.3.3 A zero age main sequence model

At an age of $6.934\,96 \times 10^7$ yr, the last of the three models presented in this section is well ensconced on the hydrogen-burning main sequence in the sense that the energy released by hydrogen-burning reactions is overwhelmingly the dominant mode of energy production, with $-L_{\mathrm{grav}} = 1.58 \times 10^{-3}\,L_\odot \ll L_{\mathrm{nuc}} = 0.704\,L_\odot$. Because less than 0.2% of the initial hydrogen has been converted into helium at the model center, the model also qualifies as a "zero-age" main sequence model star. Located in the HR diagram at $\log(L) = -0.1534$ and $\log(T_e) = 3.7508$, the model has a radius of $R = 0.8833\,R_\odot$.

Structure variables are shown as functions of the mass coordinate in Fig. 9.3.18 and as functions of the radial coordinate in Fig. 9.3.19. In Fig. 9.3.20, the radiative gradient and its three ingredients, P/P_{rad}, $L(r)/M(r)$, and κ are shown, along with the adiabatic gradient and the local polytropic index $N(r)$. Once again, the product of the three ingredients, variations in each of which are quite disparate, produce a remarkably flat radiative gradient which has a mass-averaged value close to $V_{\mathrm{rad}} = 0.25$ and a mass-averaged $N(r)$ close to 3. The gravitational binding energy and the net binding energy are, respectively, $|\Omega| = 6.077\,317\,4743 \times 10^{48}$ erg and $E_{\mathrm{bind}} = 3.046\,273\,6160 \times 10^{48}$ erg, so that $E_{\mathrm{bind}}/|\Omega| = 0.501\,25$. The gravitational binding energy of the model is the same as that of an index $N = 2.876$ polytrope of the same radius. The central concentration, at

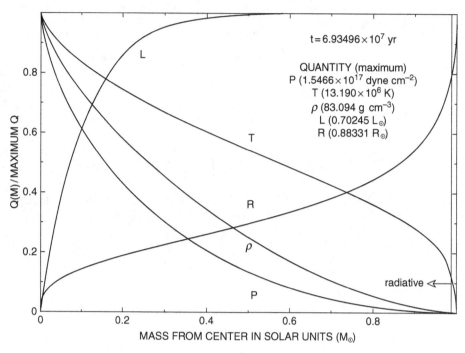

Fig. 9.3.18   Structure of a zero–age main sequence model ($M = 1M_\odot$, $Z = 0.01$, $Y = 0.25$)

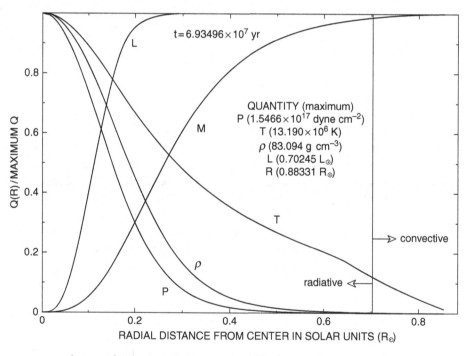

Fig. 9.3.19   Structure of a zero–age main sequence model ($M = 1M_\odot$, $Z = 0.01$, $Y = 0.25$)

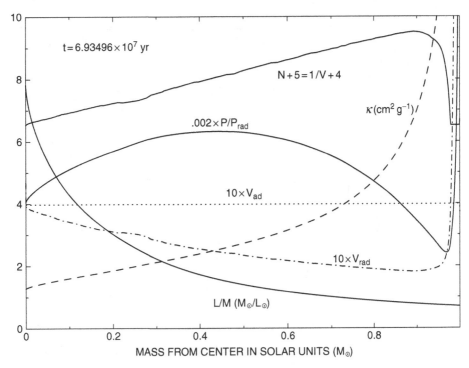

Fig. 9.3.20   Logarithmic gradients in a zero–age main sequence model (M = 1$M_\odot$, Z = 0.01, Y = 0.25)

$\rho_c/\bar{\rho} = 40.6208$, is approximately midway between the central concentration of an index $N = 2.5$ polytrope and the central concentration of an index $N = 3$ polytrope.

At the model center, matter is marginally stable against convection, as is evident from a comparison between V_{ad} and V_{rad} near the origin in Fig. 9.3.20. On the other hand, the opacity remains high in subphotospheric layers and, with a base at a distance $R_{CEB} = 0.6561\ R_\odot$ from the center and at a distance $\Delta R_{CE} = 0.2272\ R_\odot$ from the photosphere, the subsurface convective zone continues to occupy a substantial fraction (40%) of the volume of the model, although the mass of the zone is now only $\Delta M_{CE} = 0.0219\ M_\odot$.

Profiles of the number abundances of various isotopes, including those involved in the pp chains, are shown in Figs. 9.3.21–9.3.23. The effects of nuclear burning on the number abundances of the stable isotopes ^{11}B, ^{10}B, ^{9}Be, ^{7}Li, and ^{6}Li, are evident from a comparison of the abundance profiles in Fig. 9.3.15 with those in Fig. 9.3.21: for each isotope, the position of the point where the abundance is half of the maximum abundance has moved to a detectably larger mass even during the short time of ~3.6×10^7 yr that separates the two models being compared. The maximum abundance of ^{3}He has more than doubled, demonstrating that the real counterpart of the model is a potential source of new ^{3}He in the Universe.

In order to understand the distribution of ^{7}Li in the mass range 0.74–0.92 $M_\odot$, it is necessary to consider reactions involving electrons, protons, alpha particles, and the three isotopes ^{10}B, ^{7}Be, and ^{7}Li. Logarithms of the number abundances of the three isotopes are shown in Fig. 9.3.22 as functions of the mass coordinate. The reactions of relevance

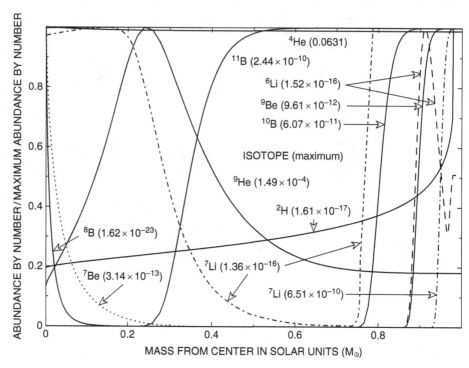

Fig. 9.3.21   Light isotope abundances in a zero–age main sequence model ($M = 1M_\odot$, $Z = 0.01$, $Y = 0.25$)

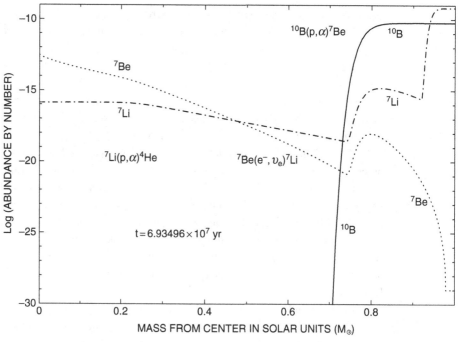

Fig. 9.3.22   Abundances of ^{7}Li, ^{7}Be, and ^{10}B in a zero–age main sequence model ($M = 1M_\odot$, $Z = 0.01$, $Y = 0.25$)

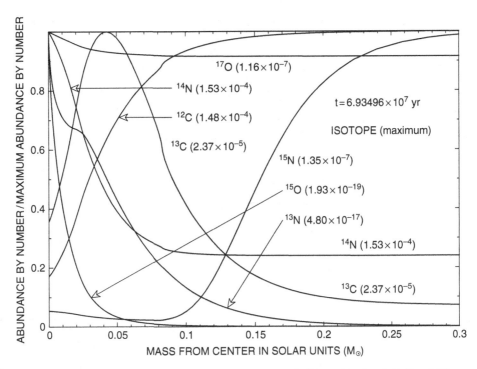

Fig. 9.3.23   Abundances of CN-cycle isotopes in a zero-age main sequence model (M = 1$M_\odot$, Z = 0.01, Y = 0.25)

are ^{10}B$(p, \alpha)^7$Be$(e^-, \nu_e)^7$Li$(p, \alpha)^3$He. The bump in the ^{7}Be abundance with a relative maximum at $M \sim 0.8\ M_\odot$ is due to ^{10}B burning, and the bump in the ^{7}Li abundance, also exhibiting a relative maximum at $M \sim 0.8\ M_\odot$, is the consequence of a balance between creation by the ^{10}B$(p, \alpha)^7$Be$(e^-, \nu_e)^7$Li reactions and destruction by the ^{7}Li$(p, \alpha)^3$He reaction. Although the reactions and the changes in light element abundances involved have absolutely no evolutionary consequences or effects on the model structure, it is nevertheless of intellectual interest to know that ^{10}B, whether it is produced in the Big Bang or by spallation reactions in the interstellar medium, is responsible for a relative maximum in the ^{7}Li abundance below the surface of a main sequence star and that ^{10}B is transformed in the stellar interior ultimately into three alpha particles.

Abundances of isotopes of carbon, nitrogen, and oxygen in the ZAMS model are presented in Fig. 9.3.23. Comparison with the abundance profiles in Fig. 9.2.16 shows that, during its brief sojourn on the main sequence, the evolving model has made progress in converting the initial abundance of ^{12}C into ^{14}N in central regions; at the center, over 80% of the initial ^{12}C has been converted into ^{14}N. However, nowhere are the abundances of ^{12}C and ^{14}N in equilibrium. That is, the rate of nuclear energy generation by CNO isotopes is controlled by the abundance of ^{12}C and the effective cross section of the ^{12}C$(p, \gamma)^{13}$N reaction and not, as in the CN cycle, by the abundance of ^{14}N and the effective cross section of the ^{14}N$(p, \gamma)^{15}$O reaction. Thus, the contribution to the luminosity by CNO isotopes is designated by $L_{C \to N}$ rather than by L_{CN}.

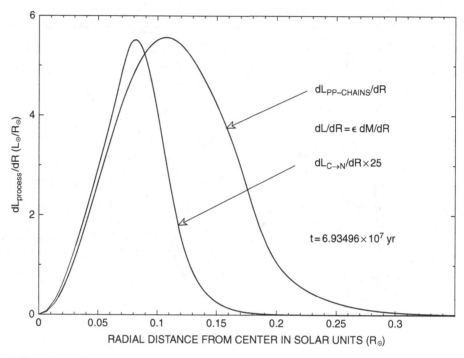

Fig. 9.3.24 Differential contributions to luminosity in a zero–age main sequence model (M = 1M$_\odot$, Z = 0.01, Y = 0.25)

At masses beyond where the maximum in the ^{13}C abundance occurs ($M \gtrsim 0.04\, M_\odot$), the abundance of ^{13}C is increasing. At smaller masses, the abundance of ^{13}C is determined by a balance between creation by the ^{12}C$(p, \gamma)^{13}$N$(e^+\nu_e)^{13}$C reactions and destruction by the ^{13}C$(p, \gamma)^{14}$N reaction.

Differential contributions to the luminosity by pp-chain reactions (L_{pp}) and by reactions involving CNO isotopes ($L_{C \to N}$) are compared in Fig. 9.3.24. The luminosity involving CNO isotopes is completely dominated by the conversion of ^{12}C into ^{14}N. The peak in $dL_{C \to N}/dR$ is closer to the center than the peak in dL_{pp}/dR simply because the ^{12}C$(p, \gamma)^{13}$N reaction is much more highly temperature dependent than is the initiating pp reaction which, by this time, is in complete control of the time scale on which the pp chains operate. The control of the pp reaction itself over the effective temperature dependence of the other reactions in the pp chains is assured by the fact that, as is evident from the abundance distributions in Fig. 9.3.21, everywhere in the region of nuclear energy production, all isotopes involved in the pp chains (other than protons, of course) are in local equilibrium with respect to creation and destruction mechanisms.

Integration under the curves in Fig. 9.3.24 shows that the pp chains produce energy at a rate 40 times larger than the rate at which C $\to$ N reactions produce energy. The center of the hydrogen-burning shell is at $M_{0.5} = 0.072\,365\, M_\odot$, where $P = 1.0598 \times 10^{17}$ dyne cm^{-2}, $T_6 = 11.448$, $L = 0.349\,21\, L_\odot$, and $R = 0.112\,14\, R_\odot$. Not surprisingly, the radius at the shell center defined in this way coincides with the location of the peak in

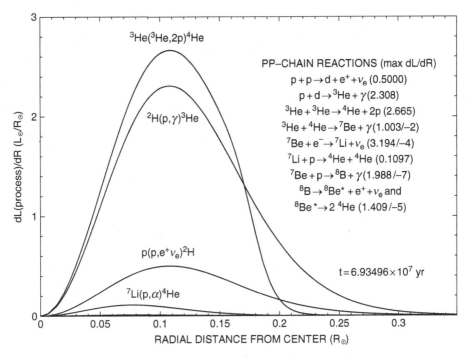

Fig. 9.3.25   pp–chain contributions to luminosity in a zero–age main sequence model (M = 1M$_\odot$, Z = 0.01, Y = 0.25)

dL_{PP}/dR in Fig. 9.3.24. With $M_{0.1} = 0.009\,88\ M_\odot$ and $M_{0.9} = 0.2304\ M_\odot$, the burning shell has a thickness in mass of 0.2205 $M_\odot$. This corresponds to about 0.8% of the volume of the model.

Contributions to the surface luminosity by individual reactions in the pp chains are shown in Fig. 9.3.25. In the legend within the figure, the number in parentheses beside the description for each reaction gives the maximum in the value of $dL_{reaction}/dR$ for that reaction. Comparing distributions in Figs. 9.3.17 and 9.3.25, it is evident that, apart from the additional structure occasioned by mixing in a convective core in the younger model, the overall features of the two distributions are quite similar, both qualitatively and quantitatively. That is, the two reactions which dominate in nuclear energy production, ^{3}He(^{3}He,$2p$)^{4}He and ^{2}H(p, γ)^{3}He, contribute comparably to the global luminosity, and the reactions ^{2}H(p, γ)^{3}He and ^{7}Li(p, α)^{3}He are minor players insofar as global energy production is concerned.

Contributions to the surface luminosity by individual reactions involving CNO isotopes are shown in Fig. 9.3.26. The two peaks in the distribution $dL_{^{15}N(p,\alpha)^{12}C}/dR$ may be accounted for by the fact that the rate of the ^{15}N(p, α)^{12}C reaction is proportional to the abundance of ^{15}N, which increases outward (see Fig. 9.3.23), and is also proportional to the density, which increases inward, and to a high power of the temperature, which also increases inward. The temperature and density variations are responsible for the peak closer to the center of the model and the variation in the ^{15}N abundance is responsible for the peak closer to the surface of the model.

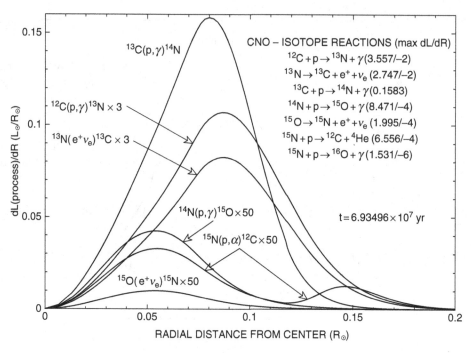

CNO-ISOTOPE REACTIONS (max dL/dR)
$^{12}C + p \rightarrow ^{13}N + \gamma (3.557/-2)$
$^{13}N \rightarrow ^{13}C + e^+ + \nu_e (2.747/-2)$
$^{13}C + p \rightarrow ^{14}N + \gamma (0.1583)$
$^{14}N + p \rightarrow ^{15}O + \gamma (8.471/-4)$
$^{15}O \rightarrow ^{15}N + e^+ + \nu_e (1.995/-4)$
$^{15}N + p \rightarrow ^{12}C + ^4He (6.556/-4)$
$^{15}N + p \rightarrow ^{16}O + \gamma (1.531/-6)$

$t = 6.93496 \times 10^7$ yr

Fig. 9.3.26   CNO-isotope contributions to luminosity in a zero–age main sequence model ($M = 1M_\odot$, $Z = 0.01$, $Y = 0.25$)

Although the rate of release of nuclear energy dwarfs the rate of release of gravothermal energy in the ZAMS model, the balance between the work done by or against gravity and the work done by compression or expansion is completely unaffected by this disparity. The rate at which gravity does work ($\mathbf{g} \cdot \mathbf{v} > 0$) and the rate at which work is done against gravity ($\mathbf{g} \cdot \mathbf{v} < 0$) are shown in Fig. 9.3.27, along with the components g and v of these work rates. Also shown are the rates at which work is being done by compression $\epsilon_{\text{compression}} > 0$ and by expansion $\epsilon_{\text{compression}} < 0$. The quantities $\int \epsilon_{\text{work}} \, dM$ and $\int \epsilon_{\text{compression}} \, dM$ are identical.

In contrast to all previously described phases, the velocity is now negative only in regions in and near the region where nuclear energy is being copiously released and it is negative there primarily because the mean molecular weight is increasing locally due to the transformation of four protons and four electrons into one helium nucleus and two electrons. Now, for the first time, velocities are positive over much of the model (in regions outside of the region where nuclear transformations are occurring at a high rate) and work is being done to push matter outward *against* gravity ($\epsilon_{\text{work}} < 0$).

The contrast is reinforced by a comparison of the three quantities $\epsilon_{\text{compression}}$, $\epsilon_{\text{internal}}$, and $\epsilon_{\text{gravothermal}}$ for the ZAMS model (Fig. 9.3.28) with the same three quantities for the model settling onto the main sequence (Fig. 9.3.12). In the model settling onto the main sequence, dU/dt is everywhere positive; in regions near the center, nuclear energy is converted into thermal energy and, in the contracting envelope, the work of compression is converted into thermal energy. In the ZAMS model, dU/dt is everywhere negative. In that part of the ZAMS model which is moving outward, thermal energy is contributing to the

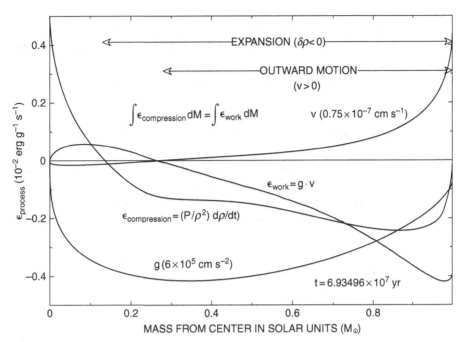

Fig. 9.3.27 Compressional energy–generation rate and gravitational work rate in a zero–age main sequence model ($M = 1M_\odot$, $Z = 0.01, Y = 0.25$)

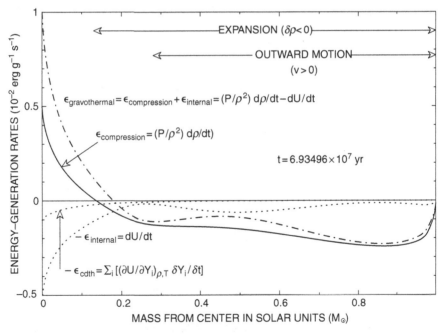

Fig. 9.3.28 Gravothermal energy–generation rates in a zero–age main sequence model ($M = 1M_\odot, Z = 0.01, Y = 0.25$)

work being done against gravity. In the region where hydrogen is being actively converted into helium, the increase in the molecular weight accounts for only part ($-\epsilon_{cdth}$) of the rate of decrease in the internal energy; the bulk of the rate of decrease must simply be accepted as a fact.

In the model settling onto the main sequence, $\epsilon_{compression}$ is everywhere positive because matter is everywhere moving inward; in the ZAMS model, matter in that part of the envelope which is moving outward is cooling and, everywhere that matter is expanding, the compressional work is negative. The net result is that, in the model settling onto the main sequence, $\epsilon_{gravothermal}$ is negative in the region where nuclear reactions are increasing in importance and positive in the region where the energy released by gravitational contraction is still important. In the ZAMS model, $\epsilon_{gravothermal}$ is positive where nuclear transformations are occurring, matter is contracting and, for whatever reason, cooling; it is negative over the entire region where matter is moving outward and over most of the region where expansion is occurring.

The complexity of the gravothermal energy-generation rate in the ZAMS model is highlighted in Fig. 9.3.29 which shows that, in the radial coordinate representation, there are two peaks in the differential contribution to the luminosity due to the rate of change in the internal energy. The peak at the larger distance from the center can be understood as the consequence of a contribution of thermal energy to the work of expansion. One way of understanding the first peak is to suppose that, during the transition from predominantly gravothermal to predominantly nuclear energy generation in central regions, a temperature

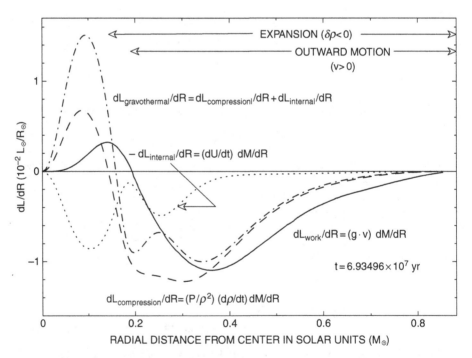

Fig. 9.3.29   Contributions to gravothermal luminosities in a zero–age main sequence model ($M = 1M_{\odot}, Z = 0.01, Y = 0.25$)

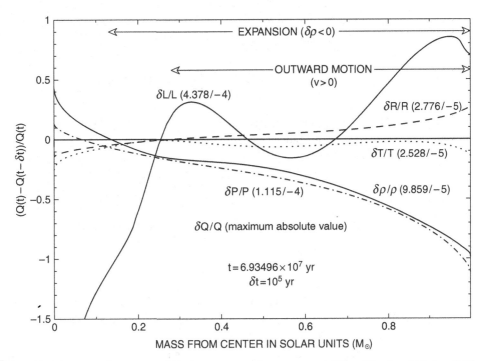

Fig. 9.3.30 Logarithmic increments in structure variables in a zero–age main sequence model ($M = 1M_\odot, Z = 0.01, Y = 0.25$)

overshoot has occurred. That is, one may suppose that nuclear energy has been injected sufficiently rapidly to lead to an excessive increase in the thermal energy content in central regions and that the observed rate of decrease in thermal energy near the center of the ZAMS model is the consequence of thermal relaxation.

For completeness, logarithmic increments in structure variables in the ZAMS model are shown in Fig. 9.3.30. The signs of the changes in temperature, density, and radius have already been noted and their consequences have been explored. The one new feature is that the surface luminosity is increasing with time, in consequence of an increase in the rate of nuclear energy generation in the interior.

9.4 Evolution of a 5 $M_\odot$ population I model to the main sequence: gravitational contraction, C → N burning, CN-cycle burning, and properties of a zero age main sequence model

The choice of an initial abundance distribution for a model is necessarily rather arbitrary. Since the lifetime of a 5 $M_\odot$ star is short compared with a Hubble time, if one of the objectives is to be able to compare model properties with those of observed stars in our Galaxy or in, say, the Magellanic Clouds, an appropriate choice for the metallicity is that of a typical population I star. For definiteness, assume that the initial abundances of elements heavier than helium are related to the hydrogen abundance as in Section 9.2. Conservation of mass means that

$$X \left(1 + 4\,Y_4' + \sum_j A_j Y_j' \right) = X + Y + Z = 1, \tag{9.4.1}$$

where Y_4' is the abundance by number of helium relative to hydrogen, Y_i' is the abundance by number of an isotope heavier than helium relative to hydrogen, and X, Y, and Z are the abundances by mass of hydrogen, helium, and elements heavier than helium. Thus,

$$Y = 4\,X\,Y_4' \text{ and } Z = X \sum_j A_j Y_j'. \tag{9.4.2}$$

Adopting the abundance parameters Y_j' from Table 9.2.1, one has

$$\sum_i A_j Y_i' = 0.0135. \tag{9.4.3}$$

With the convention that this sum is appropriate for the choice $Z = 0.01$, it follows from eq. (9.4.2) that $X = 0.74$, and $Y = 0.25$. This is the abundance chosen for the models described in Sections 9.2 and 9.3.

One could just as well have chosen Y and found the corresponding values of X and Z. Studies of the evolution with time of mean abundances in galaxies suggest that, to first approximation,

$$Y = 0.23 + 3Z, \tag{9.4.4}$$

where $Y = 0.23$ is an observation-based estimate of the helium abundance by mass produced in the Big Bang. Conservation of mass requires that

$$X = 1 - (0.23 + 4Z). \tag{9.4.5}$$

Thus, $Z = 0.01$ may be associated with $Y = 0.26$ and $X = 0.73$, and $Z = 0.02$ may be associated with $Y = 0.29$ and $X = 0.69$. In Chapter 10, it is argued that consistency with results of solar neutrino experiments can be achieved if, for the Sun, $Z \sim 0.015$. For this value of Z, eqs. (9.4.4) and (9.4.5) give $Y = 0.275$ and $X = 0.71$. In this section and the next, these values for initial X, Y, and Z are adopted as representative of the initial abundance of a typical population I star. The initial abundances relative to hydrogen of isotopes of elements heavier than helium are taken to be 1.5 times the values of Y_j' in Table 9.2.1.

Although nuclear transformations alter the composition in evolving models, only changes in the abundances of hydrogen and helium are taken into account in estimating the opacity. For example, even though large variations in the distribution of CNO isotopes occur, the opacity is calculated as if the distribution of elements heavier than helium remains constant. Put another way, the opacity is assumed to be characterized by X, Y, and Z, with Z fixed at the initial value of $Z = 0.015$.

Instead of constructing an initial 5 $M_\odot$ model high along a fully convective Hayashi track, as was done in Section 9.2 for a 1 $M_\odot$ model, in this section a 5 $M_\odot$ model is created

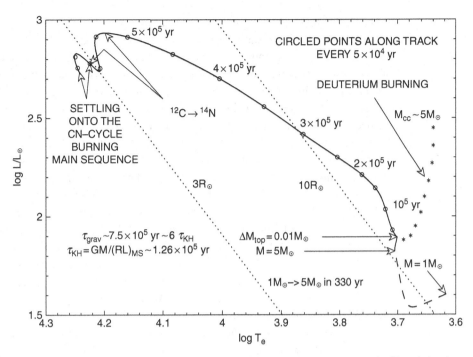

Fig. 9.4.1 Evolutionary path of a 5 $M_\odot$ model of composition $Z = 0.015$, $Y = 0.275$ contracting onto the CN–cycle burning main sequence

by artificially increasing the mass of the 1 $M_\odot$ starting model. Beginning with an initial pair of models of arbitrary mass, a pair of models of any other mass may be obtained by using the evolutionary program as a mass-changing tool. Before taking a time step, the mass of every shell of the initial model pair is increased by the same factor $1 + df$, where df is small compared with 1. This procedure is repeated in subsequent time steps until the mass of a model pair reaches the desired final value. During a time step, the mass of the static envelope is maintained at a fixed value, but new static surfaces which take into account the changed mass of the interior are constructed. After the mass-enhancement episode has been completed, subsequent evolution at constant mass produces models which become increasingly independent of the details of the mass-changing episode.

In the present study, the initial model is the gravitationally contracting 1 $M_\odot$ model with a completely convective interior which follows the evolutionary track in the HR diagram shown in Fig. 9.2.11. At the start of the calculation, the abundances in the model are changed abruptly to those appropriate for $Z = 0.015$ and $Y = 0.275$, time steps are fixed at $\Delta t = 10$ yr, and the mass-changing parameter is taken as $df = 0.05$. In $N \sim \log 5 / \log(1 + 0.05) \sim 33$ steps, or 330 yr, the model achieves a mass just short of the desired value of 5 $M_\odot$. The parameter df is reduced so that the model achieves the mass 5 $M_\odot$ in the next time step and evolution is continued at constant mass for another 16 models. The evolutionary track during the mass-changing phase is described by the heavy dashed curve in the lower right hand corner of Fig. 9.4.1.

During the mass-changing episode, the radius of the model does not vary appreciably. Since pressure balance requires that central temperature and radius be related approximately by $T_c \propto M/R$, interior temperatures in the final 5 $M_\odot$ model are of the order of five times larger than in the starting 1 $M_\odot$ model. The heat generated in the artificial mass-changing process increases interior temperatures to the extent that deuterium is destroyed over most of the interior. Another consequence of the increase in temperatures is a decrease in interior opacities to the extent that the inner 60% of the mass of the model is in radiative equilibrium. After reducing the mass of the static surface from 0.02 $M_\odot$ to 0.01 $M_\odot$ (to enlarge the fraction of the model in which gravothermal energy is properly treated), and increasing the time step size to 100 yr, evolution is continued. The model becomes steadily brighter and bluer on a characteristic time scale of several times 10^5 yr. The travel time between circled points along the evolutionary track in Fig. 9.4.1 is 5×10^4 yr.

The variations with time of several central and global characteristics of the 5 $M_\odot$ model are shown in Fig. 9.4.2. As is evident from the curve labeled $M_{RC}/5M_\odot$, the mass of the radiative core increases until, after $\sim 10^5$ yr of evolution, energy is carried outward by radiation throughout the entire model. Comparing with the history of the radiative core along the evolutionary track of the 1 $M_\odot$ model in Fig. 9.2.11, one may infer that the 5 $M_\odot$ model has been constructed at a point along the Hayashi track near the minimum in luminosity during the transition between the fully convective phase at high luminosity and low surface temperature and the phase characterized by a radiative core of increasing mass.

The curve defined by asterisks (stars) in Fig. 9.4.1 is a freehand estimate of a portion of the evolutionary track of a model which is initially fully convective. The real analogue of an isolated 5 $M_\odot$ model attains its final mass through accretion of matter at its surface, and may first appear at some point along the starred track. Where deuterium burning occurs along this track can be estimated from the fact that, in the middle of its deuterium-burning phase, the 1 $M_\odot$ model of Section 9.2 is characterized by $\log L \sim 1.0$ and $\log T_e \sim 3.64$ (see Fig. 9.2.11). A completely convective 5 $M_\odot$ model of the same central temperature would have a radius approximately five times larger than that of the 1 $M_\odot$ model. On the other hand, its central density would be approximately 25 times smaller. If the 5 $M_\odot$ model had approximately the same surface temperature as the 1 $M_\odot$ model, its luminosity would be 25 times larger than the luminosity of the 1 $M_\odot$ model, placing it in the HR diagram at $\log L \sim 2.4$ and $\log T_e \sim 3.64$. Taking into account that densities in the 5 $M_\odot$ model at this position are ~ 25 times smaller than in the 1 $M_\odot$ model, the center of the deuterium-burning phase in the 5 $M_\odot$ model will be at a lower luminosity along the starred track, at a position indicated schematically in Fig. 9.4.1.

From the discussion which leads to eq. (9.2.50), an upper limit to the time for quasistatic evolution along the starred portion of the track to the luminosity minimum (estimated to be at $L \sim 80\,L_\odot$ and $R \sim 10.7R_\odot$) is of the order of

$$\tau_{\text{to min luminosity}} \sim 4.456 \times 10^6 \text{ yr } \frac{R_\odot}{R} \frac{L_\odot}{L} \left(\frac{M}{M_\odot}\right)^2$$

$$\sim 4.456 \times 10^6 \text{ yr } \frac{1}{10.7} \frac{1}{80} 5^2 = 1.3 \times 10^5 \text{ yr}. \qquad (9.4.6)$$

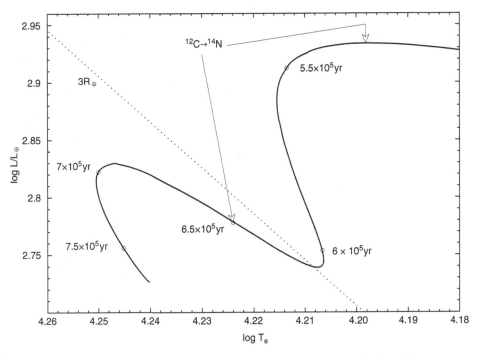

Fig. 9.4.1a
Evolutionary path of a 5 $M_\odot$ model of composition $Z = 0.015$, $Y = 0.275$ approaching the CN–cycle burning main sequence

Since it is not known where a real 5 $M_\odot$ analogue first appears along a quasistatic evolutionary track, the times posted along the calculated (heavy solid) track in Fig. 9.4.1 are meaningful only relative to one another, with the starting time relative to that of a real analogue being imprecise by of the order of 10^5 yr.

For approximately 5×10^5 yr following the formal start of model evolution, the release of gravitational potential energy is essentially the only source of surface luminosity. As contraction proceeds, the central temperature T_c increases steadily, slowly at first but ultimately by a factor of about 6 before nuclear burning begins to affect structure. Nuclear energy released by the conversion of ^{12}C into ^{14}N then begins to compete with the release of gravitational potential energy, leading to the distinctive drop in surface luminosity followed by an increase in surface luminosity shown at the very left of the track in the HR diagram in Fig. 9.4.1, and more clearly on an expanded scale in Fig. 9.4.1a.

The histories of various internal and global characteristics of the model are described in Fig. 9.4.2 as functions of the logarithm of time. The linear times noted just above the horizontal baseline in this figure permit a correspondence to be made between the location in the figure and the locations in the HR diagrams of Figs. 9.4.1 and 9.4.1a. For example, the maximum in luminosity in Fig. 9.4.1a occurs at $t \sim 5.2 \times 10^5$ yr and coincides with the formation of a convective core indicated by the dashed curve in Fig. 9.4.2 beginning at the the far right of the figure at $\log t\,(\mathrm{s}) = 13.215$. The formation and growth of the convective core is driven by the energy released by the conversion of ^{12}C into ^{14}N.

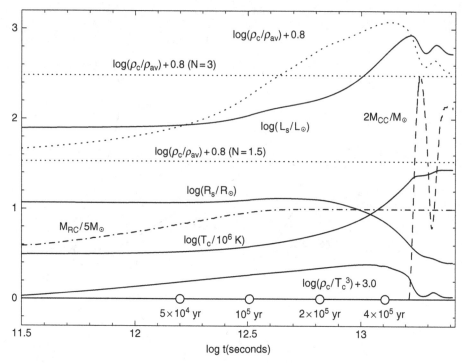

Fig. 9.4.2 Central and global characteristics of a 5 $M_\odot$ model ($Z = 0.015$, $Y = 0.275$) evolving onto the CN–cycle burning main sequence

As shown by the lowermost curve in Fig 9.4.2, the ratio of central density ρ_c to T_c^3 remains nearly constant, consistent with the expectation that $\rho_c / T_c^3 \overset{\sim}{\propto} 1/M^2$. The central concentration, as measured by ρ_c/ρ_{av}, where ρ_{av} is the average density, begins at a value close to that of an $N = 1.5$ polytrope but becomes considerably larger than one appropriate for an $N = 3$ polytrope before dropping to a value similar to that of an $N = 3$ polytrope. The drop in the effective polytropic index coincides with the development of the convective core for which the energy flux due to the C $\rightarrow$ N reactions is responsible (note that the effective polytropic index given by the heavy dotted curve is approximately anticorrelated with the mass of the convective core given by the heavy dashed curve labeled $2M_{CC}/M_\odot$).

Variations with time in the surface luminosity L_{surf}, in the net rate of release of gravothermal energy L_{grav}, and in the rates of release of nuclear energy by the conversion of ^{12}C into ^{14}N ($L_{C\rightarrow N}$), by the pp chains (L_{pp}), and by equilibrium CN-cycle reactions (L_{CN}) are shown in Fig. 9.4.3. The abundances by number of ^{12}C and ^{14}N in the convective core are shown by the dotted curves labeled $Y(^{12}C)$ and $Y(^{14}N)$, respectively. The mass of the convective core is indicated by the dashed curve.

The rate of energy release by nuclear burning exceeds the net rate of gravothermal energy release during two distinct phases separated by a short lived phase during which the rate of gravothermal energy release is the major contributor to the surface luminosity.

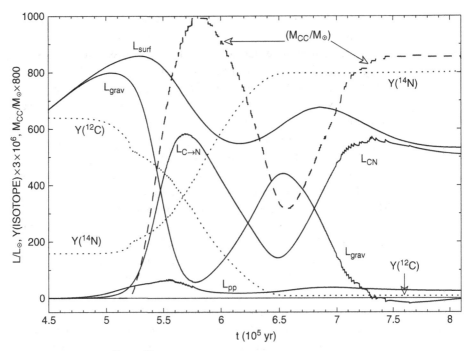

Fig. 9.4.3 Global luminosities, central ^{12}C and ^{14}N abundances, and convective core mass for a 5 $M_\odot$ model ($Z = 0.015$, $Y = 0.275$) settling onto the main sequence

The characteristics of the first nuclear burning phase are determined by the facts that (1) carbon is initially more abundant than nitrogen and (2) at a given mean temperature and density in the convective core, the effective cross section for the reaction ^{12}C$(p, \gamma)^{13}$N is approximately 40 times larger than the effective cross section for the ^{14}N$(p, \gamma)^{15}$O reaction. Thus, the conversion of ^{12}C into ^{14}N by the reactions ^{12}C$(p, \gamma)^{13}$N$(e^+\nu_e)^{13}$C $(p, \gamma)^{14}$N is not followed immediately by the burning of ^{14}N. The conversion of the original ^{12}C into ^{14}N in the convective core lasts for approximately 1.2×10^5 yr. The rate $L_{C \to N}$ at which energy is released due to the conversion reaches a maximum when about half of the original ^{12}C has been converted into ^{14}N.

As the ratio $Y(^{12}\text{C})/Y(^{14}\text{N})$ drops towards the CN-cycle equilibrium value of $1/40$ in the convective core, $L_{C \to N}$ decreases. The net rate of gravothermal energy production increases and exceeds the rate of nuclear energy production for $\sim 0.5 \times 10^5$ yr. As further contraction and heating occur, the release of nuclear energy by CN-cycle isotopes in local or global equilibrium in the convective core supplants the release of gravitational potential energy as the dominant source of surface luminosity. The rate of release of nuclear energy by CNO isotopes in the convective core is this time controlled by the abundance of ^{14}N and the effective cross section for the ^{14}N$(p, \gamma)^{15}$O reaction. The luminosity L_{CN} shown in Fig. 9.4.3 is produced predominantly in the convective core by the full set of CN-cycle reactions ^{14}N$(p, \gamma)^{15}$O$(e^+\nu_e)^{15}$N$(p, \alpha)^{12}$C$(p, \gamma)^{13}$N$(e^+\nu_e)^{13}$C$(p, \gamma)^{14}$N, where all abundances are in equilibrium.

It would be nice if the structure of the evolutionary track in Fig. 9.4.1a could be confirmed by an analysis of the distribution in the HR diagram of stars in a real open cluster in which the most massive stars are of intermediate mass. The analysis would involve examining the density of stars as a function of position in the HR diagram as this density is related to the rate of evolution described in Figs. 9.4.3 and 9.4.1a.

In the next nineteen graphs, the structure, composition, and energy-generation characterisitics of three models are explored. The first model, of age 5.7×10^5 yr, is approximately midway in the conversion of ^{12}C into ^{14}N in a convective core which is approaching its maximum mass, and $L_{C \rightarrow N}$ is at its maximum. Although the net rate of gravothermal energy generation is near a relative minimum, a substantial fraction of the surface luminosity is due to the release of gravothermal energy in still contracting outer layers of the model. The net gravothermal energy-generation rate is small since a large fraction of the nuclear energy released is absorbed by matter in regions near the center which are expanding and heating; expansion and heating are, of course, due to the energy absorbed. In the second model, of age 6.9×10^5 yr, all CN-cycle isotopes are in equilibrium in a re-growing convective core, but the conversion of ^{12}C into ^{14}N outside the core contributes approximately 10% of the surface luminosity. Gravothermal energy liberated from the still contracting envelope contributes nearly 40% of the surface luminosity.

The third model, of age 8.1×10^5 yr, is basically a ZAMS model, defined here as a configuration in which the rate of energy generation by equilibrium CN-cycle nuclear reactions is the dominant source of surface luminosity, the conversion of hydrogen into helium has just begun, and the rate at which gravitational potential energy is released is completely controlled by nuclear transformations. From this point onward, until hydrogen is nearly exhausted at the model center, the rate at which gravity does work in nuclear energy-generating regions is just sufficient to maintain pressure balance in response to molecular weight changes wrought by nuclear transformations. Outside of the nuclear energy-generating region, matter expands and moves outward. The energy required to move matter outward against gravity is supplied by absorption from the ambient flow of energy produced by nuclear reactions in central regions.

Structure variables in the 5.7×10^5 yr old model are shown as functions of the mass coordinate in Fig. 9.4.4. The location of the outer edge of the convective core at $M_{CC} \sim 1.20$ $M_\odot$ is indicated by the vertical line labeled M_{CC}. Matter to the left of the dotted vertical line at $M \sim 1.65$ $M_\odot$ is expanding and matter to the right of this line is contracting. Matter to the left of the vertical line segment at $M \sim 2.556$ $M_\odot$ is moving outward and matter to the right of this line is moving inward. Finally, to the left of the vertical line labeled watershed, at $M \sim 1.7485$ $M_\odot$, the gravothermal energy-generation rate is negative, and to the right of this line it is positive.

The density, temperature, pressure, and radius profiles in Fig. 9.4.4 are morphologically quite similar to those encountered in the 1 $M_\odot$ models in Section 9.3, as, for example, exemplified by the profiles in Fig. 9.3.9. The luminosity profile in Fig. 9.4.4 is more complex. The rise in $L(M)$ from the center outward is due to the release of nuclear energy, modified by the absorption of energy required to expand and heat matter in the region of nuclear energy production. The following dip in $L(M)$ is due to energy used up in heating and expanding matter outside the primary region of nuclear energy production, and the

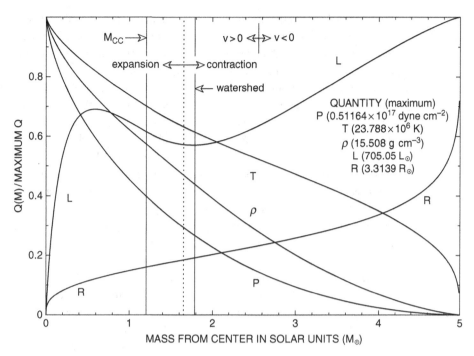

Fig. 9.4.4 Structure of a 5 $M_\odot$ model converting ^{12}C into ^{14}N ($Z = 0.015$, $Y = 0.275$, and AGE $= 5.7 \times 10^5$ yr)

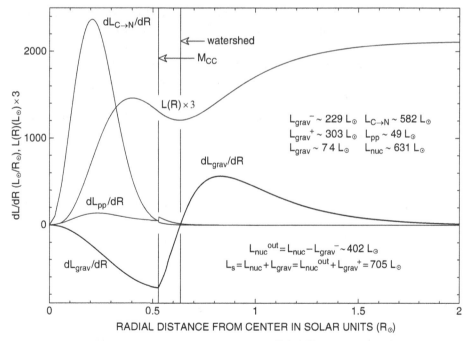

Fig. 9.4.5 Contributions to the surface luminosity in a 5 $M_\odot$ model converting ^{12}C into ^{14}N ($Z = 0.015$, $Y = 0.275$, and AGE $= 5.7 \times 10^5$ yr)

final rise in $L(M)$ is due to the release of gravitational potential energy in still contracting matter. The contribution to the surface luminosity of this final rise is $(L_{grav})^+ \sim 303\ L_\odot$, over four times larger than the net rate of gravothermal energy production which, as given by the curve for L_{grav} in Fig. 9.4.3, is $L_{grav} \sim 74\ L_\odot$.

Differential contributions to the luminosity by the two major sources of nuclear energy and by the gravothermal energy source are shown in Fig. 9.4.5 as functions of radial distance from the center. The location of the outer edge of the convective core coincides with the discontinuity in slope at $R \sim 0.5273\ R_\odot$ along the curve dL_{grav}/dR and the radial location of the watershed at $R \sim 0.6348\ R_\odot$ coincides, by definition, with the radius where $dL_{grav}/dR = 0$. The luminosity $L(R)$, which is the sum of the integrals under the differential energy curves, is also shown. Results of integrations under the complete nuclear energy-generation curves and results of integrating under portions of the gravothermal energy-generation curve on either side of the watershed are given in the middle right hand portion of Fig. 9.4.5. Results of manipulating the different luminosity components are given in the lower right hand portion of the figure.

The total rate of nuclear energy production is $L_{nuc} = L_{C \to N} + L_{pp} \sim 582\ L_\odot + 49\ L_\odot = 631\ L_\odot$. Thus, up to the relative minimum in luminosity (at $R \sim 0.6348\ R_\odot$ and $M(R) \sim 1.7848\ M_\odot$), the luminosity profiles in Figs. 9.4.4 and 9.4.5 are due to an injection of nuclear energy at the rate $L_{nuc} \sim 631\ L_\odot$, diminished by gravothermal absorption at the rate $(L_{grav})^- \sim (L_{grav})^+ - L_{grav} = (303 - 74)\ L_\odot = 229\ L_\odot$, leaving a total contribution of nuclear energy to the surface luminosity of $L_{nuc}^{out} = (631 - 229)\ L_\odot = 402\ L_\odot$. The contribution to the surface luminosity of gravitationally contracting matter in the region to the right of the watershed is related to the contribution of nuclear energy to the surface luminosity by $(L_{grav})^+ / L_{nuc}^{out} \sim 303/402 = 0.754$, considerably larger than $L_{grav}/L_{nuc} \sim 74/631 = 0.117$, which expresses the net rate of release of gravothermal energy relative to the total rate of release of nuclear energy.

Ingredients and components of the gravothermal energy-generation rate are shown in Fig. 9.4.6 as functions of distance from the center. The watershed where $dL_{grav}/dR = dL_{compression}/dR + dL_{internal}/dR = 0$ is located at the point where the curves for $-dL_{internal}/dR$ and $dL_{compression}/dR$ intersect. In most of the region to the left of the watershed, $dL_{compression}/dR = \epsilon_{compression}\ dM/dR < 0$, demonstrating that the injection of nuclear energy leads to expansion. The fact that $dL_{internal}/dR = \epsilon_{internal}\ dM/dR < 0$ in this inner region demonstrates that the injection of nuclear energy also leads to heating. The outer boundary of the region within which matter moves outward in response to the absorption of energy occurs at $R = 0.7696\ R_\odot$, at the intersection of the curves for the velocity v and the rate dL_{work}/dR at which gravity does work. By inspection, the theorem $\int dL_{compression}/dR\ dR = \int dL_{work}/dR\ dR$ holds.

Number abundances of CNO isotopes in the 5.7×10^5 yr old model are shown in Fig. 9.4.7. The outer edge of the convective core is marked by discontinuities in abundances at $M(R) \sim 1.2\ M_\odot$. In the convective core, isotopes which react with protons do so on time scales long compared with the mixing time in the core, which is of the order of one month, much shorter than a computational time step of 100 yr. With the exception of ^{14}C, which has a half life of 5730 yr, isotopes that beta decay are everywhere in local equilibrium. The fact that $Y^{12}(C)/Y^{13}(C) \sim 4$ everywhere in the convective core shows that ^{13}C is in local as well as in global equilibrium in the core.

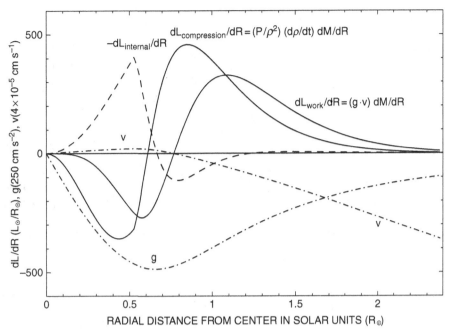

Fig. 9.4.6 Ingredients of gravothermal energy–generation rates in a 5 $M_\odot$ model converting ^{12}C into ^{14}N ($Z = 0.015$, $Y = 0.275$, and AGE $= 5.7 \times 10^5$ yr)

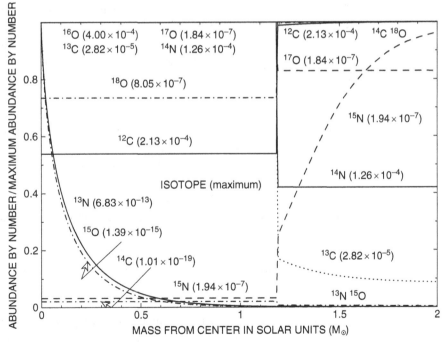

Fig. 9.4.7 CNO isotope abundances in a 5 $M_\odot$ model converting ^{12}C into ^{14}N ($Z = 0.015$, $Y = 0.275$, and AGE $= 5.7 \times 10^5$ yr)

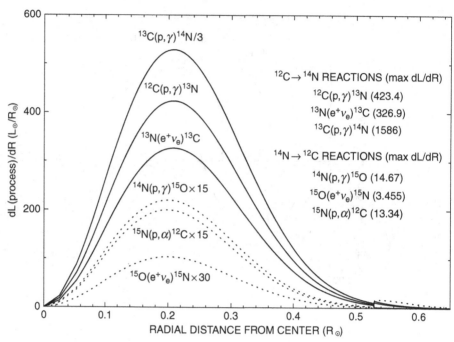

Fig. 9.4.8   CNO isotope contributions to luminosity in a 5 $M_\odot$ model converting ^{12}C into ^{14}N ($Z = 0.015$, $Y = 0.275$, and AGE $= 5.7 \times 10^5$ yr)

Differential energy-generation rates associated with reactions involving several CNO isotopes are shown in Fig. 9.4.8. Energy production is essentially confined to the convective core (outer edge at $R \sim 0.527\ R_\odot$). Interestingly, the distributions of individual rates with respect to radius are almost symmetric about a location slightly to the left of the center of the core. The most important distinction between the reaction rates depicted is that the rate of energy production associated with the conversion of ^{12}C into ^{14}N far exceeds the rate of energy production associated with the completion of the CN cycle by the conversion of ^{14}N into ^{12}C. The reasons for this are worth repeating: (1) the abundances of ^{12}C and ^{14}N in the convective core are almost identical, but (2) the effective cross section for the ^{12}C$(p, \gamma)^{13}$N reaction is approximately 40 times larger than the effective cross section for the ^{14}N$(p, \gamma)^{15}$O reaction.

The abundance curves $Y(^{12}$C$)$ and $Y(^{14}$N$)$ in Fig. 9.4.3 show that, in the convective core, ^{12}C and ^{14}N reach equlibrium with respect to transformations into one another at $t \sim 6.5 \times 10^5$ yr. At this time, the rate of energy release by CNO isotopes (now via the complete CN cycle in the convective core) is at a relative minimum and, due to the diminished flux of energy, the mass of the convective core is near a relative minimum. In response to a decline in nuclear energy production that accompanies the decrease in the ^{12}C abundance in the convective core, central regions contract and heat. The net rate of gravitational energy production reaches a relative maximum at about the same time that the rate of nuclear energy production by CNO isotopes reaches a relative minimum.

As core temperatures and densities continue to increase, the rate of energy release by CN-cycle reactions increases. Conditions in the model at time $t = 6.9 \times 10^5$ yr are

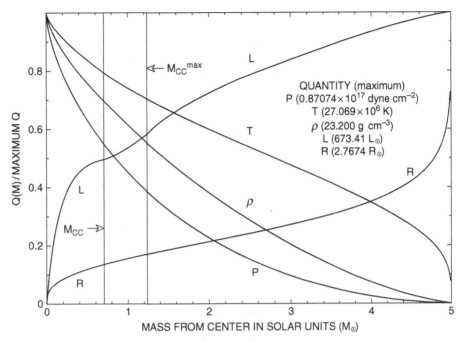

Fig. 9.4.9   Structure of a 5 $M_\odot$ model approaching the main sequence ($Z = 0.015$, $Y = 0.275$, and AGE $= 6.9 \times 10^5$ yr)

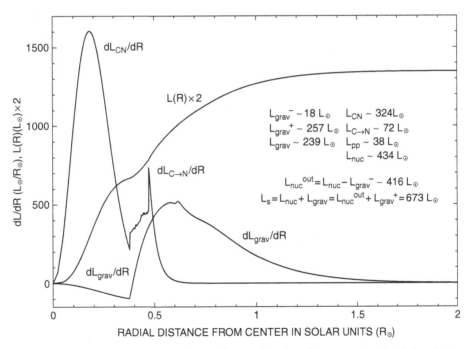

Fig. 9.4.10   Contributions to the surface luminosity in a 5 $M_\odot$ model approaching the main sequence ($Z = 0.015$, $Y = 0.275$, AGE $= 6.9 \times 10^5$ yr)

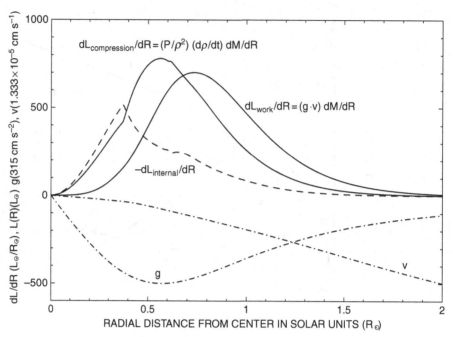

Fig. 9.4.11 Ingredients of gravothermal energy–generation rates in a 5 $M_\odot$ model approaching the main sequence ($Z = 0.015$, $Y = 0.275$, AGE $= 6.9 \times 10^5$ yr)

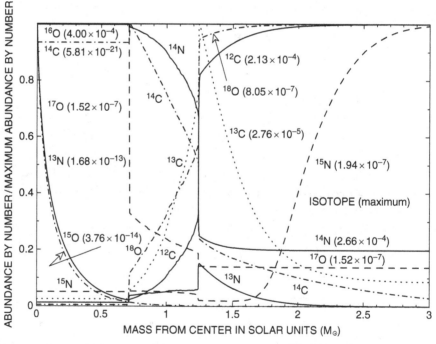

Fig. 9.4.12 CNO isotope abundances in a 5$M_\odot$ model approaching the main sequence ($Z = 0.015$, $Y = 0.275$, and AGE $= 6.9 \times 10^5$ yr)

described in Figs. 9.4.9–9.4.13. The convective core has a mass of about $M_{CC} \sim 0.71 \, M_\odot$, 0.53 $M_\odot$ shy of the maximum mass $M_{CC}^{max} \sim 1.24 \, M_\odot$ achieved at the height of the phase of core C $\rightarrow$ N burning. These convective core masses are indicated by appropriately labeled vertical lines in Fig. 9.4.9, where structure variables are given as functions of mass. All parts of the model are contracting and moving inward.

The variation of luminosity with spatial distance from the center is shown in Fig. 9.4.10, along with the differential contributions to the luminosity by nuclear reactions involving CNO isotopes and by gravothermal energy. Results of integrating under the differential distributions are shown in the middle right hand portion of the figure. The radial locations of the outer edge of the current convective core and of the convective core at its maximum earlier extent coincide, respectively, with the two largest vertical discontinuities in $dL_{C \rightarrow N}/dR$ at $R \sim 0.37 \, R_\odot$ and at $R \sim 0.47 \, R_\odot$. The watershed where $dL_{grav}/dR = 0$ is very close to the outer edge of the current convective core at $R \sim 0.38 \, R_\odot$.

The small discontinuites along the C $\rightarrow$ N distribution in the region between the two large discontinuities are the consequence of discontinuities in abundances left at the outer edge of the convective core as core mass declines from its maximum during the second half of the main ^{12}C $\rightarrow$ ^{14}N conversion phase. Abundance changes in the convective core also occur in finite steps, producing the discontinuous changes in M_{CC} evident in Fig. 9.4.3.

In the convective core of the 6.9×10^5 yr old model, energy production involving CNO isotopes is via the complete CN cycle. All participating isotopes which beta decay are in local equilibrium and those which involve proton capture are in global equilibrium in the convective core. Outside of the convective core, the conversion of ^{12}C into ^{14}N continues. As indicated in Fig. 9.4.10, the contributions of CNO isotopes are $L_{CN} \sim 324 \, L_\odot$ and $L_{C \rightarrow N} \sim 72 \, L_\odot$. The contribution of pp-chain reactions to the total rate of nuclear energy generation (not shown in the figure) is $L_{pp} \sim 38 \, L_\odot$. Thus, C $\rightarrow$ N reactions outside of the convective core produce almost twice as much energy as do pp-chain reactions throughout the model.

Components of the differential gravothermal energy-generation rate are shown in Fig. 9.4.11, along with the differential rate at which gravity does work and the two key ingredients of this rate: g and v. The fact that v is everywhere negative demonstrates that all parts of the model are moving inward and the fact that $dL_{compression}/dR$ is everywhere positive demonstrates that all parts of the model are contracting. The fact that the differential rate $dL_{internal}/dR$ at which the internal energy increases is everywhere negative demonstrates that all parts of the model are heating.

It is interesting to contrast and compare the distributions in Figs. 9.4.10 and 9.4.11 with those in Figs. 9.4.5 and 9.4.6. In both models, heating occurs throughout the region of nuclear energy generation. However, in response to the rapid injection of nuclear energy by C $\rightarrow$ N reactions, matter in central regions of the younger model expands outward and work is done against gravity. In contrast, in nuclear energy-generating regions of the older model, which relies primarily on the intrinsically weaker CN-cycle reactions, the matter in nuclear burning regions contracts and gravity continues to do positive work. The total gravothermal absorption rate, as measured by $\left(L_{grav} \right)^-$, is smaller by almost a factor of 30 in the older model than it is in the younger model. On the other hand, inward velocities in outer regions of both models are quite similar and the contributions of gravothermal energy

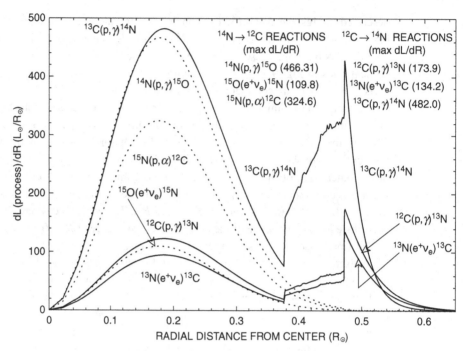

Fig. 9.4.13 CN–cycle and C → N contributions to luminosity in a 5 $M_\odot$ model approaching the main sequence ($Z = 0.015$, $Y = 0.275$, and AGE $= 6.9 \times 10^5$ yr)

to surface luminosity by both models are also quite similar, being $(L_{\text{grav}})^+ \sim 303\ L_\odot$ in the younger model and $(L_{\text{grav}})^+ \sim 257\ L_\odot$ in the older model. Thus, in outer layers, both models remain seriously engaged in gravothermal energy production, but the inner, nuclear energy-generating region of the older model is approaching a stable structure in which the rate of gravothermal energy production is quite small relative to the rate of nuclear energy production.

Abundances by number of CNO isotopes in the 6.9×10^5 year old model are shown in Fig. 9.4.12 as functions of the mass coordinate, and differential energy-generation rates due to several reactions involving these isotopes are shown in Fig. 9.4.13 as functions of the radial coordinate. The locations of the outer edge of the convective core in the 6.9×10^5 yr old model and the edge of this core at its earlier maximum extent are well delineated by large discontinuities in both figures: discontinuities in abundances in Fig. 9.4.12 and discontinuities in $dL(\text{process})/dR$ in Fig. 9.4.13. In both figures, the jaggedness along curves in the region between the two sets of discontinuities reflects the unevenness of abundances left behind at the outer edge of the convective core as its mass declines from its absolute maximum during the core C → N-burning phase.

An interesting feature of the distributions in Fig. 9.4.13 is the fact that the maxima in the distributions for the three C → N reactions in the convective core are comparable in magnitude with the maxima for the corresponding reactions in the region where a net conversion of ^{12}C to ^{14}N is still taking place. A second interesting feature is that, in the convective core where the complete CN cycle prevails, the maxima in the distributions for the reactions $^{14}\text{N} \rightarrow {}^{12}\text{C}$ which complete the CN cycle (dotted curves) are comparable in magnitude with the maxima in the distributions for the reactions $^{12}\text{C} \rightarrow {}^{14}\text{N}$ which initiate the CN cycle (solid curves).

As the model evolves between $t \sim 7 \times 10^5$ yr and $t \sim 7.7 \times 10^5$ yr, the outer edge of the convective core moves outward through the region where abundances vary jaggedly, producing the discontinuous changes evident along the curves for M_{CC}, L_{CN}, and L_{grav} in Fig. 9.4.3. The discontinuous changes are correlated, occasioned by ingestion into the convective core of discontinuously varying abundances. From the point of view of mathematics, the departures from smoothness are manifestions of numerical noise which could be reduced and perhaps eliminated by more clever zoning and choice of time step. From the point of view of physics, convection is a stochastic process characterized by fluctuations, and the jaggedness of the numerical results is perhaps a more realistic description of real analogues of the models than is a mathematically smooth description.

As evolution progresses, more and more of the model interior becomes stabilized against gravitation-induced contraction. The model eventually settles into a configuration in which the release of gravitational potential energy in central regions is completely controlled by changes in molecular weight due to nuclear transformations. Instead of contracting and releasing gravothermal energy, as during the approach to the main sequence, outer layers of the model expand. The increase in the potential energy of envelope matter is accomplished by absorption of energy from the energy flux produced by nuclear reactions in central regions. The net rate of gravothermal energy production, L_{grav}, becomes negative and adopts a slowly varying (local steady state) value which is only a fraction of a percent of L_{nuc}.

Somewhat arbitrarily, a model of age 8.1×10^5 yr is here designated as the zero age main sequence (ZAMS) model. Some of its characteristics are described in Figs. 9.4.14–9.4.22. As indicated in Fig. 9.4.14, its luminosity is $\sim$533 $L_\odot$ and its radius is $\sim$2.55 $R_\odot$, with approximately one fourth of the radius contained in a static envelope of mass 0.01 $M_\odot$. Once more, the most interesting profile among structure variables is the luminosity profile, which is made up of three segments: (1) a segment extending from the center to the outer edge of the convective core at $M_{\text{CC}} \sim 1.07\ M_\odot$, (2) a segment between the outer edge of the convective core and the outer edge of the convective core at its maximum extent $M_{\text{CC}}^{\text{max}} \sim 1.24\ M_\odot$ during the core $\text{C} \rightarrow \text{N}$ episode, and (3) a third segment extending beyond the first two. The derivatives dL/dR along the three segments differ because of differences in abundance and nuclear energy-generation characteristics. Gravothermal energy generation contributes imperceptibly to the luminosity distribution. Nevertheless, the differential distribution of its contribution to the luminosity provides the quintessential clue to the character of model evolution. To the left of the vertical line labeled watershed, the rate of gravothermal energy production is positive. To the right, it is negative.

Differential contributions to the luminosity profile are described as functions of the radius coordinate in Fig. 9.4.15. The three regions corresponding to those selected in the mass representation of Fig. 9.4.14 are identified by discontinuities at $R \sim 0.448\ R_\odot$ and at $R \sim 0.475\ R_\odot$. The three regions are defined by $R \leq 0.448\ R_\odot$, 0.448

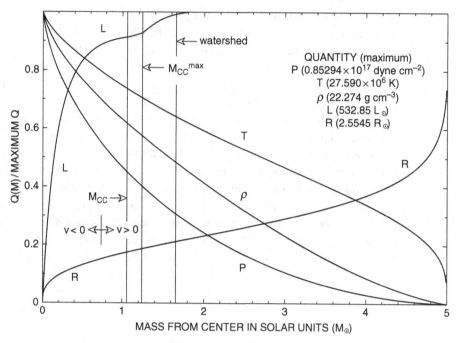

Fig. 9.4.14   Structure of a 5$M_\odot$ model on the zero age main sequence ($Z = 0.015$, $Y = 0.275$, and AGE $= 8.1 \times 10^5$ yr)

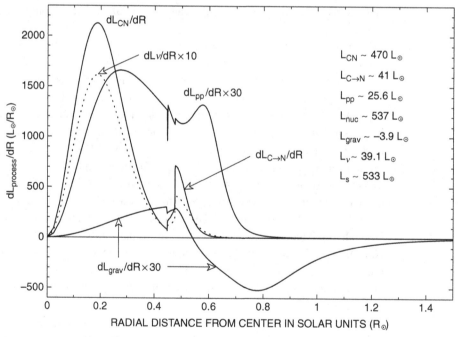

Fig. 9.4.15   Differential contributions to luminosity in a zero age main sequence model ($M = 5M_\odot$, $Z = 0.015$, $Y = 0.275$, and AGE $= 8.1 \times 10^5$ yr)

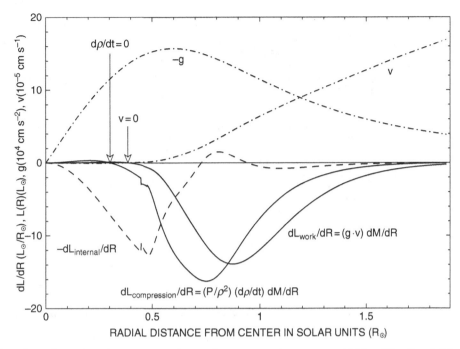

Fig. 9.4.16 Ingredients of gravothermal energy–generation rates in a zero age main sequence model (M = 5M$_\odot$, Z = 0.015, Y = 0.275, and AGE = 8.1 × 10^5 yr)

$R_\odot < R \le 0.475\ R_\odot$, and $R > 0.475\ R_\odot$. In the first, or convective core region, the isotopes ^{12}C, ^{13}C, and ^{14}N are in global equilibrium and the isotopes ^{13}N and ^{15}O are everywhere in local equilibrium. In the second and third regions, ^{12}C continues to be converted into ^{14}N. As detailed in the upper right hand portion of the figure, of the total nuclear energy-production rate, CN-cycle reactions contribute ∼87.5%, C → N reactions contribute ∼7.6%, and pp reactions contribute ∼4.8%. The rate of energy loss by electron neutrinos in beta-decay reactions is about 7% of the rate at which nuclear reactions inject heat into the model. The net gravothermal energy-production rate is only 0.7% of the total nuclear energy-production rate.

Ingredients of the gravothermal energy-generation rate are shown in Fig. 9.4.16. As demonstrated by the fact that $dL_{internal}/dR$ is positive everywhere except in a tiny region about $R \sim 0.8\ R_\odot$, matter is cooling over most of the interior of the ZAMS model. As is shown in Fig. 9.3.29, this is also a characteristic of the 1 $M_\odot$ ZAMS model. In both cases, cooling must be a consequence of "overshoot", since, as is described in Chapters 10 and 11, during the bulk of the main sequence phase, the increase in molecular weight in regions where hydrogen is being converted into helium forces an increase in both density and temperature in nuclear energy-generating regions. On the other hand, another major characteristic of the main sequence phase in full swing is expansion and outward motion in regions outside of the region of nuclear energy production, and this behavior is well developed in the 5 $M_\odot$ ZAMS model, as described by the curves for v and $dL_{compression}/dR$ in Fig. 9.4.16.

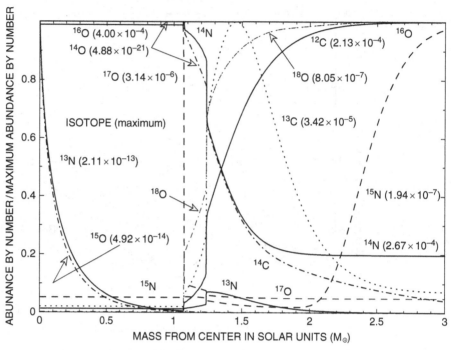

Fig. 9.4.17   CNO isotope abundances in a 5M$_\odot$ zero age main sequence model (Z $=$ 0.015, Y $=$ 0.275, and AGE $=$ 8.1 $\times$ 10^5 yr)

Number abundance parameters for CNO isotopes in the 5 $M_\odot$ ZAMS model are shown in Fig. 9.4.17 and differential contributions to the luminosity by reactions involving these isotopes are shown in Fig. 9.4.18. Comparison of the distributions in Figs. 9.4.12 and 9.4.17 shows that the outer edge of the convective core has moved outward in mass, aproaching the location of the edge at its earlier maximum extent outward in mass. In the zone of decreasing mass between the two edges, the jaggedness in abundance distributions has been noticeably reduced. As is evident from a comparison between Figs. 9.4.13 and 9.4.18, the reduction in jaggedness in abundances translates into a significant reduction in jaggedness along the profile for the differential contribution to the luminosity of the ^{13}C$(p, \gamma)^{13}$N reaction outside the growing convective core. The reductions in jaggedness are due to numerical smoothing during rezoning.

Integration under the abundance profiles for ^{13}C in Figs. 9.4.12 and 9.4.17 shows that the amount of fresh ^{13}C has increased by roughly 70% in the 1.2 $\times$ 10^5 yr separating the two models. On the other hand, integrations under the differential luminosity profiles for the ^{13}C$(p, \gamma)^{13}$N reaction in Figs. 9.4.18 and 9.4.13 shows that the contribution to the luminosity of this reaction outside the convective core relative to its contribution in the convective core has decreased, from approximately a third to a fifth. As detailed in Figs. 9.4.10 and 9.4.15, the contribution of C $\rightarrow$ N reactions outside the convective core relative to the contribution of CN-cycle reactions in the convective core has dropped from about 0.22 to about 0.09.

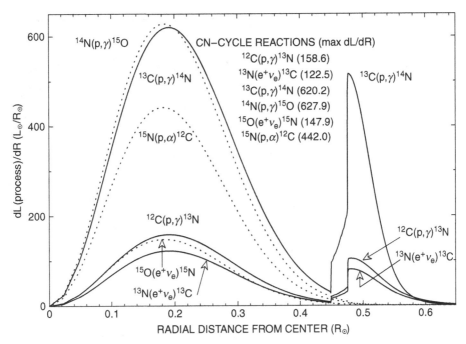

Fig. 9.4.18 CN–cycle and C $\rightarrow$ N contributions to luminosity in a zero–age main sequence model (M = 5$M_\odot$, Z = 0.015, Y = 0.275, AGE = 8.1 × 10⁵ yr)

The ^{16}O profile in Fig. 9.4.17 shows that a slight conversion of ^{16}O into ^{14}N has taken place in the convective core. The differential contributions to the luminosity by reactions involved in this conversion are shown in Fig. 9.4.19. The smallness of these contributions relative to the differential contributions by CN-cycle reactions, shown in Fig. 9.4.18, demonstrates that the rate at which ^{16}O is converted into ^{14}N in the convective core is very slow compared with the CN-cycling rate. Clearly, the CNO bi-cycle is far from being established.

Light isotope abundance distributions in the 5 $M_\odot$ ZAMS model are shown in Fig. 9.4.20 and differential contributions to the luminosity by pp reactions are shown in Fig. 9.4.21. These distributions are worth comparing with the distributions in the 1 $M_\odot$ ZAMS model given in Figs. 9.3.21 and 9.3.25, respectively. In the convective core of the 5 $M_\odot$ model, ^{7}Be and ^{3}He are in global equilibrium and the overall rate at which ^{3}He reacts with ^{4}He to form ^{7}Be competes with the rate at which it experiences the reaction ^{3}He(^{3}He, $2p)^4$He. Both the proton-capture and electron-capture reactions on ^{7}Be produce very little energy, but they occur at comparable rates, so that both the ^{7}Li($p, \alpha)^4$He reactions and the ^{8}B($e^+ \nu_e)^8$Be$^* \rightarrow 2\alpha$ reactions make significant contributions to the luminosity, comparable, respectively, with the contributions of the ^{3}He(^{3}He, $2p)^4$He and ^{2}H($p, \gamma)^3$He reactions. In contrast, in the radiative core of the 1 $M_\odot$ ZAMS model, the contributions to the luminosity of reactions which depend on the formation of ^{7}Be are unimportant.

The shape of the differential contribution to the luminosity of the ^{7}Li($p, \alpha)^4$He reaction in the convective core is due to the fact that the ^{7}Be($e^-, \nu_e)^7$Li rate is a slowly

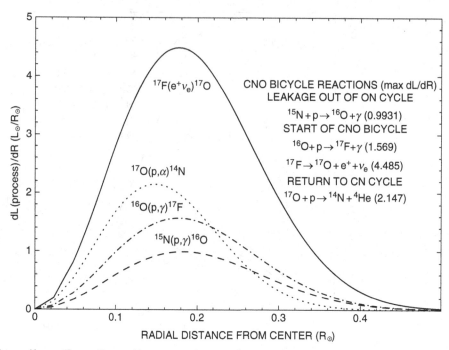

Fig. 9.4.19 $^{15}N \rightarrow \, ^{16}O \rightarrow \, ^{17}F \rightarrow \, ^{17}O \rightarrow \, ^{14}N$ contributions to luminosity in a zero age main sequence model (M $= 5 M_\odot$, Z $= 0.015$, Y $= 0.275$, and AGE $= 8.1 \times 10^5$ yr)

varying function of position in the convective core, whereas the $^7\mathrm{Be}(p, \gamma)^8\mathrm{B}$ rate, being very temperature sensitive, decreases rapidly outward. Thus, the differential contribution of the $^7\mathrm{Li}(p, \alpha)^4\mathrm{He}$ reaction increases monotonically outward through the convective core. Outside of the convective core, by far the dominant contribution to the luminosity is the $^3\mathrm{He}(^3\mathrm{He}, 2p)^4\mathrm{He}$ reaction. The half maximum in the number abundance of $^3\mathrm{He}$ in Fig. 9.4.20 occurs at $M \sim 2 \ M_\odot$. The relationship between mass and radius in Fig. 9.4.14 shows that this location corresponds to $R \sim 0.59 \ R_\odot$, which nearly coincides with the location of the maximum in dL_{33}/dR for the $^3\mathrm{He}(^3\mathrm{He}, 2p)^4\mathrm{He}$ reaction in Fig. 9.4.21. It follows that the $^3\mathrm{He}$ number-abundance profile is correlated with the dL_{33}/dR profile.

Going through the same exercise for the 1 $M_\odot$ ZAMS model, Figs. 9.3.18, 9.3.21, and 9.3.25 may be used to demonstrate that the maximum in the dL_{33}/dR profile at $R \sim 0.11$ $R_\odot$ in Fig. 9.3.25 nearly coincides with the location of the half maximum of $^3\mathrm{He}$ to the left of the maximum at $M \sim 0.24 \ M_\odot$ in Fig. 9.4.21. The difference between the two cases is that, during evolution to the main sequence, the 1 $M_\odot$ model has had time to create $^3\mathrm{He}$ at a maximum number abundance of $Y_3 \sim 1.49 \times 10^{-4}$, whereas the 5 $M_\odot$ model, while having converted its initial deuterium into $^3\mathrm{He}$ at a number abundance of $Y_3 \sim 2.58 \times 10^{-5}$, has managed only to destroy $^3\mathrm{He}$ over the inner approximately 40% of its mass. Creation of new $^3\mathrm{He}$ occurs during the ensuing main sequence phase.

In understanding the nature of energy flow in a model, it is useful to examine the variations of opacity and of the radiative and adiabatic temperature gradients in the model interior. These quantities are shown for the 5 $M_\odot$ ZAMS model in Fig. 9.4.22. Near the center

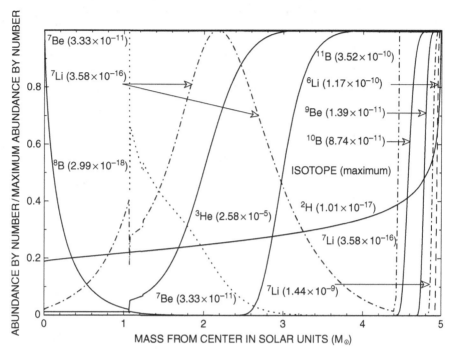

Fig. 9.4.20 Light isotope abundances in a zero–age main sequence model (M = 5$M_\odot$, Z = 0.015, Y = 0.275, and AGE = 8.1 × 10^5 yr)

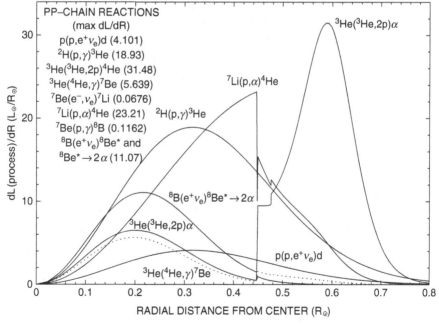

Fig. 9.4.21 pp–chain contributions to luminosity in a zero–age main sequence model (M = 5$M_\odot$, Z = 0.015, Y = 0.275, and AGE = 8.1 × 10^5 yr)

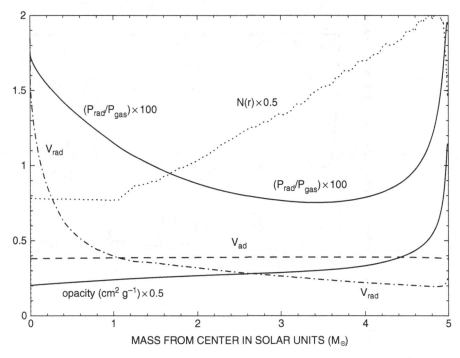

Fig. 9.4.22   Interesting variables in a 5 $M_\odot$ ZAMS model ($Z = 0.015$, $Y = 0.275$, and AGE $= 8.1 \times 10^5$ yr)

of the model, the opacity is only modestly larger than would be the case if electron scattering were the only contributer to the opacity. The opacity increases monotonically outward, reaching a value approximately four times larger than given by the electron-scattering contribution. The increase is primarily due to free–free absorption.

The radiative gradient V_{rad} and the adiabatic gradient V_{ad} in Fig. 9.4.22 intersect, of course, at the outer edge of the convective core. Within the convective core, the radiative gradient increases rapidly inward because the energy-generation rate due to CN-cycle reactions increases rapidly inward, causing the factor $\bar{\epsilon}(r) = L(r)/M(r)$ in the expression for V_{rad} given by eq. (9.2.62) to increase rapidly inward.

It is significant that the adiabatic gradient in the model, at $V_{ad} \sim 0.38$, is several percent smaller than the value $V_{ad} = 0.4$ which would hold for a perfect gas in the absence of radiation pressure. As shown by the curve for P_{rad}/P_{gas} in Fig. 9.4.22, radiation pressure contributes to the total pressure at the 1–2% level, with a mass-averaged value of the ratio of radiation pressure to gas pressure being ~ 0.01. It is evident that, because of the reduction of V_{ad}, the mass of the convective core is larger than it would have been in the absence of radiation pressure. Although the effect is small in the case of the 5 $M_\odot$ model, it becomes an important effect in more massive models, as demonstrated in the case of a 25 $M_\odot$ model in Section 9.5.

Following the development in Section 3.3, another estimate of the relative importance of radiation pressure can be obtained by making use of global energetics. For the 5 $M_\odot$ ZAMS model, direct integration gives

$$E_{gas} = 3.014\,247\,3855 \times 10^{49} \text{ erg,} \tag{9.4.7}$$

$$E_{thermal} = 3.092\,690\,1638 \times 10^{49} \text{ erg,} \tag{9.4.8}$$

and

$$\Omega = -6.106\,937\,5493 \times 10^{49} \text{ erg,} \tag{9.4.9}$$

where E_{gas} is the total thermal energy of material particles, $E_{thermal}$ is the total thermal energy, and $|\Omega|$ is the gravitational binding energy. Using eqs. (9.4.7) and (9.4.9) in eq. (3.3.11), one has that

$$\tilde{\beta} = 2\,\frac{E_{gas}}{|\Omega|} = 0.987\,155, \tag{9.4.10}$$

where $\beta = P_{gas}/P$, $P = P_{gas} + P_{rad}$, and $\tilde{\beta}$ is the reciprocal of an average of $1/\beta$ given by eq. (3.3.10). From eq. (3.2.18),

$$\frac{\beta}{1-\beta} = \frac{P_{rad}}{P_{gas}}. \tag{9.4.11}$$

Setting $\beta = \tilde{\beta}$ in eq. (9.4.11) gives a convoluted average estimate of

$$\left\langle \frac{P_{rad}}{P_{gas}} \right\rangle \sim 0.013, \tag{9.4.12}$$

only modestly larger than the mass average of P_{rad}/P_{gas} in Fig. 9.4.22.

A local polytropic index defined by

$$N(r) = \left(\frac{d\log P(r)}{d\log \rho(r)} - 1 \right)^{-1} \tag{9.4.13}$$

is described by the dotted curve in Fig. 9.4.22. The fact that, in the convective core, the index is a few percent larger than $N = 1.5$ is a consequence of the fact that radiation pressure contributes at the 1–2% level to the total pressure. Beyond the core, the index increases monotonically to about 4 before dropping near the surface to about 3. A mass-averaged value of the local index outside of the convective core is $\langle N \rangle \sim 2.5$. On the other hand, the central concentration of the 5 $M_\odot$ ZAMS model is $\rho_c/\rho_{av} = 52.6742$, quite close to 54.185, the central concentration of an index $N = 3$ polytrope, but quite far from 23.407, the central concentration of an index $N = 2.5$ polytrope. The difference between the estimates demonstrates, once again, that, although there is considerable pedagogic value in comparing realistic stellar models with polytropic models, the comparison is primarily conceptually useful.

9.5 Evolution of a 25 $M_\odot$ gravitationally contracting population I model through deuterium burning and two C $\rightarrow$ N burning phases, and properties of a CN-cycle burning zero age main sequence model

To avoid short circuiting the deuterium-burning phase as in Section 9.4, the initial 25 $M_\odot$ model is constructed from the initial 1 $M_\odot$ Hayashi-band model of Section 9.2 by multiplying the mass and radius at every point in the 1 $M_\odot$ model by a factor of 25 while maintaining the temperature in every mass shell fixed. This means that the density in every mass shell is reduced by a factor of $25^2 = 625$ and the pressure in every shell is decreased by the same factor. Because the ratio of density to temperature has been reduced significantly, the ratio of particle pressure to radiation pressure has been similarly reduced. Therefore, the state variables in the initial 25 $M_\odot$ model are not related exactly as prescribed by an equation of state. The correct relationships are restored in the course of evolving the model. Insisting that the surface temperatures of the 25 $M_\odot$ model and of the 1 $M_\odot$ model be the same, the surface luminosity in the 25 $M_\odot$ model must be chosen larger than the luminosity of the 1 $M_\odot$ model by a factor of 625. Supposing that the effective energy-generation law is proportional to the temperature, as expected in a fully convective model, the luminosity at every point in the 25 $M_\odot$ model is chosen to be larger than the luminosity at the corresponding point in the 1 $M_\odot$ model by a factor of 625.

After effecting these transformations, a first attempt to evolve the 25 $M_\odot$ model fails because, in the process of iterating for a solution, temperatures and densities called for lie outside of the range for which extrapolations of tabular opacities are meaningful. However, by choosing the analytic opacity described in Section 7.12, convergence is achieved. The resultant evolutionary track in the HR diagram is described by the solid curve in Fig. 9.5.1 bounded by the points labeled "ANALYTIC OPACITY" and "SWITCH TO TABULAR OPACITY". Note that the initial evolutionary model, in which all state variables obey a correct equation of state, lies very close to the initial artificially constructed model to which it is connected by a light dotted line. As evolution with the analytic opacity progresses, the model develops a large radiative core.

On switching to the tabular opacity, changes in density and temperature called for in iterating for a solution lie within permissible bounds, but changes in interior variables are effectively discontinuous in the next three time steps. The locations of the model in the HR diagram at the ends of these time steps are shown by the three asterisks at the lower right in Fig. 9.5.1. The reduction in model luminosity achieved in the three time steps is due to the fact that the tabular opacities are somewhat larger than given by the analytic opacity. Following the three step adjustment, the evolutionary path of the model continues smoothly. The discontinuous changes introduced by changing the opacity abruptly are displayed again by asterisks in Fig. 9.5.2, where global luminosities are given as functions of time. Evolution on either side of the discontinuous segment is quite smooth.

The most noteworthy event during the first portion of the gravitationally contracting phase is a brief period of deuterium burning. The evolutionary track in Fig. 9.5.1 and

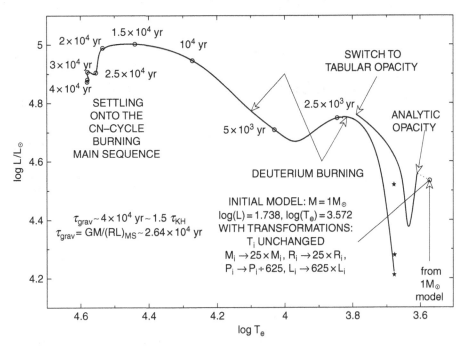

Fig. 9.5.1 Evolutionary path of a 25 $M_\odot$ model of composition $Z = 0.015$, $Y = 0.275$ contracting onto the CN–cycle burning main sequence

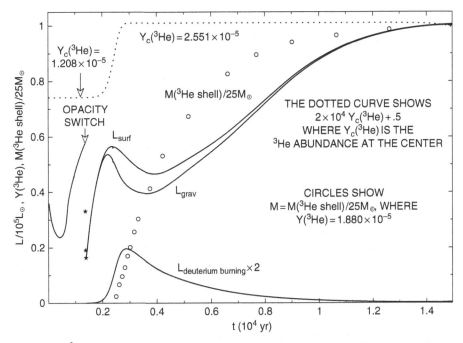

Fig. 9.5.2 Luminosities and ^{3}He in a 25 $M_\odot$ model ($Z = 0.015$, $Y = 0.275$) during early gravitational contraction and deuterium–burning phases

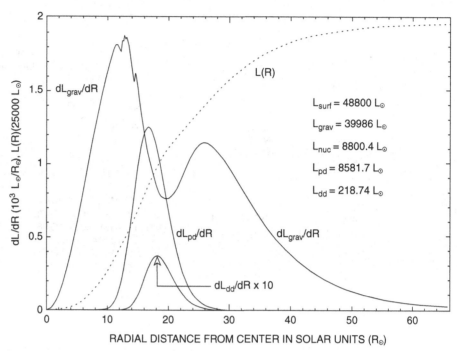

Fig. 9.5.3   Contributions to the surface luminosity in a 25 $M_\odot$ model burning deuterium ($Z = 0.015$, $Y = 0.275$, AGE $= 3.32 \times 10^3$ yr)

the luminosity curves in Fig. 9.5.2 show that the major consequence of deuterium burning is to cause a temporary drop in the surface luminosity. The total amount of energy liberated by deuterium burning over the period $(0.2 \rightarrow 0.7) \times 10^4$ yr is only ~ 10 % of the energy emitted at the surface and, thus, the time scale for evolutionary changes is largely unaffected by deuterium burning. In contrast with the 1 $M_\odot$ model, in which deuterium drops at the same rate in all parts of the well mixed interior, the 25 $M_\odot$ model is in radiative equilibrium throughout most of the interior, so deuterium first disappears at the center and burning then continues in a shell moving outward in mass. As deuterium is converted into ^{3}He, the number abundance of ^{3}He increases from an initial value of $Y(^3\text{He}) = 1.208 \times 10^{-5}$ to $Y(^3\text{He}) = 2.551 \times 10^{-5}$. The dotted curve in Fig. 9.5.2 plots the quantity $2 \times 10^4 \, Y_\text{C}(^3\text{He}) + 0.5$, where $Y_\text{C}(^3\text{He})$ is the abundance of ^{3}He at the center.

The center of the deuterium-burning shell may be defined as the place where the deuterium abundance is half its initial abundance, or, equivalently, as the place where the ^{3}He abundance is halfway between its initial and final abundances, namely at $Y(^3\text{He}) = 1.880 \times 10^{-5}$. The location in mass of the center of the deuterium-burning shell defined in this way, relative to the total mass of the model, is described as a function of time by the open circles in Fig. 9.5.2.

Differential contributions to the surface luminosity of nuclear reactions and of the gravothermal energy source are compared in Fig. 9.5.3 at a time $t \sim 3.32 \times 10^3$ yr when L_grav in Fig. 9.5.2 is near a relative minimum, just before the relative minimum in L_surf near

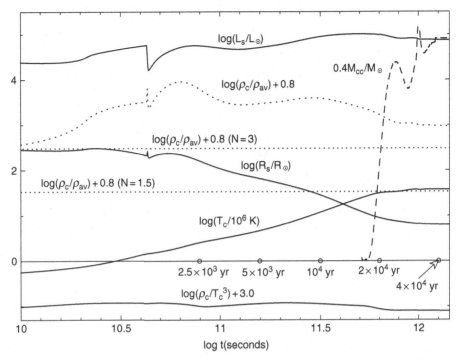

Fig. 9.5.4   Central and global characteristics of a 25 $M_\odot$ model ($Z = 0.015$, $Y = 0.275$) evolving onto the CN–cycle burning main sequence

$\log T_e \sim 4$ in Fig. 9.5.1. By far the major contribution to the nuclear burning luminosity is energy liberated in the reaction $p(d, \gamma)^3$He. Nonetheless, although they provide energy at a rate which is only about 2.5% of the total rate of nuclear energy generation, the contributions of the $d(d, n)^3$He and $d(d, p)^3$H($e^- \bar{\nu}_e)^3$He reactions to the surface luminosity is many (219) times larger than the Sun's luminosity.

Since dL_{grav}/dR is everywhere positive in the deuterium-burning model, it is evident that all of the gravitational potential energy liberated in the interior as well as all of the nuclear energy liberated reaches the surface. The main effect of nuclear burning is to modestly decrease the rate of release of gravitational potential energy. This contrasts with the much larger influence of deuterium burning on the contraction rate of the 1 $M_\odot$ model in the fully convective Hayashi band, as described in Figs. 9.2.10 and 9.2.11.

The time evolution of various internal and global characteristics of the 25 $M_\odot$ model is shown in Fig. 9.5.4. Guidance in relating location in this figure with location in the HR diagram of Fig. 9.5.1 is provided by the five locations marked by circles along the horizontal axis in Fig. 9.5.4. The evolution of characteristics in Fig. 9.5.4 may be compared with the evolution of these same characteristics for the 5 $M_\odot$ model in Fig. 9.4.2 and for the 1 $M_\odot$ model in Fig. 9.2.16. One characteristic of the 25 $M_\odot$ model which distinguishes it from the less massive models is the near constancy of ρ_c/T_c^3. A feature common to all three sets of models is the development of a convective core driven by nuclear burning which signals the beginning of the end of the gravitational contraction phase.

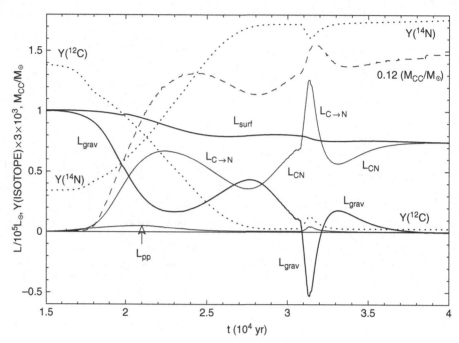

Fig. 9.5.5   Global luminosities, central ^{12}C and ^{14}N abundances, and convective core mass for a 25 M$_\odot$ model ($Z = 0.015$, $Y = 0.275$) settling onto the main sequence

Time dependences of global and interior characteristics of the 25 $M_\odot$ model during the transitional phase from almost exclusive reliance on gravitational potential energy to almost exclusive reliance on nuclear energy as the source of surface luminosity are shown in Fig. 9.5.5. These time dependences may be compared with the time dependences during the same transitional phase for the 5 $M_\odot$ model shown in Fig. 9.4.3. An interesting difference between the two cases is in the variation in the mass of the convective core in response to changes in the nature of nuclear burning. In both cases, after the convective core has reached a relative maximum in mass while ^{12}C is being converted into ^{14}N, the mass of the core declines to a relative minimum and then again increases as the flux of energy due to equilibrium CN-cycle burning increases with increasing temperatures. In the less massive model, the mass of the convective core does not become as large as during the peak of the C $\rightarrow$ N episode and the rate of release of gravitational potential energy declines gradually as the model settles onto the ZAMS. In the more massive model, the mass of the convective core grows to exceed its value during the height of the C $\rightarrow$ N episode; injection of fresh ^{12}C at an abundance several ($\sim$5) times larger than the CN-cycle equilibrium abundance leads to a second ^{12}C $\rightarrow$ ^{14}N episode during which much of the energy produced by nuclear burning is used up in expanding matter, with a consequent delay in the final approach to the ZAMS.

The effect of each C $\rightarrow$ N episode is to cause a decrease in the surface luminosity. The second decrease is barely detectible in Figs. 9.5.1 and 9.5.5, but is much more apparent in a blow up of the relevant portion of the HR diagram presented in Fig. 9.5.6. The shape of

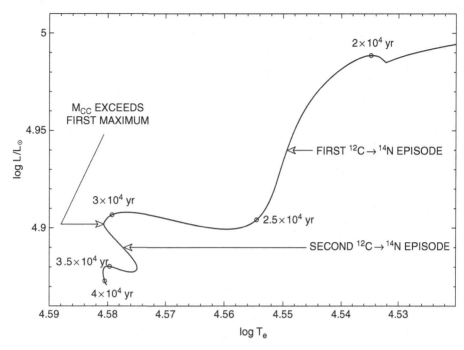

Fig. 9.5.6 Evolutionary path of a 25 $M_\odot$ model of composition $Z = 0.015$, $Y = 0.275$ approaching the CN–cycle burning main sequence

the evolutionary track during the first C → N episode in the 25 $M_\odot$ case is qualitatively different from the shape of the evolutionary track of the 5 $M_\odot$ model during its only C → N episode (see Fig. 9.4.1a), but the 25 $M_\odot$ track during and following the second C → N episode is a miniature version of the evolutionary track of the 5 $M_\odot$ track during and following its only central C → N episode.

The curves for M_{CC}, L_{CN}, and L_{grav} versus time are much smoother in the 25 $M_\odot$ case, Fig. 9.5.5, than in the 5 $M_\odot$ case, Fig. 9.4.3. The difference is due to the fact that the evolution of the more massive model has been calculated twice, with mass zones in the second calculation being selected in anticipation of subsequent evolutionary changes rather than in response to changes as they occur. In particular, in the second calculation, zoning in the region $9 \gtrsim M/M_\odot \gtrsim 12$ has been predetermined and not adjusted as the mass at the outer edge of the convective core passes through the region. On average, the 25 $M_\odot$ models have twice as many mass zones as do the 5 $M_\odot$ models. The distribution of zones in a 25 $M_\odot$ model of age 4.562×10^4 yr shown in Fig. 9.5.7 is typical of the zoning during the period when $M_{CC} > 9\ M_\odot$.

Structural characteristics of the 4.562×10^4 yr old model are shown in Fig. 9.5.8 and differential contributions to the surface luminosity of the model by nuclear energy sources and by the gravothermal energy source are shown in Fig. 9.5.9. The fact that the gravothermal energy-generation rate is everywhere negative means that the gravitational potential energy of the model is being increased by absorption of energy from the ambient luminosity for

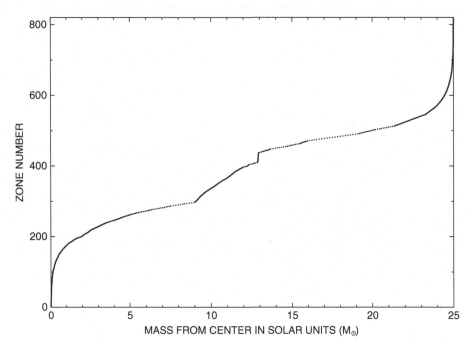

Fig. 9.5.7   Zoning in a 25 $M_\odot$ model near the zero age main sequence ($Z = 0.015$, $Y = 0.275$, and AGE = 4.562×10^4 yr)

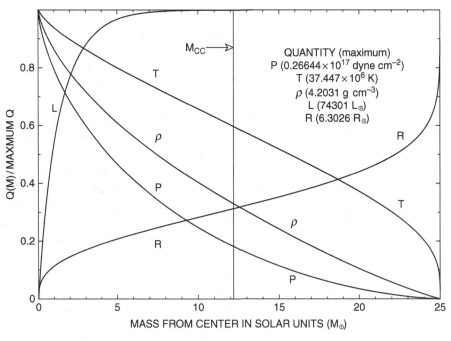

Fig. 9.5.8   Structure of a 25 $M_\odot$ ZAMS model ($Z = 0.015$, $Y = 0.275$, and AGE = 4.562×10^4 yr)

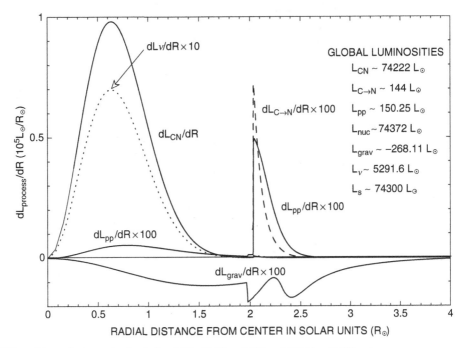

Fig. 9.5.9 Differential contributions to luminosity in a 25 $M_\odot$ ZAMS model ($Z = 0.015$, $Y = 0.275$, and AGE $= 4.562 \times 10^4$ yr)

which nuclear energy sources are responsible. The facts that the total gravothermal absorption rate $-L_{grav}$ is a small (<0.4 %) fraction of the surface luminosity and that only a tiny fraction ($<0.15\%$) of hydrogen has been converted into helium in the convective core qualify the model to be designated as a ZAMS model.

The most obvious distinction between the distributions in the 25 $M_\odot$ ZAMS model shown in Fig. 9.5.8 and those in the 5 $M_\odot$ ZAMS model shown in Fig. 9.4.14 is that, in the 5 $M_\odot$ model, a discernable ($\sim7\%$) fraction of the surface luminosity is produced outside of the convective core, whereas, in the 25 $M_\odot$ model, essentially all of the surface luminosity is produced in the convective core of mass $\sim12M_\odot$. Another distinction is that all of the matter in the more massive model is moving outward and expanding, while matter near the center of the lower mass model continues to contract and move inward (see Fig. 9.4.16).

Differential contributions to the surface luminosity of the 25 $M_\odot$ ZAMS model are shown in Fig. 9.5.9. Curiously, by far the dominant contribution of reactions which involve CNO isotopes is confined to the convective core, whereas the dominant contribution of pp-chain reactions occurs outside of the convective core. The gravothermal energy source contributes negatively everywhere and the pp-chain reactions, the C $\rightarrow$ N reactions outside of the convective core, and gravothermal energy sources make contributions to the surface luminosity quite small compared with the contribution of CN-cycle reactions in the convective core. The rate at which energy is carried away by neutrinos is roughly 7% of the rate at which nuclear reactions inject thermal energy into the model.

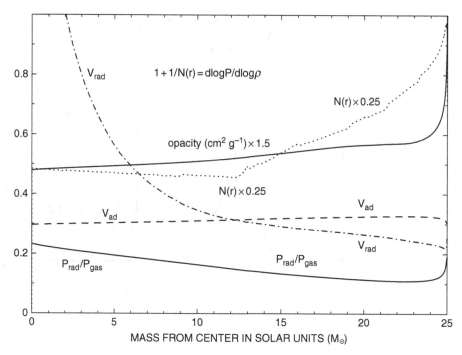

Fig. 9.5.10   Interesting variables in a 25 $M_\odot$ ZAMS model ($Z = 0.015$, $Y = 0.275$, and AGE $= 4.562 \times 10^4$ yr)

Additional interior characteristics of the 25 $M_\odot$ ZAMS model are shown in Fig. 9.5.10. Particularly interesting is the fact that the ratio of radiation pressure P_{rad} to gas pressure P_{gas} is everywhere larger than 10% and, near the center and near the outer edge of the model, the ratio is of the order of 20%. More precisely, $P_{rad}/P_{gas} \sim 0.234$ at the center and $P_{rad}/P_{gas} \sim 0.191$ in the last interior zone of the model. At the halfway point in mass, $P_{rad}/P_{gas} \sim 0.147$, and the mass-averaged ratio is slightly larger than this. From the definition of the quantity β in eq. (3.2.18), one has that

$$\beta = \frac{P_{gas}}{P} = \left(1 + \frac{P_{rad}}{P_{gas}}\right)^{-1}. \tag{9.5.1}$$

At the center of the model, $\beta \sim 0.81$. In the middle of the model, at $M(r) = 12.5 M_\odot$, $\beta \sim 0.87$, and at the outer edge of the interior, $\beta \sim 0.84$. The mean of these values is $\langle\beta\rangle = 0.84$. Equation (3.2.20) defines an average value of β of a model of homogeneous composition in terms of the mass and composition of the model. Setting $\bar{\alpha} = 0.25$, $\bar{\mu} = 1$, $M = 25 M_\odot$, and $\beta_c = 0.81$ in this expression gives $\bar{\beta} \sim 0.84$.

Another meaningful measure of β follows from the development in Section 3.3 which makes use of global energies of a model. The total thermal energy, gravitational binding energy, and net binding energy of the 25 $M_\odot$ ZAMS model are, respectively,

$$E_{thermal} = 2.962\,17 \times 10^{50} \text{ erg,} \tag{9.5.2}$$

$$|\Omega| = 5.177\,73 \times 10^{50} \text{ erg,} \tag{9.5.3}$$

and

$$E_{\text{bind}} = 2.21556 \times 10^{50} \text{ erg}. \qquad (9.5.4)$$

As shown in Section 3.3, the kinetic energy of material particles, E_{gas}, is equal to the net binding energy. Therefore, the total energy in photons is

$$E_{\text{rad}} = E_{\text{thermal}} - E_{\text{gas}} = E_{\text{thermal}} - E_{\text{bind}} = 0.74661 \times 10^{50} \text{ erg} = 0.337 \, E_{\text{gas}}. \quad (9.5.5)$$

From this, it follows that the kinetic energy of photons makes up approximately one quarter of the thermal energy in the model, and the kinetic energy of material particles makes up about three quarters of the thermal energy. From eqs. (3.3.10) and (3.3.11), the reciprocal of a gas-pressure weighted average of $1/\beta$ is

$$\tilde{\beta} = 2 \frac{E_{\text{bind}}}{|\Omega|} = 0.858, \qquad (9.5.6)$$

close to the average values of β given by the two previously invoked algorithms.

Related to the fact that radiation pressure is an important component of the total pressure is the fact that, at $V_{\text{ad}} \sim 0.3$, the adiabatic gradient is everywhere significantly smaller than the value $V_{\text{ad}} = 0.4$ which would obtain if radiation pressure were negligible. Comparing with the radiative gradient V_{rad} in Fig. 9.5.10, it is evident that radiation pressure is responsible for a convective core mass considerably larger than would be the case in the absence of radiation pressure ($\sim 12 \, M_\odot$ rather than $\sim 8 \, M_\odot$).

The local polytropic index defined by eq. (9.4.13) is described by the dotted curve in Fig. 9.5.10. The fact that, at $N \sim 1.9$, the index in the convective core is significantly larger than $N = 1.5$ is due to the importance of radiation pressure. The mass-weighted average over the interior is $\langle N(r) \rangle \sim 2.3$. The central concentration of the model, at $\rho_c / \rho_{\text{av}} = 29.8566$, lies closer to the central concentration of an $N = 2.5$ polytrope (23.407) than to that of an $N = 3$ polytrope (54.185).

Insight into the nature of energy flow in the 25 $M_\odot$ model is provided by the variation in opacity and by the radiative and adiabatic gradients in Fig. 9.5.10. Over most of the interior, the opacity is nearly constant, varying from $\kappa \sim 0.32 \text{ cm}^2 \text{ g}^{-1}$ at the center to $\kappa \sim 0.38 \text{ cm}^2$ g^{-1} at $M = 22.5 \, M_\odot$. This demonstrates that, over most of the interior, the primary source of opacity is electron scattering, as approximated, e.g., by eq. (3.4.19). The relatively small densities and large temperatures in the 25 $M_\odot$ ZAMS model are responsible for this result.

From the definition of V_{rad} given by eq. (3.4.28) and described qualitatively by eq. (9.2.62), coupled with the facts that κ is nearly constant and $P/T^4 \propto P/P_{\text{rad}} = 1 + P_{\text{gas}}/P_{\text{rad}}$ decreases only slowly inward through the convective core, it is evident that, as in the case of the 5 $M_\odot$ model of Section 9.4, it is the increase in $\bar{\epsilon}(r) = L(r)/M(r)$, the average energy generation rate per unit mass interior to $M(r)$, that is responsible for the increase in V_{rad} inward through the core and, as in the 5 $M_\odot$ model, the increase inward in $\bar{\epsilon}(r)$ is due to the large temperature sensitivity of the CN-cycle energy-generation rate.

The number abundances of CNO isotopes in the 25 $M_\odot$ ZAMS model displayed in Fig. 9.5.11 may be compared with the abundances of these same isotopes in the 5 $M_\odot$ ZAMS model displayed in Fig. 9.4.17. One striking difference is that, in central regions of the convective core of the more massive model, the beta-unstable isotopes ^{13}N and ^{15}O

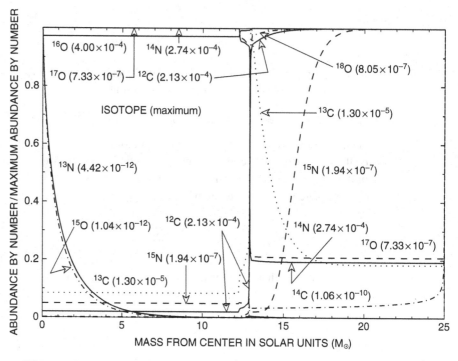

Fig. 9.5.11   CNO isotope abundances in a 25 $M_\odot$ ZAMS model ($Z = 0.015$, $Y = 0.275$, and AGE $= 4.562 \times 10^4$ yr)

are approximately 20 times more abundant than their counterparts in central regions of the convective core of the less massive model. The difference occurs because, while their decay rates are fixed, the rates at which the beta-unstable isotopes are produced increase rapidly with temperature, and, in central regions, temperatures in the more massive model are ~30% larger than in the less massive model.

On the other hand, the stable isotope ^{17}O is about four times more abundant in the convective core of the less massive model than in the more massive model. The difference is due to the fact that the rate at which ^{17}O is created by the ^{16}O$(p, \gamma)^{17}$F$(e^+\nu_e)^{17}$O reactions increases with temperature less rapidly than does the rate at which ^{17}O is destroyed by the reaction ^{17}O$(p, \alpha)^{14}$N.

Differential contributions to the surface luminosity by CN-cycle and C $\rightarrow$ N reactions in the 25 $M_\odot$ ZAMS model are shown in Fig. 9.5.12. The outer edge of the convective core occurs at $R \sim 1.975\ R_\odot$. Contributions by CN-cycle reactions are confined to the convective core and contributions by C $\rightarrow$ N reactions occur just outside the core. The C $\rightarrow$ N contributions are completely dwarfed by the CN-cycle contributions. This is in marked contrast with what occurs in the 5 $M_\odot$ ZAMS model (see Fig. 9.4.18) where C $\rightarrow$ N burning outside the convective core makes a substantial (~7%) contribution to the surface luminosity. Actually, the absolute contributions of C $\rightarrow$ N reactions to the surface luminosity are approximately the same in both models (~30–40 $L_\odot$), demonstrating that conditions just outside the convective cores of both models are approximately the same. The difference in the relative contributions of CN-cycle and C $\rightarrow$ N reactions in the two

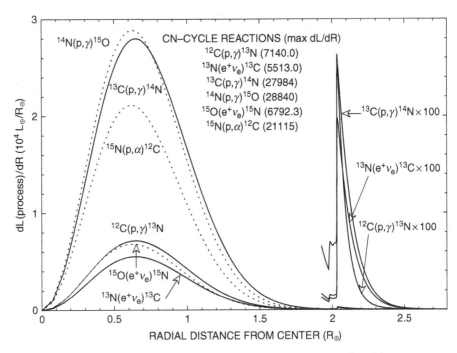

Fig. 9.5.12 CN–cycle and C $\rightarrow$ N contributions to luminosity in a 25 $M_\odot$ ZAMS model ($Z = 0.015, Y = 0.275$, AGE $= 4.562 \times 10^4$ yr)

models is due to the significantly larger temperatures in the convective core of the more massive model.

The differential contributions to the surface luminosity of isotopes participating in the CNO-bi-cycle in the 25 $M_\odot$ ZAMS model are shown in Fig. 9.5.13. Comparison with the distributions in Fig. 9.4.19 for the 5 $M_\odot$ ZAMS model shows that, relative to one another, the contributions of the reactions $^{16}O(p, \gamma)^{17}F$, and $^{17}F(e^+\nu_e)^{17}O$ are exactly the same in the two models, demonstrating that ^{17}F is everywhere in local equilibrium in both models. The stable isotope ^{17}O is in global equilibrium in the convective core of both models, with the ^{17}O number abundance in the convective core of the 25 $M_\odot$ model being about four times smaller than in the convective core of the 5 $M_\odot$ model. Because of the different temperature dependences of the reaction rates, the overall rate of the $^{17}O(p, \alpha)^{14}N$ relative to the rate of the $^{16}O(p, \gamma)^{17}F$ reaction in the convective core is $\sim$70% larger in the more massive model. Comparison of Fig. 9.5.12 with Fig. 9.5.13 and of Fig. 9.4.18 with Fig. 9.4.19 reveals that, in both models, CNO-bi-cycle contributions to the surface luminosity are about 1% of the CN-cycle contributions.

Number abundances of light isotopes in the region outside the convective core of the 25 $M_\odot$ ZAMS model are shown in Fig. 9.5.14. The abundance distributions are qualitatively similar to the abundance distributions outside the convective core of the 5 $M_\odot$ ZAMS model shown in Fig. 9.4.20. In the convective core of the 25 $M_\odot$ model, the ^8B abundance varies from 3.290×10^{-18} at the center to 4.788×10^{-21} at the edge and the ^7Li abundance varies from 1.187×10^{-19} at the center to 1.420×10^{-17} at the edge. The abundances of ^7Be

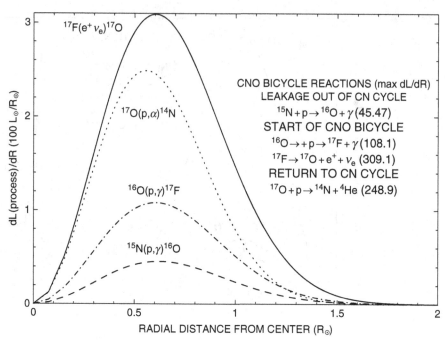

Fig. 9.5.13 $^{15}N \rightarrow {}^{16}O \rightarrow {}^{17}F \rightarrow {}^{17}O \rightarrow {}^{14}N$ contributions to luminosity in a 25 $M_\odot$ ZAMS model ($Z = 0.015$, $Y = 0.275$, and AGE $= 4.562 \times 10^4$ yr)

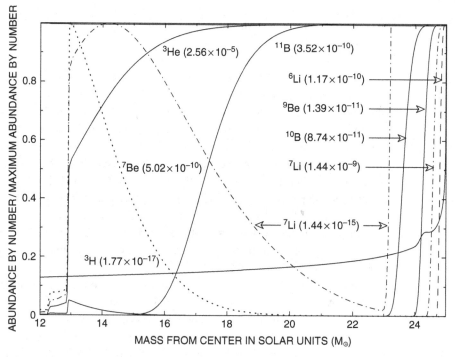

Fig. 9.5.14 Light isotope abundances in a 25 $M_\odot$ ZAMS model ($Z = 0.015$, $Y = 0.275$, and AGE $= 4.562 \times 10^4$ yr)

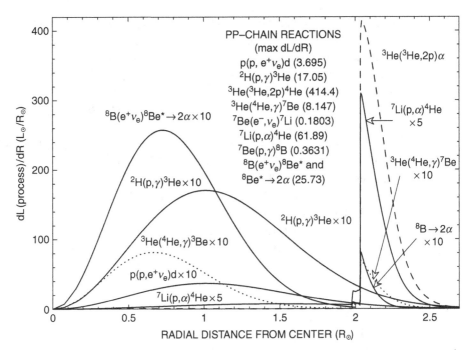

Fig. 9.5.15 pp–chain contributions to luminosity in a 25 $M_\odot$ ZAMS model ($Z = 0.015$, $Y = 0.275$, and AGE $= 4.562 \times 10^4$ yr)

and ^{3}He are constant at 8.638×10^{-12} and 2.056×10^{-8}, respectively. In the 5 $M_\odot$ ZAMS model, the corresponding number abundances are: ^{8}B $= 2.992 \times 10^{-18} \rightarrow 6.147 \times 10^{-20}$; ^{7}Li $= 7.054 \times 10^{-18} \rightarrow 1.409 \times 10^{-16}$; ^{7}Be $= 3.332 \times 10^{-11}$; and ^{3}He $= 2.987 \times 10^{-7}$.

Differential pp-chain contributions to the surface luminosity in the 25 $M_\odot$ ZAMS model are shown in Fig. 9.5.15. From Fig. 9.5.9, it is clear that approximately two-thirds of the total contribution of pp-chain reactions comes from outside the convective core and Fig. 9.5.15 shows that most of this latter contribution is from the ^{3}He(^{3}He, $2p$)α reaction. As shown by Figs. 9.4.15 and 9.4.21, in the case of the 5 $M_\odot$ ZAMS model, approximately two thirds of the contribution of pp chains to the surface luminosity comes from the convective core, but of the one third contribution from the region outside the core, the large majority comes from the ^{3}He(^{3}He, $2p$)α reaction.

Bibliography and references

D. R. Alexander & J. W. Ferguson. *ApJ*, **437**, 879, 1994.

C. Angulo, M. Arnould, M. Rayet, *et al.*, *Nucl. Phys. A*, **656**, 3, 1999.

I. Appenzeller & W. Tscharnuter, *A&Ap*, **40**, 397, 1975.

M. Asplund, N. Grevesse & A. J. Sauval, in *Cosmic Abundances as Records of Stellar Evolution and Nucleosynthesis*, eds. F. N. Bash & T. G. Barnes, ASP Conf. Ser., Vol. **30**, 2005.

S. Basu & Telemachos Ch. Mouschovias, *ApJ*, **432**, 720B, 1994; **452**, 386B, 1995a; **453**, 271B, 1995b.

Peter Bodenheimer, *ApJ*, **144**, 103, 1966a; **144**, 709, 1966b.

Peter Bodenheimer & Alan Sweigert, *ApJ*, **152**, 515, 1968.

W. B. Bonner, *MNRAS*, **116**, 351, 1956.

C. Chiappini, A. Renda, & Franka Matteucci, *A&A*, **395**, 789, 2002.

André Louis Cholesky, *Bull. Geodésiques*, **1**, 159, 1922; **2**, 67, 1924.

G.E. Ciolek & T. Ch. Mouschovias, *ApJ*, **418**, 774, 1993; **425**, 142, 1994; **454**, 194, 1995.

Martin Cohen & Leonard V. Kuhi, *ApJS*, **41**, 743, 1979.

R. Ebert, *Zeits. für Ap*, **27**, 217, 1955; **43**, 263, 1957.

R. A. Fiedler & Telemachos Ch. Mouschovias, *ApJ*, **391**, 199, 1992; **418**, 744, 1993.

Brian D. Fields, *ApJ*, **456**, 478, 1996.

John E. Gaustad, *ApJ*, **138**, 1050, 1963.

J. Geiss and G. Gloeckler, *Space Sci. Rev.*, **84**, 239 & 250, 1998.

Chushiro Hayashi, *Publ. Astron. Soc. Japan*, **13**, 450, 1961.

Chushiro Hayashi, R. Hoshi, & Dai'ichiro Sugimoto, *Progr. Theoret. Phys. Suppl.*, **22**, 1, 1962.

George H. Herbig, *ApJS*, **4**, 337, 1960.

Chris Hunter, *ApJ*, **218**, 834, 1977.

Icko Iben, Jr., *ApJ*, **141**, 993, 1965.

Carlos A. Iglesias & Forrest J. Rogers, *ApJ*, **464**, 943, 1996.

James H. Jeans, *Astronomy and Cosmology* (Cambridge: Cambridge University Press), 1928; (New York: Dover), 1961.

Leonard V. Kuhi, *J. Roy. Astr. Soc. Can.*, **61**, 1, 1966.

Richard B. Larson, *MNRAS*, **145**, 271, 1969a; **145**, 297, 1969b.

Leslie Looney, Lee G. Mundy, & W. J. Welch, *ApJ*, **592**, 255, 2003.

Leon Mestel & Lyman J. Spitzer, *MNRAS*, **116**, 503, 1956.

J. A. Morgan, in *Astronomical Data Analysis Software and Systems IV*, ed. R. A. Shaw, H. E. Payne, & J. J. E. Hayes, PASP Conf. Ser. **77**, 129, 1995.

Telemachos Ch. Mouschovias, *ApJ*, **206**, 753, 1976a; **207**, 141, 1976b; **211**, 147, 1977; **228**, 159, 1979.

Telemachos Ch. Mouschovias & Lyman J. Spitzer, Jr., *ApJ*, **210**, 326, 1976.

Francesco Palla & Steven W. Stahler, *ApJ*, **375**, 288, 1991; **392**, 667, 1992; **418**, 414, 1993.

M. V. Penston, *Roy. Obs. Bull.*, No. **117**, 299, 1966.

F. J. Rogers & C. A. Iglesias, *ApJ*, **401**, 361, 1992.

R. A. Shaw, H. E. Payne, & J. J. E. Hayes, *PASP Conf. Ser.*, **77**, 129, 1995.

Frank H. Shu, *ApJ*, **214**, 488, 1977a; **214**, 488, 1977b.

Steven W. Stahler, *ApJ*, **332**, 804, 1988.

Steven W. Stahler, Frank H. Shu, & Ronald E. Taam, *ApJ*, **241**, 637, 1980a; **242**, 226, 1980b; **248**, 727, 1981.

Konstantinos Tassis and Telemachos Ch. Mouschovias, *ApJ*, **618**, 769, 2005a; **618**, 783, 2005b.

10 Solar structure and neutrino physics

In the last four decades of the twentieth century, the detection of neutrinos from the Sun became a reality. Using a detection scheme beginning with a chloroethylene-filled tank in the Homestake mine in South Dakota, Raymond Davis and his collaborators (Davis, Harmer, & Hoffman, 1968) established upper limits on the fluxes of neutrinos made in the Sun by the reactions $^8\text{B} \rightarrow \,^8\text{Be}^* + e^+ + \nu_e$ and $^7\text{Be} + e^- \rightarrow \,^7\text{Li} + \nu_e$ and reaching the Earth as electron-flavor neutrinos. The experiment relied on the reaction $^{37}\text{Cl}(\nu_e, e^-)^{37}\text{A}$, which has a threshold (at 0.814 MeV) approximately twice as large as the 0.43 MeV maximum energy of the neutrino emitted in the pp reaction, slightly smaller than the 0.861 MeV energy of the neutrino emitted in the $^7\text{Be} + e^- \rightarrow \,^7\text{Li} + \nu_e$ reaction, and much smaller than the maximum energy of the neutrino emitted in the $^8\text{B}(e^+\nu_e)^8\text{Be}^*$ reaction.

The limits established by Davis *et al.* were an order of magnitude smaller than fluxes which had been predicted on the basis of solar models that incorporated the then best guesses as to the appropriate input physics. Among other consequences, the discrepancy, commonly referred to as the solar neutrino problem, led to a re-examination of available data on relevant nuclear cross sections, revisions and new measurements of these cross sections, and to a refinement over time in the solar models. Ultimately, positive detections were achieved, not only by the Davis group, but by groups in Italy, Russia, Japan, and Canada, and, despite occasional biasing in the choice of input physics in ways to minimize the discrepancy, fluxes of electron-flavor neutrinos produced by conventional solar models continued to exceed observed fluxes of electron-flavor neutrinos, typically by factors of the order of 2 to 3.

In the mid 1980s, the possibility that neutrinos might not retain their birth flavor on traveling from their point of origin in the Sun to the Earth became seriously considered as an explanation for the observed discrepancies. Two ideas gained currency, both requiring that at least one of the three known flavors of neutrino has a mass or acquires an effective mass, with the consequence that oscillations between neutrino flavors can occur. The first idea was that, even in a vacuum, at least one neutrino has an intrinsic mass. The second idea was that, due to an interaction between an electron neutrino and electrons in the dense solar environment, a neutrino born near the solar center as an electron neutrino can acquire an effective mass in consequence of which it can transform into, say, a muon neutrino as it makes its way out of the Sun.

Experiments involving antineutrinos from terrestial nuclear reactors have now demonstrated that, even in a vacuum, electron antineutrinos and muon antineutrinos oscillate with large amplitude between flavor types in ways that imply that at least two of the three classical neutrino types – electron, muon, and tau – have intrinsic masses and that all neutrino states are linear combinations of mass eigenstates. The magnitude of the discrepancy between fluxes from the Sun predicted on the assumption of no flavor mixing and observed fluxes of high energy electron neutrinos is evidence that, on interacting with electrons on its passage through the Sun, a high energy electron neutrino acquires an effective mass comparable to or larger than the difference in mass between electron and muon neutrinos in vacuum.

Studies of the abundances of elements heavier than hydrogen in the solar photosphere have suggested values for the metallicity parameter ranging from $Z \sim 0.0122$ to $Z \sim 0.019$ (see Section 2.1). There is no reason to suspect the integrity of any of the studies, but the estimate of $Z \sim 0.0122$ involves a three-dimensional hydrodynamical analysis and might be expected to yield more definitive results than the more usual hydrostatic analyses. However, even if the surface metallicity were known precisely, thanks to chemical diffusion across the base of the convective envelope, surface abundances are not the same as those in the deep interior. So, for comparison with results of solar neutrino experiments, it is prudent to be equipped with models with at least two choices of interior heavy element abundances. Given that readily available opacity tables exist with state of the art opacities for $Z = 0.01$ and $Z = 0.02$ and a variety of hydrogen abundances (see Section 7.15), it is convenient to construct models with these two metallicities having the Sun's mass, luminosity, radius, and age. Fitting to results of neutrino experiments can be accomplished by interpolating between the two models.

The construction of models with $Z = 0.01$ and $Z = 0.02$ are described in the first parts of Sections 10.1 and 10.2, respectively. Initial models are in the last phase of pre-main sequence gravitational contraction and, for each choice of Z, a set of solar mass models covering a range of Y (helium abundance by mass) values is constructed to permit an estimate of $Y_\odot(Z)$, the initial helium abundance by mass of a model which reaches the Sun's luminosity after approximately 4.6×10^9 yr of evolution. The solar radius does not act as a strong constraint on the model since the model radius can be adjusted by varying the mixing length to scale height ratio. The results are $Y(Z_\odot = 0.01) \sim 0.237$ and $Y_\odot(Z = 0.02) \sim 0.296$. The structure and abundance characteristics and the nuclear and gravothermal energy-generation rates of the ~ 4.6 billion year old models with $Z = 0.01$ and $Z = 0.02$ are described in the second parts of Sections 10.1 and 10.2, respectively. Contributions of various nuclear reactions to the photon luminosity and to the neutrino luminosity for the two solar-like models are described in Section 10.3.

In Section 10.4, the neutrino fluxes produced by the solar models are compared with neutrino fluxes estimated in four different experiments. The total flux of neutrinos produced by the $^8B \rightarrow {}^8Be^* + e^+ + \nu_e$ reaction in the Sun measured by the one experiment which is sensitive to high energy neutrinos of all flavors corresponds to the total flux produced by a solar model characterized by the interior composition parameters $Z \sim 0.0166$ and $Y \sim 0.28$, these numbers being obtained by interpolating between the $Z = 0.01$ and $Z = 0.02$ models.

In Section 10.5, the formalism for the propagation through a vacuum of neutrinos with mass differences is developed and the results of two sets of experiments which estimate mass squared differences and mixing angles for reactor-produced antineutrinos and for muon neutrinos produced by cosmic rays impinging upon the Earth's atmosphere are summarized. Assuming that the mixing angle and mass squared difference found experimentally for electron and muon antineutrinos are the same as these parameters for electron and muon neutrinos, comparison of solar model neutrino flux characteristics with the results of experiments described in Section 10.4 suggests that low energy electron neutrinos created in the pp reaction and in the $^7\text{Be}(e^-, \nu_e)^7\text{Li}$ reaction are not affected by interaction with solar electrons by nearly as much as are the generally much higher energy neutrinos created in the $^8\text{B} \rightarrow {}^8\text{Be}^* + e^+ + \nu_e$ reaction.

The formalism for calculating the flavor wave function of a neutrino as it travels through the Sun is presented in Section 10.6 and results of numerical solutions for neutrinos which pass through a simple exponential electron-density distribution are described in Section 10.7. In Section 10.8, the contributions to neutrino fluxes at the Earth by reactions occurring at different distances from the center of realistic solar models are presented. An important result is that, although the flavor combination of a neutrino emerging from the Sun depends upon its energy and upon its distance from the solar center at its point of origin, this combination does not depend upon the particular trajectory through the Sun. That is, the emergent combination is the same for all neutrinos of a given energy which are produced in a spherical shell centered at the Sun's center, regardless of point of origin in the shell.

Another important result is that, for any given choices for the vacuum mixing angle and for the difference between squares of the vacuum masses of electron and muon neutrinos, the larger the neutrino energy, the smaller is the fraction of time which a neutrino arriving at the Earth manifests the electron flavor. In particular, with the choices $\Delta^2 = 0.8 \times 10^{-4}$ (eV)^2 and $\sin^2(\theta) = 0.8$, consistent with experimental results for reactor and cosmic ray neutrinos, this time fraction is of the order of 0.3–0.4 for typical neutrinos from the decay of ^{8}B and of the order of 0.6 for typical neutrinos produced by the pp, pep, and ^{7}Be electron-capture reactions. These choices thus suggest roughly a factor of two difference between the production rate and the detection rate of neutrinos made by pp, pep, and ^{7}Be electron-capture reactions and roughly a factor of three difference between the production rate and the detection rate of neutrinos made by ^{8}B decay.

In Section 10.9, a summary is followed by the conclusion that a solar neutrino problem no longer exists: consistency between the results of terrestial neutrino experiments and solar models can be achieved by assuming that (1) the difference between electron and muon neutrino squared masses is the same as the experimentally determined difference between squared masses of electron and muon antineutrinos and that (2) an interaction between electrons and electron-flavor neutrinos predicted by the current–current interaction theory of Feynman and Gell-Mann operates in the Sun with a coupling constant which is identical with the vector coupling constant in ordinary beta decay.

10.1 Construction and properties of a $Z = 0.01$ solar-like model with the Sun's luminosity, radius, and estimated age

For any choice of metallicity Z and attendant opacity, iterations are required to determine $Y_\odot(Z)$, the initial helium abundance by mass of a model which achieves the Sun's

Table 10.1.1 Relationships between age and initial helium abundance Y for models of near solar luminosity when $Z = 0.01$ and rmixs $= 0.6$

Y	t_9	$L/L_\odot$	$R/R_\odot$
0.25	3.35	0.999666	0.973417
0.24	4.32	1.000001	0.985827
0.23	5.36	1.000136	0.998386

luminosity in 4.6×10^9 yr of evolution after the pre-main sequence deuterium-burning phase. For the model to also have the Sun's radius requires additional iterations to determine an appropriate ratio of mixing length to pressure scale height.

Adopting the $Z = 0.01$ opacity given by the OPAL website referred to in Section 7.15, but exaggerating the contributions of carbon and oxygen to the opacity by a factor of two, the required iterations produce $Y_\odot(0.01) = 0.25$ and

$$\text{rmixs} = \frac{l_{\text{mix}}}{2H_{\text{P}}} = 0.6. \qquad (10.1.1)$$

The initial number abundance parameters for all isotopes in the initial model used to obtain these results are listed in the next to last column of Table 9.2.1 and the evolutionary track of the model during the gravitational contraction phase is given in Fig. 9.2.11.

Using OPAL opacities without carbon and oxygen enhancements, a model with $Y = 0.25$ and rmixs $= 0.6$ attains the Sun's luminosity after only 3.35 billion years of evolution and has a radius smaller than the Sun's by several percent, as recorded in the first row of Table 10.1.1. The basic reason for the difference in ages is that, in regions where the temperature and density are such that the contributions of carbon and oxygen to the opacity are large, the opacity in the calculation reported in this section is smaller than the opacity in the calculation reported in Section 9.2. All other things being equal, a smaller opacity means a larger luminosity; hence, in the present calculation, the Sun's luminosity is reached in less time than in the Section 9.2 calculation.

The time to reach the solar luminosity may be increased by decreasing the initial helium abundance by mass. Decreasing Y means increasing the particle density at a given mass density and therefore decreasing the temperature at a given pressure. Thus, the luminosity at the beginning of the nuclear burning phase of evolution is reduced, and more time is required to reach the Sun's luminosity. Results of choosing $Y = 0.24$ and $Y = 0.23$ are shown, respectively, in the second and third rows of Table 10.1.1. In the corresponding initial models, the values for the number abundance parameters Y_{H} and Y_{He} listed in the first and third row of the next to last column in Table 9.2.1 have been changed from $(Y_{\text{H}}, Y_{\text{He}}) = (0.74, 0.0625)$ when $Y = 0.25$ to $(Y_{\text{H}}, Y_{\text{He}}) = (0.75, 0.0600)$ when $Y = 0.24$ and to $(Y_{\text{H}}, Y_{\text{He}}) = (0.76, 0.0575)$ when $Y = 0.23$. No other initial number abundance parameters have been changed, and so the abundances by mass of elements heavier than helium are the same for all three models.

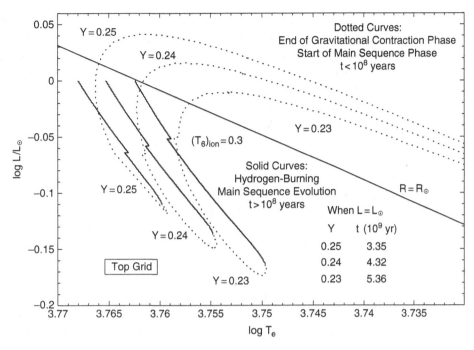

Fig. 10.1.1   Evolution in the HR diagram of 1 $M_\odot$ models with $Z = 0.01$, $Y = 0.25, 0.24$, and 0.23, and rmixs $= 0.60$

Portions of the evolutionary tracks in the HR diagram for the three models are displayed in Fig. 10.1.1. The dotted tracks describe evolution during the last part of the gravitational contraction phase (beginning at a time $t \sim 2 \times 10^7$ yr after the deuterium-burning phase) and the very first portion of the main sequence nuclear burning phase (up to a time $t \sim 10^8$ yr after the deuterium-burning phase). The solid tracks describe evolution during core hydrogen burning. Time steps during this latter phase have been chosen for convenience as $\Delta t = 10^7$ yr, even though the criteria for time zoning described in Section 8.12 would allow for much larger time steps. The entries in Table 10.1.1 are the consequence of selecting the model with luminosity closest to the solar luminosity.

A prominent glitch is evident along each of the solid tracks and along two of the dotted tracks at $-0.08 < \log(L/L_\odot) < -0.05$ and $3.763 > \log(T_e) > 3.758$. The glitches are associated with one or more of three discontinuities in the evolutionary program which has been used to generate the tracks. Static surfaces are constructed with $\log(L/L_\odot)$ and $\log(T_e)$ specified at points located at four corners of a rectangle, an example of which is the rectangle labeled "Top Grid" in the lower left hand corner of Fig. 10.1.1. Interpolation within each grid is linear in $\log(L/L_\odot)$ and $\log(T_e)$, but discontinuities in numerical derivatives occur when, in the course of iterations for a model, the required values of L and/or T_e pass between adjacent grid rectangles. If these discontinuities were the only source of the observed glitches, the locations of the glitches should change when the size and normalization of the grids are randomly altered. In the case of the $Y = 0.24$ models, neither decreasing the grid area by a factor of four nor arbitrarily changing the grid normalization, at fixed grid size, changes the locations of the glitches.

The second source of numerical noise is a discontinuity in the equation of state that migrates from one zone to another as evolution progresses. In the evolutionary program that leads to the tracks in Fig. 10.1.1, a model is divided into two parts according to the value assigned to a temperature $(T_6)_{ion}$. During all iterations for a model, if T_6 in a mass zone satisfies $T_6 \leq (T_6)_{ion}$ at the start of iterations, the equation of state for matter is that of a partially ionized perfect gas, with the degree of ionization of hydrogen and helium being determined by relevant Saha equations. If initially $T_6 \geq (T_6)_{ion}$, it is assumed for all iterations that the gas is completely ionized and an equation of state appropriate for completely ionized matter is employed. Since abundances of elements heavier than helium are different in surface and interior solutions, the equation of state on both sides of the temperature divider are different in surface and interior solutions. Experiments show that the locations of the observed discontinuities in the evolutionary tracks can sometimes be changed by changing $(T_6)_{ion}$. The fact that glitch locations nevertheless occur near surface grid boundaries suggests that both of the first two sources of numerical noise contribute to the appearance of some glitches in evolutionary tracks.

A third source of numerical noise is rezoning. Since the gradients of all structure and composition variables change as evolution progresses, by adhering rigorously to the demands of an algorithm for rezoning as described in Section 8.12, every now and again some shells are merged and others halved. Each change leads to a glitch in the rate of gravitational energy production in a finite region centered on the location of the merger/splitting process. The discontinuous changes in temperature and density associated with rezoning lead to discontinuous changes in the rate of nuclear energy production which can also in principle contribute to the occurrence of glitches in observable properties.

Glitches can be dealt with by smoothing computational results to hide the glitches or by going to the trouble of preparing an equation of state which makes a smooth transition between completely ionized and partially ionized matter, making static surfaces with as many elements as are in the interior, and making bicubic interpolations in surface solutions. A third alternative, adopted here, is to accept the occurrence of glitches as a fact of life that can contribute to an understanding of the relationship between the input physics used in constructing models and observable properties of models. In the case of glitches for which rezoning is responsible, one course of action is to analyze the results of the first calculation of an evolutionary phase during which rezoning has been employed, to rezone the initial model in a way which takes into account the zoning requirements encountered, and to then repeat the calculation with no rezoning. In the early days of stellar evolution calculations, such a course of action was not practical because of the limited memory and slow speed of available computers.

Linear interpolation between rows 2 and 3 in Table 10.1.1 suggests that the choice $Y_\odot(Z = 0.01) \sim 0.2373$ should produce a model of solar luminosity at $t \sim 4.6 \times 10^9$ yr after the deuterium-burning phase. It is evident from Table 10.1.1 (and from the termination of the solid tracks in Fig. 10.1.1) that, the larger Y is, the smaller is the radius (and the larger is the surface temperature) of the model of solar luminosity. By decreasing rmixs from 0.6, which characterizes the models which produce the tracks in Fig. 10.1.1, to rmixs = 0.55, and repeating the last 1.5×10^9 yr of evolution of the model with $Y = 0.24$, the final

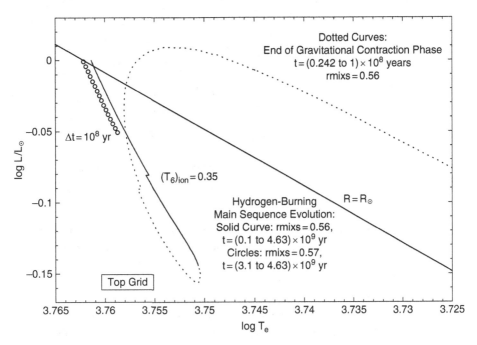

Evolution to the Sun's position in the HR diagram; $M = 1M_\odot$, $Z = 0.01$, $Y = 0.2373$, and rmixs $= 0.56$ and 0.57

radius of the model is increased from $\sim 0.986\ R_\odot$ to $\sim 1.000\ R_\odot$, while the final luminosity remains at $\sim 1.000\ L_\odot$.

These results indicate that the choices $Y_\odot(Z = 0.01) \sim 0.2373$ and rmixs ~ 0.56 should produce a model which has both the Sun's luminosity and the Sun's radius after approximately 4.6×10^9 yr of evolution. With the objective of reducing the effect of the discontinuity in the equation of state in the evolution program, $(T_6)_{ion}$ has been increased from 0.30 to 0.35 in order to decrease the degree of ionization of hydrogen and helium at the discontinuity.

A calculation with these three changes produces the dotted and solid curves in Fig. 10.1.2 and the entries in the first two rows of Table 10.1.2. For $t \sim 4.60 \times 10^9$ yr, the luminosity of the model is just shy of the Sun's luminosity (by $\sim 0.24\%$) and its radius is a touch larger than the Sun's (by $\sim 0.3\%$). A further 3×10^7 yr of evolution produces a model with the Sun's luminosity but with a radius larger than the Sun's by another 0.1%. The attempt to eliminate the glitches along the evolutionary tracks succeeds only in decreasing the luminosity at which the glitches appear, from $\log(L) \sim -0.06 \pm 0.05$ for the $Y = 0.24$ model described in Fig. 10.1.1 to $\log(L) \sim -0.085 \pm 0.05$ for the $Y = 0.2373$ model described in Fig. 10.1.2.

Apart from the single glitch, the solid track in Fig. 10.1.2 is extremely smooth. In contrast, all three solid tracks in Fig. 10.1.1 exhibit a staircase-like structure. The difference is entirely due to roundoff error. In the case of the staircase-like tracks, the graphics routine has been provided with values of $\log(L)$ and $\log(T_e)$ that have four significant figures to

Table 10.1.2 Properties of models of near solar luminosity, radius, and age when $Y \sim Y_\odot (Z = 0.01) \sim 0.2373$

t_9	$L/L_\odot$	rmixs	$R/R_\odot$
4.60	0.997650	0.56	1.002799
4.63	1.000058	0.56	1.003737
4.60	0.998076	0.57	0.999118
4.63	1.000508	0.57	1.000066

the right of the decimal place (i.e., $\pm 0.xxxx$ and $3.yyyy$); in the case of the smooth track, values of $\log(L)$ and $\log(T_e)$ provided have five significant figures to the right of the decimal place (i.e., $\pm 0.xxxxx$ and $3.yyyyy$). The difference in appearance makes statements both about the resolution of which the human eye is capable and about the accuracy of the numerical calculation.

Repeating the last 1.5×10^9 yr of the evolution of the $Y = 0.2373$ model with rmixs = 0.57 produces the set of circles to the left of the solid track in Fig. 10.1.2 and the entries in the last two rows of Table 10.1.2. For $t \sim 4.60 \times 10^9$ yr, the luminosity is still short of the Sun's luminosity (by $\sim 0.2\%$) and the radius is still slightly larger than the Sun's (by $\sim 0.1\%$). A further 3×10^7 yr of evolution essentially reproduces the Sun's estimated luminosity and radius.

Obviously, one could iterate further to match the solar radius and luminosity at exactly 4.6 billion years. On the other hand, the Sun's age, secular mean bolometric luminosity, and secular mean photospheric radius are not known to better than several tenths of a percent, if that well. More importantly, no element of the input physics employed in constructing models is known precisely at even the one percent level. It is therefore not very rewarding to pursue a closer match between the luminosity, radius, and age of imprecise models of the Sun and estimates of these same quantities for the real Sun.

The model with characteristics given in the last row of Table 10.1.2, hereinafter called model A, is adopted for comparison with a corresponding model for $Z = 0.02$ (model B) described in Section 10.2. Structure variables for model A are shown in Figs. 10.1.3 and 10.1.4 as functions, respectively, of mass and radius. Variables in both figures are plotted in such a way that the maximum value (given in parentheses in the top right hand quadrant of the figures) is unity. Although the two figures contain the same information, the different perspectives which they provide contribute complementary insights. Regarding relationships between most variables, the same understanding can be achieved by inspection of either figure.

For example, from both figures it is readily apparent that approximately half of the mass of the model is contained within a sphere of radius $\sim 0.26\ R_\odot$ and that approximately 90% of the mass is contained within a sphere of radius $\sim 0.5\ R_\odot$. Similarly, a quick inspection of either figure reveals that 90% of the luminosity of the model is produced in the inner $\sim 0.29\ M_\odot$ of the model and that this 90% is produced in a sphere of radius $\sim 0.18\ R_\odot$.

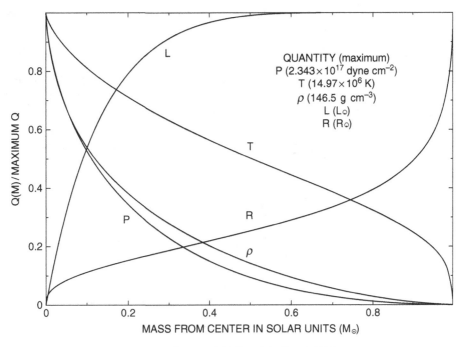

Fig. 10.1.3   Structure of solar model A ($Z = 0.01$, $Y = 0.2373$)

Not so trivially identical in both figures is information regarding the location of and properties of matter at the base of the convective envelope. In Fig. 10.1.4, a clear change in the slope of the temperature versus radius profile is evident at a distance ~ 0.76 $R_{\odot}$ from the center where the temperature and density are, respectively, $\sim 1.71 \times 10^6$ K and $\rho \sim 0.0939$ g cm^{-3}. This change identifies the location of the base of the convective envelope which, from the mass versus radius profile, is seen to have a mass of ~ 0.012 $M_{\odot}$. The temperature versus radius profile terminates before the surface at 1 $R_{\odot}$ is reached because the display of structure does not extend into the static envelope. Although, as Fig. 10.1.4 demonstrates, the static envelope extends from ~ 0.943 $R_{\odot}$ to 1.0 $R_{\odot}$, it contains very little mass (actually, only ~ 0.0003 $M_{\odot}$).

At the base of the static envelope, where the temperature and density are $T_6 \sim 0.30$ K and $\rho \sim 7.87 \times 10^{-3}$ g cm^{-3}, respectively, hydrogen and helium are not completely ionized ($\sim 1.5\%$ of hydrogen nuclei retain an electron and $\sim 7\%$ of helium nuclei retain an electron). The fact that hydrogen and helium are still not completely ionized at $T_6 = (T_6)_{\text{ion}} = 0.35$, where $\rho \sim 1.82 \times 10^{-2}$ g cm^{-3}, contributes to the occurrence of the glitches in the evolutionary tracks in the HR diagram.

Abundances by number of light isotopes involved in the pp-chain reactions are plotted in Fig. 10.1.5 as functions of the mass coordinate. The scale for each isotope extends from zero to unity, with the actual maximum in abundance for each isotope being given in parentheses next to the symbol for the isotope. The neutron abundance is shown primarily for amusement, as neutron-production and -capture reactions are quite rare indeed. The source of neutrons is the alpha-capture reaction ^{13}C$(\alpha, n)^{16}$O and neutron capture on

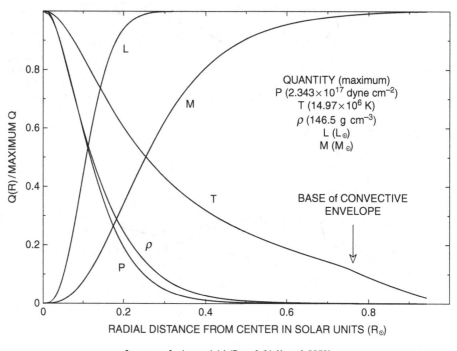

Fig. 10.1.4

Structure of solar model A ($Z = 0.01, Y = 0.2373$)

protons and light elements with multiple isotopes in the set shown in Table 9.2.1 leads to the tiny equilibrium abundances reported in the figure.

All of the deuterium in the precursor Hayashi-track model (initial abundance $Y(^2H) = 1.48 \times 10^{-5}$) has been converted into ^{3}He (initial abundance $Y(^3He) = 1.26 \times 10^{-5}$), and this is reflected in the abundance of ^{3}He near the surface of the solar model where $Y(^3He) = 2.7 \times 10^{-5}$. In the current solar model, the abundance of deuterium is determined by a balance between the rate of the pp reaction and the rate of the $d(p, \gamma)^3$He reaction; this equilibrium abundance is between twelve and thirteen orders of magnitude smaller than the abundance in the initial model.

Most of the ^{3}He in the middle part of the model is a consequence of the burning of protons through the $p(p, e^+\nu_e)d(p, \gamma)^3$He reactions, with the maximum abundance, $Y(^3He) \sim 1.29 \times 10^{-3}$ at $M \sim 0.55\ M_\odot$, being 47 times larger than the sum of the initial abundances of deuterium and ^{3}He. The fact that essentially all of the ^{3}He in this peak is eventually ejected in a planetary nebula event demonstrates that low mass stars are major sources of ^{3}He in the Universe. Outwards (to the right) of the peak, ^{3}He is still being created by the $p(p, e^+\nu_e)d(p, \gamma)^3$He reactions. Inwards of the peak, the abundance of ^{3}He is determined by a balance between the rate of the $d(p, \gamma)^3$He reaction and the rate of the ^{3}He(^{3}He, $2p)^4$He reaction.

The isotope ^{6}Li (initial abundance $\sim 8.1 \times 10^{-11}$) has been been basically destroyed over the entire model (final maximum abundance $\sim 6.56 \times 10^{-15}$) and ^{7}Li (initial abundance $\sim 10^{-9}$) has been effectively destroyed over most of the radiative interior of the model (the maximum abundance in the deep interior is at the center where $Y(^7Li) = 2.57 \times 10^{-16}$).

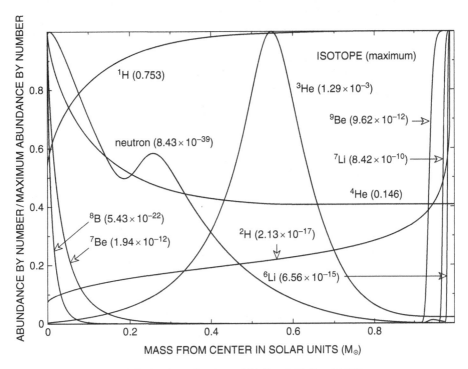

Fig. 10.1.5

pp-chain abundances in solar model A ($Z = 0.01, Y = 0.2373$)

However, in the convective envelope, the abundance of ^{7}Li is only 16% smaller than in the original model. Just below the base of the convective envelope, there is a finite region of mass $\Delta M \sim 0.02$ $M_\odot$ over which the abundance of ^{7}Li declines approximately linearly with mass to a nominal value. Given the fact that the observed lithium to hydrogen ratio in the Sun's photosphere is two orders of magnitude smaller than in the Solar System distribution (compare the entries in columns 4 and 5 with those in columns 6 and 7 for Li in Table 2.1.1), it is likely that, in the real analogue of the model, chemical diffusion operates efficiently over this region during the pre-main sequence gravitational contraction phase to siphon ^{7}Li out of the convective envelope into hotter regions where it can be destroyed. Diffusion is discussed at length in Chapter 12 of Volume 2 of this book.

The abundances of ^{7}Be and ^{8}B in the deep interior of the model are equilibrium abundances representing an exact balance between creation and destruction mechanisms, with, e.g., ^{7}Be being made by the ^{3}He(α, γ)^{7}Be reaction and being destroyed primarily by capture of free electrons.

Abundances of isotopes involved in CN-cycle reactions are shown as functions of the mass coordinate in Fig. 10.1.6. Once again, the scale for each isotope extends from zero to unity, with the actual maximum in abundance for each isotope being given in parentheses next to the symbol for the isotope. From the perspective of element building in the Universe, the most interesting features of the abundance profiles are that, in the interior of a solar-like star, ^{12}C and ^{15}N are being destroyed while ^{13}C and ^{14}N are being created.

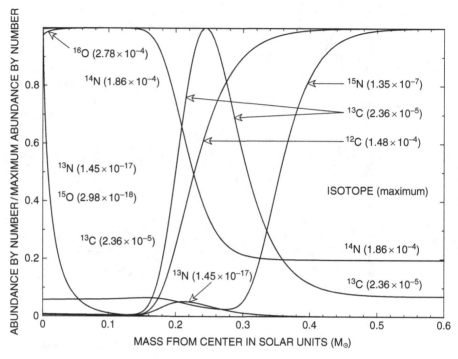

Fig. 10.1.6   CN–cycle abundances in solar model A ($Z = 0.01, Y = 0.2373$)

The abundance at the peak of the ^{13}C abundance profile is ∼14 times larger than the initial number abundance of ∼1.65×10^{-6}. Inwards (to the left) of the peak, ^{13}C and ^{12}C are at their equilibrium abundances. To the right of the peak, ^{13}C is still being produced by the ^{12}C$(p, \gamma)^{13}$N$(e^{+}\nu_{e})^{13}$C reactions more rapidly than it is being destroyed by the ^{13}C$(p, \gamma)^{14}$N reaction.

The fact that the peak in the neutron abundance extending from $M \sim 0.19$ $M_{\odot}$ to ∼0.32 $M_{\odot}$ in Fig. 10.1.5 coincides with the peak in the ^{13}C abundance curve in Fig. 10.1.6 solidifies the identification of the ^{13}C$(\alpha, n)^{15}$O reaction as the primary neutron source. Since protons are far more abundant than CNO elements and since the cross section for neutron capture by protons is of the order of 100 times larger than the cross section for neutron capture by ^{12}C and several thousand times larger than the cross sections for neutron capture by ^{14}N and ^{16}O, the neutron abundance is determined by the balance between the ^{13}C$(\alpha, n)^{16}$O and $n(p, \gamma)d$ reaction rates.

As described in Chapter 11, when a model leaves the main sequence and evolves to the red giant branch, the base of the convective envelope extends inward into the region where ^{13}C has been created, leading to a decrease in the surface ratio of ^{12}C to ^{13}C. From the information in Fig. 10.1.6, it is possible to anticipate the minimum degree of enhancement in the carbon-isotope ratio at the red giant tip of a solar mass star. By inspection, the number of fresh ^{13}C nuclei in solar model A is ∼$2.4 \times 10^{-6}N_{\odot}$, where $N_{\odot}$ is the number of nucleons in the Sun, whereas the number of original ^{13}C nuclei remaining is ∼$1.4 \times 10^{-6}N_{\odot}$ and the number of ^{12}C nuclei remaining is ∼$110 \times 10^{-6}N_{\odot}$. Thus, if the base of the convective envelope in the daughter red giant encompasses of the order of 80% of the stellar mass, the carbon isotopic ratio at the surface becomes

$$\left(\frac{^{12}C}{^{13}C}\right)_{\text{red giant}} \sim \frac{110}{2.4 + 1.4} \sim 29, \tag{10.1.2}$$

which is to be compared with the adopted initial ratio of

$$\left(\frac{^{12}C}{^{13}C}\right)_{\text{interstellar medium}} = 90. \tag{10.1.3}$$

Given the fact that, during the rest of the main sequence phase, the conversion of ^{12}C into ^{13}C continues, it is evident that the estimate given by eq. (10.1.2) is an upper limit and, since most of the matter in the envelope of a red giant is ultimately returned to the interstellar medium in a planetary nebula ejection event, it is clear that the burning of ^{12}C in main sequence stars is a major source of ^{13}C in the Universe.

Over the inner $\sim 25\%$ of the mass of the solar model, ^{12}C has been converted into ^{14}N, creating an abundance of ^{14}N that exceeds the initial abundance by a factor of 5. As evolution continues during the main sequence phase, the transition region between ^{12}C-rich and ^{14}N-rich matter moves outward in mass. Therefore, for the same reason that they develop an enhancement of the surface ratio of ^{13}C to ^{12}C, the red giant progeny of main sequence stars develop an enhanced $^{14}N/^{12}C$ ratio (Iben, 1964). The conversion of ^{12}C into ^{14}N in stars on the main sequence is the primary mechanism for the production of nitrogen in the Universe.

As is demonstrated by the slight dip in the ^{16}O number abundance in the upper left hand corner of Fig. 10.1.6, temperatures in model A have not become large enough for all but a tiny fraction of ^{16}O near the very center to have have experienced the $^{16}O(p, \alpha)^{12}C$ reaction to become catalysts in the CN cycle.

Energy-generation rates for four different processes are shown in Fig. 10.1.7 as functions of mass. For each process, the maximum local rate and the integrated rate (luminosity) are shown in the upper right hand quadrant of the figure, where the processes are ordered according to their importance in the energy budget. It is interesting that the rate at which energy leaves the star as neutrino energy is greater by over a factor of 5 than the integrated rate at which CN-cycle reactions deposit energy locally. About half of the neutrino energy loss comes from the pp reaction and about half comes from the electron capture reaction on ^7Be.

The location of the secondary peak in the CN-cycle energy-generation rate centered at $M \sim 0.21\ M_\odot$ in Fig. 10.1.7 coincides roughly with the location of the peak in the number abundance curve for ^{13}N in Fig. 10.1.6. The facts that, at a given temperature and density, the cross sections for proton capture on ^{12}C and ^{13}C are, respectively, approximately 100 times and 400 times larger than the cross section for proton capture on ^{14}N, coupled with the fact that the abundances of the carbon isotopes are much larger than if they were in local equilibrium, are responsible for the peak.

As is evident in Fig. 10.1.7, gravothermal energy production is positive in the inner $0.3\ M_\odot$ of the model. As demonstrated by the hydrogen-abundance profile in Fig. 10.1.5,

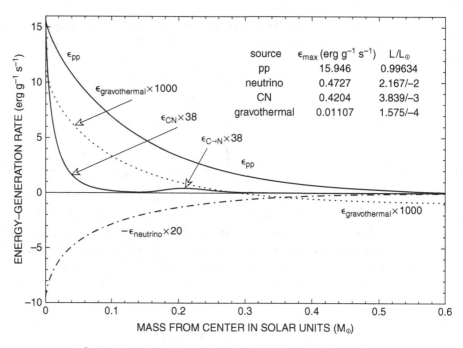

Fig. 10.1.7   Energy-generation rates versus mass in model A ($Z = 0.01$, $Y = 0.2373$)

nuclear reactions are active in this region in reducing the number of particles per unit mass and the dominant contribution to $\epsilon_{\text{gravothermal}}$ is due to the necessity for compression to help maintain pressure balance. The tendency for the temperature to increase in response to a reduction in the number of particles assists in maintaining pressure balance. In regions where nuclear transformations are not important, the gravothermal energy-generation rate is negative for non-trivial reasons requiring detailed examination to understand properly.

Energy-generation rates in the radius-coordinate representation are obtained by multiplying energy-generation rates per unit mass by the derivative

$$\frac{dM(r)}{dr} = 4\pi r^2 \, \rho(r), \tag{10.1.4}$$

which is shown for model A in Fig. 10.1.8, along with $M(r)$. The peak in the derivative occurs at $r \sim 0.21 \, R_\odot$, and half of the mass is contained within a sphere of radius $\sim 0.26 \, R_\odot$. The mass distribution described by the derivative looks very much like the mass distribution shown in Fig. 5.2.5 for the index $N = 3$ polytrope.

In the expression

$$\frac{dL_i(r)}{dr} = \frac{dL_i(r)}{dM(r)} \frac{dM(r)}{dr} = \epsilon_i(r) \frac{dM(r)}{dr}, \tag{10.1.5}$$

the quantity

$$\epsilon_i(r) = \frac{dL_i(r)}{dM(r)}, \tag{10.1.6}$$

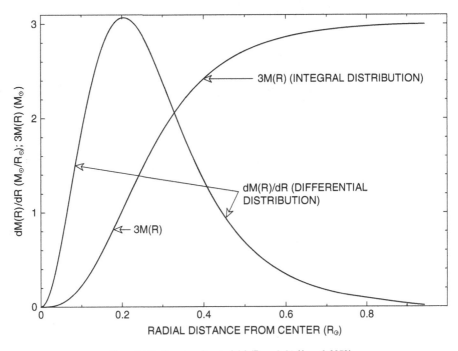

Fig. 10.1.8   Mass distributions in solar model A ($Z = 0.01$, $Y = 0.2373$)

is the energy-generation rate per unit mass of the source of type i and $dL_i(r)$ is the contribution to the luminosity by this energy source in matter in a spherical shell of radius r and mass thickness $dM(r)$. In Fig. 10.1.9 are shown the differential contributions to the radiative luminosity by pp-chain and CN-cycle reactions, as well as the differential contribution to the neutrino luminosity by all neutrino-emission processes. The peak in the contribution of pp-chain reactions to the surface luminosity occurs at $r \sim 0.095$ $R_\odot$. The peak in the contribution to the neutrino luminosity is displaced to a slightly smaller radius $r \sim 0.085$ $R_\odot$, due to the fact that the rate of production of ^{8}B in spherical shells of unit thickness peaks at a radius smaller than does the rate of production of ^{7}Be in these shells. The primary peak in the contribution of CN-cycle reactions occurs at $r \sim 0.047$. The secondary peak, which occurs at $r \sim 0.155$ $R_\odot$, is labeled $dL_{C \to N}/dR$: to the right of the peak, ^{13}C has not reached its equilibrium abundance with respect to ^{12}C and, to the left of the peak, ^{14}N has not reached its equilibrium abundance with respect to the carbon isotopes.

Understanding the nature of gravothermal energy generation begins with an examination of the rates at which structure variables change with time. In Fig. 10.1.10, structure variables in model A are compared with structure variables in the immediately preceding model, which is younger than model A by $\delta t = 10^7$ yr. The logarithmic increments $\delta Q/Q$ in structure variables are plotted in such a way that the maximum in the absolute magnitude of each increment is unity (the actual maximum in absolute magnitude is given in parentheses next to each increment).

As may be inferred from the curves for $\delta R/R$ and $\delta \rho/\rho$, matter moving inward in space can be either contracting or expanding, whereas all matter moving outward in space is

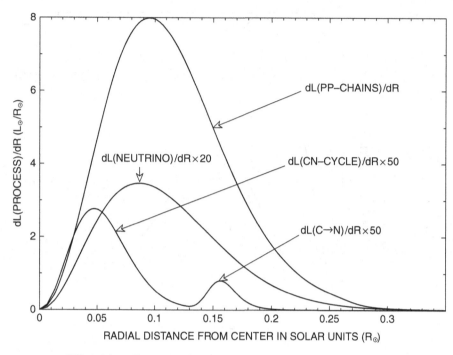

Fig. 10.1.9   Differential contributions to luminosity in solar model A (Z = 0.01, Y = 0.2373)

expanding. Although approximately 70% of the matter in the model is moving inward, only the inner 45% of the mass of the model is contracting. On the other hand, although the outermost 30% of the mass is moving outward, approximately 55% of the mass of the model is expanding. On comparing the luminosity profile in Fig. 10.1.3 with $\delta\rho/\rho$ in Fig. 10.1.10, it is evident that contraction is confined to the region within which nuclear transformations are causing an increase in molecular weight. In the outer region which is both expanding and moving outward, the mean temperature is increasing; absorption of energy from the ambient flow of energy from the interior is used to supply the increase in thermal energy and to do the work required to move matter outward against gravity.

Energy-generation rates in the mass-coordinate representation and in the radius-coordinate representation are shown in Figs. 10.1.11 and 10.1.12, respectively. The scalar values, v and g, of the two components of the work done by gravity are shown in Fig. 10.1.12 as functions of radius. In Fig. 10.1.11, the gravothermal energy-generation rate ϵ_{grav} given by the dash-dot curve is negative over the outer $\sim$70% of the mass of the model. The dotted curves give the total rate dU/dt at which the internal energy is increasing and the contribution $-\epsilon_{cdth} = (dU/dt)_{\rho,T}$ which abundance changes make to dU/dt. The $-\epsilon_{cdth}$ profile reflects the fact that, the nearer the center, the larger are the rates of nuclear transformations which decrease the number of particles per unit mass. As may be deduced from the hydrogen number-abundance profile in Fig. 10.1.5, these transformations become negligible for $M \gtrsim 0.6\ M_{\odot}$.

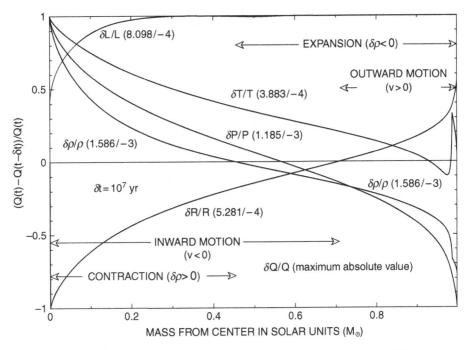

Fig. 10.1.10   Logarithmic increments in structure variables in solar model A ($Z = 0.01$, $Y = 0.2373$)

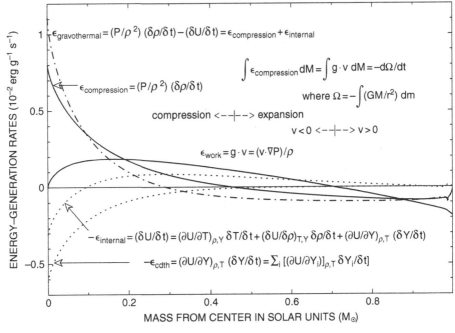

Fig. 10.1.11   Gravothermal energy generation rates in solar model A ($Z = 0.01$, $Y = 0.2373$)

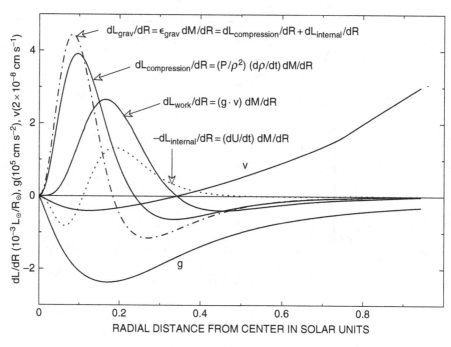

Fig. 10.1.12   Differential contributions to gravothermal luminosities in solar model A ($Z = 0.01$, $Y = 0.2373$)

The two solid curves in Fig. 10.1.11 labeled $\epsilon_{compression}$ and ϵ_{work} describe, respectively, the rate at which work is done locally by compression, $(P/\rho^2)\,(d\rho/dt)$, and the local contribution to the rate at which work is done by gravity, $\mathbf{g} \cdot \mathbf{v} = (1/\rho)\,(\mathbf{v} \cdot \nabla P)$. Integration under the two curves shows that

$$\int \epsilon_{compression}\,dM = \int \frac{P}{\rho^2} \frac{d\rho}{dt}\,dM$$
$$= \frac{dE_{work}}{dt} = \int (\mathbf{g} \cdot \mathbf{v})\,dM = 3.171 \times 10^{-4} L_\odot. \qquad (10.1.7)$$

The fact that the two curves have completely different shapes and yet, when integrated over mass, produce the same result demonstrates once again that (1) gravitational potential energy is not a local property and that (2) the release of gravitational potential energy is communicated by pressure-gradient forces and is manifested locally as compressional work.

The rate at which work is done by gravity and the rate at which compressional work is done are more transparently related to one another in the radial coordinate representation of Fig. 10.1.12 than in the mass coordinate representation of Fig. 10.1.11. As in Fig. 10.1.11, the thick solid curves in Fig. 10.1.12 describe the rate at which work is done by the action of gravity,

$$\frac{dL_{work}}{dr} = (\mathbf{g} \cdot \mathbf{v}) \frac{dM}{dr}, \qquad (10.1.8)$$

and the rate at which, concomitantly, pressure-gradient forces convert this action into compressional work,

$$\frac{dL_{compression}}{dr} = \frac{P}{\rho^2}\frac{d\rho}{dt}\frac{dM}{dr}. \tag{10.1.9}$$

The dotted curve gives

$$-\frac{dL_{internal}}{dr} = \frac{dU}{dt}\frac{dM}{dr}, \tag{10.1.10}$$

the rate at which the internal energy increases, and the dash-dot curve gives

$$\frac{dL_{gravothermal}}{dr} = \frac{dL_{compression}}{dr} + \frac{dL_{internal}}{dr}, \tag{10.1.11}$$

the rate at which gravothermal energy contributes to the ambient flow of energy in various ways. The peak in the dL_{work}/dr distribution occurs at $r \sim 0.16\ R_\odot$, whereas the peak in the $dL_{compression}/dr$ distribution occurs at $r \sim 0.09\ R_\odot$, or at about half the distance from the center as the peak in the gravitational work distribution.

Extrapolation to the model surface of the velocity curve in Fig. 10.1.12 shows that the surface of the model is expanding at the rate

$$\frac{dR_{surface}}{dt} \sim 6.8 \times 10^{-8}\ \text{cm s}^{-1} = 2.15\ \text{cm yr}^{-1} \sim 0.85\ \text{inch per year}$$

$$\sim \frac{5.7\ \text{feet}}{80\ \text{years}} \sim \frac{\text{average human height}}{\text{average human lifetime}}. \tag{10.1.12}$$

Over its final 3 million years of evolution, the model's radius has increased by about one hundredth of the Earth's radius, or by about 40 miles.

In regions where $\epsilon_{nuc} \gg \epsilon_{grav}$, the details of ϵ_{grav} are of essentially no consequence for the structure of the model; the values of state variables are controlled by the pressure balance equation and the balance between the rate of nuclear energy generation and the outward radiative flux, almost completely independent of ϵ_{grav}, let alone of the details of the variation in ϵ_{grav} with position in the star. It is nevertheless of extraordinary interest to explore these details, as it is in these details that lies the key to understanding that, despite the dominance of the nuclear energy source, gravity does the indispensable work necessary to compress and heat matter in the deep interior. The change in composition brought about by nuclear transformations is, of course, the trigger for compression; in order that pressure balance be maintained, the decrease in the number abundances of pressure-producing particles must be offset by an increase in density and/or in temperature. But it is the work done by gravity that supplies the energy necessary for this increase. Comparison of $\delta\rho/\rho$ in Fig. 10.1.10 with ϵ_{cdth} in Fig. 10.1.11 shows that compression is confined to the region in which the conversion of eight particles (four protons and four electrons) into three particles (one alpha particle and two electrons) is occurring.

It is instructive to examine the gravothermal energy-generation rate in the approximation that the only particles present are protons, alpha particles, and electrons and that all particles obey the perfect gas law. Designating the number abundance parameters of these particles as Y_1 (protons), Y_4 (alpha particles), and Y_e (electrons), mass conservation gives

$Y_1 + 4Y_4 \sim 1$ and number conservation gives $Y_e = Y_1 + 2Y_4$. The sum of number abundance parameters is therefore

$$Y_e + Y_1 + Y_4 = \frac{5Y_1 + 3}{4}. \tag{10.1.13}$$

The pressure is

$$P = N_A \, kT \, \frac{5Y_1 + 3}{4} \rho \tag{10.1.14}$$

and the energy density per unit mass is

$$U = N_A \, \frac{3}{2} \, kT \, \frac{5Y_1 + 3}{4}. \tag{10.1.15}$$

From these relationships,

$$\delta U = U \left[\frac{\delta T}{T} + \frac{5\delta Y_1}{5Y_1 + 3} \right] = N_A \, kT \left[\frac{5Y_1 + 3}{4} \frac{3}{2} \frac{\delta T}{T} + \frac{15}{8} \delta Y_1 \right], \tag{10.1.16}$$

$$\frac{P}{\rho^2} \delta \rho = N_A \, kT \, \frac{5Y_1 + 3}{4} \frac{\delta \rho}{\rho}, \tag{10.1.17}$$

and

$$\epsilon_{\mathrm{grav}} = \frac{P}{\rho^2} \frac{\delta \rho}{\delta t} - \frac{\delta U}{\delta t} \tag{10.1.18}$$

$$= \frac{N_A \, kT}{\delta t} \left[\frac{5Y_1 + 3}{4} \left(\frac{\delta \rho}{\rho} - \frac{3}{2} \frac{\delta T}{T} \right) - \frac{15}{8} \delta Y_1 \right]. \tag{10.1.19}$$

The magnitude of the time step which produces the increments in Fig. 10.1.11 is $\delta t = 3.156 \times 10^{14}$ s, so, for solar model A,

$$\epsilon_{\mathrm{grav}} = 0.2635 \ \mathrm{erg \ g^{-1} \ s^{-1}} \ T_6 \left[\frac{5Y_1 + 3}{4} \left(\frac{\delta \rho}{\rho} - \frac{3}{2} \frac{\delta T}{T} \right) - \frac{15}{8} \delta Y_1 \right]. \tag{10.1.20}$$

At the very center, $T_6 = 14.97$, $\delta T/T = 3.883 \times 10^{-4}$, $\delta \rho/\rho = 1.586 \times 10^{-3}$, $Y_1 = 0.323$, and $\delta Y_1 = -1.674 \times 10^{-3}$. Thus, at the center,

$$\epsilon_{\mathrm{grav}} = 3.945 \times 10^{-3} \ \mathrm{erg \ g^{-1} \ s^{-1}} \ [1.259 \ (1.586 - 1.5 \times 0.3883) - 1.875 \times (-0.807)]$$

$$= 3.945 \times 10^{-3} \ \mathrm{erg \ s^{-1}} \ [1.264 + 1.5131] = 1.095 \times 10^{-2} \ \mathrm{erg \ g^{-1} \ s^{-1}}, \tag{10.1.21}$$

which compares with the model result of $\epsilon_{\mathrm{grav}} = 1.107 \times 10^{-2} \ \mathrm{erg \ g^{-1} \ s^{-1}}$.

Comparisons between ϵ_{grav} in Fig. 10.1.11 and ϵ_{grav} given by eq. 10.1.20 are particularly straightforward where $\delta T/T = 0.0$ and $\delta \rho/\rho = 0.0$. At $M \sim 0.452 \ M_\odot$, where $\delta \rho/\rho \sim 0.0$, $T_6 \sim 7.96$, $\delta T/T \sim 1.46 \times 10^{-4}$, $Y_1 \sim 0.744$, and $\delta Y_1 \sim -5 \times 10^{-5}$, one has

$$\epsilon_{\mathrm{grav}} \sim 2.097 \ \mathrm{erg \ s^{-1}} \left[1.680 \left(-1.5 \times 1.46 \times 10^{-4} \right) + 5 \times 10^{-5} \right]$$

$$\sim -0.67 \times 10^{-3} \ \mathrm{erg \ g^{-1} \ s^{-1}}, \tag{10.1.22}$$

compared with the model result of $\epsilon_{grav} \sim -0.68 \times 10^{-3}$ erg g^{-1} s^{-1}. Although the reduction in the local abundance of particles continues to make a positive contribution to ϵ_{grav}, energy absorbed from the ambient flow of energy from interior regions dominates in producing a net negative ϵ_{grav} and makes the major contribution to the increase in the local temperature.

At $M \sim 0.9845 \; M_{\odot}$, where $\delta T/T \sim 0.0$, $T_6 \sim 1.931$, $\delta\rho/\rho \sim -8.86 \times 10^{-4}$, $Y_1 \sim 0.753$, and $\delta Y_1 \sim 0.0$, one has

$$\epsilon_{grav} \sim 0.5088 \text{ erg g}^{-1} \text{ s}^{-1} \left[1.691 \, (-8.86 \times 10^{-4}) \right]$$
$$\sim -0.762 \times 10^{-3} \text{ erg g}^{-1} \text{ s}^{-1}, \tag{10.1.23}$$

compared with the model result of $\epsilon_{grav} = -0.763 \times 10^{-3}$ erg g^{-1} s^{-1}. In this instance, energy absorbed from the ambient flow of energy from the interior does the work required to expand matter locally.

Global gravothermal characteristics are of interest. In model A,

$$\Omega = -\int \frac{GM}{r^2} \, dM = -6.247\,756\,0642 \times 10^{48} \text{ erg} = -1.644 \frac{GM_{\odot}^2}{R_{\odot}} \tag{10.1.24}$$

and, in its immediate precursor, $\Omega = -6.247\,369\,8283 \times 10^{48}$ erg. Thus,

$$-\frac{\delta\Omega}{\delta t} = 1.2238 \times 10^{30} \text{ erg s}^{-1} = 3.171 \times 10^{-4} L_{\odot}, \tag{10.1.25}$$

which agrees exactly with $\int (\mathbf{g} \cdot \mathbf{v}) \, dM$ and with $\int (P/\rho^2) \, (d\rho/dt) \, dM$ as given by eq. (10.1.7).

As an aside, it is interesting that a polytrope of index $N = 3.175$ (see eq. (5.2.30)) produces the same relationship between Ω and $GM_{\odot}^2/R_{\odot}$ as given by eq. (10.1.24). On the other hand, the central concentration of solar model A is

$$\frac{\rho_{central}}{\rho_{mean}} = 103.6, \tag{10.1.26}$$

which lies between the central concentrations (see Table 5.2.3) of 88.15 for an $N = 3.25$ polytrope and 152.88 for an $N = 3.5$ polytrope, indicating an effective polytropic index of $N \sim 3.31$. Given the fact that the solar model is not of homogeneous composition, it is remarkable that two different indicators of effective polytropic index give almost the same result: $N = 3.24 \pm 0.07$.

The total internal energy of model A is

$$E_{internal} = \int U \, dM = 3.117\,174\,2144 \times 10^{48} \text{ erg}, \tag{10.1.27}$$

and that of its immediate precursor is $E_{internal} = 3.116\,979\,6684 \times 10^{48}$ erg, so that, in both model A and its precursor,

$$\frac{E_{internal}}{\Omega} = -0.499 \sim -0.5, \tag{10.1.28}$$

as expected from the virial theorem when radiation pressure is negligible and the equation of state is that of a non-relativistic perfect gas (see Chapter 3). On the other hand,

$$\frac{\delta E_{internal}}{\delta t} = 0.6164 \times 10^{30} \text{ erg s}^{-1} = 1.597 \times 10^{-4} L_\odot = -0.504 \frac{\delta \Omega}{\delta t}. \quad (10.1.29)$$

The 1% difference between the numbers 0.499 and 0.504 in eqs. (10.1.28) and (10.1.29) may be ascribed partially to deviations from a perfect gas equation of state and partially to discontinuities in the evolutionary program.

In model A, the total kinetic energy of bulk motions for which gravitational acceleration is responsible is

$$E_{kin} = \int \frac{1}{2} v^2 \, dM = 8.138 \times 10^{16} \text{ erg} = 2.611 \times 10^{-32} \, E_{internal}. \quad (10.1.30)$$

As discussed in Section 9.2, the appropriate measure of the accuracy of the quasistatic approximation is not the ratio of E_{kin} to $E_{internal}$, but rather the ratio

$$\int \left(\mathbf{v} \cdot \frac{d\mathbf{v}}{dt} \right) \, dM \div \int (\mathbf{v} \cdot \mathbf{g}) \, dM$$

$$= \frac{dE_{kin}}{dt} \div \frac{dE_{work}}{dt} = \frac{dE_{kin}}{dE_{work}} = -\frac{dE_{kin}}{d\Omega}. \quad (10.1.31)$$

The third equality follows from eq. (8.3.26). From solar model A and its precursor, one has

$$\frac{dE_{kin}}{dE_{work}} = -\frac{dE_{kin}}{d\Omega} = \frac{5.868 \times 10^{14} \text{ erg}}{3.862 \times 10^{44} \text{ erg}} \sim 1.52 \times 10^{-30}. \quad (10.1.32)$$

Comparing with eq. (9.2.40), it appears that, for the Sun, the quasistatic approximation during the core hydrogen-burning phase is a better approximation than during the deuterium burning phase by about twelve orders of magnitude.

10.2 Construction and properties of a $Z = 0.02$ solar-like model with the Sun's luminosity, radius, and estimated age, and comparisons with the $Z = 0.01$ solar-like model

For $Z = 0.02$, the initial number abundances relative to hydrogen of all isotopes more massive than ^{4}He are larger by a factor of 2 than given in Table 9.2.1, but the relative number abundances of deuterium and ^{3}He are the same. The convection parameter has been increased to rmixs = 0.7 to take into account that, all other things being equal, an increase in the opacity means an increase in model radius, whereas, for models with convective envelopes, an increase in rmixs leads to a decrease in model radius. The discontinuity in the equation of state has been set at $(T_6)_{ion} = 0.35$.

Following the procedures outlined in Section 10.1, the entries in Table 10.2.1 and the evolutionary tracks in Fig. 10.2.1 result. Interestingly, no glitches appear along the dotted portions of the evolutionary tracks and the glitches along the solid portions of the tracks

Table 10.2.1 Relationships between age and initial helium abundance Y for models of near solar luminosity when $Z = 0.02$ and rmixs $= 0.7$

Y	t_9	$L/L_\odot$	$R/R_\odot$
0.30	4.18	0.999878	0.983521
0.29	5.24	1.000230	0.997909
0.28	6.38	1.000155	1.012933

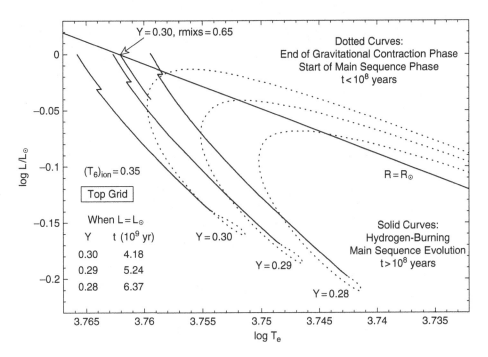

Fig. 10.2.1   Evolution in the HR diagram of 1M$_\odot$ models with $Z = 0.02$, $Y = 0.28, 0.29, 0.30$, and rmixs $= 0.7$

occur at larger luminosities ($\Delta \log(L) \sim 0.04$) than in the $Z = 0.01$ case. A recalculation with rmixs $= 0.65$ of the last 1.5 billion yr of evolution of the model with $Y = 0.30$ produces a track which achieves the solar luminosity and radius and does not display a glitch (see the solid track segment between the $Y = 0.29$ and $Y = 0.28$ solid tracks in Fig. 10.2.1). The absence of a glitch demonstrates that convection in the model envelope plays a role in determining if, when, and how the discontinuities built into the evolution program are manifested.

Interpolation in Table 10.2.1 suggests that the choice $Y \sim 0.296$ will lead to a model of solar luminosity in approximately 4.6 billion yr, and the effect of changing rmixs from 0.7 to 0.65 in the $Y = 0.30$ case suggests that the choice rmixs ~ 0.67 will produce a model of near solar radius. The results of a calculation with these choices are shown in Fig. 10.2.2

Table 10.2.2 Properties of models of near solar luminosity, radius, and age when $Y = Y_\odot(Z = 0.02) \sim 0.296$			
t_9	$L/L_\odot$	rmixs	$R/R_\odot$
4.60	0.998502	0.67	0.998711
4.62	1.000087	0.67	0.999303

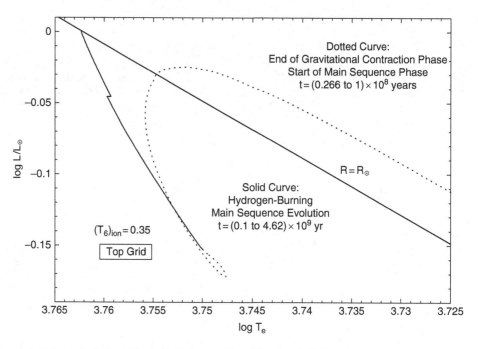

Fig. 10.2.2
Evolution to the Sun's position in the HR diagram; $M = 1M_\odot$, $Z = 0.02$, $Y = 0.296$, and rmixs $= 0.67$

and Table 10.2.2. In fact, the solar luminosity is achieved after 4.62 billion yr of evolution and the solar luminosity could be obtained by a fraction of a percent decrease in rmixs.

For estimating neutrino fluxes, it is more important for the model to have nearly the Sun's luminosity than to have an age precisely equal to 4.6×10^9 yr, so the model of age 4.62×10^9 yr is chosen as model B to compare with model A described in Section 10.1. Structure variables for model B are shown in Figs. 10.2.3 and 10.2.4 as functions, respectively, of mass and radius, and may be compared with structure variables for model A displayed in Figs. 10.1.3 and 10.1.4. From casual inspection, the two sets of structure variables, normalized as they are to maximum variable values, differ almost imperceptibly from one another. Without referring to the maximum values of the various quantities given in the upper right hand quadrants, the only clearcut distinction, which nevertheless requires a second glance to detect, is the location of the base of the convective envelope, being deeper in model B than in model A.

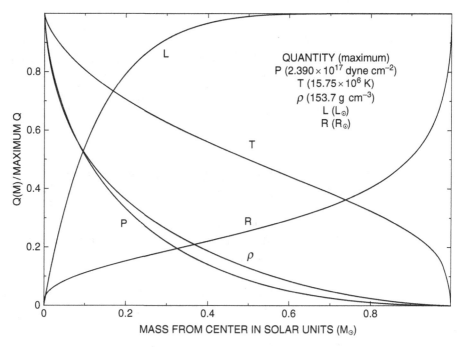

Fig. 10.2.3

Structure of solar model B ($Z = 0.02$, Y = 0.296)

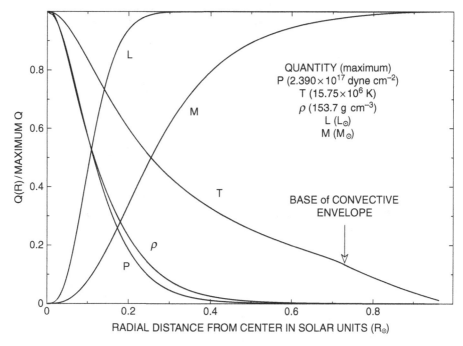

Fig. 10.2.4

Structure of solar model B ($Z = 0.02$, Y = 0.296)

		T_c 10^6 K	ρ_c g cm^{-3}	P_c 10^{17} dyne cm^{-2}	$Y_c - Y$	$\frac{L_{CN}}{L_{pp}}$	M_{BCE} $M_\odot$	R_{BCE} $R_\odot$	T_{BCE} 10^6 K	ρ_{BCE} g cm^{-3}
Z	Y									

Table 10.2.3 Comparison of properties of models A and B

Z	Y	T_c 10^6 K	ρ_c g cm^{-3}	P_c 10^{17} dyne cm^{-2}	$Y_c - Y$	$\frac{L_{CN}}{L_{pp}}$	M_{BCE} $M_\odot$	R_{BCE} $R_\odot$	T_{BCE} 10^6 K	ρ_{BCE} g cm^{-3}
0.01	0.237	14.97	146.5	2.343	0.346	3.85/−3	1.17/−2	0.762	1.71	0.0939
0.02	0.296	15.75	153.7	2.390	0.351	1.56/−2	2.00/−2	0.730	2.13	0.153

Quantitative distinctions are more readily apparent in Table 10.2.3 which compares structural quantities at the center, the ratio of CN cycle to pp-chain contributions to the luminosity, and properties of the convective envelope. The quantity $Y_c - Y$ is the amount by which the helium abundance by mass at the center is larger than the helium abundance by mass at the surface. It is interesting that, while the central temperature and density are each larger by approximately 5% in model B than in model A, the central pressure and the amount of fuel consumed at the center are larger by only ∼2% in model B. At the center of model A, the number abundance parameter for hydrogen is $Y_1 = 0.4070$ and, at the center of model B, $Y_1 = 0.3336$. Using eq. (10.1.12) as an approximation, it follows that the number of free particles at the center is approximately 7.5% smaller in model B than in model A, and this accounts adequately for the fact that the pressure in model B is only 2% larger than in model A while the product of density and temperature is 10% larger.

The ratio L_{CN}/L_{pp} is approximately four times larger in model B than in model A. This difference may be understood as a consequence of the facts that (1) the abundances of beta-stable CNO isotopes are twice as large in model B as in model A and (2) temperatures in model B are larger by 5%, leading to an enhancement in CN-cycle to pp reaction rates by an additional factor of ∼$(1.05)^{17-4} \sim 1.9$.

The fact that the convective envelope in model B is more massive than in model A may be ascribed to the larger opacity in model B, forcing the radiative gradient to be larger than the adiabatic gradient further into the interior. The larger temperatures and densities at the base of the convective envelope in model B account for the fact that more ^{7}Li is burned up in the convective envelope of this model, as is detailed in Fig. 10.2.5, which shows the distribution of abundances of light isotopes, most of which are involved in the pp chains. The number-abundance parameter for ^{7}Li in the convective envelope is five times smaller than the initial value of ∼2×10^{-9}. The factor of five reduction is still far short of the two orders of magnitude reduction anticipated from a comparison of the observed solar abundance of lithium with the abundance of lithium in Solar System material, so diffusion of lithium through the base of the convective envelope at some point prior to the present must still be invoked to understand the observations.

The ^{9}Be profile and the peak in the abundance of ^{6}Li in Fig. 10.2.5, which is a consequence of the reaction ^{9}Be$(p, \alpha)^6$Li, have moved further towards the surface than at the beginning of the main sequence phase, as may be inferred by comparing with the abundances of ^{6}Li and ^{9}Be in Fig. 9.3.15.

Abundances of isotopes involved in CN-cycle reactions are shown as functions of the mass coordinate in Fig. 10.2.6. In the inner 0.2 $M_\odot$, beta-unstable isotopes in local

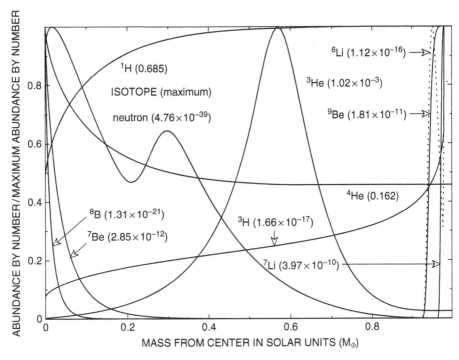

Fig. 10.2.5   pp-chain abundances in solar model B ($Z = 0.02$, $Y = 0.296$)

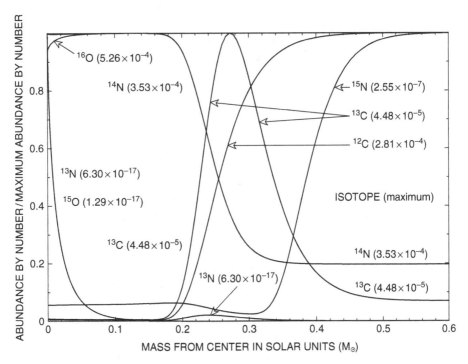

Fig. 10.2.6   CN–cycle abundances in solar model B ($Z = 0.02$, $Y = 0.296$)

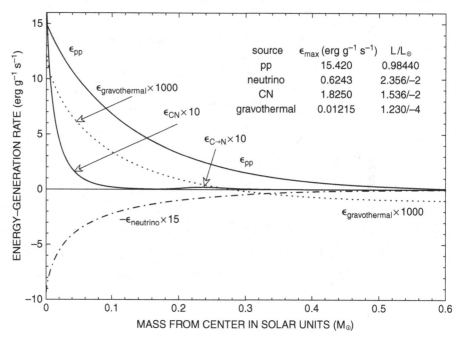

Fig. 10.2.7   Energy generation rates versus mass in model B (Z = 0.02, Y = 0.296)

equilibrium in the CN cycle are approximately four times more abundant than these same isotopes in model A described in Fig. 10.1.6, consistent with the difference in temperatures and the differences in initial abundances of beta-stable CN-cycle isotopes. Due to the larger temperatures in model B, the peak in the ^{13}C abundance distribution and the profiles of ^{14}N, ^{12}C, and ^{15}N are all shifted outwards by about 0.025–0.03 $M_\odot$ relative to these same quantities in Fig. 10.1.6 for model A. Repeating for model B the calculation described by eq. (10.1.2) produces

$$\left(\frac{^{12}C}{^{13}C}\right)_{\text{red giant}} \sim \frac{210}{4.6 + 2.5} \sim 30, \tag{10.2.1}$$

essentially the same result as for model A (see eq. (10.1.2)).

Model B analogues of Figs. 10.1.7–10.1.12 for model A are presented as Figs. 10.2.7–10.2.12. Comparing the luminosities for different processes listed in the upper right hand quadrants of Figs. 10.2.7 and 10.1.7, one sees that the neutrino luminosity in model B is about 10% larger than in model A and the gravothermal contribution to the radiant luminosity is about 20% smaller. In going from model A to model B, the ~ -0.0119 $L_\odot$ decrease in the luminosity due to the pp chains is essentially compensated for by the increase of ~ 0.0115 $L_\odot$ in the luminosity due to CN-cycle reactions. That the CN-cycle luminosity increases from being 5.5 times smaller than the neutrino luminosity in model A to becoming comparable with the neutrino luminosity in model B is a point brought home by comparing Figs. 10.2.9 and 10.1.9.

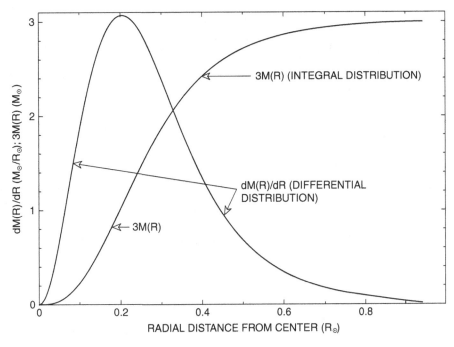

Fig. 10.2.8   Mass distributions in solar model B ($Z = 0.02$, $Y = 0.296$)

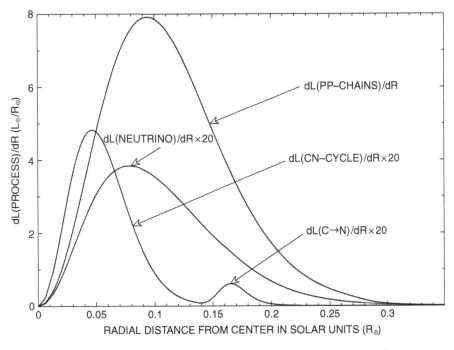

Fig. 10.2.9   Differential contributions to luminosity in solar model B ($Z = 0.02$, $Y = 0.296$)

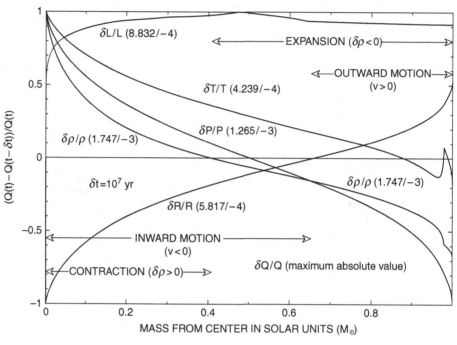

Fig. 10.2.10

Logarithmic increments in structure variables in solar model B ($Z = 0.02$, $Y = 0.296$)

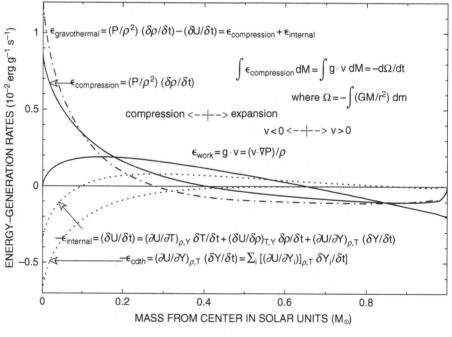

Fig. 10.2.11

Gravothermal energy generation rates in solar model B ($Z = 0.02$, $Y = 0.296$)

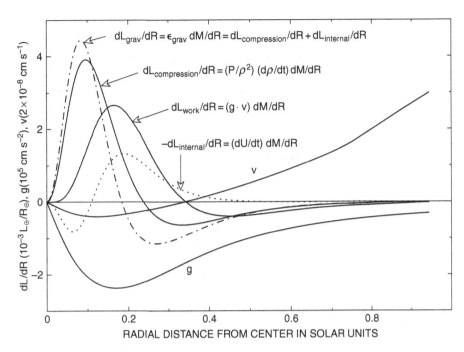

Fig. 10.2.12   Differential contributions to gravothermal luminosities in solar model B ($Z = 0.02, Y = 0.296$)

Comparison of Figs. 10.2.10 and 10.1.10 shows that the absolute magnitudes of the rates of change of all structure variables are roughly 10% larger in model B than in model A. From this difference, one might surmise that the gravothermal luminosity of model B is larger than that of model A. Just the reverse is true: the gravitational luminosity of model B is about 20% smaller than that of model A. Comparison of Figs. 10.2.11 and 10.1.11 shows that, for whatever reason, less of the mass of model B is being compressed than is the case in model A and the rate at which work is done to expand matter outward is larger in model B than in model A, with the consequence that the net rate at which gravothermal energy is converted into escaping energy is smaller in model B. The same conclusion follows from a careful comparison of Figs. 10.2.12 and 10.1.12, even though the differences in gravothermal behavior in the two models are very subtle.

A final check on the accuracy of the calculation for model B is provided by a consideration of global energy characteristics. By direct integration, the gravitational binding energy of model B is

$$\Omega = -\int \frac{GM}{r^2}\, dM = -6.175\,873\,1975 \times 10^{48} \text{ erg} = -1.625 \frac{GM_\odot^2}{R_\odot} \qquad (10.2.2)$$

and that of its immediate precursor, younger by 10^7 yr, is $\Omega = -6.175\,570\,3468 \times 10^{48}$ erg. Thus,

$$-\frac{\delta\Omega}{\delta t} = 0.9596 \times 10^{30} \text{ erg s}^{-1} = 2.486 \times 10^{-4} L_\odot, \qquad (10.2.3)$$

compared with

$$\int (\mathbf{g} \cdot \mathbf{v}) \, dM = \int \left(\frac{P}{\rho^2}\right) \frac{d\rho}{dt} \, dM = 2.487 \times 10^{-4} \, L_\odot, \tag{10.2.4}$$

also given by direct integration. The total internal energy of model B is

$$E_{internal} = \int U \, dM = 3.081\,607\,8816 \times 10^{48} \, \text{erg} \tag{10.2.5}$$

and that of its immediate precursor is $E_{internal} = 3.081\,454\,6648 \times 10^{48}$ erg, so that,

$$\frac{\delta E_{internal}}{\delta t} = 0.4943 \times 10^{30} \, \text{erg s}^{-1} = 1.281 \times 10^{-4} L_\odot = -0.506 \, \frac{\delta \Omega}{\delta t}. \tag{10.2.6}$$

In both model B and its immediate precursor,

$$\frac{E_{internal}}{\Omega} = -0.499. \tag{10.2.7}$$

The difference beteen the factors 0.499 and 0.506 in eqs. (10.2.6) and (10.2.7) is a measure of departures from a perfect gas equation of state and of inaccuracies in the calculation.

Comparing Ωs given by eqs. (10.2.2) and (5.2.30), one infers an effective polytropic index of $N = 3.154$ for model B compared with an index $N = 3.175$ inferred from the same comparison for model A. The central concentration of model B is

$$\frac{\rho_{central}}{\rho_{mean}} = 108.8, \tag{10.2.8}$$

which is 5% larger than the central concentration of model A. The effective polytropic index suggested by this indicator is $N \sim 3.33$, differing by less than 1% from the index $N \sim 3.31$ suggested by the same indicator for model A.

10.3 Contributions to photon and neutrino luminosities and to neutrino fluxes at the Earth

The contributions of various reactions in the pp chains to the surface luminosity are shown in Fig. 10.3.1 for model A and in Fig. 10.3.2 for model B. An important lesson from these figures is the fact, emphasized in earlier sections, that although the $p(p, e^+ \nu_e)d$ reaction controls the whole set of processes in the pp chains, approximately ten times more energy is provided by subsequent reactions. Another lesson is that by far the most frequent way in which ^{3}He nuclei disappear is by reactions with other ^{3}He nuclei rather than by reactions with ^{4}He nuclei, with the consequence that the rate of production of energetic neutrinos by reactions beginning with ^{7}Be is much smaller than the rate of production of the very low energy pp-reaction neutrinos. Even so, the contribution of the ^{7}Li$(p, \alpha)^4$He reaction to surface luminosity is comparable to that of the pp reaction.

Shown in Figs. 10.3.3 and 10.3.4 are the contributions of the ^{7}Be$(e^-, \nu_e)^7$Li reaction, the ^{7}Be$(p, \gamma)^8$B reaction, and the ^{8}B$(e^+ \nu_e)^8$Be$^*(2\alpha)$ reactions. It is ironic that the neutrinos from these reaction channels, which play such insignificant roles in the dynamics of the

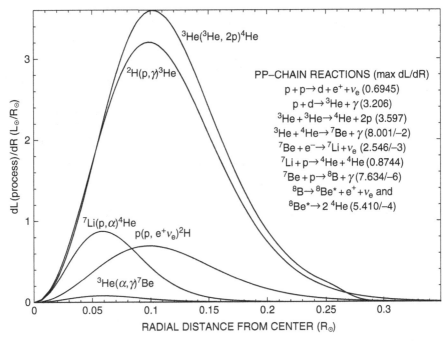

Fig. 10.3.1 pp-chain contributions to luminosity in solar model A ($Z = 0.01$, $Y = 0.2373$)

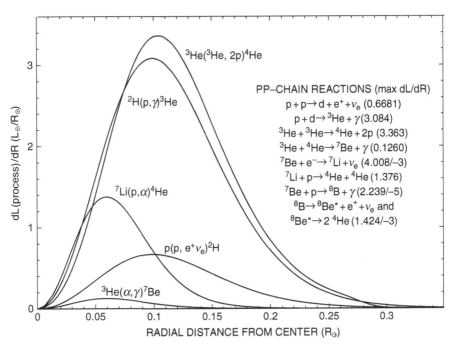

Fig. 10.3.2 pp–chain contributions to luminosity in solar model B ($Z = 0.02$, $Y = 0.296$)

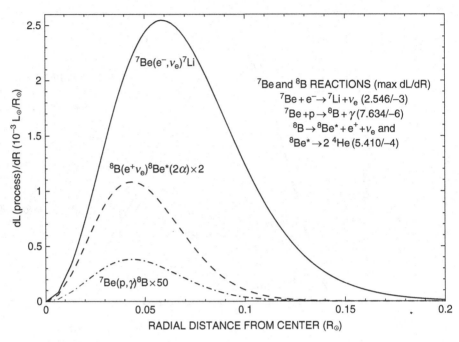

Fig. 10.3.3 ^{7}Be and ^{8}B contributions to luminosity in solar model A ($Z = 0.01$, $Y = 0.2373$)

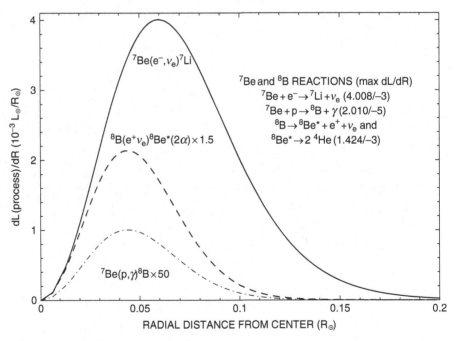

Fig. 10.3.4 ^{7}Be and ^{8}B contributions to luminosity in solar model B ($Z = 0.02$, $Y = 0.296$)

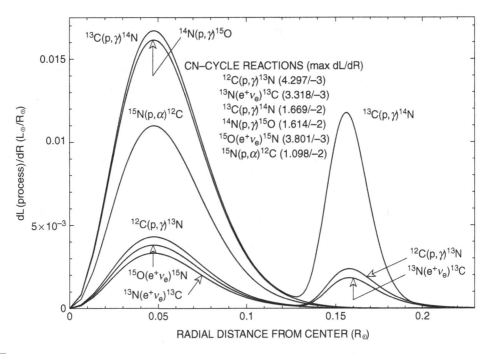

Fig. 10.3.5   CN–cycle contributions to luminosity in solar model A ($Z = 0.01, Y = 0.2373$)

Sun, have played such important roles in the development of mankind's understanding of the neutrino.

Differential contributions to the surface luminosity from CN-cycle reactions are shown in Figs. 10.3.5 and 10.3.6 for models A and B, respectively. There are two sets of peaks in the distributions. All of the distributions making up the peaks closest to the center scale perfectly with one another, showing that the CN-cycle is in full swing, with all associated abundances in equilibrium everywhere. The secondary set of peaks show where pristine ^{12}C is being converted into ^{14}N. The fact that the three peaks scale with one another indicates that the two intermediate isotopes, ^{13}C and ^{13}N, are at their equilibrium abundances.

In Figs. 10.3.7 and 10.3.8 are shown differential contributions to the surface luminosity of four reactions, three of which are of interest because they are responsible for the slow conversion of initial ^{16}O into ^{14}N, thus increasing the pool of isotopes participating in the CN cycle. The fourth reaction, $^{15}N(p, \gamma)^{16}O$, occurs approximately 2200 times less frequently than does the $^{15}N(p, \alpha)^{12}C$ reaction, and is thus just a temporary leakage out of the CN cycle that is repaired by the other three reactions. The energy released in one $^{15}N(p, \gamma)^{16}O$ reaction is 20.5 times larger than the energy released in one $^{16}O(p, \gamma)^{17}F$ reaction (see Table 6.2.3), so, despite the fact that the rate of energy generation by the reactions which create ^{16}O is approximately 1.6 times larger than the rate of energy generation by the reactions which destroy ^{16}O, ^{16}O is actually being destroyed approximately 13 times faster than it is being created. The conversion of the ^{16}O with which the Sun was born into the set of isotopes participating in the CN-cycle is just barely discernable

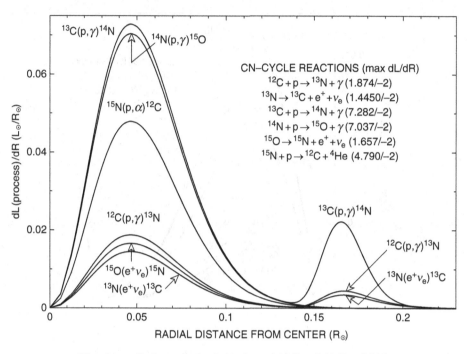

Fig. 10.3.6   CN–cycle contributions to luminosity in solar model B ($Z = 0.02, Y = 0.296$)

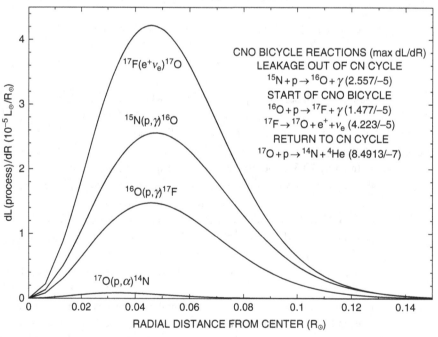

Fig. 10.3.7 $^{15}N \rightarrow {}^{16}O \rightarrow {}^{17}F \rightarrow {}^{17}O \rightarrow {}^{14}N$ contributions to luminosity in solar model A ($Z = 0.01, Y = 0.2373$)

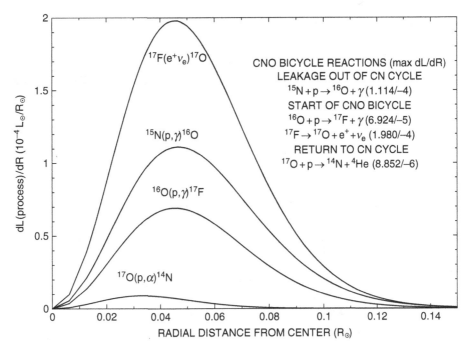

Fig. 10.3.8 ^{15}N $\to$ ^{16}O $\to$ ^{17}F $\to$ ^{17}O $\to$ ^{14}N contributions to luminosity in solar model B ($Z = 0.02$, $Y = 0.296$)

in the ^{16}O abundance profiles near the center of the solar models in the upper left hand portions of Figs. 10.1.6 and 10.2.6.

In Figs. 10.3.9 and 10.3.10 are shown contributions of the various neutrino-loss mechanisms to escaping energy. Neutrinos from the pp reaction account for a loss rate of $\sim$0.02 $L_\odot$ for both models A and B. Losses due to ^{7}Be electron capture account for $\sim$0.0026 $L_\odot$ for model A and $\sim$0.004 $L_\odot$ in model B. Losses due to ^{8}B neutrinos account for $\sim$0.000 023 $L_\odot$ in model A and $\sim$0.000 06 $L_\odot$ in model B. It is incredible that ^{8}B neutrinos, which have played such an important role in clarifying the nature of the neutrino, make a contribution to the total neutrino-loss rate which is so tiny compared with losses by other processes.

10.4 The solar neutrino problem

In Table 10.4.1 are the neutrino fluxes at the Earth produced by the solar models described in Section 10.1. It has been assumed that the flux $\phi(\text{pep})$ of neutrinos from the infrequent $p + p + e^- \to d + \nu_e$ reaction (Bahcall & May, 1968), with a minimum neutrino energy of 1.443 MeV, is related to the flux $\phi(\text{pp})$ of pp neutrinos by $\phi(\text{pep}) = 0.0025 \times \phi(\text{pp})$ (Abraham & Iben, 1971).

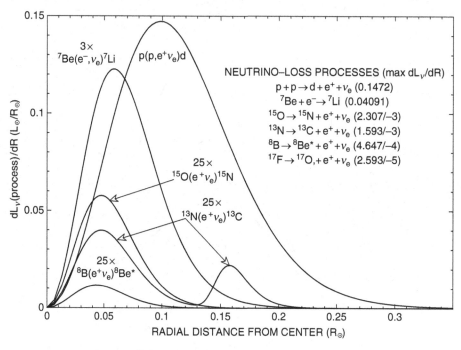

Fig. 10.3.9   Contributions to neutrino luminosity in solar model A ($Z = 0.01$, $Y = 0.2373$)

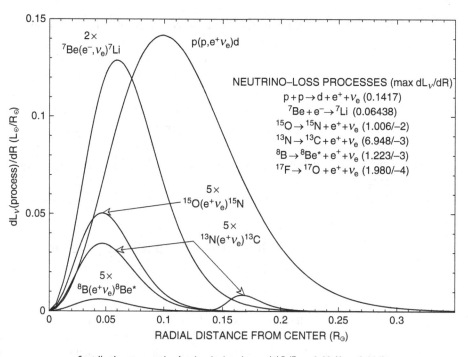

Fig. 10.3.10   Contributions to neutrino luminosity in solar model B ($Z = 0.02$, $Y = 0.296$)

Table 10.4.1 Neutrino fluxes at the Earth (in units of $cm^{-2} s^{-1}$) when $Z = 0.01$									
Y	t_9	$\phi(pp)$	$\phi(pep)$	$\phi(^7Be)$	$\phi(^8B)$	$\phi(^{13}N)$	$\phi(^{15}O)$	$\phi(^{17}F)$	$\phi(^{18}F)$
0.23	5.36	6.13/10	1.53/8	3.62/9	3.08/6	1.67/8	1.32/8	1.43/6	2.10/4
0.24	4.32	6.19/10	1.55/8	3.09/9	2.34/6	1.45/8	1.07/8	1.13/6	1.26/4
0.25	3.35	6.24/10	1.56/8	2.64/9	1.77/6	1.31/8	8.69/7	9.01/5	7.58/3
0.2373	4.60	6.16/10	1.54/8	3.21/9	2.49/6	1.49/8	1.12/8	1.19/6	1.42/4
0.2373	4.63	6.17/10	1.54/8	3.25/9	2.56/6	1.52/8	1.14/8	1.21/6	1.48/4
0.2373	4.60	6.16/10	1.54/8	3.21/9	2.50/6	1.49/8	1.12/8	1.19/6	1.42/4
0.2373	4.63	6.17/10	1.54/8	3.26/9	2.56/6	1.52/8	1.14/8	1.22/6	1.49/4
0.25*	4.60	6.11/10	1.53/8	3.46/9	2.87/6	1.65/8	1.27/8	1.36/6	1.82/4

Entries in the first three rows of Table 10.4.1 are for the three models whose properties are described in Table 10.1.1 and which have been used to interpolate for the parameters which produce models of near solar age. Entries in the next four rows are for the models with properties described in Table 10.1.2, with entries in the seventh row describing neutrino fluxes from model A. Entries in the last row are for the model of solar age with solar-like global characteristics, but with an opacity given by factors of two enhancements in the carbon and oxygen abundances. Hereinafter, this last model, with $Z = 0.01$ and $y = 0.25$, is designated as model A*.

The most important lesson from Table 10.4.1 is that, for any choice of the initial helium abundance or of the opacity, as long as the model has the solar luminosity and age, the fluxes of pp and pep neutrinos are effectively fixed, the flux of ^8B neutrinos is determined to within a factor of two, and the fluxes of neutrinos due to other sources is determined to within 40% or less. The second most important lesson is that, for a fixed age, a factor of two variation in the abundance of carbon and oxygen adopted in the opacity introduces only a $\sim 10\%$ variation in the flux of ^8B neutrinos. The third lesson is that, for a fixed opacity algorithm and a fixed composition, a variation in age by 1% has a negligible effect on the most important neutrino fluxes.

In Table 10.4.2 are the neutrino fluxes at the Earth produced by the solar models presented in Section 10.2. Entries in the first three rows of the table are for models with properties listed in Table 10.2.1, entries in the fourth row are for model B, with properties listed in the second row of Table 10.2.2, and entries in the last row are for a model of solar age with solar-like global characteristics, but with an opacity appropriate for factors of two enhancements in the carbon and oxygen abundances. This latter model is hereinafter designated as model B*.

Neutrino fluxes at the Earth from model A (row 5 in Table 10.4.1) and Model B (row 4 in Table 10.4.2) are given in Table 10.4.3.

The Homestake mine experiment to detect solar neutrinos relies on the reaction

$$\nu_e + {}^{37}Cl \rightarrow {}^{37}Ar + e^-. \tag{10.4.1}$$

				Table 10.4.2 Neutrino fluxes at the Earth (in units of cm^{-2} s^{-1}) when $Z = 0.02$					
Y	t_9	$\phi(pp)$	$\phi(pep)$	$\phi(^7Be)$	$\phi(^8B)$	$\phi(^{13}N)$	$\phi(^{15}O)$	$\phi(^{17}F)$	$\phi(^{18}F)$
0.28	6.38	5.75/10	1.44/8	6.36/9	9.86/6	7.78/8	7.16/8	8.41/6	3.66/5
0.29	5.24	5.86/10	1.47/8	5.58/9	7.80/6	6.28/8	5.62/8	6.51/6	2.15/5
0.30	4.18	5.95/10	1.49/8	4.84/9	6.05/6	5.20/8	4.46/8	5.07/6	1.26/5
0.296	4.62	5.91/10	1.48/8	5.15/9	6.75/6	5.62/8	4.92/8	5.63/6	1.58/5
0.309*	4.6	5.82/10	1.46/8	5.45/9	7.54/6	6.13/8	5.43/8	6.27/6	1.94/5

				Table 10.4.3 Neutrino fluxes at the Earth (in units of cm^{-2} s^{-1}) for models A and B					
Model	Z	Y	$\phi(pp)$	$\phi(pep)$	$\phi(^7Be)$	$\phi(^8B)$	$\phi(^{13}N)$	$\phi(^{15}O)$	$\phi(^{17}F)$	$\phi(^{18}F)$
A	0.01	0.2373	6.17/10	1.54/8	3.25/9	2.56/6	1.51/8	1.14/8	1.21/6	1.48/4
B	0.02	0.296	5.91/10	1.48/8	5.15/9	6.75/6	5.62/8	4.92/8	5.63/6	1.58/5

The threshold neutrino energy for detection is 0.814 MeV, approximately twice the maximum energy of a neutrino emitted in the pp reaction, but smaller than the 0.861 MeV energy of the neutrino emitted when ^{7}Be captures an electron in a transition to the ground state of ^{7}Li.

Assuming no flavor change following creation, the theoretical detection rate is

$$\sum^{Cl} \sigma_i \phi_i \text{ (SNU)}$$
$$= \frac{\phi(pep)}{6.25 \times 10^8} + \frac{\phi(^7Be)}{4.17 \times 10^9} + \frac{\phi(^8B)}{9.00 \times 10^5} + \frac{\phi(^{13}N)}{5.88 \times 10^9} + \frac{\phi(^{15}O)}{1.47 \times 10^9}, \quad (10.4.2)$$

where the ϕs are total neutrino fluxes (in particles cm^{-2} s^{-1}) reaching the Earth and

$$1 \text{ SNU} = 10^{-36} \text{ reaction s}^{-1} \text{ per } ^{37}\text{Cl nucleus.} \quad (10.4.3)$$

The fluxes in the denominators of terms in eq. (10.4.2) are based on cross sections σ_i for the capture of electron-flavor neutrinos by a ^{37}Cl nucleus taken, with one exception, from J. N. Bahcall (*Neutrino Astrophysics*, 1989). The exception is $\sigma(^8B) = (1.11 \pm 0.08) \times 10^{-46}$ cm^2 determined by Aufderheide, Bloom, Resler, & Goodman (1994). It is interesting that the fluxes in the denominators of eq. (10.4.2) are larger by, respectively, 10%, 19%, 23%, 26%, and 16% than follow from the earliest calculations of detection cross sections (Bahcall, 1964a, 1964b, 1966).

Inserting the fluxes given in the first row of Table 10.4.3, one obtains for model A

$$\left(\sum^{Cl} \sigma_i \phi_i \right)_A \text{ (SNU)}$$

$$= \frac{1.54 \times 10^8}{6.25 \times 10^9} + \frac{3.25 \times 10^9}{4.17 \times 10^9} + \frac{2.56 \times 10^6}{9.00 \times 10^5} + \frac{1.51 \times 10^8}{5.88 \times 10^9} + \frac{1.14 \times 10^8}{1.47 \times 10^9}$$

$$= 0.0246 + 0.779 + 2.84 + 0.0257 + 0.0776 = 3.75. \tag{10.4.4}$$

The same calculation for the $Z = 0.01$, $Y = 0.25$ model with carbon- and oxygen-enhanced opacities (fluxes in the eighth row of Table 10.4.1) gives $\left(\sum^{Cl} \sigma_i \phi_i \right)_{A*} = 4.15$ SNU.

For model B, the $Z = 0.02$ solar model with fluxes given in the second row of Table 10.4.3, one obtains

$$\left(\sum^{Cl} \sigma_i \phi_i \right)_B = (0.024 + 1.24 + 7.50 + 0.0956 + 0.335) \text{ SNU} = 9.19 \text{ SNU}, \tag{10.4.5}$$

compared with $\left(\sum^{Cl} \sigma_i \phi_i \right)_{B*} = 10.10$ SNU for the $Z = 0.02$ solar model with carbon- and oxygen-enhanced opacities (fluxes in the last row of Table 10.4.2).

Assuming that the theoretical detection rate and Z are related logarithmically, it follows from models A and B that $\sum^{Cl} \sigma_i \phi_i \sim 4.85$ SNU for $Z = 0.0122$ and $\sum^{Cl} \sigma_i \phi_i \sim 8.60$ SNU for $Z = 0.019$. From the models with C- and O-enhanced opacity, it follows that $\sum^{Cl} \sigma_i \phi_i \sim 5.36$ SNU for $Z = 0.0122$ and $\sum^{Cl} \sigma_i \phi_i \sim 9.45$ SNU for $Z = 0.019$.

These results are not inconsistent with results from models constructed in the 1960s with the then commonly adopted nuclear cross section factors. For example, at fixed Y, the theoretical flux of ^{8}B neutrinos is related to cross section factors and central temperature T_c approximately by (Iben, 1969)

$$\phi(^8\text{B}) \propto \frac{S_{34}}{(S_{11} \times S_{33})^{0.5}} \times \frac{S_{17}}{S_{e7}} \times T_c^{20.3}, \tag{10.4.6}$$

where the first factor on the right hand side of the equation comes from assuming that ^{3}He is in local equilibrium, the second factor comes from the competition between proton capture and electron capture on ^{7}Be, and the third factor comes from the temperature dependences of the reactions at temperatures near 14–16×10^6 K. The constraint that luminosity must remain fixed, coupled with the temperature dependence of the pp reaction rate, gives $\rho_c T_c^{4.35} \sim$ constant (see Table 6.4.6 in Section 6.4 for the relevant temperature dependence of the pp reaction) and the fact that, to the extent that solar models approximate index $N = 3$ polytropes, $\rho_c \overset{\sim}{\propto} T_c^3$, means that, roughly, $T_c \propto S_{11}^{-1/7.35}$. Inserting this last relationship into eq. (10.4.6) gives, approximately,

$$\phi(^8\text{B}) \propto S_{34} \times \frac{1}{S_{33}^{0.5}} \times S_{17} \times \frac{1}{S_{e7}} \times \frac{1}{S_{11}^{3.26}}. \tag{10.4.7}$$

From this expression it follows that, at fixed Y, the ^{8}B flux produced by models constructed here and the flux produced by models constructed in 1969 by Iben with a particular choice of cross section factors ($S_{ij} = S_{ij}^0$) are related approximately by

$$\frac{\phi(^8B)_{2005}}{\phi(^8B)_{1969}} \sim \frac{0.54}{0.6} \times \frac{0.021}{0.03} \times \left(\frac{5180}{6500}\right)^{0.5} \times \frac{1.0}{1.2} \times \left(\frac{3.5}{3.94}\right)^{3.26} \times 0.81$$

$$= 0.9 \times 0.7 \times 0.89 \times 0.83 \times 0.68 \times 0.81 \sim \frac{1}{4}, \tag{10.4.8}$$

where the final factor of 0.81 in the product of factors is a consequence of the reduction in the adopted cross section for ^8B neutrino capture by ^{37}Cl. Major contributors to the factor of four decrease in the theoretical rate is an approximately 50% decrease in the estimate of the $^7Be(p, \gamma)^8B$ cross section and a 20% decrease in the experimental estimate of the half life of the neutron from $\sim$770 s in the middle 1960s to $\sim$614 s in the first decade of the twenty-first century, leading to a corresponding increase in the square of the effective beta decay coupling constant to which S_{11} is directly proportional.

The Homestake mine experiment began with 520 tons of liquid C_2Cl_4 (tetrachloroethylene) in a horizontal cylindrical tank 4850 feet underground in the Homestake gold mine in Lead, South Dakota. After 25 years of operation, involving 108 extractions of argon, the experiment yielded an ^{37}Ar production rate of (Cleveland *et al.*, 1998)

$$\sum \sigma_i \phi_i \sim 2.56 \text{ SNU}, \tag{10.4.9}$$

which is only 15% smaller than the initial upper limit of 3 SNU. Both systematic and statistical errors are estimated to be $\pm$6%. In the year 2002, Ray Davis shared the Nobel prize for physics for his pioneering role in the experiment.

The observed detection rate of electron neutrinos of energy greater than 0.814 MeV is, respectively, 1.9 and 3.4 times smaller than the detection rate predicted by solar models A and B for $Z = 0.0122$ and $Z = 0.019$ on the assumption of no change in neutrino flavor after birth. Given the history of the changes in the choice of nuclear reaction cross sections over the years and the fact that the cross sections of critical reactions remain guesses based on extrapolation beyond where experimental data exists, an interpretation of the meaning of the discrepancy cannot be entirely unambiguous.

Two experiments in Gran Sasso, Italy (GALLEX, gallium experiment and GNO, gallium neutrino observatory) and one at the foot of Mt. Elbrus in the Northern Caucasus (SAGE, Soviet–American gallium experiment) make use of the reaction

$$\nu_e + {}_{31}Ga_{40} \rightarrow {}_{32}Ge_{39} + e^- \tag{10.4.10}$$

followed by the electron-capture decay of ^{71}Ge (half life = 11.43 days). Detection of pp-reaction neutrinos and of ^7Be electron-capture neutrinos are in principle possible because neutrino-capture transitions from the ground state of ^{71}Ga to the ground state of ^{71}Ge, with a threshold of 0.233 MeV, and transitions to the first two excited states of ^{71}Ge, with threshold energies of 0.408 MeV and 0.733 MeV, respectively, can occur. Detection of ^8B, CNO, and pep neutrinos involve also transitions to more highly excited states. Taking cross sections σ_i for the capture of electron-flavor neutrinos by a ^{71}Ga nucleus from J. N. Bahcall (1996), the theoretically expected counting rate (if creation flavor is conserved) is

$$\sum^{\text{Ga}} \sigma_i \phi_i \; (\text{SNU}) = \frac{\phi(\text{pp})}{8.53 \times 10^8} + \frac{\phi(\text{pep})}{4.90 \times 10^7} + \frac{\phi(^7\text{Be})}{1.39 \times 10^8}$$

$$+ \frac{\phi(^8\text{B})}{4.17 \times 10^5} + \frac{\phi(^{13}\text{N})}{1.66 \times 10^8} + \frac{\phi(^{15}\text{O})}{8.80 \times 10^7} + \frac{\phi(^{17}\text{F})}{8.80 \times 10^7}. \quad (10.4.11)$$

Inserting fluxes from Table 10.4.3 produced by the $Z = 0.01$ model A, the predicted detection rate is

$$\left(\sum^{\text{Ga}} \sigma_i \phi_i\right)_A (\text{SNU}) = \frac{6.17 \times 10^{10}}{8.53 \times 10^8} + \frac{1.54 \times 10^8}{4.90 \times 10^7} + \frac{3.25 \times 10^9}{1.39 \times 10^8}$$

$$+ \frac{2.56 \times 10^6}{4.17 \times 10^5} + \frac{1.51 \times 10^8}{1.66 \times 10^8} + \frac{1.14 \times 10^8}{8.80 \times 10^7} + \frac{1.21 \times 10^6}{8.80 \times 10^7}$$

$$= 72.3 + 3.14 + 23.4 + 6.14 + 0.910 + 1.30 + 0.0138 = 107.2, \quad (10.4.12)$$

with 92% of the predicted rate produced by pp, pep, and ^{7}Be neutrinos. Inserting fluxes from the last row of Table 10.4.1 produced by model A*, the $Z = 0.01$ model made with C- and O-enhanced opacity, one obtains $\left(\sum^{\text{Ga}} \sigma_i \phi_i\right)_{A*} = 108.9$ SNU.

Inserting fluxes from the second row in Table 10.4.3 for the $Z = 0.02$ model B,

$$\left(\sum^{\text{Ga}} \sigma_i \phi_i\right)_B (\text{SNU}) = 69.3 + 3.02 + 37.1 + 16.2 + 3.39 + 5.59 + 0.064 = 134.7, \quad (10.4.13)$$

with 81% of the predicted rate produced by pp, pep, and ^{7}Be neutrinos. Using neutrino fluxes given by model B*, the $Z = 0.02$ model made with C- and O-enhanced opacities, gives $\left(\sum^{\text{Ga}} \sigma_i \phi_i\right)_{B*} = (68.2 + 2.98 + 39.2 + 18.1 + 3.69 + 6.17 + 0.071)$ SNU = 138.4 SNU.

Interpolating logarithmically between the detection rates predicted by using neutrino fluxes given by models A and B, one finds that the theoretical detection rate varies from 115.5 SNU for $Z = 0.0122$ to 132.6 SNU for $Z = 0.019$. Using neutrino fluxes given by the models constructed with C- and O-enhanced opacities, one finds 116.6 SNU for $Z = 0.0122$ and 136.1 SNU for $Z = 0.019$.

The GALLEX experiment was terminated in 1998 after a total of 65 runs of 3–4 weeks duration each, a total of 1594 days of exposure, and a total number of 300 measured ^{71}Ge electron-capture events (W. Hampel et al., 1999). The stated counting rate is $\sum \sigma_i \phi_i \sim$ 77.5 SNU, with a quoted error of about 10%. Early results of the successor GNO experiment, when folded with the GALLEX data, is stated by M. Altmann et al. (2000) to be $\sum \sigma_i \phi_i \sim 74.1$ SNU, with a quoted 7.3% error.

After 12 years of operation, analysis of 92 runs and 158 individual data sets, and attribution of 1723 events to electron-capture by ^{71}Ge, the SAGE experiment determines (J. N. Abdurashitov *et al.*, 2002)

$$\overset{\text{Ga}}{\sum} \sigma_i \phi_i \sim 70.8 \text{ SNU}, \tag{10.4.14}$$

with a quoted 7.5% error.

Once again, the theoretical detection rates based on the assumption of no flavor changes are larger than experimentally observed rates, this time by factors of the order of 1.5–1.8, compared with the factors of 1.9–3.4 for the ^{37}Cl experiment. Since the ^{71}Ga experiment is sensitive primarily to low energy neutrinos from pp, pep, and ^{7}Be electron-capture reactions and the ^{37}Cl experiment is sensitive primarily to the high energy neutrinos from ^{8}B decay, one might tentatively infer that, when compared with theoretical predictions without neutrino-flavor changes: (1) the diminution in electron-flavor fluxes for low energy neutrinos might be due primarily to vacuum oscillations whereas (2) interactions in the Sun may play an important role in the diminution in electron-flavor fluxes for high energy neutrinos.

The Super-Kamiokande detector and its predecessor, the Kamiokande detector, are located in the Mozumi mine of the Kamiokande Mining and Smelting Company in central Honchu, Japan. The Kamiokande detector consisted of a tank of 3000 tons of purified ordinary water (H_2O) viewed by 1000 photomultiplier tubes which detect the Cherenkov radiation emitted by atomic electrons elastically scattered by neutrinos in the reaction

$$\nu_x + e^- \rightarrow \nu_x + e^-. \tag{10.4.15}$$

The detection of neutrinos from Supernova 1987a by this device was responsible for the award to M. Koshiba of a share of the Nobel prize for physics in 2002.

The Super-Kamiokande (SK) detector consists of a vertical cylindrical tank containing 50 000 tons of purified water viewed initially by over 11 000 photomultiplier tubes (an accident in the year 2001 destroyed roughly half of the photomultiplier tubes, which have since been reconstituted, and to a corresponding reduction in the effective volume of water). Analysis of 1258 days of data collected over a five year period led to an estimate of 18 464 scattering events in 22 500 tons of water, involving neutrinos of energies in the range 5–20 MeV, and to an estimate of a corresponding flux of neutrinos from the Sun of (S. Fukuda *et al.*, 2001)

$$\phi_{ES} = 2.32 \times 10^6 \text{ cm}^{-2} \text{ s}^{-1}, \tag{10.4.16}$$

with a stated accuracy of about 3.4%. Most impressively, the experiment measures a 7% seasonal variation in flux corresponding to the 3.4% seasonal variation in the distance between the Earth and the Sun.

The observed scattered electron energy spectrum is consistent, within roughly 10%, with an undistorted terrestial ^{8}B neutrino energy spectrum (Ortiz *et al.*, 2000), suggesting that, whatever combination of flavors is impinging on the detector, the interaction between electron neutrinos and electrons in the Sun produces roughly the same combination of flavors for all neutrinos of energy in the range 5–14 MeV. The SK experiment also determines an upper limit on the flux of neutrinos due to the reaction $^3\text{He} + p \to {}^4\text{He} + e^+ + \nu_\text{e}$ (minimum neutrino energy of 18.77 MeV) of $\phi(^3\text{He}p) = 4 \times 10^3 \text{ cm}^{-2} \text{ s}^{-1}$.

The Sudbury Neutrino Observatory (SNO) is located in the International Nickel Company Creighton mine in Ontario. The detector consists of 1000 tons of nearly pure heavy water (D_2O) in a 12 meter diameter acrylic sphere viewed by 9456 photomultiplier tubes. Experiments utilize the electron-scattering (ES) reaction, eq. (10.4.15), as well as the reactions

$$\nu_\text{e} + d \to p + p + e^- \tag{10.4.17}$$

and

$$\nu_x + d \to p + n + \nu_x. \tag{10.4.18}$$

The charged current (CC) reaction, eq. (10.4.17), is sensitive only to electron-flavor neutrinos whereas the neutral current (NC) reaction, eq. (10.4.18), is sensitive to all flavors. The theoretical threshold for the CC reaction is

$$E(\nu_\text{e}) = (M_\text{p} - M_\text{n})c^2 + E_\text{bind} + m_\text{e}c^2$$
$$= -1.293 \text{ MeV} + 2.225 \text{ MeV} + 0.511 \text{ MeV} = 1.443 \text{ MeV}, \tag{10.4.19}$$

where $E_\text{bind} = 2.225$ MeV is the binding energy of the deuteron, and the theoretical threshold for the NC reaction is just the binding energy of the deuteron. Thus, the current-current reactions in the SNO experiment are not under any circumstance sensitive to pp and ^7Be neutrinos. Because of background noise, the effective threshold for both the CC and NC reactions is about 5 MeV.

The results of 254 days of observation involving 3055 events in the years 2001–2002 are stated as (S.N. Ahmed *et al.*, 2004)

$$\phi_\text{CC} = 1.59 \times 10^6 \text{ cm}^{-2} \text{ s}^{-1}, \tag{10.4.20}$$

$$\phi_\text{ES} = 2.21 \times 10^6 \text{ cm}^{-2} \text{ s}^{-1}, \tag{10.4.21}$$

and

$$\phi_\text{NC} = 5.21 \times 10^6 \text{ cm}^{-2} \text{ s}^{-1}, \tag{10.4.22}$$

with quoted errors of ± 5 %, ± 13 %, and ± 5 %, respectively. It is encouraging that the values of ϕ_ES found by the SK and SNO experiments differ by only 5%.

The quantity ϕ_NC presumably measures the total flux of neutrinos originating from ^8B decay in the Sun, regardless of the combination of flavors impinging upon the Earth, whereas the quantity ϕ_CC measures only the flux of ^8B-produced neutrinos arriving at the Earth as electron-flavor neutrinos. The fact that

$$\frac{\phi_\text{CC}}{\phi_\text{NC}} = \frac{1.59}{5.21} = \frac{1}{3.28} = 0.305 \tag{10.4.23}$$

may be taken as evidence that, in the Sun, electron-flavor neutrinos interact with electrons to acquire an effective mass and that the degree of mixing between flavors that this interaction induces increases with the energy of the neutrino. This theme is developed in Section 10.5.

Even at this juncture, the SNO estimate of the flux of ^{8}B neutrinos can be used to make a quantitative statement about the likely values of Z and Y appropriate for the solar interior. Logarithmic interpolation in the ^{8}B neutrino fluxes for models A and B given in Table 10.4.3 shows that a value of $\phi_{NC} = 5.21 \times 10^6$ cm^{-2} s^{-1} corresponds to interior composition parameters

$$Z \sim 0.0166 \text{ and } Y \sim 0.279. \tag{10.4.24}$$

Using, instead, ^{8}B neutrino fluxes given by the models made with C- and O-enhanced opacity (column 8 in Table 10.4.1 and column 5 in Table 10.4.2), one finds $Z \sim 0.0153$ and $Y \sim 0.285$.

Studies of galactic nucleosynthesis suggest that interstellar metallicity and helium abundances are related by $dY/dZ \sim 3$. Big Bang nucleosynthesis calculations, as well as estimates of the helium abundance in extreme population II objects, suggest $Y \sim 0.23$ when $Z = 0$, implying $Y \sim 0.28$ when $Z \sim 0.0166$. A possible interpretation of the solar interior Z and Y given by eq. (10.4.24), in conjunction with the estimate of $Z = 0.0122$ for the photospheric metallicity of the Sun, is that diffusion of elements heavier than helium out of the convective envelope has led to a reduction of the surface metallicity by about 25%, from an initial value of $Z \sim 0.0166$ to a current value of $Z \sim 0.0122$, but to a minimal diffusion inward of helium. Other interpretations include the possibilities that further estimates of the surface metallicity of the Sun are appropriate and that a more sophisticated treatment of the opacity in the solar interior which takes into account the transformations of isotope abundances by nuclear reactions is called for.

10.5   Neutrino oscillations in vacuum

The crucial ingredient of the resolution of the solar neutrino problem is the supposition that neutrinos are not massless and that, because they are composites of mass eigenstates, neutrinos born in the electron-flavor state spend part of their lives in the muon-flavor state and are therefore detectable only part of the time by a device which is sensitive only to neutrinos in the electron-flavor state.

Oscillations between flavor states even in a vacuum are inevitable consequences of intrinsic mass differences, with the time scale for an oscillation being directly proportional to the energy of the neutrino and inversely proportional to the difference in the squares of mass eigenvalues. At the large electron-densites in the solar interior, an interaction with electrons causes the electron-flavor state to acquire an effective mass proportional to the product of the Fermi weak interaction constant and the electron density, and, even if intrinsic mass differences did not exist, the induced mass differences can, under the right circumstances, cause an electron neutrino to evolve into a muon-like neutrino before leaving the Sun. Since the flavor expressed most of the time by a muon neutrino cannot, by definition, be the electron flavor, the probabily of detection by a device sensitive only to neutrinos in the electron-flavor state is reduced even further than if an interaction between electrons and neutrinos in the electron-flavor state did not occur.

The simplest possibility to explore is oscillations in vacuum between only two flavor states, say, electron flavor and muon flavor. Let $v_e(t)$ and $v_\mu(t)$ denote, respectively, the wave functions for electron-flavor and muon-flavor states, and let $v_1(t)$ and $v_2(t)$ denote, respectively, the wave functions for mass eigenstates of these two kinds of neutrinos. A unitary transformation which connects the two sets of wave functions can be written as

$$\begin{pmatrix} v_e(t) \\ v_\mu(t) \end{pmatrix} = \begin{pmatrix} \cos\theta & \sin\theta \\ -\sin\theta & \cos\theta \end{pmatrix} \begin{pmatrix} v_1(t) \\ v_2(t) \end{pmatrix}, \tag{10.5.1}$$

where θ is a vacuum mixing angle to be found by experiment. Inverting this equation gives

$$\begin{pmatrix} v_1(t) \\ v_2(t) \end{pmatrix} = \begin{pmatrix} \cos\theta & -\sin\theta \\ \sin\theta & \cos\theta \end{pmatrix} \begin{pmatrix} v_e(t) \\ v_\mu(t) \end{pmatrix}. \tag{10.5.2}$$

For the mass eigenstates, one may adopt wave functions of the form

$$v_{1,2}(t) = A_{1,2} \, \exp\left(-i \, \frac{E_{1,2}}{\hbar} t \right), \tag{10.5.3}$$

where

$$E_{1,2} = \sqrt{p_{1,2}^2 + m_{1,2}^2}, \tag{10.5.4}$$

p stands for momentum times c (the speed of light in vacuum), and m stands for mass times c^2. Choosing as initial conditions

$$v_e(0) = 1 \text{ and } v_\mu(0) = 0, \tag{10.5.5}$$

eqs. (10.5.2) and (10.5.3) give

$$A_1 = v_1(0) = \cos\theta \text{ and } A_2 = v_2(0) = \sin\theta. \tag{10.5.6}$$

Thus,

$$v_e(t) = \cos^2\theta \, \exp\left(-i \, \frac{E_1}{\hbar} t \right) + \sin^2\theta \, \exp\left(-i \, \frac{E_2}{\hbar} t \right) \tag{10.5.7}$$

and

$$v_\mu(t) = \cos\theta \, \sin\theta \left[-\exp\left(-i \, \frac{E_1}{\hbar} t \right) + \exp\left(-i \, \frac{E_2}{\hbar} t \right) \right]. \tag{10.5.8}$$

Choosing as initial conditions

$$v_\mu'(0) = 1 \text{ and } v_e'(0) = 0, \tag{10.5.9}$$

then

$$A_1 = v_1'(0) = -\sin\theta \text{ and } A_2 = v_2'(0) = \cos\theta, \tag{10.5.10}$$

where the prime distinguishes the wave functions which describe a neutrino born in a pure muon-flavor state from those which describe a neutrino born in a pure electron-flavor state. The resulting time-dependent wave functions,

$$
\nu'_e(t) = -\cos\theta \, \sin\theta \left[\exp\left(-i\frac{E_1}{\hbar}t\right) - \exp\left(-i\frac{E_2}{\hbar}t\right)\right] \tag{10.5.11}
$$

and

$$
\nu'_\mu(t) = \sin^2\theta \, \exp\left(-i\frac{E_1}{\hbar}t\right) + \cos^2\theta \, \exp\left(-i\frac{E_2}{\hbar}t\right), \tag{10.5.12}
$$

clearly differ from those that originate from a neutrino born in a pure electron-flavor state.

The difference translates into a dependence on origin of the fraction of time that a neutrino of either type spends in a particular flavor state. The probability that a neutrino born in an electron-flavor state is in a muon-flavor state at time t is

$$
P_{\nu_\mu}(t) = |\nu_\mu(t)|^2 \tag{10.5.13}
$$

$$
= \cos^2\theta \, \sin^2\theta \left[2 - \exp\left(i\frac{E_1 - E_2}{\hbar}t\right) - \exp\left(-i\frac{E_1 - E_2}{\hbar}t\right)\right]
$$

$$
= 2 \cos^2\theta \, \sin^2\theta \left[1 - \cos\left(\frac{E_1 - E_2}{\hbar}t\right)\right]
$$

$$
= \sin^2(2\theta) \, \sin^2\left(\frac{E_1 - E_2}{2\hbar}t\right), \tag{10.5.14}
$$

where, in the final transformation, use has been made of the facts that $(1 - \cos\phi) = 2\sin^2\phi/2$ and $2\cos\theta\sin\theta = \sin 2\theta$. Since $|\nu_e(t)|^2 + |\nu_\mu(t)|^2 = 1$, it follows that the probability that the neutrino born in an electron-flavor state is in an electron-flavor state after a time t is

$$
P_{\nu_e}(t) = |\nu_e(t)|^2 = 1 - \sin^2(2\theta) \, \sin^2\left(\frac{E_1 - E_2}{2\hbar}t\right). \tag{10.5.15}
$$

The probabilities are exactly reversed when the neutrino is born in a pure muon-flavor state. Then,

$$
P'_{\nu_e}(t) = |\nu'_e(t)|^2 = \sin^2(2\theta) \, \sin^2\left(\frac{E_1 - E_2}{2\hbar}t\right) \tag{10.5.16}
$$

and

$$
P'_{\nu_\mu}(t) = |\nu'_\mu(t)|^2 = 1 - \sin^2(2\theta) \, \sin^2\left(\frac{E_1 - E_2}{2\hbar}t\right). \tag{10.5.17}
$$

To distinguish precisely between the terms "neutrino flavor" and "neutrino type", one may decree that, on average, a neutrino born in the electron-flavor state spends a fraction

$$
\langle P_{\nu_e}(t)\rangle = 1 - \frac{1}{2} \, \sin^2(2\theta) \tag{10.5.18}
$$

of its time expressing the electron flavor and a fraction

$$
\langle P_{\nu_\mu}(t)\rangle = \frac{1}{2} \, \sin^2(2\theta) \tag{10.5.19}
$$

of its time expressing the muon flavor. An electron neutrino (type) may be defined as a neutrino which expresses the two flavor states with these mean probabilities, with the additional proviso that P_{ν_e} oscillates between unity and something larger than zero. Similarly, a muon neutrino (type) may be defined as a neutrino which expresses the two flavor states with just the reverse mean probabilities, with the additional proviso that P_{ν_μ} oscillates between unity and something larger than zero. Thus, the term "electron neutrino" refers to a neutrino which spends a larger fraction of its time expressing the electron flavor and the term "muon neutrino" refers to a neutrino which spends a larger fraction of its time expressing the muon flavor.

The full amplitude of the oscillation between neutrino types is given by $\sin^2(2\theta)$ and the time-averaged probability of expressing either flavor state depends on the original flavor state. With the restriction that $\theta \leq \pi/4$, the larger the angle, the larger is the full oscillation amplitude. Only if $\theta = 0$ and $E_1 - E_2 = 0$ would there be such a thing as a neutrino continuously in a pure flavor state. If $\theta = \pi/4$, the two neutrino types are, for all intents and purposes, indistinguishable insofar as detectability is concerned.

It is instructive to examine an analogy with chameleons. Just as a neutrino expresses flavors with different probabilities, a chameleon is never only red or only blue. Consider two types of chameleon, C_A and C_B, and two types of chameleon detectors, D_A and D_B. Type C_A chameleons are genetically and environmentally programmed to be red 80% of the time and blue 20% of the time, whereas type C_B chameleons are programmed to be blue 80% of the time and red 20% of the time. Type D_A detectors are capable of detecting only red chameleons and type D_B detectors are capable of detecting only blue chameleons. Suppose that a steady stream of chameleons, equally divided into two parts, flow toward the detectors from opposite directions. Together, the two detectors will detect all of the chameleons, but being direction sensitive, one detector claims that four times as many chameleons are coming from one direction than from the other while the other detector claims just the reverse. In another experiment, put detector D_A in place and send N chameleons of type C_A towards it. The detector will count $N_A = 0.8N$ chameleons. Put detector D_B in place and send the same N chameleons of type C_A towards it. The detector will count $N_B = 0.2N$ chameleons. And so forth.

In the context of neutrino experiments, the distance over which a complete oscillation takes place is of prime relevance. A vacuum oscillation time t_{vac} and a vacuum oscillation length L_{vac} may be defined by

$$\frac{|E_1 - E_2|}{2\,\hbar}\, t_{\text{vac}} = \frac{|E_1 - E_2|}{2\,\hbar}\, \frac{L_{\text{vac}}}{c} = \pi. \qquad (10.5.20)$$

Turning a blind eye to the fact that linear momentum and energy cannot be conserved simultaneously in the transformation of one mass eigenstate into the other, one may adopt $p_1 = p_2 = p$ and, assuming that $(m_1, m_2) \ll p$, one has that

$$E_{1,2} \sim p\left(1 + \frac{1}{2}\frac{m_{1,2}^2}{p^2}\right). \qquad (10.5.21)$$

With these choices,

$$L_{vac} = 4\pi\,\hbar c\,\frac{p}{\Delta^2} = 248\,\frac{E(\text{MeV})}{\Delta^2(\text{eV})^2}\,\text{cm}\qquad(10.5.22)$$

where

$$\Delta^2 = \left|\,m_1^2 - m_2^2\,\right|\qquad(10.5.23)$$

and the approximation $E \sim p$ has been adopted.

If, for example, $\Delta^2 \sim 10^{-4}$ $(\text{eV})^2$ and $\sin^2(2\theta) \sim 0.8$, a 10 MeV neutrino born in the electron-flavor state expresses the electron flavor for 60% of the time in traveling over a distance $L_{vac} \sim 250$ km and expresses the muon flavor for 40% of the time. With the same choice of parameters, a 1 MeV neutrino is characterized by $L_{vac} \sim 25$ km. The fact that the oscillation phenomenon appeared first as a theoretical prediction rather than as the result of laboratory experiments may be ascribed to the fact that typical values of L_{vac} are large relative to normal laboratory dimensions.

In interpreting the results of an experiment to detect neutrinos from the Sun, it is necessary to compare L_{vac} with two other distances: (1) the characteristic distance L_{source} over which the neutrino of a given type is formed in the Sun, and (2) the distance L_{exp} by which the distance between the Earth and the Sun changes over the course of the experiment. If either of these two distances is large compared with L_{vac}, then the detector registers the time-averaged flux of the particular flavor of neutrino to which it is sensitive.

The Earth–Sun distance is given approximately by

$$r_{ES} \sim \langle r_{ES}\rangle\,(1 - e\cos(\omega_{orb}\,t)) \sim 1.5 \times 10^{13}\,\text{cm}\left[1 - 0.017\,\cos\left(\frac{2\pi}{365\,\text{days}}t\right)\right],\qquad(10.5.24)$$

where $e = 0.017$ is the eccentricity of the Earth's orbit, and the root mean square change in this distance over a time δt is

$$\delta r_{ES} = e\,r_{ES}\,\sqrt{\langle\sin^2(\omega_{orb}\,t)\rangle}\,\delta t = 4.39 \times 10^4\,\text{km}\,\sqrt{\langle\sin^2(\omega_{orb}\,t)\rangle}\,\frac{\delta t}{\text{day}},\qquad(10.5.25)$$

where $\langle\sin^2(\omega_{orb}\,t)\rangle$ is an average over time δt. Thus, an experiment lasting several months any time of the year would result in

$$L_{exp} \sim 4 \times 10^4\,\text{km}\qquad(10.5.26)$$

and meet the criterion $L_{exp} \gg L_{vac}$.

As an example of L_{source}, consider ^{8}B neutrinos which are produced in the inner $\sim$4% of the Sun's mass. Roughly,

$$150\,\text{g cm}^{-3} \times \frac{4\pi}{3}\,R_{B8}^3 \sim 0.04\,M_\odot = 0.04 \times 1.4\,\text{g cm}^{-3} \times \frac{4\pi}{3}\,R_\odot^3,\qquad(10.5.27)$$

or

$$L_{source} = R_{B8} \sim 0.072\,R_\odot \sim 5 \times 10^4\,\text{km}.\qquad(10.5.28)$$

Thus, independent of its duration, an experiment will detect time-averaged fluxes for all types of solar neutrino as long as Δ^2 is larger than, say, $\sim 3 \times 10^{-6}$ $(\text{eV})^2$.

Vacuum oscillations were proposed by Bruno Pontecorvo (1967). Experiments carried out in Japan with the Super-Kamiokande detector demonstrate that cosmic-ray-produced muon neutrinos mix with tau neutrinos and experiments with the KamLAND liquid scintillator detector located at the site of the first Kamiokande detector demonstrate that reactor-produced electron antineutrinos mix with muon antineutrinos. The liquid-scintillator experiment is particularly important for the interpretation of solar neutrino experiments since it is reasonable to suppose that the parameters Δ^2 and $\sin(2\theta)$ for neutrinos are identical with the same parameters for antineutrinos.

The first experiment of the Super-Kamiokande group showed that muon neutrinos created by cosmic rays impinging on the Earth's atmosphere mix with tau neutrinos (Fukuda, 2000). Further experiments showed that Δ^2, the difference between the squared masses associated with muon and tau neutrinos, satisfies (Ashie *et al.*, 2004)

$$1.9 \times 10^{-3} \ (\text{eV})^2 < \Delta^2 < 3.0 \times 10^{-3} \ (\text{eV})^2 \tag{10.5.29}$$

and confirmed the dependence on energy predicted by the analogues of eqs. (10.5.14)–(10.5.17). The mixing angle θ between tau and muon neutrinos satisfies (Ashie *et al.*, 2005)

$$\sin^2(2\theta) > 0.92. \tag{10.5.30}$$

The results are a consequence of the observation of over 15 000 events in a five year period, with neutrino energies ranging from 100 MeV to 10 TeV and neutrino flight lengths varying from 10 km to 13 000 km.

The KamLAND experiment (Kamioka Liquid Scintillator Anti-Neutrino Detector) counts antineutrinos from 53 surrounding nuclear reactors using various consequences of the reaction

$$\bar{\nu}_e + p \rightarrow n + e^+. \tag{10.5.31}$$

The delayed 2.225 MeV photon from the capture of a neutron by a proton is useful in reducing background noise. The prompt scintillation light from the positron (including twice the rest mass energy plus the kinetic energy of the positron) gives an estimate of the energy of the antineutrino. A kiloton of liquid scintillator is contained in a transparent spherical balloon 13 meters in diameter suspended in non-scintillating oil and viewed by an array of nearly 2000 photomultipliers. The reactors vary in distance from the detector from 88 km to over 295 km, with approximately 80% of the antineutrino flux coming from 26 reactors at distances between 138 km and 214 km.

The initial analysis of 145 days of counting over a 19 month interval by the KamLAND collaboration yielded 54 $\bar{\nu}_e$ capture events versus the 87 events expected if no flavor oscillations had occurred, with most of the deficiency in flux attributable to a deficiency of antineutrinos from the 26 reactors at an average distance of 180 km (Eguchi *et al.*, 2003). These results were interpreted to imply, tentatively, $\Delta^2 \sim 0.55 \times 10^{-4}$ (eV)^2 and $\sin^2 2\theta \sim 0.83$.

After upgrading the detector and collecting data over an additional 22 months, the KamLAND collaboration (Araki *et al.*, 2005) observed 258 events versus the 365 events expected in the absence of oscillations. Combining with data from the first experimental run, the collaboration concluded that

$$\Delta^2 = (0.79 \pm 0.06) \times 10^{-4} \ (\text{eV})^2 \tag{10.5.32}$$

and

$$\sin^2 (2\theta) \sim 0.82. \tag{10.5.33}$$

The discussion in Sections 10.6 and 10.7 shows that the interaction between electron-flavor neutrinos and electrons in the Sun has a relatively minor effect on neutrinos having energies as small as those emitted in the pp, pep, and $^7\text{Be}(e^-, \nu_e)$ reactions. This means that the average flux of low energy electron neutrinos arriving at the Earth from the Sun should be smaller than the total flux of such neutrinos predicted by solar models by roughly the probability factor given by eq. (10.5.18) for oscillations in vacuum.

Assuming that the mixing angle between flavors is the same for neutrinos as for antineutrinos, the contribution of low energy neutrinos to the solar-model-predicted counting rate for the GALLEX, GNO, and SAGE experiments should be smaller than the flux predicted in the absence of vacuum oscillations by the factor

$$\langle P_{\nu_e} \rangle \sim 1 - \frac{1}{2} \sin^2 (2\theta) = 1 - \frac{1}{2} 0.82 = 0.59. \tag{10.5.34}$$

Thus, lumping together all neutrino sources other than the ^8B source as low energy neutrinos, the predicted contribution of low energy neutrinos to a gallium detector counting rate is in the range

$$\left(\sum^{\text{Ga}} \sigma_i \phi_i \right)^{\text{OSC}}_{\text{low energy}} (\text{SNU}) = 60 \ (Z = 0.01) - 70 \ (Z = 0.02). \tag{10.5.35}$$

This result follows from eqs. (10.4.12) and (10.4.13) for models A and B as well as from similar equations for the models made with C- and O-enhanced opacity.

The SNO experiment demonstrates that high energy ^8B neutrinos reaching the Earth manifest the electron flavor only for a fraction 0.305 of the time (eq. (10.4.23)), compared with the fraction 0.59 expected if only vacuum oscillations were involved in determining the degree of mixing. Hence, it is necessary to invoke another process such as the MSW effect discussed in Section 10.6. Given the SNO result, and attributing the result to the operation of the MSW process, the predicted contribution of ^8B neutrinos to a gallium detector counting rate, as given by models A and B, is in the range

$$\left(\sum^{\text{Ga}} \sigma_i \phi_i \right)^{\text{OSC+MSW}}_{\text{high energy}} (\text{SNU}) = 1.9 \ (Z = 0.01) - 4.9 \ (Z = 0.02). \tag{10.5.36}$$

Altogether, the predicted gallium detector counting rate is in the range

$$\left(\sum^{\text{Ga}} \sigma_i \phi_i \right)^{\text{OSC+MSW}}_{\text{all energies}} (\text{SNU}) = 62 \ (Z = 0.01) - 75 \ (Z = 0.02). \tag{10.5.37}$$

Adopting the metallicity $Z = 0.0166$ of the interpolated solar model which produces a total ^{8}B neutrino flux consistent with the SNO estimate of the total flux (eq. (10.4.22)) and interpolating logarithmically in eq. (10.5.37) gives

$$\left(\sum^{\text{Ga}} \sigma_i \phi_i \right)^{\text{OSC+MSW}}_{Z=0.0166,\ Y=0.279} = 71 \text{ SNU}, \tag{10.5.38}$$

which is identical with the SAGE experimental estimate (eq. (10.4.14)).

Performing the same exercise with the predicted counting rates for the ^{37}Cl experiment, one obtains from models A and B

$$\left(\sum^{\text{Cl}} \sigma_i \phi_i \right)^{\text{OSC+MSW}}_{\text{low energy}} (\text{SNU}) = 0.535\ (Z = 0.01) - 0.866\ (Z = 0.02), \tag{10.5.39}$$

$$\left(\sum^{\text{Cl}} \sigma_i \phi_i \right)^{\text{OSC+MSW}}_{\text{high energy}} (\text{SNU}) = 0.9998\ (Z = 0.01) - 2.288\ (Z = 0.02), \tag{10.5.40}$$

and

$$\left(\sum^{\text{Cl}} \sigma_i \phi_i \right)^{\text{OSC+MSW}}_{\text{all energies}} (\text{SNU}) = 1.401\ (Z = 0.01) - 3.287\ (Z = 0.02). \tag{10.5.41}$$

Interpolating logarithmically in Z, the observed counting rate of 2.56 SNU (eq. (10.4.9)), is achieved for

$$Z = 0.0163, \tag{10.5.42}$$

which is essentially the same value as obtained (eq. (10.4.23)) by insisting that the total ^{8}B neutrino flux given by $\phi(NC)$ found in the SNO experiments is the same as the model ^{8}B neutrino flux when flavor mixing is discounted.

The task remaining is to show that the interaction between electrons and electron neutrinos of energy in the range 5–14 MeV produces neutrinos that, on emerging from the Sun, express the electron flavor for about 30% of the time and that neutrinos of energy less than 1 MeV manifest the electron flavor with essentially the same probability as do electron neutrinos of this energy in vacuum.

10.6 The MSW effect

Lincoln Wolfenstein (1978) showed that the same term in the Feynman–Gell-Mann weak interaction Hamiltonian that leads to the conversion of an electron and a positron into an electron neutrino–antineutrino pair (see Chapter 14 in Volume 2) also leads to an interaction between electrons and electron-flavor neutrinos that may be interpreted as giving the electron-flavor neutrino an effective mass proportional to the product of the electron density and the weak coupling constant. This interaction implies that the electron neutrino, which at small electron densities is created predominantly in the lower mass eigenstate ($\cos \theta \sim 0.72$ when $\sin^2 (2\theta) \sim 0.8$), is, at large electron densities, created with an effective mass larger than that of the higher mass vacuum eigenstate and, as it travels through regions of lower and lower electron density on its passage out of the Sun, evolves towards being a muon neutrino. If the vacuum mixing angle were small, this mechanism would be very effective in reducing the flux of electron neutrinos arriving at the Earth.

The small mixing angle solution was proposed as a solution to the solar neutrino problem by S. P. Mikheyev and A. Yu. Smirnov (1986) and the entire process is referred to as the Mikheyev–Smirnov–Wolfenstein (MSW) mechanism. Studies by H. A. Bethe (1986), S. P. Rosen & J. M. Gelb (1986), S. J. Parke (1986), and many others have contributed to the elucidation of the mechanism.

To see how the counting rate in an experiment to detect neutrinos in the electron-flavor state is affected by the MSW effect, consider neutrinos for which the oscillation length is small compared with the variation in the Earth–Sun distance over the duration of the experiment. If a neutrino created in the Sun in the electron-flavor state emerges from the Sun as an electron neutrino, the time averaged probability for its detection at the Earth is given by eq. (10.5.18) and the detected neutrino flux ϕ_{observed} is related to the originating flux ϕ_{origin} by the factor

$$R_{\text{vacuum}} = \frac{\phi_{\text{origin}}}{\phi_{\text{observed}}} = \frac{1}{1 - 0.5 \, \sin^2 (2\theta)} = \frac{2}{2 - \sin^2 (2\theta)}. \tag{10.6.1}$$

If a neutrino created in the Sun as an electron neutrino emerges from the Sun as a muon neutrino, the time-averaged probability for its detection at the Earth in the electron-flavor state is given by the time average of eq. (10.5.16) and the detected neutrino flux is smaller than it would have been had the transformation into a muon neutrino not occurred by the factor

$$R_{\text{MSW}} = \frac{\phi_{\text{origin}}}{\phi_{\text{observed}}} = \frac{2}{\sin^2 (2\theta)}. \tag{10.6.2}$$

If, for example, $\sin^2 (2\theta) = 0.8$, then $R_{\text{vacuum}} = 5/3 \sim 1.67$ and $R_{\text{MSW}} = 5/2 = 2.5$. Thus, in this example, the reduction factor is increased by 50% over that prevailing in the absence of the MSW effect.

Before finding the additional mass term due to the weak interaction and exploring the transition from electron to muon neutrino in the Sun, it is instructive to examine the mass matrix in vacuum in the mass and flavor representations. A mass operator $\mathcal{M}_{\text{op}}$ and a mass matrix $\mathcal{M}^{\text{mass}}$ may be defined in such a way that

$$\mathcal{M}_{\text{op}} \begin{pmatrix} \nu_1 \\ \nu_2 \end{pmatrix} = \mathcal{M}^{\text{mass}} \begin{pmatrix} \nu_1 \\ \nu_2 \end{pmatrix} = \begin{pmatrix} m_1^2 & 0 \\ 0 & m_2^2 \end{pmatrix} \begin{pmatrix} \nu_1 \\ \nu_2 \end{pmatrix} = \begin{pmatrix} m_1^2 \, \nu_1 \\ m_2^2 \, \nu_2 \end{pmatrix}. \tag{10.6.3}$$

Using eq. (10.6.3) and the relationships between mass and flavor states given by eq. (10.5.1), elements of the vacuum mass matrix in the flavor representation can be found. For example,

$$M_{\nu_e,\nu_\mu}^{\text{flavor}} = M_{\nu_\mu,\nu_e}^{\text{flavor}} = \langle (\cos \theta \, \nu_1 + \sin \theta \, \nu_2)^* \, | \, \mathcal{M} \, | \, (-\sin \theta \, \nu_1 + \cos \theta \, \nu_2) \rangle$$

$$= \left(m_2^2 - m_1^2 \right) \cos \theta \, \sin \theta = \frac{1}{2} \left(m_2^2 - m_1^2 \right) \sin (2\theta), \tag{10.6.4}$$

and

$$M_{\nu_e,\nu_e}^{\text{flavor}} = \cos^2\theta \; m_1^2 + \sin^2\theta \; m_2^2 = \frac{1}{2}\left(m_1^2 + m_2^2\right) - \frac{1}{2}\left(m_2^2 - m_1^2\right)\left(\cos^2\theta - \sin^2\theta\right)$$

$$= \frac{1}{2}\left(m_1^2 + m_2^2\right) - \frac{1}{2}\left(m_2^2 - m_1^2\right)\cos(2\theta). \tag{10.6.5}$$

The matrix element $M_{\nu_\mu,\nu_\mu}^{\text{flavor}}$ is the same as $M_{\nu_e,\nu_e}^{\text{flavor}}$ except that the sign of the second term is reversed. Altogether,

$$\mathcal{M}^{\text{flavor}} = \frac{1}{2}\left(m_1^2 + m_2^2\right)\begin{pmatrix} 1 & 0 \\ 0 & 1 \end{pmatrix} + \frac{1}{2}\Delta^2 \begin{pmatrix} -\cos(2\theta) & +\sin(2\theta) \\ +\sin(2\theta) & +\cos(2\theta) \end{pmatrix}, \tag{10.6.6}$$

where Δ^2 is given by eq. (10.5.23).

The relevant term in the weak interaction Hamiltonian, as given by eq. (15.5.9) in Chapter 15 of Volume 2, is, in the notation of this section,

$$H_{\text{int}} = \frac{g_{\text{weak}}}{\sqrt{2}}\left[\bar{\nu}_e \, \gamma_\mu \, (1 + i\gamma_5)\, \nu_e\right]\left[\bar{\psi}_e \, \gamma_\mu \, (1 + i\gamma_5)\, \psi_e\right], \tag{10.6.7}$$

where g_{weak} is the axial vector weak interaction constant, ν_e and ψ_e are, respectively, electron-flavor neutrino and electron plane-wave functions which look basically like $(1/\sqrt{V})\, e^{(i/\hbar)\,(\mathbf{p}\cdot\mathbf{r}-Et)}$, and the γs are 4×4 matrices described in Section 15.4. A summation over the index $\mu = 1, 4$ is implied. Using the techniques developed in Section 15.5, and summing up over all electrons in the volume V, it follows that, for electrons at rest, the only term in the second square bracket which does not vanish is

$$\left[\bar{\psi}_e \, \gamma_4 \, (1 + i\gamma_5)\, \psi_e\right] = \bar{\psi}_e\psi_e = n_e, \tag{10.6.8}$$

where n_e is the electron number density (in cm^{-3}). The one relevant term in the first square bracket of eq. (10.6.7) is

$$\left[\bar{\nu}_e \, \gamma_4 \, (1 + i\gamma_5)\, \nu_e\right] = \frac{2}{V}. \tag{10.6.9}$$

Thus,

$$H_{\text{int}} = \frac{g_{\text{weak}}}{\sqrt{2}}\frac{2}{V}\, n_e \tag{10.6.10}$$

and

$$E_{\text{int}} = \int H_{\text{int}}\, dV = \sqrt{2}\, g_{\text{weak}}\, n_e. \tag{10.6.11}$$

The quantity E_{int} is the interaction energy between one electron-flavor neutrino and all surrounding electrons and it may be interpreted as a potential energy in the equation

$$p^2 + m^2 = (E - E_{\text{int}})^2 = E^2 - 2\,E\,E_{\text{int}} + E_{\text{int}}^2 \sim E^2 - 2E\,E_{\text{int}}. \tag{10.6.12}$$

Thus,

$$E^2 \sim p^2 + (m^2 + 2E\,E_{\text{int}}) = p^2 + \left(m^2 + 2E\,\sqrt{2}\, g_{\text{weak}}\, n_e\right), \tag{10.6.13}$$

and one may envision that the mass matrix in the flavor representation is to be augmented by

$$M' = \begin{pmatrix} A_{int} & 0 \\ 0 & 0 \end{pmatrix}, \tag{10.6.14}$$

where

$$A_{int} = 2E \sqrt{2}\, g_{weak}\, n_e. \tag{10.6.15}$$

With the choice $g_{weak} = 1.42 \times 10^{-49}$ erg cm^3,

$$A_{int} = 1.504 \times 10^{-7}\ (\text{eV})^2\ E(\text{MeV})\ \frac{\rho(\text{g cm}^{-3})}{\mu_e}, \tag{10.6.16}$$

where μ_e is the electron molecular weight and ρ is the matter density in g cm^{-3}.

Adding the matrix in eq. (10.6.14) to the matrix in eq. (10.6.6) gives

$$\mathcal{M}^{\text{flavor}} = \frac{1}{2} \left(m_1^2 + m_2^2 + A_{int} \right) \begin{pmatrix} 1 & 0 \\ 0 & 1 \end{pmatrix}$$

$$+ \frac{1}{2} \begin{pmatrix} A_{int} - \Delta^2 \cos(2\theta) & \Delta^2 \sin(2\theta) \\ \Delta^2 \sin(2\theta) & -A_{int} + \Delta^2 \cos(2\theta) \end{pmatrix}. \tag{10.6.17}$$

The eigenvalues of $\mathcal{M}^{\text{flavor}}$ are

$$m_\nu^2 = \frac{1}{2} \left(m_1^2 + m_2^2 + A_{int} \right) \pm \frac{1}{2} \left[(A_{int} - \Delta^2 \cos(2\theta))^2 + \Delta^2 \sin^2(2\theta) \right]^{1/2}. \tag{10.6.18}$$

The branch of the solution which follows from the choice of the plus sign in eq. (10.6.18) gives

$$m_+^2 = m_1^2 + \frac{A_{int}}{2}\, [1 + \cos(2\theta)] \tag{10.6.19}$$

when $A_{int} \ll \Delta^2 \cos(2\theta)$ and

$$m_+^2 = m_1^2 + A_{int} \tag{10.6.20}$$

when $A_{int} \gg \Delta^2 \cos(2\theta)$. The branch of the solution which follows from the choice of the minus sign gives

$$m_-^2 = m_2^2 + \frac{A_{int}}{2}\, [1 - \cos(2\theta)] \tag{10.6.21}$$

when $A_{int} \ll \Delta^2 \cos(2\theta)$, and

$$m_-^2 = m_2^2 \tag{10.6.22}$$

when $A_{int} \gg \Delta^2 \cos(2\theta)$.

When $A_{\text{int}} \neq 0$, the mixing angle θ_m that relates eigenfunctions of the mass matrix to flavor eigenfunctions is not the same as the vacuum mixing angle θ. In eq. (10.5.2), replacing θ by θ_m and replacing ν_1 and ν_2 by, respectively, ν_+ and ν_-, it follows that

$$\nu_+(t) = \cos\theta_m \, \nu_e(t) - \sin\theta_m \, \nu_\mu(t) \tag{10.6.23}$$

and

$$\nu_-(t) = \sin\theta_m \, \nu_e(t) + \cos\theta_m \, \nu_\mu(t). \tag{10.6.24}$$

Since the off-diagonal matrix elements of $\mathcal{M}$ vanish in the mass representation, the relationship between θ_m and θ can be found by forming and setting to zero the quantity

$$\begin{aligned}
\langle \nu_+ | \mathcal{M} | \nu_- \rangle &= \langle (\cos\theta_m \, \nu_e - \sin\theta_m \, \nu_\mu)^* | \mathcal{M} | (\sin\theta_m \, \nu_e + \cos\theta_m \, \nu_\mu) \rangle \\
&= \cos\theta_m \sin\theta_m \, \mathcal{M}_{ee}^{\text{flavor}} - \sin^2\theta_m \, \mathcal{M}_{\mu e}^{\text{flavor}} \\
&\quad + \cos^2\theta_m \, \mathcal{M}_{e\mu}^{\text{flavor}} - \sin\theta_m \cos\theta_m \, \mathcal{M}_{\mu\mu}^{\text{flavor}} \\
&= 2\cos\theta_m \sin\theta_m \left[A_{\text{int}} - \Delta^2 \cos(2\theta) \right] + \left[\cos^2\theta_m - \sin^2\theta_m \right] \Delta^2 \sin(2\theta) \\
&= \sin(2\theta_m) \left[A_{\text{int}} - \Delta^2 \cos(2\theta) \right] - \cos(2\theta_m) \Delta^2 \sin(2\theta) = 0. \tag{10.6.25}
\end{aligned}$$

The last identity in eq. (10.6.25) can be manipulated to yield

$$\sin(2\theta_m) = \frac{\sin(2\theta)}{\sqrt{\sin^2(2\theta) + \left[(A_{\text{int}}/\Delta^2) - \cos(2\theta) \right]^2}}. \tag{10.6.26}$$

If the electron neutrino is created in a region where

$$A_{\text{int}} \sim \Delta^2 \cos(2\theta), \tag{10.6.27}$$

then

$$\sin(2\theta_m) = 1, \ \theta_m = \frac{\pi}{4}, \ \text{and}$$

$$\nu_+(0) \sim \nu_-(0) = \frac{1}{\sqrt{2}}. \tag{10.6.28}$$

Thus, if at the point of creation the resonance condition given by eq. (10.6.27) is met, mixing between mass eigenstates is maximal. If the electron neutrino is created in a region where

$$A_{\text{int}} \gg \Delta^2 \cos(2\theta), \tag{10.6.29}$$

then

$$\sin(2\theta_m) = 0, \ \theta_m = 0, \ \nu_+(0) = 1 \ \text{and} \ \nu_-(0) = 0, \tag{10.6.30}$$

so the electron-flavor neutrino is completely in the more massive mass eigenstate.

How its combination of flavor states varies as a neutrino passes through a varying density environment on its way out of the Sun requires careful scrutiny. If the width of the resonance region is large compared with the vacuum mixing length, the neutrino will emerge from the resonance region predominantly in the muon flavor state. However, if the mixing angle in vacuum is near $\pi/4$ and the mixing length in vacuum is small compared with the variation in the Earth–Sun distance, whether or not the neutrino leaves the Sun as an electron neutrino or as a muon neutrino makes essentially no difference. For most neutrino energies of interest, the MSW effect is important only if the vacuum mixing angle is such that $\sin^2(2\theta)$ is smaller than unity by a non-trivial amount (see eqs. (10.6.1) and (10.6.2)).

In order to follow the time development of the neutrino wave functions when the interaction between electrons and the neutrino in the electron-flavor state is taken into account, it is convenient to recast the treatment of the time development in vacuum. In the energy representation, the Hamiltonian matrix is

$$H^{\mathrm{mass}} = \begin{pmatrix} E_1 & 0 \\ 0 & E_2 \end{pmatrix} \sim \begin{pmatrix} p + \frac{m_1^2}{2p} & 0 \\ 0 & p + \frac{m_2^2}{2p} \end{pmatrix}. \tag{10.6.31}$$

The matrix elements of H in the flavor representation may be obtained from those in the mass representation by using eq. (10.5.1). The results are

$$H_{\mathrm{ee}}^{\mathrm{flavor}} = \langle \nu_{\mathrm{e}}|H|\nu_{\mathrm{e}} \rangle = H_{\mu\mu}^{\mathrm{flavor}} = \langle \nu_{\mu}|H|\nu_{\mu} \rangle$$

$$= \langle (\cos\theta\, \nu_1 + \sin\theta\, \nu_2)^* |H| (\cos\theta\, \nu_1 + \sin\theta\, \nu_2) \rangle$$

$$= \cos^2\theta \left(p + \frac{m_1^2}{2p} \right) + \sin^2\theta \left(p + \frac{m_2^2}{2p} \right) = p + \cos^2\theta\, \frac{m_1^2}{2p} + \sin^2\theta\, \frac{m_2^2}{2p} \tag{10.6.32}$$

and

$$H_{\mathrm{e}\mu}^{\mathrm{flavor}} = \langle \nu_{\mathrm{e}}|H|\nu_{\mu} \rangle = H_{\mu \mathrm{e}}^{\mathrm{flavor}} = \langle \nu_{\mu}|H|\nu_{\mathrm{e}} \rangle = \sin\theta\, \cos\theta\, \frac{\Delta^2}{2p}. \tag{10.6.33}$$

Thus,

$$H^{\mathrm{flavor}} = \begin{pmatrix} p & 0 \\ 0 & p \end{pmatrix} + \frac{1}{2p} \begin{pmatrix} m_1^2 \cos^2\theta + m_2^2 \sin^2\theta & \sin\theta\, \cos\theta\, \Delta^2 \\ \sin\theta\, \cos\theta\, \Delta^2 & m_1^2 \sin^2\theta + m_2^2 \cos^2\theta \end{pmatrix}, \tag{10.6.34}$$

which can be transformed into

$$H^{\mathrm{flavor}} = \frac{1}{2} \left(2p + \frac{m_1^2 + m_2^2}{2p} \right) \begin{pmatrix} 1 & 0 \\ 0 & 1 \end{pmatrix} + \frac{\Delta^2}{4p} \begin{pmatrix} -\cos(2\theta) & \sin(2\theta) \\ \sin(2\theta) & \cos(2\theta) \end{pmatrix}. \tag{10.6.35}$$

Note the similarities and differences between the Hamiltonian matrix and the mass matrix given by eq. (10.6.6).

The interaction between electrons and a neutrino in the electron-flavor state is introduced by replacing m_1^2 in the first term of eq. (10.6.35) by $m_1^2 + A_{\mathrm{int}}$. This is equivalent to adding to the Hamiltonian matrix of eq. (10.6.35) the matrix or matrices

$$H_{\text{add}}^{\text{flavor}} = \begin{pmatrix} \frac{A_{\text{int}}}{2p} & 0 \\ 0 & 0 \end{pmatrix} = \begin{pmatrix} \frac{A_{\text{int}}}{4p} & 0 \\ 0 & \frac{A_{\text{int}}}{4p} \end{pmatrix} + \begin{pmatrix} \frac{A_{\text{int}}}{4p} & 0 \\ 0 & -\frac{A_{\text{int}}}{4p} \end{pmatrix}, \tag{10.6.36}$$

so that

$$H^{\text{flavor}} = \left(p + \frac{m_1^2 + A_{\text{int}} + m_2^2}{4p} \right) \begin{pmatrix} 1 & 0 \\ 0 & 1 \end{pmatrix}$$

$$+ \frac{1}{4p} \begin{pmatrix} A_{\text{int}} - \Delta^2 \cos(2\theta) & \Delta^2 \sin(2\theta) \\ \Delta^2 \sin(2\theta) & -A_{\text{int}} + \Delta^2 \cos(2\theta) \end{pmatrix}. \tag{10.6.37}$$

Note again the similarities and differences between the Hamiltonian matrix and the mass matrix as given by eq. (10.6.17). Combining the two matrices in eq. (10.6.37), it is evident that the off-diagonal elements are equal and that only the upper left hand element is a function of position. Given its small rest mass energy compared with its total energy, a typical neutrino travels essentially at the speed of light in vacuum, and, knowing its trajectory through the density profile, A_{int} can be converted into a function only of time. The Hamiltonian matrix may therefore be written in the flavor representation as

$$H^{\text{flavor}} = \begin{pmatrix} \alpha(t) & \beta \\ \beta & \gamma \end{pmatrix}, \tag{10.6.38}$$

where

$$\alpha(t) = \left(p + \frac{m_1^2 + m_2^2}{4p} \right) + \frac{\Delta^2}{4p} \left(\frac{2 A_{\text{int}}(t)}{\Delta^2} - \cos(2\theta) \right), \tag{10.6.39}$$

$$\beta = \frac{\Delta^2}{4p} \sin(2\theta), \tag{10.6.40}$$

and

$$\gamma = \left(p + \frac{m_1^2 + m_2^2}{4p} \right) + \frac{\Delta^2}{4p} \cos(2\theta). \tag{10.6.41}$$

Following the logic of Wolfenstein (1978), one may postulate that the time development of the flavor states is given by the solution of

$$-\frac{\hbar}{i} \frac{\partial}{\partial t} \begin{pmatrix} \nu_e \\ \nu_\mu \end{pmatrix} = H^{\text{flavor}} \begin{pmatrix} \nu_e \\ \nu_\mu \end{pmatrix}, \tag{10.6.42}$$

or

$$-\frac{\hbar}{i} \frac{d}{dt} \begin{pmatrix} \nu_e \\ \nu_\mu \end{pmatrix} = \begin{pmatrix} \alpha & \beta \\ \beta & \gamma \end{pmatrix} \begin{pmatrix} \nu_e \\ \nu_\mu \end{pmatrix}. \tag{10.6.43}$$

By a differentiation with respect to time, followed by several substitutions, eq. (10.6.43) can be converted into a second order differential equation for ν_e:

$$-\frac{\hbar}{i}\frac{d^2\nu_e}{dt^2} = \alpha \frac{d\nu_e}{dt} + \nu_e \frac{d\alpha}{dt} + \beta \frac{d\nu_\mu}{dt}$$

$$= -\frac{i}{\hbar}\alpha\left(\alpha\,\nu_e + \beta\,\nu_\mu\right) + \nu_e\frac{d\alpha}{dt} - \frac{i}{\hbar}\beta\left(\beta\,\nu_e + \gamma\,\nu_\mu\right)$$

$$= \alpha\frac{d\nu_e}{dt} - \frac{i}{\hbar}\left(\alpha^2 + \beta^2\right)\nu_e + \nu_e\frac{d\alpha}{dt} - \frac{i}{\hbar}\beta\left(\alpha + \gamma\right)\nu_\mu$$

$$= \alpha\frac{d\nu_e}{dt} - \frac{i}{\hbar}\left(\alpha^2 + \beta^2\right)\nu_e + \nu_e\frac{d\alpha}{dt} - \frac{i}{\hbar}\beta\left(\alpha + \gamma\right)\frac{1}{\beta}\left(-\frac{\hbar}{i}\frac{d\nu_e}{dt} - \alpha\,\nu_e\right)$$

$$= \alpha\frac{d\nu_e}{dt} - \frac{i}{\hbar}\left(\beta^2 - \gamma\,\alpha\right)\nu_e + \nu_e\frac{d\alpha}{dt} + \left(\alpha + \gamma\right)\frac{d\nu_e}{dt}.$$

$$= \left(2\,\alpha + \gamma\right)\frac{d\nu_e}{dt} - \frac{i}{\hbar}\left(\beta^2 - \gamma\,\alpha\right)\nu_e + \nu_e\frac{d\alpha}{dt}, \tag{10.6.44}$$

or

$$\frac{d^2\nu_e}{dt^2} - \frac{\hbar}{i}\left(2\,\alpha + \gamma\right)\frac{d\nu_e}{dt} + \left(\beta^2 - \gamma\,\alpha\right)\nu_e = -\frac{i}{\hbar}\nu_e\frac{d\alpha}{dt}. \tag{10.6.45}$$

Following Rosen and Gelb (1986), one may set

$$\nu_e(t) = z(t)\,e^{-i\phi(t)}, \tag{10.6.46}$$

where $z(t)$ is an amplitude to be found and the phase factor $\phi(t)$ is given by

$$\phi(t) = \frac{1}{\hbar}\int_0^t \alpha(t')\,dt'. \tag{10.6.47}$$

Using

$$\frac{d}{dt}\,e^{-i\phi(t)} = -\frac{i}{\hbar}\,\alpha(t)\,e^{-i\phi(t)}, \tag{10.6.48}$$

it follows that

$$\frac{d\nu_e}{dt} = \left(\frac{dz(t)}{dt} - \frac{i}{\hbar}\alpha(t)\,z(t)\right)e^{-i\phi(t)} \tag{10.6.49}$$

and

$$\frac{d^2\nu_e}{dt^2} = \left(\frac{d^2z(t)}{dt^2} - \frac{i}{\hbar}\,2\,\alpha(t)\frac{dz(t)}{dt} - \frac{\alpha^2(t)}{\hbar^2}\,z(t) - \frac{i}{\hbar}\frac{d\alpha(t)}{dt}\right)e^{-i\phi(t)}. \tag{10.6.50}$$

Using eqs. (10.6.46), (10.6.49), and (10.6.50) in eq. (10.6.45), one obtains

$$\frac{d^2z(t)}{dt^2} - \frac{i}{\hbar}\left(\alpha - \gamma\right)\frac{dz(t)}{dt} + \frac{\beta^2}{\hbar^2}\,z(t) = 0. \tag{10.6.51}$$

Combining with eqs. (10.6.39)–(10.6.41), one gets

$$\frac{d^2 z(t)}{dt^2} - 2\frac{i}{\hbar}\frac{\Delta^2}{4p}\left(\frac{A_{\text{int}}(t)}{\Delta^2} - \cos(2\theta)\right)\frac{dz(t)}{dt} + \frac{1}{\hbar^2}\left(\frac{\Delta^2}{4p}\right)^2 \sin^2(2\theta)\, z(t) = 0.$$

(10.6.52)

To complete the story, it follows from eqs. (10.6.39) and (10.6.47) that

$$\phi(t) = \frac{1}{\hbar}\int_0^t \left[\left(p + \frac{m_1^2 + m_2^2}{4p}\right) + \frac{\Delta^2}{4p}\left(\frac{2\,A_{\text{int}}(t)}{\Delta^2} - \cos(2\theta)\right)\right] dt'. \qquad (10.6.53)$$

When $A_{\text{int}}(t) = 0$, setting

$$z(t) = e^{i\omega t} \qquad (10.6.54)$$

in eq. (10.6.52) gives

$$(\hbar\omega)^2 + 2\frac{\Delta^2}{4p}\cos(2\theta)\,\hbar\omega - \left(\frac{\Delta^2}{4p}\sin(2\theta)\right)^2 = 0, \qquad (10.6.55)$$

which has the solutions

$$\hbar\omega = -\frac{\Delta^2}{4p}\cos(2\theta) \pm \frac{\Delta^2}{4p}. \qquad (10.6.56)$$

The general solution in vacuum for $z(t)$ is then

$$z(t) = \exp\left(-\frac{i}{\hbar}\frac{\Delta^2}{4p}\cos(2\theta)\,t\right)\left[N_+ \exp\left(+\frac{i}{\hbar}\frac{\Delta^2}{4p}\,t\right) + N_- \exp\left(-\frac{i}{\hbar}\frac{\Delta^2}{4p}\,t\right)\right],$$

(10.6.57)

where N_+ and N_- are constants associated, respectively, with the frequencies ω_+ and ω_-, where

$$\hbar\omega_+ = \frac{\Delta^2}{4p}\left(1 - \cos(2\theta)\right) \qquad (10.6.58)$$

and

$$\hbar\omega_- = -\frac{\Delta^2}{4p}\left(1 + \cos(2\theta)\right). \qquad (10.6.59)$$

From eq. (10.6.53) one has

$$\phi(t) = \frac{1}{\hbar}\left[\left(p + \frac{m_1^2 + m_2^2}{4p}\right) - \frac{\Delta^2}{4p}\cos(2\theta)\right] t. \qquad (10.6.60)$$

Note that, in the product $z(t)\,e^{-i\phi(t)}$, the cosine term drops out. Note further that

$$\exp\left(-\frac{i}{\hbar}\frac{m_1^2+m_2^2}{4p}t\right)\exp\left(+\frac{i}{\hbar}\frac{\Delta^2}{4p}t\right)=\exp\left(-\frac{i}{\hbar}\frac{m_1^2}{2p}t\right) \qquad (10.6.61)$$

and that

$$\exp\left(-\frac{i}{\hbar}\frac{m_1^2+m_2^2}{4p}t\right)\exp\left(-\frac{i}{\hbar}\frac{\Delta^2}{4p}t\right)=\exp\left(-\frac{i}{\hbar}\frac{m_2^2}{2p}t\right). \qquad (10.6.62)$$

Putting all the pieces together, the general solution for the electron neutrino in vacuum is

$$\nu_e(t)=N_+\,\exp\left(-i\frac{E_1}{\hbar}t\right)+N_-\,\exp\left(-i\frac{E_2}{\hbar}t\right). \qquad (10.6.63)$$

Comparing with eq. (10.5.7) it is evident that, if the initial state is the electron-flavor state, the appropriate choices for the coefficients are

$$N_+=\cos^2\theta=\frac{1}{2}\left(1+\cos\left(2\theta\right)\right)\ \text{and}\ N_-=\sin^2\theta=\frac{1}{2}\left(1-\cos\left(2\theta\right)\right). \qquad (10.6.64)$$

If the initial state is the muon-flavor state, eq. (10.5.11) shows that the appropriate choices for the coefficients are

$$N_+=-\sin\theta\cos\theta=-\frac{1}{2}\sin\left(2\theta\right)\ \text{and}\ N_-=+\sin\theta\cos\theta=\frac{1}{2}\sin\left(2\theta\right). \qquad (10.6.65)$$

Defining a vacuum frequency as

$$\omega_{\text{vac}}=\frac{|E_2-E_1|}{2\hbar}\sim\frac{1}{\hbar}\frac{\Delta^2}{4p}, \qquad (10.6.66)$$

eq. (10.6.57) can be written as

$$z(t)=e^{-i\cos\left(2\theta\right)\,\omega_{\text{vac}}t}\left[N_+\,e^{+i\omega_{\text{vac}}t}+N_-\,e^{-i\omega_{\text{vac}}t}\right] \qquad (10.6.67)$$

and the amplitudes for the electron-flavor wave functions of electron and muon neutrinos in vacuum may be derived, respectively, from

$$z_e(t)=e^{-i\cos\left(2\theta\right)\,\omega_{\text{vac}}t}\left[\cos\omega_{\text{vac}}t+i\cos\left(2\theta\right)\sin\omega_{\text{vac}}t\right] \qquad (10.6.68)$$

and

$$z_e'(t)=e^{-i\cos\left(2\theta\right)\,\omega_{\text{vac}}t}\left[-i\sin\left(2\theta\right)\sin\omega_{\text{vac}}t\right]. \qquad (10.6.69)$$

When multiplied by $e^{-i\phi(t)}$ the z_es in eqs. (10.6.68) and (10.6.69) become, respectively, the ν_es of eqs. (10.5.7) and (10.5.11).

In general, a neutrino formed in the deep interior of the Sun does not emerge as either an electron or a muon neutrino. Assuming that the electron-flavor component of any neutrino in vacuum is a linear combination of the electron-flavor wave functions, one may write

$$z(t) = a\, z_{\mathrm{e}}(t) + b\, z'_{\mathrm{e}}(t), \tag{10.6.70}$$

$$|z(t)|^2 = |a|^2 \left(1 - \sin^2(2\theta)\, \sin^2 \omega_{\mathrm{vac}} t\right) + |b|^2 \sin^2(2\theta)\, \sin^2 \omega_{\mathrm{vac}} t$$

$$- \frac{1}{2}(a^*b + b^*a)\, \sin(4\theta)\, \sin^2 \omega_{\mathrm{vac}} t, \tag{10.6.71}$$

and, averaged over time,

$$\langle |z(t)|^2 \rangle = |a|^2 \left(1 - \frac{1}{2}\sin^2(2\theta)\right) + |b|^2 \frac{1}{2}\sin^2(2\theta) - \frac{1}{4}(a^*b + b^*a)\, \sin(4\theta). \tag{10.6.72}$$

Comparison with eqs. (10.5.15) and (10.5.16) shows that

$$\langle |z(t)|^2 \rangle = |a|^2 \langle P_{\nu_{\mathrm{e}}}(t)\rangle + |b|^2 \langle P'_{\nu_{\mathrm{e}}}(t)\rangle - \frac{1}{4}(a^*b + b^*a)\, \sin(4\theta), \tag{10.6.73}$$

where $P_{\nu_{\mathrm{e}}}(t)$ and $P'_{\nu_{\mathrm{e}}}(t)$ are the probabilities for the electron flavor to be expressed by, respectively, an electron neutrino and a muon neutrino. Presumably, $|a|^2 + |b|^2 = 1$, and the third term in eq. (10.6.73) shows that, after having passed through a variable matter-density profile, the time-averaged probabilities with which the hybrid neutrinos emerging from the Sun express the electron flavor could be less (or larger) than the time-averaged probabilities with which either electron or muon neutrinos express this flavor.

10.7 Numerical solutions for neutrinos produced at the center of a simple solar model

When the trajectory of a neutrino passes through matter, solutions in general require numerical integration. As a first step, eq. (10.6.52) may be converted into two first order differential equations:

$$\frac{dz(t)}{dt} = y(t) \tag{10.7.1}$$

and

$$\frac{dy(t)}{dt} = 2\frac{i}{\hbar}\frac{\Delta^2}{4p}\left(\frac{A_{\mathrm{int}}(t)}{\Delta^2} - \cos(2\theta)\right) y(t) - \frac{1}{\hbar^2}\left(\frac{\Delta^2}{4p}\right)^2 \sin^2(2\theta)\, z(t). \tag{10.7.2}$$

From eqs. (10.6.46) and (10.6.47),

$$\frac{dz(t)}{dt} = \frac{d\nu_{\mathrm{e}}(t)}{dt}\, e^{i\phi(t)} + \frac{i}{\hbar}\alpha(t)\, z(t) \tag{10.7.3}$$

and, from eq. (10.6.43),

$$\frac{dv_e(t)}{dt} = -\frac{i}{\hbar} \left[\alpha(t) \, v_e(t) + \beta \, v_\mu(t) \right].$$

(10.7.4)

Since, by assumption, neutrinos emitted in nuclear beta-decay reactions are in the pure electron-flavor state,

$$v_e(0) = 1 \text{ and } v_\mu(0) = 0.$$

(10.7.5)

From eq. (10.6.47), $\phi(0) = 0$. So, from eqs. (10.7.3)–(10.7.5), initial conditions for z and y are

$$z(0) = 1 \text{ and } y(0) = \left(\frac{dz(t)}{dt} \right)_{t=0} = 0.$$

(10.7.6)

In numerical integrations, the time step Δt may be controlled by insisting that $\omega_{max} \Delta t \ll 1$, where $\omega_{max} = \max(|\omega_+|, |\omega_-|)$ and

$$\hbar \omega_\pm = \frac{\Delta^2}{4p} \left(\frac{A_{int}(t)}{\Delta^2} - \cos(2\theta) \right) \pm \frac{\Delta^2}{4p} \sqrt{\left(\frac{A_{int}(t)}{\Delta^2} - \cos(2\theta) \right)^2 + \sin^2(2\theta)}.$$

(10.7.7)

In calculations which produce the results described hereinafter, the third fourth order Runge-Kutta routine described in Section 8.5 has been used, with typically 200 time steps per oscillation period.

In an illustrative example, consider a 10 MeV neutrino produced at the center of a simple solar model in which the matter density drops off exponentially with distance from the center according to

$$\rho = \rho_0 \, \exp\left(-\frac{r}{R_{sc}} \right),$$

(10.7.8)

where ρ_0 is the central density in g cm^{-3} and $R_{sc} \ll R_\odot$. Conservation of mass requires that

$$M_\odot = \frac{4\pi}{3} \bar{\rho} \, R_\odot^3 = 4\pi \rho_0 \int r^2 \, e^{-r/R_{sc}} \, dr \sim 8\pi \rho_0 \, R_{sc}^3,$$

(10.7.9)

or

$$\rho_0 \sim \frac{\bar{\rho}}{6} \left(\frac{R_\odot}{R_{sc}} \right)^3,$$

(10.7.10)

where $\bar{\rho} = 1.41$ g cm^{-3}. Choosing

$$R_{sc} = \frac{1}{10} R_\odot,$$

(10.7.11)

it follows that that $\rho_0 \sim 233$ g cm^{-3}. Assuming that the average electron molecular weight in central regions is $\mu_e \sim 1.5$, the electron-density profile may be approximated roughly by

$$\rho_e \sim 150 \text{ g cm}^{-3} \exp\left(-\frac{r}{10R_\odot}\right). \tag{10.7.12}$$

With the choices

$$\Delta^2 = 0.8 \times 10^{-4} \text{ (eV)}^2 \tag{10.7.13}$$

and

$$\sin^2(2\theta) = 0.8, \tag{10.7.14}$$

eq. (10.6.16) gives that, at the model center, the matter-induced mass squared of the electron-flavor neutrino is related to the mass squared difference in vacuum by

$$\frac{A_{\text{int}}}{\Delta^2} = 0.282 \, E(\text{MeV}). \tag{10.7.15}$$

For a 10 MeV neutrino, the mixing parameter in matter, as given by eq. (10.6.26), is

$$\sin^2(2\theta_m) = 0.1244, \tag{10.7.16}$$

and the matter-induced oscillation frequency, as given by eq. (10.7.7), is

$$\omega_+ = 14\,915 \text{ s}^{-1}. \tag{10.7.17}$$

If A_{int} were constant, the square of the amplitude of the solution would be exactly

$$|z|^2 = 1 - \sin^2(2\theta_m) \sin^2 \omega_{\text{osc}} \, t, \tag{10.7.18}$$

where

$$\omega_{\text{osc}} = \frac{\omega_+}{2} \sim 7458 \text{ s}^{-1}, \tag{10.7.19}$$

and the neutrino would express the electron-neutrino flavor 93.78 % of the time.

Near the model center, $A_{\text{int}}(t)$ varies sufficiently slowly with time that, for the first few oscillation cycles, the numerical solution is very well fitted by eq. (10.7.18), with parameters given by eqs. (10.7.16) and (10.7.19). As the neutrino travels outward, the oscillation frequency becomes smaller and both the term $C(t)$ and the coefficient $\sin^2(2\theta_m)$ in the equation

$$|z(t)|^2 = C(t) - \sin^2(2\theta_m) \sin^2 \omega_{\text{osc}} \, (t - t_0(t)) \tag{10.7.20}$$

decrease, $\sin^2(2\theta_m)$ at first decreasing more rapidly than $C(t)$. The parameter $t_0(t)$ has been introduced because $|z(t)|^2$ and $\sin^2 \omega_{\text{osc}} t$ are slightly out of phase.

By the time the neutrino has traversed half the radius of the model, $C(t)$ has become nearly independent of time at ~ 0.449 and $\omega_{\text{osc}} \sim \omega_{\text{vac}}$, where

$$\omega_{\text{vac}} \sim 3040 \text{ s}^{-1} \tag{10.7.21}$$

Table 10.7.1 Characteristics of emergent neutrinos as a function of energy when $\sin^2(2\theta) = 0.8$, $\Delta^2 = 0.8 \times 10^{-4}(eV)^2$, ρ_e as given by eq. (10.7.12)

E (MeV)	ω_+ (s^{-1})	ω_{vac} (s^{-1})	A_{int}/Δ^2	$\|z\|^2_{min}$	$\|z\|^2_{max}$	$\|z\|^2_{av}$
15	15537	2026	4.230	0.1795	0.3852	0.2825
10	14915	3040	2.820	0.1330	0.4485	0.2908
5	13837	6077	1.410	0.0318	0.6404	0.3362
2.5	14447	12154	0.705	0.0084	0.8676	0.4378
2	15478	15193	0.564	0.0276	0.9143	0.4706
1	32657	30385	0.282	0.1010	0.9799	0.5399
0.5	76061	60771	0.141	0.1656	0.9971	0.5808
0.25	167745	121542	0.0705	0.1754	0.9972	0.5856

is the oscillation frequency in vacuum. On emerging from the model, the square of the amplitude of the solution can be fitted by

$$|z(t)|^2 = 0.4485 - 0.3155 \, \sin^2 \omega_{vac}(t - t_0), \tag{10.7.22}$$

where time t_0 has been chosen so that the minimum in $\sin^2 \omega_{vac}(t - t_0)$ coincides with the maximum in $|z(t)|^2$, as is the case at the center of the model.

The emerging neutrino is neither an electron neutrino, for which, in eq. (10.6.73), $|a|^2 = 1$ and $|b|^2 = 0$, nor a muon neutrino, for which, in this same equation, $|a|^2 = 0$ and $|b|^2 = 1$. Nevertheless, however one wishes to assign the mixture of flavors, the bottom line is independent of this assignment: the time-averaged square of the electron flavor amplitude changes from $\langle|z|^2\rangle = 0.938$ at the center of the model to $\langle|z|^2\rangle = 0.291$ as the neutrino emerges from the model. Note that $\langle|z|^2\rangle$ in vacuum is smaller than the value of 0.4 which would result if the emerging neutrino were a muon neutrino.

In Table 10.7.1 the characteristics of emergent neutrinos created at the model center are shown as a function of energy. In this table, $|z|^2_{max}$, $|z|^2_{min}$, and $|z|^2_{av}$ are, respectively the maximum, minimum, and average values of the square of the amplitude of the electron-flavor wave function of the emergent neutrino. As the neutrino energy tends toward zero, the interaction between the neutrino and electrons in the model goes to zero and not only does the probability for expressing the electron flavor approach the value for an electron neutrino in vacuum, but the maximum and minimum values of the square of the electron-flavor amplitude approach the values appropriate for an electron neutrino in vacuum.

The most important lessons from Table 10.7.1 are that, assuming that the mixing angle and mass-squared difference for electron and muon neutrinos are similar to the values found by experiment to characterize electron and muon antineutrinos, on emerging from the model after being created at the model center, (1) pp neutrinos express the electron-neutrino flavor with essentially the same probability with which electron neutrinos of the same energy in vacuum express the electron-neutrino flavor and (2) ^{8}B neutrinos with energies in the range 5–15 MeV express the electron-neutrino flavor with approximately half the probability with which electron neutrinos of the same energy in vacuum express the electron-neutrino flavor.

The particular choice for the scale length R_{sc} in eqs. (10.7.11) and (10.7.12) is somewhat arbitrary. For example, the detailed solar model characterized by $(Z, Y) = (0.02, 0.309)$ has a central matter density $\rho_0 \sim 154$ g cm^{-3} which implies, through eq. (10.7.10), a scale length $R_{sc} \sim (1/9) R_\odot$ and a central electron density $\rho_e \sim 102.5$ g cm^{-3}. On the other hand, as long as $R_{sc} \ll R_\odot$, the properties of emergent neutrinos are independent of R_{sc}. That is, for given choices of Δ^2 and $\sin^2 2\theta$, the properties of emergent neutrinos are completely independent of the gradient in the electron density, and depend solely on the electron density at the point of origin. More precisely, characteristics of solutions are functions of only two parameters, $\sin^2 2\theta$ and

$$\frac{A_{int}}{\Delta^2} = \frac{2E \sqrt{2} \, g_{weak} \, n_e}{\Delta^2}$$

$$= 1.880 \times 10^{-3} \, E(\text{MeV}) \, \frac{0.8 \times 10^{-4} \, (\text{eV})^2}{\Delta^2} \, \rho_e(\text{g cm}^{-3}). \qquad (10.7.23)$$

10.8 Solutions for neutrinos produced in realistic solar models

Although the rate per unit volume $\dot{\nu}$ at which neutrinos of any given type are formed in a solar model peaks at the center, because the contribution to the total rate $\dot{N}_\nu$ by a spherical shell of radius r and radial thickness dr is

$$d\dot{N}_\nu = \dot{\nu} \, 4\pi r^2 dr, \qquad (10.8.1)$$

the contribution peaks at a position off center. It is therefore necessary to examine trajectories and flavor changes experienced by neutrinos created away from the model center. From any point in the model, neutrinos stream outward with the same probability in every direction, but those reaching the Earth pass along trajectories which are essentially parallel to each other. This means that every neutrino created in the hemispherical shell of radius r which faces the Earth passes through the same total subsequent change in electron density as any other neutrino so created. The only difference is that, the greater the angle which the neutrino's initial radius vector **r** makes with the line joining the centers of the Earth and the Sun, the slower does the decrease in density occur as the neutrino passes to the outer edge of the Sun. Just as the characteristics of emergent neutrinos produced at the center are independent of the scale length R_{sc}, as long as $R_{sc} \ll R_\odot$, so the flavor mix of an Earth-directed emergent neutrino originating anywhere on the hemisphere is independent of the initial location on the hemisphere.

Neutrinos which are born in the hemispherical shell of radius r which faces away from the Earth encounter an increasing electron density until they cross the disk which separates the two hemispheres and, as they approach the facing hemispherical shell, they encounter a decreasing electron density which is just the reverse of the prior increase. Intuition suggests that, in passing from the backward facing shell to the forward facing shell, there is

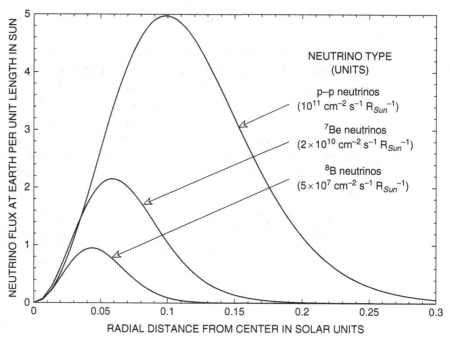

Fig. 10.8.1   Neutrino fluxes from solar model A ($Z = 0.01$, $Y = 0.2373$)

no net change in the flavor mixture. That is, an electron neutrino born on the back facing hemispherical shell and heading toward the Earth reaches the forward facing hemispherical shell as an electron neutrino and thereafter experiences the same changes as does the neutrino which is born on the facing hemispherical shell and heads toward the Earth. For all cases examined here, the intuitive expectation is born out.

These considerations allow the construction of differential distributions which describe the contributions of spherical shells in realistic solar models to the fluxes of neutrinos reaching the Earth. Figures 10.8.1 and 10.8.2 describe contributions to the flux of neutrinos of the three most relevant types from solar models characterized, respectively, by $Z = 0.01$ and $Z = 0.02$.

Although, as shown by Figs. 10.8.1 and 10.8.2 and by Table 10.4.3, the total fluxes of all neutrino types other than pp and pep neutrinos depend sensitively on the choice of solar metallicity, the shapes of the differential distributions in the models are much less sensitive to metallicity. Comparing the differential distributions for pp, ^{7}Be, and ^{8}B neutrinos in Fig. 10.8.1 with those in Fig. 10.8.2, it is evident that the radial locations of the peaks in the distributions are essentially independent of metallicity and each distribution for $Z = 0.01$ is related by a simple scaling factor to the corresponding distribution for $Z = 0.02$. That is, for the three types of neutrino, the ratios of fluxes in the two rows of Table 10.4.3 are effectively identical with the ratios of maximum $d\dot{N}_\nu/dr$ as given in Figs. 10.8.1 and 10.8.2.

The total flux of any neutrino type is obtained by integrating the differential distribution for that type. To estimate the detectability of this flux is a much more complicated process. The first step is to determine how the probability with which an emergent neutrino

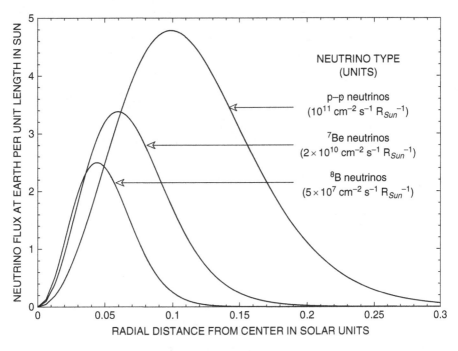

Fig. 10.8.2   Neutrino fluxes from solar model B ($Z = 0.02$, $Y = 0.296$)

expresses the electron flavor depends on birth position and energy. The electron-density distributions required for making this determination are displayed in Fig. 10.8.3 for the two choices of metallicity. Interestingly, the distributions are almost independent of metallicity and can approximated to sufficient accuracy by the function

$$\rho_e = 102.5 \text{ g cm}^{-3} \ \exp\left(-\left(\frac{r}{0.16373}\right)^2\right) \tag{10.8.2}$$

for $r \le 0.1363 R_\odot$, and by the function

$$\rho_e = 51.25 \text{ g cm}^{-3} \ \exp\left(-\frac{r - 0.1363}{0.100176}\right) \tag{10.8.3}$$

for $r > 0.1363 R_\odot$. In these equations, r is in units of $R_\odot$. The functions match at the fitting point at $r = 0.1363 \ R_\odot$ and their slopes there are nearly identical. The analytic approximation of eq. (10.8.2) fits the $Z = 0.01$ electron-density distribution slightly better than the $Z = 0.02$ distribution for much of the range $r \le 0.1363$, and the approximation given by eq. (10.8.3) fits the $Z = 0.02$ distribution better at larger r.

Coupling the near metallicity independence of the electron-density distribution with the fact that the shapes of the differential neutrino-flux distributions are also almost metallicity independent, it follows that the results of numerical experiments with the analytic approximation to ρ_e are effectively as applicable to the $Z = 0.01$ model as to the $Z = 0.02$ model, thus justifying the interpolations between models made in Sections 10.4 and 10.5.

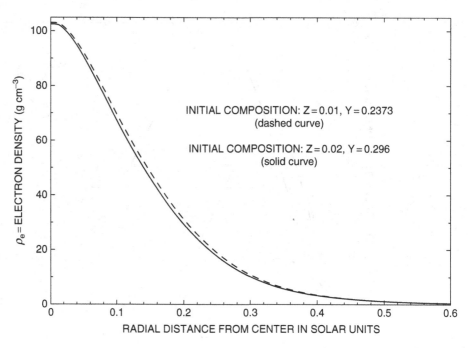

Fig. 10.8.3   Electron density versus radius in two solar models

Consider first the characteristics of emergent ^{8}B neutrinos with energies large enough to be detectable by the SNO experiment, namely, $E \geq 5$ MeV. From Figs. 10.8.1 and 10.8.2, it is evident that each peak in the distributions for ^{8}B neutrinos occurs at $r \sim 0.045~R_\odot$, and that approximately equal contributions to the emergent flux are made from regions on either side of the spherical shell of radius $r \sim 0.05~R_\odot$. Hence, it is sufficient to explore birth radii in the range $r = 0.0$ to $r = 0.1$. Several results when $\sin^2(2\theta) = 0.8$ and $\Delta^2 = 0.8 \times 10^{-4}(\text{eV})^2$ are given in Table 10.8.1. In a zeroth order approximation, one might assume that a detection probability is influenced most strongly by the properties of neutrinos emitted at the peak in the creation distribution function $d\dot{N}_\nu/dr = \dot{v}~4\pi r^2$ and at neutrino energies near the center of the distribution in energy space. The entry in Table 10.8.1 which corresponds most closely to these criteria is the one for $E = 7.5$ MeV and $r = 0.05$, suggesting $\langle P_{\nu_e} \rangle = |z|^2_{\text{av}} \sim 0.34$, approximately 10% larger than given by the SNO experiment (eq. (10.4.23)).

For comparison, characteristics of emergent low energy neutrinos are given in Table 10.8.2 as functions of birth radius when $\sin^2(2\theta) = 0.8$ and $\Delta^2 = 0.8 \times 10^{-4}(\text{eV})^2$. The range in r has been increased to reflect the larger effective volumes over which pp and ^{7}Be neutrinos are emitted.

The last set of entries in Table 10.8.2 demonstrates that the MSW effect has an essentially negligible influence on the character of the emergent neutrino created by the pp reaction. Not only is the probability of expressing the electron flavor very close to that of an electron-flavor neutrino in vacuum, but the minimum amplitude squared is close to that of a pure electron-flavor neutrino. In the case of the 0.861 MeV ^{7}Be electron-capture neutrinos,

Table 10.8.1 Characteristics of emergent high energy neutrinos as functions of energy and starting position when $\sin^2(2\theta) = 0.8$, $\Delta^2 = 0.8 \times 10^{-4}$(eV)^2, ρ_e as given by eqs. (10.8.2) and (10.8.3)

E (MeV)	r ($R_\odot$)	ρ_e ($\rho_e(0)$)	ω_+ (s^{-1})	ω_{vac} (s^{-1})	A_{int}/Δ^2	$|z|^2_{min}$	$|z|^2_{max}$	$|z|^2_{av}$
15	0.0	1	10220	2026	2.891	0.136	0.444	0.290
	0.05	0.91	9212		2.633	0.124	0.462	0.293
	0.1	0.69	6740		1.991	0.082	0.531	0.307
12.5	0.0	1	10009	2431	2.409	0.111	0.482	0.297
	0.05	0.91	9108		2.194	0.097	0.505	0.301
	0.1	0.69	6606		1.659	0.054	0.586	0.320
10	0.0	1	9750	3039	1.927	0.077	0.540	0.308
	0.05	0.91	8790		1.755	0.063	0.568	0.316
	0.1	0.69	6486		1.327	0.024	0.662	0.343
7.5	0.0	1	9473	4051	1.445	0.035	0.632	0.333
	0.05	0.91	8575		1.317	0.023	0.665	0.344
	0.1	0.69	6470		0.995	0.002	0.764	0.383
5	0.0	1	9413	6077	0.964	0.001	0.775	0.388
	0.05	0.91	8648		0.878	0.000	0.806	0.403
	0.1	0.69	6907		0.644	0.013	0.882	0.448

the minimum amplitude squared is definitely smaller than that of a pure electron-flavor neutrino, but the probability of manifesting the electron flavor is only 5 ± 2 % smaller than that of an electron-flavor neutrino in vacuum.

An enhanced feeling for the dependence on various parameters of the probability $\langle P_{\nu_e} \rangle = |z|^2_{av}$ with which an emergent neutrino expresses the electron flavor can be achieved graphically. In Fig. 10.8.4, $\langle P_{\nu_e} \rangle$ is shown as a function of A_{int}/Δ^2 for three choices of $\sin^2(2\theta)$. The energies E placed near the top border of the figure are appropriate for $\Delta^2 = 0.8 \times 10^{-4}$ (eV)^2, the electron-density distribution given by eqs. (10.8.2) and (10.8.3), and the radial distances r ($R_\odot$) = 0.0, 0.05, and 0.10.

The contribution from any point r in a solar model to the flux of neutrinos of any given provenance detected by a terrestial experiment is an integral over energy of the product of four factors. The first factor is the height along the appropriate differential neutrino-flux distribution in the solar model, e.g., the distributions for ^{8}B in Figs. 10.8.1 and 10.8.2. The second factor is the distribution with respect to energy of neutrinos of the selected type. For example, the spectral distribution of ^{8}B neutrinos at birth has, in first approximation, the allowed shape

$$f(E)\, dE \sim E^2 p_e E_e dE = E^2 \left(E_0 + m_e c^2 - E\right) \left[\left(E_0 + m_e c^2 - E\right)^2 - (m_e c^2)^2\right]^{1/2} dE,$$

$$(10.8.4)$$

Table 10.8.2 Characteristics of low energy emergent neutrinos as functions of energy and starting position when $\sin^2(2\theta) = 0.8$, $\Delta^2 = 0.8 \times 10^{-4}(\mathrm{eV})^2$, ρ_e as given by eqs. (10.8.2) and (10.2.3)

E (MeV)	r ($R_\odot$)	ω_+ (s^{-1})	ω_{vac} (s^{-1})	A_{int}/Δ^2	$|z|^2_{min}$	$|z|^2_{max}$	$|z|^2_{av}$
2.5	0.0	11299	12154	0.482	0.045	0.938	0.491
	0.1	12364		0.332	0.085	0.972	0.528
	0.2	15430		0.128	0.154	0.996	0.575
2	0.0	14560	15193	0.385	0.069	0.961	0.515
	0.1	16629		0.265	0.106	0.983	0.544
	0.2	19810		0.102	0.163	0.998	0.580
1	0.0	35990	30385	0.193	0.131	0.991	0.561
	0.1	38365		0.133	0.152	0.996	0.573
	0.2	41763		0.051	0.182	0.999	0.590
0.5	0.0	71790	60771	0.095	0.166	0.997	0.581
	0.1	82223		0.064	0.176	0.998	0.587
	0.2	85721		0.026	0.191	0.999	0.595
0.25	0.0	167538	121542	0.048	0.183	0.998	0.590
	0.1	170115		0.033	0.189	0.998	0.591
	0.2	173661		0.013	0.196	0.999	0.597

where $E_0 \sim 13.92$ MeV. Deviations from this differential distribution occur at the lower end of the spectrum due to the Coulomb interaction between the nucleus and the emitted positron (this distortion is negligible for ^{8}B neutrinos detected by SNO) and at the upper end of the spectrum due to the ~ 1.5 MeV width of the first excited state of ^{8}Be to which the beta decay normally occurs (e.g., Ortiz *et al.*, 2000). The third factor is the probability, as a function of neutrino energy, that the neutrino will manifest the electron flavor on emerging from the Sun. For example, for ^{8}B neutrinos born at $r = 0.05$, the third multiplicative factor, as given by Table 10.8.1, is 0.403 at $E = 5$ MeV, 0.316 at 10 MeV, and 0.293 at 15 MeV. The final factor is the cross section at the detector. For neutrino capture reactions of the form $\nu_e + {}_Z A_N \rightarrow {}_{Z+1} B_{N-1} + e^+$, the cross section for allowed transitions depends on the energy of the emitted electron as

$$\sigma_{capture} \propto (cp_e)\, E_e = \left[E_e^2 - (m_e c^2)^2\right]^{1/2} E_e \rightarrow E_e^2, \qquad (10.8.5)$$

where $E_e = E - E_{threshold}$ and the approximate E_e^2 dependence occurs when $E_e \gg m_e c^2$. Since, for most experiments, $E_{max} \sim 14$ MeV $\gg E_{threshold}$, the peak in the product of eqs. (10.8.4) and (10.8.5) occurs at several MeV above the ~ 7.5 MeV peak in the distribution of eq. (10.8.4).

Folding all of the factors together, it is evident that a reasonable first approximation to a predicted experimental ^{8}B neutrino counting rate (using a detector sensitive only to the

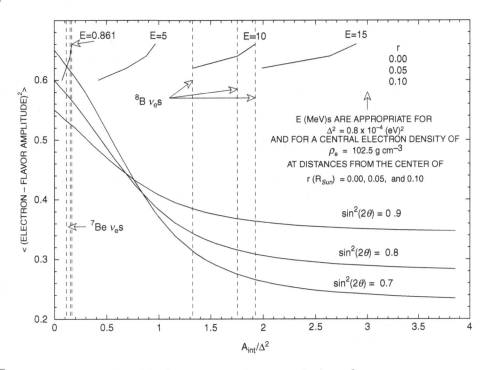

Fig. 10.8.4   Probability that emergent neutrino expresses the electron flavor

electron flavor) is given by considering neutrinos originating within a distance $\Delta r = \pm 0.05$ of the peak in the differential ^{8}B neutrino-flux distribution near $r \sim 0.05$, and at energies where the product of eqs. (10.8.4) and (10.8.5) has its maximum; namely, near $E \sim 10$ MeV. The intersection with the $\langle P_{\nu_e} \rangle$ vs A_{int}/Δ^2 curves in Fig. 10.8.4 of the three vertical dashed lines labeled ^{8}B ν_es define the electron-flavor characteristics of the emergent neutrinos which dominate in determining an experimental counting rate. The SNO result of $\langle P_{\nu_e} \rangle \sim 0.305$ suggests that, with $\Delta^2 \sim 0.8 \times 10^{-4}$ (eV)^2, the vacuum mixing angle satisfies $\sin^2(2\theta) \leq 0.8$. If a larger or smaller value of Δ^2 is chosen, the dashed curves are shifted to the left or to the right, leading to a reduction or increase in the value of $\sin^2(2\theta)$ estimated by comparison with, say, the SNO experiment. The discussion in Section 10.4 suggests that the choices $\Delta^2 \sim 0.8 \times 10^{-4}$ (eV)^2 and $\sin^2(2\theta) \sim 0.8$ are good first approximations.

The intersection with the $\langle P_{\nu_e} \rangle$ vs A_{int}/Δ^2 curves in Fig. 10.8.4 of the three vertical dashed lines labeled ^{7}Be ν_es define the electron-flavor characteristics of the emergent neutrinos which dominate in determining an experimental counting rate of ^{7}Be neutrinos. On comparison with results of the ^{37}Cl and gallium experiments, additional constraints can be put on estimates of the value of $\sin^2(2\theta)$ and Δ^2. Once again, as has already been explored in Section 10.5, the comparison reinforces the choices $\Delta^2 \sim 0.8 \times 10^{-4}$ (eV)^2 and $\sin^2(2\theta) \sim 0.8$.

10.9 Summary and conclusions

Matching the ^{8}B neutrino flux produced by solar models (Table 10.4.3) with the Sudbury
Neutrino Observatory (SNO) estimate of this flux based on consequences of a decomposition of the deuteron mediated by a neutral current (eq. (10.4.18)), leads to the estimate of
$Z = 0.0166$ and $Y = 0.28$ for the Sun's composition below the convective envelope. Comparing the rate of the deuteron decomposition reaction mediated by a charged current (eq.
(10.4.17)) with the rate of decomposition mediated by the neutral current, the SNO experiment reveals that approximately 0.3 of the ^{8}B neutrinos produced in the Sun reach the
Earth as electron-flavor neutrinos. Experiments by the Super-Kamiokande group show that
the propagation properties of antineutrinos can be explained by choosing the mass squared
difference between muon and electron antineutrinos in vacuum as $\Delta^2 \sim 0.8 \times 10^{-4}$ (eV)^2
and a mixing angle described by $\sin^2(2\theta) \sim 0.8$.

Adopting the Feynman–Gell-Mann current–current formulation of weak interaction theory, one can estimate the effective mass acquired by a neutrino interacting with electrons as it travels out of the Sun. Using the electron density given by the solar models and assuming that vacuum values of $\sin^2(2\theta)$ and Δ^2 for neutrinos are the same as
those for antineutrinos, it is found (Table 10.8.1) that neutrinos of energy 5–15 MeV
produced in the Sun as electron-flavor neutrinos reach the Earth manifesting the electron flavor a fraction $\sim$0.4–0.3 of the time, compared with the SNO estimate of 0.305.
Low energy neutrinos produced by the pp, pep, and ^{7}Be electron-capture reactions reach
the Earth manifesting the electron flavor approximately 59% of the time (Table 10.8.2).
Using these diminution factors, it is shown in Section 10.5 that the results of the gallium
detector experiments (GALLEX, GNO, and SAGE) can be matched by neutrino fluxes
from a solar model characterized by $Z \sim 0.0166$ and $Y \sim 0.279$ and that the results of the
^{37}Cl experiment can also be fitted using fluxes from a solar model of essentially the same
composition.

The rough comparisons which have been made in this chapter could certainly be refined
by varying the parameters Δ^2 and $\sin^2(2\theta)$. However, the persistent uncertainties in pertinent nuclear cross sections and in other relevant input physics prevent results of such calculations from being definitive. It seems wiser to leave the determination of the two critical
parameters for electron and muon neutrinos to terrestial long baseline experiments such
as those undertaken between CERN and Gran Sasso and to use the experimentally determined parameters to choose among models of the current Sun, as described in Sections 10.4
and 10.5.

Although inevitable uncertainties which can never be completely eliminated remain, to
all intents and purposes, the solar neutrino problem is no longer a serious problem. Consistency between the results of terrestial neutrino experiments and the characteristics of solar
models can be achieved by assuming that (1) the mass squared difference between electron
and muon neutrinos is the same as the experimentally determined mass squared difference
between electron and muon antineutrinos and that (2) an interaction between electrons and

electron-flavor neutrinos predicted by the current–current interaction theory of Feynman and Gell-Mann operates in the Sun and the coupling constant for this interaction is identical with the vector coupling constant in ordinary beta decay.

10.10 Postscript

References incorporated in the preceding text are based on a literature search ending in early 2005. Subsequent experimental and theoretical results of relevance are discussed in reviews by K. Nakamura and Boris Kayser, both reviews appearing in the compilation by C. Amsler *et al.* (2008). In §5 of the Nakamura review, it is stated that the best fit parameters for a two neutrino analysis of SNO data (B. Aharmim *et al.*, 2005) and KamLAND data (T. Araki *et al.*, 2005) results in $\Delta^2 = 8.0^{+0.6}_{-0.4} \times 10^{-5}$ (eV)^2 and $\tan^2(\theta) = 0.45^{+0.09}_{-0.07}$, which corresponds to $\sin^2(2\theta) = 0.86^{+0.05}_{-0.06}$. These values are consistent with the choices adopted in §10.7 (eqs. (10.7.12) and (10.7.14)). The reviews state further that a three neutrino analysis of the same data sets (A. B. Balantekin *et al.*, 2005; A. Strumia & F. Vissani, 2005; and G. L. Fogli. *et al.*, 2006) gives results consistent with a two neutrino analysis.

A discussion of neutrino experiments in progress and envisioned in the twenty first century and the manifold ramifications for astrophysics of anticipated results is given by G. B. Gelmini, A. Kusenko, and T. J. Weiler (2010). Results of analyses of ongoing experiments at two different times by G. L. Fogli and colaborators (Fogli *et al.*, 2006; Fogli *et al.*, 2011) are presented in Table 10.10.1 where the θ_{ij} are mixing angles between electron and tau neutrinos ($ij = 13$), electron and muon neutrinos ($ij = 12$), and muon and tau neutrinos ($ij = 23$). The mass-squared difference for electron and muon neutrinos is given by δm^2_{12} and the mass-squared difference for muon and tau neutrinos is given by $\Delta m^2 = m^2_3 - (m^2_1 + m^2_2)/2$.

Direct estimates of the electron-tau mixing angle obtained in reactor experiments have been reported by three collaborations: (1) $\sin^2(2\theta_{13}) = 0.086 \pm 0.04 (\sin^2(\theta_{13}) = 0.022)$ by the Double (two detectors) Choos (nuclear power plant in France) collaboration (Y. Abe, *et al.*, 2012); (2) $\sin^2(2\theta_{13}) = 0.113 \pm 0.013 (\sin^2(\theta_{13}) = 0.029)$ by the RENO (reactor experiment for neutrino oscillations) collaboration which detects electron antineutrinos from the Yonggwang nuclear power plant in South Korea (Soo-Bong Kim, *et al.*, 2012);

Table 10.10.1 Parameter Estimates by Fogli *et al.* (2006) and by Fogli *et al.* (2011)

quantity	$\sin^2(\theta_{13})$	δm^2_{12}	$\sin^2(\theta_{12})$	Δm^2	$\sin^2(\theta_{23})$
date/units	10^{-2}	10^{-5} (eV)^2	1	10^{-3} (eV)^2	1
2006	$0.9\left(1^{+2.3}_{-0.9}\right)$	$7.92\left(1^{+0.04}_{-0.04}\right)$	$0.314\left(1^{+0.18}_{-0.15}\right)$	$2.4\left(1^{+0.21}_{-0.26}\right)$	$0.44\left(1^{+0.11}_{-0.22}\right)$
2011	2.1–2.5	7.58	0.306–0.312	2.35	0.42

and (3) $\sin^2(2\theta_{13}) = 0.092 \pm 0.016(\sin^2(\theta_{13}) = 0.024)$ by a collaboration using the nuclear reactors in Daya Bay near Hong Kong, China (F. P. An, *et al.*, 2012). The smallness of the estimates of θ_{13} relative to the estimates of θ_{12} suggests that the two neutrino analysis of Solar neutrino experiments that has been conducted in this chapter provides a reasonably adequate interpretation of the facts.

Evidence for day–night variations in detected Solar neutrino fluxes has been reported. Such variations have been interpreted to mean that passage through the earth leads to a change in the flavor mix of neutrinos, but this interpretation contradicts the results of the experiments presented in §10.8 of this chapter which show that an electron neutrino born in a Solar spherical shell at a point on the side further from the earth and headed toward the earth leaves the Sun in the same flavor mix as does an electron neutrino born in the same Solar spherical shell but at a point on the side closer to the Earth.

Bibliography and references

J. N. Abdurashitov, E. P. Veretenkin, V. M. Vermul, *et al., J. Exp. Theor. Phy.*, **95**, No. 2, 181, 2002.

Y. Abe, C. Aberle, T. Akiri, *et al., Phys. Rev. Lett.*, **108**, 131801, 2012.

Zulema Abraham & Icko Iben, Jr., *ApJ*, **170**, 157, 1971.

B. Aharmim, S. N. Ahmed, A. E. Anthony, *et al., Phys. Rev.*, **C72**, 055502, 2005; **C81**, 055505, 2010.

Q. R. Ahmad, R. C. Allen, T. C. Anderson, *et al., Phys. Rev. Lett.*, **89**, 011301, 2002.

S. N. Ahmed, A. E. Anthony, E. W. Beier, *et al., Phys. Rev. Lett.*, **92**, 181301, 2004.

M. Altmann, M. Balata, P. Belli, *et al., Phys. Lett. B*, **490**, 16, 2000.

C. Amsler, M. Doser, M. Antonelli, *et al., Phys. Lett.*, **B667**, 1, 2008.

F. P. An, J. Z. Bai, A. B. Balantakin, *et al., Phys. Rev. Lett.*, **108**, 171803, 2012.

T. Araki, K. Eguchi, S. Enomoto, *et al., Phys. Rev. Lett.*, **94**, 081801, 2005.

Y. Ashie, J. Hosaka, K. Ishihara, *et al., Phys. Rev. Lett.*, **93**, 101801, 2004.

Y. Ashie, J. Hosaka, K. Ishihara, *et al., Phys. Rev.*, **D71**, 112005, 2005.

M. B. Aufderheide, S. B. Bloom, D. A. Resler, & C. D. Goodman, *Phys. Rev.*, **C49**, 678, 1994.

John N. Bahcall, *Phys. Rev.*, **B137**, 135, 1964a; **C56**, 3391, 1996.

John N. Bahcall, *Phys. Rev. Lett.*, **12**, 300, 1964b; **17**, 398, 1966.

John N. Bahcall, *Neutrino Astrophysics* (Cambridge, Cambridge University Press), 1989.

John N. Bahcall & R. M. May, *ApJL*, **152**, L17, 1968.

John N. Bahcall, A. M. Serenelli, & S. Basu, *ApJL*, **621**, L85, 2005.

A. B. Balantekin, V. Barger, D. Marfatia, S. Pakvasa, & H. Yuksel, *Phys. Lett.*, **B613**, 61, 2005.

Hans A. Bethe, *Phys. Rev. Lett.*, **56**, 1305, 1986.

B. T. Cleveland, T. Daily, R. Davis, Jr., *et al., ApJ*, **496**, 505, 1998.

Raymond Davis, Jr., D. S. Harmer, & K. C. Hoffman, *Phys. Rev. Lett.*, **20**, 1205, 1968.

K. Eguchi, S. Enomoto, K. Furuno, *et al., Phys. Rev. Lett.*, **90**, 021802, 2003.

G. L. Fogli. E. Lisi, A. Marrone, & A. Palazzo, *Prog. Part. Nucl. Phys.*, **57**, 742, 2006.

G. L. Fogli, E. Lisi, A. Marrone, A. Palazzo, & D. Montanino, *Phys. Rev. D.*, **66**, 053010, 2002.

G. L. Fogli, E. Lisi, A. Marrone, A. Palazzo, & A. M. Rotunno, *Phys. Rev. D.*, **84**, 053007, 2011.

S. Fukuda, *Phys. Rev. Lett.*, **85**, 3999, 2000.

S. Fukuda, Y. Fukuda, S. Ishitsuka, *et al., Phys. Rev. Lett.*, **86**, 5656, 2001.

Graciela B. Gelmini, Alexander Kusenko, and Thomas J. Weiler, *Scientific American*, **302**, No. 5, 38, May 2010.

W. Hampel, J. Handt, G. Heusser, *et al., Phys. Lett. B*, **447**, 127, 1999.

Icko Iben, Jr., *ApJL*, **140**, 1631, 1964.

Icko Iben, Jr., *Ann. Phys.*, **54**, 164, 1969.

Boris Kayser, in C. Amsler *et al., Phys. Lett.*, **B667**, 1, 2008, pdg.lbl.gov.

S. P. Mikheyev and A. Yu. Smirnov, *Nuovo Cimento C*, **9**, 17, No. 1, 1986.

K. Nakamura, in C. Amsler *et al., Phys. Lett.*, **B667**, 1, 2008, pdg.lbl.gov.

Soo-Bong Kim, for the RENO collaboration, arXiv: 1204. 062612, 2012.

C. E. Ortiz, A. Garcia, R. A. Waltz, M. Battacharya, & A. K. Komives, *Phys. Rev. Lett.*, **85**, 2909, 2000.

S. J. Parke, *Phys. Rev. Lett.*, **57**, 1275, 1986.

Bruno Pontecorvo, *Zh. Eksp. Teor. Fiz.*, **53**, 1771, 1967; translated in *Sov. Phys., JETP*, **26**, 984, 1968.

S. P. Rosen & J. M. Gelb, *Phys. Rev. D*, **34**, 4, 1986.

A. Strumia & F. Vissani, *Nucl. Phys.*, **B726**, 294, 2005.

Lincoln Wolfenstein, *Phys. Rev. D*, **17**, issue 9, 2369, 1978.

Evolution through hydrogen-burning phases of models of mass 1, 5, and 25 $M_\odot$

In this chapter, models of mass 1, 5, and 25 $M_\odot$ and of population I composition ($Z = 0.015$ and $Y = 0.275$) are evolved through all phases of hydrogen burning up to the point when helium is ignited in a hydrogen-exhausted core. The model masses have been chosen with the aim of representing three broad classes of stars. Models of mass less than 0.5 $M_\odot$ have been excluded from consideration not only because they evolve on a time scale much longer than a Hubble time, but, because they remain completely convective throughout their nuclear burning lives, they can be described adequately by a sequence of polytropes, as discussed in Section 5.6, without invoking the elaboration of an evolutionary calculation. The 1 $M_\odot$ model is representative of a class of stars which evolve in less than a Hubble time into red giants with an electron-degenerate helium core and ignite helium in a semiexplosive fashion. These stars, of mass extending to $\sim$2.25 $M_\odot$, eventually become AGB stars with a carbon–oxygen core and, after ejecting a nebular shell, evolve into CO white dwarfs of mass $\sim$0.55 $M_\odot$. The 5 $M_\odot$ model is representative of stars in the approximate mass range $2.25 < M/M_\odot < 10.5$ which ignite helium under non-electron-degenerate conditions, but during and following the quiescent helium-burning phase evolve in a fashion similar to the evolution of lower mass stars during and after the quiescent helium-burning phase. Stars in this class which are initially less massive than $\sim$8.5 $M_\odot$ become CO white dwarfs of mass less $\sim$1.1 $M_\odot$. Those of initial mass in the approximate range $8.5 < M/M_\odot < 10.5$ develop an electron-degenerate core composed of oxygen and neon before becoming AGB stars and eventually evolve into ONe white dwarfs of mass in the approximate range $1.1 < M/M_\odot < 1.37$. The 25 $M_\odot$ model is representative of stars which burn both helium and carbon under non-electron-degenerate conditions, experience core collapse and evolve into neutron stars or black holes after experiencing a type II supernova explosion during which they eject all of the matter above the collapsing core.

While hydrogen remains at the center, the 1 $M_\odot$ model burns hydrogen primarily by the pp-chain reactions and therefore does not possess a convective core. After exhausting hydrogen at the center, the model burns hydrogen in a thick shell above a nearly isothermal core and the rate of producing energy by CN-cycle reactions becomes increasingly competitive with the rate of producing energy by the pp chains. When the center of the hydrogen-burning shell approaches a mass of approximately 0.12–0.13 $M_\odot$ and the ratio of the density at the center to the density at the middle of the shell approaches a value of approximately 32, the hydrogen-exhausted core contracts at an accelerated rate and develops a noticeable temperature gradient; as electron degeneracy in the core becomes important, energy transport by electon conduction reduces the temperature gradient. During this phase, the model evolves in the HR diagram at nearly constant luminosity along the subgiant branch, the base of the convective envelope extends ever more deeply inward in mass, and the mass width of the hydrogen-burning shell narrows significantly.

As the mass between the hydrogen-burning shell and the base of the convective envelope continues to decrease, the effects of H^- opacity in the cooling and radiative surface layer lead to an upturn in the HR diagram and the model evolves into a red giant which is characterized by a compact electron-degenerate core and an expanding convective envelope, the core being capped by a very narrow hydrogen-burning shell and the burning shell being separated from the base of the convective envelope by a transition zone in which all structure variables change by an order of magnitude or more in a relatively small mass interval. During the red giant branch phase of evolution, hydrogen burning in the shell adds mass to the electron-degenerate core which grows to a mass of ~ 0.47 $M_\odot$ before helium is ignited in a semi dynamical fashion near the model center. Atlhough the phase is relatively shortlived ($t_{evo} \sim 5 \times 10^8$ yr), evolution in the HR diagram is quite extensive, with the luminosity of the model increasing by over a factor of 100. Subsequent evolution of the 1 $M_\odot$ model is described in Chapter 17 in Volume 2 of this book.

While on the main sequence, the 5 $M_\odot$ and 25 $M_\odot$ models burn hydrogen primarily by the CN-cycle reactions in large convective cores which slowly shrink in mass as evolution progresses. As it exhausts hydrogen in its convective core, the 5 $M_\odot$ model develops a thick hydrogen-burning shell which becomes the primary source of nuclear energy. The hydrogen-exhausted core becomes nearly isothermal and contracts until, when the central density becomes approximately 32 times larger than the density in the middle of the hydrogen-burning shell (located at a mass which is approximately 11% of the model mass), the model becomes gravothermally unstable. The core contracts at an accelerated rate and heats, developing a strong temperature gradient. The hydrogen-burning shell moves spatially inward, becomes narrower in mass, and, for a time, decreases in strength. The envelope of the model expands outward and, because of the work required for movement against gravity, absorbs a substantial fraction of the nuclear energy produced by the hydrogen-burning shell.

Due to increasing opacities in the expanding and cooling envelope, convection extends inward from the surface and the 5 $M_\odot$ model evolves into a red giant with a contracting and heating helium core. The rate of nuclear energy generation by shell hydrogen burning increases and absorption of energy from the ambient flow, being effectively confined to the radiative portion of the envelope, decreases; the model climbs upward in the HR diagram. When temperatures in the core become high enough for helium to burn, electrons in the core are only mildly degenerate, so helium is ignited gently and burns quiescently. The energy produced by helium burning replaces compressional energy produced during the phase of rapid core contraction as the primary source of energy in the hydrogen-exhausted core, and the gravothermally unstable phase is terminated. Evolution continues on a time scale determined by the rate of core helium burning, as described in Chapter 18 in Volume 2.

The 25 $M_\odot$ model differs from the 5 $M_\odot$ model by having a much larger ratio of radiation pressure to gas pressure (resulting in a smaller adiabatic gradient) and an opacity which (being dominated by electron scattering and free–free contributions) depends approximately linearly on the abundance of hydrogen. The consequence is that, in the 25 $M_\odot$ model, convection competes with radiation as the primary mode of energy transfer over an extended region outside the convective core, leading to a hydrogen-abundance distribution such that the adiabatic and radiative gradients are nearly the same. The process whereby equilibration of gradients is achieved is called semiconvection. Hydrogen burning in the convective core remains the primary source of surface luminosity until the hydrogen abundance in the core drops below $Y_H \sim 10^{-4}$, after which an increasing fraction of the total rate of nuclear energy production takes place outside the convective core.

Once the central hydrogen abundance drops to $Y_H \sim 10^{-5}$, approximately 40% of the original hydrogen in the model has been converted into helium and the mass of the convective core in which hydrogen has been severely depleted is approximately 25% of the mass of the model. Gravothermal energy delivered by the rapidly contracting core has become the primary source of energy flux sustaining convection about the center and sustaining semiconvection in regions outside the convective core.

The further development of the model depends sensitively on the choice of evolutionary time step as well as on the hydrogen-abundance distribution built up during prior evolution, the latter distribution being a function of the choices of spatial and temporal zoning during prior evolution. When time steps are chosen to be too small, spurious mixing occasioned by an artificially large release of gravothermal energy results in two additional episodes of core hydrogen burning. When time steps are chosen to be larger than the time scale on which hydrogen burns at the center, hydrogen is exhausted in central regions and hydrogen burning continues exclusively in a thick shell above the core. Matter in the region below the shell contracts and heats on a thermal time scale and the envelope expands as the nuclear burning shell gains in strength. Helium is ignited at the center before the model becomes a red giant and further evolution proceeds on a helium-burning timescale in a region in the HR diagram not far from the observational main sequence, as described in Chapter 20 in Volume 2.

The hydrogen-burning evolution of the 1 $M_\odot$, 5 $M_\odot$, and 25 $M_\odot$ constant mass models is described in Sections 11.1, 11.2, and 11.3, respectively. In Section 11.4, global properties of the three models are compared and an estimate is made of the extent of mass loss from the surface of real counterparts of the models.

11.1   Evolution of a 1 $M_\odot$ model during core and shell hydrogen burning on the main sequence, formation of an electron-degenerate core, and core growth during shell hydrogen burning on the red giant branch

The initial ZAMS (zero age main sequence) model for the 1 $M_\odot$ study is obtained by repeating the exercises described in Sections 9.2 and 9.3, but for the initial composition $Z = 0.015$ and $Y = 0.275$. The resultant evolutionary track is shown in Fig. 11.1.1. Times

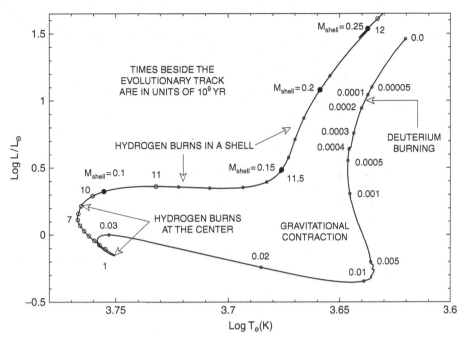

Fig. 11.1.1 Evolutionary track of a 1M$_\odot$ model (Z = 0.015, Y = 0.275) during gravitational contraction and central and shell hydrogen-burning phases

recorded beside the track are given in units of 10^9 yr. The lifetime of the deuterium-burning phase which occurs at $L \sim 10\ L_\odot$ along the Hayashi portion of the evolutionary track is measured in hundreds of thousands of years. Evolution through the gravitational contraction phase to the beginning of the main sequence occurs in roughly 30 million years. Hydrogen burns at the center (and in a large region beyond the center) for approximately 8.5 billion years and, for another several billion years, hydrogen continues to burn in a thick, but narrowing, shell above a contracting hydrogen-exhausted core which becomes increasingly isothermal until the mass of the hydrogen-free core approaches ~ 0.12–0.13 $M_\odot$. A temperature gradient develops in the core as the model evolves along the nearly horizontal portion of the evolutionary track in the HR diagram called the subgiant branch for approximately 5×10^8 yr. Because of rapidly mounting densities in the contracting core, electrons in the core become increasingly degenerate and electron conduction inhibits the increase in core temperatures, causing the core to revert again to a nearly isothermal state. In response to increasing opacities in the expanding envelope, the outer convective zone in the envelope moves inward in mass and the influence of H$^-$ opacity in surface layers causes the evolutionary track to evolve upward in the HR diagram along a track which is approximately parallel to the track during the Hayashi phase. This latter, or red giant branch portion of the track lasts for approximately 500 million years, as the hydrogen-burning shell grows to about 0.47 $M_\odot$.

The time dependences of several interior features are shown in Figs. 11.1.2–11.1.4. The bumpy nature of some of the curves in the first two of these figures is not a consequence

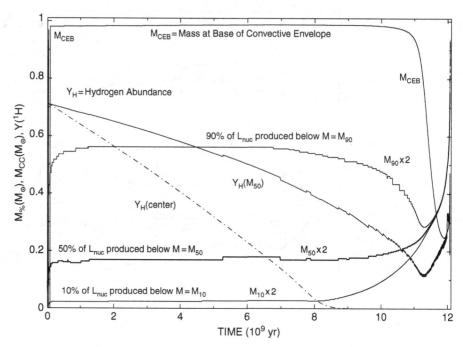

Fig. 11.1.2   Characteristics of the hydrogen-burning zone and location of the base of the convective envelope in a 1$M_\odot$ model ($Z = 0.015, Y = 0.275$)

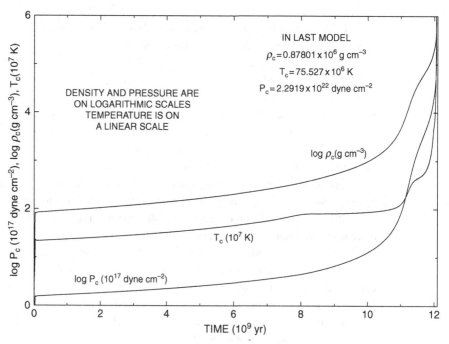

Fig. 11.1.3   State variables at the center of a 1$M_\odot$ model ($Z = 0.015, Y = 0.275$) burning hydrogen on the main sequence, subgiant, and red giant branches

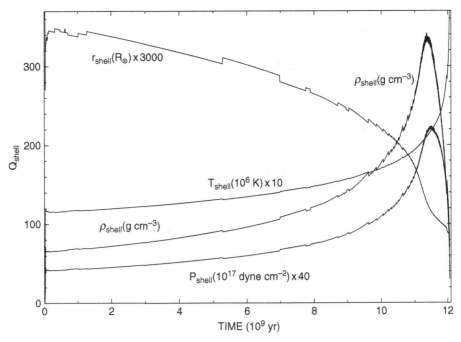

Characteristics at the middle of the hydrogen-burning region during main sequence and giant branch phases in a 1$M_\odot$ model ($Z = 0.015$, $Y = 0.275$)

of inaccurate calculation, but rather a consequence of the fact that the evolution program has not been instructed to obtain exact estimates of the associated quantities. For example, having available the total nuclear burning luminosity at the end of every time step, the program has been instructed to find and record the zone at the outer edge of which the nuclear-burning luminosity is larger than or equal to 50% of the total value; it has not been asked to interpolate within the model mesh to obtain a more precise estimate at half maximum. As a result, if no rezoning occurs between models, the mass associated with the center of the hydrogen-burning shell remains constant or increases discontinuously with each evolutionary time step. On the other hand, if rezoning occurs, the mass assigned to the center of the shell can decrease in consequence of zone splitting.

The curves labeled M_{10} and M_{90} in Fig. 11.1.2 outline the region within which 80% of the nuclear energy generation occurs and the curve labeled Y_H(center) shows the central hydrogen abundance. From these curves, it is evident that the region where most of the hydrogen burning occurs remains approximately fixed in the mass coordinate until the hydrogen abundance at the center decreases below Y_H(center) ~ 0.02 at $t \sim 8 \times 10^9$ yr. Hydrogen effectively vanishes at the center at $t \sim 9 \times 10^9$ yr and the thickness of the hydrogen-burning shell (defined as the region within which 80% of the nuclear energy generation occurs) decreases steadily as evolution progresses. The abundance of hydrogen at the middle of the hydrogen-burning region, defined as the place below which 50% of the nuclear luminosity is produced, is approximated by the curve in Fig. 11.1.2 labeled $Y_H(M_{50})$. This abundance also decreases steadily with time until $M_{\text{shell}} \sim 0.1236\ M_\odot$ at $t \sim 11.20 \times 10^9$ yr.

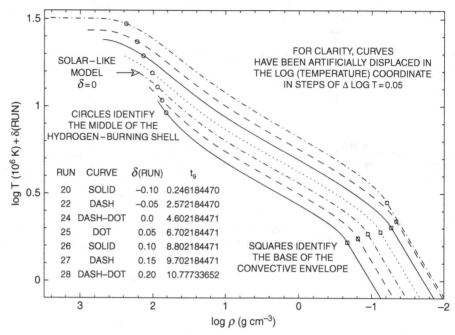

Fig. 11.1.5   Temperature versus density in a $1M_\odot$ model ($Z = 0.015$, $Y = 0.275$) during the main sequence phase and the early subgiant phase

    The time dependences of state variables at the center of the model are given in Fig. 11.1.3 and conditions at the middle of the hydrogen-burning region are given as functions of time in Fig. 11.1.4. The radius at the middle of the hydrogen-burning region, depicted by the curve labeled r_{shell} in Fig. 11.1.4, decreases monotonically with time, whereas the temperature at the middle of the region, shown by the curve labeled by T_{shell}, increases monotonically with time. The density at the middle of the region, shown by the curve labeled ρ_{shell}, increases with time until $t \sim 11.35 \times 10^9$ yr and then decreases with time. The density at the model center, shown by the curve labeled $\log \rho_c$ in Fig. 11.1.3, increases monotonically with time, consistent with the fact that r_{shell} decreases steadily with time and the fact that M_{shell} (M_{50} in Fig. 11.1.2) is essentially constant while hydrogen is present at the center and then increases monotonically after hydrogen vanishes at the center. On the other hand, whereas T_{shell} increases monotonically with time over most of the evolutionary history of the model, the central temperature, as given by the curve labeled T_c in Fig. 11.1.3, increases monotonically until the hydrogen abundance at the center decreases to $Y_H \sim 0.02$ at $t \sim 8 \times 10^9$ yr, but then remains nearly constant at $T_c \sim$ 19-21 $\times 10^6$ K for the next approximately 2.5×10^9 yr. Over this 2.5 billion year time period, T_{shell} increases from $\sim 13 \times 10^6$ K to $\sim 20 \times 10^6$ K. In short, over this phase of evolution, the region between the center of the model and the hydrogen-burning shell becomes nearly isothermal.

    The approach to isothermality in the hydrogen-exhausted core is demonstrated explicitly in the temperature-density diagram of Fig. 11.1.5 by the relationships between temperature and density for the three oldest models represented in the figure. That these three

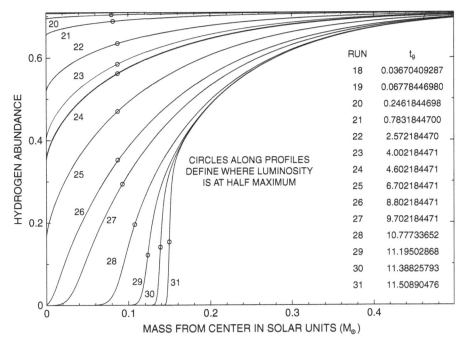

Fig. 11.1.6 Hydrogen–abundance profiles in a 1M$_\odot$ model ($Z = 0.015$, $Y = 0.275$) during the main sequence and subgiant branch phases

models have no hydrogen at the center is corroborated by the hydrogen-abundance profiles in Fig. 11.1.6. In both figures, open circles identify the location of the middle of the hydrogen-burning region where $M(r) = M_{50} = M_{shell}$. In Fig. 11.1.5, open squares identify the location of the base of the convective envelope.

In Fig. 11.1.5, the temperature–density profiles have been artificially displaced from one another in steps of $\Delta \log T = 0.05$. Since, at any given density, the separation between adjacent curves is of the order of $\delta \log T = 0.05$, it is evident that, where they overlap in density, the actual temperature–density profiles for all of the models are, with one exception, almost identical. The exception is that, in the region near and beyond the base of the convective envelope in the two most evolved models, the actual temperature–density curve for the most evolved model crosses over to be below the actual curve for its less evolved precursor. Given that, in regions of overlap at high densities, temperature–density distributions are almost identical, one has the picture that, as evolution progresses, the center of the hydrogen-burning zone moves inward to higher densities and temperatures along a common trajectory and that the very center of the model moves even more rapidly inward in density along this trajectory, with the difference in temperature between the center of the model and the middle of the hydrogen-burning zone decreasing in magnitude.

An additional consequence of artificially separating the temperature-density curves in Fig. 11.1.5 is that, instead of increasing with time as suggested by the open squares in the figure, the temperature at the base of the convective envelope actually decreases with time

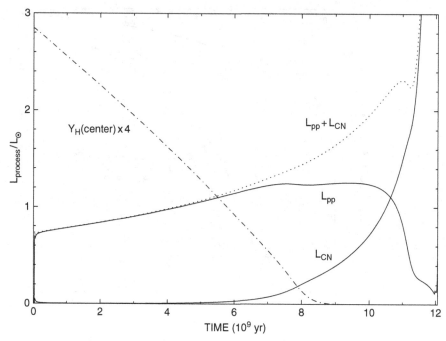

Fig. 11.1.7 pp–chain and CN–cycle contributions to the luminosity in a 1$M_\odot$ model (Z = 0.015, Y = 0.275) during main
sequence and subgiant phases

during the main sequence phase. On the other hand, the open square along the curve for the
most evolved model in Fig. 11.1.5 lies above the open square of the second most evolved
model by more than $\Delta \log T = 0.05$, indicating that, in the transitional model between the
main sequence and the subgiant branch, the base of the convective envelope is moving
inward in mass and becoming hotter.

Contributions to surface luminosity by the two major nuclear burning sources are shown
as functions of time in Fig. 11.1.7. Over the first half of the evolution shown, pp-chain
reactions completely dominate CN-cycle reactions as contributors to surface luminosity.
Thereafter, the contributions of pp-chain reactions level off to an approximately constant
value which is maintained for about 3 billion years, while the contributions of CN-cycle
reactions increase steadily. The pp-chain reaction luminosity begins to decline at about
$t \sim 10^{10}$ yr, and the two sources achieve parity at $t \sim 10.7 \times 10^9$ yr. As evolution progresses still further, CN-cycle reactions become by far the dominant contributors to surface
luminosity and the contributions of pp-chain reactions drop to a nominal value somewhat
smaller than $\sim 0.2\ L_\odot$.

Having explored evolutionary changes of several global characteristics and internal features of the hydrogen-burning model over a long time scale, it is of interest to inquire how
closely the model comes to acquiring solar-like characteristics after 4.6 billion years of
evolution. Following its formal creation on the Hayashi track at $t = 0$ and $L \sim 30\ L_\odot$, the
model attains the Sun's luminosity at $t \sim 4.3 \times 10^9$ yr. At an age of $t = 4.602 \times 10^9$ yr, its
luminosity exceeds the Sun's by about 3.3% and its radius exceeds the Sun's by about 1%.

Table 11.1.1 Neutrino fluxes at the Earth (in units of $cm^{-2} s^{-1}$) for solar-like models										
Model	Z	Y	$\phi(pp)$	$\phi(pep)$	$\phi(^7Be)$	$\phi(^8B)$	$\phi(^{13}N)$	$\phi(^{15}O)$	$\phi(^{17}F)$	$\phi(^{18}F)$
$t_9=4.3$	0.015	0.275	6.10/10	1.53/8	4.14/9	4.26/6	3.07/8	2.51/8	2.78/6	5.09/4
$t_9=4.6$	0.015	0.275	6.20/10	1.55/8	4.74/9	5.56/6	3.70/8	3.15/8	3.55/6	8.17/4
C*	0.015	0.272	6.08/10	1.50/8	4.32/9	4.63/6	3.13/8	2.73/8	3.06/6	6.89/4
A	0.01	0.2373	6.17/10	1.54/8	3.25/9	2.56/6	1.51/8	1.14/8	1.21/6	1.48/4
B	0.02	0.296	5.91/10	1.48/8	5.15/9	6.75/6	5.62/8	4.92/8	5.63/6	1.58/5
C	0.015	0.270	6.02/10	1.51/8	4.25/9	4.51/6	3.26/8	2.68/8	2.97/6	5.91/4

From Tables 10.1.1 and 10.2.1, one has that, for a given choice of Z, a change in the age of a model of solar luminosity is related to a change in the initial helium abundance by

$$-\frac{\Delta t_9}{\Delta Y} = 105 \pm 5, \qquad (11.1.1)$$

where Δt_9 is the age change in units of 10^9 yr. The smaller value of the derivative holds for $Z = 0.01$ and the larger value holds for $Z = 0.02$. Thus, for the choice $Z = 0.015$, to obtain a model of the Sun's luminosity at an age 4.6×10^9 yr requires a change in initial Y by $\Delta Y = -0.3/105 \sim -0.003$. At this point, making the estimated change in initial composition and repeating the calculations would distract from the main objective of the chapter and would offer very little in the way of additional enlightenment. Suffice it to demonstrate that the neutrino fluxes for a solar model with $(Z, Y) = (0.015, 0.272)$ are consistent with those which can be estimated from the model results presented in Chapter 10.

The neutrino fluxes at the Earth's distance from the Sun in the $t_9 = 4.3$ and $t_9 = 4.6$ models constructed here are given in the first two rows of Table 11.1.1. By interpolation in Tables 10.4.1 and 10.4.2, one can estimate the neutrino fluxes expected from a solar model with $Y(Z = 0.015) \sim 0.272$. The results are given in the third row (model C*) of Table 11.1.1. For comparison, the fluxes from solar models A and B given in Table 10.4.3 are reproduced in the fourth and fifth rows of Table 11.1.1. In the sixth row (model C) are the results of interpolating logarithmically between models A and B to estimate the fluxes from a solar model specified by $Z = 0.015$ and $Y(Z = 0.015) \sim 0.270$. Given the crudeness of the interpolations, it is encouraging that the neutrino fluxes in the interpolated models C and C* agree to within a few percent, with the exception of the flux of neutrinos from the beta decay of ^{18}F which differs by 17% in the two estimates. Fortunately, this flux contributes very little to an estimated counting rate in solar neutrino experiments.

Several structure and composition characteristics of the 4.602 billion year old model are shown in the mass and radius representations in Figs. 11.1.8 and 11.1.9, respectively. The structure variables may be compared with those of models A in Figs. 10.1.3 and 10.1.4 and with those for model B in Figs. 10.2.3 and 10.2.4. The distributions of the few composition variables shown in Figs. 11.1.8 and 11.1.9 may be compared with distributions of these same variables in Figs. 10.1.5, 10.1.6, 10.2.5, and 10.2.6.

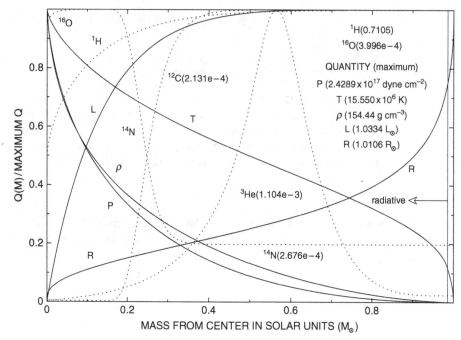

Fig. 11.1.8 Structure and composition variables in a $1M_\odot$ model ($Z = 0.15$, $Y = 0.275$) with solar-like characteristics ($t_g = 4.602184471$)

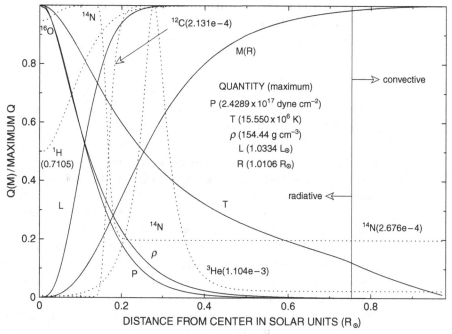

Fig. 11.1.9 Structure and composition variables in a $1M_\odot$ model ($Z = 0.15$, $Y = 0.275$) with solar-like characteristics ($t_g = 4.602184471$)

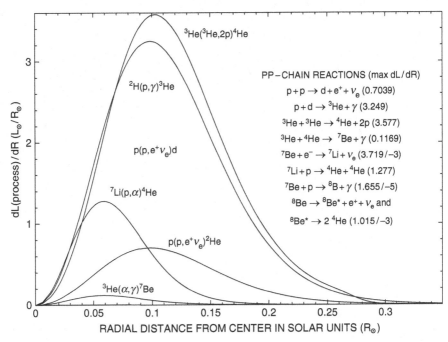

pp-chain contributions to luminosity in a 1M$_\odot$ model (Z $= 0.015$, Y $= 0.275$) with solar-like characteristics (t$_g$ $= 4.602184471$)

Contributions to the surface luminosity by various reactions in the 4.602 billion year old model are shown in the radius representation in Figs. 11.1.10–11.1.12. These may be compared with corresponding distributions in model A in Figs. 10.3.1, 10.3.3, and 10.3.5 and with those in model B in Figs. 10.3.2, 10.3.4, and 10.3.6. It is instructive that the maxima in the distributions for CN-cycle reactions in Fig. 11.1.12 are all larger by the same amount (∼4.4%) than the average of the maxima in the distributions in Figs. 10.3.5 and 10.3.6. That they are larger is consistent with the fact that the luminosity of the $Z = 0.015$ model is larger (by ∼3.3%) than that of models A and B. Given the similarity in distribution shapes for each reaction type, it means further that, for models of solar luminosity and age, the total contribution to the luminosity by CN-cycle reactions of a given type is almost exactly proportional to the metallicity Z, despite the differences in the temperature and density distributions in the models of different metallicity.

Returning now to a general description of the evolution of the 1 $M_\odot$ ($Z = 0.015$, $Y = 0.275$) model, an interesting question is how to define the upper boundary of the main sequence phase. Considerations involved in defining phases and boundaries between phases include the rate at which internal and/or global characteristics change with time, morphology in the HR diagram, and basic differences in the structure of the interior. Given the fact that there are no dramatic changes in the time scales on which either internal or external characteristics change when hydrogen vanishes at the center, it seems inappropriate to confine the definition of the main sequence to models that burn hydrogen at the center.

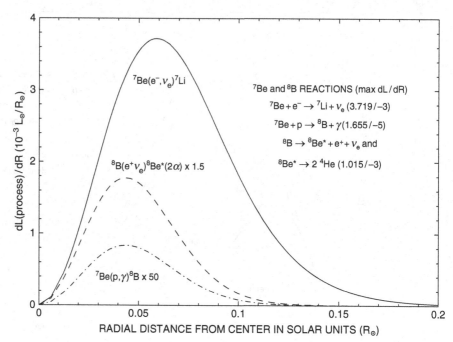

Fig. 11.1.11 ^{7}Be and ^{8}B contributions to luminosity in a 1$M_\odot$ model ($Z = 0.015$, $Y = 0.275$) with solar-like characteristics ($t_g = 4.602184471$)

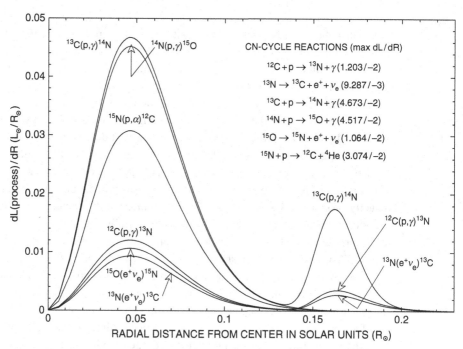

Fig. 11.1.12 CN-cycle contributions to luminosity in a 1$M_\odot$ model ($Z = 0.015$, $Y = 0.275$) with solar-like characteristics ($t_g = 4.602184471$)

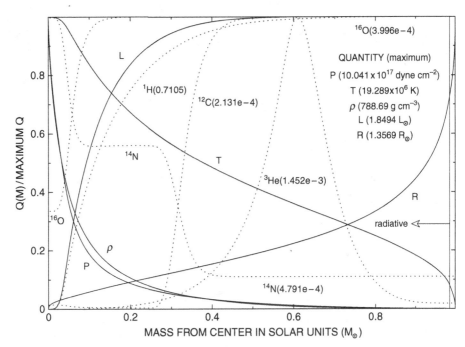

Fig. 11.1.13 Structure and composition variables in a $1M_\odot$ model ($Z = 0.15$, $Y = 0.275$) near the end of the main sequence phase ($t_g = 9.702184471$)

A better definition of the main sequence includes that part of the shell hydrogen-burning phase during which the hydrogen-exhausted core becomes and remains nearly isothermal. Distributions of structure and abundance characteristics in two models which fall into this category are shown in Figs. 11.1.13–11.1.16. One problem with the inclusion of both of these models is that, in the HR diagram (Fig. 11.1.1), the second model lies along the first portion of the nearly horizontal segment of the evolutionary track that, in the context of globular clusters, has traditionally been called the subgiant branch. As a compromise, this second model is here designated as being in a transitional phase between the main sequence and subgiant phases.

Inspection of Fig. 11.1.7 shows that, in the transitional model, the contributions to the total luminosity of pp-chain reactions and CN-cycle reactions are nearly identical. This near identity is demonstrated further by energy-generation profiles in the mass and radius representations in Figs. 11.1.17 and 11.1.18, respectively. From the luminosity curves also shown in the two figures, one may determine that the center of the hydrogen-burning shell (defined as the point where the luminosity is at half maximum) is located where $M_{shell} \sim 0.107 \ M_\odot$ and $r_{shell} \sim 0.063 \ R_\odot$. These estimates agree with those obtainable from Figs. 11.1.2 and 11.1.4, respectively. The additional insight provided by Figs. 11.1.17 and 11.1.18 over that provided by Figs. 11.1.2 and 11.1.4 is that the centroids of the two major sources of nuclear energy are at two distinct locations separated in the two representations by amounts ($\Delta M \sim 0.04 \ M_\odot$ and $\Delta r \sim 0.02 \ R_\odot$, respectively) which are

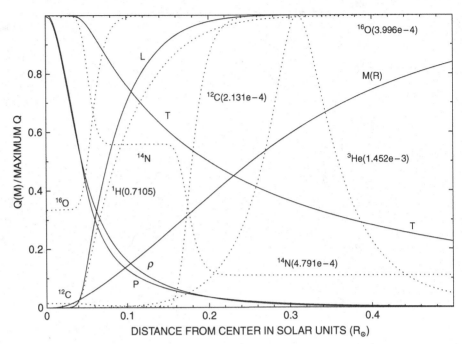

Fig. 11.1.14   Structure and composition variables in a 1$M_\odot$ model ($Z = 0.15$, $Y = 0.275$) near the end of the main sequence phase ($t_g = 9.702184471$)

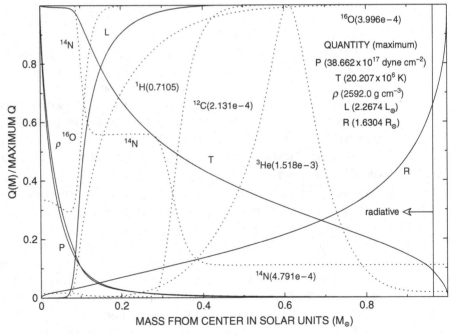

Fig. 11.1.15   Structure and composition variables in a 1$M_\odot$ model ($Z = 0.15$, $Y = 0.275$) in transition from the main sequence to the subgiant phase ($t_g = 10.77733652$)

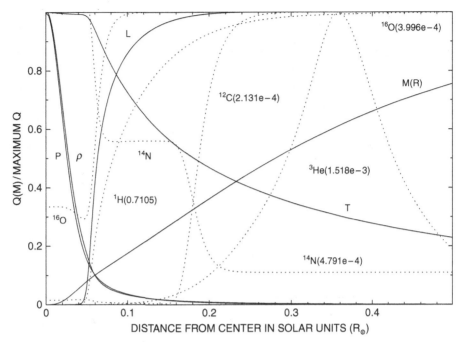

Fig. 11.1.16 Structure and composition variables in a 1$M_\odot$ model ($Z = 0.15$, $Y = 0.275$) in transition from the main sequence to the subgiant phase ($t_g = 10.77733652$)

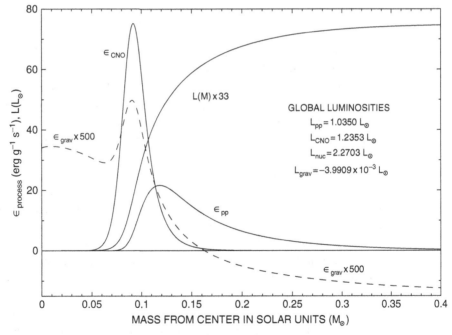

Fig. 11.1.17 Contributions to the luminosity in a 1$M_\odot$ model ($Z = 0.015$, $Y = 0.275$) in transition from the main sequence to the subgiant branch ($t_g = 10.77733652$)

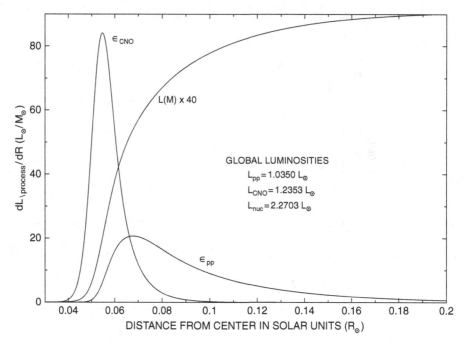

Fig. 11.1.18 Differential contributions to the luminosity in a $1M_\odot$ model ($Z = 0.015$, $Y = 0.275$) in transition from the main sequence to the subgiant branch ($t_9 = 10.77733652$)

approximately one third of the widths of the entire nuclear burning region in the two representations.

Relevant to the association of the transitional model with a named phase of evolution is the fact that, as is evident from a comparison of the curve labeled ϵ_{grav} with the curves for the nuclear energy-generation rates in Fig. 11.1.17, the contribution to the surface luminosity of gravothermal energy is inconsequential compared with the contribution of nuclear energy. That is, evolution is occurring on a nuclear burning time scale despite the fact that gravothermal processes are influencing the direction of evolution in the HR diagram.

Structure and composition characteristics in two models along the subgiant branch proper are presented in Figs. 11.1.19 and 11.1.20. In the first model, of age $t_9 = 11.195$, the mass at the hydrogen-burning shell is $M_{shell} = 0.12359\ M_\odot$, and in the second model, of age $t_9 = 11.388$, $M_{shell} = 0.13843\ M_\odot$. Several features distinguish the two subgiant models from models which are near the end of the main sequence phase or are in transition between the main sequence and the subgiant branch: the densities in the core have increased markedly, the hydrogen-exhausted core has developed a pronounced temperature gradient, and the base of the convective envelope has moved inward to the extent that ^{3}He is being dredged upward into surface layers. From the (barely) discernable rise in luminosity through the model cores, it is evident that gravothermal energy is being released by the contracting cores. In both models, the central concentration has become so large that meaningful information about density and pressure variations beyond the core and the

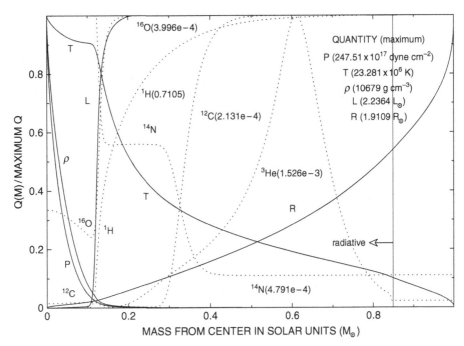

Fig. 11.1.19 Structure of and composition in a 1$M_\odot$ model ($Z = 0.15$, $Y = 0.275$) midway in the subgiant phase ($t_g = 11.19502868$)

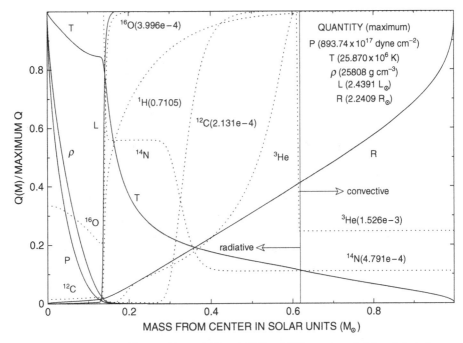

Fig. 11.1.20 Structure and composition variables in a 1$M_\odot$ model ($Z = 0.15$, $Y = 0.275$) near the end of the subgiant phase ($t_g = 11.38825793$)

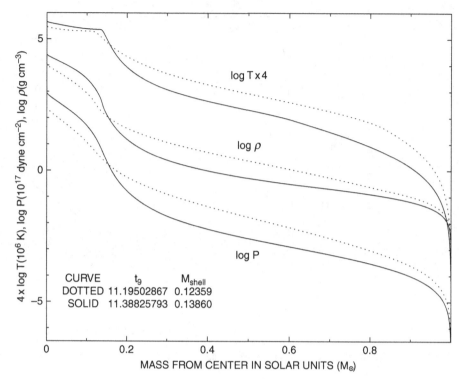

Fig. 11.1.21 Structure variables in 1M$_\odot$ models (Z $=$ 0.015, Y $=$ 0.275) in the middle (t$_g$ $=$ 11.19502868) and near the end (t$_g$ $=$ 11.38825793) of the subgiant phase

nuclear burning shell requires plotting on a logarithmic scale, as is done in Fig. 11.1.21. To retain adequate resolution in temperature, $\log T$ in this figure has been multiplied by a factor of 4.

Results of two classical studies have bearing on the changes in the character of evolution and structure which occur on the subgiant branch. The first study, by Mario Schönberg and Subrahmanyan Chandrasekhar (1942), examines the structure of a static model consisting of a hydrogen-exhausted isothermal core in which both electrons and nuclei obey a perfect gas equation of state and an envelope of homogeneous composition in which energy flow is by radiation controlled by a Kramers opacity and at the base of which the flux of energy is due to a shell source of infinitesimal extent powered by CN-cycle reactions. The variable parameters are the mass of the isothermal core relative to the total mass of the model and the molecular weight in the envelope relative to the molecular weight of core material. They find that no static solutions exist for mass ratios larger than a critical value which has come to be known as the Schönberg–Chandrasekhar limit. The limit increases as the molecular weight ratio increases. For molecular weight ratios now thought to be appropriate, the S–C limit in the case of a 1 $M_\odot$ model is $M_{core}/M_{model} \sim 0.13$.

The second study, by Donald Lynden-Bell & Roger Wood (1968), examines the properties of a spherically symmetric distribution of point particles (stars) treated as a perfect gas in a hypothetical container bounded by a heat reservoir at constant temperature from which

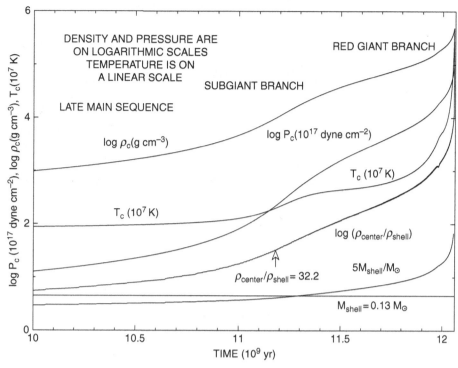

State variables at the center of a 1$M_\odot$ model ($Z = 0.015$, $Y = 0.275$) during late main sequence, subgiant, and giant branch phases

the particles are specularly reflected. When the density at the center relative to the density at the boundary reaches a critical value given by $\rho_{center}/\rho_{boundary} \sim 32.2$, thermal equilibrium (static) solutions cease to exist and the gas begins to contract at an ever accelerating rate in a process described by them as a "gravothermal catastrophe". The same critical ratio has been demonstrated by Masayuki Fujimoto & Iben (1991) to be relevant for understanding the behavior of stellar models with helium-exhausted cores and helium-burning shells and by Iben (1993) to be relevant for understanding stellar models of intermediate mass with hydrogen-exhausted cores and hydrogen-burning shells.

The two critical ratios are related and, in the present context, serve as convenient markers. But, because the equation of state in the core of the subgiant departs significantly from that of a perfect gas, the ratios remain primarily as convenient markers and not as indicators of subsequent catastrophic behavior. That they are related is demonstrated in Fig. 11.1.22, where the logarithm of $\rho_{center}/\rho_{shell}$ and the quantity $M_{shell}/M_\odot$ are plotted against time. The solid horizontal line indicates where $M_{shell}/M_\odot = 0.13$. It is evident that the time when the ratio $\rho_{center}/\rho_{shell} \sim 32.2$ and the time when $M_{shell} = 0.13\ M_\odot$ nearly coincide. However, the only evident changes which occur in central characteristics are slight, temporary increases in the rates of change of central density and temperature. From the time dependences of shell characteristics shown in Fig. 11.1.23, it is evident that passage of core characteristics through the markers does not influence the characteristics at the middle of

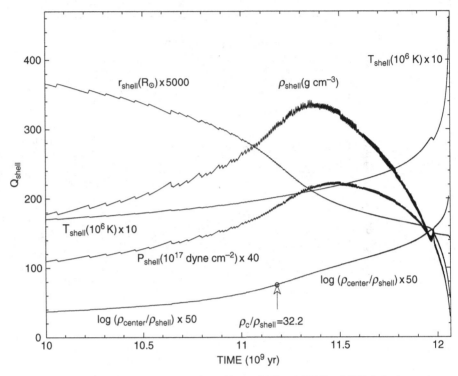

Q_{shell}

$r_{shell}(R_\odot) \times 5000$

$\rho_{shell}(g\ cm^{-3})$

$T_{shell}(10^6\ K) \times 10$

$T_{shell}(10^6\ K) \times 10$

$P_{shell}(10^{17}\ dyne\ cm^{-2}) \times 40$

$\log(\rho_{center}/\rho_{shell}) \times 50$

$\log(\rho_{center}/\rho_{shell}) \times 50$

$\rho_c/\rho_{shell} = 32.2$

TIME (10^9 yr)

Fig. 11.1.23 Characteristics at the middle of the H–burning shell in a 1$M_\odot$ model ($Z = 0.015$, $Y = 0.275$) during late main sequence, subgiant, and giant branch phases

the hydrogen-burning shell. The shell serves as a boundary condition for the core but is not itself obviously affected by changes in core characteristics.

In the current context, exceeding the critical quantities predicted by the classical experiments does not lead to dramatic changes in core temperatures because only a small fraction of the compressional energy released by contraction in the hydrogen-exhausted core is converted into thermal kinetic energy, and gravothermal energy released locally in the core is carried away efficiently by electron conduction. Both of these occurrences are due to the increase in the degree of electron degeneracy in the core.

Responsible for the increase in the degree of electron degeneracy is the fact that the density in the core increases more rapidly with time than does the temperature, a feature highlighted in Fig. 11.1.22 where the central temperature is plotted on a linear scale but central density is plotted on a logarithmic scale. Density and temperature at the model center during the subgiant phase and the first part of the red giant phase are each plotted on a linear scale in Fig. 11.1.24 by the dashed lines labeled ρ_{c4} and $T_{c6}/2$, respectively. The increase with time in the degree of electron degeneracy at model center is demonstrated in Fig. 11.1.24 by the related quantities $P_e/n_e kT$ (with values at the ends of calculational runs given by open circles) and ϵ_F/kT (values given along the heavy solid curve). Here, P_e is the electron pressure, n_e is the electron number density, and ϵ_F is the electron Fermi energy. Electron energies are not relativistic and the relationship between quantities are as

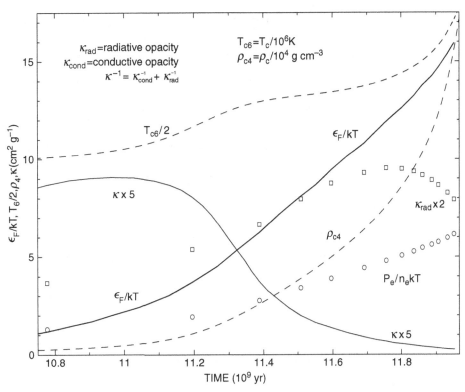

Fig. 11.1.24 Physical characteristics at the center of a 1$M_\odot$ model ($Z = 0.015$, $Y = 0.275$) during the subgiant and early red giant branch phases

described in general terms in Sections 4.6 and 4.7 and in quantitative detail in columns 2 and 4 of Table 4.7.1. In first approximation, $P_e \sim (2/5)\, n_e \epsilon_F$, so

$$\frac{\epsilon_F}{kT} \sim 2.5\, \frac{P_e}{n_e kT}. \tag{11.1.2}$$

In Fig. 11.1.24, the total opacity κ at the model center is defined by the curve labeled $\kappa \times 5$ and the radiative opacity κ_{rad} at the center is defined by the open squares. The conductive opacity κ_{cond} can be deduced from the relationship $1/\kappa = 1/\kappa_{cond} + 1/\kappa_{rad}$. Even though the radiative opacity increases with time over most of the time period shown in the figure, the total opacity decreases significantly over most of this time, indicating that the conductive opacity decreases significantly as well. For example, at the center of the 11.388 billion year old model on the subgiant branch, $\kappa_{rad} = 3.33$ cm^2 g^{-1}, $\kappa = 0.748$ cm^2 g^{-1}, and $\kappa_{cond} = 0.965$ cm^2 g^{-1}, showing that conduction is over three times more effective than radiation in transporting energy.

Gravothermal characteristics in the deep interior of the 11.388 billion year old subgiant model are shown in Fig. 11.1.25. The peak in the curve ϵ_{grav} ($\epsilon_{gravothermal}$) identifies where the nuclear shell source is centered. All matter lying between the model center and the place beyond the nuclear burning shell where the curve for $\epsilon_{work} = \mathbf{g} \cdot \mathbf{v}$ intersects the

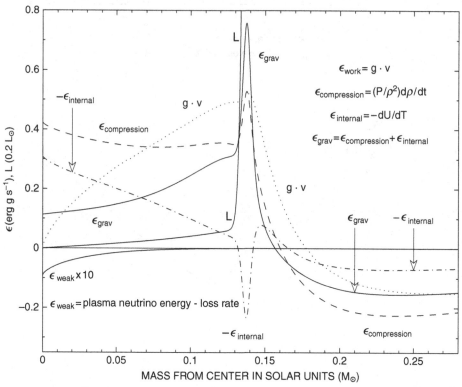

Fig. 11.1.25 Gravothermal energy-generation rates in the deep interior of a $1M_\odot$ model ($Z = 0.015$, $Y = 0.275$) at the base of the red giant branch ($t_g = 11.38825793$)

x-axis is moving inward. All matter between the center and where the curve for $\epsilon_{compression}$ intersects the x-axis is increasing in density. This latter intersection also lies beyond the nuclear burning shell. Thus all matter in the hydrogen-exhausted core and in the hydrogen-burning shell is moving toward the center and contracting. By the same reasoning, one may infer that all matter in the envelope of the model is both expanding and moving outward.

From the curves for $\epsilon_{compression}$, $\epsilon_{internal}$, and $\epsilon_{grav} = \epsilon_{compression} + \epsilon_{internal}$, one may deduce that, in the core, approximately half of the compressional energy released is used up in increasing the internal energy, with the remainder sustaining an outward flow of energy which produces the increase in the luminosity L through the core. The same three curves show that, in the displayed portions of the envelope, approximately one third of the energy used locally for expansion is supplied by heat given up by cooling matter; the other two thirds is abstracted from the ambient flow of energy. In the nuclear burning shell, a substantial fraction of the rate of decrease of internal energy ($-dU/dt = \epsilon_{internal} > 0$) is due to the decrease in particle numbers brought about by nuclear transformations.

The gravothermal characteristics displayed in Fig. 11.1.25 in the core region are shown again in Fig. 11.1.26, but on an expanded scale, and a curve labeled $-\epsilon_{ions} \times 10$ has been added. The quantity

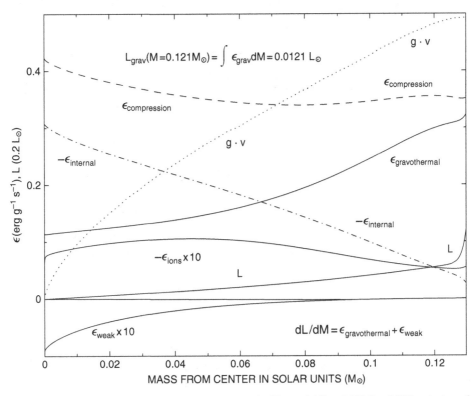

Fig. 11.1.26 Gravothermal energy-generation rates in the contracting core of a 1$M_\odot$ model ($Z = 0.015$, $Y = 0.275$) at the base of the red giant branch ($t_g = 11.38825793$)

$$-\epsilon_{ions} = \frac{dU_{ions}}{dt} = N_A \, Y_{nuc} \, \frac{d}{dt} \left(\frac{3}{2} \, kT \right), \tag{11.1.3}$$

where N_A is Avogadro's number and $Y_{nuc} = \sum_i Y_i$, where Y_i is the number abundance of the ith nucleus, describes the rate per gram per second at which the kinetic energy of nuclei increases. In the non-relativistic approximation, the kinetic energy density of electrons, given by eqs. (4.6.41) and (4.6.43), is (in erg g^{-1})

$$U'_e = \frac{3}{5} \, N_A \, Y_e \, \epsilon_{F0} \left(1 + \frac{5\pi^2}{12} \left(\frac{kT}{\epsilon_{F0}} \right)^2 - \frac{\pi^4}{16} \left(\frac{kT}{\epsilon_{F0}} \right)^4 + \cdots \right), \tag{11.1.4}$$

where ϵ_{F0} is the electron Fermi energy at zero temperature given by eq. (4.6.27) and $Y_e = \sum_i \frac{Z_i}{A_i} Y_i$ is the electron number abundance. Since ϵ_{F0} is a function only of density,

$$\left(\frac{dU_e}{dt} \right)_\rho = \frac{3}{5} \, N_A \, Y_e \, \frac{5\pi^2}{6} \, \frac{kT}{\epsilon_{F0}} \, \frac{d}{dt} \, (kT) \left(1 - \frac{3\pi^2}{10} \left(\frac{kT}{\epsilon_{F0}} \right)^2 + \cdots \right), \tag{11.1.5}$$

or

$$\left(\frac{dU_e}{dt}\right)_\rho \sim \frac{\pi^2}{3} \frac{kT}{\epsilon_{F0}} N_A Y_e \frac{d}{dt}\left(\frac{3}{2}kT\right). \tag{11.1.6}$$

In the hydrogen-exhausted core, $Y_e \sim 0.5$ and $Y_{nuc} \sim 0.25$. At the center of the 11.338 billion year old subgiant model, $kT/\epsilon_{F0} \sim 1/6.25$, so

$$\left(\frac{dU_e}{dt}\right)_\rho \sim 0.263 \, N_A \frac{d}{dt}\left(\frac{3}{2}kT\right), \tag{11.1.7}$$

compared with

$$\frac{dU_{ions}}{dt} = 0.25 \, N_A \frac{d}{dt}\left(\frac{3}{2}kT\right). \tag{11.1.8}$$

Thus, at the center, the rate of increase of the thermal energy of electrons is comparable with the rate of increase of the thermal energy of nuclei. This is true over most of the rest of the core as well.

Since, as shown in Fig. 11.1.26, over most of the core,

$$\frac{dU_{ions}}{dt} \ll \frac{dU_{internal}}{dt}, \tag{11.1.9}$$

it is clear that most of the increase in internal energy in the core goes into increasing the kinetic energy of degenerate electrons rather than into the thermal energy of either nuclei or electrons. It is this circumstance that best explains why the rate of temperature increase in the core continues to be modest after the Schönberg–Chandrasekhar limit has been reached.

The quantity ϵ_{weak} plotted in Figs. 11.1.25 and 11.1.26 is the rate at which energy is converted into escaping neutrinos by the plasma process mediated by the weak interaction, a process described in Chapter 15 of Volume 2. At the center of the model, $\epsilon_{weak} \sim \epsilon_{ions}$. Taking into account the fact that, at the center, the rate of heating electrons near the top of the Fermi sea is comparable with the rate of heating of ions, it is evident that heating occurs at approximately twice the rate of cooling. As evolution progresses, the sense of this inequalty is maintained at the center, with the consequence that matter at the center continues to become hotter. On the other hand, at positions progressively further from the center, heating rapidly outpaces cooling by neutrino losses. As evolution progresses, this differential behavior persists and, ultimately, the maximum temperature in the core moves away from the model center.

Taking into account the relative magnitudes of the various gravothermal energy-generation rates in the core, it is evident that, along the subgiant branch at least, the dominant source of energy is work done by the gravitational field that is converted into the release of compressional energy, approximately half of which contributes to an out-going flux of radiant energy. Because the choice of 0.2 $L_\odot$ as a unit for the luminosity is somewhat unorthodox, it is worthwhile to examine carefully the relationship between $\epsilon_{gravothermal}$ and L in Fig. 11.1.26. A sharp increase in the slope of the curve for L at $M \gtrsim 0.125 \, M_\odot$ indicates that nuclear energy release is contributing to the luminosity. Integration under the curve for $\epsilon_{gravothermal}$ from the center out to a point just before this sharp increase gives an estimate of the gravothermal luminosity which can be compared with the luminosity given by the curve labeled L. For example, $L_{\text{gravothermal}}(M = 0.121\,M_\odot) = \int_0^{0.121M_\odot} \epsilon_{\text{gravothermal}}\, dM \sim 0.0121\, L_\odot$. At $M = 0.121\,M_\odot$ along the curve for L, one finds that $L \sim 0.0605 \times 0.2\, L_\odot = 0.0121\, L_\odot$. Thus, all is well with the choice of units. A further integration under the curve for ϵ_{grav} out to $M \sim 0.16\,M_\odot$ reveals that the total positive contribution to L_{grav} is $L_{\text{grav}}^+ \sim 0.0161\, L_\odot$. Between $\sim 0.16\,M_\odot$ and the base of the convective envelope at $\sim 0.62\,M_\odot$, ϵ_{grav} remains negative. Over the convective envelope, $\epsilon_{\text{grav}} \sim 0$ and an integration shows that the total negative contribution to L_{grav} is $L_{\text{grav}}^- \sim 0.0219\, L_\odot$. Altogether, $L_{\text{grav}} = L_{\text{grav}}^+ - L_{\text{grav}}^- \sim -0.0058\, L_\odot$, which says that slightly more gravothermal energy is used up in expanding matter in the envelope outward against gravity than is produced by compression in the core and in the hydrogen-burning shell.

An important feature of subgiant branch evolution is that, because of increasing opacities in the expanding and cooling outer layers of the model envelope, the base of the convective envelope moves inward in the mass coordinate, as demonstrated in Figs. 11.1.19 and 11.1.20. One obvious consequence of this inward movement is that the surface abundances of isotopes which are easily destroyed over most of the interior during the main sequence phase are dramatically reduced. At the same time, surface abundances of isotopes such as ^{3}He and ^{13}C which are produced in envelope regions are increased. The process of changing surface abundances due to mixing outward consequences of prior nuclear processing in the interior is known as the first dredge-up. The increase in the surface ^{3}He abundance brought about as the base of the convective envelope sweeps into the region where the ^{3}He abundance has been built up during the main sequence phase is demonstrated by the profiles in Figs. 11.1.19 and 11.1.20, which prevail, respectively, prior to dredge-up and after dredge-up has begun.

Another important consequence of the inward extension of convection is that, because of the shrinkage in mass of the radiative region between the hydrogen-burning shell and the base of the convective envelope, the opacity in radiative surface layers, where H^- absorption has become the dominant source of opacity, plays a major role in determining the relationship between surface luminosity and surface temperature. The fact that the opacity in surface layers decreases with decreasing surface temperature is responsible for the upturn in the HR diagram from the subgiant branch to the red giant branch.

Following the phase of evolution to the red at nearly constant luminosity along the subgiant branch, the subsequent evolution to higher luminosity at ever lower surface temperatures is essentially the inverse of the evolution during the pre-main sequence phase of gravitational contraction downward and blueward in the HR diagram. During the pre-main sequence phase, when the interior is completely convective, the H^- opacity in surface layers drives evolution downward in the HR diagram. During the second phase, when the core is radiative and grows in mass, the decrease in interior opacity with increasing interior temperatures drives evolution upward in the HR diagram.

In Fig. 11.1.27, the mass M_{CEB} at the base of the convective envelope is compared with masses M_{10}, M_{50}, and M_{90} describing the location of the hydrogen-burning shell. Shown also are $\Delta M_{\text{shell}} = M_{90} - M_{10}$, the width in mass of the hydrogen-burning shell, and $Y_{\text{H}}(M_{50})$, the abundance of hydrogen at the middle of the hydrogen-burning shell. At $t_9 \sim 11.89$, when the base of the convective envelope reaches its maximum inward extent,

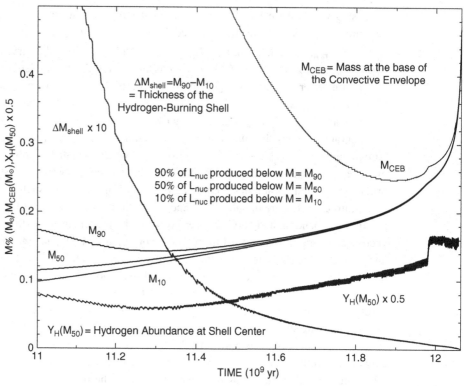

Fig. 11.1.27 Hydrogen–burning shell properties and convective envelope base location in a 1$M_\odot$ model ($Z = 0.015, Y = 0.275$) on the subgiant and giant branches

$M_{CEB} \sim 0.246\ M_\odot$, approximately 0.04 $M_\odot$ above the middle of the hydrogen-burning shell. Thereafter, the base of the envelope convective zone recedes outward in mass. This, then, marks the end of the first dredge-up episode. Abundances in column 2 of Table 11.1.4 are those in the model prior to the pre-main sequence deuterium-burning phase and those in column 3 are those in the model at the end of the first dredge-up episode. In the case of the isotopes ^{1}H, ^{4}He, ^{12}C, ^{13}C, and ^{14}N, surface abundances at the beginning of the dredge-up episode are the same as those prior to deuterium burning. Conversion of initial deuterium into ^{3}He increases the abundance of ^{3}He to 2.601×10^{-5} just prior to the start of the first dredge-up episode. In the case of the isotopes ^{1}H, ^{4}He, ^{12}C, ^{13}C, and ^{14}N, ratios are between abundances before and after the first dredge-up episode. In the case of ^{3}He, the ratio between abundances before and after the first dredge-up episode is 18.0 rather than 38.4. Table 11.1.4 is located near the end of this section.

The variation with time of $Y_H(M_{50})$ in Fig. 11.1.27 can be understood by an examination of the hydrogen-abundance profiles in Fig. 11.1.28. In Fig. 11.1.28, the profile labeled 30 describes the distribution of hydrogen in the 11.388 billion year old subgiant model which lies in the HR diagram more or less at the intersection of the subgiant branch and the red giant branch. The outer part of this profile is identical with the profile prevailing at the

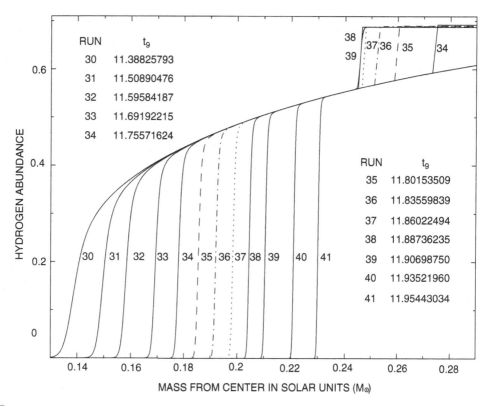

Fig. 11.1.28 Hydrogen-abundance profiles in a 1M$_\odot$ model ($Z = 0.015, Y = 0.275$) during early ascent of the red giant branch

end of the main sequence phase. As the model climbs the red giant branch, the base of the hydrogen profile in the hydrogen-burning shell steepens and the abundance at the center of the profile increases, explaining the more or less monotonic increase with time in the curve for $Y_H(M_{50})$ in Fig. 11.1.27 for $11.3 \lesssim t_9 \lesssim 11.98$.

While approaching its maximum inward extent, the base of the convective envelope moves into regions within which substantial conversions of hydrogen into helium have occurred during the main sequence phase. Hydrogen is convected inward, producing the hydrogen-abundance profiles in Fig. 11.1.28 at masses larger than 0.24 $M_\odot$. The near discontinuities in the profiles mark the location of the base of the convective envelope. At $t_9 \sim 11.89$, when the base of the convective envelope reverses direction in the mass coordinate, the hydrogen profile (roughly that of the model labeled 38) consists of four segments: a steep inner segment defining the location of the hydrogen-burning shell, a slowly and smoothly varying segment of width about 0.04 $M_\odot$ which is what remains of the hydrogen profile established on the main sequence, another steep segment which is the near discontinuity left behind by the receding base of the convective envelope, and, finally, an outer segment of constant abundance. The smoothing of the outer near discontinuity which is evident along the profile labeled 39 is a consequence of numerical diffusion associated with rezoning.

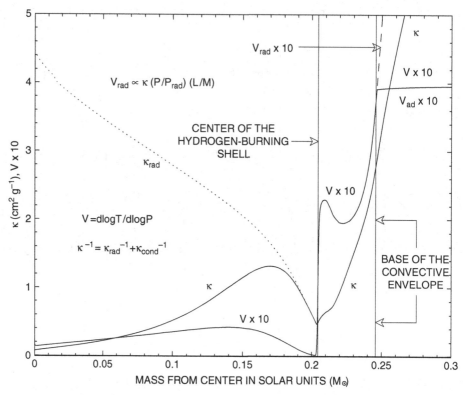

Fig. 11.1.29 Opacities and temperature gradients in the $1M_\odot$ red giant model ($Z = 0.015$, $Y = 0.275$) with the most massive convective envelope ($t_g = 11.88736235$)

In Fig. 11.1.29, the logarithmic temperature gradient $V = d\log T/d\log P$ is shown as a function of mass in the red giant model in which the base of the convective envelope is near its minimum value of $M_{CEB} \sim 0.246\ M_\odot$ (labeled model 38 in Fig. 11.1.28). In the convective envelope, $V = V_{ad} \cong 0.4$. Elsewhere, $V = V_{rad}$, where

$$V_{rad} = \frac{1}{16\pi Gc}\ \kappa\ \frac{P}{P_{rad}}\ \frac{L}{M}. \tag{11.1.10}$$

Over much of the region between the hydrogen-burning shell and the base of the convective envelope, $V_{rad} \sim 0.23 \pm 0.03$. In the hydrogen-exhausted core, $V_{rad} < V_{rad}^{max}$, where $V_{rad}^{max} \sim 0.04$ is the relative maximum at $M \sim 0.145\ M_\odot$. Just below the base of the hydrogen-burning shell, V_{rad} becomes much smaller than V_{rad}^{max}.

In the hydrogen-exhausted core, of the three variable factors in eq. (11.1.10) which enter into the determination of V_{rad}, the opacity κ is the one most responsible for the decrease inward from the relative maximum at $M \sim 0.145\ M_\odot$. In Fig. 11.1.29, the total opacity κ is shown by a solid curve and the radiative opacity κ_{rad} is shown by the dotted curve. From approximately $0.19\ M_\odot$ outward, $\kappa \sim \kappa_{rad}$. Progressing inward from this point into the core, electron conductivity becomes ever more effective in reducing κ. For example, at

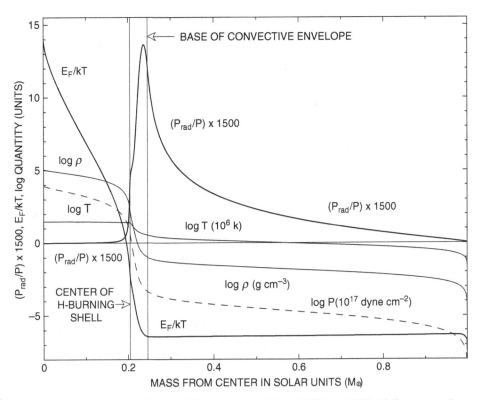

Fig. 11.1.30 E_F/kT, P_{rad}/P, and structure variables in the 1M$_\odot$ red giant model ($Z = 0.015$, $Y = 0.275$) with the most massive convective envelope ($t_g = 11.88736235$)

$M = 0.15\ M_\odot$, $\kappa_{rad} \sim \kappa_{cond} \sim 2\kappa$. At $M = 0.1\ M_\odot$, $\kappa_{rad} \sim 2.8$, $\kappa \sim 0.52$ and $\kappa_{cond} \sim 0.64$, all in units of cm^2 g^{-1}. Progressing further inward, κ rapidly approaches κ_{cond}.

On progressing inward, the switch from energy transfer primarily by radiation to energy transfer primarily by conduction is a consequence of the increase in the degree of electron degeneracy. An indicator of the degree of electron degeneracy is the function ϵ_F/kT shown in Fig. 11.1.30. Comparing κ and κ_{rad} in Fig. 11.1.29 with ϵ_F/kT shows that κ_{cond} decreases below κ_{rad} once $\epsilon_F/kT \gtrsim 4$.

To understand the behavior of $V = V_{rad}$ just below the hydrogen-burning shell requires an examination of the factors P/P_{rad} and L/M in eq. (11.1.10). The inverse of the first of these factors is shown as a function of mass in Fig. 11.1.30, along with the main structural variables ρ, T, and P which determine its value. In the electron-degenerate core, radiation pressure is quite small compared with the total pressure; proceeding outward through the core, P_{rad}/P does not become noticeable on the scale of Fig. 11.1.30 until just below the hydrogen-burning shell as the Fermi energy becomes negative. Conversely, the ratio P/P_{rad} is large in the electron-degenerate core and drops to relatively nominal values just below the hydrogen-burning shell. This is demonstrated in Fig. 11.1.31 where the ratio, normalized to the ratio at the center, becomes undetectable beyond the hydrogen-burning shell.

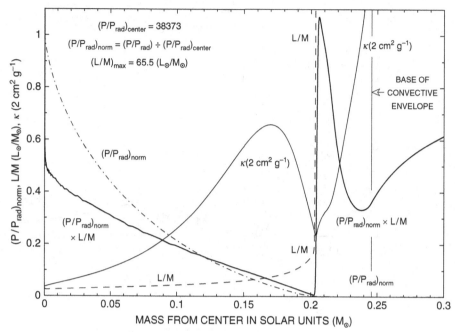

Fig. 11.1.31   Ingredients of V_{rad} in the $1M_\odot$ red giant model ($Z = 0.015$, $Y = 0.275$) with the most massive convective envelope ($t_g = 11.88736235$)

The curve labeled L/M in Fig. 11.1.31 describes the final ingredient for determining V_{rad}. It is small in the core, reaching a value of $L/M \sim 0.17 \, L_\odot/M_\odot$ at the base of the hydrogen-burning shell, and then climbs precipitously to $L/M \sim 65.5 \, L_\odot/M_\odot$ at the outer edge of the burning shell. The product of the normalized ratio P/P_{rad} and the ratio L/M is given by the heavy solid curve in Fig. 11.1.31. Putting all three factors together, it is evident that the primary responsibility for the minimum in V_{rad} at the very base of the hydrogen-burning shell is the relatively large value of P_{rad}/P at the base of the shell. In Fig. 11.1.30, the base of the shell occurs at the mass which coincides with the location of the inflection point along the curve for P_{rad}/P.

At time $t_9 \sim 11.98$, the middle of the hydrogen-burning shell reaches the near discontinuity left behind by the convective envelope at its maximum extent inward, and, as shown in Fig. 11.1.27, the value of $Y_H(M_{50})$ increases discontinuously. The consequence for the evolutionary track in the HR diagram of Fig. 11.1.1 is a barely discernable drop in surface luminosity just below the uppermost filled circle labeled $M_{shell} = 0.25 \, (M_\odot)$. Thereafter, $Y_H(M_{50})$ remains nearly constant at slightly less than half of the surface abundance of hydrogen, which by now has dropped to $Y_H \sim 0.68$, and the base of the convective envelope recedes outward, remaining slightly above the nuclear burning shell.

Evolution in the HR diagram during and after these interesting changes in the distribution of hydrogen in the interior is shown in Fig. 11.1.32. The two large glitches in the evolutionary track as M_{shell} increases from $0.35 \, M_\odot$ to $0.45 \, M_\odot$ are consequences of increasing the mass in the static envelope in an effort to reduce computational time by decreasing

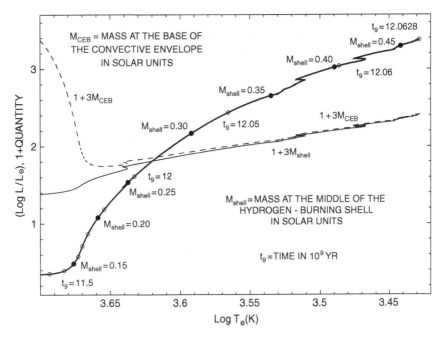

Fig. 11.1.32 Evolutionary track of a 1M$_\odot$ model ($Z = 0.015$, $Y = 0.275$) as a red giant and locations of the H-burning shell and the base of the convective envelope

the number of mass points in the interior model. As is evident from the curve for ΔM_{shell} in Fig. 11.1.27, the thickness of the hydrogen-burning shell decreases significantly with time; the consequence is an increase in the number of mass zones in the neighborhood of the burning shell required to resolve the shell. The increase in luminosity with time, coupled with a decrease in the mass of the hydrogen-burning shell, means that the rate of change in the number abundances of isotopes in the hydrogen-burning shell increases with time, necessitating a decrease in the size of evolutionary time steps in order to prevent excessively large fractional changes in abundances in a time step. The net result is a huge expenditure of computer time to reach the termination of the red giant branch phase and an incentive to sacrifice information about gravothermal changes in the outer part of the model by reducing the mass of the interior model relative to the mass of the static envelope.

Fortunately, the glitches in surface relationships occasioned by reallocation of mass between the quasistatic interior and the static surface do not appear in the relationship between surface luminosity and hydrogen-burning shell mass shown in Fig. 11.1.33. The one bump in the relationship which begins at about $\log L/L_\odot \sim 1.5$ and $\log M_{shell}/M_\odot \sim -0.615$ ($M_{shell} \sim 0.243 \, M_\odot$) is a legitimate reaction to the advance of the hydrogen-burning shell into the hydrogen profile deposited at the base of the convective envelope at its maximum inward extent.

A red giant branch model can be thought of as consisting of four main parts: (1) a compact, electron-degenerate, hot white-dwarf-like core; (2) a very narrow hydrogen-burning shell in which the composition and luminosity change almost discontinuously; (4) a very

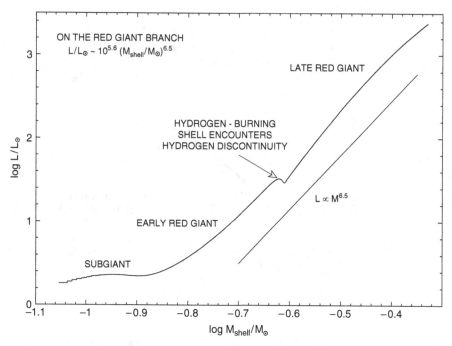

Fig. 11.1.33 Relationship between luminosity and hydrogen-burning shell mass for a $1M_\odot$ model ($Z = 0.015$, $Y = 0.275$) on the subgiant and red giant branches

extended and cool convective envelope; and (3) a region of very rapid changes in all structure variables between the outer edge of the burning shell and the base of the convective envelope (i.e. between regions (2) and (4)).

In the next five figures, some large-scale characteristics of four red giant branch models are examined and compared with those of the two subgiant models which have already been examined in some detail. The four red giant branch models include a model with two near discontinuities in the hydrogen abundance associated, respectively, with the hydrogen-burning shell and the base of the convective envelope at its maximum inward extent, and three, more luminous, models in which the only near discontinuity in the hydrogen abundance is associated with the hydrogen-burning shell.

In Fig. 11.1.34 are temperature profiles of the six models in the mass representation. In each case, the location of the center of the hydrogen-burning shell is indicated by an open circle and, for each of the four red giant branch models, the location of the base of the convective envelope is shown by an open square. The base of the convective envelope in the two subgiant models lies outside of the figure (see Figs. 11.1.19 and 11.1.20). In good approximation, the temperature at the center of the hydrogen-burning shell for the two subgiant models and for the three most luminous red giant models is related to the mass at the center of the shell by

$$T_{\text{shell}} = 10 \times 10^6 K + 84 \times 10^6 K \left(\frac{M_{\text{shell}}}{M_\odot} \right). \qquad (11.1.11)$$

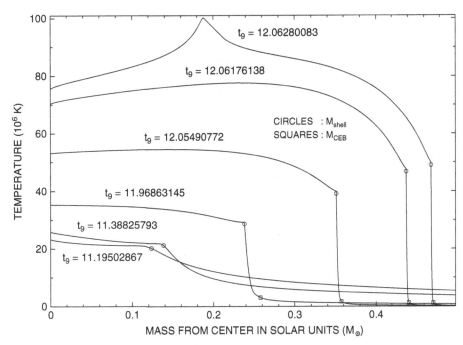

Fig. 11.1.34 Temperature profiles in the core of a 1M$_\odot$ model (Z = 0.015, Y = 0.275) burning hydrogen in a shell as an evolved subgiant and as a red giant

Equation (11.1.11) overestimates the shell temperature of the least luminous red giant model by only $\sim 10^6$ K.

Several features in Fig. 11.1.34 stand out. First, whereas the temperature gradient dT/dM in the two subgiant models is everywhere relatively modest, in the four red giant branch models, the gradient just outside of the hydrogen-burning shell is huge compared with both the gradient in the hydrogen-exhausted core and the gradient in the convective envelope. Second, in the red giant branch models, the separation in mass between the hydrogen-burning shell and the base of the convective envelope is small compared with the mass in the hydrogen-exhausted core below the burning shell and the extremely large temperature gradient is confined to the region between the shell and the base of the convective envelope. Third, in the three most luminous red giant branch models, the maximum temperature is not at the center, indicating that energy flows inward from the point of maximum temperature. Energy loss due to the plasma neutrino process is responsible for this phenomenon. Fourth, as time passes, the temperature at the center of the core lags progressively further behind the maximum temperature in the core. This is because, as core densities increase, the rate of energy loss due to the plasma neutrino process increases. Finally, in the most luminous red giant branch model, the injection of energy from helium burning has led to the formation of a temperature spike. The flux produced by helium burning has become large enough to engender a small convective shell, the presence of which is manifested by a temperature gradient of constant slope in a region extending outward by about 0.03 $M_\odot$ from the point of maximum temperature. Helium-burning reactions and energy-generation rates are

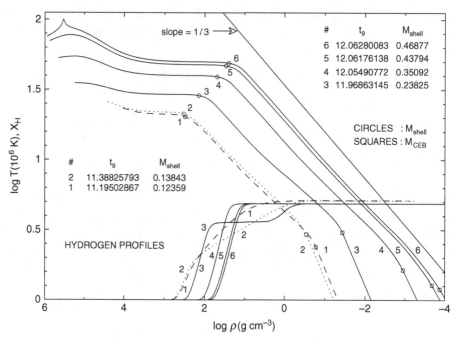

Fig. 11.1.35 Temperature vs density in the core of a 1$M_\odot$ model ($Z = 0.015, Y = 0.275$) burning hydrogen in a shell as an evolved subgiant and as a red giant

discussed in Chapter 16 of Volume 2 and the helium-burning evolution of the 1 $M_\odot$ mass model is discussed in Chapter 17 of Volume 2.

The relationship between temperature and density in the high temperature interior of each model is shown in Fig. 11.1.35. Along each temperature–density curve, the location of M_{shell} is indicated by an open circle and the location of M_{CEB} is indicated by an open square. Hydrogen abundance profiles for each of the models are shown at the bottom of the figure. The temperature–density relationships in the hydrogen-exhausted cores are quite shallow, consistent with the shallowness of the temperature–mass relationships in Fig. 11.1.34. In the regions between M_{shell} and M_{CEB}, the temperature–density gradients in the three most luminous models are slightly less steep than a line of slope $d\log T / d\log \rho = 1/3$.

The fact that the logarithmic temperature–density gradients in the two subgiant models are nearly constant in the regions between M_{shell} and M_{CEB}, but somewhat shallower than the corresponding gradients for the three most luminous models, can be understood from the fact that the hydrogen abundance in the subgiant models increases monotonically in the region between the boundary masses M_{shell} and M_{CEB} rather than being constant as in the most luminous red giants. In the least luminous red giant model, there is a jump in the hydrogen abundance over a density range small compared with the entire density range between boundary masses and this results in an inflection in the slope of the temperature–density profile which coincides with the jump in the hydrogen abundance.

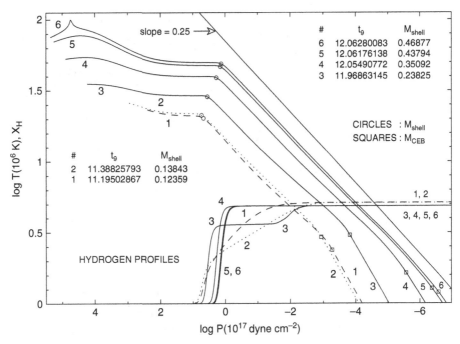

Fig. 11.1.36 Temperature vs pressure in the core of a 1$M_\odot$ model ($Z = 0.015$, $Y = 0.275$) burning hydrogen in a shell as an evolved subgiant and as a red giant

In the convective envelopes of the three least luminous models where it is measurable in Fig. 11.1.35, d $\log T/$d $\log \rho \sim 2/3$, as expected.

The hydrogen-abundance profiles in Fig. 11.1.35 show that, in each case, density drops by approximately a factor of three through the hydrogen-burning shell, with a corresponding factor of three reduction in relevant reaction rates at constant temperature and composition. Comparing with the temperature–density profiles, this range in density corresponds to an approximately 7% drop in temperature through the burning shell. Supposing that the relevant reaction rates depend on about the 17th power of the temperature, this drop in temperature contributes, at constant composition, to an approximate factor of three drop in reaction rates. Thus, density and temperature decreases through the burning shell contribute comparably to the determination of the thickness of the shell.

The relationship between temperature and pressure in each of the six models is given in Fig. 11.1.36, along with relationships between the hydrogen abundance and pressure. Again, in each case, the location of the center of the hydrogen-burning shell is indicated by an open circle and the base of the convective envelope is indicated by an open square. Note that, in each case, the logarithmic temperature gradient d $\log T/$d $\log P = V$ is effectively zero over an extended region below the hydrogen-burning shell, consistent with the lesson provided by Fig. 11.1.29. In the region between M_{shell} and M_{CEB}, d $\log T/$d $\log P \sim 1/4$. In the convective envelopes, d $\log T/$d $\log P \sim 2/5$.

The mass distributions and the hydrogen-abundance profiles in Fig. 11.1.37 and the mass distributions in Fig. 11.1.38 demonstrate that, although essentially the entire mass of a red

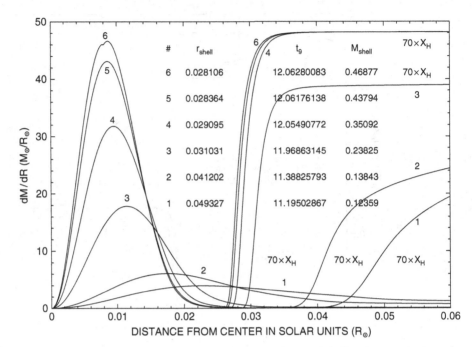

Fig. 11.1.37   Mass profiles in the cores and hydrogen-abundance profiles outside of the cores of $1M_\odot$ subgiant and red giant models ($Z = 0.015$, $Y = 0.275$)

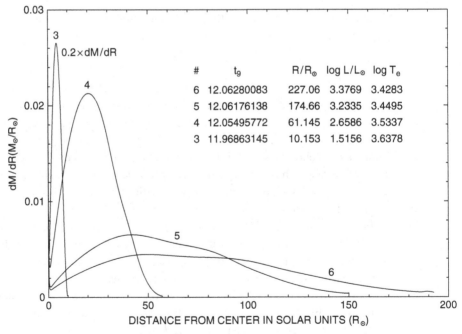

Fig. 11.1.38   Mass profiles in the envelope of a $1M_\odot$ model ($Z = 0.015$, $Y = 0.275$) burning hydrogen in a shell during ascent of the red giant branch

giant is concentrated into two distinct configurations, a core of white dwarf-like dimensions and a halo that occupies a volume orders of magnitude larger than the core, the hydrogen-burning shell and, by extension, the region between the burning shell and the base of the halo convective zone occupy much more volume than does the white-dwarf-like structure forming the nucleus of the hydrogen-exhausted core. In all four of the red giants represented in the figures (curves labeled 3 through 6), the peak in the nuclear core distribution (Fig. 11.1.37) is at a radius comparable with the Earth's radius, whereas the peak in the halo distribution (Fig. 11.1.38) varies from several times the Sun's radius to approximately a quarter of the distance from the Sun to the Earth. The mass distributions in the hydrogen-exhausted cores of the two subgiant models shown by curves labeled 1 and 2 in Fig. 11.1.37 are significantly more diffuse than those in the cores of the red giants and the envelopes of the subgiants are too small to display effectively on the scale of Fig. 11.1.38.

The hydrogen-abundance profiles in the subgiants are relatively shallow and extend into the tails of the core mass distributions. In contrast, the hydrogen-abundance profiles in the red giants are steep and lie distinctly outside the core mass distributions. The steep portions of the profiles essentially outline the locations of the hydrogen-burning shells. It is interesting that, in the red giant models, the radius at the center of the hydrogen-burning shell decreases by only ~10% as the core mass increases by a factor of two. Exaggerating somewhat, one may characterize the burning shell as a transformer frozen in space which acquires matter from an expanding halo, converts the hydrogen in this matter into helium, and deposits the helium ash into a container of nearly constant volume.

The next eleven figures provide additional information about the four basic structure zones in three of the four red giant branch models selected for detailed study. The distributions of structure variables in Figs. 11.1.39–11.1.41 describe conditions in the least luminous ($L = 32.8\,L_\odot$) of the four red giant models. Figure 11.1.39 gives a broad brush picture of structure variables for the entire model. The two vertical lines mark, respectively, the locations of the middle of the hydrogen-burning shell and the base of the convective envelope. It is evident that, in the region between these lines, the rates of change of all structure variables are significantly larger than on either side of this region. It is this circumstance that motivates calling the region between the hydrogen-burning shell and the base of the convective envelope a transition zone between the hydrogen-burning shell and the red giant halo.

Structural variations in the transition zone are shown on an expanded scale in Fig. 11.1.40. The two vertical lines labeled $M_{10\%}$ and $M_{90\%}$ identify the region in which 80% of the nuclear energy is produced and thus provide a measure of the size of the hydrogen-burning shell, the center of which is identified by the vertical line labeled $M_{50\%}$. An important lesson from Fig. 11.1.40 is that the size of the hydrogen-burning shell is small compared with the size of the transition zone.

Through the transition zone, radius increases by a factor of ~27. Temperature, density, and pressure decrease, respectively, by factors of ~9, ~2280, and ~17 000. The fact that the drop in pressure is ~20% smaller than the product of the drops in temperature and density is related to the fact that the hydrogen abundance in the transition zone is variable in two places (compare the curve for the hydrogen abundance in model 3 in Fig. 11.1.35 with the curve for temperature versus density in model 3).

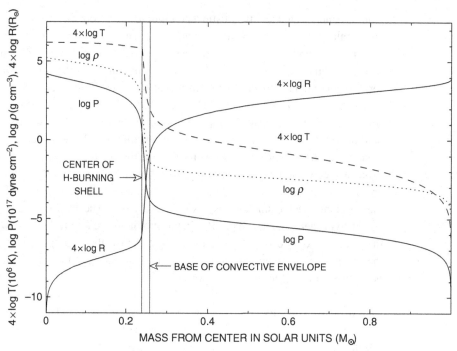

Fig. 11.1.39 Structure variables in a 1$M_\odot$ model ($Z = 0.015$, $Y = 0.275$) burning hydrogen in a shell as a red giant ($t_g = 11.96863145$)

Small though it is compared with the other three structure zones, the hydrogen-burning shell is the engine that controls red giant branch evolution. Conditions in the vicinity of the hydrogen-burning shell are shown in Fig. 11.1.41. Except for density, which is plotted on a logarithmic scale, all variables are plotted on a linear scale. Variables in addition to temperature and density are the reciprocal of the radius coordinate, the luminosity, the hydrogen abundance, and the nuclear energy-generation rate. It is evident that, with appropriate normalization, the hydrogen-abundance distribution and the luminosity curve are almost identical. It is also interesting that, in the direction of increasing mass, the shallower the hydrogen-abundance profile becomes, the closer does the curve for the reciprocal of the radius come to being parallel to the curve for the temperature.

The distributions of structure variables in Figs. 11.1.42–11.1.44 describe conditions in the next most luminous ($L = 456\,L_\odot$) red giant model. In this model there is no variation in the hydrogen abundance beyond the variation in and near the hydrogen-burning shell. The two vertical lines in Fig. 11.1.42 have the same significance as do those in Fig. 11.1.39. Structure variables in the vicinity of these lines are shown in Fig. 11.1.43 on exactly the same scale used in Fig. 11.1.40, facilitating direct comparisons between the two models. Even though the transition zone in the more luminous model is approximately three times narrower in the mass coordinate than the transition zone in the less luminous model, changes in structure variables through the transition zone in the more luminous model are significantly larger than in the less luminous model. Through the transition zone of the

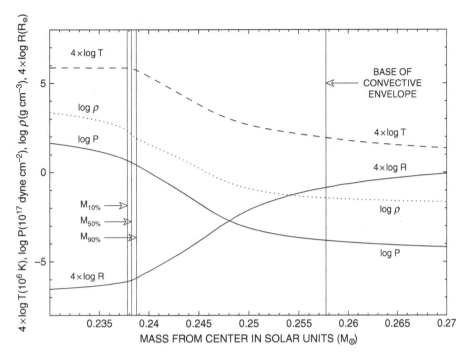

Fig. 11.1.40 Structure variables between the H-burning shell and the base of the convective envelope in a 1$M_\odot$ red giant model ($Z = 0.015$, $Y = 0.275$, $t_9 = 11.96863145$)

more luminous model, radius increases by a factor of $\sim$38 while temperature, density, and pressure decrease, respectively, by factors of $\sim$22, $\sim$23 000, and $\sim$456 000. The fact that the drop in pressure is only $\sim$10% smaller than the product of the drops in temperature and density is related to the fact that the only change in the hydrogen abundance in the transition zone occurs in and near the hydrogen-burning shell (compare the curve for the hydrogen abundance in model 4 in Fig. 11.1.35 with the curve for temperature versus density in model 4).

The quantities plotted in Fig. 11.1.44 in the vicinity of the hydrogen-burning shell in the more luminous model are the same as those in Fig. 11.1.41 in the less luminous model. Although the scaling in the mass coordinate and the scaling in the vertical coordinates are quite different in the two figures, variations in all of the displayed variables are morphologically equivalent. Of particular interest are the facts that, when appropriately normalized, the luminosity and the hydrogen abundance share a common profile and that, in the region where the hydrogen abundance is nearly constant, the curves for the temperature and for the reciprocal of the radius are basically parallel.

The distributions of structure variables in Figs. 11.1.45–11.1.47 describe conditions in the model of the penultimate luminosity ($L = 1712\,L_\odot$) in the suite of selected red giant branch models. The distributions of variables in these figures are qualitatively identical with those for the two less luminous models which are shown in Figs. 11.1.39–11.1.41 and Figs. 11.1.42–11.1.44, respectively, except that gravothermal energy-generation rates are

Table 11.1.2 Characteristics at boundaries of the transition zone between the hydrogen-burning shell and the base of the convective envelope of a 1 $M_\odot$ model red giant

$\dfrac{M_{se}}{M_\odot}$	$\dfrac{M_b}{M_\odot}$	$\dfrac{L_{se}}{L_\odot}$	$\dfrac{\kappa_{se}}{cm^2\,g^{-1}}$	$\dfrac{\kappa_b}{cm^2\,g^{-1}}$	R_{se}/R_b	T_{se}/T_b	ρ_{se}/ρ_b	P_{se}/P_b
0.23874	0.25770	32.8	0.448	2.30	27	9	2.28/3	1.70/4
0.35107	0.35728	456	0.326	0.980	38	22	2.30/4	4.56/5
0.43803	0.44123	1712	0.195	0.650	52	33	6.75/4	1.82/6

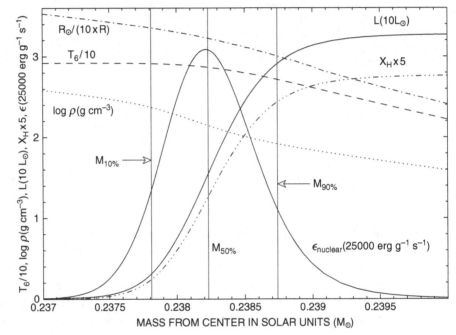

Fig. 11.1.41 Structure variables, nuclear energy-generation rate, and H abundance in the H-burning shell of a $1M_\odot$ red giant model ($Z = 0.015$, $Y = 0.275$, $t_g = 11.96863145$)

shown additionally in Fig 1.1.47. At the same time, quantitative differences are large. The factors by which structural variables change in the transition zones for the three models are summarized in Table 11.1.2, where the subscript se denotes shell edge and the subscript b denotes base of the convective envelope.

The morphological similarities in structural characteristics of the red giant models displayed by the temperature–density and temperature–pressure relationships in the region between the edge of the burning shell and the base of the convective envelope in Figs. 11.1.35 and 11.1.36 and displayed by model characteristics in the structure variable versus mass relationships in Figs. 11.1.39–11.1.47 suggest that it might be possible to construct simple analytical approximations to help understand the factors which are

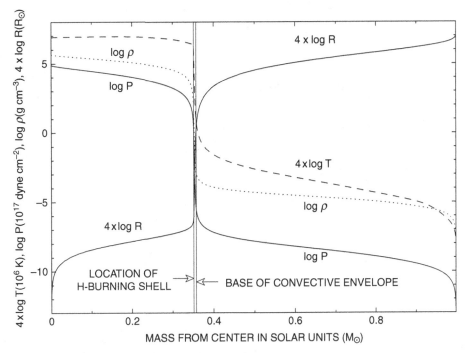

Fig. 11.1.42 Structure variables in a 1$M_\odot$ model (Z = 0.015, Y = 0.275) burning hydrogen in a shell as a red giant (t_g = 12.05490772)

primarily responsible for the changes which occur in the transition zone. Between the boundary masses, the differential relationship between temperature and pressure is given by eq. (11.1.10), which can also be written as

$$V_{\rm rad} = \frac{P}{T}\frac{dT}{dP} = \frac{3}{16\pi acG}\frac{\kappa P}{T^4}\frac{L}{M}. \tag{11.1.12}$$

To the extent that the variation in L/M may be neglected,

$$T^4 - T^4_{\rm shell} = \alpha(P - P_{\rm edge}), \tag{11.1.13}$$

where

$$\alpha = \frac{3\langle\kappa\rangle}{4\pi cGa}\frac{L}{M_{\rm shell}}, \tag{11.1.14}$$

$\langle\kappa\rangle$ is an average of the opacity with respect to pressure, and $L/M_{\rm shell}$ is the luminosity–mass ratio just outside of the hydrogen-burning shell. Equation (11.1.13) suggests that

$$\frac{T^4}{P} = \frac{T^4_{\rm edge}}{P_{\rm edge}} = \alpha. \tag{11.1.15}$$

In the perfect gas approximation this is equivalent to

$$\frac{T^3}{\rho} \sim \frac{3}{4\pi c\,G\,a}\frac{k}{M_{\rm H}}\frac{\langle\kappa\rangle}{\mu}\frac{L}{M_{\rm shell}}. \tag{11.1.16}$$

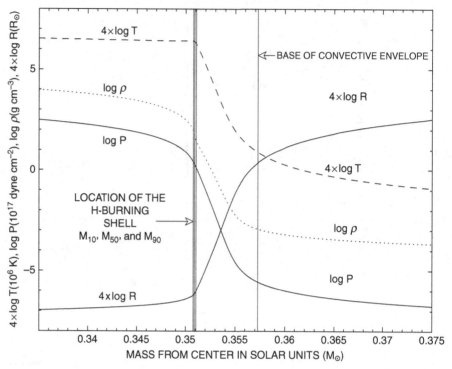

Fig. 11.1.43 Structure variables between the H-burning shell and the base of the convective envelope in a 1$M_\odot$ red giant model ($Z = 0.015$, $Y = 0.275$, $t_9 = 12.05490772$)

From the radiative transfer equation,

$$\frac{\mathrm{d}T}{\mathrm{d}r} = -\frac{3}{4a\,c}\,\kappa\,\frac{\rho}{T^3}\,\frac{L}{4\pi r^2} = \frac{3}{4a\,c}\,\kappa\,\frac{\rho}{T^3}\,\frac{L}{4\pi}\,\frac{\mathrm{d}}{\mathrm{d}r}\left(\frac{1}{r}\right). \qquad (11.1.17)$$

Using eq. (11.1.16) to replace ρ/T^3 in eq. (11.1.17), and approximating κ by $\langle\kappa\rangle$, one has

$$T \sim \frac{1}{4}\,\frac{G\,M_{\mathrm{H}}\,M_{\mathrm{shell}}}{k}\,\frac{\mu}{r}, \qquad (11.1.18)$$

where the constant of integration has been ignored. Comparing the curve for T with the curve for $1/r$ in Figs. 11.1.41, 11.1.44, and 11.1.47, it is apparent that, just outside of the hydrogen-burning shell, the T and $1/r$ profiles can be normalized to match one another reasonably well. However, this does not guarantee that the normalization given by eq. (11.1.18) is the correct choice and, in fact, as the base of the convective envelope is approached, it overestimates the temperature by approximately a factor of 2.

Combining eqs. (11.1.16) and (11.1.18) gives

$$\rho \sim \left(\frac{1}{4}\,\frac{GM_{\mathrm{H}}M_{\mathrm{shell}}}{k}\,\frac{\mu}{r}\right)^3\,\frac{4\pi\,cGa}{3}\,\frac{M_{\mathrm{H}}}{k}\,\frac{\mu}{\langle\kappa\rangle}\,\frac{M_{\mathrm{shell}}}{L}. \qquad (11.1.19)$$

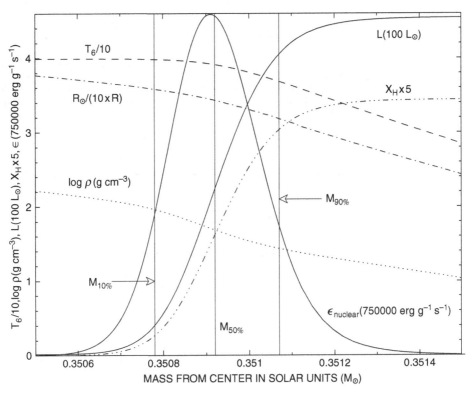

Structure variables, nuclear energy-generation rate, and H abundance in the H-burning shell of a 1$M_\odot$ red giant model ($Z = 0.015$, $Y = 0.275$, $t_9 = 12.05490772$)

The mass between any two radii in the transition zone is given by

$$m(r) - m(r_0) = \int_{r_0}^{r} 4\pi r^2 \rho \, dr. \tag{11.1.20}$$

Using eq. (11.1.19) to replace ρ in eq. (11.1.20) gives

$$m(r) - m(r_0) \sim \frac{(4\pi)^2}{4^3\,3} \left(\frac{GM_{\mathrm{H}}M_{\mathrm{shell}}}{k}\right)^4 \frac{ca}{L} \int_{r_0}^{r} \left(\frac{\mu}{r}\right)^3 \frac{\mu}{\langle \kappa \rangle} \, r^2 dr \tag{11.1.21}$$

or

$$m(r) - m(r_0) \sim \frac{\pi^2}{12} \left(\frac{GM_{\mathrm{H}}M_{\mathrm{shell}}}{k}\right)^4 \frac{ca}{L} \frac{\mu^4}{\langle \kappa \rangle} \int_{r_0}^{r} \frac{dr}{r}. \tag{11.1.22}$$

Doing the integration and evaluating the numerical coefficient, one has finally that

$$\log_e(r/r_0) \sim 0.0187 \frac{\langle \kappa \rangle}{\mu^4} \frac{L_{\mathrm{shell}}}{M_{\mathrm{shell}}^3} \frac{m(r) - m(r_0)}{M_{\mathrm{shell}}}, \tag{11.1.23}$$

where luminosity and mass are in solar units. Thus, the analytical approximation predicts that the radius in the transition zone increases exponentially with mass.

Table 11.1.3 Gravothermal energy-generation rates in the core of a model at the base of and in the core of a model near the tip of the red giant branch

$M_{core}/M_\odot$	$L_s/L_\odot$	$L_{grav}^{core}/L_\odot$	L_{grav}^{core}/L_s	$\bar\epsilon_{comp}^{core}$	$\dot{E}_{comp}^{core}/L_\odot$	$\dot{E}_{comp}^{core}/L_s$	$L_{grav}/L_\odot$
0.1384	2.44	0.0121	0.014	0.36	0.024	0.0010	−0.0058
0.4376	1712	7.59	0.016	90	20.3	0.0012	18.09

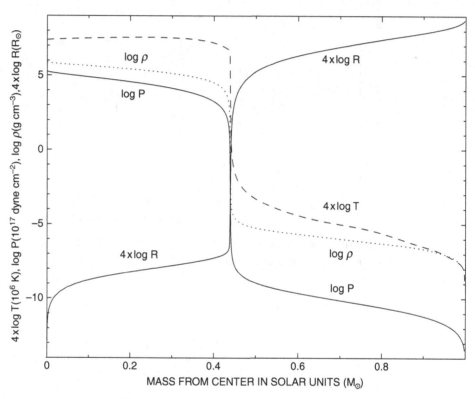

Fig. 11.1.45 Structure variables in a 1$M_\odot$ model ($Z = 0.015$, $Y = 0.275$) near the tip of the red giant branch ($t_g = 12.06176138$)

For the most luminous model in the set of red giant branch models selected for detailed scrutiny, eq. (11.1.23) becomes

$$\log_e(r/r_0) \sim 0.0187\,\langle\kappa\rangle(1.596)^4\,\frac{1712}{(0.438\,03)^3}\,\frac{0.0032}{0.438\,03} = 18.07\,\langle\kappa\rangle. \qquad (11.1.24)$$

As recorded in Table 11.1.3, at $M_{90\%} = 0.438\,03\ M_\odot$, $\kappa = 0.195$ cm^2 g^{-1} and if opacity were constant all the way to the base of the convective shell, $\log_e r/r_0 = 3.53$ or $r/r_0 \sim 34$. Actually, at the base of the convective envelope, $\kappa \sim 0.65$ cm^2 g^{-1}, and one might adopt $\langle\kappa\rangle \sim 0.42$, leading to the estimate $\log_e r/r_0 = 7.65$ or $r/r_0 \sim 2100$, far larger than

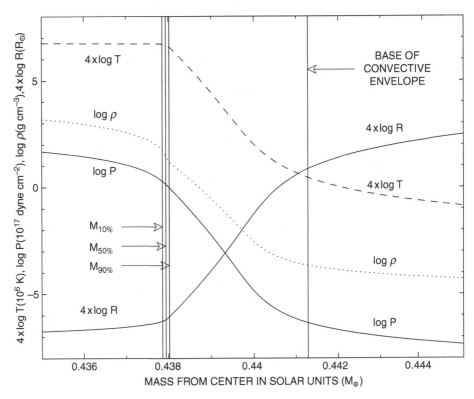

Fig. 11.1.46 Structure variables between the H-burning shell and the CEB in a 1$M_\odot$ model near the red giant branch tip ($Z = 0.015$, $Y = 0.275$, $t_g = 12.06176138$)

the ratio of 52 recorded in Table 11.1.2. To achieve the ratio found in the model requires that $\langle \kappa \rangle \sim 0.219$ cm^2 g^{-1}.

Given the fact that the pressure drops by a factor of over a million in the transition zone, it is not unreasonable that an average of κ over pressure is heavily weighted toward the high pressures just outside the hydrogen-burning shell. Nevertheless, the weighting by a factor of forty required to achieve agreement of the analytic estimate with the one obtained by straightforward model calculations suggests that attempts to construct simple analytic approximations are not always rewarding. At some point, it is necessary to concede that solutions to some problems are basically numerical. One of the flaws in the mathematical description presented here is that, although relationships such as eq. (11.1.18) may be useful over some parts of the transition zone, they fail badly over other parts.

Figure 11.1.47 is distinguished from its counterparts, Figs. 11.1.44 and 11.1.41, by showing abundances of the beta-stable CNO isotopes as well as the gravothermal energy-generation rate and the creation–destruction component of this rate. It is evident that, over most of the hydrogen-burning shell, all of the alpha-particle CNO isotopes are in local equilibrium and that the full CNO-bicyle is operating. It is also evident that, insofar as they contribute to the luminosity and therefore to the gradients in structure variables in and

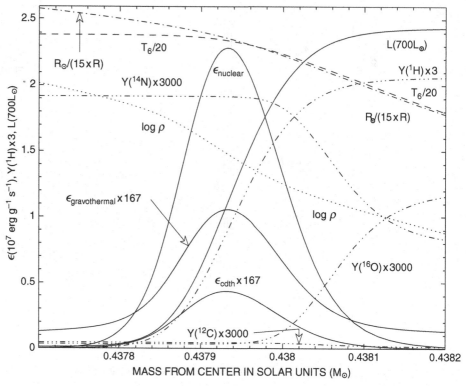

Fig. 11.1.47 Energy-generation, abundance, and structural characteristics of the H-burning shell in a $1M_\odot$ model near the red giant tip ($Z = 0.015$, $Y = 0.275$, $t_g = 12.06176138$)

above the hydrogen-burning shell, gravothermal energy-generation rates are of negligible importance relative to the nuclear energy-generation rate.

However, in Fig. 11.1.48, where the nuclear energy-generation rate is presented from the perspective of the gravothermal rates, gravothermal energy-generation characteristics take center stage. In this second perspective, it is dramatically clear that the profiles of the gravothermal energy-generation rates are serenely independent of the details of the nuclear energy-generation rate. Despite the overwhelming quantitative dominance of the nuclear energy-generation rate relative to the gravothermal rates, changes in local structural characteristics (temperature, density, and radial position) are basically a consequence of gravothermal changes related to the rate of transformation of particle numbers in the main peak of the energy-generation profile rather than to the rate at which energy is released in conjunction with these transformations.

The profile of the nuclear energy-generation rate has three peaks, only one of which is discernable in Fig. 11.1.47. The main peak is due to equilibrium CNO-bicycle burning. As shown in Fig. 11.1.48, the energy-generation rates in the other two peaks, although quite small relative to that in the main peak, are comparable with the gravothermal energy generation rate $\epsilon_{\text{gravothermal}}$ over a region which extends significantly beyond what has heretofore been identified as the hydrogen-burning shell. The first subsidiary peak, at

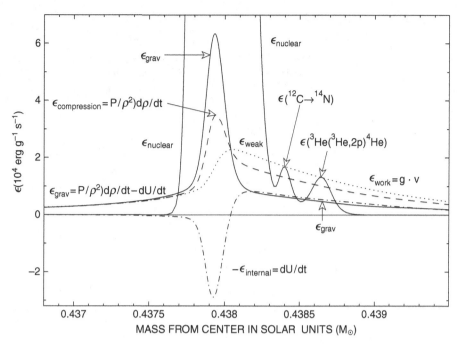

Fig. 11.1.48 Gravothermal and nuclear energy-generation rates in and near the H-burning shell of a 1$M_\odot$ model near the red giant tip ($Z = 0.015$, $Y = 0.275$, $t_g = 12.06176138$)

$M \sim 0.4384$ $M_\odot$, is due to the reactions involved in the conversion of ^{12}C into ^{14}N and the second subsidiary peak at ~ 0.43863 $M_\odot$ is due to the reaction ^{3}He(^{3}He, $2p$)^{4}He.

From the perspective of Fig. 11.1.47, the nuclear burning shell appears to be confined to the mass range $0.4378 \rightarrow 0.4381$ $M_\odot$. From the perspective of the gravothermal rates in Fig. 11.1.48, the main peak in the nuclear burning shell extends over the mass range $0.4377 \rightarrow 0.4383$ $M_\odot$, twice the size of the first estimate. Further, energy-generation rates responsible for the two subsidiary peaks are larger than $\epsilon_{\text{gravothermal}}$ out to $M \sim 0.4387$ $M_\odot$. Thus, depending upon the perspective, the width of the hydrogen-burning shell varies from $\Delta M_{\text{shell}} = 0.0003$ $M_\odot$ to $\Delta M_{\text{shell}} = 0.0010$ $M_\odot$, a difference of over a factor of 3.

Conditions in the hydrogen-exhausted core of the red giant model of penultimate luminosity are shown in Fig. 11.1.49. Comparisons with the conditions in the subgiant model at the base of the red giant branch shown in Fig. 11.1.26 are instructive. The first difference to note is that, in the core of the red giant model, the rate at which gravity does work, as reflected in $\epsilon_{\text{compression}}$, is only slightly larger than the rate at which the kinetic energy of degenerate electrons is increased, as reflected in $-\epsilon_{\text{internal}}$, so that only a small fraction of the energy released by compression is converted into the rate $\epsilon_{\text{gravothermal}}$ which contributes to the outward flow of energy. In the core of the subgiant model, on average almost half of the rate of compressional energy release appears in $\epsilon_{\text{gravothermal}}$ and therefore contributes substantially to the outward flow of energy.

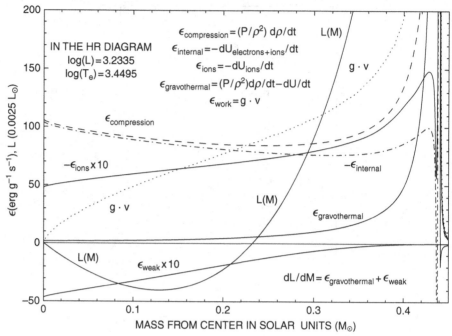

Fig. 11.1.49 Gravothermal and neutrino energy-generation rates in the core of a 1$M_\odot$ model approaching the red giant tip ($Z = 0.015$, $Y = 0.275$, $t_g = 12.06176138$)

In both figures, the quantity $-\epsilon_{ions} = dU_{ions}/dt$ is the rate at which the thermal energy of heavy ions increases and the quantity $-\epsilon_{weak}$ is the rate at which energy is lost due to the plasma neutrino process. In both cases, at the center of the model, $\epsilon_{weak} \sim \epsilon_{ions}$. The discussion of eqs. (11.1.2)–(11.1.7) shows that, in the electron-degenerate regions of the helium core, the rate of heating electrons near the top of the Fermi sea is related to the rate of heating of ions by

$$\left(\frac{dU_e}{dt}\right)_\rho \sim \frac{2\pi^2}{3}\frac{kT}{\epsilon_{F0}}\frac{dU_{ions}}{dt}. \tag{11.1.25}$$

At the center of the subgiant model, $\epsilon_{Fermi}/kT = 6.25$, so $\left(dU_e/dt\right)_\rho \sim dU_{ions}/dt$, and therefore heating is about twice as effective as cooling. In the red giant model, $\epsilon_{Fermi}/kT = 19.975$, so $\left(dU_e/dt\right)_\rho \sim (1/3) dU_{ions}/dt$, and heating is only about 30% more efficient than cooling. In both cases, at positions progressively further from the center, heating gains in importance relative to cooling by neutrino losses, so temperatures increase more rapidly at points off center than at the center. However, the contrast between the rate of heating at the center and the rate of heating where $\epsilon_{weak} \ll \epsilon_{ions}$ is greater in the red giant model than in the subgiant, and this accounts for the fact that the maximum temperature in the core of the red giant model is not at the center.

The direction of energy flow in the core is determined by the sum of the gravothermal energy-generation rate and the neutrino-loss rate. That is,

$$\frac{dL}{dM} = \epsilon_{\text{gravothermal}} + \epsilon_{\text{weak}}. \tag{11.1.26}$$

In the core of the subgiant model near the base of the red giant branch, $\epsilon_{\text{gravothermal}}$ is large compared with ϵ_{weak}, so the luminosity increases positively outward everywhere in the core. In the core of the highly evolved red giant model, where $\epsilon_{\text{compression}} \sim -\epsilon_{\text{internal}}$, and over the inner part of the core where ϵ_{weak} is large, $\epsilon_{\text{gravothermal}}$ is small compared with $-\epsilon_{\text{weak}}$. The net result is that, in the luminous red giant model, energy flows inward from the point where $\epsilon_{\text{gravothermal}} = -\epsilon_{\text{weak}}$.

In all models on the red giant branch, from the base of the branch to its tip, the gravothermal response of the hydrogen-exhausted core to the rate at which the hydrogen-burning shell processes matter is very stable. This is demonstrated in Table 11.1.3, where $L_{\text{grav}}^{\text{core}}$ is the rate at which the gravothermal source delivers energy at the base of the hydrogen-burning shell, L_{s} is the rate at which the hydrogen-burning shell produces energy, $\bar{\epsilon}_{\text{comp}}^{\text{core}}$ (in units of erg g^{-1} s^{-1}) is the average local rate at which compressional energy is being released in the core, and $\dot{E}_{\text{comp}}^{\text{core}}$ is the total rate of compressional energy generation in the core. It is evident that, to a good approximation, both $L_{\text{grav}}^{\text{core}}$ and $\dot{E}_{\text{comp}}^{\text{core}}$ are proportional to L_{s} and therefore proportional to the rate at which mass is being added to the core.

There is a basic difference between the subgiant model and the red giant model with regard to the gravothermal energetics outside of the core. In the subgiant model, there is a large region between the hydrogen-burning shell and the base of the convective envelope in which energy is absorbed from the ambient flow and the total rate of absorption L_{grav}^{-} is actually larger than the rate of production of gravothermal energy by the core and by the hydrogen-burning shell, leading to the negative value for L_{grav} recorded in the last column of Table 11.1.3. In the red giant, the base of the convective envelope lies just above the hydrogen-burning shell and, in the region between the shell and the base of the convective envelope, matter flows inward and gravity does work. The total rate at which gravothermal energy is produced from the base of the hydrogen-burning shell to the base of the convective envelope is $L_{\text{grav}}^{\text{shell}} \sim 10.5\, L_\odot$. There is essentially no absorption of energy from the ambient flow in the convective envelope. The total rate of gravothermal energy production is therefore $L_{\text{grav}} = L_{\text{grav}}^{\text{core}} + L_{\text{grav}}^{\text{shell}} \sim 18.09\, L_\odot$.

Further insight into the gravothermal behavior of 1 $M_\odot$ models is given by the curve of binding energy versus time in Fig. 11.1.50 and by the slope of this curve, dE_{bind}/dt, which is of course identical with L_{grav}. During main sequence evolution, up to $t_9 \sim 8$, the binding energy increases with time as hydrogen is converted into helium in central regions and the model becomes centrally more compact in response to a decrease in the number of particles. Over the next 2 billion years, centered at $t_9 \sim 9$ when the abundance of hydrogen at the center vanishes, E_{bind} remains nearly constant, with the gravothermal energy required to expand envelope regions being balanced by gravothermal emission from the hydrogen-depleted core and the hydrogen-burning shell. The binding energy then decreases at an accelerating rate as the model evolves onto and along the subgiant branch. Along the

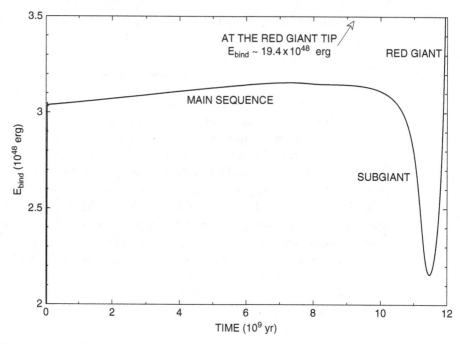

Fig. 11.1.50 Binding energy in a 1$M_\odot$ model (Z = 0.015, Y = 0.275) during the main sequence, subgiant, and early red giant branch phases

subgiant branch, more gravothermal energy is needed to expand envelope material than is supplied by the contracting core. At $t_9 \sim 11.2$, as the magic ratio $\rho_c/\rho_{shell} \sim 32.2$ is traversed, an increase in the core contraction rate moderates the rate of decline in binding energy and, at $t_9 \sim 11.4$, the rate of gravothermal energy release from the core exceeds the rate at which gravothermal energy is needed to expand envelope material. From then on, the binding energy increases with time.

Along the red giant branch proper, almost all of the envelope matter above the hydrogen-burning shell is in convective equilibrium and the energy required to expand envelope material is supplied locally by cooling. Thus the change in binding energy becomes essentially the change in the binding energy of the hydrogen-exhausted core. The relationship between the total binding energy and the binding energy of the core is worth exploring. In the last model, the total binding energy is $E_{bind} \sim 19.4 \times 10^{48}$ erg. From eq. (5.2.31) in Section 5.2, the binding energy of a polytrope of index N in which radiation pressure is negligible is

$$E_{bind}(N) = \frac{1}{2} \frac{3}{5-N} G \frac{M^2}{R} = 1.90 \times 10^{48} \text{ erg} \frac{3}{5-N} \frac{\bar{M}^2}{\bar{R}}, \quad (11.1.27)$$

where $\bar{M}$ and $\bar{R}$ are, respectively, mass and radius in solar units. At the center of the last model, the average energy of an electron is approximately $\sim$15% larger than the rest mass of an electron, which means that the degenerate electrons in the core are only modestly

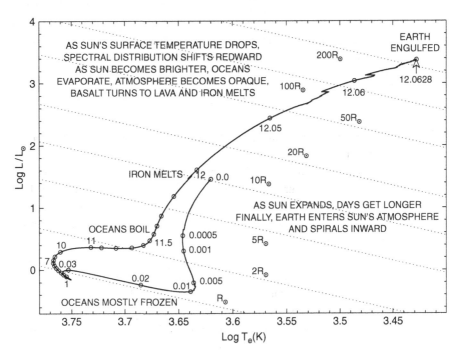

Impact on Earth of the evolution of the Sun

relativistic and that the core can be approximated by a polytrope of index only modestly larger than $N = 1.5$. Thus, a lower limit on the binding energy of the core is

$$E_{\rm bind}^{\rm core} \gtrsim 1.63 \times 10^{48} \ {\rm erg} \ \frac{\bar{M}_{\rm core}^2}{\bar{R}_{\rm core}}. \tag{11.1.28}$$

In the last model, $M_{\rm core} \sim M_{\rm shell} = 0.468\,77 \ M_\odot$, and, from an examination of the mass distribution for the last model in Fig. 11.1.37, $r_{\rm core} \sim 0.02 \ R_\odot$, compared with the radius of the shell, $r_{\rm shell} = 0.028\,106 \ R_\odot$. Inserting these values for $M_{\rm core}$ and $r_{\rm core}$ in eq. (11.1.28) gives

$$E_{\rm bind}^{\rm core} \gtrsim 1.79 \times 10^{49} \ {\rm erg}, \tag{11.1.29}$$

compared with the total binding energy of

$$E_{\rm bind} \sim 1.94 \times 10^{49} \ {\rm erg}. \tag{11.1.30}$$

A modest increase in the effective polytropic index for the core from $N = 1.5$ to $N \sim 1.77$ leads to $E_{\rm bind}^{\rm core} = E_{\rm bind}$. Thus, the binding energy of a highly evolved red giant of low mass is essentially the binding energy of its white-dwarf-like core.

Given the fact that, after $\sim 4.6 \times 10^9$ years of evolution, the global characteristics of the $1 \ M_\odot$ model chosen for study in this section approximate our Sun's current global properties, it is pertinent to examine some of the possible effects on conditions at the Earth's surface of changes in the global characteristics of the model over the entire span of evolution examined. The complete track of the model in the HR diagram is shown in Fig. 11.1.51, along with lines of constant radius and comments about changing conditions at the Earth's surface. Temperatures at the Earth's surface are approximated by assuming that the Earth absorbs and radiates as a black body, so that

$$4\pi R_{\text{earth}}^2 \, \sigma T_{\text{earth}}^4 = \pi R_{\text{earth}}^2 \, \frac{L_{\text{sun}}}{4\pi (\text{AU})^2}, \tag{11.1.31}$$

where 1 AU is the distance between the Earth and the Sun. In this approximation, all complications associated with the fact that an atmosphere can modify the simple relationship both by reflecting some of the energy impinging from the Sun and by blocking loss of energy from the surface are ignored. Also ignored is the fact that the Earth retains heat from its formative years, that this heat flows outward continuously toward the Earth's surface, and that volcanic eruptions sporadically modify the Earth's atmosphere and the heat balance at the Earth's surface.

Despite these caveats, the situation depicted in Fig. 11.1.51 is not excluded. For a billion or so years after the Sun's gravitational contraction phase had been completed, average temperatures at the Earth's surface hovered near the freezing point of water. The ramifications of this fact for life on Earth depended, of course, on when water became a prominent feature at the Earth's surface and on the extent to which emerging life forms could exist below the Earth's surface, utilizing heat flowing out from the Earth's interior.

As the Sun becomes brighter, the situation becomes less tentative. There is no question but that, as the model evolves along the subgiant branch, the flux of energy reaching the Earth from the Sun will shift significantly to the red, affecting the visual acuity of all life forms, and water will be near the boiling point. Oceans will eventually evaporate, significantly affecting the properties of the Earth's atmosphere and certainly making the existence of life extremely problematical. As the Sun climbs upward along the red giant branch, the basalt in ocean basins will melt, and then solids such as iron will melt. As the Sun's radius increases, daylight hours on the Earth will increase, and, most important of all, the Earth may find itself entering the Sun's atmosphere. There, under the influence of drag forces, it will spiral inward, all the time evaporating material that becomes a part of the Sun's envelope.

The prediction that the Earth may be engulfed by the Sun prior to the ignition of helium at the Sun's center is not a rock solid prediction. If the metallicity of the Sun is significantly smaller than that given by the choice $Z = 0.015$, the radius at the red giant tip of a 1 $M_\odot$ model could be smaller than 1 AU, so that the incorporation of the Earth into the Sun could be delayed until a later stage when the Sun has evolved into an asymptotic giant branch star with an electron-degenerate carbon–oxygen core. On the other hand, real low mass red giants are known to lose mass at rather modest rates via an evaporative wind characterized by speeds of the order of 10–20 km s^{-1}. An analysis of observational evidence by D. Reimers (1975) suggests

$$\dot{M}_{\text{wind}} \left(M_\odot \, \text{yr}^{-1} \right) \sim 4 \times 10^{-13} \, \eta \, \frac{L}{L_\odot} \, \frac{(g \, R)_\odot}{g \, R} = 4 \times 10^{-13} \, \eta \, \frac{L \, R}{M} \left(\frac{M}{L \, R} \right)_\odot, \tag{11.1.32}$$

Table 11.1.4 Abundances at the surface of the model at the red giant tip relative to those in the starting model

Isotope	Y_{end}	Y_{start}	Ratio
^{1}H	0.6877	0.7105	0.968
^{2}H	1.380/−17	1.410/−5	1/−12
^{3}He	4.679/−4	1.219/−5	38.4
^{4}He	0.07414	0.06880	1.078
^{6}Li	1.267/−17	1.167/−10	1/−7
^{7}Li	1.351/−11	1.439/−9	0.00939
^{9}Be	9.740/−13	1.380/−11	0.0703
^{10}B	1.207/−11	8.739/−11	0.138
^{11}B	1.730/−10	3.517/−10	0.492
^{12}C	1.838/−4	2.132/−4	0.862
^{13}C	6.175/−6	2.372/−6	2.603
^{14}N	7.840/−5	5.275/−5	1.486
^{15}N	1.419/−7	1.940/−7	0.731

where $1/3 < \eta < 3$, g is the surface gravity, L is the surface luminosity, and R is the radius. This expression predicts a mass loss along the giant branch of the order of 0.1–0.2 $M_\odot$. Model calculations show that, at a given luminosity along the giant branch, a smaller mass leads to a larger radius. For example, Allen V. Sweigart and Peter G. Gross (1978) find that, for $Z = 0.01$ models at the red giant tip, a model of mass 0.9 $M_\odot$ is approximately 5.4% larger than a model of mass 1.1 $M_\odot$. So, even with a smaller metallicity, because of mass loss, the Sun's radius might exceed 1 AU before the Sun reaches the tip of the red giant branch.

Yet another caveat can be identified. In response to a decreasing gravitational force from a mass-losing Sun, the Earth's orbit must expand. In first approximation, the product of the semimajor axis and the mass of the Sun remains constant. Thus a 20% decrease in the mass of the Sun means a 20% increase in the diameter of the Earth's orbit about the Sun, and this could be the factor which delays the possible engulfment of the Earth to the time when the Sun has evolved into a TPAGB star on its way to ejecting most of its remaining hydrogen-rich surface layers and passing through the planetary nebula phase.

Whatever fate the Sun has in store for the Earth, the changes in abundances that have occurred in the interior and contributed to the abundances in the convective envelope of the model at the red giant tip will make their way into the interstellar medium, initially in consequence of mass loss during the first ascent of the red giant branch and finally in consequence of the ejection of most of the hydrogen-rich envelope during the TPAGB phase. Surface abundances of a selection of isotopes in the final red giant branch model and in the initial luminous pre main sequence model are shown in the second and third columns of Table 11.1.4, respectively. The ratios $Y_{\text{end}}/Y_{\text{start}}$ are shown in the fourth column.

By far the most significantly enhanced abundance is that of ^{3}He. Deuterium is converted completely into ^{3}He over the entire model during the pre-main sequence phase so that the number abundance of ^{3}He in the initial ZAMS model is $Y(^3\text{He}) = 2.598 \times 10^{-5}$, slightly more than twice the number abundance in the initial model. Processing in the model interior leads to a final number abundance of ^{3}He in the convective envelope which is 17 times larger than the sum of the number abundances of ^{3}He and deuterium in the matter out of which the starting model was constructed. One infers that 1 $M_{\odot}$ and other low mass stars are almost certainly the major contributors of ^{3}He to the interstellar medium. Also enhanced are ^{13}C and ^{14}N, both at the expense of ^{12}C. Although the enhancements are modest, they are expected to occur as a consequence of main sequence burning in most stars and it is probable that main sequence transformations of ^{12}C into ^{13}C and ^{14}N are responsible for most of the ^{13}C and ^{14}N in the Universe.

Both deuterium and ^{6}Li are essentially totally absent in the final red giant. Deuterium destruction occurs early on at high luminosity and low surface temperature during the pre-main sequence gravitationally contracting phase, while ^{6}Li destruction occurs at the base of the shrinking convective envelope as luminosity and surface temperature increase during the final approach to the main sequence. The ^{7}Li number abundance near the surface is reduced by a factor of 1.8 during the final approach to the main sequence, remains constant during the main sequence phase, and begins to decrease as the mass of the convective envelope increases during the subgiant and red giant branch phases, with the result that, when the model reaches the tip of the red giant branch, the number abundance of ^{7}Li at the surface has been reduced by a factor of about 60 relative to the surface abundance prevailing during the main sequence phase. The isotopes ^{9}Be and ^{10}B are reduced by factors of order 10, and the abundance of ^{11}B is halved. Similar reductions are expected in other low mass main sequence stars and this supports the view that, in interstellar space, isotopes such as these are primarily products of spallation reactions occurring in the interstellar medium between cosmic rays and the CNO isotopes composed of alpha particles.

Changes in the abundances of ^{1}H and ^{4}He, though seemingly small, are sufficiently large to allow one to infer that ejection of matter from stars which have experienced hydrogen burning is responsible for the increase in helium in the Universe that has occurred since the Big Bang.

11.2   Evolution of a 5 $M_{\odot}$ model during core hydrogen burning, development of a thick hydrogen-burning shell, and shell hydrogen-burning evolution up to the onset of core helium burning as a red giant

The ZAMS model for the 5 $M_{\odot}$ study is available from the work described in Section 9.4. Subsequent evolution of the model in the HR diagram during the core hydrogen-burning phase and during the first portion of the shell hydrogen-burning phase is shown by the heavy solid curve in Fig. 11.2.1. The times to reach various points along the track are

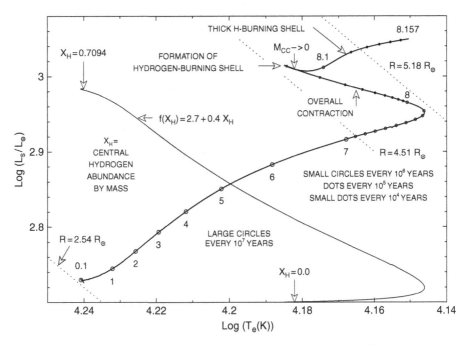

Fig. 11.2.1 Evolution in the HR-diagram of a 5M$_\odot$ model (Z = 0.015, Y = 0.275) during core hydrogen burning, overall contraction, and thick shell hydrogen burning

indicated by circles and dots, larger circles being separated by 10^7 yr, smaller circles by 10^6 yr, and dots by 10^5 yr. Numbers beside points give the total elapsed time in units of 10^7 yr. For further orientation, three lines of constant radius are shown in the figure as dotted lines; the line characterized by $R = 2.54 \ R_\odot$ passes near the location of the starting ZAMS model, while the lines for $R = 4.51 \ R_\odot$ and $R = 5.18 \ R_\odot$ bracket a region in the HR diagram traversed by the model near the end of the main sequence phase and during the first part of the thick shell hydrogen-burning phase.

Along that portion of the track which extends from the point of lowest surface temperature at $t \sim 7.9 \times 10^7$ yr to the following relative maximum surface temperature point near $t \sim 8.09 \times 10^7$ yr, the slope of the track, $d \log(L)/d \log(T_e)$, is less steep than the slope of a line of constant radius, demonstrating that model radius is decreasing. Along this portion of the track, matter everywhere in the interior of the model is contracting, and the model is said to be in the overall contraction phase.

The abundance by mass of hydrogen at the model center, denoted by X_H, is described in the figure by the function $f(X_H) = 2.7 + 0.4 X_H$ and varies from an initial value of $X_H = 0.71$ on the ZAMS to $X_H = 0$ just after the termination of the overall contraction phase. Whether the term main sequence is used to describe the phase of evolution between the ZAMS and the start of the overall contraction phase, between the ZAMS and the end of the overall contraction phase, or between the ZAMS and the time at which most of the nuclear energy generated by the model switches from a convective core to an off-center shell is a matter of taste.

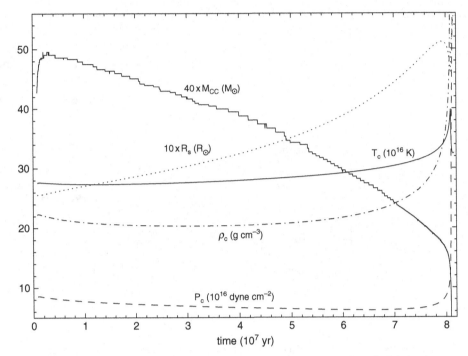

Fig. 11.2.2 Radius, convective core mass, and central characteristics of a 5$M_\odot$ model ($Z = 0.015$, $Y = 0.275$) during the core hydrogen-burning phase

The evolution with time of the radius, mass of the convective core, and state variables at the center of the model are shown in Fig. 11.2.2. The evolution with time of nuclear luminosites and of number abundances at the model center of hydrogen and of the primary isotopes of oxygen and nitrogen are shown in Fig. 11.2.3. The reciprocal of the molecular weight at the center in Fig. 11.2.3 is obtained by using the approximation

$$\left(\frac{1}{\mu}\right)_c = 2Y(^1\mathrm{H}) + 3Y(^4\mathrm{He}) + \frac{Z}{2}, \tag{11.2.1}$$

where $Y(^1\mathrm{H})$ and $Y(^4\mathrm{He})$ are the abundances by number at the center of ^{1}H and ^{4}He, respectively. During the evolution shown, the equation of state is essentially that of a perfect gas, so the central pressure P_c is proportional to the product $\rho_c T_c / \mu_c$.

With the exception of the convective core mass M_{CC}, all characteristics described in Figs. 11.2.2 and 11.2.3 vary smoothly with time. The stairstep variation with time of M_{CC} is a consequence of the fact that the position where the radiative and adiabatic gradients are the same does not coincide with the boundary of any given zone. In every mass zone, both the radiative gradient and the adiabatic gradient are defined at the center of the zone and M_{CC} is defined as the mass at the outer edge of the outermost mass zone in which the radiative gradient is larger than the adiabatic gradient. The rezoning algorithm, which is activated after every third time step, ensures only that the number of mass zones in the neighborhood of the outer boundary satisfies a predetermined criterion; it does not attempt

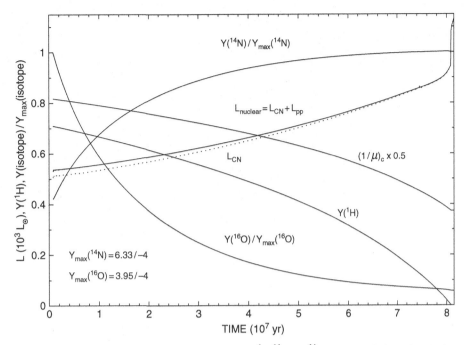

Fig. 11.2.3 Nuclear burning luminosities and central number abundances of ^{1}H, ^{16}O, and ^{14}N during core hydrogen burning in a 5$M_\odot$ model ($Z = 0.015$, $Y = 0.275$)

to identify the precise position where $V_{ad} = V_{rad}$ and insert a zone boundary there. Thus, even though the radiative and adiabatic gradients in any mass shell vary smoothly with time, the outer boundary of the convective zone must change discontinuously. No attempt has been made to fine tune the zoning in either space or time to produce a variation in M_{CC} smoother than the one shown in Fig. 11.2.2. Over the course of the evolution shown, between 500 and 750 mass zones have been employed and over 3500 time steps have been taken.

During the first part of main sequence evolution, the mass of the convective core increases from $M_{CC} \sim 1.07\,M_\odot$ in the ZAMS model to a maximum of $M_{CC} \sim 1.24\,M_\odot$, which is identical with the maximum value achieved during the C$\rightarrow$N phase of pre main sequence evolution. Remaining near its maximum for approximately 10^6 yr until $t \sim 3 \times 10^6$ yr, M_{CC} declines essentially monotonically for the remainder of the core hydrogen-burning phase.

During the bulk of the main sequence phase, the central temperature and density change only slightly. Over the first seven-eighths of the phase, the increase in central temperature is less than 10% and the variation in the central density about the mean is of the order of 10%. Over the same time period, the central pressure decreases by about 30%, with most of the decrease being a consequence of an approximately 40% decrease in the number of free particles per unit mass, as measured by the reciprocal of the molecular weight. Global features experience larger changes, with luminosity increasing by $\sim$50% and radius increasing by over 70% during the first seven-eighths of the main sequence phase. The largest changes of all are in the relative number abundances of hydrogen, helium, oxygen, and nitrogen in

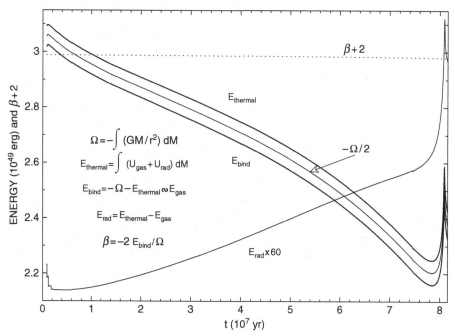

Fig. 11.2.4 Global gravothermal energies of a 5$M_\odot$ model ($Z = 0.015$, $Y = 0.275$) during the main sequence phase and the start of shell hydrogen burning

central regions. At the center, the number abundance of ^{1}H goes from 0.71 to nearly 0.0, the number abundance of ^{16}O decreases by a factor of $\sim$17, and the number abundances of ^{4}He and ^{14}N increase by factors of $\sim$3.6 and $\sim$2.4, respectively. The curves for L_{nuclear} and L_{CN} in Fig. 11.2.3 show that, as evolution progresses, the contribution to the nuclear burning luminosity of pp-chain reactions relative to the contribution of CN-cycle reactions becomes progressively smaller, consistent with the increase in central temperature shown in Fig. 11.2.2 and the fact that CN-cycle reactions are much more sensitive to temperature changes than are pp-chain reactions.

The variations with time of three main gravothermal characteristics of the model – the gravitational binding energy $|\Omega|$, the total thermal energy content $E_{\text{thermal}} = E_{\text{gas}} + E_{\text{rad}}$, and the net binding energy E_{bind} – are shown in Fig. 11.2.4. The definition of these characteristics and several relationships between them, as developed in Section 3.3, are noted in the left hand part of the figure. The fact that $E_{\text{bind}} \stackrel{\sim}{=} E_{\text{gas}}$, where E_{gas} is the global kinetic energy of the material particles in the model, is a consequence of the fact that the mean velocities of the material particles are essentially non-relativistic. Also shown explicitly in Fig. 11.2.4 is the total energy E_{rad} of photons in the model. From eqs. (3.3.16) and (3.3.17), one has that

$$E_{\text{rad}} = 2 \left(\frac{1}{2} |\Omega| - E_{\text{bind}} \right) = 2 \left(E_{\text{thermal}} - \frac{1}{2} |\Omega| \right), \qquad (11.2.2)$$

corroborating the visual impression from Fig. 11.2.4 that the curve for $|\Omega|/2$ is exactly midway between the curve for E_{thermal} and the curve for E_{bind}.

During most of the core hydrogen-burning phase, until central hydrogen decreases to $Y(^1H) \sim 0.05$ at $t \sim 7.9 \times 10^7$ yr, all of the global energy-measuring characteristics decrease. The decrease in gravitational binding energy can be understood as a consequence of an expansion outward of much of the matter outside the nuclear burning region, an expansion which is indicated by the steady increase in model radius. This expansion requires an expenditure of energy that translates into a decrease in gravitational binding energy. The total thermal energy content of the model decreases because (1) in regions outside the nuclear burning region, the average temperature decreases with time, and (2) in the region of active nuclear burning where temperatures are increasing, the decrease in the number of material particles per gram more than offsets the increase in temperatures. The binding energy decreases for the same reason that the thermal energy decreases.

In the current context, the changes in gravothermal energies are much smaller than energies released in nuclear reactions. It is a marvel that, in spite of this disparity, the relationships between gravothermal quantities maintain an integrity which is independent of the nature and strength of the nuclear energy sources. That is, independent of the magnitude of the contribution to the ambient flow of energy due to nuclear reactions in the deep interior, the rate at which kinetic energy is released in consequence of the destruction of material particles in the convective core and in consequence of the decrease in temperatures outside of the core exactly matches the rate of decrease in the net binding energy and, to the extent that radiation pressure is small compared with the gas pressure, the decrease in the kinetic energy of material particles contributes precisely half of the energy required to push most of the matter in the model outward against gravity.

Equation (3.3.14) defines a quantity $\tilde{\beta} = 2E_{gas}/|\Omega|$, the value of which can in principle be inferred from the dotted curve in Fig. 11.2.4. It is evident that $\tilde{\beta}$ remains close to unity and, at first glance, appears to be nearly constant. However, the ratio of global photon energy to global particle kinetic energy is given by (see eq. (3.3.19))

$$\frac{E_{rad}}{E_{gas}} = 2 \frac{1 - \tilde{\beta}}{\tilde{\beta}}. \tag{11.2.3}$$

Noting from Fig. 11.2.4 that the fractional increase in radiation energy is comparable with the fractional decrease in the binding energy, it is evident that, during the time period shown, $1 - \tilde{\beta}$ varies by approximately 50%, from a minimum of ~ 0.013 through a maximum of ~ 0.021.

Interestingly, and in contrast with the behavior of the 1 $M_\odot$ model, all portions of the 5 $M_\odot$ model expand during the first half of the main sequence phase. Structure variables at a representative point during the overall expansion phase, at $t = 3 \times 10^7$ yr, are shown in Fig. 11.2.5. Two features distinguish the structure variables in the evolved model from the structure variables in the 5 $M_\odot$ ZAMS model described in Fig. 9.4.14. First, in the evolved model, essentially all of the surface luminosity is produced in the convective core, with no significant contribution from outside of the core, as is the case in the ZAMS model. This means that, in the evolved model, the surface luminosity is due almost entirely to equilibrium CN-cycle burning in the core, with an insignificant contribution from C$\rightarrow$N burning outside of the core. The time at which the rate of energy generation by C$\rightarrow$N reactions becomes negligible compared with the rate of energy generation by equilibrium

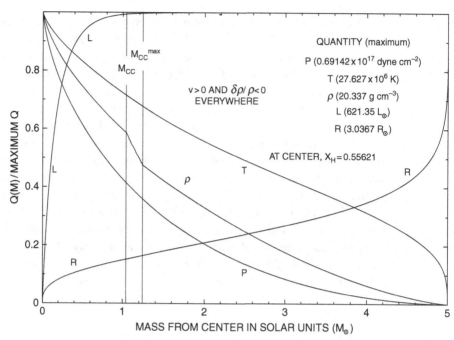

Fig. 11.2.5 Structure of a 5 $M_\odot$ model ($Z = 0.015$, $Y = 0.275$) after completion of three-eighths of the main sequence phase (AGE $= 3.00 \times 10^7$ yr)

CN-cycle reactions can be identified with the time at which M_{CC} during the main sequence phase reaches and thereafter declines from the maximum achieved during the first pre-main sequence C$\rightarrow$ N phase. From Fig. 11.2.2, this occurs at $t \sim 3 \times 10^6$ yr.

Second, in the evolved model, in the region between the location of the edge of the convective core at its maximum mass and the location of the current edge of the convective core, the slope of the density profile is noticeably steeper than on either side of this region. It is precisely in the region of the steeper density profile that the number of free particles per unit mass decreases inward, reflecting the hydrogen/helium ratio left behind as the outer edge of the convective core moves inward in mass. The correspondence between the steeper density profile and the molecular weight profile is made clear in Fig. 11.2.6, where number abundances of the most abundant isotopes in the model are shown. Since the pressure distribution in the region of strongly variable molecular weight is determined by the balance between the gravitational force and the spatial gradient of $\rho \, T/\mu$ and since the profiles of both temperature and pressure exhibit no discontinuous changes in slope at the boundaries of the region, it is evident that the product ρ/μ has no discontinuity in slope at the boundaries of the region and varies through the region in approximately the same way that density would vary in the absence of a molecular weight gradient.

The half maximum in the abundance of ^{12}C in Fig. 11.2.6 occurs at $M \sim 2.5 \, M_\odot$, and this coincides approximately with the location of the peak in the number abundance of ^{13}C (not shown). From Fig. 11.2.5 one sees that the temperature at $M = 2.5 \, M_\odot$ is only about half of the central temperature, and this accounts for the fact that C$\rightarrow$ N reactions make no

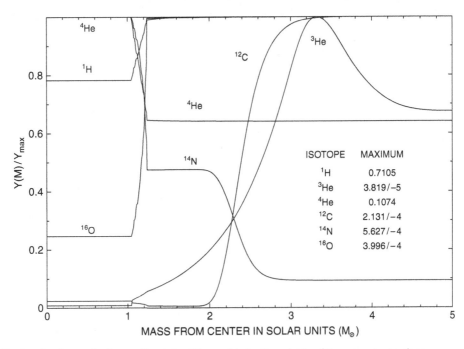

Fig. 11.2.6 Number abundances of major constituents in a 5$M_\odot$ model after three eighths of the main sequence phase ($Z = 0.015$, $Y = 0.275$, AGE $= 3.00 \times 10^7$ yr)

perceptible contribution to the surface luminosity relative to the contribution of CN-cycle reactions in the convective core.

Gravothermal characteristics of the 3×10^7 year old model are shown in Figs. 11.2.7–11.2.10. The explanation already given for the decline in $E_{\text{thermal}} \sim E_{\text{gas}}$ in Fig. 11.2.4 is made explicit in Fig. 11.2.7. In the convective core, the decrease in particle number per unit mass is responsible for a rate of decrease in thermal energy, expressed by $(dU/dt)_{\rho,\,T} = -\epsilon_{\text{cdth}} < 0$, which is far larger than the rate of increase in thermal energy there due to a modest rate of increase in temperatures over the inner two thirds of the core's mass, where $(dU/dt)_Y = -\epsilon_{\text{th}} > 0$. Outside the convective core, except for a very small region just above the outer edge of the core, the local particle number abundance remains constant in time and cooling in the expanding envelope leads to $(dU/dt)_Y = -\epsilon_{\text{th}} < 0$. Integration under the curves in Fig. 11.2.7 gives

$$L_{\text{cdth}} = \int \epsilon_{\text{cdth}} \, dM = 0.384 \, L_\odot \tag{11.2.4}$$

as the global rate at which thermal energy is produced by a change in particle numbers at constant density and temperature and

$$L_{\text{th}} = \int \epsilon_{\text{th}} \, dM = 0.261 \, L_\odot \tag{11.2.5}$$

as the global rate at which thermal energy is changed by a change in temperature when the particle number densities are held constant In order to avoid possible confusion between

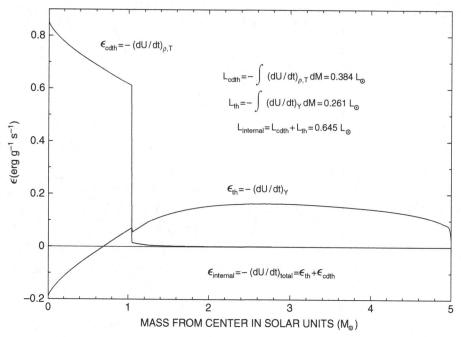

Fig. 11.2.7   Components of dU/dt in a 5M$_\odot$ model after three eighths of the main sequence phase ($Z = 0.015$, $Y = 0.275$, AGE $= 3.00 \times 10^7$ yr)

the meanings of the labels "th" and "thermal," the word internal has been used instead of the word thermal to label the local rate of change of thermal energy as

$$\epsilon_{\text{internal}} = \epsilon_{\text{cdth}} + \epsilon_{\text{th}}. \tag{11.2.6}$$

The total global rate of change of the internal (thermal) energy is given by

$$L_{\text{internal}} = \int \epsilon_{\text{internal}} \, dM = L_{\text{cdth}} + L_{\text{th}}, \tag{11.2.7}$$

which in the current instance is $L_{\text{internal}} = 0.645 \, L_\odot$. The same result may be obtained by evaluating the negative of the slope of the curve of E_{thermal} versus t in Fig. 11.2.4 at $t = 3 \times 10^7$ yr. That is, $L_{\text{internal}} = L_{\text{thermal}} = -dE_{\text{thermal}}/dt \sim 0.65 \, L_\odot$.

The sum of the two contributions to the local rate of change of internal energy is shown in Fig. 11.2.8 by the dashed curve labeled $\epsilon_{\text{internal}}$. Reflections about the horizontal axis of the solid curves in Fig. 11.2.8 labeled $\epsilon_{\text{compression}}$ and ϵ_{work} give, respectively, the rate at which energy is required locally to expand matter and the rate at which work is being done locally against gravity. By inspection, the integrals under the two curves are identical. The same information is given in Fig. 11.2.9 in the radius representation. The total rate at which gravothermal energy is changing locally is given by the dotted curves labeled ϵ_{grav} in Figs. 11.2.8 and 11.2.9. Positive values of ϵ_{grav} and dL_{grav}/dR mean that gravothermal energy is contributing locally to the outward flow of energy (because of a decrease in internal energy and/or because of work being done by compression), negative values mean that energy is being absorbed from the ambient flow (either because of an increase in

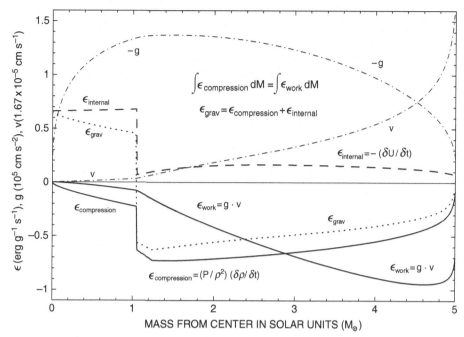

Gravothermal energy-generation rates in a 5$M_\odot$ model after three eighths of the main sequence phase ($Z = 0.015$, $Y = 0.275$, AGE $= 3.00 \times 10^7$ yr)

internal energy and/or because of the work used to expand matter). When it is positive, L_{grav} represents the rate at which a global decrease in gravothermal energy contributes to the surface luminosity. Although gravothermal sources can locally add or subtract from the ambient flow of energy, when L_{grav} is negative, $-L_{grav}$ is the net or global rate at which energy is absorbed from the ambient flow and converted into gravothermal energy, and nuclear energy sources are entirely responsible for the amount of energy absorbed.

The two variables v and g whose product gives the rate at which work is done against gravity are shown as functions of mass in Fig. 11.2.8 and as functions of radius in Fig. 11.2.9. As recorded at the lower right in Fig. 11.2.9, the global rate at which energy due to nuclear sources is being converted into gravothermal energy ($-L_{grav} \sim$ 0.67 $L_\odot$) is slightly larger than the rate at which internal thermal energy is being released ($L_{internal} \sim 0.65\ L_\odot$). The difference is due to the fact that the total energy in photons is increasing, and can be understood by recognising that, to the extent that kinetic energies of material (gas) particles are not relativistic,

$$L_{grav} = \frac{dE_{gas}}{dt} = \frac{d(E_{internal} - E_{rad})}{dt} = \frac{dE_{internal}}{dt} - \frac{dE_{rad}}{dt} = -L_{internal} - \frac{dE_{rad}}{dt}.$$
(11.2.8)

Since, as shown in Fig. 11.2.4, E_{rad} increases with time during the main sequence phase, it is clear that, when $L_{grav} < 0$, it is also true that $-L_{grav} > L_{internal}$.

Differential contributions to the surface luminosity by the two major nuclear energy sources and by the gravothermal source are shown in the radius representation in

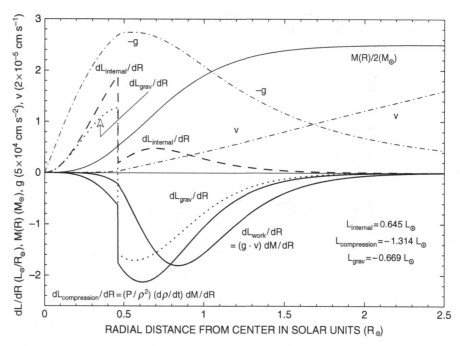

Fig. 11.2.9 Gravothermal energy-generation rates in a 5$M_\odot$ model after three eighths of the main sequence phase ($Z = 0.015$, $Y = 0.275$, AGE $= 3.00 \times 10^7$ yr)

Fig. 11.2.10. Also shown is the differential rate of energy loss by electron neutrinos emitted primarily by ^{13}N and ^{15}O. As sources of surface luminosity, CN-cycle reactions are roughly forty times more important than pp reactions and, despite its intriguing aspects, gravothermal energy production plays an exceedingly minor role in influencing the surface luminosity.

For comparison with the distributions with respect to the mass coordinate that have already been given, the mass coordinate $M(R)$ and its derivative with respect to radius dM/dR are also shown in Fig. 11.2.10. Prior to reaching its maximum at $R \sim 0.65\ R_\odot$, the mass gradient passes through a region in which it decreases sharply. This region, which lies between $R \sim 0.45\ R_\odot$ and $R \sim 0.5\ R_\odot$ coincides with the region bounded by M_{CC} and M_{CC}^{max} in Fig. 11.2.5, which in turn coincides with the region in Fig. 11.2.6 through which hydrogen increases sharply outward. If density were to behave according to $\rho \propto 1/r^2$, the mass gradient would be a constant with respect to r; the fact that the density gradient in the region of strongly variable molecular weight decreases means that density decreases outward through this region more rapidly than the inverse square of r.

The variations with mass of the logarithmic radiative and adiabatic temperature gradients in the 3×10^7 year old model are shown in Fig. 11.2.11, along with the ingredients of the radiative gradient and with the local polytropic index $N \overset{\sim}{=} 1/V - 1$, where $V = \min(V_{ad}, V_{rad})$. Radiation pressure is responsible for the facts that the adiabatic gradient, at $V_{ad} \sim 0.38$, is slightly smaller than 0.4 and that the local polytropic index in the convective core, at $N \sim 1.65$, is slightly larger than would be the case in the absence of

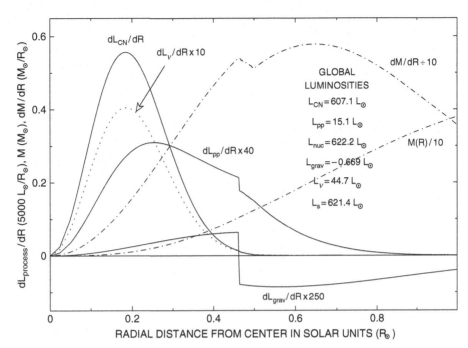

Fig. 11.2.10 Differential contributions to luminosity in a 5$M_\odot$ model after three eighths of the main sequence phase ($Z = 0.015$, $Y = 0.275$, AGE $= 3.00 \times 10^7$ yr)

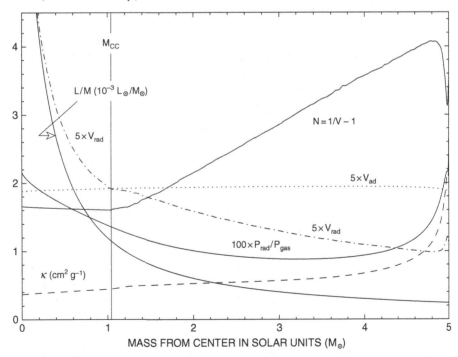

Fig. 11.2.11 Logarithmic gradients in a 5$M_\odot$ model after three eighths of the main sequence phase ($Z = 0.015$, $Y = 0.275$, AGE $= 3.00 \times 10^7$ yr)

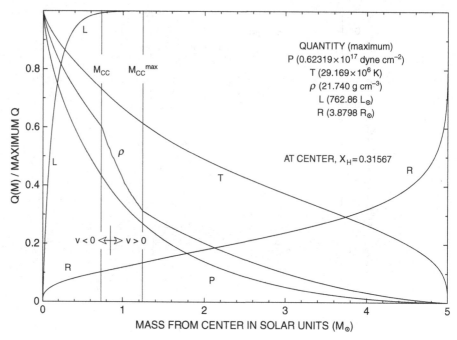

Structure of a 5 $M_\odot$ model after completing three quarters of the main sequence phase ($Z = 0.015$, $Y = 0.275$, AGE $= 6.00 \times 10^7$ yr)

radiation. The large value of L/M is the primary factor in producing $V_{rad} > V_{ad}$ in central regions where κ is essentially constant in space and $P/P_{rad} \sim P_{gas}/P_{rad}$ increases by about 50%. It is instructive that the mass weighted value of the local polytropic index, at $\langle N \rangle \sim 2.5$, is not far from Eddington's favorite value of $N = 3$.

Characteristics of the 5 $M_\odot$ model at a representative point during the phase of core contraction and envelope expansion, at time $t = 6 \times 10^7$ yr, are shown in Figs. 11.2.12–11.2.18. These figures are, respectively, counterparts of Figs. 11.2.5–11.2.11 describing the 5 $M_\odot$ model during the earlier phase of overall expansion at time $t = 3 \times 10^7$ yr. As indicated in Fig. 11.2.12, the dividing line in the more evolved model between where matter moves inward and matter moves outward ($v = 0$) occurs slightly above the convective core at $M \sim 0.851$ $M_\odot$. The division between matter which is contracting and matter which is expanding (not shown explicitly in Fig. 11.2.12) lies just below the outer edge of the convective core (at $M \sim 0.691$ $M_\odot$).

The profiles of structure variables in the 6×10^7 year old model shown in Fig. 11.2.12 are qualitatively quite similar to the profiles in the 3×10^7 model shown in Fig. 11.2.5. The most obvious difference between the two models is due to the fact that the convective core in the older model is distinctly smaller than in the younger model, leading to a larger region $\Delta M = M_{CC} - M_{CC}^{max}$ in the older model where the density profile is at its steepest.

The central temperature and central density in the older model are larger than in the younger model, while the central pressure is smaller, consistent with the fact that the number of particles per gram in the convective core of the older model is significantly smaller

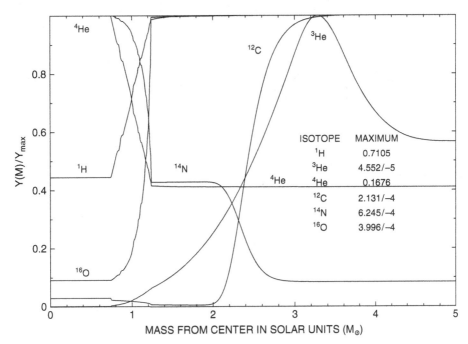

Fig. 11.2.13 Number abundances of major constituents in a 5$M_\odot$ model after three fourths of the main sequence phase ($Z = 0.015$, $Y = 0.275$, AGE $= 6.00 \times 10^7$ yr)

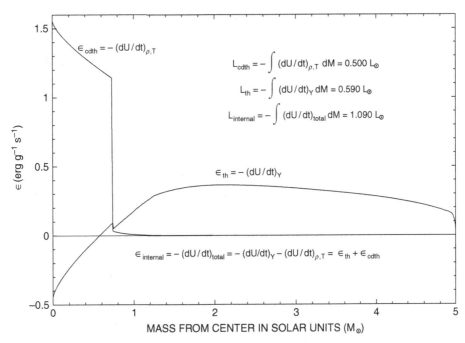

Fig. 11.2.14 Components of dU/dt in a 5$M_\odot$ model after three fourths of the main sequence phase ($Z = 0.015$, $Y = 0.275$, AGE $= 6.00 \times 10^7$ yr)

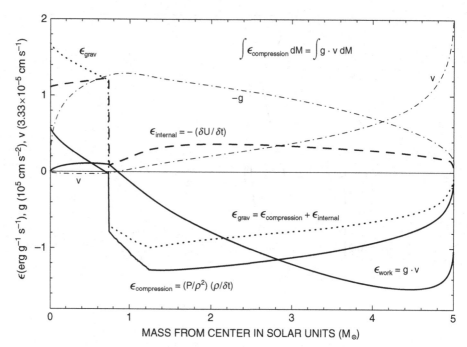

Fig. 11.2.15   Gravothermal energy-generation rates in a $5M_\odot$ model after three fourths of the main sequence phase ($Z = 0.015$, $Y = 0.275$, AGE $= 6.00 \times 10^7$ yr)

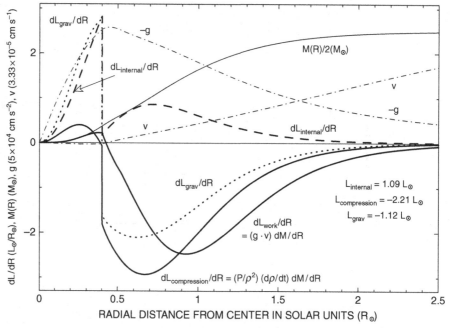

Fig. 11.2.16   Gravothermal energy-generation rates in a $5M_\odot$ model after three fourths of the main sequence phase ($Z = 0.015$, $Y = 0.275$, AGE $= 6.00 \times 10^7$ yr)

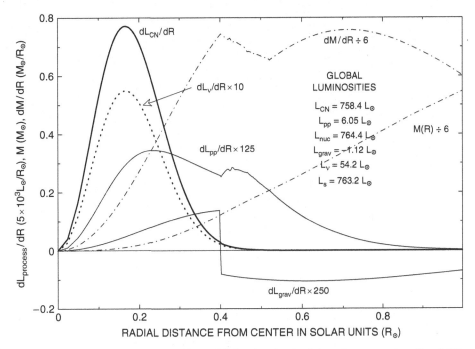

Fig. 11.2.17 Differential contributions to luminosity in a 5$M_\odot$ model after three fourths of the main sequence phase ($Z = 0.015$, $Y = 0.275$, AGE $= 6.00 \times 10^7$ yr)

than in the younger model. Comparison of number abundances in Figs. 11.2.13 and 11.2.6 reveals that, although in the region $M \gtrsim M_{CC}^{max}$ the number abundances of hydrogen and helium are essentially the same in the two models, the number abundance of hydrogen in the convective core has decreased from $Y_H \sim 0.554$ to $Y_H \sim 0.313$, resulting in a decrease in the reciprocal of the molecular weight from $1/\mu \sim 1.42$ to $1/\mu \sim 1.14$.

The conversion of ^{16}O into ^{14}N in the convective core has neared an asymptotic value, with the number abundance of ^{16}O declining from slightly less than a quarter of its primordial value in the younger model to slightly less than a tenth of its primordial value in the older model. Outside the convective core, the only major isotope which has changed its number abundance significantly is ^{3}He, with the peak number abundance increasing from $Y(^3\text{He}) \sim 3.8 \times 10^{-5}$ to $Y(^3\text{He}) \sim 4.6 \times 10^{-5}$.

Comparison between Figs. 11.2.14 and 11.2.7 shows that the manner in which the local thermal energy content changes is qualitatively the same in both models: in the convective core, the thermal energy per unit mass decreases primarily because of a reduction in the particle number densities there; in the region above the convective core, the thermal energy per unit mass decreases primarily because of a decrease in temperature. In both models, the global rates at which the two sources of internal energy change are comparable in magnitude. On the other hand, the global rates of change of both sources are larger in the older model than in the younger one, consistent with the fact that the global rate at which nuclear energy is being produced is increasing with time. This result can also be deduced

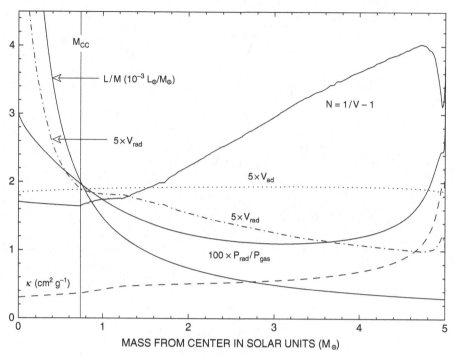

Fig. 11.2.18 Logarithmic gradients and ingredients of V_{rad} in a 5$M_\odot$ model after three fourths of the main sequence phase
($Z = 0.015$, $Y = 0.275$, AGE $= 6.00 \times 10^7$ yr)

from the fact that the slope of the curve for $E_{thermal}$ versus t in Fig. (11.2.4) is larger in
absolute value at $t = 6 \times 10^7$ yr than at $t = 3 \times 10^7$ yr.

Gravothermal energy-generation rates and the ingredients of the gravitational work rate
in the 6×10^7 year old model are shown in Fig. 11.2.15. The only qualitative difference
from these quantities in the 3×10^7 year old model shown in Fig. 11.2.8 is related to the
fact that matter in the convective core of the older model is moving inward and contracting,
with the consequence that the rate of compressional work is positive rather than negative,
in contrast with the younger model which is expanding everywhere. The same inferences
follow from comparisons of gravothermal energy-generation rates shown in the radius representations of Figs. 11.2.16 and 11.2.9.

Comparing quantities in Figs. 11.2.17 and 11.2.10, one sees that, in the 3×10^7 years of
evolution separating the two models, the contribution of pp-chain reactions to the surface
luminosity has decreased by a factor of about 2.5, while the contribution of CN-cycle
reactions has increased by about 25%. Most of the decline in L_{pp} is due to the fact that the
hydrogen abundance in the convective core has decreased by a factor of $\sim$1.77 whereas
the temperature at the center has increased by only $\sim$5.6%, with the result that ϵ_{pp} at the
center decreases by a factor of $\sim(1.77)^2/(1.056)^4 \sim 2.5$. The increase in L_{CN} is due to the
fact that the high sensitivity of CN-cycle reaction rates to temperature more than offsets
the decrease in the number abundance of hydrogen in the convective core.

The distinctive behavior of the mass gradient through the region $R \sim 0.4\,R_\odot$ to $R \sim 0.52$ $R_\odot$ in Fig. 11.2.17 identifies the region in Fig. 11.2.12 bounded by M_{CC} and M_{CC}^{max} which coincides with the region in Fig. 11.2.13 where the abundances of hydrogen and helium vary most strongly. Just as in the case of the 3×10^7 year old model, the fact that the slopes of the curves for P and T in Fig. 11.2.12 do not change discontinuously at the boundaries of the identified region means that the slope of the quantity ρ/μ also does not change discontinuously at the boundaries of this region. And, just as in the case of the 3×10^7 year old model, in the region below the maximum in dM/dR where this derivative decreases outward in mass, density in the 6×10^7 year old model drops off with distance from the center more steeply than $1/r^2$.

Comparing quantities in Figs. 11.2.18 and 11.2.11, and using the fact that $V_{rad} \propto \kappa(P/P_{rad})(L/M)$, one can determine what is responsible for the decrease with time in the mass of the convective core. The opacity in central regions is smaller in the convective core of the older model because the number of electrons per unit mass is smaller, reducing the contributions to the opacity by both electron scattering and free–free absorption. Over the mass range defined by the convective core in the older model, the decrease in opacity varies from $\sim 20\%$ at the center to $\sim 10\%$ at the outer edge of the convective core in the older model. In the mass range $M < M_{CC}^{max}$, the ratio $P/P_{rad} \sim P_{gas}/P_{rad}$ is over 25% smaller in the older model than in the younger model. At the outer edge of the convective core of the older model, the average energy-generation rate $\epsilon_{average}(r) = L(r)/M(r)$ is $\sim 25\%$ larger in the older model than in the younger one. Thus, a reduction in opacity and an increase in the relative importance of radiation pressure more than offset the increase in the local average energy-generation rate in central regions to cause a decrease in V_{rad} everywhere in the convective core, with the result that V_{rad} falls below V_{ad} at smaller and smaller masses as time advances.

In Fig. 11.2.19, the number abundances of beta-stable CNO isotopes are shown as functions of the mass coordinate in the 6×10^7 year old model. In the region of the model beyond $1.2\,M_\odot$, the total amount of ^{12}C has been decreased from a primordial value of $N_{C12} \sim 8.1 \times 10^{-4}$ (in units of Avogadro's number times the solar mass) to $N_{C12} \sim 5.54 \times 10^{-4}$ and the total amount of ^{13}C in this region has been increased from a primordial value of $N_{C13} \sim 0.94 \times 10^{-5}$ to $N_{C13} \sim 4.07 \times 10^{-5}$. One may anticipate that, when the model becomes a red giant, if the base of the convective envelope extends inward to a mass of the order of $M_{CE} \sim 1.2\,M_\odot$, the ratio of carbon isotopes will become of the order of $N_{C12}/N_{C13} \sim 14$ compared with a primordial ratio of $N_{C12}/N_{C13} \sim 90$. The amount of ^{14}N in the 6×10^6 year model above $1.2\,M_\odot$ is proportional to $N_{N14} \sim 4.2 \times 10^{-4}$, compared with a primordial value in this region proportional to $N_{N14} \sim 2.1 \times 10^{-4}$. Thus, the ratio of ^{14}N to ^{12}C in the red giant progeny may be expected to reach $N_{N14}/N_{C12} \sim 0.77$, over three times larger than the primordial ratio of $N_{N14}/N_{C12} \sim 0.25$.

In Fig. 11.2.20, the number abundances of light isotopes are shown as functions of mass for the 6×10^7 year old model. One of the most interesting number abundance distributions is that of ^3He. In the region beyond $M \sim 1.0\,M_\odot$, which eventually becomes incorporated into the convective envelope of the daughter red giant, the total amount of ^3He is not much different from the amount of this isotope in the ZAMS precursor. Thus, although prior to the main sequence phase the 5 $M_\odot$ model converts all of the deuterium

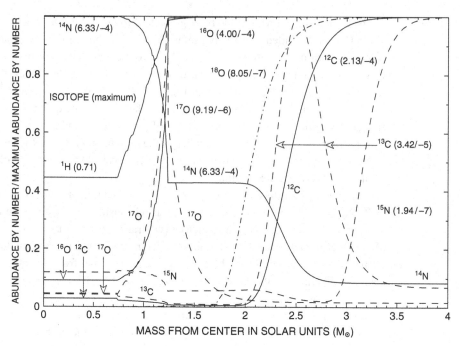

Fig. 11.2.19 β–stable CNO–isotope abundances in a 5 $M_\odot$ model after three fourths of the main sequence phase ($Z = 0.015$, $Y = 0.275$, age $= 6.00 \times 10^7$ yr)

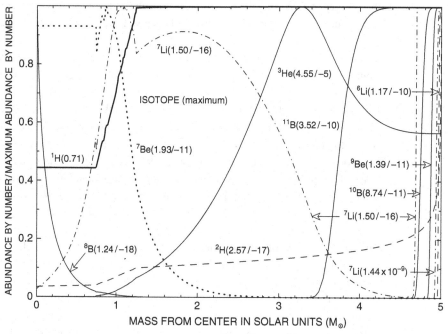

Fig. 11.2.20   Light isotope abundances in a 5$M_\odot$ model after three fourths of the main sequence phase ($Z = 0.015$, $Y = 0.275$, AGE $= 6.00 \times 10^7$ yr)

which it incorporates from the interstellar medium at birth into ^{3}He, in the region that eventually becomes incorporated into a red giant convective envelope, the model destroys almost as much ^{3}He by ^{3}He(^{4}He, γ)^{7}Be and ^{3}He(^{3}He, $2p$)^{4}He reactions as it creates by the $p(p, e^+\nu_e)d(p, \gamma)^3$He reactions. One infers tentatively that population I stars of inter- mediate mass may contribute to the enrichment of ^{3}He in the interstellar medium only to the extent that they convert their initial store of deuterium into ^{3}He during the pre-main sequence phase.

Lithium isotopes are also of interest. They are destroyed over most of the interior by (p, α) reactions, but ^{7}Li can be made by the reactions ^{3}He(^{4}He, γ)^{7}Be(e^-, ν_e)^{7}Li. The number abundance of ^{6}Li is shown in Fig. 11.2.20 by the dashed curve which appears to be almost vertical. A blowup of the region near the surface shows that the half maxi- mum along the ^{6}Li profile lies approximately 0.030 $M_\odot$ below the surface and the width of the profile is approximately 0.01 $M_\odot$. The number abundance of ^{7}Li is plotted on two different scales by dash-dot curves. Near the surface, the ^{7}Li number-abundance profile rises seemingly vertically from zero to a maximum value of 1.17×10^{-10}. The half max- imum along the profile occurs at a mass $\sim$0.058 $M_\odot$ below the surface and the width of the profile is $\sim$0.025 $M_\odot$. Over the entire region below $M \sim 4.7\ M_\odot$, ^{7}Li is in local equi- librium with respect to creating and destroying reactions, but the maximum number abun- dance of ^{7}Li is only 1.50×10^{-16}, totally negligible with respect to the number abundance in layers close to the surface. Their abundances being significantly reduced over all but very small regions near the surface, both isotopes experience large reductions in their sur- face abundances as the base of the convective envelope moves inward during the red giant phase.

When the number abundance of hydrogen in its convective core declines sufficiently, the evolving main sequence model embarks on a phase of overall contraction and heating. Increases in the mean temperature and density in the convective core accelerate to compen- sate for the decrease in the molecular weight in maintaining pressure balance. Increases in density and temperature also help maintain a high rate of nuclear energy generation in the core in spite of the decline in the hydrogen abundance. Thanks to the increases in temper- atures and densities in the region beyond the outer edge of the convective core where the abundance of hydrogen remains relatively high, the contribution to the surface luminosity by nuclear reactions outside of the convective core increases in importance relative to the contribution from nuclear reactions in the convective core.

Figures 11.2.21–11.2.25 show characteristics of the model near the start of the overall contraction phase at $t = 7.977 \times 10^7$ yr, when the hydrogen abundance at the center is $Y_H = 0.025\,77$, and Figs. 11.2.26–11.2.30 show characteristics of the model midway in the overall contraction phase at $t = 8.067 \times 10^7$ yr, when the hydrogen abundance at the center is about 4.5 times smaller at $Y_H = 0.005\,739$. From the luminosity profiles in Figs. 11.2.21 and 11.2.26, one sees that, over the 0.9×10^6 yr which separates the two contracting and heating models, the fraction of the surface luminosity produced outside of the convective core increases from $\sim$3.5% to $\sim$15.5%. Despite the factor of 4.5 difference in the hydrogen number abundance in the convective cores of the models, the luminosity of the convective core in the more evolved model is nearly the same ($\sim$93% as large) as that of the less evolved model. In the more evolved model, there is a detectable contribution to surface

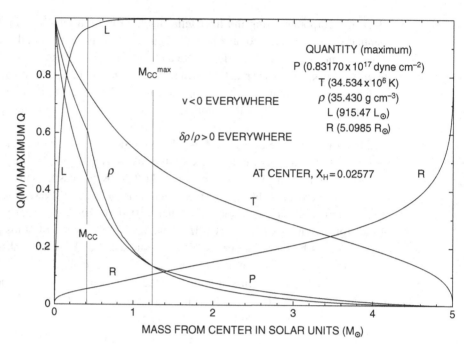

Fig. 11.2.21   Structure of a 5$M_\odot$ model near the start of the overall contraction phase ($Z = 0.015$, $Y = 0.275$, AGE $= 7.977 \times 10^7$ yr)

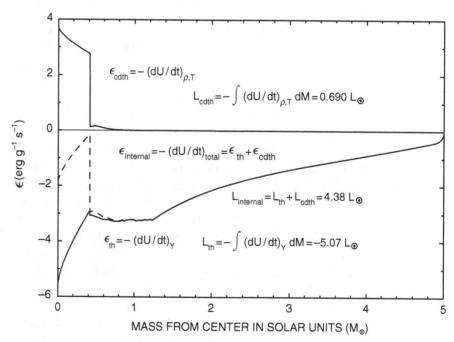

Fig. 11.2.22   Components of dU/dt in a 5$M_\odot$ model near the start of the overall contraction phase ($Z = 0.015$, $Y = 0.275$, AGE $= 7.977 \times 10^7$ yr)

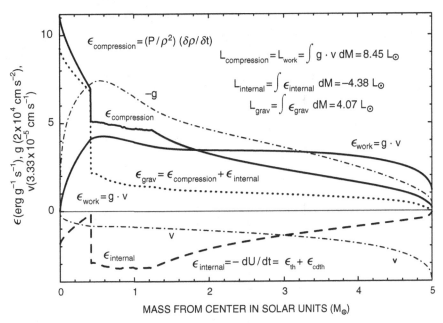

Fig. 11.2.23 Gravothermal energy-generation rates in a 5$M_\odot$ model at the start of the overall contraction phase ($Z = 0.015$, $Y = 0.275$, AGE = 7.977×10^7 yr)

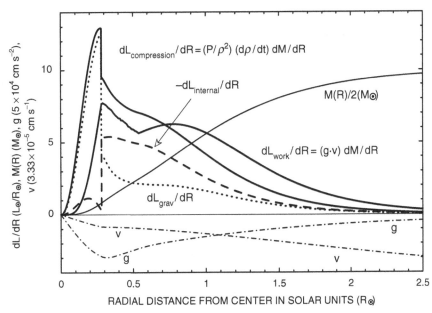

Fig. 11.2.24 Gravothermal energy-generation rates in a 5$M_\odot$ model at the start of the overall contraction phase ($Z = 0.015$, $Y = 0.275$, AGE = 7.977×10^7 yr)

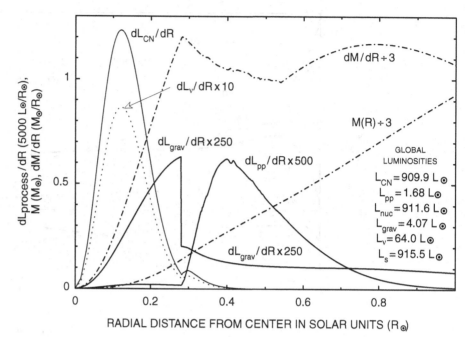

Fig. 11.2.25 Differential contributions to luminosity in a $5M_\odot$ model at the start of the overall contraction phase ($Z = 0.015$, $Y = 0.275$, AGE $= 7.977 \times 10^7$ yr)

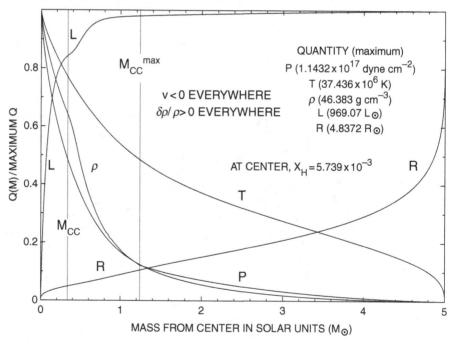

Fig. 11.2.26 Structure of a $5M_\odot$ model midway in the overall contraction phase ($Z = 0.015$, $Y = 0.275$, AGE $= 8.067 \times 10^7$ yr)

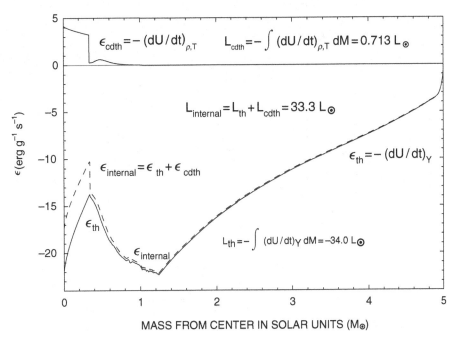

Fig. 11.2.27 Components of dU/dt in a 5M$_\odot$ model midway in the overall contraction phase (Z = 0.015, Y = 0.275, AGE = 8.067×10^7 yr)

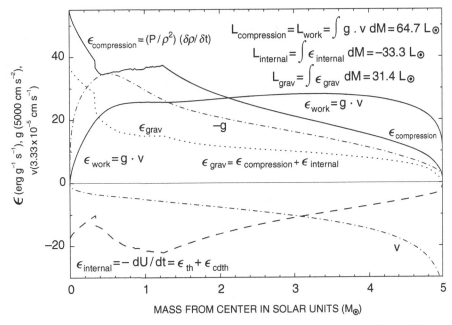

Fig. 11.2.28 Gravothermal energy-generation rates in a 5M$_\odot$ model midway in the overall contraction phase (Z = 0.015, Y = 0.275, AGE = 8.067×10^7 yr)

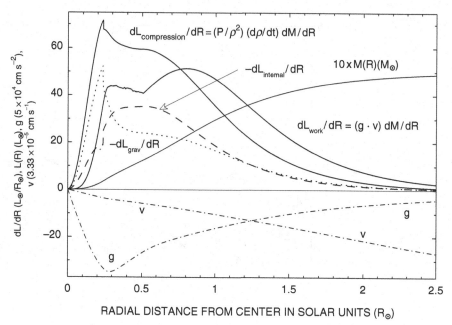

Fig. 11.2.29   Gravothermal energy-generation rates in a 5$M_\odot$ model midway in the overall contraction phase ($Z = 0.015$, $Y = 0.275$, AGE $= 8.067 \times 10^7$ yr)

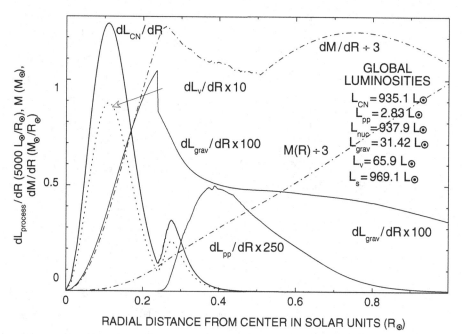

Fig. 11.2.30   Differential contributions to luminosity in a 5$M_\odot$ model midway in the overall contraction phase ($Z = 0.015$, $Y = 0.275$, AGE $= 8.067 \times 10^7$ yr)·

t (yr)	L_{cdth} ($L_\odot$)	L_{nuc} ($L_\odot$)	$\langle T \rangle$ (10^6 K)	$T_{L(r)=L_s/2}$ (10^6 K)
3×10^7	0.384	632	26.5	26.5
6×10^7	0.500	764	28.5	28.8
7.977×10^7	0.690	912	33.0	33.3
8.067×10^7	0.713	938	33.2	33.4
8.157×10^7	0.859	1130	33.2	33.5

Table 11.2.1 Average temperature in the hydrogen-burning region of a 5 $M_\odot$ model estimated from L_{cdth} and L_{nuc} and at $L(r) = L_{\text{surface}}/2$

luminosity due to the release of gravothermal energy by matter outside of the region of nuclear energy production.

Figures 11.2.22 and 11.2.27 show the contributions to the gravothermal energy source of changes in the local and global thermal energy content of the two models. The differential contributions of the particle creation–destruction component ϵ_{cdth} in the two models are qualitatively the same. They are also qualitatively the same as the contributions of this component in the two younger models at ages 3×10^7 yr and 6×10^7 yr shown in Figs. 11.2.7 and 11.2.14, respectively. Values of L_{cdth} for the four models are shown in the second column of the first four rows of Table 11.2.1. Since the rate of nuclear energy production may be approximated by $L_{\text{nuc}} \propto dN/dt$, where dN/dt is the rate of change of a typical isotope involved in CN-cycle nuclear energy production, and since $L_{\text{cdth}} \propto \langle T \rangle \, dN/dt$, where $\langle T \rangle$ is a mean temperature in the nuclear energy producing region, one expects that

$$\langle T \rangle = \alpha \, \frac{L_{\text{cdth}}}{L_{\text{nuc}}}, \tag{11.2.9}$$

where α is a constant. Values of L_{nuc} and $\langle T \rangle$ when $\alpha = 4.3615 \times 10^{10}$ K are given for the four models in the third and fourth columns, respectively, of Table 11.2.1. For comparison, the temperature in the models where the luminosity is one half of the surface luminosity is shown in the last column of the table. The agreement between the two estimates for the least evolved model has been forced by the choice of α. It is remarkable that the two temperatures in each of the other four models do not differ by more than $\sim 1\%$.

What dramatically differentiates the two more evolved models from the two less evolved models (of the set described in the first four rows of Table 11.2.1) is the nature and magnitude of the purely temperature-change-related component of the gravothermal energy-generation rate $\epsilon_{\text{th}} = -(dU/dt)_Y$. Outside the convective cores, not only does this component in the two more evolved models have the opposite sign of the component in the two less evolved models, but it is much, much larger in absolute value in the two more evolved models. In the two less evolved models, ϵ_{th} is negative over only a portion of the convective core, but in the two more evolved models, it is emphatically negative over the entire convective core, as well as being large and negative above the convective core. Progressing from the least evolved to the most evolved of the four models, the global internal (thermal) energy-generation rate progresses through $L_{internal} = L_{thermal} = 0.261\ L_\odot$, $0.590\ L_\odot$, $-5.07\ L_\odot$, and $-34.0\ L_\odot$.

Gravothermal energy-generation rates and the ingredients of the gravitational work rate in the model near the beginning of the overall contraction phase are shown in Fig. 11.2.23 in the mass representation and in Fig. 11.2.24 in the radius representation. In both representations, the location of the outer edge of the convective core coincides with discontinuities in the curves for the gravothermal energy-generation rates. The rates and ingredients in the two representations are shown for the model midway in the overall contraction phase in Figs. 11.2.28 and 11.2.29, respectively. Comparison between these figures and the corresponding figures for the two younger models – Figs. 11.2.8 and 11.2.9 for the 3×10^7 year old model and Figs. 11.2.15 and 11.2.16 for the 6×10^7 year old model – are instructive. For example, the differences between L_{grav} and $L_{internal}$ ($L_{thermal}$) documented in Figs. 11.2.23, 11.2.28, 11.2.9, and 11.2.16 show that, whereas $|L_{grav}|$ and $|L_{internal}|$ in the two younger models differ by only $\sim$0.02–0.03 $L_\odot$, they differ in the models of age 7.977×10^7 yr and 8.067×10^7 yr by 0.31 $L_\odot$ and 1.9 $L_\odot$, respectively. The differences are due to differences in the rate of change with time of the global radiation energy content E_{rad} and are explicable quantitatively by invoking eq. (11.2.8) and finding dE_{rad}/dt from the curve for E_{rad} versus t in Fig. 11.2.4.

The differential contributions of nuclear and gravothermal energy sources in the radius representation are compared in Fig. 11.2.25 for the 7.977×10^7 model and those for the 8.067×10^7 model are compared in Fig. 11.2.30. The global rate of gravothermal energy generation (due to compression throughout the model modified primarily by cooling above the convective core) is almost eight times larger in the more evolved model. The secondary peak above the convective cores in the distributions dL_{CN}/dR is noticeably larger in the more evolved model, demonstrating that a nuclear energy source distributed in a radiative shell located above the convective core is building in strength relative to the strength of a weakening convective core nuclear energy source.

Ultimately, hydrogen effectively vanishes at the center, core convection ceases, and nuclear energy generation becomes concentrated in a thick shell centered at a point near the base of the hydrogen profile located above the outer edge of the shrinking convective core. The transition between core hydrogen burning and shell hydrogen burning is remarkably abrupt and involves dramatic changes in the interior.

Some insight into the transition from core hydrogen burning to thick shell hydrogen burning is provided by the temporal variations of various quantities in Figs. 11.2.31– 11.2.33. The global rates of the nuclear, gravothermal, and surface luminosities are shown in Fig. 11.2.31. During the first part of the transition from core to shell burning, between $\sim$8.06 $\times 10^7$ yr and $\sim$8.12 $\times 10^7$ yr, the global rate of gravothermal energy generation rises gradually to $\sim +8\%$ of the surface luminosity, and then drops abruptly to $\sim -7\%$ of the surface luminosity before rising gradually to a locally asymptotic value of the order of -1% of the surface luminosity.

The mass of the convective core is shown in both Fig. 11.2.31 and Fig. 11.2.32. The number abundance of hydrogen in the core is shown in Fig. 11.2.31, but the hydrogen

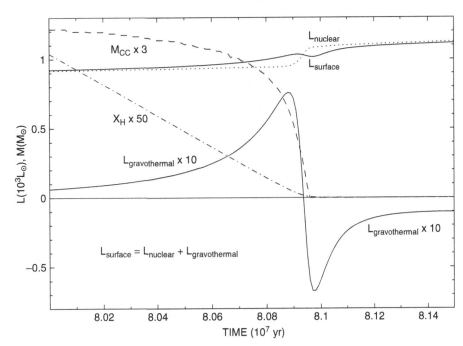

Fig. 11.2.31 Luminosities, central ^{1}H abundance, and convective core mass during the transition from core to shell H-burning in a 5$M_\odot$ model (Z = 0.015, Y = 0.275)

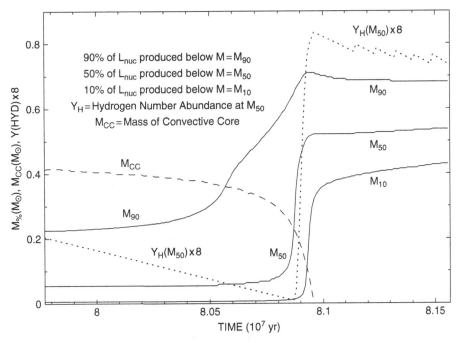

Fig. 11.2.32 Mass and ^{1}H abundance at the middle of the H-burning zone as core H-burning ends and shell H-burning begins in a 5$M_\odot$ model (Z = 0.015, Y = 0.275)

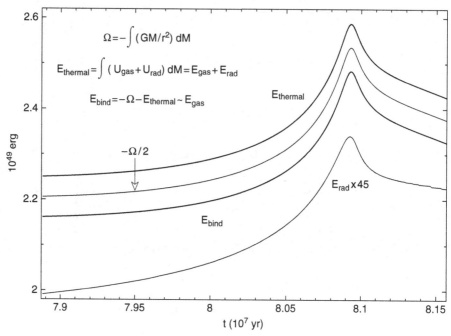

$$\Omega = -\int (GM/r^2)\, dM$$

$$E_{thermal} = \int (U_{gas} + U_{rad})\, dM = E_{gas} + E_{rad}$$

$$E_{bind} = -\Omega - E_{thermal} \sim E_{gas}$$

Fig. 11.2.33 Global gravothermal energies during the transition from core to shell hydrogen burning in a 5$M_\odot$ model ($Z = 0.015$, $Y = 0.275$)

number abundance shown in Fig. 11.2.32 is that at the point within which 50% of the nuclear energy generation occurs. This latter number abundance does not vanish as the convective core vanishes, but increases to reach a value of the order of $Y_H = 0.1$ in the fully developed hydrogen-burning shell.

The masses within which 90%, 50%, and 10% of the nuclear energy are being generated are described in Fig. 11.2.32 by the curves labeled M_{90}, M_{50}, and M_{10}, respectively. The midway point in the transition between core hydrogen burning and shell hydrogen burning may be defined as the time when half of the rate of nuclear energy release is contributed by reactions taking place in the convective core and half is contributed by reactions taking place outside the core. This time is identified by the point where the curves for M_{CC} and M_{50} in Fig. 11.2.32 intersect. The intersection occurs at $t \sim 8.087 \times 10^7$ yr, when the mass of the convective core is approximately $M_{CC} \sim 0.22\ M_\odot$ and the number abundance of hydrogen in the core is about $Y(^1H) \sim 0.001$. In $\delta t \sim 0.9 \times 10^5$ additional years, convection has died out at the model center and the middle of the hydrogen-burning shell, defined as the point below which half of the nuclear burning luminosity is produced, has moved outward in mass to $\sim 0.53\ M_\odot$, where the abundance by number of hydrogen is $Y_H \sim 0.1$.

The gravothermal luminosity is also the time derivative of the net binding energy of the model, as may be verified by comparing $L_{gravothermal}$ in Fig. 11.2.31 with the slope of the curve of E_{bind} versus t which is barely distinguishable among the crowded curves on the right hand side of Fig. 11.2.4, but is easily identified in Fig. 11.2.33, which is an

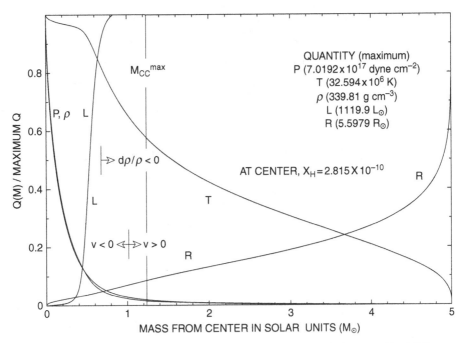

Q(M) / MAXIMUM Q

MASS FROM CENTER IN SOLAR UNITS ($M_\odot$)

QUANTITY (maximum)
P (7.0192×10^{17} dyne cm^{-2})
T (32.594×10^6 K)
ρ (339.81 g cm^{-3})
L (1119.9 $L_\odot$)
R (5.5979 $R_\odot$)

AT CENTER, $X_H = 2.815 \times 10^{-10}$

Fig. 11.2.34 Structure of a 5$M_\odot$ model after the establishment of a thick hydrogen-burning shell ($Z = 0.015, Y = 0.275$, AGE $= 8.157 \times 10^7$ yr)

expanded version of the rightmost portion of Fig. 11.2.4. It is evident, for example, that the location of the maximum of the binding energy curve in Fig. 11.2.33 at $t \sim 8.093 \times 10^7$ yr coincides with the location of the zero point in $L_{\text{gravothermal}}$ in Fig. 11.2.31. As the transition between core and shell hydrogen burning takes place in the interior, the model evolves along the track in the HR diagram of Fig. 11.2.1 between the dots at 8.08 and 8.11, reversing direction in the surface temperature coordinate when the model radius reaches a minimum slightly larger than $R = 4.51$ $R_\odot$.

Structure variables in a model at $t = 8.157 \times 10^7$ yr, near the beginning of the thick shell hydrogen-burning phase, are shown in Fig. 11.2.34. Comparing the characteristics in this figure with those of core hydrogen-burning models described in Figs. 11.2.5, 11.2.12, 11.2.21, and 11.2.26, it is evident that, in the central region where hydrogen has effectively vanished, the steep temperature gradient associated with a nuclear burning energy source is replaced by a relatively shallow temperature gradient maintained by a much weaker gravothermal energy source. The reduction in the temperature gradient is accomplished by cooling in the hydrogen-vacated region; the central temperature in the thick shell hydrogen-burning model is about 15% smaller than in the core hydrogen-burning model described by Fig. 11.2.26. On the other hand, the central density in the thick shell hydrogen-burning model is over seven times larger. Thus, in the central, hydrogen-exhausted region, although the gravothermal release of energy during the transition from core to shell hydrogen burning is partially a consequence of cooling, the major source of released gravothermal energy is compressional work.

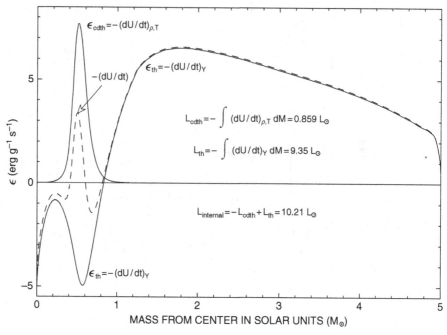

Fig. 11.2.35 Components of dU/dt in a 5 $M_\odot$ model with a thick hydrogen-burning shell ($Z = 0.015$, $Y = 0.275$, AGE $= 8.157 \times 10^7$ yr)

The net local rate at which internal thermal energy decreases in the 8.157×10^7 year old model is given by the dashed curve in Fig. 11.2.35 labeled $-dU/dt$, and the two solid curves describe its components, $\epsilon_{cdth} = -(dU/dt)_{\rho,\, T}$ and $\epsilon_{th} = -(dU/dt)_Y$. Since it is proportional to the rate at which material particles are being destroyed, ϵ_{cdth} provides a precise outline of where in the model active nuclear burning is taking place: a region of half width $\Delta M_{1/2} \sim 0.22\ M_\odot$ centered at $M \sim 0.53\ M_\odot$. From the curves for M_{10} and M_{90} in Fig. 11.2.32, one may deduce that 80% of the nuclear energy is released in a region of mass $\Delta M_{80\%} \sim 0.25\ M_\odot$.

Over the inner $\sim 0.9\ M_\odot$ of the model, the net rate at which the internal energy changes averages out to nearly zero. That is, in this inner region, the rate of increase in the thermal energy per particle is almost exactly balanced by the rate of decrease in the number density of particles times the thermal energy of a particle. Another way of putting it is that, even though temperatures in the inner 0.9 $M_\odot$ are increasing, the number of particles is decreasing at such a rate that the net heat content is not changing. Thus, effectively the entire contribution to $L_{internal}$, the global rate of change of internal energy, is produced outside of the active nuclear burning region by cooling in the region extending from $\sim 0.9\ M_\odot$ to the surface. Quantitatively,

$$\int_0^{5M_\odot} \epsilon_{internal}\, dM \sim \int_{0.9M_\odot}^{5M_\odot} \epsilon_{th}\, dM \sim 10.2\ L_\odot. \qquad (11.2.10)$$

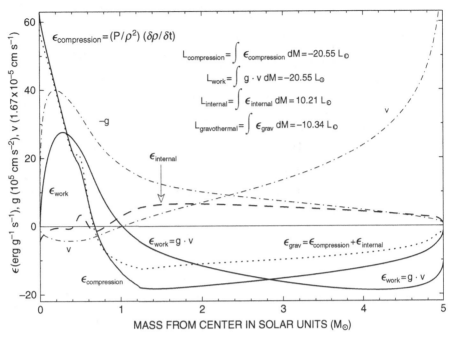

Fig. 11.2.36 Gravothermal energy-generation rates in a 5M$_\odot$ model with a thick hydrogen-burning shell (Z = 0.015, Y = 0.275, AGE = 8.157 × 10^7 yr)

The dominance of compressional energy release in central regions is made explicit in Fig. 11.2.36, where the curves for $\epsilon_{\mathrm{gravothermal}}$ (dotted curve) and $\epsilon_{\mathrm{compression}}$ (solid curve) are almost identical over the inner ~0.8 $M_\odot$ of the model and, as already noted, $\epsilon_{\mathrm{internal}}$ (dashed curve) averages out to about zero. The rate at which gravity does work is shown by another solid curve and the velocity and gravitational acceleration ingredients of this rate are shown by dash-dot curves. Figure 11.2.37 contains the same information regarding gravothermal energy-generation rates as does Fig. 11.2.36, but in the radius representation. The dashed curve labeled $10 \times M(R)$ describes the mass coordinate as a function of the radius coordinate, facilitating comparison between quantities in the two figures.

The position in Fig. 11.2.37 where $dL_{\mathrm{compression}}/dR = 0$ coincides with the position in Fig. 11.2.34 where $\delta\rho/\rho = 0$. It is evident that, in the hydrogen-exhausted core and over most of the nuclear energy-producing region, as delineated by the luminosity profile in Fig. 11.2.34, matter is being compressed and the rate of compressional energy release far exceeds the rate at which internal energy is either released or absorbed. On the other hand, in the region extending from the outer edge of the luminosity profile to the surface, matter is expanding and, over most of this region, it is also cooling. Thus, in this latter region, the release of internal energy assists in doing the work required to expand and to move matter outward against gravity. In spite of this, as anticipated in eqs. (8.1.68) and (8.3.24), an exact global balance exists between the total rate at which expansional work is done and the total rate at which work is done against gravity. This means that the compressional

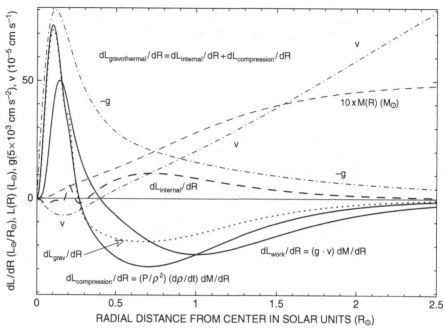

Fig. 11.2.37 Gravothermal energy-generation rate components in a $5M_\odot$ model with a thick H-burning shell ($Z = 0.015$, $Y = 0.275$, AGE $= 8.157 \times 10^7$ yr)

energy produced in the core of the model contributes to the expansional energy used up in the envelope of the model.

Integrating under the curve for $\epsilon_{\text{compression}}$ from $M = 0$ to $M \sim 0.7\, M_\odot$, where $\epsilon_{\text{compression}}$ vanishes, one finds that the core contributes to the outward flow of radiation at the rate $L^{\text{core}}_{\text{compression}} \sim +10\, L_\odot$. Integration from $M \sim 0.7\, M_\odot$ to $M = 5\, M_\odot$ shows that envelope expansion requires energy at the rate $L^{\text{envelope}}_{\text{expansion}} = -L^{\text{envelope}}_{\text{compression}} \sim 31\, L_\odot$. Since cooling of envelope material supplies energy at the rate $L^{\text{envelope}}_{\text{internal}} \sim 10\, L_\odot$, energy must be absorbed from the ambient flow of radiant energy at the rate $L^{\text{envelope}}_{\text{absorption}} \sim 21\, L_\odot$. Approximately half of the absorption rate may be ascribed to the release of compressional energy in the core, and the rest may be ascribed to nuclear energy released in the core. The fact that compressional energy released in the core contributes in a delayed fashion to the expansion of the envelope is reflected in a similar relationship between the rate at which work is done by gravity in the inner $1\, M_\odot$ of the model where $v < 0$ and the rate at which work is done against gravity in the outer $4\, M_\odot$ of the model. Rough integrations give $L^{v<0}_{\text{work}} \sim 9\, L_\odot$ and $-L^{v>0}_{\text{work}} \sim 30\, L_\odot$.

In Fig. (11.2.38), differential contributions by nuclear energy sources are compared with the contribution by the gravothermal source. It is interesting that the total rate at which gravothermal energy is released in the contracting core ($L_{\text{grav}}(\dot\rho > 0) \sim 10\, L_\odot$) exceeds by over a factor of three the rate at which pp reactions contribute energy in a region extending outward from near the centroid of the profile defined by the primary CN-cycle energy source. In the expanding region extending from the outer edge of the luminosity

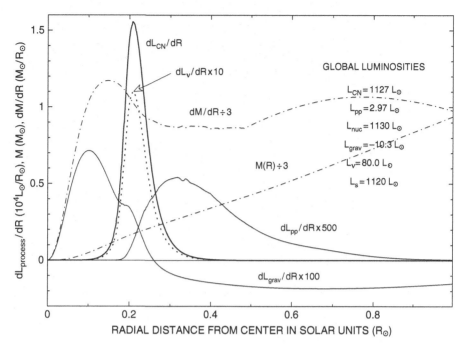

Fig. 11.2.38 Differential contributions to luminosity in a 5$M_\odot$ model with a thick hydrogen-burning shell ($Z = 0.015$, $Y = 0.275$, AGE $= 8.157 \times 10^7$ yr)

profile to the surface, the total rate of absorption by the gravothermal energy source is $-L_{\mathrm{grav}}(\dot{\rho} < 0) \sim 20\,L_\odot$, so the net rate of gravothermal energy generation is $L_{\mathrm{gravothermal}} \sim -10\,L_\odot$, about -1% of the total rate of nuclear energy generation.

Shown also in Fig. 11.2.38 as functions of the radial coordinate are the mass coordinate $M(R)$ and the radial derivative of the mass coordinate. In the outer half of the region extending from the center to $\sim 0.5\,R_\odot$ (that is, from about the maximum in the $\mathrm{d}L_{\mathrm{CN}}/\mathrm{d}R$ profile at $\sim 0.2\,R_\odot$ to a distance comparable with the distance from the model center to the maximum in the profile), $\mathrm{d}M/\mathrm{d}R$ is nearly constant. This means that, over a region comparable in radial extent with that of the hydrogen-exhausted core, the density falls off almost exactly in proportion to the inverse square of the distance from the center.

To complete the description of the representative thick shell hydrogen-burning model, number abundance distributions are shown in Figs. 11.2.39–11.2.41. Comparison of the distributions of CNO isotopes in Figs. 11.2.39 and 11.2.40 with those in Fig. 11.2.19 shows that, in the outer 4 $M_\odot$ of the two models, the distributions of β-stable CNO isotopes are essentially identical. In Fig. 11.2.39, the profiles of the beta-unstable isotopes nicely outline the location of the CN-cycle driven hydrogen-burning shell and the near identity of the profile for ^{17}F with those for ^{13}N and ^{15}O demonstrates that, in the main nuclear burning region, CNO-bi-cycle reactions are in equilibrium.

Light isotope abundances in the thick shell hydrogen-burning model are shown in Fig. 11.2.41 and are to be compared with those for the 6×10^7 year old model in

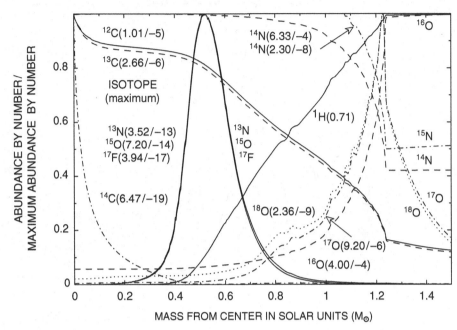

Fig. 11.2.39   CNO isotope abundances in a 5 $M_\odot$ model with a thick H-burning shell ($Z = 0.015$, $Y = 0.275$, AGE $= 8.157 \times 10^7$ yr)

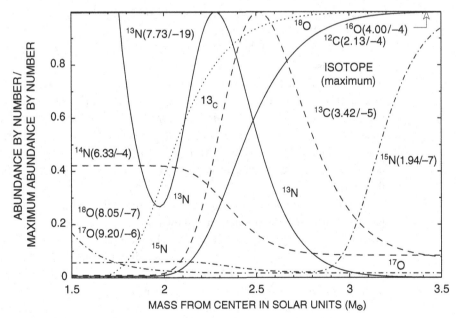

Fig. 11.2.40   CNO isotope abundances in a 5 $M_\odot$ model with a thick H-burning shell ($Z = 0.015$, $Y = 0.275$, AGE $= 8.157 \times 10^7$ yr)

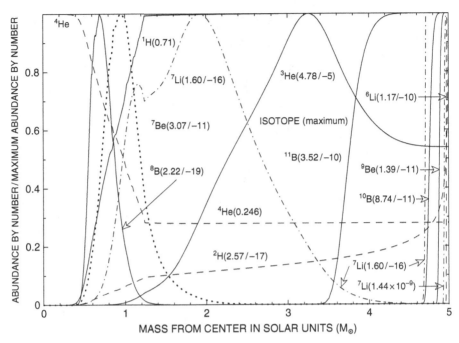

Fig. 11.2.41 Light isotope abundances in a 5$M_\odot$ model with a thick H-burning shell ($Z = 0.015$, $Y = 0.275$, AGE $= 8.157 \times 10^7$ yr)

Fig. 11.2.20. In the $\sim 2 \times 10^7$ yr separating the two models, there have been essentially no changes in the number abundance distributions of ^{3}He, ^{6}Li, ^{7}Li, ^{9}Be, ^{10}B, and ^{11}B in the outer 1.5 $M_\odot$. There has been a very slight increase in the abundance of ^{3}He in the middle 2 $M_\odot$ of the model. The largest change has, of course, been the almost complete destruction of hydrogen over the inner $\sim 10\%$ of the mass of the model. Interestingly, the maximum abundances of two of the trace isotopes involved in the pp chains, ^{7}Li and ^{7}Be, are quite similar in the two models, and the maximum abundance of ^{8}B differs by less than an order of magnitude in the two models, being approximately six times smaller in the more evolved model which sports higher temperatures.

Evolution in the HR diagram during shell hydrogen burning, up to the ignition of helium in the core, is shown by the solid curve in Fig. 11.2.42. The passage of the evolutionary track across dotted lines of ever larger radius demonstrates that the model becomes a red giant, and the evolutionary ages given beside dots along the track (in units of 10^7 yr) demonstrate that the rate of evolution accelerates rapidly after the maximim in luminosity along the track labeled "onset of gravothermal instability" at $t \sim 8.25 \times 10^7$ yr is reached. The rate of evolution slows noticeably again only when the luminosity along the red giant branch reaches a value comparable with that at the onset of the gravothermal instability. At this juncture, helium burning in the core has terminated the instability.

State variables at the center of the model are shown as functions of time in Fig. 11.2.43. During the phase of hydrogen burning in a thick shell, the central temperature remains

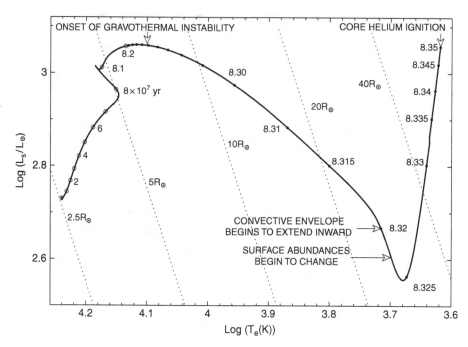

Fig. 11.2.42 Evolution in the HR–diagram of a 5$M_\odot$ model (Z = 0.015, Y = 0.275) through core and shell hydrogen burning to the onset of core helium burning

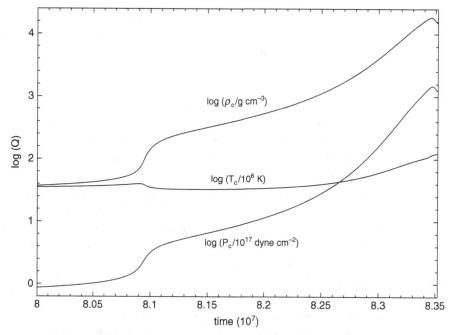

Fig. 11.2.43 State variables at the center of a 5$M_\odot$ model (Z = 0.015, Y = 0.275) from the end of core hydrogen burning to the start of core helium burning

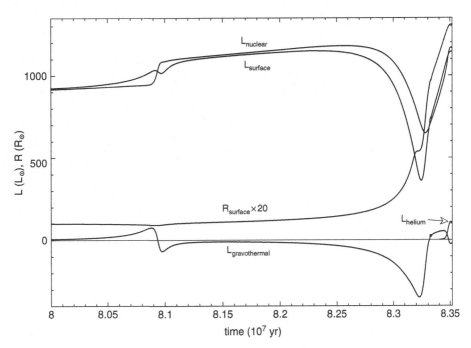

Fig. 11.2.44 Global luminosities and radius of a 5$M_\odot$ model ($Z = 0.015$, $Y = 0.275$) from the end of core hydrogen burning to the start of core helium burning

nearly constant over an extended period, not reaching the maximum of $\sim 40 \times 10^6$ K achieved during the overall contraction phase at $t \sim 8.09 \times 10^7$ yr until $t \sim 8.25 \times 10^7$ yr.

Model radius and global luminosities are shown as functions of time in Fig. 11.2.44. During the time shown, there are two phases when the rate of gravothermal energy production, $L_{\mathrm{gravothermal}} = L_{\mathrm{surface}} - L_{\mathrm{nuclear}}$, becomes noticeable relative to the rate of nuclear energy production. Modestly large changes in $L_{\mathrm{gravothermal}}$ occur during the overall contraction phase between $t \sim 8.07 \times 10^7$ yr and $\sim 8.11 \times 10^7$ yr, with maximum values of $|L_{\mathrm{gravothermal}}|$ being of the order of 10% of L_{nuclear}. Changes in $L_{\mathrm{gravothermal}}$ over the time period ~ 8.25–8.35×10^7 yr are manifestations of the gravothermal instability marking the transition between hydrogen–burning in a thick shell and core helium burning. The maximum in $|L_{\mathrm{gravothermal}}|$ during the transition is approximately five times larger than maximum values during the preceeding overall contraction phase. The minimum in the gravothermal luminosity, $L_{\mathrm{gravothermal}}^{\mathrm{min}} \sim -350\, L_\odot$, is reached at $t \sim 8.322 \times 10^7$ yr, shortly before the nuclear burning luminosity reaches a minimum of $L_{\mathrm{nuclear}}^{\mathrm{min}} \sim 640\, L_\odot$ at $t \sim 8.328 \times 10^7$ yr. At $t \sim 8.322 \times 10^7$ yr, $L_{\mathrm{gravothermal}} \sim -1/2\, L_{\mathrm{nuclear}}$.

The abrupt changes in sign of the slopes of the global luminosity curves follow the formation of a convective envelope, the base of which, at $t \sim 8.32 \times 10^7$ yr, begins to penetrate inward in mass (see a later figure, Fig. 11.2.57). In the convective portion of the envelope, absorption of energy from the ambient flow is negligible, and it is the decrease with time in the mass of the radiative portion of the envelope between the outer edge of the

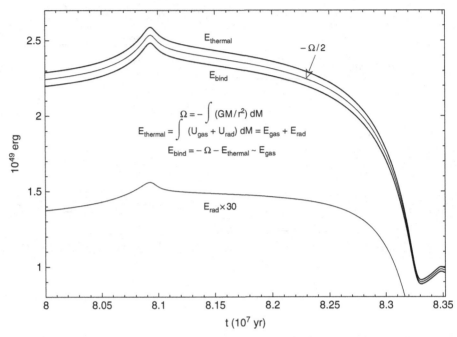

Fig. 11.2.45   Global gravothermal energies of a $5M_\odot$ model ($Z = 0.015$, $Y = 0.275$) from the end of core hydrogen burning to the start of core helium burning

hydrogen-burning shell and the base of the convective envelope that is responsible for the decrease with time of $|L_{\text{gravothermal}}|$ after the minimum in $L_{\text{gravothermal}}$ has been reached.

Global changes in thermal and gravothermal energies are shown as functions of time in Fig. 11.2.45. Two quantities in Figs. 11.2.44 and 11.2.45 are related by the fact that $L_{\text{gravothermal}} = dL_{\text{bind}}/dt$. The upturns in the quantities E_{thermal}, $|\Omega|/2$, and E_{bind} on the very far right in Fig. 11.2.45 are consequences of the extension inward of the base of the convective envelope. In doing work against gravity, rather than relying on absorption from the ambient flow of energy, as in the radiative portion of the envelope, the convective envelope locally converts thermal energy into the energy of expansion (with the result that $\epsilon_{\text{grav}} = 0$).

Characteristics of the hydrogen-burning shell are shown as functions of time in Figs. 11.2.46 and 11.2.47. As revealed in Fig. 11.2.46, during the thick shell-burning phase prior to the onset of the gravothermal instability, the mass within which 80% of the nuclear energy is produced decreases from $\sim 0.25\ M_\odot$ to $\sim 0.17\ M_\odot$. During the phase of rapid core contraction and envelope expansion, this mass decreases to $\sim 0.04\ M_\odot$, or less than 1% of the mass of the model. Figure 11.2.47 shows that, while the temperature in the middle of the hydrogen-burning shell (where $M = M_{50}$) increases only modestly (from $T_{\text{shell}} \sim 30 \times 10^6$ K to $T_{\text{shell}} \sim 35 \times 10^6$ K), the density there more than doubles (from $\rho_{\text{shell}} \sim 30$ g cm^{-3} to $\rho_{\text{shell}} \sim 65$ g cm^{-3}) before stabilizing at $\rho_{\text{shell}} \sim 50$ g cm^{-3} when helium burning in the core takes hold. The radius of the sphere defined by the middle of the hydrogen-burning shell declines monotonically to $R_{\text{shell}} \sim 0.1\ R_\odot$.

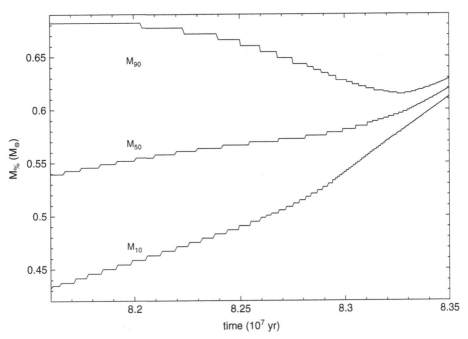

Fig. 11.2.46 Mass characteristics of the hydrogen–burning shell in a 5$M_\odot$ model (Z = 0.015, Y = 0.275) evolving from the main sequence to the giant branch

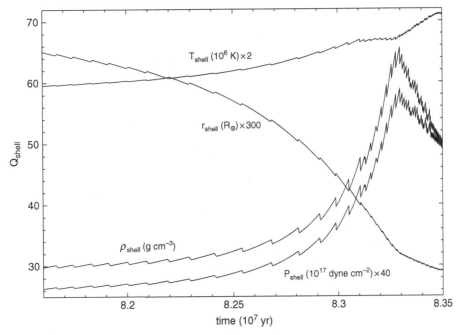

Fig. 11.2.47 Radius and state variables at the center of the hydrogen-burning shell in a 5$M_\odot$ model (Z = 0.015, Y = 0.275) during the transition to the giant branch

The gravothermal instability uncovered by Lynden-Bell & Wood (1968) in the context of a stellar cluster and the upper limit found by Schönberg and Chandrasekhar (1942) on the mass of an isothermal core capped by a hydrogen-burning shell were shown in Section 11.1 to have a bearing on the evolution of a low mass stellar model, even though the high degree of electron degeneracy in the low mass model violates a basic condition (perfect gas equation of state) on which the instability and the limit have been predicated. In the 5 $M_\odot$ model under consideration here, electron degeneracy is not an important factor, so the instability and the limit should be more transparently relevant than in the 1 $M_\odot$ model.

In the Lynden-Bell and Wood experiments, when the density at the center relative to the density at the boundary reaches the critical value $\rho_{center}/\rho_{boundary} \sim 32.2$, thermal equilibrium solutions cease to exist and a "gravothermal catastrophe" ensues. From the curve for $\log(\rho_c)$ in Fig. 11.2.43 and the curve for ρ_{shell} in Fig. 11.2.47, one finds that ρ_c/ρ_{shell} reaches the critical value at $t_9 \gtrsim 8.25 \times 10^7$ yr. From Fig. 11.2.46, one finds that, at $t_9 = 8.25 \times 10^7$ yr, $M_{shell} \sim 0.57 M_\odot$, or $M_{shell}/M_{model} = 0.114$, close to the Schönberg–Chandrasekhar limit. Finally, the position of maximum luminosity along the evolutionary track in Fig. 11.2.42 occurs at $t_9 \gtrsim 8.25 \times 10^7$ yr and the rate of evolution along the track after this time obviously accelerates. The simultaneous closeness of the model ratios to the critical ratios in the Lynden-Bell–Wood and Schönberg–Chandrasekhar studies, along with the acceleration in the evolution rate in the HR diagram, all at $t_9 \sim 8.25 \times 10^7$ yr, motivates the identification of the position of maximum luminosity as the point at which the gravothermal instability sets in.

The instability is manifested in Fig. 11.2.43 by the fact that the rate at which the central density increases with time accelerates with time. Its effects are even more apparent in the acceleration in the rate of increase of $-L_{gravothermal}$ in Fig. 11.2.44. Paradoxically, an even more dramatic evidence of the instability is the acceleration in the rate of decrease in the binding energy displayed in Fig. 11.2.45. The decrease in binding energy is due to the expansion of the model envelope (see $R_{surface}$ in Fig. 11.2.44) which requires absorption of energy from the ambient flow (see $L_{surface}$ and $L_{nuclear}$ in Fig. 11.2.44).

Given the fact that the phase of rapid core contraction and envelope expansion is halted once core helium burning takes hold, it is somewhat hyperbolic to call the phase a gravothermal catastrophe. A more appropriate description of the phenomenon is that it is a normal gravothermal adjustment bridging two phases of core nuclear burning. During phases when the rate of nuclear burning controls the evolutionary time scale, gravothermal energy generation is essentially negligible relative to the rate of nuclear energy generation. During the transition between stable core nuclear burning phases, core contraction and heating and envelope expansion and cooling take center stage, with the evolutionary time scale being controlled by the gravothermal changes. Cause and effect during evaluation from the onset of the gravothermal instability to the red giant branch and along the red giant branch are discussed at length from a philosophical perspective by Iben (1993).

In the case of the 5 $M_\odot$ model described here, the duration of the thick shell hydrogen-burning phase is comparable to the duration of the overall contraction phase and the lengths of the evolutionary tracks in the HR diagram for the two phases are also comparable (see

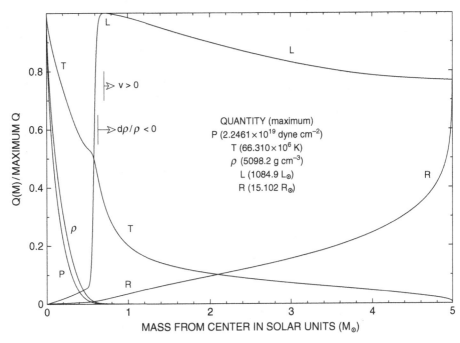

Fig. 11.2.48 Structure of a 5$M_\odot$ model in the Hertzsprung gap, midway between the main sequence and the giant branch
($Z = 0.015, Y = 0.275$, age $= 8.30693 \times 10^7$ yr)

Figs. 11.2.1 and 11.2.42). The fact that the evolutionary tracks have the opposite slopes suggest that, near the upper end of the main sequence in the color magnitude diagram defined by stars in an open cluster, the distribution of stars might exhibit interesting structure. It turns out that this is in fact the case, with the cluster M67 exhibiting a gap in the distribution of stars at the upper end of the main sequence (see Fig. 3 in Johnson and Sandage (1955)). The age of M67 is about 5×10^9 yr, so the masses of stars near the upper end of the main sequence are of the order of 1.25 $M_\odot$ (see Fig. 11.4.1) and the appearance of a prominent gap is most likely due to the fact that the ratio of the timescales for the thick shell hydrogen-burning and the overall contraction phases increases with decreasing stellar mass. In the models discussed by Iben (1967), the ratio increases from $\sim$0.6 for a 5 $M_\odot$ model to $\sim$5.7 for a 1.25 $M_\odot$ model.

Structure variables in the interiors of several 5 $M_\odot$ models at various stages of the gravothermal adjustment phase are shown in Figs. 11.2.48–11.2.52. The most transparently instructive features in the figures are the luminosity and temperature profiles, coupled with the maximum values along structure profiles listed in the inserts headed by the label QUANTITY (maximum). In each case, the outer edge of the hydrogen-exhausted core is located near where the luminosity profile undergoes an abrupt increase in slope (at $M \sim 0.55$–0.6 $M_\odot$). From the densities and temperatures listed, it is evident that the cores are contracting and heating. In each case, the location of the outer edge of the hydrogen-burning shell, which may also be called the base of the model envelope, coincides with the location of the maximum in the luminosity profile. This maximum nearly

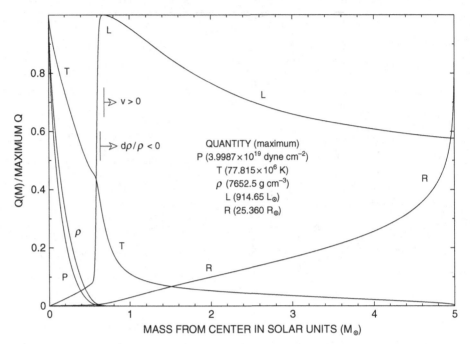

Fig. 11.2.49 Structure of a 5$M_\odot$ model nearing the giant branch, about to develop a convective envelope ($Z = 0.015$, $Y = 0.275$, AGE $= 8.31824 \times 10^7$ yr)

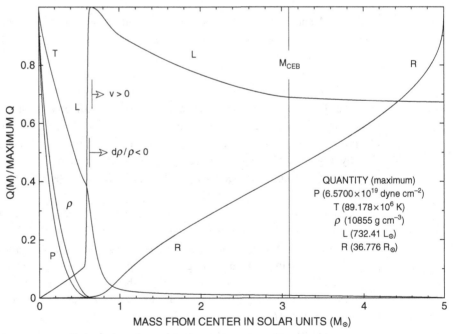

Fig. 11.2.50 Structure of a 5$M_\odot$ model ascending the red giant branch with a deepening convective envelope ($Z = 0.015$, $Y = 0.275$, AGE $= 8.32785 \times 10^7$ yr)

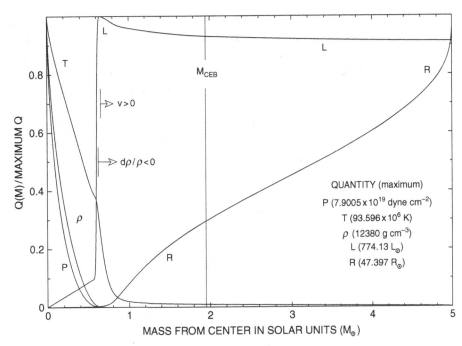

Fig. 11.2.51   Structure of a 5$M_\odot$ model ascending the red giant branch with a deepening convective envelope ($Z = 0.015$, $Y = 0.275$, AGE $= 8.33168 \times 10^7$ yr)

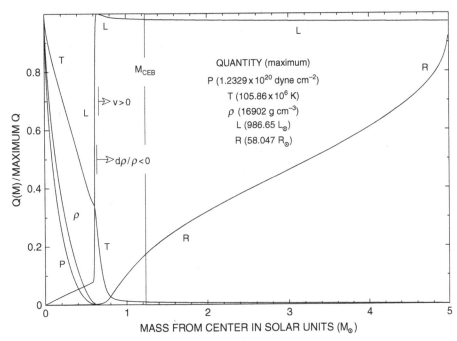

Fig. 11.2.52   Structure of a 5$M_\odot$ model approaching the tip of the red giant branch ($Z = 0.015$, $Y = 0.275$, AGE $= 8.3415 \times 10^7$ yr)

coincides with the vertical line (labeled $\rightarrow v > 0$) to the right of which all matter is moving outward and to the left of which all matter is moving inward. The division between contracting and expanding matter is marked by another vertical line (labeled $\rightarrow \delta\rho/\rho < 0$) which lies just below the outer edge of the hydrogen-burning shell. To the right of this second line, all matter is expanding and, to the left of the line, all matter is contracting.

The models all have in common that the contracting and heating core contributes to the ambient flow of energy at the base of the hydrogen-burning shell at a rate of the order of $L_{core} \sim 65$–$85\ L_\odot$. Models in Figs. 11.2.48–11.2.51 also have in common that a substantial fraction of the luminosity produced by the contracting core and the hydrogen-burning shell is absorbed from the ambient flow in the radiative portion of the expanding and cooling envelope above the hydrogen-burning shell. On the other hand, as illustrated by Figs. 11.2.50–11.2.52, very little energy is absorbed from the ambient flow by matter in the convective portion of the expanding envelope, which means that, in the convective envelope, thermal energy provides locally most of the energy required to expand matter. This contrasts with the situation in gravitationally contracting models during the completely convective Hayashi phase when approximately half of the energy released by compression is converted locally into heat and half contributes to the ambient flow of energy.

The base of the convective envelope begins to penetrate inward at $t \sim 8.32 \times 10^7$ yr and contains about half of the mass of the model star at $t \sim 8.329 \times 10^7$ yr (as shown explicitly in Fig. 11.2.59). Figure 11.2.44 shows that, over this time interval, the global gravothermal luminosity increases from a minimum value of $L_{gravothermal} \sim -360\ L_\odot$ to $L_{gravothermal} = 0\ L_\odot$ and then begins to increase with time. Figure 11.2.45 shows that the point at which the gravothermal energies $E_{thermal}$, $|\Omega|$, and E_{bind} stop decreasing and begin to increase with time coincides with the inward penetration of convection. Both changes may be ascribed to the decreasing rate at which the expanding envelope abstracts energy from the ambient flow. The reversal in the direction in which global characteristics change with time is reflected also in changes in the direction and rates of change of characteristics in the hydrogen-burning shell. Figure 11.2.47 shows that, as the base of the convective envelope reaches $M_{CEB} \sim 2.5\ M_\odot$ at $t \sim 8.329 \times 10^7$ yr, the density and pressure at the middle of the hydrogen-burning shell switch from increasing to decreasing with time and the rate of increase of temperature at the middle of the shell increases abruptly.

Of the five models described in Figs. 11.2.48–11.2.52, the models described by Figs. 11.2.49 and 11.2.50 most closely bracket the region surrounding the base of the red giant branch defined by the evolutionary track in Fig. 11.2.42, the one just prior to the formation of a convective envelope (at $\log L_s \sim 2.72$, $\log T_e \sim 3.73$), the other after convection has penetrated in mass almost half way from the surface to the outer edge of the hydrogen-burning shell (at $\log L_s \sim 2.69$, $\log T_e \sim 3.66$). The surface luminosities of the two models are almost identical at $L_s \sim 500\ L_\odot$ ($\log L_s \sim 2.7$). Local gravothermal and nuclear energy-generation rates in the first model, of age $t = 8.318\,24 \times 10^7$ yr, are described in Figs. 11.2.53 and 11.2.54. The model is approaching the red giant branch, but its envelope, being entirely in radiative equilibrium, absorbs over 40% of the energy produced by the contracting core and the hydrogen-burning shell. Figure 11.2.53 shows local gravothermal energy-generation rates throughout the model as functions of the mass coordinate and Fig. 11.2.54 compares the differential contributions to the luminosity of

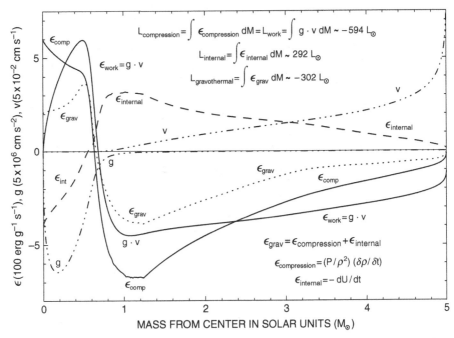

Fig. 11.2.53 Gravothermal energy-generation rates in a 5$M_\odot$ model approaching the red giant branch ($Z = 0.015$, $Y = 0.275$, AGE = 8.31824×10^7 yr)

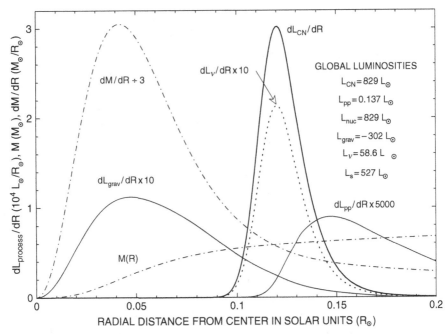

Fig. 11.2.54 Differential contributions to luminosity in a 5$M_\odot$ model approaching the red giant branch ($Z = 0.015$, $Y = 0.275$, AGE = 8.31824×10^7 yr)

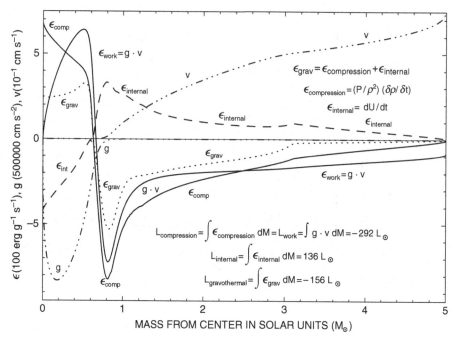

Fig. 11.2.55 Gravothermal energy-generation rates in a 5$M_\odot$ model ascending the red giant branch (Z $= 0.015$, Y $= 0.275$, Age $= 8.32785 \times 10^7$ yr)

gravothermal energy generation and nuclear energy generation by hydrogen-burning reactions between the model center and the outer edge of the hydrogen-burning shell as functions of the radius coordinate.

Profitable comparisons may be made between these figures and the corresponding figures for a model in the thick hydrogen-burning shell phase, Figs. 11.2.37 and 11.2.38. Whereas gravothermal energy generation rates are a relatively minor perturbation prior to the onset of the gravothermal instability, they become major players during the instability, at maximum being approximately 30 times more powerful contributors to the energy budget than in the thick shell-burning phase.

Integration of ϵ_{grav} in Fig. 11.2.53 and integration of dL_{grav}/dR in Fig. 11.2.54 both give $L_{\mathrm{grav}}^{\mathrm{core}} \sim 85\ L_\odot$. Additional integrations in Fig. 11.2.53 give $L_{\mathrm{compression}}^{\mathrm{core}} \sim 145\ L_\odot$ and $L_{\mathrm{internal}}^{\mathrm{core}} \sim -60\ L_\odot$, from which one infers that $L_{\mathrm{grav}}^{\mathrm{core}} = L_{\mathrm{compression}}^{\mathrm{core}} + L_{\mathrm{internal}}^{\mathrm{core}} \sim 85\ L_\odot$, confirming the two direct integrations for the core gravothermal luminosity. Since the global gravothermal luminosity of the model is $L_{\mathrm{gravothermal}} = -302\ L_\odot$, the envelope of the model absorbs energy from the ambient flow at the rate $L_{\mathrm{absorption}}^{\mathrm{envelope}} = -L_{\mathrm{gravothermal}} + L_{\mathrm{gravothermal}}^{\mathrm{core}} \sim (302 + 85)\ L_\odot = 387\ L_\odot$, which is 42.5% of the rate $L_{\mathrm{envelope\ base}} = 914.7\ L_\odot$ at which energy flows into the envelope.

For the model ascending the red giant branch with a deep convective envelope and with structure variables given by Fig. 11.2.50, Figs. 11.2.55 and 11.2.56 provide the same kind of information as provided by Figs. 11.2.53 and 11.2.54 for the model approaching the red giant branch with a completely radiative envelope. Integrations of the relevant

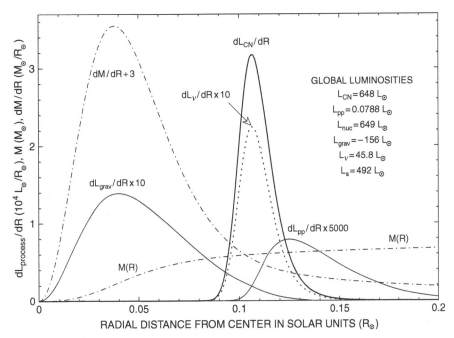

Differential contributions to luminosity in a 5$M_\odot$ model ascending the red giant branch ($Z = 0.015$, $Y = 0.275$, AGE $= 8.32785 \times 10^7$ yr)

quantities give $L_{\text{grav}}^{\text{core}} \sim 90 \ L_\odot$. Since the global gravothermal luminosity of the model is $L_{\text{gravothermal}} = -156 \ L_\odot$, the envelope of the model absorbs energy from the ambient flow at the rate $L_{\text{absorption}}^{\text{envelope}} \sim (156 + 90) \ L_\odot = 246 \ L_\odot$, which is 33.6% of the rate $L_{\text{envelope base}} = 732.4 \ L_\odot$ at which energy flows into the envelope. It is revealing that, whereas both the rate of nuclear energy generation and the rate of absorption by the envelope decrease substantially, the rate at which the core produces gravothermal energy remains fairly constant. This reinforces the interpretation of the transition between thick hydrogen-shell burning and core helium burning as being driven by an instability in the hydrogen-exhausted core, with the remainder of the model star being forced to accomodate itself to the evolution of the core.

The decline in the energy-absorption rate in the envelope can be understood as a consequence of the behavior of the convective portion of the envelope. It is evident from Fig. 11.2.55 that, in the convective envelope which extends from $M \sim 3.1 \ M_\odot$ to the surface, the local rate $-\epsilon_{\text{compression}}$ at which energy is used up in expanding matter is almost the same as the local rate $\epsilon_{\text{internal}}$ at which thermal energy is decreasing, with the consequence that $\epsilon_{\text{gravothermal}} = \epsilon_{\text{internal}} + \epsilon_{\text{compression}}$ is small compared with either component.

Differential contributions to the luminosity in the model are shown in Fig. 11.2.56. The dotted curve labeled dL_ν/dR describes the differential rate at which energy is carried off by electron neutrinos produced in the CN cycle. That it appears everywhere to be precisely proportional to the curve for dL_{CN}/dR demonstrates that CN-cycle reactions are

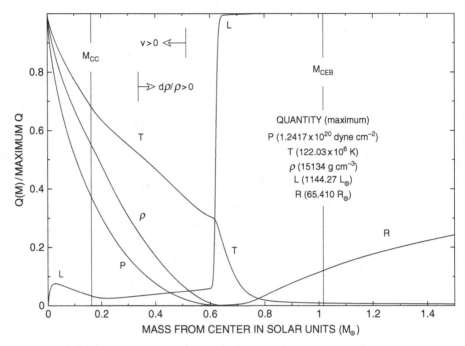

Fig. 11.2.57 Structure of a 5$M_\odot$ model at the start of the core helium-burning phase ($Z = 0.015$, $Y = 0.275$, AGE = 8.3515×10^7 yr)

everywhere in local equilibrium in the energy-generating region. The areas under the two curves are listed in the figure under the heading GLOBAL LUMINOSITIES and the proportionality constant is given by the ratio $L_\nu/L_{CN} = 0.07$, which is the same as the ratio of energy-generation rates $\epsilon_\nu/\epsilon_{CN}$ derived for the equilibrium CN cycle in Section 6.10 (see eq. (6.10.10)).

Figure 11.2.57 shows structure variables in a model in which energy generation by helium burning has effectively replaced gravothermal energy generation in the hydrogen-exhausted core, terminating the phase of rapid core contraction and heating. Differential contributions of various energy sources to the luminosity in the core and in the hydrogen-burning shell are shown in Fig. 11.2.58. The luminosities shown in the latter figure demonstrate that the luminosity at the base of the hydrogen-burning shell, given by $L_{\text{base of shell}} \sim L_{\text{He}} + L_{\text{grav}} \sim (106 - 23)\, L_\odot = 83\, L_\odot$ is essentially the same as the base luminosity supplied by the contracting core during the transition phase. Discussion of the ensuing core helium-burning phase is deferred until Chapter 18 in Volume 2, after helium-burning reactions have been described in Chapter 16.

The story of 5 $M_\odot$ evolution along the red giant branch up to the onset of core helium burning would be incomplete without a description of the quantitative changes in surface abundances brought about by the inward penetration of the convective envelope. Figure 11.2.59 shows the mass at the base of the convective envelope as a function of time and the surface abundances of several isotopes which change as a consequence of

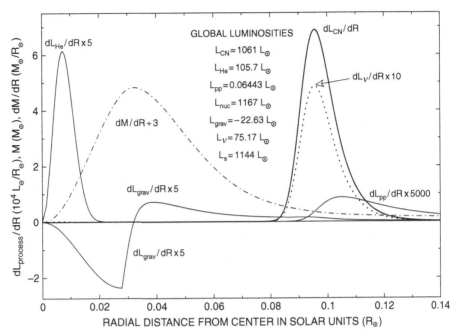

Fig. 11.2.58 Differential contributions to luminosity in a 5$M_\odot$ model at the start of core helium burning ($Z = 0.015$, $Y = 0.275$, AGE $= 8.35153 \times 10^7$ yr)

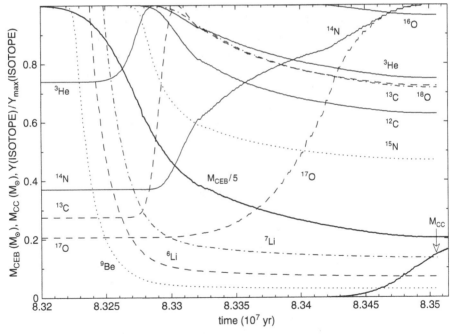

Fig. 11.2.59 Mass at base of convective envelope, convective core mass, and surface abundances in a 5$M_\odot$ model ($Z = 0.015$, $Y = 0.275$) on the giant branch

Table 11.2.2 Number abundances at the surface of a $5\,M_\odot$ model during the first ascent of the red giant branch

Isotope	start	maximum	minimun	end	end/start
^{1}H	0.71048	0.71048	0.70398	0.70398	0.99085
^{3}He	2.5782/−5	3.4873/−5	2.5782/−5	2.6099/−5	1.0123
^{4}He	0.068797	0.070389	0.068797	0.070389	1.0231
^{6}Li	1.1673/−10	1.1673/−10	8.1264/−13	8.1264/−13	0.0069617
^{7}Li	1.4387/−9	1.4387/−9	1.9337/−11	1.9337/−11	0.013441
^{9}Be	1.3855/−11	1.3855/−11	3.9595/−13	3.9595/−13	0.028578
^{10}B	8.7390/−11	8.7930/−11	5.1172/−12	5.1172/−12	0.05856
^{11}B	3.5169/−10	3.5169/−10	1.0564/−10	1.0564/−10	0.30038
^{12}C	2.1315/−4	2.1315/−4	1.3375/−4	1.3375/−4	0.62749
^{13}C	2.3716/−6	8.6235/−6	2.3716/−6	6.2354/−6	2.6292
^{14}N	5.2754/−5	1.4225/−4	5.2754/−5	1.4225/−4	2.6965
^{15}N	1.9396/−7	1.9396/−7	9.1028/−8	9.1028/−8	0.46931
^{16}O	3.9965/−4	3.9965/−4	3.8544/−4	3.8544/−4	0.96444
^{17}O	1.5240/−7	7.3712/−7	1.5240/−7	7.3712/−7	4.8367
^{18}O	8.0498/−7	8.0498/−7	5.7704/−7	5.7704/−7	0.71684
^{19}F	1.6093/−8	1.6093/−8	1.5189/−8	1.5189/−8	0.94383
^{22}Ne	5.3287/−6	5.3287/−6	4.7962/−6	4.7962/−6	0.90007

what is called the first dredge-up episode to distinguish it from a second dredge-up episode which occurs during a second ascent of the giant branch after the exhaustion of helium at the center and during a series of (third) dredge-up episodes which occur during the asymptotic giant branch phase of evolution. Initial, maximum, and final surface abundances of isotopes which have been followed are given in Table 11.2.2.

The time dependences of the abundance changes shown in Fig. 11.2.59 could be anticipated by folding the information concerning M_{CEB} in this figure with the abundance profiles in Figs. 11.2.39–11.2.41 which obtain near the beginning of the thick shell hydrogen-burning phase. In particular, the first isotopes to decrease in abundance at the surface are those which have been effectively destroyed over all but a narrow region near the surface of the thick shell hydrogen-burning model. From Fig. 11.2.41, these include ^{6}Li, ^{7}Li, ^{9}Be, and ^{10}B. By the time the base of the convective envelope has reached its minimum value of $M_{CEB}^{min} \sim 1\,M_\odot$, the abundances of ^{6}Li, ^{7}Li, ^{9}Be, ^{10}B have decreased, respectively, by factors of 144, 74, 35, and 17 relative to their initial values.

Among the isotopes which experience abundance increases at the surface are the intermediate mass isotopes ^{13}C, ^{14}N, and ^{17}O; their abundances increase, respectively, by factors of 2.6, 2.7, and 4.8. The initial increases in the three isotopes occur successively later in time as the base of the convective envelope reaches the regions where they have been produced, respectively, by the reactions $^{12}C(p, \gamma)^{13}N(\beta^+, \nu_e)^{13}C$, $^{13}C(p, \gamma)^{14}N$ (following $^{12}C(p, \gamma)^{13}N(\beta^+, \nu_e)^{13}C$), and $^{16}O(p, \gamma)^{17}F(\beta^+, \nu_e)^{17}O$ (see Figs. 11.2.40 and 11.2.39).

The final surface ratio of ^{12}C to ^{13}C is ^{12}C/^{13}C ~ 21.5, compared with the initial ratio of ^{12}C/^{13}C ~ 90. The final surface ratio of ^{16}O to ^{17}O is ^{16}O/^{17}O ~ 523, compared with the initial ratio of ^{16}O/^{17}O ~ 2620. On the other hand, the surface ratio of ^{16}O to ^{18}O increases from an initial value of ^{16}O/^{18}O ~ 496, to a final value of ^{16}O/^{18}O ~ 668, larger than the final surface ratio of ^{16}O to ^{17}O.

Although the bulk of the increase in ^{14}N is at the expense of ^{12}C, there is a secondary contribution from primary ^{16}O as the base of the convective envelope extends into the region where, during the main sequence phase, substantial amounts of ^{16}O have been converted in the convective core into ^{14}N by the reactions ^{16}O$(p, \gamma)^{17}$F$(\beta^+, \nu_e)^{17}$O$(p, \alpha)^{14}$N. The receding convective core leaves behind a profile of ^{14}N below the mass $M_{CC}^{max} \sim 1.24 M_\odot$ defining the maximum outer extent of the convective core (see, e.g., Fig. 11.2.13) and, as M_{CEB} reaches the outer edge of this profile and extends inward (to a minimum of $M_{CEB}^{min} \sim 1\,M_\odot$), the ^{14}N surface abundance increases at an accelerated rate before tapering off to an asymptotic value. The final surface ratio of ^{14}N to ^{12}C is ^{14}N/^{12}C ~ 1.06, corresponding to a factor of 4.3 increase over the initial ratio.

In a discussion of ^{3}He abundances in the 6×10^7 year old main sequence model described in Fig. 11.2.13, it was anticipated that, although the 5 $M_\odot$ model succeeds in converting deuterium inherited from the interstellar medium into ^{3}He, during the main sequence phase it produces only a very tiny additional amount of ^{3}He that is brought into the convective envelope during the first dredge-up. This anticipation is confirmed by the variation in the surface abundance of ^{3}He in Fig. 11.2.59. Although $Y(^3$He$)$ at first increases as M_{CEB} sweeps into the region $M \sim 4.4 \to 2.4\ M_\odot$ where fresh ^{3}He has been produced by the $p(p, e^+\nu_e)d(p, \gamma)^3$He reactions, it then proceeds into the region in which ^{3}He has been destroyed by reactions with itself and ^{4}He. At the completion of the first dredge-up phase, the final surface abundance of ^{3}He is only 1% larger than the sum of the abundances of initial deuterium and ^{3}He.

Changes in the number abundances of protons and alpha particles, the dominant constituents of the model star, although small, reflect the fact that, despite the dramatic changes in abundances of such elements as lithium and nitrogen, the major overall driver of evolution remains the conversion of hydrogen into helium in the interior. At the surface of the final model, the hydrogen number abundance is smaller by $-\Delta Y(^1$H$) = 0.006\,50$ than at the beginning of the dredge-up phase. The surface number abundances of helium and nitrogen are larger at the end of the dredge-up episode than at the beginning by $\Delta Y(^4$He$) = 0.001\,592$ and $\Delta Y(^{14}$N$) = 0.000\,895$, respectively, and the surface abundance of oxygen is smaller by $-\Delta Y(^{16}$O$) = 0.000\,014$. In first approximation, the associated decrease in the number abundance of hydrogen is $-\Delta Y(^1$H$) \sim 4 \times \Delta Y(^4He) + 2 \times \Delta Y(^{14}N) + 2 \times \Delta Y(^{16}O) = 0.006\,52$, where the term involving oxygen takes into account that only two protons are involved in the production of α particles via the reactions ^{16}O$(p, \gamma)^{17}$F$(p, \alpha)^{14}$N.

In order of magnitude, the changes in abundances in other intermediate mass models experiencing the first dredge-up episode are quite similar to the changes found in the 5 $M_\odot$ model and some of the predicted changes can be tested observationally. Though fascinating to explore, the changes in the abundances of hydrogen and helium are too small to be observed directly. On the other hand, the expected changes in the ratios N/C and ^{12}C/^{13}C and, most especially, the change in the lithium abundance are susceptible to observational check. One of the most felicitous cases in point is the abundance of lithium in the two components of Capella A and B, first explored by George Wallerstein (1964). Both components are of nearly identical mass ($\sim$2.5 $M_\odot$, as determined by A. H. Batten, G. Hill, & W. Lu, 1991) and the abundances of lithium in the two components differ by a factor of the order of 100 (C. A. Pilachowski & J. R. Sowell, 1992), with the larger abundance being similar to abundances found at the surfaces of less evolved population I stars.

Unfortunately, the masses of the two stars are so nearly alike that one cannot claim an observationally based demonstration that the initially more massive star is the more evolved internally. Further, the fact that mass loss from the surface can decrease the layer of unburned lithium prior to the first dredge-up makes the interpretation of the observed lithium abundances in terms of the behavior of constant mass models ambiguous (Iben, 1965). On the other hand, however much one quibbles about the details, it is incontrovertible that the observational facts are extremely strong evidence that the first dredge-up occurs in real stars and that the component of Capella with the smaller lithium abundance has experienced the first dredge-up, whereas the component with the larger lithium abundance has not.

11.3 Evolution of a 25 $M_\odot$ model without mass loss during core hydrogen burning on the main sequence and during shell hydrogen burning near the main sequence up to the onset of core helium burning as a blue giant

Real stars of very large mass lose a substantial fraction of their mass during hydrogen-burning phases by way of a radiative wind. As described in Section 11.4, a crude theoretical estimate of the mass-loss rate can be constructed which indicates that the fraction of mass lost during the main sequence phase increases with initial mass, amounting to about 17% for a star of initial mass 25 $M_\odot$. Recognizing that an estimated mass-loss rate is uncertain by at least a factor of 2, in this section, the evolution of a model of mass 25 $M_\odot$ is followed without mass loss through hydrogen-burning phases up to the ignition of helium in the core. One way of viewing the results of the calculation is that they describe approximately the evolution of a model of average mass 25 $M_\odot$.

The initial model is the final model described in Section 9.5. Evolution in the HR diagram is shown in Fig. 11.3.1. The radius of the model at any point along the track can be gauged by comparing with the dotted lines of constant radius, adjacent ones of which are separated by a factor of $\sqrt{2}$ in radius. Large open circles along the evolutionary track are placed at intervals of a million years, small open circles occur every 10^5 yr, and small solid dots are placed at intervals of 10^4 yr. At several points along the track, the age of the model is noted explicitly in millions of years.

Occurrences in the interior are also noted at several points along the track. Core hydrogen burning lasts for approximately 6.35×10^6 yr and evolution from the point where

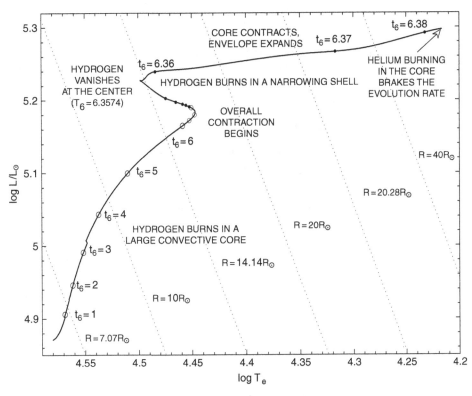

Fig. 11.3.1 Evolutionary path of a 25M$_\odot$ model through core hydrogen burning to the onset of core helium burning ($Z = 0.015, Y = 0.275$)

hydrogen (effectively) vanishes from the center to the point where core helium burning begins to control the rate of evolution lasts for approximately 2×10^4 yr or only about 0.3% of the core hydrogen-burning lifetime. This contrasts with the 5 $M_\odot$ model described in Section 11.2 for which the phase of rapid core contraction and envelope expansion leading to the ignition of helium does not begin until some time after the effective disappearance of hydrogen at the center and lasts for $\sim 10^7$ yr, or about 1.25% of the core hydrogen-burning lifetime. The reason for the difference is the fact that the initial mass of the hydrogen-exhausted core relative to the model mass is very close to the Schönberg–Chandrasekhar limit in the less massive model but is considerably larger than this limit in the more massive model.

The time dependences of three global characteristics of the 25 $M_\odot$ model are shown in Fig. 11.3.2 and the time dependences of three structure variables at the model center are shown in Fig. 11.3.3. The time dependences of the hydrogen number abundance and of the number abundances of the main beta-stable CNO isotopes at the model center are shown in Fig. 11.3.4. Comparing number abundances in Fig. 11.3.4 with those in Fig. 11.2.3, it is evident that the conversion of ^{16}O into ^{14}N in the convective core proceeds approximately twenty times more rapidly in the 25 $M_\odot$ model than it does in the 5 $M_\odot$ model, with the number abundance of ^{16}O at the center of the 25 $M_\odot$ model being halved in the first $\sim 10\%$

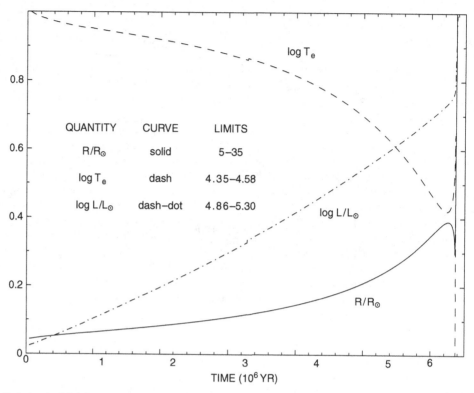

QUANTITY	CURVE	LIMITS
$R/R_\odot$	solid	5–35
$\log T_e$	dash	4.35–4.58
$\log L/L_\odot$	dash–dot	4.86–5.30

Fig. 11.3.2 Variations in global characteristics of a 25$M_\odot$ model evolving through core hydrogen burning to the onset of core helium burning ($Z = 0.015, Y = 0.275$)

of the main sequence lifetime of $\sim6.3 \times 10^6$ yr, compared with a halving of the ^{16}O number abundance at the center of the 5 $M_\odot$ model in the first $\sim17\%$ of the $\sim8.1 \times 10^7$ yr main sequence lifetime.

Comparing the hydrogen number-abundance profile in Fig. 11.3.4 with the radius profile in Fig. 11.3.2, it is evident that, near the end of the main phase of core hydrogen burning, the radius of the 25 $M_\odot$ model shrinks after the hydrogen number abundance at the center drops below a value of the order of $Y(^1\mathrm{H}) \sim 0.02$. Comparison between the hydrogen number-abundance profile in Fig. 11.2.3 and the model radius profile in Fig. 11.2.2 shows that, also in the 5 $M_\odot$ case, radial shrinkage begins when the number abundance of hydrogen at the center is reduced to $Y(^1\mathrm{H}) \sim 0.02$. In both cases, all parts of the model contract and the phases of radial shrinkage are therefore dubbed phases of overall contraction, as explicitly noted in Figs. 11.3.1 and 11.2.1.

Conditions in the 25 $M_\odot$ model approximately halfway through the phase of core hydrogen burning are described in Figs. 11.3.5–11.3.7. An important feature to notice is that, over a distance in mass of about 2.5 $M_\odot$ extending from the location of the outer edge of the convective core (indicated by the vertical line labeled M_{CC} in Fig. 11.3.5) to slightly beyond the initial location of this edge in the ZAMS precursor (indicated by the vertical line labeled M_{CC}(start)), the hydrogen number-abundance profile Y_{H} is such that the

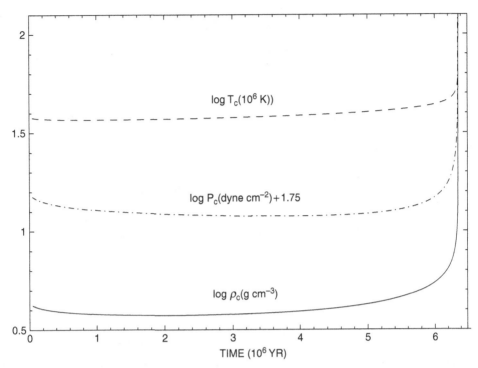

Fig. 11.3.3 Evolution of central state variables in a 25$M_\odot$ model through core hydrogen burning to the onset of core helium burning ($Z = 0.015$, $Y = 0.275$)

adiabatic and radiative temperature gradients, respectively V_{ad} and V_{rad} in Fig. 11.3.6, are nearly identical. This is the classical condition characterizing a semiconvective zone.

No attempt has been made to adjust the hydrogen abundance distribution to achieve the near identity of the adiabatic and radiative gradients. Instead, it has been assumed that mixing of abundances in adjacent zones for which $V_{rad} > V_{ad}$ is instantaneous, consistent with the fact that mixing time scales are small compared with adopted evolutionary time steps; the near equality of the radiative and adiabatic gradients in the region of variable hydrogen abundance is a consequence of implementing this procedure. Thus, although the details of the resultant stairstep-like hydrogen number-abundance profile vary as evolution progresses and although the exact form of the profile at any time is sensitive to the temporal and spatial zoning, the fact that $V_{rad} \sim V_{ad}$ in the region of variable hydrogen abundance is a consequence of the adopted mixing algorithm and not a consequence of adjusting the Y_H profile to achieve $V_{rad} \sim V_{ad}$. In other words, the formation of a semiconvective region in which radiative and adiabatic gradients are nearly identical is a natural consequence of implementing the Schwarzschild criterion for convective instability rather than a consequence of assuming the equality of the two temperature–pressure gradients.

Due to the substantial ratio of radiation pressure to gas pressure, the adiabatic gradient, at $V_{ad} \sim 0.3$, is significantly smaller than the value $V_{ad} = 0.4$ which would obtain if radiation pressure were not important, with the consequence that the mass of the convective core is roughly 50% larger than it would have been in the absence of radiation pressure.

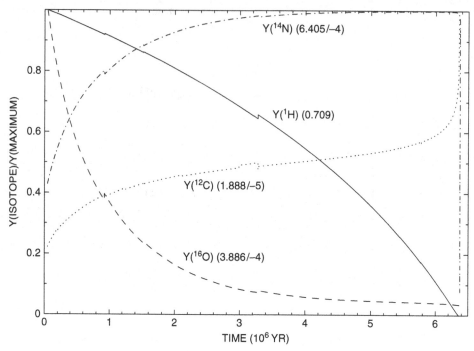

Fig. 11.3.4 Hydrogen and CNO abundances at the center of a $25M_\odot$ model through core hydrogen burning to the onset of core helium burning ($Z = 0.015$, $Y = 0.275$)

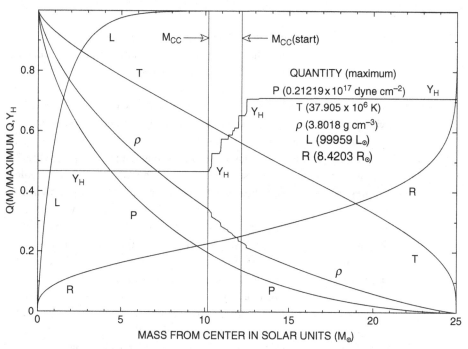

Fig. 11.3.5 Structure of a $25M_\odot$ model midway in the core hydrogen-burning phase ($Z = 0.015$, $Y = 0.275$, and AGE $= 3.18032 \times 10^6$ yr)

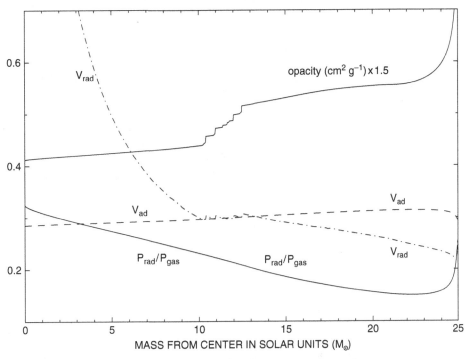

Interesting variables in a 25M$_\odot$ model midway in the core hydrogen-burning phase ($Z = 0.015$, $Y = 0.275$, and AGE $= 3.18032 \times 10^6$ yr)

Comparison of the opacity profile in Fig. 11.3.6 with the hydrogen number-abundance pro-file in Fig. 11.3.5, coupled with the fact that, when $Y_H = 0.6 \pm 0.1$, the electron-scattering opacity is $\kappa = 0.32 \pm 0.02$ cm^2 g^{-1}, demonstrates that electron scattering is a major con-tributor to the opacity and that the consequent nearly linear dependence of the radiative gradient on the hydrogen number abundance plays a major role in the establishment of a hydrogen profile such that $V_{rad} \sim V_{ad}$.

Differential contributions to the luminosity in the model are shown in Fig. 11.3.7. Just as in the 5 $M_\odot$ models described in Figs. 11.2.54, 11.2.56 and 11.2.58, the ratio of dL_ν/d$R = \epsilon_\nu$ to dL_{CN}/d$R = \epsilon_{CN}$ is precisely constant at 0.07, accounting for the fact that the global ratio is given by $L_\nu/L_{CN} = 0.07$. The neutrino energy-loss rate is due to beta decays in CN-cycle reactions and the fact that the ratio of the neutrino-loss rate and the nuclear energy-generation rate is everywhere the same locally demonstrates that CN-cycle reac-tions are everywhere in local equilibrium.

One of three additional lessons from Fig. 11.3.7 is that temperatures in the 25 $M_\odot$ main sequence model are so large that the contribution of CN-cycle reactions to the luminosity is almost four thousand times larger than the contribution of pp-chain reactions. A second lesson is that, although the rate of gravothermal energy generation in the convective core is very small relative to the rate of nuclear energy generation, the fact that the rate is positive demonstrates that the reduction in particle number abundances in the core is responsible for a slow contraction of the core. A third lesson is that the rate of gravothermal energy generation outside the convective core is negative, implying that energy is being abstracted

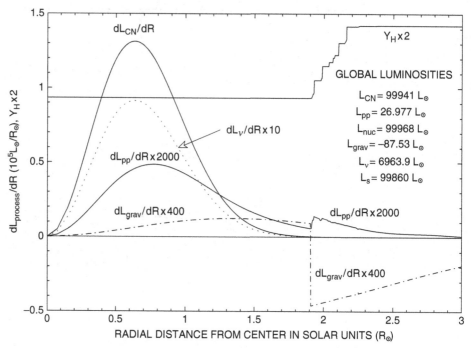

Fig. 11.3.7 Differential contributions to the luminosity in a 25$M_\odot$ model midway in the core hydrogen-burning phase ($Z = 0.015$, $Y = 0.275$, and AGE = 3.18032×10^6 yr)

from the ambient energy flow to do work against gravity, and this accounts for the fact that the radius of the model increases with time.

Conditions in the 25 $M_\odot$ model as it approaches the overall contraction phase are shown in Figs. 11.3.8–11.3.10. Comparing the hydrogen number-abundance profile Y_H in Fig. 11.3.8 with the gradients V_{ad} and V_{rad} in Fig. 11.3.9, one sees that semiconvective activity extends outward continuously from the outer edge of the convective core in the current model (M_{CC}) to approximately 2.5 $M_\odot$ beyond the outer edge of the convective core in the ZAMS model (M_{CC}(start)). Comparing with the same quantities in the less evolved model described in Figs. 11.3.5 and 11.3.6, it is evident that, during evolution between the models, semiconvective activity has modified the hydrogen profile existing in the region in the less evolved model between M_{CC} and M_{CC}(start). This is in marked contrast with the situation in the 5 $M_\odot$ model in which that portion of the hydrogen-abundance profile found outside of the convective core at any given time does not change as the mass of the convective core continues to decrease.

As shown in Fig. 11.3.9, the opacity increases outward through the convective core, which extends from the center to ~7 $M_\odot$, demonstrating that the density- and temperature-dependent free–free contribution to the opacity in the core is not negligible. The same holds true for the opacity in the basically convective region extending from ~11.3 $M_\odot$ to ~14 $M_\odot$.

Integrating under the curves for dL_{grav}/dR in Figs. 11.3.10 and 11.3.7, it is evident that the rate at which gravothermal energy is released in the convective core is over ten times larger in the more evolved model than in the less evolved model, increasing from

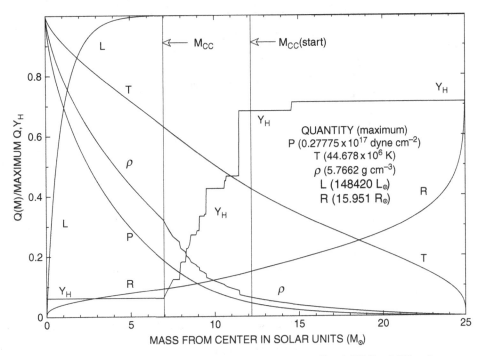

Fig. 11.3.8   Structure of a 25$M_{\odot}$ model near the end of the core hydrogen-burning phase ($Z = 0.015$, $Y = 0.275$, and AGE $= 6.07982 \times 10^6$ yr)

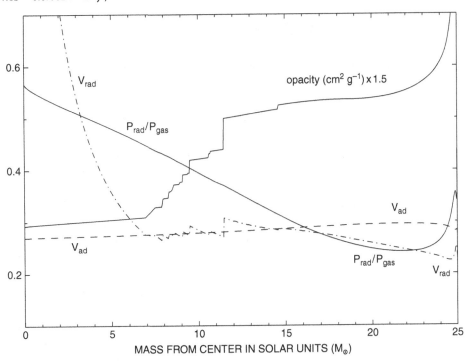

Fig. 11.3.9   Interesting variables in a 25$M_{\odot}$ model near the end of the core hydrogen-burning phase ($Z = 0.015$, $Y = 0.275$, and AGE $= 6.07982 \times 10^6$ yr)

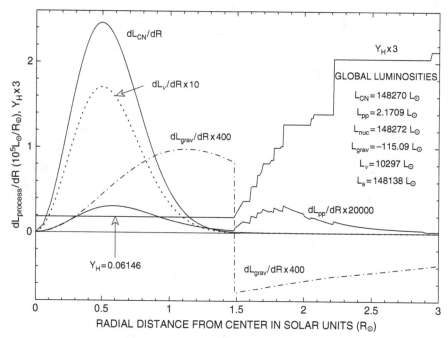

Fig. 11.3.10 Differential contributions to the luminosity in a 25$M_\odot$ model near the end of the core hydrogen-burning phase ($Z = 0.015$, $Y = 0.275$, and AGE = 6.07982×10^6 yr)

~40 $L_\odot$ to ~475 $L_\odot$. The increase is related to the factor of ~7 decrease in the hydrogen abundance in the core from $Y_H \sim 0.41$ to $Y_H \sim 0.061$. In order for pressure balance to be maintained, the decrease in particles per unit mass must be compensated for by an increase in density, leading to a release of compressional energy. Simultaneously, the rate of absorption of gravothermal energy outside the core has increased from ~130 $L_\odot$ to ~590 $L_\odot$, an increase by over a factor of 4 which is related to the increase in luminosity emanating from the core.

Conditions in the 25 $M_\odot$ model near the very end of the overall contraction phase are shown in Figs. 11.3.11–11.3.13. From the hydrogen number-abundance profile in Fig. 11.3.11, one might conclude that, for all intents and purposes, hydrogen has been exhausted over a region extending from the center to a position slightly more than 2 $M_\odot$ beyond the outer edge of the convective core, located at $M_{CC} \sim 4\ M_\odot$, or over a mass which is roughly 25% of the mass of the model. The ratio of the mass of the hydrogen-depleted region to the model mass is roughly twice the relevant Schönberg–Chandrasekhar limit, and one might expect that the hydrogen-depleted region is contracting very rapidly and that the source of energy producing the luminosity profile in Fig. 11.3.11 out to $M \sim 6M_\odot$ is primarily gravothermal. These expectations are confirmed by the profiles for dL_{grav}/dR, dL_{CN}/dR, and $L(R)$ in Fig. 11.3.13. At the outer edge of the convective core, the luminosity is ~$10^5\ L_\odot$, of which only ~$0.175 \times 10^5\ L_\odot$ is contributed by CN-cycle burning. Near the base of the hydrogen-burning shell, the luminosity is $L(R) \sim 1.2 \times 10^5$ $L_\odot$, so the total rate at which gravothermal energy is being produced by the contracting

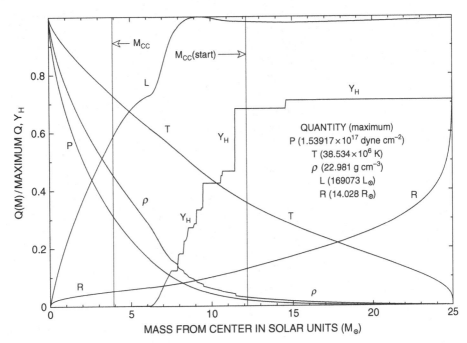

Fig. 11.3.11   Structure of a 25$M_\odot$ model near the very end of the overall contraction phase ($Z = 0.015$, $Y = 0.275$, and AGE $= 6.356674 \times 10^6$ yr)

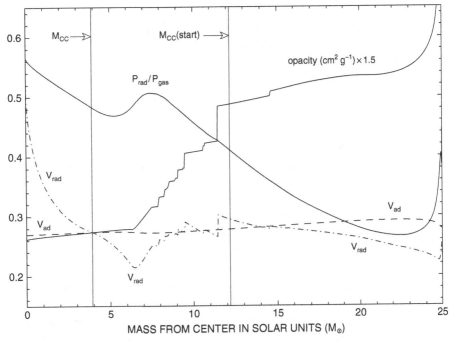

Fig. 11.3.12   Interesting variables in a 25$M_\odot$ model near the very end of the overall contraction phase ($Z = 0.015$, $Y = 0.275$, and AGE $= 6.356674 \times 10^6$ yr)

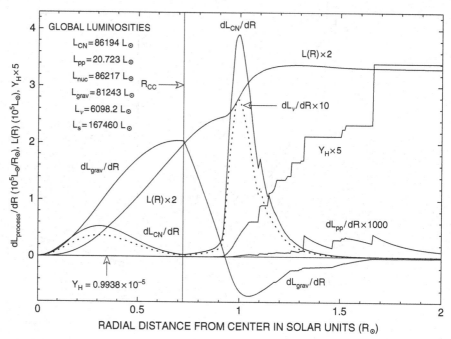

Fig. 11.3.13 Contributions to the luminosity in a 25$M_\odot$ model near the very end of the overall contraction phase ($Z = 0.015$, $Y = 0.275$, and age $= 6.356674 \times 10^6$ yr)

core below the base of the hydrogen-burning shell is $L^+_{\mathrm{grav}} \sim 10^5 \, L_\odot$. Put another way, over 80% of the luminosity at the base of the CN-cycle burning shell is due to the release of gravothermal energy.

It is remarkable that, even though the hydrogen number abundance in the convective core has been reduced to less than $Y_H = 10^{-5}$, the core contributes $\sim$20% of the total nuclear burning rate, which is $L_{\mathrm{nuc, total}} \sim 8.62 \times 10^4 \, L_\odot$. The other 80% is contributed by CN-cycle burning in a shell extending from $R \sim 0.9 \, R_\odot$ to $R \sim 1.3 \, R_\odot$ (corresponding to a mass extent from $M \sim 6 \, M_\odot$ to $\sim$9.4 $M_\odot$).

In the construction of models leading up to the model described in Figs. 11.3.11–11.3.13, the choice of evolutionary time steps has been controlled by the criterion that the rate at which the hydrogen number abundance may change locally in a time step at any point in the absence of convection is less than 10%. This means, of course, a change per time step in the average abundance of hydrogen in the fully convective core much smaller than 10%. The time step for the model described in Figs. 11.3.11–11.3.13 is slightly smaller than three years.

If the 10% time-step limiting criterion were maintained, and the convective core were to remain large even as the contribution of hydrogen burning to the maintainance of convection becomes smaller and smaller, ever smaller time steps would ensue. Clearly, some other criterion must be adopted for choosing time steps. Several experiments have been performed to explore subsequent evolutionary behavior as a function of time-step size when this size is held fixed at some predetermined value.

If the evolutionary time step is fixed at 5 years or less, the immediately ensuing evolution is quite bizarre and can best be described by saying that mixing zones of variable size and duration spring up over most of the model interior, from the center to very close to the surface. As evolution progresses, the convective core incorporates zones with successively larger hydrogen abundances and the model embarks on another episode of core hydrogen burning much like the first episode. Results of experiments with fixed time steps of 5 years and 1 year show that the mass and hydrogen abundance in the resurrected core increase with a decrease in the size of the time step. Following a second overall contraction phase a third extended phase of core burning ensues, followed by a final phase of overall contraction and evolution to the red giant branch where, finally, helium is ignited in the core.

These results are spurious. They are mentioned only as a warning that a tremendous amount of time can be spent pursuing a beguiling scenario before examining first why the recurrence of core hydrogen burning is tied to a choice of time step size below a critical threshold. Suspicious behavior appears almost immediately, with the luminosity profile in the model exhibiting wild gyrations, peaking at values one to two orders of magnitude larger than the surface luminosity and even becoming negative. Astonishingly, model convergence is achieved in every time step (equations in each grid point are satisfied to better than one part in 10^4), but the fact that as many as 15 iterations are required to achieve convergence to this accuracy raises the suspicion that the models are non-physical.

The luminosity peaks, which are correlated with discontinuities in the hydrogen-abundance profile, are gravothermal responses to sudden changes in the hydrogen number-abundance distribution. The large gravothermal luminosities are, in turn, responsible for the occurrence of extended convective regions. Each time the abundance profile is changed, the changes in density and pressure required to ensure mass conservation and pressure balance are to a large extent independent of the time step chosen. However, the gravothermal response to these changes is inversely proportional to the time step size. This helps explain why gravothermal energy production and absorption rates in the models made with 1 year time steps are as much as two orders of magnitude larger than those obtained when the time step is chosen as 100 years and why mixing all the way to the center becomes more likely as time step size is decreased. For small enough time steps, large fluxes due to artificially enhanced gravothermal energy-generation rates spawn artificial convectives zones and the associated artificially large and abrupt changes in the distribution of composition variables act as jolts which elicit artificially large gravothermal responses.

When the time-step size is forced to be an order of magnitude larger than five years, hydrogen is effectively destroyed over central regions in a very few time steps. The core contracts at an accelerated rate, matter in the hydrogen-burning shell heats as it is drawn inward, and the increased rate at which energy is injected into the envelope causes matter in the envelope to expand on a time scale controlled by the rate of core contraction. For a time-step size of 100 yr, the path in the HR diagram during this phase of core contraction and envelope expansion is the one shown in Fig. (11.3.1), beginning where hydrogen effectively vanishes at the center (between the solid dot at $t_6 = 6.35$ and the solid dot at $t_6 = 6.36$), and extending to the point where the rate of helium burning in central regions begins to brake the rate of core contraction (near the solid dot at $t_6 = 6.38$). Subsequent

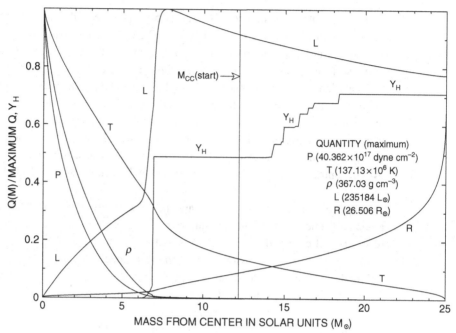

QUANTITY (maximum)
P (40.362×10^{17} dyne cm^{-2})
T (137.13×10^6 K)
ρ (367.03 g cm^{-3})
L (235184 $L_\odot$)
R (26.506 $R_\odot$)

MASS FROM CENTER IN SOLAR UNITS ($M_\odot$)

Fig. 11.3.14 Structure of a 25$M_\odot$ model with a contracting core and expanding envelope after exhausting hydrogen at the center ($Z = 0.015$, $Y = 0.275$, and AGE $= 6.367763 \times 10^6$ yr)

evolution during helium-burning phases is discussed in Chapter 20 of Volume 2, after a description of helium-burning reaction rates and energy-generation rates (in Chapter 16 of Volume 2).

Characteristics of the model approximately two thirds of the way between the solid dots at $t_6 = 6.36$ and $t_6 = 6.37$ along the evolutionary track in Fig. 11.3.1, at $\log T_e \sim 4.355$ and $\log L \sim 5.261$ ($R \sim 26.5$), are shown in Figs. 11.3.14–11.3.16. From the curves V_{rad} and V_{ad} in Fig. 11.3.15 and the luminosity and hydrogen-abundance profiles in Fig. 11.3.14, it is clear that the flux of energy due to the combination of both gravothermal and nuclear sources maintains a large convective shell extending from $\sim$7 $M_\odot$ to $\sim$13.5 $M_\odot$, and that energy flow in the hydrogen-depleted region between the center and the inner edge of the convective shell is carried by radiation.

In the region between $\sim$14 $M_\odot$ and $\sim$18 $M_\odot$, V_{rad} and V_{ad} are nearly equal and the hydrogen abundance varies relatively smoothly, demonstrating that the hydrogen profile is the consequence of semiconvective activity. Comparing the hydrogen-abundance profiles in Figs. 11.3.14 and 11.3.11, it is evident that, during evolution between the two models, approximately 0.8 $M_\odot$ of hydrogen from the region between 11.5 $M_\odot$ and 18 $M_\odot$ in the less evolved model has been transferred by semiconvection into the convective shell of the more evolved model.

In the luminosity profile of Fig. 11.3.14 one can distinguish four segments, the origins of which are elucidated by the differential luminosity distributions shown in Fig. 11.3.16. The first segment, extending from the center to $M \sim 6.2$ $M_\odot$, is a consequence of gravothermal

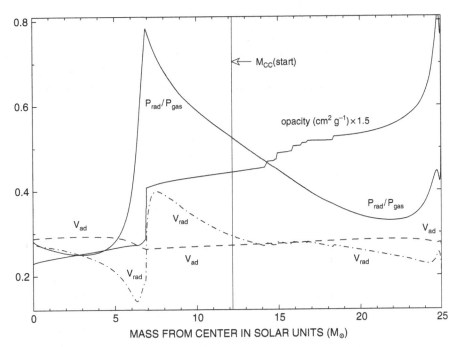

Fig. 11.3.15 Variables in a 25$M_\odot$ model with a contracting core and expanding envelope after exhausting hydrogen at the center ($Z = 0.015$, $Y = 0.275$, and AGE $= 6.367763 \times 10^6$ yr)

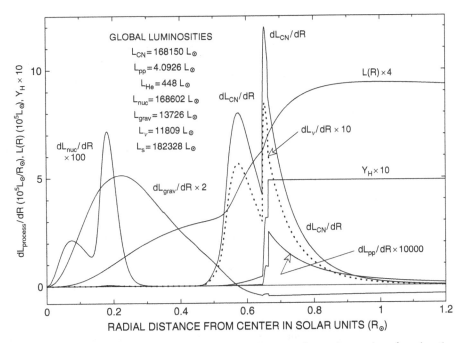

Fig. 11.3.16 Contributions to the luminosity in a 25$M_\odot$ model with a contracting core and expanding envelope after exhausting hydrogen at the center ($Z = 0.015$, $Y = 0.275$, and AGE $= 6.367763 \times 10^6$ yr)

energy delivered by the rapidly contracting, highly hydrogen-depleted core. CN-cycle reactions are responsible for the second segment which extends through a radiative region from ~6.2 $M_\odot$ to the base of a fully convective shell at ~6.9 $M_\odot$. The convective shell extends out to ~14 $M_\odot$. The third segment is contributed by CN-cycle reactions in a region near the base of the convective shell between $M \sim 6.9$ $M_\odot$ and ~7.5 $M_\odot$. The final segment, which extends from ~7.5 $M_\odot$ to the surface, is one in which energy from the ambient flow and from the thermal content of the envelope is being used to expand matter against gravity.

Nuclear burning near the model center makes a minor contribution to the energy balance there, as demonstrated by the distribution labeled dL_{nuc}/dR in Fig. 11.3.16, but the double peak in the distribution deserves comment. The peak at smaller radius is due to helium-burning reactions, which is not surprising given the large temperatures and densities near the center ($T_c \sim 137 \times 10^6$ K and $\rho_c \sim 367$ g cm^{-3}). The peak at larger radius is due to assorted p-capture and beta-decay reactions which persist despite the extremely small abundance of hydrogen in the contributing regions ($Y_H \sim 7 \times 10^{-11}$ near the second peak in the distribution). Since, strictly speaking, the abundance of hydrogen everywhere in the model is finite (at the center, $Y_H \sim 10^{-19}$), the statement that hydrogen vanishes at the center is actually an exaggeration and has prompted the use of the adjective hydrogen-depleted in place of hydrogen-exhausted to describe the core.

The hydrogen-depleted core continues to contract and heat until helium burning takes over from the gravothermal source as the dominant source of energy in the core while the model is still near the observational main sequence in the HR diagram. Conditions in the interior of the model at this point are as shown in Figs. 11.3.17–11.3.19 and they may be profitably compared with conditions in the interior of the model described in Figs. 11.3.14–11.3.16.

The hydrogen-abundance distributions in the two models are almost identical except near the base of the hydrogen-burning shell where the abundance profile has steepened in the course of evolution. The largest difference in characteristics occurs in the hydrogen-depleted core. In the more evolved model, the only gravothermal energy generation of any consequence takes place over a broad region in the outer portion of the core below the hydrogen-burning shell where matter is being compressed by the flow of fuel-exhausted material from the hydrogen-burning shell, as shown by the curve dL_{grav}/dR in Fig. 11.3.19.

The large differences in the luminosity profiles shown in Figs. 11.3.14 and 11.3.17 demonstrate that the development of a quiescent helium-burning energy source in central regions dramatically reduces the strength of gravothermal energy-generation rates in regions on both sides of the hydrogen-burning shell. In the hydrogen-depleted core of the less evolved model, the luminosity increases steadily outward throughout the entire core, demonstrating that all parts of the core are rapidly contracting and releasing compressional energy; in the more evolved model, the profile in the core has a very steep slope over a relatively small mass near the center, consistent with a highly temperature- and density-dependent nuclear energy source, whereas, over the remainder of the core, the luminosity is nearly constant, consistent with a distributed gravothermal source which at any point is characterized by an energy-generation rate that is approximately one and a half orders of magnitude smaller than the nuclear energy-generation rate in central regions. In the region above the hydrogen-burning shell in the less evolved model, the pronounced decrease

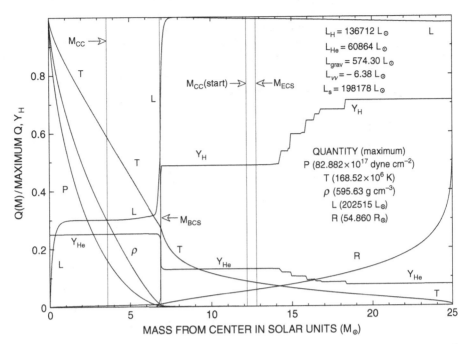

Fig. 11.3.17 Structure of a 25$M_\odot$ shell hydrogen-burning model stabilized by core helium burning ($Z = 0.015$, $Y = 0.275$, and AGE $= 6.377601 \times 10^6$ yr)

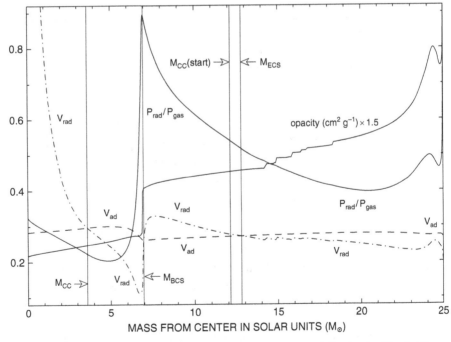

Fig. 11.3.18 Interesting variables in a 25$M_\odot$ shell hydrogen-burning model stabilized by core helium burning ($Z = 0.015$, $Y = 0.275$, and AGE $= 6.377601 \times 10^6$ yr)

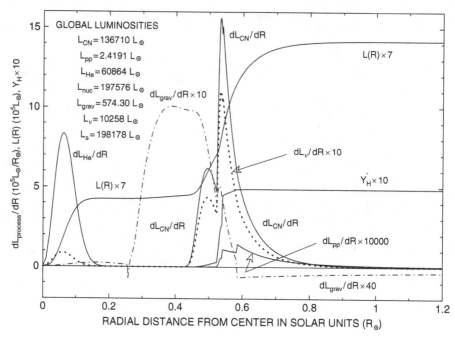

Fig. 11.3.19 Contributions to the luminosity in a 25$M_\odot$ shell hydrogen-burning model stabilized by core helium burning ($Z = 0.015$, $Y = 0.275$, and AGE $= 6.377601 \times 10^6$ yr)

outward in the luminosity profile is due to absorption from the ambient flow of energy necessary to sustain the rapid expansion of the envelope in response to the rapid contraction of the core. In the more evolved model, the ascendancy of quiescent helium burning in central regions has dramatically reduced the rate of core contraction, resulting in a correspondingly dramatic decrease in the rate of expansion of the envelope.

11.4 Global properties of main sequence models as functions of model mass and estimates of surface mass loss during pure hydrogen-burning phases

11.4.1 Global main sequence properties

Several properties of the three main sequence models described in previous sections are compared in Table 11.4.1. In the third column is the logarithm of the ratio of the maximum mass of the convective core, M_{CC}^{max}, to the mass of the model. The last column gives the logarithm of the main sequence lifetime, t_{MS}. In the fourth and fifth columns of the table, luminosity L, mass M, and radius R are in solar units. The values for luminosity and radius are approximations to the values prevailing midway in the main sequence phase at time $t \sim t_{MS}/2$.

Table 11.4.1 Average global properties of population I main sequence models					
$M/M_\odot$	$\log M/M_\odot$	$\log M_{CC}^{max}/M$	$\log L/M$	$\log R/M$	$\log t_{MS}(yr)$
1	0.0	—	0.05	0.045	10.04
5	0.699	−0.611	2.121	−0.192	7.90
25	1.398	−0.312	3.612	−0.469	6.80

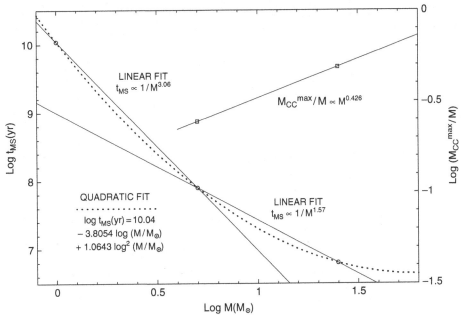

Fig. 11.4.1 Main sequence lifetime and maximum mass of the convective core versus model mass for models of mass 1, 5, and 25 $M_\odot$ ($Z = 0.015$, $Y = 0.275$)

In Fig. 11.4.1, circles identify the logarithms of t_{MS} for all three models as functions of the logarithm of M. The dotted line of negative slope is a quadratic fit to the circles. The two squares in the figure identify the logarithms of M_{CC}^{max}/M for the 5 $M_\odot$ and 25 $M_\odot$ models and the solid straight line of positive slope passes through the squares. In Fig. 11.4.2, circles and squares identify the logarithms of L/M and R/M, respectively, as functions of $\log M$ for all three models, with the dotted and dashed curves being quadratic fits to the data points.

More informative in some respects than the quadratic fits in Figs. 11.4.1 and 11.4.2 are the linear fits to pairs of points shown as solid straight lines in the same figures. In Fig. 11.4.1, the solid line labeled

$$t_{MS} \propto \frac{1}{M^{3.06}} \qquad (11.4.1)$$

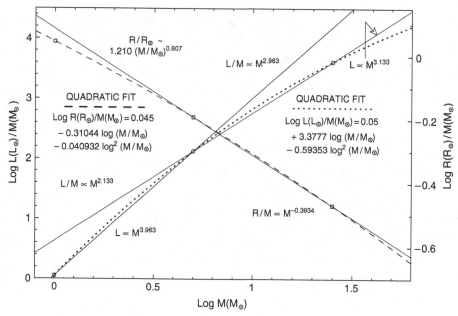

Fig. 11.4.2 Luminosity/mass versus mass and radius/mass versus mass for models of mass 1, 5, and 25 $M_\odot$ ($Z = 0.015$, $Y = 0.275$)

is a straight line fit to the 1 $M_\odot$ and 5 $M_\odot$ model results and the solid line labeled

$$L \propto M^{3.963} \tag{11.4.2}$$

in Fig. 11.4.2 is a fit to the results for the same two models. What these lines demonstrate is that, for small enough model masses,

$$t_{MS} \overset{\sim}{\propto} \frac{M}{L}, \tag{11.4.3}$$

showing that the mass of fuel consumed during the main sequence phase is proportional to the mass of the model. The fact that $L \overset{\sim}{\propto} M^4$ demonstrates that model results are consistent with the mass–luminosity relationship defined by real stars of mass in the range given by $-0.2 < \log M/M_\odot < 1.1$, as shown in Fig. 2.3.1.

The solid line in Fig. 11.4.2 labeled

$$L \propto M^{3.133} \tag{11.4.4}$$

is a straight line fit to the 5 $M_\odot$ and 25 $M_\odot$ model results. Since one has from Fig. 11.4.1 that, for the same two models,

$$M_{CC}^{max} \overset{\sim}{\propto} M^{1.426}, \tag{11.4.5}$$

it follows that

$$\frac{M_{CC}^{max}}{L} \overset{\sim}{\propto} \frac{1}{M^{1.707}}. \tag{11.4.6}$$

The solid line through the t_{MS} points for the 5 $M_\odot$ and 25 $M_\odot$ models in Fig. 11.4.1 shows that

$$t_{MS} \stackrel{\sim}{\propto} \frac{1}{M^{1.57}}. \tag{11.4.7}$$

The coefficients of M in eqs. (11.4.6) and (11.4.7) differ by only $\sim 9\%$, so one can also write

$$t_{MS} \stackrel{\sim}{\propto} \frac{M_{CC}^{max}}{L}, \tag{11.4.8}$$

which says that the amount of fuel consumed during the main sequence phase by models with large convective cores is approximately proportional to the maximum mass of the convective core.

The straight line fits to the data for the 5 $M_\odot$ and 25 $M_\odot$ models in Figs. 11.4.1 and 11.4.2 can be written as

$$\frac{L/M}{L_\odot/M_\odot} = 4.27 \left(\frac{M}{M_\odot} \right)^{2.13} \tag{11.4.9}$$

and

$$\left(\frac{R/M}{R_\odot/M_\odot} \right)^{1/2} = 1.10 \left(\frac{M}{M_\odot} \right)^{-0.20}. \tag{11.4.10}$$

Eliminating mass from eqs. (11.4.9) and (11.4.10) gives

$$\frac{R}{R_\odot} = 0.92 \left(\frac{L}{L_\odot} \right)^{0.19}. \tag{11.4.11}$$

The dashed line in the HR diagram of Fig. 2.2.1 which passes through the locations of the stars α Cen A and Regulus describes reasonably well the mean main sequence defined by population I stars in our Galaxy. This line may be approximated by

$$\log T_e = 3.73 + 0.14 \, (\log L/L_\odot). \tag{11.4.12}$$

Since

$$\frac{L/L_\odot}{(R/R_\odot)^2} = \left(\frac{T_e}{5760} \right)^4, \tag{11.4.13}$$

it follows that

$$\frac{R}{R_\odot} = 1.15 \left(\frac{L}{L_\odot} \right)^{0.22}. \tag{11.4.14}$$

Only a slight adjustment in the zero point and the coefficent of $\log L$ in eq. (11.4.14) is required to reproduce the relationship given by eq. (11.4.11) which is, in any case, a very rough fit to constant mass models of intermediate to high mass. Thus, one may argue that the models are in reasonable agreement with the observations in regard to the mean radius–luminosity relationship along the main sequence.

11.4.2 Mass loss during hydrogen-burning phases

Mass loss from the surface has not been incorporated in the construction of models described in previous sections, but real stars lose mass. It is therefore of interest to examine to what extent the neglect of mass loss constitutes a serious error and in what ways mass loss affects quantities such as the main sequence lifetime and changes in surface abundances along the red giant branch as determined on the assumption of no mass loss. Qualitatively, it is clear that, for any given initial mass, the main sequence lifetime will be increased and, for elements that are destroyed over all but the outer few percent of a model's mass, mass loss from the surface will decrease the masses of the surface layers in which the elements survive and therefore increase the extent to which the surface abundances of these elements are depleted along the giant branch.

The Sun loses mass at an average rate of the order of 10^{-14} $M_\odot$ yr^{-1} by an evaporative wind. The wind, which acquires speeds of $\sim$400 km s^{-1} at large distances from the Sun, is thought to be driven by the energy released by reconnection of magnetic fields in a turbulently convective envelope. The current mass-loss rate is so small that one might reasonably assume that the Sun and other single low mass main sequence stars with convective envelopes may be adequately approximated by constant mass models.

On the other hand, real stars rotate and possess global magnetic fields, two additional phenomena not taken into account in the models. The fact that the rotating magnetic field of a real star exerts a torque on wind particles implies a loss of angular momentum by the wind-emitting star which accounts qualitatively for the fact that the average rotational velocity of low mass stars in young clusters such as the Pleiades and the Hyades decreases with the age of the cluster and for the fact that the considerably older Sun is rotating much more slowly than solar mass stars in the young clusters. Thus it is that wind mass loss through a magnetic field during the early part of the main sequence phase makes it possible to neglect both rotation and the magnetic field during the rest of the main sequence phase of low mass models.

This is not to say that rotation and magnetic fields in solar mass models are always of secondary importance. In the close binary stars called cataclysmic variables, which consist of a white dwarf and a low mass main sequence companion, the angular momentum lost by the stellar wind from the main sequence star forces the separation of the components to decrease until the main sequence star fills its Roche lobe and begins to transfer mass to the white dwarf. Tidal forces ensure that the donor star rotates in synchronism with the orbital rotation rate and angular momentum loss from the donor star translates into orbital angular momentum loss. When donor mass is or becomes less than accretor mass, the donor maintains contact with its Roche lobe and mass transfer continues on a non-dynamical time scale. In a typical system, whenever the mass of hydrogen-rich material accreted by the white dwarf component reaches a critical value of about 10^{-5} $M_\odot$, hydrogen is ignited explosively in the accreted material and the ensuing eruption is observed as a nova. In summary, while a low mass main sequence star which does not have a close stellar companion can be adequately described by a constant mass model, the result of mass loss from such a star in a close binary can have spectacular results.

Low mass stars along the red giant branch lose mass at the rate approximated by eq. (11.1.32). For the most luminous model constructed in Section 11.1, with $L \sim 2.51 \times 10^3$ $L_\odot$, $R \sim 210\ R_\odot$, and age = 12.0628 billion yr, this expression predicts a mass-loss rate of $\sim 2.11 \times 10^{-7}\ \eta\ M_\odot\ \mathrm{yr}^{-1}$, where $1/3 < \eta < 3$. For a model of age 12.05 billion yr, with a luminosity ten times smaller and radius about five times smaller, the expression yields $\dot{M} \sim 4 \times 10^{-9}\ \eta\ M_\odot\ \mathrm{yr}^{-1}$. It is apparent that, over the course of its lifetime as a red giant, a typical low mass star may lose of the order of one to two tenths of a solar mass.

In galactic globular clusters, the distribution in the HR diagram of horizontal branch stars, which are the core helium-burning progeny of red giant branch stars, can be understood as a consequence of mass loss along the giant branch at rates which vary from one progenitor star to another. That is, it is necessary to assume a spread in masses among horizontal branch stars of the order of 0.1–0.2 $M_\odot$ (Iben & Robert T. Rood, 1970). Coupled with the empirical evidence for wind mass-loss rates that imply a total mass loss on the red giant branch measured in tenths of a solar mass, the requirement of a mass spread of the same order reinforces the empirical estimates of mass-loss rate and, at the same time, demonstrates that the mass-loss rate is the consequence of a stochastic mass-loss process that can vary from one star to another as a function of individual characteristics such as rotation rate and magnetic field configuration.

Massive main sequence stars are known to lose mass at rates which are roughly proportional to the luminosity of the star (see, e.g., Peter S. Conti, 1978). Given that such stars do not have convective envelopes, observed mass-loss rates must be due to radiative winds and a first estimate can be made by assuming that the wind momentum is equal to the momentum in the radiation field. Thus,

$$\frac{L}{c} \stackrel{\sim}{=} \dot{M}_{\mathrm{wind}}\, v_\infty, \tag{11.4.15}$$

where L/c is the momentum of the radiation from the surface, $\dot{M}_{\mathrm{wind}}$ is the mass-loss rate, and v_∞ is the wind speed above the acceleration zone. Rough agreement with rates estimated from empirical evidence can be achieved by adopting

$$v_\infty \sim 3\, v_{\mathrm{escape}}, \tag{11.4.16}$$

where v_{escape} is the escape velocity at the stellar surface given by

$$\frac{1}{2}\, v_{\mathrm{escape}}^2 = \frac{GM}{R}. \tag{11.4.17}$$

Putting in numbers,

$$v_{\mathrm{escape}} = 618\ \mathrm{km\ s}^{-1} \left(\frac{M}{R}\frac{R_\odot}{M_\odot} \right)^{1/2}, \tag{11.4.18}$$

and

$$\dot{M}_{\mathrm{wind}} \left(M_\odot\ \mathrm{yr}^{-1} \right) \sim 1.1 \times 10^{-11}\, \frac{L}{L_\odot} \left(\frac{R}{R_\odot}\frac{M_\odot}{M} \right)^{1/2}. \tag{11.4.19}$$

From the path of the 5 $M_\odot$ model in Fig. 11.2.1, representative values of radius and luminosity during the main sequence phase are, respectively, $R/R_\odot \sim 2.5\sqrt{2}$ and $L/L_\odot \sim 700$. Inserting these numbers in eq. (11.4.19) gives $\dot{M}_{\rm wind} \sim 0.65 \times 10^{-8}\ M_\odot\ {\rm yr}^{-1}$. Thus, over its $\sim 8 \times 10^7$ yr main sequence lifetime, a star of initial mass 5 $M_\odot$ may lose of the order of 0.52 $M_\odot$, or $\sim 10\%$ of its initial mass. From the path of the 25 $M_\odot$ model in Fig. 11.3.1, one has that at a midway point in its main sequence evolution, the radius and the luminosity of the model are, respectively, $R/R_\odot \sim 8.5$ and $L/L_\odot \sim 1.05 \times 10^5$. Inserting these numbers in eq. (11.4.19) gives $\dot{M}_{\rm wind} \sim 0.67 \times 10^{-6}\ M_\odot\ {\rm yr}^{-1}$. Thus, over its ~ 6.3 million year main sequence lifetime, a star of initial mass 25 $M_\odot$ may lose approximately 4.2 $M_\odot$, or $\sim 17\%$ of its initial mass. As already noted, this means that the results of evolutionary calculations of constant mass models are best interpreted as describing the evolution of models of average mass equal to that of the constant mass models.

As initial mass is increased still further, the product of the estimated mass-loss rate and the estimated main sequence lifetime continues to increase. A very rough estimate of the mass dependence can be obtained by extrapolating on the basis of the straight line fits to the data for the 5 $M_\odot$ and 25 $M_\odot$ models in Figs. 11.4.1 and 11.4.2. For the main sequence lifetime, the fit is given by

$$t_{\rm MS}({\rm yr}) = 10^9 \left(\frac{M}{M_\odot}\right)^{-1.57}. \tag{11.4.20}$$

Coupling this equation and eqs. (10.4.9) and (10.4.10) with eq. (11.4.19) gives

$$\frac{\Delta M}{M} = \frac{\dot{M}\, t_{\rm MS}}{M} = 0.052 \left(\frac{M}{M_\odot}\right)^{0.36}, \tag{11.4.21}$$

where ΔM is the total amount of mass lost during time $t_{\rm MS}$. This equation yields 9.3% and 16.6% for the fractional mass loss by the 5 $M_\odot$ and 25 $M_\odot$ models, respectively, close to the initial estimates of 10% and 17% . For a model of mass 63 $M_\odot$ ($\log M/M_\odot = 1.8$), it predicts a fractional mass loss of 23%. However, comparison of the quadratic fits to the ingredients of eq. (11.4.21) with the linear fits to these ingredients suggests that this is a considerable underestimate. Relative to what is given by the quadratic fits, the product of the luminosity and the square root of the radius given by the linear fits is overestimated by a factor of about 1.8, but the main sequence lifetime is underestimated by a factor of about 3, suggesting that the fractional mass loss is more like 38%.

This degree of mass loss helps account for the prevalence in our galaxy of luminous Wolf-Rayet stars which exhibit highly evolved surface abundances, yet lie in a region which encompasses the main sequence. W. Schmutz, W. R. Hamann, & U. Wessolowski (1989) examine 20% of the WR stars known at the time, concentrating on the WN stars with helium-rich and nitrogen-rich surface abundances. Surface temperatures lie in the range $4.5 \gtrsim \log T_{\rm e} \gtrsim 5.0$ and the WN stars break into two groups, with the higher-temperature WN stars having luminosities given by $\log L/L_\odot = 5 - 5.5$ and and the lower-temperature WN stars having luminosities given by $\log L/L_\odot = 5.5 - 6$. Mass-loss rates, at $\dot{M} \sim 10^{-5.3} - 10^{-3.9}\ M_\odot\ {\rm yr}^{-1}$, are substantially larger than mass-loss rates of OB stars of comparable luminosity and it is thought that the region of wind acceleration lies below

Table 11.4.2 Exposure of layers containing matter enriched by products of fresh hydrogen burning

$M/M_\odot$	M_{CC}^{max}/M	$\Delta M/M$	$M_{terminal}/M$	$\Delta M_{exp}/M$	$\log L_{MS}/L_\odot$
5	0.24	0.09	0.91	-0.67	2.82
25	0.49	0.17	0.83	-0.34	5.00
63	0.72	0.38	0.62	0.10	5.95

the photosphere (see, e.g., Mariko Kato & Iben, 1992), in contrast with the situation in OB stars where the relationship described by eq. (11.4.5) indicates that the region of acceleration lies above the photosphere.

The development of highly evolved surface abundances in Wolf-Rayet stars requires that, as a consequence of mass loss, matter which has experienced extensive hydrogen burning in the interior of the Wolf-Rayet precursor becomes exposed at the surface. For initial masses of 5 and 25 $M_\odot$, M_{CC}^{max}/M is 0.24 and 0.49, respectively. The proportionality constant on the right hand side of eq. (11.4.5) is 0.123, so, for $M = 63$ $M_\odot$, a straight line logarithmic extrapolation gives $M_{CC}^{max}/M \sim 0.72$. The entries in the second column of Table 11.4.2 give an estimate of the fraction of the initial mass within which some degree of helium burning has taken place. The third column gives $\Delta M/M$, the fractional mass lost, the fourth column gives $M_{terminal}/M = 1 - \Delta M/M$ as an estimate of the mass of the star at the end of the main sequence phase relative to the initial mass of the star. The fifth column gives $\Delta M_{exp}/M = (M_{CC}^{max} - M_{terminal})/M$ and the sixth column gives the average main sequence luminosity.

At the end of the main sequence phase, the postulated observational counterpart of the model of constant mass 25 $M_\odot$ has a mass of $\sim$20.75 $M_\odot$. Inspection of Fig. 11.3.14 shows that, in the constant mass model, semiconvection has led to a depletion of hydrogen over a region out to $\sim$18.5 $M_\odot$, considerably larger than $M_{CC}^{max} \sim 12.5$ $M_\odot$. Even so, the final edge of the postulated observational counterpart does not extend inward to the region where fresh products of hydrogen burning have appeared. However, in the postulated observational counterpart of a constant mass model of 63 $M_\odot$, the exposed layer extends inward beyond M_{CC}^{max} by over 6 $M_\odot$ and, thanks to the operation of semiconvection, it is to be expected that the surface of the observational counterpart exhibits a helium to hydrogen ratio perhaps considerably larger than unity and thus appears as a Wolf-Rayet star.

From an examination of over 1000 supergiants and O-stars in the galaxy and the Magellanic Clouds, Roberta M. Humphreys (1978, 1979) and Humphreys & Kris Davidson (1978) find that in the HR diagram there is an upper bound to the luminosity that increases from $\log (L/L_\odot) \sim 5.7$ to $\log (L/L_\odot) \sim 6.6$ as surface temperature increases from a value given by $\log T_e \sim 4.2$ to a value given by $\log T_e \sim 4.8$. In the entire region in which $\log T_e \gtrsim 4.2$, there are fewer than a handful of stars more luminous than given by $\log (L/L_\odot) \sim 5.7$, in stark contrast with what one would expect if evolution beyond the main sequence were that predicted by constant mass models in which shell hydrogen burning continues to play an important role, carrying the model to the red, followed by core carbon burning at an even redder location.

Although it is a simple matter to construct evolutionary models with mass loss at an arbitrarily chosen rate, the manner in which the evolutionary path in the HR diagram for a mass-losing model deviates from that of a constant mass model and how surface abundances evolve depend on the adopted mass-loss rate. Progress requires a sophisticated understanding of how the rate of mass loss depends on global stellar characteristics and on extensive comparisons between the properties of real stars and the properties of stellar models constructed with various theoretically-motivated mass-loss rates normalized to observationally based estimates of mass-loss rates.

As an example of what can be learned by comparing the properties of real stars with those of theoretical models, consider the surface Li abundance predicted by theoretical models with and without mass loss. In the case of a star which arrives on the main sequence with a mass of 5 $M_\odot$, the theoretical analogue with no mass loss, as described in Section 11.2, destroys ^{7}Li over all but the outer $\sim$0.06 $M_\odot$ of the model and destroys ^{6}Li over all but the outer $\sim$0.03 $M_\odot$, with the result that, as the model climbs the red giant branch and the mass of the convective envelope increases to $\sim$4 $M_\odot$, the total Li abundance at the surface is reduced by a factor of the order of 70. If, as suggested by the estimate in this section, $\sim$10% of the mass of the real analogue is lost during the main sequence phase, all of the Li in surface layers is lost prior to the red giant phase and the surface Li abundance in the real red giant analogue is zero.

However, as also discussed in Section 11.2, the number abundances of Li at the surfaces of both of the $\sim$2.5 $M_\odot$ components of the binary Capella are measurable, although differing by approximately two orders of magnitude. In a theoretical constant-mass model of 3 $M_\odot$ (Iben, 1965), ^{7}Li is destroyed over all but the outer $\sim$0.04 $M_\odot$ during the main sequence phase and, as the base of the convective envelope extends inward until convection prevails over the outer $\sim$2.46 $M_\odot$ of the model, the surface Li abundance is reduced by a factor of $\sim$60. A reduction by another factor of 2, to bring the prediction into line with the observed Li abundances, would be accomplished if mass loss during the main sequence phase and/or during the transition from the main sequence to the base of the giant branch reduced the mass of the subsurface Li layer established on the main sequence by a factor of 2, but by no more than this, prior to dredge-up along the giant branch, suggesting that mass loss of a star of initial mass $\sim$2.5 − 3.0 $M_\odot$ is much less than predicted by eq. (11.4.19). Thus, even for relatively early phases of evolution, a proper incorporation of all of the relevant physics required for the construction of stellar models the evolution of which faithfully mimics the evolution of real stars is far from complete. Much work remains to be done.

In this volume, some of the basic physical processes which occur in stars have been presented and transformations in the internal characteristics of stars brought about by the operation of these processes during evolution onto the main sequence and through hydrogen-burning phases up to the ignition of helium in a hydrogen-exhausted core have been examined. In the companion Volume 2, physical processes that play important roles during more advanced evolutionary phases are described and the properties of quasistatic models in helium-burning phases of evolution and beyond are examined.

Bibliography and references

A. H. Batten, G. Hill, & W. Lu, *PASP*, **103**, 623, 1991.

Peter S. Conti, *ARAA*, **16**, 371, 1978.

Masayuki Fujimoto & Icko Iben, Jr., *ApJ*, **374**, 63, 1991.

Roberta M. Humphreys, *ApJS*, **38**, 309, 1978; **39**, 389, 1979.

Roberta M. Humphreys & Kris Davidson, *ApJ*, **233**, 409, 1978.

Icko Iben, Jr., *ApJ*, **142**, 1447, 1965; **415**, 767, 1993.

Icko Iben, Jr. & Robert T. Rood, *ApJ*, **161**, 587, 1970.

Icko Iben, Jr., Ann. Rev. Ast. Ap., 5, 571, 1967.

Harold L. Johnson, & Allan R. Sandage, ApJ, 121, 616, 1955.

Mariko Kato & Icko Iben, Jr., *ApJ*, **194**, 305, 1992.

Donald Lynden-Bell & Roger Wood, *MNRAS*, **138**, 495, 1968.

W. H. McCrea, *QJRAS*, **3**, 63, 1962.

Andre Maeder, & George Meynet, *Ast. Ap. Supp.*, **76**, 411, 1988; *AstAp*, **210**, 155, 1989.

C. A. Pilachowski & J. R. Sowell, *AJ*, **103**, 1668, 1992.

D. Reimers, *Mem. Roy. Soc. Liège*, 6e Ser., **8**, 369, 1975.

W. Schmutz, W. R. Hamann, & U. Wessolowski, *A&A*, **164**, 86, 1989.

Mario Schönberg & Subrahmanyan Chandrasekhar, *ApJ*, **96**, 161, 1942.

Richard B. Stothers, *MNRAS*, **151**, 65, 1970.

Richard B. Stothers & Chao-Wen Chin, *ApJ*, **292**, 222, 1985.

Allen V. Sweigart and Peter G. Gross, *ApJS*, **36**, 405, 1978.

George Wallerstein, *Nature*, **204**, 367, 1964.

Index